AF323262

DOE/TIC-22800

TRANSURANIC ELEMENTS IN THE ENVIRONMENT

A Summary of Environmental Research on Transuranium
Radionuclides Funded by the U. S. Department of Energy
Through Calendar Year 1979

Wayne C. Hanson, Editor
Pacific Northwest Laboratory

Prepared for the U. S. Department of Energy
Assistant Secretary for Environment
Office of Health and Environmental Research

1980

Published by
Technical Information Center/U. S. Department of Energy

Library of Congress Cataloging in Publication Data
Main entry under title:

Transuranic elements in the environment.

"DOE/TIC-22800."
Includes bibliographical references and index
1. Transuranium elements—Environmental aspects. 2. Radioecology.
3. Radioactive pollution. I. Hanson, Wayne C. II. United States. Dept. of Energy. Office of Health and Environmental Research.
QH545.T74T73. 574.5'222 80-607069
ISBN 0-87079-119-2

Available as DOE/TIC-22800 for $18.50 from

National Technical Information Service
U. S. Department of Commerce
Springfield, Virginia 22161

DOE Distribution Category UC-11

Printed in the United States of America

April 1980

Foreword

Before 1973 environmental research into the behavior of the transuranium elements was conducted on an ad hoc basis. It was usually prompted by some contamination event, such as the loss of nuclear material in the military aircraft accidents at Palomares, Spain, and Thule, Greenland, or the discovery of plutonium concentrations that exceeded fallout levels at such locations as the Rocky Flats Plant near Golden, Colo., and the Nevada Test Site. These research activities were usually aimed at describing the distribution of plutonium and appraising the health hazard at the individual site. Because this information was gathered at specific sites, it was not sufficient for generalized statements about environmental movement. In about 1970 the Nevada Applied Ecology Group began an integrated program at the Nevada Test Site in an attempt to provide a broader information base on transuranium elements. This program, however, was applicable primarily to desert environments. Some experimental studies at other locations were concerned with the uptake of transuranium elements by vegetation, but most of these dealt with western soils of high pH. No concerted effort was made to study transuranic radionuclide behavior in the marine environment except for studies at Thule, Greenland, and the Pacific Testing Grounds in the Marshall Islands.

In 1973 the U. S. Atomic Energy Commission, Division of Biomedical and Environmental Research (BER) (now U. S. Department of Energy, Office of Health and Environmental Research), performed an intensive study of its research efforts in support of the development of nuclear power with special emphasis on the Liquid Metal Fast Breeder Reactor (LMFBR). The environmental team reviewed information gathered up to that time on transuranic cycling in various environments and concluded that a comprehensive description of the environmental hazards of plutonium and the other transuranium elements relative to the LMFBR could not be made with the available data nor would it be forthcoming with the established research by BER contractors. It was obvious that too much of the past research had been centered on studies in the western regions, which were arid or semiarid, and essentially no studies had been made of soil movement and plant uptake in the humid eastern regions where fuel reprocessing plants were scheduled to operate. In addition, very little information was available on the cycling of plutonium through aquatic food webs inclusive of the marine studies in Greenland and the Marshall Islands. Essentially no research on the environmental behavior of transplutonic elements was under way, and the question of the long-term behavior and fate of the transuranium elements had not been addressed in any effective way. Further, the question of biological modification of the transuranium elements, which might lead to increased mobilization in the environment and possible underestimations of the dose to man, was not addressed.

The conclusions of the environmental team prompted AEC to develop a research program which would develop the information that was not available and which would be as comprehensive as possible for future assessments on the impacts of transuranic radionuclides from all stages in the nuclear fuel cycle. The program was designed to take advantage of the high-quality research that was already under way at the Health and Safety Laboratory, the Nevada Test Site, Lawrence Livermore Laboratory, Pacific Northwest Laboratory, the University of California at Los Angeles, Woods Hole Oceanographic Institute, and the University of Washington and to complement this research with research projects in other geographies and climates. Research activities were selected to cover all aspects of environmental transport from soil processes to ecosystem cycling. The objectives of the research program were to understand the cycling behavior of the transuranium elements in our environment and to determine to what degree these elements would be transported to us through food chains and aerial pathways. A further objective was to develop a satisfactory description of the degree to which the transuranium elements persist in the environment as a first step in assessing the potential hazard of these species on a historical and geological time scale.

To achieve the broad objectives of this program, we must answer many questions. If the transport of the transuranium elements is to be described, we must know to what degree these elements can be mobilized in the soils and aquatic sediments where they reside. A compendium of concentration ratios for plant uptake into food crops on various agricultural soils must be assembled. The transport through aquatic and terrestrial food chains must be quantified and appraised for the potential for human ingestion. Changes in the availability of transuranic elements due to resuspension from soils also must be better understood. Although the thrust of this work is on environmental transport, all the research scientists are alert for unusual concentration processes that might lead to radiological effects within environmental systems.

The areas of research just mentioned are of immediate concern, but beyond these near-term considerations are those related to the possible long-term persistence of the transuranic elements in available form on the scale of hundreds and thousands of years. Such considerations are very difficult to address adequately with contemporary research. However, two approaches are under way which may provide reasonable first approximations to the prediction of long-term behavior. One is the theoretical approach to studying the chemical and physical processes in soil of these radionuclides with the objective of developing good thermodynamic data. We need information on the equilibrium concentrations of the various oxidation states in different environments, on complexation processes, and on diffusion coefficients for various species. We can then apply this type of information to predictive modeling. An empirical approach would be to study the distribution and environmental behavior of naturally occurring elements that have properties analogous to those of the transuranium elements. For instance, the availability for plant uptake of the rare earth neodymium, which has been subjected to weathering for thousands of years, may provide a basis for predicting the uptake of americium after long periods of time because americium and neodymium have quite similar chemical and physical properties. Other rare earths, uranium, and thorium are also candidates as analogs for some of the transuranic elements in environmental studies.

The general areas of this program are outlined in Fig. 1. Research has now been conducted for periods of time ranging from 2 yr for new work to 6 yr for work that preceded this comprehensive program. Some of the results have been published in

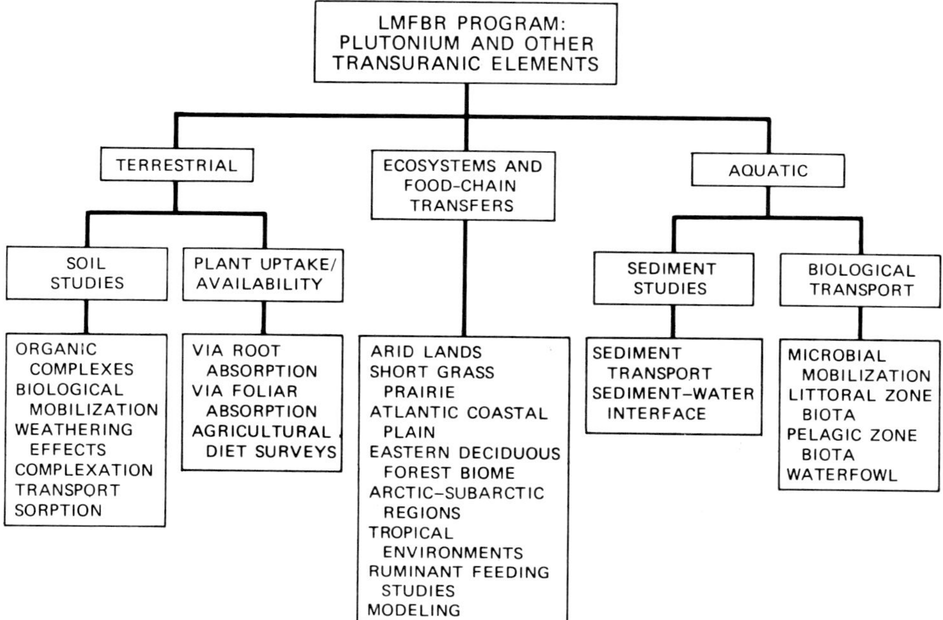

Fig. 1 United States Department of Energy, Office of Health and Environmental Research, transuranium elements research program.

laboratory reports, government documents, and refereed journals. This volume is an attempt to assemble the accumulated information as a synthesis document to provide an up-to-date interpretation of the environmental behavior of the transuranium elements.

R. L. Watters
Office of Health and Environmental Research
U. S. Department of Energy

Preface

This book evolved from recommendations made at the *Second Workshop on Environmental Research for Transuranic Elements* held at Seattle, Wash., Nov. 12–14, 1975 (proceedings available as ERDA-76/134), under the sponsorship of the Environmental Programs Branch of the U. S. Energy Research and Development Administration, Division of Biomedical and Environmental Research. A sixfold expansion of research on the environmental aspects of transuranic elements had occurred since the first plutonium workshop was held at Estes Park, Colo., July 11–12, 1974, and a need for greater communication of research results was identified. It was felt that investigators would be encouraged to publish their work in open literature following the publishing of a single publication that summarized the available information from the several Environmental Programs Branch investigations.

The objectives of this book are to assemble the available information on the behavior of transuranic nuclides in the environment following their release from a variety of source terms, their translocation by physical and biological transport phenomena, and the interpretation of the consequences of such concentrations as might be found in higher trophic levels of food webs. All such studies require attention to sampling design to provide accurate data, much of which may be destined as input to computer simulation models that are capable of treating the several variables and long half-lives that are involved in projecting the long-term environmental consequences of the nuclear fuel cycle of the future.

The authors were asked to emphasize both similarities and differences in transuranic nuclide behavior in various environmental settings and to identify research needs as they perceived them. Topics were assigned to scientists who were considered to be best qualified to address particular areas of research, often as members of a team. The cooperation of several of the authors in making this arrangement function properly was most gratifying and has magnified the application of their research to a better understanding of a difficult class of elements.

The outline of the volume and an initial evaluation of many of the manuscripts were presented at the *Third Workshop on Environmental Research for the Transuranic Elements* held at Woods Hole, Mass., Apr. 18–22, 1977 (proceedings available as CONF-770429). The hope that it would provide a comprehensive review and editorial comment of a rough draft of this book was only partially realized. That it survived at all is due in large part to the perseverance of the session chairmen.

The importance of transuranic nuclides in terms of their long physical half-lives, chemical toxicities, the effective linear energy transfer of their radiations, and the appreciable public concern about their release to the environment has prompted several symposia on this subject in recent years. As a result, the reader may wonder: "Why another publication on transuranic nuclides in the environment?" It is our intention that this book should not be "just another pretty face" but that it should uniquely represent an integration and synthesis of research results from a balanced program of studies with a common funding source. How effective we have been in achieving our goal is left to the reader to judge.

Wayne C. Hanson
Editor

Acknowledgments

The encouragement and support of R. L. Watters, W. O. Forster, and H. M. McCammon of the U. S. Department of Energy, Office of Health and Environmental Research, in the planning and executing of this publication are greatly appreciated. The assistance of D. N. Edgington, T. E. Hakonson, M. H. Smith, R. L. Watters, F. W. Whicker, and R. E. Wildung in providing the comprehensive synthesis part of this volume and for initial evaluation of various manuscripts is gratefully acknowledged. Many of the authors also served as reviewers of articles in their areas of expertise. Appreciation is expressed for additional advice on manuscripts obtained from the following persons:

A. J. Ahlquist, Los Alamos Scientific Laboratory
G. Choppin, Florida State University
J. M. Cleveland, U. S. Geological Survey, Lakewood, Colorado
P. B. Dunaway, U. S. Department of Energy, Las Vegas, Nevada
P. W. Durbin, Lawrence Berkeley Laboratory
S. W. Fowler, International Laboratory of Marine Radioactivity, Monaco
R. Fukai, International Laboratory of Marine Radioactivity, Monaco
D. W. Gillette, National Center for Atmospheric Research
J. A. Hayden, Rockwell International, Rocky Flats
J. W. Healy, Los Alamos Scientific Laboratory
P. W. Krey, U. S. Department of Energy, Environmental Measurements Laboratory
C. L. Osterberg, International Laboratory of Marine Radioactivity, Monaco
J. Pentreath, Ministry of Agriculture, Food and Fisheries, Lowestoft, England
V. Schultz, Washington State University
L. C. Schwendiman, Battelle Pacific Northwest Laboratory
F. B. Turner, University of California, Los Angeles
D. C. Wolf, University of Arkansas
D. S. Woodhead, International Laboratory of Marine Radioactivity, Monaco

Thanks are extended to many of the authors who promptly responded to the several deadlines that came and went during the long and tedious effort to produce this volume.

We greatly appreciate the secretarial support of V. J. McCabe and M. A. Rosenthal of the Los Alamos Scientific Laboratory and L. V. Kupinski and K. A. Tallent of Battelle Pacific Northwest Laboratory. Financial support was supplied by DOE contracts W-7405-ENG-36 and EY-76-C-06-1830 at Los Alamos Scientific Laboratory and Battelle Pacific Northwest Laboratory, respectively.

M. C. Fox, Jean S. Smith, and other members of the DOE Technical Information Center provided expert technical editing and other assistance in the publication of this volume. R. F. Pigeon of the DOE Office of Information Services rendered helpful advice for achieving the editorial goals. M. L. Merritt of Sandia Laboratories gave me the benefit of his experience in editing the DOE publication *The Environment of Amchitka Island, Alaska,* and otherwise provided suggestions.

Wayne C. Hanson
Editor

Contributors

D. C. Adriano
Savannah River Ecology Laboratory, Aiken, South Carolina
T. M. Beasley
Oregon State University, Newport, Oregon
S. G. Bloom
Battelle Columbus Laboratories, Columbus, Ohio
E. A. Bondietti
Oak Ridge National Laboratory, Oak Ridge, Tennessee
A. L. Boni
E. I. du Pont de Nemours and Company, Aiken, South Carolina
D. A. Cataldo
Battelle Pacific Northwest Laboratory, Richland, Washington
J. F. Cline
Battelle Pacific Northwest Laboratory, Richland, Washington
J. C. Corey
E. I. du Pont de Nemours and Company, Aiken, South Carolina
F. A. Cross
National Marine Fisheries Service, Beaufort, South Carolina
R. C. Dahlman
U. S. Department of Energy, Washington, D. C.
T. J. Dobry
U. S. Department of Energy, Washington, D. C.
P. G. Doctor
Battelle Pacific Northwest Laboratory, Richland, Washington
L. L. Eberhardt
Battelle Pacific Northwest Laboratory, Richland, Washington
D. N. Edgington
University of Wisconsin, Milwaukee, Wisconsin
D. R. Elle
U. S. Department of Energy, Richland, Washington
R. M. Emery
Battelle Pacific Northwest Laboratory, Richland, Washington
L. D. Eyman
Oak Ridge National Laboratory, Oak Ridge, Tennessee
G. C. Facer
U. S. Department of Energy, Washington, D. C.

R. H. Gardner
 Oak Ridge National Laboratory, Oak Ridge, Tennessee
T. R. Garland
 Battelle Pacific Northwest Laboratory, Richland, Washington
C. T. Garten, Jr.
 Oak Ridge National Laboratory, Oak Ridge, Tennessee
R. O. Gilbert
 Battelle Pacific Northwest Laboratory, Richland, Washington
T. E. Hakonson
 Los Alamos Scientific Laboratory, Los Alamos, New Mexico
W. R. Hansen
 Los Alamos Scientific Laboratory, Los Alamos, New Mexico
W. C. Hanson
 Battelle Pacific Northwest Laboratory, Richland, Washington
D. W. Hayes
 E. I. du Pont de Nemours and Company, Aiken, South Carolina
J. W. Healy
 Los Alamos Scientific Laboratory, Los Alamos, New Mexico
J. H. Horton
 E. I. du Pont de Nemours and Company, Aiken, South Carolina
D. C. Klopfer
 Battelle Pacific Northwest Laboratory, Richland, Washington
M. R. Kreiter
 Battelle Pacific Northwest Laboratory, Richland, Washington
C. A. Little
 Oak Ridge National Laboratory, Oak Ridge, Tennessee
F. G. Lowman
 Environmental Protection Agency, Narragansett, Rhode Island
R. W. McKee
 Battelle Pacific Northwest Laboratory, Richland, Washington
M. C. McShane
 Battelle Pacific Northwest Laboratory, Richland, Washington
R. P. Marshall
 University of Washington, Seattle Washington
W. E. Martin
 Battelle Columbus Laboratories, Columbus, Ohio
J. E. Mendel
 Battelle Pacific Northwest Laboratory, Richland, Washington
V. E. Noshkin
 Lawrence Livermore Laboratory, Livermore, California
J. W. Nyhan
 Los Alamos Scientific Laboratory, Los Alamos, New Mexico
C. R. Olsen
 Lamont-Doherty Geological Observatory, Palisades, New York
D. Paine
 Rockwell International, Richland, Washington
R. W. Perkins
 Battelle Pacific Northwest Laboratory, Richland, Washington

J. E. Pinder III
Savannah River Ecology Laboratory, Aiken, South Carolina

J. A. Robbins
University of Michigan, Ann Arbor, Michigan

E. M. Romney
University of California, Los Angeles, California

S. M. Sanders, Jr.
E. I. du Pont de Nemours and Company, Aiken, South Carolina

W. R. Schell
University of Washington, Seattle, Washington

R. G. Schreckhise
Battelle Pacific Northwest Laboratory, Richland, Washington

G. A. Sehmel
Battelle Pacific Northwest Laboratory, Richland, Washington

H. J. Simpson
Lamont-Doherty Geological Observatory, Palisades, New York

M. H. Smith
Savannah River Ecology Laboratory, Aiken, South Carolina

T. Tamura
Oak Ridge National Laboratory, Oak Ridge, Tennessee

W. L. Templeton
Battelle Pacific Northwest Laboratory, Richland, Washington

C. W. Thomas
Battelle Pacific Northwest Laboratory, Richland, Washington

R. C. Thompson
Battelle Pacific Northwest Laboratory, Richland, Washington

J. R. Trabalka
Oak Ridge National Laboratory, Oak Ridge, Tennessee

R. M. Trier
Lamont-Doherty Geological Observatory, Palisades, New York

B. E. Vaughan
Battelle Pacific Northwest Laboratory, Richland, Washington

B. W. Wachholz
U. S. Department of Energy, Washington, D. C.

M. A. Wahlgren
Argonne National Laboratory, Argonne, Illinois

A. Wallace
University of California, Los Angeles, California

R. L. Watters
U. S. Department of Energy, Washington, D. C.

F. W. Whicker
Colorado State University, Fort Collins, Colorado

R. E. Wildung
Battelle Pacific Northwest Laboratory, Richland, Washington

Contents

TERRESTRIAL ECOSYSTEMS

Experimental Studies

Field Studies

Models

AQUATIC ECOSYSTEMS

Marine Studies

Freshwater

BIOLOGICAL EFFECTS

Synthesis of the Research Literature

**R. L. WATTERS, D. N. EDGINGTON, T. E. HAKONSON, W. C. HANSON,
M. H. SMITH, F. W. WHICKER, and R. E. WILDUNG**

This book provides a compendium of environmental research related to transuranium elements; this research has developed greatly over the last 5 yr. The individual chapters describe studies that deal with mobility and transport in various environmental media and physiographic provinces. The intent of this synthesis is to develop, from the information in this book and other publications, unifying ideas and generalizations about movement of these elements through the environment to the human population.

Chemical, physical, and biotic processes control movement of transuranic elements within ecosystems. As illustrated by the conceptual model in Fig. 1, transport processes are driven by wind, water, biotic, and mechanical activity. For example, wind, water, or mechanical resuspension of soil and sediment can result in contamination of plant and animal surfaces, and the diet of consumer organisms may therefore contain this surficially deposited material. Examples of biotic transport include the movement of soil contaminants associated with a grazing animal and the subsequent redistribution of this material through defecation and/or death. Burrowing and grooming activities, which result in contamination of the animal, are additional examples of biotic transport.

Examples of chemical transport are the passage of soluble contaminants from soil through plant roots or across physiological membranes (e.g., lungs or gut wall) and the vertical leaching of soluble contaminants through the soil, although these processes may involve biochemical and physical parameters.

To predict the behavior of transuranic elements in the environment, one must understand (1) the ecological relationships in contaminated ecosystems, including the content and size of compartments and the exchange of materials between compartments, and (2) the pathways, rates, and mechanisms of transport through the ecosystems.

The behavior of transuranic elements in the environment must be described, at present, in terms of data obtained from direct sampling of sites with different contamination histories, sources, and ecological features. This information, together with data from laboratory studies defining rates and mechanisms, provides the framework for consideration of environmental fates and effects.

The results of this synthesis are, in many cases, tentative conclusions—as one would expect in any process involving inductive reasoning. Most of the inferences must be drawn from data pertaining to plutonium, which has been more intensively studied than other transuranic elements now under investigation. Some of these conclusions will become well established as more evidence accumulates; others may require modification to emphasize exceptions.

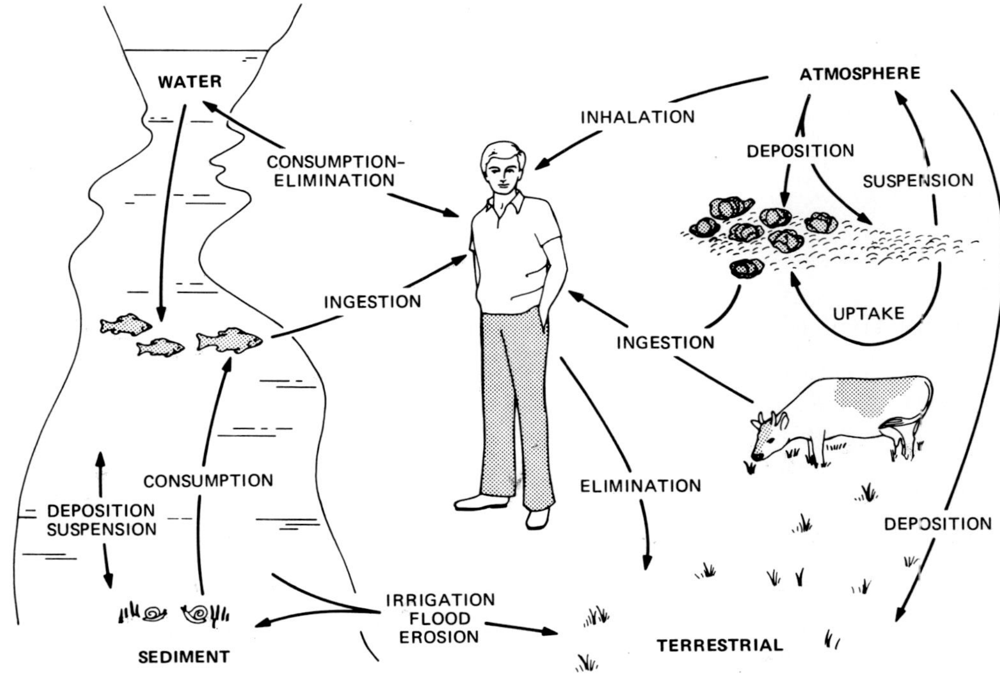

Fig. 1 Movement of transuranic elements to man from atmosphere, aquatic, and terrestrial components of the biosphere.

This synthesis is intended to clarify the present status of knowledge about transuranic elements and should stimulate further research that will define more clearly the environmental behavior of the transuranic elements and foster further analysis of new data as they become available.

Distribution and Inventory

Sources

Atomic weapons testing has been the major source of transuranic elements in the general environment. A portion of the debris from these tests was transferred into the stratosphere and then slowly returned to the Earth's surface. Because the transuranic elements were exposed to high temperatures, it was assumed that they were formed as high-fired oxides. It was further assumed that neither plutonium nor americium would move readily into biological systems because high-fired oxide particles would not dissolve in natural waters or, if they did, would form insoluble polymeric hydrated oxides. Further, it has been suggested that the behavior of transuranic elements in the environment is a function of source and, for plutonium, isotopic composition.

If these assumptions are accepted, transuranic elements are unlike any other element in the periodic table. However, experimental evidence relating to plutonium and americium in a wide variety of environments does not bear out these assumptions. It is known that a significant fraction of the plutonium deposited on the surface of the earth as fallout was produced by an (n,γ) reaction with ^{238}U and the subsequent decay of ^{239}U through ^{239}Np to ^{239}Pu (Joseph et al., 1971).* Thus a proportion of the plutonium in the environment was formed as single atoms long after the explosion and was never involved in a reaction to form a high-fired oxide.

The chemical form of transuranic elements released in small quantity during nuclear fuel reprocessing and fabrication may range from relatively insoluble oxide particles, which are of different composition than fallout (Sanders, 1977), to relatively soluble inorganic salts and organic complexes, which may be present in solid and liquid wastes. A generalized representation of the chemical forms of the transuranic elements released to the environment is given in Table 1.

TABLE 1 Major Sources and Initial Forms of Transuranic Elements Entering the Environment

Source	Form*
Nuclear weapons testing (global fallout, debris)	$TuO_x \cdot MO_x$ $TuO_x \cdot U_3O_8$
Nuclear fuel reprocessing/fabrication	$TuO_x \cdot nH_2O$ $TuO_x \cdot MO_x$ $TuNO_3$ Tu organic complexes

*Tu denotes plutonium and possibly americium, curium, and neptunium. M represents metal impurities, dust, and gaseous condensation products.

The potential movement of a transuranic pollutant from several sources, as illustrated for plutonium in Fig. 1, can be classified according to expected initial solubility in surface waters, interstitial waters of soils and sediments, and perhaps even on lung surfaces. Initially, particulate oxides of transuranic elements may be largely insoluble in solution. Ultimately, solubility will be a function of the chemical and physical properties of the particle and the matrix in which the particle is deposited. Oxide particles of the highest specific activity and containing the highest concentrations of impurities in the crystal lattice will exhibit the greatest solubility. The combination of configuration and equivalent diameter, as reflected in surface area exposed to solution, will also influence the rate of oxide solubility. Once dissolved, transuranic elements will be subject to chemical reactions governing dissolved salts. Hydrolyzable transuranic elements entering the environment in solutions sufficiently acidic to maintain soluble ions and in concentrations exceeding that of natural complexing agents will be rapidly hydrolyzed on dilution and subsequently precipitated on particle surfaces. These include Pu(III, IV, and likely VI), Am(III), Cm(III), and Np(IV and VI), although the rates of hydrolysis will vary between oxidation states. Conversely, chemical species of transuranic elements not subject to marked hydrolysis, such as Pu(V) and Np(V), initially may be more soluble

---

*As stated in the reference, approximately 1×10^{28} atoms of ^{239}Pu were generated by the explosion of thermonuclear weapons. This amounts to 4000 kg, or about two-thirds of the total plutonium estimated to have been deposited on the surface of the earth.

than the above species. Immobilization of those chemical species may occur through cation exchange reactions with particle surfaces or through redox reactions to hydrolyzable forms that become insoluble.

Transuranic elements entering the environment as stable organocomplexes, as might occur in the vicinity of a spent-fuel reprocessing facility, may be highly soluble initially (Wildung and Garland, 1975). The duration of solubility and mobility will be a function of the stability of the complex to substitution by major competing ions, such as Ca^{2+} and H^+ (Lahav and Hochberg, 1976; Stevenson and Ardakani, 1972; Norvell, 1972), the competition of other ligands forming more stable compounds, and the resistance of the organic ligand to chemical and microbial decomposition (Wildung and Garland, 1975). Disruption of the complex may lead to marked reduction in solubility through hydrolysis and precipitation reactions, as described for acid solutions on dilution. A portion of the ions released may react with other, perhaps more stable, ligands. The mobility of the intact complexes, in turn, will be principally a function of their chemical and microbiological stability and the charge on the complex, which will govern the degree of sorption on particles.

Initial chemical reactions and tendencies to remain soluble after release to the environment apparently depend on the initial chemical form of transuranic elements. However, the original source characteristics become less important as times goes on and weathering and aging processes proceed. From a consideration of the known distribution of transuranic elements in the environment and expected solubilities in the presence of particle surfaces, it is clear that their behavior will be markedly influenced by their individual chemistries and their chemical interactions in soils and sediments.

The effect of source and the immediate environment on the distribution of transuranic elements can be illustrated by comparison of the concentration of plutonium resulting from global fallout with that from more localized sources in soils, sediments, and waters (Table 2). For nuclear weapons testing, highest concentrations of plutonium in soils occur at the test locations. However, after stratospheric dispersion, concentrations are relatively low [<0.1 (d/min)/g] in surface soils, fresh water, and marine sediments. The lower concentrations of plutonium in marine sediments relative to those in soils reflect the longer residence times in the water column. Where nuclear processing facilities are known to provide a source of plutonium, soil and sediment concentrations range from fallout levels to several thousand disintegrations per minute per gram in controlled areas. Of major significance from the standpoint of environmental behavior is the fact that concentrations of plutonium in soils and sediments generally exceed those in water and other media by many orders of magnitude.

Terrestrial Ecosystems

One way of examining the distribution of any element within an ecosystem is through the use of an inventory ratio (IR). Two types of data are needed to calculate IR's: the weight (W) of each ecosystem compartment and the concentration (C) of the element within each compartment. The IR differs from the concentration ratio (CR) in that it takes into account the size of the compartments. For our discussion the compartments are soil, vegetation, litter, and animals.

The IR is calculated as

$$IR = \frac{A_c}{A_t}$$

where A_c is the amount of the chemical in a compartment and A_t is the total amount in the ecosystem. Thus IR can be dramatically changed by the size of the soil compartment, which is a direct function of the depth used to calculate this parameter. The amount (A) of the chemical of interest is simply

$$A = WC$$

where W and C are given in dry-weight units. These calculations do not require information about transfer rates or pathways within the ecosystem.

TABLE 2 Plutonium in Soils, Sediments, and Waters

Source and locations	Concentration $(^{239,240}\text{Pu})$*	Reference
Soils and sediments,† (d/min)/g		
Nuclear weapons testing		
Global fallout (soil)	0.01–0.05	Hardy, 1974
Debris (NTS, soil)	180.00–1.1 × 10⁵	Romney et al., 1976
Bikini Atoll (soil)	1.10–800.00	Nevissi, Shell, and Nelson, 1976
Lake Michigan (sediment)	0.20–0.90	Edgington, Wahlgren, and Marshall, 1976
North Atlantic (sediment, 5597 to 5968 m)	0.00–0.017	Bowen, Livingston, and Burke, 1976
Chemical processing		
Savannah River, S. C. (soil)	0.05–28.00	Adriano et al., 1975
Rocky Flats, Colo. (soil)	0.01–150.00	Krey and Hardy, 1970
Hanford, Wash. (soil)	0.01–1.50	Corley, Robertson, and Brauer, 1971
Hanford Z-9 trench, Wash. (soil, subsurface waste site)	1.50 × 10⁴–1.5 × 10⁹	Smith, 1973
Maxey Flats, Ky. (soil, waste site)	0.90–9.00	Meyer, 1976
Irish Sea (sediment)	0.70–105.00	Hetherington et al., 1976
Waters, (d/min)/liter		
Nuclear weapons testing		
Enewetak Atoll (groundwater)	4 × 10⁻⁴–1.50	Noshkin et al., 1976
Lake Michigan	5 × 10⁻⁴–7 × 10⁻⁴	Wahlgren et al., 1976
North and South Atlantic	3 × 10⁻⁴–4 × 10⁻³	Bowen, Wong, and Noshkin, 1971
North and South Pacific	1.30–9.4 × 10⁻⁴	Miyake and Sugimura, 1976
Chemical processing, runoff		
Newport, S. C., Estuary	5 × 10⁻⁴–5.6 × 10⁻³	Hayes, LeRoy, and Cross, 1976
Savannah River, S. C. (freshwater pretreatment)	5 × 10⁻³	Corey and Boni, 1976
Savannah River, S. C. (treated drinking water)	2 × 10⁻⁴	Corey and Boni, 1976
Irish Sea	0.10–1.0	Hetherington et al., 1976

*Values should not be considered representative but rather as examples of values obtained through various studies in specific localities. The original literature should be consulted before use of this information for other purposes.

†Expressed sediments on a dry-weight basis.

**TABLE 3   Range in Inventory Ratios for Plutonium in
Major Ecosystem Compartments**

Compartment	Range in inventory ratio*	Reference
Soils	0.998 to 0.986	Romney and Wallace, 1977; Dahlman, Garten, and Hakonson, this volume; Little, this volume
Vegetation	2×10^{-2} to 3×10^{-7}	Hanson, 1975; Romney et al., 1976; Dahlman, Garten, and Hakonson, this volume; Little, this volume; Pinder et al., 1979
Litter	3×10^{-2} to 2×10^{-4}	Dahlman, Garten, and Hakonson, this volume; Little, this volume
Animals	7×10^{-3} to 6×10^{-11}	Dahlman, Garten, and Hakonson, this volume; Little, this volume

*The proportion of the total plutonium in the ecosystem that is found in
each major compartment.

More than 99% of the plutonium inventory is found in the soil compartment of most
ecosystems (Table 3), and most of the contamination occurs near the soil surface
(Francis, 1973; Little and Whicker, 1977). Notable exceptions occur in arctic systems
where lichens intercept and retain fallout for long periods of time and in ecosystems that
are still receiving aerial depositions from nuclear processing facilities. However, even in
these special cases soil will be the eventual repository after deposition ceases (Hanson,
this volume; Holm and Persson, 1975; Dahlman and McLeod, 1977).

At some sites, a considerable amount of water has percolated into the soil since
the initial deposition (e.g., Savannah River Plant), but still the major inventory of
plutonium is in the top few centimeters of soil. The concentration of plutonium is only
occasionally higher in the subsoils below 10 cm than in the surface materials (Essington et
al., 1976). Plutonium is found at depths greater than 20 cm but usually in very low
concentrations unless soil or sediment mixing is actively in progress. Such mixing can
occur in steep canyons and delta regions of running-water ecosystems (Nyhan, Miera, and
Peters, 1975) and in terrestrial sites where natural biotic or human activities have mixed
or buried the plutonium.

Transuranic radionuclides can often be buried and remain immobile after deposition.
The exact distribution in the soil profile has an important influence on the availability of
transuranic elements for resuspension and root uptake.

The proportion of plutonium associated with biotic components of the ecosystem can
be as small as 0.1% (e.g., southeastern forests) (Dahlman, Garten, and Hakonson, this
volume). This fact reflects generally lower concentrations in biota but more importantly
the small mass of biota relative to soil. Even if transuranic elements were randomly
distributed among ecosystem components, the majority would still be associated with
soil.

The amount of plutonium associated with vegetation is greater than that associated
with animals (Table 3) but ranges over five orders of magnitude. Inventory ratios for
animals range over eight orders of magnitude. Most of this variation is probably due to
the amount of surface contamination on samples and not to internal concentrations of
plutonium in plants and animals (Dahlman and McLeod, 1977).

Autoradiographs of leaf tissues show that plutonium occurs primarily in discrete particles of suspendible size on the surface (Romney and Wallace, 1977). Washing plant materials to remove surface contamination reduces the concentration of plutonium in subsequent analyses (Dahlman and McLeod, 1977).

In some cases (e.g., inadequately cleaned vegetables) soil can be ingested by humans. Knowledge of the surface contamination of plant material is certainly important in determining the amount of plutonium ingested by both animals and humans.

Analyses of animal pelts, gastrointestinal tracts, and lungs give higher concentrations than those of tissues not exposed directly to surface contamination (Bradley, Moor, and Naegle, 1977). Humans do not normally eat tissues exposed to direct surface contamination to the extent that other carnivores do. Thus IR's calculated using concentration values determined from animal samples with natural levels of surface contamination may be more relevant in the assessment of potential environmental problems with food chains up to, but not including, humans.

High IR values for plant litter (Table 3) are principally due to surface contamination because smaller soil particles are impossible to remove (Romney and Wallace, 1977). The same is true for northern lichen-dominated communities where most dust particles are intercepted before they reach the surface of the ground (Hanson, 1966; Holm and Persson, 1975). Around transuranic processing facilities, aerial deposition of transuranic-bearing particles is probably the dominant form of contamination of plants (Pinder et al., 1979). Resuspension contributes importantly to surface deposition of contaminants and increases plant concentration values even in relatively moist environments (Dahlman, Garten, and Hakonson, this volume; Pinder et al., 1979).

There are few estimates of biomass of higher carnivores relative to that of vegetation. In addition, contaminated areas are generally limited in size and frequently include only parts of the ranges of a few individuals. Hence reliable IR's are not available for upper trophic levels, and animals are considered here as one compartment within the ecosystem. Inventory ratios may have characteristic values for certain ecosystems, and identification of ecosystem attributes allowing prediction of IR's would aid in assessing hazards. Knowledge of the relative biomass of ecosystem components will always be useful in modeling the long-term distribution of most contaminants, including transuranic elements. Future research on IR's should emphasize the establishment of predictable relationships and the identification of variables affecting IR values.

Aquatic Ecosystems

Experimental studies in the Great Lakes (Edgington and Robbins, 1975), Buzzards Bay (Livingston and Bowen, 1976), Irish Sea (Hetherington, Jefferies, and Lovett, 1975; Hetherington, 1978), and Trombay Harbor (Pillai and Mathew, 1976) have shown that, in comparatively shallow bodies of water, more than 96% of the total plutonium released to these environments is rapidly transferred to sediments. However, Bowen, Wong, and Noshkin (1971) estimated that as of 1969 10 to 20% of the total plutonium in deep oceans had been deposited in the sediments and that this would increase to only 30% by 1980.

In those parts of Lake Michigan and Lake Erie where sedimentation rates are greater than 5 mm/yr, a detailed analysis of plutonium and ^{137}Cs profiles in sediment cores clearly reflects the worldwide fallout maximums in 1959 and 1963 (Fig. 2). Similarly, it has been shown that americium and plutonium profiles in sediments from the Irish Sea

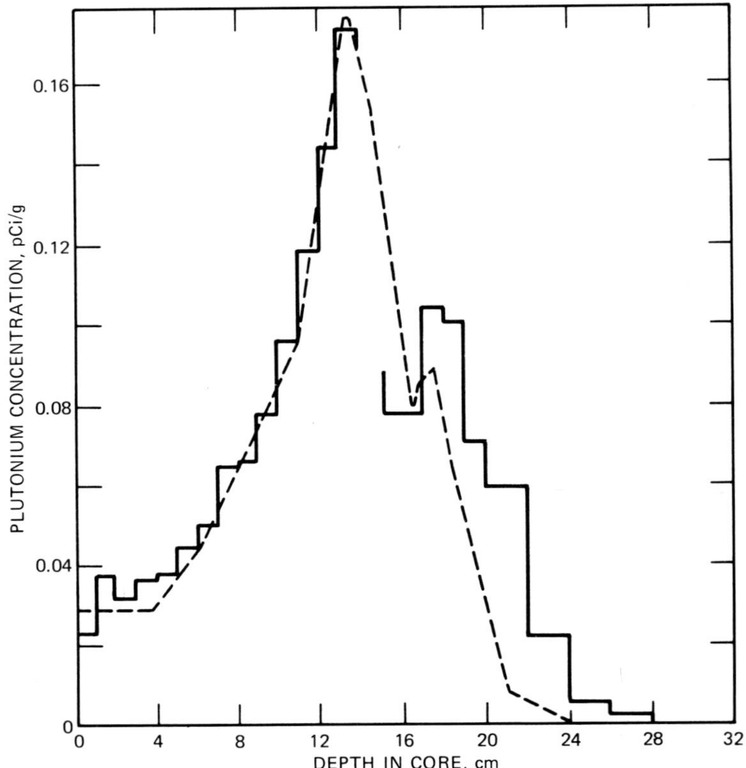

Fig. 2   Histogram of the distribution of plutonium in a sediment core from Lake Erie. The two peaks correspond to the material deposited during years of maximum fallout, 1963 and 1959, respectively. The dashed line represents the predicted distribution based on a sedimentation rate of 1.15 cm/yr at the surface and a mixing depth due to bioturbation of 4 cm.

reflect the history of releases from Windscale (Hetherington, Jefferies, and Lovett, 1975; Hetherington, 1978). Profiles measured in sediments from the Santa Barbara channel show a continuing input of plutonium due to erosion of California soils and direct input from fallout (Koide, Griffin, and Goldberg, 1975).

In other cores from Lake Michigan and Lake Erie, it is possible, by comparing plutonium and cesium profiles with those for ^{210}Pb, to estimate the effects of biotic activity on lake sediments (Robbins and Edgington, 1975) and to identify massive disturbances in sediments of the Great Lakes due to large storms (Edgington and Robbins, 1976).

Some sediment cores from Lake Ontario have plutonium concentration profiles exhibiting subsurface maximums similar to those found in Lake Michigan and Lake Erie (Bowen, 1976; Edgington and Robbins, 1976). However, in other cores from Lake Ontario and Buzzards Bay, the profiles show no subsurface maximums (Livingston and Bowen, 1976; Bowen, Livingston, and Burke, 1976). In these cores it is clear that there must be mixing downward by physical or biological processes. Repeated coring in Buzzards Bay from 1964 to 1973 showed that there was a small net loss of plutonium in the sediments. This loss was interpreted as a direct return to the water column rather than physical redistribution of sediments.

These studies indicate that sediments will be continually reworked by physical and biological processes. New plutonium will be continually added to the Great Lakes and coastal waters by wind erosion and transport of sediments down river. Because of the dynamics of the system, the inventory and distribution of plutonium will continue to change (Edgington and Robbins, 1975).

Solubility and Chemistry

Theoretical Considerations

The transuranic elements, starting with neptunium (atomic number 93), are a subset of the actinide series. This series is similar to the lanthanide series in that electrons are added successively to the 5f orbitals in a manner similar to the filling of the 4f orbitals. However, the shielding of the 5f electrons by outer electrons is less effective than that of 4f electrons; thus the chemical properties of the actinides are more complicated than those of the lanthanides. Although the latter exist primarily in the III oxidation state and exhibit ionic bonding, the actinides (through plutonium) can exist in multiple oxidation

TABLE 4 Comparison of Oxidation States for the Actinide Elements in Solution*

f =	1	2	3	4	5	6	7	8	9	10
	Ac	Th	Pa	U	Np	Pu	Am	Cm	Bk	Cf
							2			2
	3		3	3	3	3	3	3	3	3
		4	4	4	4	4	4	4		
			5	5	5	5	5			
				6	6	6	6			
					7	7				

*The solid lines bound the most likely oxidation states in aqueous solution.

states (Table 4). Because of their extreme reactivity, the II and VII oxidation states are not likely to be encountered in the environment. The oxidation—reduction behavior of the triad U—Np—Pu is complicated, and multiple oxidation states can coexist in solution. Actinides with atomic numbers exceeding that of plutonium behave similarly to the lanthanides.

The complex interactions between the various oxidation states of neptunium and plutonium are partly governed by their total concentration in solution. When concentrations are sufficient, disproportionation reactions between oxidation states are common. However, such concentrations are unlikely to be found in the environment, and the stable oxidation states will be a function of the chemical environment, e.g., the presence of oxidizing or reducing agents and complexing ligands. Thus the transuranic elements neptunium and plutonium can exist in more than one oxidation state in the environment, whereas the transplutonium elements will be 3+ cations.

Because of their very similar electronic structures and ionic radii, transuranic elements of a given oxidation state behave similarly chemically. Thus under most conditions Pu(III) is similar to Am(III) or rare earths, such as La(III); Pu(IV) is similar to Th(IV); and Pu(VI) is similar to U(VI) (Wahlgren et al., 1976). These differences in oxidation states of the transuranic elements and the ability of the elements to form complexes with natural ligands will greatly affect their availability for transfer in the biosphere (Dahlman, Bondietti, and Eyman, 1976).

Standard oxidation—reduction potentials can be used to predict the possible oxidation states of actinide elements in solution under environmental conditions. This, however, is an equilibrium prediction of the relative thermodynamic stability of various species and does not consider the effect of the kinetics of reaction or other factors, such as complexation, which affect the redox couple. Because measurements have been made of these potentials in near-neutral solution ($5 < pH < 9$), their magnitude must be calculated from the known hydrolytic behavior of the various ions, their respective formation constants, and standard potentials measured in acid solution (Kraus, 1949; Connick, 1949). The resulting oxidation—reduction potentials estimated for plutonium in neutral solution are:

$$Pu^{3+} \xrightarrow{-0.63} Pu(OH)_4 \cdot yH_2O_{(s)} \xrightarrow[\overbrace{}^{0.94}]{1.11} PuO_2 \xrightarrow{0.77} PuO_2(OH)_2$$

Since the potential for the oxygen couple in neutral solution $2H_2O \rightarrow O_2 + 4H^+(10^{-7}M) + 4_e$ is +0.815 volt, the oxidized species in any oxidation—reduction couple with a higher positive potential than this is thermodynamically unstable in water, although in practice a considerable overpotential exists (Pourbaix, 1966) which results in extremely slow reaction rates. The formation of complexes that drive the equilibrium potential to more thermodynamically stable values becomes extremely important.

With values of the oxidation potential (Eh) relative to the standard hydrogen electrode calculated for the reactions of transuranic elements in solution, it is possible to construct Eh—pH diagrams that delineate the regions of stability of ionic and solid species as a function of pH and soluble actinide concentrations. Earlier efforts at constructing these diagrams and phase relationships between plutonium species have been summarized (Bondietti and Sweeton, 1977). A comparison of Eh—pH diagrams for Pu(III) → Pu(IV) and Fe(II) → Fe(III) suggested that, under environmental conditions where ferric ion is reduced to ferrous ion, Pu(IV) may also be reduced to Pu(III) (Bondietti and Sweeton, 1977).

An Eh—pH diagram that was constructed with published values of E^0 (Pourbaix, 1966) is presented in Fig. 3. The III, IV, and VI oxidation states of plutonium were included. However, recent evidence suggests that Pu(V) can exist in aerobic environments (vide infra). The diagram shows the effect of changes in the concentration of plutonium in solution on the regions of stability of each oxidation state. Because of the tendency to form insoluble hydrolytic species, free Pu^{4+} ions can exist principally under strongly oxidizing acid conditions (region I). In the normal range of pH and plutonium concentrations encountered in the environment, plutonium could be present as PuO_2^{2+}, and this form will slowly come into equilibrium with solid $Pu(OH)_4$ (regions II and III).

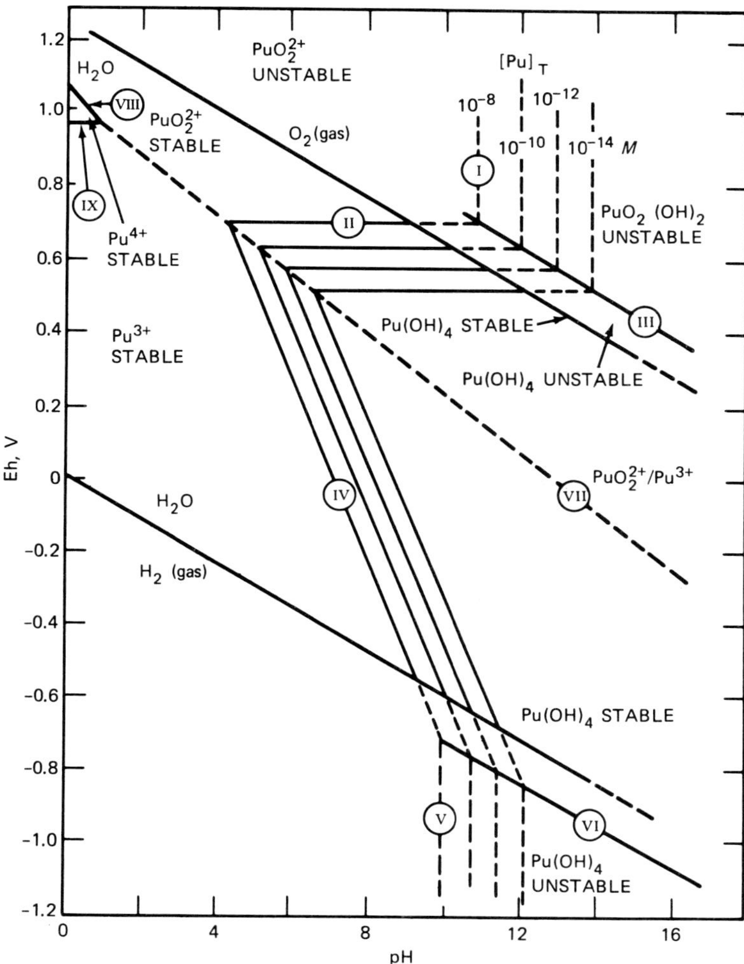

Fig. 3 Eh–pH diagram of stability fields for various plutonium species. Circled numbers represent lines of transition from one oxidation state to another and approximate the line of equilibrium between the regions in which plutonium may be susceptible to change by changing Eh or pH under the conditions specified. Reactions that are not considered may also occur and contain kinetic parameters of great importance. (See also Rai and Serne, 1977.)

Because the H_2O–O_2 couple is relatively insensitive to small changes in the partial pressure of oxygen, PuO_2^{2+} is thermodynamically stable in solution until the concentration of dissolved oxygen is essentially zero. This diagram is different from those presented by Polzer (1971), which indicated that PuO_2^+ is stable. Such differences reflect the uncertainty in many of the values of the relevant equilibrium constants and the choice of complexes considered. However, as Bondietti and Sweeton (1977) and Pourbaix (1966) have stressed, the results of these calculations are only valid for the stated conditions. In the formation of complexes with the OH^- ion and other natural ligands, the effect of insoluble compounds (such as phosphates) and the presence of natural reducing agents must be considered.

Similar Eh–pH diagrams can be constructed for americium (Pourbaix, 1966) which indicate that the standard potential for Am(III) → Am(IV) is much greater than that for Pu(III) → Pu(IV). Hence the range of stability of Am(OH)$_4$ moves to higher values of pH, and it appears unlikely that Am(IV) can exist in solution under normal environmental conditions.

The formation of complexes can also strongly affect oxidation–reduction potentials in solution and, depending on the relative values of stability constants, can stabilize different oxidation states in solution. Complex-ion formation in solution has been extensively studied because of the need to understand the behavior of transuranic elements in ion exchange, solvent extraction, and precipitation reactions. The general tendency for complex formation depends on such factors as ionic radius and charge. The order of stability constants is $M^{IV} > MO_2^{II} > M^{III} > MO_2^{I}$, and for anions it is generally $CO_3^{2-} >$ oxalate$^{2-} > SO_4^{2-}$ for divalent ions and $F^- > NO_3^- > Cl^- > ClO_4^-$ for monovalent ions. At relatively high concentrations of metal ions, hydrolysis reactions as acid solutions are neutralized lead to the formation of low-molecular-weight hydrolytic polymers, which can be described in some cases by simple equilibria, and high-molecular-weight polymers, which are not in equilibrium with the ions in solution.

Stability constants for some ligands present in natural waters have been summarized (Rai and Serne, 1977). For oxidizing conditions at pH 8, the conclusion was that the dominant species are $PuO_2CO_3OH^-$ and PuO_2 in the solution and solid phases, respectively. Unfortunately, the values of the stability constants for many of the hydroxo- and carbonato- complexes of plutonium, particularly Pu(IV), are not known, and the values given in the literature are suspect (Cleveland, 1970; 1978).

Since the effective charges of the metal ions in UO_2^{2+} and PuO_2^{2+} ions are so similar, the formation constants of complexes would be expected to be essentially the same for each ligand. Woods, Mitchell, and Sullivan (1978) measured the stability constants of complexes of PuO_2^{2+} with carbonate ions and found that at pH 8, where the HCO_3^- ion predominates, there is a 1 : 1 complex (Pu : HCO_3^-) with a formation constant of about 4×10^2. At pH 11, where the CO_3^{2-} predominates, there is a 1 : 3 complex, i.e., $PuO_2(CO_3)_3^{4-}$. In contrast, Langmuir (1978) indicated that, for UO_2^{2+} in waters at pH 8 in equilibrium with partial pressures of CO_2 much higher than atmospheric concentration, the predominant complex is $UO_2(CO_3)_3^{4-}$. He also suggested that at pH <7.5 $UO_2(HPO_4)_2^{2-}$ is the dominant species in natural waters with a phosphate concentration of $10^{-6}M$.

Although the effect of complexing in solution is to increase the total concentration of metal, it is not clear if such reactions will make them more or less available for bioaccumulation in the water column. Some of the smaller complexes may be readily assimilated; larger ones may not. The known distribution of the transuranic elements in the environment and their expected solubilities in the presence of particle surfaces indicate that their biological availability also will be markedly influenced by their chemistry/biochemistry in soils and sediments.

Terrestrial Ecosystems

Most studies indicate that plutonium is associated primarily with the solid phase in soils and sediments (Tamura, 1976; Garland and Wildung, 1977; Edgington, Wahlgren, and Marshall, 1976). Even in experiments where micromolar concentrations of Pu(NO$_3$)$_4$ are added to soil, the water-extractable and nonfilterable (<0.01 membrane filter) portion exists principally as hydrous oxide particles with a diffusion coefficient of approximately

10^{-7} (Garland and Wildung, 1977). Diffusion coefficients for total plutonium in soil are on the order of 10^{-9} (Relyea and Brown, 1975), which indicates little mobility by this mechanism. However, field studies have shown that plutonium may penetrate up to 30 cm in arid soil (Nyhan, Miera, and Neher, 1976) and that plutonium is more mobile through biological (e.g., root uptake and transport) and physicochemical mechanisms than would be predicted on the basis of diffusion alone. Furthermore, a fraction of plutonium in soils is readily dissolved. Studies of soils that had contained plutonium for over 30 yr indicated that up to 13% of the plutonium was extractable with chelating resins (Bondietti, Reynolds, and Shanks, 1976). This plutonium was probably available for plant uptake and, in the case of perennials, may continue to be available for several croppings, as demonstrated for clover (Romney, Mork, and Larson, 1970) and alfalfa (Wildung et al., 1977). The chemical/biological phenomena controlling the quantity and form of mobile plutonium are the key to predicting long-term implications of plutonium in the environment.

Under aerobic conditions the ultimate behavior of plutonium in soils and sediments, regardless of the chemical form of the source material, will be governed by processes that influence hydrolysis and sorption on the solid phase and formation of soluble complexes with organic or inorganic ligands (Fig. 4). Initially, sorption and precipitation processes predominate when Pu(IV) is added as the soluble nitrate and account for 98% of the total plutonium a few hours after Pu(IV) has been added to a neutral silt loam soil (Wildung and Garland, 1975). The addition of Pu(IV) as the DTPA complex results in nearly 100% plutonium solubility before a gradual reduction in solubility occurs by processes described earlier. From a thermodynamic standpoint, the formation of Pu(V and VI) in soil solution is theoretically possible. However, studies of the interactions of Pu(VI) with organic ligands representing a range of common soil metabolites (Wildung, Garland, and Cataldo, 1979), humic substances, and reducing sugars (Bondietti, Reynolds, and Shanks, 1976) suggest that Pu(VI) will be readily reduced to Pu(III) + (IV) in aerobic surface soils. The presence of Fe(II), a reductant, may further promote the formation of reduced plutonium under most soil conditions except highly alkaline soils.

Because Pu(IV) readily forms insoluble hydrolysis products, the interaction of these species with mineral and organic surfaces results in the relative immobility of plutonium in soils and sediments. Hydrolysis products sorb on the solid phase by mechanisms other than ion exchange, and attempts to extract exchangeable plutonium from soils using $MgCl_2$ (Muller, 1978) and resins (Bondietti, Reynolds, and Shanks, 1976) resulted in the removal of relatively small quantities (<13%) of the total plutonium. The major portion of plutonium associated with the solid phase in soils and sediments (Muller, 1978; Edgington, Wahlgren, and Marshall, 1976) was extractable with citrate—dithionite, but with citrate alone much less was extracted, which suggests the association of plutonium with the reductant-soluble iron on the surfaces of soil/sediment particles (Wildung, Schmidt, and Routson, 1977) or with iron in the original particles that were deposited.

The importance of hydrolysis in governing plutonium behavior extends to the soluble fraction, at least over the short term (months). Almost all soluble and diffusible plutonium on soil has been shown to be $Pu(OH)_n$ (Wildung et al., 1977). The small quantity remaining in soil/sediment solutions is probably present as the Pu^{4+} ion stabilized against hydrolysis by interaction with a predominant anion (CO_3^{2-} or HCO_3^-, depending on pH and ionic composition) and organic ligands (Wildung, Garland, and Cataldo, 1979). Concentration of low-molecular-weight organic ligands, bicarbonate ion, and carbonate ion are directly related to microbial metabolism and decomposition of

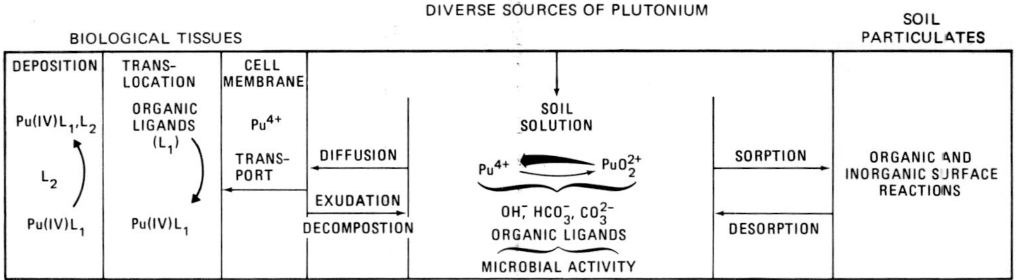

Fig. 4 Model for plutonium chemistry in the ingestion pathway. Regardless of the form of plutonium entering soils, sediments, or water, plutonium is predominantly converted (exceptions given in aquatic section) to Pu(IV), which is largely insoluble and associated with the solid phase of soils and sediments. Soluble plutonium is also present primarily as Pu(IV) stabilized by complexation with inorganic and organic ligands. The Pu(IV) complexes are largely dissociated at the cell surface with Pu^{4+} ion transport across the cell membrane. Mobility in biological tissues is facilitated by formation of secondary complexes.

organic materials. There is direct evidence that plutonium forms complexes with microbial metabolites and considerable indirect evidence supporting microbial influence on plutonium solubility in soil (Wildung and Garland, 1977; Wildung, Garland, and Cataldo, 1979). The complexed Pu(IV) is probably the only plutonium that is available for plant uptake (Fig. 4). The formation and the delivery of these complexes to roots are the rate-limiting processes in the ingestion pathway.

Chemical properties of other transuranic elements (americium, curium, and neptunium) in the environment have not been established. Laboratory studies have been limited to studies of (1) the soil sorption of americium and neptunium, which indicate sorption in the order Pu $\gg$ Am > Np (Routson, Jansen, and Robinson, 1977); (2) the sorption of Cm(III) and Np(V) on soil clay, which indicate sorption in the order Cm $\gg$ Np with an apparent association of neptunium with organic matter and amorphous iron (Bondietti and Tamura, this volume); and (3) the solution behavior of ^{244}Cm in a freshwater lake, which shows that soluble curium (50% of total) was largely anionic (Dahlman, Bondietti, and Eyman, 1976). Field studies have been limited by relatively low concentrations in the environment and lack of sensitive analytical methods for certain nuclides of importance (e.g., $^{241,243}Am$, ^{244}Cm, and ^{237}Np). The aqueous chemistry of these elements has been fairly well established (Katz and Seaborg, 1957; Keller, 1971) and allows some predictions of behavior in soils and sediments. Major differences in their environmental behavior as compared with plutonium would be expected, and sorption on solid surfaces may be a function of the predominant valence state and its tendency to hydrolyze (Dahlman, Bondietti, and Eyman, 1976). The only stable ions of americium and curium in aqueous solutions are the trivalent cations. Their chemistry in soils and sediments is similar if they are present in similar mass concentrations. Hydrolysis reactions may be a primary factor governing the behavior of americium and curium, but greater mobility and biological availability can be predicted because of greater solubility of their hydroxides in comparison with $Pu(OH)_4$. For neptunium, NpO_2^+ is the most stable species in aqueous solution and should not be subject to significant hydrolysis at environmental pH values (Burney and Harbour, 1974). Of the transuranic elements, the environmental behavior of neptunium has been least studied, but, because of its chemical characteristics, it is the most soluble in soils and may be the most available to biota. Plant uptake studies (Schreckhise and Cline, this volume) and experimental feeding studies (Sullivan, 1979) indicate that this is likely.

Aquatic Systems

Freshwater Ecosystems.   The behavior of plutonium and americium has been studied in a wide range of freshwater systems (Table 5). Some contaminated areas have been small, such as the U-pond on the Hanford Reservation in Washington (Emery and Klopfer, 1976), White Oak Lake in Tennessee (Dahlman, 1976), Rocky Flats ponds (Rees, Cleveland, and Gottschall, 1978), and the ponds and canals at the Mound Laboratory in Ohio (Bartelt et al., 1977). More extensive studies have been carried out in the Hudson River in New York (Simpson, Trier, and Olsen, this volume) and in the Great Lakes system (Wahlgren, Robbins, and Edgington, this volume; Bowen, 1976). The concentration of plutonium in these systems varied by more than four orders of magnitude (Table 6). If concentrations are calculated assuming that the ^{239,240}Pu is essentially all ^{239}Pu, concentrations in water vary between $3 \times 10^{-17} M$ in the Great Lakes and $3 \times 10^{-13} M$ in contaminated systems like U-pond on the Hanford reservation. These concentrations are low relative to concentrations of many other trace elements. For example, the concentration of thorium in Lake Michigan is $\leqslant 4 \times 10^{-12} M$ (0.1 fCi/liter) (Wahlgren et al., 1977a).

Processes controlling the solubility of plutonium in natural waters clearly are more complex than can be explained by a simple solubility product. For example, the concentration measured in Lake Michigan is higher than that predicted for $Pu(OH)_4$ and lower than that for $PuO_2(OH)_2$. These differences have been attributed to the formation of hydroxyl complexes, such as $Pu(OH)_5^-$ (Bondietti and Sweeton, 1977), carbonate complexes, such as PuO_2CO_3 or $PuO_2(CO_3)_2^{2-}$ (Moskvin and Gel'man, 1958), or $PuO_2CO_3OH^-$ (Rai and Serne, 1977).

A recent investigation of the reduction of Pu(IV) and (VI) by natural organic compounds showed that up to 15% of Pu(IV) was reduced to Pu(III) at pH 4.0 and up to 75% of Pu(VI) was reduced to Pu(IV) by fulvic acid at pH 8. However, the Pu(VI) was more stable in the presence of carbonate (Bondietti, Reynolds, and Shanks, 1976).

Wahlgren et al. (1977a) studied the behavior of plutonium in the water column in the Great Lakes and other smaller freshwater lakes to determine whether differences in chemical characteristics of lake water affect the chemical properties of plutonium (Table 7). The concentration of plutonium in waters of these lakes varied almost 100-fold. The highest concentrations of plutonium were observed in the lakes (ELA 241 and ELA 661) with low pH, a lake with a very high concentration of sulfate (Little Manitou), and the acidic southeastern United States lakes.

Using techniques to separate Pu(III) + Pu(IV) from Pu(V) + Pu(VI), Nelson and Lovett (1978) showed that in the Irish Sea plutonium was predominantly in the Pu(V) + Pu(VI) states. A similar observation was made in Lake Michigan waters (Wahlgren et al., 1977b). Because Pu(V) was thought to disproportionate at lower pH than does Pu(VI) (Pourbaix, 1966), this fraction was referred to as Pu(VI).* In all other lake waters, Pu(III) + (IV) apparently predominated.

---

*Very recent experiments at Argonne National Laboratory have shown that this assumption is not correct (D. M. Nelson and K. A. Orlandini, Argonne National Laboratory, 1979, personal communication). Techniques have been developed after the method of Inoui and Tochiyama (1977) for separation of Np(V) from Np(VI) to distinguish Pu(V) from Pu(VI) in water samples. Preliminary results indicate that all the plutonium in the higher oxidation state is present as Pu(V).

TABLE 5   Studies of Transuranic Elements in Freshwater Environments

Location	Aquatic system	Source terms and nuclides studied	Inventory	Sampling techniques for water
Hanford reservation, Wash.	U-pond Ultraeutrophic	Effluents from reprocessing plant since 1944: ^{239,240}Pu, ^{238}Pu, and ^{241}Am	~1 Ci(Pu)	Filtration through 0.45-μm filters; Battelle large volume water sampler
Rocky Flats, Colo.	A and B ponds Walnut Creek Great Western, mower reservoirs Standley Lake	Effluents from process waste treatment plant: ^{239,240}Pu	N/A*	Filtered through Millipore filters—pore size not specified
Oak Ridge, Tenn.	White Oak Lake	Effluents from waste processing streams: ^{239,240}Pu and ^{238}Pu	N/A*	
Miamisburg, Ohio	Canals and ponds Miami River	Effluents from Mound Laboratory: ^{238}Pu	Annual transport, about 30 mCi	Same as for Great Lakes
Great Lakes	Lake Michigan Lake Huron Lake Erie Lake Ontario	Atmospheric fallout: ^{239,240}Pu and ^{238}Pu	120 Ci 60 Ci	Filtration through 0.45-μm filters Fractionation using ion exchange resins, ultrafiltration, etc.
New York	Hudson River	Atmospheric fallout: ^{239,240}Pu Effluents from Indian Point	Annual transport, about 6 mCi	Continuous centrifugation, filtration through 0.45-μm filters
Savannah River Laboratory	Savannah River	Effluents from processing plant	N/A*	

*Not applicable.

TABLE 6 Behavior of Plutonium and Americium in Various Aquatic Systems

Aquatic system	^{239,240}Pu			^{238}Pu			^{241}Am		
	Water, fCi/liter	Surface sediment, fCi/g	K_D ($\times 10^{-4}$)	Water, fCi/liter	Surface sediment, fCi/g	K_D ($\times 10^{-4}$)	Water, fCi/liter	Surface sediment, fCi/g	K_D ($\times 10^{-4}$)
U-pond*									
(average values)	3.3†	177‡	5.4	5.1†	169‡	3.3	4.2†	81‡	1.9
Rocky Flats									
Pond A1 (1973)	2.0†,§	20‡	1.0						
B1 (1973)	2.2†,§	1000‡	45.0						
B2 (1973)	2.1†,§	150‡	6.8						
B3 (1974)	2.1†,§	60‡	3.0						
White Oak Lake	9.8 to 2.1								
Miamisburg									
Pond–N				310	19.5‡	6.4			
–S				540	24.8	4.6			
Canal–5				2900	230	7.9			
Great Lakes									
Michigan	0.6	200	33	<0.1	8	>8	<0.1	50	>50
Erie	0.15	100¶	15						
Hudson River	0.3	20.0	6.7						

*Sediment average of all depths of sampling.

†pCi/liter.

‡pCi/g.

§Data corrected for particulate material content (Paine, this volume).

¶Estimated from data for two cores and average ^{137}Cs concentration in sediments.

TABLE 7 Behavior of Plutonium in a Selected Group of Canadian and United States Freshwater Systems

Lake	pH	Alkalinity, mg/liter as CaCO$_3$	SO$_4^{2-}$, µg/liter	Pu, moles/liter ($\times 10^{17}$)	$\dfrac{\text{Pu(V and VI)}}{\text{Pu(IV)}}$	Fraction anionic	Ultrafiltrate
Great Lakes							
Michigan	8.2	113	16	2.1	5.0*	~1.00	0
Huron	8.0	79	17	2.3			
Erie	8.1	92	26	0.5			
Ontario	7.9	93	29	1.0			
Superior	7.8	52	4	2.2			
Canadian Lakes							
Clear (M)†	8.0	190		4.6	5.0*		
Last Mountain (S)†	8.4	305		4.5	1.0*		
Katherine (M)	8.5	180		2.8	0.7		
Lake of the Woods (O)†	7.1	50	6	2.4	0.3		
ELA 885 (M)	8.4	355		7.4	0.2		
ELA 661 (O)	4.8		Low	44	0.2		
ELA 241,‡ surface	5.9		Low	29	0.3		
ELA 241,‡ 9 m	6.0			48			
ELA 241,‡ 11 m	6.1			⩾220	⩽0.14		
Little Manito (S)	8.0	770	10^5	47	0.08		
Great Slave Lake							
Main basin	8.2	80		1.5			
Christie Bay	8.5	65	30	2.9			
McLeod Bay	7.5	15	Low	3.7			
Southeastern United States							
White Oak Lake				6	⩽0.1	~1.00	0
Banks (Ga.)							
May				20		0.21	0.70
December	3.9			5		0.39	0.53
Okefenokee (Fla.)	5.0			15.4		0.08	0.89
Par Pond (Ga.)	8.0	15		0.5			
Precipitation (Argonne National Laboratory)	4.8			20.0		0.23	0.72

*These lakes are highly oligotrophic.
†M, Manitoba; S, Saskatchewan; O, Ontario.
‡This lake is meromictic and is very reducing in the bottom water.

The Eh–pH diagram (Fig. 3) shows that under normal environmental conditions Pu(III) and Pu(VI) can coexist and that the ratio of the two states will depend on the oxidizing conditions and pH in the system. Therefore the relatively high concentrations of plutonium in ELA lakes (other than 885) and lakes in the southeastern United States could be due to the reduction of Pu(IV) to Pu(III) as well as to complexation of Pu(IV) by organic ligands. The high concentration of plutonium and the very low fraction of Pu(VI) in Little Manitou Lake, which contains high sulfate concentrations, could also be due to the formation of sulfate complexes, which stabilize the Pu(IV) state.

Similar measurements have evaluated the relative concentration of Pu(III + IV) and Pu(V + VI) in White Oak Lake water (Bondietti and Sweeton, 1977) and indicated that Pu(IV) was the dominant oxidation state present. Plutonium(IV) rather than Pu(III) was suggested because another study indicated that Pu(III) was unstable toward oxidation to Pu(IV) at pH $\geqslant 5$ (Bondietti, 1977).

Charge characteristics of plutonium in Lake Michigan water indicate that the element is not associated with colloidal matter in the size range $0.003 < x < 0.45$ μm and that it is almost quantitatively absorbed by anion exchange resins. In water samples from acidic lakes, the majority of the plutonium behaves like cationic or uncharged species. These results and the differences in oxidation state discussed earlier strongly suggest that the solubility of plutonium is governed by different complexing agents. In waters of high pH, the concentration of CO_3^{2-} and HCO_3^- ions is relatively high, and carbonate complexes can form. In waters of low pH, such complexes cannot exist, and the solubility must be due to complexing with other ligands, such as natural organic compounds.

In addition to measuring concentrations in water columns, most investigators have measured the concentration of plutonium in surficial sediments. In a few cases measurements have been made of plutonium in suspended particulate material. Table 6 shows that there is some relationship between concentrations in water and concentrations in surface sediments.

If there is mixing of the surficial sediment with the water column and a true equilibrium between the water and particulate matter or sediment, the distribution constant, K_D, for the reaction between filtered water ($\leqslant 0.45$ μm) and sediment is

$$K_D = \frac{\text{Concentration per kilogram of sediment}}{\text{Concentration per liter of water}}$$

Values of K_D vary from 10^4 to 5×10^5, but most values do not vary more than fivefold. Considering the wide variety of sediment types involved and the differences in source terms and sizes of aquatic environments, the small range in values strongly suggests a commonality in the behavior of plutonium in these systems.

Sediment characteristics affect the uptake of radionuclides, and fivefold to tenfold variations in distribution coefficients can be explained solely in terms of differences in distributions of particle sizes in sediments (Duursma and Bosch, 1970). Little information is given on sediment characteristics, but sediments from small ponds and rivers probably are generally coarser than those from deep waters of the Great Lakes.

More recent experiments have shown the distribution of plutonium between solution and solid phases to be a true equilibrium. Sediments labeled with ^{238}Pu from the Miami River were equilibrated with Lake Michigan water. Kinetic studies indicated that equilibrium was reached in 1 day or less. Moreover, the ratio of oxidation states in water from this experiment is the same as that observed for ^{239}Pu in Lake Michigan (D. M.

Nelson and D. N. Edgington, Argonne National Laboratory, personal communication). A few measurements have been made of the oxidation state of plutonium adsorbed on particulate matter (Nelson and Lovett, 1978). Samples of surface sediment from the Great Lakes and Miami River indicate that plutonium absorbed by sediment particles is predominantly in the (III) and (IV) states. On reequilibration of sediment with water, it has been shown that there is a conversion of Pu(III) + Pu(IV) back to Pu(V) or (VI) (D. M. Nelson, Argonne National Laboratory, personal communication).

TABLE 8   Values of the Distribution Coefficients (K_D) and Concentration Ratios (CR) for Phytoplankton for the Actinide and Lanthanide Elements in Lake Michigan

Element	Log K_D*	Log CR†
La(III)	5.2	3.0
Th(IV)	$\geqslant$6.5	
U(VI)	~4.0	2.2
Plutonium	5.5	4.0
Pu(IV)	6.5	4.8
Pu(VI)	4.1	~4.2
Americium	>6.0	~4.2

*Values taken from Wahlgren et al., 1976.

$$\dagger CR = \frac{\text{pCi/kg wet tissue}}{\text{pCi/kg water}}.$$

Therefore the concentration of plutonium in many freshwater lakes and streams apparently is controlled by an equilibrium between water and sediment. From the data it is possible to calculate values of K_D for Pu(III) and Pu(IV) and Pu(VI). These values are given in Table 8 and are compared with those for stable-element homologs, such as La(III), Th(IV), and U(VI). As would be expected, results for ^{238}Pu from U-pond and ponds and canals in Miamisburg (Table 6) show little or no difference in behavior of the plutonium owing to isotopic composition.

Finally, leaching experiments with sediments have shown that a major fraction of fallout plutonium can be removed with extractants, such as dilute acids or complexing agents (Alberts, Muller, and Orlandini, 1976). Furthermore, studies of plutonium in natural waters have shown low but measurable concentrations of plutonium in true solution. In Lake Michigan the plutonium concentration was essentially constant over the whole lake (Wahlgren and Nelson, 1975). The measured concentration of 0.5 fCi/liter of ^{239,240}Pu corresponds to a chemical concentration of $3 \times 10^{-17}M$, or 20,000 atoms/ml. Hence molecular collision theory implies that the formation of polymeric plutonium species in the lake (i.e., many plutonium atoms linked as $-$Pu$-$Pu or as $-$Pu$-$O$-$Pu) is unlikely. Even the possibility of the formation of dimers is vanishingly small.

Marine Ecosystems.   Studies of transuranic elements in marine and estuarine ecosystems have encompassed a wide range of sources: worldwide fallout as a result of the testing of nuclear weapons, lagoons in tropical atolls where many tests were performed, and direct discharges from nuclear processing plants into coastal zones. The study areas and characteristics of the sources are given in Table 9.

Plutonium has been measured in samples of ocean waters collected since 1963, shortly after the peak of activity in testing weapons. The concentration has decreased from 2 to 3 fCi/liter in samples of water from the northeastern Pacific in 1964 (Pillai, Smith, and Folsom, 1964) to about 0.2 to 0.9 fCi/liter in samples collected between 1968 and 1973 (Miyake and Sugimura, 1976). The major point in studying plutonium in the water column of oceans is to use variations in the concentration of this element to explain movements of water and pollutants. To this end, comparisons have been made of the movement of plutonium relative to ^{90}Sr and ^{137}Cs in the Pacific Ocean and Atlantic Ocean (Bowen, Wong, and Noshkin, 1971; Miyake and Sugimura, 1976). The Pu/^{90}Sr and Pu/^{137}Cs ratios in surface seawater are far lower than those found on land, which indicates that the residence time of plutonium in the water column is less than that of ^{137}Cs and ^{90}Sr. As early as 1968, from 10 to 20% of the total plutonium deposited over the ocean was in deep-sea sediments at water depths of about 4000 m. The depletion of plutonium from surface waters may be modeled in terms of settling rates of particulate matter in the water column. The observed distribution of total plutonium in the water column to 5000 m is explained in terms of a distribution of particles with the majority settling at an average velocity of 195 m/yr (Bowen, Wong, and Noshkin, 1971).

Another series of water samples was collected from the Pacific Ocean in 1973 as a part of the Geosecs Program. Analyses of these samples showed that there is a maximum in the concentration of plutonium at a depth of 300 to 700 m across the Pacific Ocean and that the concentration of ^{239}Pu in this stratum has not changed by more than 20% over a period of 5 yr (Bowen, 1977). This behavior can be explained by assuming a rapid transfer of plutonium to about 400 m by biogenic debris (e.g., fecal pellets) (Beasley and Cross, this volume) where the plutonium returns to a soluble species that can migrate upward or downward by diffusion. Experiments in the Irish Sea have shown that plutonium is in solution predominantly as Pu(VI) and on particles as Pu(III) + Pu(IV) (Nelson and Lovett, 1978). The rapid movement of plutonium to 700 m may be associated with particles that sink to that depth where they dissolve and the plutonium reoxidizes to Pu(VI). The higher ^{241}Am/^{239,240}Pu ratio from 1000 m relative to that in surface waters supports this hypothesis. Since the K_D for Pu(VI) is about 1000 times lower than that for Pu(III), Pu(IV), or by inference Am(III), any ^{241}Am that is released would be preferentially taken up by any remaining particles.

Plutonium in oceans occurs in solution over a wide range of concentrations (Table 10). A more surprising result is that distribution coefficients between water and suspended sediments are very similar to those in the Great Lakes and elsewhere.

The distribution of ^{239,240}Pu and ^{238}Pu in waters and sediments of Enewetak Atoll has been studied in detail (Noshkin, this volume). In 1976 the total inventory of ^{239,240}Pu in water and sediments was 1.24 and 249 Ci, respectively. The K_D for plutonium in these sediments has been independently measured in the laboratory as 1.8×10^5. A simple model can be constructed to predict the average concentration of plutonium in the lagoon by assuming this equilibrium constant. This model predicts the concentration of plutonium to be 32 fCi/liter. The average concentration measured in 1976 was 16 fCi/liter. Furthermore, there is no indication of preferential dissolution of ^{238}Pu in this lagoon because the isotopic ratios of ^{239,240}Pu and ^{238}Pu are identical in water and sediments. Similar K_D values for sediments from the Irish Sea and Enewetak Atoll suggest that similar chemical reactions are occurring in all oceans. These results are

TABLE 9   Studies of Transuranic Elements in Marine Environments

Location	Aquatic system	Source and nuclides studied	Inventory	Group responsible
Oceans	Atlantic	Atmospheric fallout: ^{239,240}Pu, ^{238}Pu, ^{241}Am, and ^{241}Pu	~150 kCi	V. T. Bowen, Woods Hole Oceanographic Institute
	Pacific			Y. Miyake, Geochemistry Research Association, Japan
	Pacific			T. R. Folsom, Scripps Oceanographic Institute
	All			E. D. Goldberg, Scripps Oceanographic Institute
Coastal zones of USA	Hudson River	Atmospheric fallout: ^{239,240}Pu, ^{238}Pu, ^{241}Am, and ^{241}Pu	Annual transport, about 6 mCi	H. J. Simpson, Columbia University
	Savannah River	Effluents from processing plants: ^{238,239,240,241}Pu and ^{241}Am	Fallout + 0.3 Ci from SRP	D. W. Hayes, Savannah River Laboratory (du Pont)
India	Trombay	Effluents from processing plants: ^{238,239,240,241}Pu and ^{241}Am	?	K. C. Pillai, Bhabha Atomic Research Center
France	English Channel Cap de la Hague	Effluents from processing plants: ^{238,239,240,241}Pu and ^{241}Am	?	CEA, Centre de la Hague, France
United Kingdom	Irish Sea (Windscale)	Effluents from processing plants: ^{238,239,240,241}Pu and ^{241}Am	Has been up to 100 Ci per month for plutonium and >200 Ci per month for americium	Fisheries Radiobiological Laboratory, Lowestoft, England; Argonne National Laboratory
United Kingdom and West Germany	North Sea	Atmospheric fallout and reprocessing wastes from Irish Sea and Cap de la Hague	In 1974	
	Mediterranean	Atmospheric fallout		International Atomic Energy Agency, Monaco; Woods Hole Oceanographic Institute
Test sites at Enewetak		Debris from weapons testing and immediate fallout: ^{238,239,240,241}Pu, ^{241}Am, and possibly ^{237}Np and the higher actinides	1200 Ci ^{239,240}Pu, 475 Ci ^{241}Am	Lawrence Livermore Laboratory
Bikini			1470 Ci ^{239,240}Pu, 1140 Ci ^{241}Am	University of Washington

TABLE 10 Behavior of Plutonium in Various Marine Ecosystems

| Ecosystem | ^{239,240}Pu | | | ^{238}Pu | | ^{241}Am | Reference |
	Water, fCi/liter	Surface sediments, fCi/g	K_D ($\times 10^{-4}$)	Water, fCi/liter	Surface sediments, fCi/g	Surface sediments, fCi/g	
Hudson River	0.25–1.18						Simpson, Trier, and Olsen, this volume
Savannah River (southern rivers)	0.24–2.4	14–100	4.2–41.2				Hayes and Horton, this volume
Trombay	4.0–20.0	400–2940	4.8–13.0				Pillai and Mathew, 1976
Irish Sea							
Distance from Windscale							Hetherington, Jefferies, and Lovett, 1975
Outfall, km							
1	460	36*	7.0			56*	Hetherington et al., 1976
4.5	370	80*	22.0			49–113*	
9	310–1020	9–29*	2.0–9.0			15–47*	
75	35	0.22*	0.6				
110	30	0.60*	2.0			0–7*	
Enewetak	16–35† 2–75	492†	2–170*	4.7–3.6	16†	81†	Noshkin, 1972; this volume
Bikini	20–100 16–500‡ (bottom)	4–110*			0.05–2.1*	1–80*	Schell, Lowman, and Marshall, this volume

*pCi/g.
†Average values for 1972 and 1976, respectively.
‡It is not clear as to whether these samples were filtered.

in marked contrast to the data for Bikini Atoll (Schell, Lowman, and Marshall, this volume) where preferential mobilization of ^{238}Pu was inferred from increased ratios of ^{238}Pu to ^{239,240}Pu between nonfilterable and filterable fractions from lagoon water. These findings were postulated to result from recoil damage from high-specific-activity ^{238}Pu, which caused increased solubility. However, such an explanation must assume that ^{239,240}Pu and ^{238}Pu are in separate particles originally.

There is little evidence to suggest that differing sources of transuranic elements affect their chemical properties when the elements are moderately well dispersed in aquatic systems. Transuranic elements are soluble, to a limited extent, in both freshwater and marine systems and are therefore available for transfer across biological membranes. Plutonium apparently behaves similarly in oceans and in the Great Lakes, as shown by values of K_D and chemical speciation. These systems can be considered oligotrophic with their chemical properties largely controlled by their respective carbonate cycles. Hence the similarities in values of K_D and ratios of Pu(VI)/Pu(IV) are expected.

Because the pH of the ocean is well buffered, plutonium apparently cannot exist except as Pu(III) or Pu(VI) in solution in the water column or as Pu(IV) in sediments if the relationships shown in Fig. 4 hold. However, in freshwater lakes large variations in composition are possible, and the pH can be relatively low (about 4). Under these conditions dramatic changes in concentrations of plutonium are observed and can be explained by the presence of Pu(III) or Pu(IV) as complexes.

Environmental studies show the danger of using the results of laboratory experiments with moderately concentrated solutions ($10^{-6}M$) to predict the behavior of plutonium in the environment, where the maximum observed concentration has not exceeded $10^{-13}M$. Somewhere within the concentration range of 10^{-13} to $10^{-6}M$, plutonium ceases to exhibit the properties of simple ions, and the possible formation of polymeric species must be considered.

Transport

Terrestrial

Most environmental plutonium exists in a strongly adsorbed state on surface soils. Hence most investigators have concluded that the transport of this element, at least over the last 30 yr, has been governed by processes regulating the distribution and transport of soil (Essington et al., 1976; Hakonson, 1975; Hakonson, Nyhan, and Purtymun, 1976; Hakonson and Nyhan, this volume; Hayes and Horton, this volume; Romney and Wallace, 1977; and Sprugel and Bartelt, 1978). In natural systems soil-erosion processes are mainly driven by wind and water.

Wind Erosion. Wind transport of plutonium in soil can be documented anywhere that appropriate soil, vegetation, and climatic conditions exist. These conditions exist when soil is loose, dry, and of optimum particle size; the soil surface is relatively smooth; vegetation cover is sparse; and winds are sufficiently strong to initiate soil movement (Beasley, 1972).

Wind redistributes plutonium in soil, as inferred from sampling of contaminated sites (Little, 1976; this volume; Markham, Puphal, and Filer, 1978; Romney and Wallace, 1977) and from studies focused specifically on wind transport of plutonium (Gallegos, 1978; Sehmel, 1978; Anspaugh, Shinn, and Wilson, 1974; Anspaugh, Shinn, and Phelps, 1974; 1975). These observations and field studies, primarily in arid regions, imply that wind transport of soil is highly seasonal and is relatively more important in dry, sparsely vegetated areas than in mesic, heavily vegetated areas.

In the arid western United States, wind erosion of soil occurs primarily in the spring and late summer months, coinciding with periods of high wind and low surface soil moisture. Studies in the humid southeast United States suggest that wind is a minor cause of transport of plutonium in soil (Dahlman, Bondietti, and Eyman, 1976) because of the low incidence of high winds and the heavy cover of vegetation.

Soil particle sizes and plutonium concentrations in soil affect the importance of wind as a plutonium transport vector. Plutonium concentrations of various soil size fractions can differ by several orders of magnitude and, depending on source characteristics, are generally highest in the smaller size fractions (Nyhan, Miera, and Neher, 1976; Nyhan, Miera, and Peters, 1976; Tamura, 1975; Little and Whicker, 1977). Furthermore, wind preferentially moves certain sizes of soil particles, depending on the physical characteristics of soil, the wind speed, and the soil moisture (Beasley, 1972). The relationship of some of these factors to plutonium transport by wind is illustrated for a 1-month sampling period at two locations in the fallout zone at Trinity Site, New Mexico, in Table 11. Within 1 km of ground zero, very little of the plutonium activity was present in

TABLE 11 Mass and Plutonium Content of Dust and Soil Samples from Two Locations in the Trinity Fallout Zone for the Period 7-14-76 to 8-10-76

Distance from crater, km	Sample mass, g	Dust*		Soil	
		% mass $<53\ \mu$m	% Pu in $<53\ \mu$m	% mass $<53\ \mu$m	% Pu in $<53\ \mu$m
1	4.02	7	2	9	0.8
45	2.10	54	45	36	73

*Saltated dust collected in the zone 0 to 15 cm above the ground surface with accumulative Bagnold dust sampler.

the silt–clay ($<53\ \mu$m) size fraction of dust or soil samples. However, about 45 km from the crater along the fallout pathway, a much higher percentage of the plutonium in dust and soil samples was present in this size fraction. These differences demonstrate the potential importance of the relationship of soil particle sizes to plutonium concentration in understanding plutonium transport within ecosystems. Silt–clay particles may be transported farther and are more likely to remain attached to biological surfaces than are larger size particles (Romney and Wallace, 1977; Romney et al., 1963; Little, 1976; this volume).

Plutonium suspended by wind can be redeposited on soil or intercepted by biological surfaces. Redeposition of plutonium on soils can lead to major changes in the distribution of the element within an ecosystem, as shown by work at the Nevada Test Site (Romney et al., 1963; Essington et al., 1976). These studies showed that plutonium associated with blown sand accumulates around the bases of shrubs where many of the desert life processes function (Romney and Wallace, 1977). Our understanding of soil plutonium in other areas and climates is limited. However, the accumulation of plutonium around vegetation clumps (or other natural or man-made obstacles) may be common to all regions where wind is a major soil-erosion agent.

The deposition of plutonium on biological surfaces can be inferred directly from concentration ratios

$$CR = \frac{\text{Transuranic concentration in receptor}}{\text{Transuranic concentration in donor}}$$

based on field data (Hakonson, 1975; Little, 1976; this volume; Romney and Wallace, 1977; Dahlman, Bondietti, and Eyman, 1976). Such ratios are much higher than those derived from greenhouse studies (Francis, 1973; Price, 1973; Schulz, 1977) and imply that root uptake cannot account for concentrations measured in field samples. Physical processes are evidently more important than chemical processes in transporting plutonium to vegetation.

Wind is apparently more important in contaminating vegetation in dry regions than it is in humid regions, as shown by plutonium CR's. These ratios decrease from about 10^{-1} in United States deserts (Hakonson, 1975; Little, 1976; Romney and Wallace, 1977) to 10^{-3} in mesic ecosystems of the southeast (Dahlman, Bondietti, and Eyman, 1976).

Additional observations implicate wind-driven processes in contaminating vegetation with plutonium. For example, plutonium concentrations are inversely correlated with height of plants above the ground (Hakonson and Johnson, 1974; Dahlman, Garten, and Hakonson, this volume). Thus low-growth forms, such as grasses, forbs, lichens, and mosses, generally exhibit higher plutonium concentrations than shrubs or trees. This pattern is consistent with soil flux—height relationships, which show that most of the soil mass transported by wind is within 1 m of the ground surface (Sehmel, 1978; Gillette, Blifford, and Fenster, 1972; Phelps and Anspaugh, 1977).

Water Erosion.   Physical transport of transuranic elements by raindrop splash or surface runoff has received little attention in terrestrial ecosystems, although these processes certainly occur (Romney and Wallace, 1977; Hakonson, Nyhan, and Purtymun, 1976; Sprugel and Bartelt, 1978; Muller, Sprugel, and Kohn, 1978). For example, Beasley (1972) has shown that a 5-cm rainstorm causes disaggregation of 200 metric tons of soil per hectare by raindrop splash and surface-water runoff. The importance of soil splash up from raindrops in contaminating vegetation is unknown, although the process certainly occurs.

In certain cases (e.g., intermittent streams) water movement of sediments may be the dominant mechanism of plutonium transport (Hakonson, Nyhan, and Purtymun, 1976). The process is primarily the physical transport of plutonium sorbed on soil particles rather than movement of dissolved plutonium (Hakonson, Nyhan, and Purtymun, 1976; Muller, Sprugel, and Kohn, 1978).

The relationship of plutonium concentration to soil particle size is also important in assessing transport because water movement preferentially sorts soil according to particle size (Hakonson, Nyhan, and Purtymun, 1976; Muller, Sprugel, and Kohn, 1978). For example, as water velocity decreases, successively smaller soil size fractions remain in suspension. Hence silt—clay fractions, which usually contain higher concentrations of plutonium, are probably carried greater distances than larger size fractions.

Water transport of soil across landscapes redeposits plutonium within local watershed soils and stream channel sediments (Hakonson, Nyhan, and Purtymun, 1976), downstream ponds (Muller, Sprugel, and Kohn, 1978), rivers (Hayes and Horton, this volume; Sprugel and Bartelt, 1978), lakes, and oceans. Studies of intermittent streams at Los

Alamos showed that stream-bank soils are a repository of effluent plutonium and serve as a source of the element to stream-bank biota (Hakonson et al., 1979).

It has been estimated that rivers contribute about 150 to 500 Ci/yr of ^{239}Pu to oceans (Simpson, Trier, and Olsen, this volume).

Studies of the Savannah River showed that about 0.005% of the total plutonium in the watershed is lost to the coastal zone annually (Hayes and Horton, this volume). In contrast, the annual loss for a typical midwestern river, the Miami (Muller, Sprugel, and Kohn, 1978; Sprugel and Bartelt, 1978), and for the Hudson (Simpson, Trier, and Olsen, this volume) is an order of magnitude greater. These results indicate a residence time of between 10^3 and 2×10^4 yr. In each river the majority of the plutonium is transported with suspended matter. The differences in loss rates are probably related to differences in watershed morphology. A major fraction of the plutonium transported in the Savannah River is probably held up in impoundments at the upper reaches.

Biotic Activity and Mechanical Disturbance.   Plutonium concentrations in animal tissues demonstrate the dominance of physical processes in transporting plutonium to animals in natural ecosystems. In addition to the gastrointestinal tract, highest concentrations of plutonium are associated with the pelt and, to a lesser degree, lung tissue as a result of interaction with plutonium on soil particles (Hakonson, 1975; Little, 1976; this volume; Hakonson and Nyhan, this volume; Bradley, Moor, and Naegle, 1977).

Work at the Nevada Test Site with cattle (Smith, 1977) shows that considerable amounts of soil are routinely ingested by grazing herbivores. Cattle ingest several hundred grams of soil daily under normal range conditions. Transport of plutonium occurs when these animals move to other areas with subsequent deposition through defecation and/or death of the animal (Arthur and Alldredge, 1979). The amount of plutonium transported in this manner is considered small. Foraging by herbivores, such as cattle, deer, rodents, and insects, may subject a substantial amount of the plutonium in soil to digestion processes over prolonged grazing histories. Whether the chemical form of ingested plutonium is altered as it passes through the gastrointestinal tracts is not known, but in vitro studies indicate changes in solubility in an artificial rumen, simulated abomasal, and intestinal fluid procedure (Barth, 1977).

Mechanical disturbances, such as soil tilling and construction, can transport large amounts of plutonium on a local scale. Plowing enhances mixing of plutonium with the soil profile and also can cause large increases in airborne soil particles. Soil tilling activities at the Savannah River Plant increased local air concentration of plutonium 100-fold (Milham et al., 1976). Mechanical harvesting of agricultural crops also results in surface contamination of edible grains (Adriano et al., 1975).

Existing data pertaining to plutonium distribution in natural ecosystems suggest that physical processes driven by wind will become less important as plutonium migrates into the soil profile. Contemporary data from fallout areas contaminated in 1945 (Hakonson and Nyhan, this volume) show that less than 50% of the soil column inventories of plutonium occurs in the surface 2.5 cm of soil. Similar relationships have been observed in a Los Alamos intermittent stream initially contaminated in 1963 (Hakonson et al., 1979) and in the grassland study site at Rocky Flats (Little, this volume). A change in physical transport of plutonium would probably change the relative importance of chemical and biological transport processes. For example, long-term cropping studies of Romney and Davis (1972) and Schreckhise and Cline (this volume) suggest that migration of plutonium into soil may create conditions more favorable for uptake by deeper rooted plants.

Plant Uptake. Transuranic elements in terrestrial environments can enter plants by foliar absorption and root uptake. The route of entry into plants will depend on the nature of the source; climatic conditions affecting deposition, retention, and chemistry of particles on leaf surfaces; the foliar surface area exposed; and soil conditions affecting resuspension and solubility.

The root is the major ion-absorbing organ of the plant. Somatic cells in the leaf possess the same potential for absorption; however, they are protected by a waxy cuticle. Foliar absorption is an efficient route of entry for nutrients (Bukovac and Wittwer, 1957; Wittwer, Bukovac, and Tukey, 1963), fission products, and activation products (Tukey, Wittwer, and Bukovac, 1961; Athalye and Mistry, 1972). Foliar absorption of ^{238}Pu and ^{241}Am can occur and is dependent on chemical form and environmental conditions with up to 10^{-5} and 10^{-4} of the foliar deposits absorbed and translocated to seeds and roots (Cataldo and Vaughan, this volume; Cataldo, Garland, and Wildung, 1978). About one one-millionth of the plutonium applied as oxide was absorbed by leaves; availability was dependent on particle size. The availability of americium applied as the oxide was two to five times as great as that of the less soluble plutonium oxide at comparable particle size. Thus foliar uptake appeared to be related to transuranic solubility.

Ion uptake by plant roots is apparently a metabolically mediated process in which ions are transported across the cell membrane. The process is concentration dependent over a broad range (10^{-6} to $10^{-2}M$), exhibits a degree of ion selectivity, and may allow for accumulation of ions against a concentration gradient (Nissen, 1973). Although the transport process is selective, plants accumulate nonnutrient ions. Processes leading to the delivery of soluble transuranic species to root membranes have been described. It is critical to determine if discrimination occurs at the membrane level because this would limit transuranic uptake by plants and incorporation into food chains.

Because of the relatively low uptake of plutonium and americium from soil by plants (CR values of 10^{-8} to 10^{-3}), it has been generally assumed that marked discrimination occurs. Evidence is increasing that solubility in soil rather than plant discrimination at the membrane level limits transuranic uptake by plants. As expected from their respective aqueous chemistries, transuranic elements are sorbed by soil in the order Pu > Am ≈ Cm ≫ Np (Table 12). Uptake of these elements is apparently inversely related to soil sorption. The addition of complexing agents, which markedly increases transuranic solubility in soil (Wildung and Garland, 1975), also increases plant uptake (10- to 10,000-fold) (Energy Research and Development Administration, 1976). Thus indirect evidence supports soil solubility as the primary factor governing transuranic availability to plants.

Experiments with plants grown in hydroponic solutions containing plutonium aid in distinguishing soil sorption and plant root discrimination when uptake is compared with uptake by plants grown in soils. When hydroponically grown soybeans *(Glycine max)* were placed in μM ^{238}Pu–DTPA solutions and permitted to accumulate plutonium for up to 49 hr (Wildung et al., 1977), CR's [(μCi/g dry plant) per (μCi/ml nutrient solution)] for shoot tissues were 6×10^{-3} and 3×10^{-1} after 1 and 24 hr, respectively. The Pu–DTPA complex supplied in the growth medium was not detected in the exudates. Similarly, leaves of bush beans *(Phaseolus vulgaris)* exhibited CR's in nutrient solution of 0.8 and 5.1 for Pu(IV) and Pu(VI), respectively. Thus plants can accumulate soluble plutonium effectively; much of the apparent discrimination found in soil–plant studies results from the effect of soil sorption in reducing the quantity of soluble plutonium available to the plant.

**TABLE 12   Distribution Coefficients (K_D) for Soil
Sorption and Relative Plant Uptake of the
Transuranic Elements**

Element	Log K_D*	Relative uptake†
Plutonium(IV)	4.0	4
Americium(III)	1.8	35
Curium(III)	2	39
Neptunium(V)	0	3×10^4

*Plutonium (Prout, 1958); americium and neptunium (Routson, Jansen, and Robinson, 1977); curium (Routson, 1978, personal communication).

†pCi/g barley per mCi/container (~0.1 mCi/container); transuranic element may not have been uniformly distributed in container (Schreckhise and Cline, this volume).

Plutonium is probably transported across biological membranes in the Pu(IV) state, particularly in plant roots (Fig. 4). Plutonium(IV) has been identified in the plant xylem of plants grown in a solution containing predominantly Pu(VI) (Delaney and Francis, 1978).

Once in the root plutonium is probably translocated downward in the root and upward in the xylem stream to shoot tissues as Pu(IV). Simple organic acids typical of microbial and plant metabolites quantitatively reduce Pu(VI) to Pu(IV) (Wildung et al., 1977). Plutonium(IV) dominates in the plant xylem regardless of the oxidation state supplied in nutrient solution (Delaney and Francis, 1978). The low solubility of Pu(IV) limits translocation in plants unless complexed, and several anionic and cationic complexes of plutonium have been identified in the xylem stream of plants supplied Pu(IV) and Pu(VI) (Wildung et al., 1977). During growth a fraction of the plutonium is lost from the root with other inorganic and organic exudates and by decomposition of sloughed cells (Fig. 4). The plutonium associated with this material in the rhizosphere may be subject to recycling into the plant, subsequent modification, leaching, and diffusion.

Translocation in plants can also serve as a primary factor governing entrance of transuranic nuclides into foodstuffs. Plutonium was mobile in barley and soybeans but was not uniformly distributed in the plant (Garland et al., 1974). In general, concentrations of plutonium in the leaves of soybeans were 5 to 10 times as high as those in stems. The lowest plutonium concentrations were observed in barley and soybean seeds, which minimized the amount of plutonium ingested with these edible tissues.

Animal Uptake.   The gut absorption of plutonium by mammals requires the presence of soluble forms, and solubility is governed by chemical reactions similar to those previously discussed (hydrolysis and complexation). When large amounts of plutonium (>1 mg) are introduced into the gut as Pu(VI) in the absence of foodstuffs (starved animals) and in the presence of large excesses of a holding oxidant, Pu(VI) is absorbed in significant quantities (Weeks et al., 1956). In Chicago plutonium is present largely as Pu(VI) in chlorinated drinking water (Larsen and Oldham, 1978). However, the reducing potential in the gut seems to be sufficient to reduce very low concentrations of Pu(VI) to Pu(IV), which would limit uptake in the absence of the holding oxidant. This conversion would be particularly pronounced if reducing substances, such as food residues, were present (Sullivan et al., 1979). In contrast, when plutonium is present as Pu(IV) complexes, such as in microbial and plant tissues, preliminary studies indicate that gut absorption is somewhat increased (Sullivan and Garland, 1977; Ballou et al., 1978). This probably occurs because of increased Pu(IV) solubility in the gut and transport across the gut wall as Pu(IV) ion or a low-molecular-weight plutonium complex. For elements that are not readily hydrolyzed in the pH range of the digestive tract, e.g., Np(V), incorporation in plant tissues may reduce gut transport relative to direct absorption from administered solutions (Sullivan, 1979).

The absorption of transuranic elements other than plutonium, under similar conditions in the gut, may be related to solubility following principles outlined in the previous section. If this is true, gut absorption in the presence of foodstuffs will follow the order Np(V) > Cm(III) $\simeq$ Am(III) > Pu(IV) $\simeq$ Pu(VI).

Gut absorption has not been studied in terrestrial invertebrates; thus comparisons with the observations reported for marked uptake in marine invertebrates (vide infra) cannot be made.

Terrestrial Food Webs.   The CR is used to assess the degree of bioaccumulation. Extensive reviews of CR data based on greenhouse and field studies are presented elsewhere (Francis, 1973; Schulz, 1977; Energy Research and Development Administration, 1976; Price, 1973). Table 13 summarizes transuranic-element CR's based on laboratory studies, and Table 14 summarizes those based on field studies. These transuranic elements are

TABLE 13   Transuranic-Element Concentration Ratios Based on Experimental Studies

Element	Agricultural crops	Native plants	Reference
^{238,239,240}Pu	10^{-10}–10^{-3}	10^{-5}–10^{-4}	Francis, 1973; Schulz, 1977; Brown, 1976; Price, 1973
^{241}Am	10^{-7}–10^{1}	10^{-5}–10^{-2}	Francis, 1973; Schulz, 1977; Price, 1973
^{244}Cm		10^{-4}–10^{-1}	Schreckhise and Cline, this volume; Price, 1972; 1973
^{237}Np		10^{-2}–10^{-1}	Schreckhise and Cline, this volume; Price, 1972; 1973

TABLE 14   Transuranic-Element Concentration Ratios Based on Field Studies

Element	Agricultural crops	Native plants	Native animals*	Reference
^{238}Pu		10^{-4}–10^{0}		Hakonson and Nyhan, this volume; Dahlman, Bondietti, and Eyman, 1976
^{239,240}Pu	10^{-5}–10^{-1}	10^{-4}–10^{0}	10^{-4}–10^{-1}	Hakonson and Nyhan, this volume; Dahlman, Bondietti, and Eyman, 1976; Little, 1976; Durbin, 1975
^{241}Am	10^{-3}–10^{0}	10^{-3}–10^{0}		Durbin, 1975

*Based on whole-body burdens.

generally not concentrated (i.e., CR < 1) by terrestrial plants and animals. On the basis of laboratory studies, neptunium may be an exception (Schreckhise and Cline, this volume; Price, 1972). However, there are no data from which to judge the behavior of this element under field conditions, particularly in acid soils where Fe(II) would reduce Np(V).

Tables 13 and 14 show that CR's based on greenhouse studies are much lower than those derived from field data. The higher CR's based on field data are likely due to surficial contamination of plants with small soil particles, whereas CR's based on greenhouse studies generally reflect only root uptake.

Agricultural plant species accumulate transuranic elements to about the same degree as native plants. The concentrations of transuranic elements in fruits and grains are 10^{-3} to 10^{-1} times lower than those in vegetative parts (Schulz, 1977).

Field data from a number of study sites containing up to several hundred picocuries per gram of soil show that plutonium transfer to native and domestic animals is also very small (Little, 1976; this volume; Hakonson and Nyhan, this volume; Bradley, Moor, and Naegle, 1977; Smith, 1977). Concentration ratios in internal tissues of rodents are comparable to those observed in internal tissues of vegetation. Concentrations of plutonium in internal tissues (i.e., liver, muscle, and bone) can seldom be measured owing to the low gut availability of this element (Durbin, 1975).

Aquatic Food Webs. Transuranic elements can enter aquatic environments at a number of points in complicated food chains encompassing all trophic levels from microbes to vertebrates. Summaries of trophic-level studies in freshwater and marine environments (Dahlman, Bondietti, and Eyman, 1976; Hetherington et al., 1976) indicate that plutonium CR's relative to water generally decrease at higher trophic levels.

Marine benthic invertebrates and invertebrate predators feeding on them exhibit the highest levels of plutonium in coastal fauna (Noshkin, 1972; Pillai and Mathew, 1976). Although these observations generally correlated with the high fraction of the plutonium inventory found in sediments, experimental studies show that marine invertebrates have remarkably high assimilation efficiencies relative to terrestrial mammals (Beasley and Cross, 1979).

No clear correlation between sources of transuranic elements and marine fish concentrations can be made at this time because of limited data. The evidence from both field and experimental studies shows variations in uptake which can be attributed to the element under study, the chemical species, and the type of fish. Studies with ^{237}Pu show that plaice can absorb plutonium as Pu(VI) by direct uptake from seawater, but absorption across the gut from labeled food or sediment is very low (Pentreath, 1978a). Elasmobranch fish, such as the thornback ray, however, do appear to absorb plutonium across the gut wall relatively easily (Pentreath, 1978b). Environmental observations indicate that americium is relatively more available to plaice than is plutonium (Pentreath and Lovett, 1978).

Except for high CR's for plutonium in phytoplankton relative to water, which appear to result from a surface-adsorption phenomenon (Beasley and Cross, this volume), and observations of a fourfold increase in the concentrations of plutonium in starfish relative to those in the mussels on which they feed (Noshkin et al., 1971), no apparent biomagnification has been observed in aquatic systems (Dahlman, Bondietti, and Eyman, 1976).

Further research should be conducted to establish the possibility of biomagnification of plutonium in invertebrate food chains and to determine the magnitude of the uptake of other transuranic nuclides.

In freshwater systems, such as the Great Lakes, food webs are moderately simple compared to those in oceans. Studies in the Great Lakes (Edgington, Wahlgren, and Marshall, 1976; Bowen, 1976) indicate that, although conspicuous biomagnification of plutonium occurs between water and phytoplankton, there is a net decrease of an order of magnitude for each higher trophic level in the food chain. Results from studies at Hanford U-pond, Rocky Flats, and the Miami River in Ohio are comparable to those from Lake Michigan. The CR's (Table 15) reflect not only biological variation but also variations in the concentration or chemical form of plutonium or americium in the water column.

TABLE 15   Accumulation of Plutonium by Aquatic Organisms Leading to Man (for Fish and Shellfish Given for Muscle Only)

Aquatic organism	Freshwater (concentration ratio)					Marine (concentration ratio)	
	Great Lakes*	Miami River†	Rocky Flats‡	U-pond §	White Oak Lake ¶	Atlantic/ Pacific**	Irish Sea
Phytoplankton							
Mixed	5700			7000		260–3500	
Cladophora	3800	220–2900 3000 1600		2000			
Macrophytes				500–5000			
Zooplankton	350		1400–1700				
Benthic organisms							
Mysis							
Pontoporeia	760–1600						
Mytilus						260–490	2000††
Worms						4000	
Crustacea							
Crayfish			600–1300				
Crab							
Fish							
Benthic	250			600	3		<1‡‡
Planktivores	14–37		ND§§				
Piscivores	1–7				0.04	1–13	
Aquatic birds				1.0			

*Edgington, Wahlgren, and Marshall, 1976.
†Wayman, Bartelt, and Alberts, 1977.
‡Paine, this volume.
§Emery and Klopfer, 1976.
¶Eyman and Trabalka, this volume.
**Noshkin, 1972.
††Hetherington et al., 1976.
§§ND, not detected in flesh.

The transfer to humans seems limited because the transuranic elements are not significantly enriched in fresh edible fish (Edgington, Wahlgren, and Marshall, 1976; Dahlman, Bondietti, and Eyman, 1976; Eyman and Trabalka, 1977; Pentreath and Lovett, 1976; 1978; Pentreath et al., 1979).

Prediction of Long-Term Behavior

The long half-lives of several isotopes of the transuranic elements necessitate the estimation of their behavior and effects over thousands of years. The behavior of transuranic elements over a 30-yr interval may not properly represent behavior over more extended periods. Uncertainties arise principally from effects of physical and biogeochemical processes on the redistribution and form of transuranic elements in the environment and from effects of these changes on biological availability.

Several research approaches have been taken to estimate the long-term behavior of transuranic elements. These include (1) basic studies of environmental influences and mechanisms that may alter distribution and biological availability over time; (2) investigations of the behavior of naturally occurring elements that have been in the environment over geologic time and may exhibit analogous behavior; and (3) investigations of the distribution and behavior of transuranic elements presently in the environment as a result of defense activities. These approaches have developed information highly useful for predictive purposes, but considerable research remains to be done before a reliable model can be developed. It is essential to understand factors influencing the chemical speciation of plutonium in the vicinity of biological membranes prior to uptake and how these chemical changes influence the transfer within organisms and between trophic levels. This will require more refined mechanistic studies (approach 1) using studies of analog elements and plutonium distribution from fallout (approaches 2 and 3) to verify predictions. For example, the behavior of plutonium is largely governed by the chemistry of its lower oxidation states, Pu(III) and Pu(IV). However, Pu(V and VI) may be present in highly oligotrophic lakes. Thus, under conditions in which the valence state controls plutonium chemistry, the behavior of naturally occurring Th(IV) and U(VI) may serve as analogs of Pu(III + IV) and Pu(VI) in tests of predictions with respect to matrix and environmental factors (e.g., pH, Eh, and ionic composition). The results of investigations to define the distribution of plutonium from defense activities (approach 3) can be used in a similar manner.

The complexity of the environmental chemistry of plutonium has required a major basic research effort. Unless the mechanisms responsible for the behavior of plutonium are known, it is not possible to develop or validate predictive models. For example, to determine the validity of certain analog elements of plutonium, one must first determine the predominant plutonium valence states and the conditions under which these valences exist. Only then can comparisons be made with naturally occurring elements with similar valences. Americium and neptunium, which have more than one oxidation state, must be studied in this manner. The chemistry of these elements is less complex than that of plutonium, and more rapid progress can be expected. Curium has only one oxidation state, Cu(III), and analog chemistry should be straightforward.

Perhaps the most important factor limiting our ability to predict the transport of transuranic elements in ecosystems is our knowledge of ecosystem structure and function. Prediction of the behavior of transuranic elements in the environment requires information as to the concentrations of these elements in important ecosystem compartments, the physical size of the compartments, and the processes and rates controlling movements of materials between compartments. Knowledge of ecosystem structure and function is useful in predicting the transport of any insoluble element where physical and biotic transport processes dominate. Quantification of physical and biotic processes controlling soil and sediment transport provides a basis for predicting the behavior of elements tightly bound to them.

Ecological Effects of Transuranic Elements in the Environment

The main purpose of the studies described in this volume is to provide information that can be used to help predict the consequences of transuranium elements in the environment. Such consequences include possible harmful effects on man and other species from current and potential levels of these elements in the biosphere. The prediction of consequences requires detailed knowledge of source terms, environmental transport, biological uptake, and biological effects expected from uptake. This review emphasizes effects that might be expressed in species populations, such as mortality and natality, and resulting perturbations in population density and community composition.

It is clear that effects induced by transuranic elements at the population or community level have not been measured directly because environmental levels have not been sufficient to produce obvious changes. Subtle changes in populations or communities are readily masked by natural variations, and ecologists are ordinarily unable to measure small perturbations and identify their causes. However, indirect calculations and extrapolations to low doses can be used to infer ecological consequences of transuranic elements presently in the environment. The task of predicting ecological impacts of a given level of transuranic elements in a particular environment is not simple. How the contaminant will behave; i.e., how it will be distributed among the various ecosystem components; how this distribution will change with time; and what physical, chemical, and biological factors will affect the distribution, must be understood. There is also the question of doses to critical biological tissues. Most transuranic elements are alpha emitters that exhibit generally heterogeneous distributions in tissues, and this makes the calculation of effective doses difficult. Finally, the relation of effective doses to biologic effects must be understood.

The bank of data from which ecological impacts of the transuranic elements can be predicted is limited. For instance, the relationships of tissue concentrations of transuranic elements to concentrations in soil, air, or water are accurately known for only a few ecosystems. Current data pertain to plutonium and americium; research on other transuranic elements only recently has been initiated. Our experience with transuranic elements in the environment has been too brief to allow us to predict long-term behavior confidently. Another problem is that the microdosimetry of transuranic elements has been well studied in only a few laboratory animals. Finally, predictable dose–effect relationships exist mainly for plutonium in laboratory animals. Thus there is considerable uncertainty as to dose–effect relationships for all the transuranic elements in plants and aquatic organisms.

To measure direct relationships between amounts of transuranic elements and effects would require purposely contaminating ecosystems at levels permitting direct observations of biological effects. In practice, however, this approach is not feasible for ecosystem-scale investigations, and such studies have not been done. Ecosystems have been contaminated with transuranic elements through mishaps or experiments for other

purposes, but the levels have generally been orders of magnitude below those presumably required to cause detectable ecological changes. Aquatic and terrestrial organisms exposed to locally high levels of transuranic elements have been studied, but no evidence of transuranic-related effects deleterious to a population has been reported (Bradley, Moor, and Naegle, 1977).

Numerous investigators have directly assayed plutonium and a few other transuranic elements in tissues of a variety of environmentally exposed aquatic and terrestrial organisms, including humans. In these experiments the tissue burdens and resulting radiation dose rates have generally been less than dose rates experienced during evolutionary time from natural sources of radiation. This is due to the low levels of transuranic elements in the environment and also to their low solubility and biological mobility. At such low levels of radiation exposure, ecological changes would be undetectable. Laboratory studies of a variety of aquatic and terrestrial species have shown that radiation dose rates several orders of magnitude higher than those resulting from natural background sources are necessary to produce gross changes in mortality or natality. This is true even for the more radiosensitive stages of comparatively sensitive organisms.

Although gross ecological effects from transuranic elements are not likely to be demonstrated at the levels hitherto experienced in the environment, there is reason to expect a statistically determined incidence of biological effects, such as tumors and genetic alterations. In the absence of sufficient data to the contrary, a linear dose—effect relationship is generally assumed for cancer induction and genetic mutations at low doses. If this assumption is correct, then any dose, however low, imposes some risk. Since concern for most plants and animals is generally for populations rather than for individuals, modest increases in genetic or somatic effects are not expected to have measurable consequences. A different attitude prevails for humans, however, where there is concern for individual organisms.

Dose rates can be calculated and compared with natural background or with literature on dose—effect relationships. Table 16 lists dose rates calculated from measured tissue concentration of plutonium in a variety of organisms exposed to elevated environmental levels and in humans exposed to fallout. The data from Windscale, Rocky Flats, and the Nevada Test Site apparently represent some of the organisms exposed to the highest doses of transuranic elements studied. Even in those highly localized cases, calculated dose rates are about the same as or less than those for natural background, and measurable population-level changes are not expected (National Academy of Sciences—National Research Council, 1972). Doses to humans exposed to fallout plutonium have been so low that specific biological effects cannot be demonstrated (Thompson and Wachholz, this volume).

Levels and distributions of transuranium nuclides in water, sediments, and selected biota, particularly in locally contaminated freshwater and marine sites, have been examined extensively. However, there are few data pertaining to biologic effects. In fact, there have been no good opportunities to observe effects of transuranic elements in natural aquatic ecosystems. Reported water concentrations of plutonium and other transuranic elements in natural environments have been 1 pCi/liter or less, and dose rates appear to be three to eight orders of magnitude less than dose rates required to produce detectable effects (Templeton, this volume; Till, Kaye, and Trabalka, 1976). Present data suggest that aquatic systems can receive several orders of magnitude more transuranic activity than experienced in the past before ecological changes will be detectable.

TABLE 16   Examples of Dose Rates Calculated from Measured Tissue Concentrations of Plutonium in Environmentally Exposed Organisms

Organism	Tissue	Environment/Location	Dose rate, mrad/day	Reference
Mussels	Viscera	Windscale/Irish Sea	1.6*	Hetherington et al., 1976
Crab	Gill	Windscale/Irish Sea	3.1*	Hetherington et al., 1976
Plaice	Bone	Windscale/Irish Sea	0.04*	Hetherington et al., 1976
	Liver	Windscale/Irish Sea	0.17*	Hetherington et al., 1976
	Kidney	Windscale/Irish Sea	0.15*	Hetherington et al., 1976
Fish	Embryos	Windscale, White Oak Lake, U-pond, Enewetak lagoon, Lake Michigan	10^{-6} to 10^{-3}	Till, Kaye, and Trabalka, 1976; Till and Franks, 1977
Small mammals	Whole	Rocky Flats	1.7	Little, 1976
	Carcass	Nevada Test Site	3.3	Bradley, Moor, and Naegle, 1977
Arthropods	Whole	Rocky Flats	0.9	Little, 1976
Cotton rats	Carcass	Savannah River Plant	0.007	McLendon et al., 1976
Humans	Bone	United States	10^{-4}	McInroy et al., 1977
	Lymph nodes	United States	10^{-3}	McInroy et al., 1977

*Includes dose from ^{241}Am.

Opportunities to observe and quantify ecological changes resulting from transuranic contamination of terrestrial environments have also been extremely limited. Terrestrial ecosystems contaminated with plutonium at levels of 10 to 1000 μCi/m^2 have been examined carefully but without demonstrable effects (Whicker, this volume). Assays of plutonium in plants and animal tissues from such contaminated areas reveal levels of plutonium generally less than 10 pCi/g. Dose rates from such plutonium concentrations are a few millirad per day. Chronic dose rates of at least a few rad per day are generally required to cause detectable ecological changes (Whicker and Fraley, 1974; Turner, 1975). Calculations based on a substantial body of information suggest that man could occupy and derive sustenance from land containing 20 to 200 μCi ^{239}Pu/m^2 without exceeding the nonoccupational maximum permissible dose to the lung or other critical organs (Healy, 1974; Martin and Bloom, 1976). Other calculations suggest that ^{239}Pu levels of 1 to 1000 mCi/m^2 would be required to cause significant mortality in plant and animal populations. Mammals would probably show mortality at lower levels than plants (Whicker, this volume).

Summary

The preceding discussion leads to a number of generalizations that can be summarized as follows:

1. The nature of the source for release to the environment is important in the initial deposition and distribution of transuranic elements. However, as environmental factors, such as erosion, chemical weathering, and biological processes, proceed, the original chemical and physical properties are altered and source influence diminishes.

2. The major repositories of plutonium and americium are soils and sediments.

3. Suspended particles in air and water act as vectors for the physical movement of plutonium and americium, and erosional processes are the principal means of translational movement in the environment.

4. In spite of the large fraction of plutonium and americium residing in soils and sediments, chemical and biological processes produce a very small fraction of soluble species in terrestrial and aquatic environments. These species are incorporated in biological tissue, but the concentrations in biota have not produced demonstrable deleterious radiation effects.

5. An increase or decrease in the soluble fraction of plutonium over long weathering times cannot be demonstrated at this time. However, preliminary observations of naturally occurring analog elements indicate that plant uptake and transfer of plutonium and americium through food chains would not be expected to change appreciably over time.

6. Concentrations of plutonium do not increase from one trophic level to the next in natural food webs except for sorption by phytoplankton and one observation of starfish feeding on mussels.

7. The environmental chemistry of transuranic elements in marine and in oligotrophic freshwater systems is similar in a number of ways. However, significant differences in chemical species exist in many lakes where chemical conditions, such as pH and ligand concentration (both organic and inorganic), may be different.

8. Present levels of transuranium elements in our environment have not produced discernible ecological effects.

Important reservations are implicit in the above generalizations mainly because of insufficient information on fundamental processes and lack of data pertaining to transuranic elements other than plutonium. Three lines of investigation are necessary in future studies:

1. Develop process and dose models as a framework to identify specific research areas where important data are lacking.

2. Expand research related to neptunium, americium, and curium to provide a broader base of information about the environmental behavior of the transuranic elements.

3. Investigate the kinetics of the Pu(IV) → Pu(V and VI) oxidation and the factors controlling this valence distribution since increasing evidence suggests that oxidation mechanisms occur that make plutonium more soluble than predicted in some environmental media.

References

Adriano, D. C., et al., 1975, A Field Study to Determine Pu Contents of Wheat and Soil in a Warm, Humid Area, in *Annual Meeting of the Society of Agronomy,* Knoxville, Tenn., Aug. 24–29, 1975.

Alberts, J. J., R. N. Muller, and K. A. Orlandini, 1976, Particle Size and Chemical Phase Distribution of Plutonium in an Estuarine Sediment, in *Radiological and Environmental Research Division Annual Report, January–December 1976,* ERDA Report ANL-76-88(Pt. 3), pp. 34-36, Argonne National Laboratory, NTIS.

Anspaugh, L. R., J. H. Shinn, and P. L. Phelps, 1975, *Resuspension and Distribution of Plutonium in Soils,* USAEC Report UCRL-76419, University of California Radiation Laboratory, NTIS.

——, J. H. Shinn, and D. W. Wilson, 1974, Evaluation of the Resuspension Pathway Toward Protective Guidelines for Soil Contamination with Radioactivity, in *Population Dose Evaluation and Standards for Man and His Environment,* Symposium Proceedings, Yugoslavia, May 20–24, 1974, pp. 513-524, STI/PUB/375, International Atomic Energy Agency, Vienna.

——, J. H. Shinn, P. L. Phelps, and N. C. Kennedy, 1975, Resuspension and Redistribution of Plutonium in Soils, in Proceedings of the Second Annual Life Sciences Symposium on

Plutonium—Health Implications for Man, Los Alamos, N. Mex., May 22–24, 1974, *Health Phys.*, 29(4): 571-582.

Arthur, W. J., III, and A. W. Alldredge, 1979, Soil Ingestion by Mule Deer in Northcentral Colorado, *J. Range Manage.*, 32(1): 67-71.

Athalye, V. V., and K. B. Mistry, 1972, Foliar Retention, Transport and Leaching of Polonium-210 and Lead-210, *Radiat. Bot.*, 12: 287-290.

Ballou, J. E., K. R. Price, R. A. Gies, and P. G. Doctor, 1978, The Influence of DTPA on the Biological Availability of Transuranics, *Health Phys.*, 34: 445-450.

Bartelt, G. E., C. W. Wayman, S. E. Groves, and J. J. Alberts, 1977, ^{238}Pu and ^{239,240}Pu Distribution in Fish and Invertebrates from the Great Miami River, Ohio, in *Transuranics in Natural Environments*, Symposium Proceedings, Gatlinburg, Tenn., Oct. 5–7, 1976, M. G. White and P. B. Dunaway (Eds.), ERDA Report NVO-178, pp. 517-530, Nevada Operations Office, NTIS.

Barth, J., 1977, Application of the Artificial Rumen and Simulated Bovine Gastrointestinal Fluids Procedure in the Study of the Bioavailability of Transuranics, in *Transuranics in Natural Environments*, Symposium Proceedings, Gatlinburg, Tenn., Oct. 5–7, 1976, M. G. White and P. B. Dunaway (Eds.), ERDA Report NVO-178, pp. 419-433, Nevada Operations Office, NTIS.

Beasley, R. P., 1972, *Erosion and Sediment Pollution Control*, Iowa State University Press, Ames, Iowa.

Bondietti, E. A., 1978, Actinide Elements in Aquatic and Terrestrial Environments, in *Environmental Sciences Division Annual Progress Report for Period Ending Sept. 30, 1977*, ERDA Report ORNL-5365, pp. 58-74, Oak Ridge National Laboratory, NTIS.

——, and F. H. Sweeton, 1977, Transuranic Speciation in the Environment, in *Transuranics in Natural Environments*, Symposium Proceedings, Gatlinburg, Tenn., Oct. 5–7, 1976, M. G. White and P. B. Dunaway (Eds.), ERDA Report NVO-178, pp. 449-476, Nevada Operations Office, NTIS.

——, S. A. Reynolds, and M. H. Shanks, 1976, Interaction of Plutonium with Complexing Substances in Soils and Natural Waters, in *Transuranium Nuclides in the Environment*, Symposium Proceedings, San Francisco, 1975, pp. 273-287, STI/PUB/410, International Atomic Energy Agency, Vienna.

Bowen, V. T., 1976, *Plutonium and Americium Concentration Along Fresh Water Chains of the Great Lakes, U.S.A.*, General Summary of Progress, USAEC Report COO-3568-13, Woods Hole Oceanographic Institution, NTIS.

——, 1977, *Radioelement Studies in the Oceans, Progress Report January–December 1977*, Report COO-3563-71, Woods Hole Oceanographic Institution.

——, H. D. Livingston, and J. C. Burke, 1976, Distribution of Transuranium Nuclides in Sediment and Biota of the North Atlantic Ocean, in *Transuranium Nuclides in the Environment*, Symposium Proceedings, San Francisco, 1975, pp. 107-120, STI/PUB/410, International Atomic Energy Agency, Vienna.

——, K. M. Wong, and V. E. Noshkin, 1971, ^{239}Pu in and over the Atlantic Ocean, *J. Mar. Res.*, 29(1).

Bradley, W. G., K. S. Moor, and S. R. Naegle, 1977, Plutonium and Other Transuranics in Small Vertebrates: A Review, in *Transuranics in Natural Environments*, Symposium Proceedings, Gatlinburg, Tenn., Oct. 5–7, 1976, ERDA Report NVO-178, pp. 385-406, Nevada Operations Office, NTIS.

Brown, K. W., 1976, Americium—Its Behavior in Soil and Plant Systems, U. S. Environmental Protection Agency Ecological Research Series, Report EPA 600/3-76-005, pp. 1-10.

Bukovac, M. J., and S. H. Wittwer, 1957, Absorption and Mobility of Foliar Applied Nutrients, *Plant Physiol.*, 32: 428-435.

Burney, G. A., and R. M. Harbour, 1974, *Radiochemistry of Neptunium*, USAEC Report NAS-NS-3060, National Academy of Sciences–National Research Council, NTIS.

Cataldo, D. A., T. R. Garland, and R. E. Wildung, 1978, Foliar Absorption of Transuranic Elements: Effects of Chemical Form and Environmental Factors, *J. Environ. Qual.* (in preparation).

Cleveland, J. M., 1970, *The Chemistry of Plutonium*, Gordon and Breach, Science Publishers, Inc., New York.

——, 1979, A Critical Review of Plutonium Equilibria of Environmental Concern, in *Chemical Modeling in Aqueous Systems*, E. A. Jenne (Ed.), A.C.S. Symposium Series No. 93, pp. 331-338, American Chemical Society, Washington, D. C.

Connick, R. E., 1949, Oxidation States, Potentials, Equilibria and Oxidation–Reduction Reactions of Plutonium, in *The Actinide Elements,* National Nuclear Energy Series, Div. IV, Vol. 14-A, G. T. Seaborg, J. J. Katz, and W. M. Manning (Eds.), pp. 221-300, McGraw Hill Book Company, New York.

Corey, J. C., and A. L. Boni, 1976, Removal of Pu from Drinking Water by Community Water Treatment Facilities, in *Transuranium Nuclides in the Environment,* Symposium Proceedings, San Francisco, 1975, pp. 401-408, STI/PUB/410, International Atomic Energy Agency, Vienna.

Corley, J. P., D. M. Robertson, and F. P. Brauer, 1971, Plutonium in Surface Soil in the Hanford Plant Environs, in *Proceedings of Environmental Plutonium Symposium,* Los Alamos, N. Mex., Aug. 4–5, 1971, USAEC Report LA-4756, p. 119, Los Alamos Scientific Laboratory, NTIS.

Dahlman, R. C., 1976, *Transuranium Elements in Aquatic and Terrestrial Environments,* Environmental Sciences Division Annual Progress Report, 1976, USAEC Report ORNL-5257, pp. 131-147, Oak Ridge National Laboratory, NTIS.

——, and K. W. McLeod, 1977, Foliar and Root Pathways of Plutonium Contamination of Vegetation, in *Transuranics in Natural Environments,* Symposium Proceedings, Gatlinburg, Tenn., Oct. 5–7, 1976, M. G. White and P. B. Dunaway (Eds.), ERDA Report NVO-178, pp. 303-320, Nevada Operations Office, NTIS.

——, E. A. Bondietti, and L. D. Eyman, 1976, Biological Pathways and Chemical Behavior of Plutonium and Other Actinides in the Environment, in *Actinides in the Environment,* ACS Symposium Series No. 35, pp. 47-80, American Chemical Society.

Delaney, M. S., and C. W. Francis, 1978, The Relative Uptake of ^{237}Pu(IV) and Pu(VI) Oxidation States from Water by Bushbeans, *Health Phys.,* 34: 492-494.

Durbin, P., 1975, Plutonium in Mammals: Influence of Plutonium Chemistry, Route of Administration and Physiological Status of the Animal on Initial Distribution and Long Term Metabolism, *Health Phys.,* 29: 495-510.

Duursma, E. K., and C. J. Bosch, 1970, Theoretical, Experimental and Field Studies Concerning Diffusion of Radioisotopes in Sediments and Suspended Solid Particles of the Sea, Part B, Methods and Experiments, *Neth. J. Sea Res.,* 4: 395-469.

Edgington, D. N., and J. A. Robbins, 1975, The Behavior of Plutonium and Other Long-Lived Radionuclides in Lake Michigan. II. Patterns of Deposition in the Sediments, in *Impacts of Nuclear Releases into the Aquatic Environment,* Symposium Proceedings, Otaniemi, Finland, 1975, pp. 245-260, STI/PUB/406, International Atomic Energy Agency, Vienna.

——, and J. A. Robbins, 1976, Records of Lead Deposition in Lake Michigan Sediments Since 1830, *Environ. Sci. Technol.,* 10: 266-274.

——, M. A. Wahlgren, and J. S. Marshall, 1976, The Behavior of Plutonium in Aquatic Ecosystems: A Summary of Studies on the Great Lakes, in *Environmental Toxicity of Aquatic Radionuclides: Models and Mechanisms,* M. W. Miller and J. N. Stannard (Eds.), pp. 45-80, Ann Arbor Science Publishers, Ann Arbor, Mich.

Emery, R. M., and D. C. Klopfer, 1976, The Distribution of Transuranic Elements in a Freshwater Pond Ecosystem, in *Environmental Toxicity of Aquatic Radionuclides: Models and Mechanisms,* M. W. Miller and J. N. Stannard (Eds.), pp. 269-286, Ann Arbor Science Publishers, Ann Arbor, Mich.

Energy Research and Development Administration, 1976, Environmental Research for Transuranic Elements, in *Workshop on Environmental Research for Transuranic Elements,* Proceedings of the Workshop, Battelle Seattle Research Center, Seattle, Wash., Nov. 12–14, 1975, ERDA Report ERDA-76/134, NTIS.

Essington, E. H., E. B. Fowler, R. O. Gilbert, and L. L. Eberhardt, 1976, Plutonium, Americium, and Uranium Concentration in Nevada Test Site Soil Profiles, in *Transuranium Nuclides in the Environment,* Symposium Proceedings, San Francisco, 1975, pp. 157-173, STI/PUB/410, International Atomic Energy Agency, Vienna.

Eyman, L. D., and J. R. Trabalka, 1977, Distribution Patterns and Transport of Plutonium in Freshwater Environments with Emphasis on Primary Producers, in *Transuranics in Natural Environments,* Symposium Proceedings, Gatlinburg, Tenn., Oct. 5–7, 1976, M. G. White and P. B. Dunaway (Eds.), ERDA Report NVO-178, pp. 477-488, Nevada Operations Office, NTIS.

Francis, C. W., 1973, Plutonium Mobility in Soil and Uptake in Plants, *J. Environ. Qual.,* 2: 67-70.

Gallegos, A. F., 1978, Preliminary Model of Plutonium Transport by Wind at Trinity Site, in *Selected Environmental Plutonium Research Reports of the NAEG,* M. G. White and P. B. Dunaway (Eds.), USDOE Report NVO-192(Vol.2), pp. 681-695, Nevada Operations Office, NTIS.

Garland, T. R., and R. E. Wildung, 1977, Physiochemical Characterization of Mobile Plutonium Species in Soils, in *Biological Implications of Metals in the Environment*, DOE Symposium Series, Richland, Wash., Sept. 29–Oct. 1, 1975, H. Drucker and R. E. Wildung (Eds.), pp. 254-263, CONF-750929, NTIS.

——, R. E. Wildung, A. J. Scott, and A. V. Robinson, 1974, in *Pacific Northwest Laboratory Annual Report for 1974 to the USAEC Division of Biomedical and Environmental Research*, USAEC Report BNWL-1950(Pt. 2), pp. 44-49, Battelle, Pacific Northwest Laboratory, NTIS.

Gillette, D. A., I. H. Blifford, Jr., and C. R. Fenster, 1972, Measurement of Aerosol Size Distributions and Vertical Fluxes of Aerosols on Land Subject to Wind Erosion, *J. Appl. Meteorol.*, 11: 977-987.

Hakonson, T. E., 1975, Environmental Pathways of Plutonium into Terrestrial Plants and Animals, *Health Phys.*, 29: 583-588.

——, and L. J. Johnson, 1974, Distribution of Environmental Plutonium in the Trinity Site Ecosystem After 27 Years, in *Proceedings of the Third International Radiation Protection Association Meeting*, Washington, D.C., Sept. 9–14, 1973, W. S. Snyder (Ed.), USAEC Report CONF-730907, pp. 242-247, International Radiation Protection Association, NTIS.

——, J. W. Nyhan, and W. D. Purtymun, 1976, Accumulation and Transport of Soil Plutonium in Liquid Waste Discharge Areas at Los Alamos, in *Transuranium Nuclides in the Environment*, Symposium Proceedings, San Francisco, 1975, pp. 175-189, STI/PUB/410, International Atomic Energy Agency, Vienna.

——, G. C. White, E. S. Gladney, and M. Dreicer, 1979, Comparative Distribution of Mercury, Cesium[137], and Plutonium in an Intermittent Stream at Los Alamos, *J. Environ. Qual.* (in press).

Hanson, W. C., 1975, Studies of Transuranic Elements in Arctic Ecosystems, in *Radioecology & Energy Resources*, Proceedings of the Fourth National Symposium on Radioecology, Oregon State University, May 12–14, 1975, C. E. Cushing Jr. (Ed.), Ecological Society of America Special Publication Series No. 1, pp. 29-39, Academic Press, Inc., New York.

Hardy, E. P., 1974, Depth Distributions of Global Fallout Sr^{90}, Cs^{137}, and $Pu^{239,240}$ in Sandy Loam Soil, in *Fallout Program Quarterly Summary Report*, June 1, 1974–Sept. 1, 1974, USAEC Report HASL-286, pp. I-2 to I-10, Health and Safety Laboratory, NTIS.

Hayes, D. W., J. H. LeRoy, and F. A. Cross, 1976, Plutonium in Atlantic Coastal Estuaries in the Southeastern United States, in *Transuranium Nuclides in the Environment*, Symposium Proceedings, San Francisco, 1975, pp. 79-88, STI/PUB/410, International Atomic Energy Agency, Vienna.

Healy, J. W., 1974, *Proposed Interim Standard for Plutonium in Soils*, USAEC Report LA-5483-MS, Los Alamos Scientific Laboratory, NTIS.

Hetherington, J. A., 1978, The Uptake of Plutonium Nuclides by Marine Sediments, *Marine Sci. Commun.*, 4(3): 239-274.

——, D. F. Jefferies, and M. B. Lovett, 1975, Some Investigations into the Behavior of Plutonium in the Marine Environment, in *Impacts of Nuclear Releases into the Aquatic Environment*, Symposium Proceedings, Otaniemi, Finland, 1975, pp. 193-212, STI/PUB/406, International Atomic Energy Agency, Vienna.

——, D. F. Jefferies, N. T. Michell, R. J. Pentreath, and D. S. Woodhead, 1976, Environmental and Public Health Consequences of the Controlled Disposal of Transuranic Elements to the Marine Environment, in *Transuranium Nuclides in the Environment*, Symposium Proceedings, San Francisco, 1975, pp. 139-156, STI/PUB/410, International Atomic Energy Agency, Vienna.

Holm, E., and R. B. R. Persson, 1975, Fallout Plutonium in Swedish Reindeer Lichens, *Health Phys.*, 29: 43-51.

Inoui, Y., and O. Tochiyama, 1977, Determination of the Oxidation States of Neptunium at Tracer Concentrations by Adsorption on Silica Gel and Barium Sulfate, *J. Inorg. Nucl. Chem.*, 39: 1443-1447.

Joseph, A. B., P. F. Gustafson, I. R. Russel, E. A. Schubert, H. L. Volchok, and A. Tamplin, 1971, Sources of Radioactivity and Their Characteristics, in *Radioactivity in the Marine Environment*, pp. 6-14, National Academy of Sciences.

Katz, J. J., and G. T. Seaborg, 1957, *The Chemistry of the Actinide Elements*, John Wiley & Sons, Inc., New York.

Keller, C., 1971, *The Chemistry of the Transuranium Elements*, Verlag Chemie GmbH, Weinheim/Bergstr., West Germany.

Koide, M., J. J. Griffin, and E. D. Goldberg, 1975, Records of Plutonium Fallout in Marine and Terrestrial Samples, *J. Geophys. Res.*, 80: 4153-4162.

Kraus, K. A., 1949, Oxidation–Reduction Potentials of Plutonium Couples as a Function of pH, in *The Transuranium Elements,* National Nuclear Energy Series, Div. IV, Vol. 14-B, G. T. Seaborg, J. J. Katz, and W. M. Manning (Eds.), pp. 241 and 267, McGraw Hill Book Company, New York.

Krey, P. W., and E. P. Hardy, 1970, *Plutonium in Soil in and Around Rocky Flats Plant,* USAEC Report HASL-235, Health and Safety Laboratory, NTIS.

Lahav, N., and M. Hochberg, 1976, A Simple Technique for Characterizing the Stability of Metal Chelates in the Soil, *Soil Sci.,* 121: 58-59.

Langmuir, D., 1978, Uranium Solution–Mineral Equilibria at Low Temperatures with Applications to Sedimentary Ore Deposits, *Geochim. Cosmochim. Acta,* 42: 547-570.

Larsen, R. P., and R. D. Oldham, 1978, Plutonium in Drinking Water: Effects of Chlorinations on Its Maximum Permissible Concentration, *Science,* 201: 1008-1009.

Little, C. A., 1976, *Plutonium in a Grassland Ecosystem,* Ph.D. Dissertation, Colorado State University, Fort Collins, Colorado, ERDA Report COO-1156-83, NTIS.

——, and F. W. Whicker, 1977, Plutonium Distribution in Rocky Flats Soil, *Health Phys.,* 34: 451-457.

Livingston, H. L., and V. T. Bowen, 1976, Americium in the Marine Environment—Relationships to Plutonium, in *Environmental Toxicity of Aquatic Radionuclides: Models and Mechanisms,* M. W. Miller and J. N. Stannard (Eds.), pp. 107-130, Ann Arbor Science Publishers, Ann Arbor, Mich.

McInroy, J. F., H. A. Boyd, B. C. Eustler, M. W. Stewart, and G. L. Tietgen, 1977, Determination of Plutonium in Man, in *Biomedical and Environmental Research Program of the LASL Health Division,* DOE Report LA-6898-PR, Los Alamos Scientific Laboratory, NTIS.

McLendon, H. R., O. M. Stewart, A. L. Boni, J. C. Corey, K. W. McLeod, and J. E. Pinder, 1976, Relationships Among Plutonium Contents of Soil Vegetation and Animals Collected on and Adjacent to an Integrated Nuclear Complex in the Humid Southeastern United States of America, in *Transuranium Nuclides in the Environment,* Symposium Proceedings, San Francisco, 1975, pp. 347-361, STI/PUB/410, International Atomic Energy Agency, Vienna.

Meyer, G. L., 1976, Preliminary Data on the Occurrence of Transuranium Nuclides in the Environment at the Radioactive Waste Burial Site, Maxey Flats, Kentucky, in *Transuranium Nuclides in the Environment,* Symposium Proceedings, San Francisco, 1975, pp. 231-270, STI/PUB/410, International Atomic Energy Agency, Vienna.

Milham, R. C., J. F. Schubert, J. R. Watts, A. L. Boni, and J. C. Corey, 1976, Measured Plutonium Resuspension and Resulting Dose from Agricultural Operations on an Old Field at the Savannah River Plant in the Southeastern United States of America, in *Transuranium Nuclides in the Environment,* Symposium Proceedings, San Francisco, 1975, pp. 409-420, STI/PUB/410, International Atomic Energy Agency, Vienna.

Miyake, Y., and Y. Sugimura, 1976, The Plutonium Content of Pacific Ocean Waters, in *Transuranium Nuclides in the Environment,* Symposium Proceedings, San Francisco, 1975, pp. 91-104, STI/PUB/410, International Atomic Energy Agency, Vienna.

Moskvin, A. I., and A. D. Gel'man, 1958, Determination of the Composition and Instability-Constants of Oxalate and Carbonate Complexes of Plutonium (IV), *Z. Neorg. Khim.,* 3(4): 198.

Muller, R. N., 1978, Chemical Characterization of Local and Stratospheric Plutonium in Ohio Soils, *Soil Sci.,* 125(3): 131-136.

——, D. G. Sprugel, and B. Kohn, 1978, Erosional Transport and Deposition of Plutonium and Cesium in Two Small Midwestern Watersheds, *J. Environ. Qual.,* 7(2): 171-174.

National Academy of Sciences–National Research Council, 1972, *The Effects on Populations of Exposure to Low Levels of Ionizing Radiation,* Advisory Committee on the Biological Effects of Ionizing Radiation.

Nelson, D. M., and M. B. Lovett, 1978, The Oxidation State of Plutonium in the Irish Sea, *Nature (London),* 276: 599-601.

Nevissi, A., W. R. Schell, and V. A. Nelson, 1976, Plutonium and Americium in Soils of Bikini Atoll, in *Transuranium Nuclides in the Environment,* Symposium Proceedings, San Francisco, 1975, pp. 691-701, STI/PUB/410, International Atomic Energy Agency.

Nissen, P., 1973, Multiphasic Ion Uptake in Roots, in *Ion Transport in Plants,* pp. 539-553, W. P. Anderson (Ed.), Academic Press, Inc., New York.

Norvell, W. A., 1972, Equilibria of Metal Chelates in Soil Solutions, in *Micronutrients in Agriculture,* J. J. Mortvedt, P. M. Giordano, and W. L. Lindsay (Eds.), pp. 115-138, Soil Science Society of America.

Noshkin, V. E., 1972, Ecological Aspects of Plutonium Dissemination in Aquatic Environments, *Health Phys.*, 22: 537-539.

——, V. T. Bowen, K. M. Wong, and J. C. Burke, 1971, Plutonium in North Atlantic Ocean Organisms: Ecological Relationships, in *Radionuclides in Ecosystems*, Proceedings of the Third National Symposium on Radioecology, Oak Ridge, Tenn., May 10–12, 1971, D. J. Nelson (Ed.), USAEC Report CONF-710501-P2, pp. 681-688, Oak Ridge National Laboratory, NTIS.

——, K. M. Wong, K. March, R. Eagle, G. Holladay, and R. W. Buddemeier, 1976, Plutonium Radionuclides in the Groundwaters at Enewetak Atoll, in *Transuranium Nuclides in the Environment*, Symposium Proceedings, San Francisco, 1975, pp. 517-543, STI/PUB/410, International Atomic Energy Agency, Vienna.

Nyhan, J. W., F. R. Miera, Jr., and R. E. Neher, 1976, Distribution of Plutonium in Trinity Soils After 28 Years, *J. Environ. Qual.*, 5(4): 431-437.

——, F. R. Miera, Jr., and R. J. Peters, 1975, The Distribution of Plutonium and Cesium in Alluvial Soils of the Los Alamos Environs, in *Radioecology and Energy Resources*, Fourth National Symposium on Radioecology, C. E. Cushing, Jr. (Ed.), pp. 49-57, Dowden, Hutchinson and Ross, Inc., Stroudsburg, Pa.

——, F. R. Miera, Jr., and R. J. Peters, 1976, Distribution of Plutonium in Soil Particle Size Fractions in Liquid Effluent-Receiving Areas at Los Alamos, *J. Environ. Qual.*, 5: 50-56.

Pentreath, R. J., 1978a, Plutonium 237 Experiments with the Plaice, *Pleuronectes platessa, Marine Biol.*, 48: 327-335.

——, 1978b, Plutonium 237 Experiments with the Thornback Ray, *Raja clavata, Marine Biol.*, 48: 337-342.

——, and M. B. Lovett, 1976, Occurrence of Plutonium and Americium in Plaice from the Northeastern Irish Sea, *Nature (London)*, 262: 814-816.

——, and M. B. Lovett, 1978, Transuranic Nuclides in Plaice, *Pleuronectes platessa* from the Northeastern Irish Sea, *Marine Biol.*, 48: 19-26.

——, M. B. Lovett, B. R. Harvey, and R. D. Ibbett, 1979, Alpha-Emitting Nuclides in Commercial Fish Species Caught in the Vicinity of Windscale, UK, and Their Radiological Significance to Man, in *Biological Implications of Radionuclides Released from Nuclear Industries*, Symposium Proceedings, Vienna, Mar. 26–30, 1978, International Atomic Energy Agency, Vienna.

Phelps, P. L., and L. R. Anspaugh, 1977, Development of Specialized Instruments and Techniques for Resuspension Studies, in *Transuranics in Natural Environments*, Symposium Proceedings, Gatlinburg, Tenn., Oct. 5–7, 1976, M. G. White and P. B. Dunaway (Eds.), ERDA Report NVO-178, pp. 273-286, Nevada Operations Office, NTIS.

Pillai, K. C., and E. Mathew, 1976, Plutonium in the Aquatic Environment, Its Behavior, Distribution and Significance, in *Transuranium Nuclides in the Environment*, Symposium Proceedings, San Francisco, 1975, pp. 25-45, STI/PUB/410, International Atomic Energy Agency, Vienna.

——, A. C. Smith, and J. R. Folsom, 1964, Plutonium in the Marine Environment, *Nature (London)*, 203: 581-586.

Pinder, J. R., III, M. H. Smith, A. L. Boni, J. C. Corey, and J. H. Horton, 1979, Plutonium Inventories in Two Old-Field Ecosystems in the Vicinity of a Nuclear-Fuel Processing Facility, *Ecology* (in press).

Polzer, W. L., 1971, Solubility of Plutonium in Soil/Water Environments, in *Proceedings of the Rocky Flats Symposium on Safety in Plutonium Handling Facilities*, Apr. 13–16, 1971, USAEC Report CONF-710401, pp. 411-429, NTIS.

Pourbaix, M., 1966, *Atlas of Electrochemical Equilibria*, pp. 198-212, Pergamon Press, Inc., New York, 1977.

Price, K. R., 1972, Uptake of ^{237}Np, ^{239}Pu, ^{241}Am and ^{244}Cm from Soil by Tumbleweed and Cheatgrass, USAEC Report BNWL-1688, pp. 1-14, Battelle, Pacific Northwest Laboratory, NTIS.

——, 1973, A Review of Transuranic Elements in Soil, Plants, and Animals, *J. Environ. Qual.*, 2: 62-66.

Prout, W. E., 1958, Adsorption of Radioactive Waste by Savannah River Plant Soil, *Soil Sci.*, 86: 13-17.

Rai, E., and R. J. Serne, 1977, Plutonium Activities in Soil Solutions and the Stability and Formation of Selected Minerals, *J. Environ. Qual.*, 6(1): 89-95.

Rees, T. F., J. M. Cleveland, and W. C. Gottschall, 1978, Dispersion of Plutonium from Contaminated Pond Sediments, *Environ. Sci. Technol.*, 12(9): 1085-1087.

Relyea, F. J., and D. A. Brown, 1975, The Diffusion of Pu-238 in Aqueous Solutions and Soil Systems, in *Annual Meeting of the Society of Agronomy*, p. 124, Knoxville, Tenn., Aug. 24–30, 1975.

Robbins, J. A., and D. N. Edgington, 1975, Determination of Recent Sedimentation Rates in Lake Michigan Using ^{210}Pb and ^{137}Cs, *Geochim. Cosmochim. Acta*, 39: 285-304.

Romney, E. M., and J. J. Davis, 1972, Ecological Dissemination of Plutonium in the Environment, *Health Phys.*, 21: 551-

——, and A. Wallace, 1977, Plutonium Contamination of Vegetation in Dusty Field Environments, in *Transuranics in Natural Environments*, Symposium Proceedings, Gatlinburg, Tenn., Oct. 5–7, 1976, M. G. White and P. B. Dunaway (Eds.), ERDA Report NVO-178, pp. 287-302, Nevada Operations Office, NTIS.

——, H. M. Mork, and K. H. Larson, 1970, Persistence of Plutonium in Soil, Plants, and Small Mammals, *Health Phys.*, 18: 487-491.

——, A. Wallace, R. O. Gilbert, and J. E. Kinnear, 1976, ^{239,240}Pu and ^{241}Am Contamination of Vegetation in Aged Fall-Out Areas, in *Transuranium Nuclides in the Environment*, Symposium Proceedings, San Francisco, 1975, pp. 479-490, STI/PUB/410, International Atomic Energy Agency, Vienna.

——, R. G. Lindberg, H. A. Hawthorne, B. G. Bystrom, and K. H. Larson, 1963, Contamination of Plant Foliage with Radioactive Fallout, *Ecology*, 44: 343-349.

Routson, R. C., G. Jansen, and A. V. Robinson, 1977, ^{99}Tc, ^{237}Np and ^{241}Am Sorption on Two United States Subsoils from Differing Weathering Intensity Areas, *Health Phys.*, 33: 311-317.

Sanders, S. M., Jr., 1977, Characterization of Airborne Plutonium-Bearing Particles from a Nuclear Fuel Reprocessing Plant, DOE Report DP-1470, Savannah River Laboratory, NTIS.

Schulz, R. K., 1977, Root Uptake of Transuranic Elements, in *Transuranics in Natural Environments*, Symposium Proceedings, Gatlinburg, Tenn., Oct. 5–7, 1976, M. G. White and P. B. Dunaway (Eds.), ERDA Report NVO-178, pp. 321-330, Nevada Operations Office, NTIS.

Sehmel, G. A., 1978, *Transuranic and Tracer Simulant Resuspension*, DOE Report BNWL-SA-6236, Battelle, Pacific Northwest Laboratory, NTIS.

Smith, A. E. (Comp.), 1973, Nuclear Reactivity Evaluation of 216-Z-9 Enclosed Trench, USAEC Report ARH-2915, Atlantic Richfield Hanford Company, NTIS.

Smith, D. D., 1977, Review of Grazing Studies of Plutonium Contaminated Rangelands, in *Transuranics in Natural Environments*, M. G. White and P. B. Dunaway (Eds.), ERDA Report NVO-178, pp. 407-417, Nevada Operations Office, NTIS.

Sprugel, D. G., and G. E. Bartelt, 1978, Erosional Removal of Fallout Plutonium from a Large Midwestern Watershed, *J. Environ. Qual.*, 7(2): 175-177.

Stevenson, F. J., and M. S. Ardakani, 1972, Organic Matter Reactions Involving Micronutrients in Agriculture, J. J. Mortvedt, P. M. Giordano, and W. L. Lindsay (Eds.), pp. 79-114, Soil Science Society of America, Madison, Wisconsin.

Sullivan, M. F., 1979, Absorption of Actinide Elements from the Gastrointestinal Tract of Neonatal Animals, *Health Phys.* (in press).

——, and T. R. Garland, 1977, Gastrointestinal Absorption of Alfalfa-Bound Plutonium-238 by Rats and Guinea Pigs, in *Pacific Northwest Laboratory Annual Report for 1976 to the ERDA Assistant Administrator for Environment and Safety*, ERDA Report BNWL-2100(Pt. 1), Battelle, Pacific Northwest Laboratory, NTIS.

——, J. L. Ryan, L. L. Gorham, and K. M. McFadden, 1979, The Influence of Oxidation State on Maximum Permissible Concentration in Drinking Water, *Radiat. Res.* (in press).

Tamura, T., 1975, Distribution and Chemical Characterization of Plutonium in Soils from the Nevada Test Site, *J. Environ. Qual.*, 4: 350-355.

——, 1976, Physical and Chemical Characteristics of Plutonium in Existing Contaminated Soils and Sediments, in *Transuranium Nuclides in the Environment*, Symposium Proceedings, San Francisco, 1975, pp. 213-230, STI/PUB/410, International Atomic Energy Agency, Vienna.

Till, J. E., and M. L. Franks, 1977, Bioaccumulation, Distribution and Dose of ^{241}Am, ^{244}Am and ^{238}Pu in Developing Fish Embryos, in *Proceedings of the Fourth International Congress of the International Radiation Protection Association*, Paris, Vol. 2, pp. 645-648.

——, S. V. Kaye, and J. R. Trabalka, 1976, *Toxicity of Uranium and Plutonium to the Developing Embryos of Fish*, ERDA Report ORNL-5160, Oak Ridge National Laboratory, NTIS.

Tukey, H. B., S. H. Wittwer, and M. J. Bukovac, 1961, Absorption of Radionuclides by Aboveground Plant Parts and Movement Within the Plant, *J. Agric. Food Chem.*, 9: 106-113.

Turner, F. B., 1975, Effects of Continuous Irradiation on Animal Populations, in *Advances in Radiation Biology*, J. T. Lett and H. Adler (Eds.), Vol. 5, pp. 83-144, Academic Press, Inc., New York.

Wahlgren, M. A., and D. M. Nelson, 1975, Plutonium in the Laurentian Great Lakes: Comparison of Surface Waters, *Verh. Int. Verein. Limnol.*, 19: 317-322.

——, J. J. Alberts, D. M. Nelson, and K. A. Orlandini, 1976, Study of the Behaviour of Transuranics and Possible Chemical Homologues in Lake Michigan Water and Biota, in *Transuranium Nuclides in the Environment,* Symposium Proceedings, San Francisco, 1975, pp. 9-24, STI/PUB/410, International Atomic Energy Agency, Vienna.

——, J. J. Alberts, K. A. Orlandini, and E. T. Kucera, 1977a, A Comparison of the Concentrations of Fallout-Derived Plutonium in a Series of Freshwater Lakes, in *Radiological and Environmental Research Division Annual Report, January—December 1977,* ERDA Report ANL-77-65(Pt. 3), pp. 92-94, Argonne National Laboratory, NTIS.

——, J. J. Alberts, D. M. Nelson, K. A. Orlandini, and E. T. Kucera, 1977b, Study of the Occurrence of Multiple Oxidation States of Plutonium in Natural Water Systems, in *Radiological and Environmental Research Division Annual Report, January—December 1977,* ERDA Report ANL-77-65(Pt. 3), pp. 95-98, Argonne National Laboratory, NTIS.

Wayman, C. W., G. E. Bartelt, and J. J. Alberts, 1977, Distribution of ^{238}Pu and ^{239,240}Pu in Aquatic Macrophytes from a Midwestern Watershed, in *Transuranics in Natural Environments,* Symposium Proceedings, Gatlinburg, Tenn., Oct. 5—7, 1976, M. G. White and P. B. Dunaway (Eds.), ERDA Report NVO-178, pp. 505-516, Nevada Operations Office, NTIS.

Weeks, M. H., et al., 1956, Further Studies on the Gastrointestinal Absorption of Plutonium, *Radiat. Res.*, 4: 339-347.

Whicker, F. W., and L. Fraley, Jr., 1974, Effects of Ionizing Radiation on Terrestrial Plant Communities, in *Advances in Radiation Biology*, J. T. Lett, H. Adler, and M. R. Zelle (Eds.), Vol. 4, pp. 317-366, Academic Press, Inc., New York.

Wildung, R. E., and T. R. Garland, 1975, Relative Solubility of Inorganic and Complexed Forms of Plutonium-238 and Plutonium-239 in Soil, in *Pacific Northwest Laboratory Annual Report for 1974 to the USAEC Division of Biomedical and Environmental Research,* USAEC Report BNWL-1950(Pt. 2), Battelle, Pacific Northwest Laboratory, NTIS.

——, and T. R. Garland, 1977, *The Relationship of Microbial Processes to the Fate and Behavior of Transuranic Elements in Soils, Plants, and Animals,* DOE Report BNWL-2416, Battelle, Pacific Northwest Laboratory, NTIS.

——, T. R. Garland, and D. A. Cataldo, 1979, Environmental Processes Leading to the Presence of Organically-Bound Plutonium in Plant Tissues Consumed by Animals, in *Biological Implications of Radionuclides Released from Nuclear Industries,* Symposium Proceedings, Vienna, Mar. 26—30, 1978, International Atomic Energy Agency, Vienna.

——, R. L. Schmidt, and R. C. Routson, 1977, Phosphorus Status of Eutrophic Lake Sediments as Related to Changes in Limnological Conditions—Phosphorus Mineral Components, *J. Environ. Qual.*, 6: 100-104.

——, T. R. Garland, K. M. McFadden, D. A. Cataldo, H. Drucker, and M. Sullivan, 1977, Transuranic Complexation in Soil and Uptake by Plants, in *Pacific Northwest Laboratory Annual Report for 1976 to the ERDA Assistant Administrator for Environment and Safety,* ERDA Report BNWL-2100(Pt. 2), pp. 4.6-4.10, Battelle, Pacific Northwest Laboratory, NTIS.

Wittwer, S. H., M. J. Bukovac, and H. B. Tukey, 1963, Advances in Foliar Feeding of Plant Nutrients, in *Fertilizer Technology and Usage,* M. H. McVickor, G. L. Bridger, and L. B. Nelson (Eds.), pp. 429-455, American Society of Agronomy.

Woods, M., M. G. Mitchell, and J. C. Sullivan, 1978, Determination of the Stability Quotient for the Formation of a Tris Carbonato Plutonate (VI) Complex in Aqueous Solution, *Inorg. Nucl. Chem. Lett.*, 14: 465-467.

Radiological Assessments, Environmental Monitoring, and Study Design

WAYNE R. HANSEN and DONALD R. ELLE

Studies of the behavior of transuranic elements in the environment form the basic data for applied programs in radiological assessment, environmental monitoring, derivation of radiation-protection standards, and environmental impact statements. This chapter introduces some of the major information requirements of these applications of transuranic research data. Characteristics of the source terms from nuclear activities usually are needed for an analysis of environmental pathways or deployment of monitoring systems. Major inhalation and ingestion pathways are considered in assessments of hazards from transuranics and are discussed from the viewpoint of research needed.

In conducting radiological assessments, writing environmental impact statements, attempting to derive standards, or designing monitoring programs for transuranic elements, one must rely on data from existing studies of transuranic elements in the environment. In each of these types of assessments, man is usually the major receptor considered for a variety of pathways. The data used to estimate the radiological impact of transuranics on man derive from the results of research carried out with a variety of objectives. The objectives may have been limited to the assessment of a specific pathway at a specific site. Data obtained for a particular pathway or portion of a pathway and geographical area often are applied, with modifying assumptions, to other geographical areas for lack of data specific to the area of interest.

The design of environmental monitoring programs for estimating radiological effects on man must include consideration of a large number of factors. The following discussion reiterates some of these factors in study design and analysis for use in radiological assessments. A brief discussion of the nontechnical influences on radiological assessments and thus study design is also included. Statistical considerations and modeling are considered elsewhere but will be referred to as necessary. Hopefully, a statistician will always be included in the design phase of any study or monitoring system.

Prior to the projected growth of the nuclear-power industry, efforts to study the environmental behavior of transuranic elements were centered around dispersal by nuclear weapons testing programs. Early radiological assessments of the behavior of transuranium elements in the environment relied on conservative assumptions owing to the lack of empirical data and concentrated on plutonium, which was the major transuranic element in weapons manufacture and testing. Weapons plutonium is still the major transuranic element available for study. Data are less available or nonexistent for curium, americium, and neptunium, but assessments should still include these elements.

The need for additional data on plutonium dispersal and behavior in the event of an accident with weapons components was recognized early. Examples of early studies designed to provide data for radiological assessments are Operation Plumbbob in 1957 at the Nevada Test Site and Operation Roller Coaster in 1963 at the Tonopah Test Range. Jordan (1971) described the objectives of these tests as being primarily concerned with obtaining data about the dispersion characteristics and biomedical impacts on animals exposed to the airborne plutonium and with evaluating instrumentation and decontamination methods. During the design of the experiments, the personnel involved decided not to attempt measurements of resuspension because of the complex nature of the process. Analysis of the data from these tests provided the experience and guidance required to deal with the dispersion of plutonium from nuclear weapons accidents in Palomares, Spain, and Thule, Greenland. In addition, the test areas in Nevada provided research areas for the ongoing study and evaluation of the long-term environmental behavior of the residual plutonium in a desert ecosystem.

Environmental Assessments, Impact Statements

With the growth in the number of light water reactors (LWR), the projected recycle of plutonium in LWR, and the projected liquid-metal fast breeder reactor (LMFBR), the detailed radiological assessments of transuranics increased to keep pace with planned fuel fabrication and fuel reprocessing facilities. Beginning in 1970 the National Environmental Policy Act (NEPA) of 1969 (U. S. Congress, 1970) required that prior to implementing "major Federal actions significantly affecting the quality of the human environment, a detailed statement" must be prepared which addresses the environmental impact, any adverse environmental effects that cannot be avoided, alternatives, relationships between short-term uses and long-term productivity of the environment, and any irreversible and irretrievable commitments of resources. The Council on Environmental Quality (1976) issued guidelines for the preparation of environmental impact statements in 1976. Actions taken in the past by the U. S. Atomic Energy Commission and presently by the U. S. Nuclear Regulatory Commission and the U. S. Department of Energy have been considered major actions that significantly affect the human environment. Both the environmental impact statement on the LMFBR program by the U. S. Energy Research and Development Administration (1975) and the environmental statement by the U. S. Nuclear Regulatory Commission (1976) on the use of recycle plutonium in mixed-oxide fuel in light-water-cooled reactors (GESMO) contain radiological assessments that estimate the radiation doses from transuranics. In each case estimates of the radiological impacts on man were made from data available at the time of the preparation of the environmental impact statement. In most cases the limited data available required that conservative assumptions and extrapolations be used in predicting the impacts as required by NEPA and the CEQ guidelines. These estimates become decision-making tools and are the subject of debate in hearings or litigation for licensing of facilities.

Generic environmental statements and modeling efforts, such as those carried out by Bloom and Martin (1976), have been based on hypothetical individuals who obtain air, food, and water from the area of maximum transuranic concentration. The estimated radiation dose for such broad studies is usually for transuranics from a postulated source. Existing facilities usually have accumulated some data that describe the source of transuranics. Existing facilities carry out environmental monitoring programs that are designed to detect changes in transuranics and other radionuclides in environmental media, such as air, water, soil, vegetation, and animals, and thus verify the results of emission and effluent monitoring programs from which the radiological impacts from existing facilities are estimated. The design of the surveillance programs and the radiological assessments performed, however, are still heavily dependent on data provided by the studies of transuranics in different ecosystems.

General Aspects of Environmental Monitoring

The general design of networks and of programs for the measurement of radioactive materials in the environment has been described by the International Commission on Radiological Protection (1965) and by the World Health Organization (1968). The International Atomic Energy Agency (1966; 1975) has published two guides for environmental monitoring, and more recently, the National Council on Radiation Protection and Measurements (1976) published a report entitled *Environmental Radia- . tion Measurements*. In addition to the recommendations of scientific bodies such as ICRP and NCRP, more specific guidance for environmental monitoring is provided by government agencies (U.S. Atomic Energy Commission, 1974). The Nuclear Regulatory Commission issues general guidance as regulatory guides. Regulatory Guide 4.5, issued in 1974, for example, deals with the sampling and analytical procedures for plutonium in soil. The regulatory guides are not regulations but represent methods that are acceptable for licensing actions or compliance with regulations for operating facilities. The Department of Energy relies on surveillance programs tailored to specific sites and problems. Its contractors issue annual reports of the methods and results of the surveillance programs. The DOE follows *A Guide for Environmental Radiological Surveillance at ERDA Installations* (Corley et al., 1977), which is based on a state-of-the-art review of environmental monitoring practices. The specific objective of the guide was to develop guidance for achieving comparable, high-quality, environmental monitoring and reporting programs at DOE installations, which encompass a wide variety of nuclear activities, i.e., plutonium production, reactor operation, and research studies. Environmental monitoring systems for all types of facilities have many similarities. Although somewhat specific to reactors, the *Environmental Radioactivity Surveillance Guide* issued by the U. S. Environmental Protection Agency (1972) contains general information on sampling methods and frequencies that can be applied.

Monitoring systems are usually designed to verify that a facility is operating within limits specified as safe by a scientific body. These limits become law where incorporated into state or federal codes, such as *Code of Federal Regulations*, Title 10, Part 20 (*Federal Register*, 1976). Each monitoring program should, prior to deployment, identify the pathways to man for transuranic elements in addition to the normal pathways requiring monitoring by regulation. Figures 1 and 2 are examples of simplified pathway diagrams for the movement of radionuclides to man. Not all the pathways in the diagrams will be present for a given site. For a specific facility and location, however, all pathways should be identified and analyzed for their contribution of transuranic elements to the total radionuclide uptake by man or biota. As pathways are identified and analyzed, the number of pathways requiring routine monitoring will be reduced. In the analysis of pathways, the short-term and long-term aspects of accumulation and movement of transuranics should be kept in mind.

The considerations that are included in the identification and analysis of the pathways are the many aspects of environmental studies described in other chapters of this book, starting with the source term. Characteristics of the source term are important for the selection of monitoring methods and instrumentation. Some of the general characteristics of the transuranics to be considered are

1. Quantity.
2. Rate of release.
3. Chemical form.
4. Physical characteristics, such as particle size distribution or ionic state.
5. The presence of radionuclides.
6. The presence of nonradioactive chemicals.

For routine operation of such facilities as plutonium fabrication, reactor-fuel fabrication, and reactor-fuel reprocessing, the emission and effluent monitoring systems will provide the information about quantities and rates of release. The chemical form may be inferred from the operations being carried out by the facility. The physical characteristics of the emissions or effluents can be identified by a specific study or inferred from the operating parameters for the waste-treatment system. The transuranic isotopic composition is usually well defined by criticality and safeguards calculations or by analysis of actual effluent samples. The presence of other radionuclides, such as fission products, may interfere in monitoring transuranics or may be helpful by serving as tracers from which ratios with respect to transuranics can be determined. The presence of nonradioactive chemicals combined with the transuranics in emission or effluents may alter the original chemical forms introduced into the waste-treatment systems. For normal operation of nuclear facilities handling transuranics, the quantities released usually are small. Monitoring for the transuranics in the environment from normal operations of facilities fabricating plutonium metal, heat sources, or reactor fuel or reprocessing reactor fuel become oriented to long-term buildups. The source-term parameters for unexpected or

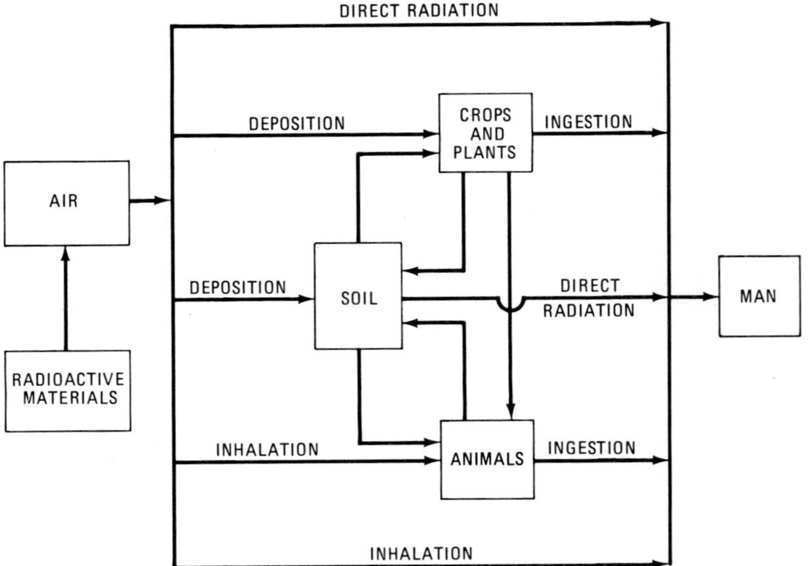

Fig. 1 Simplified pathways between radioactive materials released to atmosphere and man. [From International Commission on Radiological Protection (1965).]

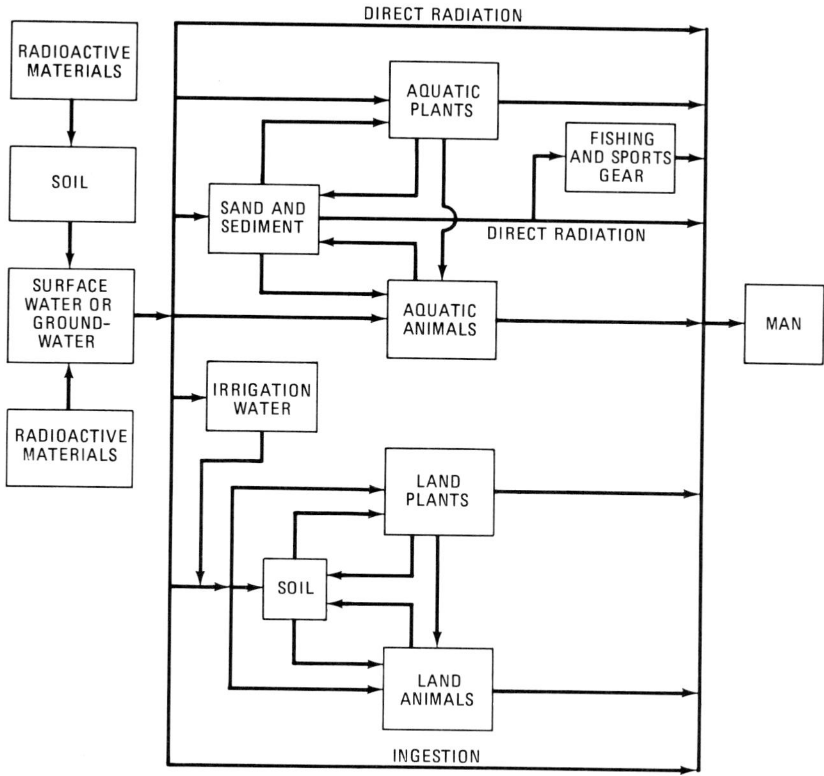

Fig. 2 Simplified pathways between radioactive materials released to groundwater or surface water (including oceans) and man. [From International Commission on Radiological Protection (1965).]

sudden releases, such as may occur from accidents, are identified after an occurrence. Although facilities incorporate engineered protection against such releases, environmental surveillance programs must be prepared to trace the movement of transuranics released to the environment after the initial assessments of the emission or effluent monitoring results. Existing areas of transuranics in the environment also present a challenge to the design of an adequate monitoring system. Once the transuranics are deposited in the environment, however, the pathway considerations for routine and accidental sources are much the same at a given site. To predict the radiological impact from areas of potential future contamination requires a knowledge of the parameters used to estimate the dose from transuranic pathways.

The methods for calculating dose use ICRP models that predict the metabolic fate of radionuclides on the basis of the chemical and physical characteristics of the mode of intake. The parameters needed for dose calculations should be considered during the analysis of pathways for the design of monitoring systems so that the number of assumptions needed for the dose estimations can be minimized. Details of the dose-estimation methods and consequences are discussed in many other publications. At the risk of being redundant, some of the methods are discussed below in relation to pathway information.

Pathways and Inhalation

The inhalation pathways for a number of exposure modes for man have been identified by ICRP. In addition to the direct inhalation of airborne transuranics released from an operating facility, other secondary pathways are possible and have been discussed by Healy (1974). These secondary pathways are the primary inhalation pathways from existing areas of transuranics in the environment. Resuspension by wind is discussed in another chapter. Other means of generating airborne transuranics, such as agricultural activities (including home gardening), carriage into homes on clothes and pets, children's play, and movement by vehicular activity, need further quantification.

Estimates of the doses from inhalation of the different airborne transuranics use the ICRP II lung model (International Commission on Radiological Protection, 1960), which categorizes the inhaled material as soluble or insoluble. The Task Group on Lung Dynamics model (International Commission on Radiological Protection, 1966) has three categories· of solubility. Additionally, the deposition of particulates in three respiratory regions is dependent on particle size in the Task Group model. Residence time in the lung is dependent on solubility classification in both models. After transuranics have resided in the environment for months or years, the solubility and particle sizes may change. Data are needed which evaluate the changes of solubility and particle-size distribution with time and weathering. Although some work has been started for plutonium, more long-term studies of the solubility and particle-size changes are needed. Data for neptunium, americium, and curium is sparse to nonexistent. The derivation or collection of some of this information could be incorporated into the design of present environmental monitoring programs. More detail about the association of the transuranics with soil particles, particle-size distribution, and chemical changes with time and weathering is gathered in separate projects that are usually beyond the scope and budget of monitoring programs.

Pathways and Ingestion

Although inhalation is the primary pathway for human exposure, ingestion also plays an important role. Transuranics can be ingested through numerous pathways. These pathways are strongly influenced by local water-use practices, agricultural systems, sport fisheries and wildlife use, and estuarine and marine fisheries, and by the amount of soil directly ingested on contaminated plants and from the hands. The deliberate ingestion of soil by children is a special pathway for evaluation identified by Healy (1974). Dose assessments for transuranics have generally indicated that inhalation is the dominant pathway. The limited data available for such transuranics as neptunium, americium, and curium, however, indicate that these elements are possibly more available than plutonium for plant uptake, as indicated by an examination of the data available for plant uptake of transuranics. Thomas and Healy (1976) concluded that current information is inadequate for accurate dose assessments. The research needs discussed in the following chapters on physical and biological transport mechanisms are many of those which would reduce the great number of uncertainties contained in present dose assessments for transuranics from ingestion. Pathways in particular that need further quantification are the contamination of food plants by resuspension and the consequences of incorporation of transuranics into organic molecules.

The identification of local food webs is one of the more important aspects of dose assessment and design of monitoring programs. Regional diets throughout the United

States and the world vary widely according to the local agricultural, sport fisheries and wildlife, commercial fisheries, and water-use practices. Food imported or exported may serve to reduce or increase the total dose of transuranics. Local gardens or small farms may lead to maximization of dose to an individual through the food pathway. Dose assessments have usually been calculated for a "maximum individual," i.e., one who obtains all sustenance at a facility site boundary. For an actual facility where food chains to man could be identified, the use of the "fence-post man" would appear to contain numerous conservatisms.

Sampling for transuranics in soils, plants, or other media in areas with existing levels of radionuclides must be designed according to the particular needs of the study and must include consideration of existing levels of transuranics in the environment. Methods for inventory sampling may not always be adequate for defining resuspendible material leading to inhalation or plant contamination. Studies must be carried out which provide information about inventory mobility and the consequential effect on human health. General comments throughout this section have referred to the general considerations of monitoring program design with little reference to specific guidance. The selection of sampling methods and measurement techniques for environmental media, including those used to calculate radiation dose to people but also commonly used as trend indicators, have been discussed in several review articles and publications; NCRP 50, the DOE guide, ICRP-7, and the NRC regulatory guides are examples. As already discussed, the proper selection of media samples, based on the detailed pathway analysis, is important in assessing the dose to people. It would not be possible to detail these considerations here for all media; therefore discussion of soil contamination is presented as an example.

As guidance for ingestion and inhalation of transuranics in the environment was being developed, consideration of soil concentration limits and resulting problems in applicability of a soil limit led to the recognition that dose limits to the lung and bone are most important. The calculation of this dose can be based on air measurements, in which case more information on particle size and the physical parameters of the transuranics is necessary than is normally developed in environmental surveillance programs. Thus the need for research input and cooperation with environmental programs becomes important.

Although the air pathway is of primary concern, the environmental measurements eventually must be or will be translated into soil concentrations. Thus the usual inventory measurements and sample techniques will not provide adequate information on the resuspendible and respirable fraction of transuranics in the environment. Definitions of such things as resuspendible surface, sample collection methods, sample preparation, and particle-size determinations are all factors in the radiological assessment of transuranic-contaminated soils. Much of the necessary data is yet to be determined. Sampling techniques have been reviewed by several researchers, including Bernhardt (1976). Differences in techniques exist which necessitate evaluation and verification of methodology used in assessing radiological impacts of transuranics in the environment.

References

Bernhardt, D. E., 1976, *Evaluation of Sample Collection and Analysis Techniques for Environmental Plutonium*, Technical Note ORP/lv-76-5, U. S. Environmental Protection Agency, NTIS.

Bloom, S. G., and W. E. Martin, 1976, *A Model to Predict the Environmental Impact of the Release of Long-Lived Radionuclides*, Final Report to the U. S. Environmental Protection Agency, Battelle, Columbus Laboratories, NTIS.

Corley, J. P., D. H. Denham, D. E. Michels, A. R. Olsen, and D. L. White, 1977, *Guide for Environmental Radiological Surveillance at ERDA Installations*, ERDA Report ERDA-77-24, Battelle, Pacific Northwest Laboratories, NTIS.

Council on Environmental Quality, 1976, *Environmental Quality*, The Fifth Annual Report of the Council on Environmental Quality, Appendix D, Preparation of Environmental Impact Statements: Guidelines, pp. 506-522, GPO.

Federal Register, 1976, Special Edition, *Code of Federal Regulations*, Title 10, Energy, GPO.

Healy, J. W., 1974, *A Proposed Interim Standard for Plutonium in Soils*, USAEC Report LA-5483-MS, Los Alamos Scientific Laboratory, NTIS.

International Atomic Energy Agency, 1966, *Manual on Environmental Monitoring in Normal Operation*, Safety Series No. 16, STI/PUB/98, International Atomic Energy Agency, Vienna.

——, 1975, *Objectives and Design of Environmental Monitoring Programmes for Radioactive Contaminants*, Safety Series No. 41, STI/PUB/385, International Atomic Energy Agency, Vienna.

International Commission on Radiological Protection, 1960, Report of Committee II on Permissible Dose for Internal Radiation, *Health Phys.*, 3: 27-39.

——, 1965, *Principles of Environmental Monitoring Related to the Handling of Radioactive Material*, ICRP Publication 7, Pergamon Press, Inc., New York.

——, 1966, Task Groups on Lung Dynamics. Deposition and Retention Models for Internal Dosimetry of the Human Respiratory Tract, *Health Phys.*, 12: 173-207.

Jordan, H. S., 1971, Distribution of Plutonium from Accidents and Field Experiments, in *Proceedings of Environmental Plutonium Symposium*, Los Alamos, N. Mex., Aug. 4–5, 1971, USAEC Report LA-4756, pp. 21-24, NTIS.

National Council on Radiation Protection and Measurements, 1976, *Environmental Radiation Measurements*, NCRP Report No. 50, NCRP Publications, Washington, D. C.

Thomas, R., and J. W. Healy, 1976, *An Appraisal of Available Information on Uptake by Plants of Transplutonium Elements and Neptunium*, ERDA Report LA-6460-MS, Los Alamos Scientific Laboratory, NTIS.

U. S. Atomic Energy Commission, 1974, *Measurements of Radionuclides in the Environment: Sampling and Analysis of Plutonium in Soil*, Office of Standards Development, NTIS.

U. S. Congress, 1970, Public Law 91-100, *National Environmental Policy Act of 1969*, 91st Congress, S. 1075.

U. S. Energy Research and Development Administration, 1975, *Final Environmental Statement on the Liquid Metal Fast Breeder Reactor Program*.

U. S. Environmental Protection Agency, 1972, *Environmental Radioactivity Surveillance Guide*, Report ORP/SID-72-2, Office of Radiation Programs.

U. S. Nuclear Regulatory Commission, 1976, *Final Generic Environmental Statement on the Use of Recycle Plutonium in Mixed Oxide Fuel in Light Water Cooled Reactors, Health Safety and Environment*, Office of Nuclear Material Safety and Safeguards, NTIS.

World Health Organization, 1968, *Routine Surveillance for Radionuclides in Air and Water*.

Worldwide Fallout

R. W. PERKINS and C. W. THOMAS

Since the first nuclear weapons test at Alamogordo, N. Mex., on July 16, 1945, approximately 360,000 Ci (360 kCi) of $^{239,240}Pu$ has been injected into the atmosphere. In addition, 17,000 Ci (17 kCi) of ^{238}Pu entered the atmosphere in April 1964 as a result of the high-altitude burnup of a SNAP-9 satellite power source. Since most of the plutonium from nuclear weapons testing, as well as that from the SNAP-9 burnup, entered the stratosphere, fallout has been worldwide. The deposition is influenced by meteorological conditions and also by topographical features of the earth's surface. Residence time in the stratosphere is about 10 to 11 months; however, because of the high-altitude burnup of the SNAP-9 device, it was 2 yr before significant amounts of this debris reached the earth's surface.

In addition to plutonium, substantial amounts of ^{241}Am are formed from the decay of the weak beta emitter ^{241}Pu and are an important constituent of fallout.

The majority of radioactivity entering the stratosphere during this past decade has been a result of the Chinese nuclear weapons testing. The ratio of plutonium to ^{137}Cs has been relatively constant throughout the nuclear weapons period, and thus a measurement of ^{137}Cs permits a reasonable estimate of the plutonium deposition. The ratio of transuranic elements in fallout is substantially different from that in power reactor wastes, which contain far more americium and curium relative to plutonium. Fresh fallout from thermal nuclear weapons contains large amounts of short-lived ^{237}U and ^{239}Np, and these may contribute substantially to the radiation exposure at the earth's surface.

The first significant injection of transuranium elements into the atmosphere occurred as the result of the nuclear weapons testing in Alamogordo, N. Mex., on July 16, 1945. Between then and 1952 further nuclear detonations resulted in additional injections to the atmosphere; however, because of their relatively low yield, most of this debris was confined to the troposphere. On Nov. 1, 1952, the first thermonuclear device was detonated. This 14-Mt explosion injected large amounts of debris into the stratosphere. The relatively high energy yield of this fusion device, together with a much higher integrated neutron flux, greatly increased the production of the transuranium elements. The majority of the transuranium elements and other nuclear debris which has been injected into the atmosphere was produced during the 1961 and 1962 United States (U. S.) and Union of Soviet Socialist Republics (U.S.S.R.) nuclear testing programs. A nuclear weapons test-ban agreement between the United States, United Kingdom, and Soviet Union in early 1963 suspended atmospheric testing. However, in late 1964 the Chinese exploded their first atmospheric nuclear test, and since that time they have continued testing in the northern hemisphere. France was not a member of the test-ban agreement, and in mid-1966 they began atmospheric testing in the southern hemisphere. The test-ban agreement in 1963 did not rule out underground tests, which do not vent to the atmosphere. Since 1963 major underground testing, which included several hundred underground nuclear devices, has been conducted but has in only a few cases released radioactivity to the atmosphere.

Most of the studies of transuranium elements from nuclear weapons testing have been concerned with the measurement of ^{239,240}Pu isotopes. However, the complete decay of the accompanying short-lived ^{241}Pu (15 yr) results in the formation of a quantity of ^{241}Am which approaches that of ^{239,240}Pu. A major injection of ^{238}Pu into the atmosphere occurred in April 1964 when a navigational satellite failed to achieve a stable orbit and disintegrated on reentry into the atmosphere. The 17 kCi of ^{238}Pu that was added to the atmosphere was relatively small compared with the 360 kCi of ^{239,240}Pu that has been added by nuclear weapons testing. However, it greatly increased the worldwide ^{238}Pu deposition, and its point-source injection has been useful in developing models describing global atmospheric mixing.

Other incidents have added to the environmental distribution of transuranic elements but not on a worldwide scale. An aerial refueling explosion involving a B-52 bomber carrying four plutonium-bearing nuclear weapons and a KC-135 tanker occurred on Jan. 16, 1966, 28,000 ft above the Mediterranean coastline near the Spanish village of Palomares, Spain. The high-explosive component part of two weapons exploded on impact, releasing the weapons plutonium inventory over the hillside outside the village. On Jan. 21, 1968, a B-52 with four plutonium-bearing nuclear weapons on board attempted an emergency landing at Thule Air Force Base. At 9000 ft over the base, the crew bailed out, and the abandoned plane crashed on the ice of North Star Bay. The high-explosive components of all weapons detonated, and the plutonium inventory was scattered over the ice.

The operation of nuclear reactors also results in the production of transuranium elements, and the potential exists for release of some of these to the atmosphere during reactor operation and subsequent fuel processing. The modern nuclear power plants, which are designed for the generation of electric energy, use very long fuel exposure periods and may in the future recycle the fuel to burn the resulting ^{239}Pu. This results in successive neutron capture of the transuranium elements and production of very substantial quantities of higher mass elements. It has been estimated that approximately 2×10^9 Ci of transuranium elements may be produced as radioactive waste through the year 2000. Whereas ^{239,240}Pu and ^{241}Am are the main transuranium alpha activities from nuclear weapons testing, ^{238}Pu, ^{241}Am, and the curium isotopes will be the principal alpha activities from nuclear reactor operations. Accidental releases of transuranium elements to the atmosphere have occurred both from nuclear plant operation and from the transport of nuclear weapons. The total amounts released to the atmosphere by these processes have been relatively minor; however, such accidents may have rather significant local effects.

Distribution of Transuranium Elements from Nuclear Explosions

The amounts of transuranium elements from nuclear testing distributed over the world surfaces have been estimated on the basis of the nuclear tests of all nations (Hardy, 1964; United Kingdom Atomic Energy Authority, 1972; 1973; 1974; 1975; Nakahara et al., 1975). Tables 1 to 5* are summaries of the individual tests performed by each

---

*Publication of this book does not constitute a DOE endorsement of the accuracy or completeness of the list of alleged tests contained in these tables.

(Text continues on page 59.)

TABLE 1   United States Nuclear Detonations

Date	Name	Height of burst, ft	Type	Yield	Cloud top, ft	Location
Trinity						
July 16, 1945	Trinity	100	Tower	19 kt	35,000	Alamogordo, **N. Mex.**
World War II						
Aug. 5, 1945	World War II	~1,850	Air	20 kt		Hiroshima, Japan
Aug. 9, 1945	World War II	~1,850	Air	20 kt		Nagasaki, Japan
Crossroads						
June 30, 1946	Able	520	Air	20 kt	35,000	Bikini Atoll
July 24, 1946	Baker	−90	Underwater	20 kt	8,000	Bikini Atoll
Sandstone						
Apr. 14, 1948	X-ray	200	Tower	37 kt	56,000	Enewetak Atoll
Apr. 30, 1948	Yoke	200	Tower	49 kt	55,000	Enewetak Atoll
May 14, 1948	Zebra	200	Tower	18 kt	28,000	Enewetak Atoll
Ranger						
Jan. 27, 1951	Able	1,060	Air	1 kt	17,000	Nevada Test Site
Jan. 28, 1951	Baker	1,080	Air	8 kt	35,000	Nevada Test Site
Feb. 1, 1951	Easy	1,080	Air	1 kt	12,000	Nevada Test Site
Feb. 2, 1951	Baker-2	1,100	Air	8 kt	36,000	Nevada Test Site
Feb. 6, 1951	Fox	1,435	Air	22 kt	42,000	Nevada Test Site
Greenhouse						
Apr. 7, 1951	Dog	300	Tower			Enewetak Atoll
Apr. 20, 1951	Easy	300	Tower	47 kt	40,000	Enewetak Atoll
May 8, 1951	George	200	Tower			Enewetak Atoll
May 24, 1951	Item	200	Tower			Enewetak Atoll
Buster-Jangle						
Oct. 22, 1951	Able	100	Tower	<0.1 kt	8,000	Nevada Test Site
Oct. 28, 1951	Baker	1,118	Air	3.5 kt	29,000	Nevada Test Site
Oct. 30, 1951	Charlie	1,132	Air	14 kt	40,000	Nevada Test Site
Nov. 1, 1951	Dog	1,417	Air	21 kt	40,000	Nevada Test Site
Nov. 5, 1951	Easy	1,314	Air	31 kt	45,000	Nevada Test Site
Nov. 19, 1951	Sugar	4	Surface	1.2 kt	16,000	Nevada Test Site
Nov. 29, 1951	Uncle	−17	Underground	1.2 kt	11,000	Nevada Test Site
Tumbler-Snapper						
Apr. 1, 1952	Able	793	Air	1 kt	16,000	Nevada Test Site
Apr. 15, 1952	Baker	1,050	Air	1 kt	16,000	Nevada Test Site
Apr. 22, 1952	Charlie	3,447	Air	31 kt	42,000	Nevada Test Site
May 1, 1952	Dog	1,040	Air	19 kt	42,000	Nevada Test Site
May 7, 1952	Easy	300	Tower	12 kt	34,000	Nevada Test Site
May 25, 1952	Fox	300	Tower	11 kt	41,000	Nevada Test Site
June 1, 1952	George	300	Tower	15 kt	37,000	Nevada Test Site
June 5, 1952	How	300	Tower	14 kt	41,000	Nevada Test Site
Ivy						
Oct. 31, 1952	Mike		Surface	14 Mt	~100,000	Enewetak Atoll
Nov. 15, 1952	King	1,480	Air	High yield	~70,000	Enewetak Atoll
Upshot-Knothole						
Mar. 17, 1953	Annie	300	Tower	16 kt	41,000	Nevada Test Site
Mar. 24, 1953	Nancy	300	Tower	24 kt	42,000	Nevada Test Site
Mar. 31, 1953	Ruth	300	Tower	0.2 kt	14,000	Nevada Test Site
Apr. 6, 1953	Dixie	6,020	Air	11 kt	43,000	Nevada Test Site
Apr. 11, 1953	Ray	100	Tower	0.2 kt	13,000	Nevada Test Site
Apr. 18, 1953	Badger	300	Tower	23 kt	35,000	Nevada Test Site
Apr. 25, 1953	Simon	300	Tower	43 kt	45,000	Nevada Test Site
May 8, 1953	Encore	2,425	Air	27 kt	41,000	Nevada Test Site
May 19, 1953	Harry	300	Tower	32 kt	43,000	Nevada Test Site
May 25, 1953	Grable	524	Gun	15 kt	38,000	Nevada Test Site
June 4, 1953	Climax	1,334	Air	61 kt	43,000	Nevada Test Site

(Table continues on the next page.)

TABLE 1   (Continued)

Date	Name	Height of burst, ft	Type	Yield	Cloud top, ft	Location
Castle						
Feb. 28, 1954	Bravo		Surface	15 Mt	114,000	Bikini Atoll
Mar. 26, 1954	Romeo		Barge			Bikini Atoll
Apr. 6, 1954	Koon		Surface	100 kt		Bikini Atoll
Apr. 25, 1954	Union		Barge			Bikini Atoll
May 4, 1954	Yankee		Barge			Bikini Atoll
May 13, 1954	Nectar		Barge			Enewetak Atoll
Teapot						
Feb. 18, 1955	Wasp	762	Air	1 kt	22,000	Nevada Test Site
Feb. 22, 1955	Moth	300	Tower	2 kt	25,000	Nevada Test Site
Mar. 1, 1955	Tesla	300	Tower	7 kt	30,000	Nevada Test Site
Mar. 7, 1955	Turk	500	Tower	43 kt	44,000	Nevada Test Site
Mar. 12, 1955	Hornet	300	Tower	4 kt	35,000	Nevada Test Site
Mar. 22, 1955	Bee	500	Tower	8 kt	40,000	Nevada Test Site
Mar. 23, 1955	Ess	−67	Underground	1 kt	12,000	Nevada Test Site
Mar. 29, 1955	Apple I	500	Tower	14 kt	32,000	Nevada Test Site
Mar. 29, 1955	Wasp Prime	740	Air	3 kt	32,000	Nevada Test Site
Apr. 6, 1955	HA	36,620 (mean sea level)	Air	3 kt	55,000	Nevada Test Site
Apr. 9, 1955	Post	300	Tower	2 kt	16,000	Nevada Test Site
Apr. 15, 1955	Met	400	Tower	22 kt	40,000	Nevada Test Site
May 5, 1955	Apple II	500	Tower	29 kt	43,000	Nevada Test Site
May 15, 1955	Zucchini	500	Tower	28 kt	35,000	Nevada Test Site
Wigwam						
May 14, 1955	Wigwam	−2,000	Underwater	30 kt		29° N 126° W
Safety experiment						
Jan. 18, 1956			Surface			Nevada Test Site
Redwing						
May 4, 1956	Lacrosse		Surface	20 kt		Enewetak Atoll
May 20, 1956	Cherokee	4,320	Air	Several megatons		Bikini Atoll
May 27, 1956	Zuni		Surface			Bikini Atoll
May 30, 1956	Erie	300	Tower			Enewetak Atoll
June 6, 1956	Seminole		Surface			Enewetak Atoll
June 11, 1956	Flathead		Barge			Bikini Atoll
June 11, 1956	Blackfoot	200	Tower			Enewetak Atoll
June 16, 1956	Osage	680	Air			Enewetak Atoll
June 25, 1956	Dakota		Barge			Bikini Atoll
July 8, 1956	Apache		Barge			Enewetak Atoll
July 10, 1956	Navajo		Barge			Bikini Atoll
July 20, 1956	Tewa		Barge			Bikini Atoll
July 21, 1956	Huron		Barge			Enewetak Atoll
Plumbbob						
May 28, 1957	Boltzmann	500	Tower	12 kt	33,000	Nevada Test Site
June 2, 1957	Franklin	300	Tower	140 kt	17,000	Nevada Test Site
June 5, 1957	Lassen	500	Balloon	0.5 kt	7,000	Nevada Test Site
June 18, 1957	Wilson	500	Balloon	10 kt	35,000	Nevada Test Site
June 24, 1957	Priscilla	700	Balloon	37 kt	43,000	Nevada Test Site
July 5, 1957	Hood	1,500	Balloon	74 kt	48,000	Nevada Test Site
July 15, 1957	Diablo	500	Tower	17 kt	32,000	Nevada Test Site
July 19, 1957	John	20,000 (mean sea level)	Rocket	~2 kt	44,000	Nevada Test Site
July 24, 1957	Kepler	500	Tower	10 kt	28,000	Nevada Test Site
July 25, 1957	Owens	500	Balloon	9.7 kt	35,000	Nevada Test Site
July 26, 1957	Pascal A		Underground	Slight	6,000	Nevada Test Site
Aug. 7, 1957	Stokes	1,500	Balloon	19 kt	37,000	Nevada Test Site
Aug. 18, 1957	Shasta	500	Tower	17 kt	32,000	Nevada Test Site

TABLE 1 (Continued)

Date	Name	Height of burst, ft	Type	Yield	Cloud top, ft	Location
Plumbbob (Continued)						
Aug. 23, 1957	Doppler	1,500	Balloon	11 kt	38,000	Nevada Test Site
Aug. 30, 1957	Franklin Prime	750	Balloon	4.7 kt	32,000	Nevada Test Site
Aug. 31, 1957	Smoky	700	Tower	44 kt	38,000	Nevada Test Site
Sept. 2, 1957	Galileo	500	Tower	11 kt	37,000	Nevada Test Site
Sept. 6, 1957	Wheeler	500	Balloon	197 tons	17,000	Nevada Test Site
Sept. 6, 1957	Coulomb B		Surface	0.3 kt	18,000	Nevada Test Site
Sept. 8, 1957	Laplace	750	Balloon	1 kt	20,000	Nevada Test Site
Sept. 14, 1957	Fizeau	500	Tower	11 kt	40,000	Nevada Test Site
Sept. 16, 1957	Newton	1,500	Balloon	12 kt	32,000	Nevada Test Site
Sept. 23, 1957	Whitney	500	Tower	19 kt	30,000	Nevada Test Site
Sept. 28, 1957	Charleston	1,500	Balloon	12 kt	32,000	Nevada Test Site
Oct. 7, 1957	Morgan	500	Balloon	8 kt	40,000	Nevada Test Site
Safety experiment						
Dec. 9, 1957	Coulomb C		Surface	0.5 kt		Nevada Test Site
Hardtack — Phase I						
Apr. 28, 1958	Yucca	86,000	Balloon			12°37′N
May 5, 1958	Cactus		Surface			163°01′E
May 11, 1958	**Fir**		Barge			Enewetak Atoll
May 11, 1958	Butternut		Barge			Bikini Atoll
May 12, 1958	Koa		Surface			Enewetak Atoll
May 16, 1958	Wahoo	−500	Underwater			Enewetak Atoll
May 20, 1958	Holly		Barge			Enewetak Atoll
May 21, 1958	Nutmeg		Barge			Bikini Atoll
May 26, 1958	Yellowwood		Barge			Enewetak Atoll
May 26, 1958	Magnolia		Barge			Enewetak Atoll
May 30, 1958	Tobacco		Barge			Enewetak Atoll
May 31, 1958	Sycamore		Barge			Bikini Atoll
June 2, 1958	Rose		Barge			Enewetak Atoll
June 8, 1958	Umbrella	−150	Underwater			Enewetak Atoll
June 10, 1958	Maple		Barge			Bikini Atoll
June 14, 1958	Aspen		Barge			Bikini Atoll
June 14, 1958	Walnut		Barge			Enewetak Atoll
June 18, 1958	Linden		Barge			Enewetak Atoll
June 27, 1958	Redwood		Barge			Bikini Atoll
June 27, 1958	Elder		Barge			Enewetak Atoll
June 28, 1958	Oak		Barge			Enewetak Atoll
June 29, 1958	Hickory		Barge			Bikini Atoll
July 1, 1958	Sequoia		Barge			Enewetak Atoll
July 2, 1958	Cedar		Barge			Bikini Atoll
July 5, 1958	Dogwood		Barge			Enewetak Atoll
July 12, 1958	Poplar		Barge			Bikini Atoll
July 22, 1958	Olive		Barge			Enewetak Atoll
July 26, 1958	Pine		Barge			Enewetak Atoll
Aug. 1, 1958	Teak	252,000	Rocket	Megaton range		Johnston Island
Aug. 12, 1958	Orange	141,000	Rocket	Megaton range		Johnston Island
Hardtack — Phase II						
Sept. 12, 1958	Otero	−480	Underground	38 tons	9,000	Nevada Test Site
Sept. 17, 1958	Bernalillo	−456	Underground	15 tons	7,500	Nevada Test Site
Sept. 19, 1958	Eddy	500	Balloon	83 tons	11,000	Nevada Test Site
Sept. 21, 1958	Luna	−484	Underground	1.5 tons	Low diffuse cloud	Nevada Test Site
Sept. 26, 1958	Valencia	−484	Underground	2 tons	5,500	Nevada Test Site
Sept. 28, 1958	Mars		Underground	13 tons	Low diffuse cloud	Nevada Test Site
Sept. 29, 1958	Mora	1,500	Balloon	2 kt	18,500	Nevada Test Site
Oct. 5, 1958	Hidalgo	377	Balloon	77 tons	12,000	Nevada Test Site
Oct. 5, 1958	Colfax	−350	Underground	5.5 tons	5,500	Nevada Test Site
Oct. 8, 1958	Tamalpais	−330	Underground	72 tons	Low diffuse cloud	Nevada Test Site

(Table continues on the next page.)

TABLE 1   (Continued)

Date	Name	Height of burst, ft	Type	Yield	Cloud top, ft	Location
Hardtack—Phase II (Continued)						
Oct. 10, 1958	Quay	100	Tower	79 tons	10,000	Nevada Test Site
Oct. 13, 1958	Lea	1,500	Balloon	1.4 kt	17,000	Nevada Test Site
Oct. 14, 1958	Neptune	−98.5	Underground	115 tons	11,000	Nevada Test Site
Oct. 15, 1958	Hamilton	50	Tower	1.2 tons	6,000	Nevada Test Site
Oct. 16, 1958	Dona Ana	450	Balloon	37 tons	11,000	Nevada Test Site
Oct. 17, 1958	Vesta		Surface	24 tons	10,000	Nevada Test Site
Oct. 18, 1958	Rio Arriba	72.5	Tower	90 tons	13,500	Nevada Test Site
Oct. 22, 1958	Socorro	1,450	Balloon	6 kt	26,000	Nevada Test Site
Oct. 22, 1958	Wrangell	1,500	Balloon	115 tons	10,000	Nevada Test Site
Oct. 22, 1958	Rushmore	500	Balloon	188 tons	11,500	Nevada Test Site
Oct. 24, 1958	Catron	72.5	Tower	21 tons	8,500	Nevada Test Site
Oct. 24, 1958	Juno		Surface	1.7 tons	5,500	Nevada Test Site
Oct. 26, 1958	Ceres	25	Tower	0.7 tons	6,000	Nevada Test Site
Oct. 26, 1958	Sanford	1,500	Balloon	4.9 tons	26,000	Nevada Test Site
Oct. 26, 1958	De Baca	1,500	Balloon	2.2 kt	17,500	Nevada Test Site
Oct. 27, 1958	Chavez	52.5	Tower	0.6 tons	6,500	Nevada Test Site
Oct. 29, 1958	Evans	−848	Underground	55 tons		Nevada Test Site
Oct. 29, 1958	Humboldt	25	Tower	7.8 tons	7,500	Nevada Test Site
Oct. 30, 1958	Santa Fe	1,500	Balloon	1.3 kt	18,000	Nevada Test Site
Oct. 30, 1958	Blanca	−835	Underground	19 kt	7,700	Nevada Test Site
Oct. 30, 1958	Titania	25	Tower	0.2 tons	6,000	Nevada Test Site
Argus						
Aug. 27, 1958	Argus-1	~300 miles	Rocket	1-2 kt		38°S 12°W
Aug. 30, 1958	Argus-2	~300 miles	Rocket	1-2 kt		50°S 8°W
Sept. 6, 1958	Argus-3	~300 miles	Rocket	1-2 kt		50°S 10°W
Continental (I)						
Sept. 15, 1961	Antler	−1,319	Underground	2.4 kt	Low diffuse cloud	Nevada Test Site
Sept. 16, 1961	Shrew		Underground	Low	Low diffuse cloud	Nevada Test Site
Oct. 10, 1961	Chena		Underground	Low	Low diffuse cloud	Nevada Test Site
Oct. 29, 1961	Mink		Underground	Low	Low diffuse cloud	Nevada Test Site
Dec. 3, 1961	Fisher	−1,193	Underground	13.5 kt	Low diffuse cloud	Nevada Test Site
Dec. 10, 1961	Gnome	−1,184	Underground	3 ± 1 kt	Low diffuse cloud	Nevada Test Site
Dec. 13, 1961	Mad	−594	Underground	430 tons	Low diffuse cloud	Nevada Test Site
Dec. 17, 1961	Ringtail		Underground	Low	Low diffuse cloud	Nevada Test Site
Dec. 22, 1961	Feather		Underground	Low	Low diffuse cloud	Nevada Test Site
Continental (II)						
Jan. 9, 1962	Stoat		Underground	4.5 kt	Low diffuse cloud	Nevada Test Site
Jan. 30, 1962	Doormouse		Underground	Low	Low diffuse cloud	Nevada Test Site
Feb. 9, 1962	Armadillo	−786	Underground	6.6 kt	Low diffuse cloud	Nevada Test Site
Feb. 15, 1962	Hardhat	−950	Underground	5.9 kt	Low diffuse cloud	Nevada Test Site
Feb. 19, 1962	Chinchilla	−504	Underground	1.8 kt	Low diffuse cloud	Nevada Test Site
Feb. 24, 1962	Platypus		Underground	Low	Low diffuse cloud	Nevada Test Site
Mar. 5, 1962	Danny Boy	−110	Underground	430 tons	Low diffuse cloud	Nevada Test Site
Mar. 6, 1962	Ermine		Underground	Low	Low diffuse cloud	Nevada Test Site
Mar. 8, 1962	Brazos		Underground	7.8 kt	Low diffuse cloud	Nevada Test Site
Mar. 31, 1962	Chinchilla II		Underground	Low	Low diffuse cloud	Nevada Test Site
Apr. 14, 1962	Platte		Underground	1.7 kt	Low diffuse cloud	Nevada Test Site
May 12, 1962	Aardvark	−1,444	Underground	37 kt	Low diffuse cloud	Nevada Test Site
May 19, 1962	Eel		Underground	Low	Low diffuse cloud	Nevada Test Site
June 6, 1962	Packrat		Underground	Low	Low diffuse cloud	Nevada Test Site
June 13, 1962	Des Moines		Underground	Low	Low diffuse cloud	Nevada Test Site
June 21, 1962	Daman I		Underground	Low	Low diffuse cloud	Nevada Test Site
June 27, 1962	Haymaker		Underground	56 kt	Low diffuse cloud	Nevada Test Site
June 28, 1962	Marshmallow		Underground	Low	Low diffuse cloud	Nevada Test Site
July 6, 1962	Sedan	−635	Underground	100 kt	12,000	Nevada Test Site
July 7, 1962	Little Feller I	Slightly above ground	Surface	Low	8,000	Nevada Test Site
July 11, 1962	Johnie Boy	Shallow depth	Underground	500 tons	11,000	Nevada Test Site

TABLE 2   United Kingdom Nuclear Detonations

Date	Name	Type	Yield	Location
Oct. 3, 1952	Hurricane	Ship	Kiloton range	Monte Bello Islands
Oct. 14, 1953	Totem	Tower	Kiloton range	Tests held at Emu Field,
Oct. 26, 1953	Totem	Tower	Kiloton range	300 miles northwest of Woomera
May 16, 1956	Mosaic	Tower	Kiloton range	Monte Bello Islands
June 19, 1956	Mosaic	Tower	Kiloton range	Monte Bello Islands
Sept. 27, 1956	Buffalo	Tower	Kiloton range	Maralinga
Oct. 4, 1956	Buffalo	Surface	Low yield	Maralinga
Oct. 11, 1956	Buffalo	Air drop	Low yield	Maralinga
Oct. 22, 1956	Buffalo	Tower	Kiloton range	Maralinga
May 15, 1957	Grapple	Air drop	Megaton range	Christmas Island Area
May 31, 1957	Grapple	Air drop	Megaton range	Christmas Island Area
June 19, 1957	Grapple	Air drop	Megaton range	Christmas Island Area
Sept. 14, 1957	Antler	Tower	Low yield	Maralinga
Sept. 25, 1957	Antler	Tower	Kiloton range	Maralinga
Oct. 9, 1957	Antler	Balloon	Kiloton range	Maralinga
Nov. 8, 1957	Grapple	Air drop	Megaton range	Christmas Island Area
Apr. 28, 1958	Grapple	Air drop	Megaton range	Christmas Island Area
Aug. 22, 1958	Grapple	Balloon	Kiloton range	Christmas Island Area
Sept. 2, 1958	Grapple	Air drop	Megaton range	Christmas Island Area
Sept. 11, 1958	Grapple	Air drop	Megaton range	Christmas Island Area
Sept. 23, 1958	Grapple	Balloon	Kiloton range	Christmas Island Area
Mar. 1, 1962	Pampas	Underground	Low	Nevada Test Site
July 17, 1964		Underground	Low	Nevada Test Site
Sept. 10, 1965		Underground	Low to intermediate	Nevada Test Site

nation and include the yield of each device. During the course of nuclear weapons testing from 1945 through 1976, it has been estimated by Harley (1975) and updated by using announced nuclear tests that approximately 230 Mt of fission yield were introduced into the atmosphere, which produced approximately 360 kCi of ^{239,240}Pu and lesser amounts of other transuranic elements.

Prior to the detonation of the first thermonuclear device (Mike) in 1952, atmospheric injections were confined mainly to the troposphere, and the mass of most of the transuranic isotopes was lower than about 243. In debris from the Mike, which was detonated at the Enewetak Atoll on Nov. 1, 1952, transuranium elements with masses through 255 were observed. The much higher neutron yield of the Mike and subsequent fusion devices than that of earlier fission devices permitted very substantial multiple neutron capture by uranium, which allowed production of the very heavy elements. In the detonation process, multiple neutron capture by ^{238}U results in the production of extremely neutron-rich products, the beta decay of which produces nuclides along the line of greatest stability. Table 6 shows the relative abundance of the transuranium isotopes that were formed in the Mike test (Diamond et al., 1961) as well as those measured in fallout debris. The isobars of significant half-life are shown together with

TABLE 3 Union of Soviet Socialist Republics Nuclear Detonations

Date	Type	Yield	Cloud top, ft	Location
Aug. 29, 1949				U.S.S.R.
Oct. 3, 1951*				U.S.S.R.
Oct. 22, 1951*				U.S.S.R.
Aug. 12, 1953		Thermonuclear		U.S.S.R.
Aug. 23, 1953		Fission		U.S.S.R.
Oct. 26, 1954*				U.S.S.R.
Aug. 4, 1955*				U.S.S.R.
Sept. 24, 1955*				U.S.S.R.
Nov. 10, 1955*				U.S.S.R.
Nov. 23, 1955*	Air	Megaton range		U.S.S.R.
Mar. 21, 1956*				U.S.S.R.
Apr. 2, 1956*				U.S.S.R.
Aug. 24, 1956		<1 Mt		Siberia
Aug. 30, 1956		Large		Siberia
Sept. 2, 1956				U.S.S.R.
Sept. 10, 1956				U.S.S.R.
Nov. 17, 1956		Large		U.S.S.R.
Jan. 19, 1957				U.S.S.R.
Mar. 8, 1957				U.S.S.R.
Apr. 3, 1957				U.S.S.R.
Apr. 6, 1957				U.S.S.R.
Apr. 10, 1957		Large		U.S.S.R.
Apr. 12, 1957				U.S.S.R.
Apr. 16, 1957		Large		Siberia
Aug. 22, 1957		Substantial size		Siberia
Sept. 9, 1957*		Moderate intensity		Siberia
Sept. 24, 1957		Megaton range		Arctic
Oct. 6, 1957		Thermonuclear		U.S.S.R.
Oct. 10, 1957		Small explosion		Arctic
Dec. 28, 1957				Siberia
Feb. 23, 1958		Megaton range		Arctic
Feb. 27, 1958		Megaton range		Arctic
Feb. 27, 1958		Large		Arctic
Mar. 14, 1958		Below megaton range		Arctic
Mar. 14, 1958		Below megaton range		Siberia
Mar. 15, 1958		Below megaton range		Siberia
Mar. 20, 1958		Small range		Siberia
Mar. 21, 1958				Arctic
Mar. 22, 1958		Medium range		Arctic
Sept. 30, 1958		Moderate to high		Arctic
Sept. 30, 1958		Moderate to high		Arctic
Oct. 2, 1958		Moderate		Arctic
Oct. 5, 1958				Arctic
Oct. 10, 1958		Relatively large†		Arctic
Oct. 12, 1958		Large†		Arctic
Oct. 15, 1958		Large†		Arctic
Oct. 18, 1958		Large†		Arctic
Oct. 19, 1958		Small		Arctic

TABLE 3   (Continued)

Date	Type	Yield	Cloud top, ft	Location
Oct. 20, 1958		Large†		Arctic
Oct. 22, 1958		Large†		Arctic
Oct. 24, 1958		Large†		Arctic
Oct. 25, 1958		Relatively large		Arctic
Nov. 1, 1958		Relatively low		Siberia
Nov. 3, 1958		Relatively low		Siberia
Sept. 1, 1961	Atmospheric	Intermediate		Semipalatinsk
Sept. 4, 1961	Atmospheric	Low		Semipalatinsk
Sept. 5, 1961	Atmospheric	Low to intermediate		Semipalatinsk
Sept. 6, 1961	Atmospheric	Low to intermediate		East of Stalingrad
Sept. 10, 1961	Atmospheric	Several megatons		Novaya Zemlya
Sept. 10, 1961	Atmospheric	Low to intermediate		Novaya Zemlya
Sept. 12, 1961	Atmospheric	Several megatons		Novaya Zemlya
Sept. 13, 1961	Atmospheric	Low to intermediate		Semipalatinsk
Sept. 13, 1961	Atmospheric	Low to intermediate		Novaya Zemlya
Sept. 14, 1961	Atmospheric	Several megatons		Novaya Zemlya
Sept. 16, 1961	Atmospheric	Order of a megaton		Novaya Zemlya
Sept. 17, 1961	Atmospheric	Intermediate		Semipalatinsk
Sept. 18, 1961	Atmospheric	Order of a megaton		Novaya Zemlya
Sept. 20, 1961	Atmospheric	Order of a megaton		Novaya Zemlya
Sept. 22, 1961	Atmospheric	Order of a megaton		Novaya Zemlya
Oct. 2, 1961	Atmospheric	Order of a megaton		Novaya Zemlya
Oct. 4, 1961	Atmospheric	Several megatons		Novaya Zemlya
Oct. 6, 1961	Atmospheric	Several megatons		Novaya Zemlya
Oct. 8, 1961	Atmospheric	Low		Novaya Zemlya
Oct. 12, 1961	Atmospheric	Low to intermediate		Semipalatinsk
Oct. 20, 1961	Atmospheric	Several megatons	>12,000	Novaya Zemlya
Oct. 23, 1961	Atmospheric	About 25 Mt		Novaya Zemlya
Oct. 23, 1961	Underwater	Low		South of Novaya Zemlya
Oct. 25, 1961	Atmospheric	Intermediate to high		Novaya Zemlya
Oct. 27, 1961	Atmospheric	Low to intermediate		Novaya Zemlya
Oct. 30, 1961	Atmospheric	55 to 60 Mt	>12,000	Novaya Zemlya
Oct. 31, 1961	Atmospheric	Several megatons		Novaya Zemlya
Oct. 31, 1961	Atmospheric	Intermediate to high		Novaya Zemlya
Nov. 2, 1961	Atmospheric	Low to intermediate		Novaya Zemlya
Nov. 2, 1961	Atmospheric	Low to intermediate		Novaya Zemlya
Nov. 4, 1961	Atmospheric	Several megatons		Novaya Zemlya
Feb. 2, 1962	Underground			Semipalatinsk
Aug. 5, 1962	Atmospheric	30 Mt		Novaya Zemlya
Aug. 7, 1962	Atmospheric	Low		Central Siberia
Aug. 10, 1962	Atmospheric	<1 Mt		Novaya Zemlya
Aug. 20, 1962	Atmospheric	Order of several megatons		Novaya Zemlya
Aug. 22, 1962	Atmospheric	Low megaton		Novaya Zemlya
Aug. 25, 1962	Atmospheric	Order of several megatons		Novaya Zemlya
Aug. 25, 1962	Atmospheric	Low		Semipalatinsk
Aug. 27, 1962	Atmospheric	Several megatons		Novaya Zemlya
Sept. 2, 1962	Atmospheric	Intermediate		Novaya Zemlya
Sept. 8, 1962	Atmospheric	Megaton		Novaya Zemlya
Sept. 15, 1962	Atmospheric	Several megatons		Novaya Zemlya
Sept. 16, 1962	Atmospheric	Several megatons		Novaya Zemlya

(Table continues on the next page.)

TABLE 3   (Continued)

Date	Type	Yield	Cloud top, ft	Location
Sept. 18, 1962	Atmospheric	Few megatons		Novaya Zemlya
Sept. 19, 1962	Atmospheric	Multimegatons		Novaya Zemlya
Sept. 21, 1962	Atmospheric	Few megatons		Novaya Zemlya
Sept. 25, 1962	Atmospheric	Multimegatons		Novaya Zemlya
Sept. 27, 1962	Atmospheric	<30 Mt		Novaya Zemlya
Oct. 7, 1962	Atmospheric	Intermediate		Novaya Zemlya
Oct. 14, 1962	Atmospheric	Low		Semipalatinsk
Oct. 22, 1962	High altitude	Few hundred kilotons		Central Asia
Oct. 22, 1962	Atmospheric	Several megatons		Novaya Zemlya
Oct. 27, 1962	Atmospheric	Intermediate		Novaya Zemlya
Oct. 28, 1962	High altitude	Intermediate		Central Asia
Oct. 28, 1962	Atmospheric	Low		Semipalatinsk
Oct. 29, 1962	Atmospheric	Intermediate		Novaya Zemlya
Oct. 30, 1962	Atmospheric	Intermediate		Novaya Zemlya
Nov. 1, 1962	High altitude	Intermediate		Central Asia
Nov. 1, 1962	Atmospheric	Intermediate		Novaya Zemlya
Nov. 3, 1962	Atmospheric	Intermediate		Novaya Zemlya
Nov. 3, 1962	Atmospheric	Intermediate		Novaya Zemlya
Nov. 4, 1962	Atmospheric	Intermediate		Semipalatinsk
Nov. 17, 1962	Atmospheric	Low		Semipalatinsk
Dec. 18, 1962	Atmospheric	Intermediate		Novaya Zemlya
Dec. 18, 1962	Atmospheric	Intermediate		Novaya Zemlya
Dec. 20, 1962	Atmospheric	Low		Novaya Zemlya
Dec. 22, 1962	Atmospheric	Intermediate		Novaya Zemlya
Dec. 24, 1962	Atmospheric	About 20 Mt		Novaya Zemlya
Dec. 25, 1962	Atmospheric	Few megatons		Novaya Zemlya

*Date of announcement not necessarily shot date.

†Mr. McCone on Oct. 24, 1958, announced that these seven tests had a high yield, meaning that each had an explosive power equal to millions of tons of TNT.

their decay properties and half-lives. Their total activities are normalized to ^{239}Pu to permit comparison of their relative production rates. The radioisotope ^{241}Pu (a beta-decay isotope with a 14.7-yr half-life), which decays to ^{241}Am, is the most abundant activity.

The distribution on the earth's surface of transuranic elements produced during nuclear weapons testing depends on whether the debris is contained in the stratosphere or troposphere. Such partitioning is dependent on many things, including yield of the device, the "burst" height, and the height of the troposphere. Figure 1 shows the percent of debris in the troposphere as a function of the yield of a nuclear device (Ferber, 1964). From these data it can readily be seen that devices in the low-kiloton range place most of the debris in the troposphere, whereas weapons in the megaton range inject most of the debris into the stratosphere. Prior to 1952 all the nuclear explosions were in the low-kiloton range; the residence time for this debris is about 20 to 40 days (Stewart, Crooks, and Fisher, 1955; United Nations, 1964; Krey and Krajewski, 1970a). After 1952 numerous multimegaton tests took place in which most of the debris was injected into the lower stratosphere where the residence half-time is about 1 yr (Thomas et al., 1970).

TABLE 4 Republic of France Nuclear Detonations, 1960 to 1971

Date of detonation	Type	Yield	Location
Feb. 13, 1960	Tower	60 to 70 kt	Reggan, Algeria
Apr. 1, 1960	Surface	Small	Reggan, Algeria
Dec. 27, 1960	Tower	Small	Reggan, Algeria
Apr. 25, 1961	Tower	Small	Reggan, Algeria
Nov. 7, 1961	Underground	Weak	Sahara Desert
May 1, 1962	Underground	Middle	Sahara Desert
Mar. 18, 1963	Underground	Weak	Sahara Desert
Mar. 30, 1963	Underground	Weak	Sahara Desert
Oct. 20, 1963	Underground	Middle	Sahara Desert
Feb. 14, 1964	Underground	Weak	Sahara Desert
June 15, 1964	Underground	Weak	Sahara Desert
Nov. 28, 1964	Underground	Weak	Sahara Desert
Feb. 27, 1965	Underground	Middle	Sahara Desert
May 30, 1965	Underground	Weak	Sahara Desert
Oct. 1, 1965	Underground	Weak	Sahara Desert
Dec. 1, 1965	Underground	Weak	Sahara Desert
Feb. 16, 1966	Underground	Weak	Sahara Desert
July 2, 1966	Barge	Small	Mururoa Island
July 19, 1966	Air	Small	Mururoa Island
Sept. 11, 1966	Balloon	Small	Mururoa Island
Sept. 24, 1966	Barge	Small	Fangataufa Island
Oct. 4, 1966	Barge	200 to 300 kt	Mururoa Island
June 5, 1967	Balloon	Small	Mururoa Island
June 27, 1967	Balloon	Small	Mururoa Island
July 2, 1967	Balloon	Small	Mururoa Island
July 7, 1968	Balloon	Small	Mururoa Island
July 15, 1968	Balloon	0.5 Mt	Mururoa Island
Aug. 3, 1968	Balloon	Low to intermediate	Mururoa Island
Aug. 24, 1968	Balloon	Low megaton (first H bomb)	Fangataufa Island
Sept. 8, 1968	Balloon	Low megaton	Mururoa Island
May 15, 1970	Balloon	Low	Mururoa Island
May 22, 1970	Balloon	Intermediate	Mururoa Island
May 30, 1970	Balloon	Intermediate (megaton range)	Fangataufa Island
June 24, 1970	Balloon	Low	Mururoa Island
July 3, 1970	Balloon	Intermediate (1 Mt)	Mururoa Island
July 27, 1970	Balloon	Low	Mururoa Island
Aug. 2, 1970	Balloon	Low to intermediate	Fangataufa Island
Aug. 6, 1970	Balloon	Intermediate	Mururoa Island
June 5, 1971	Balloon	Low	Mururoa Island
June 12, 1971	Balloon	Intermediate	Mururoa Island
July 4, 1971	Balloon	Low	Mururoa Island
Aug. 8, 1971	Balloon	Low	Mururoa Island
Aug. 14, 1971	Balloon	Intermediate	Mururoa Island
June 25, 1972		Low	South Pacific
July 1, 1972		Low	South Pacific
July 29, 1972		Low	South Pacific

(Table continues on the next page.)

TABLE 4   (Continued)

Date of detonation	Type	Yield	Location
July 21, 1973	Low		South Pacific
July 28, 1973	Low		South Pacific
Aug. 18, 1973	Low		South Pacific
Aug. 24, 1973	Low		South Pacific
Aug. 28, 1973	Low		South Pacific
June 16, 1974	Low		South Pacific
July 7, 1974	Unknown		South Pacific
July 17, 1974	Unknown		South Pacific
July 25, 1974	Unknown		South Pacific
Aug. 15, 1974	Unknown		South Pacific
Aug. 24, 1974	Unknown		South Pacific
Sept. 15, 1974	Low		South Pacific

TABLE 5   People's Republic of China Nuclear Detonations

Date of detonation	Type	Yield	Location
Oct. 16, 1964	Tower	~20 kt	Lop Nor
May 14, 1965	Air drop	>20 kt	
May 9, 1966	Air drop	200 to 500 kt	
Oct. 27, 1966	Missile, not HH§*	<20 kt	
Dec. 28, 1966	Tower	300 kt	
June 17, 1967	Air drop	3 Mt	
Dec. 24, 1967	Air drop	15 to 25 kt	
Dec. 27, 1968	Air drop	3 Mt	
Sept. 22, 1969	Underground	~25 kt	
Sept. 29, 1969	Air drop	3 Mt	
Oct. 14, 1970	Air drop	3 Mt	
Nov. 18, 1971	Tower	~20 kt	
Jan. 7, 1972	Atmospheric	<20 kt	
Mar. 18, 1972	Atmospheric	20 to 200 kt	
June 27, 1973	Missile?	1 to 3 Mt?	
June 17, 1974	Atmospheric	1 Mt	
Oct. 27, 1975	Underground	20 kt	
Jan. 23, 1976	Atmospheric	<20 kt	
Sept. 25, 1976	Atmospheric	200 kt	
Oct. 27, 1976	Underground	200 kt	
Nov. 17, 1976	Atmospheric	4 Mt	

*HH§ stands for launching by missile to high altitude.

TABLE 6   Relative Abundance of Heavy Elements Produced During Mike Test Compared with That Measured in Worldwide Fallout (mass abundances at time = 0)

Mass No.	Isobar	Type decay	$t_{1/2}$, yr	Relative abundance, atoms		Relative abundance, activity	
				Mike	Fallout	Mike	Fallout
239	Plutonium	α	2.44×10^4	1.0	1.0	1.0	1.0
240	Plutonium	α	6.54×10^3	0.363	0.18	1.35	0.669
241	Plutonium	$-\beta$	15	0.039	0.013	63	21
242	Plutonium	α	3.87×10^5	1.9×10^{-2}	0.004	1.2×10^{-3}	2.53×10^{-4}
243	Americium	α	7.37×10^3	2.1×10^{-3}		6.9×10^{-3}	
244	Plutonium	α	8.3×10^7	1.2×10^{-3}		3.5×10^{-7}	
245	Curium	α	8.5×10^3	1.2×10^{-4}		3.6×10^{-4}	
246	Curium	α	4.76×10^3	4.8×10^{-5}		2.4×10^{-4}	
247	Curium	α	1.54×10^7	3.9×10^{-6}		6.2×10^{-9}	
248	Curium	α	3.5×10^5	1.2×10^{-6}		8.4×10^{-8}	
249	Berkelium	$-\beta$	0.852	1.1×10^{-7}		3.2×10^{-3}	
250	Curium	SF*	1.13×10^4	$\sim 3 \times 11^{-8}$		$\sim 6.5 \times 10^{-8}$	
251	Californium	α	9.0×10^2	$\sim 1.4 \times 10^{-8}$		$\sim 3.8 \times 10^{-7}$	
252	Californium	α	2.63	1.0×10^{-9}		9.3×10^{-6}	
253	Californium	$-\beta$	0.049	5.0×10^{-10}		2.5×10^{-4}	
254	Californium	$-\beta$	0.164	5.0×10^{-11}		7.4×10^{-6}	
255	Einsteinium	$-\beta$	0.107	4.0×10^{-11}		9.1×10^{-6}	

*SF, spontaneous fission.

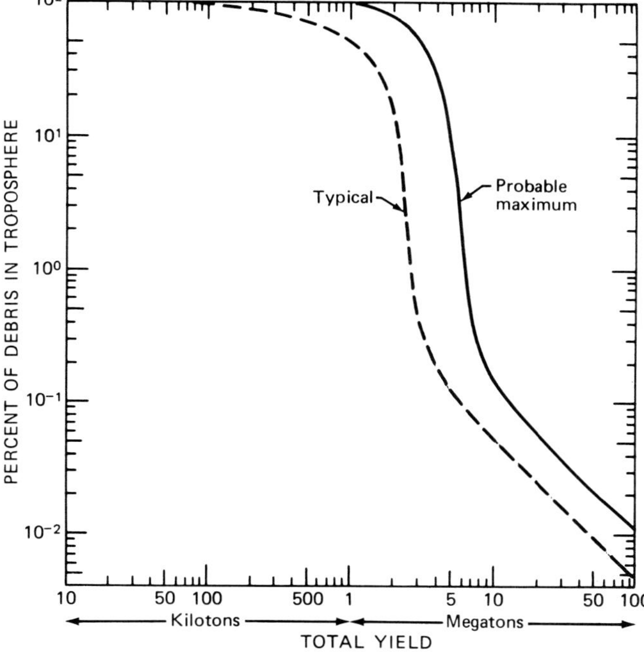

Fig. 1   Percent of total activity initially injected in the troposphere as a function of total yield for air bursts in a tropical atmosphere.

Figure 2 shows the distribution of radioactive fallout (in millicuries per 100 square miles) between 2 and 35 days following the Mike detonation explosion in the Marshall Islands (Machta, 1964). These results represent only that fractional amount of the debris which was contained in the troposphere. Since the residence time in the troposphere is on the order of 20 to 40 days, the deposition rate was quite rapid and was confined mainly to the hemisphere where the test took place; the higher concentration was near the latitude where the explosion occurred.

The distribution of debris of stratospheric origin is considerably different from that of tropospheric origin. Most of the debris leaving the stratosphere does so through the tropopause discontinuity, which occurs near the midlatitudes and is almost independent of the latitude of the detonation. Empirical box models (Krey and Krajewski, 1970b), which describe the movement of radioactivity from the upper to the lower stratosphere, between hemispheres, from the stratosphere to the troposphere, and the deposition rate on the earth's surface, have been developed and appear to be reasonably satisfactory.

The movement of radioactive debris in the troposphere is influenced by all the forces of the weather. Rain and snow will scavenge radioactive particles, which will cause the debris to be distributed unevenly on the earth's surface. Recently, measurements have shown that the scavenging of radionuclides by cirrus cloud ice particles resulted in major depletion of radionuclides from atmospheric layers of 1.3 to 2.8 km thick at about 10 km (Young, Wendell, and Wogman, 1975). The mixing of air masses as they move west to east across the United States and are orographically lifted over mountain ranges can increase the ground-level concentration of radionuclides on the downwind side. This effect is presumably due to the downwind mixing of high-level air, which contains higher concentrations of both cosmogenic and nuclear-weapons-produced radionuclides. This effect is shown in Fig. 3 where the atmospheric concentrations of ^{7}Be and ^{137}Cs for a period of $1\frac{1}{2}$ yr for Quillayute, Wash. (48°N, 125°W), Richland, Wash. (46°N, 119°W), and Rocky Flats, Colo. (40°N, 106°W), are compared. Storm systems originating in the Aleutians move air masses over the Quillayute sampling site which are orographically lifted several thousand feet by the Cascade Mountain range before they descend to the Richland site. The air mass is again lifted by the Rocky Mountain range before it descends to the Rocky Flats sampling site. The average annual air concentrations of ^{7}Be and ^{137}Cs during 1973 through early 1975 were 2.1 to 2.4 times as great at Richland and 2.9 to 3.1 times as great at Rocky Flats as those at Quillayute (Thomas, 1972).

Production and Characteristics of Individual Transuranic Elements

Because of their methods of production, the relative abundances of the transuranium elements are considerably different in nuclear detonations than in reactor operations. In nuclear detonations neutron capture occurs in extremely rapid succession, producing uranium or plutonium isotopes of very high mass which rapidly decay to form a spectrum of transuranium elements. In this case there is no opportunity for the decay of the various uranium isotopes, which could break the chain of successive neutron capture. In the reactor production of radionuclides, the neutrons are captured only one at a time, and the resulting product may decay before additional neutron capture. In Table 7 the amounts of the various transuranium elements resulting from the Mike nuclear test are compared with those which result from nuclear power generation. It is immediately evident that the transuranium elements resulting from nuclear energy generation are much higher relative to ^{239}Pu, particularly in the region just below and above 239 than

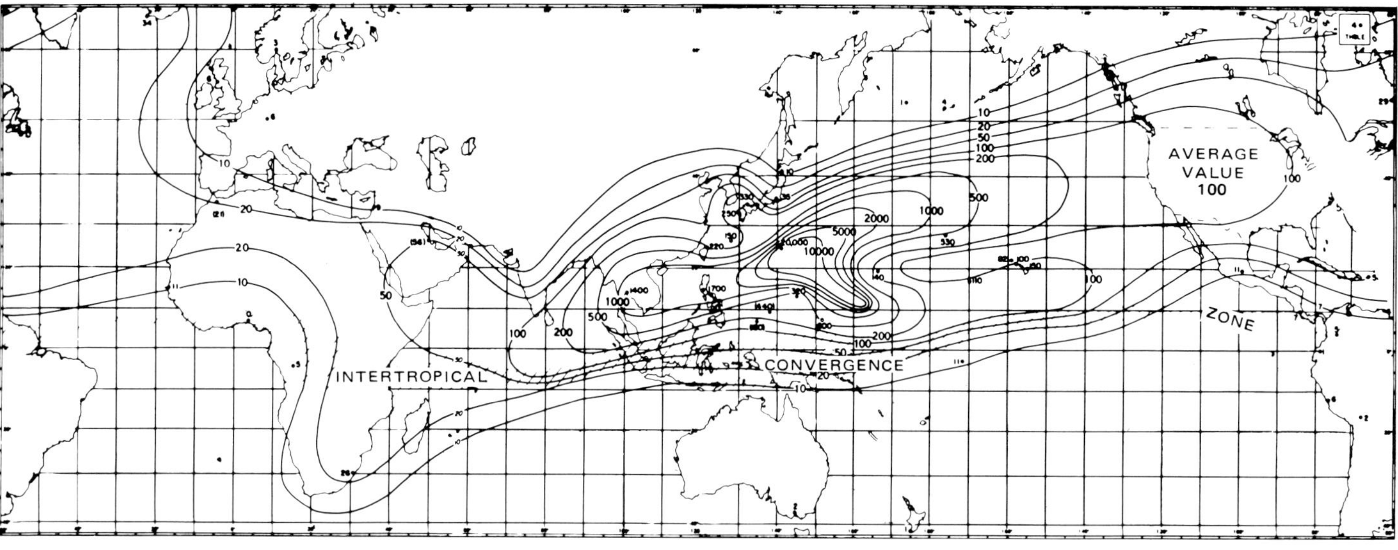

Fig. 2   Radioactive fallout from a multimegaton thermonuclear explosion in the Marshall Islands in November 1952. The values shown are millicuries per 100 square miles between 2 and 35 days after the explosion. [From L. Machta, R. J. List, and L. F. Hubert, World-Wide Travel of Atomic Debris, *Science*, 124: 474-477 (Sept. 14, 1956).]

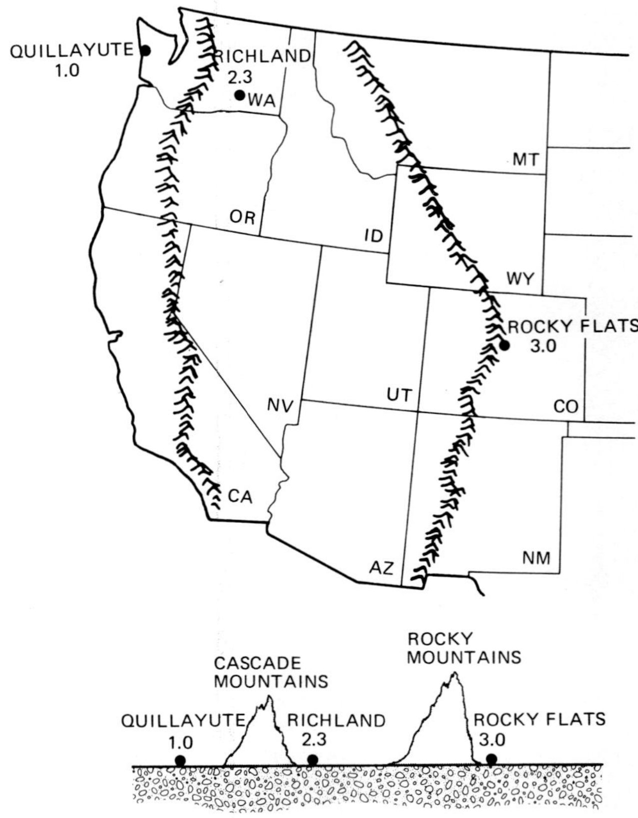

Fig. 3   Relative concentrations of [7]Be and ^{137}Cs in air from late 1973 through early 1975 normalized to 1 for Quillayute.

those from nuclear weapons testing. Comparable data for the production of isotopes of mass greater than 246 by the nuclear industry were not available, but higher concentrations relative to ^{239}Pu would be expected through perhaps mass 252.

Reported releases of transuranium elements from nuclear plant operations indicate that, in general, these have been very small. Loss of material around the Rocky Flats plant (Krey and Hardy, 1970) has resulted in some environmental contamination, and elevated atmospheric concentrations have been observed through resuspension. Accidents involving aircraft carrying nuclear weapons (Langham, 1970) have resulted in the spread of plutonium over limited areas; however, these appear to have resulted in rather minor injections of plutonium into the atmosphere.

Since the testing of the first thermonuclear device in 1952, substantial amounts of fission products, as well as transuranium elements, have entered the stratosphere. These injections result in the long-term, relatively slow deposition of radioactivity over the entire surface of the earth. As a first approximation, the transuranium elements appear to behave in their atmospheric transport in essentially the same manner as other fission products. Figure 4 shows, for example, that the ratio of ^{137}Cs to ^{239,240}Pu has been

**TABLE 7   Relative Compositions of Transuranium
Elements from Power Reactors and Mike Shot
(values normalized to ^{239}Pu)**

Mass No.	Isotope	Power reactors*	Mike shot
238	Plutonium	44.1	0.015
239	Plutonium	1	1
240	Plutonium	1.65	1.35
241	Plutonium	306	63
	Americium	95	
242	Plutonium	4.3×10^{-3}	1.2×10^{-3}
	Americium	5.64	
243	Plutonium		6.9×10^{-3}
	Americium	11.2	
244	Curium	15.11	3.5×10^{-7}
245	Curium	0.22	3.6×10^{-4}
246	Curium	4.5×10^{-2}	2.4×10^{-4}
247	Curium		6.2×10^{-9}
248	Curium		8.4×10^{-8}
249	Berkelium		3.2×10^{-3}
250	Curium		$\sim 6.5 \times 10^{-8}$
251	Californium		$\sim 3.8 \times 10^{-7}$
252	Californium		9.3×10^{-6}
253	Californium		2.5×10^{-4}
254	Californium		7.4×10^{-6}
255	Einsteinium		9.1×10^{-6}

*Assuming 30,000 Mwd/ton exposure (Schneider, 1974).

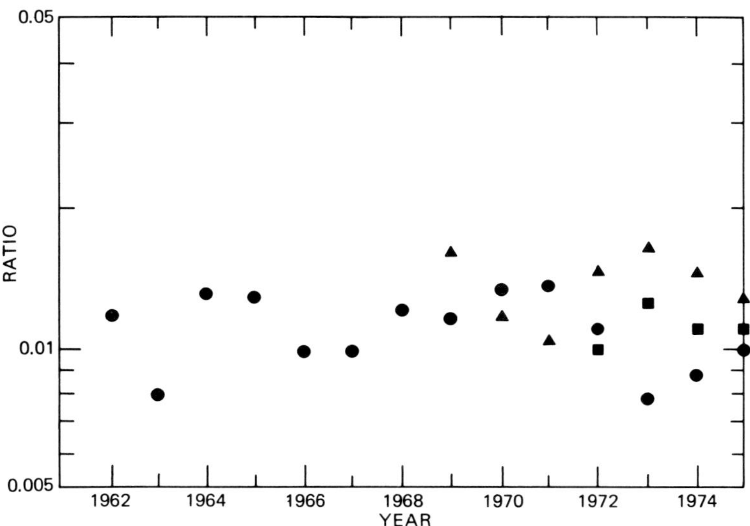

Fig. 4   Activity ratio of ^{239,240}Pu/^{137}Cs in tropospheric air. ●, Richland, Wash. ▲, New York, N. Y. ■, Harwell, England.

reasonably constant in fallout since the U. S. and U.S.S.R. nuclear tests of 1961 and 1962. These data are based on measurements beginning in 1962 at Richland, Wash. (46°N), and in later periods in New York, N. Y. (41°N), and Harwell, England (52°N). The ^{239,240}Pu/^{137}Cs ratio appears to be constant in most cases within the accuracy of the measurements and thus tends to indicate a general constancy of the transuranium production and atmospheric behavior relative to that of the fission products.

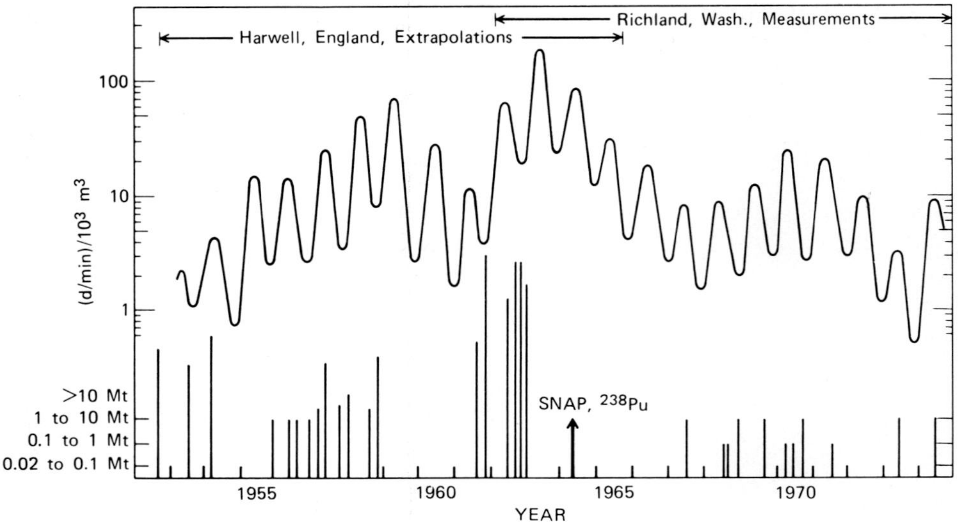

Fig. 5 Concentrations of ^{137}Cs in surface air at 46°N latitude since 1953. The concentrations prior to 1962 were estimated by normalizing concentrations measured at Harwell, England.

On the basis of observed ^{137}Cs concentrations at 46°N latitude since 1962 and an extrapolation back to 1953 by normalizing Harwell, England, to Richland, Wash., ^{137}Cs air concentrations during the period 1962 to 1964 (as indicated in Fig. 5), it should be possible to obtain a good estimate of the airborne plutonium concentrations during this entire period. Such extrapolations are, of course, subject to some uncertainty.

An atmospheric sampling program using high altitude aircraft has been conducted since 1959 (Hardy, 1973). Sampling aircraft normally operate at four latitudes—70°N, 35°N, 10°N, and 40°S. Sampling altitudes normally range from 15,000 to 70,000 ft. Figure 6 shows the ratios of ^{240}Pu to ^{239}Pu, ^{241}Pu to ^{239}Pu, and ^{242}Pu to ^{239}Pu in air at 70°N latitude as a function of time. It is evident that there is considerable variation in these ratios which is undoubtedly associated with the type and energy of the weapon responsible for the plutonium isotope production. There is a substantial increase in the heavy-to-light plutonium isotopes immediately following the 1961 and 1962 U.S.– U.S.S.R. test series.

Figure 7 shows the concentrations of ^{238}Pu and ^{239,240}Pu from 1962 to the present. These measurements, which were made near Richland, Wash., show that seasonal variations in the ^{238,239,240}Pu were similar to those of other nuclear-weapons-produced radionuclides of stratospheric origin; maximum concentrations occur in the late spring,

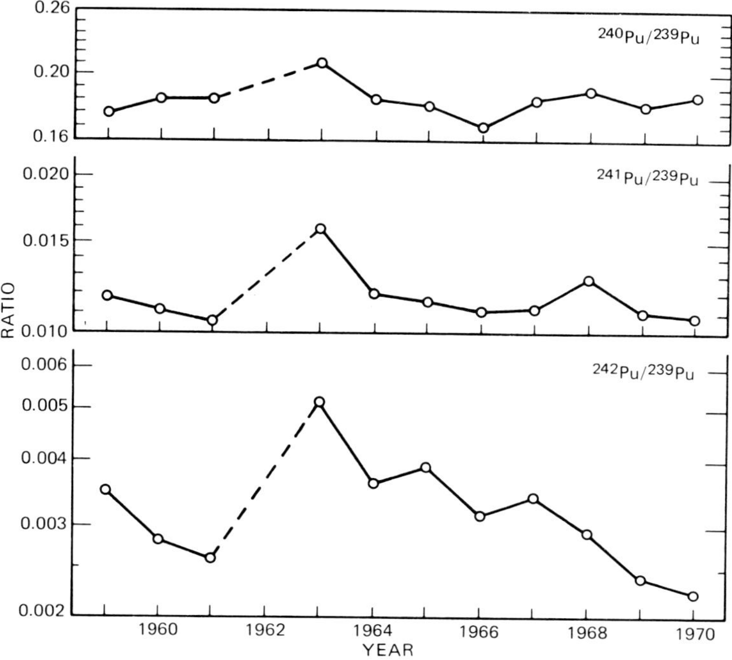

Fig. 6 Atom ratios of plutonium isotopes from an air column (15,000 to 17,000 ft high) at 70°N latitude.

and minimum concentrations occur in the winter. The rate of decrease in the ^{239,240}Pu concentrations from 1963 through 1967 corresponded to a stratospheric half-residence time of 10 to 11 months, which is similar to the half-residence times calculated from measurements of other radionuclides of stratospheric origin. The ^{239,240}Pu concentrations remained fairly constant from 1967 to 1972, primarily because of yearly injections of plutonium by thermonuclear tests conducted by the Chinese at Lop Nor (44°N); the contribution from the French tests in the South Pacific (23°S) may also have significance.

From 1962 through 1965 the ^{238}Pu and ^{239,240}Pu in surface air at Richland, Wash., came primarily from the 1961 and 1962 U.S. and U.S.S.R. series. The ^{238}Pu/^{239,240}Pu activity ratio averaged about 0.020 in 1964. The activity ratio stayed almost the same in 1965, but, by the spring of 1966, it had increased to 0.042, which suggests that ^{238}Pu from the SNAP-9A burnup was present. The amount of SNAP-9A ^{238}Pu present was determined from the ^{238}Pu concentrations and the ^{238}Pu/^{239,240}Pu activity ratios; the activity ratio in debris from nuclear weapons tests was assumed to be 0.020. These considerations indicate that the ^{238}Pu in Richland air from 1967 to 1971 came largely from SNAP-9A. From 1967 through 1969, the concentrations of SNAP-9A plutonium at Richland remained fairly constant, which indicates that the ^{238}Pu was being transferred into the northern hemispheric lower stratosphere at a rate comparable to the rate at which ^{238}Pu was being deposited on the earth's surface. This suggests that a substantial amount of ^{238}Pu was retained in the upper stratosphere, and its slow movement into the lower stratosphere maintained a nearly constant level for about 2 yr. The fact that the ^{238}Pu concentrations showed the usual seasonal variations typical of radionuclides of

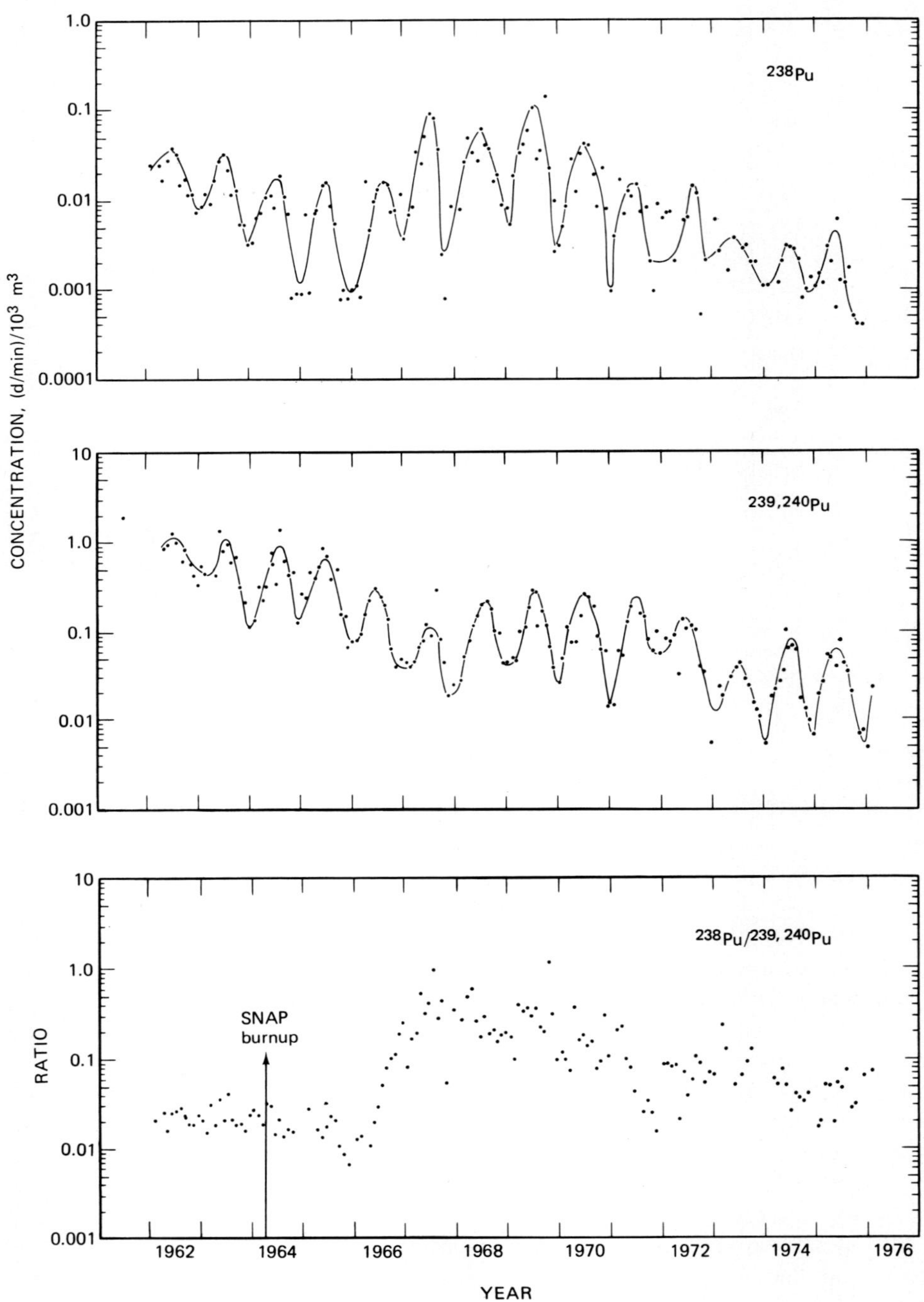

Fig. 7   Concentrations of [238]Pu and [239,240]Pu in surface air at Richland, Wash.

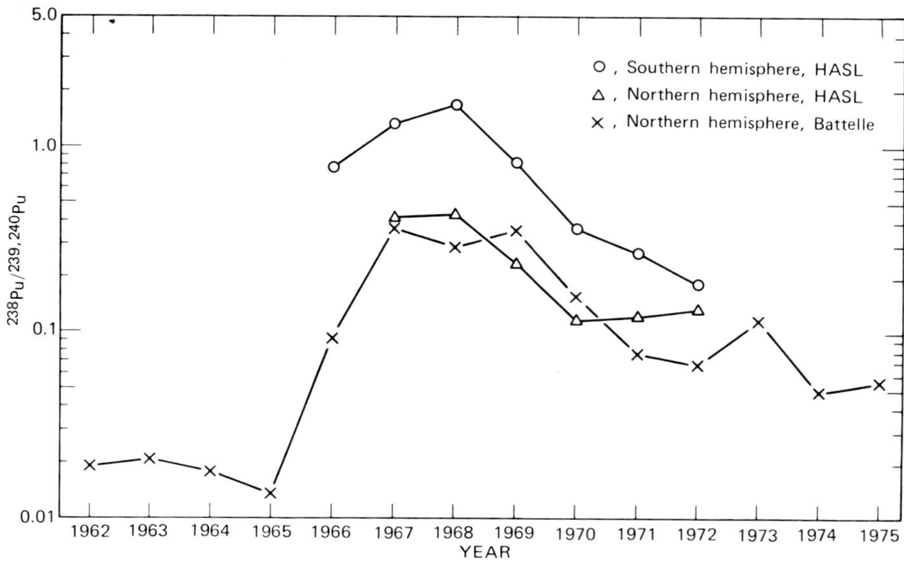

Fig. 8   Average yearly activity ^{238}Pu/^{239,240}Pu ratios in surface air.

stratospheric origin indicates that the transfer involved movement into the northern stratosphere and then to the troposphere. The hemispheric yearly averages compared with the yearly average at Richland, Wash., are shown in Fig. 8. Concentrations of SNAP-9A ^{238}Pu in the northern hemisphere and at Richland have decreased rapidly since 1968 and 1969, respectively.

Similar changes in ground-level air concentrations were observed at other locations in the northern and southern hemispheres (Hardy, 1976). These results indicate that the stratospheric debris injected into the high stratosphere may not produce high concentrations of the debris in ground-level air until 2 yr later. These ground-level concentrations may in some regions remain nearly constant for about 2 yr before they begin to decrease.

Although not formed directly in the nuclear weapons detonation, considerable amounts of ^{241}Am are present in fallout debris. This, of course, results from the decay of ^{241}Pu. On the basis of the amount of ^{239,240}Pu in the atmosphere and the ratio of ^{239}Pu to ^{241}Pu observed in the Mike test, one can calculate the ^{241}Am as a function of time in the atmosphere. The ^{241}Pu and the ^{241}Am ratios are plotted in Fig. 9 together with the observed concentrations of ^{241}Am as measured from samples taken at a monitoring station in Richland, Wash. It is evident that the airborne concentrations are in reasonably good agreement with those calculated. Also, the ratio of ^{241}Am to ^{239,240}Pu does increase, as would be expected, as the debris ages. On the basis of the yields of transuranium elements, which were observed in the Mike tests, and the ^{239,240}Pu updated inventory established by the Environmental Measurements Laboratory (EML), the total amounts of the other transuranium elements that have entered the atmosphere can be estimated. These values are shown in Table 8. For isotopes of mass greater than 244, the total atmospheric injections are in the range of hundredths to tens of curies, and the total alpha-decay activity of all the transuranium elements of mass greater than 241 is only about 1% of the ^{239,240}Pu.

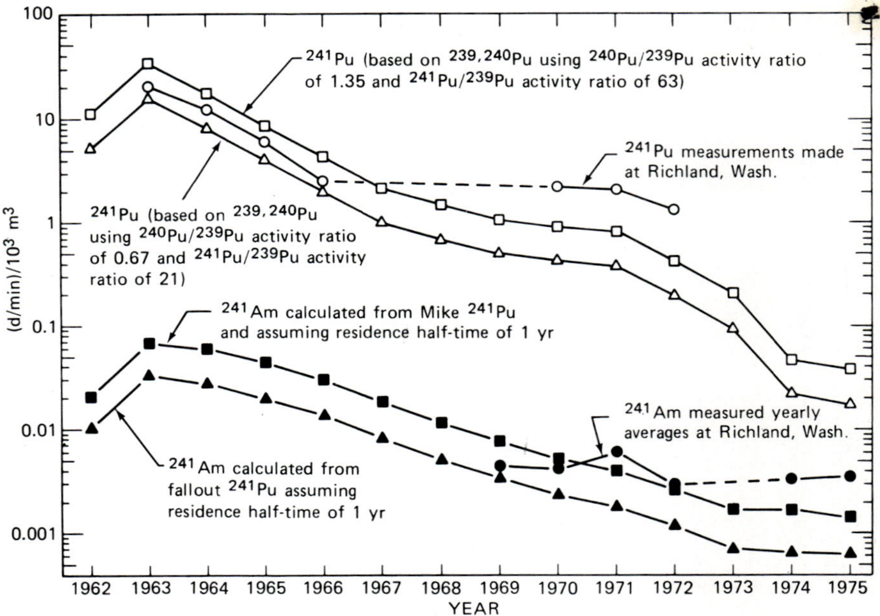

Fig. 9   Estimated concentrations of ^{241}Pu and ^{241}Am as compared with the measured concentration of ^{241}Am in surface air since 1962 at Richland, Wash.

TABLE 8   Relative Abundance and Estimated
Amounts of Transuranium Elements That
Have Been Injected into the Atmosphere

	Activity abundance	Total injection,* kCi
^{239}Pu	1	154
^{240}Pu	1.35	209
^{241}Pu	63	9720
^{241}Am		336†
^{242}Pu	1.2×10^{-3}	0.19
^{243}Am	6.9×10^{-3}	1.07
^{244}Pu	3.5×10^{-7}	5.2×10^{-5}
^{245}Cm	3.6×10^{-4}	5.6×10^{-2}
^{246}Cm	2.4×10^{-4}	3.7×10^{-2}
^{247}Cm	6.2×10^{-9}	8.8×10^{-7}
^{248}Cm	8.4×10^{-8}	1.3×10^{-5}
^{249}Bk	3.2×10^{-3}	0.49
^{250}Cm	6.5×10^{-8}	8.8×10^{-6}
^{251}Cf	3.8×10^{-7}	5.8×10^{-5}
^{252}Cf	9.3×10^{-6}	1.4×10^{-3}
^{253}Cf	2.5×10^{-4}	3.9×10^{-2}
^{254}Cf	7.4×10^{-6}	1.1×10^{-3}
^{255}Es	9.1×10^{-6}	1.4×10^{-3}

*Assumes 360,000 Ci (360 kCi) ^{239,240}Pu atmospheric injection with Mike ratios.

†Americium-241 formed on total decay of ^{241}Pu.

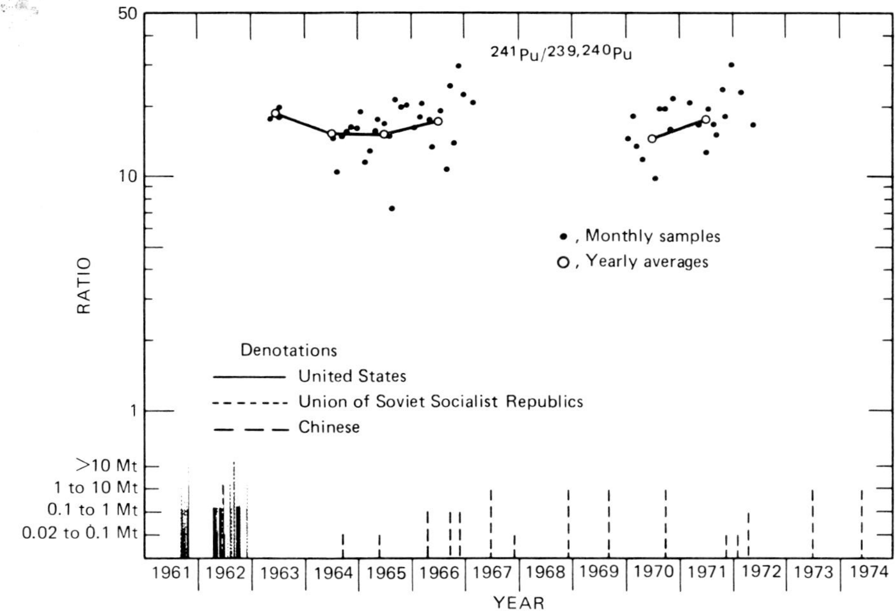

Fig. 10   Concentration ratio of ^{241}Pu and ^{239}Pu in air at Richland, Wash.

The most abundant plutonium isotope produced during nuclear detonation is the weak beta-emitting ^{241}Pu. The atmospheric concentrations observed in air at Richland, Wash., from 1963 to 1972 ranged from 20 (d/min)/10^3 m^3 at standard temperature and pressure in 1963 to 0.7 (d/min)/10^3 m^3 in 1972, whereas the ratio of ^{241}Pu/^{239,240}Pu was about 15 (Thomas and Perkins, 1974). These data are summarized in Fig. 10.

Americium-241, which is the daughter of ^{241}Pu, in global fallout can be estimated from the plutonium isotopic composition data. Americium-241 is an alpha emitter with a half-life of 433 yr. It is a bone seeker, and, on the basis of the International Commission on Radiological Protection (ICRP) maximum permissible concentrations in air and water, its toxicity is comparable to that of ^{239}Pu. In the nuclear power industry, ^{241}Am is particularly important because it is a relatively large contributor to the total alpha activity of high-burnup nuclear fuel (Thomas and Perkins, 1974). Like ^{240}Pu, most of the ^{241}Pu in large nuclear weapons test debris is produced in the detonation. By making appropriate weighting and radioactive decay corrections, including in-growth from parent—daughter relationships, the ^{241}Am/^{239,240}Pu activity ratio of integrated global fallout in February 1974 was estimated to be 0.22 (Krey et al., 1976). Two soil samples analyzed for ^{241}Am fallout (Krey et al., 1976) gave ^{241}Am/^{239,240}Pu ratios of 0.25 and 0.22.

Further calculations of the parent—daughter relationships indicate that the ^{241}Am content of present integrated fallout in soil will peak in 2037 and will represent 42% of the ^{239,240}Pu activity.

The distribution of the ^{240}Pu/^{239}Pu atom ratio in soil is shown in Fig. 11. Similar patterns emerge from the ^{241}Pu and ^{242}Pu data. There is a marked reduction in the ratio in the southwestern United States and along the west coast of South America owing to

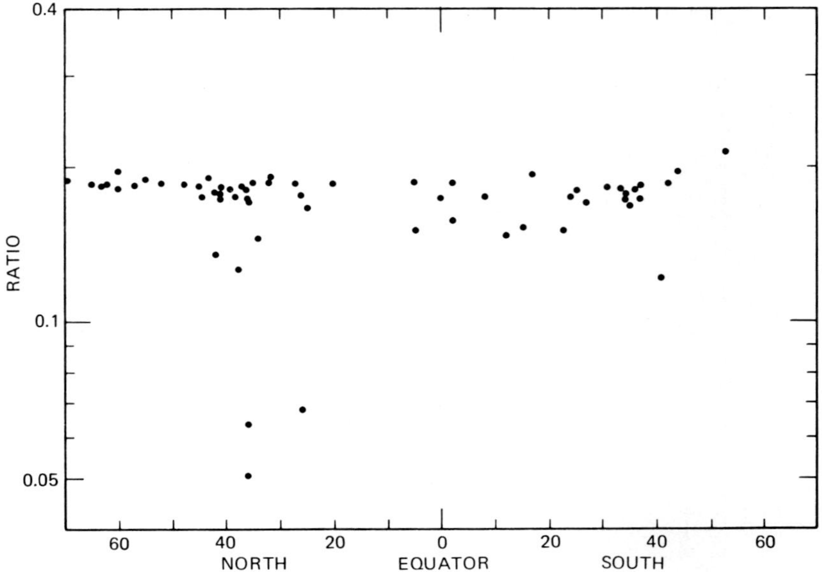

Fig. 11   Atom ratio of ^{240}Pu/^{239}Pu in worldwide fallout soil sample.

the relatively higher deposition of atypical debris from the Nevada Test Site and from the French testing site at Mururoa Atoll, respectively.

Figure 11 also shows a slightly reduced ^{240}Pu/^{239}Pu mass ratio in the equatorial region between 30°N and 30°S. This observation can be explained by the following considerations. The neutrons generated in a nuclear detonation increase with the yield. Therefore it seems reasonable that the ^{240}Pu/^{239}Pu atom ratio will also increase with the size of the nuclear detonation. Most of the nuclear test sites are located within the 30°N to 30°S region where the tropopause height is at its maximum and only the debris clouds from the larger yield tests in this region had sufficient momentum to penetrate the stratosphere. Debris from these tests, which have an elevated ^{240}Pu/^{239}Pu ratio, which entered the stratosphere is largely deposited in the middle latitudes because of the greater transfer rates from the stratosphere to the troposphere at these latitudes. By contrast, debris from the smaller yield tests, which have lower ^{240}Pu/^{239}Pu mass ratios, remained within the troposphere and are deposited on the earth's surface predominantly at the latitude of the detonation. Therefore a relatively greater amount of fallout plutonium from low-yield detonations is deposited in the equatorial regions, which gives rise to the reduced ^{240}Pu/^{239}Pu mass ratio (Hardy, Krey, and Volchok, 1973). The average atom ratios for ^{240}Pu/^{239}Pu, ^{241}Pu/^{239}Pu, and ^{242}Pu/^{239}Pu corrected to January 1971 were 0.179 ± 0.014, 0.0083 ± 0.0017, and 0.0036 ± 0.0011, respectively. Samples containing obvious contributions from the Nevada Test Site were excluded in the determining of these ratios.

The concentration of ^{238}Pu from the SNAP-9A device had been measured in surface monthly air samples from northern to southern latitudes (Hardy, 1977). The yearly average of these data is shown in Fig. 12. At most latitudes the ^{238}Pu from the SNAP-9A incident was barely detectable in 1966 but built up to a maximum in 1967 and 1968. The

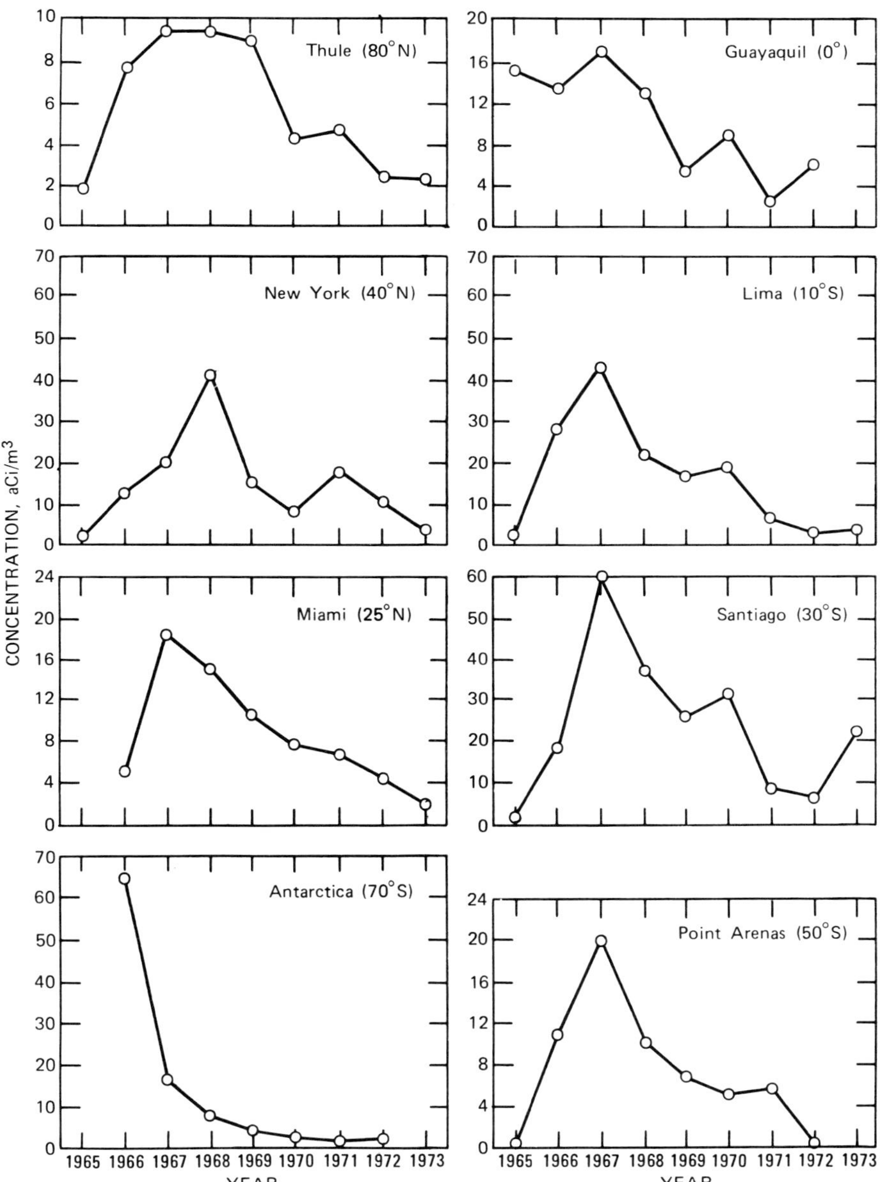

Fig. 12 SNAP-9A ^{239}Pu tropospheric air concentration.

notable exception to this trend was the Antarctica sample (70°S), where 1966 represented the maximum concentration. It has been theorized that air movement above 21 km involves an ascent of air over the summer pole, a mesospheric meridional flow from the summer to the winter hemisphere, and a descent of air over the winter pole. This, coupled with the residence half-time at this altitude of 6 months (Thomas et al., 1970) and reentry burnup of the SNAP-9A generator occurring at 46 km in the southern hemisphere over the Indian Ocean, produced a condition where the meridional flow from the northern hemisphere to the southern hemisphere had either begun or was about to begin. All these factors resulted in a disproportionation of ^{238}Pu deposition, 73% in the southern hemisphere and 27% in the northern hemisphere, and could have transferred debris over the south pole, resulting in a more rapid movement downward at the pole than at other southern or northern latitudes. Whatever the mechanism is for distribution of SNAP-9A debris throughout the hemisphere, these data show that debris injected in April 1964 at 46 km altitude reached a maximum in ground-level air about 2 or 3 yr later, depending on the latitude, and by 1971 it was largely depleted from the atmosphere.

From October 1970 until January 1971, soils were collected by EML (Thomas and Perkins, 1974) from undisturbed areas at 65 sites around the world to determine the total deposition of plutonium. Each sample consisted of ten 8.9-cm-diameter cores taken to 30-cm depth, which represented a surface area of 622 cm^2. The measured ^{238}Pu included that from weapons testing plus the SNAP-9A contribution. The ^{239,240}Pu was assumed to be entirely derived from weapons testing. The EML estimated the weapons ^{238}Pu contribution by multiplying the ^{239,240}Pu values by 0.024, their average weapons ^{238}Pu/^{239,240}Pu ratio found for six soils collected before fallout from the SNAP-9A (Krey et al., 1976). These soils were selected to cover a range of latitudes from 71°N to 35°S. The SNAP-9A ^{238}Pu was simply the difference between the total measured ^{238}Pu and the weapons ^{238}Pu. The EML deposition sites were grouped into ten-degree latitude bands and the deposition values averaged as shown in Table 9. The average activities of ^{238}Pu per square kilometer, ^{239,240}Pu per square kilometer, and the ^{238}Pu/^{239,240}Pu ratios in each ten-degree latitude band are shown in Figs. 13 to 15, respectively.

The distribution pattern for weapons plutonium shows heaviest deposition in the northern hemisphere temperate latitudes and a minimum in the equatorial region. The rise in the southern hemisphere temperate zone is, at its peak, about one-fifth of that in the northern hemisphere maximum. The SNAP-9A ^{238}Pu has an entirely different distribution pattern. Most of the SNAP debris was deposited in the southern hemisphere where the total fallout is 2.5 times as great as that in the northern hemisphere.

Short-Lived Transuranic Radionuclides in Fallout from Nuclear Weapons Testing

Nuclear debris from the past several Chinese tests has been examined to estimate the radiation exposure resulting from individual short-lived radioisotopes (Thomas, 1979a, 1979b, 1979c; Thomas and Jenkins, 1974; Thomas, Jenkins, and Perkins, 1976; Thomas et al., 1976a, 1976b). Table 10 shows the ratios of the concentrations of ^{239}Np and ^{237}U relative to ^{140}Ba. It is evident that the ratios of each of these transuranium elements to the major fission product, ^{140}Ba, are rather high. This fact becomes important when one calculates the radiation exposure from a submersion dose or from ground shine. In fallout debris from such a test, the radiation exposure from these short-lived transuranic radionuclides makes up a significant portion of the total exposure of fresh fallout debris.

TABLE 9   Average Latitudinal Distributions of Cumulative ^{239,240}Pu and ^{238}Pu Fallout*

Hemisphere	Latitude band, degrees	^{239,240}Pu	^{238}Pu Weapons	^{238}Pu SNAP-9A	$\dfrac{^{238}\text{Pu}}{^{239,240}\text{Pu}}$
			Millicuries per square kilometer		
Northern	90–80	(0.10 ± 0.04)	(0.002 ± 0.001)	(<0.001)	0.020
	80–70	0.36 ± 0.05	0.009 ± 0.001	<0.001	0.025
	70–60	1.6 ± 1.0	0.038 ± 0.025	0.026 ± 0.015	0.040
	60–50	1.3 ± 0.2	0.031 ± 0.004	0.013 ± 0.004	0.034
	50–40	2.2 ± 0.5	0.053 ± 0.011	0.026 ± 0.011	0.036
	40–30	1.8 ± 0.6	0.042 ± 0.014	0.025 ± 0.015	0.037
	30–20	0.96 ± 0.07	0.023 ± 0.002	0.011 ± 0.004	0.035
	20–10	0.24 ± 0.10	0.006 ± 0.002	0.003 ± 0.002	0.038
	10–0	0.13 ± 0.06	0.003 ± 0.001	<0.001	0.023
					$\bar{x} = 0.032 ± 0.0073$
Southern	0–10	0.30 ± 0.20	0.007 ± 0.005	0.010 ± 0.007	0.057
	10–20	0.18 ± 0.05	0.004 ± 0.001	0.036 ± 0.021	0.222
	20–30	0.39 ± 0.16	0.009 ± 0.004	0.070 ± 0.042	0.203
	30–40	0.40 ± 0.12	0.009 ± 0.003	0.061 ± 0.020	0.175
	40–50	0.35 ± 0.21	0.008 ± 0.005	0.069 ± 0.038	0.220
	50–60	(0.20 ± 0.09)	(0.005 ± 0.002)	(0.044 ± 0.023)	0.245
	60–70	(0.10 ± 0.04)	(0.002 ± 0.001)	(0.022 ± 0.012)	0.240
	70–80	(0.03 ± 0.01)	(0.001 ± 0.001)	(0.008 ± 0.005)	0.300
	80–90	(0.01 ± 0.004)	(<0.001)	(0.004 ± 0.002)	0.400
					$\bar{x} = 0.229 ± 0.092$
			Kilocuries deposited (through 1971)		
Northern		256 ± 33	6.1 ± 0.8	3.1 ± 0.8	0.036
Southern		69 ± 14	1.6 ± 0.3	10.8 ± 2.1	0.180
Global		325 ± 36	7.7 ± 0.9	13.9 ± 2.2	0.066

*Results in parentheses were derived by extrapolation; error terms are standard deviations.

TABLE 10   Ratio of ^{237}U and ^{239}Np Activities to ^{140}Ba Chinese Test Debris Collected in Surface Air Samples at Richland, Wash.

Date	$^{237}\text{U}/^{140}\text{Ba}$	$^{239}\text{Np}/^{140}\text{Ba}$
May 9, 1966	4.52	31.3
Dec. 27, 1968	3.97	29.4
Sept. 29, 1969	4.76	34.8
Oct. 15, 1970	3.70	27.2
June 26, 1973	7.55	46.5
June 17, 1974	9.71	44.2

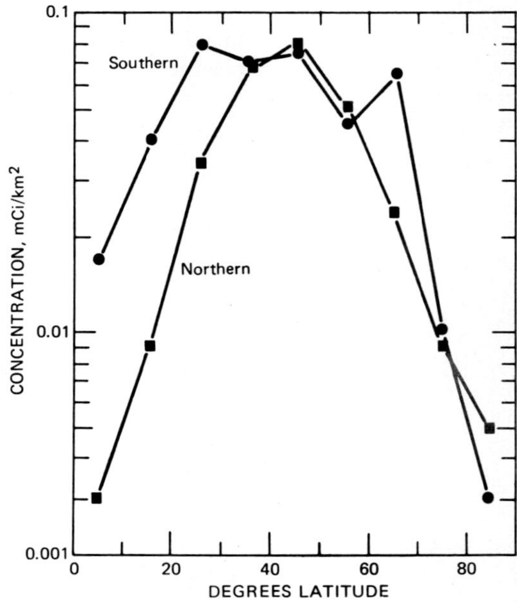

Fig. 13   Concentration of ^{238}Pu as a function of latitude in integrated soil sample.

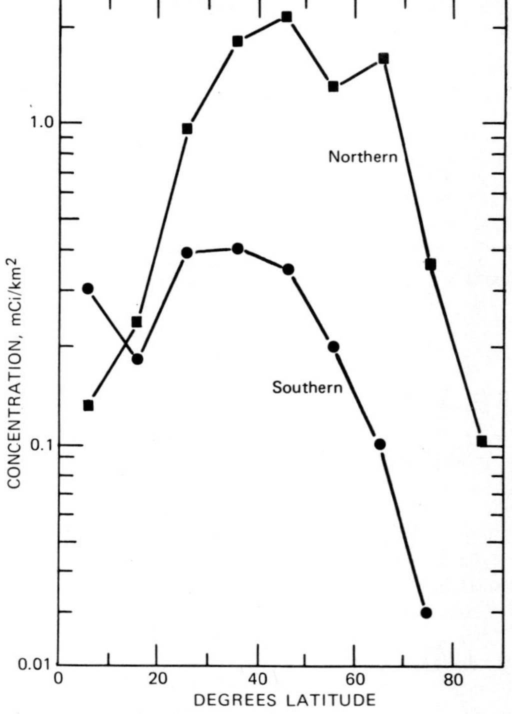

Fig. 14   Concentration of ^{239,240}Pu as a function of latitude in integrated soil sample.

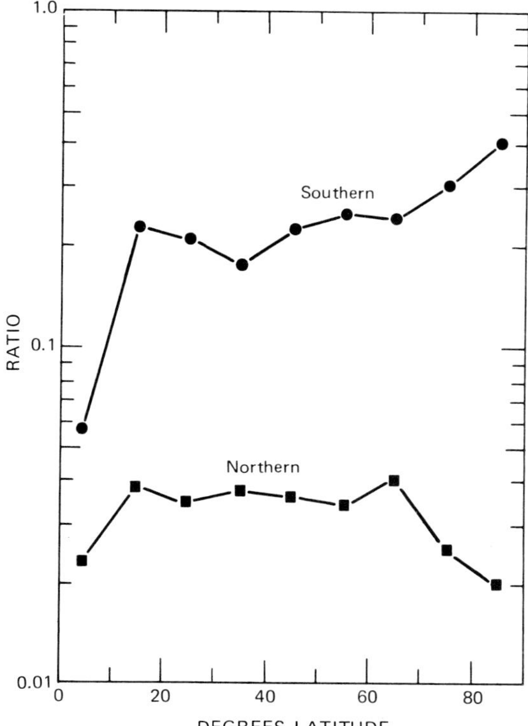

Fig. 15   Activity ratio of ^{238}Pu/^{239,240}Pu as a function of latitude in integrated soil sample.

References

Diamond, H., et al., 1961, Heavy Isotopes Abundances in Mike Thermonuclear Device, *Phys. Rev.*, 119(6): 2000-2004.

Ferber, Gilbert J., 1964, Distribution of Radioactivity with Heights in Nuclear Clouds, in *Proceedings of the Second Conference on Radioactive Fallout from Nuclear Weapons Tests*, Germantown, Md., Nov. 3–6, 1964, USAEC Report CONF-765, pp. 629-645, NTIS.

Hardy, E. P., Jr., 1964, *Fallout Program, Quarterly Summary Report, Jan. 1, 1963, Through Dec. 1, 1964*, USAEC Report HASL-142, p. 225, Health and Safety Laboratory, NTIS.

——, 1973, *Fallout Program, Quarterly Summary Report, Dec. 1, 1972, to Mar. 1, 1973*, USAEC Report HASL-273, pp. 111-112, Health and Safety Laboratory, NTIS.

——, 1976, *Health and Safety Laboratory Environmental Quarterly, Nov. 1, 1975, to Mar. 1, 1976*, USAEC Report HASL-302, pp. B111-119, Health and Safety Laboratory, NTIS.

——, 1977, *Health and Safety Laboratory Environmental Quarterly, Sept. 1, 1976, to Dec. 1, 1976*, USAEC Report HASL-315, p. B-118, Health and Safety Laboratory, NTIS.

——, P. W. Krey, and H. L. Volchok, 1973, Global Inventory and Distribution of Fallout Plutonium, *Nature*, 241(5390): 444-445.

Harley, J. H., 1975, *Transuranium Elements on Land*, paper presented at the American Nuclear Society Winter Meeting 1974, ERDA Report HASL-291, pp. I/105-109, Health and Safety Laboratory, NTIS.

Krey, P. W., and E. P. Hardy, 1970, *Plutonium in Soil Around the Rocky Flats Plant*, USAEC Report HASL-235, Health and Safety Laboratory, NTIS.

——, and B. Krajewski, 1970a, Tropospheric Scavenging of ^{90}Sr and ^{3}H, in *Precipitation Scavenging (1970)*, DOE Symposium Series, Richland, Wash., June 1–5, 1970, R. J. Engelmann and W. G. N. Slinn (Coordinators), pp. 447-463, CONF-700601, NTIS.

——, and B. Krajewski, 1970b, Comparison of Atmospheric Transport Model Calculations with Observations of Radioactive Debris, *J. Geophys. Res.*, 75(15): 2901-2908.

——, E. P. Hardy, C. Pachucki, F. Rourke, J. Coluzza, and W. K. Benson, 1976, Mass Isotopic Composition of Global Fallout Plutonium in Soil, in *Transuranium Nuclides in the Environment*, Symposium Proceedings, San Francisco, 1975, pp. 671-678, STI/PUB/410, International Atomic Energy Agency, Vienna.

Langham, W., 1970, USAF, *Nucl. Safety*, 65: 36.

Machta, L., 1964, Status of Global Radioactive Fallout Predictions, in *Proceedings of the Second Conference on Radioactive Fallout from Nuclear Weapons Tests*, Germantown, Md., Nov. 3–6, 1964, USAEC Report CONF-765, NTIS.

Nakahara, H., T. Sotobayashi, O. Nitho, T. Suzuki, S. Koyama, and S. Tonouchi, 1975, Fallout from the 15th Chinese Nuclear Test, *Health Phys.*, 29(2): 291-300.

Schneider, K. J. (Ed.), 1974, *High-Level Radioactive Waste Management Alternatives*, ERDA Report BNWL-1900 (Vol. 1), Battelle, Pacific Northwest Laboratory, NTIS.

Stewart, N.G., R. N. Crooks, and E. M. R. Fisher, 1955, The Radiological Dose to Persons in the U. K. due to Debris from Nuclear Test Explosions, in *Report for the M.R.P. Committee on the Medical Aspects of Nuclear Radiation*, Report AERE-HP/R-1701, Atomic Energy Research Establishment.

Thomas, C. W., 1972, *Comparison of Atmospheric Radionuclide Concentrations at Neah Bay and Richland, Washington*, USAEC Report BNWL-1651(Pt. 2), Battelle Pacific Northwest Laboratories, NTIS.

——, 1979a, *Radioactive Fallout from Chinese Nuclear Test, December 14, 1978*, DOE Report PNL-2912, Battelle, Pacific Northwest Laboratory, NTIS.

——, 1979b, *Radioactive Fallout from Chinese Nuclear Weapons Test, March 15, 1978*, DOE Report PNL-2913, Battelle, Pacific Northwest Laboratory, NTIS.

——, 1979c, *Short-Lived Debris from the Chinese Nuclear Test of September 17, 1977*, DOE Report PNL-2914, Battelle, Pacific Northwest Laboratory, NTIS.

——, and C. E. Jenkins, 1974, *Radioactive Debris from the Chinese Nuclear Test of June 19, 1974*, ERDA Report BNWL-B-367, Battelle, Pacific Northwest Laboratory, NTIS.

——, and R. W. Perkins, 1974, Transuranium Elements in the Atmosphere, in *Proceedings of the Winter Meeting of the American Nuclear Society on Environmental Levels of the Transuranium Elements*, Washington, D.C., Oct. 30, 1974, American Nuclear Society.

——, C. E. Jenkins, and R. W. Perkins, 1976, *Behavior and Characteristics of Radioactive Debris from the Chinese Nuclear Weapons Test of June 26, 1973*, ERDA Report BNWL-2160, Battelle, Pacific Northwest Laboratory, NTIS.

——, W. B. Silker, C. E. Jenkins, and R. W. Perkins, 1976a, *Behavior and Characteristics of Radioactive Debris from the Chinese Nuclear Weapons Tests of January 7, 1972, and March 18, 1972*, ERDA Report BNWL-2161, Battelle, Pacific Northwest Laboratory, NTIS.

——, J. K. Soldat, W. B. Silker, and R. W. Perkins, 1976b, *Radioactive Fallout from Chinese Nuclear Weapons Test, September 26, 1976*, ERDA Report BNWL-SA-6054, Battelle, Pacific Northwest Laboratory, NTIS.

——, J. A. Young, N. A. Wogman, and R. W. Perkins, 1970, The Measurement and Behavior of Airborne Radionuclides Since 1962, in *Radionuclides in the Environment*, Advances in Chemistry Series, No. 93, American Chemical Society.

United Kingdom Atomic Energy Authority, 1972, *Radioactive Fallout in Air and Rain, Results to the Middle of 1972*, Report AERE-R-7245.

——, 1973, *Radioactive Fallout in Air and Rain, Results to the Middle of 1973*, Report AERE-R-7540.

——, 1974, *Radioactive Fallout in Air and Rain, Results to the Middle of 1974*, Report AERE-R-7832.

——, 1975, *Radioactive Fallout in Air and Rain, Results to the End of 1975*, Report AERE-R-8267.

United Nations, New York, 1964, *Report of the United Nations Scientific Committee on the Effects of Atomic Radiation*, General Assembly, Supplement No. 14 (A/5814).

Young, J. A., L. Wendell, and N. A. Wogman, 1975, *Cirrus Scavenging as a Mechanism for the Production of Radionuclide Concentration Minimums Between 6 and 9 km*, USAEC Report BNWL-1950(Pt. 3), p. 136, Battelle, Pacific Northwest Laboratories, NTIS.

Transuranic Elements in Space Nuclear Power Systems

THADDEUS J. DOBRY, JR.

In the 20 years of the space age, the U. S. Department of Energy and its predecessors, the U. S. Energy Research and Development Administration and the U. S. Atomic Energy Commission, have had a growing role in our country's exploration and exploitation of space. From a few early earth orbital missions through lunar landings to long-term outer-planetary journeys, the safety, compactness, reliability, and life of nuclear isotope power supplies have been essential to mission success. Technology improvements are continuing to virtually eliminate the release of radioactive fuel during normal operations and accident situations.

Between June 1961 and December 1976 the United States launched 19 spacecraft designed with electrical systems powered by the transuranic element plutonium, which contained approximately 80% ^{238}Pu and 17% ^{239}Pu by weight. Of these 19 systems, 7 were U. S. Department of Defense (DOD) satellites and 12 were National Aeronautics and Space Administration (NASA) scientific spacecraft (2 weather satellites, 6 Apollo lunar experiments, 2 Pioneer interplanetary probes, and 2 Viking Mars landing vehicles). Table 1 lists these launchings and gives the status of the systems.

Of the 941,600 Ci of ^{238}Pu (700 Ci of ^{239}Pu) launched to date, 379,200 Ci (282 Ci of ^{239}Pu), approximately 40%, is in long-term earth orbit. There is 222,500 Ci of ^{238}Pu (165 Ci of ^{239}Pu), or approximately 24% of the total, on the lunar surface; 160,000 Ci (119 Ci of ^{239}Pu), approximately 17%, has been ejected from our solar system, and 84,000 Ci (63 Ci of ^{239}Pu), 9% of the total, is on the surface of the planet Mars. The remaining 10%, 95,900 Ci (71 Ci of ^{239}Pu) was involved in three in-flight vehicle aborts.

These aborts did not result in nuclear accidents and were of the following three types:

1. The DOD Transit satellite 5 BN-3 launched in April 1964 from the Western Test Range reentered the atmosphere, and the 17,000 Ci of ^{238}Pu (13 Ci of ^{239}Pu) promptly burned up at high altitude over the Mozambique channel. Prior to 1967 plutonium metal was used as a fuel, and burnup with subsequent atmospheric dilution was a design and safety requirement. Since 1967 progress has been made in virtually eliminating reentry burnup of the fuel in the event of an in-flight abort and in minimizing the probability of releases of radioactive fuels from launch aborts. An intact reentry—intact impact philosophy has been invoked to counter aborts leading to uncontrolled random worldwide land or sea impacts.

2. The NASA Nimbus B-1 weather satellite launched in May 1968 from the Western Test Range was a range safety destruct action that resulted in the intact impact of 34,400 Ci of ^{238}Pu (25 Ci of ^{239}Pu) in the Santa Barbara channel. The fuel remained intact in two containers, which were subsequently recovered.

3. After being launched in April 1970, the Apollo 13 was aborted in flight, and 44,500 Ci of ^{238}Pu (33 Ci of ^{239}Pu) attached to the lunar landing vehicle was deliberately disposed of in the Pacific Ocean near the Tonga Trench as a preplanned flight contingency.

All the above aborts were identified and their probabilities and consequences were analyzed in the risk assessments submitted for the preflight presidential approval actions.

The space program has handled approximately 1 MCi of plutonium, and only 17,000 Ci has been released to the environment as worldwide fallout. The activity disposed of in the deep ocean is contained in corrosion-resistant materials that should prevent the release of the fuel for a period of time equivalent to 10 half-lives of the ^{238}Pu fuel. Safety design requirements have been established which limit the release of fuel to the environment. Since accidents are not expected to occur each time a system is launched

TABLE 1   Summary of Launched Space Nuclear Power Systems

Mission	Launch date	Fuel form	Activity, Ci ^{238}Pu	^{239}Pu	Disposition
Transit 4-A	6/61	Plutonium metal	1,800	1.3	In >1000-yr earth orbit
Transit 4-B	11/61	Plutonium metal	1,800	1.3	In >1000-yr earth orbit
Transit 5-BN-1	9/63	Plutonium metal	17,000	13	In >1000-yr earth orbit
Transit 5-BN-2	12/63	Plutonium metal	17,000	13	In >1000-yr earth orbit
Transit 5-BN-3	4/64	Plutonium metal	17,000	13	Aborted, burned up on reentry
Nimbus B-1	5/68	PuO_2 microspheres	34,400	25	Aborted, containers recovered
Nimbus 111	4/69	PuO_2 microspheres	37,600	28	In ~3000-yr earth orbit
Apollo 12	11/69	PuO_2 microspheres	44,500	33	On lunar surface
Apollo 13	4/70	PuO_2 microspheres	44,500	33	Aborted, intact in Pacific Ocean
Apollo 14	1/71	PuO_2 microspheres	44,500	33	On lunar surface
Apollo 15	7/71	PuO_2 microspheres	44,500	33	On lunar surface
Apollo 16	1/72	PuO_2 microspheres	44,500	33	On lunar surface
Pioneer F	3/72	Plutonium molybdenum cermet	80,000	59.5	Ejected from solar system
Transit	9/72	Plutonium molybdenum cermet	24,000	18	In <1000-yr earth orbit
Apollo 17	12/72	PuO_2 microspheres	44,500	33	On lunar surface
Pioneer G	4/73	Plutonium molybdenum cermet	80,000	59.5	Ejected from solar system
Viking-1	8/75	Plutonium molybdenum cermet	42,000	31	On Mars surface
Viking-2	9/75	Plutonium molybdenum cermet	42,000	31	On Mars surface
Les 8/9	3/76	Pressed PuO_2	280,000	208.4	In >100,000-yr earth orbit
		Total	941,600	700	

and since the fuel inventory required in the system could be contained in multiple structures, the extent of release per container is controlled within the heat-source design by a probabilistic scaling factor, where the release probability, including the occurrence probability of the accident, is inversely proportional to the quantity of fuel released. Risk assessments of current space systems have indicated that a source term to the biosphere of 1 to 10 Ci of respirable fuel might be expected with a probability of from 10^{-8} to 10^{-5}. This source term could be either an atmosphere release at altitude or a ground point-source release. Releases to the hydrosphere are controlled by the integrity of the fuel containers and the dissolution rate of the fuel form. On the basis of release-rate experiments, plutonium concentrations in seawater are generally in the picocurie per milliliter range at distances of 10 m or more from the point source.

In summary, progress has been made in the space program to virtually eliminate the release of radioactive fuel during normal operations and launch aborts. As future trends require larger systems with higher electrical power levels and larger fuel inventories, more-stringent system safety requirements and more-sophisticated analytical and test methods to improve the quality or risk assessments and source-term evaluations are being developed and enforced.

Quantities of Transuranic Elements in the Environment from Operations Relating to Nuclear Weapons

GORDON FACER

Only nuclear explosions near or above the earth's surface or under water have contributed substantial amounts of transuranic materials to the world bioenvironment. The amounts of transuranics placed in the environment through underground test ventings, accidents involving U. S. nuclear weapons, and releases during weapon production operations have been negligible in comparison with those from atmospheric testing of nuclear explosives. On the order of 10^5 Ci of plutonium has been dispersed within our environment from about 400 nuclear explosive tests, including those by the United States, Great Britain, and Russia, between 1945 and 1963, plus more recent nuclear explosive tests in the atmosphere by China, India, and France.

The main source of transuranic material, particularly plutonium, presently in the human environment, other than that which occurs in nature (Meyers and Lindner, 1971), is nuclear weapons.* Weapons testing in the atmosphere since 1945 has distributed by far the largest part of the existing transuranic inventory throughout the world. However, smaller amounts of transuranic materials have reached the environment as the result of accidents, both real and simulated, with nuclear weapons and of releases of transuranic materials during weapon development and fabrication operations. It must be assumed that other countries have had releases of transuranic materials comparable to those for which the United States was responsible.

All U. S. weapons explosions in atmospheric or near-surface (ground or water) environments took place between 1945 and September 1963. The United States, Great Britain, and Russia joined in terminating atmospheric testing when the Limited Test Ban Treaty was established in September 1963. Only China, India, and France (not parties to the Limited Test Ban Treaty) have continued testing nuclear explosives in the atmosphere since the 1963 date.

The quantities of transuranics released to the environment from nuclear testing are somewhat uncertain. First, the amounts of transuranics that have been placed within test devices and the numbers of such devices that have been tested are topics that have been closely held by the respective testing countries. Second, even if we knew the amount of materials in each specific test device, there would be no accurate means for determining the amount of material that may have reached the environment from the detonation of those devices. Some undisclosed amount of the transuranic material was expended in the

---

*For this discussion, the term ''nuclear weapons'' is used to mean all nuclear explosives, including some designed for peaceful applications.

fission process; in addition, an unknown amount became environmentally inaccessible because of the circumstances under which the tests were done. However, even with these limitations, certain approximations can be made.

About 195 (Glasstone, 1962) U. S. nuclear tests have been conducted in emplacement locations from which transuranic materials might have reached the environment, including all atmospheric and most underwater and cratering tests.* Allowing for as many nuclear tests by other countries as the United States has conducted brings us to an approximate worldwide total of about 400 tests. [Carter and Moghissi (1977) reported 389 such tests through June 1975.]

Some of the devices tested have been of a pure fission design. Many others, however, probably reflected a variety of designs involving combinations of fission and fusion processes. The transuranic material released to the environment per test certainly has varied considerably through the range of tests that have been done. For this discussion I have assumed that, as an order of magnitude, more than 100 Ci and less than 1000 Ci of transuranic source material was residual to each test. Hence a residual of between 4×10^4 and 4×10^5 Ci (1 Ci of ^{239}Pu = 16 g) of transuranic materials might remain environmentally available from worldwide nuclear weapons testing in the atmosphere; however, only a small part of the real total may remain accessible to the human environment today.

The nature of individual test emplacements has a considerable influence on the amount of the residual transuranic material that actually becomes environmentally available. Of the 195 U. S. tests (Glasstone, 1962), 60 were fired at or near the earth's surface and 45 were fired atop steel towers. About 90 others were detonated in a way that would make most residual material environmentally available, e.g., devices emplaced on tethered balloons and those positioned by airdrop, rocket, or gun.

The fraction of transuranic material released to the world environment from nuclear tests fired very close to the surface has probably been relatively small (Glasstone and Dolan, 1977). Plutonium particles in particular have a strong tendency to attach to other materials; hence most of the residual plutonium from a near-surface explosion would become attached to the enormous amount of earthen material disturbed by the explosion. Most of these plutonium-laden earthen materials, which were in the form of large particles, remained fairly close to the detonation point after the test explosion. (A different but somewhat comparable situation exists with the near-surface explosions on moored barges in the shallow lagoons of Bikini and Enewetak.) Part of the work that has been done by the Nevada Applied Ecology Group† (NAEG) (Dunaway and White, 1974; White et al., 1975; 1976; 1977) at Nevada on the behavior of this earth-entrained plutonium has been aimed at defining the nature of these distributions.

The nuclear explosives detonated on steel towers represent an intermediate situation wherein relatively little surface material was disturbed and the fraction of transuranic residue that became associated with earthen materials was much smaller. However, the towers used in these tests furnished materials that probably influenced the behavior of those transuranics. The typical tower was made of heavy structural steel with an open

---

*For this discussion, there would be no point in including the contained nuclear tests from which plutonium does not become environmentally available.

†The Nevada Applied Ecology Group was established in 1970 within the Nevada Operations Office of the U. S. Atomic Energy Commission to design a comprehensive studies program looking into specific environmental problems that might already exist or that might arise in connection with nuclear weapons test activities.

square framework about 22 by 22 ft (7 by 7 m) in cross section and extending high enough that the explosion fireball did not reach the ground surface. This meant that most towers were between 300 and 500 ft (90 and 150 m) high (Glasstone, 1962); the tallest was 700 ft (213 m). Steel guy wires gave these towers lateral support. When a nuclear test was fired, the great heat vaporized much of the tower steel and carried it upward within the rapidly rising cloud. As the vaporized tower material cooled, it condensed, and most of it tended to fall nearby, carrying along with it some of the residual transuranic materials from the tested device which were condensing at the same time.

Balloon-suspended nuclear tests and airdropped nuclear tests, which were detonated well above the ground, contributed very little material to which the residual transuranic materials might become attached. Hence probably a fairly large fraction of the transuranic material from those tests became widely dispersed throughout the world.

Much of the material that was released into the environment from nuclear testing has by today become relatively inaccessible to the environment. That which fell on the lakes, oceans, and seas is in the process of sinking to the bottom or already has reached the bottom sediments (Edgington, Wahlgren, and Marshall, 1976), and much of that which fell on land areas will soon be beneath the immediate surface layer (Essington and Fowler, 1976). Because of these factors and because of the fairly large uncertainties as to the amounts of transuranic materials that originally had been injected into the world environment from nuclear weapons testing, quantitative (source-term) estimates based on weapons testing history may not be as useful in handling specific localized problems as are regional estimates of the materials present based on local samplings.

On the other hand, regional estimates may not extrapolate well as a means of determining the world source term. Environmental releases of transuranics from nuclear weapons have varied considerably by latitude and have been far greater in the northern hemisphere than in the southern hemisphere. If we should assume that the U. S. plutonium distribution level (Harley, 1971) in the surface soils of about 1 mCi/km^2 persists worldwide (about 4.8×10^8 km^2), there would be 480,000 Ci, or about 8,000 kg. Hardy, Krey, and Volchok (1973) have estimated that worldwide there was 325,000 Ci, or about 5200 kg, of ^{239}Pu and ^{240}Pu in weapons-fallout debris. Although quantities of transuranics in this range are entirely credible, my review of atmospheric nuclear testing indicates that those estimates may be somewhat high. All things considered, a 1×10^5 Ci source term for environmentally available ^{239}Pu, although very approximate, appears conservatively suitable.

As mentioned, there is some transuranic material in the environment as a result of accidents with nuclear weapons. A certain amount of this transuranic dispersal took place as the result of deliberate tests of the behavior of weapons under accident conditions.* Several such accident-simulation tests were done at and near the Nevada Test Site and at the Tonopah Test Range near Tonopah, Nev. Although there was on the order of 10^3 to 10^4 Ci of plutonium (total) involved in those tests, some of the material was recovered and removed from the environment by personnel manually searching for and picking up the scattered metal pieces. On the basis of site-intensive inventories at the locations of these safety tests (White and Dunaway, 1975) conducted under the NAEG studies program, about 160 Ci of ^{239}Pu and ^{240}Pu remain environmentally available (in the

---

*U. S. nuclear weapons are designed so that no accident can create the nuclear circumstances necessary to deliver *nuclear yield*. However, certain accident conditions could cause the materials associated with a nuclear device to burn or detonate and thus disperse transuranic materials.

top 5 cm of soil) in the immediate vicinity of those accident-simulation test sites; probably there are on the order of a few curies from the same tests dispersed outside the immediate vicinity but within a few miles of those locations. No such inventory yet has been feasible for the sites of U. S. atmospheric tests.

Very few accidents with U. S. plutonium weapons have placed transuranic materials in the environment. The following weapons accidents are noted (U. S. Atomic Energy Commission, 1974). One in 1960 involved the burning of a missile on its launcher at Maguire Air Force Base, N. J. That fire completely melted plutonium in the nuclear warhead but apparently did not lead to any appreciable dispersal. A second accident in January 1966 resulted from the collision of a B-52 bomber with a tanker aircraft over Palomares, Spain. Two nuclear weapons on board the bomber fell to the ground; the impact detonated the high explosive in the weapons, and the contained plutonium was dispersed nearby, mainly within an area totaling about 500 acres (200 ha). As the result of extensive cleanup by the U. S. Air Force, the amount of residual plutonium that is environmentally available from that accident has been estimated to be quite small. A third accident (U. S. Air Force, 1970) involved the in-flight fire and crash of a B-52 bomber on the ice of an Arctic bay at Thule, Greenland, in January 1968. When the plane impacted, there was a large explosion and intense fire. After that accident considerable plutonium was found associated with the black crustation of burned jet fuel distributed over about 30 acres (12 ha) of the snow-covered surface of the bay ice. More of the material was found on the aircraft wreckage. Almost all this plutonium was removed through cleanup operations. An estimated 25 Ci probably went to nearby soils and bottom sediments; of the remainder, only a small fraction of the total plutonium in the accident appears to have been dispersed via the atmosphere away from the crash location.

During Operation Hardtack II at Johnston Island in 1962, four THOR missiles being used in connection with the atmospheric nuclear tests failed to perform properly.* Three of these missiles had to be destroyed in flight; transuranic materials from the attached nuclear explosives were scattered over nearby ocean areas. A fourth THOR caught fire on the launch pad on July 25, 1962, and high explosive, associated with the nuclear explosive, was detonated as a protective measure. Plutonium metal and plutonium oxide were scattered as a result. There was extensive decontamination of the launch pad following this specific incident; the more accessible plutonium was removed, and some less-accessible material was painted over or paved over.

From the record of these accidents, it appears that from 10 to 100 Ci of plutonium may remain available in the environment from that type happening involving U. S. nuclear weapons.

Finally, transuranic materials have been released in the course of operations at the laboratories and plants where nuclear weapons are designed and built. An amount estimated at between 10 and 100 Ci of plutonium was released to the soil through leakage from stored waste at the Rocky Flats Plant over a period of several years. About 300 Ci of ^{238}Pu was released at Mound Laboratory in 1969 owing to a break in a waste transfer line (U. S. Atomic Energy Commission, 1975); this material, however, was not weapons related. The total from all such releases at all weapons laboratories and facilities probably

---

*This information is based on personal communications with Layton O'Neill, Nevada Operations Office. Although the individual incidents were made public through press releases in 1962, no comprehensive account has been located in published literature.

does not exceed 400 Ci, of which a small percentage may have reached the general off-site environments.

Something should be said with regard to the ingrowth of americium as a result of the decay of plutonium. During the production of weapons-grade plutonium in a reactor, a small fraction of the material produced is ^{241}Pu, which has a 13-yr half-life for decay by beta emission to ^{241}Am. As this decay progresses, the americium activity grows to approach the long-term level. Through this ingrowth mechanism the inventory of environmentally available transuranic material has substantially increased. Because the percent ^{241}Pu produced depends on the conditions under which the material was produced in the reactor, the amount of ^{241}Am that will result from the ingrowth process is not a fixed fraction that can be predicted. For plutonium more than 20 or 30 yr after its production, however, americium activity at several percent of the curie level of the host plutonium is to be expected.

In summary, on the order of 10^5 Ci of weapons plutonium has probably been broadly distributed within the world environment from all sources. Of that amount, between 10^3 and 10^4 Ci is probably concentrated in surface soils around the U. S. test sites. Plutonium that remains dispersed and environmentally available from actual weapon accidents may be on the order of 10^2 Ci with perhaps an additional 10^2 Ci dispersed and environmentally available as a result of weapon accident tests. On the order of 10^2 Ci of plutonium is probably accessible in the environment owing to spills and releases at laboratories and plants.

References

Carter, M. W., and A. A. Moghissi, 1977, Three Decades of Nuclear Testing, *Health Phys.*, 33: 55-71.

Dunaway, P. B., and M. G. White, 1974, *The Dynamics of Plutonium in Desert Environments*, Nevada Applied Ecology Group Progress Report as of January 1974, USAEC Report NVO-142, Nevada Operations Office, NTIS.

Edgington, D. N., M. A. Wahlgren, and J. S. Marshall, 1976, The Behavior of Plutonium in Aquatic Ecosystem: A Summary of Studies on the Great Lakes, in *Environmental Toxicity of Aquatic Radionuclides: Models and Mechanisms*, Proceedings of the 8th Rochester International Conference on Environmental Toxicology, Rochester, N. Y., June 2–4, 1975, M. W. Miller and J. N. Stannard (Eds.), pp. 45-79, Ann Arbor Science Publishers, Inc., Ann Arbor, Mich.

Essington, E. H., and E. B. Fowler, 1976, Distribution of Transuranic Radionuclides in Soils, in *Transuranics in Natural Environments*, Symposium Proceedings, Gatlinburg, Tenn., Oct. 5–7, 1976, M. G. White and P. B. Dunaway (Eds.), ERDA Report NVO-178, p. 41, Nevada Operations Office, NTIS.

Glasstone, Samuel, 1962, *The Effects of Nuclear Weapons*, Appendix B, U. S. Atomic Energy Commission, GPO.

——, and Philip J. Dolan, 1977, *The Effects of Nuclear Weapons*, Sec. 9.50, U. S. Department of Defense and U. S. Department of Energy, GPO.

Hardy, E. P., P. W. Krey, and H. L. Volchok, 1973, Global Inventories and Distribution of Fallout Plutonium, *Nature*, 241: 444.

Harley, J. H., 1971, Worldwide Plutonium Fallout from Weapons Tests, in *Environmental Plutonium Symposium Proceedings*, Los Alamos, New Mexico, USAEC Report LA-4756, Los Alamos Scientific Laboratory, NTIS.

Meyers, Wm. A., and M. Lindner, 1971, Precise Determination of the Natural Abundance of ^{237}Np and ^{239}Pu in Katanga Pitchblende, *J. Inorg. Nucl. Chem.*, 33: 3233-3238.

U. S. Air Force, 1970, *USAF Nucl. Saf.*, 65(2), No. 1, Special Edition, Crested Ice.

U. S. Atomic Energy Commission, 1974, *Plutonium and Other Transuranium Elements*, USAEC Report WASH-1359, NTIS.

——, 1975, *Investigation of the Circumstances Associated with the Appearance of Plutonium-238 Contamination in Waterways Adjacent to Mound Laboratory,* Albuquerque Operations Office, unnumbered report.

White, M. G., and P. B. Dunaway (Eds.), 1975, *The Radioecology of Plutonium and Other Transuranics in Desert Environments,* Nevada Applied Ecology Group Progress Report, ERDA Report NVO-153, Nevada Operations Office, NTIS.

——, and P. B. Dunaway (Eds.), 1976, *Studies of Environmental Plutonium and Other Transuranics in Desert Ecosystems,* Nevada Applied Ecology Group Progress Report, ERDA Report NVO-159, Nevada Operations Office, NTIS.

——, P. B. Dunaway, and W. A. Howard (Eds.), 1977, *Environmental Plutonium on the Nevada Test Site and Environs,* ERDA Report NVO-171, Nevada Operations Office, NTIS.

Transuranic Wastes from the Commercial Light-Water-Reactor Cycle

M. R. KREITER, J. E. MENDEL, and R. W. McKEE

Airborne and transuranic-contaminated wastes generated in postfission activities are identified by quantity and radioactivity for the case in which spent fuel is declared waste (once-through cycle) and that in which spent fuel is reprocessed and the recovered uranium and plutonium are recycled. Because no standard defining transuranic wastes is available at this time, in this chapter the waste source is used as the basis for such a definition.

Radioactive wastes are generally treated to reduce their volume and/or mobility. For convenience the radioactive wastes discussed are categorized according to the treatment they require. A selected treatment process as well as the final treated volume is presented for each of the seven categories of waste.

In addition to wastes generated during the operation of fuel reprocessing and mixed-oxide-fuel fabrication plants, transuranic wastes resulting from activities associated with decommissioning postfission fuel-cycle facilities are identified. Dismantlement is the mode assumed for decommissioning the facilities.

A listing of projected nuclear power growth is presented both for the Organization for Economic Cooperation and Development nations and for the United States to provide perspective regarding the quantities of waste generated.

Radioactive wastes result from the fissioning of nuclear fuels used in producing energy at nuclear power plants. In this chapter radioactive wastes are defined as all materials actually or potentially contaminated with radioactivity and subsequently disposed of when worn out, defective, or of no further use.

Radioactive wastes can be categorized as transuranic contaminated or nontransuranic contaminated. Currently, there is no standard or criterion defining a commercially generated transuranium waste. A proposed rule making would consign to licensed burial grounds wastes that have been contaminated with no more than 10 nCi of transuranic elements per gram of waste [*Fed. Regist. (Washington, D.C.)*, 39: 32922 (Sept. 12, 1974)]. Ten nanocuries per gram was chosen as representing the upper range of concentration of radium in the earth's crust. This proposal would imply that wastes exceeding 10 nCi/g can be classed as transuranic. Studies are in progress to assess the numerical validity of the transuranic limit in this proposed rule making (Adam and Rogers, 1978). For this chapter we will assume that reactor operations and spent-fuel storage-basin operations do not normally produce transuranic wastes [*Fed. Regist. (Washington, D.C.)*, 39: 32922 (Sept. 12, 1974)]. Transuranic wastes would include spent fuel if it is declared waste, high-level waste, cladding hulls, and others that will be identified later in this chapter.

Both as-generated (untreated) and treated wastes are, in most cases, addressed in the following text. Untreated wastes are generally exposed to some form of treatment to reduce activity levels, to decrease the volume to be handled, and/or to decrease the mobility of the waste. Treatment processes can be as uncomplicated as a simple compaction scheme or technically quite sophisticated, as with high-level waste vitrification. The technology for a variety of alternative waste-treatment processes is at this time commercially available or under active development (Energy Research and Development Administration, 1976; U.S. Department of Energy, 1979).

Attention is focused here on the quantities and radioactivities of those transuranic-contaminated wastes generated in postfission activities involved in the light-water-reactor (LWR) fuel cycle for commercial power production only. A description is given of wastes resulting from operating and decommissioning of the fuel-cycle facilities. The projected waste characteristics are often necessarily conjectural owing to lack of hard data from plant operating experience.

Nuclear Power Growth

A nuclear power generation forecast to the year 2000 is presented for the OECD nations* to provide a frame of reference for the magnitude of the worldwide generation of wastes (excluding those nations with centrally planned economies). Forecasts of installed nuclear capacity have been declining as a result of a downward trend in projected electric power requirements from all energy sources. For example, in the 3-yr period between 1973 and 1976, the high and low nuclear power growth estimates for OECD nations decreased 20% and 33%, respectively (Müda, Häussermann, and Mankin, 1977). Such reductions generally reflect uncertainties due to lower than expected growth in energy use and greater than anticipated delays because of concerns about safety and the environment.

Table 1 shows the results of a 1976 estimate of the nuclear power growth for OECD nations. Since 1976 an additional downward revision of from 5 to 10% in the low growth estimate has been proposed (letter from R. Gene Clark, Chief, Nuclear Energy Analysis Division of DOE to M. W. Shupe, DOE, Richland Operations Office, July 12, 1978). This table also provides a recent projection of the nuclear power growth for the United States (U.S. Department of Energy, 1978).

Waste Descriptions and Classifications

For the LWR fuel cycle, there are two generic operating modes to be considered, that with and that without spent-fuel reprocessing. In the nonreprocessing mode, currently referred to as the once-through cycle, energy values contained in irradiated fuel removed from a nuclear power plant are not recovered by reprocessing and recycling. Irradiated fuel is considered a radioactive waste and after storage for some period is sent to disposal.

The reprocessing mode includes different alternatives, for example, the recycle of uranium only or the recycle of uranium and plutonium. In this chapter we are discussing the transuranium wastes resulting from the recycling of uranium and plutonium. In this recycle uranium and plutonium are chemically recovered from the irradiated fuel and then purified and formed into fresh fuel for generating electricity in a nuclear power plant. Recycling uranium and plutonium requires a plant for processing irradiated fuels

---

*The OECD (Organization for Economic Cooperation and Development) nations are Australia, New Zealand, Canada, Japan, United States, and Western Europe.

TABLE 1   Nuclear Power Growth Estimates, GW(e)

	Year				
	1980	1985	1990	1995	2000
OECD					
Minimum	147	293	467	681	829
Maximum	172	389	680	1090	1640
USA only	61	127	195	283	380

and a plant for fabricating fresh fuel elements containing the recycled plutonium. The plants are commonly referred to as fuels reprocessing plants and mixed-oxide fuels fabrication plants, respectively. In this chapter these two plants are considered to be the only sources of transuranic-contaminated wastes for the reprocessing fuel-cycle mode. It is assumed that the recovered uranium and plutonium are converted to UF_6 and PuO_2, respectively, at the fuels reprocessing plant. The wastes from these conversion processes are included with the fuels reprocessing plant wastes.

Primary wastes are untreated initial wastes issuing from a fuel-cycle facility. The primary wastes are processed to form treated wastes. Only treated wastes are allowed to leave the confines of the originating plant and are always under careful control. Treated wastes are of two types: (1) those which are treated to reduce their activity levels so that they can be released to the environment without harm to man and (2) those which are conditioned for long-term containment so that their radioactivity will remain confined and out of contact with man's environment. The latter are covered in this chapter.

Secondary wastes are generated in treating primary wastes and in the subsequent handling of treated wastes. Thus secondary wastes are generated not only from the initial waste-processing steps but also from the storage, transportation, and disposal steps. The amount of secondary wastes is generally small compared with that of primary wastes. Nevertheless, an assessment of waste management is not complete until the effects of secondary wastes are included. Secondary wastes are of the same general classifications as primary wastes and require the same treatments. Most can be recycled to incoming primary-waste streams for treatment. In remote locations without primary-waste-treatment facilities (for instance, isolation sites), special facilities must be provided for treating the secondary wastes.

Many methods of classifying radioactive wastes are in use, e.g., the kind of radioactivity contained, the amount of radioactivity contained, the untreated physical form, and the treated physical form. It is convenient to classify the primary and secondary wastes into categories according to the treatment they require; i.e., all wastes requiring a similar treatment are included in one category. The categories and a brief generic description of each are given in Table 2. The first three waste categories listed in Table 2 are generated to some degree in almost any facility in which radioactive materials are processed, treated, or handled. Thus both primary and secondary wastes in these categories are found throughout the postfission LWR fuel cycles. The remaining four waste categories are specific to certain fuel cycles. Spent fuel as a waste is specific only to the once-through cycle, and high-level liquid waste and hulls, only to the fuel cycles that use fuels reprocessing.

**TABLE 2 Classification of Primary Transuranic Wastes
from the Postfission LWR Fuel Cycle**

Waste category	General description
Gaseous	Mainly two types: (1) large volumes of ventilation air, potentially containing particulate activity, and (2) smaller volumes of vessel vent and process off gas, potentially containing volatile radioisotopes in addition to particulate activity.
Compactible trash and combustible wastes	Miscellaneous wastes, including paper, cloth, plastic, rubber, and filters. Wide range of activity levels dependent on source of waste.
Concentrated liquids, wet wastes, and particulate solids	Miscellaneous wastes, including evaporator bottoms, filter sludges, resins, etc. Wide range of activity levels dependent on source of waste.
Failed equipment and non-compactible, noncombustible wastes	Miscellaneous metal or glass wastes, including massive process vessels. Wide range of activity levels dependent on source of waste.
Spent UO_2 fuel	Irradiated PWR and BWR fuel assemblies containing fission products and actinides in ceramic UO_2 pellets sealed in Zircaloy tubes.
High-level liquid waste	Concentrated solution containing over 99% of the fission products and actinides, except uranium and plutonium, in the spent fuel. Contains about 0.5% of the uranium and plutonium in the spent fuel.
Hulls and assembly hardware	Residue remaining after UO_2 has been dissolved out of spent fuel. Includes short segment of Zircaloy tubing (hulls) and stainless-steel assembly hardware. Activity levels are next highest to high-level liquid wastes.

A large plant can have many sources of wastes belonging to the same category. Thus the generation rate (waste volume/time) and radioactive content of the wastes from each source must be analyzed and summed to obtain the overall description of wastes in each category. This has been done for the main generic plant components of the postfission LWR fuel cycle in Tables 3 to 7. The waste volumes and activities are given per GW(e)-yr; 1 GW(e)-yr corresponds to the annual electric power needs of about 500,000 people in the United States. For the generic LWR fuel cycle, on which this chapter is based, 38 metric tons of UO_2 fuel must pass through the cycle to generate 1 GW(e)-yr. The radioactivity contributions of important isotopes present in the wastes are given individually in the tables along with total radioactivities.

Besides the wastes generated from operation of the main plant components of the postfission LWR fuel cycle, wastes will be generated from the decommissioning of these facilities (see Table 8), and miscellaneous secondary wastes will be generated in the ancillary activities of the postfission LWR fuel cycle, such as transportation and geologic isolation. The volume of these secondary wastes is minor, and their radioactivity content is insignificant compared with that of the primary wastes. These ancillary secondary wastes will be treated at the main plant waste-treatment facilities where possible. Special waste-treatment facilities, which are scaled-down versions of the large-plant waste-treatment facilities, will be installed where required, e.g., at geologic repository receiving stations.

TABLE 3   Spent Fuel as a Primary Waste (TRU)

Weight, MTHM/GW(e)-yr*	38
Radionuclide content,† Ci/GW(e)-yr	
Important activation products	
^{14}C	4.2
^{55}Fe	1.6×10^5
^{60}Co	1.6×10^5
^{95}Zr	3.1×10^3
Total activation products	4.8×10^5
Important fission products	
^{3}H	1.6×10^4
^{85}Kr	3.4×10^5
^{90}Sr	2.5×10^6
^{106}Ru	6.5×10^6
^{129}I	1.3
^{137}Cs	3.5×10^6
Total fission products	5.3×10^7
Important actinides	
Np	5.4×10^2
Pu	4.3×10^6
Am	1.5×10^4
Cm	1.9×10^5
Total actinides	4.6×10^6

*Weight is in metric tons heavy metal.
†Content 1.5 yr after removal from reactor.

Nuclear-Power-Plant Wastes

For this chapter the general wastes issuing from nuclear power plants are not considered to be transuranic contaminated. However, for the once-through cycle, irradiated fuel issuing from a nuclear power plant would be considered a transuranic (TRU) waste. The characteristics of spent fuel as a waste are given in Table 3. For the purposes of the generic treatment used here, a reference LWR fuel assembly has been defined as a composite with properties characterized as between those of PWR and BWR fuel assemblies. Each assembly weighs about 430 kg (80 wt.% core and 20 wt.% Zircaloy cladding and stainless-steel hardware) and is slightly over 5 m long.

After a period of storage in water basins, the spent fuel, if declared waste, will be placed in a container, which will be subsequently filled with helium or a metal with high thermal conductivity and sealed for ultimate disposal.

Fuel Reprocessing Plant Wastes

The fuel reprocessing plant TRU primary wastes are described in Tables 4 to 6. The characteristics of each waste after treatment are also shown in the tables.

The waste treatments shown in the tables are those defined as references for this chapter. Other waste treatments could result not only in differing characteristics in the treated wastes but also in differing amounts and types of secondary wastes. The effects of secondary-waste management are included in the tables. The volumes of primary waste shown have been increased appropriately to reflect the recycle of secondary wastes.

The general-operations wastes from the fuel reprocessing plant are described in Table 4. The three categories of solid and liquid general-operations waste include all the miscellaneous wastes; the volumes and activities can vary widely depending on plant operation. The wastes are all packaged and transported to off-site geologic isolation after treatment. Combustible wastes are incinerated to ensure that the treated wastes sent to the geologic repository are nonflammable. High-efficiency particulate aerosol (HEPA) filters are included in the "compactible trash and combustible wastes" category. The filter cartridges are punched out and packaged with compaction; the combustible filter frames are incinerated. The incinerator ashes are immobilized cement, as are all concentrated liquids, wet wastes, and particulate solids. The remaining failed equipment and noncompactible, noncombustible wastes are disassembled and packaged. Large equipment is disassembled to fit into boxes 1.2 by 1.8 by 1.8 m. All remaining general-operations wastes are packaged in 55- or 80-gal drums.

High-level liquid waste is described in Table 5. Although of relatively small volume, particularly after treatment, high-level liquid waste initially contains over 100 times more radioactivity than the rest of the wastes combined. The first solvent-extraction battery in the Purex process separates plutonium and uranium from the remaining radionuclides. The plutonium and uranium are extracted into an immiscible organic fluid and separated from the starting aqueous solution, which still contains over 99% of the nonvolatile fission products and actinides other than plutonium and uranium. After concentration this aqueous solution becomes high-level liquid waste. Since the extraction of plutonium and uranium is not perfect, some is left behind as "waste losses." The amount of plutonium and uranium present in high-level liquid waste can vary owing to waste loss. It is assumed that 0.5% of the plutonium and uranium in the spent fuel ends up in the high-level liquid waste.

The reference treatment for high-level liquid waste used for Table 5 calculations is vitrification. This treatment encapsulates the waste in a durable, temperature- and radiation-resistant glass that is cast in stainless-steel canisters. The canisters are hermetically sealed before they leave the fuel reprocessing plant.

The second most radioactive waste from the fuel reprocessing plant includes the hulls and assembly hardware (Table 6). This solid waste results from the feed-preparation step in the fuel reprocessing plant. In the feed-preparation step, the fuel assemblies are mechanically chopped to expose the fuel so that it can be dissolved in nitric acid. The hulls are the chopped segments of Zircaloy tubing from which the UO_2 fuel has been dissolved. The assembly hardware consists mainly of the fuel-assembly end fittings, which are removed before the fuel is dissolved.

The activity in hulls and assembly hardware wastes is from (1) neutron activation of Zircaloy and stainless steel; (2) neutron activation of trace materials, such as uranium in the cladding metal; and (3) residual fission products and actinides, which were either not dissolved or had diffused into the metal surfaces so that they could not be dissolved.

Gaseous streams originating in a fuels reprocessing plant must be treated to remove airborne radioactive materials before they are discharged to the atmosphere. The principal gas streams include dissolver off-gas, process vessel off-gases, ventilation air, vaporized excess water, and off-gases from the uranium fluorination process. The transuranic elements contained in the untreated gaseous streams are a small fraction of the airborne radioactive materials, and most are transported by entrainment.

Estimated transuranic activities discharged to the atmosphere following two different treatments are given in Table 7. The treatment in case 1 (Fig. 1) consists of directing the

TABLE 4   Fuel Reprocessing Plant General-Operations Wastes (TRU)

	Gaseous wastes	Compactible trash and combustible wastes	Concentrated liquids, wet wastes, and particulate solids	Failed equipment and noncompactible, noncombustible wastes
Primary waste				
Volume,* m^3/GW(e)-yr	1.2×10^8	69	6	27
Radionuclide content,† Ci/GW(e)-yr				
Important activation product				
^{14}C	2.1×10^1	2.4×10^{-4}	2.1×10^{-4}	4.2×10^{-5}
Total activation products	2.1×10^1	2.4×10^{-4}	2.1×10^{-4}	4.2×10^{-5}
Important fission products				
^{3}H	1.3×10^4	1.8×10^{-1}	1.6×10^1	3.2×10^{-2}
^{85}Kr	3.2×10^5	0	0	0
^{90}Sr	4.6×10^{-1}	2.6×10^1	2.3×10^1	4.6
^{106}Ru	1.5×10^3	1.6×10^2	7.3×10^3	1.5×10^1
^{129}I	1.3	1.7×10^{-2}	4.0×10^{-3}	2.6×10^{-5}
^{137}Cs	7.2×10^{-1}	4.0×10^1	3.5×10^1	7.2
Total fission products	3.3×10^5	6.7×10^2	8.1×10^3	1.1×10^2
Important actinides				
Np	3.6×10^{-4}	2.2×10^{-2}	1.8×10^{-2}	3.6×10^{-3}
Pu	1.4	1.8×10^4	7.1×10^3	1.4×10^3
Am	6.5×10^{-3}	3.9×10^{-1}	3.2×10^{-1}	6.5×10^{-2}
Cm	1.3×10^{-1}	7.9	6.5	1.3
Total actinides	1.6	1.8×10^4	7.3×10^3	1.4×10^3
Treated waste				
Treatment	Filtration of particulates, cryogenic isolation of ^{85}Kr, adsorption of ^{129}I on Ag–zeolite, adsorption of Ru on silica gel, conversion of ^{14}C to $CaCO_3$	Incineration, with ash and scrub solution immobilized with cement. Filters punched out, media compressed, frames incinerated	Immobilization with cement	Package with minimum treatment

Volume, m^3/GW(e)-yr	6.1 filters,‡ 0.12 ^{85}Kr, 0.01 Ru on silica gel, 0.13 ^{129}I on Ag–zeolite, 0.04 ^{14}C as $CaCO_3$	13.7	10.1	26
Containers/GW(e)-yr	2.8 cylinders of ^{85}Kr gas, each 42.5 liters, steel 0.05 55-gal drums of Ru-loaded silica gel 0.61 55-gal drums of I-loaded Ag–zeolite 0.18 55-gal drums of ^{14}C as $CaCO_3$	65.6 55-gal drums	48 55-gal drums	94 55-gal drums 1.4 containers, each 76 by 300 cm 1.1 boxes, each 1.2 by 1.8 by 1.8 m

*38 metric tons heavy metal must pass through the cycle to generate 1 GW(e)-yr.

†Content 1.5 yr after removal from reactor.

‡Untreated volume, included as primary waste in compactible-trash and combustible-wastes column.

TABLE 5 High-Level Liquid Waste (TRU)

Primary waste	
Volume,* m^3/GW(e)-yr	23
Radionuclide content,† Ci/GW(e)-yr	
Important fission products	
^{3}H	1.3×10^3
^{85}Kr	0
^{90}Sr	2.3×10^6
^{106}Ru	7.3×10^6
^{129}I	6.7×10^{-3}
^{137}Cs	3.6×10^6
Total fission products	5.3×10^7
Important actinides	
Np	1.8×10^3
Pu	3.5×10^4
Am	3.2×10^4
Cm	6.5×10^5
Total actinides	7.2×10^5
Treated waste	
Treatment	Vitrification
Volume, m^3/GW(e)-yr	2.8
Containers/GW(e)-yr	
(stainless-steel canister, 30 by	
300 cm)	12.6

*38 metric tons heavy metal must pass through
the cycle to generate 1 GW(e)-yr.
†Content 1.5 yr after removal from reactor.

dissolver and vessel off-gases and the ventilation air through an atmospheric protection system (APS). This system is comprised of a Group III prefilter followed by a single bank of HEPA filters. After treatment the gas is released through a stack to the atmosphere. Vaporized excess water and the off-gases arising from UF_6 conversion are sent directly to the stack without treatment.

The treatment in case 2 (Fig. 1) directs the dissolver off-gas through iodine, carbon, and krypton removal systems and the vessel off-gas through a filter, an iodine recovery system, and a nitrogen oxide removal system before entry into the APS. The remaining streams are handled in the same manner as in case 1. In the reference processes, iodine is removed by reaction with a silver-loaded adsorbent; carbon is captured in the form of CO_2 on a zeolite (molecular sieve) and converted to calcium carbonate; and krypton is cryogenically liquefied along with xenon and argon, followed by fractionation. Removing iodine, carbon, and krypton in conjunction with the APS results in about twice the reduction in transuranic activity discharged as with the APS alone.

Mixed-Oxide-Fuel Fabrication Wastes

The wastes from mixed-oxide-fuel fabrication are shown in Table 8. It is assumed that about 20% of the fuel is mixed oxide; thus 7.6 metric tons of mixed oxide is equivalent to 1 GW(e)-yr for the reference system.

There are no unique waste streams from the mixed-oxide-fuel fabrication plant such as those which occur at the fuel reprocessing plant. All are general-operations wastes and

TABLE 6 Hulls and Assembly Hardware (TRU)

Primary waste	
Volume,* m³/GW(e)-yr	12.3
Radionuclide content,† Ci/GW(e)-yr	
Important activation products	
^{14}C	4.6
^{55}Fe	1.5×10^5
^{60}Co	1.5×10^5
^{95}Zr	3.1×10^3
Total activation products	4.8×10^5
Important fission products	
^{3}H	2.4×10^3
^{90}Sr	1.2×10^3
^{106}Ru	3.6×10^3
^{137}Cs	1.8×10^3
Total fission products	2.9×10^4
Important actinides	
Np	9×10^{-1}
Pu	3.5×10^3
Am	1.6×10^1
Cm	3.2×10^2
Total actinides	3.9×10^3
Treated waste	
Treatment	Package without compaction
Volume, m³/GW(e)-yr	12.8
Containers/GW(e)-yr	9.2
(canister, 76 by 300 cm)	

*38 metric tons heavy metal must pass through the cycle to generate 1 GW(e)-yr.

†Content 1.5 yr after removal from reactor.

TABLE 7 Release of Airborne Transuranics from a Fuels Reprocessing Plant

	Activity,* Ci/GW(e)-yr		
	Into atmospheric protection system	Into stack	
Element		Case 1	Case 2
Neptunium	3.6×10^{-4}	3.6×10^{-8}	2×10^{-8}
Plutonium	1.4	3×10^{-4}	2×10^{-4}
Americium	6.5×10^{-3}	6.5×10^{-7}	3×10^{-7}
Curium	1.3×10^{-1}	1.3×10^{-5}	6.5×10^{-6}

*Activity 1.5 yr after removal from reactor; 38 metric tons heavy metal must pass through the cycle to generate 1 GW(e)-yr.

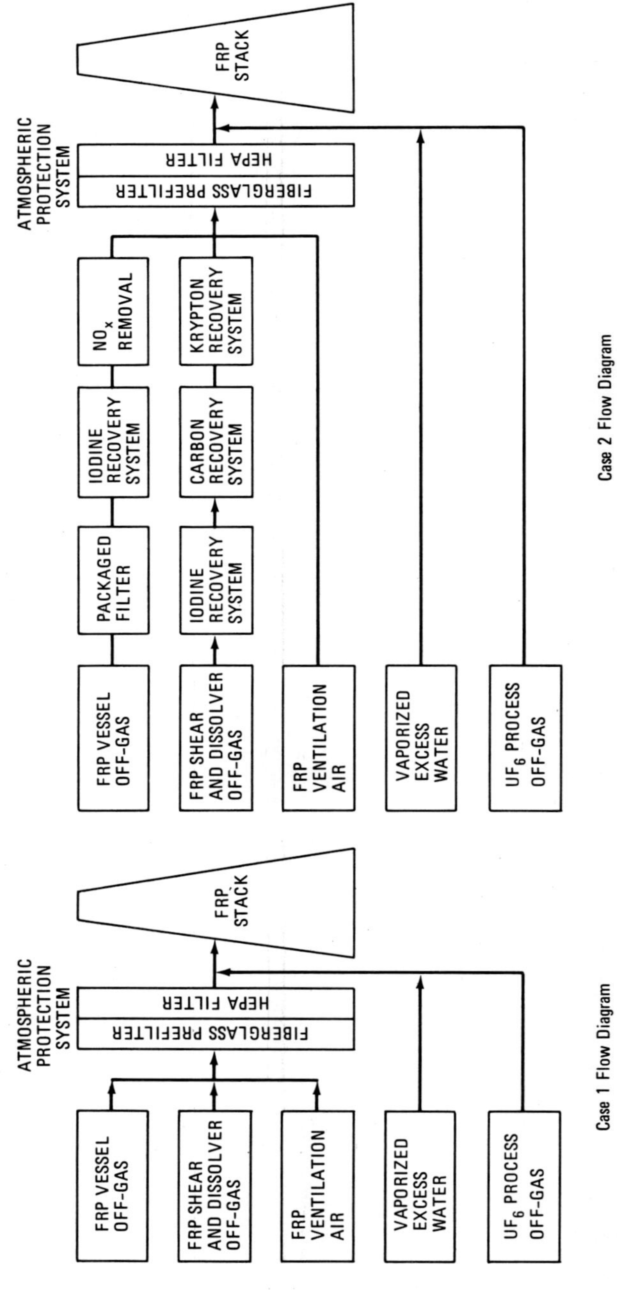

Fig. 1   Fuel reprocessing plant gaseous waste streams and atmospheric protection system.

TABLE 8 Mixed-Oxide-Fuel Fabrication Wastes (TRU)

	Gaseous wastes	Compactible trash and combustible wastes	Concentrated liquids, wet wastes, and particulate solids	Failed equipment and noncompactible, noncombustible wastes
Primary waste				
Volume,* m^3/GW(e)-yr	8.4×10^6	5	3	3
Radionuclide content (important actinides),† Ci/GW(e)-yr				
Np	3.7×10^{-10}	1.2×10^{-6}	1.3×10^{-7}	1.3×10^{-7}
Pu	1.4	4.6×10^3	5×10^2	5×10^2
Am	2.2×10^{-3}	7.3	1.5×10^2	0.8
Cm				
Total actinides	1.4	4.6×10^3	6.6×10^2	5.1
Treated waste				
Treatment	Filtration of particulates	Incineration, with ash and blowdown immobilized with cement. Filters compacted separately.	Immobilized with cement	Package with minimum treatment
Volume, m^3/GW(e)-yr	0.8 (filters)‡	4.2	2.3	3
Containers/GW(e)-yr	0	21 (55-gal drums)	11 (55-gal drums)	7.5 (55-gal drums) 0.4 (boxes, each 1.2 by 1.8 by 1.8 m)

*7.6 metric tons of mixed-oxide fuel produced per 1 GW(e)-yr.
† Based on fuel fabrication 1 yr after reprocessing (2.5 yr out of reactor).
‡ Included as primary waste in compactible-trash and combustible-wastes column.

as such can vary considerably in volume and radioactive content, depending on the day-to-day variables of plant operation. Since there are no volatiles in the gaseous wastes, filtration to remove particulates is sufficient treatment. As in the fuel reprocessing plant, excess wastewater is vaporized with the gaseous waste; thus there are no releases of liquid waste to the environment. The highly contaminated liquid wastes are immobilized in concrete. The combustible wastes are incinerated and immobilized with cement.

After immobilization and packaging, the mixed-oxide-fuel fabrication wastes are sent to geologic disposal in either 55- or 80-gal drums or in 1.2- by 1.8- by 1.8-m boxes.

A filtration system comprised of a prefilter and two banks of HEPA filters was used as the reference for removing contaminated airborne particulates from mixed-oxide-fuel fabrication-plant ventilation air discharged to the atmosphere. Vaporized excess water from processing is not filtered but is sent directly to the stack. Estimated activities of the major transuranic elements directed into the filtration system and subsequently discharged to the atmosphere are given in Table 9.

Decommissioning Wastes

Nuclear power plants and postfission fuel-cycle facilities become contaminated during power-production, fuel-cycle, and waste-treatment operations. On retirement these facilities become a waste that requires management, commonly termed decommissioning. Various alternatives are available for decommissioning these retired facilities. Three basic decommissioning modes are considered.

TABLE 9 Release of Airborne Transuranics from a Mixed-Oxide-Fuel Fabrication-Plant Center

	Activity,* Ci/GW(e)-yr	
Element	Into filter system	Into stack
Neptunium	3.7×10^{-10}	2×10^{-15}
Plutonium	1.4	5×10^{-6}
Americium	2.2×10^{-3}	8×10^{-7}

*Activity 2.5 yr after discharge; 7.6 metric tons of mixed-oxide fuel produced per 1 GW(e)-yr.

Protective storage. At shutdown the facility is prepared to be left in place for an extended period. Temporary physical barriers are constructed between the environment and radioactive contamination in the facility. Continuing surveillance is required after the facility has been placed in protective storage. Surveillance continues until all radioactivity in the facility has decayed or further decommissioning activities are carried out.

Entombment. At shutdown the facility is prepared to be left in place until all radioactivity has decayed to nonhazardous levels. Permanent physical barriers are constructed between the environment and the radioactive contamination in the facility. Minimal surveillance is required at an entombed facility.

Dismantlement. At shutdown all potentially hazardous amounts of radioactive contamination are removed from the facility to an approved disposal site. Plans for future use of the site dictate which noncontaminated portions of the facility remaining after dismantlement will be demolished and removed.

Combinations of these basic modes can also be used to decommission a retired facility. For example, a facility can be placed in protective storage at shutdown and dismantled after radioactive decay has reduced radiation levels in the facility.

The decommissioning mode assumed here for each fuel-cycle facility is dismantlement. Decommissioning by dismantlement requires that all potentially hazardous amounts of radioactivity be packaged and removed from the site to an approved disposal location. Uncontaminated portions of the facility can be reclaimed for other uses or demolished and removed. In either case there would be no restrictions on subsequent use of the site; no residual from its use in the nuclear fuel cycle would remain.

For mixed-oxide-fuel fabrication plants, immediate dismantlement after a 30-yr useful life is assumed. For nuclear power plants and fuel reprocessing plants, a 30-yr useful life is assumed, but dismantlement is preceded by 50- and 30-yr periods of protective storage, respectively, to allow short-lived activity to decay. During protective storage of nuclear power plants and fuel reprocessing plants, the radioactivity is consolidated in portions of the facility with relatively high contamination levels. Appropriate security measures are established, and a surveillance and monitoring program is maintained. Because most wastes generated in preparing for protective storage will be stored on site, major shipments of wastes will occur only at the final dismantlement.

Wastes generated during decommissioning are listed in Table 10. The wastes shown in Table 10 must be packaged and shipped from the site by truck or rail to an approved disposal location. There will also be atmospheric releases of gases and releases of water to

TABLE 10 Wastes from Decommissioning of Retired Facilities

| Waste type | Source facility | Volume of primary waste,* m^3/GW(e)-yr | | Radionuclide content,* Ci/GW(e)-yr | | | | | | | | |
| | | | | Major fission products | | | Total fission | Major actinides | | | | Total |
		Untreated	Treated	^{90}Sr	^{137}Cs	^{151}Sm	products	Np	Pu	Am	Cm	actinides
Compactible trash and combustible wastes (TRU)	Generic FRP†	0.41	0.72	2×10^{-1}	2×10^{-1}	5×10^{-3}	6×10^{-1}	1×10^{-5}	0.35	2.5×10^{-3}	3×10^{-2}	4×10^{-1}
	Generic MOX-FFP‡	0.08	0.05					7×10^{-6}	6×10^{1}	2		6×10^{1}
Concentrated liquids, wet wastes, and particulate solids (TRU)	Generic FRP	0.15	0.24	2×10^{-1}	1×10^{-1}	3×10^{-3}	6×10^{-1}	3×10^{-6}	9×10^{-2}	7×10^{-4}	8×10^{-3}	1×10^{-1}
	Generic MOX-FFP	0.20	0.13					3×10^{-6}	2×10^{1}	7×10^{-1}		2×10^{1}
Failed equipment and noncompactible, noncombustible wastes (TRU)	Generic FRP	1.4	1.4	9×10^{-1}	7×10^{-1}	2×10^{-2}	2	2×10^{-5}	5×10^{-1}	4×10^{-1}	4×10^{-2}	6×10^{-1}
	Generic MOX-FFP	1.6	1.6					7×10^{-6}	6×10^{1}	2		6×10^{1}

*To obtain volume of waste per source facility, multiply m^3/GW(e)-yr by 1600. Use the same factor to convert Ci/GW(e)-yr to total activity present in the waste of each facility at time of decommissioning.

†Fuels reprocessing plant (FRP) assumed to operate for 30 yr and decommissioned 30 yr after shutdown.

‡Mixed-oxide-fuel fabrication plant (MOX-FFP) assumed to operate for 30 yr and decommissioned at shutdown.

the ground. The gases will be subjected to the standard off-gas treatments normally used in the plants, and the atmospheric releases will be controlled to below those when the plants were operating. Water releases will also be below established limits. Water inventories from cooling basins, for instance, will be released to the ground in the location of the site but only after suitable analyses are made to ensure that the activity contained in the basin water is below established limits.

References

Adam, J. A., and V. L. Rogers, 1978, *A Classification System for Radioactive Waste Disposal—What Waste Goes Where?* Report NUREG-0456 FBDU-224-10, Office of Nuclear Material Safety and Safeguards, U.S. Nuclear Regulatory Commission.

Energy Research and Development Administration, 1976, *Alternatives for Managing Wastes from Reactors and Post-Fission Operations in the LWR Fuel Cycle,* ERDA Report ERDA-76-43, NTIS.

Miida, J., W. Häussermann, and S. Mankin, 1977, Nuclear Power Programmes and Medium-Term Projections in the OECD Area, in *Nuclear Power and Its Fuel Cycle,* Symposium Proceedings, Salzburg, Austria, 1977, pp. 213-232, STI/PUB/465, Vol. 1, International Atomic Energy Agency, Vienna.

U. S. Department of Energy, 1978, *Report of Task Force for Review of Nuclear Waste Management: Draft,* DOE/ER-0004/D.

——, 1979, *Technology for Commercial Radioactive Waste Management,* Report DOE/ET-0028, NTIS.

The Detection and Study of Plutonium-Bearing Particles Following the Reprocessing of Reactor Fuel

S. MARSHALL SANDERS, JR., and ALBERT L. BONI

A method has been developed to identify and study individual airborne particles containing ^{239}Pu from fission-fragment and alpha-particle tracks produced by them in a polycarbonate film with a nuclear-track-emulsion coating. Membrane filters, used to collect the particles from atmospheric effluents, are cast into films composed of a polycarbonate matrix containing the particles. When a particle is located, the amount of ^{239}Pu in it is determined by counting the tracks, a small portion of the film containing the particle is isolated, the emulsion removed, the polycarbonate dissolved, the track replicas oxidized, and the elemental composition of the ^{239}Pu-bearing particle determined by electron-microprobe analysis. The elemental compositions, sizes, structures, and ^{239}Pu contents were determined for 558 plutonium-bearing particles isolated from various locations in the exhaust from a nuclear processing facility at the Savannah River Plant. These data were compared with data from natural aerosol particles.

Nuclear fuel reprocessing facilities at the Savannah River Plant release to the atmosphere minute quantities (<1 mCi/yr) of ^{239}Pu in particulate form. These particles have been isolated and studied as to size, elemental composition, and radioactive properties with autoradiographic techniques.

Leary (1951) first used an autoradiographic technique to measure particle size-frequency distributions of radioactive aerosols in 1950. With his procedure, aerosols fed to and discharged from a decontamination pilot plant at Los Alamos Scientific Laboratory were collected on filter paper. A sample of the filter paper was then placed in contact with nuclear track emulsion for various exposure times. Assuming that the aerosol particles contained no nonradioactive material and that the isotopic composition of the radioactive compounds was known, Leary determined the size of each radioactive particle by counting the number of alpha-ray tracks produced by it in the emulsion for a given exposure time. This method distinguished between the radioactive and inert particles and thus was particularly useful for aerosols in which the abundance of these radioactive particles was low relative to atmospheric dust. It had, however, two serious deficiencies: (1) the actual particles were never observed and (2) plutonium could not be distinguished from uranium. In spite of these deficiencies, this method was the basis for other techniques for more than a decade.

Quan (1959) overcame the first of these deficiencies in 1959 by permanently bonding the aerosol particles between the nuclear track emulsion and the Millipore filter used in their collection so that the particles were not separated from the tracks they produced in the emulsion. With Quan's method, the Millipore filter with the contaminated surface upward was cemented with collodion to a stainless-steel frame. The upper surface was then covered with a thin collodion membrane and a gelatin bonding medium. A square of Kodak autoradiographic permeable base film was then floated onto the surface of the gelatin with the sensitive emulsion upward. After exposure and development, the Millipore filter was made transparent with a collodion solution.

About the same time Moss, Hyatt, and Schulte (1961) developed a simpler method of relating the particles with the tracks they produce. These investigators collected airborne plutonium samples on membrane filters in a processing plant at Los Alamos Scientific Laboratory where plutonium metal was handled in glove boxes. The membrane filters were pressed on a glass slide covered with nuclear track emulsion with the contaminated side in contact with the emulsion. When excess water was blotted from the slide, the emulsion, which had been softened by submerging it in water for several minutes before using it, became tacky. The filter was then separated from the emulsion, and the particles were left embedded in the emulsion.

To calculate the particle size, Moss, Hyatt, and Schulte (1961) assumed that the particles were spheres of pure $^{239}PuO_2$. However, some particles appeared larger than calculated, which led the group to speculate that, "when plutonium dust settles on surfaces and is resuspended during cleaning, the resuspended particles are much larger and are only partly composed of plutonium."

Andersen (1964) made a particle size study of plutonium aerosols in employee work areas at Hanford Laboratories in 1963. Here samples were collected on membrane filters and Hollingsworth and Vose type 70 filter paper. The filters were contact exposed to nuclear track emulsions by a method similar to that of Leary (1951). The filters were mounted in X-ray exposure holders altered so that the nuclear track film could be positioned reproducibly. These and the use of a microscope-stage micrometer permitted sufficient reproducibility in film location that a particle in question could be readily relocated. He found the plutonium particles to be small with a geometric mean diameter less than 0.04 to 0.1 μm. He assumed that plutonium was not attached to dust particles since copious quantities of plutonium were generally involved during particle formation and the effect of foreign particles was not extensive.

An autoradiographic technique for the location and examination of alpha active dust particles collected on glass-fiber paper, developed by Stevens (1963), was used by Sherwood and Stevens (1963; 1965) to analyze laboratory air samples taken at the Atomic Energy Research Establishment at Harwell, England, in 1963. Each filter was mounted in an Araldite (epoxy resin) mixture, which renders the filter transparent, and was covered with autoradiographic stripping film. After exposure and development, the sample was viewed with a high-powered optical microscope. The particles that were identified as radioactive by the alpha-particle tracks emanating from them were sized, and their radioactivity was determined by counting the number of tracks. Stevens and Sherwood found that relatively few of the particles collected were pure plutonium or plutonium compounds. Most of the particles were large inert particles contaminated with plutonium.

As late as 1965 Kirchner (1966) used the contact-exposed method of Leary (1951) to analyze air samples obtained from work areas in plutonium chemistry and fabrication plants at Rocky Flats. Although this procedure did not permit examination of inert particles, Kirchner believed that the autoradiographs indicated that agglomeration with inert or other active particles was rare. He also believed that, despite good agreement in the activity median diameters reported at Harwell (Sherwood and Stevens, 1965), lung retention patterns from accidental exposures at Rocky Flats support the premise that aerosols there are primarily pure compounds of plutonium.

Ettinger, Moss, and Johnson (1971) developed a technique in 1971 to measure ^{238}Pu particles, which was a modification of the one used earlier by Moss, Hyatt, and Schulte (1961), in which the particles were embedded in the emulsion. The samples were collected from several ^{238}PuO$_2$ glove-box areas at Los Alamos Scientific Laboratory by two-stage (cyclone-filter) air samplers and on gross filter samplers using Millipore filters. Their modified technique permitted multiple exposures of aerosols to additional nuclear track plates. Additional plates were placed against the plate containing the embedded aerosols for shorter periods of time and thus increased the range of sensitivity by providing both long and short exposures to the same particles. Ettinger felt that the alpha tracks in the emulsion being symmetrical suggested that the plutonium was not attached to inert aerosols.

Hayden (1976), at Rocky Flats, was the first to combine alpha-particle and fission-fragment tracks to isolate ^{239}Pu-bearing particles from other fissile material in 1974. The sample was placed in intimate contact with a 10-μm-thick polycarbonate film. A cellulose nitrate film was then placed on the polycarbonate film. The package was allowed to set for a predetermined exposure time. The cellulose nitrate was removed, and the remaining package was irradiated in a reactor for a desired neutron fluence. Both films were then etched and examined for tracks. The fission-fragment tracks appeared in the polycarbonate film and the alpha-particle tracks in the cellulose nitrate film. The presence of fission-fragment and alpha-particle tracks indicated that the particle contained ^{239}Pu. The presence of fission-fragment tracks alone indicated that the particle contained uranium. The Murri (Hayden, Murri, and Baker, 1972) equation was used to calculate particle size from the fission-fragment track count, and the Leary (Leary, 1951) equation was used to calculate particle size from the alpha-particle track count. However, the sample could not be mounted with precise positioning of reference symbols so that a specific particle could be evaluated. Thus the actual size of the particle or the presence or composition of inert material could not be determined.

Only two previous studies of plant effluents have been made, neither of which has used autoradiographic techniques. The first was of effluent aerosols downstream from high efficiency particulate air (HEPA) filters undertaken by Mishima and Schwendiman (1972) in 1971 at the plutonium finishing plant at the Hanford site. Filter and cascade impactor samples were taken of the stack gases and various exhaust systems of the plant to determine the aerodynamic characteristics and distribution of plutonium-bearing particles with their associated radioactivity. They found that the overall efficiency of the exhaust system was high, that little, if any, of the alpha radioactivity leaving the stack was being recycled back into the ventilation system, and that the plutonium present appeared to be attached to large, nonradioactive particles.

Systematic measurements and analyses of plutonium-bearing particles in off-gas were also made by Elder, Gonzales, and Ettinger (1974) and Seefeldt, Mecham, and Steindler (1976) at Los Alamos Scientific Laboratory in 1972. They collected samples from five operations—two research and development, two fabrication, and one recovery—where isotopes of plutonium were handled at Rocky Flats, Mound Laboratory, and Los Alamos Scientific Laboratory. The sampling stations were upstream from the HEPA filters, where aerosol concentrations were adequately high. Particle size characteristics were determined by radiometric analyses of the material deposited on each of the eight stages of Andersen impactors and material deposited on a backup membrane filter that collected particles that passed through the impactor. The material at each impactor stage had been characterized as being within a certain particle size range. Since the stage at which a particle is deposited is a complex function of actual particle size, shape, and density, the unit of size measurement used by them was the activity median aerodynamic diameter (AMAD), which is the diameter of a unit-density sphere with the same settling velocity as a plutonium-bearing particle in a population divided so that the radioactivity of all the larger particles equals that of the smaller ones.

They found that the two fabrication facilities produced the largest AMAD (4.0 and 2.7 μm) and the recovery facility produced the smallest AMAD (0.3 μm). The two research and development facilities produced intermediate size particles.

In 1975 Sanders (1976; 1977; 1978; 1979) began a study of plutonium-bearing particles in various parts of the chemical separations process exhaust system at the Savannah River Plant using autoradiographic techniques to record both fission-fragment and alpha-particle tracks.

Methods and Materials

Particle Collection

Particles are collected by drawing a fraction of exhaust air through membrane filters. These filters are polycarbonate films that are 47 mm in diameter and 5 μm thick with 3×10^8 0.1-μm-diameter pores per square centimeter, which gives a filter porosity of 0.024. The filters are supported in a polycarbonate aerosol holder.* Air is drawn through the holder by a small diaphragm pump with a Viton† diaphragm at a rate of 4 liters/min to give a face velocity at the filter of 3.8 cm/sec. At this flow the total efficiency for particle collection by the processes of impaction, diffusion, and interception, calculated according to Spurny et al. (1969) is 100% for all particles with diameters of 0.001 μm (the diameter of gas molecules) or larger.

Arrangement of the air-sampling system is shown in Fig. 1. As particles accumulate on the membrane filters, membrane porosity and airflow are reduced. Integrated airflow is measured with a dry-type test meter‡ in series with the diaphragm pump to determine the fraction of the exhaust sampled. When nitrogen dioxide is present, exhaust gas is passed through two gas-drying towers between the filter and the pump. The first tower contains indicating Drierite§ to remove moisture from the air and save the Ascarite¶ in the second tower. The self-indicating Ascarite, in turn, absorbs nitrogen dioxide to protect the pump and the dry test meter. A small flowmeter is mounted on the exhaust side of the dry test meter to give an indication of the instantaneous flow rate through the system. Air from the meter is fed back into the exhaust system to prevent its release to the service area.

Film Preparation

Figure 2 shows the procedure for converting the particle-containing filter membrane to a polycarbonate film. After air has been sampled, the radioactivity retained on each filter is

---

*The aerosol holders and membrane filters were produced by Nuclepore Corporation, Pleasanton, Calif., and obtained from them or Bio-Rad Laboratories, Richmond, Calif.

†Trademark of E. I. du Pont de Nemours & Company, Inc.

‡Manufactured by the American Meter Division of Singer.

§Trademark of W. S. Hammond Drierite Company.

¶Trademark of Arthur H. Thomas Company.

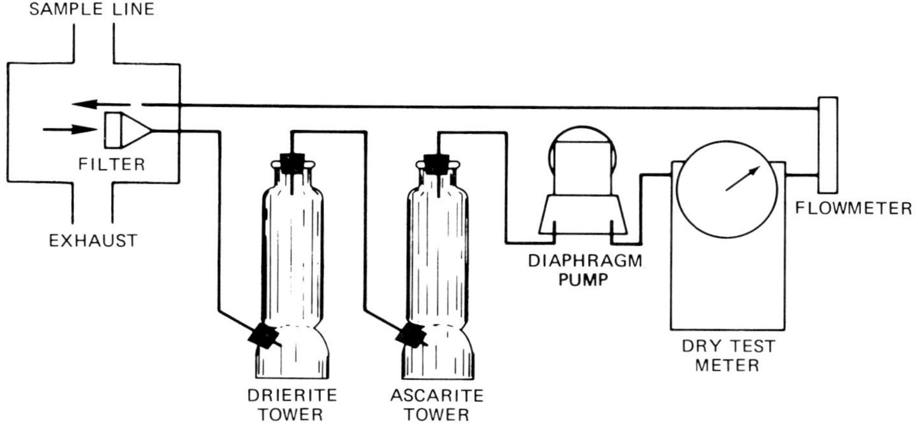

Fig. 1 Arrangement of sample collection equipment.

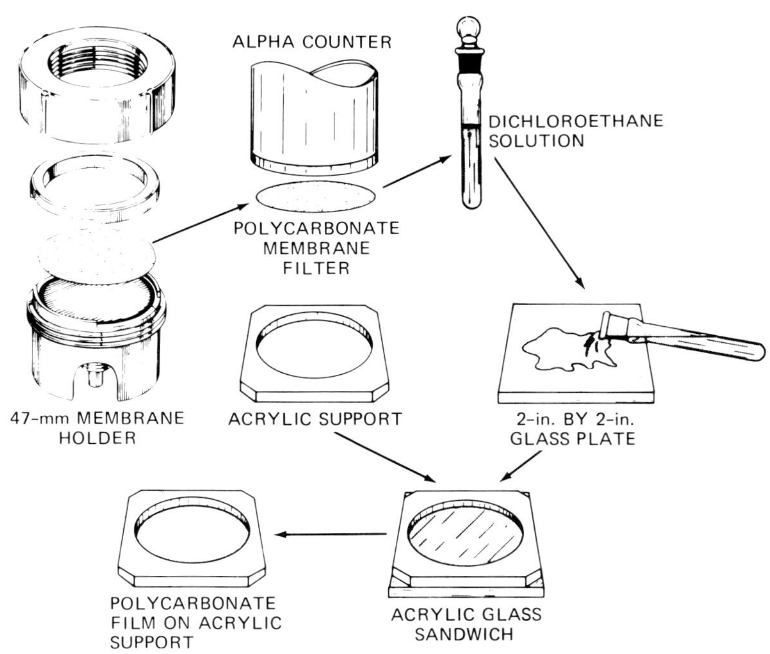

Fig. 2 Procedure for preparing polycarbonate films.

measured before it is handled in the laboratory. Each filter is then dissolved in a 40% (vol./vol.) solution of 1,2-dichloroethane in dichloromethane. The filters are folded, and each is placed in a 1-ml volumetric flask. A second clean, unused filter is placed in the same flask to give sufficient polycarbonate to form a 50-mm^2 film. Volume of the dichloroethane solution in the flask is adjusted to about ¾ ml. This mixture is stirred until the polycarbonate filters dissolve. The flasks are stoppered and allowed to stand for 30 min to allow trapped air bubbles to rise to the surface.

The clear polycarbonate solution containing the particles is poured onto a clean, 50-mm (2-in.) square glass plate (see Fig. 2). One edge of a second 50-mm^2 glass plate is used to spread the solution evenly over the surface of the first plate. The solution is stirred continuously with the second plate for about half a minute while the solution thickens. A 50-mm^2 1.6-mm-thick acrylic support with a 45-mm-diameter hole is placed on top of the wet film. The support and plate combinations are placed in covered petri dishes for 16 hr while the films continue to dry.

The glass plates are then removed by dipping the support and plate combinations in distilled water and prying the supports from the glass with tweezers.

Film Irradiation

The cast film is irradiated in a thermal neutron fluence of about 9×10^{14} neutrons/cm^2 to produce fission-fragment tracks in the polycarbonate film by which particles containing fissionable material can be identified. Films are arranged for irradiation by stacking the supports on top of each other, thus sandwiching each film between two supports. Included in the stack are blank films that are prepared in the same way as the sample films from clean unused filters. The assembled stack is wrapped with cellophane tape. Wrapped with each stack are preweighed 25.4-mm-diameter 0.25-mm-thick type 302 stainless-steel disks. The induced radioactivity from 27-day ^{51}Cr in these disks is later measured to determine the thermal neutron fluence to which the particles are exposed.

The packaged stacks are irradiated in a 3-in.-diameter hole in a light-water-cooled enriched-uranium-fueled standard pile with graphite reflectors (Axtmann et al., 1953). Following irradiation, the induced radioactivity of the stacks is allowed to decay several days before the packaged stacks are returned to the laboratory.

Film Etching

The polycarbonate film is etched for 10 min in 6N NaOH at 52 to 55°C to make the fission-fragment tracks visible with an optical microscope. During this etching process, a portion of all polycarbonate surfaces is dissolved: the outer surface of the cast film, the surface around the particle, and especially that along the fission-fragment tracks.

Emulsion Coating

For the identification of the fissionable material in each particle, the alpha-particle emission rate is measured by coating the polycarbonate film with a photographic emulsion, which is developed after a predetermined exposure time.

Kodak type NTB nuclear track emulsion is used to coat irradiated films. Under darkroom lighting (No. 2 Wratten-filtered) a 4-oz. jar of emulsion is partly immersed in a water bath that is maintained at 40°C until the emulsion melts (between 15 and 20 min). Slightly over half the molten emulsion is carefully poured into a narrow polyethylene container in the water bath. The molten emulsion is tested by dipping a clean glass microscope slide into it and examining the coat on the glass under a safelight to determine whether bubbles are present. If bubbles are present, they are scooped from the surface of the molten emulsion with a porcelain spoon.

The polycarbonate films are coated with emulsion by holding the supports containing the films vertically by one corner and dipping them into the clear molten emulsion for about 1 sec. The films are kept vertical until the excess emulsion has drained off. The coated films are then placed horizontally in a TH-Junior temperature–humidity test chamber* that is maintained at 28°C and about 80% relative humidity until the emulsion cools and gels (about 30 min).

Exposure

The polycarbonate films are exposed for 1 week before being developed to determine the alpha-particle emission rate for each aerosol particle. The films are stored in spun aluminum *Desicoolers†* containing 60 g of indicating *Drierite* during this exposure of the emulsion to the particles. The *Desicoolers* are sealed with black adhesive tape and stored in a refrigerator between 4 and 5°C for the duration of the exposure.

Emulsion Processing

At the end of the exposure period, the alpha-particle tracks in the emulsion are developed, and all substances other than tracks are removed from the emulsion. The emulsion is developed in a 1 : 1 solution of Dektol‡ developer for 3 min at 17°C (Eastman Kodak Company, 1976; Boyd, 1955; Kopriwa and Leblond, 1962).

Immediately following development the film is rinsed in 28% (vol./vol.) acetic acid for 10 sec. The high acid concentration is used to prevent reticulation of the emulsion and its separation from the supporting polycarbonate film.

The rinsed emulsion is fixed by placing it for 5 min in a 1 : 3 dilution of Kodak rapid fixer concentrate§ containing 2.8% (vol./vol.) hardener concentrate.

A batch process is used to wash all chemicals except the metallic silver from the emulsion. The emulsion-covered film is placed in distilled water, and the chemicals in the emulsion and wash water are allowed to approach equilibrium for 2 min. The emulsion is then placed in a second container of distilled water while the water in the first container is being changed. This process is repeated a total of eight times. After the water wash, the emulsion-coated polycarbonate films are placed in racks and allowed to dry in a dust-free atmosphere.

Track Counting

The film is prepared for track counting by placing the acrylic support on a 50-mm-square, 1.0-mm-thick polycarbonate block.

Particles with tracks are located under a Bausch and Lomb zoom stereomicroscope by using transmitted light and a magnification of 105 ×. Each particle with tracks is circled with a felt-tip marking pen. After a particle has been marked, the support and block holding the film are moved to a Zeiss photomicroscope where the fission-fragment and

---

*Manufactured by Tenney Engineering, Inc., Union, New Jersey.
†Trademark of Fisher Scientific Company.
‡Trademark of Eastman Kodak Company.
§Kodak Photographic Products catalog numbers 146 4106 or 146 4114.

alpha-particle tracks are counted by using transmitted light and a magnification of 1000 ×. Epiplan, flat-field objectives are used because they are corrected for uncovered specimens and do not require cover glasses. The numbers of alpha-particle tracks in the lower and upper emulsions are added to give the total number of alpha tracks observed.

Three Polaroid pictures of tracks from a single particle—one with the focal plane in the lower emulsion, one in the polycarbonate film, and one in the upper emulsion—are shown in Fig. 3.

Identification of Fissionable Materials

Table 1 gives the theoretical ratios of alpha-particle to fission-fragment tracks which would be produced from particles irradiated with a fluence of 8.64×10^{14} thermal neutrons/cm^2 when there is a 7-day interval between film casting and etching and during exposure to nuclear track emulsion. The stipulation that etching follows film casting by 7 days is included because spontaneous fissions will add to the number of fission-fragment tracks during this period.

The atom percents in the uranium mixtures in Table 1 are given in Table 2, and those in the plutonium mixtures are given in Table 3.

The isotopic mixtures of plutonium contain ^{241}Pu. All but 0.0023% of this nuclide decays by beta emission to ^{241}Am, which has a 433-yr half-life. The amount of this americium nuclide in a mixture will reach a maximum of 0.887 of the initial ^{241}Pu atom percent in 74.6 yr. Americium-241 will add additional alpha tracks to those from plutonium. Thus two ratios are given in Table 1 for each plutonium mixture, one for freshly purified plutonium and one for 75-yr-old plutonium containing the maximum ^{241}Am activity. In two of the three mixtures, this caused an increase in alpha-particle-to-fission-fragment ratios. However, with heat-source plutonium, the decrease in ^{238}Pu activity was not compensated for by the increase in ^{241}Am activity.

This identification procedure can be used to distinguish particle-bound plutonium from uranium. Table 2 shows that, of the six isotopic mixtures of uranium, only the highly enriched uranium mixture will give a number of fission-fragment tracks comparable to that of the plutonium mixtures. Even if there should be enough uranium to produce fission-fragment tracks, mixtures of these isotopes would not produce alpha-particle tracks.

This procedure can be used not only to identify plutonium but also to identify the plutonium isotopic composition in a particle. For example, a particle having 10 fission-fragment tracks would also have 5 alpha-particle tracks if the mixture were low-irradiation plutonium, 640 alpha-particle tracks if it were high-irradiation plutonium, and 5080 alpha-particle tracks if it were heat-source plutonium.

Table 1 includes (in addition to uranium and plutonium track data) a number of curium and californium nuclides that could mimic the plutonium mixtures. Some of these nuclides decay by spontaneous fission. The polycarbonate film should be allowed to stand several weeks after casting and then be etched both before and after thermal neutron irradiation to detect spontaneous fissioning. Under these conditions tracks due to spontaneous fissioning will appear in unirradiated films.

Measurement of Plutonium and Uranium Ratios

For a demonstration of the effectiveness of this identification method in distinguishing between plutonium and uranium, samples of particles were obtained from two sources of

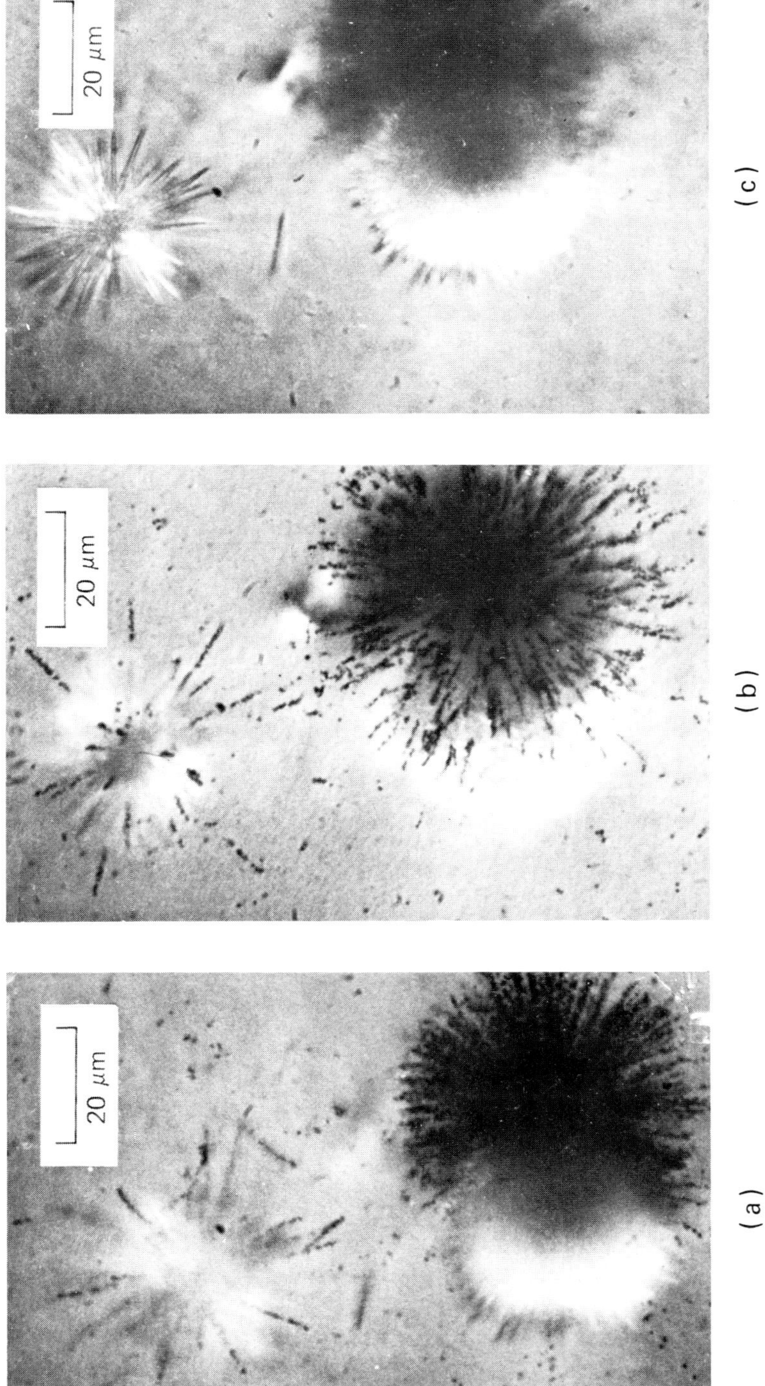

Fig. 3 Alpha-particle and fission-fragment tracks in the lower photographic emulsion (a), the polycarbonate film (b), and the upper photographic emulsion (c).

TABLE 1   Theoretical Number of Fission-Fragment and Alpha-Particle Tracks from 10^{10} Atoms and the Ratio of Alpha-Particle to Fission-Fragment Tracks

Nuclides	Fission-fragment tracks per 10^{10} atoms	Alpha-particle tracks per 10^{10} atoms	Ratio of alpha-particle tracks to fission-fragment tracks
Power reactor fuel			
(4% ^{235}U)	3.99×10^2	2.14×10^{-1}	5.36×10^{-4}
Low-burnup uranium			
(2.5% ^{235}U)	2.50×10^2	1.50×10^{-1}	6.00×10^{-4}
Highly enriched uranium			
(90% ^{235}U)	8.98×10^3	6.54	7.29×10^{-4}
Natural uranium			
(99% ^{238}U)	7.18×10^1	5.74×10^{-2}	7.99×10^{-4}
Depleted uranium			
($\sim$100% ^{238}U)	2.49×10^1	3.60×10^{-2}	1.44×10^{-3}
High-burnup uranium			
(1% ^{235}U)	8.88×10^1	1.44×10^{-1}	1.62×10^{-3}
^{247}Cm	1.25×10^3	8.10	6.48×10^{-3}
^{242m}Am	1.31×10^5	4.20×10^3	3.19×10^{-2}
^{233}U	9.06×10^3	8.20×10^2	9.05×10^{-2}
^{245}Cm	3.73×10^4	1.56×10^4	4.17×10^{-1}
Low-irradiation plutonium			
(94% ^{239}Pu)	1.20×10^4	(6.48 to 7.52) $\times 10^3$	(5.40 to 6.27) $\times 10^{-1}$
^{247}Cm (SF)	<6.64	8.10	>1.22
^{251}Cf	8.30×10^4	1.48×10^5	1.78
^{248}Cm (SF)	5.78×10^1	3.37×10^2	5.84
High-irradiation plutonium			
(40% ^{239}Pu)	8.02×10^3	(5.13 to 9.31) $\times 10^4$	(6.40 to 1.16) $\times 10^1$
^{249}Cf	2.88×10^4	3.79×10^5	1.32×10^1
^{252}Cf (SF)	3.13×10^6	5.02×10^7	1.61×10^1
^{243}Cm	1.19×10^4	4.43×10^6	3.71×10^2
Heat-source plutonium			
(80% ^{238}Pu)	2.40×10^3	(1.22 to 1.07) $\times 10^6$	(5.08 to 4.46) $\times 10^2$
^{250}Cf (SF)	1.56×10^4	1.02×10^7	6.50×10^2
^{246}Cf (SF)	1.56×10^1	2.76×10^4	1.76×10^3
^{241}Am	5.43×10^1	3.07×10^5	5.66×10^3
^{252}Cf	5.53×10^2	5.02×10^7	9.08×10^4
^{244}Cm (SF)	2.03×10^1	7.34×10^6	3.62×10^5

known nuclide mixtures, one of low-irradiation plutonium and one of highly enriched uranium. Polycarbonate films were prepared containing particles from either one or the other source. The films were irradiated and coated with emulsion, and the emulsion was exposed and developed by this procedure. The number of alpha-particle and fission-fragment tracks with each particle were counted.

The data from 315 particles containing low-irradiation plutonium are given in Table 4 and those from 350 particles containing highly enriched uranium are given in Table 5. The data were ranked according to the number of observed fission-fragment tracks per particle to determine whether the number of tracks influenced the measured ratios. The mean and standard deviation of the ratios in each track interval are also given.

TABLE 2   Atom Percents in Six Typical Mixtures of Uranium

Nuclide	Natural uranium	Power reactor fuel	Low-burnup uranium	High-burnup uranium	Highly enriched uranium	Depleted uranium
^{234}U	0.005	0.030	0.018	0.017	1.160	0.001
^{235}U	0.720	4.000	2.506	0.890	90.000	0.250
^{236}U		0.320	0.363	0.393	2.770	0.014
^{238}U	99.275	95.650	97.113	98.700	6.070	99.735

TABLE 3   Atom Percents in Three Typical Mixtures of Plutonium

Nuclide	Low-irradiation plutonium	High-irradiation plutonium	Heat-source plutonium
^{236}Pu		5×10^{-6}	1×10^{-4}
^{238}Pu	0.0115	2.9	80.3
^{239}Pu	93.6	39.6	15.87
^{240}Pu	5.9	25.6	3.00
^{241}Pu	0.4	16.8	0.72
^{242}Pu	0.013	15.0	
^{244}Pu	0.02		

From Tables 4 and 5 the mean ratio (alpha-particle-to-fission-fragment tracks) for low-irradiation plutonium is 9.1×10^{-1}, and that for highly enriched uranium is 1.8×10^{-3}. Thus ^{239}Pu can clearly be distinguished from ^{235}U by this procedure if there is a sufficient number of tracks. However, these ratios are 1.7 and 2.5 times the theoretical ratios given in Table 1. For highly enriched uranium, all alpha-particle tracks were observed as single tracks only, some of which may have been due to background radiation; this would explain the higher mean ratio for uranium. With plutonium the higher observed ratios are probably due to the geometry of the media in which the tracks are formed.

Quantitative Radiographic Analysis

Alpha-particle and fission-fragment track counts will provide not only a ratio from which the fissionable material carried on the particles can be identified but also an estimate of the quantity of the radioactive nuclides present. One femtocurie ($1 \text{ fCi} = 10^{-15} \text{ Ci}$) of ^{239}Pu will produce about 22 alpha particles in a week, and, when irradiated with a fluence of 8.64×10^{14} thermal neutrons/cm^2, it will produce about 40 fission fragments. In a mixture of low-irradiation plutonium, the number of fission fragments produced will be increased to 53 with between 28 and 33 alpha particles, depending on the age of the mixture. Only about half of these particles will produce tracks; yet this radiographic technique is much more sensitive than electron-microprobe analysis, which is not sensitive to less than 10 fCi (Sanders, 1976).

TABLE 4   Analyses of Particles Containing ^{239}Pu

Fission-fragment tracks	Number of particles	Total fission-fragment tracks	Total alpha-particle tracks	Ratio of alpha-particle tracks to fission-fragment tracks	Standard deviation
3–4	33	125	108	0.86	0.68
5–9	80	528	358	0.68	0.49
10–14	65	783	569	0.73	0.44
15–19	36	607	489	0.81	0.45
20–24	26	570	432	0.76	0.38
25–29	11	294	275	0.94	0.50
30–34	13	408	417	1.02	0.38
35–39	7	262	287	1.10	0.32
40–44	16	672	819	1.22	0.44
45–49	9	423	398	0.94	0.45
50–54	4	212	302	1.42	0.72
55–59	1	58	41	0.71	
60–64	6	371	362	0.98	0.30
65–69	2	136	124	0.91	0.08
80–84	2	160	172	1.08	0.46
85–89	1	86	86	1.00	
90–94	2	182	147	0.81	0.18
100–104	1	104	56	0.54	
Total	315	5981	5442	0.91	

TABLE 5   Analyses of Particles Containing ^{235}U

Fission-fragment tracks	Number of particles	Total fission-fragment tracks	Total alpha-particle tracks	Ratio of alpha-particle tracks to fission-fragment tracks
3–4	124	435	3	0.0069
5–9	146	935	2	0.0021
10–14	39	460	0	0.0000
15–19	18	293	0	0.0000
20–24	10	214	0	0.0000
25–29	3	82	0	0.0000
30–34	4	126	0	0.0000
35–39	3	110	0	0.0000
40–44	3	125	0	0.0000
Total	350	2780	5	0.0018

Particle Isolation

After a particle has been identified and photographed and the tracks have been counted, it is excised from the film in a polycarbonate square. For this the support and block holding the film are returned to the stereomicroscope. In transmitted illumination and at a magnification of 105 ×, two parallel cuts are made through the emulsion-coated film on either side of the particle with an ultra microlance. The film is then rotated through 90°, and two more cuts are made; the cut film then forms a square [Fig. 4(a)]. The cut square is then probed in one corner by a 15-mm-long, electrolytically sharpened tungsten needle (made by placing a pair of 0.52-mm-diameter tungsten wires in a 3N NaOH solution and applying a 60-Hz 10-volt potential between them for 10 to 15 min). With this needle the cut square containing the particle is lifted from the film and placed on a

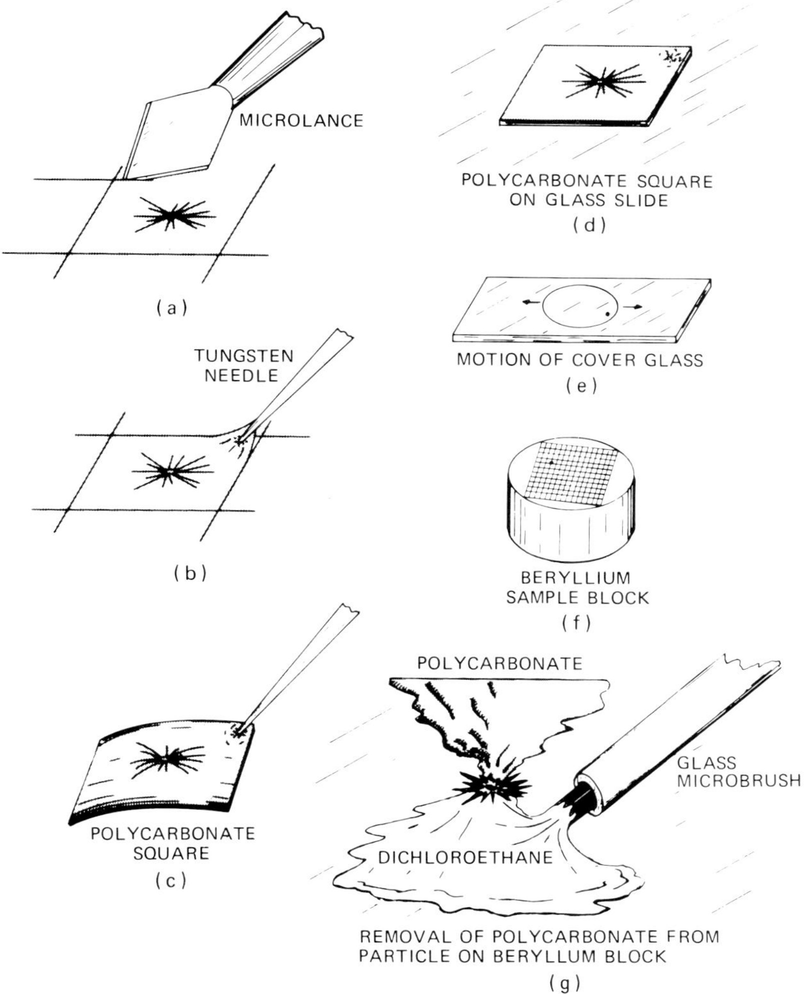

Fig. 4 Procedure for mounting particle for fissionable material identification. See text for explanation of (a) etc.

glass microscope slide [Figs. 4(b), 4(c), and 4(d)]. The polycarbonate square is freed from the needle by rotating it so that the corner of the square opposite that stuck by the needle strikes the slide and causes the square to rotate and fall.

The emulsion layers are then removed from the polycarbonate square by placing a cover glass on top of the square. Water is introduced between the cover glass and slide with a glass microbrush made from a 20-μl glass disposable pipet. (A 0.025-mm-diameter tungsten wire doubled and threaded through the lumen forms a loop at one end. A small amount of glass wool is placed through this loop and is then drawn into the end of the pipet. The glass fibers are then cut off about 2 mm from the end of the pipet.) The microbrush is dipped in water, and the glass fibers are touched to the edge of the cover glass to allow the water to flow from the brush to between the slide and the cover glass.

The emulsion is then removed by gently moving the cover glass a few millimeters from side to side [Fig. 4(e)]; this rolls the swollen emulsion from the surface of the film but not from the fission-fragment tracks themselves. The cover glass is carefully lifted from the glass microscope slide, taking care not to lose the polycarbonate square containing the particle.

Particle Mounting

To mount a particle, the polycarbonate square is placed in a selected grid location on a beryllium sample mounting block* [Fig. 4(f)]. These sample mounting blocks are 25 mm in diameter and 13 mm thick and fit the standard electron-microprobe sample holders, which grip the sides and provide the necessary electrical contact. The top surface of the block is highly polished and contains a grid network of 1-mm squares inscribed on the surface. The squares are numbered in mirror-image fashion both vertically and horizontally through the center.

With coaxial (reflected light) illumination and 15 × magnification under a stereo-microscope, the polycarbonate squares are moved from the microscope slide to the beryllium block with an electrolytically sharpened tungsten needle.

The polycarbonate square is then dissolved and washed back from the particle with dichloroethane, leaving the particle usually connected to the main body of polycarbonate by a thin isthmus of plastic. This connection does not seriously affect the microprobe analysis and aids in later locating the particles and holding them on the beryllium block. A glass microbrush is rinsed in dichloroethane to remove any foreign material and is filled by immersing the bristled end in a second beaker of dichloroethane. The magnification was increased to 105 ×. Dichloroethane from the brush is dispensed on the beryllium block just in front of the polycarbonate square until the square is engulfed in the solution. The microbrush is then used to push the solution back from the particle. Gelatin replicas of the fission-fragment tracks remained with the particles.

The beryllium block is returned to the photomicroscope where a second Polaroid picture of each particle is made at a magnification of 556 × to identify the particles after the gelatin has been removed.

The gelatin with each particle is oxidized by exposure to an oxygen plasma for 3 hr in a low-temperature asher.† In this asher a gas plasma is generated in oxygen with the energy of electrons in the gas. Power is supplied to electrons at 13.56 MHz by a

---

*Walter C. McCrone Associates, Inc., catalog number XIII-403-3.
†Manufactured by International Plasma Corporation.

radiofrequency generator. Since the energy to do this with a low-temperature asher is provided through the electrons instead of heat energy, high-temperature degradation, volatilization, or fusion of the inorganic constituents of the particles is eliminated.

Figure 5 illustrates the last three stages in the preparation of one particle. Part a is the particle in the polycarbonate film with emulsion stripped off. Part b is the same particle with the polycarbonate removed showing the gelatin replicas of the fission-fragment tracks. Part c is a scanning electron micrograph of the particle after oxidation of the gelatin. In this photograph traces of the gelatin replicas and silver grains can be seen. Here what had appeared to be a single particle is actually a conglomeration of at least five and possibly ten smaller particles.

Particle Sizing

For control of particles after the gelatin track replicas have been oxidized, the beryllium sample block is returned to the photomicroscope where each particle is located and photographed again under reflected light by using Polaroid film and a magnification of 556 X. An arrow pointing to the particle is marked on the film so that there will be no mistake in what is intended for analysis.

The size of each particle is estimated from these Polaroid pictures taken after oxidation of the completely denuded particles. An average of the smallest and largest dimensions of the photographed particle is measured in micrometers and divided by the magnification.

Elemental Analysis

For the determination of elemental composition of the particles, the particles are analyzed on a Cameca MS46 electron microprobe, equipped with four crystal, wave-length-dispersive spectrometers (take-off angle, $18°$) and an EDAX 701/MICROEDIT* energy-dispersive analyzer. X-ray intensities resulting from the electron bombardment of the particles and particle sizes and shapes are estimated. These estimates, along with estimated average densities, are used in the FRAME program (Yakowitz, Myklebust, and Heinrich, 1973) as modified for particles work by Armstrong and Buseck (1975) on a UNIVAC† 1110 computer. This calculation gives the particle composition in elemental weight percents. Ratios of the elemental weight percents are used to calculate enrichment factors, explained in the appendix, which are used to compare the composition of these particles with that of other aerosols.

Sampling Locations

Particles were collected from air in both exhaust systems in nuclear fuel reprocessing facilities at the Savannah River Plant. A schematic diagram of these systems is given in Fig. 6. System I takes room air from inside wet cabinets (where plutonium is in solution) and from work areas and exhausts it via the JB-Line stack (Sanders, 1976). System II takes air from the mechanical line (where plutonium is handled in metallic form) and exhausts it via the 291-F stack. In System I samples were taken of unfiltered cabinet air from the fifth and sixth levels (sampling points 29 and 30, respectively), of filtered air from both locations (sampling point 27), unfiltered room air from the fifth level

---

*Trademark of EDAX Internation, Inc.
†Trademark of Sperry Rand Corporation.

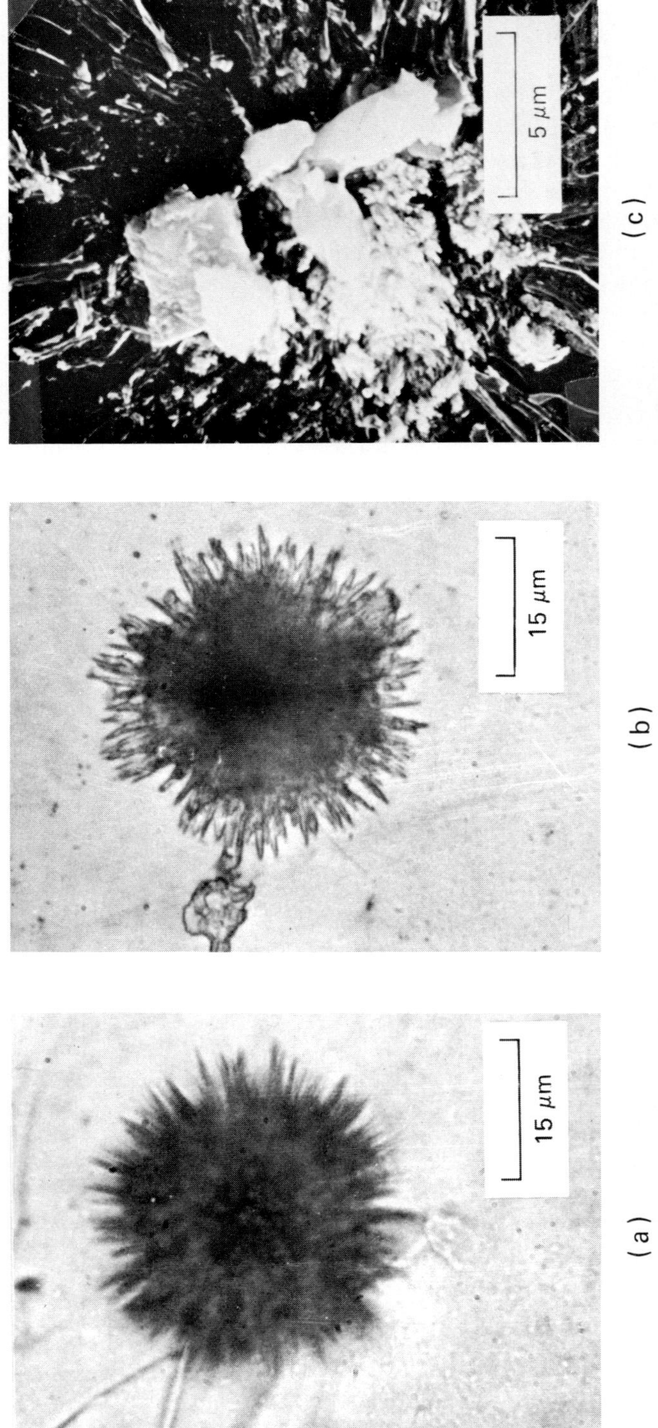

Fig. 5   Plutonium-bearing particles in last stages of mounting. (a) Particle in polycarbonate film with emulsion stripped off. (b) Same particle with polycarbonate removed showing the gelatin replicas of the fission-fragment tracks. (c) Scanning electron micrograph of the particle after oxidation of the gelatin.

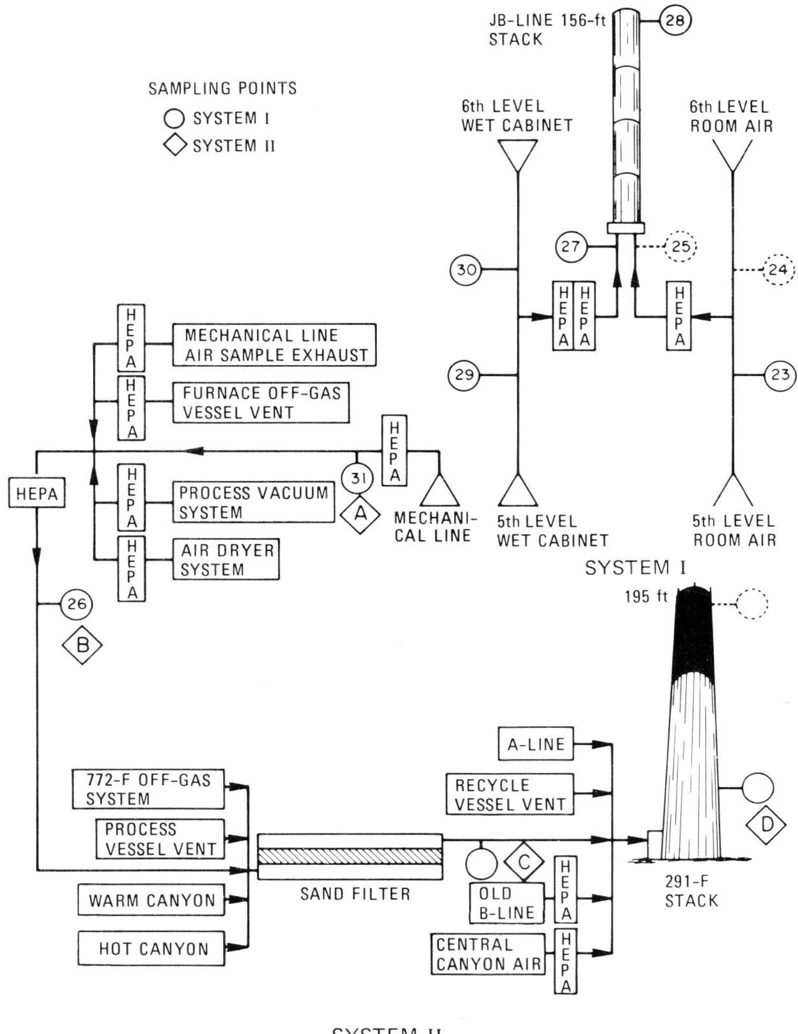

Fig. 6   Savannah River Nuclear Fuel Reprocessing Facility exhaust systems.

(sampling point 23), and of air at the 156-ft level of the JB-Line stack (sampling point 28). In System II samples were taken of mechanical line air from just beyond the first HEPA filters located in back of the cabinets (sampling point A or 31); of the combined air from the mechanical line, air sample exhaust, furnace off-gas vessel vent, process vacuum system, and air-dryer system after the second HEPA filter (sampling point B or 26); of the air leaving the sand filter, which also contained air from the support laboratory off-gas system of building 772-F, the fuel dissolving and extraction process vessel vent system, and building 221-F canyons containing the process vessels (sampling point C); and of air from the 50-ft level in the 291-F stack where air from the sand filter mingles with that from the uranium recovery A-line and other sources (sampling point D).

A total of 121 particles were analyzed from System I (16 from sampling point 23, 67 from point 29, and 38 from point 30) and 417 from System II (125 from sampling point A, 107 from point B, 114 from point C, and 71 from point D). These figures do not include 20 particles that contained no elements with atomic numbers greater than 9 and were assumed to be organic.

Grouping of Data by Enrichment Factors

The results were expressed in terms of "enrichment factors" (dimensionless ratios of elemental concentrations), which enabled the intercomparison of the compositions of plutonium-bearing particles with other atmospheric aerosols and the intracomparison among particles collected from different sampling points. A definition of enrichment factors and an explanation of their development and application in this work are given in the appendix.

For a comparison of the chemical composition of the particles collected from Systems I and II with each other and with the average for global crustal aerosol, the particle analyses were grouped according to the level of the enrichment factors. Four groups were established for each element by using the elemental concentration data in Table A.1 of the appendix. The first group contained particles with no detectable amounts of the element sought. The second group contained detectable amounts with enrichment factors less than one standard deviation below the geometric mean enrichment factor, $\overline{EF}_g/s_g$. The third group contained particles with enrichment factors between the lower and upper limits of one standard deviation from the geometric mean enrichment factor, $\overline{EF}_g/s_g$ and $\overline{EF}_g \times s_g$, respectively. The fourth group contained enrichment factors greater than one standard geometric mean enrichment factor, $\overline{EF}_g \times s_g$. The third column of Table 6 gives the percent of the particles analyzed which gave positive analyses for each element. The fourth, fifth, and sixth columns of Table 6 contain the percent of those having positive analyses which had enrichment factors less than, between, and more than the lower and upper limits of the geometric standard deviation.

For a comparison of the chemical composition of particles collected at the various sample points in System II with each other and with global crustal aerosol (Table A.1), this process was repeated, and the results are listed in Table 7.

Particles with no detectable amounts of an element were not counted with those with enrichment factors less than the lower limit for the geometric standard deviation (s_g) because there can be no zero or negative concentration of enrichment-factor values in log-normal frequency distributions. Thus the size of the three groups is expressed as the percent of the particles giving positive analyses rather than the percent of the total number of particles.

Particle Evaluation by Size

In this study particles were selected for analysis on the basis of the number of observed fission-fragment tracks. Since there were many more particles than could be analyzed, those having 3 or 4 tracks were generally passed over in favor of those surrounded by 50 or more tracks. The selection of particles for analysis, however, was not biased by

**TABLE 6 Comparison of Analyses of Particles
from Systems I and II**

			% of positive analyses*		
Element	System	Positive analyses, %	Less than†	Within‡	Greater than§
Silicon	I	100	47	24	29
	II	99	29	30	41
Aluminum	I	84	0	100	0
	II	88	0	100	0
Iron	I	93	14	35	51
	II	79	36	33	31
Calcium	I	70	53	30	17
	II	52	41	40	19
Sodium	I	70	13	72	15
	II	54	8	81	10
Potassium	I	90	56	30	14
	II	63	35	41	24
Magnesium	I	51	24	59	17
	II	39	38	52	10
Titanium	I	74	20	17	65
	II	31	12	13	76
Phosphorus	II	1	0	17	83
Manganese	I	10	0	0	100
	II	12	4	8	88
Barium	II	0.5	0	0	100
Sulfur	I	17	47	47	5
	II	70	28	60	13
Chlorine	I	34	13	67	21
	II	40	2	82	16
Chromium	I	53	0	18	82
	II	29	0	9	91
Nickel	I	56	2	25	73
	II	9	0	3	97
Zinc	I	64	4	41	55
	II	45	5	52	43
Cobalt	II	1	0	0	100
Scandium	II	0.2	0	0	100
Copper	I	36	12	37	51
	II	7	6	29	65
Tungsten	I	1	0	0	100
	II	0.5	0	0	100
Cadmium	II	0.2	0	0	100

*The percent of the positive analyses less than, within, and greater than one geometric standard deviation of the global geometric mean enrichment factor.

†$EF < \overline{EF}_g/s_g$.

‡$\overline{EF}_g/s_g \leqslant EF \leqslant \overline{EF}_g\,s_g$.

§$EF > \overline{EF}_g\,s_g$.

TABLE 7   **Comparison of Analyses of Particles from Sampling Points A, B, C, and D of System II**

			% of positive analyses*		
Element	Sampling point	Positive analyses, %	Less than†	Within‡	Greater than §
Silicon	A	99	35	33	32
	B	98	8	24	69
	C	100	36	31	33
	D	99	40	31	29
Aluminum	A	79	0	100	0
	B	94	0	100	0
	C	89	0	100	0
	D	96	0	100	0
Iron	A	98	31	22	46
	B	100	40	39	21
	C	58	33	41	21
	D	49	46	34	20
Calcium	A	56	20	44	36
	B	77	54	38	9
	C	41	45	40	15
	D	27	53	32	16
Sodium	A	55	18	60	22
	B	90	4	94	2
	C	39	5	82	14
	D	24	6	94	0
Potassium	A	76	20	48	31
	B	73	46	44	10
	C	55	35	32	33
	D	37	54	27	19
Magnesium	A	63	33	58	9
	B	48	35	55	10
	C	17	47	42	11
	D	21	60	27	13
Titanium	A	42	12	6	83
	B	27	10	24	66
	C	30	14	11	74
	D	15	9	18	73
Phosphorus	A	2	0	50	50
	C	3	0	0	100
	D	1	0	0	100
Manganese	A	7	13	0	88
	B	30	3	9	88
	C	5	0	0	100
	D	6	0	25	75
Barium	A	1	0	0	100
	B	1	0	0	100
Sulfur	A	58	30	54	17
	B	93	21	69	10
	C	66	24	61	15
	D	61	47	44	9

TABLE 7 (Continued)

Element	Sampling point	Positive analyses, %	% of positive analyses*		
			Less than†	Within‡	Greater than §
Chlorine	A	43	4	47	49
	B	72	1	99	0
	C	27	0	97	3
	D	10	14	86	0
Chromium	A	27	0	6	94
	B	58	0	15	85
	C	13	0	0	100
	D	14	0	0	100
Nickel	B	27	0	3	97
	C	4	0	0	100
	D	7	0	0	100
Zinc	A	53	14	35	51
	B	88	0	69	31
	C	22	4	24	72
	D	6	0	75	25
Cobalt	B	5	0	0	100
Scandium	C	1	0	0	100
Copper	A	22	7	33	59
	B	3	0	0	100
	C	1	0	0	100
Tungsten	A	1	0	0	100
	B	1	0	0	100
Cadmium	D	1	0	0	100

*The percent of the positive analyses less than, within, and greater than one geometric standard deviation of the global geometric mean enrichment factor.

†$EF < \overline{EF}_g/s_g$.

‡$\overline{EF}_g/s_g \leqslant EF \leqslant \overline{EF}_g\, s_g$.

§$EF > \overline{EF}_g\, s_g$.

physical size. The size of the particles was not measured until after the particles had been mounted and the polycarbonate film containing the tracks dissolved. Thus the size distribution of the analyzed particles is indicative of the size distribution of particles in the aerosol carrying most of the plutonium.

Cumulative frequency plots were constructed for particles from Systems I and II. Particles in each system were first ranked in order of their approximate diameter (in micrometers) from the smallest to the largest. A list of the number of particles with successively larger diameters was made. A cumulative total of the number of particles at increasing diameter segments was calculated and then normalized by dividing by the total number of particles from each system. This gave the fraction of the particles having a diameter equal to or smaller than any particular diameter. Table 8 lists the particle

TABLE 8   Comparison of Size Distributions of Particles from Systems I and II with Natural Aerosols*

Diameter (D), μm	Fraction with diameter $\leqslant$ D					
	System I	Sampling point A	Sampling point B	Sampling point C	Sampling point D	Natural aerosol
0.4	0.03			0.01	0.01	
0.5	0.04			0.02		
0.9	0.07			0.04	0.06	
1.1				0.09	0.08	0.25
1.2				0.10		0.42
1.4				0.11		0.64
1.7				0.14	0.11	
1.8	0.28	0.02		0.17		0.83
2.2	0.30			0.32	0.21	0.91
2.5				0.34	0.23	
2.7	0.35	0.04	0.01	0.37	0.24	0.949
3.0				0.42	0.31	
3.3				0.43	0.32	
3.6	0.53	0.10	0.06	0.46	0.34	0.979
3.9		0.11				0.983
4.0			0.07	0.51	0.38	
4.4				0.52	0.41	
4.5	0.54	0.13	0.10			0.989
5.0			0.11	0.55	0.48	
5.4	0.62	0.20	0.18	0.60		0.994
5.8			0.20	0.59		
6.1			0.22	0.61	0.56	
6.3		0.23	0.26			0.996
6.7			0.27	0.64	0.58	
7.0				0.67		
7.2	0.67	0.33	0.35	0.68		0.997
7.4				0.68		
7.8			0.36	0.70		
8.0	0.68	0.34	0.37	0.71	0.59	0.998
8.6			0.39		0.63	
9.0	0.75	0.40	0.46	0.74	0.69	0.999
10.0		0.41	0.48	0.78	0.75	
10.8	0.79	0.44	0.60	0.81		
11.7		0.46	0.61	0.82	0.79	
12.6	0.83	0.54	0.62	0.83	0.80	1.000
13.5		0.55	0.66		0.82	
14.4	0.84	0.58		0.89		
14.9			0.67		0.83	
16.2	0.88	0.61	0.71			
17.1		0.62		0.89	0.89	
18.0	0.92	0.64	0.75	0.90		
20.7		0.70	0.81	0.91	0.92	
21.6	0.93	0.71	0.82			
23.4		0.74	0.85	0.930	0.930	
24.3		0.75	0.86			

TABLE 8   (Continued)

Diameter (D), μm	System I	Sampling point A	Sampling point B	Sampling point C	Sampling point D	Natural aerosol
			Fraction with diameter ≤ D			
25.2	0.94	0.78	0.87		0.972	
26.9	0.95	0.81	0.91	0.939		
27.9		0.83	0.92			
28.8			0.93	0.956	0.986	
30.6		0.86	0.935			
31.5		0.87		0.974		
32.4		0.88	0.944			
33.5		0.89	0.963			
34.2	0.959	0.94	0.972	0.982		
35.1				1.000	1.000	
36.0	0.975	0.944	0.981			
39.6		0.968	0.991			
41.4	0.983		1.000			
50.4	0.992					
53.9		0.992				
59.4	1.000					
62.9		1.000				

*The percent of the positive analyses less than, within, and greater than one geometric standard deviation of the global geometric mean enrichment factor.

diameters (in micrometers); and, in columns 2, 3, 4, 5, and 6, the fraction of the particles having diameters equal to or less than each diameter measured in System I and sampling points A, B, C, and D in System II, respectively. These fractions are also plotted on the logarithmic probability graph given in Fig. 7.

For comparison a cumulative frequency plot was also made of the size distribution of particles in natural atmospheric aerosols. A very simple function that has been used extensively in atmospheric research to express particle size distribution in both natural and polluted atmospheres is

$$\frac{dN}{dD} = aD^{-b} \tag{1}$$

where N is the number concentration or total number of particles per unit volume having diameters from the lower limit of definition of aerosols up to diameter D (in micrometers). From the relationships

$$dD = D\,d(\ln D) \tag{2}$$

and

$$\ln D = \ln 10 \log D \tag{3}$$

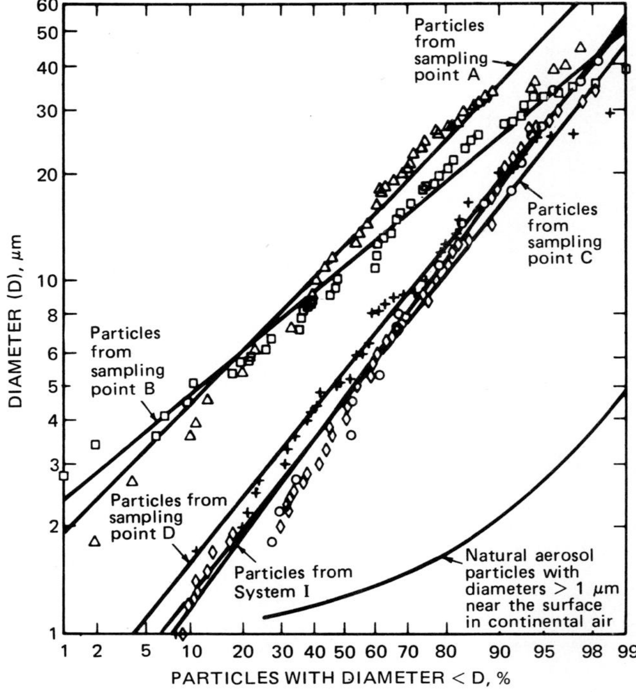

Fig. 7   Size distribution plots for natural and collected particles.

the more useful expression

$$\frac{dN}{d(\log D)} = (\ln 10)aD^{-c}$$

(4)

is obtained, where $c = b - 1$ and $dN/d(\log D)$ is the number distribution. Junge (1965) found c to be about 3 over the size range $-0.7 < \log D < 1.5$ or $0.2 < D < 32$ µm. Integrating the first equation between D_0 and D ($D_0 < D$) gives

$$N = \left.\frac{aD^{-c}}{c}\right]_D^{D_0} = \frac{a}{3}\left(\frac{1}{D_0^3} - \frac{1}{D^3}\right)$$

(5)

Instead of expressing the distribution as the number of particles per unit volume, it can be expressed as a fraction, F, of the total number of particles, or

$$F = \frac{N}{N_T} = 1 - \left(\frac{D_0}{D}\right)^3$$

(6)

where N_T is the total number of particles when $D = \infty$ and $N_T = a/3D_0^3$. For a reasonable distribution, only those particles which could be easily seen with an optical microscope were included. Thus D_0 was assumed to be 1 µm, and Eq. 6 can be expressed as

$$F = 1 - \frac{1}{D^3}$$

(7)

The frequency distribution for natural aerosols with particle diameters between 1 μm and D, calculated from this expression, is given in column 7 of Table 8 and plotted in Fig. 7.

To see how closely the distribution of particle diameters resembles a log-normal distribution, we assumed that the observed diameters represented a sample of a population having a log-normal distribution. The geometric mean diameter, $\overline{D}_g$, and geometric standard deviation, s_g, were calculated from these data by using equations similar to those given earlier for the geometric mean enrichment factor and geometric standard deviation. These values are given in Table 9. Values for the upper 68.27% limit for the diameters were calculated from the product of $\overline{D}_g$ and s_g. The best-fit log-normal probability curves were plotted on the logarithmic probability graph in Fig. 7 by drawing straight lines through coordinates for $\overline{D}_g$ and $\overline{D}_g \times s_g$ on the 50.00 and 84.14* cumulative percent abscissae, respectively.

TABLE 9 Distribution of Particle Diameters in Systems I and II

System	Sample location	Data points, N	Geometric mean diameter $(\overline{D}_g)$	Geometric standard deviation (s_g)	Skewness (SK)
I		121	4.64	2.92	0.71
II	A	125	12.27	2.24	0.04
II	B	107	10.82	1.93	0.34
II	C	114	4.48	2.75	0.37
II	D	71	5.43	2.69	0.23

To determine the degree of asymmetry, we calculated the skewness (SK) of these frequency distributions by using the relationship

$$SK = 3 \left(\frac{\ln \overline{D}_g - \ln D_{med}}{\ln s_g} \right) \tag{8}$$

where D_{med} is the median diameter. A perfect log-normal distribution has a skewness of zero. If a distribution has a higher tail to the right than to the left, it is positively skewed. Most of the distributions encountered here are negatively skewed, i.e., have higher left-hand tails. Calculated skewness values are given in Table 9.

Particle Evaluation by Plutonium Content

Another characteristic studied was the distribution of plutonium among the particles as indicated by the observed number of fission-fragment tracks in the surrounding polycarbonate.

The track distribution among particles from both systems was evaluated in the same way as the particle diameters. The fraction of the particles with the number of tracks

---

*$50.00 + \dfrac{68.27}{2}$.

equal to or less than a selected number, T, are given for sampling points A, B, C, and D in Table 10. Figure 8 is a logarithmic probability plot of cumulative percent of particles from each of these sampling points. Figure 9 is a similar plot for particles from four locations in System I. The calculated geometric mean for the number of fission-fragment tracks per particle, the geometric standard deviation, and the skewness for particles from each sampling point are given in Table 11. Best-fit log-normal probability curves for each distribution are plotted in Figs. 8 and 9. For comparison of the track distributions for particles from the various sampling points in System I with those from System II, the probability curve for the track distribution for particles from sampling point A in System II is plotted with the distributions from System I in Fig. 9.

Table 10   Distribution of Fission-Fragment Tracks Among Plutonium-Bearing Particles Collected from Sampling Points A, B, C, and D

Number of tracks	Fraction with tracks $\leqslant$ T			
	Sampling point A	Sampling point B	Sampling point C	Sampling point D
1	0.04			
2	0.05			
3	0.09			
4	0.13			0.01
5	0.15		0.06	0.03
6	0.19	0.01	0.09	
7	0.21		0.11	
8	0.26		0.13	0.06
9	0.31	0.03	0.15	0.10
10	0.34	0.04	0.20	0.13
11	0.36		0.24	0.21
12	0.38		0.26	0.25
13	0.40	0.05	0.32	0.28
14	0.44	0.07	0.38	0.32
15	0.45	0.07	0.44	0.42
16	0.47	0.08	0.48	0.46
17	0.50	0.11	0.49	0.58
18	0.51	0.15	0.54	0.63
19	0.54	0.17	0.56	0.69
20	0.59	0.21	0.59	0.70
21	0.60	0.22	0.63	0.72
22	0.63	0.26	0.66	0.75
23	0.64	0.31	0.70	0.77
24	0.68	0.37	0.72	0.82
25	0.70		0.75	
26	0.72	0.39	0.78	
27	0.74	0.40	0.82	
28	0.75	0.44	0.87	0.83
29	0.76	0.45	0.89	0.85
30	0.78	0.48	0.90	0.86
31	0.79	0.50	0.92	0.89
32	0.81	0.52	0.93	
33	0.82	0.56	0.947	0.92
34	0.84	0.58		0.93
35	0.85	0.59		0.944

Number of tracks	Fraction with tracks $\leqslant$ T			
	Sampling point A	Sampling point B	Sampling point C	Sampling point D
36	0.86	0.60	0.956	
37		0.61	0.965	0.958
38	0.87	0.62		
39	0.88			
40	0.89	0.63		0.972
41		0.65		
42	0.90	0.66		
43		0.68		
44	0.91	0.69		
46	0.91			
47		0.73		
48	0.92	0.75		
49	0.93	0.76		
50	0.945	0.77		0.986
51	0.950			
52		0.78		
54	0.955		0.991	
55		0.80		
57		0.81		
58		0.84		
59		0.85		
60	0.960	0.89		
63		0.90		
65			1.000	
68		0.91		
70		0.92		
72	0.965			
73		0.93		
75	0.970			
80	0.980	0.93		1.000
82		0.953		
84		0.972		
98		0.981		
100	0.990	0.991		
150	0.995	1.000		
200	1.000			

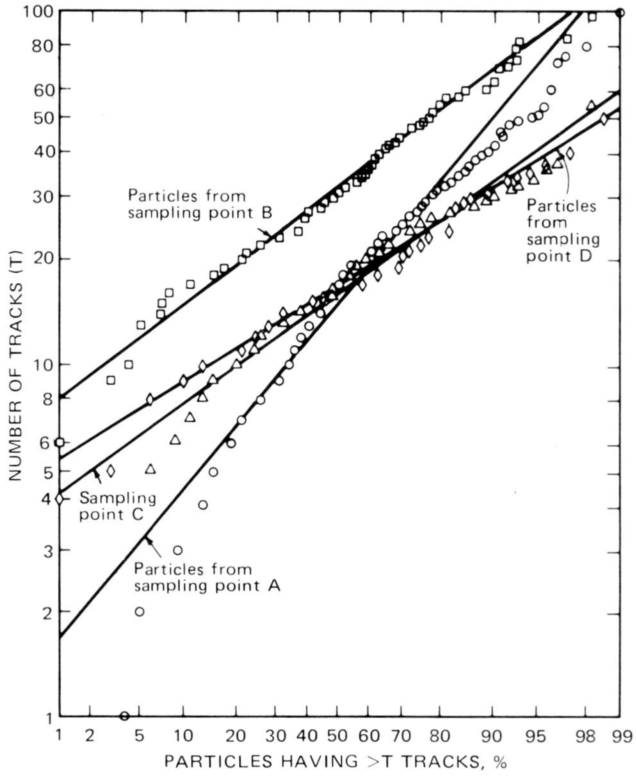

Fig. 8 Distribution of the number of tracks per particle for particles collected from sampling points A, B, C, and D in System II.

Discussion

The most abundant elements in average crustal rock (and soil) are oxygen (46.60%), silicon (27.72%), aluminum (8.13%), iron (5.00%), calcium (3.63%), sodium (2.83%), potassium (2.59%), magnesium (2.09%), and titanium (0.44%) (Mason, 1966). Except for oxygen, which was not detected by electron-microprobe analyses, these elements are also found in most inorganic particles (Tables 6 and A.1). This supports the idea that most plutonium-bearing particles are airborne crustal material to which minute quantities of plutonium have become attached.

Of particular interest is the quantity of ^{239}Pu contained on these particles. One femtocurie of ^{239}Pu irradiated under the conditions described here should produce 41 fission-fragment tracks. The minimum detection limit for electron-microprobe analysis of plutonium is about 0.2 pg, or about 10 fCi, of ^{239}Pu (Sanders, 1976), which is equivalent to 410 fission-fragment tracks. Because of this relatively low sensitivity of electron-microprobe analysis, plutonium could be detected by this method in only 1 of the 558 particles selected for analysis, even though all the particles produced fission-fragment tracks. This single particle was a small, 1-μm-diameter particle collected from unfiltered wet-cabinet exhaust. It contained 73% PuO_2 by weight (equivalent to 170 fCi of ^{239}Pu) in combination with Fe_2O_3 and mica.

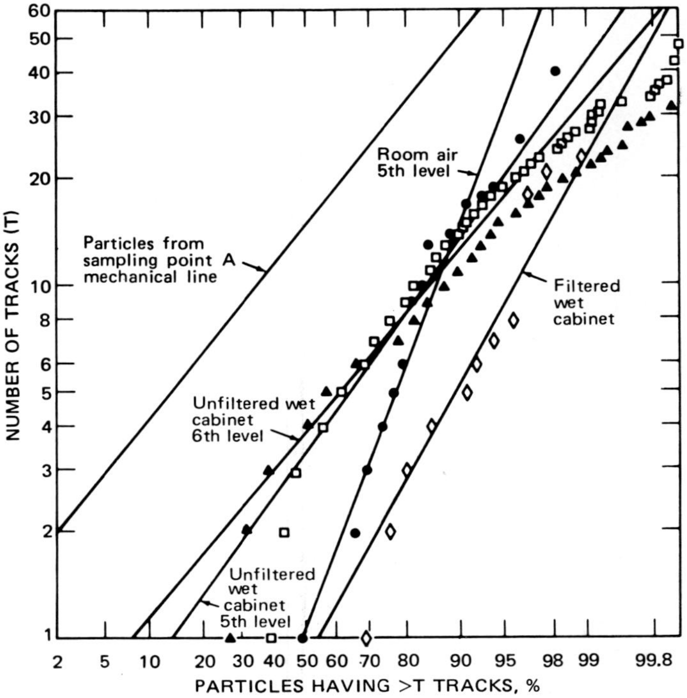

Fig. 9 Distribution of the number of tracks per particle for particles collected from the mechanical line (sampling point A), wet cabinets, and room air.

TABLE 11 Distribution of Fission-Fragment Tracks Among Plutonium-Bearing Particles from Various Sources in Systems I and II

Source	Data points, N	Geometric mean of number of fission tracks $(\bar{T}_g)$	Geometric standard deviation (s_g)	Skewness (SK)	Geometric mean activity particle, fCi
		System I			
Unfiltered fifth-level wet-cabinet air	15,987	3.76	2.56	−0.20	0.09
Unfiltered sixth-level wet-cabinet air	7,042	3.32	2.99	−0.51	0.08
Fifth-level room air*	53	1.00	8.40	−0.98	0.02
Filtered wet-cabinet air*	98	0.87	4.14	−0.29	0.02
		System II			
Sampling point A air	200	14.74	2.69	−0.43	0.36
Sampling point B air	107	32.38	1.78	0.23	0.79
Sampling point C air	114	16.50	1.75	−0.22	0.40
Sampling point D air	71	17.01	1.65	0.24	0.41

*Values determined graphically.

Of the major crustal elements listed in Table 6, silicon and iron were the most ubiquitous, being found in most particles. The enrichment-factor distribution for these elements, however, does not fall within the log-normal distribution for crustal material. For the enrichment factors of an element to match the log-normal distribution of crustal material in aerosols, there should be about 16% of the enrichment factors of less than one geometric standard deviation, 68% within one geometric standard deviation of the mean, and another 16% above one geometric standard deviation. This lack of conformity may result from the low values for the geometric standard deviations of the enrichment factors for these elements in aerosols.

Only the enrichment factors for sodium and chlorine fall within the log-normal distribution for crustal material. This may be due to the relatively high solubility of compounds of these elements and, in the case of chlorine, the high value for the geometric standard deviation.

Particles from System I contain a greater variety of elements than those from System II, and thus all but four elements are contained on a higher proportion of particles from System I than from System II. The most striking example was nickel. Although 56% of the particles from System I contained nickel, only 9% of those in System II did. The major crustal elements (those in Table A.1 comprising 0.4% or more of crustal material) are contained on over half the particles from System I and, except for magnesium in particles from sampling points C and D and titanium, are also contained on over half the particles from System II. Some of the minor elements (those comprising 0.1% or less of crustal material) are present in over half the particles, viz, nickel, chromium, and zinc in particles from System I and sulfur, chromium, and zinc in particles from sampling point B of System II. The chromium and nickel may have come from the 304L stainless-steel alloy of cabinets and exhaust ducts or the *Hastelloy*-C* alloy in the wet cabinets. However, few of the particles contained the proper ratio of chromium to nickel found in either alloy. Also, if *Hastelloy-C* contributed the nickel in the particles, some molybdenum should also have been detected.

Of the elements that are present on less than 10% of the particles, all but copper on particles from System II have high enrichment factors. This indicates that the minor constituents of crustal material are not uniformly distributed among particles but are concentrated on a few particles where they represent a major constituent.

The plutonium-bearing particles were larger than natural aerosol particles collected at relatively low altitude (<2.3 km), as shown in Fig. 7. Particles collected from sampling points A and B of System II were larger than those from System I, with geometric mean diameters two or three times as great as those of particles from other locations.

The size of about 95% of the plutonium-bearing particles ranges between 0.4 and 37 μm in diameter. Morrow (1964) estimated that with normal respiration all particles in a monodispersed aerosol of unit-density spheres 37 μm in diameter will be deposited in the nasopharyngeal region of the respiratory tract. [With larger (>37 μm) particles the fraction deposited rapidly decreases.] As the diameter decreases, the fraction deposited in the respiratory tract decreases until a minimum of 20% deposition is reached for particles that are around 0.1 to 0.2 μm in diameter, where the particles tend to remain airborne. As the diameters decrease below 37 μm, a larger fraction is deposited in the tracheobronchial region until 70% of the particles 5 μm in diameter are deposited in the tracheobronchial region and only 5% in the nasopharyngeal and 5% in the alveolar

---

*Trademark of Cabot Corporation, Boston, Mass.

regions. With still smaller particles, the fraction deposited in the tracheobronchial region decreases until at 0.2 μm diameter only 10% is deposited in the tracheobronchial and 10% in the alveolar regions. For dust particles with a density of around 2.5, this distribution will be shifted toward smaller diameters so that 100% deposition in the nasopharyngeal region occurs around 5 μm.

Particles from all parts of System II also contained, on the average, more plutonium per particle than those from System I. As shown in Table 11, the geometric mean number of tracks per particle from unfiltered wet-cabinet air was just over three for both fifth- and sixth-level cabinets (averaging about 0.08 fCi per particle), whereas that for filtered wet-cabinet air was about one-third of this, or almost the same for room air (averaging about 0.02 fCi/particle).

A comparison of the mean diameters of particles collected from different sampling points, given in Table 9, with the mean number of fission-fragment tracks for particles from the same location, given in Table 11, indicates a possible relationship between particle size and plutonium content. Correlation coefficients between the cube of the particle diameter and the number of fission-fragment tracks from each particle from sampling points B, C, and D were calculated. These are given in Table 12. These coefficients differ significantly from that expected from a random sample from a population of paired variables having a correlation coefficient of zero. Thus, even though

TABLE 12   Correlation and Coefficient of Alienation for the Cube of the Diameter and the Number of Fission-Fragment Tracks for Particles from Sampling Points B, C, and D of System II

Sampling point	Number of particles	Correlation coefficient
B	107	0.69
C	114	0.29
D	71	0.36

the points on a plot of particle diameter cubed vs. number of fission-fragment tracks appear scattered, there is a significant correlation between the quantity of plutonium in particles collected from sampling points B, C, and D in System II and the particle volume. (Tracks with particles collected at other sampling points, where only ^{239}Pu could be found, were counted but not recorded for each particle. Only where a ratio of alpha-particle to fission-fragment tracks was needed to distinguish plutonium-bearing particles from those having other fissionable materials were the track counts recorded.)

Summary and Conclusions

The elemental compositions, sizes, structures, and ^{239}Pu contents were determined for 558 plutonium-bearing particles collected from various locations in the exhaust from a reactor fuel reprocessing facility. Airborne particles were collected on polycarbonate membrane filters. Particles containing ^{239}Pu were identified by fission-fragment and alpha-particle tracks produced by them in a polycarbonate film with a nuclear-track-emulsion coating. When located, the amount of ^{239}Pu in each particle was determined by counting the tracks, a small portion of the film containing the particle was isolated, the emulsion removed, the polycarbonate dissolved, the track replicas oxidized, and the elemental composition of the ^{239}Pu-bearing particle determined by electron-microprobe analysis. These data were compared with data from natural aerosol particles.

Most of the collected particles were composed of aggregates of crustal materials. Of the particles, 3.6% was organic and 1.7% was metallic, viz., iron, chromium, and nickel. High enrichment factors for titanium, manganese, chromium, nickel, zinc, and copper were evidence of the anthropogenic nature of some of the particles. The amount of plutonium in most particles was very small (less than 1 fCi of ^{239}Pu). Thus plutonium concentrations had to be determined by the fission-track counting method. Only one particle contained sufficient plutonium for detection by electron-microprobe analysis. This was a 1-μm-diameter particle containing 73% PuO_2 by weight (estimated to be 170 fCi of ^{239}Pu) in combination with Fe_2O_3 and mica. The plutonium-bearing particles were generally larger than natural aerosols. The geometric mean diameter of those collected from the mechanical line exhaust was larger than that of particles collected from the wet-cabinet exhaust (12.3 μm vs. 4.6 μm). Particles from the mechanical line also contained more plutonium per particle than those from the wet cabinets. The amount of plutonium per particle decreased with the distance of each sampling point from the mechanical line.

The size and ^{239}Pu content distribution among particles collected from the sand filter effluent and at the 50-ft level of the 291-F stack were almost the same. The geometric mean and standard deviation of the diameter of ^{239}Pu-bearing particles at the 50-ft level was 5.43 ± 2.69 μm. The relatively large size of these particles is believed to be due to coagulation of submicrometer particles by thermal and turbulent mechanisms to form larger agglomerates. The elemental composition of these particles, which contain very small amounts of plutonium in combination with crustal elements not used in the recovery process, supports this assumption. Scanning electron micrographs, such as Fig. 5(c), also show these particles to be agglomerates of smaller dissimilar particles.

Fleischer and Raabe (1977) have observed alpha-decay-induced fragmentation of $^{239}PuO_2$ particles probably caused by the heavy recoiling nuclei. When suspended in water, these particles produce fragments, or subparticles, which contain from 50 to 10,000 ^{239}Pu atoms, the abundance of which follows a power-law relation with the largest particles being the least abundant. The possibility exists that PuO_2 particles, large enough to be trapped on HEPA filters, fragment owing to alpha decay. The small fragments then pass through the filters where they coagulate with dust composed of crustal elements. The larger dust particles may not have passed through the HEPA filters but entered the exhaust system through leaks in the ducts, as illustrated in Fig. 10. Such leaks might remain undetected as long as the exhaust system remained under negative pressure with respect to the atmosphere.

The geometric mean and standard deviation of the number of fission-fragment tracks per ^{239}Pu-bearing particle collected from the 50-ft level during July, August, and September 1977 was 17.01 ± 1.65 tracks. One femtocurie of ^{239}Pu in a mixture of low-irradiation plutonium will produce 52.6 fission fragments when irradiated with a fluence of 8.64 x 10^{14} thermal neutrons/cm^2. Only about half, or 26.3, of the fragments will produce tracks in the polycarbonate film. Thus the calculated geometric mean radioactivity on the ^{239}Pu-bearing particles leaving the stack is 0.65 fCi/particle. During these 3 months a total of 82 μCi of ^{239}Pu was discharged to the environment. This amounts to an average of 0.62 nCi/min. Thus during this period about 10^6 ^{239}Pu-bearing particles per minute were discharged from the 291-F stack to the environment. With a flow rate in the stack of 2×10^5 cfm, the average ^{239}Pu-bearing particle concentration in stack air was 5 particles/ft^3.

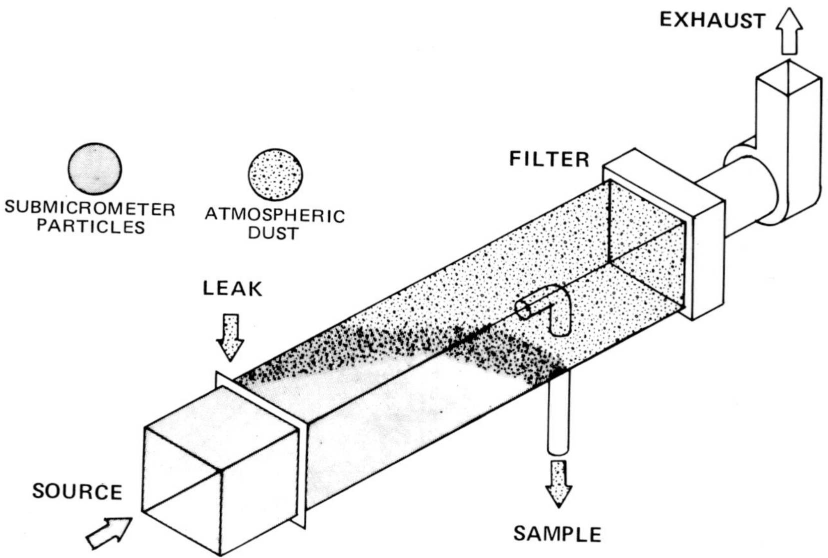

Fig. 10   Formation of plutonium-bearing particles in exhaust systems by the coagulation of submicrometer plutonium particles with atmospheric dust.

Acknowledgment

I gratefully acknowledge the assistance of E. F. Holdsworth and J. T. Armstrong of the Chemistry Department of the Arizona State University, Tempe, Ariz., who performed the electron-microprobe analyses.

Appendix: Use of Elemental Enrichment Factors to Express Particle Compositions

Background

Two recent developments in aerosol studies have provided valuable tools for the analysis of particle composition data. The first is the use of ratios of elemental concentrations called "enrichment factors" to compare aerosol compositions. Begun in the early seventies, this technique has gained wide acceptance in the last few years (Rahn, 1971; 1976; Zoller, Gladney, and Duce, 1974; Duce, Hoffman, and Zoller, 1975; Neustadter, Fordyce, and King, 1976). The second development is the availability of data on the composition of natural aerosols. In the last few years, Rahn (1976) published a compilation of 104 data sets of trace elements in aerosols along with the geometric mean and geometric standard deviation of the enrichment factors for each of the elements. These data sets were from sampling sites ranging from highly industrialized temperature zones to the tropics and poles and represent all continents except South America as well as various marine locations. As a framework from which to view much of the order in atmospheric aerosols, Rahn used the concept of aerosol-crust enrichment factors for the elements. This concept has been applied to analyzing data collected in this study to provide for (1) the intercomparison of the compositions of plutonium-bearing particles with atmospheric aerosols compiled by Rahn and (2) the intracomparison among particles collected from different sampling points.

Microprobe Analyses of Particles

For comparison, results of microprobe analyses must be expressed as elemental ratios because not all elements that may be present in an aerosol are detected by microprobe analysis. The microprobe used in this study is quantitative only for elements with atomic numbers greater than 10. It is only semiquantitative for oxygen (the most-abundant element in crustal material) as well as for other major elements of low atomic number, such as hydrogen, fluorine, and carbon. Atmospheric aerosols are known to contain, in addition to elements and oxides, carbonaceous material, such as sooty carbon and organics, and water-soluble ionic material, such as sulfate, nitrate, and ammonium ions. Thus elemental weight percents, normalized to 100 on the basis of the elements detected, cannot be compared. Even the addition of a hypothetical oxygen concentration, calculated on the supposition that all elements are present as oxides of known valence, will still not account for the organic fraction of particles. However, a ratio of the concentrations of one element to another will normally be relatively unaffected by the concentrations of other elements that may be present and thus can be used for comparisons even when a complete analysis of all the elements in an aerosol or single particle is not available.

Enrichment Factors

A dimensionless ratio of elemental concentrations, called the enrichment factor, has been defined as

$$EF(X) = \frac{(X/Ref)_{aerosol}}{(X/Ref)_{source}} \qquad (A.1)$$

where $EF(X)$ is the enrichment factor of element X in an aerosol relative to some source material and X/Ref is the ratio of the concentration of element X to the concentration of the reference element, Ref, in both the aerosol and the source material.

Source Material

Elemental ratios in aerosols or in single particles are normalized by dividing them by ratios of the same elements in a standard source material to obtain the enrichment factors. If a particle is composed of the same material as the source, the enrichment factor will be 1.00 for all elements. If the ratio of an element to the reference element is greater or less than the same ratio in the source material, the enrichment factors will be greater or less than 1.00, and the particle is said to be either enriched or depleted, respectively, in that element.

The most commonly used crustal source material for continental enrichment-factor calculations is globally averaged crustal rock. (For marine enrichment-factor calculations, sea salt is used.) The selection of rock may seem strange because there is little doubt that soil rather than rock is the precursor to the crustal aerosol. Some 93% of the earth's continental surface is covered by soils (Kothny, 1973). Many of these soils are in states of loose aggregation which can easily be made airborne by the wind. Chemically, however, Rahn (1976) has found that the composition of the crustal aerosol is not unambiguously that of soil. Elements in natural aerosols with rock-like enrichment factors include silicon, iron, calcium, potassium, and chromium; those with soil-like enrichment factors are titanium and barium. One would expect natural aerosols to be, like soil, depleted in the more-soluble elements. Except for glacial activity and to a lesser extent in deserts, physical weathering processes, which ultimately produce small particles from boulders, are very slow and are accompanied at all stages by intense chemical weathering. Thus large masses of physically pulverized rock which have not been chemically weathered are not available for aerosol production.

Rahn (1976) speculates that remote continental aerosols are never as depleted in the soluble elements (e.g., sodium, potassium, calcium, and magnesium) as they should be relative to rock (if natural aerosols were purely soil derived) because of the presence of small amounts of marine aerosol. Soluble elements, especially sodium and magnesium, are abundant in the marine aerosol; thus only small amounts of this aerosol in remote continental areas would noticeably raise the proportions of soluble elements in an aerosol collected there.

In addition to the similarity in the elemental composition of aerosol and crustal rock, available analytical data are much less numerous and less reliable for soils, especially for several interesting trace elements that are enriched in aerosols.

For these reasons the majority of investigators who calculate aerosol-crust enrichment factors have chosen one of the several available tables of elemental abundances in average crustal rock. Because the composition of plutonium-bearing particles are compared with data reported by Rahn (1976), the same crustal-rock composition used by him [that reported by Mason (1966)] was selected as the source material composition for this work. Column 2 of Table A.1 gives the elemental concentrations in globally averaged crustal rock for those elements found in plutonium-bearing particles.

Reference Element

Of the various elements that seem to be reliably crust derived in aerosols, aluminum, silicon, and iron are generally considered to be the most suitable reference elements. (When sea salt is the source material, the nearly universal choice is sodium.) An acceptable crustal reference element should have high concentrations in rock and soil, very low pollution potential, ease of determination by a number of analytical techniques, and freedom from contamination during sampling. Iron has markedly higher pollution potential than aluminum and so is less suited for use with urban or rural aerosols. Silicon is probably the most unambiguous elemental indicator of crustal material. Unfortunately, silicon has been determined in so few aerosol samples that it cannot be used as the reference element where comparisons are to be made. Aluminum is a major element (81,300 ppm in rock), well determined by a variety of analytical techniques, and has a minimum of specific pollution sources.

Thus for this work enrichment factors for element X in most particles were calculated using

$$EF(X) = \frac{(X/Al)_{particle}}{(X/Al)_{rock}} \tag{A.2}$$

TABLE A.1   Elemental Concentrations in Average Crustal Rock and Geometric Mean Enrichment Factors of Various Aerosols

| | | Geometric mean enrichment factors | | | | | |
| | | Global | | | Remote marine | Remote continental | Urban |
Element	Concentration, ppm	$\overline{EF}_g/s_g$	$\overline{EF}_g$	$\overline{EF}_g\, s_g$	$\overline{EF}_g$	$\overline{EF}_g$	$\overline{EF}_g$
Silicon	277,200	0.62	0.79	1.01	0.7	0.7	0.79
Aluminum	81,300	1.00	1.00	1.00	1.0	1.0	1.00
Iron	50,000	1.05	2.06	4.06	2.5	1.5	2.2
Calcium	36,000	1.15	2.84	7.04	8	1.5	2.9
Sodium	28,300	0.64	4.44	30.8	$10^2 - 10^3$	0.4	1.81
Potassium	25,900	0.99	1.98	3.98	6	1.5	1.63
Magnesium	20,900	0.64	2.38	8.90	$10^1 - 10^2$	0.7	2.0
Titanium	4,400	1.01	1.39	1.92	1.2	1.2	1.63
Phosphorus	1,050	0.79	2.63	8.71			2.6
Manganese	950	1.45	3.91	10.5	3	2	3.2
Barium	425	2.61	5.50	11.6		~2	4.8
Sulfur	260	228	608	1620			490
Chlorine	130	100	740	5470	$10^4 - 10^5$	40	300
Chromium	100	2.50	8.11	26.3	20	6	6.2
Nickel	75	8.74	31.9	116	100	50	10.8
Zinc	70	79.7	257	832	400	80	300
Copper	55	34.0	102	304	150	20	149
Tungsten	1.5	4.89	19.1	74.3			11.0
Cadmium	0.2	274	1920	13400	5000	2000	940

with aluminum as the reference element and average crustal rock as the source material. However, 18 particles from System I and 37 from System II contained no aluminum. Thus the enrichment factors had to be based on silicon rather than on aluminum, where

$$EF(X) = \frac{(X/Si)_{particle}}{(X/Si)_{rock}} \frac{(\overline{Si}/Al)_{g\ aerosol}}{(Si/Al)_{rock}} = 0.79 \frac{(X/Si)_{particle}}{(X/Si)_{rock}} \qquad (A.3)$$

(The second set of ratios is the geometric mean of the global aerosol-crust enrichment factor explained in the next section.)

With the use of these two relationships, the enrichment factors were calculated from the elemental weight percents obtained for 115 particles in System I and 156 particles in System II. Six small (0.5 to 3.6 μm in diameter) iron particles in System I and two particles [~15 μm in diameter and containing potassium, chromium, and iron (1 : 3 : 3)] from sample point A of System II contained neither aluminum nor silicon and were thus not included in the study.

Comparative Aerosol Data

For a comparison of the elemental composition of plutonium-bearing particles with that of atmospheric aerosols, enrichment factors calculated for elements in these particles were grouped according to data supplied by Rahn (1976) for aerosols. In his report trace-element concentrations in aerosols from 104 published and unpublished data sets were used to calculate enrichment factors. From the enrichment factors in each data set, the geometric mean enrichment factor ($\overline{EF}_g$) and geometric standard deviation (s_g) of the logarithmic frequency distributions of enrichment factors were calculated for each element by the following formulas:

$$\overline{EF}_g = \exp\left(\frac{1}{N}\sum_{i=1}^{N} \ln EF_i\right) \qquad (A.4)$$

and

$$s_g = \exp\left\{\left[\frac{1}{N-1}\sum_{i=1}^{N}(\ln EF_i - \ln \overline{EF}_g)^2\right]^{\frac{1}{2}}\right\}$$

where N is the number of data points and EF_i is the enrichment factor of the *ith* point.

The geometric mean enrichment factors obtained by Rahn (1976) for 19 elements are given in Table A.1 for global, remote marine, remote continental, and urban aerosols. The geometric means of the global aerosol enrichment factors include data from all points and may be weighted too heavily toward cities, but they can serve as a useful first approximation to a general aerosol. The urban enrichment factors are geometric means for 29 cities. The enrichment factors for remote continental and remote marine areas were read from the enrichment-factor plots and are therefore somewhat subjective.

Values for $\overline{EF}_g/s_g$ and $\overline{EF}_g \times s_g$, respectively, were calculated with the use of global values to obtain the lower and upper limits for 68.27% of the enrichment factors closest to the geometric mean. (When describing concentrations at selected statistical levels remote from a mean, the s_g is a multiplier or divider of the $\overline{EF}_g$, whereas its counterpart, Gaussian standard deviation, functions as an increment to the arithmetic mean. This is a consequence of the fact that multiplying and dividing values are equivalent to adding and subtracting their logarithms.) The results from these calculations are also given in Table A.1.

References

Andersen, B. V., 1964, Plutonium Aerosol Particle Size Distributions in Room Air, *Health Phys.*, 10: 899-907.

Armstrong, John T., and Peter R. Buseck, 1975, Quantitative Chemical Analysis of Individual Microparticles Using the Electron Microprobe: Theoretical, *Anal. Chem.*, 47: 2178-2192.

Axtmann, R. C., L. A. Heinrich, R. C. Robinson, O. A. Towler, and J. W. Wade, 1953, *Initial Operation of the Standard Pile*, USAEC Report DP-32, E. I. du Pont de Nemours and Co., Savannah River Laboratory, NTIS.

Boyd, George A., 1955, *Autoradiography in Biology and Medicine*, Academic Press, Inc., New York.

Duce, Robert A., Gerald L. Hoffman, and William H. Zoller, 1975, Atmospheric Trace Metals at Remote Northern and Southern Hemisphere Sites: Pollution or Natural? *Science*, 187: 59.

Eastman Kodak Company, 1976, *Kodak Materials for Nuclear Physics and Autoradiography*, Kodak Technical Pamphlet No. P-64.

Elder, John C., Manuel Gonzales, and Harry J. Ettinger, 1974, Plutonium Aerosol Size Characteristics, *Health Phys.*, 27: 45-53.

Ettinger, Harry J., William D. Moss, and Lamar J. Johnson, 1971, Size Selective Sampling for Plutonium-238, *Health Phys.*, 23: 41-46.

Fleischer, Robert L., and Otto G. Raabe, 1977, Fragmentation of Respirable PuO_2 Particles in Water by Alpha Decay—A Mode of Dissolution, *Health Phys.*, 32: 253-257.

Hayden, J. A., 1976, Characterization of Environmental Plutonium by Nuclear Track Techniques, in *Atmosphere–Surface Exchange of Particulate and Gaseous Pollutants (1974)*, ERDA Symposium Series, No. 38, Richland, Wash., Sept. 4–6, 1974, R. J. Engelmann and G. A. Sehmel (Coordinators), pp 648-660, CONF-740921, NTIS.

——, R. L. Murri, and H. M. Baker, 1972, *The Application of Fission Tracks to Study Environmental Samples at Rocky Flats*, USAEC Report RFP-1922, Dow Chemical Company, NTIS.

Junge, C. E., 1965, *Handbook of Geophysics and Space Environments*, pp. 5-23, S. L. Valley (Ed.), Air Force Cambridge Research Laboratories.

Kirchner, R. A., 1966, A Plutonium Particle Size Study in Production Areas at Rocky Flats, *Am. Ind. Hyg. Assoc. J.*, 27(4): 396-401.

Kopriwa, Beatrix Markus, and C. P. Leblond, 1962, Improvements in the Coating Technique of Radioautography, *J. Histochem. Cytochem.*, 10(3): 269.

Kothny, E. L., 1973, The Three-Phase Equilibrium of Mercury in Nature, in *Trace Elements in the Environment*, Advances in Chemistry Series, No. 123, pp. 48-80, American Chemical Society, Washington, D. C.

Leary, J. A., 1951, Particle-Size Determination in Radioactive Aerosols by Radioautograph, *Anal. Chem.*, 23(6): 850-853.

Mason, B., 1966, *Principles of Geochemistry*, 3rd ed., p. 45, John Wiley & Sons, Inc., New York.

Mishima, J., and Lysle C. Schwendiman, 1972, Characterization of Radioactive Particles in a Plutonium Processing Plant Exhaust System, in *Pacific Northwest Laboratory Annual Report for 1971 to the USAEC Division of Biology and Medicine*, Volume II: Physical Sciences, Part 1. Atmospheric Sciences, USAEC Report BNWL-1651(Pt.1), pp. 88-90, Battelle, Pacific Northwest Laboratories, NTIS.

Morrow, P. E., 1964, Evaluation of Inhalation Hazards Based Upon the Respirable Dust Concept and the Philosophy and Application of Selective Sampling, *Am. Ind. Hyg. Assoc. J.*, 25(2): 213.

Moss, W. D., E. C. Hyatt, and H. F. Schulte, 1961, Particle Size Studies on Plutonium Aerosols, *Health Phys.*, 5: 212-218.

Neustadter, H. E., J. S. Fordyce, and R. B. King, 1976, Elemental Composition of Airborne Particulates and Source Identification: Data Analysis Techniques, *J. Air Pollut. Control Assoc.*, 26(11): 1079.

Quan, J. T., 1959, A Technique for the Radioautography of Alpha-Active Aerosol Particles on Millipore Filters, *Am. Ind. Hyg. Assoc. J.*, 20: 61-65.

Rahn, Kenneth A., 1971, *Sources of Trace Elements in Aerosols—An Approach to Clean Air*, Ph.D. Thesis, University of Michigan, Ann Arbor, Mich., Xerox University Microfilms (Order No. 72-4956).

——, 1976, *The Chemical Composition of the Atmospheric Aerosol*, Technical Report, Graduate School of Oceanography, University of Rhode Island.

Sanders, S. Marshall, Jr., 1976, *Composition of Airborne Plutonium-Bearing Particles from a Plutonium Finishing Operation*, ERDA Report DP-1445, E. I. du Pont de Nemours and Co., Savannah River Laboratory, NTIS.

——, 1977, *Characterization of Airborne Plutonium-Bearing Particles from a Nuclear Fuel Reprocessing Plant*, ERDA Report DP-1470, E. I. du Pont de Nemours and Co., Savannah River Laboratory, NTIS.

——, 1978, Methodology for the Isolation and Characterization of Individual Plutonium-Bearing Particles in Atmospheric Effluents from a Nuclear Processing Plant, in *Nuclear Methods in Environmental and Energy Research*, Proceedings of the Third International Conference, Oct. 10–13, 1977, ERDA Report CONF-771072, pp. 425-440, NTIS.

——, 1979, Isolation and Characterization of Individual Plutonium-Bearing Particles in Atmospheric Effluents from a Nuclear Processing Plant, *Health Phys.*, 36: 381-385.

Seefeldt, W. B., W. J. Mecham, and M. J. Steindler, 1976, *Characterization of Particulate Plutonium Released in Fuel Cycle Operations*, ERDA Report ANL-75-78, pp. 35-51, Argonne National Laboratory, NTIS.

Sherwood, R. J., and D. C. Stevens, 1963, *Some Observations on Nature and Particle Size of Airborne Plutonium in the Radiochemical Laboratories, Harwell*, Report AERE-R4672, Atomic Energy Research Establishment.

——, and D. C. Stevens, 1965, Some Observations of the Nature and Particle Size of Airborne Plutonium in the Radiochemical Laboratories, Harwell, *Ann. Occup. Hyg.*, 8: 93.

Spurny, Kvetoslav R., James P. Lodge, Jr., Evelyn R. Frank, and David C. Sheesley, 1969, Aerosol Filtration by Means of Nuclepore Filters: Structural and Filtration Properties, *Environ. Sci. Technol.,* 3(5): 453-464.

Stevens, D. C., 1963, Location and Examination of Alpha Active Airborne Dust Particles Collected on Glass Fibre Filter Paper, *Ann. Occup. Hyg.,* 6: 1.

Yakowitz, H., R. L. Myklebust, and K. F. J. Heinrich, 1973, *FRAME: An On-Line Correction Procedure for Quantitative Electron Probe Microanalysis,* National Bureau of Standards, Technical Note 796, GPO.

Zoller, William H., E. S. Gladney, and Robert A. Duce, 1974, Atmospheric Concentrations and Sources of Trace Metals at the South Pole, *Science,* 183: 198.

Physicochemical Associations of Plutonium and Other Actinides in Soils

E. A. BONDIETTI and T. TAMURA

Soil physicochemical behavior of plutonium and other actinides is discussed with primary emphasis on the behavior of plutonium in actual contaminated soil and the importance of actinide speciation in interpreting laboratory results. The behavior of actinides in soil is strongly influenced by physical form and/or oxidation state. The chemistry of plutonium, americium, curium, and neptunium is reviewed, particularly with respect to the oxidation states likely to control their behavior in most soils. Several aspects of sorption to soils are discussed, particularly those for plutonium. The comparative behavior of plutonium, thorium, and uranium in soil is also illustrated to provide a perspective for evaluating long-term environmental behavior. The relative hazard associated with plutonium-contaminated soil is evaluated and the importance of both physicochemical form of the plutonium and the soil particle-size association is emphasized.

The exposure of internal organs to ionizing radiation is the major potential hazard associated with the production and release of actinide alpha emitters to the environment. Two predominant pathways of exposure from this environmental contamination are inhalation and ingestion.

Inhalation of discrete radioactive particles (i.e., PuO_2) and carriers (i.e., contaminated soil particles) can result in exposure of the lung and of other body organs following transport of particles or ions through the lung to other organs. Ingestion of radionuclides present in biologically assimilated forms or as surface contamination also serves as a source of possible exposure to critical organs.

The relative importance of inhalation and ingestion as pathways for human exposure depends on many environmental parameters exclusive of the physicochemical associations of the radioelement in soils. However, the physicochemical properties of the radioelement in soils strongly influence the pathway and magnitude of transport. This is illustrated by the $\sim 10^3$ greater assimilation by plants of plutonium added in monomeric form as compared with PuO_2 microspheres (Adams et al., 1975).

This chapter examines two aspects of the physicochemical associations of plutonium and other actinides in contaminated soils: (1) the case in which plutonium is monomerically distributed in soil (i.e., the potentially most biologically available form) and (2) the case in which plutonium exists or its origin is traced to a discrete particulate source. However, it is also important to evaluate general principles of actinide–soil interactions. For this purpose we also explore several aspects of sorption behavior to soil colloids.

General

*Chemical Characteristics of the Transuranium Elements (Np, Pu, Am, and Cm)
and Related Actinides (U and Th)*

The alpha-emitting elements of significance in the uranium and thorium fuel cycles
include Th, U, Np, Pu, Am, and Cm. These radioelements can be released to the biosphere
by fuel reprocessing, radioactive waste handling and disposal, and fuel fabrication. Many
of these elements have already been dispersed through defense-related activities.

Although the geochemistries of U and Th are reasonably well understood, the largely
man-made elements (Pu, Np, Am, and Cm) are only now under study. Fortunately the
chemical characteristics of the transuranic elements are very similar to the naturally
occurring rare earths (oxidation state III), Th (oxidation state IV), and U (oxidation
states IV and VI). Indeed, early studies on the solution and solid-phase chemistry of the
transuranic elements used these analogies (Hindman, 1954; Connick, 1954; Cunningham
and Hindman, 1954; Thompson et al., 1949).

The complexities of the chemistry of any element usually depend on the number of
oxidation states that the element can exhibit. For the actinide elements under discussion,
these oxidation states are represented by the M^{3+}, M^{4+}, MO_2^+(V), and MO_2^{2+}(VI) species.
Uranium is a classic example of the influence of oxidation state on environmental
chemistry. Both U and Th in the tetravalent state are extremely resistant to leaching.
However, the oxidation of U(IV) to U(VI) results in much higher U mobilities in the
environment (Adams, Osmond, and Rogers, 1959). Likewise, although uranyl ion is
relatively stable in seawater as the uranyl carbonate complex, ^{230}Th formed from the
radiodecay of ^{234}U rapidly becomes depleted with respect to ^{234}U (and ^{238}U) (Cherry
and Shannon, 1974).

Environmentally Important Oxidation States of the Transuranium Elements

Curium is trivalent in solution and probably will be present in air-ignited oxides as
Cm_2O_3. Americium is also likely to be trivalent in environmental solutions, although the
IV, V, and VI oxidation states are known in the laboratory. The dioxide, AmO_2, can be
formed on ignition. Plutonium can exhibit valences of III, IV, V, and VI in solution and
the IV state as the dioxide. Neptunium is probably tetravalent or pentavalent in
environmental solutions and tetravalent in the oxide. Since limited experimental
information is available, these valence assignments for environmental systems are subject
to revision.

Environmental conditions, such as pH and Eh, will control the oxidation-state
distribution, although the kinetics of the redox reactions are unknown. Experimental
determinations of Pu and Np oxidation states in an environmental context have been
undertaken (Bondietti and Reynolds, 1976; Bondietti and Sweeton, 1977; Bondietti,
Reynolds, and Shanks, 1976; Bondietti, 1976). A number of speciation diagrams have
been constructed (Polzer, 1971; Andelman and Rozzell, 1970; Rai and Serne, 1977)
which attempt to evaluate the oxidation-state distribution of Pu in environmental
solutions. These investigators have recognized that understanding the environmental
speciation of Pu is critical to evaluating its biogeochemistry. The oxidation states of Pu in
natural water have actually been determined (Bondietti and Sweeton, 1977).

**TABLE 1 Effect of Clay Treatment on Adsorption of
Actinide Elements to Miami Silt Loam Clay**

Treatment*	Cation exchange capacity, meq/100 g	Percent adsorbed[†]			
		^{234}Th(IV)	^{244}Cm(III)	^{233}U(IV)	^{237}Np(V)
Intact clay	17	99.7	98.6($\pm$0.2)[‡]	95.6($\pm$0.2)[‡]	61.8
Organic matter removed	11	99.8	99.6($\pm$0.2)[‡]	96.4($\pm$0.4)[‡]	49.7
Fe and organic matter removed	9.9	99.7	95.6($\pm$0.5)[‡]	99.1($\pm$0.5)[‡]	18.2

*Organic matter removed with NaOCl; Fe removed with sodium dithionite.

[†]pH 6.5; 5mM Ca(NO$_3$)$_2$; solution/clay ratio of 400/1; 48 hr equilibration; U and Np at <microgram per gram levels; Th and Cm at <nanogram per gram levels.

[‡]Mean $\pm$ standard deviation.

Interactions with Environmental Colloids

Effect of Oxidation State on Sorption. The partitioning of the actinides between solid and solution phases may be dependent on the charge characteristics of the element, the physicochemical characteristics of the solid, and the composition of the solution. Complexation by OH$^-$ (hydrolysis) and other ligands affects sorption because all four common oxidation states (III, IV, V, and VI) form complexes of varying stabilities. For example, the competition between hydrolysis and complexation by carbonate dominates the sorption behavior of uranyl ion in natural solutions. Above pH 7.5 (and in equilibrium with atmospheric CO$_2$) soluble uranyl carbonate complexes can predominate; below this pH sorption to particulates readily occurs (Starik and Kolyadin, 1957). Another example is Pu(IV), which is extensively hydrolyzed in near-neutral solutions and is probably not adsorbed by normal ion exchange mechanisms (Tamura, 1972). Oxidation–reduction reactions, since they affect oxidation state, also influence sorption. Thus NpO$_2^+$ shows poor adsorption to soil, but reduction to Np(IV) increases sorption (Bondietti, 1976).

The different actinide oxidation states with respect to sorption are compared in Table 1. The distribution of Th(IV), U(VI), Cm(III), and Np(V) between a soil–clay fraction and a 5mM Ca(NO$_3$)$_2$ solution showed that relative sorption followed the oxidation-state order IV > III > VI > V under the specified conditions. The organic matter and free iron oxide (Fe) coatings were removed to evaluate the effects of colloid surface constituents [and thus cation exchange capacity (CEC)]. Only Np(V) sorption was strongly influenced by these treatments. The removal of organic matter decreased the CEC by 35% and Np sorption by 20%.

Removal of organic matter and Fe did not further affect the CEC, but the Np(V) sorption value decreased to 29% of the intact clay value. This observation of an apparent surface-dependent sorption mechanism suggests that, even for the MO$_2^+$ oxidation-state species, which is largely unhydrolyzed at environmental pH's, mass-action relationships may not describe adsorption equilibria.

Mass-action expressions have been used to describe ion-exchange equilibria. The exchange of NpO$_2^+$ on a sodium-saturated clay can be expressed as

$$NpO_2^+ + NaC \rightleftharpoons NpO_2C + Na^+ \tag{1}$$

and the equilibrium expression as

$$K = \frac{(NpO_2 C)\,(Na)}{(NpO_2^+)\,(NaC)} \tag{2}$$

where K is the equilibrium constant, C is clay, and the parentheses denote activity.

In a system in which neptunium is present in trace quantities and the clay is essentially sodium saturated, the activity coefficients of the sodium clay, neptunium clay, and neptunium and sodium ions can be considered to be a constant. Equation 2 can be rewritten in the form

$$K' = \frac{[NpO_2 C]\,[Na]}{[NpO_2^+]\,[NaC]} \tag{3}$$

where K' incorporates the constant activity coefficients and the brackets denote concentrations. By definition, the distribution coefficient (K_d) for neptunium is

$$K_d = \frac{[NpO_2 C]}{[NpO_2^+]} \tag{4}$$

and Eqs. 3 and 4 can be combined to form

$$K_d = K' \frac{[NaC]}{[Na]} \tag{5}$$

Since the clay remains essentially sodium saturated, NaC is a constant, and a log K_d vs. log [Na] plot should yield a straight line. The slope of the line is a function of the exponent of Na^+; in this case the slope is -1. If the clay is calcium saturated, it can be written as $Ca_{0.5}C$ in Eq. 1. The final expression of Eq. 5 would contain the exponent of 0.5, and thus the slope would be -0.5.

Data reported by Routson, Jansen, and Robinson (1975) on neptunium sorption by two soils at different sodium and calcium ion concentrations are shown in Fig. 1. The notable feature in the plot is the absence of any effect of sodium on neptunium sorption. Thus the slope of the NpO_2^+ K_d vs. Na^+ concentration plot approaches 0 rather than -1. Calcium exerted a more pronounced effect on Np sorption, but even here the Np K_d vs. Ca^{2+} concentration plot had a slope of about -0.3 rather than -0.5.

The data of Routson, Jansen, and Robinson (1975) and the effect of clay surface treatment on NpO_2^+ sorption (Fig. 1) indicate that electrostatic interactions alone do not explain NpO_2^+ sorption. One surface component, organic matter, appears to have an influence. In general, the stability of Np(V) chelates is comparable with divalent cation chelates (Zn^{2+}, Ca^{2+}, etc.). The interactions of Np(V) with soil humic acids reflect this. Figure 2 represents the observed distribution of Zn^{2+}, Cd^{2+}, Ca^{2+}, Sr^{2+}, and NpO_2^+ between complexed and free forms in the presence of soil humic acids. As is illustrated in the figure, Zn and Cd form stronger complexes than Ca or Sr, which is expected. The NpO_2^+ cation forms complexes that are slightly stronger than Ca. It should be apparent from the above samples that even the least hydrolytic actinide oxidation state interacts with soil constituents in complicated ways.

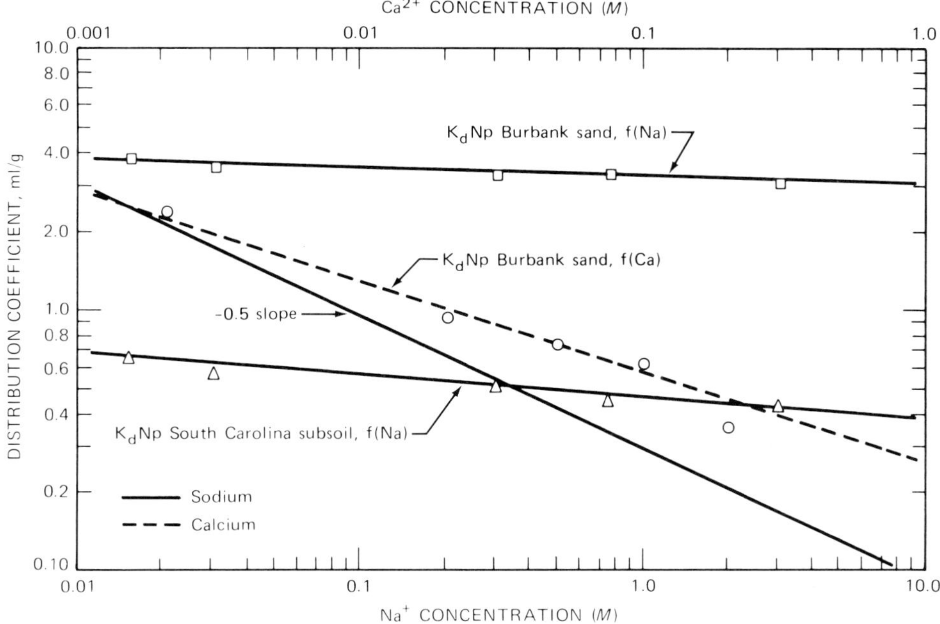

Fig. 1   Distribution coefficients of neptunium in selected soils. (Modified after Routson, Jansen, and Robinson, 1975.)

Plutonium Sorption.   Plutonium in oxidation state IV is very insoluble in water in the absence of soluble complexers. Given a solubility product (Ksp) for $Pu(OH)_4$ of $\sim 10^{-55}$ (Coleman, 1965), soluble monomeric Pu(IV) species should be difficult to assay in near-neutral solutions. Considering the various hydrolytic species $[Pu(OH)_3^+, Pu(OH)_2^{2+},$ etc.], concentrations of soluble Pu in equilibrium with crystalline PuO_2 might approach those depicted in Fig. 3. This figure was plotted using the hydrolysis constants evaluated by Baes and Mesmer (1976). By analogy to U(IV), a negatively charged pentahydroxy species, $Pu(OH)_5^-$, was postulated to exist in their analysis of hydrolytic constants.

Studies on the effect of pH on Pu(IV) sorption by soils have shown that, in the pH range 2 to 8, 99+% of the added Pu is lost from solution (Rhodes, 1957; Rogers, 1975). Rogers (1975) showed that maximum sorption was at about pH 5.5; sorption was less at lower and at higher pH's. Above pH 8, Rhodes (1957), Rogers (1975), Prout (1958), and Nishita (1978) observed substantial increases in the concentration of Pu in the supernatant (Fig. 4). Rogers (1975) attributed this behavior to dispersed soil colloids that failed to sediment during centrifugation. Rhodes (1957) and Rogers (1975) observed that this decrease in sorption might also have been due to the dispersal of Pu polymer or hydroxy species.

Prout (1958) observed a decrease in adsorption for three Pu oxidation states (III, IV, and VI) above pH 7 to 8 which might argue against polymer dispersion. In addition, he found that radiostrontium and radiocesium in low-ionic-strength solutions also showed a decrease in adsorption above pH 7 to 8. In higher ionic-strength solutions, the adsorption increased with increasing pH. Figure 5 illustrates this ionic-strength effect using selected adsorption curves reported by Prout (1958). The concentration of strontium used was

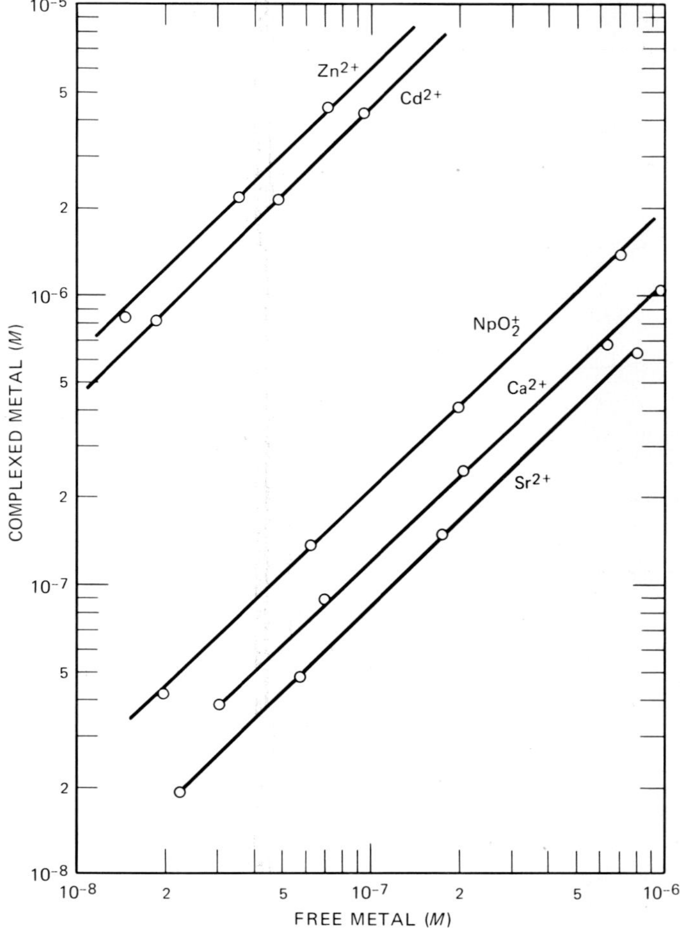

Fig. 2 Comparison of the relative complexing of Zn, Cd, Np(V), Ca, and Sr by soil humic acids (pH 7.0).

$5 \times 10^{-8}M$; if no sodium nitrate were added, strontium adsorption increased to a maximum around pH 7 and then decreased. When 1% $NaNO_3$ was added, no decrease was observed with increasing pH. Cesium showed the same effect: tracer Cs ($5 \times 10^{-8}M$) displayed maximum adsorption around pH 7.7, but increasing the Cs concentration to $5 \times 10^{-4}M$ removed this effect. These results suggest that the decrease above pH 7 to 8 observed with $\sim 10^{-6}M$ Pu was due to dispersion of clay-size particles containing sorbed ions. As the ionic strength was increased in the Sr and Cs studies, the clay remained flocculated and the observed K_d increased.

Plutonium represents an element in which the simultaneous presence of more than one oxidation-state species in solution can influence the observed adsorption behavior. Determination of the extent of this problem has unfortunately been ignored in most Pu adsorption experiments reported in the literature. Bondietti and Reynolds (1976),

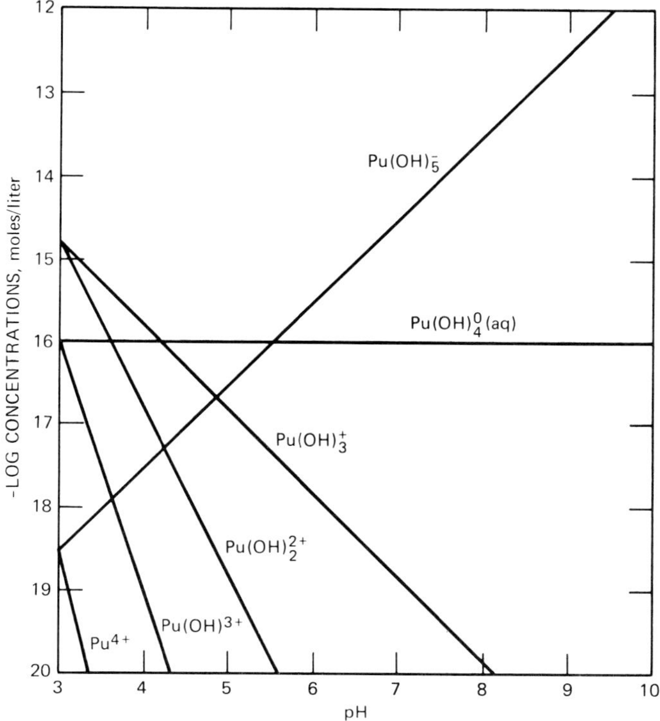

Fig. 3 Estimated concentrations of hydrolyzed Pu(IV) species in solutions saturated with crystalline PuO_2. (Adapted from C. F. Baes and R. E. Mesmer, *The Hydrolysis of Cations*, John Wiley & Sons, Inc., New York, 1976.)

however, reported the presence (and problem) of multiple oxidation states in Pu–clay equilibrations. These data are summarized in Table 2.

When ^{238}Pu(IV) (as the nitrate) was added to treated clays, the organic-matter removal treatment showed the least Pu sorption at 3 weeks (61.5%). With time, sorption values approached the other two treatments (i.e., 99+%). The reduced adsorption was not directly caused by the removal of organic matter but was influenced by disproportionation of Pu(IV), yielding Pu(III) and Pu(V) and/or Pu(VI) during the tracer addition. The low initial adsorption in the organic-matter removal treatment was apparently due to the fact that the clays were "oxidized" from the NaOCl treatment used to remove organic matter and the PuO_2^+ and/or PuO_2^{2+}, resulting from disproportionation, became stabilized. For the Fe removal treatment, the clays were in a reduced state, which minimized the presence of Pu(V + VI). The memory of these treatments is observable 2 yr after treatment (Table 2). For the intact clay, 79% of the soluble Pu was present as Pu(III + IV) (Bondietti and Reynolds, 1976). The largest amount of oxidized Pu [Pu(V) + Pu(VI)] was found in the organic removal treatment (35%). Very little oxidized Pu was found in the iron treatment (7%). The presence of more than one oxidation state in the control solutions provides support for an initial disproportionation reaction since oxidized or reduced clay surfaces were not present.

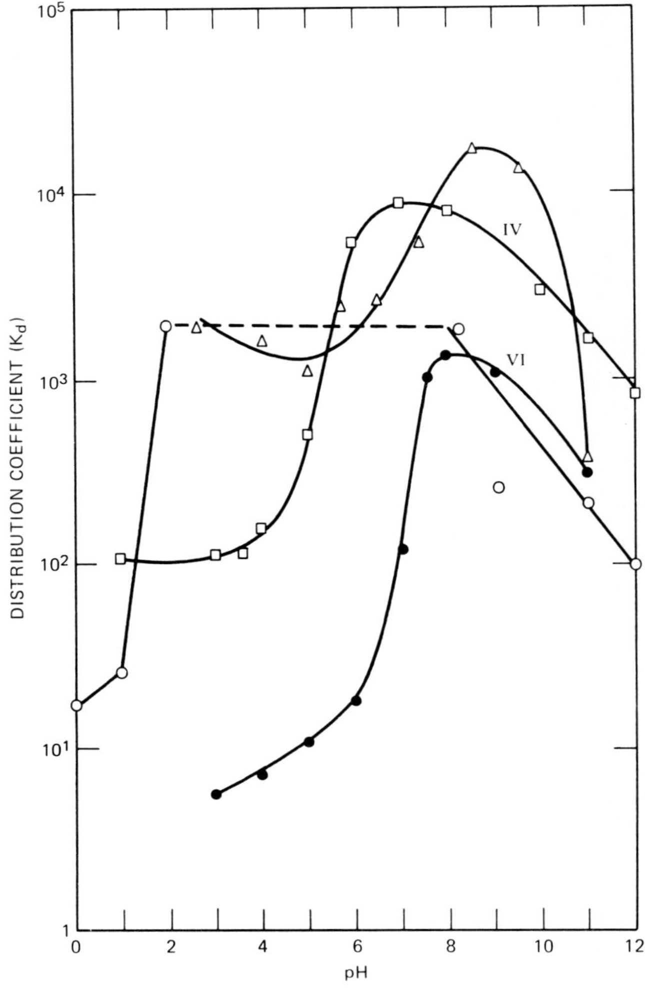

Fig. 4 Literature observations on the effect of pH on Pu sorption to soils. △, soil/solution ratio, 1/12.5 (Nishita, 1976). ●, soil/solution ratio, 1/10 (Prout, 1958). □, soil/solution ratio, 1/10 (Prout, 1958). ○, soil/solution ratio, 1/20 (Rhodes, 1957).

It is significant to note that, in the results described above, the initial ^{238}Pu concentration was about 3 ng/ml ($1.3 \times 10^{-8}M$). Experiments described by Jacobson and Overstreet (1948) and Prout (1958) indicated that Pu(III) was adsorbed more readily than Pu(IV), which was itself adsorbed more readily than Pu(VI). However, the molar concentrations of Pu used by these investigators [Jacobson and Overstreet (1948), $7 \times 10^{-7}M$; Prout (1958), $10^{-6}M$] were greater than those used here. Consequently the reason the Pu(IV) sorption was intermediate between Pu(III) and Pu(VI) was probably due to an initial disproportionation of the added Pu(IV). The true order is probably IV > III > VI, as shown in Tables 1 and 2 (intact clay). Additional support for this concept is provided in data by Bondietti, Reynolds, and Shanks (1976), where the K_d for

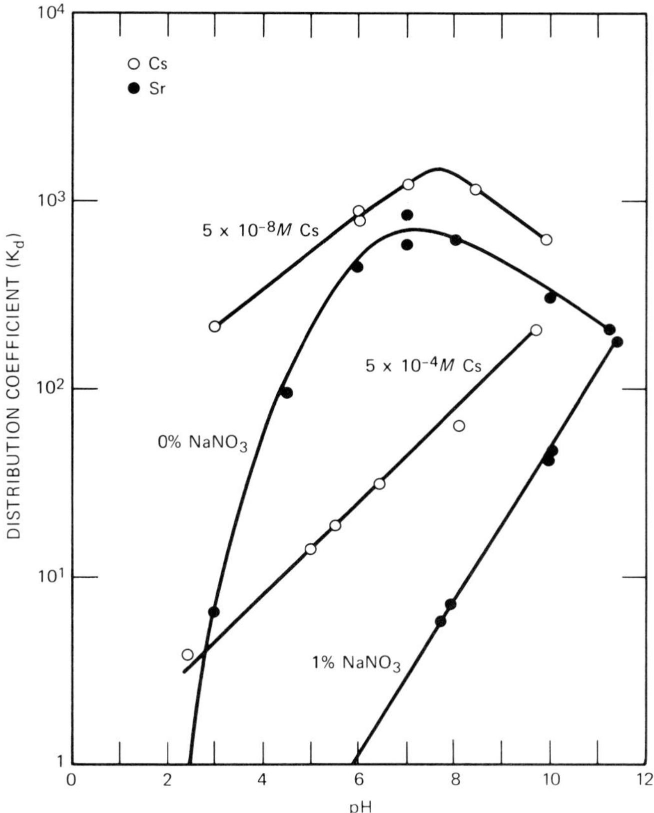

Fig. 5 Effect of ionic strength on sorption of radiostrontium and radiocesium by soil. [Modified from E. E. Prout, Adsorption of Radioactive Wastes by Savannah River Plant Soil, *Soil Science*, 86: p. 15 (1958).]

TABLE 2 Adsorption and Solution-Phase Characterization of ^{238}Pu(IV) Added to Miami Silt Loam Clay (pH 4.0)

Treatment*	Percent adsorbed† following indicated equilibration time (weeks)				Soluble phase characterization‡ (104 weeks)	
	3	18	52	104	Percent Pu(III + IV)	Percent Pu(V + VI)
Intact clay	99.9	99.8	99.9	99.9	79	19
Organic matter removed	50.0	61.5	99.8	99.4	65	35
Organic matter and Fe removed	99.8	99.9	99.8	99.9	93	7
Control §	71.0	71.7	78.6	82.1	80	20

*Organic matter removed with NaOCl; Fe removed with sodium dithionite.

†pH 4.0; 5m*M* Ca(NO$_3$)$_2$; solution/clay ratio of 400/1.

‡See Bondietti and Reynolds (1976) for methodology.

§No clay present.

Th was 20 times as high as that for Pu(IV) when both elements were equilibrated with montmorillonite clay. No valences were determined, but the concentration of ^{239}Pu used undoubtedly resulted in some initial disproportionation.

This discussion is meant to illustrate that the adsorption of Pu to soils is more complex than the simple distribution of a radioelement between a solid phase and a solution. For Pu the problem of mixed-oxidation-state species is significant. Attempts to delineate the soil/sediment chemistry of Pu must consider that more than one oxidation state may be present in stock solutions or may be formed during the experiment. Attempts to correlate Pu adsorption with soil type may be confounded by the complex interplay between soil components and the stability of various Pu oxidation-state species. Thus Glover, Miner, and Polzer (1976) and Polzer and Miner (1976) observed that the adsorption of Am(III) by various soils was as great as or greater than that of Pu(IV). They also noted that the variability in Am(III) sorption was much less than that in Pu. At the Pu concentrations used (10^{-8} to $10^{-6}M$), disproportionation may have been the cause of this variability. A small proportion of Pu(V) or Pu(VI) with their correspondingly lower sorption tendencies would provide erroneous sorption values for "Pu(IV)."

Plutonium at Contaminated Sites

General

Contaminated field sites provide the best situations for studies to understand the behavior of plutonium in environmental systems. Full appreciation of the behavior of Pu in these sites requires knowledge of the initial character of the contaminating event(s). From this information conclusions can be drawn regarding Pu behavior up to the time of sampling and potential behavior extrapolated for different source terms.

Nevada Test Site. One of the largest contaminated areas in the United States is the Nevada Test Site (NTS), which serves as the test area for nuclear detonations. Within the NTS several sites were used for safety shot evaluation; these sites, which have been declassified, contain dispersed plutonium from a series of high-explosive detonations simulating an accidental detonation of a subcritical atomic device. The detonation would be expected to produce a wide range of particles. Since plutonium metal is relatively reactive, the oxide form would be expected to be produced (Cunningham, 1954). Although a size distribution as a function of distance from ground zero (GZ) might be expected, this relationship is difficult to establish since considerable cleanup took place after the test. Some indication of decreasing size with increasing distance from GZ has been reported by Tamura (1975). Samples taken from 500 to 6700 ft from GZ showed that at 500 ft the 125- to 50-μm soil-size fraction contained 25% of the activity; at 6700 ft this fraction contributed less than 2%.

The solubility of plutonium oxides decreases with increasing ignition of the oxide (Cunningham, 1954). The plutonium in the safety shot sites was not subjected to fission temperatures but to explosion temperatures. The lower solubility of plutonium at NTS has been reported by Tamura (1976), who subjected contaminated soils from NTS to $8M$ nitric acid extraction at room temperatures. Compared with samples from Oak Ridge National Laboratory (ORNL) and Mound Laboratory (ML), the NTS samples were only one-fifth to one-eighth as soluble.

Rocky Flats. The contamination at the Rocky Flats (RF) plant in Colorado was caused by leaking barrels of Pu-contaminated cutting oil (Krey and Hardy, 1971). Before containment the plutonium-bearing oil was filtered to remove the large particles (Navratil and Baldwin, 1977). The plutonium would not oxidize in the oil; however, once the plutonium was in the soil and the protective oil was leached, the oxide form would predominate. With room-temperature extraction using 8M nitric acid, the RF samples showed 15 to 20% dissolution; this can be compared with the 10 to 15% observed in the NTS samples. This suggests that both RF and NTS plutonium are similar in character; the higher solubility of the RF vs. the NTS samples in the mineral acid may be ascribed to the smaller size of the plutonium or to the lower temperature of ignition of the RF samples.

Mound Laboratory. Contamination in the canal at ML occurred in 1969 through sedimentation of eroded soil particles contaminated with ^{238}Pu. Initially, the plutonium was in an acidic solution as plutonium nitrate. During transfer the pipeline ruptured and the plutonium was sorbed on soil particles. The sorbed plutonium was eroded during cleanup operations by intense rains (Rogers, 1975). Thus, unlike the metallic origin of the plutonium at the NTS and RF locations, the ML contamination was originally in a soluble form. That the character of the plutonium differed from the NTS and RF samples is exemplified by the higher solubility (80 to 85%) in cold 8M nitric acid. The contamination in the canal should be differentiated from the soil contamination reported by Muller and Sprugel (1977). They reported that the source of the ^{238}Pu in the soil 1 mile east of the Laboratory was aerial emissions from stacks. These emissions were not characterized physically or chemically (Muller and Sprugel, 1977).

Oak Ridge National Laboratory. The ORNL site involves two different contaminating situations. One of the sites was formerly a holdup pond of wastewater for radionuclide retention. After the pond was drained in 1944, the bottom sediment was exposed, and a young forest developed on the floodplain. The depth distribution of the plutonium suggests that the plutonium was sorbed on particles that settled to the bottom. The 8M nitric acid extraction revealed that 60 to 75% was soluble at room temperature and 1-hr extraction. This relatively high extraction suggests a monomeric hydrolyzed form of plutonium.

The second site of contamination was the bottom sediment of a pond that served initially as a waste-receiving pond for ORNL liquid waste. With an improved waste-management system, the pond then served as a secondary settling pond for effluent from a low-level wastewater treatment plant (Tamura, Sealand, and Duguid, 1977). The higher activity level in the pond and its close proximity to ORNL suggest that the pond served as the primary waste retention system before overflowing into the White Oak Creek. The 8M nitric acid extraction revealed that 90% of the Pu in this sediment was soluble.

Citric acid extractability of the NTS, ML, and ORNL (floodplain) samples has been published (Tamura, 1976). The increasing order of extraction by the citrate was: NTS, 1%; ORNL, 25%; and ML, 50%. The time of contact of the citric acid with the solids was 30 min at room temperature. Unpublished data by these investigators show that citric acid treatment of the RF sample extracted approximately 10% of the plutonium.

The percentage range in Pu extractability by the mineral acid and citric acid from the different site samples reveals the differences in the plutonium at these sites. Although not established quantitatively, these differences should also be reflected in the uptake coefficients of vegetation grown at the sites. The extraction data presented further suggest that plutonium derived from metallic sources, such as at NTS and RF, is less soluble than that derived from an initially solubilized form, such as at ML and ORNL.

Chemical Characterization

Plutonium Behavior in ORNL Soil

We have conducted a partial characterization of Pu behavior in an alluvial soil contaminated in 1944 at ORNL. The nature of the Pu source material appears to have been low-level waste solutions in which the Pu was probably present in monomeric forms. This soil probably represents the oldest contamination event since the first milligram and gram amounts of Pu were extracted from the Oak Ridge graphite reactor in January and February 1944. The study area was used for low-level radioactive-waste retention from March 1944 to September 1944 and was subsequently abandoned.

Because the mode of release was soluble Pu rather than calcined PuO_2, the site is proving invaluable for developing concepts on the biogeochemistry of Pu. Plutonium distributions in the contaminated floodplain soil and biota are available (Dahlman, Gartin, and Hakonson, this volume). Total soil Pu has been determined by hot $8M$ HNO_3 leaching. These values are the same as those for $HF-HNO_3$ leaches.

Comparative Behavior of Pu, Th, and U.   Comparative extractions have been conducted to compare the extractabilities of Pu, U, and Th. Two extractants are used, $1M$ HNO_3 and 10% sodium carbonate and 5% sodium bicarbonate. The former represents a weak acid extraction, and the latter is used for U extraction in ore processing. Table 3

**Table 3   Extraction of U, Th, and Pu from
Floodplain Soil Using Mild Extractants**

	Percent extracted*		
Extractant	U	Th	Pu
$1M$ HNO_3	73	7.9	7.7
10% $Na_2CO_3-5\%$ $NaHCO_3$	71	45	54

*Based on $8M$ HNO_3 extractable U (8.18 $\mu g/g$), Th (16.8 $\mu g/g$), and Pu (135 dpm/g).

illustrates that Pu and Th are extracted similarly, whereas U is much more readily extracted. Although U extraction changes little between the acid and basic systems, Th and Pu are extracted more readily with carbonate. In the case of carbonate extraction, considerable organic matter (humic material) was solubilized; dialysis studies in carbonate indicate that the carbonate-extracted Pu was not bound to the solubilized organic matter but was present as a diffusible carbonate complex. As is discussed later, however, some of the Pu was associated with humic acids in the soil.

Thorium-234 and plutonium-236 were added to replicate 10-g samples containing 20 ml of $0.5M$ nitric acid to further characterize the relative behavior of Pu and Th in the floodplain soil. The samples were equilibrated for 0.5, 1, 4, and 24 hr. The amount of indigenous $^{239,240}Pu$ and ^{232}Th in solution and the amount of added isotopes in solution were assayed. The results are plotted in Fig. 6. The hot $8M$ nitric acid extractable ^{232}Th and ^{239}Pu were used as the basis for inventorying the total indigenous elements; thus, within 0.5 hr for Th and 1 hr for Pu, both indigenous isotopes achieved the same soil-to-solution distribution as the added isotopes. The sluggishness of the ^{236}Pu toward

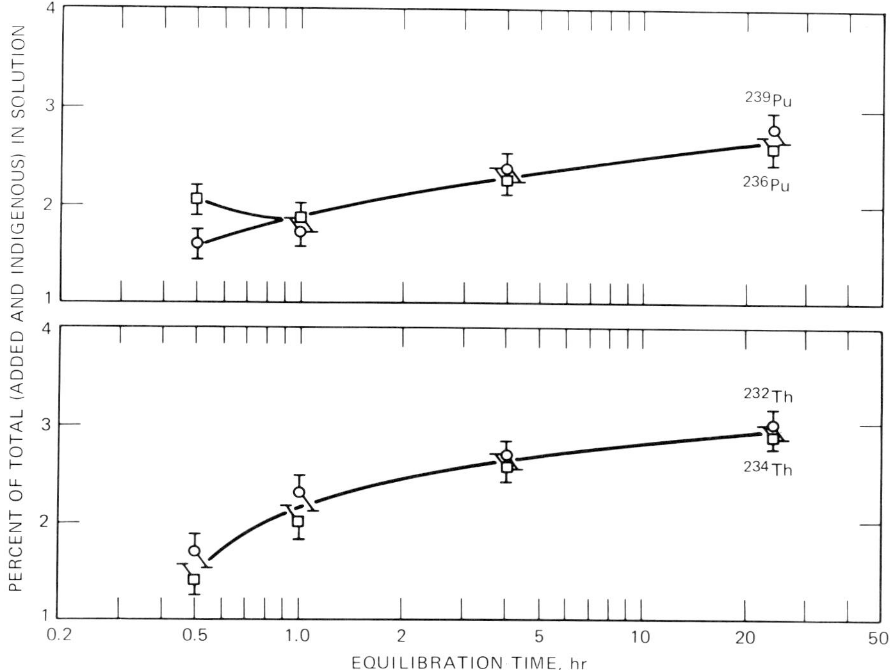

Fig. 6 Comparative behavior of added ^{236}Pu, ^{234}Th, and indigenous ^{239,240}Pu and ^{232}Th during equilibration with 0.5M HNO$_3$. Soil was contaminated with ^{239,240}Pu in 1944.

equilibration during the first 0.5 hr is not yet understood. It may have been due to the presence of a small amount of Pu(V and VI) at the start of the equilibration. It is apparent, however, that the 8M nitric acid soluble isotopes rapidly redistribute in the same manner as the added isotopes and that the solution-phase indigenous isotopes are in equilibrium with soil-bound elements. This behavior strongly suggests that both Pu and 8M nitric acid soluble Th are surface sorbed rather than occluded or trapped in mineral matrices. The strong carbonate-extraction results further support this concept since only a few elements are soluble in this extractant. Iron, for example, is not solubilized.

Other Observations on Physicochemical Associations. Bondietti, Reynolds, and Shanks (1976) discussed the probable association of part of the Pu with soil organic matter. This conclusion was based on the solubilization of part of the soil humic acids using chelating resin (Na form). The resin was prebuffered to the soil pH (6.5); the subsequent decalcification of soil solubilized 15% of the soil organic C. These soluble humates contained 5% of the soil Pu. In addition to the humic-associated Pu, 13% of the Pu was associated with the resin itself. Assuming equal distribution of Pu with soil organic C (which is not likely) and assuming that the resin-associated Pu was organically bound initially, 55% of the Pu at most was associated with organic matter. In reality, the fraction was probably substantially less. Repeated treatments of the soil with NaOCl to destroy organic matter did, however, remove 82% of the soil Pu. The bleach treatment, which was conducted at pH 9.5, minimized inorganic mineral destruction. Removal of most of the Pu with bleach also suggests that the Pu was surface sorbed.

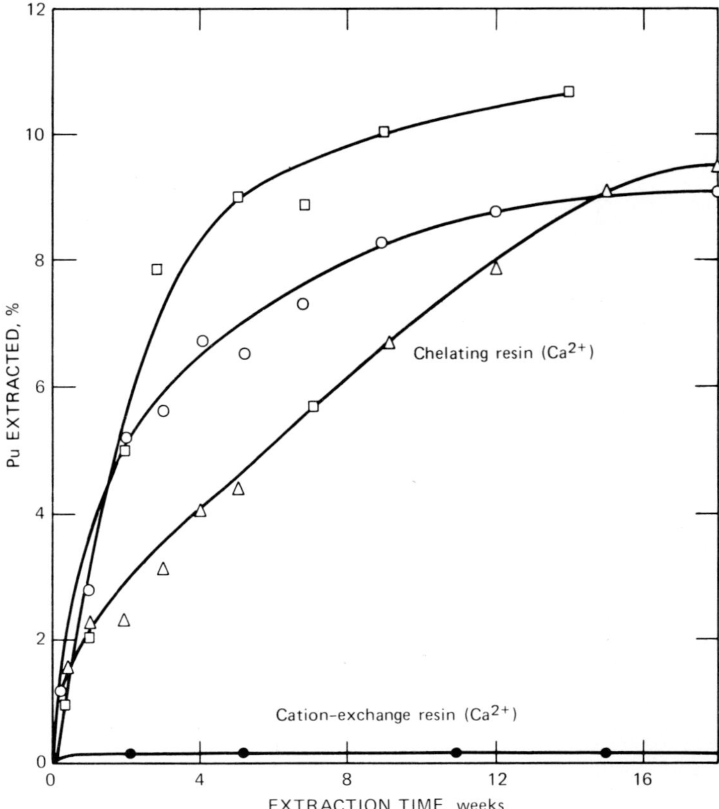

Fig. 7 Extraction of Pu from contaminated soil by chelating and cation-exchange resins.

Mild Extractants. Bondietti, Reynolds, and Shanks (1976) reported on the transfer rates of this soil-bound Pu to chelating resin. The objective was to evaluate what fraction of the soil Pu would transfer from the soil solid phase to a resin solid phase as an approximation of soil–root interactions. For three soil samples, the transfer of Pu was characterized by an initial period of rapid Pu transfer followed by much slower rates (Fig. 7). Between 9 and 11% of the Pu desorbed from the soil in 14 weeks. The soil-bound Pu did not transfer to a cation-exchange resin (Dowex-50) (Fig. 7). These observations indicate that there is a form of Pu present that will transfer from the soil to resin, but it is not bound by cation exchange. Also, the mobile species are exhausted rather quickly, which suggests that only Pu on the surface of soil peds is involved in the redistribution.

Tamura (1976) reported that 7% of the Pu desorbed when equilibrated with sodium citrate (pH 7.3), which suggests a fraction similar to the resin-extractable Pu. Extraction of the soil with citric acid (pH 3) removed 23% of the Pu in that study. Bondietti, Reynolds, and Shanks (1976), using DTPA (pH 6.5) to extract Pu from this soil, found that 28% was removed with no difference between 4 and 20 hr of equilibration. In addition to the extractions using organic chelating agents, the soil was equilibrated for 18 hr with $10^{-2} M$ sodium bicarbonate. At a 4/1 solution/soil ratio, 0.08% of the Pu was

TABLE 4 Concentration Ratios for U, Th, and Pu
of Plants Grown on Contaminated Soil*

Plant	Concentration ratio ($\times 10^{-3}$)		
	^{238}U	^{232}Th	^{239,240}Pu
Leaves and stems			
Snapbean (2)	9.3	2.1	1.9 ± 0.7
Soybean (3)	17 ± 1.7†	4.6 ± 0.35†	2.0 ± 0.04†
Millet	17 ± 0.7†	0.1 ± 0.5†	0.2 ± 0.07†
Tomato (2)	23 ± 1.4†	5.8 ± 0.4†	6.0 ± 0.5†
Garden beet	8.3	2.5	3.4
Fruit, seed, and storage organ			
Soybean (2)	0.55 ± 0.12†	0.1 ± 0.05†	$\leqslant 0.05 \pm 0.01$†
Squash			
Whole	1.9	0.45	0.081
Peeled	$\leqslant 3.0$	$\leqslant 0.18$	$\leqslant 0.15$
Irish potato			
Whole (4)	9.9 ± 0.01†	2.0 ± 0.03†	1.4 ± 0.08†
Peeled (3)	0.9 ± 0.03†	$\leqslant 0.12 \pm 0.02$†	0.18 ± 0.02†
Snapbean (pod)	1.0	0.3	0.1
Millet	1.0	2.0	0.2
Tomato (4)	1.5 ± 0.07†	Not determined	0.1 ± 0.03†
Tomato (1)	0.7	0.17	Not determined
Beet (peeled)	0.3	0.3	0.6

*Values are ± standard errors of replicate analysis; for single samples, U and Th analyses are typically ±10%; Pu analysis is ±50% (counting error). Data set represents plants cleaned by washing. Edible tissues were peeled or cleaned as if being prepared for cooking; several analyses (not included) showed high and similar CR's for three elements, indicating soil contamination; in no case, however, was Th and/or Pu significantly higher than U.

†Mean ± standard deviation.

found in the aqueous phase after ultracentrifugation (9000 $\times g$ for 1 hr). This corresponds to a desorption K_d of 5×10^3 ml/g.

Extraction by Growing Plants. Like chemical extractants, plants grown on this floodplain soil reveal that U is extracted more readily than Pu or Th. Table 4 illustrates this for both whole plants and reproductive and storage organs. Both Th and Pu appear in vegetative tissue in similar concentrations relative to 8M nitric acid extractable soil values. This observation suggests that the Pu is probably present in the III or IV oxidation state rather than in the V or VI. The similarity to Th suggests that the IV state dominates.

Conclusions on Chemical Associations of Pu in an Alluvial Soil

Tamura (1976) concluded that the Pu in the floodplain soil was most likely present in monomer rather than polymer form. This conclusion was reached because the Pu was highly extractable in citrate and leached in cold 8M HNO$_3$. The extraction of the Pu by strong carbonate solution and the extractability with diethylenetriaminepentaacetic acid also suggest that monomeric forms are present as do the isotopic-exchange results.

The Pu associated with this soil appears to represent surface-sorbed associations, with humic materials representing one, but not the only, site. About 7 to 11% of the Pu is readily mobilized by citrate or resin equilibrations where the pH is maintained at natural values. A small fraction (0.08%) is soluble in dilute sodium bicarbonate. On the basis of chemical extractions with dilute nitric acid and strong carbonate solutions and by plant-extraction data, the Th and Pu appear to behave similarly, which suggests similar chemical associations in the soil. Since the soil itself is largely alluvium, the extractable Th is probably associated with colloidal surfaces of secondary minerals. The intrusion of Pu may have resulted in similar associations.

For this soil and for the type of contamination that was believed to have occurred, it appears reasonable to conclude that 30+ yr after deposition in soil Pu is not assimilated by vegetation to a greater extent than natural Th. To the extent that Th is recognized as an element which does not readily transfer in biological food chains, a similar behavior for Pu should be observed many centuries after its release to the biosphere.

Physical Characterization

Observations of Particle-Size Association of Pu in Contaminated Sites

As noted in an earlier section, plutonium can actually be released into the environment by several modes. In addition, the size and character of the plutonium as it is being introduced are subject to change as it interacts with environmental material. After the interaction, some of the properties of the plutonium can be controlled by the matrix properties.

Studies of the association of plutonium with soil and sediment particles have been reported by Tamura (1976) for several contaminated sites. He reported that, in the safety shot sites of NTS, the plutonium found in the soils surrounding the site was primarily associated with coarse-silt (50 to 20 μm) and fine-sand size (125 to 50 μm) fractions. The activity/size ratio of the coarse-silt fraction of two samples reported by Tamura (1976) was approximately 7.7 and of the clay size was approximately 0.5. Since the size segregation was based on the density of silicate particles (2.65 g/cm^3 density), it could not be ascertained whether the plutonium particles were of the designated sizes or whether the plutonium particles were of a finer size and attached to the silicate surfaces. These size associations are consistent with the findings of Mork (1970), who reported in his studies of NTS soils that the major portion of the activity was associated with particles larger than 44 μm.

In contrast to the NTS soils, the Pu in the bottom sediment originating from a waste-transfer line leak at ML was primarily associated with particles less than 2 μm in diameter (Tamura, 1976). Interestingly, Muller and Sprugel (1977) reported that small amounts of plutonium released from stacks at ML and absorbed by soils were also primarily associated with the <2-μm soil particles. They also found that fallout plutonium in the environs of ML showed the same pattern of concentration in the various size fractions.

A sample from the floodplain soil at ORNL showed that the plutonium distribution followed the soil-particle size distribution (Tamura, 1976). This distribution would indicate that the plutonium in the floodplain was part of settling sediment particles that had reacted with the plutonium farther upstream. As noted earlier, upstream of the floodplain is a waste pond that contains plutonium in the bottom sediment; the overflow from this pond empties into the creek flowing to the floodplain. The activity/size ratio of the floodplain sample was 1.40 in the clay size and 0.97 in the coarse-silt fractions; in comparison, the ratio for the two sizes in the ML samples were 2.34 and 0.29 in the sediment sample and 2.42 and 0.60 in the soil sample. These results show that the ML samples are enriched in the fine-clay size fraction.

Tamura (1976) reported on the size association of plutonium in a RF soil taken from the 5- to 10-cm depth near the spill site. The activity/size ratio was 2.26 in the clay fraction and 1.26 in the coarse-silt fraction. In another sample taken 1 km east of the spill site, the ratios of the two sizes were 3.20 and 0.98, respectively (Tamura, 1977a).

The size associations of the plutonium show that in the safety shot sample of NTS the clay-size fraction is not enriched in plutonium; at the other sites the clay size is relatively enriched. However, the association with clay in the enriched samples does not necessarily mean high acid solubility. The 8M nitric acid extraction at room temperature revealed that the ML and ORNL samples are quite soluble (over 60%); the RF sample was less soluble (15 to 20%). The difference in the acid solubility is likely due to the initial soluble form in the ML and ORNL releases and the metallic nature of the RF sample.

Implication of Particle Size Association

Tamura (1977b) attempted to evaluate the significance of the size association of plutonium on soil particles in terms of potential hazard due to resuspension and inhalation of contaminated particles. He considered the soil particle size association of the plutonium, the depositional character of the different particle sizes in the pulmonary compartment of the lung, and the fraction of activity in the resuspendible fraction in the soil. This initial attempt did not include considerations of soil erodibility, vegetation, field size, and surface-roughness factors, which are important in wind erosion of soils (Skidmore, 1976).

Table 5 shows the soil plutonium indexes calculated from the three factors for the four contaminated sites. The less than 125-μm size is considered to be the resuspendible fraction; others have suggested the less than 100-μm sizes (Chepil, 1945; Healy, 1974), but the available data are given for the slightly larger size. The soil activity factor is defined as the activity per unit weight of mass for each size fraction. This factor is derived by dividing the activity portion of a given size by the mass contribution of that size; it therefore weights the activity in the different potentially inhalable sizes.

Table 5 also gives the depositional fraction derived by the Task Group on Lung Dynamics (1966) and the depositional factor derived as a product of the soil activity factor and the depositional fraction. The depositional percentage of the larger resuspendible sizes is relatively low; most of these particles are filtered by the upper respiratory tract and have a short biological half-life (Task Group on Lung Dynamics, 1966). Also included in Table 5 is the fraction of the activity found in the < 125-μm sizes. The high percentage results in a small effect on the final soil factor. The activity distribution was determined by using water suspension and either ultrasonic treatment or chemical dispersant (ML sample). Thus the actual association of the plutonium in the soil may be different and should be evaluated.

The final soil index shows a range of 0.52 for the NTS sample to 1.26 for the RF sample. This implies that the plutonium in the soil at the RF site is potentially about 2.4 times as hazardous in terms of the inhalation pathway. It should be emphasized that the number of samples is limited, and the factors may change with more information. Furthermore, the erodibility and other factors of the soils were not evaluated; hence, until these factors are evaluated and larger numbers of samples are investigated, the indexes are only tentative.

A soil-level standard of 2 (d/min) g^{-1} of soil was tentatively set for the state of Colorado (Johnson, Tidball, and Severson, 1976). The value of 2 (d/min) g^{-1} did not specify the soil size fraction. If this value were used for the RF sample, then the equivalent standard for the NTS would be 5 (d/min) g^{-1} (1.26/0.52 × 2 = 5). Similar calculations for ML and ORNL give 2 and 3 (d/min) g^{-1}, respectively.

Healy (1974) suggested 500 (d/min) g^{-1} for bare soil without reference to any particular site; thus this suggested level might be interpreted as a general guide. The soil factors in Table 5 show that NTS has the lowest value of 0.52. If 500 (d/min) g^{-1} were used as a general guide, the allowable concentration for NTS would be 960 (d/min) g^{-1} (1/0.52 × 500 = 960). Similarly, for RF, ML, and ORNL, the values would be 395, 425, and 615 (d/min) g^{-1}, respectively.

The importance of establishing a soil standard is related to long-term health risk for exposed populations. For example, the U. S. Environmental Protection Agency (1977) proposed a soil-screening level of 0.2 μCi/m^2 (surface 1-cm depth). This value was established as a reasonable soil level whereby the resulting lung and bone doses to the critical segment of the exposed population would be below proposed limits.

TABLE 5   Soil Factor Calculated from Soil Activity Factor, Depositional Factor, and Resuspendible Fraction

Size, μm	Soil fraction	Activity fraction	Soil activity factor	Depositional fraction	Depositional factor	Resuspendible fraction	Soil Pu index
			Nevada Test Site (Area 13)				
<2	0.04	0.03	0.75	0.40	0.30		
2 to 5	0.03	0.04	1.33	0.12	0.16		
5 to 125	0.43	0.92	2.19	0.03	0.07		
	0.50	0.99	4.27		0.53	0.99	0.52
			Rocky Flats				
<2	0.12	0.28	2.33	0.40	0.93		
2 to 5	0.04	0.14	3.50	0.12	0.42		
5 to 125	0.34	0.49	1.44	0.03	0.04		
	0.50	0.91	7.27		1.39	0.91	1.26
			Mound Laboratory*				
<2	0.19	0.46	2.42	0.40	0.97		
2 to 4	0.09	0.14	1.56	0.12	0.19		
4 to 125†	0.72	0.40	0.56	0.03	0.02		
	1.00	1.00	4.54		1.18	1.00	1.18
			Oak Ridge National Laboratory				
<2	0.29	0.40	1.38	0.40	0.55		
2 to 5	0.10	0.09	0.90	0.12	0.11		
5 to 125	0.59	0.51	0.86	0.03	0.03		
	0.98	1.00	3.14		0.69	1.00	0.69

*Data from Muller and Sprugel, 1976.
†Assumes particles greater than 45 μm to be less than 125 μm.

References

Adams, J. A. S., J. K. Osmond, and J. J. W. Rogers, 1959, The Geochemistry of Thorium and Uranium, in *Physics and Chemistry of the Earth,* Vol. 3, L. H. Ahrens, F. Press, K. Rankama, and S. K. Runcorn (Eds.), Pergamon Press, Inc., New York.

Adams, W. H., et al., 1975, *Studies of Plutonium, Americium, and Uranium in Environmental Matrices,* USAEC Report LA-5661, Los Alamos Scientific Laboratory, NTIS.

Andelman, J. B., and T. C. Rozzell, 1970, Plutonium in the Water Environment, in *Radionuclides in the Environment,* Advances in Chemistry Series, No. 93, pp. 118-137, American Chemical Society.

Baes, C. F., and R. E. Mesmer, 1976, *The Hydrolysis of Cations,* John Wiley & Sons, Inc., New York.

Bondietti, E. A., 1976, Environmental Chemistry of Neptunium: Valence Stability and Soil Reactions, *Agronomy Abstracts,* p. 126, American Society of Agronomy, Madison, Wisc.

——, and S. A. Reynolds, 1976, Field and Laboratory Observations on Plutonium Oxidation States, in *Proceedings of an Actinide–Sediment Reactions Working Meeting,* Seattle, Wash., Feb. 10–11, 1976, L. L. Ames (Ed.), USAEC Report BNWL-2117, Battelle, Pacific Northwest Laboratories, NTIS.

——, S. A. Reynolds, and M. H. Shanks, 1976, Interaction of Plutonium with Complexing Substances in Soils and Natural Waters, in *Transuranium Nuclides in the Environment,* Symposium Proceedings, San Francisco, Nov. 17–21, 1975, pp. 273-287, STI/PUB/410, International Atomic Energy Agency, Vienna.

——, and F. H. Sweeton, 1977, Transuranic Speciation in the Environment, in *Transuranics in Natural Environments,* Symposium Proceedings, Gatlinburg, Tenn., Oct. 5–7, 1976, M. G. White and P. B. Dunaway (Eds.), ERDA Report NVO-178, pp. 449-476, Nevada Operations Office, NTIS.

Chepil, W. S., 1945, Dynamics of Wind Erosion: II. Initiation of Soil Movement, *Soil Sci.,* 60: 397-411.

Cherry, R. D., and L. V. Shannon, 1974, The Alpha Radioactivity of Marine Organisms, *At. Energy Rev.,* 12(1): 3-46.

Coleman, G. H., 1965, *The Radiochemistry of Plutonium,* USAEC Report NAS-NS-3058, National Academy of Sciences–National Research Council, NTIS.

Connick, R. E., 1954, Oxidation States, Potential, Equilibria and Oxidation–Reduction Reactions of Plutonium, in *The Actinide Elements,* Chap. 8, G. T. Seaborg and J. J. Katz (Eds.), McGraw-Hill Book Company, New York.

Cunningham, B. B., 1954, Preparation and Properties of the Compounds of Plutonium, in *The Actinide Elements,* Chap. 10, G. T. Seaborg and J. J. Katz (Eds.), McGraw-Hill Book Company, New York.

——, and J. C. Hindman, 1954, The Chemistry of Neptunium, in *The Actinide Elements,* Chap. 12, G. T. Seaborg and J. J. Katz (Eds.), McGraw-Hill Book Company, New York.

Glover, P. A., F. S. Miner, and W. L. Polzer, 1976, Plutonium and Americium Behavior in the Soil/Water Environment, in *Proceedings of an Actinide–Sediment Reactions Working Meeting,* Seattle, Wash., Feb. 10–11, 1976, L. L. Ames (Ed.), USAEC Report BNWL-2117, Battelle, Pacific Northwest Laboratories, NTIS.

Healy, J. W., 1974, *Proposed Interim Standard for Plutonium in Soils,* USAEC Report LA-5483-MS, pp. 1-100, Los Alamos Scientific Laboratory, NTIS.

Hindman, J. C., 1954, Ionic and Molecular Species of Plutonium in Solution, in *The Actinide Elements,* Chap. 9, G. T. Seaborg and J. J. Katz (Eds.), McGraw-Hill Book Company, New York.

Jacobson, L., and R. Overstreet, 1948, The Uptake of Plants and Plutonium and Some Products of Nuclear Fission Absorbed to Soil Colloids, *Soil Sci.,* 65: 129-134.

Johnson, C. J., R. R. Tidball, and R. C. Severson, 1976, Plutonium Hazard in Respirable Dust on the Surface of Soil, *Science,* 193: 488-490.

Krey, P. W., and E. P. Hardy, Jr., 1971, *Plutonium in Soil Around the Rocky Flats Plant,* USAEC Report HASL-235, Health and Safety Laboratory, NTIS.

Mork, H. M., 1970, *Redistribution of Plutonium in the Environs of the Nevada Test Site,* USAEC Report UCLA-12,590, pp. 1-29, University of California, NTIS.

Muller, R. N., and D. G. Sprugel, 1977, Distribution of Local and Stratospheric Plutonium in Ohio Soils, *Health Phys.,* 33(5): 411-416.

Navratil, J. D., and C. E. Baldwin, 1977, *Recovery Studies for Plutonium Machining Oil Coolant,* ERDA Report RFP-2579, University of Utah, NTIS.

Nishita, H., 1978, Extractability of ^{239}Pu and ^{242}Cm from a Contaminated Soil As a Function of pH and Certain Soil Components. $CH_3COOH-NH_4$ System, in *Environmental Chemistry and Cycling Processes,* DOE Symposium Series, No. 45, Augusta, Ga., Apr. 28–May 1, 1976, D. C. Adriano and I. Lehr Brisbin, Jr. (Eds.), pp. 403-416, CONF-760429, NTIS.

Polzer, W. L., 1971, Solubility of Plutonium in Soil/Water Environments, in *Proceedings of the Rocky Flats Symposium on Safety in Plutonium Handling Facilities,* Golden, Colo., Apr. 13–16, 1971, USAEC Report CONF-710401, pp. 411-429, Dow Chemical Company, NTIS.

——, and F. J. Miner, 1976, The Effect of Selected Chemical and Physical Characteristics of Aqueous Plutonium and Americium and Their Sorption by Soils, in *Proceedings of an Actinide–Sediment Reactions Working Meeting,* Seattle, Wash., Feb. 10–11, 1976, L. L. Ames (Ed.), USAEC Report BNWL-2117, Battelle, Pacific Northwest Laboratories, NTIS.

Prout, W. E., 1958, Adsorption of Radioactive Wastes by Savannah River Plant Soil, *Soil Sci.,* 86: 13-17.

Rai, D., and R. J. Serne, 1977, Plutonium Activities in Soil Solutions and the Stability and Formation of Selected Minerals, *J. Environ. Qual.,* 6(1): 89-95.

Rhodes, D. W., 1957, Adsorption of Plutonium by Soil, *Soil Sci.,* 84: 465-469.

Rogers, D. R., 1975, Mound Laboratory Environmental Plutonium Study, 1974, USAEC Report MLM-2249, pp. 1-141, Mound Laboratory, NTIS.

Routson, R. C., G. Jansen, and A. V. Robinson, 1975, *Sorption of ^{99}Tc, ^{237}Np and ^{241}Am on Two Subsoils from Differing Weathering Intensity Areas,* USAEC Report BNWL-1889, Battelle, Pacific Northwest Laboratories, NTIS.

Skidmore, E. L., 1976, A Wind Erosion Equation: Development, Application, and Limitation, in *Atmosphere–Surface Exchange of Particulate and Gaseous Pollutants (1974),* ERDA Symposium Series, No. 38, Richland, Wash., Sept. 4–6, 1974, R. J. Engelmann and G. A. Sehmel (Coordinators), pp. 452-465, CONF-740921, NTIS.

Starik, I. E., and L. B. Kolyadin, 1957, Mode of Occurrence of Uranium in Ocean Water, *Geokhimiya,* No. 3.

Tamura, T., 1972, Sorption Phenomena Significant in Radioactive Waste Disposal, in *Underground Waste Management and Environmental Implications,* T. D. Cook (Ed.), American Association of Petroleum Geologists, Memoir 18.

——, 1975, Characteristic of Plutonium in Surface Soils from Area 13 of the New Test Site, in *Radioecology of Plutonium and Other Transuranics in Desert Environments,* M. G. White and P. B. Dunaway (Eds.), USAEC Report NVO-153, pp. 27-41, Nevada Operations Office, NTIS.

——, 1976, Physical and Chemical Characteristics of Plutonium in Existing Contaminated Soils and Sediments, in *Transuranium Nuclides in the Environment,* Symposium Proceedings, San Francisco, Nov. 17–21, 1975, pp. 213-229, STI/PUB/410, International Atomic Energy Agency, Vienna.

——, 1977a, Effect of Pretreatments on the Size Distribution of Plutonium in Surface Soil from Rocky Flats, in *Transuranics in Desert Ecosystems,* M. G. White, P. B. Dunaway, and D. L. Wireman (Eds.), USDOE Report NVO-181, pp. 173-186, Nevada Operations Office, NTIS.

——, 1977b, Plutonium Association in Soils, in *Transuranics in Natural Environments,* Symposium Proceedings, Gatlinburg, Tenn., Oct. 5–7, 1976, M. G. White and P. B. Dunaway (Eds.), ERDA Report NVO-178, pp. 97-114, Nevada Operations Office, NTIS.

——, O. M. Sealand, and J. O. Duguid, 1977, *Preliminary Inventory of ^{239,240}Pu, ^{90}Sr, and ^{137}Cs in Waste Pond No. 2 (3513),* ERDA Report ORNL-TM-5802, pp. 1-31, Oak Ridge National Laboratory, NTIS.

Task Group on Lung Dynamics, 1966, Deposition and Retention Models for Internal Dosimetry of the Human Respiratory Tract, *Health Phys.,* 12: 173-207.

Thompson, S. B., L. O. Morgan, R. A. James, and I. Perlman, 1949, The Trace Chemistry of Americium and Curium in Aqueous Solutions, in *The Transuranium Elements,* Part II, paper 191, G. T. Seaborg, J. J. Katz, and W. M. Manning (Eds.), McGraw-Hill Book Company, New York.

U. S. Environmental Protection Agency, 1977, *Proposed Guidance on Dose Limits for Persons Exposed to Transuranium Elements in the General Environment,* Report EPA 520/4-77-016.

Sources of Variation in Soil Plutonium Concentrations

JOHN E. PINDER III and DONALD PAINE

Variations in ^{238}Pu and $^{239,240}Pu$ concentrations in surface soil near a nuclear-fuel reprocessing facility were attributed to distance from the point of aerial release; microtopographical heterogeneity in deposition; and sampling error, which included aliquoting and analytical errors. Distance from the point of release accounted for approximately 75% of the variation in concentrations of both nuclides, whereas sampling error accounted for less than 5% of the variation. Microtopographical heterogeneity accounted for approximately 20% of the variation in $^{239,240}Pu$ concentrations but only 5% of the variation in ^{238}Pu concentrations. This difference may be due to different histories of deposition of the nuclides at the site. Other sources of variation, errors in the statistical models, and the implications for future sampling are discussed.

Concentrations of radionuclides in soils, plants, and animals are usually highly variable, with coefficients of variation (standard deviation/mean) usually exceeding 1.0 (Eberhardt, 1964; Remmenga and Whicker, 1967; Pinder and Smith, 1975; Shanks and De Selm, 1963). This is especially true of the isotopes of plutonium. Large coefficients of variation in soil plutonium concentrations have been reported for plutonium contamination resulting from weapons testing (Nyhan, Miera, and Neher, 1976; Romney et al., 1976), "safety-shots," i.e., chemical explosions of nuclear weapons material (Gilbert et al., 1976), aqueous discharges from industrial facilities that handle plutonium (Hakonson and Nyhan, this volume), and deposition of aerial releases from reprocessing facilities (Adriano, Corey, and Dahlman, this volume; Adriano and Pinder, 1977; McLendon, 1975; McLendon et al., 1976). A portion of this large variability may be due to the release of plutonium in particulate form and the analytical errors caused by including various amounts of these particles in a sample (Doctor et al., this volume; Adriano, Wallace, and Romney, this volume); however, there must be a greater understanding of the causes of this variation before the cycling processes of plutonium can be fully understood or efficient sampling programs can be designed to estimate plutonium concentrations or inventories. The purpose of this study was to evaluate the relative importance of several potential sources of variation in soil concentrations of ^{238}Pu and $^{239,240}Pu$ that had been released to the atmosphere from a reprocessing facility at the U. S. Department of Energy's Savannah River Plant (SRP) near Aiken, South Carolina.

We hypothesized three main components of variation in plutonium concentrations. First, we expected the distance from the point of release to be important because soil concentrations have been shown to decrease with increasing distance from the point of release (McLendon, 1975; McLendon et al., 1976). Soil concentrations were negligibly affected by reprocessing facilities at distances exceeding 10 km from the point of release (McLendon et al., 1976). Second, we anticipated that plutonium concentrations in soils collected in close proximity (i.e., within 2 m) to one another might vary greatly owing to microtopographical heterogeneity in the deposition of plutonium particles. The occurrence of a single large particle at a site could greatly elevate the soil plutonium concentration. Soil concentrations could also be elevated by an accumulation of resuspended plutonium particles due to some feature of terrain, vegetation, or surface roughness. The third source of variation was termed "sampling error." This is the variation in plutonium concentrations observed among aliquots of a well-mixed soil and includes both errors introduced in taking an aliquot for analysis and the analytical errors associated with determining plutonium concentration in the aliquot.

Large plutonium particles may affect both the microtopographical heterogeneity and the sampling error components of variation. If plutonium deposition from aerial releases occurs as large particles that are resistant to weathering, the particle deposition will influence the sampling error term because the particle will occur in one aliquot and not in the others. If, however, the deposition occurs as large particles that are easily weathered into smaller independent particles which can be easily homogenized in the sample, then aliquots of the soil will have similar concentrations. In that case, the effect of particulate deposition would be expressed as large differences in concentrations among closely spaced soils, i.e., as microtopographical heterogeneity. A mixture of particle sizes, or the confounding effects of resuspension and redistribution of particles, may produce results that are intermediate between the above two extremes.

Methods

To partition the total variation in soil plutonium concentration into components that are due to distance from the source, microtopographical heterogeneity, and sampling error, we collected three soils from each of five independent transects located from 183 to 436 m to the northwest of the point of release (a 62-m stack that exhausts filtered air from the internal atmosphere of the reprocessing facility). Each transect was 2 m long, and soil was collected at distances of 0, 1 and 2 m along the transect. Soil was obtained from the upper 5 cm of the profile with a soil auger. For each point on each transect, we homogenized the soil by vigorous, manual shaking in a cardboard ice-cream carton for 1 min and divided the soil into two samples of equal volume. The auger was cleaned between sampling points, and a separate ice-cream carton was used at each point to prevent cross-contamination. The samples were collected in the vicinity of H-area at the SRP on a field that was being used for long-term studies of the transport and fate of transuranic elements in agricultural ecosystems. The samples were collected before the soil had been disturbed for agricultural purposes in November 1974.

The concentrations of ^{238}Pu and ^{239,240}Pu were determined by alpha spectrometry under the direction of A. L. Boni at the Savannah River Laboratory (operated by E. I. du Pont de Nemours & Co.). Ten-gram aliquots of soil were ashed and leached with HCl. A triisooctylamine, 200- to 800-mesh solid ion-exchange resin was used to remove the plutonium from the leachate by liquid ion exchange. Plutonium was leached from the resin with H_2SO_3, electroplated on platinum disks, and counted. Plutonium-236 was used as an internal standard in estimating recoveries of plutonium from the soil. Large soil particles (>3 mm in diameter) were removed from the samples before aliquots were drawn.

Mathematical Formulation

Comparisons of concentrations between samples at each transect point provide an estimate of sampling error. Comparisons of samples among points on each transect provide an estimate of microtopographical heterogeneity in soil concentrations, and comparisons among transects provide an estimate of the importance of distance from the point of release.

Let Y_{ijk} be the plutonium concentration in the kth sample ($k = 1,2$) drawn from the jth transect point ($j = 1,2,3$) in the ith transect ($i = 1,2, \ldots ,5$). Y_{ijk} may be partitioned into components according to the statistical model,

$$Y_{ijk} = \mu + D_i + M_{ij} + e_{ijk} \tag{1}$$

where μ = grand mean of concentrations
   D_i = effect of the ith transect location
   M_{ij} = effect of the jth position in the ith transect
   e_{ijk} = sampling error

We assume that D_i, M_{ij}, and e_{ijk} are normally distributed random variables with $\mu_D = \mu_M = \mu_e = 0$ and variances σ_D^2, σ_M^2, and σ_e^2 and that D_i, M_{ij}, and e_{ijk} are independent or, in other words, that σ_D^2, σ_M^2, and σ_e^2 are constant for all combinations of D and M. Later in this chapter we will test the validity of these assumptions and discuss the inaccuracies introduced by the failure of the data to meet them. Random samples of Y have expected mean μ and expected variance $\sigma^2 = \sigma_D^2 + \sigma_M^2 + \sigma_e^2$. The parameters σ_D^2, σ_M^2, and σ_e^2 are termed the variance components of σ^2. Equation 1 represents a two-way nested analysis of variance with random effects. Nested analyses and random-effect models are discussed in greater detail by Scheffé (1959) and Searle (1971), who also give procedures for estimating σ_D^2, σ_M^2, and σ_e^2 and calculating confidence intervals about the estimates. The relative importances of distance, microtopographical heterogeneity, and sampling error are given by the intraclass correlation coefficients, $\rho_D = \sigma_D^2/\sigma^2$, $\rho_M = \sigma_M^2/\sigma^2$, and $\rho_e = \sigma_e^2/\sigma^2$, respectively (Scheffé, 1959). The estimated ρ_a for the ath effect is given by $\rho_a = \hat{\sigma}_a^2/\hat{\sigma}^2$, where $\hat{\sigma}^2$ is the sum of the estimates of the individual variance components. Scheffé (1959) also gives procedures for testing the statistical null hypothesis, $H_0 : \sigma_a^2 = 0$, versus the alternative hypothesis, $H_A : \sigma_a^2 > 0$. The formulas and procedures outlined by Scheffé (1959) were used in the following analyses. The Statistical Analysis System was used for the computations (Barr et al., 1976).

Results

Estimates of σ_D^2, σ_M^2, and σ_e^2; 95% confidence intervals about the estimates; and estimates of intraclass correlation coefficients for the concentrations of ^{238}Pu and ^{239,240}Pu are given in Table 1. All the variance components for ^{239,240}Pu were statistically greater ($P < 0.05$) than the corresponding variance components for ^{238}Pu. The $\hat{\sigma}^2$ for ^{239,240}Pu was 11.853, whereas $\hat{\sigma}^2$ for ^{238}Pu was only 0.129. Mean concentrations were 2.23 pCi/g for ^{239,240}Pu and 0.481 pCi/g for ^{238}Pu.

For both radionuclides, sampling error accounted for less than 5% of the total variance, and $\hat{\sigma}_D^2$ was the largest component of the total variance for both nuclides. The major difference between the radionuclides occurred in the relative importance of $\hat{\sigma}_M^2$. Microtopographical heterogeneity was an important component of the variation in

^{239,240}Pu concentrations and accounted for approximately 20% of the variation. The 95% confidence interval about $\hat{\sigma}^2_M$ for ^{238}Pu concentrations includes 0 and suggests that σ^2_M may not be greater than 0; however, the inclusion of 0 is the result of a slight inaccuracy in confidence intervals calculated according to the formulas of Scheffé (1959, pp. 231-235). An F-test of the null hypothesis, $H_0 : \sigma^2_M = 0$, indicated that σ^2_M was significantly greater than 0 (F = 2.99; df = 10,15; P < 0.05).

Alternate Statistical Models

If each transect is considered separately, a one-way analysis of variance procedure can be used to estimate σ^2_M and σ^2_e (Scheffé, 1959, pp. 221-224). These estimates, symbolized $\hat{\sigma}^2_{M,i}$ and $\hat{\sigma}^2_{e,i}$, can be compared among transects to evaluate the assumptions of independence of D_i, M_{ij}, and e_{ijk}. The $\hat{\sigma}^2_{M,i}$ and $\hat{\sigma}^2_{e,i}$ for ^{238}Pu and ^{239,240}Pu concentrations for each transect are compared in Table 2.

The $\hat{\sigma}^2_{M,i}$ for ^{238}Pu concentrations range from −0.0003333 to 0.01674. Although negative $\sigma^2_{M,i}$ are impossible, negative $\hat{\sigma}^2_{M,i}$ can occur owing to the sampling variances of mean-square terms (Scheffé, 1959, pp. 228-229; Searle, 1971, pp. 406-408). The 95% confidence intervals for all the $\hat{\sigma}^2_{M,i}$ overlapped and indicated that the estimates could have been drawn from a common value of $\sigma^2_{M,i}$ at all transects. The $\hat{\sigma}^2_{e,i}$ differed significantly among transects ($F_{max.}$ = 85.6; df = 5,3; P < 0.05) (Kirk, 1968) and tended to decrease as the mean concentration decreased.

The $\hat{\sigma}^2_{M,i}$ for ^{239,240}Pu concentrations ranged from −0.004167 to 12.16 and were significantly greater than 0 for three of the five transects. Again, the 95% confidence intervals for the $\hat{\sigma}^2_{M,i}$ overlapped. The broad confidence bands around the $\hat{\sigma}^2_{M,i}$ in Table 2 for both ^{238}Pu and ^{239,240}Pu concentrations were due to the small number of degrees

TABLE 1   Estimated Variance Components, 95% Confidence Intervals About the Estimates, and Intraclass Correlation Coefficients for the Model $\sigma^2 = \sigma^2_D + \sigma^2_M + \sigma^2_e$, Which Partitions the Total Variance of Soil Plutonium Concentrations (σ^2) into Components due to Distance from the Source (σ^2_D), Microtopographical Heterogeneity (σ^2_M), and Sampling Error (σ^2_e)*

Component	Estimated variance component	Symbol	Degrees of freedom	95% Confidence interval		Estimated intraclass correlation coefficients
				Lower bound	Upper bound	
			^{238}Pu			
Distance	0.1168	$\hat{\sigma}^2_D$	4	0.03947	0.9830	0.904
Microtopographical heterogeneity	0.006185	$\hat{\sigma}^2_M$	10	−0.0001446	0.02555	0.048
Sampling error	0.006220	$\hat{\sigma}^2_e$	15	0.003393	0.01490	0.048
			^{239,240}Pu			
Distance	9.000	$\hat{\sigma}^2_D$	4	2.689	80.91	0.759
Microtopographical heterogeneity	2.575	$\hat{\sigma}^2_M$	10	1.175	8.220	0.217
Sampling error	0.2771	$\hat{\sigma}^2_e$	15	0.1511	0.6639	0.023

*Estimates were computed from 30 determinations of ^{238}Pu and ^{239,240}Pu. Concentrations are in picocuries per gram.

TABLE 2   Distances from the Source of Plutonium Release, Mean Plutonium Concentrations in Soil, and Estimates of the Variance Components for the Model $\sigma_i^2 = \sigma_{M,i}^2 + \sigma_{e,i}^2$ for Each Transect ($i = 1,2, \ldots ,5$)*

Transect	Distance from source, m	Mean concentration, pCi/g	$\hat{\sigma}_{M,i}^2$	95% Confidence interval Lower bound	95% Confidence interval Upper bound	$\hat{\sigma}_{e,i}^2$
			^{238}Pu			
1	183	1.02	0.009167	−0.196	0.664	0.1568
2	214	0.43	0.004742	−0.189	0.422	0.01218
3	275	0.55	0.01674	0.00388	0.675	0.0007167
4	406	0.19	0.0006083	−0.00054	0.0276	0.0001833
5	436	0.20	−0.0003333	−0.0881	0.0318	0.002333
			^{239,240}Pu			
1	183	7.81	12.16	0.986	505.7	1.324
2	214	0.40	−0.004167	−0.0882	1.34	0.01515
3	275	1.56	0.6259	0.155	25.1	0.01910
4	406	0.79	0.04060	−0.145	2.07	0.02415
5	436	0.61	0.05770	0.0132	2.33	0.002550

*Each $\hat{\sigma}_{M,i}^2$ has df = 2, and each $\hat{\sigma}_{e,i}^2$ has df = 3. Mean concentrations are computed from all six samples at each transect.

of freedom associated with the $\hat{\sigma}_{M,i}^2$ and $\hat{\sigma}_{e,i}^2$. The $\hat{\sigma}_{e,i}^2$ for ^{239,240}Pu concentrations also differed significantly among transects ($F_{max.}$ = 519.3; df = 5,3; P < 0.01) and tended to decrease as the mean ^{239,240}Pu concentrations decreased.

Although the $\hat{\sigma}_{M,i}^2$ did not differ significantly among transects, there appeared to be a positive correlation between $\hat{\sigma}_{M,i}^2$ and mean concentration for both ^{238}Pu and ^{239,240}Pu concentrations. The apparent correlations of $\hat{\sigma}_{M,i}^2$ and $\hat{\sigma}_{e,i}^2$ with mean concentration suggested that $\sigma_{M,i}^2$ and $\sigma_{e,i}^2$ were proportional to mean concentration. Proportional relationships between means and variances can be expected because concentrations varied over a broad range and analytical error was controlled to plus or minus a percentage of the measured value. The components $\sigma_{M,i}^2$ and $\sigma_{e,i}^2$ may also vary because the statistical model was inappropriate (e.g., a linear model for a nonlinear process) or failed to contain important causes of variation, such as soil type or disturbance. The variation in $\hat{\sigma}_{M,i}^2$ and $\hat{\sigma}_{e,i}^2$ recorded in Table 2 probably resulted from a combination of factors, including the proportionality between analytical error and concentration, the nonlinear relationship between concentration and distance from the point of release, a change in soil type between transects 3 and 4, and soil disturbances. The impacts of the first three of these factors on our interpretations of the relative importance of microtopographical heterogeneity and sampling error were evaluated by comparing the interpretations resulting from applying more-complex statistical models to the original arithmetic data as well as logarithmic transformations of the data. None of the results for these alternative models affected our conclusions that sampling error was a small fraction of the total variation or that microtopographical heterogeneity was more important for ^{239,240}Pu than for ^{238}Pu. Because some readers may be interested in specific alternate interpretations, we have added the raw data as Table 3.

**TABLE 3 Concentrations of ^{238}Pu and ^{239,240}Pu
for Each Replicate at Each Transect**

Transect*	Sample location†	Replicate	Concentration,‡ pCi/g ^{238}Pu	^{239,240}Pu
1	0	1	1.13 ± 0.02	12.8 ± 0.07
		2	0.99 ± 0.02	10.7 ± 0.06
	1	1	1.02 ± 0.02	6.09 ± 0.04
		2	1.26 ± 0.03	7.73 ± 0.08
	2	1	0.95 ± 0.01	5.22 ± 0.03
		2	0.82 ± 0.02	4.30 ± 0.04
2	0	1	0.41 ± 0.01	0.37 ± 0.01
		2	0.62 ± 0.02	0.56 ± 0.01
	1	1	0.52 ± 0.15	0.43 ± 0.01
		2	0.41 ± 0.01	0.35 ± 0.01
	2	1	0.38 ± 0.01	0.46 ± 0.01
		2	0.25 ± 0.01	0.24 ± 0.01
3	0	1	0.49 ± 0.02	0.66 ± 0.02
		2	0.46 ± 0.01	0.62 ± 0.02
	1	1	0.49 ± 0.02	1.89 ± 0.03
		2	0.52 ± 0.01	2.20 ± 0.03
	2	1	0.69 ± 0.02	1.93 ± 0.03
		2	0.74 ± 0.01	2.06 ± 0.02
4	0	1	0.22 ± 0.01	0.71 ± 0.02
		2	0.19 ± 0.01	0.92 ± 0.02
	1	1	0.15 ± 0.01	0.41 ± 0.03
		2	0.16 ± 0.01	0.67 ± 0.01
	2	1	0.19 ± 0.01	0.91 ± 0.02
		2	0.20 ± 0.01	1.08 ± 0.03
5	0	1	0.26 ± 0.01	0.81 ± 0.01
		2	0.20 ± 0.01	0.77 ± 0.02
	1	1	0.17 ± 0.01	0.35 ± 0.01
		2	0.19 ± 0.01	0.31 ± 0.01
	2	1	0.23 ± 0.01	0.75 ± 0.01
		2	0.13 ± 0.01	0.64 ± 0.01

*Distances from the point of plutonium release to each transect are given in the text.

†Numbers are the distance (in meters) from the start of the transect.

‡Concentration ± counting error (1σ).

Evidences of soil disturbance due to man's activities were present in the area, and the importance of disturbance was suggested by the data from transect 2 (Table 2), where plutonium concentrations were lower than might be expected for the proximity to the stack. The low concentrations may have been caused by a disturbance that moved relatively uncontaminated subsurface material to the soil surface. This illustrates a problem in our interpretation of σ_D^2 as a distance effect. The among-transect comparisons used to estimate σ_D^2 contain not only distance effects but also disturbance, soil type, tree cover, and other uncontrolled effects. The relative importances of distance and the other effects could be estimated by sampling two or more transects at each distance. Numerous other more-complex sampling designs could be used to further subdivide σ^2, but we believe that the design used provided basic data about the sources of variation and that more-complex designs involving a greater number of samples may not be worth the increased cost.

Discussion

The data for ^{239,240}Pu indicated that distance from the source and microtopographical heterogeneity in deposition were important causes of variation in concentrations and that sampling error was a minor component of the total variation. The data for ^{238}Pu concentrations also indicated that distance from the source was an important cause of variation but suggested a lesser effect for microtopographical heterogeneity. Sampling error was only a small part of the total variation in ^{238}Pu concentrations. It would be interesting to compare our variance component estimates with similar estimates for different ecosystems and plutonium source terms, but comparable data are unavailable.

The difference in the importance of microtopographical heterogeneity between ^{238}Pu and ^{239,240}Pu may be due to the history of plutonium deposition at the site. Approximately 0.44 Ci of ^{239,240}Pu was released from the H-area facility before the installation of high-efficiency filters on the exhaust air systems in 1955. From 1956 through 1966, ^{239,240}Pu releases averaged 4 mCi/yr. From 1967 through 1974, normal plutonium releases averaged 14 mCi/yr; approximately 85% of the released plutonium was ^{238}Pu. The unfiltered releases of ^{239,240}Pu probably involved large-size plutonium-containing particles, but no data on particle sizes are available from the periods of ^{239,240}Pu releases. The more-recent filtered releases of ^{238}Pu have probably involved smaller particle sizes. An additional release of 0.6 Ci of ^{238}Pu occurred during a several-day period in 1969 following the accidental failure of the filtering mechanism. This release may have included some large particles of ^{238}Pu, but the low soil concentrations of ^{238}Pu suggest that relatively little of this accidental release was deposited on our study area.

The larger $\hat{\sigma}_M^2$ observed for ^{239,240}Pu probably reflects the larger particle sizes; the relatively small $\hat{\sigma}_e^2$ observed for ^{239,240}Pu, however, indicates that the particles which caused the significant $\hat{\sigma}_M^2$ have since decomposed to smaller units that can be easily homogenized within the sample. The smaller $\hat{\sigma}_M^2$ for ^{238}Pu would be expected if ^{238}Pu deposition had occurred as many small particles. Although the accumulation of resuspended plutonium particles at sites on the soil surface could have produced the large $\hat{\sigma}_M^2$ for ^{239,240}Pu, resuspension does not appear to be the important mechanism in creating microtopographical heterogeneity. If resuspension were the mechanism, a large $\hat{\sigma}_M^2$ should have occurred for ^{238}Pu because a greater proportion of the ^{238}Pu can be resuspended from this soil. Milham et al. (1976) reported that approximately 50% of the plutonium resuspended from the soil surface by a 6 m/sec wind velocity was ^{238}Pu, but only 25% of the plutonium in the upper 5 cm of soil was ^{238}Pu. The differences in patterns of ^{238}Pu and ^{239,240}Pu variation emphasize the importance of considering the form of released materials in studies of plutonium contamination.

The large $\hat{\sigma}_M^2$ and the relatively small $\hat{\sigma}_e^2$ for ^{239,240}Pu concentrations indicate that future sampling would be rendered more efficient by analyzing the radionuclide concentration in a single aliquot taken from a thoroughly mixed composite of two or more samples collected from each sampling location. If the particle problem had expressed itself as a large σ_e^2, the efficiency of future sampling would have been increased by determining concentrations in several aliquots of a single sample or in several independent samples. The small $\hat{\sigma}_M^2$ for ^{238}Pu indicates that a single sample from a location is sufficient to provide a good estimate of concentration. The relatively small $\hat{\sigma}_e^2$

for both nuclides indicates that increased analytical precision would produce only a small reduction in total variation.

Acknowledgments

This research was supported by contract EY-76-C-09-0819 between the U. S. Department of Energy and the University of Georgia. We thank R. A. Geiger, R. M. Klein, E. H. Lebetkin, and K. W. McLeod for assistance in either the field or the laboratory and J. G. Wiener, J. J. Alberts, and K. W. McLeod for their review of an early draft of the manuscript. J. Henry Horton, Jr., of the Savannah River Laboratory, kindly provided data on the history of releases from the H-area facility.

References

Adriano, D. C., and J. E. Pinder III, 1977, Aerial Deposition of Plutonium in Mixed Forest Stands from Nuclear Fuel Reprocessing, *J. Environ. Qual.*, 6: 303-307.

Barr, A. J., J. H. Goodnight, J. P. Sall, and J. T. Helwig, 1976, *A User's Guide to SAS 76*, Sparks Press, Raleigh, N. C.

Eberhardt, L. L., 1964, Variability of the Strontium-90 and Cesium-137 Burden of Native Plants and Animals, *Nature (London)*, 204: 238-240.

Gilbert, R. O., L. L. Eberhardt, E. B. Fowler, E. H. Essington, and E. M. Romney, 1976, Statistical Analyses and Design of Environmental Studies for Plutonium and Other Transuranics at NAEG "Safety Shot" Sites, in *Transuranium Nuclides in the Environment*, Symposium Proceedings, San Francisco, 1975, pp. 449-460, STI/PUB/410, International Atomic Energy Agency, Vienna.

Kirk, R. E., 1968, *Experimental Design: Procedures for the Behavioral Sciences*, Brooks/Cole Publishing Company, Belmont, Calif.

McLendon, H. R., 1975, Soil Monitoring for Plutonium at the Savannah River Plant, *Health Phys.*, 28: 347-354.

——, O. M. Stewart, A. L. Boni, J. C. Corey, K. W. McLeod, and J. E. Pinder, 1976, Relationships Among Plutonium Contents of Soil, Vegetation and Animals Collected on and Adjacent to an Integrated Nuclear Complex in the Humid Southeastern United States, in *Transuranium Nuclides in the Environment*, Symposium Proceedings, San Francisco, 1975, pp. 347-363, STI/PUB/410, International Atomic Energy Agency, Vienna.

Milham, R. C., J. F. Schubert, J. R. Watts, A. L. Boni, and J. C. Corey, 1976, Measured Plutonium Resuspension and Resulting Dose from Agricultural Operations on an Old Field at the Savannah River Plant in the Southeastern United States of America, in *Transuranium Nuclides in the Environment*, Symposium Proceedings, San Francisco, 1975, pp. 409-421, STI/PUB/410, International Atomic Energy Agency, Vienna.

Nyhan, J. W., F. R. Miera, Jr., and R. E. Neher, 1976, Distribution of Plutonium in Trinity Soils After 28 Years, *J. Environ. Qual.*, 4: 431-437.

Pinder, J. E., III, and M. H. Smith, 1975, Frequency Distributions of Radiocesium Concentrations in Soil and Biota, in *Mineral Cycling in Southeastern Ecosystems*, ERDA Symposium Series, Augusta, Ga., May 1–3, 1974, F. G. Howell, J. B. Gentry, and M. H. Smith (Eds.), pp. 107-125, CONF-740513, NTIS.

Remmenga, E. E., and F. W. Whicker, 1967, Sampling Variability in Radionuclide Concentrations in Plants Native to the Colorado Front Range, *Health Phys.*, 13: 977-983.

Romney, E. M., A. Wallace, R. O. Gilbert, and J. E. Kinnear, 1976, ^{239,240}Pu and ^{241}Am Contamination of Vegetation in Aged Fall-Out Areas, in *Transuranium Nuclides in the Environment*, Symposium Proceedings, San Francisco, 1975, pp. 479-491, STI/PUB/410, International Atomic Energy Agency, Vienna.

Scheffé, H., 1959, *The Analysis of Variance*, John Wiley & Sons, Inc., New York.

Searle, S. R., 1971, *Linear Models*, John Wiley & Sons, Inc., New York.

Shanks, R. E., and H. R. De Selm, 1963, Factors Related to Concentration of Radiocesium in Plants Growing on a Radioactive Waste Disposal Area, in *Radioecology*, Proceedings of the First National Symposium on Radioecology, Fort Collins, Colo., Sept. 10–15, 1961, V. Shultz and A. W. Klement, Jr. (Eds.), pp. 97-101, Reinhold Publishing Corporation, New York.

Statistics and Sampling in Transuranic Studies

L. L. EBERHARDT and R. O. GILBERT

The existing data on transuranics in the environment exhibit a remarkably high variability from sample to sample (coefficients of variation of 100% or greater). This chapter stresses the necessity of adequate sample size and suggests various ways to increase sampling efficiency. Objectives in sampling are regarded as being of great importance in making decisions as to sampling methodology. Four different classes of sampling methods are described: (1) descriptive sampling, (2) sampling for spatial pattern, (3) analytical sampling, and (4) sampling for modeling. A number of research needs are identified in the various sampling categories along with several problems that appear to be common to two or more such areas.

Most of the existing data on transuranic elements in the environment exhibits a remarkably high variability from sample to sample. Since analytical procedures for these elements are both complicated and expensive, many investigators use relatively few replicates. In those few cases where moderately large samples have been taken, the underlying frequency distributions generally have been badly skewed (nonsymmetrical). The use of statistical methods in the design and analysis of studies and the use of efficient sampling practices would help avoid the reporting of questionable conclusions.

This chapter identifies some sources of information on sampling and statistical methods relevant to transuranic studies and suggests further research on particular problems along these lines. We believe that too many studies of transuranic elements are currently being conducted with unrealistically small samples. In many such cases, statistical analyses are limited to reporting "counting errors"; thus the inadequacy of the sampling goes unrecognized, at least by those preparing a report on its outcome. This is, of course, not universally true but is all too often the case. As time goes on, it is to be hoped that statistical measures of the adequacy of the sampling and chemical-analysis procedures will become more widely used. The need for efficient statistical designs should then become immediately apparent to investigators and sponsors. Efficient sampling plans, however, require that rather definite objectives be specified for the study. We consider objectives and the appropriate sampling plans in this chapter.

We will direct our discussion primarily to sampling designs rather than to experimental designs since many of the "experiments" concerning transuranic elements that we have encountered thus far are not replicated or have so few replicates that statistical analysis of the results has little meaning. For purposes of this chapter, an experiment occurs when the investigator controls, through randomization, the assignment of treatments to his experimental material. Studies, by random sampling, of uncontrolled phenomena are nòt experiments in this sense.

Perhaps it is advisable to digress here to make a more detailed distinction between a sampling design and an experimental design. Consider a study of the effect of, say, four levels of concentration of some chemical substance on a specific kind of plant. A suitable experimental design might use, say, 25 individually potted plants kept in a growth chamber. Each of 5 randomly selected pots would be treated with the same concentration, and one lot of 5 plants would be a control. Another randomization would be used in assigning the 25 pots to places in the growth chamber. There would be thus 5 replicates of each type of treatment (concentration of chemical) and 5 control pots.

In contrast, a sample survey design might be used to study the concentration of, say, plutonium in a natural population of plants in the vicinity of a nuclear test site. Here we can only observe, by sampling, the results of events over which we usually have no control. There are no true replicates in the experimental sense. It is now fairly common practice, however, to speak of "replicate samples," and we would only stress the need to use the word "sample" to be precise. The investigator cannot control the physical relationships involved; he has to take specimens from an existing population of plants in the positions in which they occur. He can, of course, control the sampling process so that he obtains several individuals of the same species growing roughly the same distance from ground zero and so on.

Since science is commonly thought of as being practically synonymous with the experimental method, many people prefer to regard observations taken on some uncontrolled process as experiments. Such a view is largely immaterial and irrelevant insofar as the mathematical and computational aspects of statistical methods are concerned. Only when we begin to draw inferences from analyses of the data does the real distinction between "experiment" and "observation" become apparent; i.e., in a true experiment we can use rather homogeneous material and, by randomization, ensure that any effects due to position in the growth chamber, genetic factors, etc., are reflected by the error term in the statistical analysis. Without this element of deliberate control of the experiment, there is no assurance that unknown extraneous factors will not also influence the factors under study. Hence the analysis of data obtained by sampling is a rather more hazardous affair. However, exactly the same statistical analyses and an identical mathematical model can be used for either an experiment or a sample survey.

For the present, many of the immediate needs for statistical guidance in transuranic research programs can best be served from the sampling point of view. Many textbooks and experienced statistical practitioners are available to aid in the design and analysis of experiments. We are also interested in the design of experiments, but we have elected to concentrate on sampling in this chapter. We will not try to be explicit as to the role of source terms, but a long list of sources must, of course, be considered; e.g. (1) worldwide fallout from nuclear events; (2) localized fallout from nuclear events; (3) localized dispersion without a nuclear event, such as safety tests; (4) stack releases; (5) liquid effluents; (6) accidents involving nuclear weapons; (7) burnup of SNAP devices; (8) damage to other power sources; and (9) various kinds of eventualities concerning stored wastes, transport, etc.

We will not attempt to deal with studies of the biological effects of exposures to various substances. We have discussed the statistical problems in assessing the effects of low-level, chronic pollutants elsewhere (Eberhardt, 1975a) and can only note here that they are even more perplexing than those associated with sampling alone.

Objectives in Sampling

There are many ways in which the objectives of environmental studies can be arranged, and we have suggested several such sets (Eberhardt, 1976; Eberhardt, 1977; Eberhardt et al., 1976). In this chapter we consider four sampling methods to meet various objectives: descriptive sampling, sampling for spatial pattern, analytical sampling, and sampling for modeling. Although the four categories are neither mutually exclusive nor all-inclusive, they do seem to serve as useful devices to cover most situations. Figure 1 illustrates the different objectives using the same physical example. Since the objectives

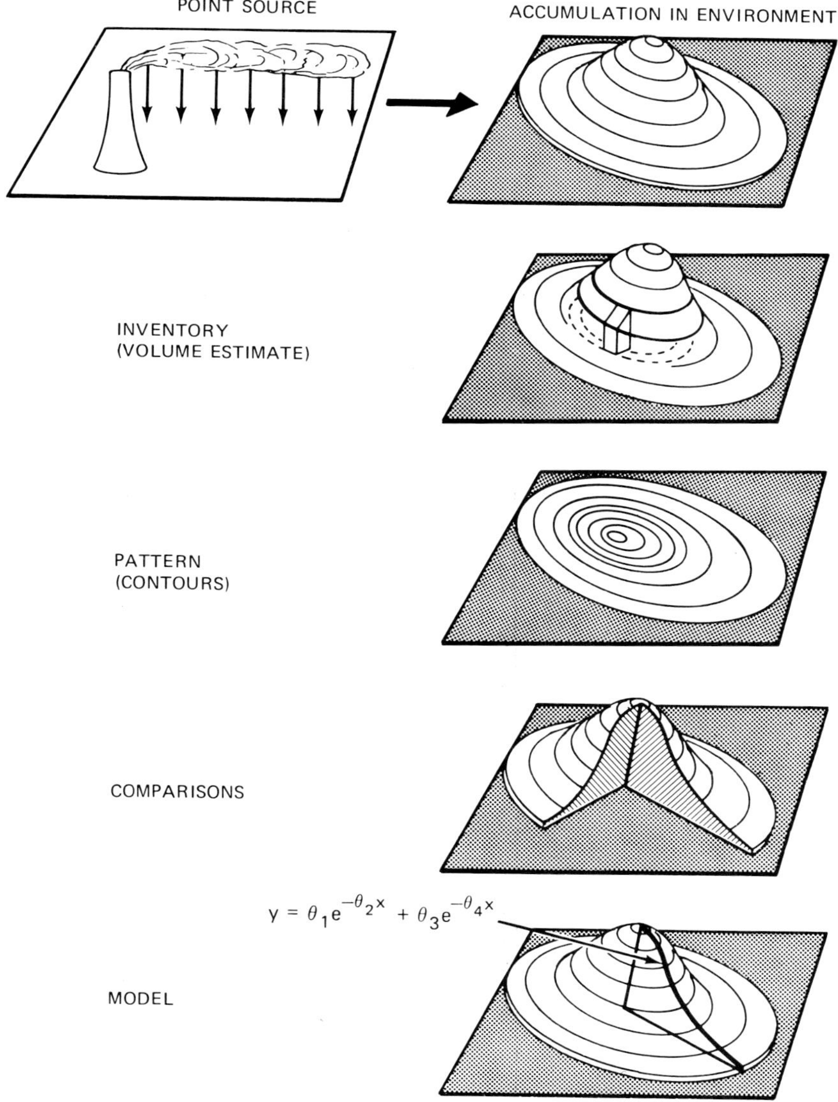

$$y = \theta_1 e^{-\theta_2 x} + \theta_3 e^{-\theta_4 x}$$

Fig. 1 Sampling objectives.

are quite diverse, illustrating them in the same example stretches the analogies somewhat and does not allow adequate coverage of the wide range of possibilities. The figure is meant to portray surface soil concentration of some contaminant emanating from a point source, such as a power plant stack or cooling tower. Figure 2 is a different summary of the four objectives.

Descriptive Sampling

Descriptive sampling is the classical approach to sampling. The objectives are to estimate a total or a mean for some variable (or set of variables) over a definite population. Textbooks on the subject, perhaps more generally known as survey sampling, have been available since the early 1950s. The textbook by Cochran (1963) is best known to biologists, but there may now be 10 or 12 books on descriptive sampling. We have become accustomed to hearing this methodology called "sampling for inventory" in the studies of plutonium in soil at the Nevada Test Site. Inventory, however, is not a good

1. INVENTORY

    COMMONLY KNOWN AS "DESCRIPTIVE" OR "SURVEY" SAMPLING

EXAMPLE

    STRATIFIED RANDOM SAMPLING

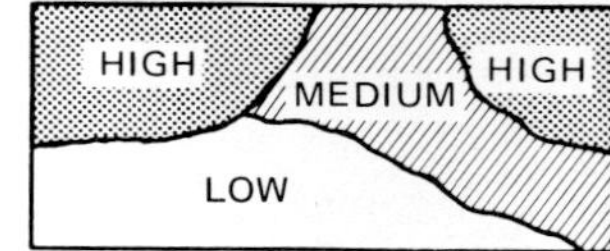

2. PATTERN

    PRESENT APPLICATIONS MAINLY IN GEOLOGY (PETROLEUM AND MINING EXPLORATION)

MAIN OUTCOME

    CONTOUR MAPS

3. COMPARISON

    LIMITED ATTENTION IN SURVEY SAMPLING TEXTS ("ANALYTICAL SAMPLING") . . . CLOSE RELATION TO ANALYSIS AND DESIGN OF EXPERIMENTS

EXAMPLES

- ANALYSIS OF VARIANCE
- ANALYSIS OF COVARIANCE

4. MODELING

    MAINLY DEVELOPED IN INDUSTRIAL EXPERIMENTATION (G.E.P. BOX AND OTHERS)

EXAMPLE

    ESTIMATE PARAMETERS IN TWO-COMPARTMENT RETENTION MODEL

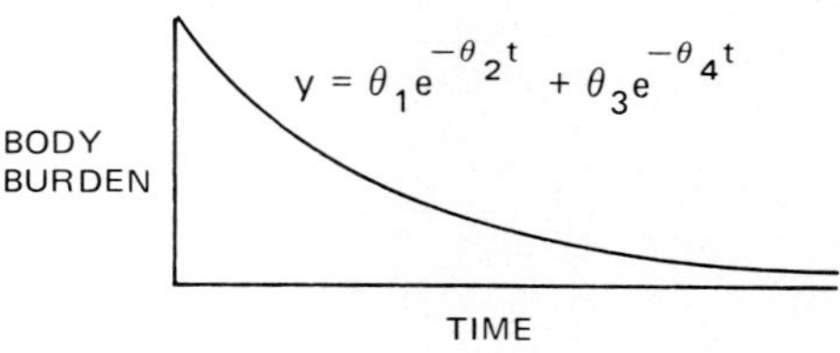

Fig. 2   Some relevant statistical methodology.

term to use here because it connotes results broader than those supplied in the usual application of descriptive sampling. Most people think of an inventory as supplying information on both quantity and location. When sampling is used, it is necessary to decide which of these attributes should be emphasized. Descriptive sampling is concerned with quantity.

In Fig. 1 descriptive sampling is illustrated by suggesting a volume-estimation (integration) process. For graphic purposes, accumulation in the environment has been shown as a surface. In fact, differential accumulation in, say, soil is, of course, reflected in changes in concentration, and a total is usually estimated by averaging. Figure 3 shows a common technique in descriptive sampling, i.e., stratification (which is described further later in this chapter).

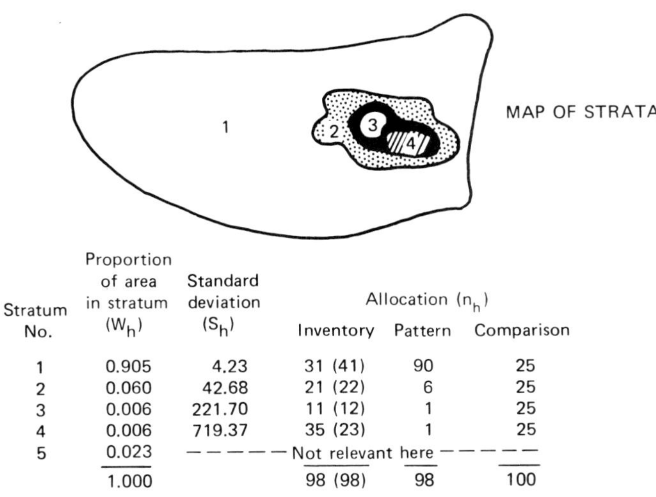

Stratum No.	Proportion of area in stratum (W_h)	Standard deviation (S_h)	Allocation (n_h)		
			Inventory	Pattern	Comparison
1	0.905	4.23	31 (41)	90	25
2	0.060	42.68	21 (22)	6	25
3	0.006	221.70	11 (12)	1	25
4	0.006	719.37	35 (23)	1	25
5	0.023	— — — — — Not relevant here — — — — —				
	1.000		98 (98)	98	100

Fig. 3 Example showing how sampling intensity differs according to objectives. Sampling for plutonium in surface soil at GMX site (area 5) at the Nevada Test Site.

Sampling for Spatial Pattern

When location is the major objective, the best sampling system may be quite different from that prescribed in the survey-sampling textbooks for estimating a total. So far most of the relevant results in this area have been produced in geology and geography and have not really begun to show up in the statistical literature or the textbooks on sampling. We (Eberhardt and Gilbert, 1976) have described some of the varied aspects of sampling for spatial pattern relative to transuranic studies. The rather lengthy discussion following our presentation in the work cited (see pp. 197-208) should be of interest in the present context. A textbook on the subject is that of Agterberg (1974).

The basic method for describing pattern is that of drawing contour lines to show regions of equal concentration (isopleths), as illustrated in Fig. 1.

Analytical Sampling

Cochran (1963, p. 4) gives a good description of analytical sampling: "Comparisons are made between different subgroups in the population, in order to discover whether differences exist among them that may enable us to form or to verify hypotheses about the forces at work in the population." Not much space is devoted to this topic in the textbooks, perhaps because it is very similar to the older body of methodology encompassed under the heading of the analysis of variance and mostly (but not necessarily) used in an experimental context. As mentioned in the introduction, the mechanical details of analysis are very much the same whether an experiment or a survey is involved. It may be, however, that survey sampling methodology has something additional to offer in the way of ideas on allocating samples to "domains of study." We have been impressed by the potential advantages of using auxiliary variables to increase efficiency, i.e., to reduce costs by use of the analysis of covariance (Eberhardt, 1975b).

One of many comparisons that can be made is illustrated in Fig. 1 where the diagram has been cut in two directions to suggest that an investigator may want to use statistical methods to determine whether there are significant differences in concentrations along transects in two (or more) directions.

Sampling for Modeling

Sampling for modeling covers a lot of ground. Perhaps one instance will serve to illustrate the topic. Consider the uptake and retention of some radionuclide in an animal. Suppose the analyses are expensive and involve sacrificing animals for each determination (as is often the case for the transuranics). Since the process is a dynamic one, the outcomes usually are represented by fitting a curve (modeling) and estimating rate constants (or half-times). Some time, trouble, and money might be saved by studying ways to select sampling times so as to obtain the "best" estimates of the parameters (rate constants). So far as we can tell, this point has not been considered in the many thousands of laboratory studies on radionuclides and other trace substances over the last 30 yr in which sampling times are either uniformly spaced or occasionally separated by geometrically increasing intervals. One must, of course, also pay attention to determining the structure of the models used and to various other points (Eberhardt et al., 1976; Eberhardt, 1978).

In Fig. 1 the basic idea is suggested by a heavy line denoting a specific model fitted to concentrations in one direction from the source. In practice one might want to fit a model to the entire surface. This might also involve comparisons to see whether a directional component is needed in the model; so analytical aspects also may be involved.

Sampling Methods and Research Needs

In this section we cite additional references and identify problems that require additional consideration. The discussions are arranged in the four categories of sampling methods previously described and are followed by a section devoted to common problems.

Descriptive Sampling

The textbooks devoted to descriptive sampling discuss extensively a rather wide variety of methods for estimating totals or means by sampling. We mention only those few methods with which we have had some experience. The main method is stratified sampling, wherein elements of the population to be sampled are assigned to one of a number of strata. The basic idea is to assign elements to strata so that the elements in each are as nearly alike as possible. If this is successful, then the variability within a given stratum is kept small, and the costs of sampling are thereby reduced. Some advance knowledge on which to base the stratification (the classification of population elements) is evidently necessary. Often this prior information can be obtained by some relatively inexpensive means of measurement, or it may be known or inferred from information about the source of contamination etc.

Once the strata have been determined, the total sample must be allocated to the several strata. Two general approaches have been used, proportional and optimum allocation. In proportional allocation the sample is distributed simply in proportion to the number of elements in each stratum (e.g., in soil sampling, to the area of the stratum). This scheme is suitable if the variances are about the same in each stratum. Variances associated with the transuranic elements, however, increase dramatically with mean concentrations. We thus recommend optimum allocation, which is based on both size of stratum and variability within the stratum.

Our initial efforts at stratification for sampling soil for plutonium at the Nevada Test Site are described by Eberhardt and Gilbert (1972). Many of the details of our subsequent experience appear in Gilbert et al. (1975). An alternative approach to stratified sampling is to use an accurate but expensive method (such as chemical analyses for plutonium) to "calibrate" a less accurate but cheaper method. Methods of this sort fall under the heading of double sampling in textbooks. Some details of an application of double sampling to sampling for plutonium appear in Gilbert and Eberhardt (1976a). The method uses ratios or regressions of rather variable quantities and thus poses some statistical problems (mentioned again later in this chapter). An important action in designing a double-sampling scheme is to use a cost function to find the combination that yields minimum cost (or that maximizes precision).

In soil sampling for inventory, sampling by depth needs further study. Much of our work with stratified sampling has been concerned chiefly with a thin surface soil layer. Since most of the plutonium is in that layer and resuspension questions focus there, this is a logical approach. Some soil profiles, however, have been taken to investigate vertical dispersion, and a detailed evaluation of allocation schemes for sampling in depth is in order. The problem in profile sampling is, of course, the analytical costs. If 10 increments per profile are taken, costs of even a modest sampling scheme become exorbitant.

In summary, descriptive (inventory) sampling has a well-known technology. Applications in any new area do, however, require statistical attention and a certain amount of research. Unfortunately, methods designed for a specific application are often used in other situations where they are not appropriate, e.g., the use of methods developed for global fallout surveys for entirely unrelated purposes (Eberhardt, 1976, pp. 201-202).

Sampling for Spatial Pattern

In a variety of situations, the main objective in sampling is to determine a geographical pattern rather than to simply estimate total quantities of a substance present in any particular area. As noted previously the objectives of a study should determine the sampling scheme. Different objectives, for example, may require very different allocations of samples to strata (see Eberhardt and Gilbert, 1976). An example of the remarkable contrast in the way samples might be assigned to several strata according to sampling plans tailored to three different objectives is shown in Fig. 3. The box labeled "allocation" gives the distribution of samples to strata appropriate under three of the objectives of Figs. 1 and 2. Two sets of figures are given for "inventory." One is the sampling pattern actually used, and the other is based on the results of the survey.

Two applications of sampling for spatial pattern are the evaluation of resuspension, where the pattern of surface contamination obviously is a controlling factor, and the difficult matter of determining what portions of an area of high concentration should be cleaned up (i.e., soil removed etc.). Wallace and Romney (1975) give a detailed review of past experience, methods, and problems in cleaning up contaminated areas.

From a statistical point of view, many of the problems in designing a sampling plan for measuring pattern remain unresolved or uncertain. The main technique for displaying the results of sampling for pattern is a contour map. Such a map is prepared by computer programs that interpolate between average concentrations assigned to points on a uniform grid. The chief problems to be resolved are those associated with the weighting processes used to transform the observed data into grid entries. At first glance, these problems may seem to be minor—fine points of statistical technique. We have now compared enough contours generated by various methods, however, to be able to demonstrate that the differences are not minor ones; see, for example, Fig. 3 of the report by Gilbert et al. (1976b, p. 457) and the series of graphs by Gilbert, Eberhardt, and Smith (1976). Some technical aspects of the problems have been discussed by Gilbert (1976).

Mapping the concentrations is only the first stage in the cleanup problem. If contaminated soils are removed, then it is usually necessary to make certain that the job is done adequately; i.e., Are there areas of unacceptably high concentration remaining? Gilbert and Eberhardt (1977) have described one possible sampling scheme for this purpose (acceptance sampling). Since cleanup operations are expensive and usually damaging to the environment, this matter needs further study.

Analytical Sampling

The statistical technology of analytical sampling is closely related to traditional methods of statistical analysis. An alternate designation that we find useful is "sampling for comparisons." The area is a broad one and includes changes in space and time, analysis of spatial patterns, etc. Some examples of statistical analysis that involve sampling follow.

One of the more difficult features in analyzing data is the problem of dealing with ratios of variable quantities. This well-known statistical problem is accentuated by the very high variability associated with the transuranic elements. A common example is the so-called "concentration factor," which is really a ratio. [The plant panel at the 1975 transuranic workshop in Seattle defined a "concentration ratio" (CR) and an "inventory ratio" (IR), thus supplying accurate names to replace the former "concentration factor" (Energy Research and Development Administration, 1976)]. The difficulties in dealing with ratios have been addressed in a number of the reports cited previously, and a recent evaluation is that of Doctor and Gilbert (1977). As these authors note, there are three well-known ways ratios of variable quantities can be estimated: (1) by averaging ratios of individual pairs of observations, (2) by summing up the x and y observations and calculating a ratio of totals, and (3) by a calculation of the type used to obtain the slope estimate in regression analysis (except that the "corrections for the means" are dropped so that the slope is appropriate to a regression calculated through the origin, i.e., the intercept is zero). They also describe two other possibilities and note that the several methods can give quite different results. Thus there is not only the problem of which method of estimation to use but also the issue of how to allocate sampling effort so as to make comparisons (of ratios) that are as meaningful as possible.

Another problem in statistical analysis is associated with interlaboratory comparisons. We have pointed out something of the uncertainties involved (Eberhardt and Gilbert,

1972) and have continued to study the issues (Gilbert and Eberhardt, 1976b). The main difficulty again has to do with variability since rather large numbers of replicate determinations are required to give reasonable assurance of detecting differences between laboratories. Proper allocation of samples (replicates) to methods, elements, and laboratories should help, but this has not been investigated in any detail.

In a statistical analysis of transuranic data, frequency distributions are often dramatically skewed (asymmetrical). Thus consideration should be given to transforming the data before analysis. We have looked into this option in some detail (Eberhardt and Gilbert, 1973; Eberhardt et al., 1976). Our recommendation is to use a logarithmic transformation of the data before doing any statistical analyses involving significance tests based on the normal distribution. A feasible alternative is to consider nonparametric (or distribution-free) methods. We have begun some limited investigations in this area. We particularly do not recommend estimating means (averages) by transforming back the mean of log-transformed data (Link and Koch, 1975). If interest is directed chiefly to estimating means on the original (untransformed) scale, we recommend use of the ordinary arithmetic average of the untransformed data. If interest is chiefly in a statistical analysis, then the results should be discussed in terms of the transformed data. The problem of how to allocate samples to accommodate both purposes, however, seems to us to need more attention.

Sampling for Modeling

Many sampling and statistical problems must be dealt with if modeling is eventually to achieve truly satisfactory status in environmental studies. Most of the present prospects for models contain a substantial number of rate functions that are little more than guesses. Rather than go into these problems, which transcend statistical and sampling issues, we will only mention some simple models. A few details of methods for finding optimum sampling times for a rather simple, but widely applicable, model are given by Eberhardt (1978).

Three categories of simple models can alternatively be described as profiles of concentration in time or space: (1) retention of some substance by an animal, (2) measuring concentrations away from a point source, and (3) studying soil profiles. Two facets of such studies may need to be considered in designing a study by sampling. One is whether the investigator's main interest is in estimating rate constants or in describing the profile itself since different sampling plans are then appropriate. A second concerns the nature of replications. In retention studies individual animals can serve as replicates, but, in the evaluation of soil profiles, the word "replicate" will not have the same meaning; so sampling results may have rather different interpretations in the two instances. Eberhardt (1978) gives some further details on sampling profiles. Essington et al. (1976) give a number of details on actual soil profiles of several transuranics.

The notion of sampling for modeling, which can also be described as sampling for curve fitting (in a more restrictive sense), appears to be new in environmental studies. As such it poses a number of problems that need further evaluation. The reader interested in technical details might well start with the review by Cochran (1973), which provides additional references. Papers by Atkinson and Hunter (1968), Box and Lucas (1959), and Box (1968; 1970; 1971) should also be consulted. As has already been noted, our attention has been focused on finding the optimum times (or depths, or distances) for sampling in the interest of obtaining a maximum amount of information for a given sampling cost.

Among the difficulties that need further review, we would like to mention the following:

1. A model has to be assumed; so it is usually desirable, or necessary, to consider several candidate models. This then brings in a need to try to decide from the data which is "the" correct model and hence statistical analysis and selection of the best sampling scheme for discriminating between models.

2. When several parameters are involved, the optimization process usually requires a computer to perform the calculations.

3. In most cases advance estimates of the parameters are required. This may be seen as a rather severe restriction, but, since most of the important models to be considered are nonlinear, experimentation is not very practical without a fair amount of advance knowledge of the system anyhow.

Some Common Problems

A number of problems common to the several kinds of sampling described here and to studies of transuranics and other trace substances need further consideration. Those associated with "counting statistics" and "counting errors" have been the subject of many investigations but continue to pose difficulties. Gilbert (1975) has prepared a review relative to counting statistics in studies of the transuranics. One common problem with low-level measurements has to do with the "below background," "not detectable," or "trace" measurements. An approach that we believe deserves further study is the use of simple nonparametric methods. When a data set contains a number of below-background measurements, simple averages are likely to be biased, the direction of the bias depending on how the questionable measurements are handled. In such cases it seems that the median may be a preferable measure of central tendency and that assessing variability may be approached through order statistics. As an example, suppose one takes 30 samples and, after analysis for a transuranic, find that 10 results are reported as "not detectable." If a mean is to be computed, one must then decide how to incorporate these 10 measurements. Should they all be assumed to contain none of the substance under study and be assigned "zero" values? Usually this is not reasonable since a longer counting time, larger sample mass, etc., would likely have turned up detectable levels in some of the 10 samples. The procedure suggested here is simply not to use the mean but to change over to the median, which does not require any decisions about the troublesome 10 "not-detectable" samples (except that their levels were, in fact, less than those for which levels were reported). Rather than calculating standard deviations, one would need to calculate measures of variability based on order statistics (Conover, 1971; Hollander and Wolfe, 1973). Random sampling is, however, also required for nonparametric methods.

Another common problem is how to deal with the practice of taking aliquots (subsamples). A related issue is the compositing of a group of samples. Both are related to the analytical (chemical) process. In the former instance sample mass is reduced to facilitate analysis; in the latter, a number of samples are combined in an effort to reduce costs. The very high sample-to-sample variability characteristic of the transuranic elements may cause trouble for the unwary investigator in either case. In both cases we believe replicate samples should be taken to provide a measure of the variability inherent in the process. Of course, doing this increases costs. Compositing samples needs some careful study to see whether it does, in fact, offer the gains that are usually assumed without any checking. A major question about compositing soil samples is whether or not they can be adequately mixed. The basic idea is that the composite sample will provide an accurate average value for the individual samples used to make up the composite. If the entire composite is used for analysis, there should be no problem with the concept. For many transuranic analyses, however, only a relatively small mass is used; so a composite itself may be subsampled (aliquoted) at the chemical-analysis stage. Whether compositing is worthwhile, then, depends on how well the sample is, or can be, mixed. The hot-particle problem in plutonium analyses, along with the common practice of using a small sample mass for analysis, suggests that compositing may not be very effective. Some further details and suggestions have been reported by Eberhardt (1976) and Eberhardt et al. (1976). A statistical evaluation of compositing (Rohde, 1976) suggests that the correct measure of variability for the average of composited samples is difficult to obtain.

Transformation of skewed data has already been mentioned; it deserves some further research. One of the difficulties has to do with transforming back; i.e., if a logarithmic transformation is used for statistical analysis, many investigators prefer to express the final results on an arithmetic scale. Simply using antilogarithms introduces a bias. In many instances it may be quite feasible to simply stay on the logarithmic scale; the consequences of doing so need to be further evaluated and explained (Agterberg, 1974, pp. 289-300; Aitchison and Brown, 1966, pp. 44-48; Heien, 1968). An interesting sidelight is that some investigators seem to believe that correlations calculated on log-transformed data are not legitimate. Whether this is true or not depends on the statistical model assumed; so that issue has to be resolved by specifying such a model. There is, in fact, some reason to argue that correlations involving transuranics should be done on log-transformed data. The usual model for correlation (bivariate normal distribution) involves a linear model with normally distributed errors (deviations). As has been pointed out several times, data on transuranics are generally not normally distributed, and relationships between different variables may be nonlinear.

A matter of substantial importance is the choice between random and systematic sampling. We have thus far largely advocated random sampling since it is the only approach generally accepted as providing unbiased estimates of population parameters. As Cochran (1963, Chap. 8) shows, systematic samples are vulnerable to unsuspected periodicities in the variable being studied, and no widely trustworthy method for estimating the variance is known. On the other hand, if adjacent samples have closely correlated values, then under random sampling two sample points that fall close together essentially duplicate the same information. Hence a systematic sample gives more for the money spent, i.e., a smaller sampling error (even though we may not get a suitable estimate of the sampling error). This point has been particularly emphasized in references on sampling for pattern. Some recent work by Barnes, Gilbert, and Delfiner (1977), using the field instrument for the determination of low-energy radiation (FIDLER) data from a safety-shot site on the Nevada Test Site, suggests that readings taken on a grid result in more precise estimates of americium contours than do comparable (FIDLER) readings taken at random within activity strata. These analyses were made with "kriging" techniques (see Delfiner and Delhomme, 1975), a method we are currently studying for potential applications in transuranic studies.

In the first few years that we worked with field sampling for plutonium, very little data on variability were available (Eberhardt and Gilbert, 1972). As more and more data collected by random sampling have become available, we have come around to the opinion that it will be worthwhile to look more carefully into systematic sampling and to begin tests of its utility in a variety of field situations. Since most of the applications will be two dimensional (area sampling), a logical approach is grid sampling. If we then use stratification, we will, in most instances, want to vary the mesh of the grid between strata. This can be done easily if mesh sizes change geometrically; i.e., for areas of the highest concentration, use a grid size of one unit, for the next lowest concentration stratum, use a grid twice as large, for the next four times as large, and so on down to the lowest concentration stratum, which will have the coarsest grid mesh. With such a scheme the problem of matching grid meshes at stratum boundaries is fairly simple. Some of the flexibility of stratified sampling, however, is thus given up, and this probably reduces its efficiency in situations where concentration gradients are known to change rapidly. Hence a variety of field trials will be needed to work out details of this and other problems.

Many other problems might be considered here, but we conclude by noting one that may go largely unnoticed, i.e., the practice of sieving soil samples and doing chemical analyses on only certain sieve fractions. Such a practice can introduce important biases, some of which are described by Gilbert et al. (1976a). Examples using plutonium data from safety tests are given in the report by Gilbert and Eberhardt (1976b, pp. 131-153).

Conclusions

We have attempted to briefly mention some of the many facets of statistical and sampling methodology which are relevant to studies of the transuranic elements. We believe that too many investigators working with these elements are not sufficiently aware of the very large component of "chance" error inherent in their data. Much of our experience has been with data on plutonium in soils, mostly that resulting from "safety shots" at the Nevada Test Site, but we have also studied data from a number of other sites (Enewetak, Los Alamos, Rocky Flats, and other locations). In other circumstances variability may be much reduced. However, if it is not, then it is likely that many "significant" findings to date are largely artifacts resulting from inadequate sampling.

References

Agterberg, F. P., 1974, *Geomathematics,* American Elsevier Publishing Co., New York.

Aitchison, J., and J. A. C. Brown, 1966, *The Lognormal Distribution,* Cambridge University Press, London.

Atkinson, A. C., and W. G. Hunter, 1968, The Design of Experiments for Parameter Estimation, *Technometrics,* 10(2): 271-289.

Barnes, M., R. O. Gilbert, and P. Delfiner, 1977, A New Statistical Tool, Kriging, and Some Applications to Area 13 Data, in *Transuranics in Desert Ecosystems,* M. G. White, P. B. Dunaway, and D. L. Wireman (Eds.), USDOE Report NVO-181, pp. 431-445, Nevada Operations Office, NTIS.

Box, M. J., 1968, The Occurrence of Replications in Optimal Designs of Experiments to Estimate Parameters in Nonlinear Models, *J. R. Stat. Soc., Ser B,* 30: 290-302.

——, 1970, Some Experiences with a Nonlinear Design Criterion, *Technometrics,* 12(3): 569-589.

——, 1971, An Experimental Design Criterion for Precise Estimation of a Subset of the Parameters in a Nonlinear Model, *Biometrika,* 58: 149-153.

——, and H. L. Lucas, 1959, Design of Experiments in Nonlinear Situations, *Biometrika,* 46: 77-90.

Cochran, W. G., 1963, *Sampling Techniques,* 2nd ed., John Wiley & Sons, Inc., New York.

——, 1973, Experiments for Nonlinear Functions, *J. Am. Stat. Assoc.,* 68(344): 771-781.

Conover, W. J., 1971, *Practical Non-Parametric Statistics,* John Wiley & Sons, Inc., New York.

Delfiner, P., and J. P. Delhomme, 1975, Optimum Interpolation by Kriging, in *Display and Analysis of Spatial Data,* pp. 96-114, J. D. Davis and M. J. McCullagh (Eds.), John Wiley & Sons, Inc., New York.

Doctor, P. G., and R. O. Gilbert, 1977, Ratio Estimation Techniques in the Analysis of Environmental Transuranic Data, in *Transuranics in Natural Environments,* Symposium Proceedings, Gatlinburg, Tenn., Oct. 5–7, 1976, M. G. White and P. B. Dunaway (Eds.), USDOE Report NVO-178, Nevada Operations Office, NTIS.

Eberhardt, L. L., 1975a, Some Problems in Measuring Ecological Effects of Chronic Low-Level Pollutants, in *Hearings Before the Subcommittee on Environment and the Atmosphere of the Committee on Science and Technology, U. S. House of Representatives, 94th Congress,* pp. 565-586, No. 49, GPO.

——, 1975b, Some Methodology for Appraising Contaminants in Aquatic Systems, *J. Fish Res. Board Can.,* 32(10): 1852-1859.

——, 1976, Sampling for Radionuclides and Other Trace Substances, in *Radiological Problems Associated with the Development of Energy Sources,* Fourth National Radioecology Symposium, C. E. Cushing (Ed.), Dowden, Hutchinson and Ross, Inc., Stroudsburg, Pa.

——, 1977, Applied Systems Ecology: Models, Data and Statistical Methods, in *New Directions in the Analysis of Ecological Systems,* pp. 43-55, G. S. Innis (Ed.), Simulation Councils Proceedings Series, 5(1), Simulations Councils, Inc.

——, 1978, Designing Ecological Studies of Trace Substances, in *Environmental Chemistry and Cycling Processes,* DOE Symposium Series, Augusta, Ga., Apr. 28–May 1, 1976, D. C. Adriano and I. Lehr Brisbin, Jr. (Eds.), pp. 8-33, CONF-760429, NTIS.

——, and R. O. Gilbert, 1972, *Statistical Analysis of Soil Plutonium Studies, Nevada Test Site,* USAEC Report BNWL-B-217, Battelle, Pacific Northwest Laboratories, NTIS.

——, and R. O. Gilbert, 1973, *Gamma and Lognormal Distributions As Models in Studying Food-Chain Kinetics,* USAEC Report BNWL-1747, Battelle, Pacific Northwest Laboratories, NTIS.

——, and R. O. Gilbert, 1976, Sampling the Environs for Contamination, in *Proceedings of the First ERDA Statistical Symposium,* Los Alamos, November 1975, ERDA Report BNWL-1986, pp. 187-196, Battelle, Pacific Northwest Laboratories, NTIS.

——, R. O. Gilbert, H. L. Hollister, and J. M. Thomas, 1976, Sampling for Contaminants in Ecological Systems, *Environ. Sci. and Tech.,* 10: 917-925.

Energy Research and Development Administration, 1976, *Workshop on Environmental Research for Transuranic Elements,* Proceedings of the Workshop, Battelle Seattle Research Center, Seattle, Wash., Nov. 12–14, 1975, USAEC Report ERDA-76/134, NTIS.

Essington, E. H., E. B. Fowler, R. O. Gilbert, and L. L. Eberhardt, 1976, Plutonium, Americium, and Uranium Concentrations in Nevada Test Site Soil Profiles, in *Transuranium Nuclides in the Environment,* Symposium Proceedings, San Francisco, 1975, pp. 157-173, STI/PUB/410, International Atomic Energy Agency, Vienna.

Gilbert, R. O., 1975, *Recommendations Concerning the Computation and Reporting of Counting Statistics for the Nevada Applied Ecology Group,* USAEC Report BNWL-B-368, Battelle, Pacific Northwest Laboratories, NTIS.

——, 1976, *Estimation of Spatial Pattern for Environmental Contaminants,* draft of paper presented at 1976 Annual Meeting of the American Statistical Association, Boston, Aug. 23–26, 1976, USAEC Report BNWL-SA-5771, Battelle, Pacific Northwest Laboratories, NTIS.

——, and L. L. Eberhardt, 1976a, An Evaluation of Double Sampling for Estimating Plutonium Inventory in Surface Soil, in *Radiological Problems Associated with the Development of Energy Sources,* Fourth National Radioecology Symposium, pp. 157-163, C. E. Cushing (Ed.), Dowden, Hutchinson and Ross, Inc., Stroudsburg, Pa.

——, and L. L. Eberhardt, 1976b, Statistical Analysis of "A Site" Data and Interlaboratory Comparisons for the Nevada Applied Ecology Group, in *Studies of Environmental Plutonium and Other Transuranics in Desert Ecosystems,* M. G. White and P. B. Dunaway (Eds.), ERDA Report NVO-159, pp. 117-154, Nevada Operations Office, NTIS.

——, and L. L. Eberhardt, 1977, Some Design Aspects of Transuranium Field Studies, in *Transuranics in Natural Environments,* Symposium Proceedings, Gatlinburg, Tenn., Oct. 5–7, 1976, M. G. White and P. B. Dunaway (Eds.), ERDA Report NVO-178, Nevada Operations Office, NTIS.

——, L. L. Eberhardt, and D. O. Smith, 1976, An Initial Synthesis of Area 13 ^{239}Pu Data and Other Statistical Analyses, USAEC Report BNWL-SA-5667, Battelle, Pacific Northwest Laboratories, NTIS.

——, L. L. Eberhardt, E. B. Fowler, and E. H. Essington, 1976a, Statistical Design Aspects of Sampling Soil for Plutonium, in *Atmosphere–Surface Exchange of Particulate and Gaseous Pollutants,* ERDA Symposium Series, Richland, Wash., Sept. 4–6, 1974, Rudolf J. Engelmann and George A. Sehmel (Coordinators), pp. 689-708, CONF-740921, NTIS.

——, L. L. Eberhardt, E. B. Fowler, E. H. Essington, and E. M. Romney, 1976b, Statistical Analysis and Design of Environmental Studies for Plutonium and Other Transuranics at NAEG "Safety-Shot" Sites, in *Transuranium Nuclides in the Environment,* Symposium Proceedings, San Francisco, 1975, pp. 449-460, STI/PUB/410, International Atomic Energy Agency, Vienna.

——, L. L. Eberhardt, E. B. Fowler, E. M. Romney, E. H. Essington, and J. E. Kinnear, 1975, Statistical Analysis of $^{239\text{-}240}$Pu and ^{241}Am Contamination of Soil and Vegetation on NAEG Study Sites, in *Radioecology of Plutonium and Other Transuranics in Desert Environments,* M. G. White and P. B. Dunaway (Eds.), USAEC Report NVO-153, pp. 339-448, Nevada Operations Office, NTIS.

Heien, D. M., 1968, A Note on Log-Linear Regression, *J. Amer. Stat. Assoc.,* 63: 1034-1038.

Hollander, M., and D. A. Wolfe, 1973, *Nonparametric Statistical Methods,* John Wiley & Sons, Inc., New York.

Link, R. F., and G. S. Koch, Jr., 1975, Some Consequences of Applying Log-Normal Theory to Pseudolognormal Distributions, *Math. Geology,* 7: 117-128.

Rohde, C. A., 1976, Composite Sampling, *Biometrics,* 32: 273-282.

Wallace, A., and E. M. Romney, 1975, Feasibility and Alternate Procedures for Decontamination and Post-Treatment Management of Pu-Contaminated Areas in Nevada, in *Radioecology of Plutonium and Other Transuranics in Desert Environments,* M. G. White and P. B. Dunaway (Eds.), USAEC Report NVO-153, pp. 251-337, Nevada Operations Office, NTIS.

Appropriate Use of Ratios in Environmental Transuranic Element Studies

P. G. DOCTOR, R. O. GILBERT, and J. E. PINDER III

This chapter discusses some statistical aspects of two types of ratios used extensively in environmental transuranic studies. The two ratios discussed, concentrations and pure ratios, have different uses and different statistical problems. A concentration gives units of numerator (Y) per unit of denominator (X), e.g., nanocuries of ^{238}Pu per gram of soil. Concentrations are viewed as raw data and are used as input for further statistical analysis. For this type of ratio, Y is assumed to be proportional to X. For environmental radionuclide concentrations, variability between aliquots for small aliquot sizes tends to become large. The choice of aliquot size permitting a reliable estimate of concentration is a major problem with this type of ratio. For a pure ratio the numerator and denominator are measured in the same units, e.g., nanocuries of ^{238}Pu over nanocuries of ^{239}Pu. In transuranic field studies both the numerator and denominator may vary considerably among aliquots in the same sample. Pure ratios often appear as a ratio of concentrations, e.g., concentration ratios and inventory ratios. However, pure ratios provide accurate information on the relationship between Y and X only when Y is proportional to X. The statistical problems of pure ratios center on an assessment of whether the multiplicative assumption is valid. Multivariate statistical techniques offer alternatives to a pure ratio for expressing the relationship between Y and X. The purpose of this chapter is not to provide a catalog of statistical methods for ratio estimates but to stimulate critical thinking about the use of ratios and to suggest approaches to the task of ratio estimation compatible with the behavior of environmental radionuclide data.

Ratios are used extensively in scientific work, particularly in the environmental and life sciences, to express the relationship between two independently measured attributes of, for example, the same animal, soil sample, plant part, or geographic locality. Examples in the field of environmental transuranic element research include ^{238}Pu activity/weight for a soil sample, the $^{239}Pu/^{241}Am$ ratio in a vegetation sample, and the ratio of $^{134-137}Cs$ activity in plant tissue to that in soil at a particular location.

A ratio is one of the simplest mathematical techniques for relating two numbers. Another approach is to compute their difference. However, both techniques have precise mathematical assumptions underlying their use. The use of a ratio implies that the relationship is multiplicative; that is, if Y is the numerator and X the denominator of the ratio, then

$$Y = \gamma X \tag{1}$$

where γ is the proportionality constant. The use of a difference implies that the relationship is additive; i.e.,

$$Y = \alpha + X$$

where α is the additive constant. Using ratios or differences without consideration for these mathematical assumptions can produce misleading results.

Since the world does not behave exactly according to mathematical principles, the decision of whether the multiplicative assumption holds is a substantive as well as a statistical question. If the data do not support the notion that the relationship is multiplicative, then other methods should be found to relate the two variables. More sophisticated mathematical and statistical techniques for relating two variables include linear and nonlinear regression, correlation and cluster analysis, and multivariate regression and analysis of variance.

There are two main types of ratios. The first, called a dimensioned ratio by Simpson, Roe, and Lewonton (1960, p. 13), expresses the amount of one variable per unit of a second variable. The second variable is usually weight or volume. Examples are ^{241}Am activity per gram of soil and ^{239}Pu activity per liter. Concentrations are so fundamental to environmental radionuclide research that they are literally viewed as raw data. However, the calculation of a concentration is a preprocessing step whose purpose is scaling; i.e., the resulting value should be independent of the size of the sample on which it was measured. The tacit assumption underlying this approach to scaling is the multiplicative one that plutonium activity in a sample is proportional to the size of the sample. Although this seems to be a reasonable assumption, it is not always true. Failure to meet this assumption can produce severe problems when low-level concentrations (picocurie and femtocurie range) and small amounts of sample material are present.

The second type of ratio is called dimensionless by Simpson, Roe, and Lewonton (1960, p. 13). This ratio is unitless because the units of the numerator and denominator are the same; so they algebraically cancel each other. Within the class of dimensionless ratios, there are two subtypes. One is a percent or fraction, e.g., the amount of ^{239}Pu expressed as a fraction of plutonium in a soil sample. This ratio is special in that its values are restricted to 0 to 100 for percents or 0.0 to 1.0 for fractions. Chayes (1971), in the context of petrology, discusses the properties of and methods for dealing with percentages. The second subtype of dimensionless ratio we call a pure ratio. The denominator does not include the numerator; so the possible values of the ratio are unconstrained. An example is the isotopic ratio, ^{238}Pu/^{239}Pu, in which the numerator and denominator are measured on the same sample. Of the two types of dimensionless ratios, pure ratios seem to be more prominently used in environmental radionuclide research.

The extent to which ratios are considered essential to environmental radionuclide research is evidenced by the compounding of ratios, i.e., a ratio of ratios. Two examples are found on pp. 23-24 of the proceedings of the November 1975 Workshop on Environmental Research for Transuranic Elements (U. S. Energy Research and Development Administration, 1976), the concentration ratio (CR), defined as

$$\frac{\text{Activity per weight of plant part}}{\text{Activity per weight of substrate or reference material}}$$

and the inventory ratio (IR), defined as

$$\frac{\text{Activity per unit area in product}}{\text{Activity per unit area in source}}$$

These compound ratios are pure ratios whose numerator and denominator are concentrations. The farther one is removed from the raw data by the compounding of ratios, the harder it is to justify theoretically and statistically the multiplicative assumption. Therefore, Simpson, Roe, and Lewonton (1960, p. 18) conclude that the compounding of ratios should be done with great care.

The calculation of a ratio is simple; however, the ramifications as they affect the statistical analyses are often complex (Sokal and Rohlf, 1969, pp. 17-19). First, ratios magnify the inaccuracies of the component variables. For example, consider the average ratio 1.0/2.0. Suppose the true measurements lie between 0.95 and 1.10 and between 1.90 and 2.10 for the numerator and denominator, respectively. There is a maximum relative error of $10\% = [(1.10 - 1.00)/1.00] \times 100$ for the numerator. However, the range for the ratio lies between $0.45 = 0.95/2.10$ and $0.58 = 1.10/1.90$, giving a maximum error of $16\% = [(0.58 - 0.50)/0.50] \times 100$ for the ratio. Moreover, the midpoint of the range of the ratio (in this case 0.52) is not the best estimate of the ratio.

Second, the frequency distribution of a ratio can be skewed or multimodal. This is particularly true if either the numerator or denominator is a discrete random variable, i.e., it can take on only a small number of possible values (Simpson, Roe, and Lewonton, 1960, pp. 15-16). An example is a low-level concentration where the number of counts is near zero. Multiplying by conversion factors and dividing by sample weight may produce numbers that appear to represent a continuum, but the number of values the ratio can take on is still small.

Third, taking ratios of two random variables does not preserve either of their distributions. For example, the ratio of two normal random variables is not a normal variable. This can present serious problems since most statistical methods require that the data be at least approximately normally distributed. The underlying probability distribution of the ratio (except in a few well-known situations) (Mielke and Flueck, 1976) cannot be inferred from the distributions of the two component variables. However, one useful exception is that the ratio of two log-normal variables is log-normally distributed. Finally, the ratio provides little information on the relationship between the component variables unless that relationship is multiplicative.

The problems of simple ratios (those composed of variables that are directly measured) are magnified when the component variables are themselves ratios, for example, CR's and IR's. Moreover, the generally unknown distributional properties of ratios make their uncritical use as input for further statistical procedures problematic. Chayes (1971) and Atchley, Gaskins, and Anderson (1976) discuss the behavior of ratios of percents and correlated normal variables, respectively, when they are used as raw data for some statistical procedures.

This chapter discusses some numerical and statistical problems encountered in using concentrations and pure ratios in environmental radionuclide research. Its purpose is not to provide a catalog of statistical methods for ratio estimates but to stimulate critical thinking about the use of ratios and to suggest approaches to the task of ratio estimation compatible with the behavior of environmental radionuclide data.

Concentrations

Recall that the purpose of the concentration is to eliminate the effect of the denominator (aliquot size) on the numerator (radionuclide activity). This implies, in theory, that the concentration can be represented as

$$\frac{Y}{X} = \gamma$$

which is obtained from Eq. 1 by dividing both sides by X. The graphic representation is given in Fig. 1, where the concentration is the same regardless of aliquot size.

The fundamental assumption that underlies the mathematical assumption of a multiplicative relationship between activity and aliquot size is that the radionuclide activity is homogeneously dispersed throughout the medium on which it is measured. This is the justification for the aliquoting procedure in which, for example, a 1-g aliquot is taken for analysis from a 100-g soil sample and the observed aliquot concentration is ascribed to the entire sample.

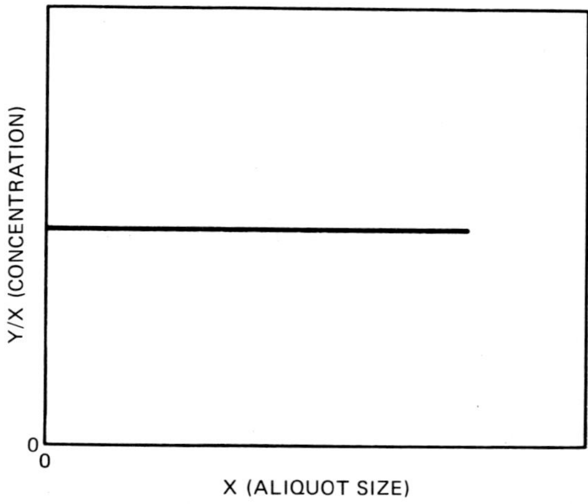

Fig. 1 Theoretical relationship between concentration and aliquot size.

The classical statistical approach to estimating a ratio from a supposedly homogeneous set of data is linear regression. (See Snedecor and Cochran, 1967, pp. 167-171, for a complete discussion.) The model is

$$Y = \gamma X + \epsilon \qquad (2)$$

where Y, X, and γ are as in Eq. 1 and ϵ is the deviation of the data from the fitted model. The estimate of γ is the ratio estimate.

An assumption underlying this approach is that Y is a random variable and X is known without error. For a radionuclide concentration, this is a reasonable assumption. The denominator is usually a weight or volume and is considered to vary through measurement error only. This error is usually negligible compared with the sampling and measurement errors associated with radionuclide activity. For that reason all the variability in the ratio is attributed to the numerator.

Another statistical assumption underlying the usual unweighted least-squares fit of Eq. 2 is that the variance of Y is the same at each value of X; i.e.,

$$\sigma^2 (Y|X) = \sigma^2$$

Both Y and X are assumed to be positive, and, from Eq. 2, the fitted line is forced to pass through the origin (Fig. 2). If both X and Y are assumed to be positive, the consequence,

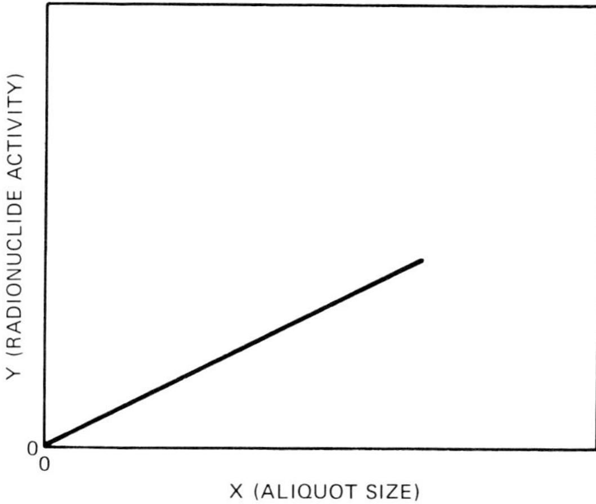

Fig. 2   **Theoretical relationship between activity and aliquot size. (Adapted from Doctor and Gilbert, 1979.)**

assuming a symmetric distribution for Y about the fitted line, is that the variance of Y must go to zero as X goes to zero. This is usually stated: the variance of Y at a particular X is some function of X; i.e.,

$$\sigma^2(Y|X) = f(X)$$

Snedecor and Cochran (1967, pp. 166-171) give methods for obtaining the weighted least-squares estimates of γ when $f(X) = kX$ and kX^2, where k is a constant. Doctor and Gilbert (1977) compared the behavior of these ratio estimates along with the sample median of the ratios and the log-normal estimate of the median ratio for three sets of transuranic data from Nevada Applied Ecology Group studies.

This classic approach to ratio estimation is often not applicable for estimating the true mean concentration from a set of environmental radionuclide data for two reasons. First, as the amount of sample material decreases, the variability in observed radionuclide activity tends to increase rather than decrease. Second, in the case in which the sample size is under the researcher's control, e.g., soil samples, the usual laboratory practice is to analyze only one size aliquot; so regression analysis is impossible. The variability problem of radionuclide concentrations is discussed in the context of two examples: first, ^{239}Pu soil concentrations in a desert environment as the result of a nuclear test and, second, $^{134-137}$Cs vegetation concentrations taken from a stream bed receiving reactor effluents.

Soil Concentrations

The data consist of twenty ^{241}Am concentrations from each of five aliquot sizes (1, 10, 25, 50, and 100 g) taken from the same composite soil sample collected near nuclear site 201 at the Nevada Test Site. The aliquoting procedure (discussed by Doctor and Gilbert, 1979) was designed to ensure as homogeneous a dispersal of the americium as possible. The concentrations are plotted in Fig. 3, where the solid lines delineate the range of the data. The variability tends to increase as aliquot size decreases; the variability of the 1-g

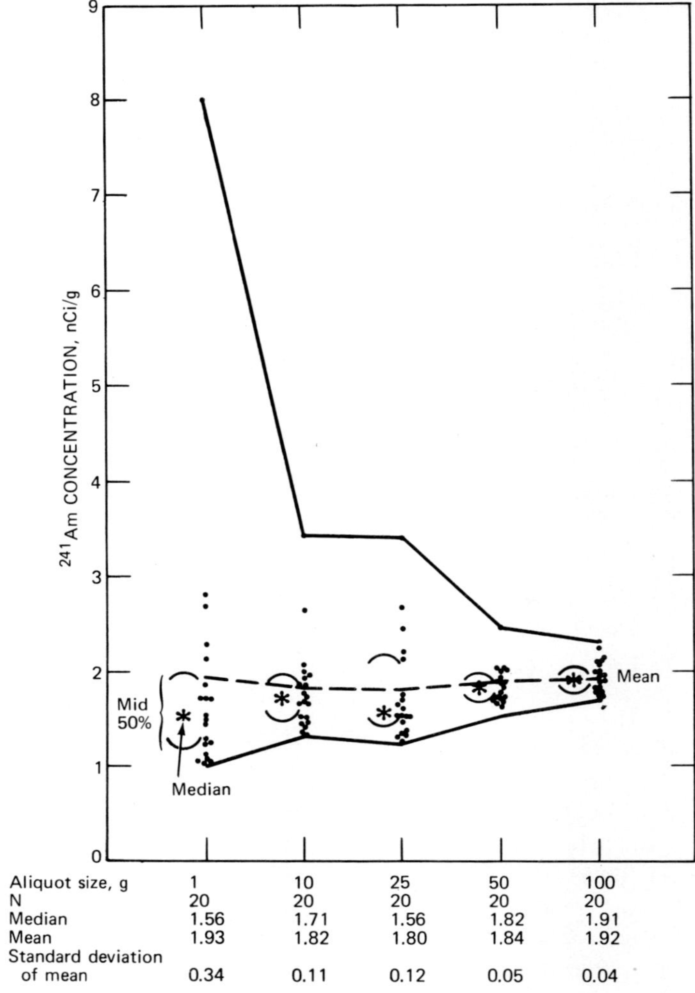

Aliquot size, g	1	10	25	50	100
N	20	20	20	20	20
Median	1.56	1.71	1.56	1.82	1.91
Mean	1.93	1.82	1.80	1.84	1.92
Standard deviation of mean	0.34	0.11	0.12	0.05	0.04

Fig. 3 Americium concentrations in soil aliquots of different sizes from nuclear site 201, Nevada Test Site.

aliquots is much larger (standard error, 0.34) than that of the 10-g aliquots (standard error, 0.11). The observed particulate nature of environmental transuranic element concentrations suggests a hypothesis for this phenomenon. Assume that the americium occurs as particles that are randomly dispersed throughout the soil. For a large (100-g) aliquot size in which the probability of sampling a particle is high, the observed concentrations will probably be reasonably consistent. However, as the aliquot size decreases, the probability of getting a particle also decreases. This tends to violate the homogeneous dispersal assumption. The result is that the observed concentrations will tend to show more variability, as evidenced in the extreme case by the 1-g aliquots.

The data show a positive skewness to high values which is most pronounced for the smaller aliquot sizes. This behavior is consistent with the hypothesis of the previous paragraph, in which the assumptions are similar to the axioms that generate the Poisson process (Parzen, 1967, p. 118). The Poisson distribution is a distribution of the number of randomly dispersed particles in a given volume. The appropriateness of this approach for describing the variability in observed radionuclide concentration as a function of aliquot size depends on the relationship between radionuclide concentration and the number of particles in a sample. We feel that the Poisson distribution as a model of radionuclide concentration within a sample deserves study. This skewness to large values is also characteristic of environmental radionuclide concentrations taken from samples at different locations; so perhaps the Poisson distribution would find use in this situation as well.

Although it is unusual to analyze 20 aliquots per sample, it is instructive to compare the effects of this variability on the observed average concentration for each aliquot size. The sample median (middle value of the observed concentrations) is denoted by an asterisk in Fig. 3. Contrary to what might be expected, the medians show more variation with aliquot size than the sample arithmetic means (connected by the broken line). Skewness affects the median more than the mean, as evidenced by the 50- and 100-g data for which skewness is least pronounced, and the mean and median are very close [see Michels (1977) for theoretical justification]. Since aliquots of all sizes are all from the same composite sample, theoretically the means should remain constant across aliquot size. This appears to be the case here.

It should be emphasized that this example illustrates within-sample variability and not variability due to different locations. However, in an environmental radionuclide study, one is faced with between-location variability as well. One value of this study is that it provides information on within-sample variability which, in turn, permits an evaluation of the amount of observed variability due to location differences. Similar studies might precede a full-scale sampling effort that encompasses a new radionuclide, a new source, or a new medium. Such studies provide a rationale for choosing an aliquot size that will minimize within-sample variability under the constraints of laboratory capability and cost (Doctor and Gilbert, 1979).

Vegetation Concentrations

The problems of obtaining reliable vegetation concentrations in the picocurie range are illustrated by data on plant uptake of $^{134-137}$Cs collected at the Savannah River Plant near Aiken, S. C. [See Sharitz et al. (1975) for a description of the site.] Here the sample or aliquot size, unlike that for soil concentrations, cannot be controlled by the researcher. The data consist of fifty-five $^{134-137}$Cs concentrations and sample weights of leaves from *Hypericum walteri* growing on the floodplain of a South Carolina stream receiving reactor effluents. Since some sampling designs may require that the sample be collected from a species at a particular location without regard to the size of the individual, the same type of data as that illustrated in Fig. 4 may result. Except for two high values at 0.4 g, the variability is dramatically increased for samples weighing <0.2 g. It appears that for these samples the assumption of uniform dispersal is seriously violated, which shows that observed concentration is not independent of sample size (compare with Fig. 1). The errors in measuring the cesium are large relative to the weight of the sample. Furthermore, if negative readings occur and are either disregarded or reanalyzed until positive values are obtained, a positive bias will be introduced, and this bias will be greater for the smaller samples. Even if the accuracy of determining radionuclide content is controlled, variation due to small sample weights may still be a problem if the range of sample weights is large.

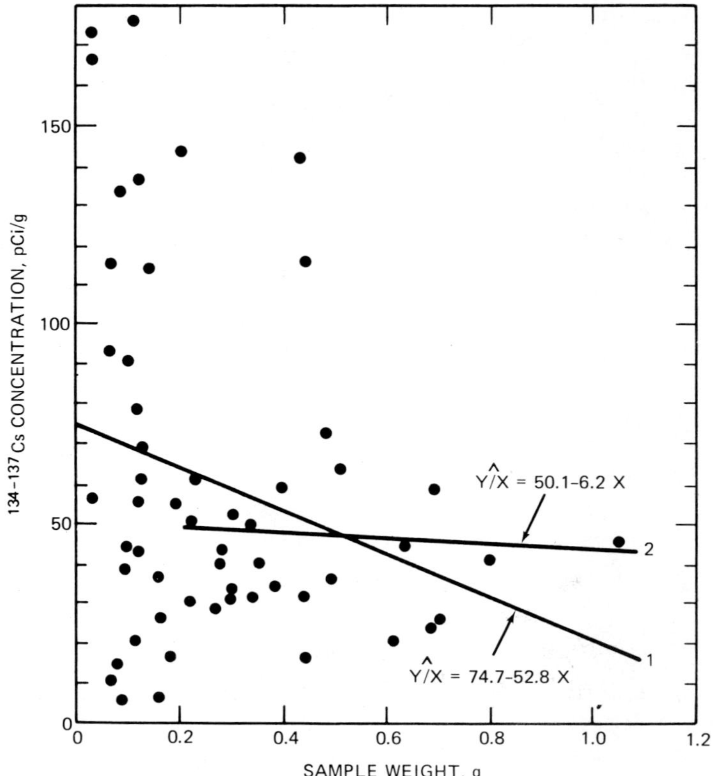

Fig. 4   Relationship between [134–137]**Cs concentration and sample weight in** *Hypericum walteri* **from a floodplain receiving reactor effluents.**

In this case we do not have the luxury of taking a larger size sample to reduce variability as we do for soil samples. Assuming that the concentration is constant, what is the best way to estimate it? The problem can be viewed as twofold: to determine (1) the minimum amount of sample required to yield a consistent estimate of the true concentration and (2) the method for combining the data to give a best estimate of the concentration.

A method of estimating the minimum sample weight required for an analysis consistent with the homogeneous dispersal assumption is illustrated in Figs. 5 and 6. The graph in Fig. 5 was obtained by arranging the samples in random order and computing the variance for an increasing number of samples starting at 10. The initially highly fluctuating variance that then decreased to a final value of about 1950 is typical of the plots obtained when this procedure is applied to positively skewed and leptokurtotic (sharp peak) data. A similar procedure was used to generate the graph in Fig. 6 except that the samples were arranged in order of decreasing weight. The form of the initial phase of Fig. 6 (sharp increase followed by a steady decrease) is similar to that of Fig. 5, and the variance appears to stabilize at approximately 750 for samples weighing >0.2 g; however, when samples weighing <0.2 g are added, the variance then increases continually until the same final value as that in Fig. 5 is obtained. This suggests that

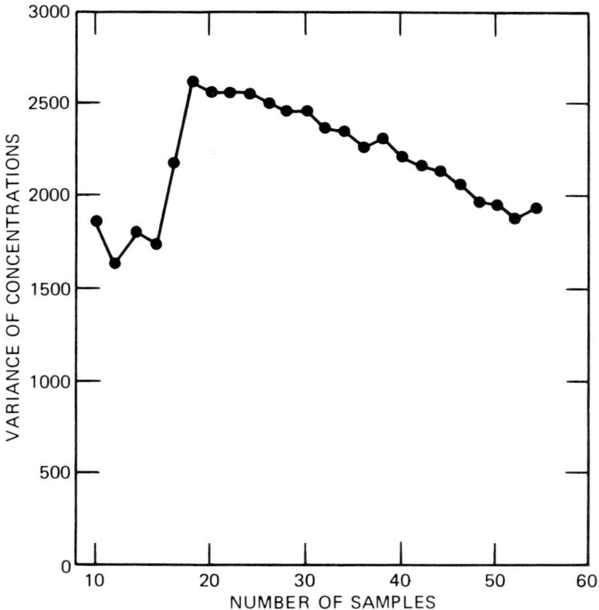

Fig. 5   Variance of $^{134-137}$Cs concentrations in *Hypericum walteri* as a function of number of samples randomly ordered by weight.

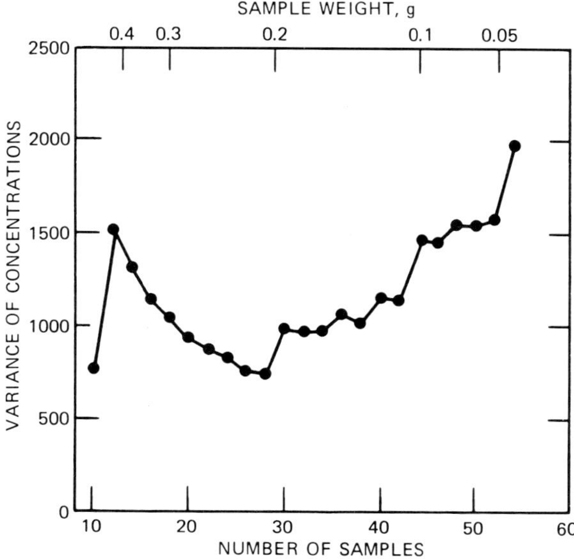

Fig. 6   Variance of $^{134-137}$Cs concentrations in *Hypericum walteri* as a function of the number of samples ordered by decreasing weight.

concentrations from samples weighing <0.2 g provide rather inaccurate estimates of the true concentration. Other sets of data may not present as distinct a choice for the minimum weight, but this way of looking at the data can provide insights into how the behavior of the data vis-a-vis the assumption of uniform dispersal affects the average concentration.

We now discuss the problem of estimating the "true" concentration from this type of data. Line 1 in Fig. 4 results when a straight line ($Y/X = \alpha + \beta X + \epsilon$) is fit to all 55 data points. Because of the large variability in concentration for the samples weighing <0.2 g, the line has a statistically significant (at the 0.05 level) negative slope ($\hat{\beta} = -52.8$; $F_{1,53}* = 4.02$). This suggests that the mean of all 55 observations, 60.1 pCi/g with a standard deviation of 44.2 pCi/g, is not a good estimate of the true concentration for this set of data. However, when samples weighing <0.2 g are excluded, leaving 28 data points, and the line is refit (line 2), the slope, although still negative, is not significantly different from zero ($\hat{\beta} = -6.2$; $F_{1,26} = 0.06$). This suggests that the mean of the 28 observations, 47.2 pCi/g, with a standard deviation of 27.1 pCi/g, is a more reasonable estimate of the true concentration.

These two examples illustrate the fact that there is no one best method for estimating a true concentration for a set of data. The importance of investigating the relationship between concentration and aliquot size before assuming a constant concentration cannot be overstated. Since the purpose of the concentration is to produce a value independent of the size of the sample on which it is measured, disregarding that relationship can produce misleading results in further statistical analyses.

Pure Ratios

Recall that for pure ratios the numerator and denominator are measured in the same units, e.g., $^{238}\text{Pu}/^{239}\text{Pu}$ both in nanocuries, and ^{238}Pu concentrations in vegetation and soil measured in picocuries per gram. In this case both numerator and denominator are random variables as compared with a concentration in which only the numerator is a random variable. This distinction complicates the statistical treatment of this type of data, but the basic assumption underlying the ratio in this situation, as for concentrations, is still proportionality. If the assumption of proportionality cannot be supported either theoretically or statistically, then other methods of relating the variables should be found. In this section we discuss some statistical problems associated with pure ratios encountered in environmental radionuclide research and illustrate the use of multivariate techniques as a substitute for and a means of testing the validity of a ratio.

First, we discuss a situation particular to radionuclide research. In contradiction to a statement in the introduction to this chapter that ratios tend to be more variable than the component variables, there is a situation where the ratio is the stable variable. An example is the ratio of ^{239}Pu to ^{241}Am observed at a safety-shot site on the Tonopah Test Range cited in Doctor and Gilbert (1977) (see Fig. 7). The ^{239}Pu and ^{241}Am values were individually quite variable, but their ratio was essentially constant.

Sokal and Rohlf (1969, p. 17) suggest that a pure ratio should be used to explain the relationship between two variables only if there is evidence that the process under study is a function of (or operates on) the ratio of the two variables and not of the variables

---

*Observed value of the F statistic with 1 and 53 degrees of freedom (Snedecor and Cochran, 1967, pp. 259-260).

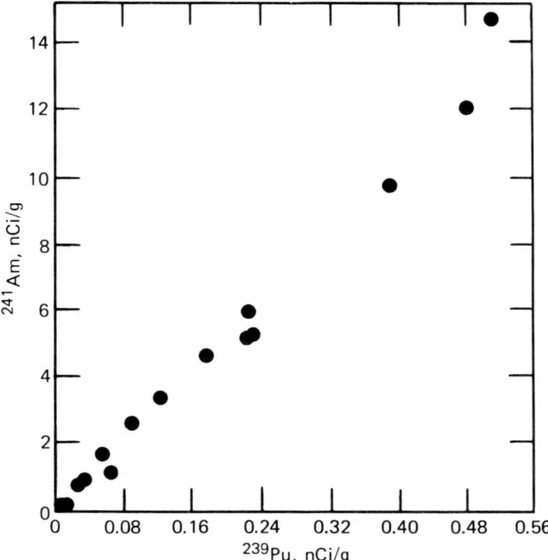

Fig. 7 Relationship between ^{239}Pu and ^{241}Am in soil at a safety shot site on the Tonopah Test Range. (From Doctor and Gilbert, 1977.)

individually. An example is provided by plant uptake of technetium, which appears to be related more to the ratio of pertechnetate to sulfate in the soil than to pertechnetate soil concentration (Cataldo, Wildung, and Garland, 1978). It would seem that this criterion would be met if the relationship between numerator and denominator is multiplicative. Two more examples include the plutonium/americium ratio just mentioned and the concentration ratio between two components of an ecosystem compartment model with a constant transfer coefficient. In these situations the reason for using ratios far outweighs their potential disadvantages. However, in some cases there may be more informative and less misleading ways to relate the two variables than by the use of a ratio.

Whether or not one can assume that the relationship between two random variables is multiplicative, the approach to these data should be a multivariate one. A first step is to plot the numerator vs. the denominator as in Fig. 7. The Pearson product moment correlation coefficient (Snedecor and Cochran, 1967, p. 172),

$$r = \frac{\sum_{i=1}^{n} (X_i - \overline{X})(Y_i - \overline{Y})}{\left[\sum_{i=1}^{n} (X_i - \overline{X})^2 \sum_{i=1}^{n} (Y_i - \overline{Y})^2 \right]^{\frac{1}{2}}}$$

measures the degree of linear association between two normally distributed random variables. Although we cannot assume that radionuclide activity is normally distributed, the correlation coefficient is still a useful piece of information. If the correlation is low, the ratio will be highly variable. The correlation coefficient provides a measure of linear association but not whether the relationship is multiplicative. That information can be gained from a regression analysis.

Since both numerator (Y) and denominator (X) are variables, the unconstrained regression of X on Y

$$X = \alpha' + \beta'Y + \epsilon'$$

(Y assumed not variable) is as valid statistically as the regression of Y on X

$$Y = \alpha + \beta X + \epsilon \tag{3}$$

where α', β', and ϵ' are different from α, β, and ϵ (Fig. 8). The two regression lines will never coincide unless there is a perfect multiplicative functional relationship (Y = βX) between Y and X. When we constrain one of the regressions, say Y on X, to go through the origin, Y = βX + ϵ, the other regression (X on Y) will not go through the origin unless

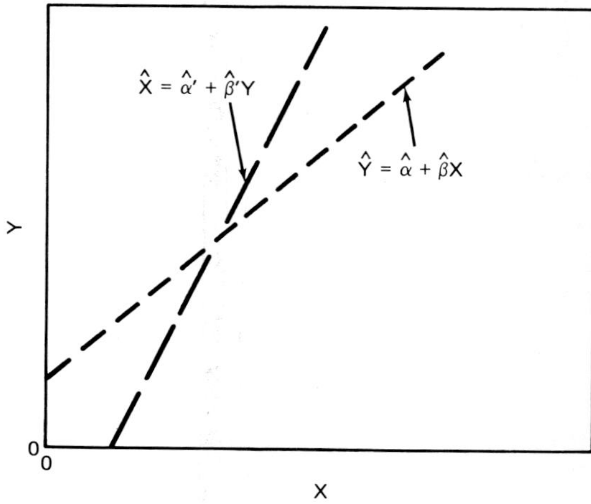

Fig. 8   Relationship between linear regression of Y on X and that of X on Y.

X is a function (without error) of Y (Snedecor and Cochran, 1967, pp. 172-181). If the multiplicative relationship seems valid, there are methods of taking a compromise slope (Ricker, 1973). If both X and Y are normally distributed, then an estimator of the slope (constant ratio), $\overline{Y}/\overline{X}$, has some nice statistical properties (Ricker, 1973; Creasy, 1956; Cochran, 1977, Chap. 6; Doctor and Gilbert, 1977). The sample median of the observed ratios is another useful estimator of the constant ratio because it is not greatly affected by extremely high or low values and because no assumption about the distribution of X and Y or their ratio (Doctor and Gilbert, 1977) is required.

Suppose that the use of a ratio is justified, e.g., the isotopic ratio ^{239}Pu/^{241}Am for the Tonopah Test Range data mentioned above (Fig. 7). This is often the first step in a series of statistical analyses; so several caveats should be mentioned. Making inferences about ratios is risky because we are usually forced to make distributional assumptions that ratios rarely fulfill. For example, a test of hypothesis regarding the difference between two samples of ratios may assume that the ratio data are normally distributed. Nonparametric techniques, such as the Wilcoxon rank sum test (Hollander and Wolfe,

1973, pp. 67-74), may alleviate that problem to a certain extent. Ratios are often used as the raw data for standard statistical methods, such as regression and analysis of variance. The appropriateness of this practice is determined by the behavior of the ratios vis-a-vis the assumptions underlying these methods.

A more general approach to a situation where a pure ratio can be used is a multivariate one. This approach is appropriate whether or not the multiplicative assumption is valid, and some multivariate techniques can be used as a check on that assumption. The previously mentioned Pearson product moment correlation coefficient, r, is a multivariate technique. The multivariate approach allows one to observe the behavior of the two random variables simultaneously. A multivariate (bivariate if the number of variables is two) variable can be most easily explained by an example. Let Y and X be, respectively, the ^{238}Pu and ^{241}Am activity at various depths in a soil profile. Both isotopes together can provide a more complete picture of the process of leaching of radionuclides in soil than either can provide separately. The joint distribution of ^{238}Pu and ^{241}Am activity at a particular profile depth may look like the two-dimensional curve in Fig. 9. The points (x and y) under the highest part of the curve correspond to the

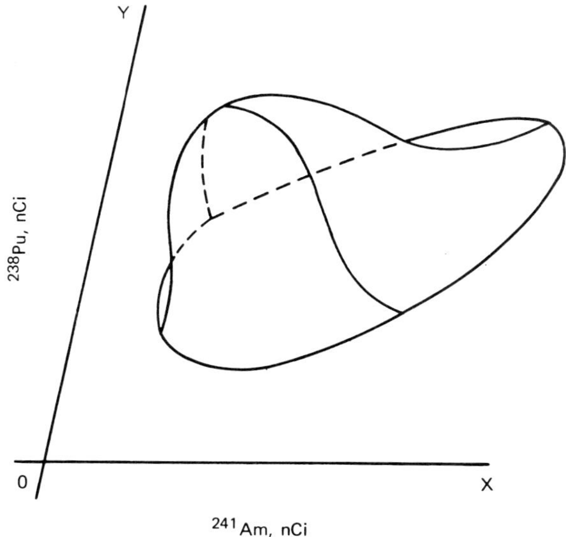

Fig. 9   Hypothetical joint distribution of ^{241}Am and ^{238}Pu.

combinations of ^{238}Pu and ^{241}Am activity most likely to occur. The relationship between the two variables is completely specified by the joint distribution. Everything that can be done statistically for single random variables—e.g., testing for differences between groups, regression, and analysis of variance—can be done for multivariate random variables (Morrison, 1967).

With this brief introduction to the rationale underlying multivariate statistical techniques, we discuss two such techniques on two sets of environmental radionuclide data. The first technique is multivariate regression and relates soil and vegetation concentrations to distance from a point source of contamination. The second technique, profile analysis, tests the hypothesis of different resuspension parameters for two ecosystems. Both techniques are based on the multivariate normal distribution. Despite the caveats mentioned earlier regarding the use of the normal distribution for analyzing environmental data, these techniques can still provide valuable insight into the behavior of the data with respect to the multiplicative assumption. Also, transforming the data by logarithms, as we have done here, tends to reduce variability and make the normality assumption more tractable.

Multivariate Regression

The data consist of ^{235}U vegetation and associated soil concentrations taken at various distances from ground zero (GZ) at a safety-shot site (A site in Area 11) on the Nevada Test Site (Gilbert and Eberhardt, 1976). The problem is to relate the soil and vegetation concentrations jointly as a function of distance from GZ. There are at least two possible approaches. We compare a univariate method that regresses the CR (vegetation/soil) on distance with a multivariate one that regresses soil and vegetation concentrations jointly on distance. A univariate regression with the CR as the dependent variable assumes that the relationship between vegetation and soil concentrations is multiplicative. Multivariate regression is not so constrained, and, because there are fewer assumptions on the relationship of Y and X, it can be used to assess whether the relationship is multiplicative.

The data shown in Fig. 10 consist of 14 pairs of soil and vegetation concentrations of ^{235}U taken from random locations within 300 ft of GZ. Figure 11 shows the ratio of vegetation to soil as a function of distance from GZ. The straight line in Fig. 11 is the least-squares fit of

$$\ln \left(\frac{Y}{X}\right) = \alpha + \beta D + \epsilon$$

where Y and X represent, respectively, vegetation and soil concentrations and D is the distance from GZ. Direction with respect to GZ does not appear to be an important factor; so it was ignored for this example. Although the fit in Fig. 11 looks reasonable by the R^2 criteria* ($R^2 = 0.56$), note that the range of the ratio for distances greater than 250 ft spans three orders of magnitude.

Compare this with the linear multivariate fit in Fig. 10 (Anderson, 1958, pp. 179-187), which is the simultaneous least-squares fit of

$$\ln (Y) = \alpha_1 + \beta_1 D + \epsilon_1 \tag{4}$$

$$\ln (X) = \alpha_2 + \beta_2 D + \epsilon_2 \tag{5}$$

$$*R^2 = 1 - \frac{\sum\limits_{i=1}^{n} (z_i - \hat{z}_i)^2}{\sum\limits_{i=1}^{n} (z_i - \bar{z})^2}$$

where $\hat{z}_i$ is the estimate of z, R^2 is a measure of the amount of variability accounted for by the model, $R^2 = 1$ implies a perfect functional relationship, and $R^2 = 0$ implies no linear relationship (Snedecor and Cochran, 1967, p. 402).

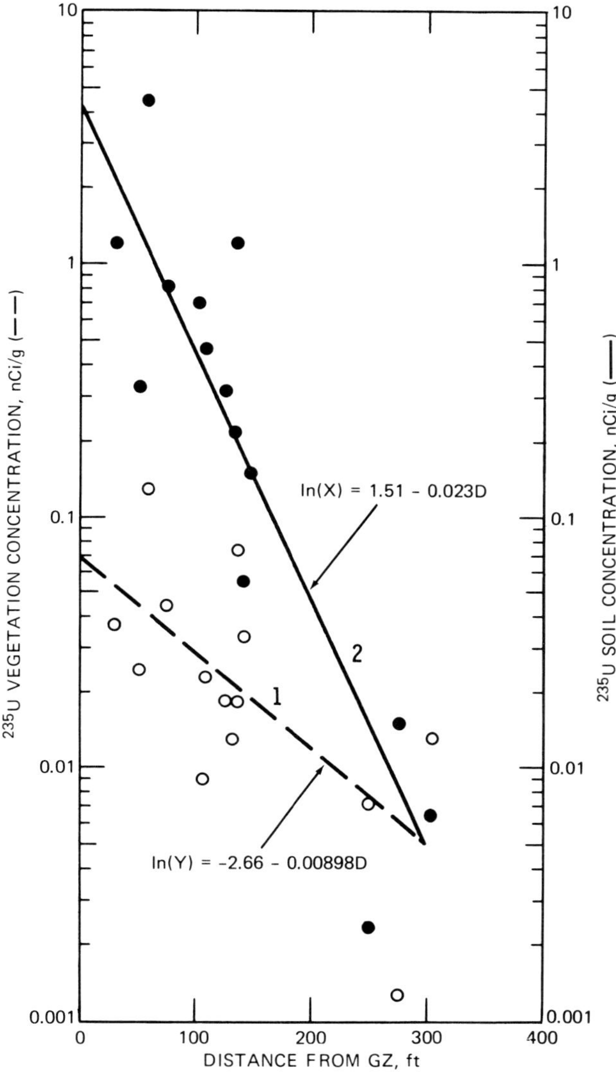

Fig. 10 Uranium-235 concentrations in vegetation and soil as a function of distance from GZ at A site, Area 11, Nevada Test Site.

where α_1, β_1, ϵ_1 and α_2, β_2, ϵ_2 represent, respectively, the parameters and error terms of the regression of ln vegetation concentration on distance and ln soil concentration on distance. Equation 4 corresponds to line 1 in Fig. 10 and Eq. 5 to line 2. This approach assumes that ϵ_1 and ϵ_2 are related; they represent, respectively, the deviations from the models given by Eqs. 4 and 5 on the same unit, in this case location. Again the straight-line fits look reasonable; for vegetation the univariate $R^2 = 0.45$ and for soil $R^2 = 0.77$.

The multivariate approach has two advantages: (1) it can be used to check the assumption of proportionality that underlies the use of the ratio and (2) it can help

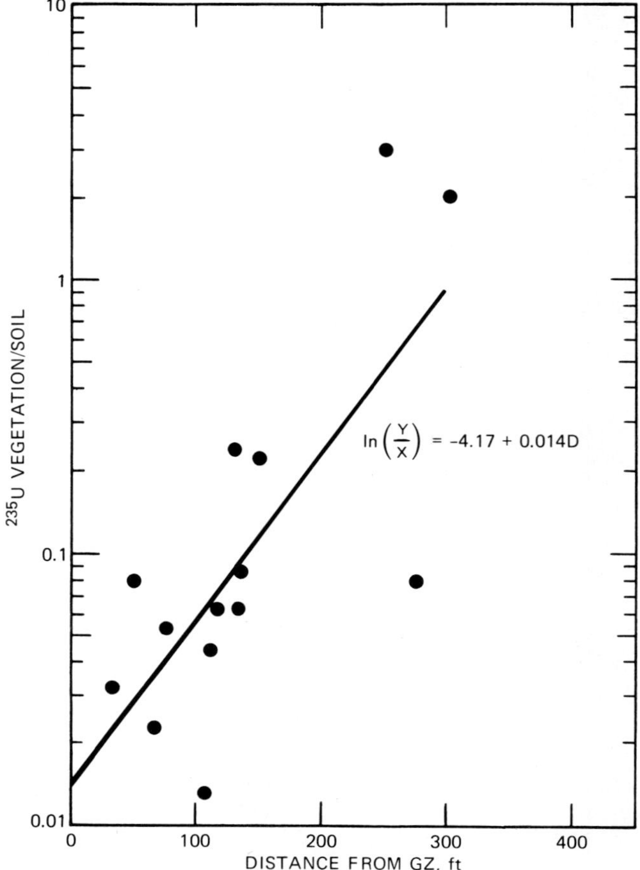

Fig. 11   **Ratio of** [235]**U concentrations in vegetation and soil as a function of distance from GZ at A site, area 11, Nevada Test Site.**

explain the observed variability in the ratios. Most important is the question of proportionality. If vegetation concentration is proportional to soil concentration, lines 1 and 2 in Fig. 10 should be parallel. A comparison of their slopes would be a test of that assumption. The slopes for the vegetation and soil concentrations are -0.00898 and -0.023, respectively.   In Fig. 10 the two regression lines do not appear to be parallel. Simultaneous 95% confidence intervals for β_1 and β_2, which are rather large because there are so few data points, are given by

$$-0.017 \leqslant \beta_1 \leqslant -0.00098$$

$$-0.032 \leqslant \beta_2 \leqslant -0.013$$

(Morrison, 1967, pp. 107-109). Since the two intervals overlap, one cannot claim that β_1 and β_2 differ significantly, although making a judgment on so few observations is risky.

The second advantage to the multivariate approach is that it can explain the variability in the ratios. A large ratio can be caused by either a large numerator or a small denominator and a small ratio by the converse. The uncritical use of ratios can obscure information. For example, the apparently strong trend of increasing ratio with increasing distance in Fig. 11 is, in part, due to the two samples at 273 and 306 ft, for which the vegetation concentration is larger than the soil concentration. Admittedly these anomalies may be traceable to the near background levels of ^{235}U at greater distances from GZ, but this example illustrates the need for caution when using ratios as input to further statistical analyses.

Profile Analysis

In this example a technique called profile analysis by Morrison (1967, pp. 186-197) is used to compare inventory ratios (IR) between ecosystem components at several sites. The elements of the profile are ^{238}Pu standing crops in the various ecosystem components at one location. Comparative studies of radionuclide inventories in ecosystems are often based on models such as the simple three-compartment one illustrated in Fig. 12. The box-and-arrow model represents the aboveground components

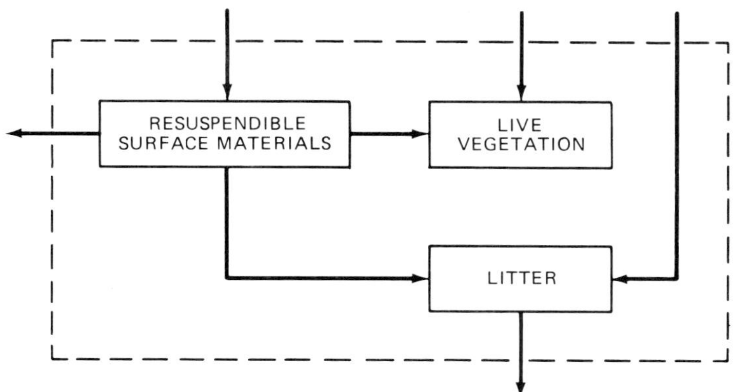

Fig. 12 Simple three-compartment model of aboveground components of herbaceous plant community.

of a herbaceous plant community on an abandoned field. The boxes denote ecosystem compartments (resuspendible surface materials, live vegetation, and litter), and the arrows denote fluxes of ^{238}Pu. Inputs to the model (arrows entering the larger dashed box) represent aerial deposition of ^{238}Pu from a reprocessing facility. Outputs (arrows exiting the dashed box) represent either wind dispersal of resuspended ^{238}Pu or ^{238}Pu incorporated into the soil. Other potentially important compartments and fluxes have been omitted for simplicity. Resuspendibles are those materials which can be resuspended into the atmosphere by a 6 m/sec wind velocity at ground level (McLendon et al., 1976).

The question often asked is whether the radionuclide distribution among compartments is the same at each site although the amount of radionuclide per unit area may differ between sites. The question can be stated another way: Are the IR's of the amount of radionuclide in one compartment to that in another the same at the various sites? Assume for ease of exposition that there are two sites and three compartments. Let ν_{ij} ($i = 1,2$; $j = 1,2,3$) represent the true radionuclide content of the jth compartment at site i. Then the IR of compartment 1 to compartment 2 at site 1 can be expressed as ν_{11}/ν_{12}. The rephrased question can be expressed as the statistical null hypothesis

$$H_0: \begin{bmatrix} \dfrac{\nu_{11}}{\nu_{12}} \\[2mm] \dfrac{\nu_{11}}{\nu_{13}} \\[2mm] \dfrac{\nu_{12}}{\nu_{13}} \end{bmatrix} = \begin{bmatrix} \dfrac{\nu_{21}}{\nu_{22}} \\[2mm] \dfrac{\nu_{21}}{\nu_{23}} \\[2mm] \dfrac{\nu_{22}}{\nu_{23}} \end{bmatrix} \tag{6}$$

Recall that $\ln (X/Y) = \ln (X) - \ln (Y)$. By defining $\mu_{ij} = \ln \nu_{ij}$, then

$$\ln \left(\frac{\nu_{i1}}{\nu_{i2}} \right) = \mu_{i1} - \mu_{i2}$$

If the data are collected so that a measurement of radionuclide content for each compartment is made at each sampling location, the hypothesis (Eq. 6) can then be expressed as

$$H_0': \begin{bmatrix} \mu_{11} - \mu_{12} \\ \mu_{12} - \mu_{13} \end{bmatrix} = \begin{bmatrix} \mu_{21} - \mu_{22} \\ \mu_{22} - \mu_{23} \end{bmatrix} \tag{7}$$

The terms $(\mu_{i,j} - \mu_{i,j+1})$ are estimated by

$$(\widehat{\mu_{i,j} - \mu_{i,j+1}}) = \frac{1}{n_i} \sum_{k=1}^{n_i} (\ln x_{i,j,k} - \ln x_{i,j+1,k})$$

where $x_{i,j,k}$ is the radionuclide content observed at the kth sampling location in the jth compartment at site i. Note that

$$\ln \left(\frac{x_{i,j,k}}{x_{i,j+1,k}} \right) = \ln x_{i,j,k} - \ln x_{i,j+1,k}$$

so consequently $(\widehat{\mu_{i,j} - \mu_{i,j+1}})$ is the mean of the logarithms of the n_i observed IR's of compartment j to compartment j + 1 for site i.

There are several reasons for doing this. First, instead of a nonlinear hypothesis on the ratios of random variables, we now have a linear hypothesis for which multivariate techniques currently exist. Second, taking logarithms tends to equalize variability and make the normality assumption more tractable; both of these conditions are assumed by the test of H_0', which is discussed below.

Note that for three compartments there are three possible IR's, which are listed in Eq. 6. For H_0' the IR's are expressed as $\mu_{i,j} - \mu_{i,j+1}$; so three compartments can be represented by only two IR's. A nice property of the following test of H_0' is that the ordering of the compartments is independent of the test, and compartment $j + 1$ need not be the source for compartment j. The hypothesis H_0' is tested by the statistic

$$T^2 = \frac{n_1 n_2}{n_1 + n_2} \left\{ [(\widehat{\mu_{1,1} - \mu_{1,2}}), (\widehat{\mu_{1,2} - \mu_{1,3}})] - [(\widehat{\mu_{2,1} - \mu_{2,2}}), (\widehat{\mu_{2,2} - \mu_{2,3}})] \right\} \underline{\underline{S}}^{-1}$$

$$\times \left\{ \begin{bmatrix} (\widehat{\mu_{1,1} - \mu_{1,2}}) \\ (\widehat{\mu_{1,2} - \mu_{1,3}}) \end{bmatrix} - \begin{bmatrix} (\widehat{\mu_{2,1} - \mu_{2,2}}) \\ (\widehat{\mu_{2,2} - \mu_{2,3}}) \end{bmatrix} \right\} \tag{8}$$

where $\underline{\underline{S}}$ is the pooled sites covariance matrix of the $(\widehat{\mu_{i,j} - \mu_{i,j+1}})$ terms. T^2 is distributed as Hotelling's T^2 (Morrison, 1967, p. 117), and

$$F = \frac{(n_1 + n_2 - 3) T^2}{2(n_1 + n_2 - 2)} \tag{9}$$

where F has the F distribution with 2 and $n_1 + n_2 - 3$ degrees of freedom if H_0' is true. Detailed procedures for testing H_0' (test of parallelism) are given by Morrison (1967, pp. 143 and 188).

The data   for this example are from unpublished data supplied by A. L. Boni, J. C. Corey, H. H. Horton, and M. H. Smith of the Savannah River Ecology Laboratory, Aiken, S. C. They consist of ^{238}Pu inventories (measured in picocuries per square meter) for the three compartments for two sites located at 0.23 km (community 1) and 0.43 km (community 2) from the point of aerial release of ^{238}Pu from a reprocessing facility at the U. S. Department of Energy Savannah River Plant. The data are given in Fig. 13, where the horizontal and vertical bars denote, respectively, the arithmetic means and 95% confidence intervals computed from 17 samples in community 1 and 12 samples in community 2. For this set of data,

$$T^2 = \frac{17 \times 12}{17 + 12} \left\{ [4.63, 1.04] - [3.44, 2.04] \right\} \begin{bmatrix} 0.578 & 0.263 \\ 0.263 & 0.773 \end{bmatrix} \left\{ \begin{bmatrix} 4.63 \\ 1.04 \end{bmatrix} - \begin{bmatrix} 3.44 \\ 2.04 \end{bmatrix} \right\}$$

$$= 6.8 \tag{10}$$

With F = 3.27 (computed using Eq. 9), the null hypothesis is rejected at the $\alpha = 0.10$ level. These results indicate that a greater proportion of the ^{238}Pu occurs in the resuspendible compartment at the more highly contaminated site (community 1), whereas a greater proportion of the ^{238}Pu occurs in the litter at the less contaminated site (community 2).

The method is easily extended to more than three compartments or more than two locations (Morrison, 1967, p. 188). It can also be applied to concentration ratios.

Assume that there are p compartments. It might appear easier to choose $p - 1$ IR's of interest (denoted by K_{ij}, $j = 1, \ldots, p - 1$) from the p compartments and test the null hypothesis

$$H_0: \begin{bmatrix} K_{1,1} \\ K_{1,2} \\ . \\ . \\ . \\ K_{1,p-1} \end{bmatrix} = \begin{bmatrix} K_{2,1} \\ K_{2,2} \\ . \\ . \\ . \\ K_{2,p-1} \end{bmatrix}$$

using the means of the observed IR's for each site. This test is, however, dependent on which $p-1$ of the possible $p(p-1)$ IR's is selected. This is not true of Eq. 10, where the T^2 value obtained is independent of the ordering of the compartments and the acceptance of H_0' implies that all the possible $p(p-1)$ IR's in the two locations have not been demonstrated to be unequal.

From these two examples we have seen that multivariate statistical techniques can be used to address such questions as the assumption of a multiplicative relationship between variables and the simultaneous test of equality of IR's, which are not easily dealt with by the use of ratios or traditional univariate techniques.

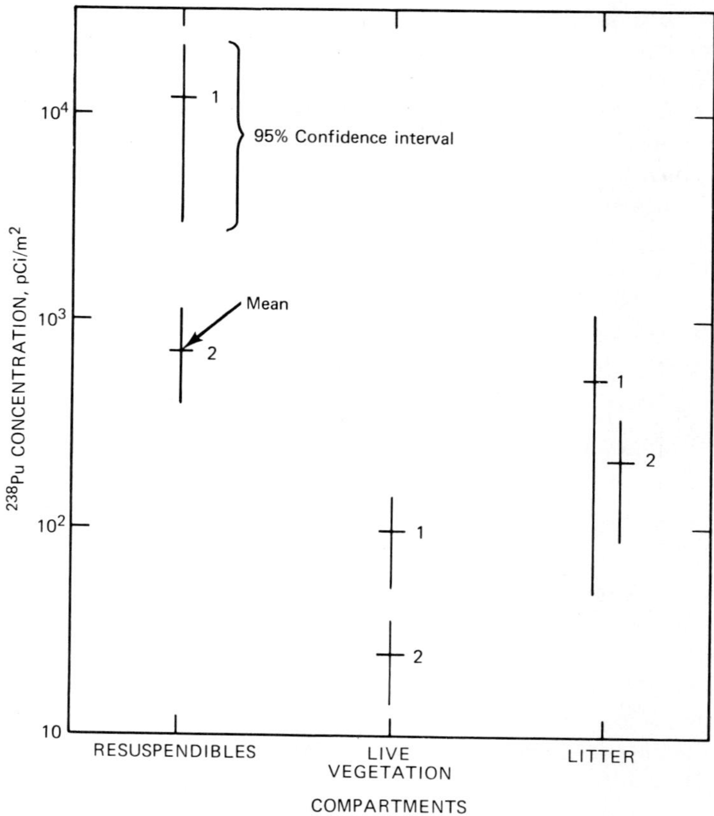

Fig. 13 Plutonium-238 standing crop in three ecosystem compartments for two sites.

Summary

The appropriate use of ratios in environmental radionuclide research demands some knowledge of the relationship between the two variables composing the ratio. Concentrations and pure ratios are two types of ratios used extensively in environmental transuranic element research. A concentration is the amount of activity per unit of weight or volume, e.g., nanocuries per gram. A pure ratio, say $^{238}Pu/^{239}Pu$ or ^{238}Pu in vegetation/^{238}Pu in soil, is used to express a relationship between the two random variables. In either case the ratio is a valid description of that relationship only if the relationship is approximately multiplicative. For concentrations a major problem (addressed in this chapter using two sets of environmental radionuclide data) is determining the aliquot size for which activity is homogeneously dispersed throughout the medium. For pure ratios the problem is assessing whether the relationship is multiplicative. If not, more meaningful methods, such as multivariate statistical techniques, must be found for relating the two variables. This chapter discusses two multivariate techniques, regression and profile analysis, which might be used as a test of the multiplicative assumption.

Acknowledgment

This paper was written under U. S. Department of Energy contract EY-76-C-06-1830.

References

Anderson, T. W., 1958, *An Introduction to Multivariate Statistical Analysis,* John Wiley & Sons, Inc., New York.

Atchley, W. R., C. T. Gaskins, and D. Anderson, 1976, Statistical Properties of Ratios. I. Empirical Results, *Syst. Zool.,* 25: 137-148.

Cataldo, D. A., R. E. Wildung, and T. R. Garland, 1978, Technetium Accumulation, Fate, and Behavior in Plants, in *Environmental Chemistry and Cycling Processes,* DOE Symposium Series, No. 45, Augusta, Ga., Apr. 28–May 1, 1976, D. C. Adriano and I. L. Brisbin, Jr. (Eds.), pp. 538-549, CONF-760429, NTIS.

Chayes, F., 1971, *Ratio Correlation,* University of Chicago Press, Chicago.

Cochran, W. G., 1977, *Sampling Techniques,* 3rd ed., John Wiley & Sons, Inc., New York.

Creasy, M. A., 1956, Confidence Limits for the Gradient in the Linear Functional Relationship, *J. R. Stat. Soc., Ser. B,* 18: 65-69.

Doctor, P. G., and R. O. Gilbert, 1977, Ratio Estimation Techniques in the Analysis of Environmental Transuranic Data, in *Transuranics in Natural Environments,* Symposium Proceedings, Gatlinburg, Tenn., Oct. 5–7, 1976, M. G. White and P. B. Dunaway (Eds.), ERDA Report NVO-178, pp. 601-619, Nevada Operations Office, NTIS.

——, and R. O. Gilbert, 1979, Two Studies in Variability for Soil Concentrations: With Aliquot Size and with Distance, in *Selected Environmental Plutonium Research Reports of the NAEG,* M. G. White and P. B. Dunaway (Eds.), USDOE Report NVO-192, Nevada Operations Office, NTIS.

Gilbert, R. O., and L. L. Eberhardt, 1976, Statistical Analysis of "A Site" Data and Interlaboratory Comparisons for the Nevada Applied Ecology Group, in *Studies of Environmental Plutonium and Other Transuranics in Desert Ecosystems,* M. G. White and P. B. Dunaway (Eds.), ERDA Report NVO-159, pp. 117-154, Nevada Operations Office, NTIS.

Hollander, M., and D. A. Wolfe, 1973, *Non-Parametric Statistical Methods,* John Wiley & Sons, Inc., New York.

McLendon, H. R., et al., 1976, Relationships Among Plutonium Contents of Soil, Vegetation and Animals Collected on and Adjacent to an Integrated Nuclear Complex in the Humid Southeastern United States of America, in *Transuranium Nuclides in the Environment,* Symposium Proceedings, San Francisco, Nov. 17–21, 1975, pp. 347-363, STI/PUB/410, International Atomic Energy Agency, Vienna.

Michels, D. E., 1977, Sample Size Effect on Geometric Average Concentrations for Log-Normally Distributed Contaminants, *Environ. Sci. Technol.,* 11: 300-302.

Mielke, P. W., Jr., and J. A. Flueck, 1976, Distribution of Ratios for Some Selected Bivariate Probability Functions, in *1976 Proceedings of Social Statistics Section,* American Statistical Association, Washington, D. C.

Morrison, D. F., 1967, *Multivariate Statistical Methods,* McGraw-Hill Book Company, New York.

Parzen, E., 1967, *Stochastic Processes,* Holden-Day, Inc., San Francisco, Calif.

Ricker, W. E., 1973, Linear Regression Fisheries Research, *J. Fish. Res. Board Can.,* 30: 409-434.

Sharitz, R. R., S. L. Scott, J. E. Pinder III, and S. K. Woods, 1975, Uptake of Radiocesium from Contaminated Floodplain Sediments by Herbaceous Plants, *Health Phys.,* 28: 23-28.

Simpson, G. G., A. Roe, and R. C. Lewonton, 1960, *Quantitative Zoology,* Harcourt Brace Jovanovich, Inc., New York.

Snedecor, G. W., and W. G. Cochran, 1967, *Statistical Methods,* 6th ed., Iowa State University Press, Ames, Iowa.

Sokal, R. R., and F. J. Rohlf, 1969, *Biometry,* W. H. Freeman and Company, San Francisco.

U. S. Energy Research and Development Administration, 1976, *Workshop on Environmental Research for Transuranic Elements,* Battelle Seattle Research Center, Seattle, Wash., Nov. 12–14, 1975, ERDA Report ERDA-76/134, NTIS.

Review of Resuspension Models

J. W. HEALY

Resuspension has been recognized as a potential mode of human exposure to contaminants in the soil for a number of years. However, methods for expressing the resulting concentrations quantitatively have been crude; in most cases, the resuspension factor was used. In this chapter we distinguish three main types of resuspension: wind-borne, mechanical disturbance, and local. The three main models for estimating concentrations (resuspension factor, resuspension rate, and mass loading) are described, and the data applicable to each are reviewed.

The studies of wind erosion of desert sands and of agricultural soils have provided much of the information available on the mechanisms involved in wind resuspension. However, such a body of evidence is not available for mechanical disturbance of the soil, although recent experiments have provided some data. Information on concentrations in the vicinity of the individual causing the disturbance is still poor.

Considerable progress has been made in the past few years on wind resuspension. For both wind and mechanical resuspension, scientists are on the verge of providing improved estimates using the resuspension rate at specific locations. However, for generic studies the mass-loading approach is recommended.

Resuspension from soils and subsequent inhalation of the resuspended material has long been considered the chief source of exposure to transuranium elements deposited in soils. In spite of the obvious importance of this pathway, research has been limited; thus it is difficult to obtain a reasonable prediction of the resulting concentrations and inhalation. In fact, Slinn (1978) indicates that he does not trust resuspension factors, rates, or velocities to within many orders of magnitude. In the following text we will review the data available and attempt to arrive at as reasonable an answer as possible.

Types of Resuspension Considered

The overall resuspension problem can be divided into three types for conceptual understanding and calculations (Healy, 1977a; 1977b): (1) wind-driven resuspension, (2) mechanical resuspension, and (3) local resuspension. For wind resuspension the energy required to dislodge the particles arises from the wind, and the particles then disperse downwind, depositing on surfaces at a rate depending on the aerodynamic properties of the particles and the nature of the terrain. Both mechanical resuspension and local resuspension result from mechanical disturbance of the soil. However, in the mechanical resuspension case the concern is with concentrations downwind following dispersion and deposition. Local resuspension, on the other hand, describes the exposure in the immediate vicinity of the individual before dispersion occurs.

Another type of resuspension that will be considered only briefly is transfer resuspension. This involves the transfer of the contaminant from its place of deposit to another place where inhalation may be more probable. Unfortunately data to describe this type of resuspension are extremely limited.

Each of these types of resuspension will be considered in deriving the best estimate of exposure to an individual in an area contaminated by the spread of transuranium elements in the soil.

Resuspension Modeling

The three basic techniques in use for resuspension modeling are (1) the resuspension factor, (2) the resuspension rate, and (3) the mass-loading approach. Each method has its strengths and its weaknesses, particularly in view of the state of the technology at this time. Each of the techniques is described, and its advantages and disadvantages are discussed.

Resuspension Factor

The resuspension factor is defined as the ratio of the concentration in the air at a reference height (usually 1 m) to the quantity of the contaminant per unit area on the surface of the ground. The usual units are meters^{-1}. Its chief advantages are its simplicity and the fact that most measurements in the past have been expressed in this form; so values are available for calculation. The chief disadvantages are that it is a completely empirical formulation, and thus it is difficult to extrapolate from one terrain to another, and it ignores the distribution of the contamination over the area and the size of the contaminated area involved. Thus the denominator contains the local quantity on the ground, and the numerator is an undefined function of the air concentration resulting from upwind contaminated areas and activities.

A problem common to this model, as well as to all others, is the uncertainty resulting from the depth to be used in assessing the quantity per unit area to be used in the denominator. For a uniform profile in the soils, the quantity per unit area will increase in direct proportion to the depth of the sample used. For nonuniform profiles the estimate of the quantity per unit area will also change, depending on the depth used. For wind resuspension it can be assumed that the appropriate depth is small, perhaps a millimeter or less, although it is likely that this depth may be variable, depending on wind speed and the degree of saltation allowed by the size of the area and the nature of the soils. For mechanical disturbance the depth will be some function of the depth to which the disturbance occurs, the function depending on the relative ease with which the particles can escape from the soil to the atmosphere.

Mishima (1964) has tabulated resuspension factors measured over a variety of conditions. It is frequently noted that these values range over eleven orders of magnitude. However, the values quoted represent both outdoor and indoor conditions, with and without mechanical disturbance, and at various times after the contaminant has been deposited. In a brief review of Mishima's table, it is noted that the values for mechanical disturbance range from about 2×10^{-6} to 7×10^{-5} m^{-1} (with one value of 10^{-3} m^{-1} based on uranium contaminant with dust stirred up and sampling at a height of 1 ft ignored). For periods of no activity, with relatively freshly deposited material, the values generally range from 10^{-8} to 2×10^{-6} m^{-1}, whereas for aged material they range from 6×10^{-10} to 10^{-13} m^{-1}. It is difficult to generalize these numbers because the exact value will depend on the degree of disturbance, the placement of the sampler, the meteorological conditions at the time, and the nature of the soils. However, many of the measurements were made in désert areas with low soil moisture where resuspension would be expected to be highest.

An apparent reduction in the resuspension factor with time was indicated by Wilson, Thomas, and Stannard (1960) from measurements at the Nevada Test Site (NTS). Here samples were taken at three different distances from the center of a safety shot starting about 1 month after the contamination occurred. The investigators noted that the data were too erratic to establish half-times for the decay of the air concentration beyond a very crude estimate. This estimate was made by plotting the medians of the data at each sampling distance, and it provided a value of 5 weeks for the concentration half-time. This value was used by Langham (1969) in assessing future hazards from plutonium contamination. A somewhat larger half-life of 45 days was used by Kathren (1968) in a study of acceptable contamination levels for plutonium in soils. Anspaugh et al. (1973) measured the decrease in air concentration with time immediately following a cratering event in Nevada and following the venting of an underground shot. For the cratering event sampling was carried out for 6 weeks following the event with a measured half-time of 38 days. The venting experiment started 3 months after the event and continued for 9 to 10 months, with the most predominant radionuclides being the isotopes of ruthenium. Here a half-life of 66 days was found.

However, the consequences of continuing such half-times over a long period were pointed out by Healy (1974) and Anspaugh (1974). Healy noted that samples taken at the same location as those taken by Wilson, Thomas, and Stannard (1960) but about 1 yr after the conclusion of the Wilson, Thomas, and Stannard study (Olafson and Larson, 1961) gave values up to several orders of magnitude greater than would be predicted by the 35-day half-life. Anspaugh (1974) indicated that the functional nature of the decrease in resuspension rate with time cannot be confidently extrapolated and that previously published models should not be applied to calculations many years after the contamination event. He also cited two sets of measurements at NTS where the area had been contaminated with plutonium by high-explosive detonations some 20 yr earlier. These studies gave values for the resuspension factor of 3×10^{-10} m^{-1} and 2×10^{-9} m^{-1}. These data indicate unequivocally that resuspension does occur after this period of time, although predictions using the short half-life following deposition would result in unmeasurable values of air concentration.

Anspaugh, Shinn, and Wilson (1974) used the available information to derive an empirical expression for the resuspension factor which allows the resuspension factor to decrease with time. In their derivation they used four constraints: (1) the apparent half-time of decrease during the first 10 weeks should approximate a value of 5 weeks, (2) this half-life should about double over the next 30 weeks, (3) the initial resuspension factor should be 10^{-4} m^{-1}, and (4) the resuspension factor 17 yr after the contaminating event should approximate 10^{-9} m^{-1}. The value for the resuspension factor in the aged source resulted from 23 individual air concentration measurements at a location contaminated with plutonium 17 yr earlier where the average resuspension factor was found to be 10^{-9} m^{-1}. An expression that approximates these constraints is given as

$$R(t) = 10^{-4} \exp [-0.15 (t)^{\frac{1}{2}}] + 10^{-9} \qquad (1)$$

where R(t) is the resuspension factor (meters^{-1}) and t is the time since the contaminating event (days).

Oksza-Chocimowski (1977) provided a "generalized" model for the change in half-time of the resuspension factor which allows the ratio of the resuspension factor at time zero to the resuspension factor at a long time $[R(0)/R(\infty)]$ to vary:

$$T_{\frac{1}{2}}(t) = A \ln \, (1 + B + Ct^D) + \frac{t \ln 2}{\ln \dfrac{R(0)}{R(\infty)}} \tag{2}$$

where A = constant coefficient, days
    B = constant
    C = constant coefficient, days^{-D}
    D = constant exponent
  $R(0)$ = initial resuspension factor, m^{-1}
 $R(\infty)$ = final resuspension factor, m^{-1}
     t = time since contaminating event, days

This expression was evaluated for a range of values of the constants with limiting values corresponding to ratios of $R(0)/R(\infty)$ of 10 and 10^{11}. However, data are not available to make adequate choices for a given area.

In both the Anspaugh, Shinn, and Wilson and the Oksza-Chocimowski models, the constants are evaluated on the basis of data from a limited number of areas and conditions. In Anspaugh's model, for example, it is apparent that the final resuspension factor after long aging is based on wind resuspension only in a desert area. Whether disturbance in the area would cause an increase is unknown, although it appears likely to do so. Sehmel and Orgill (1974) made measurements downwind from the original oil-storage pad at Rocky Flats and related the concentrations found to the 2.1 power of the wind speed. Subsequent measurements were higher, however, owing to the digging of a ditch between the oil storage area and the sampler. This work also involved increased vehicular activity. This disturbance caused about an order of magnitude increase in concentrations at the samplers which persisted after the work had been completed presumably because of a change in the character of the surface. The half-time for decrease over the next 7 months appeared to be about 9 months. Thus it is possible that disturbance could not only increase the concentration at the time that it occurs but also could result in increased wind-driven resuspension factors for some time thereafter.

Sehmel and Lloyd (1976a) and Sehmel (1977a) have also provided data on the wind-borne resuspension of a submicrometer tracer, calcium molybdate, that was applied to a test area by spraying. In this experiment, using cascade impactors arranged to operate in different wind conditions, they noted that the resuspension factor at 1.8 m height increased as the 6.5 power of the wind speed. However, this slope was determined by drawing the line through the bottom end of uneven wind-speed ranges, a procedure that could well result in an overestimate. They also noted, for this system, that there seemed to be little, if any, decrease in the resuspension over a period of 3 yr. Whether this is due to a "preaging" by the application in water is unknown.

In concluding the discussion of the resuspension factor, it is apparent that this empirical approach does not inherently incorporate many of the variables, and present estimates are relatively crude. In particular, the present estimates appear to be based on short-term experiments with little attempt to provide a factor applicable to long-term (say, 1 yr) averages.

Resuspension Rate

The resuspension rate is defined as the fraction of the contaminant present on the ground that is resuspended per unit time by either winds or mechanical disturbance. Once obtained, it can be used to describe concentrations at any point around a nonuniform contaminated area by the use of point-source dispersion and deposition equations and integration over the area. This potential use was illustrated by applying it to an area contaminated with plutonium by a safety shot (Healy, 1974) and the inverse use at the same area to obtain resuspension rates from measured air concentrations (Anspaugh et al., 1975). It was introduced for use in resuspension calculations by Healy and Fuquay (1958), although in a crude form.

Slinn (1978) has pointed out that the resuspension rate can be converted to a resuspension velocity by multiplying by the ratio of the quantity of contaminant per unit area and dividing by the volumetric concentration of the contaminant in the soil. Such a velocity is analogous to the deposition velocity with, however, a negative sign when conditions are such that net resuspension occurs.

Three techniques have been used to measure resuspension rates for a given area: (1) measurement of air concentrations resulting from a known pattern of a tracer material on the ground and inferring the resuspension rate from height profiles, which gives the total transport, or from use of dispersion equations in known meteorology; (2) measurement of air concentrations from an existing contaminated area and obtaining the resuspension rate as given in 1; and (3) measurement of natural dust fluxes and relating these fluxes to some association between the concentration of the contaminant in the soil and the dust flux. The last method has been used only for wind resuspension.

Wind Resuspension.   A detailed body of knowledge exists on the mechanisms of the movement of soils by wind through the classic studies of Bagnold (1943) on desert sands and the detailed studies of Chepil (1941; 1945a; 1945b; 1945c; 1951a; 1951b; 1956; 1957; 1960), Bisal and Hsieh (1966), Woodruff and Siddoway (1965), U. S. Department of Agriculture (1968), and Malina (1941) on agricultural soils. These studies will not be reviewed in detail since much of the information is applicable to the limited condition of erosion of erodible soils. There are, however, data and concepts applicable to the resuspension process, at least for the limited conditions of agricultural soil, and a brief review of these is in order.

The relationship between erosion and winds is complex; a large number of variables affect the outcome. Chepil (1945a) listed the most important of the factors as related to the three categories given in Table 1. In the following discussion I will briefly describe some of the more important findings applicable to the general problem of resuspension from the extensive work on soil erosion.

Soil movement across an eroding field is primarily from movement of the smaller particles, usually less than about 1 mm in size. There are three mechanisms for movement, and the particular size for each is somewhat dependent on the wind speed. The heaviest particles move by surface creep or movement along the surface. Chepil (1945a) noted that these grains were too heavy to be moved by the direct pressure of the wind but were propelled by the impacts of smaller grains moving in the second method of movement, saltation. In saltation the grains, after being rolled by the winds, suddenly leap almost vertically. Some grains rise only a short distance, whereas others, depending on the wind speed, can rise to several feet. They are then carried forward by the winds

**TABLE 1   Most Important Factors in Wind Erosion
of Agricultural Soils**

I. Air	II. Ground	III. Soil
Velocity	Roughness	Structure
Turbulence	Cover	Organic matter
Density	Obstructions	Lime content
Temperature	Temperature	Texture
Pressure	Topographic features	Specific gravity
Humidity		Moisture content
Viscosity		

*Based on data from Chepil (1945a).

and settle by gravity until they strike the ground. Chepil (1945a) attributes this sudden rise to the spinning of the grain as it rolls along the surface, with the Bernoulli effect causing a difference in pressure at the top and bottom of the grain. The third method of movement is, then, suspension of the small soil particles by the wind. In the last case the particles must be small enough to be kept airborne by the turbulent forces in the atmosphere overcoming the force of gravity. Since the turbulence varies with atmospheric stability and, to some extent, with wind speed, one would expect that larger particles would be suspended in strong turbulent winds and that these particles would settle out as the winds decrease. However, the fine particles can remain suspended for long times and can cover large distances. It is the suspension fraction, and particularly the smaller particles, that is of interest in resuspension since resuspension is defined as the suspension of a previously deposited contaminant.

The airflow characteristics over the surface of the soil are important in transmitting force to the soil grains and in determining the velocity at which they start moving. In a neutral atmosphere (i.e., temperature decrease with height is adiabatic), the velocity profile is logarithmic with height; so one can write (Prestley, 1959),

$$u_z = \frac{u_*}{K} \ln\left(\frac{z}{zo}\right) \tag{3}$$

where u_z = wind speed at a height z
    zo = height at which the wind speed is zero
    u_* = friction velocity (or drag velocity)
    K = von Kármán constant, which has been found by integration to equal 0.4

Zo is a characteristic of the surface, increasing as the roughness of the surface increases. The friction velocity is of importance in determining the force exerted on any object protruding above the laminar layer of the atmosphere and has been shown to be the characteristic wind speed that affects soil movement (Bagnold, 1943; Chepil, 1945b; 1945c).

Bagnold (1943) has shown that in severe erosion conditions the profile of wind speed with height is changed by the momentum transfer to the particles in saltation. It has been observed that the wind velocity must exceed some threshold value to induce movement of the soil. Chepil (1945b) has studied the movement of particular sizes of grains and has noted that the curve of grain diameter times the specific gravity vs. threshold friction velocity for initial movement has a minimum at about 0.15 mm (150 μm) with a friction velocity of about 0.15 m/sec; i.e., grains smaller than this size and grains larger than this size require higher friction velocities to initiate any movement. The stability of the finer grains is illustrated by the simple experiment of Bagnold (1943), in which he placed a pile of talcum powder on a smooth surface and exposed it to winds. The layer was stable at relatively high wind speeds. However, a few grains of larger particles sprinkled on the pile resulted in rapid dispersion at a wind speed much lower than would serve to disperse the particles without this added factor. It is believed that the relative stability of the small particles is due to the fact that they do not protrude above the laminar layer; thus no drag force is exerted on them. The minimum in the curve of velocity required to institute movement and particle size, then, is due to the balance between the increasing drag force as the particle increases in size and the increasing downwind force from gravity as the particle becomes larger. Above about 0.15 to 0.2 mm, the threshold velocity required to initiate movement increases as the square root of the product of the particle diameter and its specific gravity (Bagnold, 1943; Chepil, 1945a).

As a result of this threshold friction velocity, it is apparent that direct aerodynamic pickup of small particles from the soil is unlikely. Instead, the process of saltation is the key to producing the suspended fraction because the impact of these particles as they strike the ground provides the energy to propel the smaller particles above the laminar layer into the wind stream where they are transported by eddies in the wind. Thus, in the talcum powder example, the sand particles sprinkled on the talcum powder served the function of the saltating particles. In a field, knolls, ridges, sand pockets, or other areas most exposed to the wind and/or containing the easily erodible grains start to erode at a lower velocity than the rest of the field. Once the erosion starts, it spreads downwind, and the bombarding action of the particles in saltation causes movement in other parts of the field that normally would not be eroded (Chepil, 1945a). The threshold velocity of the field is therefore the threshold of the most exposed or most erosive spots in the field. Since the avalanching effect of saltation increases down the field in the direction of the wind, the length of the field is also a factor in the degree of erosion.

An important consequence of the role of saltation in the production of resuspension is that there will be no dust, or particles flowing in suspension, unless the wind speeds are great enough to produce saltation under the conditions of the field. This places a threshold condition on the wind speeds required to resuspend particles from the ground.

Bagnold (1943) measured the rate of soil flow for desert sands and found that these rates could be described by an equation of the form of

$$q = C\, u_*^3\, \frac{\rho}{g} \tag{4}$$

where q = rate of soil flow (grams per centimeter width per second)
    ρ = density of the air
    g = acceleration of gravity
    C = constant that differs for differing soils and forms of erosion

Bagnold concluded that on the desert sands the flow in suspension was small, about $\frac{1}{20}$*th* of the total flow, as compared with saltation and surface creep. Chepil (1945a; 1945b) made similar measurements on agricultural soils in a wind tunnel. His results indicated that all three methods of flow appear to follow Bagnold's law of the cube of the wind velocity, at least for the soils tested, a fine sandy loam and a heavy clay soil. The constant C for total soil flow varied widely for different soils; the range in these experiments was 1.0 to 3.1. Chepil also measured the proportion of each type of flow on four widely different soils. These results are given in Table 2.

TABLE 2   Relative Portion of Three Types of Flow on Different Soils

| | Percent of flow in | | |
Soil type	Surface creep	Saltation	Suspension
Sceptre heavy clay	24.9	71.9	3.2
Haverhill loam	7.4	54.5	38.1
Hatton fine sandy loam	12.7	54.7	32.6
Fine dune sand	15.7	67.7	16.6

*Based on data from Chepil (1945b).

It is apparent that the fraction of the total flow carried in suspension is considerably higher on agricultural soils than on desert sands (presumably because of the availability of the smaller particles). These studies showed that the logarithm of the flow plotted against height was essentially a straight line and that relative concentrations of soil particles at different heights remained the same with wind velocities ranging from 13 to 30 miles/hr (6 to 13.5 m/sec). Presumably, then, the relative flows also remained constant.

Chepil (1945c) also provided the sizes of the particles in the soils studied, between <0.1 and 0.83 mm. The relative suspension flow vs. the fraction of particles <0.1 mm is plotted in Fig. 1. The use of any other particle size range or cumulative percentages gave erratic results. This may indicate the importance of the fraction of the smaller particles in the soil in producing the suspension fraction.

An important factor in the suspension fraction is the aggregate state of the smaller particles in the soil. Particles in the submicron size range rarely exist as such in the soils because they tend to either clump together or to adhere to larger particles and thus become small aggregates. In fact, Chepil (1945b) states that particles smaller than 0.005 mm (5 μm) do not exist as such in ordinary soils because they are aggregated into larger individual grains. He adds also that single grains or aggregates 0.05 to 0.5 mm in diameter have little or no cohesive properties and are easily carried by the winds. This means that contaminants in the soils, either as fine particulates or absorbed on the surface of soil particles, will largely exist as soil aggregates and will behave in the same manner as the soil. Chepil (1957) demonstrated the aggregation of material carried in suspension at heights of 4 to 8 ft in a dust storm by sizing particles by sedimentation in CCl_4, a nonpolar solvent that tends to preserve aggregates, and then repeating in water following dispersion with sodium hexametaphosphate. The curves show the percentage of particles smaller than a given value reaching zero at about 5 μm in diameter in a 1954 storm and about 10 μm in a 1955 storm. By contrast, the dispersed samples showed 15 to 25% of the particles smaller than 5 μm.

An important factor in aggregation is the moisture content of the soil. This has been investigated by Chepil (1956) and by Bisal and Hsieh (1966). Chepil (1956) has provided a formulation for the soil flow taking into account the increased resistance to movement

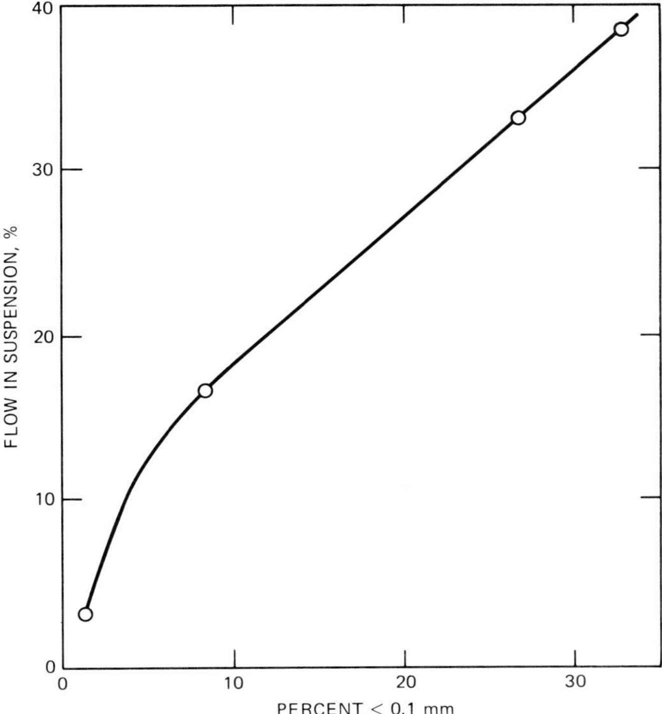

Fig. 1 Suspension flow vs. percent of grains in soil <0.1 mm.

due to the cohesion of the absorbed water films. However, it is noted that the water content of a field can decrease rapidly following a rain owing to the drying actions of the winds.

Many other factors influence the erosion of the field, such as the presence or absence of ridges, the quantity of vegetative cover, and the presence or absence of a surface crust. These have been combined to give a soil-erosion equation (Chepil, 1960; Woodruff and Siddoway, 1965)

$$E = f(I',C',K',L',V) \tag{5}$$

where E = potential average annual soil loss (tons per year)
    I' = soil and knoll erodibility
    C' = local wind-erosion climatic factor
    K' = soil ridge roughness factor
    L' = field-length factor
    V = equivalent quantity of vegetation

Mathematical relations have been established between the individual variables. The relationships, however, are so complex that the individual factors are evaluated separately in a form in which combinations of the factors can be evaluated graphically.

We will not attempt here a complete discussion of this equation along with numerical factors but will discuss each of the factors in a qualitative fashion because it appears that many of these could be of importance in any case of wind resuspension.

The soil and knoll erodibility consists of two terms. The soil-erodibility index is related to the percentage of dry aggregates greater than 0.84 mm in diameter. The higher this percentage, the lower the soil erodibility. Conversely, the higher the percentage of aggregates less than 0.84 mm in diameter, the greater the erodibility. The knoll erodibility is expressed as the percentage of the level-ground erosion that occurs at various slopes of the knolls in the field. This factor ranges from 0 for a level field to about 650% at the top of a knoll having a slope of 10% and about 360% from that portion of the windward slope where the drag velocity is the same as the top of the knoll (about the upper third of the slope). A surface crust stability factor is usually ignored because the crust disintegrates rapidly as a result of abrasion once the wind erosion starts.

The soil ridge roughness is a measure of the surface roughness other than that caused by clods or vegetation. Ridges of 2 to 4 in. in height have been found to be the most effective in controlling erosion; the erosion for ridges of this height is about 50% of that over a smooth surface.

The wind-erosion climatic factor includes the influence of wind speed and moisture. The rate of soil movement varies directly as the cube of the wind velocity. In this factor the mean annual wind velocity, corrected to a standard height of 30 ft, is used. Since atmospheric wind velocities are normally distributed, the probability of obtaining high winds is higher with higher mean velocities. The rate of soil movement varies approximately as the square of effective soil moisture. The wind-erosion climatic factor has been given for a number of locations by the U. S. Department of Agriculture (1968) for each month of the year. To illustrate the differences in this factor from one place to another, I have given rough ranges for four locations. For the State of Washington, the climatic index varies from about 1 to 50; the highest values are for the months of March through May. For eastern New Mexico, along the northeastern border, the climatic factor remains the highest in the nation throughout the year; values range from about 70 to 300. For the State of Ohio, the factor ranges from 1 to 10, and for the State of Georgia, about 1 to 5. Thus there are widely differing wind and moisture factors throughout the country; the east, in particular, has low factors as compared with the west. It could be predicted that the resuspension of contaminants from the soil will also be lower in the areas of low climatic factor for erosion.

The field-length factor again has two parts: the distance across the field and the sheltered distance. The distance across the field is measured along the prevailing wind-erosion direction. On an unprotected field, the rate of soil flow is zero on the windward edge and increases with distance downwind until, for a large field, the flow reaches a maximum that the wind of a given velocity can maintain. The sheltered distance is that distance along the prevailing wind-erosion direction that is sheltered by any barrier.

Vegetative cover is an important factor in controlling wind erosion. Three different factors are included in the equivalent quantity of vegetative cover. The first is the quantity of the vegetative cover expressed as clean, air-dried residue. The second denotes the total cross-sectional area of the vegetative material. The finer the material and the greater the surface area, the more it reduces the wind velocity and the more it reduces wind erosion. The third factor is the orientation of the cover. The more erect the vegetation, the higher it stands above the ground, the more it reduces wind velocity near the ground, and the lower the erosion.

Gillette and his collaborators (1972; 1973; 1974; 1976) have been studying the vertical flux of dust from agricultural soils under the influence of the winds. In a study of particle size distribution at 1.5 and 6.0 m along with the horizontal velocity, it was concluded that the upper limit of the ratio of the settling velocity to the friction velocity for aerosols having the potential for long-range transport is approximately 0.2 or slightly greater (Gillette and Blifford, 1974). Also, by studying the sizes of the aerosols produced at different values of the friction velocity, it was concluded that the dominant injection mechanism of soil aggregates less than 10 μm in diameter is not direct aerodynamic entrainment (Gillette and Blifford, 1974). It was noted that the size distribution of the vertical flux [expressed as dN/d (log r)] was proportional to r^2 for particle sizes greater than 2 μm. A similar particle size distribution was noted for the parent soil when the size distribution was measured with liquid Freon dispersal. Since the dielectric constant of liquid Freon is close to that of air, it tends to preserve the aggregate state of the soil. However, if the aggregates are broken up, the number of particles less than 10 μm greatly increases (Gillette and Blifford, 1974), showing once again the importance of aggregation of the smaller particles in the soil.

In a study of the vertical flux above several soils, Gillette (1974) noted that the production of a flux of particles less than 10 μm in size increased more rapidly with friction velocity on a soil that had a higher percentage of silt and clay than it did on a soil with a relatively low percentage of these materials. On both soils extrapolation of the curves to the point of intersection indicates a threshold friction velocity of about 0.18 m/sec, about the same value as that for the impact threshold discussed earlier. For the soils containing 3.5% clay (<1 μm) and either 0.5 or 1.0% silt (<25 μm), the vertical flux of particles <1 μm increased as the ratio of the friction velocity to the threshold friction velocity raised to the 5.14 power. For the soil containing 3.5% clay and 1% silt, the vertical flux increased as the 9.62 power of the ratio of friction velocity to the threshold friction velocity. This provides a model for the vertical flux of

$$F_A = \text{Const.} \ (u_*/u_{*\text{threshold}}) \qquad (6)$$

For the two soils evaluated, the constant at the point of intersection was about 6×10^{11} cm^3 sec^{-1} cm^{-2}, or, assuming an aggregate density of 2.5 g/cm^3, about 1.5×10^{-6} g sec^{-1} m^{-2}. An important conclusion from these studies was that the increase in the vertical flux at powers of the friction velocity much greater than the observed cube for the horizontal flux was due to the breakup of the aggregates by sandblasting and release of the smaller particles to suspension.

In a further study of the vertical flux resulting from eight different soils (Gillette, 1976), a regular pattern of increase with friction velocity appeared to exist with considerable spread. A curve that I fitted by eye to the bulk of the data indicated an average increase of vertical flux as the 5th power of the friction velocity with a constant at the assumed threshold friction velocity as given by the earlier study of two soils. A comparison of the size distribution of a parent sandy soil with the size distribution of the horizontal flux to a height of 1.3 cm indicated that the size distribution of the horizontal flux is very similar to the size distribution of the parent soil aggregates. The aerosol at 1 m in height showed a mode for the particles greater than 20 μm centered around 50 μm. As height increased, the particles less than 20 μm became an increasingly larger proportion of the total aerosol. It was also noted that the concentration of larger particles increased with the wind speed.

Travis (1975; 1976) has developed a model for the redistribution of wind-eroding soil-contaminant mixtures using the Gillette and Blifford (1972) and Gillette, Blifford, and Fryrear (1973) relationships for the vertical and horizontal flux. This model assumes that the contaminant is closely associated with the soil aggregates and moves in the same manner as the soil. Since it incorporates only studies from eroding agricultural soils, it should be limited to these soil conditions.

Shinn et al. (1976) measured the vertical profile of dust in the atmosphere at two sites and related these profiles to the vertical dust flux by the eddy-correlation method. The two sites were an area at NTS where plutonium contamination had occurred during a series of nonnuclear tests over 20 yr earlier (GMX Area) and in an agricultural field in west Texas. Measurements were made simultaneousiy of the dust flux by an optical particle detector and the mean wind and temperature profiles.

It was noted that the mass-concentration particle size distributions at both sites had a maximum at about 4 or 5 μm mass median aerodynamic diameter, and this decreased by an order of magnitude at 1 and 10 μm. This distribution does not agree with the data of Chepil and Woodruff (1957) or Sehmel (1976a), which show significant quantities of particles much greater than 10 μm at elevations up to 30 m above the ground. Shinn et al. (1976) indicate that this could be due to reaggregation after sampling because the sedimentation velocity of such particles would be greater than u_*. However, this remains an uncertainty that requires investigation.

The wind-profile measurements showed the roughness length (z_0) at the Texas site to be 0.044 cm and at the GMX site to be 2.0 cm. This resulted in a drag coefficient referenced to the wind speed at 2 m (u_*/u_{200}) of 0.05 at the Texas site and of 0.10 at the GMX site. It was noted that the dust profiles in this study, as well as in previous ones, fit a power law with exponents of either -0.2 or -0.35. This is due to the fact that all are nearly bare surfaces and the measurements were made in dynamic neutral atmospheric stability conditions. The dust flux calculation was parameterized by several simple relations.

$$F = K \frac{d\chi}{dz} \tag{7}$$

where F = flux
    Z = height
    χ = dust concentration
    K = eddy diffusivity

Since

$$K = u_* kz \tag{8}$$

where k is the van Kármán constant (k = 0.4). Since the dust concentration follows a power-law distribution with height,

$$\frac{d\chi}{dz} = P \frac{\chi}{z}$$

Since the power, P, is about 3 and the concentration over the heights from 0.7 to 2 m deviates only about 20% from the 1-m reference velocity, one obtains, by combining the above,

$$F \approx -0.12 \, u_* \chi_1 \tag{10}$$

The data at GMX and Texas, respectively, give values for χ_1 of

$$\chi_1 = 6.1 \, u_*^{2.09} \tag{11}$$

and

$$\chi_1 = 522 \, u_*^{6.37} \tag{12}$$

For GMX and Texas, respectively, these values then result in fluxes of

$$F = 0.73 \, u_*^{3.09} \tag{13}$$

and

$$F = 62.6 \, u_*^{7.38} \tag{14}$$

A tentative model of the upward dust flux was derived from profile and soil erosion data from a number of locations. This was titled the Gillette–Shinn model and is expressed by

$$F = -F_0 \left(\frac{u_*}{u_0} \right)^{\gamma+1} \tag{15}$$

where F is the flux, the reference wind speed is 1 m/sec, F_0 is a reference dust flux at $u_* = 0$, and γ is the power in the dust profile. From a series of experiments by Gillette and Goodwin (1974), a tentative relationship between the parameters in the Gillette– Shinn model and the soil-erosion index was derived (Fig. 2).

An initial attempt to assess resuspension from the ground was made by Healy and Fuquay (1958) and Healy (1974) using data from Hilst (1955) and Hilst and Nickola (1959) who were using zinc sulfide (ZnS) particles to estimate the effects of various surfaces on wind erosion. At this time the high rate of increase with wind speed due to breakup of aggregates was not known, and it was assumed that the rate increased as the square of the wind speed. Later, the rates were converted to a cube relation with the wind speed (Healy, 1977a), although it was noted that for these results the square appeared to give less variance in the data. This indicated a pickup rate of $5 \times 10^{-9} \, u_*^3$ per second. Since the wind speeds were measured at a 2-m height (Hilst, 1955) and the Hanford area, the site of the experiments, has a z_0 of about 2 cm, this value would be about 4×10^{-11} u_*^3 per second. This can be only a preliminary estimate, however, because of the alteration of the natural roughness feature by the change in courses.

Sehmel (1977b; 1977c) and Sehmel and Lloyd (1976a) have studied the resuspension of a tracer, submicrometer calcium molybdate, sprayed as a suspension over a lightly vegetated area with a roughness height of 3.4 cm. The area sprayed was within a circle of 29-m radius with a sampling tower in the middle. The average surface concentration on

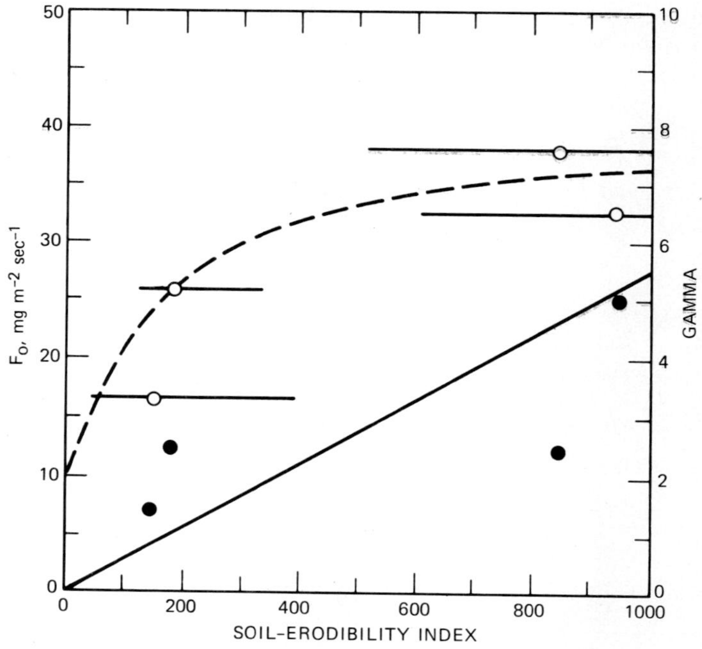

Fig. 2 Tentative relation for parameterization of dust flux for the Gillette—Shinn model.

the ground was 0.62 g of molybdenum per square meter. Measurements were made with a cowled impactor, which always faced into the wind, at heights to 6.1 m; some measurements were made at specific wind speeds as measured at a height of 2.1 m above the ground. Resuspension rates were calculated from a mass balance calculated from the profile. Those particles depositing in the cowl were considered as "nonrespirable," and those entering the impactor were considered as "respirable." The impactor separated the respirable particles into nominal diameters for unit-density spheres of 7, 3.3, 2.0, and 1.1 μm and smaller particles on the backup filters.

In these experiments the resuspension rates ranged from about 10^{-11} to 10^{-7} sec^{-1} (Sehmel, 1977a). Plots of all the data showed that the resuspension rates increased with the 1.0 to 4.8 power of the wind speed. However, these results included wind intervals with wide speed increments; thus an uncertainty as to the actual wind speed to be used existed. When these intervals were eliminated, the resuspension rates for all sizes in the impactor and the backup filter increased with wind speed to the 4.8 power. It may be noted, however, that these experiments were run over a finite period of time; so differences in stability would occur. This would cause differences in the wind profile and a change in u_*. Thus, if the majority of the lower speed winds were in the stable condition, where u_* would be lower than for the neutral condition, and if the majority of the higher speed winds were in the unstable condition, where u_* was greater than for the neutral condition, an exaggeration of the slope of the curve would occur. Since lower wind speeds frequently occur in the stable condition and higher wind speeds in the unstable condition, it can be postulated that such an effect has influenced these relationships, albeit to an unknown degree. It is noted, however, that dust loadings by the same technique appear to increase as the 2.9 power of the wind speed for the higher wind speeds (Sehmel, 1977c).

The data from the backup filters, which showed an increase as the 4.8 power of the wind speed and for which both the small interval and wide interval of sampling fitted the curve, gave a fit to

$$RR = 1.96 \times 10^{-13} \, \bar{u}^{4.82} \tag{16}$$

where RR is the resuspension rate. If one assumes that the logarithmic wind profile existed throughout the period, this becomes

$$RR = 2 \times 10^{-16} \, u_*^{4.82} \tag{17}$$

when the threshold velocity is ignored.

Sehmel (1972) also measured the resuspension rate of zinc sulfide tracer particles from an asphalt surface. Average resuspension rates were determined to range from 5×10^{-9} to 6×10^{-8} sec^{-1} for average wind speeds from 2 to 9 mph (0.9 to 4.2 m/sec). The dependency of resuspension rate on wind speed was not determined, but there was some indication that wind gusts greater than about 15 mph (7 m/sec) rapidly suspended particles.

In a similar experiment but in an area of cheatgrass, a resuspension rate of 3.4×10^{-8} sec^{-1} occurred in an area of surface roughness of 4 cm and a friction velocity of 0.52 m/sec (Sehmel, 1976b). After truck traffic that removed 0.35% of the ZnS, the surface roughness was reduced to 3 cm, and a resuspension rate of 1.25×10^{-9} sec^{-1} was measured in a friction velocity of 0.51 m/sec.

Sehmel (1975) measured the resuspension rate of 10-μm-diameter uranine particles deposited on the inner surface of an aluminum tube with a 2.93-cm inside diameter. The resuspension rates ranged from 10^{-6} to 10^{-3} sec^{-1} and were dependent on airflow rates and resuspension time. Orgill, Petersen, and Sehmel (1976) measured the resuspension of DDT from forests in the Pacific northwest. The DDT was sprayed in the early morning with low wind speeds, and sampling was conducted by an aircraft-mounted sampler for 5 days. Calculated resuspension rates on three of the sampling days were 1.0×10^{-8}, 2.5×10^{-8}, and 7.7×10^{-8}. Sehmel (1976b; 1977c) proposed a correlation between resuspension rate and the roughness height z_0 using the data from the experiments on the aluminum tube, the ZnS from the asphalt surface, the molybdenum tracer from desert soil, and the DDT from the forest. This curve shows a decrease in resuspension rate from the aluminum-tube data to the molybdenum tracer with the asphalt surface falling on the lines between these two points. However, the DDT from the forest, with a large roughness height, had an increase by 2.5 orders of magnitude from the soil data. It is not clear how these resuspension-rate data were corrected for the differing wind speeds or values of u_* that existed in each of the experiments. It is further noted that the surfaces and, possibly, mechanisms of wind pickup from the surface were different. For example, the resuspension of DDT from the forest could have been primarily a result of the mechanical movement of leaves, needles, and branches rather than the types of force found on the soil surface.

There have been many measurements of the air concentration in or near an area contaminated with radioactive materials, but most of these are not suitable for estimating resuspension rates or dependency on wind speed because of the lack of detailed meteorological data at the time of the measurement or the lack of a detailed knowledge of the configuration and contamination level of the source. Anspaugh et al. (1975; 1976) studied the resuspension of plutonium at the GMX Area in Nevada using an ultrahigh flow-rate sampler so that samples could be obtained in a relatively short time while meteorological conditions were relatively consistent. The resuspension rates measured ranged from 2.7×10^{-12} to 4.8×10^{-10} sec^{-1}. There was a strong correlation between the resuspension rate and u_*^3. Dividing the resuspension rate by u_*^3 greatly reduced the variance found in the resuspension rate itself. This result is consistent with Shinn's value for the dependence of dust flux on u_* at this location. The value of the normalized ratio to u_*^3 ranged from 1.5×10^{-11} to 10^{-9}. It is noted that the pattern of plutonium in the soil used for these derivations was determined by field instrument for the determination of low-energy radiation (FIDLER) measurements of the ^{241}Am gamma (Healy, 1974) converted to plutonium concentrations by statistical comparison of the plutonium and americium content of soil samples (Eberhardt and Gilbert, 1972). This should provide an estimate of the plutonium in the surface layers of the soil; so correction for plutonium that has migrated to some depth is not required.

The dust samples used by Shinn et al. (1976) in parameterization of the dust flux in this area were also analyzed for plutonium (Anspaugh et al., 1976). The profile of plutonium concentration at a distance of 100 m from ground zero (GZ) agreed with the shape of the dust flux to a height of about 1 m and then showed lower concentrations than expected. At the 730-m point, the deviation is smaller. The deviations were as would be expected for a limited source with rather abrupt discontinuities. However, the close fit of the plutonium concentrations to the dust profile at the lower two heights indicates that the plutonium concentration at a height less than 1 m is closely coupled to the ground concentration, even though the soil contamination is less by two orders of magnitude at the greater distance downwind.

Saltation fluxes were also measured at the GMX Area (Anspaugh et al., 1976). Values ranged from 3×10^{-7} to 8×10^{-6} g cm^{-2} sec^{-1}, which are 10^{-3} to 10^{-4} of those measured by Chepil (1945a) for wind-eroded fields. In this connection Shinn (1977) points out that NTS, the location of the GMX Area, is unique in that the natural resuspension rate owing to convective winds is very low compared with more erodible sites in the western United States. He has concluded that the natural desert shrub land, covered by a "desert pavement," or the dry lakes, covered by a crust after a rain, are not subject to wind erosion unless they are physically disturbed.

Sehmel (1977a; 1977b) has measured the resuspension of plutonium at Rocky Flats and of plutonium and cesium at the Hanford plant. However, the source areas are poorly characterized; so resuspension rates cannot be estimated. Various values of the power of the increase of concentration with wind speed ranging from unity to 9.3 have been obtained in these experiments; so it is difficult to draw conclusions from these data.

Mechanical Resuspension. Mechanical resuspension is that caused by forces other than the wind. Such forces could range from the movement of small animals on the surface, through humans walking, to the movement of heavy equipment or plows across the ground. There are several differences between mechanical resuspension and wind resuspension, chief of which is the fact that the resuspending force is independent of the wind speed (although dilution downwind will increase at higher wind speeds). Instead, the resuspension rate will depend on the magnitude of the force applied as well, perhaps, as on the nature of the force. A second difference is the depth from which resuspension can possibly occur. In the case of wind resuspension, the layer from which particles can contribute to the airstream is limited by the depth to which the saltating particles can cause ejection. For mechanical disturbance the depths over which the forces can be applied varies with the means of disturbance but could reach depths of 1 or 2 ft in plowing. Of course, the probability of resuspension is not the same at all depths, but no data are available to indicate possible variations. The extreme example would be the excavation of a hole, such as a basement, where material could be ejected into the air from considerable depths in the ground.

Another difference arises from the fact that most mechanical disturbances are a point source; i.e., the disturbance occurs over a fairly limited area at any one time. There could be multiple disturbances that could result in an approximate area source or the disturbance could move with time, which would result in an average that resembled a line source or an area source. An example of the line source would be traffic moving along a road. The average resuspension rate from the road would be the product of the resuspension per vehicle times the number of vehicles divided by the time over which this number of vehicles passed. An area source would be the average result of a farmer plowing a field and producing a resuspension rate at each point. Here the average resuspension rate would be the instantaneous resuspension rate that occurs at each point divided by the time required to plow the field. Such relationships allow the derivation of calculational methods for finding the average concentration downwind if the resuspension rate from the disturbance can be defined.

There are also similarities with wind resuspension. One would expect the dust flux to be greater in fields that contain a larger quantity of small aggregates. Many of the factors, such as moisture or vegetative cover that inhibit erosion, would be expected to minimize mechanical disturbance. Thus, when digging in contaminated soil, it is common practice to keep the soil damp to minimize resuspension. However, local areas of low saltation (vegetated strips in the field) will not affect the mechanical-resuspension rate.

A few measurements of mechanical resuspension can be used to give an order-of-magnitude estimate of the rate of resuspension under different conditions. Sehmel (1972) measured the resuspension caused by an individual walking along a 50-ft length of asphalt 10 ft wide that had been previously seeded with zinc sulfide tracer. He reports that 1×10^{-5} to 7×10^{-4} (at wind speeds of 3 to 18 mph) of the tracer was resuspended per walk-through. Assuming a walking speed of 3 mph, this would result in resuspension rates of 9×10^{-7} to 9×10^{-6} sec^{-1}. Such values have much uncertainty because of the width of the seeded area, but it is noted that they are about two orders of magnitude greater than the wind-resuspension rates with wind speeds of 2 to 9 mph.

In a continuation of these same experiments, Sehmel (1973) reports the results of driving vehicles in the adjacent lane and through the tracer material. Both a car and a three-quarter-ton truck were used. His values, reported as fractional resuspension per pass, were converted to a resuspension rate through use of the length of the seeded area and the speed of the vehicle. The resuspension rate varied from 10^{-8} to 8×10^{-3} sec^{-1}, depending primarily on the speed of the vehicle. Sehmel (1973) had reported that the resuspension was proportional to the square of the speed. The resuspension rate caused by the truck was greater than that caused by the car, presumably because of the greater turbulence from the truck. A rapid weathering of the particles was also noted. In this calculation it was assumed that there was no removal by the winds during the 30 days of the experiment; so the latter rates are the lower limit of the resuspension rates. However, the resuspension rate 30 days after application was two to three orders of magnitude lower than the initial rate. These data indicate that there will be rapid depletion of the source for materials deposited on such pavements if any significant traffic occurs.

Sehmel (1976b) also performed a similar experiment with the zinc sulfide tracer placed on a strip of cheatgrass. The course was the same size as the asphalt course. A $\frac{3}{4}$-ton truck was driven through the area at different speeds. The results indicated a relatively high resuspension rate of 3.7×10^{-5} sec^{-1} on the first pass at 2.2 m/sec. For the second pass at 6.7 m/sec, the resuspension rate had decreased to 4×10^{-6} sec^{-1}, whereas for the third pass at 13.4 m/sec, the rate increased to 8×10^{-6} sec^{-1}. With the final pass at 17.9 m/sec, the resuspension rate increased to its highest value of 9.4×10^{-5} sec^{-1}. The high resuspension rate at the low speed was caused by removing the most readily suspendible tracer from the cheatgrass. By the time the higher speeds were attained, it is likely that this more readily suspendible material was removed and the resuspension was from the soil surface.

Milham et al. (1976) described the results of air sampling during the agricultural preparation of two fields having small concentrations of plutonium accumulated 25 to 30 yr earlier as the result of a release from a nearby stack. Samples were taken at several locations during operations in the field. Healy (1977a) converted these to approximate resuspension rates by use of the field sizes and meteorological parameters given in the paper. These results are given in Table 3.

TABLE 3   Resuspension Rates from Agricultural Operations*

	Estimated resuspension rates, sec			
	North field		South field	
	7.6 m	30.5 m	7.6 m	30.5 m
Milham et al. (1976)				
Bush hogging	9×10^{-8}	2×10^{-7}	1×10^{-6}	8×10^{-8}
Disking	4×10^{-8}	6×10^{-8}		
Subsoiling	7×10^{-7}	3×10^{-8}	3×10^{-8}	3×10^{-9}
Fertilizing	2×10^{-8}	3×10^{-8}	1×10^{-8}	1×10^{-8}
Planting	1×10^{-6}	4×10^{-8}	6×10^{-7}	2×10^{-8}
Myers et al. (1976)				
Rototilling	9×10^{-8}			

*From Healy (1977a).

A somewhat similar experiment was performed by Myers et al. (1976). Here the plutonium was applied to a small field in the form of digested sewage-plant sludge containing a small amount of plutonium. The sludge was allowed to dry for 4 weeks without rain, and the area was rototilled. The rototiller was 2 m wide with the dust cover removed and was pulled behind a tractor. The sampling results were, again, converted to an approximate resuspension rate by Healy (1977a), and the result is given in Table 3.

Mass Loading

The mass-loading concept is an attempt to bypass the details of the soil characteristics and the resuspension process and to relate directly measured soil concentrations of the contaminant to the air concentration by use of the mass of soil particulates in the air.

Thus the air concentration of the contaminant is given by the product of the concentration of the contaminant in the soil and the concentration of the soil particulates in the air. If the quantity of particulates in the air is known from other data, one need, theoretically, only measure the soils in the region to provide an estimate of the air concentration of the contaminant.

Two parameters, the dust loading in the atmosphere and the appropriate concentration of the contaminant in the soil, are needed to provide estimates by this method. Healy (1974) used an average value of 120 μg/m^3 of dust in a generic analysis of limits for plutonium in the soil. This was derived from the Federal Secondary Standard for particulates in the air expressed as a geometric mean of 60 μg/m^3 assuming a geometric standard deviation of 2. Anspaugh (1974) explored a reasonable mass loading in several ways. The lower bound is quoted as about 10 μg/m^3. Examination of the data on the levels in mine atmospheres which have led to a considerable prevalence of pneumoconiosis in the workers indicates that standards on the range of 1 to 10 mg/m^3 have a very small, if any, margin of safety. Anspaugh (1974) also quotes some British data which indicate that dust levels in excess of 1 mg/m^3 could lead to considerable public health problems. He also used the data on ambient mass loading for 1966 from the National Air Surveillance Network to show that the average for urban stations ranged from 33 to 254 μg/m^3 with a mean for all nonurban locations of 38 μg/m^3. For the nonurban stations the average ranged from 9 to 79 μg/m^3. From these studies he chose an average of 100 μg/m^3 as reasonable for predictive purposes (Anspaugh, Shinn, and Wilson, 1974; Anspaugh et al., 1975). The U. S. Environmental Protection Agency (1977) also examined the data from the nonurban stations from the National Air Surveillance Network for the years of 1964 and 1965. Their map of these data indicates values ranging from 9 μg/m^3 in southern Montana to 56 μg/m^3 in western Pennsylvania, 57 μg/m^3 on the southern Oregon coast, and 59 μg/m^3 on the North Carolina coast. However, the prevalence of high values in the east would indicate the possible inclusion of industrial particulates in these samples. The U. S. Environmental Protection Agency used a value of 100 μg/m^3 in the calculation of their screening level.

Several uncertainties appear in the use of the data from the National Air Surveillance Network. The first was pointed out above in that the particulates that are collected can include a portion of those generated by industrial operations; so the values could be high. The second problem arises from the fact that the samplers are frequently in positions, such as on top of buildings; so they do not measure the air actually breathed by people. Associated with this question is the potential for people engaged in various activities to generate their own dust. This would result in local concentrations in excess of the ambient value measured by the network. However, the 100 μg/m^3 value still appears reasonable from the standpoint that it is an average over a full year, and people do not work or play in dusty operations all the time. For example, if we assume that an individual spends 8 hr per day, 5 days per week, 50 weeks per year in a concentration five times the maximum value noted in the ambient air measurements, or 300 μg/m^3, the average concentration through the year would be only 115 μg/m^3. Although some individuals, such as farmers, work longer hours during the week, their exposure to dust is limited to fewer weeks per year, and a portion of their time in the field is during periods of high moisture or vegetation in the soil when dusty conditions are limited.

Associated with the question of the concentration of soil particles in the air is the question of the origin of the particles. Once airborne, the smaller particles can travel very long distances. For example, Carlson and Prospero (1972) have reported the movement of dust from the Sahara desert over the northern equatorial Atlantic Ocean, and Clayton et al. (1972) have reported evidence of transport across the Pacific Ocean to Hawaii. Since most contaminated areas are relatively small in area, one would expect that only a fraction of the dust in the air would originate from the area. Because of preferential deposition of the larger particles from sources some distance away, this "background dust" would contain a higher percentage of the smaller particles that are more readily deposited in the lung than the dust originating from the local, contaminated area. Thus Anspaugh and Phelps (1974) report that measurements at the GMX area with Anderson high-volume cascade impactors for about 1 month indicate that the mass distribution of sizes is about 1.6 μm MMAD with a geometric standard deviation (σ_g) of about 15, and the plutonium and ^{241}Am had an activity median aerodynamic diameter of about 3 μm with a σ_g of about 7. It was also noted that the average activity of the soil was about one-third that found in the soil in close proximity to the sampler. It is noted that even higher activity was upwind.

From this we conclude that a direct comparison of the size distribution of contaminated particles in the air with those in the soil is probably valid only for very large areas. For the more usual size of contaminated area, the dilution of the total mass in the air, particularly in the smaller particle sizes, could be significant. It is noted, however, that, for resuspension by mechanical disturbance, this dilution may be of lower importance because of the frequently higher concentrations resulting from such disturbances.

The second question, that of the appropriate concentration of the contaminant in the soil, is more subtle. As was discussed earlier, soil particles that are carried in suspension are the smaller ones because the larger ones will settle rapidly. Thus, if the concentration of the contaminant in the soil fraction containing the small particles is greatly different from that in the other particle sizes, it would appear that the concentration predicted by the mass-loading approach using the total soil concentration would theoretically be low.

Tamura (1977) has analyzed the particle sizes and their associated plutonium content in samples from several existing plutonium-contaminated areas and has shown that fractionation of the plutonium content by particle size does exist. Analyses were done using water as the suspending agent, and the effect of this, as compared with the carbon tetrachloride used by Chepil or the liquid Freon used by Gillette, on the aggregate size is unknown. However, in two samples from the NTS, the aggregates less than 20 μm had plutonium concentration three and five times greater than the total soil mass. (At the NTS the bulk of the activity appears to be in the 20- to 53-μm size range.) In a bottom sediment from the canal at Mound Laboratory, the concentration in the fraction lower than 20 μm was 1.8 times as high as the total; at the floodplain at ORNL, the soil concentration in the fraction lower than 20 μm was 1.1 times as high as the total; and, in a sample from Rocky Flats, the soil concentration in the fraction less than 20 μm was about 3 times as high as the total. It is of interest that these distributions reflect both the method of contamination and the soil type. At the NTS the plutonium was mechanically dispersed by explosive material, and the particle size distribution reflects the largest amount of the plutonium in the 53- to 125-μm size range, although the highest concentration was in the smaller particle sizes. The Mound Laboratory and ORNL samples reflect the distribution expected by adsorption on the smaller particles in the sample, whereas Rocky Flats is intermediate, which reflects, perhaps, some adsorption as well as direct contamination of the larger soil particles by the plutonium-bearing oil that was the source of the contamination.

The U. S. Environmental Protection Agency (1977) proposed the use of an "enrichment factor" to include these data in resuspension calculations. This is defined as the summation of the products of g_1, the ratio of the fraction of the total activity contained within the size increment i to the fraction of the total mass in the size, and f, the fraction of the airborne mass within each increment of particle size in the air. For the distribution in soil sizes at Rocky Flats, they calculate an enrichment factor of 1.5. Tamura (1977) has defined a "soil plutonium index" which accounts for the size distribution as well as the lung deposition. This is given as

$$SI = SA \times LD \times RA \tag{18}$$

where SI = soil plutonium index
    SA = soil activity factor
    LD = lung deposition factor
    RA = resuspendible activity factor

The soil activity factor is the fraction of the activity in a given mass fraction divided by the mass fraction for particles less than 100 μm. (Tamura used 125 μm in evaluating this factor because this was the sieve size used in his analysis.) Values of this factor range from 3.14 for the ORNL sample to 7.27 for the Rocky Flats sample. The lung deposition factor is the deposition in the pulmonary region as defined by the International Commission on Radiological Protection (1966). The final factor, the resuspendible activity factor, is the fraction of the total soil plutonium index activity in the resuspendible fraction. Indexes derived from available data give 0.52 for Area 13 at NTS, 1.26 for Rocky Flats, 1.18 for Mound Laboratory, and 0.69 for ORNL.

Another approach to the use of the smaller particles is that of Johnson, Tidball, and Severson (1970). Their sampling technique was to brush the surface dust into a container. The 5-μm or smaller particle sizes were then separated from the sample after aggregates had been broken up, and plutonium analyses were performed on this fraction. They found that the concentration in these small particles was 4 to about 300 times as large as that in similar samples taken to a depth of $\frac{1}{8}$ in. Their conclusion was that these results provided a better indication of the hazard than the conventional sample, although they did not explore mechanisms of breaking down the aggregates found in the soil, which severely limit the quantities of particles of this size found in natural soils, nor did they examine pathways of this material to man.

These relations have never been actually tested to show their validity. The work on soil erosion indicates the many additional factors that will influence wind erosion and resuspension. These include the soil texture, the moisture content of the soil, the presence or absence of vegetation in vegetative residues, and the characteristic surface roughness. In the case of NTS, the desert pavement undoubtedly has more influence on either wind or mechanical resuspension than the other factors. We would believe that wind erosion, in particular, is more complex than these relations would indicate. However, it is possible that such concepts may be more applicable to mechanical disturbance.

A direct test of the mass-loading technique has been made by Anspaugh, Shinn, and Wilson (1974) and Anspaugh et al. (1975). The measured concentrations of a number of

TABLE 4   Comparison of the Predicted Concentration in Air Using a Mass Loading of 100 μg/m^3 with Measured Concentrations*

Location	Nuclide	Air concentration	
		Predicted	Measured
Nevada Test Site			
NE, Anspaugh and Phelps (1974)	^{239}Pu	7.2×10^{-3} pCi/m^3	6.6×10^{-3} pCi/m^3
GZ, Anspaugh and Phelps (1974)	^{239}Pu	0.12 pCi/m^3	0.023 pCi/m^3
Lawrence Livermore Laboratory			
Gudiksen et al. (1973a)	^{238}U	150 pg/m^3	52 pg/m^3
Gudiksen et al. (1973b)	^{238}U	150 pg/m^3	100 pg/m^3
Silver et al. (1974)	^{238}U	150 pg/m^3	86 pg/m^3
Silver et al. (1974)	^{40}K	10^{-3} pCi/m^3	9.8×10^{-4} pCi/m^3
Argonne National Laboratory			
Sedlet, Golchert, and Duffy (1973)	^{232}Th	320 pg/m^3	240 pg/m^3
Sedlet, Golchert, and Duffy (1973)	Natural uranium	215 pg/m^3	170 pg/m^3
Sutton, England			
Hamilton (1970)	Natural uranium	110 pg/m^3	62 pg/m^3

*Based on data from Anspaugh, Shinn, and Wilson (1974) and Anspaugh et al. (1975).

nuclides in the air were compared with a concentration calculated from the quantity of the nuclide in the soil assuming a mass loading of 100 μg/m^3. These results are given in Table 4.

The agreement between calculated and predicted values is good. Of course, the sources for the natural isotopes are large in area. However, the values at NTS show reasonable agreement between calculation and prediction. It is believed that the soil concentration at the point of sampling was used for the predicted values. If this were the case, the discrepancy between the two results at GZ is explainable on the basis that the concentration in the soil is highest at this point and the measured dust arose from surrounding areas of lower concentration.

Discussion

This review covered concepts and numerical values related to the resuspension problem and did not include the important conceptual and modeling studies that have been carried out by several individuals, including Amato (1971), Trevino (1972), Horst (1976), and Slinn (1978). This was done deliberately in order to focus on the nonmathematical aspects of the problem and to attempt to bring the factors of importance into focus.

It is apparent that a gratifying amount of progress has been made on the determination of resuspension in the past few years. The studies by Anspaugh, Shinn, and Wilson (1974) and Anspaugh et al. (1975) at NTS have shown the feasibility of measurement of the resuspension rate in a contaminated area, and their application of the mass-loading concept has added greatly to the understanding of this model. The work of Gillette (1974; 1976) and Shinn et al. (1976) on the dust flux has given new insights into methodology and the phenomena concerned. The studies by Sehmel (1977b; 1977c), with the tracer particles, have given values that are extremely useful for application. Slinn (1978) has provided parameterization concepts that aid in understanding.

This is not to say that additional work is not needed. Further studies of both the dust flux and resuspension rate at contaminated areas in various regions and types of soil are definitely needed along with models, such as those of Gillette (1974) and Shinn et al. (1976), which provide relationships between the resuspension and readily measured parameters that can be used to estimate resuspension rates. Concurrent studies of the dust flux and resuspension of a contaminant are badly needed, particularly in undisturbed areas apart from agricultural soils. The dust-flux model requires assumptions as to the connection between resuspension of a contaminant and the dust flux. The only checkpoint now available is the measurements at the GMX Area of the dust flux by Shinn et al. (1976) and the plutonium resuspension rate by Anspaugh et al. (1975). In particular, additional data are needed on resuspension by mechanical disturbance. Few appropriate experiments are available, and it is frequently difficult to interpret them in a manner that provides useful results. A particular area of concern for which very few data are available is the possibility of contamination while playing and working in an area with subsequent transfer to a place where inhalation is more probable. Extreme examples of this would be pulling a contaminated garment over one's head or contaminating pillows or other bed clothing. Although one could feel that this could not be a major source of exposure, we cannot tell until the experiments are done.

A primary purpose of this chapter is to choose resuspension parameters to be used in the calculation of the dose to individuals in a contaminated area. The study has reinforced our previous prejudice that the resuspension factor is not the method for use because of the failure of this method to account for many of the variables and because the conditions of the measurement are seldom described in sufficient detail to allow intelligent extrapolation to areas different from those in which the measurements were made. There is some usefulness to this technique, however, in describing the exposure of the individual causing the disturbance.

The resuspension rate has been our favorite method because of the capability of integrating over a contaminated area using accepted dispersion and deposition parameters to provide concentration isopleths around the area. For a specific situation in which the soil and meteorological parameters can be defined, this is still the preferred method, and the state of the art is rapidly approaching sufficient detail that this can be done.

However, for a generic study the mass-loading approach seems to be best. The work of Tamura (1977), the U. S. Environmental Protection Agency (1977), and Johnson, Tidball, and Severson (1970) all indicate that even this approach requires revision for the distribution of contamination in the soil. However, as has been pointed out, there are factors that tend to compensate for this, such as the size of the area, and the magnitude of the correction factor proposed by the U. S. Environmental Protection Agency (1977) and Tamura (1977) is only on the order of 1.5 to 2, a value that tends to get lost in the noise of the other uncertainties. In addition, the success shown by Anspaugh (1974) in predicting the concentrations of several nuclides in widely different climates and soil types is encouraging.

Anspaugh (1974) used a mass loading of 100 μg/m^3 in his comparison. The measured values were primarily ambient air and included no component for mechanical disturbance. In view of the agreement found in his study, we would propose the use of 200 μg/m^3 for generic studies to make allowance for these other types of exposure. This seems to be unrealistically high when compared with air-sampling results. This may well be due to the factors proposed by Tamura, with the actual mass loading of little importance as compared with the correlation found by Anspaugh using an arbitrary value of 100 $\mu g/m^3$.

References

Amato, A. J., 1971, *A Mathematical Analysis of the Effects of Wind on Redistribution of Surface Contamination*, USAEC Report WASH-1187, NTIS.

Anspaugh, L. R., 1974, The Use of NTS Data and Experience to Predict Air Concentrations of Plutonium Due to Resuspension on the Enewetak Atoll, in *The Dynamics of Plutonium in Desert Environments*, P. B. Dunaway and M. G. White (Eds.), USAEC Report NVO-142, Nevada Operations Office, NTIS.

——, and P. L. Phelps, 1974, Resuspension Element Status Report: VI. Results and Data Analysis, in *The Dynamics of Plutonium in Desert Environments*, P. B. Dunaway and M. G. White (Eds.), USAEC Report NVO-142, pp. 55-81, Nevada Operations Office, NTIS.

——, P. L. Phelps, N. C. Kennedy, and H. G. Booth, 1973, Wind-Driven Resuspension of Surface Deposited Radioactivity, in *Environmental Behaviour of Radionuclides Released in the Nuclear Industry*, Symposium Proceedings, Aix-en-Provence, 1973, STI/PUB/345, International Atomic Energy Agency, Vienna.

——, P. L. Phelps, N. C. Kennedy, J. H. Shinn, and J. M. Reichman, 1976, Experimental Studies on the Resuspension of Plutonium from Aged Sources at the Nevada Test Site, in *Atmosphere—Surface Exchange of Particulate and Gaseous Pollutants (1974)*, ERDA Symposium Series, No. 38, Richland, Wash., Sept. 4–6, 1974, R. J. Engelmann and G. A. Sehmel (Coordinators), pp. 727-743, CONF-740921, NTIS.

——, J. H. Shinn, P. L. Phelps, and N. C. Kennedy, 1975, Resuspension and Redistribution of Plutonium in Soils, *Health Phys.*, 29: 571-582.

——, J. H. Shinn, and D. W. Wilson, 1974, Evaluation of the Resuspension Pathway Toward Protective Guidelines for Soil Contamination with Radioactivity, in *Population Dose Evaluation and Standards for Man and His Environment*, Symposium Proceedings, Portoroz, 1974, STI/PUB/375, International Atomic Energy Agency, Vienna.

Bagnold, R. A., 1943, *The Physics of Blown Sand and Desert Dunes*, William Morrow and Company, New York.

Bisal, F., and J. Hsieh, 1966, Influence of Moisture on Erodibility of Soil by Wind, *Soil Sci.*, 102: 143-146.

Carlson, T. N., and J. M. Prospero, 1972, The Large-Scale Movement of Saharan Air Outbreaks over the Northern Equatorial Atlantic, *J. Appl. Meteorol.*, 11: 283-297.

Chepil, W. S., 1941, Relation of Wind Erosion to the Dry Aggregate Structure of a Soil, *Sci. Agric.*, 21: 488-507.

——, 1945a, Dynamics of Wind Erosion: I. Nature of Movement of Soil by Wind, *Soil Sci.*, 60: 305-320.

——, 1945b, Dynamics of Wind Erosion: II. Initiation of Soil Movement, *Soil Sci.*, 60: 397-411.

——, 1945c, Dynamics of Wind Erosion: III. The Transport Capacity of the Wind, *Soil Sci.*, 60: 475-480.

——, 1946, Dynamics of Wind Erosion: V. Cumulative Intensity of Soil Drifting Across Eroding Fields, *Soil Sci.*, 61: 257-263.

——, 1951a, Properties of Soil Which Influence Wind Erosion: III. Effect of Apparent Density on Erodibility, *Soil Sci.*, 71: 141-153.

——, 1951b, Properties of Soil Which Influence Wind Erosion: IV. State of Dry Aggregate Structure, *Soil Sci.*, 72: 387-401.

——, 1956, Influence of Moisture on Erodibility by Wind, *Soil Sci. Soc. Am., Proc.*, 29: 602-608.

——, 1957, Sedimentary Characteristics of Dust Storms: III. Composition of Suspended Dust, *Am. J. Sci.*, 255: 206-213.

——, 1960, Conversion of Relative Field Erodibility to Annual Soil Loss by Wind, *Soil Sci. Soc. Am., Proc.*, 24: 143-145.

——, and N. P. Woodruff, 1957, Sedimentary Characteristics of Dust Storms: II. Visibility and Dust Concentration, *Am. J. Sci.*, 255: 104-114.

Clayton, R. N., R. W. Rex, J. K. Syers, and M. L. Jackson, Oxygen Isotope Abundance in Quartz from Pacific Pelagic Sediments, *J. Geophys. Res.*, 77: 3907-3915.

Eberhardt, L. L., and R. O. Gilbert, 1972, Statistical Analysis of Soil Plutonium Studies, Nevada Test Site, USAEC Report BNWL-B-217, Battelle, Pacific Northwest Laboratories, NTIS.

Gillette, D. A., 1974, On the Production of Soil Wind Erosion Aerosols Having the Potential for Long Range Transport, *J. Rech. Atmos.*, VIII(3-4): 735-744.

——, 1976, Production of Fine Dust by Wind Erosion of Soil: Effect of Wind and Soil Texture, in *Atmosphere–Surface Exchange of Particulate and Gaseous Pollutants (1974)*, ERDA Symposium Series, No. 38, Richland, Wash., Sept. 4–6, 1974, R. J. Engelmann and G. A. Sehmel (Coordinators), pp. 591-609, CONF-740921, NTIS.

——, and I. H. Blifford, Jr., 1972, Measurements of Aerosol Size Distributions and Vertical Fluxes of Aerosols on Land Subject to Wind Erosion, *J. Appl. Meteorol.*, 11: 977-987.

——, and I. H. Blifford, Jr., 1974, The Influence of Wind Velocity on the Size Distribution of Aerosols Generated by the Wind Erosion of Soils, *J. Geophys. Res.*, 79(27): 4068-4075.

——, I. H. Blifford, Jr., and D. W. Fryrear, 1973, Studies of Airborne Soil from Wind Erosion, in *Proceedings of the 28th Annual Meeting of the Soil Conservation Society of America*, Sept. 31–Oct. 3, 1973.

——, and P. H. Goodwin, 1974, Microscale Transport of Sand-Sized Soil Aggregates Eroded by Wind, *J. Geophys. Res.*, 79(27): 4080-4084.

Gudiksen, P. H., C. L. Lindeken, C. Gatrousis, and L. R. Anspaugh, 1973a, *Environmental Levels of Radioactivity in the Vicinity of the Lawrence Livermore Laboratory, 1972 Annual Report*, USAEC Report UCRL-51333, Lawrence Livermore Laboratory, NTIS.

——, C. L. Lindeken, J. W. Meadows, and K. O. Hamby, 1973b, *Environmental Levels of Radioactivity in the Vicinity of the Lawrence Livermore Laboratory, 1972 Annual Report*, USAEC Report UCRL-51333, Lawrence Livermore Laboratory, NTIS.

Hamilton, E. I., 1970, Concentration of Uranium in Air From Contrasted Natural Environments, *Health Phys.*, 19: 511-520.

Healy, J. W., 1974, *A Proposed Interim Standard for Plutonium in Soils*, USAEC Report LA-5483-MS, Los Alamos Scientific Laboratory, NTIS.

——, 1977a, *An Examination of the Pathways from Soil to Man for Plutonium*, ERDA Report LA-6741-MS, Los Alamos Scientific Laboratory, NTIS.

——, 1977b, A Review of Resuspension Models, in *Transuranics in Natural Environments*, Symposium Proceedings, Gatlinburg, Tenn., Oct. 5–7, 1976, M. G. White and P. B. Dunaway (Eds.), ERDA Report NVO-178, pp. 211-220, Nevada Operations Office, NTIS.

——, and J. J. Fuquay, 1958, Wind Pickup of Radioactive Particles from the Ground, in *Proceedings of the Second United Nations International Conference on the Peaceful Uses of Atomic Energy, Geneva, 1958*, Vol. 18, pp. 291-295, United Nations, New York.

Hilst, G. R., 1955, *Measurements of Relative Wind Erosion of Small Particles from Various Prepared Surfaces*, USAEC Report HW-39356, Hanford Laboratories, NTIS.

——, and P. W. Nickola, 1959, On the Wind Erosion of Small Particles, *Bull. Am. Meteorol. Soc.*, 40(2): 73-77.

Horst, T. W., 1976, *The Estimation of Air Concentration Due to the Suspension of Surface Contaminants by the Wind*, USAEC Report BNWL-2047, Battelle, Pacific Northwest Laboratories, NTIS.

International Commission on Radiological Protection, Committee II, 1966, Deposition and Retention Models for Internal Dosimetry of the Human Respiratory Tract, *Health Phys.*, 12: 173-207.

Johnson, C. J., R. R. Tidball, and R. C. Severson, 1970, Plutonium Hazard in Respirable Dust on the Surface of Soil, *Science*, 193: 488.

Kathren, R. L., 1968, *Towards Interim Acceptable Surface Contamination Levels for Environmental PuO₂*, USAEC Report BNWL-SA-1510, Battelle, Pacific Northwest Laboratories, NTIS.

Langham, W. H., 1969, *Biological Considerations of Nonnuclear Incidents Involving Nuclear Warheads*, USAEC Report UCRL-50639, Lawrence Radiation Laboratory, NTIS.

Malina, F. J., 1941, Recent Development in Dynamics of Wind Erosion, *Trans. Am. Geophys. Union*, 262-284.

Milham, R. C., J. F. Schubert, J. R. Watts, A. L. Boni, and J. C. Corey, 1976, Measured Plutonium Resuspension and Resulting Dose from Agricultural Operations on an Old Field at the Savannah River Plant in the South Eastern United States, in *Transuranium Nuclides in the Environment*, Symposium Proceedings, San Francisco, Nov. 17–21, 1975, pp. 409-421, STI/PUB/410, International Atomic Energy Agency, Vienna.

Mishima, J., 1964, *A Review of Research on Plutonium Releases During Overheating and Fires*, USAEC Report HW-83668, Hanford Laboratories, NTIS.

Myers, D. S., W. J. Silver, D. G. Coles, K. C. Lamson, D. R. McIntyre, and B. Mendoza, 1976, Evaluation of the Use of Sludge Containing Plutonium as a Soil Conditioner for Food Crops, in *Transuranium Nuclides in the Environment*, Symposium Proceedings, San Francisco, Nov. 17–21, 1975, pp. 311-323, STI/PUB/410, International Atomic Energy Agency, Vienna.

Oksza-Chocimowski, G. V., 1977, *Generalized Model of the Time-Dependent Weathering Half-Life of the Resuspension Factor*, U. S. Environmental Protection Agency, Technical Note ORP/LV-77-4.

Olafson, J. H., and K. R. Larson, 1961, *Plutonium, Its Biology and Environmental Persistence*, USAEC Report UCLA-501, UCLA Laboratory of Nuclear Medicine and Radiation Biology, NTIS.

Orgill, M. M., M. R. Petersen, and G. A. Sehmel, 1976, Some Initial Measurements of DDT Resuspension and Translocation From Pacific Northwest Forests, in *Pacific Northwest Laboratory Annual Report for 1974 to the USAEC Division of Biomedical and Environmental Research*, Part 3, Atmospheric Sciences, USAEC Report BNWL-1950(Pt.3), pp. 231-236, Battelle, Pacific Northwest Laboratories, NTIS.

Prestley, C. H. B., 1959, *Turbulent Transfer in the Lower Atmosphere*, The University of Chicago Press, Chicago.

Sedlet, J., N. W. Golchert, and T. L. Duffy, 1973, *Environmental Monitoring at Argonne National Laboratory, Annual Report for 1972*, USAEC Report ANL-8007, Argonne National Laboratory, NTIS.

Sehmel, G. A., 1972, Tracer Particle Resuspension Caused by Wind Forces upon an Asphalt Surface, in *Pacific Northwest Laboratory Annual Report for 1976 to the USAEC Division of Biology and Medicine*, Part 1, Atmospheric Sciences, USAEC Report BNWL-1651(Pt.1), pp. 136-138, Battelle, Pacific Northwest Laboratories, NTIS.

——, 1973, Particle Resuspension from an Asphalt Road Caused by Car and Truck Traffic, *Atmos. Environ.*, 7: 291-309.

——, 1975, Initial Correlation of Particle Resuspension Rates As a Function of Surface Roughness Height, in *Pacific Northwest Laboratory Annual Report for 1974 to the USAEC Division of Biomedical and Environmental Research*, Part 3, Atmospheric Sciences, USAEC Report BNWL-1950(Pt. 3), pp. 209-212, Battelle, Pacific Northwest Laboratories, NTIS.

——, 1976a, The Influence of Soil Insertion on Atmospheric Particle Size Distributions, in *Pacific Northwest Laboratory Annual Report for 1975 to the USERDA Division of Biomedical and Environmental Research,* Part 3, Atmospheric Sciences, ERDA Report BNWL-2000(Pt. 3), pp. 99-101, Battelle, Pacific Northwest Laboratories, NTIS.

——, 1976b, Particle Resuspension from Truck Traffic in a Cheat Grass Area, in *Pacific Northwest Laboratory Annual Report for 1975 to the USERDA Division of Biomedical and Environmental Research,* Part 3, Atmospheric Sciences, ERDA Report BNWL-2000(Pt. 3), pp. 96-99, Battelle, Pacific Northwest Laboratories, NTIS.

——, 1976c, Particle Resuspension from an Asphalt Road Caused by Car and Truck Traffic, in *Atmosphere–Surface Exchange of Particulate and Gaseous Pollutants (1974)*, ERDA Symposium Series, No. 38, Richland, Wash., Sept. 4–6, 1974, R. J. Engelmann and G. A. Sehmel (Coordinators), pp. 859-882, CONF-740921, NTIS.

——, 1977a, Plutonium and Tracer Particle Resuspension: An Overview of Selected Battelle–Northwest Experiments, in *Transuranics in Natural Environments*, Symposium Proceedings, Gatlinburg, Tenn., Oct. 5–7, 1976, M. G. White and P. B. Dunaway (Eds.), ERDA Report NVO-178, pp. 181-210, Nevada Operations Office, NTIS.

——, 1977b, *Transuranic and Tracer Simulant Resuspension*, ERDA Report BNWL-SA-6236, Battelle, Pacific Northwest Laboratories, NTIS.

——, 1977c, Plutonium and Tracer Particle Resuspension: An Overview of Selected Battelle–Northwest Experiments, in *Transuranics in Natural Environments*, Symposium Proceedings, Gatlinburg, Tenn., Oct. 5–7, 1976, M. G. White and P. B. Dunaway (Eds.), ERDA Report NVO-128, Nevada Operations Office, NTIS.

———, 1977d, *Radioactive Particle Resuspension Research Experiments on the Hanford Reservation*, ERDA Report BNWL-2081, Battelle, Pacific Northwest Laboratories, NTIS.

———, 1978, *Plutonium Concentrations in Airborne Soil at Rocky Flats and Hanford Determined During Resuspension Experiments*, USDOE Report PNL-SA-6720, Battelle, Pacific Northwest Laboratories, NTIS.

———, and F. D. Lloyd, 1976a, Particle Resuspension Rates, in *Atmosphere–Surface Exchange of Particulate and Gaseous Pollutants*, ERDA Symposium Series, No. 38, Richland, Wash., Sept. 4–6, 1974, pp. 846-858, CONF-740921, NTIS.

———, and F. D. Lloyd, 1976b, Resuspension of Plutonium at Rocky Flats, in *Atmosphere–Surface Exchange of Particulate and Gaseous Pollutants (1974)*, ERDA Symposium Series, No. 38, Richland, Wash., Sept. 4–6, 1974, R. J. Engelmann and G. A. Sehmel (Coordinators), pp. 757-779, CONF-740921, NTIS.

———, and M. M. Orgill, 1974, Resuspension Source Charge at Rocky Flats, in *Pacific Northwest Laboratory Annual Report for 1973 to the USAEC Division of Biomedical and Environmental Research*, Part 3, Atmospheric Sciences, USAEC Report BNWL-1850(Pt. 3), pp. 212-214, Battelle, Pacific Northwest Laboratories, NTIS.

Shinn, J. H., 1977, Estimation of Aerosol Plutonium Transport by the Dust-Flux Method: A Perspective on Application of Detailed Data, in *Transuranics in Natural Environments*, Symposium Proceedings, Gatlinburg, Tenn., Oct. 5–7, 1976, M. G. White and P. B. Dunaway (Eds.), ERDA Report NVO-178, Nevada Operations Office, NTIS.

———, N. C. Kennedy, J. S. Koval, B. A. Clegg, and W. M. Porch, 1976, Observations of Dust Flux in the Surface Boundary Layer for Steady and Non-Steady Cases, in *Atmosphere–Surface Exchange of Particulate and Gaseous Pollutants*, ERDA Symposium Series, No. 38, Richland, Wash., Sept. 4–6, 1974, R. J. Engelmann and G. A. Sehmel (Coordinators), pp. 625-637, CONF-740921, NTIS.

Silver, W. J., C. J. Lindeken, J. W. Meadows, W. H. Hutchin, and D. R. McIntyre, 1974, *Environmental Levels of Radioactivity in the Vicinity of the Lawrence Livermore Laboratory*, 1973 Annual Report, USAEC Report UCRL-50547, Lawrence Livermore Laboratory, NTIS.

Slinn, W. G. N., 1978, Parameterizations for Resuspension and for Wet and Dry Deposition of Particles and Gases for use in Radiation Dose Calculations, *Nucl. Saf.*, 19(2): 205-219.

Tamura, T., 1977, Plutonium Association in Soils, in *Transuranics in Natural Environments*, Symposium Proceedings, Gatlinburg, Tenn., Oct. 5–7, 1976, M. G. White and P. B. Dunaway (Eds.), ERDA Report NVO-178, pp. 97-114, Nevada Operations Office, NTIS.

Travis, J. R., 1975, *A Model for Predicting the Redistribution of Particulate Contaminants from Soil Surface*, USAEC Report LA-6035-MS, Los Alamos Scientific Laboratory, NTIS.

———, 1976, A Model for Predicting the Redistribution of Particulate Contaminants from Soil Surfaces, in *Atmospheric–Surface Exchange of Particulate and Gaseous Pollutants (1974)*, ERDA Symposium Series, No. 38, Richland, Wash., Sept. 4–6, 1974, R. J. Engelmann and G. A. Sehmel (Coordinators), pp. 906-944, CONF-740921, NTIS.

Trevino, G., 1972, *On the Natural Distribution of Surface Contamination*, USAEC Report WASH-1225, GPO.

U. S. Department of Agriculture, Agriculture Research Service, 1968, *Wind Erosion Forces in the United States and Their Use in Predicting Soil Loss*, Agriculture Handbook No. 346, GPO.

U. S. Environmental Protection Agency, 1977, *Proposed Guidance on Dose Limits for Persons Exposed to Transuranium Elements in the General Environment*.

Wilson, R. H., R. G. Thomas, and J. N. Stannard, 1960, *Biomedical and Aerosol Studies Associated with a Field Release of Plutonium*, USAEC Report WT-1511, University of Rochester Atomic Energy Project, NTIS.

Woodruff, N. P., and F. H. Siddoway, 1965, A Wind Erosion Equation, *Soil Sci. Soc. Am., Proc.*, 29: 602-608.

Transuranic and Tracer Simulant Resuspension

G. A. SEHMEL

Plutonium resuspension results are summarized for experiments conducted at Rocky Flats, on site on the Hanford reservation, and for winds blowing from off site onto the Hanford reservation near the Prosser barricade boundary. In each case plutonium resuspension was shown by increased airborne plutonium concentrations as a function of either wind speed or as compared with fallout levels. All measured airborne concentrations were below maximum permissible concentrations.

Both plutonium and cesium concentrations on airborne soil were normalized by the quantity of airborne soil sampled. Airborne radionuclide concentrations (in microcuries per gram) were related to published values for radionuclide concentrations on surface soils. For this ratio of radionuclide concentration per gram on airborne soil divided by that for ground-surface soil, there are seven orders of magnitude uncertainty from 10^{-4} to 10^3. This uncertainty in the equality between plutonium concentrations per gram on airborne and surface soils is caused by only a fraction of the collected airborne soil being transported from off site rather than all being resuspended from each study site and also by the great variabilities in surface contamination.

Horizontal plutonium fluxes on airborne nonrespirable soils at all three sites were bracketed within the same three to four orders of magnitude from 10^{-7} to 10^{-3} $\mu Ci\, m^{-2}$ day^{-1} for ^{239}Pu and 10^{-8} to 10^{-5} $\mu Ci\, m^{-2}$ day^{-1} for ^{238}Pu. These represent the entire experimental base for nonrespirable airborne plutonium transport.

Airborne respirable ^{239}Pu concentrations increased with wind speed for a southwest wind direction coming from off site near the Hanford reservation Prosser barricade. Airborne plutonium fluxes on nonrespirable particles had isotopic ratios, $^{240}Pu/^{239+240}Pu$, similar to weapons-grade plutonium rather than to fallout plutonium.

Resuspension rates were summarized for controlled inert-particle-tracer simulant experiments. Wind resuspension rates for tracers increased with wind speed to about the fifth power. This wind-speed dependency is comparable to that measured for off-site plutonium resuspension near the Prosser barricade. However, plutonium resuspension data near the U-Pond Area showed an air concentration dependency on wind speed to the 1.5 power. There is still uncertainty in the wind-speed dependency of airborne concentrations at different sites.

The weathering half-life is the average time required for airborne concentrations from resuspension sites to decrease by one-half when airborne concentrations are averaged over all meteorological conditions. Airborne plutonium and cesium concentrations measured at Hanford as well as tracer resuspension experiments show that the weathering half-life is much greater than that usually reported in the literature: 5 months or much longer rather than only 35 to 45 days.

Resuspension rates for local mechanical resuspension of inert tracer particles caused by vehicular and pedestrian traffic are summarized.

Resuspension occurs when particles on a surface are disturbed and carried up into the air by air currents. Wind-caused resuspension is the process by which wind blows particles from a surface into the air and transports them downwind. For radionuclide-contaminated surfaces, wind might cause radionuclide particles to be resuspended and transported to other sites. Resuspension occurs at radionuclide-contaminated sites on the Hanford reservation in Washington (Sehmel, 1977c; Pacific Northwest Laboratory, September 1973–October 1974), at Rocky Flats in Colorado (Johnson, Tiball, and Severson, 1976; Krey et al., 1976b; 1976c; Sehmel, 1976a; Sehmel and Lloyd, 1976b; Volchok, Knuth, and Klemman, 1972), at the Nevada Test Site (Anspaugh et al., 1969; Wilson, Thomas, and Stannard, 1961), at the Savannah River Laboratory reservation in South Carolina (Milham et al., 1976), and at other sites (Mishima, 1964; Stewart, 1967). However, with our present knowledge (Horst, 1976; Oksza-Chocimowski, 1976), amounts of wind-caused resuspension and its effects cannot be adequately predicted.

Radioactive particles deposited on natural or man-made surfaces are resuspended by both wind and mechanical activity. Wind resuspension can occur over a wide area as well as over a local area. In contrast, mechanical-activity resuspension is usually more localized and can present an immediate inhalation problem to the worker in a contaminated zone. Although mechanical activity is frequently at a point, integration of mechanical activity over time could result in an area source. For example, an area source could be generated during the plowing of a field. In both wide-area and local resuspension of radionuclide particles, particles transported downwind could become a potential radiological concern to man. Sources for resuspended particles include radioactive fallout as well as releases from nuclear facilities. At present the significance of fallout resuspension is unknown. Data are needed to define the relative inhalation hazard of fallout-particle resuspension vs. the direct delivery of stratospheric debris.

Radioactive-particle resuspension is probably more important at nuclear facilities where the surrounding environment has been contaminated with radioactive particles. These particles can be resuspended by both wind stresses and mechanical disturbances. However, resuspension mechanisms are poorly understood, and consequently resuspension rates and potential airborne inhalation hazards cannot now be adequately predicted.

The need for such predictions is not new: for many years resuspension has been known to be occurring at nuclear sites. Some of the earliest data were obtained (Anspaugh et al., 1969; Wilson, Thomas, and Stannard, 1961) at the Nevada Test Site. Ground radioactivity contours were determined as a function of time after a test detonation. Initially, ground-surface concentrations were caused by plume deposition. Subsequent ground radioactivity contours showed (Anspaugh et al., 1969) a migration of radionuclides from the Test Site which indicated that resuspension had occurred. Similarly, aerial surveys at Hanford (Bruns, 1976) have shown transport of ^{241}Am by wind resuspension.

Resuspension is of considerable interest at the Rocky Flats nuclear plant in Colorado where ground surfaces were contaminated with plutonium from leaking storage barrels containing plutonium-contaminated cutting oil (Johnson, Tiball, and Severson, 1976; Krey et al., 1976b; 1976c; Sehmel, 1976a; Sehmel and Lloyd, 1976b; Volchok, Knuth, and Klemman, 1972). After the leakage was discovered, the barrels were removed and corrective actions were taken, but plutonium resuspension from residually contaminated soil surfaces is still occurring.

More recently, resuspension has been reported at study sites on the Hanford reservation (Sehmel, 1977c; Pacific Northwest Laboratory, September 1973—October 1974). These sites were low-level liquid-waste disposal sites.

Although environmental plutonium resuspension is receiving attention, resuspension physics is poorly understood. Resuspension was early characterized by a "resuspension factor." The resuspension factor is defined as the ratio of airborne pollutant concentration (amount per cubic meter) at breathing height divided by the ground-surface contamination level (amount per square meter). Thus the resuspension factor has units of meters^{-1}. Reported resuspension factors vary many orders of magnitude with values from 10^{-11} up to 600 m^{-1} (Mishima, 1964; Stewart, 1967; Sehmel and Lloyd, 1976a). Resuspension-factor variations have not been adequately explained as a function of experimental conditions.

Resuspension factors from about 10^{-9} to 10^{-5} m^{-1} are often used in hazard evaluations. The resuspension factor is useful since a worker's inhalation hazard is most likely related to the local resuspension caused by his work activities within a contaminated zone; however, resuspension factors are only a very rough estimate of the potential airborne contaminant concentration since resuspension factors cannot be accurately predicted. In addition to local resuspension, airborne contaminated particles can reach workers from upwind contaminated areas. Hence both local and upwind resuspension should be considered, but resuspension factors in either case cannot be used in downwind transport models.

The resuspension factor is an index of only the potential inhalation concentration and not the total resuspension release rate from a surface-contaminated area. Resuspension release rates are needed for source terms in calculating total downwind diffusion and transport of resuspended particles. Only recently have particle resuspension rates been measured (Sehmel, 1973b; 1975; 1977b; Sehmel and Lloyd, 1976a; 1976c).

The objective of this chapter is to summarize reported resuspension rates (Sehmel, 1976a; Sehmel and Lloyd, 1976b) and parameters (Sehmel, 1977b; 1977c) determined at the Pacific Northwest Laboratory between 1971 and early 1977. These include plutonium resuspension measurements at Rocky Flats and at Hanford as well as results from controlled tracer simulant source resuspension experiments.

In these experiments airborne concentrations were measured as functions of wind speed, airborne particle size, and wind direction, and the collected radionuclides or tracer simulants were determined per gram of airborne soil or solids. Particulate air samples were collected as a function of wind speed to determine whether airborne radionuclide concentrations increased at higher wind speeds, and concentrations as a function of particle size were measured to determine the distribution of radionuclide particles resuspended as individual particles or attached to host soil and solid particles. In addition, airborne radionuclides were normalized by the total amount of airborne solids to relate concentration per gram of airborne solid to concentration per gram of radionuclide on the ground.

Experiments

The experiments for measuring particle resuspension reported here have been reported in fuller detail in the following references:

- Plutonium and americium from resuspension study sites at Hanford (Sehmel, 1977c) (Fig. 1).

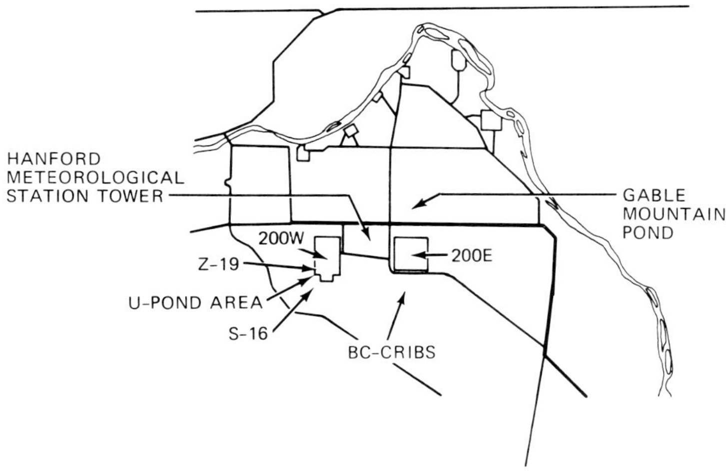

Fig. 1   Location of Hanford study sites.

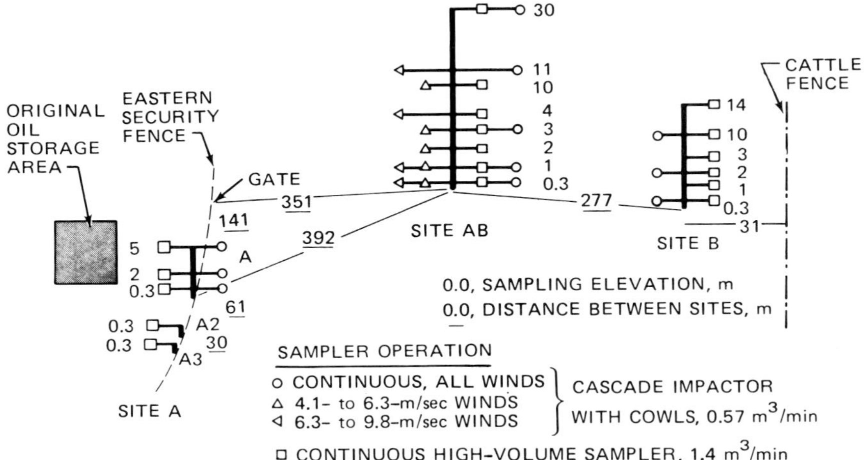

Fig. 2   Rocky Flats tower locations.

- Plutonium from contaminated environmental surfaces at Rocky Flats (Sehmel, 1976a; Sehmel and Lloyd, 1976b) (Fig. 2).
- Controlled inert simulant tracer particles from selected surfaces on the Hanford area (Sehmel, 1973b; 1975; 1976c; 1977b; Sehmel and Lloyd, 1976a; 1976c).

Plutonium resuspension results from off-site Hanford near the Prosser barricade are reported here for the first time. Most wind-caused resuspension research concerns resuspension from vegetated areas. Experiments concerning local resuspension caused by mechanical activity include tracer studies of resuspension rates for a man walking across an asphalt strip and for cars and trucks driven on asphalt or cheat grass.

Two different resuspension rates are used. For wind-caused resuspension, resuspension rates are reported as the fraction of particles resuspended per second. Thus the total wind-caused resuspension is a product of the surface contamination level, the duration of resuspension, and the resuspension rate. For local mechanical disturbances by vehicular or pedestrian traffic, resuspension was measured each time a car, $\frac{3}{4}$-ton truck, or person passed across the length of a 3-m-wide tracer-contaminated area. Thus traffic resuspension rates are reported as the fraction of particles resuspended per pass.

Particles

Resuspension was measured for several types of particles. The plutonium particle size distributions on soils at Rocky Flats and Hanford were uncontrolled. A forest-spray operation provided an opportunity to measure resuspension of DDT as tracer particles not specifically controlled for size. The controlled, inert tracer particles used were submicrometer $CaMoO_4$ particles and ZnS particles with an 8-μm mass aerodynamic equivalent diameter.

Air Samplers

Airborne resuspended particles were either sampled with total air samplers* or sized while airborne with particle cascade impactors.† Particle cascade impactors were used for plutonium and $CaMoO_4$ particles.

The particle cascade impactor for sampling respirable particles was attached to a rotating cowl, which allowed simultaneous sampling of larger nonrespirable particles. The cowl-impactor system (Sehmel, 1973a) shown in Fig. 3 was evaluated by Wedding, McFarland, and Cermak, 1977. Particles entering the 15-cm-diameter cylindrical sampler inlet of the cowl either settled on the cowl floor or were drawn up into the impactor. Particles settling on the cowl floor are called "nonrespirable" in this chapter. Respirable particles entering the particle cascade impactor were separated into nominal aerodynamic diameter ranges of 7, 3.3, 2.0, and 1.1 m, which are impactor stage 50% cutoff diameters for unit-density spheres. Smaller particles were collected on an impactor backup filter.

Results and Discussion

Airborne radionuclide concentrations were determined at transuranic resuspension study sites at Rocky Flats, Colo. (Sehmel, 1976a; Sehmel and Lloyd, 1976b) and the Hanford area in Washington (Sehmel, 1977c; Pacific Northwest Laboratory, September 1973– October 1974). In addition, some cesium resuspension data are reported for Hanford (Sehmel, 1977c). In contrast to transuranic resuspension, tracer simulants (Sehmel, 1977b) were used to determine particle resuspension rates. Results for each set of experiments are discussed separately.

Radionuclide-Particle Resuspension On Site

Airborne plutonium concentrations at Rocky Flats and Hanford were measured as a function of particle diameter, wind speed, and sampling site. Radionuclide concentrations per gram of airborne solid were determined.

---

*General Metal Works, Inc., model GMWL-2000-high-voltage air sampler with filter holder.

†Andersen 2000, Inc., model 65-100 high-volume sampler head.

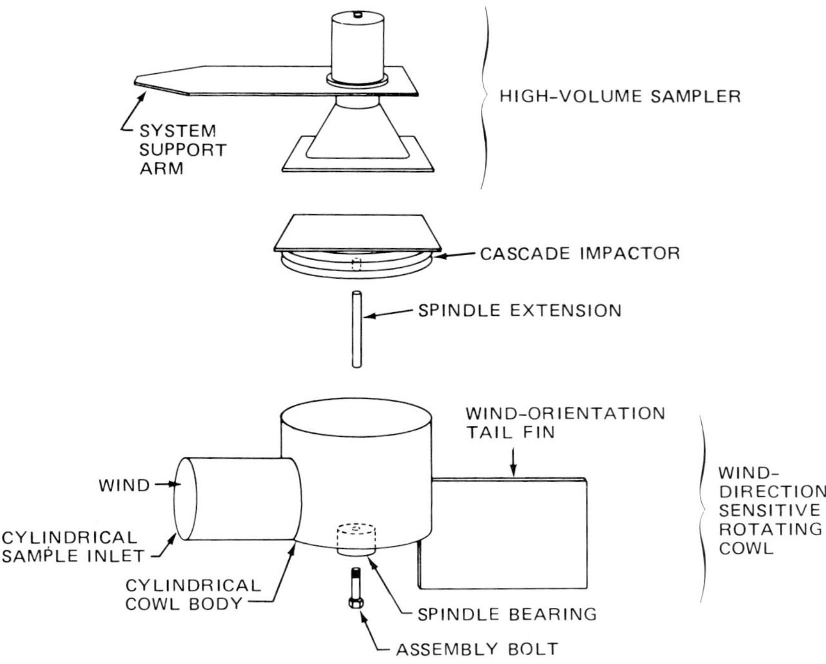

Fig. 3   Rotating cowl and impactor.

Plutonium Resuspension Research at Rocky Flats. Plutonium resuspension at Rocky Flats was investigated experimentally (Sehmel, 1976a; Sehmel and Lloyd, 1976b). In early work an empirical resuspension model was developed (Sehmel and Orgill, 1973) which was based on published weekly plutonium concentrations at Health and Safety Laboratory sampling station S-8 along the site's eastern security fence. The plutonium data were analyzed in terms of the meteorology during sampling times. Collected airborne plutonium was related to hourly average wind speeds and wind directions. Model results showed that airborne plutonium concentrations increased as the 2.1 power of wind speed. Subsequently airborne concentrations were predicted for the succeeding time period. These results showed a wide difference between predictions and experimental results. The interpretation of these differences was that the plutonium resuspension source characteristics had changed (Sehmel and Orgill, 1974).

Battelle—Northwest experimental measurements of plutonium resuspension at Rocky Flats were made in July 1973 (Sehmel, 1976a; Sehmel and Lloyd, 1976b). As shown in Fig. 2, airborne plutonium concentrations were measured at three sampling sites east of the plant. The first sampling site was along the eastern security fence. This site was called sampling site A. Sampling site B was along the eastern cattle fence, and sampling site AB was between sites A and B. The distance from site A to site AB was 227 m. Airborne plutonium at these sites was sampled and analyzed as a function of sampling height, particle size, and wind speed. For comparison, a particle cascade impactor sample was simultaneously collected at a background site 13 km west in the mountains.

Fallout levels of ^{239}Pu entering the area were estimated from the cascade impactor operated in the mountains. There was no detectable ^{239}Pu activity on the 7-, 3.3-, and 1.1-μm impactor stages, and there was no radiochemical result for the 2-μm stage. The only detectable background plutonium activity was on the backup filter that nominally collects submicrometer particles. Airborne ^{239}Pu concentration at the background station was $4 \pm 3.5 \times 10^{-18}$ μCi/cm^3, which corresponds to $0.7 \pm 0.62 \times 10^{-6}$ μCi/g of airborne soil on the backup filter. Error limits are the 2σ radiochemical counting limits.

In on-site research, airborne particles were separated in the sampling process into two main fractions. One sample contained particles collected by gravity settling in the inlet cowl section of the sampler as shown in Fig. 3. The second fraction contained those particles passing through the inlet section and collected within the high-volume cascade impactor. The smallest particles collected in the inlet cowl section were about 10 μm in diameter. This fraction was assayed for ^{238}Pu and ^{239}Pu. In some cases nonrespirable particles were sieved into smaller size fractions, and these fractions were also assayed for ^{238}Pu and ^{239}Pu. Data for respirable and nonrespirable particles are discussed separately.

Respirable Plutonium Concentrations at Rocky Flats.   Airborne ^{239}Pu concentrations at the three Rocky Flats sampling stations were reported (Sehmel and Lloyd, 1976b) in microcuries per cubic centimeter of air and microcuries per gram of airborne soil. The maximum airborne ^{239}Pu concentration was 3.7×10^{-15} μCi/cm^3. The maximum ^{239}Pu concentration on the airborne soil was 5×10^{-5} μCi/g total airborne soil and 7×10^{-5} μCi/g for the respirable fraction of airborne soil collected on the 2-μm-particle impactor stage. All airborne ^{239}Pu concentrations were significantly less than MPC's of soluble ^{239}Pu in air for occupational exposure in a 40-hr work week (2×10^{-12} μCi/cm^3) or nonoccupational exposure in a 168-hr week period (6×10^{-13} μCi/cm^3) (International Commission on Radiological Protection, 1959).

Airborne plutonium concentrations were a function of both sampling height and particle diameter. Airborne concentrations are shown for site **AB** in Fig. 4 for each particle cascade impactor stage. In contrast to simple modeling concepts, airborne concentrations did not always decrease with an increase in height. There were unexpectedly high ^{239}Pu concentrations at this site for several particle diameters and heights.

Plutonium was associated with particles collected on each particle cascade impactor stage. Since there was no plutonium in the upper stages of the impactor at the background mountain site, the ^{239}Pu found in upper stages of impactors at Rocky Flats sampling sites indicates that some plutonium was resuspended while attached to larger particles. Resuspension of submicrometer particles also occurred at Rocky Flats.

The general trend of the complete airborne ^{239}Pu concentration data is a decrease in concentration with increasing distance eastward from site A (Sehmel and Lloyd, 1976b). As might be expected, this decrease in concentration corresponded to increasing distance from the original oil storage area, which was the principal source of ground contamination. However, significant deviations did occur in concentration profiles of airborne ^{239}Pu with both distance and height. These deviations might be attributed to sampling some more-active-than-normal particles or clusters of particles. These increases in average airborne ^{239}Pu concentrations were present at both sites **AB** and **B**.

As indicated in Fig. 5 for site **AB**, which was over 392 m from the oil storage area and which was on the flat terrain, some more-active-than-normal particles or clusters of particles (hot) may have been present in the 2.0-μm size range. In this case the

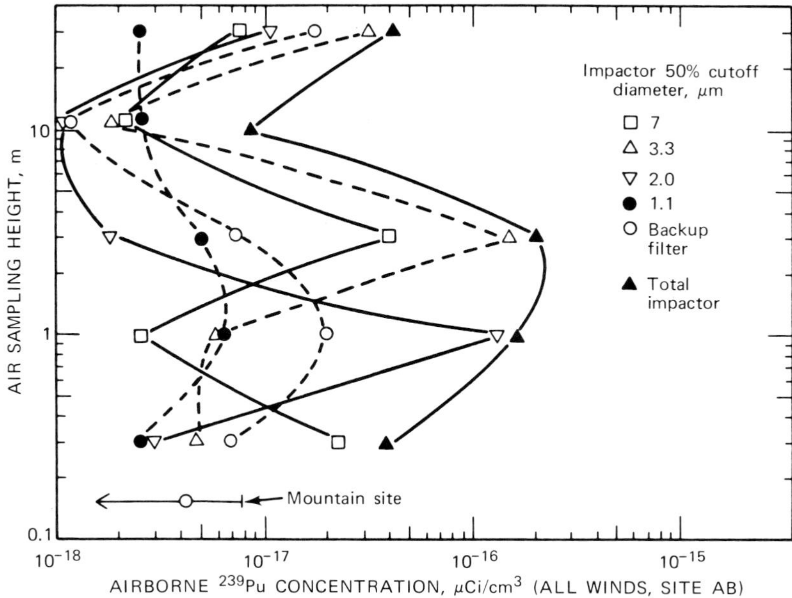

Fig. 4 Airborne ^{239}Pu concentration at site **AB** at Rocky Flats as a function of impactor collection site.

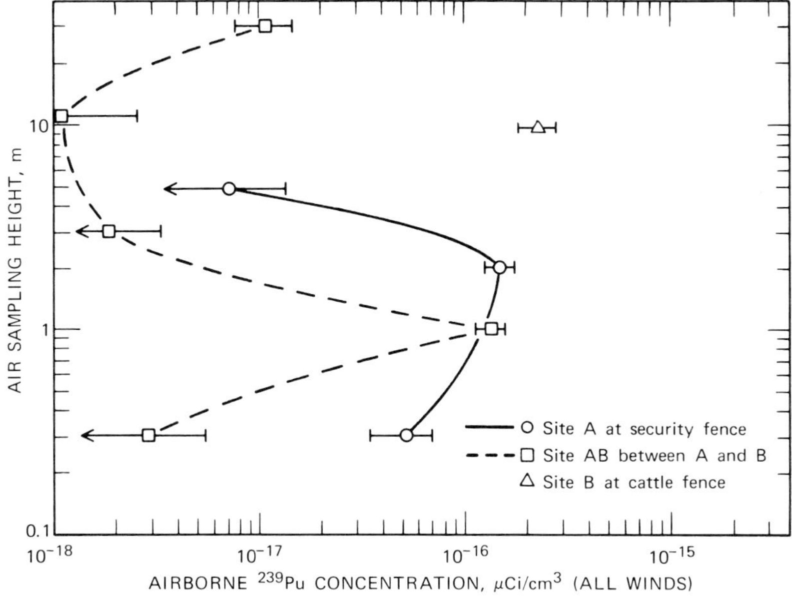

Fig. 5 Airborne ^{239}Pu concentrations from impactor 2.0-μm stage collections at Rocky Flats.

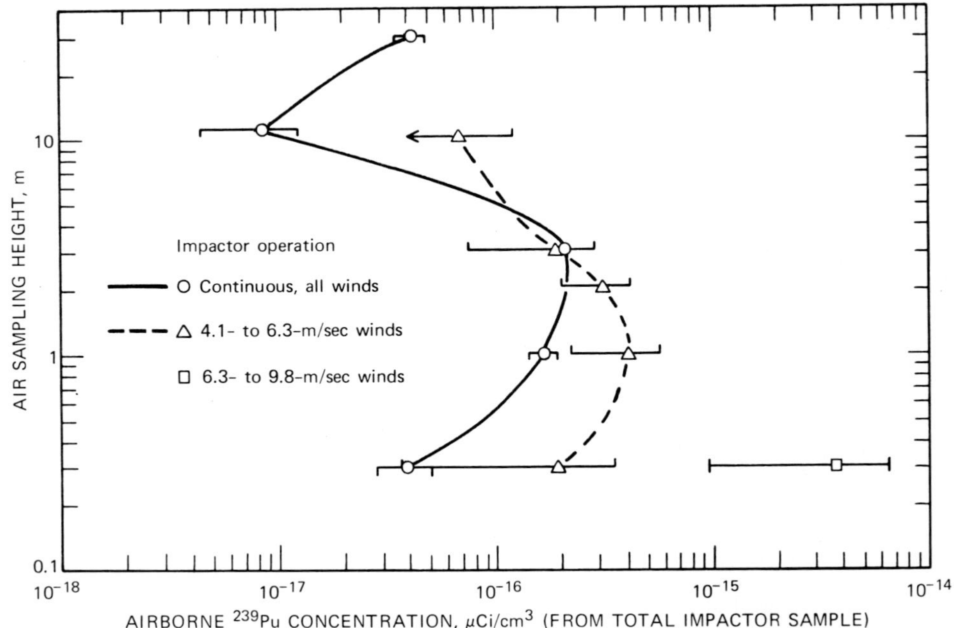

Fig. 6 Total airborne ^{239}Pu concentration at AB site at Rocky Flats as a function of wind speed.

concentration at the 1-m height of site AB is one to two orders of magnitude greater than at other heights for this site. More important to the hot particle concept is the concentration at the 10-m height of site B. This concentration of 2.3 × 10^{-16} μCi/cm^3 was the largest ^{239}Pu concentration for 2-μm particles measured at any Rocky Flats location. This relatively high concentration was unexpected since this sampling location was the most remote from both the ground and the original oil storage area. This suggests that other relatively hot particles could also be escaping from the plant boundaries; however, due caution is indicated in interpreting this hot particle concept. The total of 6 d/min collected on the 2-μm stage, or 2.3 × 10^{-16} μCi/cm^3, is much less than the MPC in air of 2 × 10^{-12} μCi/cm^3 (occupational). It is conceivable that the majority of this hot plutonium was attached to one soil particle.

The functional relationship between airborne plutonium resuspension concentrations and wind speed could not be developed as unequivocally as initially anticipated (Sehmel and Lloyd, 1976b). This was due in part to the inadvertent loss of about a fifth of the collected filter samples during radiochemical analysis. Unfortunately, most samples from the higher wind speeds were lost. Even with the limited plutonium data collected in this experiment, it was evident that airborne ^{239}Pu concentrations increased with an increase in wind speed. In Fig. 6 total airborne concentrations are shown for air sampled at all wind speeds (average wind speed of 0.9 m/sec), at wind speeds from 4.1 to 6.3 m/sec, and at wind speeds from 6.3 to 9.8 m/sec. Airborne ^{239}Pu concentrations at wind speeds from 4.1 to 6.3 m/sec are definitely larger than average airborne concentrations for continuous air sampling. However, the 2σ radiochemical counting statistics error limits are too large to determine the wind-speed dependency. Nevertheless, an attempt to approximate airborne ^{239}Pu concentrations and consequently the resuspension-rate dependency on wind speed was made for the 7-μm-diameter particles. This approximation was for the 0.3-m height at sampling site AB. For the three data points taken at the 0.3-m height, ^{239}Pu concentrations increased with the 5.9 power of wind speed. The uncertainty in this exponent is too large to make a valid comparison between airborne plutonium and soil concentrations.

The July 1973 plutonium resuspension experiment at Rocky Flats showed resuspension of both ^{238}Pu and ^{239}Pu. However, all airborne plutonium concentrations were significantly below MPC's in air. Since ^{239}Pu was collected on each particle cascade impactor stage, the suggestion is that most plutonium was attached to soil particles when the plutonium was resuspended.

Respirable Plutonium Concentrations at Hanford.   Extensive data were obtained on airborne radionuclide concentrations around resuspension sites studied (Sehmel, 1977c). These concentrations were expressed both in microcuries per cubic centimeter of filtered air and microcuries per gram of airborne solids. This report summarized ranges of data collected but did not detail data for each experiment.

Airborne plutonium concentrations for both ^{238}Pu and ^{239}Pu measured (Sehmel, 1977c) in resuspension experiments are shown in Fig. 7 and are compared with Hanford 300 Area fallout levels (Thomas, 1976) approximately 30 km distant. The data represented experiments conducted over various time periods. For each data symbol the vertical line is plotted at the mid-time of the resuspension experiment, and the experiment duration is shown by horizontal lines drawn at both maximum and minimum measured airborne concentrations. Airborne peak plutonium concentrations at resuspension study sites were significantly greater than 300 Area fallout levels (Thomas, 1976), and airborne ^{239}Pu concentrations, in general, were greater than airborne ^{238}Pu concentrations. However, although resuspension was and is still probably occurring at these sites, measured airborne concentrations were significantly less than MPC's (International Commission on Radiological Protection, 1959).

Airborne plutonium concentrations at the U-Pond Area tended to remain constant as a function of time. This constancy indicates that the weathering (or fixation) half-life for surface contamination available for resuspension at this site is on the order of years. This is much greater than the 35 to 40 days often quoted (Wilson, Thomas, and Stannard, 1961) in literature on resuspension. However, the year weathering half-life at the U-Pond could be a manifestation of some resuspension surface renewal process since this is an active waste-disposal site. The explanations are unclear for differences in weathering half-life.

The maximum airborne ^{239}Pu concentration measured was 8×10^{-15} μCi/cm^3 near the Hanford meteorological station (HMS) tower on Jan. 11, 1972. All other plutonium concentrations except one were at least one order of magnitude lower. This one exception was measured 6.1 m above ground at the U-Pond during October 1973. In comparison with other October data, the concentration for this sample was about one and one-half orders of magnitude greater than any other sample. We hypothesized that some more-active-than-normal particles or clusters of particles (hot) were resuspended and collected on this filter.

Nonrespirable Airborne Plutonium Fluxes at Rocky Flats and Hanford

Nonrespirable airborne plutonium fluxes were calculated for both ^{238}Pu and ^{239}Pu. The Rocky Flats data (Sehmel, 1976a) are shown in Figs. 8 and 9.

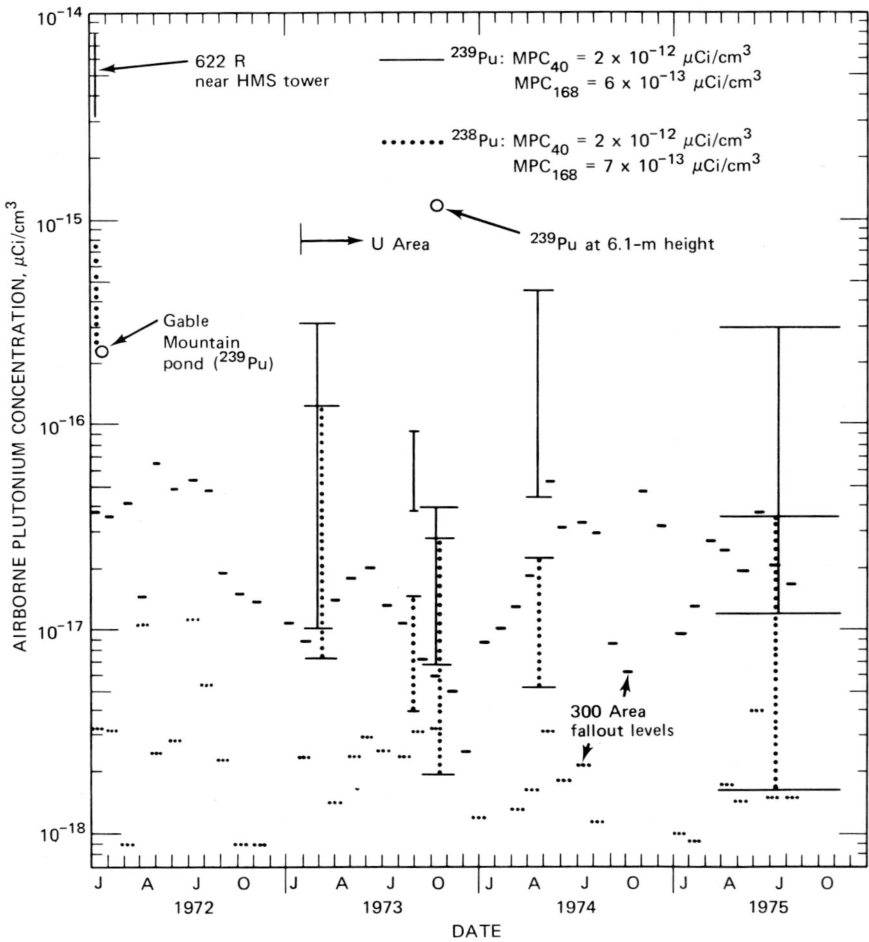

Fig. 7 Range of airborne plutonium concentrations at on-site Hanford resuspension sites compared with fallout levels.

In Fig. 8, the nonrespirable airborne ^{239}Pu horizontal flux is shown as a function of sampling distance and sampling height. As might be expected, the maximum airborne ^{239}Pu flux on nonrespirable particles was at site A near the original oil storage area. The maximum airborne ^{239}Pu flux was 6×10^{-4} μCi m^{-2} day^{-1}. The airborne ^{239}Pu flux decreased with both distance and sampling height. At site A the ^{239}Pu flux decreased over one order of magnitude as the sampling height was increased from 0.3 to 2 m above ground level. Similarly, at site AB the nonrespirable airborne ^{239}Pu flux again decreased about one order of magnitude as the sampling height was increased from 0.3 to 1 to 2 m above ground level. Airborne ^{239}Pu fluxes on nonrespirable particles decreased almost two orders of magnitude between sampling sites A and AB. However, between sampling sites AB and B, airborne ^{239}Pu fluxes on nonrespirable particles did not show a significant variation with distance.

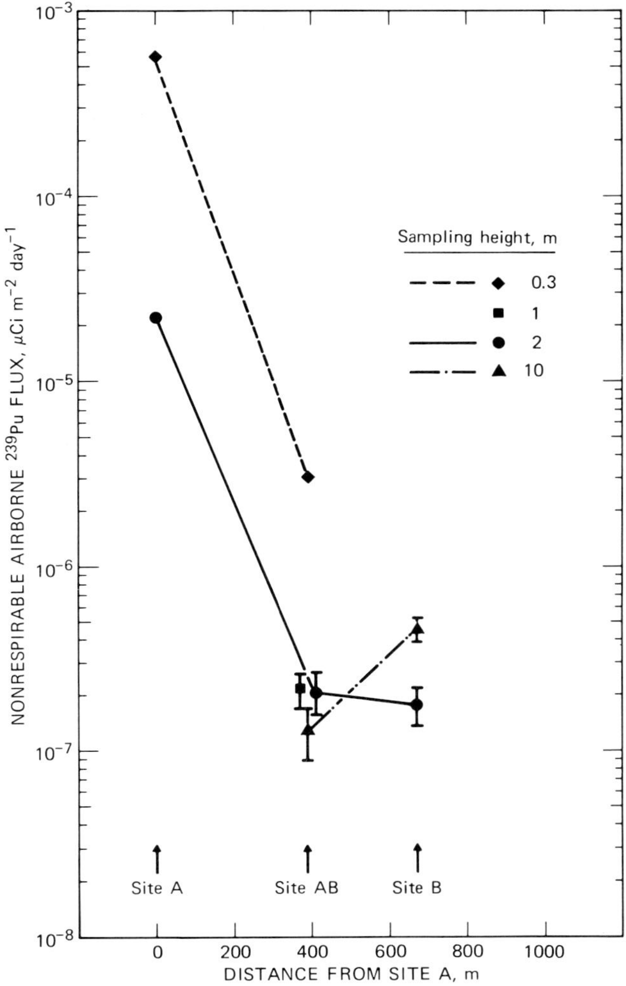

Fig. 8 Decrease with distance of total ^{239}Pu flux on nonrespirable particles at Rocky Flats.

From a comparison of the data for the three different Rocky Flats sites, the conclusion is that the airborne nonrespirable ^{239}Pu flux does not decay as a simple exponential function of distance from site A. In addition, data for sampling heights above 1 m at sites AB and B show that the airborne ^{239}Pu flux did not significantly decrease for heights greater than 1 m up to 10 m. The nonrespirable particle plume height above 10 m is unknown.

Similar results are shown in Fig. 9 for total ^{238}Pu flux on nonrespirable particles at Rocky Flats as a function of a sampling site and sampling height. The maximum nonrespirable airborne ^{238}Pu flux was 1.2×10^{-5} μCi m^{-2} day^{-1} and was at the 0.3-m sampling height at site A. Again, at site A as well as site AB, airborne ^{238}Pu fluxes decreased rapidly as the sampling height increased from 0.3 up to 10 m. However, from

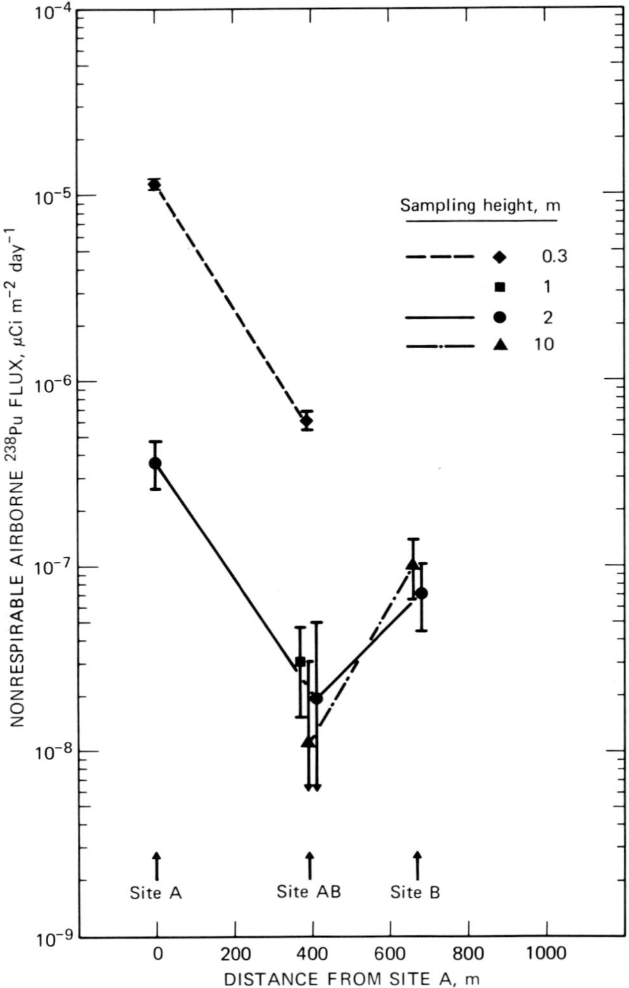

Fig. 9 Decrease with distance of total [238]**Pu flux on nonrespirable particles at Rocky Flats.**

site AB to B, an unexplained observation was made. Airborne ^{238}Pu fluxes at 2 and 10 m heights at site B were greater than those at site AB. An explanation for this increase is not apparent, but the increase is supported by comparing plutonium analyses uncertainties. Error bars for site B show a ^{238}Pu flux range significantly above error bars around site AB.

Nonrespirable airborne fluxes at Rocky Flats were greatest near the original oil storage area (source of contaminated leakage) and near ground level. Fluxes of ^{239}Pu ranged from 10^{-7} up to 10^{-3} μCi m^{-2} day^{-1}. In contrast, fluxes of ^{238}Pu ranged from 10^{-8} to 10^{-5} μCi m^{-2} day^{-1}

Nonrespirable airborne plutonium fluxes around the U-Pond within the Hanford area are shown in Fig. 10. The ^{238}Pu flux was less than the ^{239}Pu flux. This decrease is

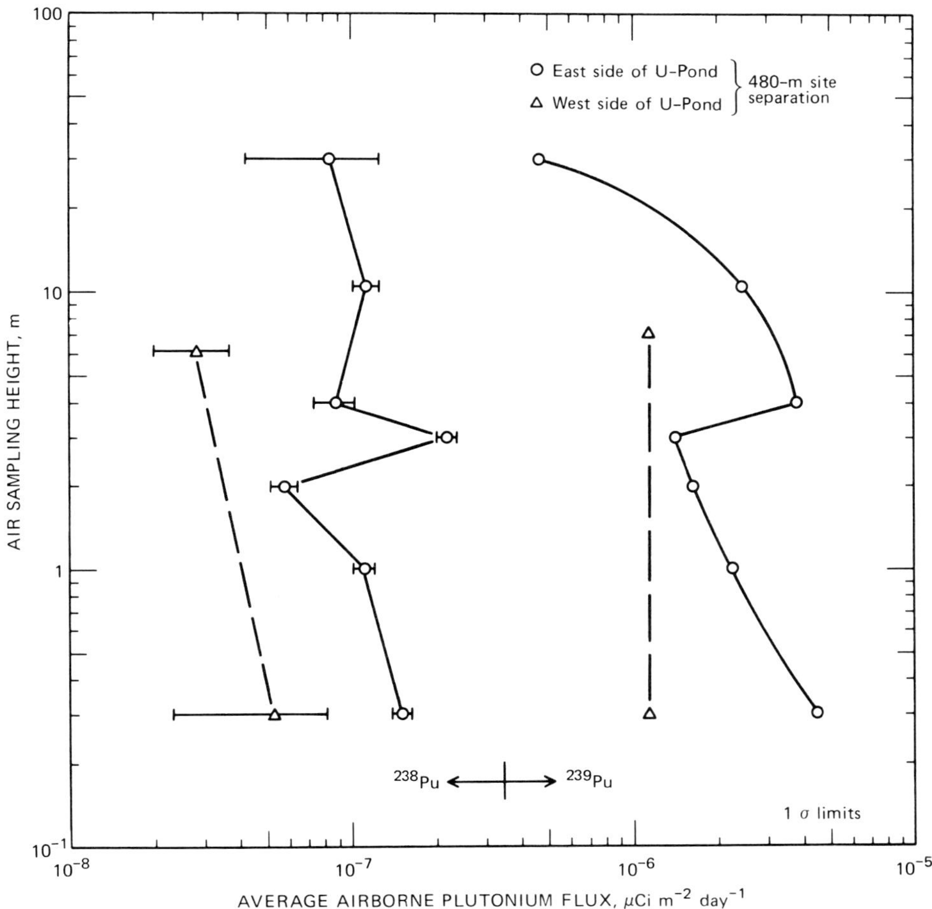

Fig. 10 Airborne ²³⁹Pu and ²³⁸Pu fluxes on nonrespirable particles at Hanford U-Pond during Feb. 27 to Nov. 10, 1975 (particles collected in cowls).

similar to the Rocky Flats data. However, the U-Pond data show that the nonrespirable plutonium flux extends at least up to 30 m above ground level. Also, there was a greater airborne plutonium flux east of the U-Pond than west of the U-Pond. This is to be expected since prevailing winds are from the west.

At the U-Pond the airborne ^{238}Pu flux ranged from 10^{-8} to 10^{-7} μCi m^{-2} day^{-1}, which is within the midrange of 10^{-8} to about 10^{-5} μCi m^{-2} day^{-1} measured at Rocky Flats. Similarly, the ^{239}Pu flux at the U-Pond ranged from about 10^{-6} to 10^{-5} μCi m^{-2} day^{-1}, which is within the 10^{-7} to 10^{-3} μCi m^{-2} day^{-1} measured at Rocky Flats. The bracketing of the nonrespirable airborne particle fluxes near the U-Pond and at Rocky Flats even within three to four orders of magnitude may be coincidental since surface sources and other factors are peculiar to each site.

The on-site data reported are the first results to quantify the range of nonrespirable airborne plutonium fluxes. Ground contamination on nonrespirable particles at both sites is poorly defined or nonuniform (Corley, Robertson, and Brauer, 1976; Krey et al., 1976b; Maxfield, 1974; Mishima, 1973; Mishima and Schwendiman, 1973; 1974; Nees and Corley, 1975); hence the data cannot at present be analyzed to reflect resuspension rates or resuspension factors for nonrespirable particles.

Plutonium Concentration per Gram of Airborne Soil

Airborne plutonium concentrations were normalized to the soil collected with the airborne plutonium. Plutonium concentrations (in microcuries per gram) were determined as a function of particle diameter as determined with both particle cascade impactors for respirable particle diameters and sieve sizes for nonrespirable particles. Resuspended plutonium is attached to nonrespirable as well as to respirable particles. Hence nonrespirable soil particles may contribute significantly to downwind airborne plutonium concentrations and represent one mechanism for transporting plutonium to surrounding land.

For Rocky Flats nonrespirable soil collected at 0.3 m above ground level was sieve sized (Sehmel, 1976a) into twelve different size increments. Each size increment was analyzed for ^{238}Pu and ^{239}Pu. Plutonium concentrations were normalized (micrograms per gram) to the grams of soil collected within each size increment. Results are shown in Fig. 11 for ^{239}Pu as a function of particle size at sites A and AB. Plutonium-239 was associated with all particle sizes. The maximum concentration was about 10^{-4} μCi/g for particle sizes between 10 and 20 μm. For larger particle diameters up to 230 μm, concentrations tended to decrease with an increase in particle diameter. Concentrations at site A were greater than those at site AB. This is expected since site A was closer to the original oil storage area at which plutonium leakage occurred.

At each site plutonium concentrations (in microcuries per gram) indicate general continuous relationships as a function of particle diameter, which might be used to infer how plutonium is attached to airborne-soil particles. For nonrespirable particle diameter ranges determined from sieve sizes, the data could be approximated by a straight line inversely proportional to particle diameter. The relationship is complicated by the collection of both contaminated on-site and uncontaminated off-site nonrespirable particles within the cowls.

For respirable particles, the ^{239}Pu microcuries per gram was nearly independent of particle diameter. This independence might suggest that plutonium attachments are volume phenomena for these respirable particles. In contrast, plutonium particle attachment to soil particles is expected to be controlled by available soil particle surface area for nonrespirable particles. Additional data are required to conclude how plutonium particles are attached to airborne particles in both respirable and nonrespirable size ranges.

Plutonium-238 concentrations on airborne soil are shown in Fig. 12. In this case only nonrespirable particle diameter ranges are shown. There was insufficient ^{238}Pu collected in the particle cascade impactor samples to yield positive results in respirable particle diameter ranges. Similar to ^{239}Pu, ^{238}Pu nonrespirable concentrations were greater at site A than at site B and also showed an inverse relationship with particle diameter. However, there is fine structure showing deviation around any apparent inverse relationship. This fine structure indicates that there is yet much to be learned about plutonium resuspension and plutonium attachment to nonrespirable as well as to respirable host particles.

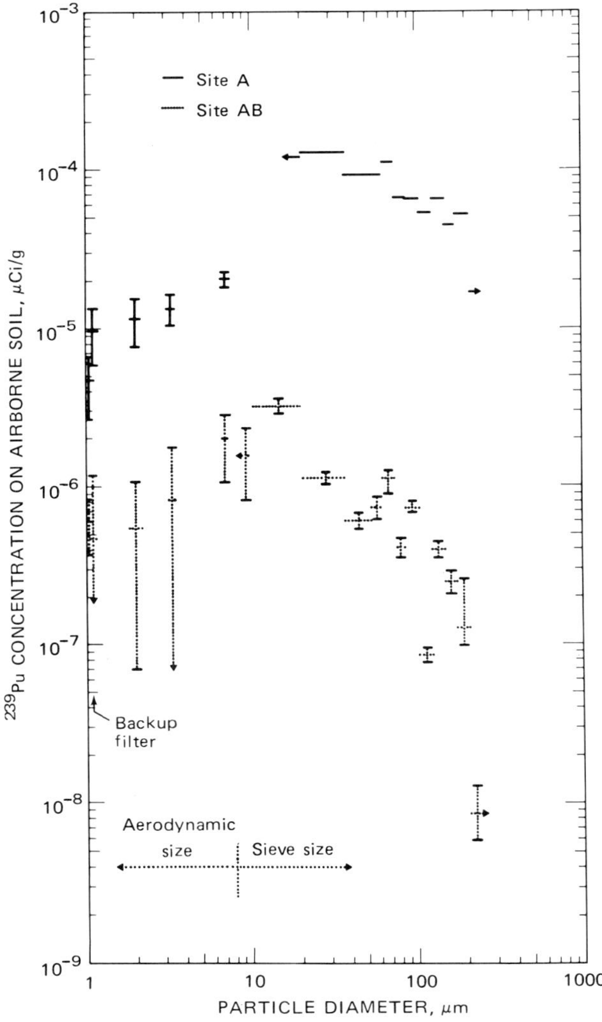

Fig. 11 Plutonium-239 concentration on airborne soil as a function of particle diameter at Rocky Flats.

Both ^{239}Pu and ^{238}Pu concentrations on airborne soil decreased from site A to site AB. Site AB/site A ratios are shown in Fig. 13. Concentrations per gram decreased by a factor of up to about 10^{-3} in the intervening 392-m distance separating sampling sites A and AB. Concentration ratios for ^{239}Pu on respirable particles were about 10^{-1} and were independent of particle size. In contrast, for nonrespirable particle sizes ^{239}Pu ratios between sites decreased nearly linearly as the particle diameter increased. Plutonium-239 concentrations on nonrespirable particles decreased at rates greater than those for the ^{238}Pu concentrations. For ^{238}Pu the concentration ratios for site AB/site A were nearly one order of magnitude greater than for ^{239}Pu. These larger ratios suggest that ^{238}Pu resuspended more readily relative to ^{239}Pu.

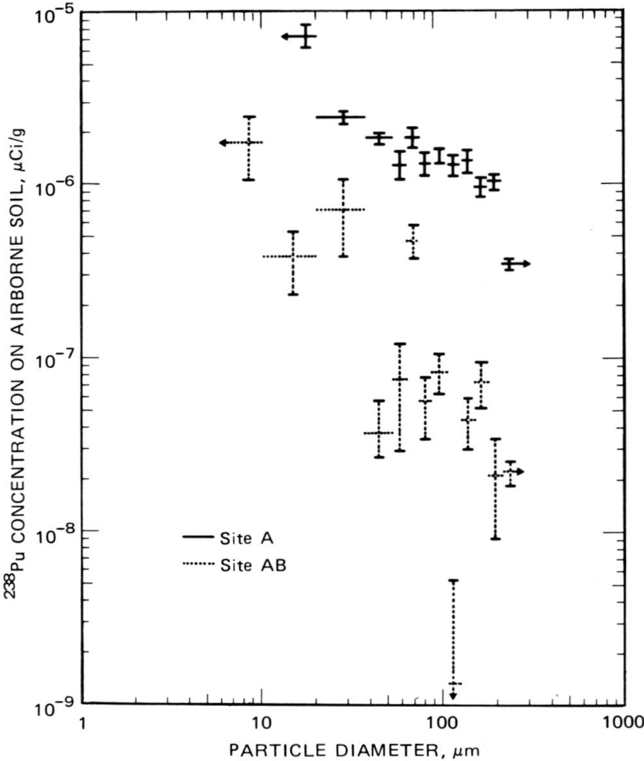

Fig. 12   Plutonium-238 concentration on airborne soil as a function of particle diameter at Rocky Flats.

Concentrations for on-site Hanford experiments are shown in Fig. 14 for collection on filters. Concentrations of ^{239}Pu were somewhat greater than those of ^{238}Pu. Plutonium concentrations on airborne solids ranged from 10^{-9} to 10^{-5} μCi/g. The only exception was the October 1973 single sample described above, for which the concentration was 6×10^{-5} μCi/g. Otherwise plutonium concentrations on airborne solids around the U-Pond appeared to be nearly independent of time.

Plutonium concentrations on airborne nonrespirable particles were also determined (Sehmel, 1977a) near the U-Pond on the Hanford reservation for sampling heights from 0.3 up to 30 m above ground. Airborne solids were sampled continuously for all wind directions at sites both east and west of the U-Pond. The distance between sampling sites was 480 m. Samples were analyzed for both ^{239}Pu and ^{238}Pu. Calculated results shown in Fig. 15 are for nonrespirable airborne solids collected within cowls. Results show that plutonium concentrations on nonrespirable airborne solids were approximately one order of magnitude higher east as compared with those west of the U-Pond. This increase is caused by prevailing west winds, which caused resuspension from this low-level liquid-waste disposal area. East of the U-Pond plutonium concentrations on nonrespirable airborne solids tended to be uniform with height up to 30 m. The plume height above 30 m is unknown. Plutonium concentrations on nonrespirable airborne solids are within the range shown in Fig. 14 for smaller particles collected on filters. In both cases ^{239}Pu

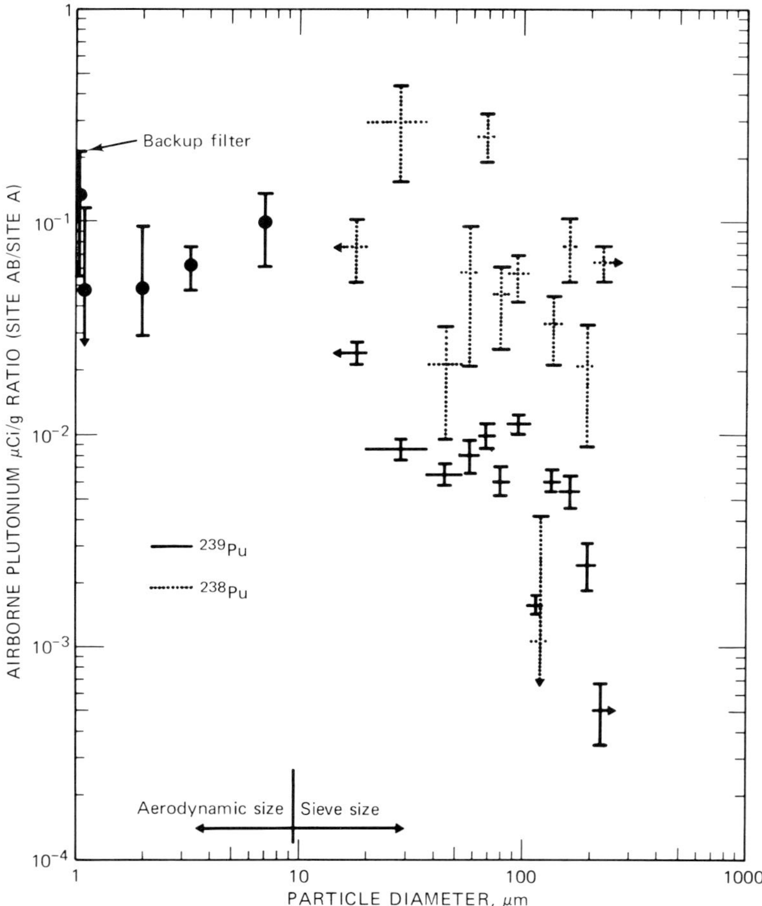

Fig. 13 Decrease in ²³⁸Pu and ²³⁹Pu concentrations on nonrespirable particles from site A to site AB at Rocky Flats.

concentrations ranged from 10^{-7} to 10^{-5} μCi/g. However, ^{238}Pu concentrations on respirable solids collected on filters tended to be greater than those on the nonrespirable particles.

Both ^{238}Pu and ^{239}Pu concentrations on nonrespirable airborne solids near the U-Pond were less than concentrations determined at Rocky Flats site A. This comparison can be seen by comparing data in Figs. 11 and 12 with data in Fig. 15. However, plutonium concentrations on nonrespirable airborne solids at Hanford's U-Pond and Rocky Flats site AB tended to be comparable.

Airborne and Ground Plutonium Ratios

There might be some relationship between plutonium concentrations on airborne soil and those on ground-surface soil if all airborne soil came from local resuspension. At Rocky Flats ground-surface soils were characterized (Krey et al., 1976b) for the same time period these nonrespirable airborne samples were collected. Ground-surface sample results

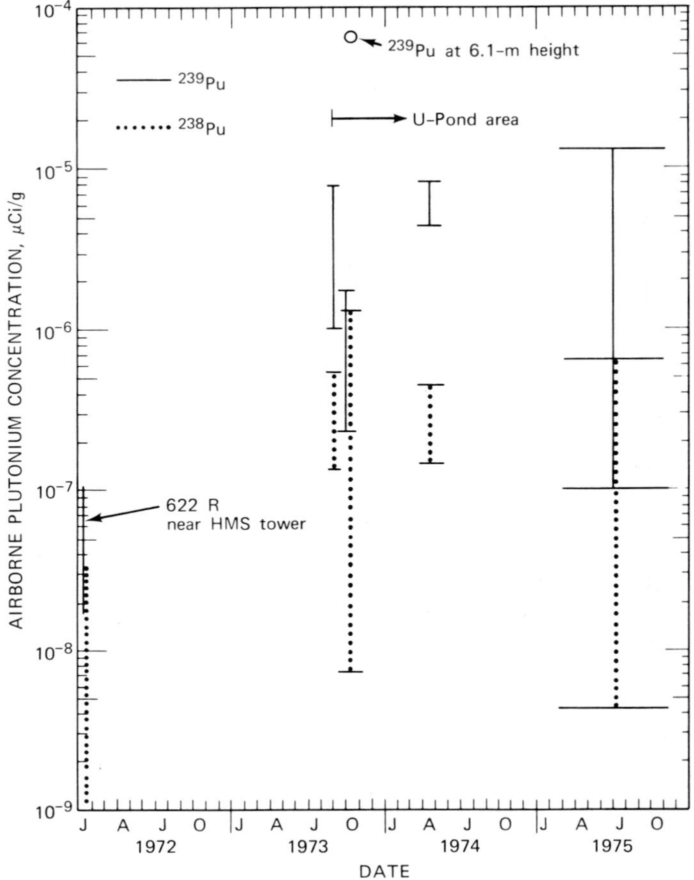

Fig. 14 Range of plutonium concentration on airborne solids at on-site Hanford resuspension sites.

for a 5-cm sampling depth are summarized in Table 1 for sampling sites near sites A, AB, and B. Ground-surface ^{238}Pu/^{239}Pu ratios are also shown.

Ground and airborne plutonium soil-sample results are compared in Table 2 for both ^{239}Pu and ^{238}Pu. Comparisons are for plutonium in ground-surface samples 5 cm deep vs. airborne nonrespirable particle concentrations in particle diameter ranges. Airborne concentrations were taken from Figs. 11 and 12 data limits. From these data activity ratios of airborne to surface concentrations (microcuries per gram) were calculated. Only the maximum ranges are reported. Maximum ranges of the ratio of airborne/ground-surface soil concentrations are shown in the last two columns. These ratios range from 1×10^{-4} up to 2. Thus in hazard evaluations (Johnson, Tiball, and Severson, 1976) one might consider maximum plutonium concentrations on airborne soil to be comparable to plutonium concentrations on ground-surface soils. This is indicated by the ratio 2. However, in most cases plutonium concentrations on airborne soil were significantly less than those on ground-surface soils.

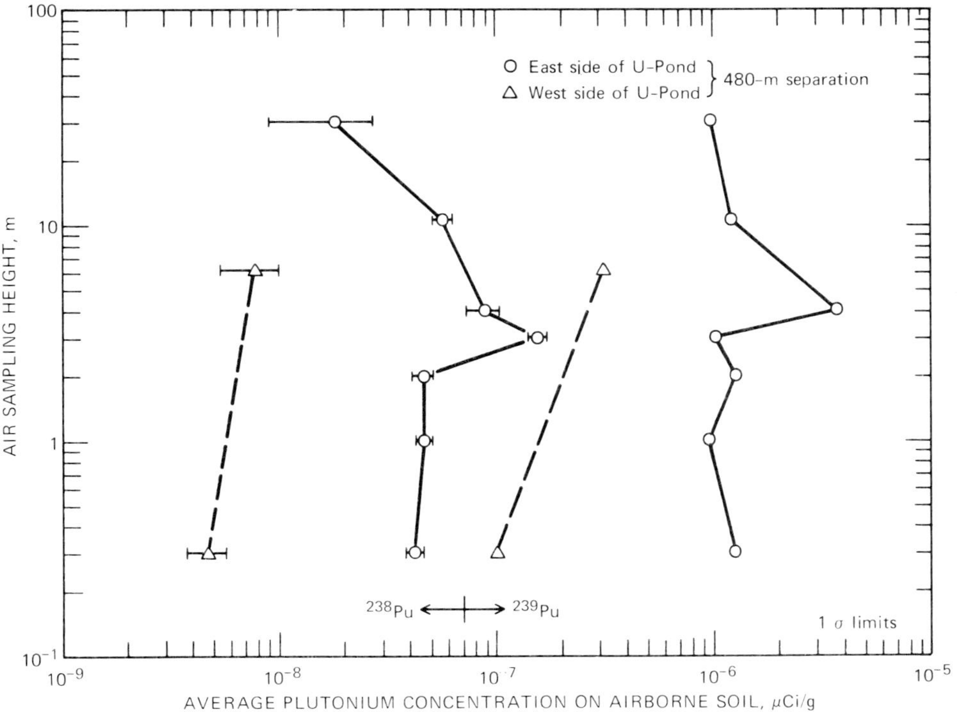

Fig. 15 Airborne nonrespirable ^{238}Pu and ^{239}Pu on nonrespirable airborne solids at Hanford U-Pond during Feb. 27 to Nov. 10, 1975 (particles collected in cowls).

TABLE 1 Results of Selected Surface Soil Samples at Rocky Flats*

Location	^{239}Pu, μCi/g	^{238}Pu, μCi/g	Sample ratio, ^{238}Pu/^{239}Pu
Near site A	3.10×10^{-3}	5.77×10^{-5}	0.019
	6.89×10^{-4}	1.24×10^{-5}	0.018
Near site AB	7.70×10^{-5}	1.56×10^{-6}	0.020
	3.86×10^{-5}	7.66×10^{-7}	0.020
Near site B	2.66×10^{-6}	5.09×10^{-8}	0.019
	3.85×10^{-5}	7.12×10^{-7}	0.018

*Microcuries per gram of dry soil ±% standard deviation; samples from 2000-cm² area at 5-cm sampling depth. Two samples per site.

Identification of relationships between sites of radionuclide concentrations on airborne solids and contaminated ground solids would be useful in establishing criteria for releasing contaminated areas for other uses. Concentrations on Hanford ground surfaces obtained from the literature (Corley, Robertson, and Brauer, 1976; Maxfield, 1974; Mishima and Schwendiman, 1973; Nees and Corley, 1975) are shown in Table 3 for

TABLE 2 Ratio of Plutonium Concentration per Gram of Airborne Soil to That per Gram of Ground-Surface Soil at Rocky Flats

| Site | Range, μCi/g | | | | Ratio range $\left(\dfrac{\mu\text{Ci/g airborne}}{\mu\text{Ci/g surface}}\right)$ | |
| | ^{239}Pu | | ^{238}Pu | | | |
	Airborne*	Surface†	Airborne‡	Surface†	^{239}Pu	^{238}Pu
A	1.3×10^{-4} to 4.6×10^{-5}	3.10×10^{-3} to 6.89×10^{-4}	7.2×10^{-6} to 3.4×10^{-7}	5.77×10^{-5} to 1.24×10^{-5}	0.04 to 0.2	0.006 to 0.6
AB	3.2×10^{-6} to 8.5×10^{-9}	7.70×10^{-5} to 3.85×10^{-5}	1.8×10^{-6} to 1.4×10^{-9}	1.57×10^{-6} to 7.66×10^{-7}	0.0001 to 0.08	0.0009 to 2

*From Fig. 11 as a function of particle diameter.
†From Table 1 for total ground-surface sample at 5-cm depth.
‡From Fig. 12 as a function of particle diameter.

TABLE 3 Concentrations of Plutonium and Cesium in Hanford Ground-Surface Solid Samples Reported in the Literature

| Location | Year | Concentration, μCi/g solids | | |
| | | ^{239}Pu | | ^{137}Cs |
		Range	Average	Range
On site (Nees and Corley, 1975)	1973	LAL* to 4×10^{-8}	9.7×10^{-7}	LAL* to 2.4×10^{-6}
Inside 200 Areas (Corley, Robertson, and Brauer, 1976)	1971	2.6×10^{-8} to 6.9×10^{-7}		Not reported
BC Area (Maxfield, 1974; Mishima and Schwendiman, 1973; Nees and Corley, 1975)	1974			2.1×10^{-5} to 2.5×10^{-2}
Within Hanford site boundary (3 to 26 km) (Corley, Robertson, and Brauer, 1976)	1971	7.2×10^{-9} to 1.2×10^{-7}		Not reported
Site perimeter (Nees and Corley, 1975)	1973	LAL* to 4×10^{-8}	7×10^{-7}	LAL* to 1.5×10^{-6}
Off site (19 to 34 km) (Corley, Robertson, and Brauer, 1976)	1971	LAL* to 7.6×10^{-8}		Not reported

*LAL, less than radiochemical analytical limit.

^{239}Pu and ^{137}Cs. The range for ^{239}Pu was from less than radiochemical analytical limits to 6.9 x 10^{-7} μCi/g of surface solids. The maximum reported ^{239}Pu concentrations for surface solids were within the 200 Areas. The minimum ^{239}Pu concentrations on surface solids reported (Corley, Robertson, and Brauer, 1976) in the literature occurred at 19 to 34 km from Hanford. Surface contamination levels were reported only out to 34 km. The ^{137}Cs concentrations ranged from 3 x 10^{-8} to 2.5 x 10^{-2} μCi/g. These contamination levels are used in Table 4.

Although there are only limited data for comparing the ratio of airborne soil to that of surface soils, these ratios were calculated from the available Hanford data summarized in Table 4. Table 4 is a summary of Hanford airborne solids concentrations (Sehmel, 1977c) for ^{239}Pu, ^{241}Am, and ^{137}Cs. Plutonium and americium concentrations were obtained from Figs. 14 and 19. Table 4 also shows ground-surface ratios of airborne solids. From these, ratios of airborne solids were determined. The last column shows maximum ratio ranges. Ratios vary from 1 x 10^{-3} to 1.5 x 10^{3}, which indicates that contamination levels on airborne solids can be either much less than or much greater than contamination levels on contaminated surface solids.

Caution should be used in interpreting these data. The ground-surface contamination data are limited in quantity and were not necessarily obtained in the same areas where resuspension experiments were performed. Airborne-particle and ground contamination levels are shown for the BC-Crib Area in the central columns of Table 4. In this case the BC-Crib Area was sampled (Mishima, 1973) in ten 1-m^2 areas. Data from these 10 squares indicated that surface contamination levels varied by about a factor of 100. Ratio ranges

TABLE 4 Ratio of Airborne to Ground-Surface Radioactivity Concentrations per Gram of Solids at Hanford

| | Concentration, μCi/g solids | | | | Maximum ratio range* |
| | Airborne solids | | Ground-surface contamination | | $\left(\dfrac{\mu\text{Ci/g air}}{\mu\text{Ci/g surface}} \right)$ |
Material	Minimum	Maximum	Minimum	Maximum	
^{239}Pu	2 x 10^{-8}	6 x 10^{-5}	LAL†	4 x 10^{-8} (Nees and Corley, 1975)	5 x 10^{-1} to 1.5 x 10^{3}
^{241}Am	1 x 10^{-7}	7 x 10^{-5}	NR†	NR†	
^{238}Pu	1 x 10^{-9}	1 x 10^{-6}	NR†	NR†	
^{137}Cs	2 x 10^{-5}	1 x 10^{-2}	NR†	NR†	
^{137}Cs at BC-Crib Area (10 1-m^2 areas) (Mishima and Schwendiman, 1973)	3 x 10^{-5}	7 x 10^{-4}	2.1 x 10^{-5}	1.2 x 10^{-3}	1 x 10^{-3}
Maximum reported value (Maxfield, 1974)				2.5 x 10^{-2}	3 x 10^{1}

*Ratios from positive reported ground contamination values.

†LAL, less than radiochemical analytical limit; NR, surface contamination levels not reported for all areas. If available, data for each area are reported separately.

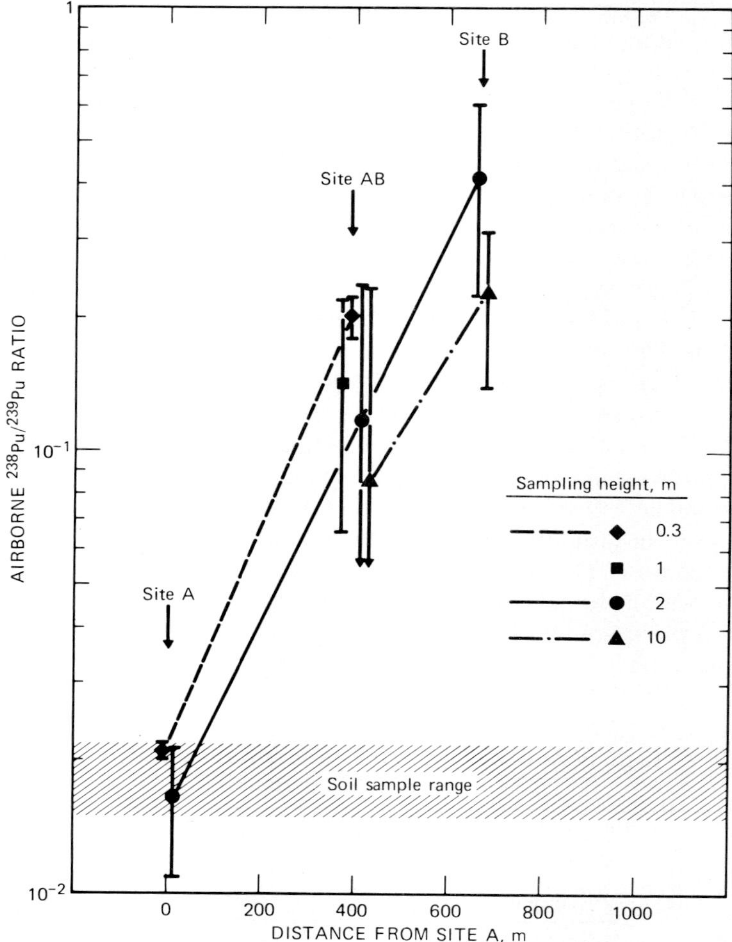

Fig. 16   Ratio of ^{238}Pu/^{239}Pu on nonrespirable airborne particles as a function of distance at Rocky Flats.

calculated from these data may be more representative than ranges calculated for ^{239}Pu. The ^{137}Cs data for the BC-Crib Area ranged from 10^{-2} to 30. The magnitude of this range is important since one might as a first approximation assume air contamination levels per gram of solids to be equal to surface contamination levels per gram of solids.

Airborne ^{238}Pu/^{239}Pu Ratios

Airborne ^{238}Pu/^{239}Pu ratios at Rocky Flats consistently changed (Sehmel, 1976a) from sampling site A to site B. Ratios shown in Fig. 16 are for total plutonium collected within each cowl at each sampling site and sampling height. At site A the ^{238}Pu/^{239}Pu ratio on airborne nonrespirable soil is comparable with ratios determined from 5-cm-deep surface-soil samples. The surface-soil range is shown as the crosshatched soil sample range.

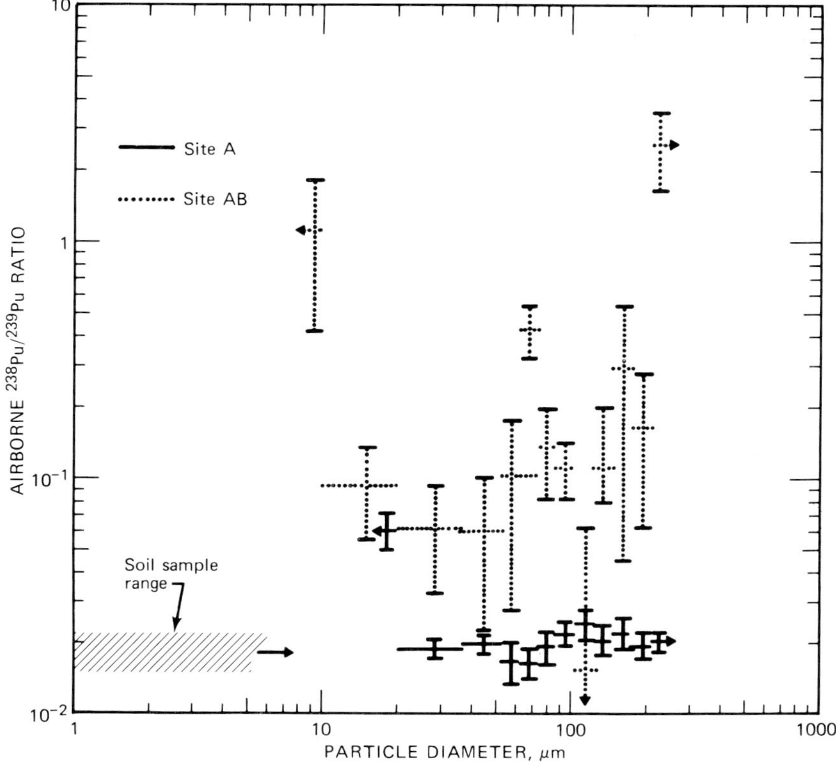

Fig. 17   Ratio of ^{238}Pu/^{239}Pu on nonrespirable airborne soil as a function of particle diameter at Rocky Flats.

This range (Krey et al., 1976b) represents surface-soil samples taken between the eastern security fence and beyond the eastern cattle fence to Indiana Avenue.

At sites AB and B, airborne ^{239}Pu/^{238}Pu ratios on nonrespirable particles are over one order of magnitude greater than similar ratios from 5-cm-deep surface-soil samples. These ratios do not appear to be significantly affected by sampling height between 0.3 and 10 m above ground level. However, considerations of the data-counting-statistics error bars at sites A and AB tend to indicate higher ^{238}Pu/^{239}Pu ratios closer to the ground.

The ^{238}Pu/^{239}Pu ratio was determined for all nonrespirable particle diameter ranges at sites A and AB at the 0.3-m sampling height. These results are shown in Fig. 17 along with the total ground-surface-soil sample range. At site A the ^{238}Pu/^{239}Pu ratio is nearly independent of particle diameter for all particle diameters above 20 μm. It is only for particles less than 20 μm that the ^{238}Pu/^{239}Pu ratio is significantly elevated at site A. The ^{238}Pu/^{239}Pu ratios on nonrespirable particles greater than 20 μm at site A are all comparable with ratios in the 5-cm-deep surface-soil samples. This similarity would be expected if there were no preferential ^{238}Pu to ^{239}Pu separation on soil surface. In contrast to site A, ^{238}Pu/^{239}Pu ratios for site AB are much different and are significantly elevated for all particle diameters above the surface-soil sample range.

One can only hypothesize as to why there should be a difference in ^{238}Pu/^{239}Pu ratios at sites AB and A as compared with soil-surface samples. One difference is that the surface activity level at site A is greater than that at site AB. As was shown in Table 1, the ground-soil-surface activity level at site A was 8 to 80 times as great as that at site AB. If plutonium particles were attacked by microorganisms (Wildung and Garland, 1977) in the soil, microorganism activity might be decreased by the increased activity level at site A. If microorganisms preferentially attacked ^{238}Pu at site AB, which had a lower plutonium contamination, ^{238}Pu on surface soils might become more readily available for resuspension. Other possibilities for increased resuspendibility of surface ^{238}Pu at site AB might be differences in soil chemistry between sites A and AB or preferential ^{238}Pu ejection (Oksza-Chocimowski, 1976) from particles during decay. Many possibilities exist, but the reasons for the elevated ^{238}Pu/^{239}Pu ratios at site AB are uncertain. Additional research is needed to determine causes of elevated ^{238}Pu/^{239}Pu ratios at site AB.

Estimation of Relative Plutonium Fluxes for Respirable and Nonrespirable Particles

Direct comparisons between airborne fluxes on respirable and nonrespirable particles were not made since respirable and nonrespirable samples were sampled differently. Respirable particles at Rocky Flats were sampled at a constant flow rate of 0.57 m^3/min, and nonrespirable particles were collected by inertial collection within cowls. To calculate the relative flux on respirable and nonrespirable particles, one needs to know (1) the average wind speed to determine the average flux on respirable particles and (2) how particles were collected within the cowl by inertial impaction. In this case inertial impaction means that particles would enter the cowl as if cowl sampling were isokinetic. Actually there is flow divergence around the cowl inlet. Consequently correction factors are needed for calculating true airborne particle fluxes for nonisokinetic sampling. However, nonisokinetic correction factors are not available. Even with these qualifications, one might still be interested in approximating the relative plutonium fluxes for respirable and nonrespirable particles. Consequently a simple calculation was made using the Rocky Flats data to illustrate the relative orders of magnitude for respirable and nonrespirable plutonium fluxes.

The horizontal plutonium flux can be estimated from airborne soil fluxes and plutonium concentrations on airborne soil. Since soil fluxes are reported (Sehmel, 1976a), plutonium concentrations on airborne soil will be discussed before plutonium fluxes. Plutonium concentrations of airborne respirable and nonrespirable soil are shown in Table 5. This table summarizes both total plutonium concentrations per gram of soil collected within cowls and respirable concentrations per gram for cases in which air was sampled continuously with particle cascade impactors. Concentrations of ^{239}Pu ranged from 2×10^{-6} to 6.2×10^{-5} μCi/g on respirable airborne soil. Concentrations of ^{238}Pu on respirable soil were less than radiochemical analytical limits. On nonrespirable soil ^{239}Pu concentrations ranged from 1×10^{-6} up to 3×10^{-4} μCi/g, and ^{238}Pu concentrations ranged from 2×10^{-7} up to 5×10^{-6} μCi/g. Respirable and nonrespirable concentrations were combined in the last two columns to estimate the average plutonium concentration on airborne soil. For this calculation isokinetic sampling was assumed. This assumption is discussed further in a later section. Total plutonium concentrations on airborne soil ranged from 1×10^{-6} up to 1.9×10^{-4} μCi/g for ^{239}Pu and from 2×10^{-7} up to 3×10^{-6} μCi/g for ^{238}Pu.

**TABLE 5 Plutonium Concentration on Airborne Soil Collected
During Continuous Air Sampling at Rocky Flats**

Sampling site	Sampling height, m	Respirable* ^{239}Pu	Nonrespirable ^{239}Pu	Nonrespirable ^{238}Pu	Total, assuming isokinetic sampling ^{239}Pu	Total, assuming isokinetic sampling ^{238}Pu
A	0.3	62.3	203	4.2	188	3.8
	2	4.6	313	5.1	98.3	1
AB	0.3	5.0	1	0.2	1.6	0.2
	1		11.5	1.6		
	2		12.4	1.6		
B	2	2.1	10.2	2.4	3.1	0.2
	10	38.2	10.0	4.2	36.4	0.2

The header spanning columns above reads: *Plutonium concentration, pCi/g (10^{-6} μCi/g)*.

*Respirable samples were collected for wind-speed increments rather than continuous sampling, ^{238}Pu was less than radiochemical analytical detection limits, 1 (d/min) g^{-1} = 0.45 pCi/g.

**TABLE 6 Average Plutonium Flux Entering Cowl-Impactor
System During Continuous Air Sampling at Rocky Flats**

Sampling site	Sampling height, m	Respirable* ^{239}Pu	Respirable* ^{238}Pu	Nonrespirable ^{239}Pu	Nonrespirable ^{238}Pu
A	0.3	4.06×10^{-7}	LAL†	1.02×10^{-5}	2.11×10^{-7}
	2	2.43×10^{-7}	LAL†	4.03×10^{-7}	6.68×10^{-9}
AB	0.3	3.22×10^{-8}	LAL†	5.66×10^{-5}	1.13×10^{-8}
B	2	1.16×10^{-8}	LAL†	8.23×10^{-9}	1.85×10^{-9}
	10	1.93×10^{-7}	LAL†	3.18×10^{-9}	1.32×10^{-9}

The header spanning columns above reads: *Plutonium flux, μCi m^{-2} day^{-1}*.

*Respirable is all material collected within a particle cascade impactor.
†LAL, less than radiochemical analytical limit.

Average airborne plutonium fluxes entering cowl inlets at Rocky Flats were calculated from airborne soil fluxes and plutonium concentrations on airborne soil. As shown in Table 6, plutonium fluxes on both respirable and nonrespirable particles were calculated. The calculation was based on collected plutonium, the sampling time, and the cross-sectional area of the cowl inlet. The cowl-inlet diameter was 15.2 cm. The maximum ^{239}Pu flux was 1×10^{-5} μCi m^{-2} day^{-1}, and the minimum flux was 3×10^{-9} μCi m^{-2} day^{-1}. On the basis of these limited data, these calculated fluxes might be used to estimate the total plutonium fluxes over larger integrated areas. However, such estimates should be made with caution since the flux variability at other sites and evaluations is unknown.

Average plutonium fluxes entering the cowl impactor system are used to estimate the respirable fraction of airborne plutonium. Estimates given (Sehmel, 1976a) in Table 7 are based on an isokinetic sampling assumption. Respirable fractions ranged from 3 to 98% of total airborne plutonium. However, one should use these numbers with caution. Plutonium collected within particle cascade impactors contained particles of 7-, 3.3-, 2.0-,

TABLE 7   Estimated Respirable Percent of Total
Airborne ^{239}Pu Flux at Rocky Flats

Sampling site	Sampling height, m	Assumed isokinetic*	Assumed correction†
A	0.3	3.8	9
	2	37.7	60
AB	0.3	36.3	59
B	2	58.5	78
	10	98.4	99

*Isokinetic = respirable/(respirable + nonrespirable).

$$\dagger\text{Correction} = \frac{(\text{respirable}) (U_a/U_s)}{(\text{respirable}) (U_a/U) + \text{nonrespirable}}$$

where respirable = (d/min) min^{-1} collected in impactor
nonrespirable = (d/min) min^{-1} collected in cowl
U_s = average flow rate through cowl inlet, 0.36 m/sec
U_a = average wind speed, 0.9 m/sec

and 1.1-μm diameter (which are impactor-stage 50% cutoff diameters for unit-density spheres) as well as smaller particles collected on the impactor backup filter. From the inhalation standpoint, particles collected on the 7-μm stage should not be included within the respirable particle size range. Only the smaller particles are usually considered respirable. However, in the present comparison between cowl-collected nonrespirable particles and impactor-collected respirable particles, there is much uncertainty in calculating the relative 7-μm particle concentration as compared with the cowl-collected particle concentration.

The better estimate of the respirable plutonium fraction at Rocky Flats is shown in the last column of Table 7. On the basis of the assumed correction factor, the fraction of respirable airborne plutonium changed from 4 to 98% for an isokinetic sampling assumption to 9 to 99% of the total airborne plutonium. True fractions of respirable plutonium should be between these limiting values.

Fractions of respirable plutonium have been reported (Volchok, Knuth, and Klemman, 1972) as 25% for plutonium collected within particle cascade impactors at Rocky Flats. Since the present results show that plutonium is also attached to particles in much larger size ranges for cowl-collected samples than for particles collected on the intial stage of an impactor, published fractions of respirable plutonium are probably indicative of maximum fractions at each sampling location rather than a true fraction.

Even if all plutonium collected in the cowl-impactor sampling systems were respirable, airborne plutonium concentrations were still below the MPC (International Commission on Radiological Protection, 1959). The $\text{MPC}_{168 \text{ hr-air}}$ for ^{239}Pu is 6×10^{-13} μCi/cm^3. The smallest fraction of respirable plutonium in Table 7 is 3.8%. This sample at 0.3 m for site A also represents the largest airborne plutonium concentration measured in our experiments. For this sample the real respirable ^{239}Pu concentration ($\leqslant$3.3 μm diameter) was 2×10^{-16} μCi/cm^3, whereas the concentration was 5×10^{-16} μCi/cm^3 if the 7-μm particles of the particle cascade impactor were

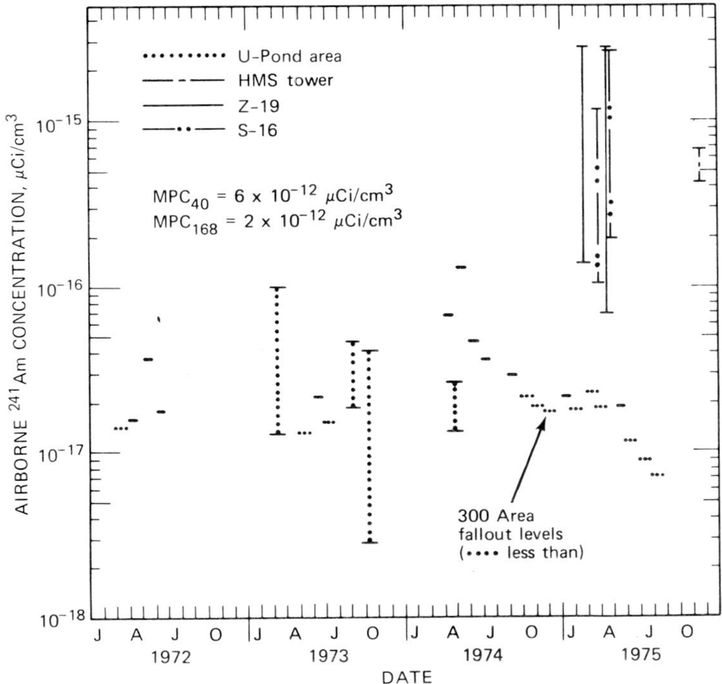

Fig. 18 Range of airborne ^{241}Am concentrations (above radiochemical detection limits) at on-site Hanford resuspension sites compared with fallout levels.

included. If all airborne plutonium is assumed respirable and soluble, a maximum calculated ^{239}Pu concentration is $(5 \times 10^{-16}/0.038)$ $1.3 \times 10^{-14}\ \mu Ci/cm^3$, which is still significantly less than the MPC of $6 \times 10^{-13}\ \mu Ci/cm^3$ for occupational exposure.

Americium Resuspension at Hanford

At Hanford airborne ^{241}Am concentrations measured (Sehmel, 1977c) at U-Pond, Z-19, S-16, and HMS tower Areas ranged from about 10^{-18} to $10^{-15}\ \mu Ci/cm^3$. However, other filter samples indicated that the total ^{241}Am collected was below radiochemical detection limits. In Fig. 18 airborne ^{241}Am concentrations are compared with the 300 Area fallout levels (Thomas, 1976). However, the comparisons are incomplete since data for ^{241}Am fallout level were not reported for all time periods. In addition, many data for fallout level were below radiochemical detection limits; these are indicated by short horizontal dotted lines. Obviously, actual fallout levels for these time periods could have been significantly less.

Airborne ^{241}Am concentrations at the U-Pond were of the same order of magnitude as reported fallout levels. In contrast, airborne concentrations at the Z-19 and S-16 Areas were significantly above fallout levels during February to May 1975. Comparisons of upwind and downwind tower air sample results at the S-16 Area showed that increased ^{241}Am airborne concentrations were from ^{241}Am resuspension from the dry S-16 Area. (This area has since been covered.) Maximum ^{241}Am concentrations measured in

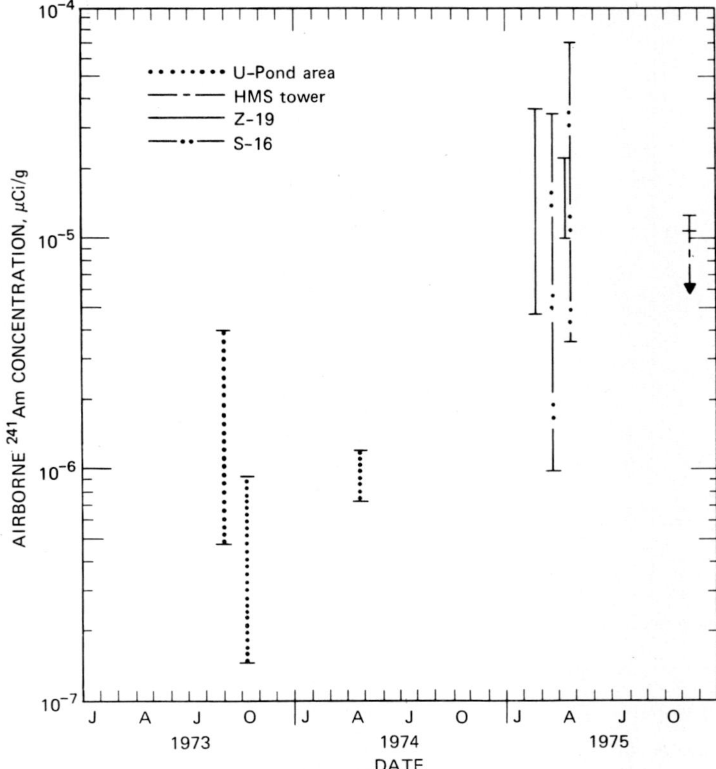

Fig. 19 Range of ^{241}Am concentrations on airborne solids at on-site Hanford resuspension sites.

November 1975 at the HMS tower are also comparable to concentrations measured at the S-16 and Z-19 Areas. However, ^{241}Am was above radiochemical detection limits on only two HMS tower air filters for this time period.

Concentrations of ^{241}Am on airborne solids at each resuspension site (shown in Fig. 19) ranged from about 10^{-7} to 10^{-4} μCi/g of airborne solids. Airborne concentrations were least for the U-Pond Area and greatest for the Z-19 and S-16 Areas. For the HMS tower data, all except two air filters collected less than radiochemical detection limits. Nevertheless, maximum ^{241}Am concentrations per gram of airborne solids at the HMS tower for these two samples were comparable to concentration ranges measured at the Z-19 and S-16 Areas.

Plutonium Resuspension from Off Site Near Hanford

Resuspension of plutonium off site near the Prosser barricade on the Hanford site was studied. The Prosser barricade is located about 19 to 20 km southeast (130 to 160°) of the fuel processing areas. Airborne solids were collected by sampling with particle cascade impactors and rotating cowl systems. Air sampling was only when wind was blowing from 190 to 260°. This range of southwest (225°) winds came from off site toward the

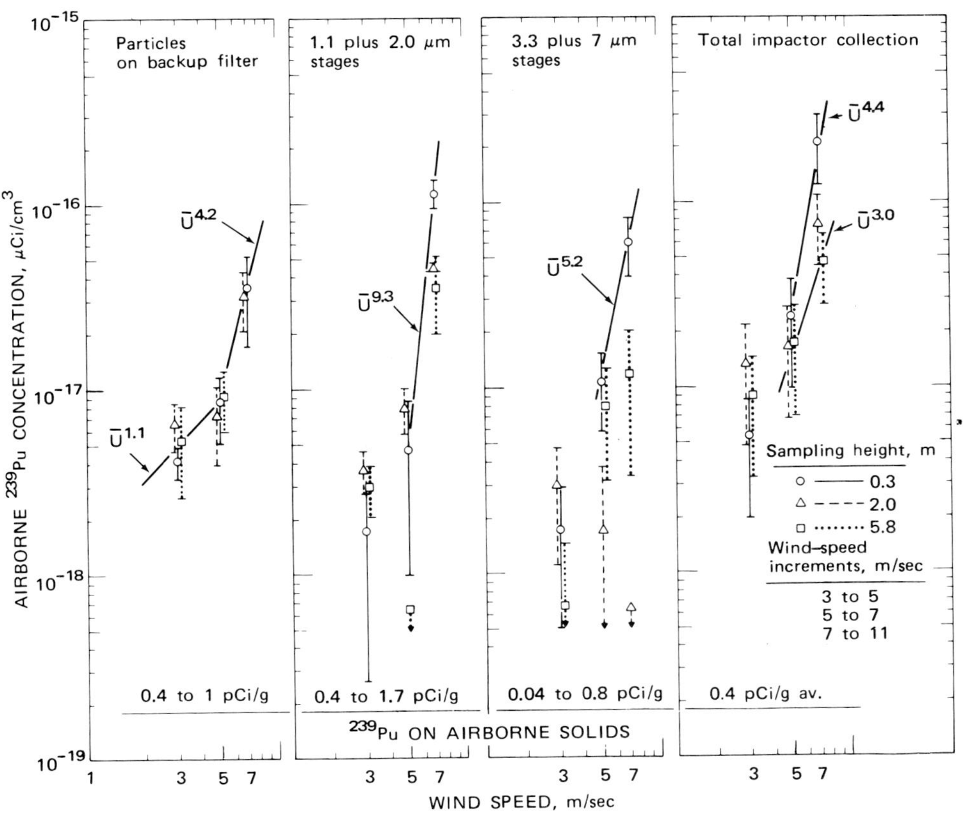

Fig. 20   Airborne ^{239}Pu concentrations near Prosser barricade at Hanford from Apr. 12 to June 29, 1976, when sampling only 190 to 270° winds.

Hanford area. All southwest winds were continuously sampled with rotating cowl systems for nonrespirable particles, whereas respirable particles were sampled with particle cascade impactors for wind-speed increments 3 to 5, 5 to 7, and 7 to 11 m/sec at a height of 1.5 m.

Airborne plutonium concentrations blowing in from off site are shown in Fig. 20 for the particle cascade impactor data. Airborne concentrations in both air and collected solid are given. Airborne plutonium concentrations determined with particle cascade impactors are shown as a function of wind-speed increments for plutonium collected on the impactor backup filter, the 1.1- plus 2.0-μm stages, and the 3.3- plus 7-μm stages.

Airborne plutonium concentrations increased with increasing wind speed. Concentrations increased up to about two orders of magnitude as wind speed increased from 3 to 5 up to 7 to 11 m/sec. Straight lines are drawn through data to direct attention to the wind-speed tendency of the data. For the plutonium collected on the cascade impactor backup filter, lines proportional to wind speed to the 1.1 and 4.2 power are shown. For wind speeds below about 5 m/sec, airborne plutonium concentrations tended to increase nearly linearly with wind speed. However, above 5 m/sec, plutonium concentrations increased with wind speed to the 4.2 power. At the present time it is unknown whether these indicated relationships would suggest a threshold wind speed of 5 m/sec for resuspension or whether a smooth curve should be drawn through all data points.

In other portions of this figure, selected straight lines suggesting wind-speed dependencies are shown only for wind speeds above 5 m/sec. For the 1.1- plus 2.0-μm impactor stages, airborne concentrations increased with wind speed to the 9.3 power. For the 3.3- plus 7-μm impactor stages, airborne concentrations increased with wind speed to the 5.2 power. For total plutonium collection within particle cascade impactors, a range of wind-speed dependency is shown on the right side of the figure. For a sampling height of 0.3 m, air concentrations increased with wind speed to the 4.4 power. However, at a sampling height of 1.8 m, airborne concentrations increased with wind speed to the 3.0 power.

These data in Fig. 20 are the first to show that plutonium is resuspended from off-site locations. In addition, airborne plutonium concentrations show a very high wind-speed dependency for this off-site plutonium resuspension. As will be discussed later, tracer wind resuspension rates suggest a wind-speed dependency to the 4.8 power. This is similar to the wind-speed dependency shown by this off-site plutonium resuspension data. However, other data on plutonium resuspension show a different wind-speed dependency. West of the U-Pond on the Hanford reservation, airborne plutonium concentrations increased (Sehmel, 1977c) with wind speed to only the 1.5 power. Reasons for these differences in the wind-speed dependency of on-site vs. off-site plutonium resuspension are unknown. Possibly a threshold wind velocity above which resuspension increases rapidly with wind speed was not exceeded at the U-Pond.

Nonrespirable airborne plutonium blowing from off site onto the Hanford reservation was also measured. In this case sampling direction was controlled by placing stops that allowed the rotating cowl (Fig. 3) inlet to rotate only with the range of 190 to 260°. Plutonium analysis was for the total nonrespirable solids collection in cowls at each height rather than as a function of particle size, as was done for Rocky Flats (see Figs. 11 and 13).

Plutonium-239 concentrations and fluxes for nonrespirable particles blowing from off site near the Hanford Prosser barricade are shown in Table 8. Plutonium concentrations on nonrespirable airborne solids ranged from 1.3×10^{-7} up to 2.1×10^{-7} μCi/g. These concentrations are similar to those shown in Fig. 15 for airborne nonrespirable particles collected west of the U-Pond.

TABLE 8   Plutonium Transport on Nonrespirable Particles from Off Site near the Prosser Barricade on the Hanford Reservation

Sampling height, m	^{239}Pu on airborne solids		Airborne ^{239}Pu nonrespirable flux, μCi m^{-2} day^{-1}	
	(d/min) g^{-1}	μCi/g	Only for 190 to 260° winds, 3 to 11 m/sec	For total time in field
0.3	0.29	1.3×10^{-7}	3.9×10^{-6}	8.0×10^{-7}
2	0.46	2.1×10^{-7}	4.0×10^{-6}	8.3×10^{-7}
5.8	0.32	1.5×10^{-7}	1.4×10^{-6}	2.8×10^{-7}

Airborne nonrespirable ^{239}Pu fluxes also were calculated for these Prosser barricade samples. Horizontal flux calculations were made for both the total time wind was between 3 and 11 m/sec and 190 to 260° and for the total time cowl air samplers were in the field. Fluxes are shown in the last two columns of Table 8. When the shorter time period (3 to 11 m/sec winds) is used for calculating the horizontal plutonium flux, fluxes range from 3.9×10^{-6} to 1.4×10^{-6} μCi m^{-2} day^{-1}. This Prosser barricade flux range is within the range measured near site A at Rocky Flats (shown in Fig. 8). However, if the total time cowl air samplers were in the field is used for calculation, Prosser barricade airborne nonrespirable off-site plutonium fluxes were lower and comparable to those measured at Rocky Flats sites AB and B (see Fig. 8).

These cross-comparison data show that there is a comparable plutonium flux on nonrespirable particles off site at Hanford, on site at Hanford U-Pond, and on site at Rocky Flats for the time periods investigated. Comparable fluxes may be caused by more soil being transported from off site at the Prosser barricade site. As shown by the range of ^{239}Pu concentrations on airborne soil from 1.3×10^{-7} up to 2.1×10^{-7} μCi/g, this range is greater than fallout levels in soil-surface samples. As shown in Table 3, reported ^{239}Pu concentrations in surface samples 19 to 34 km from Hanford had a range from 3.6×10^{-9} to 7.6×10^{-8} μCi/g. These last values are similar to a fallout level of 3.8×10^{-8} μCi/g measured (Hardy, 1974) at North Eastham, Mass.

Most plutonium collected appears not to have originated from fallout. Rather, most plutonium collected on these airborne nonrespirable particles near the Prosser barricade resembles weapons-grade plutonium (Krey, 1976; Krey et al., 1976a). Plutonium isotopic ratios (^{240}Pu/^{239,240}Pu) (in atom percent) for these nonrespirable samples were 6.10 ± 0.02 at 0.3-m height, 6.31 ± 0.02 at 2-m height, and 6.28 ± 0.03 at 5.8-m height. In comparison, the isotopic ratio determined from a sample of forest-fire smoke plume near Mt. St. Helens, Wash., was 13.82 ± 0.05. Isotopic ratios for respirable particles sampled near the Prosser barricade were not determined. Although plutonium was blowing from off site near the Prosser barricade, airborne respirable plutonium concentrations were below MPC's, as is shown in Fig. 20.

Relative Amounts of Radionuclide "Clusters" on Particles Resuspended

Data from these studies indicate that occasionally some more-radioactive-than-normal particles or clusters of radioactive particles were resuspended and collected on sampling filters. In the October 1973 plutonium data for Hanford shown in Figs. 7 and 14, one filter at 6.1-m height collected 6.5×10^{-5} μCi/g of airborne solid (1.2×10^{-15} μCi/cm^3 of filtered air). The plutonium measurement was 36 times as great as the maximum value of 1.8×10^{-6} μCi/g of airborne solid (4.0×10^{-17} μCi/cm^3 of filtered air) collected on other filters simultaneously sampling at heights of 6.1 and 0.3 m. We hypothesize that this relatively high plutonium collection on this filter was due to collection of one or more larger (more radioactive than normal usually resuspended and sampled) particles or clusters of particles.

The size of the larger particle(s) cannot be measured since the filter samples were dissolved for plutonium analysis. Nevertheless, the relative size can be estimated from the ratio of radioactivities collected for "normal" and "larger" particle sizes. Assuming that the radioactivity of a particle is proportional to its volume, then the filter with a plutonium activity 36 times as great as the next highest measured activity may have collected larger particles of plutonium activity 36 times as great as the activity of normal particles. This would correspond to a diameter for the larger, less-frequently resuspended particle(s), which may be three times the diameter of normally resuspended plutonium particles. Only since extensive air samples have been collected has the resuspension of these unusually active particles been suggested. However, the frequency of their resuspension appears to be very low.

The presence of a more-radioactive-than-normal resuspended particle is also indicated by the ^{241}Am data shown in Figs. 18 and 19. The two positive ^{241}Am air samples for the HMS tower may indicate that larger-than-normal ^{241}Am particles or clusters of particles were collected on these filters.

As discussed earlier for resuspended particles at Rocky Flats, a more-radioactive-than-average particle (or cluster of particles) was collected on the 2-μm stage of a particle cascade impactor.

Resuspension Factors at Hanford

Resuspension factors have been used to describe resuspension air concentrations. The resuspension factor (expressed in units of meters^{-1}) is defined as the airborne concentration of contaminant per cubic meter divided by the surface contamination level per square meter immediately below the point where the airborne concentration was measured. Often air concentrations for determining resuspension factors have been measured from about 1 to 1½ m above ground. However, airborne concentrations are a function of the upwind contamination level, not a contamination level immediately below the air-concentration measurement site. It is the transport from upwind contamination sites to the concentration measurement site that determines the airborne concentration.

Although the validity of resuspension factors is questionable, they were for a long time the only method for estimating air concentrations. Consequently resuspension factors were estimated from data obtained at Hanford (Sehmel, 1977c). Resuspension factors and the basis for their calculation are shown in Table 9 for both ^{239}Pu and

TABLE 9 Resuspension Factors at Hanford

Material	Area	Air concentration (χ), μCi/cm^3		Surface contamination (G), μCi/m^2		Resuspension factor* range, m^{-1}
		Minimum	Maximum	Minimum	Maximum	
^{239}Pu	Inside chemical separation areas (Corley, Robertson, and Brauer, 1976)	7×10^{-18}	8×10^{-15}	4.9×10^{-4}	1.2×10^{-2}	6×10^{-10} to 2×10^{-5}
^{137}Cs	BC (Bruns, 1976; Mishima, 1973)	2×10^{-15}	2×10^{-14}	0.29	55.4	4×10^{-11} to 7×10^{-8}

*Resuspension factor = $10^6 \chi /G$.

^{137}Cs; literature values are used as an indication of ground-surface contamination levels. Airborne concentrations (Sehmel, 1977c) are shown in Fig. 9. Resuspension factors calculated from air concentrations and ground-surface contamination levels are shown in the last column. Resuspension factors for ^{239}Pu range from 6×10^{-10} to 2×10^{-5} m^{-1}. Resuspension factors for ^{137}Cs range from 4×10^{-11} to 7×10^{-8} m^{-1}. These ranges, from 10^{-11} to 10^{-5} m^{-1}, are within ranges reported in the literature (Mishima, 1964; Sehmel and Lloyd, 1976a).

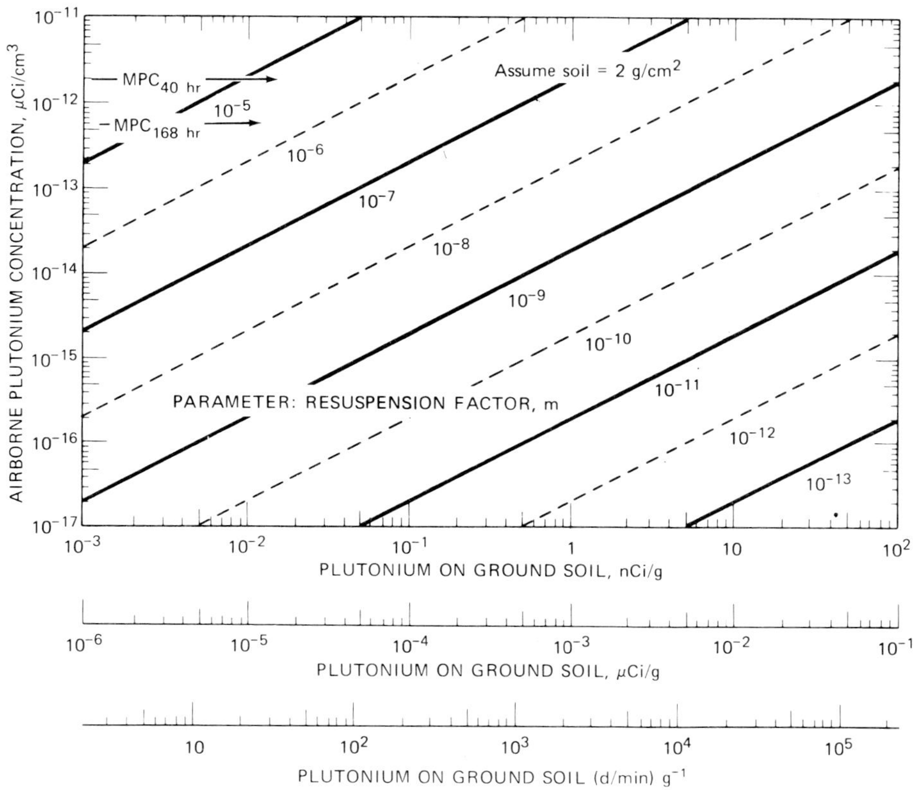

Fig. 21 Equivalency of airborne plutonium concentrations and ground-surface concentrations based on the resuspension-factor concept.

Prediction of Airborne Radionuclide Concentrations from Resuspension Factors

Airborne concentrations can be predicted from resuspension factors and surface contamination levels if both values are known. Equivalencies of airborne concentrations and ground-surface concentrations are shown in Fig. 21. The reader can estimate two of the parameters and use this figure to predict the third parameter. However, as indicated in the last section, the range of experimental resuspension factors is very large. Consequently realistically predicting the relationship between surface and airborne concentration is fraught with uncertainties.

Prediction of Airborne Radionuclide Concentrations from Airborne Solids at Hanford

Airborne concentrations can either be determined experimentally or calculated on the basis of simplifying assumptions. For example, one assumption is that radionuclide concentrations on airborne solids are equal to radionuclide concentrations per gram of ground-surface contaminated solids. As is shown in Table 4 for very limited data, this assumption is usually not valid since the ratio of radionuclide concentration per gram of airborne solids to the radionuclide concentration per gram of surface solids ranged from

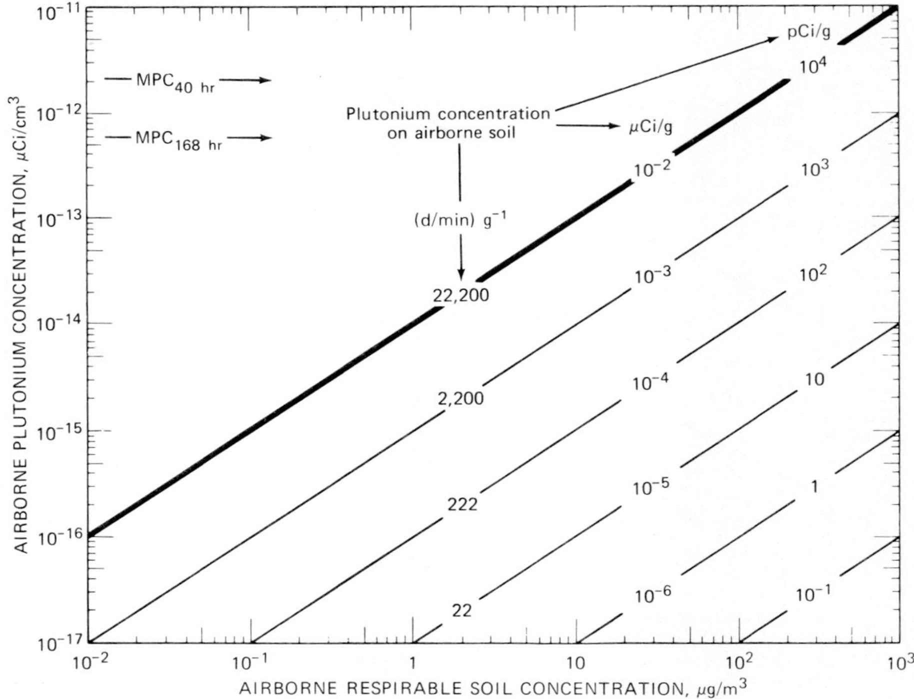

Fig. 22 Equivalencies between plutonium on airborne soil and airborne plutonium concentrations.

10^{-3} to 10^3. Nevertheless, two equality assumptions are used in this section to predict airborne radionuclide concentrations.

Airborne plutonium concentrations can be predicted by assuming both parameters of airborne soil concentration and plutonium concentration per gram on that airborne soil. In this case the equivalency between these two parameters and airborne plutonium concentrations is shown in Fig. 22. As a point of reference, airborne concentrations at Hanford are (Sehmel, 1977b) about 80 $\mu g/m^3$ for wind speeds of 5 m/sec. Hence, in using Fig. 22, a plutonium concentration on airborne soil of approximately 10^{-2} $\mu Ci/g$ would be required (assuming a surface contamination depth of 1 cm and a soil density of 2 g/cm^3) at an airborne soil concentration of 80 $\mu g/m^3$ to exceed or approach maximum permissible airborne ^{239}Pu concentrations.

As is shown in Tables 2 and 3, there is no experimental basis for adequately predicting plutonium concentration on total airborne soil since airborne soil usually consists of both uncontaminated soil blowing in from off site and the resuspended contaminated soil. At our present state of knowledge, there are only limited data for ratios of plutonium on total airborne soil compared with total surface soil. Moreover, there is no experimental resuspension data to relate plutonium concentrations on respirable surface soils vs. plutonium concentrations on respirable airborne soils.

Airborne solids concentration levels were estimated from data (Sehmel, 1976b; 1977c) shown in Fig. 23. Airborne particle volume distributions were determined at the Hanford area using both an optical particle counter and a cowl-impactor system (Sehmel,

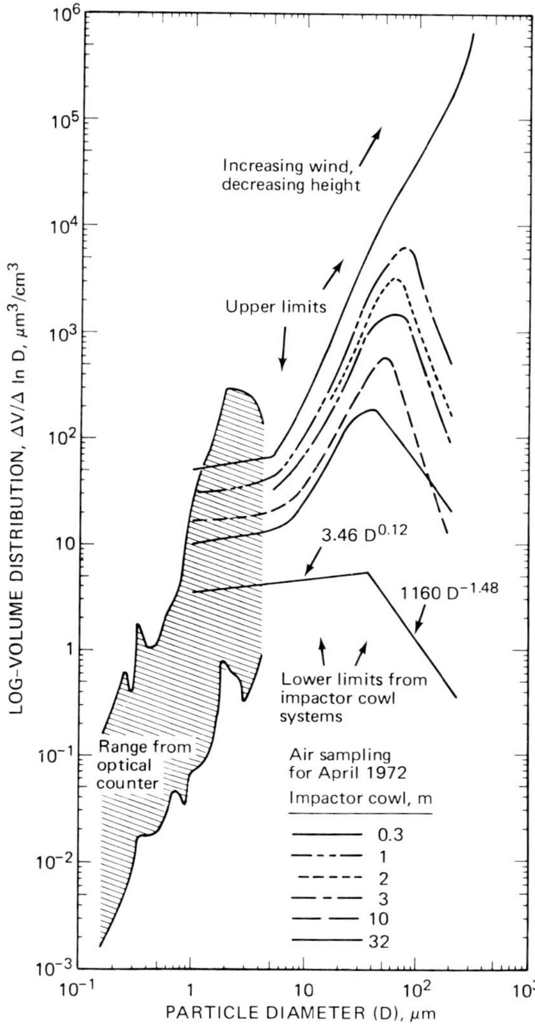

Fig. 23 Airborne particle volume distributions at Hanford.

1973a). The range obtained with the optical particle counter is shown as a crosshatched area. The cowl-impactor data have an upper limit, increasing with wind speed and decreasing height (the April 1972 data), and a lower limit for other test periods, indicated by the lines described by $3.46\,D^{0.12}$ and $1160\,D^{-1.48}$, where D is the particle diameter. The upper, or maximum, limits of the curves for any particle diameter were integrated as a function of particle diameter to determine maximum airborne mass loadings. The lower limits were also integrated as a function of particle diameter. These integrations predicted the solids mass loading per unit volume of air (shown in Table 10) as a function of four different particle diameter ranges: 0.16 to 1, 1 to 10, 10 to 100, and 100 to 230 μm.

Particles less than 10 μm in diameter are frequently considered respirable (i.e., small enough to be inhaled into the lungs), even though 3.5 μm appears to be more exact. In the following discussion, particles with diameters less than 10 μm are considered respirable and those larger than 10 μm are nonrespirable.

TABLE 10   Calculated Airborne Concentrations from Airborne Solids Concentrations
and Surface Contamination

Airborne particles	Particle diameter range, μm					
			Respirable			Nonrespirable
	0.16 to 1	1 to 10	0.16 to 10	10 to 100	100 to 230	10 to 230
Total volume, μm^3/cm^3						
Upper limit	4.58	3.5×10^2	354.6	2.22×10^4	9.37×10^4	1.16×10^5
Lower limit	0.036	3.83	3.87	9.39	0.609	10.0
Solids mass loading						
Upper limit, mg/m^3	0.00916	0.704	0.713	44.4	187.4	231.8
Lower limit, μg/m^3	0.072	7.66	7.73	18.78	1.218	20.0
Calculated* extreme maximum airborne concentration from limits of airborne solids concentrations, μCi/cm^3						
^{239}Pu, Upper limit			4.92×10^{-16}			1.60×10^{-13}
Lower limit			5.33×10^{-18}			1.38×10^{-17}
^{137}Cs, Upper limit			1.78×10^{-11}			5.79×10^{-9}
Lower limit			1.93×10^{-13}			5.00×10^{-13}

*Concentration on airborne solids assumed equal to maximum reported in Table 3 for concentration on ground
solids:

$$^{239}\text{Pu} = 6.9 \times 10^{-7}\ \mu\text{Ci/g}$$
$$^{137}\text{Cs} = 2.5 \times 10^{-2}\ \mu\text{Ci/g}$$

where

$$^{239}\text{Pu: MPC}_{40\,hr} = 2 \times 10^{-12}\ \mu\text{Ci/cm}^3$$
$$\text{MPC}_{168\,hr} = 6 \times 10^{-13}\ \mu\text{Ci/cm}^3$$
$$^{137}\text{Cs: MPC}_{40\,hr} = 6 \times 10^{-8}\ \mu\text{Ci/cm}^3$$
$$\text{MPC}_{168\,hr} = 2 \times 10^{-8}\ \mu\text{Ci/cm}^3$$

Mass loadings in these respirable and nonrespirable ranges are shown as upper limits of
0.7 mg/m^3 for respirable particles and 231.8 mg/m^3 for nonrespirable particles. The
lower limits are 7.7 μg/m^3 for respirable particles and 20 mg/m^3 for nonrespirable
particles. These mass loadings were multiplied by maximum surface contamination levels
for ^{239}Pu and ^{137}Cs as shown in Table 3. This approach yielded a predicted maximum
airborne ^{239}Pu concentration for respirable particles of 4.92×10^{-16} μCi/cm^3 and a
predicted lower limit of 5.33×10^{-18} μCi/cm^3.

Upper and lower limits of ^{137}Cs concentrations on respirable and nonrespirable
particles were calculated similarly. However, a comparison of airborne ^{137}Cs and ^{239}Pu
concentrations predicted by this method and measured airborne concentrations shows the
shortcomings of this calculational approach. The predicted respirable range of ^{239}Pu
concentrations from 5.33×10^{-18} to 4.92×10^{-16} μCi/cm^3 is within the lower two
orders of magnitude of the 10^{-18} to 10^{-14} μCi/cm^3 experimental range shown in
Fig. 7. This similarity in range is considered fortuitous when one also compares the
predicted ^{137}Cs respirable range from 1.93×10^{-13} to 1.78×10^{-11} μCi/cm^3 with the
experimental range from 2×10^{-16} to 3×10^{-13} μCi/cm^3 (Sehmel, 1977c). For ^{137}Cs
the minimum predicted airborne concentration of 1.93×10^{-13} μCi/cm^3 is 0.6 of the
maximum experimental concentration of 3×10^{-13} μCi/cm^3. However, the maximum
predicted airborne concentration for ^{137}Cs of 1.78×10^{11} μCi/cm^3 is 60 times the
maximum experimental concentration of 3×10^{-13} μCi/cm^3. These ratio comparisons

for ^{239}Pu and ^{237}Cs of predicted maximum to the experimental maximum airborne concentration show a ratio range from 0.05 for ^{239}Pu to 60 for ^{237}Cs. Thus there are at least three orders of magnitude uncertainty in using the mass-loading approach in calculating the maximum expected airborne radionuclide concentration.

This airborne-particle mass-loading approach to calculated airborne radionuclide concentrations does not distinguish the sources of airborne material. For simple wind resuspension, airborne solids included contaminated solids (lower or higher radionuclide concentrations per gram of solid) blown in from the surrounding area as well as solids resuspended from the prime resuspension study site. In contrast, airborne solids above a mechanically disturbed area are resuspended from the study site. In this case possibly such an equality assumption of radionuclide concentration per gram of solids would be appropriate for mechanical disturbances (Milham et al., 1976) of contaminated soil area.

Tracer-Particle Resuspension

Tracer particles placed on selected surfaces were used to measure (Sehmel, 1977b) resuspension rates caused by both mechanical and wind resuspension to determine particle resuspension rates. Mechanical resuspension was measured for vehicular traffic on asphalt and cheat grass areas and for pedestrian traffic on an asphalt area. Wind resuspension was measured as a function of wind speed and also as a function of respirable and nonrespirable particle diameters.

Mechanical Resuspension Rates. Mechanical resuspension includes both vehicular resuspension and pedestrian resuspension.

Vehicular Resuspension. A ¾-ton truck and a car were driven over ZnS tracer particles (8-μm mass median diameter) placed on one lane of asphalt road. Resuspended tracer was measured to determine resuspension rates (Sehmel, 1973b). Results are shown in Fig. 24 for particle resuspension rates at vehicle speeds of 5, 15, 30, and 50 mph. The resuspension rate is the fraction of particles resuspended from the tracer lane each time the vehicle was driven down the road (fraction resuspended per pass). When a car was driven through the tracer lane at speeds up to 30 mph, resuspension rates increased with the square of car speed from about 10^{-4} to 10^{-2} fraction resuspended per pass. This means that these resuspension rates were proportional to car-generated turbulence. When the car was driven on the lane adjacent to the tracer lane, resuspension rates were lower for each vehicle speed but increased with vehicle speed from about 10^{-5} to 10^{-3} fraction resuspended per pass.

Resuspension was also measured when a ¾-ton truck was driven on the tracer lane. Resuspension rates for truck passage increased from about 10^{-3} to 10^{-2} fraction resuspended per pass. Since resuspension rates were higher, truck-generated surface-stress turbulence appears to have been much greater than that for car-generated turbulence. For vehicle speeds above 20 mph, resuspension rates for car and truck passage are comparable. This similarity might be caused by tire surface-stress turbulence rather than by air turbulence.

Resuspension rates were also a function of the time tracer particles were on the asphalt road. As shown in Fig. 25, particle resuspension rates decreased as a function of time. For these data the tracer had been on the road for 4 days. Vehicle-generated resuspension rates increased from about 10^{-5} to about 10^{-3} fraction resuspended per pass as vehicle speed increased from 5 to 50 mph. For both vehicles resuspension was

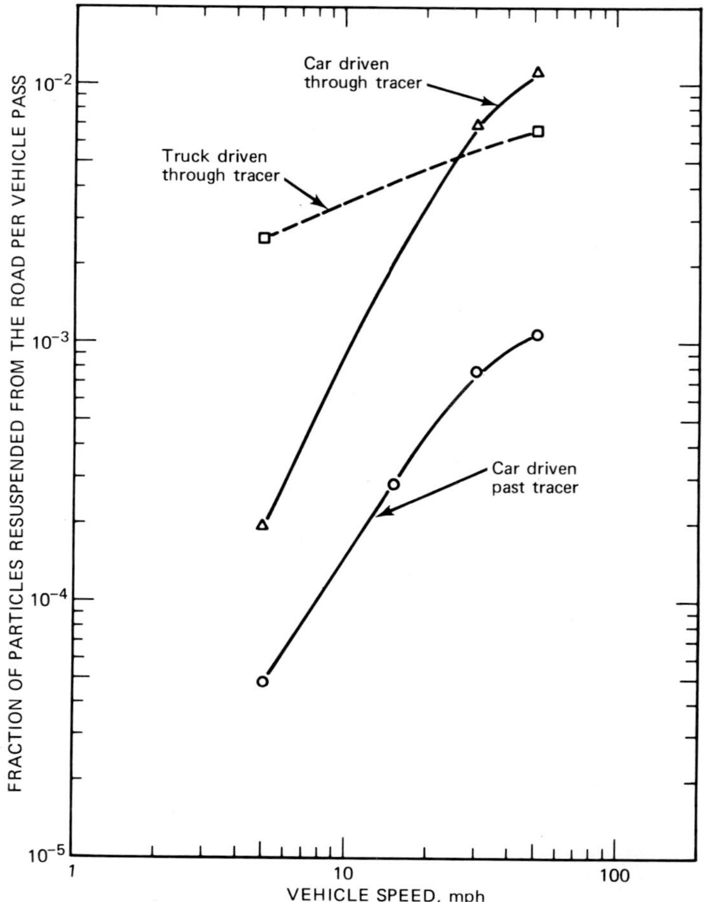

Fig. 24 Rates of tracer particle resuspension caused by vehicle passage over an asphalt road.

greater when the vehicle was driven through the tracer lane than when driven on the lane adjacent to the tracer lane.

Resuspension caused by truck passage through a cheat grass area was also measured (Sehmel, 1976c; 1977b). Results are shown in Fig. 26 along with resuspension rates from the asphalt road. Truck-caused resuspension from the cheat grass area was always less than that from the asphalt road. This decrease is attributed to the protective action of cheat grass in hindering truck-generated turbulence from reaching the ground and resuspending the tracer.

Resuspension from the cheat grass area decreased for truck speeds from 5 to 30 mph. This decrease is attributed to the sequence of experimental truck speeds. The initial truck speed was 5 mph. Apparently the relatively larger resuspension rate at 5 mph was caused by the most readily resuspended particles being removed from the cheat grass. In succeeding experiments at increasing truck speeds up to 15 mph, and possibly 30 mph, all readily resuspended tracer was removed from the cheat grass foliage. When the truck

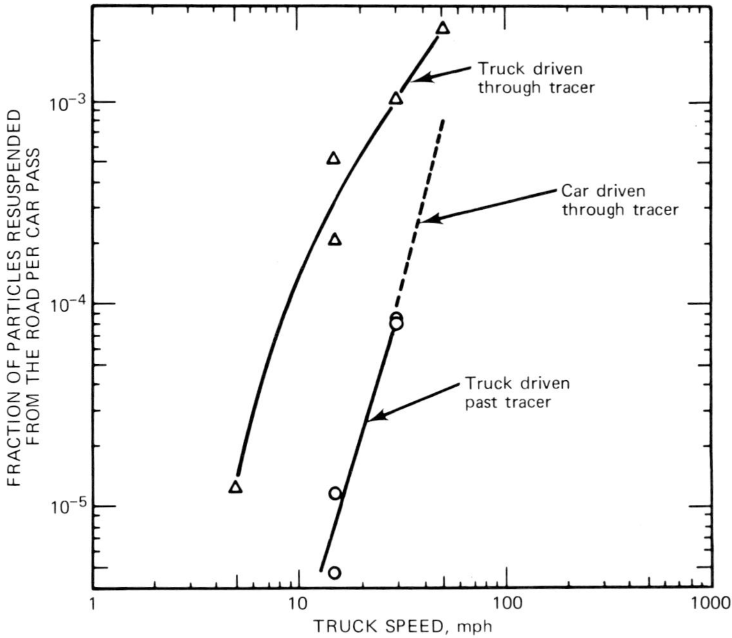

Fig. 25 Rates of tracer particle resuspension caused by vehicle passage over an asphalt road 4 days after particle deposition.

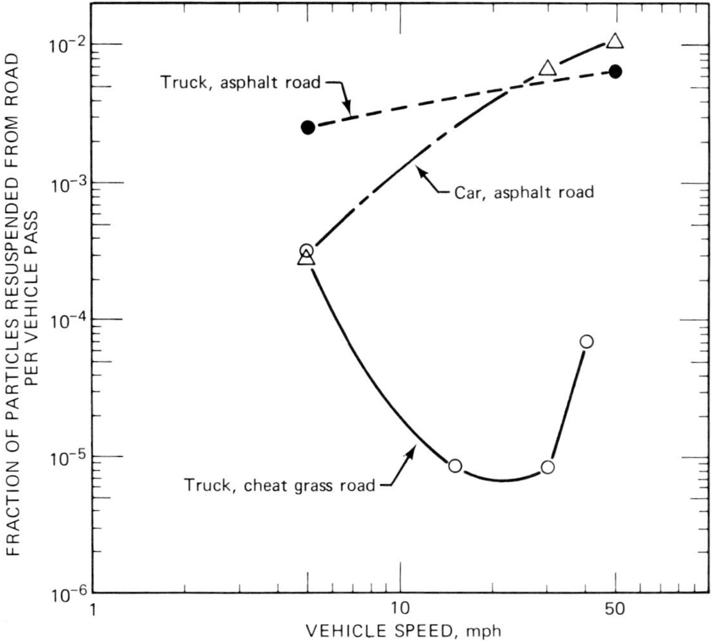

Fig. 26 Rate of tracer particle resuspension caused by vehicle passage over asphalt and cheat grass roads.

speed was subsequently increased from 30 to 40 mph, resuspension per pass also increased. Apparently increased air turbulence at the base of the cheat grass increased resuspension rates.

Pedestrian Resuspension.   Resuspension caused by a man walking along the ZnS tracer lane of the asphalt road was also measured (Sehmel, 1977b; Sehmel and Lloyd, 1972). The man walked across the tracer area in a leisurely fashion; the stride and paces per second were not measured. For tracer on a 3-m-wide road lane, the reported resuspension rate was the fraction of particles resuspended each time the person walked down the length of the tracer lane. With wind speeds of 3 to 4 m/sec, pedestrian-caused resuspension rates were from 1×10^{-5} to 7×10^{-4} fraction resuspended per pass along the tracer lane. This pedestrian-generated resuspension was greater than wind resuspension during the experiment.

Wind-Caused Resuspension.   Experimental values of wind-caused resuspension rates of tracer particles from environmental surfaces have not been experimentally determined from mass balance techniques other than for the present data (Orgill, Petersen, and Sehmel, 1976; Orgill, Sehmel, and Petersen, 1976; Sehmel, 1975; 1977b; Sehmel and Lloyd, 1972; 1976a; 1976c). Some data were initially obtained using 8-μm mass-median-diameter (MMD) ZnS particles and average wind speeds from 1 to 5 m/sec. More extensive data as a function of wind speed were obtained using submicrometer $CaMoO_4$ particles. Average resuspension rates for ZnS particles were measured for resuspension from an asphalt surface (Sehmel and Lloyd, 1972) and a cheat grass surface (Sehmel, 1976c). For average wind speeds of 1 to 4 m/sec, wind resuspension rates from an asphalt surface ranged from 5×10^{-9} to 6×10^{-8} fraction resuspended per second. For average wind speeds of 1 to 5 m/sec, wind resuspension rates from a cheat grass surface ranged from 5×10^{-9} to 6×10^{-8} fraction resuspended per second.

Wind-caused resuspension was measured for submicrometer $CaMoO_4$ particles deposited in a lightly vegetated area on the Hanford area. Tracer particles were deposited in a circular area of 23-m radius around a centrally located air-sampling tower. Resuspended particles were measured at the tower as a function of wind-speed increments for respirable particle diameters and at all wind speeds for nonrespirable particles. Respirable particles were collected within particle cascade impactors (Fig. 3), and nonrespirable particles were collected by impaction and gravity settling within cowls. Resuspension rates for each size range airborne were calculated by assuming that the entire tracer source was also in that same size range.

Wind-caused resuspension rates for the tracer–host soil particles as resuspended are shown in Fig. 27 as a function of wind speed. Resuspension rates ranged from about 10^{-11} to 10^{-7} fraction resuspended per second. Different functional dependencies of resuspension rates on wind speed can be obtained from these data, depending on which set of wind-speed increments is used. During the January to February period, air sampling was for large wind-speed increments; in subsequent experiments wind-speed increments were smaller. The straight lines shown in Fig. 27 were drawn through all data points. In these cases resuspension rates increased with the 1.0 to 4.8 power of wind speed. However, when only data points for smaller wind-speed increments were used, wind-caused resuspension rates increased with wind speed to the 4.8 power for 7-, 3.3-, 2.0-, and 1.1-μm-diameter particles as well as for the smaller particles collected on the cascade impactor backup filter.

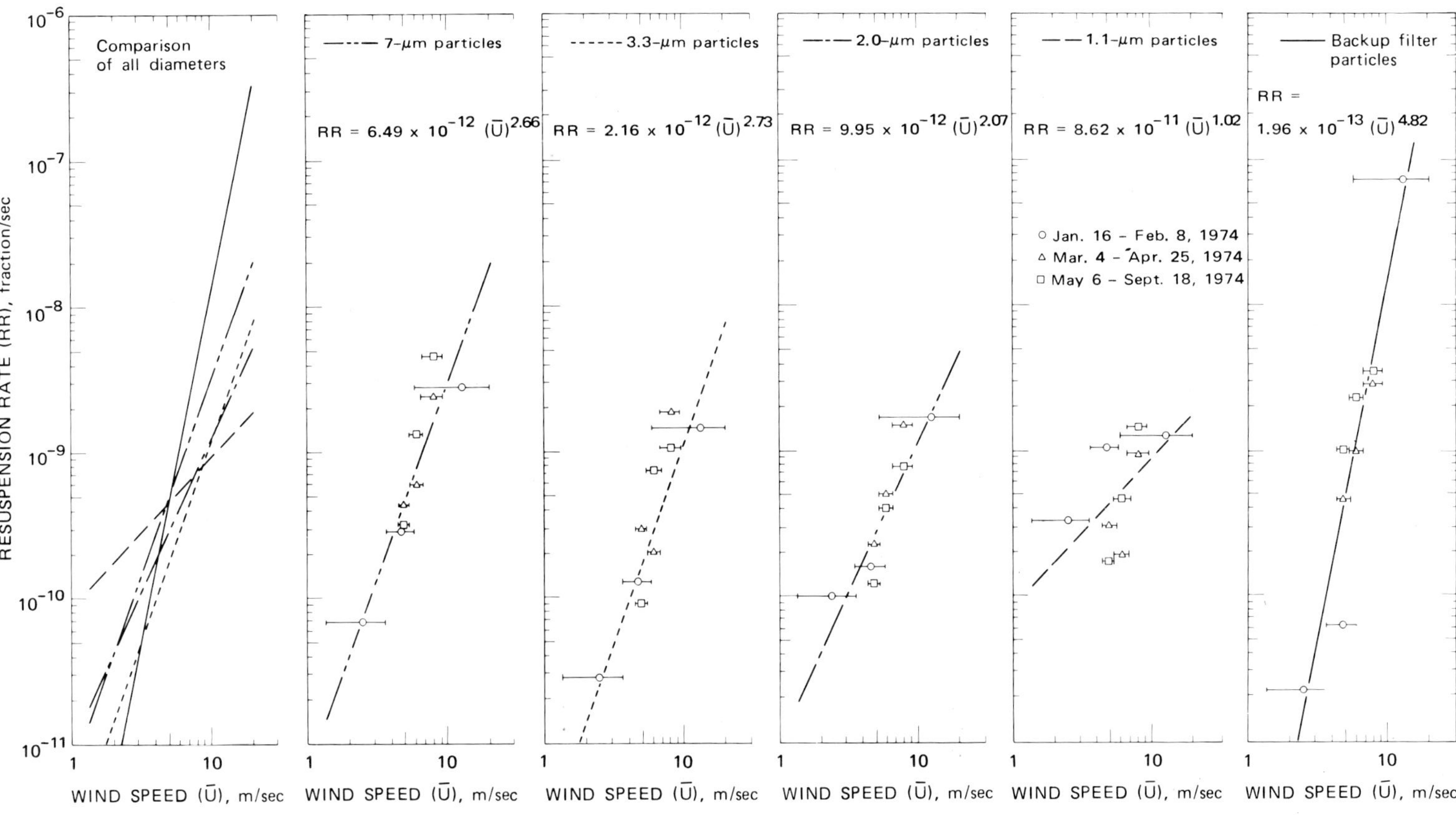

Fig. 27 Tracer resuspension rates as a function of particle diameter and wind-speed increments.

For comparable wind-speed increments, tracer resuspension rates were nearly independent of time the tracer was on the ground surface. However, it is often assumed in theoretical modeling that particles become less available for resuspension with time. The assumption in these models is that pollutant particles become fixed or attached to soil particles and subsequently "migrate" into the ground surface. This process is called weathering but is poorly understood.

The independence of wind-caused resuspension rates with time is a significantly different observation than some others have made. Literature on radioactive resuspension indicates that airborne radioactivity concentrations decrease with a weathering half-life of 35 days (Anspaugh et al., 1969; Wilson, Thomas, and Stannard, 1961). In contrast, if there is a weathering half-life for the controlled tracer experiments described above, the weathering half-life must be on the order of years. Some differences in reported weathering might be explained by how air samples were collected. In work reported by others, air concentrations were measured continuously. In contrast, in our tracer experiments air concentrations and hence resuspension rates were measured as a function of wind speed. Even these differences in determining weathering half-lives illustrate how poorly weathering is understood.

Average wind-caused tracer resuspension rates are reported for both respirable and nonrespirable particles in Fig. 28. In these cases respirable refers to all particles collected within cascade impactors and nonrespirable refers to particles collected within cowls. Nonrespirable particle resuspension rates were nearly independent of time and were of the order of 10^{-11} fraction resuspended per second.

Resuspension rates for respirable particles ranged from about 10^{-11} to 10^{-7} fraction resuspended per second. These resuspension rates did not decrease with time. For the first two sampling periods, resuspension was measured for all wind speeds. In succeeding experiments resuspension rates were measured only for wind speeds above 1 and above 4 m/sec. The upper, or solid line, portion of the respirable particle curve corresponds to resuspension rates calculated for the wind sampling periods. These periods correspond to wind speeds above 1 and above 4 m/sec. The lower limit of the respirable particle curve corresponds to resuspension rates calculated by assuming that resuspension time corresponds to the total time that cascade impactors were in the field (i.e., time included for winds less than 1 and less than 4 m/sec).

Initial Generalized Wind-Resuspension-Rate Correlation

Guidelines based on existing experimental resuspension-rate data are needed for estimating resuspension rates. An initial correlation (Sehmel, 1975; 1977b) was developed from data reported for uranine particles resuspended from a smooth surface, ZnS from an asphalt surface (Sehmel and Lloyd, 1972), submicrometer molybdenum tracer from a vegetated desert soil (Sehmel and Lloyd, 1976a), and DDT from a forest (Orgill, Sehmel, and Petersen, 1976; Orgill, Petersen, and Sehmel, 1976). Each of these surfaces has a much different estimated aerodynamic surface-roughness height, z_0, ranging from 4×10^{-3} cm for the smooth surface to 1 m for a forest. Roughness height is calculated from the log-linear velocity profile and is the height at which the extrapolated velocity profile reaches zero velocity.

Ranges of measured average resuspension rates were correlated (Sehmel, 1975; 1977b) as a function of measured or estimated surface-roughness heights (z_0) in Fig. 29. Resuspension rates range seven orders of magnitude from 10^{-10} to 10^{-3} fraction

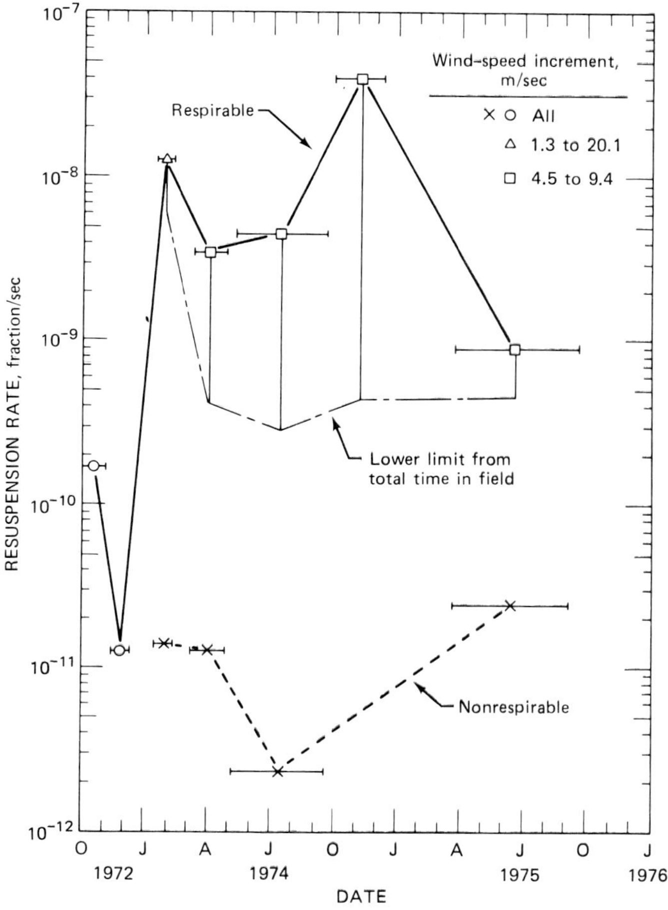

Fig. 28 "Average" wind-caused tracer resuspension rates (lightly vegetated desert on Hanford reservation).

resuspended per second. The practical significance of these numbers can be made apparent by noting that a year is 3.2×10^7 sec.

This initial resuspension-rate correlation shows that resuspension rates decrease as surface roughness increases, at least for the three smaller roughness heights. However, measured resuspension rates for DDT sprayed on a forest are two orders of magnitude greater than rates for the desert soil.

This is an unexpected and unexplained increase in resuspension rates. A possible explanation of the increase might be increased resuspension caused by tree movement in the wind. Also, various other gross differences in controlling variables and experimental factors may have influenced results. Since the data are so extremely limited, this apparent correlation should be used with extreme caution until correlations based upon several physical parameters instead of only z_0 are developed. Nevertheless, this initial correlation does give some justification for estimating resuspension rates until better correlations are developed.

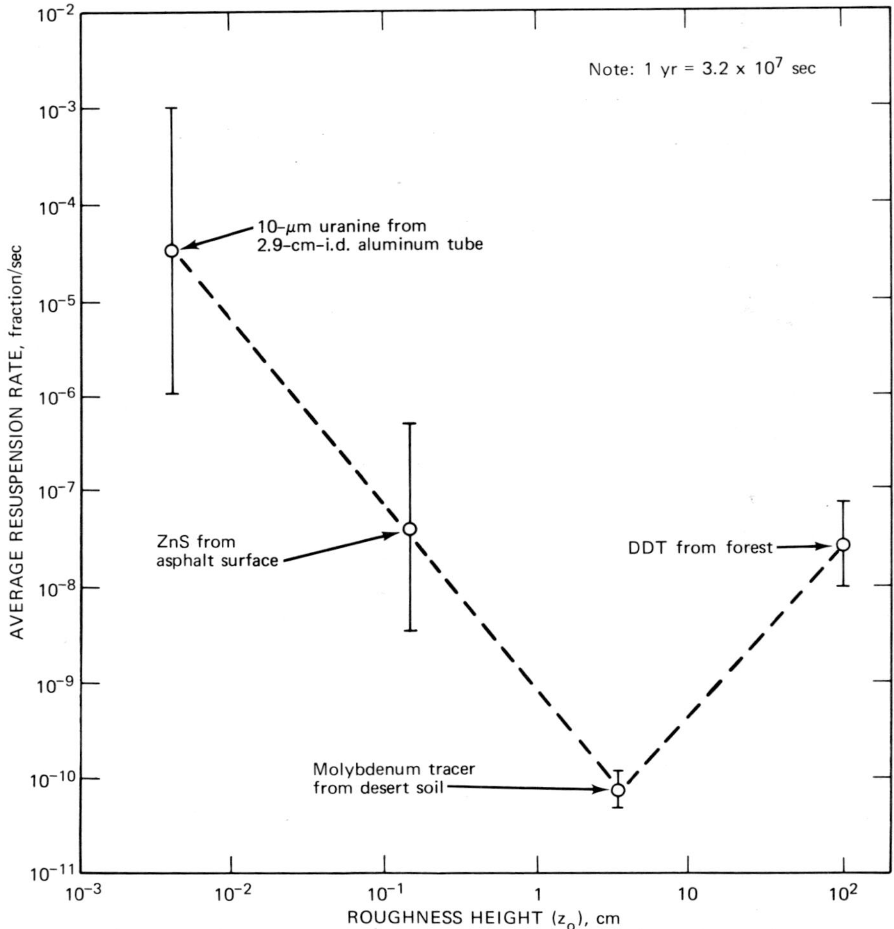

Fig. 29 Initial correlation of wind-caused tracer resuspension rates.

Conclusions

This review of 1971 to 1977 resuspension data determined at the Pacific Northwest Laboratory indicates the following problem areas:

• There are more theoretical resuspension models available for prediction than data to validate or to use in those models. Theoretical model development is limited by availability of experimental data.

• The data base discussed for relating plutonium contamination of surface soils to that of airborne soil is based on gross surface-soil and airborne-soil samples. Data have not been collected to determine any relationship between plutonium size distributions and concentrations on airborne soil and those on surface soils.

• Resuspended plutonium is transported on both respirable and nonrespirable soil particles. Data reported are the entire data base for plutonium transport on airborne nonrespirable soil. Additional data are needed to describe plutonium transport on nonrespirable particles and the subsequent degradation to a respirable resuspension source.

• One possible assumption for describing resuspension concerns the prediction of plutonium concentration per gram of airborne soil vs. plutonium concentration per gram of surface soil. However, ranges of airborne concentration per gram vs. surface-soil concentration per gram indicate a wide discrepancy between airborne vs. surface soils. This wide discrepancy shows that there seems to be no justification for assuming any equalities between plutonium concentrations on surface soils vs. airborne soils.

• The rate of change of average airborne radionuclide concentrations with time has been described by a weathering half-life. However, the weathering half-life is not well known. Data shown for radionuclides as well as for inert tracer particles indicate that the half-life is from 5 months or longer rather than the often quoted 35 to 40 days.

• At Rocky Flats the ^{238}Pu/^{239}Pu ratio can be much greater on airborne nonrespirable soils than on surface soils. Thus there may be a preferential resuspension transport of ^{238}Pu vs. ^{239}Pu. For migration within surface soils, there are data showing preferential migration with depth as well as of ^{238}Pu compared with ^{239}Pu.

Both ^{238}Pu and ^{239}Pu resuspension occurred on site at Rocky Flats and the Hanford area, but all airborne plutonium concentrations were significantly below maximum permissible concentrations (MPC's) in air. In addition, plutonium was resuspended from off site near the Hanford reservation. In all cases plutonium was deposited on each stage of particle cascade impactors, which indicates that most plutonium was resuspended while attached to larger host soil particles.

Plutonium was transported on both respirable and nonrespirable airborne-soil particles. In most resuspension research reported by other researchers as well as in air-monitoring activities, airborne concentrations of particles have been measured without sampling both respirable and all nonrespirable particle sizes present. Only respirable or near-respirable size particles are frequently measured since the usual air-sampling techniques tend to keep larger, nonrespirable particles from being collected. Consequently total airborne plutonium concentrations could be greater than normally reported using most existing sampling equipment systems. Results from those systems are a conservative estimate (high concentration) of airborne respirable plutonium concentrations. Nevertheless, plutonium transport on nonrespirable particles may be a significant factor in total plutonium transport. Larger than respirable particles are resuspended and may not travel too far downwind before redepositing again. In contrast, respirable particles remain airborne for a much longer distance. Additional research is needed to clarify the relative significance of plutonium transport on respirable as compared with nonrespirable particles.

Plutonium concentrations per gram of both respirable and nonrespirable airborne soils discussed in this chapter are summarized in Table 11. Ranges are within several orders of magnitude, from 2×10^{-8} to 6×10^{-5} μCi/g for respirable ^{239}Pu and from 1×10^{-7} to 3×10^{-4} μCi/g for nonrespirable ^{239}Pu.

In all cases plutonium concentrations per gram of soil were calculated on the basis of total soil samples. In addition, there is no proven method to predict the ratios of concentration per gram of airborne soil to concentration per gram of surface soil. As shown in Tables 2 and 4, this ratio ranges seven orders of magnitude from 10^{-4} to 10^{3}. This uncertainty range is almost as large as the uncertainty range of 10^{-11} to 600 m^{-1} for resuspension factors (Mishima, 1964; Stewart, 1967; Sehmel and Lloyd, 1976a).

TABLE 11   Summary of Plutonium Concentrations on Total Airborne Solids

| | Total airborne concentration range, μCi/g | | | |
| | ^{238}Pu | | ^{239}Pu | |
Site	Respirable	Nonrespirable	Respirable*	Nonrespirable
Rocky Flats (Sehmel, 1976a; Sehmel and Lloyd, 1976b)	NR†	2×10^{-7} to 5.1×10^{-6}	2.1×10^{-6} to 6.2×10^{-5}	1×10^{-6} to 3.1×10^{-4}
Hanford reservation				
On site (Sehmel, 1977c)	1×10^{-9} to 1×10^{-6}	4×10^{-9} to 1×10^{-7}	2×10^{-8} to 6×10^{-5}	1×10^{-7} to 4×10^{-6}
From off site near Prosser barricade	NR†	NR†	5×10^{-8} to 1×10^{-6}	1.3×10^{-7} to 2.1×10^{-7}

*Respirable as used in this chapter refers to those particles found on all filter and impactor stages as contrasted to nonrespirable particles collected by gravity in rotating cowls.

†NR, no radiochemical results.

**TABLE 12   Summary of Plutonium Total Transport
Fluxes on Nonrespirable* Particles**

| | Range of total plutonium fluxes, μCi m^{-2} day^{-1} | |
Site	^{238}Pu	^{239}Pu
Rocky Flats (Sehmel, 1976a)	1×10^{-8} to 1×10^{-5}	1×10^{-7} to 6×10^{-4}
Hanford reservation		
On site	2×10^{-8} to 2×10^{-7}	4×10^{-7} to 4×10^{-6}
From off site near Prosser barricade for 190 to 260° winds	NR†	1.4×10^{-6} to 3.9×10^{-6}

*Nonrespirable as used in this chapter refers to those particles collected within the rotating cowl shown in Fig. 3.

†NR, no radiochemical results.

Airborne plutonium transport fluxes on nonrespirable particles are summarized in Table 12. These data and the decrease of flux with distance shown in Figs. 8 and 9 constitute the present knowledge on this subject. There is surprisingly reasonable agreement within several orders of magnitude for nonrespirable airborne plutonium fluxes. However, the agreement may be caused, in part, by relatively more soil transport, with a lower plutonium-on-soil concentration being comparable to a lower soil transport and concurrent higher plutonium-on-soil concentration. Nevertheless, these ranges of horizontal plutonium fluxes on nonrespirable particles could be used in modeling efforts and hazards analysis until a greater data base is experimentally obtained.

Airborne concentrations of ^{239}Pu, ^{241}Am, ^{238}Pu, ^{137}Cs, and ^{90}Sr increased (Sehmel, 1977c) as a function of wind speed to the one to sixth power for on-site resuspension study sites. Airborne plutonium concentrations for off-site resuspension increased as a power function of wind speed. Above a wind speed of about 5 m/sec, airborne plutonium concentrations increased with wind speed to the third to fifth power. Explanations for differences in exponents are needed. Differences might be attributed to source characteristics and extent.

In contrast, for controlled source experiments, tracer-particle resuspension rates increase with about the fifth power of wind speed. This fifth power is similar to off-site plutonium resuspension.

Plutonium concentrations on collected airborne soil at Rocky Flats ranged from a maximum of twice the concentration on ground-surface soil to as low as 10^{-4} of the concentration on surface soils. The maximum reported (Krey et al., 1976b) surface-soil concentration was 3×10^{-3} μCi/g. Even at this relatively high plutonium surface-soil concentration, airborne respirable plutonium concentrations were significantly below airborne MPC's on a yearly basis. Similarly, at the Hanford resuspension study sites (Sehmel, 1977c), maximum concentrations per gram of airborne solid for transuranics were: for ^{239}Pu, 6×10^{-5} μCi/g; for ^{238}Pu, 1×10^{-6} μCi/g; and for ^{241}Am, 7×10^{-5} μCi/g.

There is much uncertainty in the relationships between radionuclide concentrations per gram of airborne solid and per gram of surface-soil solids. For the transuranic-element data reported at Rocky Flats, the ratio ranges from 10^{-4} to 2, and at Hanford the ratio ranges from 0.5 to 1.5×10^3. Uncertainties are probably complicated by spatial distributions of surface contamination.

Radionuclide particles can be resuspended either as individual particles or, more probably, attached to host soil or solid particles. An average-activity, or normal-activity, radionuclide particle distribution is usually collected on sampling filters. However, at both Rocky Flats and Hanford, one filter sample collected in each case showed significantly greater plutonium concentration than the maximum for all other samples collected during the same time period. These anomalous higher concentrations are attributed to one or more plutonium particles of unusually higher activity than those normally, or most frequently, resuspended.

There is increasing, but still conflicting, data that ^{238}Pu might be more mobile than ^{239}Pu. The isotopic-ratio data reported for airborne plutonium transport on nonrespirable particles at Rocky Flats support the greater mobility concept. This conclusion was obtained by comparing airborne ^{238}Pu/^{239}Pu ratios on nonrespirable soil near the eastern security fence and the eastern cattle fence. At the eastern security fence, airborne ^{238}Pu/^{239}Pu ratios were similar to ground surface ratios. However, at the eastern cattle fence, the ^{238}Pu/^{239}Pu ratios were significantly greater than those measured on local surface soils (see Fig. 11). Consequently an explanation is needed for the increased ^{238}Pu/^{239}Pu ratio at distances from the original oil storage area. Possible explanations include preferential biodegradation of ^{238}Pu compared with ^{239}Pu and preferential ejection of ^{238}Pu at the eastern security fence and at the eastern cattle fence. These possibilities exist but are not definitive. Further research is needed to explain the higher relative airborne ^{238}Pu/^{239}Pu concentrations near the eastern cattle fence.

As used in this chapter, the weathering half-life is the time required for airborne concentrations at a resuspension site to decrease by one-half. Weathering is probably a function of the source characteristics. However, very little is known about predicting weathering. For fallout previous literature has indicated that this half-life is between 30 and 45 days. However, results of ^{137}Cs and transuranic air concentration measurements at resuspension study sites at Hanford reported here indicate a weathering half-life of from 5 months (Sehmel, 1977c) to 1 yr or greater. Knowledge of this ill-defined half-life is important in resuspension modeling efforts to describe airborne effects of surface contamination. Changes in surface contamination availability with time must be known if models are to predict airborne concentrations. This range of from 5 months to 1 yr or greater for a weathering half-life must be considered when evaluating resuspension changes with time.

Transuranic-element resuspension rates have not been directly measured at surface contaminated sites other than inferred from nonvalidated models since published characteristics of the contaminated surface sources are not adequate for direct measurement of particle resuspension rates. Consequently resuspension rates were measured with controlled tracer-particle simulants using a uniform surface contamination source. On the basis of those measurements, the following conclusions were reached.

Particle resuspension rates are a function of at least wind speed and mechanical disturbances. Mechanical disturbances, such as vehicular traffic or a man walking, can cause high local resuspension rates. In comparison, average wind resuspension rates from a local area could be less important per unit area than local mechanical-disturbance resuspension. However, wind-caused resuspension rates apply to the entire contaminated area. In the comparison of relative resuspension from wind-caused and mechanical disturbances, one would need to know the total surface contamination area for wind resuspension vs. small localized surface contamination levels for mechanical-disturbance resuspension rates. Both mechanisms, however, do resuspend and transport potentially hazardous respirable particles.

Resuspension rates for respirable and nonrespirable particles are needed for inclusion as source terms in atmospheric diffusion and transport equations; however, model predictions are no better than the uncertainty in the source data. In the case of resuspension rates, uncertainties are very large. Much research is yet needed to develop resuspension models to predict particle resuspension rates for any situation.

Wind-caused resuspension rates from a sparsely vegetated area have only been directly measured with submicrometer tracer particles and estimated for tracer particles larger than 1 μm (Healy and Fuquay, 1958; 1959). The potential effects of different particle diameters and chemical properties on resuspension rates are unknown. It might be hypothesized that similar results would be expected for other submicrometer particles of interest since submicrometer particles are probably attached to host soil particles when particles are resuspended. If the particles of interest were much larger, it is unknown whether the particles would be resuspended attached to host soil particles or resuspended as discrete particles.

The change in airborne concentration of a pollutant as a function of time is often attributed to a weathering half-life, the fixation of the pollutant particle into the ground-surface soil. In contrast, weathering half-lives for respirable tracer particles are now estimated here as being on the order of years. Predictions using weathering half-lives of months vs. years could have a significant implication in environmental hazards evaluations. At the present time, credit for decreased airborne radioactivity from resuspension could be attributed to a weathering half-life of months. If a weathering half-life of years were applicable for transuranic elements, the potential downwind inhalation hazard from resuspended particles would be significantly increased. Additional experimental data are needed to determine what weathering half-life or variation of half-life with time should be used in hazards evaluations.

Vegetation on a resuspension surface will decrease resuspension rates. This is rather obvious but can be definitely concluded from decreased tracer-particle resuspension rates for vehicles driven on a cheat grass area compared with those for vehicles driven on an asphalt area. Since vegetation does decrease resuspension, vegetation should be retained on all areas of potential surface contamination and existing surface contaminated areas until constructive cleanup operations can be initiated.

References

Anspaugh, L. R., et al., 1969, *Bio-Medical Division Preliminary Report for Project Schooner,* USAEC Report UCRL-50718, Lawrence Radiation Laboratory, University of California, NTIS.

Bruns, L. E., 1976, Aerial Gamma Survey by Helicopter to Measure Surficial Contamination, in *Atmosphere–Surface Exchange of Particulate and Gaseous Pollutants,* (1974), ERDA Symposium Series, No. 38, Richland, Wash., Sept. 4–6, 1974, R. J. Engelmann and G. A. Sehmel (Coordinators), pp. 675-688, CONF-740921, NTIS.

Corley, J. P., D. M. Robertson, and F. P. Brauer, 1976, Plutonium in Surface Soil in the Hanford Plant Environs, in *Proceedings of Environmental Plutonium Symposium,* Los Alamos, N. Mex., Aug. 4–5, 1971, USAEC Report LA-4756, pp. 85-92, Los Alamos Scientific Laboratory, NTIS.

Hardy, E. P., 1974, *Depth Distributions of Global Fallout* $^{90}Sr,$ $^{137}Cs,$ *and* $^{239,240}Pu$ *in Sandy Loam Soil,* USAEC Report HASL-286, pp. I-2 to I-10, Health and Safety Laboratory, NTIS.

Healy, J. W., and J. J. Fuquay, 1959, Wind Pickup of Radioactive Particles from the Ground, in *Progress in Nuclear Energy, Health Physics,* Series XII, Vol. 1, pp. 427-436, Pergamon Press, Inc., New York.

Horst, T. W., 1976, *The Estimation of Air Concentrations due to the Suspension of Surface Contamination by the Wind,* USAEC Report BNWL-2047, Battelle, Pacific Northwest Laboratories, NTIS.

International Commission on Radiological Protection, 1959, *Report of Committee II on Permissible Dose for Internal Radiation,* Publication 2, p. 82, Pergamon Press, Inc., New York.

Johnson, C. J., R. R. Tiball, and R. C. Severson, 1976, Plutonium Hazard in Respirable Dust on the Surface of Soil, *Science,* 193: 488-490.

Krey, P. W., 1976, *Health Phys.,* 30(2): 209-214.

——, E. P. Hardy, C. Pachucki, R. Rourke, J. Coluzza, and W. K. Benson, 1976a, Mass Isotopic Composition of Global Fallout Plutonium in Soil, in *Transuranium Nuclides in the Environment,* Symposium Proceedings, San Francisco, Nov. 17–21, 1975, pp. 671-678, STI/PUB/410, International Atomic Energy Agency, Vienna.

——, et al., 1976b, *Plutonium and Americium Contamination in Rocky Flats Soil—1973,* ERDA Report HASL-304, Health and Safety Laboratory, NTIS.

——, R. Knuth, T. Tamura, and L. Toonkel, 1976c, Interrelations of Surface Air Concentrations and Soil Characteristics at Rocky Flats, in *Atmosphere–Surface Exchange of Particulate and Gaseous Pollutants* (1974), ERDA Symposium Series, No. 38, Richland, Wash., Sept. 4–6, 1974, R. J. Engelmann and G. A. Sehmel (Coordinators), pp. 744-756, CONF-740921, NTIS.

Maxfield, H. L., 1974, *A Preliminary Safety Analysis of the B-C Cribs Controlled Zone,* Report ARH-3088, Appendix B-4, Atlantic Richfield Hanford Company.

Milham, R. C., J. F. Schubert, J. R. Watts, A. L. Boni, and J. C. Corey, 1976, Measured Plutonium Resuspension and Resulting Dose from Agricultural Operations on an Old Field at the Savannah River Plant in the Southeastern United States of America, in *Transuranium Nuclides in the Environment,* Symposium Proceedings, San Francisco, Nov. 17–21, 1975, pp. 409-421, STI/PUB/410, International Atomic Energy Agency, Vienna.

Mishima, J., 1964, *A Review of Research on Plutonium Releases During Overheating and Fires,* USAEC Report HW-83668, Hanford Laboratories, NTIS.

——, 1973, Distribution of Cs-137 Activity During a Range Fire in the B-C Controlled Area, in *Pacific Northwest Laboratory Monthly Activities Report, September 1973: Division of Production and*

Materials Management and Hanford Plant Assistance Programs, USAEC Report BNWL-1792, p. 13, Battelle, Pacific Northwest Laboratories, NTIS.

——, and L. C. Schwendiman, 1973, Potential Airborne Release of Cs-137 Activity During a Range Fire in the B-C Controlled Area, in *Pacific Northwest Laboratory Monthly Activities Report, October 1973: Division of Production and Materials Management and Hanford Plant Assistance Programs,* USAEC Report BNWL-1799, p. 10, Battelle, Pacific Northwest Laboratories, NTIS.

——, and L. C. Schwendiman, 1974, Potential Airborne Release of Surface Contamination During a Range Fire, in *Pacific Northwest Laboratory Monthly Activities Report, January 1974: Division of Production and Materials Management and Hanford Plant Assistance Programs,* USAEC Report BNWL-1814, p. 18, Battelle, Pacific Northwest Laboratories, NTIS.

Nees, W. L., and J. P. Corley, 1975, *Environmental Status of the Hanford Reservation for CY-1973,* USAEC Report BNWL-B-336, p. 49, Battelle, Pacific Northwest Laboratories, NTIS.

Oksza-Chocimowski, George V., 1976, *Resuspension Model Review,* Report DRP/LV-76-11, U.S. Environmental Protection Agency.

Orgill, M. M., M. R. Petersen, and G. A. Sehmel, 1976, Some Initial Measurements of DDT Resuspension and Translocation from Pacific Northwest Forest, in *Atmosphere—Surface Exchange of Particulate and Gaseous Pollutants,* (1974), ERDA Symposium Series, No. 38, Richland, Wash., Sept. 4–6, 1974, R. J. Engelmann and G. A. Sehmel (Coordinators), pp. 813-834, CONF-740921, NTIS.

——, G. A. Sehmel, and M. R. Petersen, 1976, *Atmos. Environ.,* 10: 827-834.

Pacific Northwest Laboratory, *Division of Production and Materials Management and Hanford Plant Assistance Programs,* USAEC Reports BNWL-1880, pp. 6-7, October 1974; BNWL-1868, p. 7, September 1974; BNWL-1844, p. 11, June 1974; BNWL-1814, pp. 18-20, January 1974; BNWL-1808, pp. 14-15, December 1973; BNWL-1802, pp. 10-11, November 1973; BNWL-1799, p. 12, October 1973; BNWL-1792, p. 12, September 1973, NTIS.

Sehmel, G. A., 1973a, An Evaluation of a High-Volume Cascade Particle Impactor System, in *Proceedings of the Second Joint Conference on Sensing of Environmental Pollutants,* Washington, D.C., Dec. 10–12, 1973, Report ISA-JSP6671, pp. 109-115, Instrument Society of America, Pittsburgh, Pa.

——, 1973b, Particle Resuspension from an Asphalt Road Caused by Car and Truck Traffic, *Atmos. Environ.,* 7: 291-301.

——, and F. D. Lloyd, 1972, Tracer Particle Resuspension Caused by Wind Forces Upon an Asphalt Surface, in *Pacific Northwest Laboratory Annual Report for 1971 to the USAEC Division of Biology and Medicine,* Part 1, Atmospheric Sciences, USAEC Report BNWL-1651(Pt. 1), pp. 136-138, Battelle, Pacific Northwest Laboratories, NTIS.

——, and M. M. .Orgill, 1973, Resuspension by Wind at Rocky Flats, in *Pacific Northwest Laboratory Annual Report for 1972 to the USAEC Division of Biomedical and Environmental Research,* Part 1, Atmospheric Sciences, USAEC Report BNWL-1751(Pt. 1), pp. 15-22, Battelle, Pacific Northwest Laboratories, NTIS.

——, and M. M. Orgill, 1974, Resuspension Source Change at Rocky Flats, in *Pacific Northwest Laboratory Annual Report for 1973 to USAEC Division of Biomedical and Environmental Research,* Part 3, Atmospheric Sciences, USAEC Report BNWL-1850(Pt. 3), pp. 212-214, Battelle, Pacific Northwest Laboratories, NTIS.

——, 1975, Initial Correlation of Particle Resuspension Rates as a Function of Surface Roughness Height, in *Pacific Northwest Laboratory Annual Report for 1974 to USAEC Division of Biomedical and Environmental Research,* Part 3, Atmospheric Sciences, USAEC Report BNWL-1950(Pt. 3), pp. 209-212, Battelle, Pacific Northwest Laboratories, NTIS.

——, 1976a, *Airborne ^{238}Pu and ^{239}Pu Associated with the Larger than "Respirable" Resuspended Particles at Rocky Flats During July 1973,* USAEC Report BNWL-2119, Battelle, Pacific Northwest Laboratories, NTIS.

——, 1976b, The Influence of Soil Insertion on Atmospheric Particle Size Distributions, in *Pacific Northwest Laboratory Annual Report for 1975 to the USERDA Division of Biomedical and Environmental Research,* Part 3, Atmospheric Sciences, ERDA Report BNWL-2000(Pt. 3), pp. 99-101, Battelle, Pacific Northwest Laboratories, NTIS.

——, 1976c, Particle Resuspension from Truck Traffic in a Cheat Grass Area, in *Pacific Northwest Laboratory Annual Report for 1975 to the USERDA Division of Biomedical and Environmental*

Research, Part 3, Atmospheric Sciences, ERDA Report BNWL-2000(Pt. 3), pp. 96-99, Battelle, Pacific Northwest Laboratories, NTIS.

——, and F. D. Lloyd, 1976a, Particle Resuspension Rates, in *Atmosphere—Surface Exchange of Particulate and Gaseous Pollutants,* (1974), ERDA Symposium Series, No. 38, Richland, Wash., Sept. 4–6, 1974, R. J. Engelmann and G. A. Sehmel (Coordinators), pp. 846-858, CONF-740921, NTIS.

——, and F. D. Lloyd, 1976b, Resuspension of Plutonium at Rocky Flats, in *Atmosphere—Surface Exchange of Particulate and Gaseous Pollutants,* (1974), ERDA Symposium Series, No. 38, Richland, Wash., Sept. 4–6, 1974, R. J. Engelmann and G. A. Sehmel (Coordinators), pp. 757-779, CONF-740921, NTIS.

——, and F. D. Lloyd, 1976c, Resuspension Rates from a Circular Field Source, in *Pacific Northwest Laboratory Annual Report for 1975 to the USERDA Division of Biomedical and Environmental Research,* Part 3, Atmospheric Sciences, ERDA Report BNWL-2000(Pt. 3), pp. 92-93, Battelle, Pacific Northwest Laboratories, NTIS.

——, 1977a, Airborne Plutonium Transport on Nonrespirable Particles, in *Pacific Northwest Laboratory Annual Report for 1976 to USERDA Division of Biomedical and Environmental Research,* Part 3, Atmospheric Sciences, ERDA Report BNWL-2100(Pt. 3), pp. 65-73, Battelle, Pacific Northwest Laboratories, NTIS.

——, 1977b, Plutonium and Tracer Particle Resuspension: An Overview of Selected Battelle—Northwest Experiments, in *Transuranics in Natural Environments,* Symposium Proceedings, Gatlinburg, Tenn., Oct. 5–7, 1976, ERDA Report NVO-178, pp. 181-210, Nevada Operations Office, NTIS.

——, 1977c, *Radioactive Particle Resuspension Research Experiments on the Hanford Reservation,* ERDA Report BNWL-2081, Battelle, Pacific Northwest Laboratories, NTIS.

Stewart, K., 1967, The Resuspension of Particulate Material from Surfaces, in *Surface Contamination,* B. R. Fish (Ed.), Pergamon Press, Inc., New York.

Thomas, C. W., 1976, Atmospheric Fallout During 1975 at Richland, Washington, and Point Barrow, Alaska, in *Pacific Northwest Laboratory Annual Report for 1975 to the USERDA Division of Biomedical and Environmental Research,* Part 3, Atmospheric Sciences, ERDA Report BNWL-2000(Pt. 3), pp. 16-19, Battelle, Pacific Northwest Laboratories, NTIS.

Volchok, H. L., R. H. Knuth, and M. T. Klemman, 1972, *Health Phys.,* 23(3): 395-396.

Wedding, J. B., A. R. McFarland, and J. E. Cermak, 1977, *Environ. Sci. Technol.,* 11(4): 387-390.

Wildung, R. E., and T. R. Garland, 1977, *The Relationship of Microbial Processes to the Fate and Behavior of Transuranic Elements in Soils, Plants, and Animals,* ERDA Report PNL-2416, Pacific Northwest Laboratory, NTIS.

Wilson, R. H., R. G. Thomas, and J. N. Stannard, 1961, *Biomedical and Aerosol Studies Associated with a Field Release of Plutonium,* USAEC Report WT-1511, University of Rochester, Atomic Energy Project, NTIS.

Interaction of Airborne Plutonium
with Plant Foliage

D. A. CATALDO and B. E. VAUGHAN

The interaction of airborne pollutants with the foliage of terrestrial plants has been investigated from many aspects, including interception, retention, and absorption. Although interception parameters for both gaseous and particulate pollutants have been effectively modeled, the behavior and fate of pollutants, especially particulates, after foliar interception are not known. Particles with diameters of 10 to 200 µm exhibit retention half-times of 10 to 24 days. Direct and indirect data, however, suggest that submicronic particles are more effectively retained on plant foliage than are larger particles. Studies are presented to describe the retention behavior of submicron-size particles deposited on foliage of bush bean and sugar beet plants. Simulated rainfall was used to evaluate retention efficiency. These studies showed submicronic particles to be increasingly less available for leaching with increasing residence time on the leaf; e.g., more than 90% of the foliar plutonium deposits were firmly held to the leaf surface. Retention mechanisms are discussed in terms of leaf morphology and the leaching regimes used. The absorption of foliar plutonium and its subsequent translocation to seed and root tissues were dependent on a number of parameters, including chemical form and the presence or absence of a solution vector.

The behavior of the transuranic elements in the environment and their potential for transfer in the food chain have been the subject of extensive study over the past 25 years. Although there is a general understanding of many problems concerning atmospheric transport (Slinn, 1975; 1976) and of the behavior of plutonium in specific ecosystems (Nevada Test Site), little is known of the controlling mechanisms that influence the bioavailability of plutonium and the other transuranic elements and their subsequent transfers along the food web to man. With the current stratospheric depletion of fallout plutonium (Bennett, 1976), the importance of the inhalation route to man is greatly reduced. This then suggests that the major sources of transuranic elements in the future will result from resuspension of fallout-contaminated soils on a global basis, resuspension from highly contaminated local sources, accident situations, and low-level releases from nuclear facilities.

Present radiological safety estimations frequently discount foliar sorption and emphasize the soil-to-root pathway for the entry of transuranic and other radioelements to the food chain (Vaughan, Wildung, and Fuquay, 1976). Typical dose-assessment codes assume a rapidly declining exponential loss of material from leaves (Soldat, 1971). This is certainly not a general situation. It does not apply to the behavior of plutonium aerosols described here and probably applies only to very large particles and to certain gaseous radioelements, such as iodine (Markee, 1971). Our studies indicate that the fate of transuranic elements following foliar interception is influenced by particle size (mass).

The importance of the foliar-entry pathway compared to root absorption for worldwide fallout was recognized long ago (Chamberlain, 1970; Russell, 1965). In later studies of particles of probably wind-resuspended origin, 87% of the ^{90}Sr, 81% of the ^{137}Cs, and 73% of the ^{144}Ce in forage plants were derived from foliar contamination (Romney et al., 1973). This was shown by comparing plants grown inside plastic enclosures with those grown with no cover at the Nevada Test Site. Plutonium may behave similarly, but unfortunately there are no well-controlled field observations. Recently, the importance of foliar-to-root pathways in the plant was defined in the Liquid Metal Fast Breeder Reactor final environmental statement (U. S. Atomic Energy Commission, 1974). Despite inconsistencies with respect to other dose-assessment codes, risk tends to be minimized by specifying extremely conservative limits at points where radioelements actually enter the human body, i.e., air and food. As a matter of systematic practice, an improved quantitative understanding of the basic environmental processes is required. This becomes important in situations where new technology may lead to different physical (size) and chemical characteristics of the source term for release, especially in nuclear-fuel reprocessing plants, and where comparatively large increases in the handling of radioelements are projected for the future (Energy Research and Development Administration, 1975).

The following discussion reviews current knowledge regarding the retention and absorption potential of foliar surfaces and describes the fate of transuranic particles following plant-foliage interception as deduced from the extrapolation of information on the behavior of other particles and the limited information on plutonium.

The Problem of Retention of Particles on Foliar Surfaces

The retention of particles on foliar surfaces depends on many parameters associated with the foliar surface and the physical aspects of the particle. Leaf factors affecting the efficiency of particle entrapment include components of the leaf that affect roughness (Holloway, 1971), namely, venation, surface features of epidermal cells, nature of the cuticle surface, nature and frequency of trichomes, and the microstructure of surface wax. Each of the microtopographical features of the leaf may contribute to the entrainment and retention of particles. Other factors affecting retention include surface stickiness from organic and inorganic secretion, leaf wetness and charge attraction between particles, and surface waxes or components. In addition, retention is dependent on particle size, particle density, wind speed within the boundary layer, and, when a particular element comprising or contained in a particle is being considered, solubility.

Available information on foliar retention is sometimes disconcerting and contradictory when one tries to reconcile the retention and behavior of relatively insoluble particles with early fallout data on soluble or volatile fission products. Early fallout work with respect to fission products has been reviewed by Chamberlain (1970) and Russell (1965). In general, these reviews indicate a retention half-time of 10 to 14 days for soluble fission products, with losses resulting from reentrainment of carrier particles, sloughing of surface wax (Moorby and Squire, 1963), and rainfall (Middleton, 1959). Except for radioiodine (Markee, 1971), such a short retention half-time is probably characteristic only of large aerosol particles, as described later.

Several studies approached the problem of particle retention by using simulated quartz fallout material containing adsorbed fission products. Witherspoon and Taylor (1969) found that over a 33-day period 88- to 177-μm-diameter (MMD)* particles were more effectively entrained by pine foliage than by oak. Although after only 1 hr wind resuspension accounted for a 90% reduction in the number of particles in oak leaves as compared with a 10% reduction in pine, the first rainfall (t_0 + 1 day) accounted for a 15% reduction of particle activity remaining at 1 hr. Retention half-times reported for periods of 0 to 1, 1 to 7, and 7 to 37 days were 0.12, 1.4, and 25 days for oaks, and 0.26, 4.5, and 21 days for pines, respectively.

A similar study by Witherspoon and Taylor (1970) presented data for the retention of 44- to 88- and 88- to 175-μm-MMD particles by various agricultural plants. These studies indicate that average wind speeds of 0.5 mph over the initial 12-hr period following contamination are more effective in removing the smaller particles (21.1% vs. 15.8%) and that average wind speeds of 1.1 mph over a 12- to 36-hr period resulted in a higher loss of the larger particles (21.6% vs. 15.4%). Subsequent to t_0 + 6 days, varying amounts of rainfall resulted in a marked reduction in retention of particles of both size ranges. The resuspension behavior of these relatively large particles is in keeping with theoretical and empirical measurements on the inertial forces within the boundary layer required to resuspend spores from leaves (Aylor, 1975; 1976; Aylor and Parlange, 1975).

Subsequent studies by Witherspoon and Taylor (1971) with 1- to 44-μm-MMD simulated fallout particles showed longer weathering half-lives for 1- to 44-μm particles (17.9 days) than those reported earlier for 44- to 88- (15.7 days) and 88- to 175-μm (15.1 days) particles. Loss rates were also less affected by time or rainfall after a particle residence time of 7 days. This suggests that particle size does, in fact, play an important role in the extent of foliar retention.

Although these studies aid in our understanding of the interception and retention of larger particles (>10 μm) analogous to close-in fallout, questions arise as to the behavior of fallout particles of submicron size. Both Iranzo (1968) and Romney et al. (1975) reported that plutonium-containing material resuspended in field situations is difficult to remove from contaminated foliage; as much as 50% is tenaciously held on foliar surfaces. This suggests that retention is affected by factors other than the passive association of particles with relatively flat foliar surfaces where only inertial forces influence their removal or resuspension.

In studies of 6.77 ± 0.02-μm-AMAD† uranine particles, where the primary particle had an MMD in the submicron range, Wedding et al. (1975) have shown that deposition is related to the roughness of the leaf surface. By analogy, the leaf-roughness factors affecting deposition should also affect retention. The effect of wind and rainfall on foliarly deposited $PbCl_2$ particles (1- to 3-μm MMD) was evaluated by Carlson et al. (1976). These studies showed that lead particles remained fixed to leaves under controlled conditions for up to 4 weeks after fumigation; reentrainment wind speeds of up to 6.7 m/sec were ineffective in removing surface deposits. Losses due to simulated rainfall were proportional to the amount of rainfall; mists were more effective than droplets in removing lead deposited on leaf surfaces; only 15 and 5% of the foliar deposits, respectively, were leachable.

---

*Mass median diameter (MMD) assumed; particles physically measured.
†Activity median aerodynamic diameter.

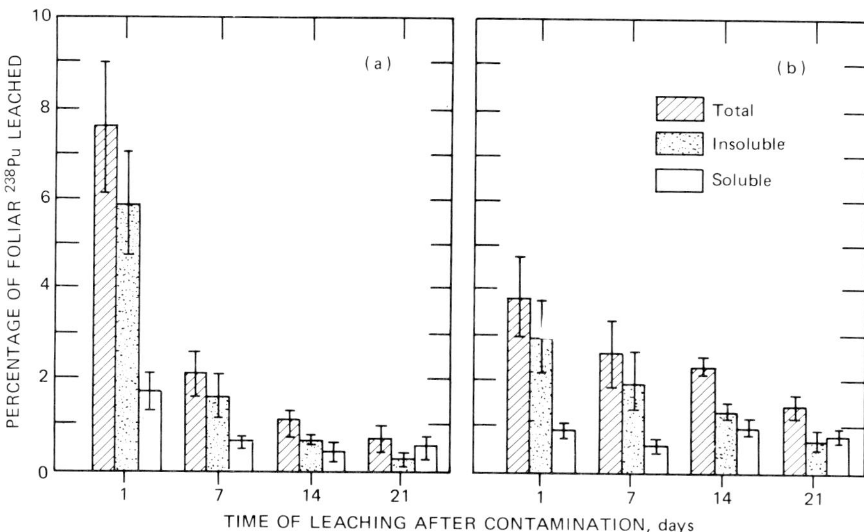

Fig. 1 Leachability of plutonium from bush bean foliage. Sets of four plants each were leached at 1, 7, 14, or 21 days after contamination; $x \pm$ SEM ($n = 4$). (a) Fresh plutonium dioxide. (b) Water-aged plutonium dioxide.

Data on the retention of plutonium by foliar surfaces are limited. The data that do exist are based on laboratory studies in which a low-windspeed exposure chamber was used to contaminate plant canopies (Cataldo, Klepper, and Craig, 1976). Figure 1 illustrates the leachability of two forms of ^{238}Pu dioxide as a function of residence time on the foliage of the bush bean following a simulated rainfall of 0.4 cm in 7 min. The particles deposited on the foliage had aerodynamic sizes (activity median aerodynamic diameter $\pm$ geometric standard deviation) of 1.274 μm $\pm$ 1.63 and 0.734 μm $\pm$ 2.16 for freshly prepared and water-aged oxides, respectively. The count modes for the log-normal distributions were 0.142 and 0.019 μm, respectively. These latter values represent the particle diameter (absolute size) with the highest frequency within the family of particles. The plutonium retained on foliage after mild leaching ranged from 92 to 99%. These data are qualitatively similar to those obtained for 1- to 3-μm lead particles (Carlson et al., 1976) but are contrary to data obtained with larger simulated fallout particles (Witherspoon and Taylor, 1969; 1970; 1971). Both the fresh and the hydrated oxide exhibit a reduced leachability with increased residence time on the leaf. The retention mechanism may be related to physical entrapment of the submicron-size particles in small fissures on the leaf surface or to charge adsorption between particles and the leaf surface. The inability to readily remove plutonium from foliar surfaces has been noted (Hanson, 1975; Romney et al., 1975; Iranzo, 1968); the mechanisms controlling retention, however, are not clear.

Little (1973) used weakly acidic solutions to study the physical processes of ion exchange involved in the retention of heavy metal particles, such as lead, on foliage. Table 1 compares the leachability of foliar plutonium by synthetic rainwater with and without 0.1% HNO_3. Leaching with acidic solution results in a moderate increase in insoluble plutonium leached from leaves contaminated with fresh PuO_2 but a substantial increase from leaves contaminated with the hydrated oxide. The large increase in the

TABLE 1   Effect of Acid Solution on Leachability of Foliar Plutonium 7 Days After Exposure of Bush Bean Plants to Fresh and Hydrated Plutonium Dioxide*

Compound	Leached component	Foliar plutonium leached,† %	
		Synthetic rainwater	Synthetic rainwater + 0.1% HNO_3 (pH 2.0)
^{238}Pu dioxide	Soluble	0.5 ± 0.1	1.0 ± 0.1
	Insoluble	1.6 ± 1.0	2.5 ± 0.6
	Total	2.1 ± 1.1	3.5 ± 0.7
^{238}Pu dioxide	Soluble	0.6 ± 0.1	4.5 ± 0.2
(hydrated)	Insoluble	2.0 ± 1.3	5.3 ± 0.5
	Total	2.6 ± 1.4	9.8 ± 0.7

*Plant foliage was exposed to polydispersed aerosols. The freshly prepared oxide had an AMAD of 1.27 μm and a GSD of 1.63; the hydrated oxide had an AMAD of 0.73 μm and a GSD of 2.16. Count modes for the aerosols were 0.140 and 0.018 μm for the fresh and hydrated oxides, respectively. Plants were leached with 200 ml of solution (equivalent to a 7-min rainfall of 0.4 cm).

†Leachability expressed as microcuries of leachate/(microcuries leached + microcuries remaining on leaves) × 100; four replicate samples, x ± SE.

soluble components may result from a solubilization of noncrystalline plutonium on the surface of the particles. The increased leachability of the hydrated oxide (0.019 μm), as compared with the fresh oxide (0.142 μm), may be related to the larger surface area available for reaction.

Even though much of the foliar-deposited plutonium is unavailable for leaching with weakly ionic pH 5.8 solution, the increased removal of both soluble and insoluble components with acidic solutions may indicate that a portion of the submicron particles intercepted by foliage may be held on the leaf surface by charge phenomena and by physical entrapment and not necessarily buried in waxy plates. Table 2 compares the leaching behavior of plutonium from two plant species with different surface roughnesses. Plants were leached with 800 ml of solution, and the leachate was collected in 50-ml fractions. Since total plutonium in the leachate decreased logarithmically (plutonium activity in the last few leachate fractions approached background levels), reported retention values represent plutonium not readily leachable.

Scanning-electron-microscope micrographs of the leaf blades show that bush bean leaves have moderate to low surface relief and sugar beet leaves are relatively flat. The difference in surface microtopography between these two species is related primarily to patterns of wax deposition and the presence of trichomes in the bush bean leaves. The surface wax of bush bean leaves is laid down in such a way as to form high longitudinally oriented ridges with deep crevasses, and the surface of the waxy plates is relatively smooth. By comparison, the surface wax of sugar beet leaves forms relatively shallow irregular convolutions, and the surface of the wax deposits is rougher than that of the bush bean. The trichomes of the bush bean leaves, which are approximately 150-μm high and spaced approximately 190 μm apart, provide additional surface relief. This microtopography and its effect on particle entrapment and leaf-surface wettability may provide a basis for understanding the processes involved in particle retention.

In general, the leaching data for sugar beets and bush beans suggest that both surface roughness and particle size affect the retention of particles on foliar surface. With the larger fresh-oxide particles (count mode approximately 0.142 μm), substantially more of the plutonium is leachable from smooth leaf surfaces under both leaching conditions. This may be the result of physical entrapment of particles in comparatively deep fissures or crevices contributing to surface roughness in the bush bean leaf, especially if it is assumed that a particle must be suspended in a water droplet to be removed from the leaf surface. Similarly, the effect of acid leachate may be in alleviating the attractive forces holding particles to leaf surfaces, particularly in the case of the sugar beet. The retention behavior of the smaller hydrated oxide particles (count mode, approximately 0.019 μm) is slightly different from that of the fresh oxide. The synthetic rainwater was about equally effective in removing particles from both the bush bean and sugar beet; the acid leach was slightly more effective with the bush bean.

Although the gross surface structure of bush bean and sugar beet leaves is obviously different, the microtopography of the surface itself may not be as different with respect to the retention of very small particles (0.02 μm). This may explain similarities in the retention behavior of plutonium deposited onto the foliage of sugar beets and bush beans. It is impossible with limited data to generalize as to mechanisms controlling the fate of particles on foliar surfaces. For the small hydrated-oxide particles, however, it appears that leachability and retention are not only dependent on particle size with respect to leaf topography and physical attraction, such as charge, but also on the ability of a water droplet to contact the particle; thus wettability and contact angle become important (Gregory, 1971), along with other environmental factors (Hull, Morton, and Wharrie, 1975) that influence the physical and chemical nature of the leaf surface.

Aside from our lack of understanding of mechanism, it is important to note that the behavior of small particles, such as that of plutonium on leaf surfaces with respect to

TABLE 2   Effect of Continuous Leaching Regimes on the Removal of Plutonium Particles from Leaves of Bush Bean and Sugar Beet*†

| | | | Plutonium retained on leaves after leaching,‡ % | |
Plant species	Leaf roughness	Plutonium form	Synthetic rainwater	Synthetic rainwater + 0.1% HNO_3 (pH 2.0)
Phaseolus vulgaris	Moderate	Fresh $^{238}PuO_2$	97.6 ± 0.9	97.0 ± 0.4
(Bush bean)		Hydrated $^{238}PuO_2$	95.5 ± 1.2	71.6 ± 6.7
Beta vulgaris	Smooth	Fresh $^{238}PuO_2$	82.0 ± 4.9	64.7 ± 9.8
(Sugar beet)		Hydrated $^{238}PuO_2$	95.7 ± 1.1	83.0 ± 3.5

*D. A. Cataldo, unpublished data.

†Plant foliage was exposed to polydispersed aerosols. Particle-size data for bush bean are given in Table 1. For sugar beet, fresh oxide had an AMAD of 1.59 μm and a GSD of 1.76; the hydrated oxide had an AMAD of 0.75 μm and a GSD of 1.84. Count modes for the aerosols were 0.130 and 0.048 μm for the fresh and hydrated oxides, respectively. Plants were leached with 800 ml of solution (equivalent to a 28-min rainfall of 1.7 cm) 7 days after exposure.

‡Four replicate samples, x ± SE.

retention half-time, differs markedly from that commonly reported for fission products and larger particles ($>10\,\mu$m). The tenacity of plutonium retention observed by Romney et al. (1975) and Iranzo (1968) tends to reinforce the laboratory studies with plutonium described here.

Availability of Foliar Deposits for Uptake and Transport to Other Plant Tissues

Foliar structures are a source of many organic and inorganic substances that either migrate to the surface by diffusion and mass flow or are actively exuded by secretory structures. This may provide a chemical environment on the leaf surface which enables readily hydrolyzable species to be complexed or chemically stabilized and therefore made more available for foliar absorption. Since foliar surfaces represent an efficient absorptive structure (Wittwer, Bukovac, and Tukey, 1963; Franke, 1967; 1971), foliar application of micronutrients to correct nutrient deficiencies is effective, especially in situations where a specific nutrient tends to be immobile and not as available for plant uptake from soil (Krantz et al., 1962; Franke, 1967). The actual mechanisms involved in foliar absorption are not totally understood. Available information indicates that, although the cuticle of the leaf is hydrophobic in nature, penetration is facilitated via intermolecular spaces (Fisher and Boyer, 1972), modification in cutin composition at anticlinal epidermal walls, and the presence of ectodesmata (Franke, 1967; 1971) and trichomes (Benzing and Burt, 1970). The role of stomates as a route of foliar penetration under normal conditions is in question and is currently considered of negligible importance (Greene and Bukovac, 1974).

The relative importance of foliar absorption as compared to root absorption as a route of entry into plant tissues depends on several factors. For soluble species that remain relatively available in soil solution, root-absorption processes are as effective as, or more effective than, foliar-absorption processes. This does not imply that foliar surfaces are not effective sites of absorption. Elements reported to be absorbed and transported from foliar surfaces include inorganic N, Rb, K, Na, Cs, P, Cl, S, Zn, Cu, B, Mn, Fe, and Mo (Wittwer, Bukovac, and Tukey, 1963). For specific nutrient deficiencies, foliar application is sometimes the method of choice (Bradford, 1966; Labanauskas, 1966). This is especially true for nutrilites that tend to hydrolyze readily in soil solution or are rapidly adsorbed to soil particles and therefore are not so available for root absorption.

The fate, with respect to foliar absorption, of relatively insoluble elements, such as plutonium, which make up or are carried on discrete particles, is in some way analogous to the behavior of micronutrients, such as iron, which tend to form relatively insoluble products in aqueous environments. If we can assume that small particles containing plutonium ($<1.0\,\mu$m) can be retained in foliar surfaces over an extended period of time, the question arises as to the absorptive capacity of foliar surfaces for available plutonium. Since absorption of a particular element is a function of the concentration of the available or soluble component, an extended residence time on plant foliage may provide the time necessary for soluble components to be chemically modified and/or absorbed by internal tissues. This may represent a more efficient route of entry than root absorption because in root absorption the same finite amount of plutonium deposited in soil may be insolubilized and adsorbed to soil particles, which, of course, reduces the concentration available for root absorption.

Absorption data from laboratory studies with bush bean plants contaminated with aerosolized plutonium are given in Tables 3 and 4; the protocol for this study was

TABLE 3 Extent of Translocation of ^{238}Pu from Contaminated Foliage of Bush Bean to Seed Tissues in the Absence and Presence of a Solution Vector*†

Time of leaching after contamination	Stage of development	Transport ratio‡			
		Pu oxide (fresh)	Pu oxide (hydrated)§	Pu citrate	Pu nitrate
No leach		$<4.5 \times 10^{-6}$	1.1×10^{-5}	5.4×10^{-6}	6.8×10^{-6}
Day 1	Preflowering	2.6×10^{-5}	4.1×10^{-5}	8.4×10^{-5}	2.6×10^{-4}
Day 7	Flowering; seed development	1.7×10^{-5}	1.8×10^{-5}	1.8×10^{-4}	1.4×10^{-3}
Day 14	Seed filling	1.8×10^{-5}	4.4×10^{-5}	3.5×10^{-5}	1.4×10^{-5}
Day 21	Seed filling completed	$<5.0 \times 10^{-6}$	4.3×10^{-5}	4.7×10^{-5}	4.0×10^{-6}
Average for leached plants		1.5×10^{-5}	3.7×10^{-5}	8.7×10^{-5}	4.2×10^{-4}

*D. A. Cataldo, unpublished data.

†All compounds were supplied from solutions at pH 5.8 to 7.0; aerosol characteristics for the fresh and hydrated oxides are given in Table 1; aerosols of plutonium citrate had an AMAD of 1.61 μm and a GSD of 1.86, and the count mode was 0.200 μm; plutonium nitrate had an AMAD of 2.29 μm and a GSD of 1.91, and the count mode was 0.152 μm. Plants were leached with 200 ml of solution (equivalent to a 7-min rainfall of 0.4 cm).

‡Transport ratio = picocuries per gram of seed tissue/picocuries per gram of contaminated leaf tissue; average of four replicate samples.

§Aged in H_2O at pH 7.0 for 10 months.

TABLE 4 Extent of Translocation of ^{238}Pu from Contaminated Foliage of Bush Bean to Root Tissues in the Absence and Presence of a Solution Vector*†

Time of leaching after contamination	Stage of development	Transport ratio‡			
		Pu oxide (fresh)	Pu oxide (hydrated)§	Pu citrate	Pu nitrate
No leach		$<6.6 \times 10^{-6}$	$<4.8 \times 10^{-5}$	7.3×10^{-6}	3.3×10^{-5}
Day 1	Preflowering	2.2×10^{-5}	1.6×10^{-5}	4.7×10^{-5}	1.1×10^{-4}
Day 7	Flowering; seed development	1.4×10^{-5}	2.2×10^{-5}	5.7×10^{-5}	4.6×10^{-4}
Day 14	Seed filling	1.1×10^{-5}	4.3×10^{-5}	7.1×10^{-5}	1.4×10^{-5}
Day 21	Seed filling completed	1.6×10^{-5}	1.9×10^{-5}	1.7×10^{-4}	3.3×10^{-5}
Average for leached plants		1.6×10^{-5}	2.5×10^{-5}	8.6×10^{-5}	1.5×10^{-4}

*D. A. Cataldo, unpublished data.

†All compounds were supplied from solutions at pH 5.8 to 7.0; aerosol characteristics are given in Table 3. Plants were leached with 200 ml of solution (equivalent to a 7-min rainfall of 0.4 cm).

‡Transport ratio = picocuries per gram of root tissue/picocuries per gram of contaminated leaf tissue; average of four replicate samples.

§Aged in H_2O at pH 7.0 for 10 months.

reported by Cataldo, Klepper, and Craig (1976). The objective was to evaluate, on the basis of chemical form supplied and the presence or absence of a solution vector (simulated rainfall), the extent of absorption and translocation of foliarly applied plutonium. All plants were exposed to plutonium at 20 days from planting (preflowering) and were held for an additional 28 days to allow time for both absorption of plutonium and seed filling. During this 28-day period, contaminated plants were either maintained in the absence of a simulated rainfall or subjected to a simulated rainfall at 1, 7, 14, or 21 days after contamination. This simulated rainfall provided a solution vector on the surfaces of contaminated leaves to enable diffusion and absorption of mobile plutonium forms. The pots containing soil and root were double bagged with polyethylene and sealed at the lower stem; the seed tissue was contained in pods that were formed after exposure. The effect of both the chemical form of the plutonium and the timing of simulated rainfall on the absorption of plutonium from foliar surfaces and its translocation was determined by analysis of uncontaminated seed and root tissues. Quantitation was by means of transport ratios: TR = picocuries of ^{238}Pu per gram (dry weight) of seed or root $\div$ picocuries of ^{238}Pu per gram (dry weight) of contaminated leaf tissue. Although there is a tendency to compare TR values with classical concentration ratios derived from plants grown on contaminated soils, this comparison is inappropriate. At harvest the dry weights of contaminated leaves and uncontaminated roots and seeds of individual plants were approximately 0.6, 0.5, and 0.3 g, respectively. Plants were contaminated with 0.25 to 1.0×10^6 d/min ^{238}Pu. Transport ratios (TR values) for root and seed tissues from plants not subjected to leaching (solution vector) ranged from $< 4.5 \times 10^{-6}$ to 3.3×10^{-5}. Application of a simulated rainfall to provide a solution vector for diffusion and absorption of soluble components on the leaf surface increased uptake and transport of plutonium to seed and root tissues for all compounds of plutonium studied (1.5×10^{-5} to 4.2×10^{-4}). Apparent differences in translocation between the various plutonium forms may result from the relative size of the soluble fraction. The fresh plutonium dioxide was truly particulate at the time of contamination, the aged oxide consisted of particles with a fractured crystal lattice (Park et al., 1974), the citrate represented a relatively stable soluble complex, and the nitrate represented an unstable complex that rapidly hydrolyzed on dilution to form colloidal hydroxides. The order of bioavailability for transport to seed and root was plutonium nitrate (hydroxide) > plutonium citrate > aged oxide > fresh oxide. (Tables 3 and 4 show average values for leaching treatment.)

An interesting aspect of these data is that maximum TR values are obtained when the simulated rainfall occurs at day 7 or 14, the time of rapid seed development. This phenomenon is of interest from the standpoint of the mobility of plutonium within the plant and the chemical form of the plutonium. It is generally accepted that materials must be transported out of mature leaves in the phloem. The entry of molecules into this transport conduit is metabolically regulated, and the loading process is highly specific for individual organic metabolites and inorganic elements (Crafts and Crisp, 1971). There is growing evidence that many inorganic nutrilites, especially multivalent cations, are transported as organic complexes in both the xylem (Tiffin, 1967; 1971; Bradfield, 1976) and phloem (van Goor and Wiersma, 1976). By analogy to the behavior of nutrilites, plutonium must be transported out of the contaminated leaves via the phloem. Similarly, it is unlikely that inorganic plutonium could remain soluble at the pH of phloem cell sap (pH 7.2 to 8.5) (Ziegler, 1975). Therefore it is possible that the mobile plutonium which was deposited in seed and root tissues may have been complexed with phloem-mobile organic species. This would explain the apparent increase in TR values seen during the time of seed development. During this period there is a significant change in both the composition and quantity of specific metabolites being produced by leaves and being exported to metabolic sinks, such as seeds and roots. This change in metabolism may increase the potential for soluble species of plutonium to become complexed with organic metabolites and subsequently to be exported to metabolic sinks. Although this is a tentative judgment and subject to substantive studies, this interpretation serves to explain the observed results on the basis of known metabolic aspects of plant function.

Conclusions

The ability of terrestrial plants to accumulate potentially hazardous elements from soils via root absorption and the relative importance of these elements in the food web to man has prompted numerous studies over the past 25 years. The majority of these investigations have been concerned with soil—plant transfer rates since the soil represents a major repository for pollutants released to the environment and since the plant root is an efficient solute-absorbing structure. Until recently the foliar portions of plants were considered to play a minor, transient role at best with respect to dose-assessment problems.

Our current understanding of the aerodynamic behavior of particles and anticipated reductions in particle-size distributions of materials such as plutonium through an expanded nuclear energy program suggests that a reevaluation of the role of plant foliage in particle interception and absorption of materials contained on airborne particles is in order.

This need is supported by both early investigations and studies currently under way. Early studies of worldwide fallout and current work on contaminated soils resuspended by wind indicate that foliar retention and foliar absorption may be as important as, and in some cases exceed, the role of root absorption with respect to food-chain transport. A critical evaluation of past literature on the leaching of foliar deposits suggests that aerosol polydispersity and large particle size (e.g., 45 μm MMD) may explain the comparatively large degree of leaching or weathering reported for comparatively large particles (Witherspoon and Taylor, 1969; 1970; 1971). This view is reinforced by data reported for well characterized particles of lead and plutonium in laboratory studies and field observations for fallout plutonium. These latter investigations indicate that a sizeable fraction ($>$80%) of submicron-size particles deposited onto foliage are tenaciously held on leaf surfaces under varied conditions (e.g., simulated rainfall and wind). Aside from the potential health implications associated with increased foliar retention, the problem of foliar absorption must be considered. In the reported studies, a substantial fraction of the foliar plutonium deposits was transported to seed and roots. Transport ratios were affected by both the presence of a solution vector (simulated rainfall) and the timing of its application with respect to the stage of plant development.

References

Aylor, D. E., 1975, Force Required to Detach Conidia of *Helminthosporium maydis, Plant Physiol.,* 56: 97-99.

——, 1976, Resuspension of Particles from Plant Surfaces by Wind, in *Atmosphere—Surface Exchange of Particulate and Gaseous Pollutants,* DOE Symposium Series, Richland, Wash., Sept. 4–6, 1974, R. J. Engelmann and G. A. Sehmel (Eds.), pp. 791-812, CONF-740921, NTIS.

——, and J. Y. Parlange, 1975, Ventilation Required to Entrain Small Particles from Leaves, *Plant Physiol.,* 56: 97-99.

Bennett, B. G., 1976, Transuranic Element Pathways to Man, in *Transuranium Nuclides in the Environment,* Symposium Proceedings, San Francisco, 1975, pp. 367-382, STI/PUB/410, International Atomic Energy Agency, Vienna.

Benzing, D. Q., and K. M. Burt, 1970, Foliar Permeability Among Twenty Species of the *Bromelioceae, Bull. Torrey Bot. Club,* 97: 269-279.

Bradfield, F. G., 1976, Calcium Complexes in the Xylem Sap of Apple Shoots, *Plant Soil,* 44: 495-499.

Bradford, G. R., 1966, Boron, in *Diagnostic Criteria for Plants and Soils,* H. D. Chapman (Ed.), pp. 33-61, Division of Agricultural Sciences, University of California.

Carlson, R. W., F. A. Bazzaz, J. J. Stukel, and J. B. Wedding, 1976, Physiological Effects, Wind Reentrainment, and Rainwash of Pb Aerosol Particulate Deposited on Plant Leaves, *Environ. Sci. Technol.,* 10: 1139-1142.

Cataldo, D. A., E. L. Klepper, and D. K. Craig, 1976, Fate of Plutonium Intercepted by Leaf Surfaces: Leachability and Translocation to Seed and Root Tissues, in *Transuranium Nuclides in the Environment,* Symposium Proceedings, San Francisco, 1975, pp. 291-301, STI/PUB/410, International Atomic Energy Agency, Vienna.

Chamberlain, A. C., 1970, Interception and Retention of Radioactive Aerosols by Vegetation, *Atmosphere Environ.,* 4: 57-58.

Crafts, A. S., and C. E. Crisp, 1971, *Phloem Transport in Plants,* pp. 85-150, W. H. Freeman and Company, San Francisco.

Energy Research and Development Administration, 1975, *A National Plan for Energy Research, Development, and Demonstration: Creating Energy Choices for the Future,* ERDA Report ERDA-48, Vol. 1, GPO.

Fisher, D. A., and D. E. Boyer, 1972, Thin Sections of Plant Cuticles Demonstration Channels and Wax Platelets, *Can. J. Bot.,* 50: 1509-1511.

Franke, W., 1967, Mechanism of Foliar Penetration of Solutions, *Annu. Rev. Plant Physiol.,* 18: 281-300.

——, 1971, The Entry of Residues into Plants via Ectodesmata (Ectocythodes), in *Residue Reviews,* Vol. 38, pp. 81-115, F. A. Gunther and J. D. Guntler (Eds.), Springer-Verlag, New York.

Greene, D. W., and M. J. Bukovac, 1974, Stomatal Penetration: Effects of Surfactants and Role in Foliar Absorption, *Am. J. Bot.,* 61: 100-106.

Gregory, P. H., 1971, The Leaf as a Spore Trap, in *Ecology of Leaf Surface Micro-organism,* T. F. Preece and C. H. Dickinson (Eds.), Academic Press, Inc., New York.

Hanson, W. C., 1975, Ecological Considerations of the Behavior of Plutonium in the Environment, *Health Phys.,* 28: 529-537.

Holloway, P. J., 1971, The Chemical and Physical Characteristics of Leaf Surfaces, in *Ecology of Leaf Surface Micro-organisms,* T. F. Preece and C. H. Dickinson (Eds.), pp. 39-53, Academic Press, Inc., New York.

Hull, H. M., H. L. Morton, and J. R. Wharrie, 1975, Environmental Influences on Cuticle Development and Resultant Foliar Penetration, *Bot. Rev.,* 41: 421-452.

Iranzo, E., 1968, First Results from the Programme of Action Following the Palomares Accident, in *Proceedings of the Symposium on Radiological Protection of the Public in Nuclear Mass Disasters,* Interlaken, Switzerland, May 27–June 1, 1968.

Krantz, B. A., A. L. Brown, D. B. Fischer, W. E. Pendury, and V. W. Brown, 1962, Foliage Sprays Correct Iron Chlorosis in Grain Sorghum, *Calif. Agric.,* 16(5): 5-6.

Labanauskas, C. K., 1966, Manganese, in *Diagnostic Criteria for Plants and Soils,* H. D. Chapman (Ed.), pp. 264-285, Division of Agricultural Sciences, University of California.

Little, P., 1973, A Study of Heavy Metal Contamination of Leaf Surfaces, *Environ. Pollut. (London),* 5: 159-172.

Markee, E. J., Jr., 1971, *Studies of the Transfer of Radioiodine Gas to and from Natural Surfaces, A Summary of Controlled Environmental Release Test Analysis (1963–1969),* Report NOAA-TM-ERL-ARL-29, National Oceanic and Atmospheric Administration, Silver Spring, Md.

Middleton, L. J., 1959, Radioactive Strontium and Cesium in the Edible Parts of Crop Plants After Foliar Contamination, *Int. J. Radiat. Biol.,* 1: 387-402.

Moorby, J., and H. M. Squire, 1963, The Loss of Radioactive Isotopes from the Leaves of Plants in Dry Conditions, *Radiat. Bot.,* 3: 163-167.

Park, J. F., D. L. Catt, D. K. Craig, R. J. Olson, and V. H. Smith, 1974, Solubility Changes of ^{238}Pu Oxide in Water Suspension and Effect on Biological Behavior After Inhalation by Beagle Dogs, in *Proceedings of the Third International Radiation Protection Association Meeting,* Wash., D. C., Sept. 9–14, 1973, W. S. Snyder (Ed.), USAEC Report CONF-730907-P1, pp. 719-724, International Radiation Protection Association, NTIS.

Romney, E. M., W. A. Rhoads, A. Wallace, and R. A. Wood, 1973, Persistence of Radionuclides in Soil, Plants and Small Mammals in Areas Contaminated with Radioactive Fallout, in *Radionuclides in Ecosystems,* Proceedings of the Third National Symposium on Radioecology, Oak Ridge, Tenn., May 10–12, 1971, D. J. Nelson (Ed.), USAEC Report CONF-710501, pp. 170-176, Oak Ridge National Laboratory, NTIS.

——, A. Wallace, R. O. Gilbert, and J. E. Kinnear, 1975, $^{239-240}Pu$ *and* ^{241}Am *Contamination of Vegetation in Aged Plutonium Fallout Areas,* ERDA Report UCLA-12-986, University of California at Los Angeles, NTIS.

Russell, R. S., 1965, An Introductory Review: Interception and Retention of Airborne Material on Plants, *Health Phys.,* 11: 1305-1315.

Slinn, W. G. N., 1975, *Wet Versus Dry Deposition Downwind of a Point Source of Pollution,* ERDA Report BNWL-SA-4800, Battelle, Pacific Northwest Laboratories, NTIS.

——, 1976, Dry Deposition and Resuspension of Aerosol Particles—A New Look at Some Old Problems, in *Atmosphere–Surface Exchange of Particulate and Gaseous Pollutants,* DOE Symposium Series, Richland, Wash., Sept. 4–6, 1974, R. J. Engelmann and G. A. Sehmel (Eds.), pp. 1-40, CONF-740921, NTIS.

Soldat, J. K., 1971, The Radiological Dose Model, in *HERMES: A Digital Computer Code for Estimating Regional Radiological Effects from the Nuclear Power Industry,* J. F. Fletcher and W. L. Dotson (Comps.), HEDL-TME-71-168, pp. 81-175, Hanford Engineering Development Laboratory, NTIS.

Tiffin, L. O., 1967, Translocation of Manganese, Iron, Cobalt and Zinc in Tomato, *Plant Physiol.,* 42: 1427-1432.

——, 1971, Translocation of Nickel in Xylem Exudate of Plants, *Plant Physiol.,* 48: 273-277.

U. S. Atomic Energy Commission, 1974, *Liquid Metal Fast Breeder Reactor Program* (Proposed Final Environmental Statement), WASH-1535, Vol. 2. (Air concens., pp. 4.3.-6, pp. 4.4-41; reprocessing, pp. 4.4-12 to 4.4-34; Pu escape, pp. 4.4-46; atmospheric dispersal/deposition, pp. II. G-13 to II.G-14, p. 6-17; soil profile movement, pp. II.G-27; food chain estimations, pp. II.G-29 to II.G-32; inhalation and ingestion, pp. II.G-5 and II.G-8), Available free from W. H. Pennington, USERDA, Office of Assistant Administrator for Environment and Safety, Washington, D. C.

van Goor, B. J., and D. Wiersma, 1976, Chemical Forms of Manganese and Zinc in Phloem Exudates, *Physiol. Plant.,* 36: 213-216.

Vaughan, B. E., R. E. Wildung, and J. J. Fuquay, 1976, Transport of Airborne Effluents to Man Via the Food Chain, in *Controlling Airborne Effluents from Fuel Cycle Plants,* ANS–AIChE Topical Meeting, Sun Valley, Idaho, Aug. 5–6, 1976, American Nuclear Society, Hinsdale, Ill.

Wedding, J. B., R. W. Carlson, J. J. Stukel, and F. A. Bazzaz, 1975, Aerosol Deposition on Plant Leaves, *Environ. Sci. Technol.,* 9: 151-153.

Witherspoon, J. P., and F. G. Taylor, 1969, Retention of Fallout Simulant Containing ^{134}Cs by Pine and Oak Trees, *Health Phys.,* 17: 825-839.

——, and F. G. Taylor, 1970, Interception and Retention of a Simulated Fallout by Agricultural Plants, *Health Phys.,* 19: 493-499.

——, and F. G. Taylor, 1971, Retention of 1-44 μm Simulant Fallout Particles by Soybean and Sorghum Plants, *Health Phys.,* 21: 673-677.

Wittwer, S. K., M. J. Bukovac, and H. E. Tukey, 1963, Advances of Foliar Feeding of Plant Nutrients, in *Fertilizer Technology and Usage,* M. H. McVicker, G. L. Bridger, and L. B. Nelson (Eds.), Soil Science Society of America, Madison, Wisc.

Ziegler, H., 1975, Nature of Transported Substance, in *Transport in Plants One Phloem Transport,* M. H. Zimmerman and J. A. Milburn (Eds.), pp. 59-100, Springer-Verlag, New York.

The Relationship of Microbial Processes
to the Fate and Behavior
of Transuranic Elements in Soils, Plants,
and Animals

R. E. WILDUNG and T. R. GARLAND

Soil physicochemical and microbial processes will influence the long-term solubility, form, and bioavailability of plutonium and other transuranic elements important in the nuclear fuel cycle. Consideration is given to the chemistry/microbiology of the transuranic elements in soil, emphasizing possible organic complexation reactions in soils and plants and the relationship of these phenomena to gastrointestinal absorption.

Initial solubility of the transuranic elements in soil is governed largely by hydrolysis with soil sorption in the order Pu > Am ≃ Cm > Np. Soluble (<0.01 µm), diffusible plutonium in soils (usually less than 0.1% of total) appears to be largely present as particulates of hydrated oxide, but several lines of evidence indicate that microorganisms influence the solubility and plant availability of plutonium and that the nonparticulate plant-available fraction is stabilized in solution by inorganic or organic ligands of limited concentration in soil. The possible role of soil microorganisms in influencing the solubility, form, and plant availability of the transuranic elements is discussed on the basis of the (1) known chemistry of organic ligands in soils; (2) effects on the soil microflora; and (3) principal microbial transformation mechanisms, including direct alteration (valence state and alkylation), indirect alteration (metabolite interactions and influence on the physicochemical environment), and cycling processes (biological uptake and release on decomposition of tissues).

The toxicity of plutonium to microorganisms depends on plutonium solubility in soil. However, soil microorganisms are generally resistant to plutonium; toxicity is due principally to radiation rather than to chemical effects. Highly resistant bacteria, fungi, and actinomycetes have been isolated from soil, and these organisms have been shown to be capable of transporting plutonium into the cell and altering its form in the cell and in solution. The resulting soluble plutonium complexes exhibit a range of mobilities in soil and tend to be of higher molecular weight than simple complexes (plutonium— diethylenetriaminepentaacetic acid) and negatively charged. The forms of plutonium complexes, although not well defined, are dependent on organism type, carbon source, and time of plutonium exposure during growth. These factors, in turn, are a function of plutonium source, soil properties, and soil environmental conditions. Knowledge of the relative influence of these factors serves as a valuable basis for predicting the long-term behavior of plutonium and other transuranic elements in soils. There is growing evidence that these phenomena also markedly influence the availability of plutonium to plants and animals.

Plutonium present in solution as an organic complex is readily assimilated by the plant in the Pu(IV) state. Evidence to date indicates that soil sorption rather than plant discrimination limits plant uptake of plutonium and that organometal complexes serve mainly to deliver plutonium to the root membrane; i.e., the ligands are not taken up by the plant stoichiometrically with the metal. After passing the root membrane, Pu(IV) is translocated to the shoots in the xylem through formation of a number of organic complexes with plant ligands. The form of plutonium differs in leaves and stems, but greater than 90% of the soluble plutonium associated with these tissues was present as complexed Pu(IV) after growth on soil to which ionic Pu(IV) had been added.

A reevaluation of plant-to-animal transfer coefficients used presently in dose assessments may be required since plutonium incorporated in plant tissues is markedly more available to animals than Pu(IV) gavaged in the inorganic plutonium solutions that were used for previous measurements. Differences in gastrointestinal transfer of plutonium in stems and leaves of alfalfa are related to differences in plutonium solubility in these tissues. Thus the form of plutonium in soils and plants may be closely related to plutonium availability to 'animals.

Although information leading to an understanding of the complex biochemical interrelationships that exist between soils, plants, and animals is rapidly developing, these phenomena are not sufficiently understood at present to be described by simple models.

The major factor governing availability of the transuranic elements to plants will be their solubility in soil since, for root uptake to occur, a soluble species must exist adjacent to the root membrane for some finite period. The form of this soluble species will have a strong influence on its stability in soil solution, on its mobility in soils, and on the rate and extent of uptake and, perhaps, on its mobility and toxicity in the plant. Furthermore, the results of preliminary studies discussed in this chapter suggest that the concentration and chemical form of the element in the plant play a major role in influencing its availability to animals on ingestion. Thus any assessment of the long-term behavior of the transuranic elements in the terrestrial environment must be based on the determination of the factors influencing solubility and on the form of soluble species in soil. These factors, illustrated in Fig. 1, include the concentration and chemical form of the element entering soil; the influence of soil properties on the elemental distribution between the solid and liquid phase; and the effect of soil processes, such as microbial activity, on the kinetics of sorption reactions, transuranic concentration, and the form of soluble and insoluble chemical species.

Portions of the soil chemical and microbiological sections of this chapter have been published elsewhere by the U. S. Department of Energy (Wildung, Drucker, and Au, 1977) and the American Society of Agronomy (Keeney and Wildung, 1977). However, at the request of the editor these have been reiterated to provide a complete treatment of the subject.

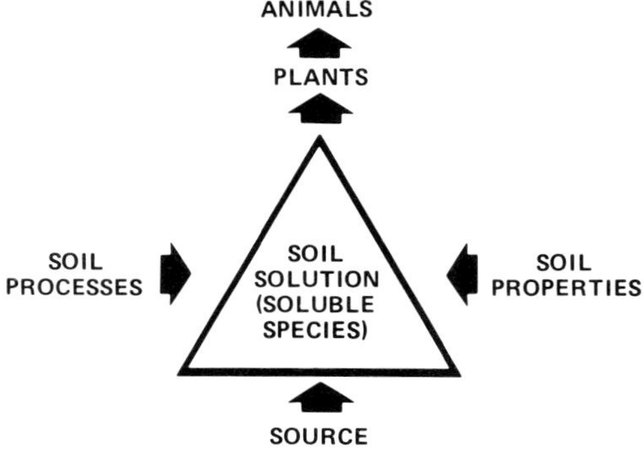

Fig. 1   Factors influencing transuranic behavior in the terrestrial environment.

Transuranic Chemistry in Soil

Sources of the Transuranic Elements

The transuranic elements of principal importance in the nuclear fuel cycle (plutonium, americium, curium, and neptunium) can enter the soil through several avenues (Vaughan, Wildung, and Fuquay, 1976), including (1) fallout from atmospheric testing, (2) possible escape of airborne particulates and liquid effluents during reprocessing of spent fuels and fuel fabrication, and (3) leaching from waste-storage facilities. The major sources of the transuranic elements can be classified according to expected initial solubility in soil:

Insoluble source terms

$$MO_x + L \rightarrow ML$$

Soluble source terms

Hydrolyzable

$$(1) \qquad M(NO_3)_x + L + H_2O \rightarrow \begin{bmatrix} MO_x \cdot nH_2O \\ or \\ M(OH)_x \end{bmatrix} \downarrow + ML$$

Nonhydrolyzable

$$(2) \qquad M(NO_3)_x + L + H_2O \rightarrow MO_x^+ + ML$$

Organic complexes

$$(3) \qquad ML_1 + L_2 + H_2O \rightarrow ML_1 + ML_2 + ML_{1,2}$$

$$ML_1 + L_2 + H_2O \rightarrow \text{as in (1) or (2)}$$

where M represents transuranic elements and L represents inorganic and organic ligands capable of reacting with transuranic elements and forming soluble or insoluble products. Particulate oxides of the transuranic elements initially can be expected to be largely insoluble in the soil solution. Ultimately, solubility is expected to be a function of the composition, configuration, and equivalent diameter of the particle as well as soil properties and processes. Oxide particles of the highest specific activity and containing the highest concentrations of impurities in the crystal lattice may exhibit the greatest solubility. The combination of configuration and equivalent diameter as reflected in surface area exposed to solution will be the other main factor governing oxide solubility. Once solubilized, the transuranic elements will be subject to the chemical reactions governing soluble salts. Hydrolyzable transuranic elements entering the soil in acid solutions sufficiently concentrated to maintain soluble ions can be expected to be rapidly insolubilized as a result of hydrolysis on dilution and subsequent precipitation on particle surfaces. These include Pu(III, IV, and VI), Am(III), Cm(III), and Np(III, IV, and VI). Conversely, transuranic elements not subject to marked hydrolysis can be initially more soluble. These include Pu(V) and Np(V). Immobilization of these chemical species (PuO_2^+

or NpO_2^+) can occur through cation-exchange reactions with particulate surfaces. Complicating this situation, disproportionation and complexation reactions may occur concurrently.

Transuranic elements entering the soil as stable organocomplexes, such as might occur in the vicinity of a spent-fuel separation facility, may be initially highly soluble (Wildung and Garland, 1975). The duration of solubility and mobility in the soil will be a function of the stability of the complex to substitution by major competing ions, such as calcium and hydrogen (Lahav and Hochberg, 1976; Lindsay, 1972; Norvell, 1972), and the stability of the organic ligand to microbial decomposition (Wildung and Garland, 1975). The disruption of the complex may lead to a marked reduction in transuranic-element solubility through hydrolysis and precipitation reactions, as described for acid solutions on dilution. A portion of the ion released may react with other, perhaps more stable, ligands in soil. The mobility of the intact complexes, in turn, will be principally a function of their chemical and microbiological stability and the charge on the complex, which will govern the degree of sorption on soil particulates.

Further generalizations of transuranic-element behavior on the basis of source terms are complicated by the overwhelming importance of soil properties and processes in influencing transuranic-element behavior on a regional and local basis. This chapter considers, in detail, the influence of soil properties and abiotic and biotic processes on the long-term solubility of the transuranic elements entering soils. Consideration is also given to the implications of these processes in terms of transuranic-element plant and animal availability. Principal emphasis is directed toward the role of soil microorganisms in this phenomenon. Microorganisms, in intimate association with soil particles, are known to play an important role in effecting solubilization of elements considered insoluble in soils strictly on the basis of their inorganic chemistry. To date studies of the microbiology of the transuranic elements have been limited principally to plutonium. This chapter will emphasize plutonium, but, where possible, the available information is used as a framework for broader discussions encompassing the long-term behavior of other transuranic elements.

Chemical Reactions Influencing Plutonium Behavior

The principal chemical reactions likely influencing plutonium behavior in soil are:

- Four oxidation states

$$Pu^{3+}, Pu^{4+}, PuO_2^{1+}, PuO_2^{2+} (Pu^{2+}, Pu^{7+})$$

- Disproportionation

$$Pu^{4+} + PuO_2^{1+} \rightleftharpoons Pu^{3+} + PuO_2^{2+}$$

- Hydrolysis

$$Pu^{4+} + 4H_2O \rightleftharpoons Pu(OH)_4 + 4H^+ (K_{sp} \approx 10^{-56})$$

- Complex formation

$$2\,Pu^{4+} + 3\,DTPA \rightleftharpoons Pu_2 DTPA_3 \quad (\log K \approx 18)$$

Plutonium ions may commonly exist in aqueous solution in valence states III, IV, $V(PuO_2^+)$, and $VI(PuO_2^{2+})$. Other valence states are known (II and VII) and predicted (VIII), but these occur under unique conditions (Cleveland, 1970). Disproportionation reactions are common, and, due to kinetic factors, plutonium is unique among the chemical elements in that it may simultaneously exist in all the common valence states. The tendency of plutonium to hydrolyze in aqueous solutions of low acidity follows the order $Pu^{4+} > PuO_2^{2+} > Pu^{3+} > PuO_2^+$ (Cleveland, 1970). Hydrolysis, which occurs in a stepwise fashion, is likely the major mechanism whereby plutonium is insolubilized in the environment. At high (grams per liter) plutonium concentrations, hydrolysis of Pu^{4+} may lead to the formation of a colloidal plutonium polymer. At these concentrations the polymer is characterized by a distinct absorption spectrum. Although the polymer has not been fully characterized, it is generally thought to be an intermediate hydrolysis product of Pu^{4+} containing oxide or hydroxide bridges with an absorption spectrum different from that of $Pu(OH)_4$. However, studies by Lloyd and Haire (1973) indicated that the polymer may be aggregates of small, discrete, amorphous or crystalline, primary particles of 5 to 20 A in diameter. It is of interest that X-ray-diffraction patterns of the polymeric plutonium and that of $Pu(OH)_4$ (Ockenden and Welch, 1956) showed a pattern characteristic of the cubic PuO_2 lattice, which suggests that the polymer and the hydroxide of Pu^{4+} may be hydrated PuO_2 with differences occurring in primary particle size and crystallinity (Lloyd and Haire, 1973). The formation of the hydrated PuO_2 is likely directly related to Pu^{4+} concentration and inversely related to the acid concentration.

Plutonium also tends to form many complexes with a range of stabilities. The strongest complexes are generally formed by reaction of organic ligands with Pu^{4+}. However, many inorganic complexes and organic complexes of all valences may be stable under appropriate conditions. The presence of organic ligands in soils likely influences the equilibrium and concentration form of plutonium in solution through complexation and subsequent inhibition of hydrolysis, polymerization, or disproportionation. These reactions, in various highly complex combinations resulting from differences in source term, soil properties, and processes, govern plutonium solubility in soil and availability to plants.

Soil chemical reactions are important in governing the behavior of the various forms of plutonium entering soil. Initially, soluble forms entering soil have the potential for undergoing a range of chemical transformations. Insoluble plutonium, such as high-fired oxide, entering soil likely will be solubilized with time, provided that soluble, stable complexes are formed. However, regardless of the form of plutonium entering soil, its ultimate solubility will be controlled by its aqueous chemistry and by soil factors. Soil physicochemical properties can be expected to have complex, interdependent effects on plutonium solubility. The long-term behavior of plutonium in soil will be a function of the kinetics of these reactions.

On the basis of research with other trace metals, recently summarized by Keeney and Wildung (1977), and limited information on the transuranic elements, it can be concluded that the soil physicochemical parameters most important in influencing the solubility of the transuranic elements include (1) solution composition, Eh and pH; (2) type and density of charge on soil colloids; and (3) reactive surface area. These phenomena will, in turn, be dependent on soil properties, including particle size distribution, organic-matter content, particle mineralogy, degree of aeration, and microbial activity. The delineation of the influence of these factors on plutonium solubility is difficult owing to the complex chemistry of plutonium.

A reasonable approach to the study of the chemistry of plutonium in soil is to direct initial attention to the factors influencing its solubility in soil. However, plutonium solubility in soil is difficult to define because solubility will depend on the method of measurement and because solubility must be arbitrarily evaluated because of the sorption of plutonium on submicron clay particles and the formation of submicron particles of hydrated plutonium oxide, which are difficult to centrifuge and may pass membrane filters. These effects can be illustrated by comparison of the differences in the solubility of plutonium in soils [100 days after amendment as $Pu(NO_3)_4$] as determined by water extraction and subsequent membrane filtration with the use of membranes of different average pore sizes (Table 1). The major fraction of the plutonium added was sorbed on

TABLE 1 Solubilities of Plutonium in Water Extracts of a Ritzville Silt Loam as Determined by Filtration with Membranes of Different Pore Sizes*

Membrane pore size, μm	Plutonium solubility,† pg/g
5	60,000
0.45	20,000
0.01	4,000
0.0015	1,000
0.0012	300
0.0010	50

*From Garland and Wildung (1977).
†Plutonium added at a level of 620,000 pg/g of soil.

the soil since a maximum of 10% of the extracted plutonium passed through the 5-μm membrane. Successive filtration through membranes with decreasing pore size resulted in decreases in plutonium concentration in the filtrate. Thus plutonium in the aqueous extract appeared to be in a wide range of particle sizes. Although membranes with pore sizes of 0.45 μm are commonly used to separate soluble matter from particulate matter, plutonium in these filtrates may be in colloidal forms. The plutonium in the 0.0010-μm filtrate appeared soluble, was stable in solution, and approximated the quantity of plutonium taken up by plants (Wildung and Garland, 1974). Of the soluble plutonium forms likely to enter soils (previous section), $Pu(NO_3)_4$ and plutonium–diethylenetriaminepentaacetic acid (DTPA) represent, in their respective chemistries, the range in soil behavior likely to occur. The water solubility (<0.01 μm) of ^{238}Pu and ^{239}Pu amended to a Ritzville silt loam (organic C content, 0.7%; pH 6.2) in the $Pu(NO_3)_4$ and Pu–DTPA forms differs markedly (Wildung and Garland, 1975). The DTPA complexes of both isotopes were water soluble in soil and appeared to be stable over the first 40 days of incubation (Fig. 2). After 7 days of incubation, the ^{238}Pu–DTPA appeared to be slightly less soluble than the ^{239}Pu–DTPA. After 95 days of incubation, both isotopes, initially added as the complex, appeared to decrease in solubility, perhaps as a result of microbial degradation of the organic moiety and the development of new chemical equilibria.

Equilibrium concentrations of soluble plutonium added as the nitrate were not obtained until 7 to 10 days. The solubility of ^{238}Pu and ^{239}Pu added to the soil as nitrates was much lower than the DTPA complexes, which likely reflects hydrolysis to

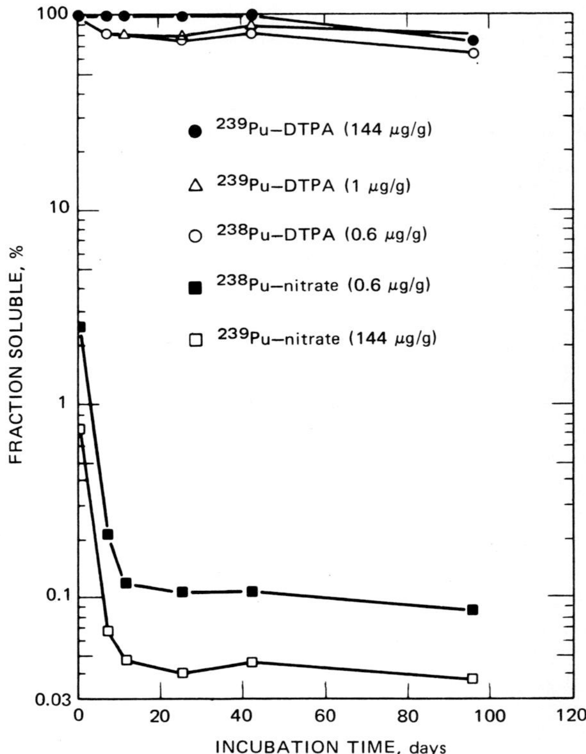

Fig. 2   Solubility of plutonium with time after addition to soil as the nitrate and the DTPA complex. [From Wildung and Garland (1975).]

the largely insoluble hydrated oxide. It is clear that organic ligands have a pronounced effect on plutonium solubility in soil. The rate of decrease in the solubility of each isotope added as the nitrate was similar. However, in contrast to the slightly lower solubility of the ^{238}Pu–DTPA compared with the ^{239}Pu–DTPA, ^{238}Pu added as the nitrate was a consistent factor of 2 to 3 times as soluble as ^{239}Pu initially added as the nitrate. This difference probably resulted from the formation of larger hydrated oxide particles at the higher plutonium concentration (^{239}Pu), but it may also have reflected the presence of soil components, such as organic ligands, which stabilized plutonium in solution but were present in limited concentrations and became important only at lower plutonium concentrations.

The water solubility of ^{238}Pu, when incorporated in relatively large plutonium oxide particles (>1 µm), would be expected to be greater than the solubility of ^{239}Pu oxide particles of similar size as a result of crystal damage and radiolysis arising from the greater specific activity of the ^{238}Pu (an approximate factor of 270). However, the behavior of the two isotopes in soil on solubilization of the oxide might be expected to follow a course similar to that exhibited by the nitrates (Fig. 2).

Equilibrium solubility after 6 days of incubation (Garland, Wildung, and Routson, 1976) of plutonium, added as Pu(NO$_3$)$_4$, in soils of different properties occurred after approximately 20 hr (Fig. 3). The quantities of plutonium soluble at equilibrium in water

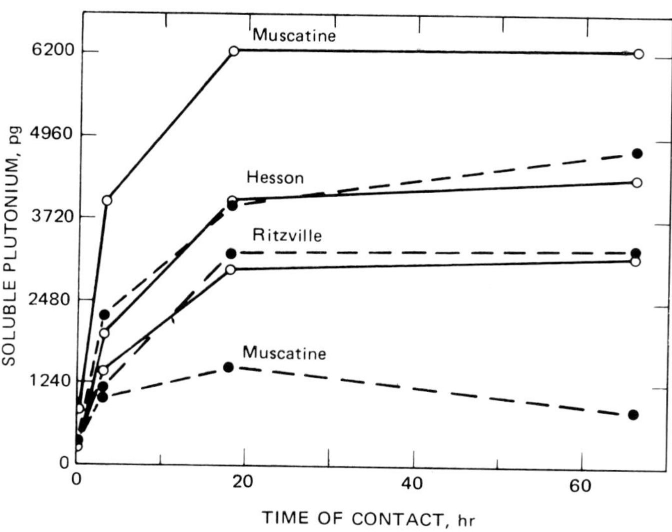

Fig. 3 Quantity of soluble plutonium removed from three soils by 0.01M CaCl$_2$. Soil : solution = 1 : 100. ○————○, distilled water. ●– – –●, 0.01M, CaCl$_2$. [From Garland, Wildung, and Routson (1976).]

and 0.01M CaCl$_2$ differed with soil type. In the CaCl$_2$ solution, solubility was lowest in the Muscatine soil, which exhibited higher silt and clay content than the other soils. Importantly, at equilibrium there was more plutonium extracted by water than by 0.01M CaCl$_2$ in the Muscatine soil. The Hesson and Ritzville soils did not exhibit this property. This may be related to a difference in the dispersibility of fine colloids in this soil and/or the presence of higher concentrations of stabilizing ligands. However, the lack of a proportional dilution effect (not shown in Fig. 3) in the water extractability of plutonium at lower solution-to-soil ratios in this soil, as compared with that in the Ritzville and Hesson soils, provided presumptive evidence for the presence of a dispersible ligand in higher concentration in the Muscatine soil.

Applying diffusion principles to characterization of mobile plutonium species in soils, Garland and Wildung (1977) estimated the concentrations and molecular weights of mobile plutonium in five surface soils representing a range in particle size distributions, pH (4.4 to 6.2), organic C (0.7 to 12.5%), and cation-exchange capacities (14 to 45 meq/100 g). Diffusion coefficients were calculated from measurements of the rate and extent of plutonium migration from soil through an agar matrix. The diffusion coefficients calculated for the most mobile species in the five soils varied from 1.5 to 3.0 × 10^{-6} cm^2/sec (Table 2). Estimated concentrations and molecular weights of the most mobile plutonium components in the five soils ranged from 9 to 55 pg/g and from 5000 to 21,000 g/mole, respectively. Thus estimated concentrations of the most mobile plutonium species were of the same order of magnitude as those observed by water extraction and subsequent ultrafiltration through the 0.0010-μm membrane (Table 1). This membrane retained Pu—DTPA (molecular weight, 1700). Hypothetical globular peptides of molecular weights of less than 500 would pass through this membrane. However, if the molecule were a hydrated PuO$_2$ sphere of similar dimensions, it would

TABLE 2   Estimated Concentrations and Molecular Weights of Mobile Plutonium in Soils from Measured Diffusion Coefficients*

	Most mobile species			Least mobile species		
Treatment	Diffusion coefficient, cm^2/sec ($\times 10^{-6}$)	Molecular weight, g/mole	Soil concentration,[†] pg/g	Diffusion coefficient, cm^2/sec ($\times 10^{-7}$)	Molecular weight, g/mole	Soil concentration,[†] pg/g
Control						
Pu–DTPA	5.8	1,700		53	1,700	
Soils						
Ritzville	3.0	5,000	24	2.3	0.9×10^6	150
Quillayute	2.5	7,200	47	2.7	0.7×10^6	1,200
Hesson	2.4	8,100	9	2.7	0.6×10^6	330
Salkum	1.5	21,000	55	2.3	0.8×10^6	340
Muscatine	1.9	13,000	36	3.1	0.5×10^6	170

*From Garland and Wildung (1977).
[†]Plutonium added at a level of 620,000 pg/g of soil.

have a molecular weight of between 10,000 and 25,000, which approximates the molecular weights of the most mobile plutonium species as determined from diffusion coefficients. This fraction, therefore, likely consisted of small particles of $Pu(OH)_4$ or hydrated oxide.

The estimated diffusion coefficients for the least mobile plutonium components ranged from 2.3 to 3.1×10^{-7} cm^2/sec with corresponding soil concentrations of 150 to 1200 pg/g (Table 2). This concentration of plutonium in soil approximated the quantity of water-soluble plutonium passing the 0.0015-μm ultrafiltration membrane (Table 1). Hypothetical globular proteins in this size range would have average molecular weights of <10,000. Particles of $Pu(OH)_4$ or hydrated oxides would have molecular weights of 200,000 to 500,000. Estimated molecular weights for these least mobile species calculated from diffusion coefficients were between 600,000 and 900,000. Thus it would appear, as in the case of the most mobile species, that the least mobile species of plutonium were particulate $Pu(OH)_4$ or hydrated oxides.

The comparison of filtration and diffusion data indicates that the mobile plutonium in incubated soils was in the form of hydrated oxide or hydroxide in a continuum of sizes. If it can be assumed that plutonium in particulate form was not available to plants, it is possible that the small fraction of plutonium taken up by plants was present in soil as reaction or dissolution products with insufficient stability and/or concentration to be detected by the methods used. Insight into this possibility was not provided by comparison of plutonium behavior in different soils, as might be expected, because the estimated concentrations and molecular weights of the mobile species were not related to the soil properties measured.

Several conclusions can be drawn from studies of the soil chemistry of plutonium which have important implications in terms of the potential role of the soil microbiota in influencing plutonium behavior in soil. The definition of plutonium solubility by filtration or diffusion alone is complicated by plutonium chemistry, but, in conjunction, the measurements suggest that mobile plutonium is largely particulate. However, a fraction of the mobile plutonium is available to plants.

This material is obviously not particulate but is present in insufficient concentration for characterization with current methods. The question remains, "What is the form of the small quantity of plutonium available to plants?" This information is essential to understanding the mechanisms whereby plutonium can be resupplied to solution from the solid phase in a range of soils and to predictions of the long-term availability of plutonium to plants. From investigations of plutonium valence state in a neutral, $0.0004M$ NH_4HCO_3 solution equilibrated with PuO_2 microspheres and in burial-ground leachates, Bondietti and Reynolds (1976) concluded that Pu(VI) may be stable in significant quantities in solution and suggested that monomeric Pu(VI) and its complexes may be important in plutonium mobilization. In the present studies, evidence was presented which suggested that plutonium ions are stabilized in soil solution by inorganic or organic ligands for subsequent uptake by the plant. Furthermore, equilibration of weathered plutonium-contaminated soil with chelating resins has been shown (Bondietti, Reynolds, and Shanks, 1976) to result in significant desorption of plutonium from the solid phase. It is known that organic ligands result in the most stable plutonium complexes. Soluble organic ligands in soil are generally derived from microbial processes.

Chemical Reactions Influencing the Behavior of Other Transuranic Elements

Other transuranic isotopes of concern in the nuclear fuel cycle include ^{241}Am, ^{243}Am, ^{242}Cm, ^{243}Cm, ^{244}Cm, and ^{237}Np. Although detailed studies of their interaction with soils are lacking, some information has become available in recent years. Furthermore, the aqueous chemistries of these elements have been fairly well established (Katz and Seaborg, 1957). The most stable ions of americium and curium in aqueous solutions are the cations (III); neptunium is most stable as the oxyion (NpO_2^+). Disproportionation is not common with these elements. Thus major differences in their environmental behavior as compared with that of plutonium would be expected. Hydrolysis reactions may still be a primary factor governing the environmental behavior of americium and curium, but greater mobility and plant availability in soils might be predicted on the basis of greater solubility of the hydroxides in comparison with $Pu(OH)_4$. The neptunium oxycation should not be subject to significant hydrolysis at environmental pH values (Burney and Harbour, 1974). Of the transuranic elements, neptunium has been the least studied, but, because of its chemical characteristics, it may be the most available to the biota. A comparison of plutonium, americium, and neptunium sorption in several soils (Routson, Jansen, and Robinson, 1975) indicated sorption in the order Pu > Am > Np. The chemistry of curium should be very similar to that of americium if present at equal mass concentrations.

Organic Complexation Reactions

Research to date on the chemistry of the transuranic elements in soil has pointed to the importance of understanding transuranic-element organic complexation reactions in soil, particularly in surface soils and aquatic sediments where organic-matter content is generally highest or in subsoils where the transuranic elements may be dispersed in conjunction with synthetic complexing agents. Very little information is available concerning the interaction of the transuranic elements with the soil organic fraction. However, despite the difficulties in characterization of soil organic complexes, much is known both theoretically and experimentally regarding the interactions of metals with functional groups of soil organic matter (Keeney and Wildung, 1977). Much of this information concerns micronutrients of greatest agronomic importance (B, Co, Cu, Fe, Mn, Mo, Se, and Zn) and has been the subject of a number of excellent reviews over the last two decades (Mitchell, 1964; 1972; Mortensen, 1963; Hodgson, 1963; Stevenson and Ardakani, 1972). Earlier studies generally emphasized metal interactions with intact soil or with the higher molecular-weight humic components of soil, whereas recent studies emphasize the more soluble components of soil.

It is practical to categorize metal complexes in soil in terms of their solubility since, in general, it is this factor, as previously noted, that most influences their mobility and plant availability. Three principal categories have been proposed (Hodgson, 1963), although the complexity of the soil system results in considerable overlap between categories. These categories include the (1) relatively high-molecular-weight humic substances containing condensed aromatic nuclei in complex polymers derived from secondary syntheses which have a high affinity for metals but are largely insoluble in soil, (2) low-molecular-weight organic acids and bases classified as nonhumic substances and derived largely from microbial cells and metabolism which demonstrate relatively high solubility in association with metals, and (3) soluble ligands that are precipitated on reaction with metals.

Humic Substances.   Humic substances are generally divided into three categories based on their solubilities (Felbeck, 1965). The humin (alkali and acid insoluble) fraction is soluble only under drastic conditions and is apparently of the highest molecular weight. The humate (alkali soluble and acid insoluble) and fulvate (alkali and acid soluble) fractions of soil may constitute up to 90% of the soil organic fraction (Kononova, 1966). The humates and fulvates are characterized, in part, by a high charge density due to acidic functional groups (Stevenson and Ardakani, 1972; Felbeck, 1965). This property leads to a high degree of reactivity, and these materials exhibit a strong pH-dependent affinity for cations in solution and are likely strongly bound to soil minerals and other organic constituents in soil (Greenland, 1965). The acidic functional groups consist principally (in general order of acidity) of carboxyl, hydroxyl (phenolic and alcoholic), enolic, and carbonyl groups (Broadbent and Bradford, 1952; Felbeck, 1965; Schnitzer, Shearer, and Wright, 1959). Total acidity has been estimated to range between 500 to 900 and 900 to 1400 meq/100 g for humic acids and fulvic acids, respectively (Stevenson and Butler, 1969). The acidic hydrogen of humic acids was differentiated by Thompson (1965) into three groups at 100 to 200, 500 to 700, and 1000 to 1200 meq/100 g using nonaqueous titration methods. Basic functional groups, likely amides and heterocyclic nitrogen compounds (Bremner, 1965), probably also contribute to retention of metals but are of much less importance than acidic groups at most soil pH values.

In batch equilibration studies (Bondietti, 1974), calcium-saturated humates removed greater than 94% of the Pu(IV) from pH 6.5 aqueous solutions (compositions not given). It is unclear whether the humates represented a surface for precipitation of hydrolyzed species or were involved in complexation of plutonium. However, in studies of plutonium desorption from humates and reference clays, citrate removed 10 to 30% of sorbed plutonium from the clays but less than 1% of that from the humic acids. Ligands forming stronger complexes with plutonium [DTPA and ethylenediaminetetraacetic acid (ETDA)] were required to remove significant quantities (up to 30%) of the plutonium from the humate complex.

Although humic and fulvic acids likely account for most of the metal immobilization attributed to the soil organic matter (e.g., Hodgson, 1963; Stevenson and Ardakani,

1972), they have the potential for formation of soluble complexes with metals, particularly in dilute solutions. Small quantities of metal fulvates, thought to be of lower molecular weight than the humates, may be present in soil solution. A nondialyzable material with infrared absorption spectra and elemental analyses similar to fulvic acids was isolated from a dilute salt (0.01M KBr) extract of a mineral soil by Geering and Hodgson (1969). The material exhibited a concentration equivalent to 2.5% of a dialyzable fraction but was more effective in complexing copper and zinc.

Nonhumic Substances with Potential for Metal Complexation. Lower molecular-weight biochemicals of recent origin have been implicated in metal complexation and solubilization in soil. These materials represent (1) components of living cells of microorganisms and plant roots and their exudates and (2) the entire spectrum of degradation products which ultimately serve as the building units of the soil humic fraction. The quantity and composition of these materials will vary with soil, vegetation, and environmental conditions (Alexander, 1961; 1971). Readily decomposable wastes disposed to soil under conditions appropriate for microbial growth may, for example, result in immediate and marked increases in organic materials identified in (1) and longer term increases of materials in (2). Conversely, toxic materials may have the opposite effects. The specific compounds produced will be dependent on the properties of the waste and soil environmental conditions after disposal (Routson and Wildung, 1969).

Although the concentration of the transuranic elements and other metals soluble in the soil solution or in mild extractants is low, often near-minimum detectable levels, the major portions of copper and zinc were shown to be associated with low-molecular-weight components. Most of the titratable acidity of this fraction was attributed (Geering and Hodgson, 1969) to aliphatic acids (<pH 7.0) and amino acids (>pH 7.0).

The production, distribution, and action of organic acids in soil were reviewed by Stevenson (1967). A wide range of organic acids is produced by microorganisms known to be present in soil. These include (1) simple acids, such as acetic, propionic, and butyric, which are produced in largest quantities by bacteria under anaerobic conditions; (2) carboxylic acids derived from monosaccharides, such as gluconic, glucuronic, and α-ketogluconic acids, which are produced by both bacteria and fungi; (3) products of the citric acid cycle, such as succinic, fumaric, malic, and citric acid, which are common metabolic excretory products of fungi; and (4) aromatic acids, such as p-hydroxybenzoic, vanillic, and syringic acids, which are thought to be fungal decomposition products of plant lignins. A variety of organic acids have also been reported in root exudates.

Amino acids are the other important group of compounds identified in significant quantities in the soil solution by Geering and Hodgson (1969) which may be expected to exhibit strong affinity for metals. The qualitative and quantitative aspects of amino acids and other nitrogenous components in soils have been reviewed by Bremner (1967). It was concluded that soil acid hydrolysates do not differ greatly in amino acid composition, but quantitative differences may occur with differences in soil, climatic, and cultural practices. A number of acidic and basic amino acids have been reported in soil. However, it appears that the major portion of amino acid-N that is present in hydrolysates is in (1) the neutral amino acids, glycine, alanine, serine, threonine, valine, leucine, isoleucine, and proline; (2) the acidic amino acids, aspartic acid and glutamic acid; and (3) the basic amino acids, lysine and arginine. Most of the amino acids detected in soil hydrolysates have also been shown to exist free in small quantities in soils with levels seldom exceeding 2 μg/g. In the soil solution (Geering and Hodgson, 1969) neutral amino acids also appeared to predominate. Basic amino acids were not detected, although two acidic amino acids (aspartic and glutamic acids) were present.

Stevenson and Ardakani (1972) concluded that organic acids and amino acids, although present only in small quantities in soil, were present in sufficient quantities in water-soluble forms to play a significant role in the solubilization of mineral matter in soil. Small quantities of a number of other complexing agents, such as nucleotide phosphates, polyphenols, phytic acid, porphyrins, and auxins, also exist in soil (pertinent references summarized by Mortensen, 1963). Complexation with biochemicals of recent origin would likely be the principal mechanism for microbial mobilization of the transuranic elements in soil.

Microbial Transformation of the Transuranic Elements in Soil

Potential Mechanisms of Transformation

From the results of limited studies of soil chemistry, microbiology, and plant availability of transuranic elements in soils and by inference from studies of complexation of other trace metals in soils, it can be concluded that the soil microflora will play a significant role in transformations governing the form, and, ultimately, the long-term solubility and behavior of transuranic elements in soil. Four general mechanisms whereby micro-organisms may alter the form of trace metals in soil (Alexander, 1961; Wood, 1974) are (1) indirect mechanisms resulting from metal interactions with microbial metabolites or changes in pH and Eh; (2) direct transformations, such as alkylation and alteration of the valence state through microbial oxidation (use of the metal as an energy source) or microbial reduction (use of the metal as an electron acceptor in the absence of oxygen); (3) immobilization by incorporation into microbial tissues; and (4) release of metals on decomposition of organic residues.

All these mechanisms may be operational in transformations of transuranic elements in soils. However, on the basis of present knowledge, it is not possible to draw conclusions as to their relative importance in affecting the long-term behavior of the transuranic elements. Since there is a paucity of information available, these mechanisms will be addressed around a framework of current information that is limited principally to plutonium.

Microbial Alteration of Solubility in Soil

To provide a preliminary assessment of the potential for microbial alteration of plutonium solubility in soil under aerobic conditions, Wildung, Garland, and Drucker (1973; 1974) measured soil respiration rate (an index of soil microbial activity), microbial types and numbers, and plutonium water solubility in sterile (gamma irradiation) and nonsterile soils that contained $10\,\mu\text{Ci}$ (Pu)/g (soil) [added as $\text{Pu(NO}_3)_4$]. Carbon dioxide evolution was used as a measure of soil respiration rate. For a measure of plutonium solubility, the soil was subsampled at intervals during incubation over a 65-day period, and the subsamples (1 g) were suspended in 1 liter of distilled water. After a 4-hr equilibration period, an aliquot of the soil suspension was filtered through 5-, 0.45-, and 0.01-μm filters. The plutonium in the 0.45- and 0.01-μm filtrates was designated water soluble, although it was recognized that plutonium likely was present as fine colloids (previous section).

Changes in the soil respiration rate and plutonium solubility during the 65-day incubation period are shown in Fig. 4(a). The lack of CO_2 evolution from the gamma-irradiated soil verified its sterility. The increased CO_2 evolution rates in the nonsterile soil over the 4- to 12-day period reflected logarithmic growth for all classes of organisms. The concentration of plutonium in the 0.01-μm filtrate during the incubation period ranged from approximately 0.04 to 0.14% of the plutonium initially applied. Solubility of plutonium was essentially identical in the sterile and nonsterile soils, decreasing with time.

In a subsequent experiment, the plutonium-containing sterile soil was inoculated with the plutonium-treated nonsterile soil (1 g) which had been previously incubated for 65 days [Fig. 4(a)], and the respiration rate and solubility of plutonium in the inoculated soil were measured for a period of 30 days [Fig. 4(b)]. When the sterile soils were inoculated with nonsterile soil, CO_2 evolution increased at a much more rapid rate without a lag phase, and this was followed by a factor of 2 increase in water solubility ($<0.01 \ \mu$m) of plutonium beginning after 5 days of incubation, which suggests the development of a microbial population in the plutonium-containing nonsterile soil that was particularly capable of alteration of plutonium solubility.

An analogous set of experiments was conducted with amended (carbon and nitrogen) sterile and nonsterile soils. The carbon and nitrogen were added to increase microbial activity and assess the effect of increased activity on plutonium solubility. The amendments markedly increased microbial activity (respiration rate, microbial types/ numbers) in the nonsterile soil but did not increase solubility in the $<0.01 \ \mu$m fraction compared to unamended soils. However, there was a significant increase in plutonium solubility in the <0.45-μm fraction of the nonsterile soil on initial incubation. As in the case of the soil that was not amended with carbon and nitrogen, reinoculation of the sterile soil with the nonsterile soil markedly increased solubility in the <0.01-μm fraction.

At least under the conditions of this study, the evidence strongly suggested that the solubility of plutonium in soil was influenced by the activity of the soil microflora. The potential mechanisms affecting the change in solubility include mechanisms (1) and (2) described on page 312, i.e., indirectly through the production of organic acids that may complex plutonium or the alteration of the solution pH and/or Eh near the soil colloid without measurable effects on the overall soil pH or directly through the reduction of plutonium to Pu(III). Oxidation to Pu(VI) would likely increase solubility, but recent evidence suggests that this does not occur in these systems. Of course, a combination of the mechanisms is possible, i.e., alteration of affinity for organic ligands through a change in valence state. If the mechanism of solubilization was indirect, the results might be applicable to other transuranic elements; e.g., from consideration of the aqueous chemistry, a reduction in pH would be expected to increase the solubility of the other transuranic elements as well as that of plutonium.

Increased water solubility of plutonium on incubation under optimum conditions for microbial activity may increase plutonium uptake by plants provided that the limiting factor is not discrimination at the root membrane. To determine if the increased solubility on incubation resulted in increased plutonium uptake by plants, the incubated soils were planted with barley and cultured by a split-root technique that allowed measurement of the uptake, sites of deposition, and chemical forms of plutonium in plant shoots and roots (Wildung and Garland, 1974). The results were compared with the results of similar plant studies in which the soils had not been incubated.

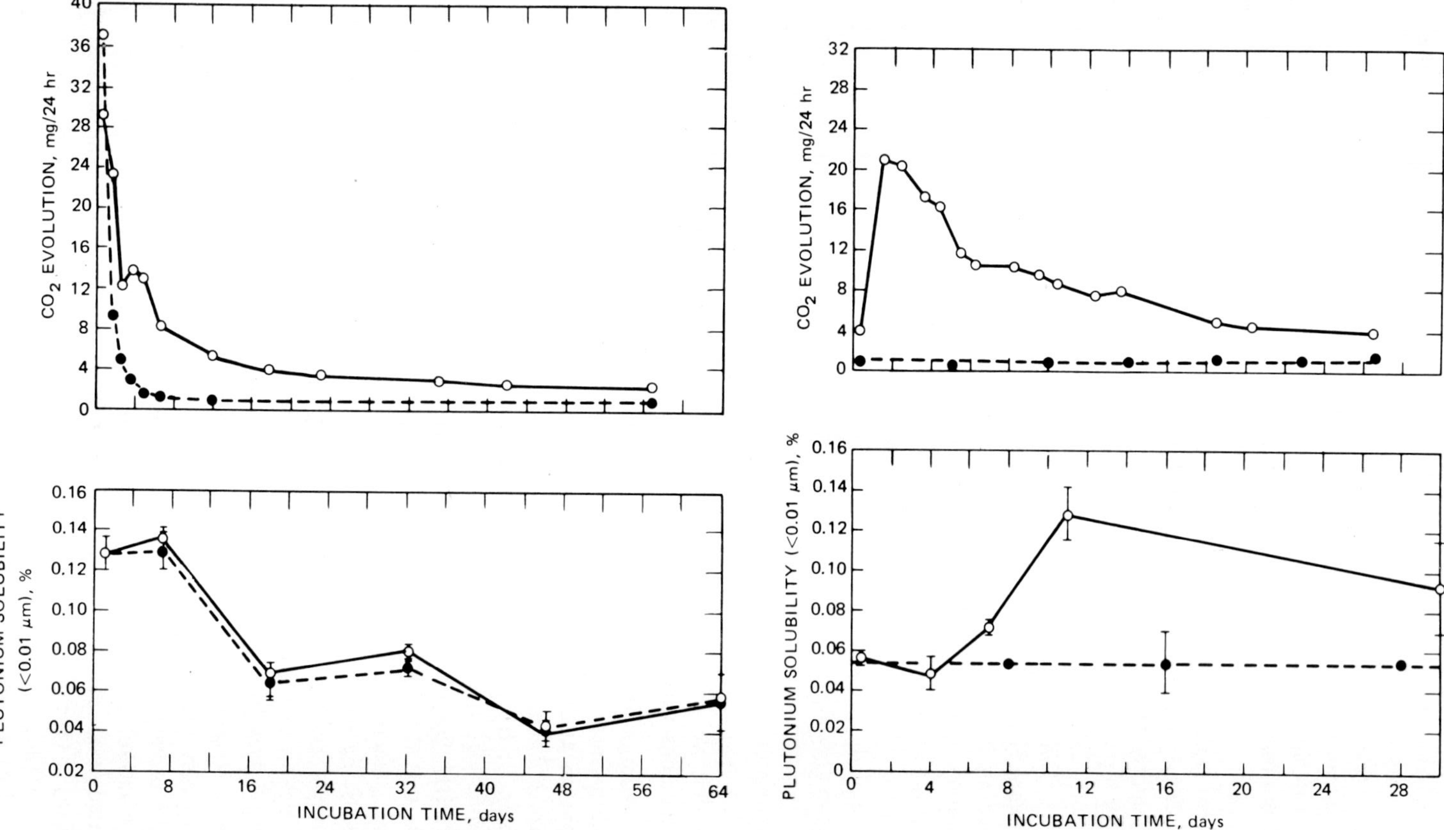

Fig. 4(a) Changes in soil respiration rate and solubility of applied plutonium with time of soil incubation in sterile (●— —●) and nonsterile (○——○) soil. [From Wildung, Garland, and Drucker (1973) and Wildung et al. (1974).] (b) Changes in soil respiration rate and solubility of plutonium in a previously sterile soil (●— —●) after inoculation with plutonium-treated nonsterile soil (○——○) (part a).

Prior incubation, which in microbial studies was shown to increase the solubility of plutonium in soil, increased plutonium and americium uptake by shoots compared with the unincubated controls (Table 3). The effect was greatly accentuated in the case of the soil-free roots, and incubation increased the soil-to-plant concentration ratios by up to 37 times relative to the unincubated control, depending on plutonium soil concentration level.

TABLE 3 Uptake of Plutonium and Americium in Barley as a Function of Prior Incubation in a Ritzville Silt Loam Soil

| | Concentration ratio* | | | |
| | Unincubated | | Incubated† | |
Plant component	Plutonium	Americium	Plutonium	Americium
Shoots	0.019	0.042	0.071	0.12
Roots (soil free)	0.060	0.13	2.2	3.6

*(Microcuries of plutonium per gram of oven-dry plant tissue per microcuries of plutonium per gram of oven-dry soil) $\times 10^{-2}$. Initial concentrations of ^{239}Pu and ^{241}Am were 0.5 μCi/g and 0.03 μCi/g, respectively, on a dry-weight basis. Mean standard errors (n = 3) were ±39 and ±10% for plutonium and americium, respectively.

†Soil was incubated 30 days after amendment with carbon and nitrogen to provide optimal microbial activity.

Effect on the Soil Microflora

Soil microorganisms may be exposed to relatively high transuranic-element concentrations even when total transuranic-element soil concentration is low. Soil organisms may be expected to be present at highest levels in the immediate vicinity of soil colloids (Alexander, 1961) where, from the aqueous chemistry of the transuranic elements and on the basis of recent information on transuranic-element chemistry in soil (previous section), the transuranic elements are likely to be concentrated. It is therefore necessary to determine the toxicity of the transuranic elements to soil microorganisms since microorganisms exhibiting resistance to the chemical effects of the transuranic elements may have the highest potential for participating in alteration of transuranic-element form. However, the transuranic-element series does not contain stable isotopes, and organisms chemically resistant to these elements must exhibit a degree of radiation resistance, which is dependent, in large part, on the radiochemistry of the isotope. Resistance to the chemical effects of transuranic elements can occur by three general mechanisms, including (1) inability of the transuranic elements to produce a toxic effect on cell metabolism at the cytoplasmic or exocytoplasmic levels; (2) inability of organisms to transport the transuranic elements; or (3) ability of the organisms to convert transuranic elements, by the direct and indirect mechanisms discussed in a previous section, to a form that is either less capable of entering the cell or is not toxic to the cell. The last mechanism is likely the most important in the alteration of transuranic-element form in soil.

Effect on Microbial Types, Numbers, and Activity. The effect of soil plutonium concentration on the soil microflora has been measured as a function of changes in microbial types and numbers and soil respiration rate (Wildung, Garland, and Drucker, 1973; Wildung et al., 1974). A noncalcareous Ritzville silt loam (pH 6.7) was amended with ^{239}Pu $(NO_3)_4$ at levels of 0.05, 0.5, and 10 μCi/g and with starch, nitrogen, and water to provide optimal microbial activity. Subsamples of soil were periodically removed to determine the changes in types and numbers of soil microflora with time. During this period soil respiration rate was monitored by continuous collection of soil-evolved CO_2.

The growth curve of fungi (Fig. 5) was generally typical of the growth response for other classes of microorganisms. Total microbial numbers were compared at the end of

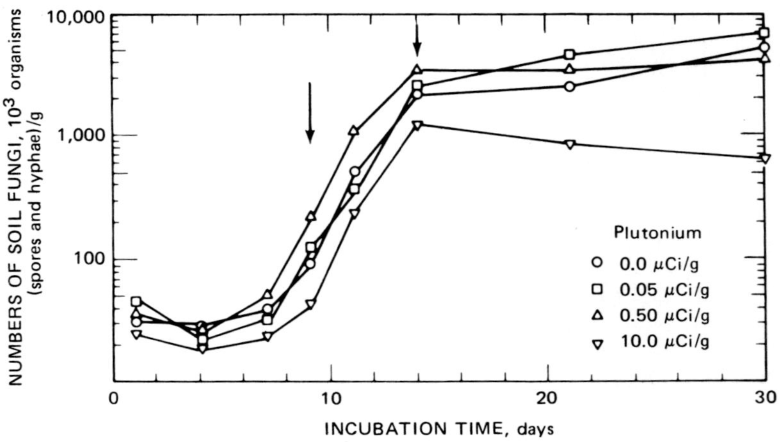

Fig. 5 Influence of plutonium concentration on the growth of fungi in soil (Wildung et al., 1974). Arrows denote time intervals at which growth rates and total numbers were compared with other microbial types (Table 4).

logarithmic growth. The organisms generally reached this stage after 8 to 14 days of incubation. Growth rates were compared over the intervals of maximum microbial growth for each organism at each plutonium concentration. The results are summarized in Table 4.

The plutonium did not generally affect the rate of growth, but it decreased the total numbers of all classes of microorganisms at levels as low as 0.05 μCi/g or 7 μg/g. The fungi were the exception, differing from the controls only at a plutonium concentration of 10 μCi/g or 144 μg/g. Thus the plutonium did not affect the maximum generation rate but rather it affected the lag period or onset of the stationary phase, which limited microbial numbers.

The accumulative CO_2 curve generally corresponded to the growth curve of the fungi. For the other classes of organisms, maximum logarithmic growth occurred before the rate of CO_2 evolution reached minimum levels. The rate of CO_2 evolution and cumulative CO_2 over the incubation period were significantly reduced only at the 144-μg/g level of plutonium amendment, although numbers of all classes of organisms except the fungi were depressed below this level (Table 4). This is in marked contrast to the results of studies with a number of other heavy metals (Drucker et al., 1973), such as silver and mercury, in which respiration rate was a sensitive measure of metal effect at levels as low as 1 μg/g in soil. Differences in the effects of the metals may be related to differences in

**TABLE 4   Effects of Plutonium at Several Soil Concentration Levels
on the Distribution of Microorganisms in Soil Relative to Controls***

| | Effect ($p < 0.05$)† of plutonium on | | | | | |
| | Growth rate at plutonium concentrations (μCi/g) of | | | Total numbers at plutonium concentrations (μCi/g) of | | |
Microbial type	0.05	0.5	10.0	0.05	0.5	10.0
Bacteria						
Aerobic and microaerophillic						
Nonspore formers	0	0	0	+	+	+
Spore formers	0	0	0	+	+	+
Anaerobic and facultative						
anaerobic						
Nonspore formers	0	+	0	+	+.	+
Spore formers	0	0	0	+	+	+
Fungi	0	0	0	0	0	+
Actinomycetes	0	0	+	+	+	+

*From Wildung, Garland, and Drucker (1973; 1974).
†Positive sign denotes significant effect. Zero indicates that there was no significant effect.

soil solubility as well as to toxicity. It should also be noted, however, that the effect on respiration rate was dependent on the magnitude of the soil respiration rate in plutonium-treated soil relative to untreated controls, which, in turn, was dependent on the initial level of microorganisms in soil. In soils exhibiting a higher CO_2 evolution rate, the reduction of respiration rate due to plutonium amendment was more pronounced. Studies of the toxicity of other transuranic elements to soil microflora have not been conducted.

Mechanism of Effect. It is important to distinguish, where possible, chemical and radiation effects of the transuranic elements on soil microorganisms to understand the long-term effects of microorganisms on transuranic-element form. Pronounced initial chemical toxicity can result in the development of special pathways of detoxification leading to alteration of transuranic-element form. The lack of chemical toxicity may imply chemical modifications of the transuranic elements through interaction with cell metabolites. In contrast, radiation resistance is associated with an enhanced ability to repair radiation damage to key macromolecules without development of new biochemical pathways leading to alteration of transuranic-element form. However, the possibilities for indirect alteration of transuranic-element form would be higher for a radiation-resistant organism than for an organism that did not exhibit either radiation or chemical resistance since, due to competitive advantage, these resistant organisms may be expected to be present in larger numbers than less-resistant organisms where transuranic elements are concentrated, such as in the vicinity of colloids.

The effects of plutonium on soil microorganisms may be due largely to radiation damage. Schneiderman et al. (1974) measured the effects of plutonium form and solubility on soil metabolic activity and on the types, numbers, and resistance of soil fungi and actinomycetes in soil separately amended with ^{239}Pu (1 to 144 μg/g) and

^{238}Pu (0.6 μg/g) in soluble nitrate and DTPA complex forms and with carbon, nitrogen, and water to provide optimal microbial activity. Subsamples of soil were removed over a 95-day aerobic incubation period to determine changes in the numbers of fungi and actinomycetes and relative water solubilities (<0.01 μm) of the plutonium forms. Comparisons of soil fungal numbers in the presence of ^{238}Pu and ^{239}Pu at common radioactivity levels, but at different mass concentrations, indicated that plutonium toxicity was due to radiation rather than to chemical effects (Fig. 6). Solubility of

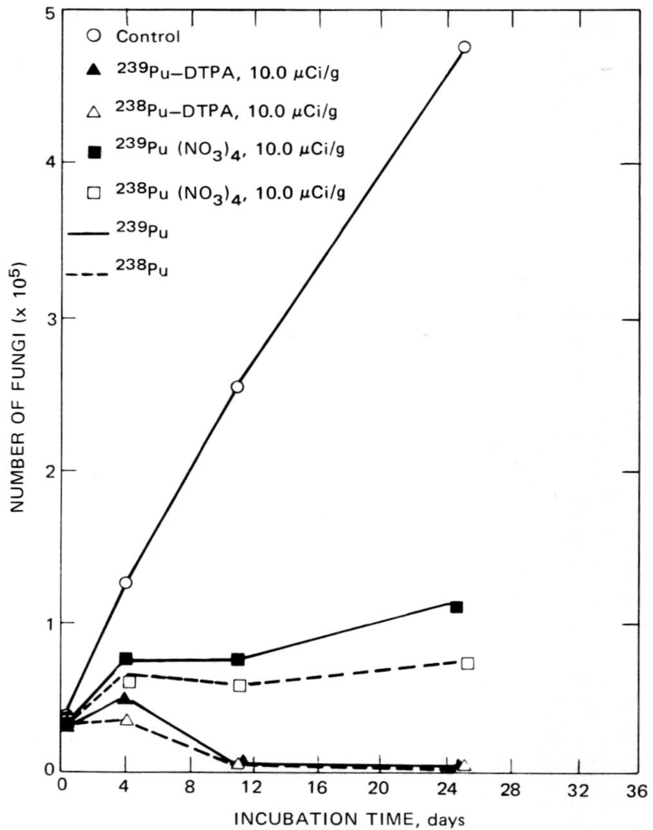

Fig. 6 Effect of different isotopes of plutonium on survival of soil fungi. [From Schneiderman et al. (1975).]

plutonium in soil influenced plutonium toxicity to microorganisms with the more-soluble Pu–DTPA forms resulting in the greatest reductions in numbers. Similar studies have not been conducted with other transuranic elements.

Isolation of Resistant Organisms. Although much information is available regarding organic ligands in soil (previous section), an organometal complex has never been isolated intact from soils. A logical approach to the study of microbial transformations of the transuranic elements is to isolate, from soil, resistant organisms most likely to alter transuranic-element form, study the transformation in vitro, and validate the results in the soil system by using techniques specifically tailored to metabolites identified from the simpler in vitro systems.

Application of enrichment techniques to the isolation of plutonium-resistant fungi, which have been demonstrated (previous section) to be the most resistant class of microorganisms, and actinomycetes from soil with the use of starch as a carbon source (Schneiderman et al., 1974) resulted in the isolation of 14 fungal cultures and 13 cultures of actinomycetes distinct in colonial morphology. Of these, 7 of the actinomycetes and 5 of the fungal isolates were capable of growth at 100 μg/ml plutonium as the soluble DTPA complex. There appeared to be a succession of actinomycete types in the soil during incubation, as indicated by the different colony morphologies obtained from enrichments after 4 and 25 days of incubation. Although this phenomenon may have resulted from changes in the soil arising from the production of metabolites or chemical degradation products, it may also have resulted from a response to the presence of plutonium. Only one actinomycete isolate was found which was common to enrichments from both incubation periods, and this organism was present at all plutonium concentrations in the media. In contrast, the fungal isolates exhibited six common morphological types regardless of incubation period.

Subsequent enrichment studies by R. A. Pelroy, Battelle-Northwest (unpublished data, 1976), have resulted in the isolation of 30 distinct cultures of bacteria from soil. Of these, 11 were resistant to plutonium at concentrations as high as 100 μg/ml. These studies also indicate that carbon source as well as soil plutonium concentration will play a role in determining the types and numbers of plutonium-resistant microorganisms present in soil, which provides presumptive evidence that microbial metabolites, which will differ with carbon source, may play a role in plutonium resistance. This subject is discussed in the next section. The presence of plutonium-resistant organisms is apparently related to factors that may be expected to vary with soil type and environmental conditions. Again similar studies have not been conducted with other transuranic elements.

Microbial Transformations

Several means exist whereby microorganisms can transform trace metals in soil. These may be generalized to (1) direct mechanisms, such as alteration in valence state or alkylation; (2) indirect mechanisms, such as interactions with normal metabolites or microbial alterations of the physicochemical environment; and (3) cycling mechanisms, such as uptake during cell growth and release on cell decomposition. In the last case, any combination of indirect and direct methods of alteration may be operational. Although there have been no studies conducted to date that would allow the unequivocal separation of these mechanisms, studies have been conducted that demonstrate the alteration of plutonium form in vitro by soil microorganisms and provide evidence for transformation of plutonium.

Direct Transformations. The potential for direct transformation of the transuranic elements through alteration of valence state or alkylation is difficult to assess. Although the transuranic elements have the potential for existing in aqueous solution in several valence states, information is not available to assess the role of soil microflora in direct alteration of valence. More information is available regarding the mechanism of metal alkylation.

Alkylation of metals involving the alkyl donor methyl cobalamine and other alkyl cobalamines has been clearly demonstrated for mercury, arsenic, and platinum (Wood,

Kennedy, and Rosen, 1968; McBride and Wolfe, 1971; Taylor and Hanna, 1977). The methyl derivative of mercury may be present in significant quantities in soils (Beckert et al., 1974). Wood (1974) suggested that methylated derivates of mercury and arsenic are important in governing their behavior in the environment. McBride and Edwards (1977) also suggested that these reactions occur abiotically. The process of biochemical methylation of metals can be described as an overlap between the chemistry of methyl cobalamine (an intermediate in methane synthesis by anaerobic bacteria and methionine synthesis in aerobic bacteria) and the chemistry of the metals. In the case of the transuranic elements, particularly plutonium, it is the complexity of the aqueous chemistry that has limited research into alkylation phenomena.

It is unknown whether an ionic species of plutonium is capable of reacting in vitro with an alkyl cobalamine. Further, if a mechanism for biological alkylation of plutonium, similar to the mercury—arsenic—platinum alkylation reaction, did exist, it would be of importance in influencing environmental behavior only if the alkylated molecule exhibited stability (Wood, 1974), i.e., a half-life in soils and sediments of hours rather than seconds. Considering the coordination chemistry of the actinides, Marks (1976) noted that U—C and Th—C linkages are formed in organic solvents and that the complexes are relatively stable thermally, although they are highly sensitive to oxygen. Meaningful microbial studies await the development of an understanding of the chemical speciation of transuranic elements in aqueous solutions at environmental concentration levels.

Indirect Transformations.   The potential for indirect transformation of the transuranic elements may be greater than that for direct transformation. The potential for plutonium interaction with microbial cells and metabolites has been demonstrated, and many of the other transuranic elements form stable complexes with known microbial metabolites.

Plutonium is taken up directly by microorganisms. Beckert and Au (1976) demonstrated the uptake of ^{238}Pu, applied initially to malt agar in nitrate, citrate, and dioxide forms, by a common soil fungus, *Aspergillus niger.* By a specialized spore collection method, the plutonium was shown to be present in the fruiting bodies. Subsequent washing to remove external contamination indicated that the major portion of the ^{238}Pu was incorporated into the spores. The order of uptake (10^{-3}) was related to pH and expected solubility of the plutonium added; plutonium in the initially soluble nitrate and citrate forms exhibited a factor of 2 to 3 greater uptake than the dioxide. The availability to microorganisms of the plutonium in citrate and nitrate might be expected to be considerably higher than that of the oxide from solubility considerations at the picocurie per milliliter level. The relatively high microbial availability of plutonium as the oxide is highly significant, and further studies are warranted to determine the mechanisms of solubilization and uptake and the significance of microorganisms in recycling processes.

The amount of literature on organic acids and bases, capable of complexing heavy elements, which are produced directly or by secondary syntheses by a variety of microorganisms, is increasing. Their concentration and form in soils will be dependent on the environmental factors influencing microbial metabolism, such as carbon source, and their residence time will be dependent on subsequent chemical and microbiological stability.

In preliminary (unpublished) studies by ourselves and others, mixed cultures of soil organisms, isolated from soil on the basis of carbon requirements and plutonium resistance, were analyzed as to their ability to transport plutonium into cells and to alter plutonium form in the cellular and exocellular media. In addition, an experiment was conducted to distinguish complexation reactions resulting from plutonium interactions with metabolites arising from normal metabolic processes and plutonium interactions with metabolites arising from plutonium resistance. For this distinction plutonium was added at the stationary growth phase of soil microorganisms isolated from soil in the absence of plutonium, and the transport and complexation were compared with microbial cultures isolated from plutonium-containing soil and grown in the presence of plutonium.

After growth for 96 hr, the cultures were separated into cellular and exocellular fractions. The cell fraction was, in turn, homogenized into intracellular soluble and cell-debris fractions. The results of studies in which plutonium was added at the stationary growth phase of cultures of fungi or bacteria grown on mixed organic acids or sugars are summarized in Table 5. These cultures, selected only on the basis of their ability to grow on either of two carbon sources, differed to a first approximation in their

TABLE 5 Distribution of Plutonium in Mixed Microbial Cultures* Exposed to Plutonium at Stationary Growth Phase and Grown on Different Carbon Sources

| | Plutonium in cultures, % | | | |
| | Fungi | | Bacteria | |
Fraction	Mixed sugars	Organic acids	Mixed sugars	Organic acids
Exocellular medium	75	42	39	89
Intracellular soluble	0.49	0.068	8.3	2
Cell debris	10	42	28	8.7

*Cultures were not replicated. Analytical precision was $< \pm 10\%$ (1 σ). Plutonium present in cell washes before homogenization is not included.

interactions with plutonium. In general, the majority of the plutonium was associated with the exocellular fraction, but significant quantities were insoluble and associated with the cell wall and membrane fractions. However, the distribution of plutonium between fractions was dependent on microorganism type and carbon source. In the case of fungi, the exocellular fraction of organisms grown on the organic acid carbon source contained less plutonium than when mixed sugars were used as a carbon source. The reverse of this relationship occurred with the bacteria.

Differences in plutonium distribution as a function of carbon source used in enrichment were also found in cultures grown in the presence of plutonium throughout incubation (Table 6). The fungal cultures grown on mixed organic acids exhibited larger concentrations of plutonium both in the exocellular fraction and bound to the cell-debris fraction; the cultures grown on mixed sugars contained a higher fraction of added plutonium in the intracellular soluble fraction. In the bacterial cultures the situation was somewhat different in that higher concentrations of plutonium occurred in the exocellular fraction of the culture grown in organic acids; less plutonium was associated with the cell-debris fraction as compared with cells grown on sugars.

In general, the continuous presence of plutonium during growth did not have pronounced effects on the distribution of plutonium in the cultures (compare Tables 5 and 6). Rather, the metabolic properties of the mixed cultures as determined by carbon source appeared to be the major factor resulting in the observed differences. Under both

TABLE 6   **Distribution of Plutonium in Mixed Microbial Cultures* Continuously Exposed to Plutonium and Grown on Different Carbon Sources**

| | Plutonium in cultures, % | | | |
| | Fungi | | Bacteria | |
Fraction	Mixed sugars	Organic acids	Mixed sugars	Organic acids
Exocellular medium	29	54	46	88
Intracellular soluble	4.2	0.24	2.7	4
Cell debris	29	39	31	3.5

*Cultures were not replicated. Analytical precision was $< \pm 10\%$ (1 σ). Plutonium present in cell washes before homogenization is not included.

sets of culture conditions, there was a high concentration of plutonium bound to the cell wall and membrane fractions and thus was insoluble. As these materials are degraded by lytic enzymes, e.g., proteases and chitinases, soluble plutonium compounds may be formed.

Preliminary characterization, using gel permeation chromatography, of the mixed culture of fungi isolated from soil and grown in sugars indicated that the plutonium form was altered during fungal growth (Fig. 7). The exocellular and intracellular soluble fractions obtained from organisms exposed to plutonium in a single exposure and in continuous exposure contained a majority of plutonium in compounds of molecular size greater than Pu–DTPA, which was used as the source of soluble plutonium. Furthermore, there appeared to be a difference in plutonium chemical form when plutonium complexes formed on simple interaction of plutonium with metabolites (single exposure) and plutonium complexes formed on interaction after continuous plutonium exposure of the culture were compared. This suggests either that the culture grown in the continuous presence of plutonium contained metabolites capable of interacting with plutonium which were different chemically from those produced by the culture grown in the absence of plutonium or that the culture grown in the presence of plutonium contained different organisms that were capable of adaptive response to the element leading to the synthesis of compounds relatively specific to detoxification of plutonium.

Further chemical characterization with the use of thin-layer chromatography and electrophoresis verified differences in plutonium form. Several solvents of different polarities and pH values were used to provide a range of chemical conditions for separation. Solvent systems included: A, butanol–pyridine, a system used in the resolution of amino acids; D, pentanol–formic acid, a system used in the separation of sugars and sugar acids; and G, water–acetic acid, a solvent used in the resolution of keto-acids and sugars. These systems were used to resolve plutonium as Pu–DTPA and plutonium in the soluble exocellular and soluble intracellular fractions of the above cultures (Fig. 8). Thin-layer chromatography with the use of solvent A indicated that the exocellular fraction contained one component of chromatographic mobility different from the added Pu–DTPA, but the complex remained present in detectable quantities. The intracellular soluble fraction contained a component of lesser chromatographic mobility, but there was no evidence of Pu–DTPA. Solvents D and G did not provide good resolution. Solvent D did not mobilize Pu–DTPA or other possible complexes; solvent G mobilized Pu–DTPA and indicated the presence of immobile plutonium components in the exocellular and intracellular fractions, but these were not resolved.

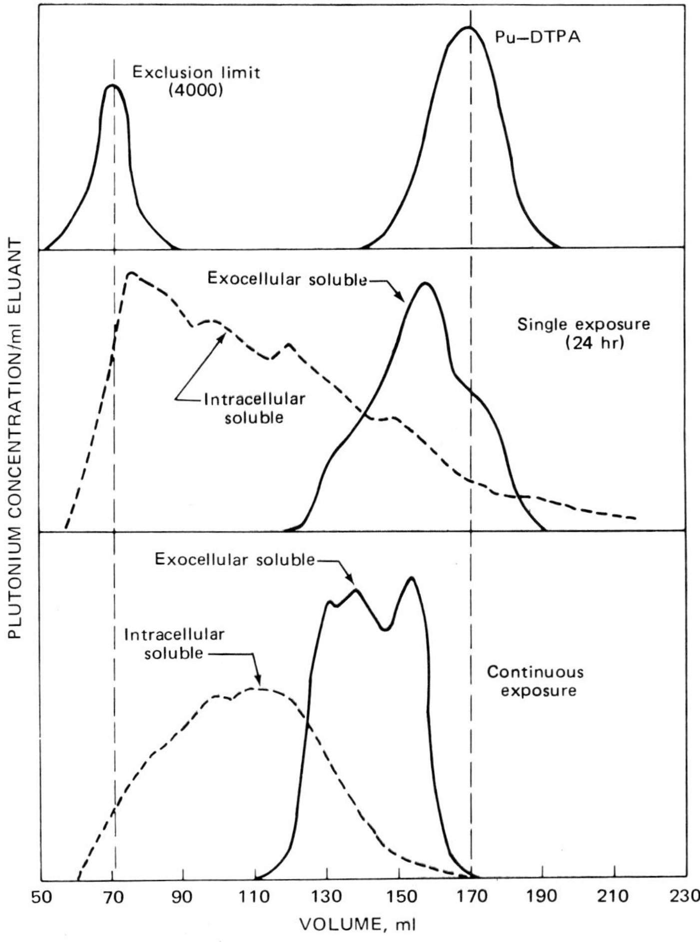

Fig. 7 Separation of soluble plutonium complexes in microbial cultures by gel permeation chromatography.

Application of thin-layer electrophoresis (pH 6.6, pyridine–acetate buffer system; cellulose support) indicated the presence (Fig. 9) of a relatively large amount of material of greater negative charge than Pu–DTPA in the exocellular fraction along with Pu–DTPA. The Pu–DTPA control contained a small quantity of plutonium, likely hydrolysis products, that did not migrate from the origin. The plutonium ligands in the intracellular fraction were either neutral in charge in this buffer system or were of a molecular size too large to migrate under the conditions of electrophoresis. Similar alterations of plutonium form by a single plutonium-resistant fungus exposed continuously to plutonium during growth have also been reported (Robinson et al., 1977).

Several phenomena may have been responsible for the observed changes in the chemical form of plutonium. The organism may have synthesized compounds that either bind Pu–DTPA or bind plutonium more tightly than DTPA, thereby successfully competing for plutonium in the presence of DTPA. Alternatively, the organism may

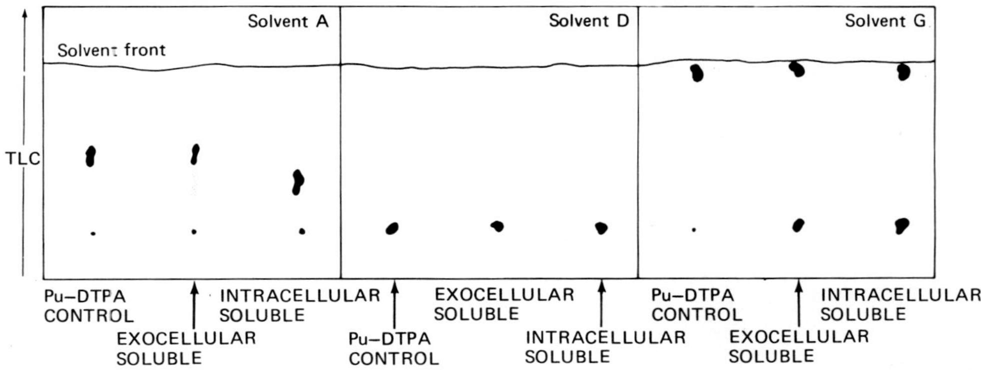

Fig. 8   Thin-layer chromatographic (TLC) behavior in three solvent systems of plutonium complexes separated by gel permeation chromatography.

degrade or modify the DTPA moiety, allowing plutonium transfer to ligands arising from microbial synthesis and degradation.

The number of known compounds with the potential to bind plutonium more strongly than DTPA appears to be quite limited, although hydroxamate derivatives (Emergy, 1974), catechol derivatives (Tait, 1975), and tetrapyrrole ring systems (Balker, 1969) may exhibit this property. If modification of the Pu—DTPA occurred prior to ligand transfer, then a myriad of microbially produced compounds, e.g., phenolic acids, peptides, and carboxylic acids, has potential for binding plutonium (see previous

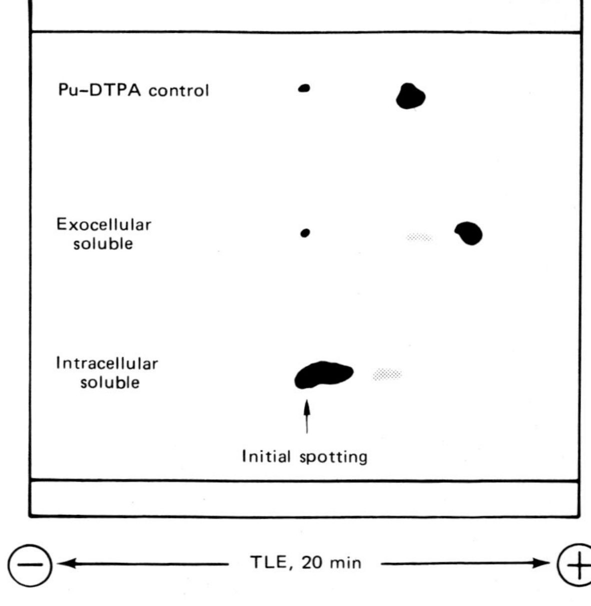

Fig. 9   Thin-layer electrophoretic (TLE) behavior of plutonium separated by gel permeation chromatography.

section; also Alexander, 1971). In either case the plutonium was not in the form initially added. Thus applications of gel-permeation chromatography, thin-layer chromatography, and thin-layer electrophoresis indicate that soil microorganisms are capable, through simple expressions of metabolic potential, of changing the chemical form of Pu—DTPA with the resulting formation of plutonium complexes exhibiting a range in chemical properties. Differences in plutonium distribution in microbial systems and in plutonium form resulted from both simple interaction with metabolites and perhaps more specific processes. These differences were dependent on organism type, metabolism, and plutonium resistance. Investigations are presently under way with pure cultures of these soil microorganisms to define complexation mechanisms. Detailed study is being directed toward those organisms ·producing exocellular metabolites which form plutonium complexes that are soluble on elution through soil.

Although published information on the transformation of transuranic elements other than plutonium is limited, it is likely that transformations similar to that of plutonium will occur. The extent of these transformations will be dependent on the solubility of the element, its availability to microorganisms, its toxicity to microorganisms, and its potential for complexation. Investigations are currently under way with pure cultures of soil microorganisms isolated in the same manner as the mixed cultures described above. Exocellular complexes mobile in soil columns are being chemically characterized for detailed study. Although microbial interactions remain to be elucidated, the solubility and potential for complexation may be preliminarily assessed from known chemistry (Table 7). It is evident that the transuranic elements form DTPA complexes with stabilities similar in magnitude to Pu—DTPA over environmental pH ranges. It can be concluded that complexation with organic ligands produced by soil microflora is highly probable, and investigations to identify and characterize the indirect processes and the ligands responsible for complexation of plutonium in soil are equally applicable to those of other transuranic elements.

Cycling During Decomposition.   A final process whereby the soil microflora may play a role in transformation of the transuranic elements involves the biological uptake (plants and microorganisms) of the elements and subsequent release on decomposition. Several studies have demonstrated plant uptake of plutonium and americium and incorporation into aboveground tissue. These tissues, deposited on soil either through litter fall or agricultural incorporation of crop residues, will be subject to microbial decomposition. Furthermore, recent studies (Wildung and Garland, 1974) have indicated that barley roots (uncontaminated with soil particles) contained three to eight times as much plutonium as the shoots. The roots of plants are in intimate contact with the soil and can be expected to decompose rapidly (weeks) under appropriate conditions of temperature and moisture even in arid regions (Wildung, Garland, and Buschbom, 1975). Relatedly, microorganisms, owing to their distribution in soil and large absorptive surface, compete efficiently with plants for ions in soil (Alexander, 1961). Studies described in a previous section demonstrated the association of plutonium with microbial cells. Growth of microbial cells, a significant portion of the soil biomass, may therefore represent an important mechanism for biological incorporation of the transuranic elements. Decomposition of microbial cells generally proceeds at a more rapid rate than that of plant tissues.

Little is known of the form of the transuranic elements in plant or microbial tissues, of the form, rate, and extent of the transuranics released on decomposition of these tissues, or of the chemical reactions governing transuranic solubility after decomposition.

TABLE 7   **Stability of DTPA Complexes with the Transuranic Elements***

Complex†	Stability constant	Stable pH range
Neptunium		
Np(III)	‡	‡
Np(IV)–DTPA	10^{24}	0.5 to 5.8
[Np(IV)]$_2$–DTPA$_3$	10^{20}	>5.8
Plutonium		
Pu(IV)–DTPA	10^{24}	1.0 to 5.8
[Pu(IV)]$_2$–DTPA$_3$	10^{18}	5.8 to 8.5
Pu(IV)–DTPA$_2$	10^{14}	>8.5
Americium		
[Am(III)]$_2$–DTPA	10^{20}	1.8 to 6
Am(III)–DTPA	10^{23}	>6

*From Hafez (1969).

†Curium can be expected to form complexes of stabilities similar to americium.

‡Unstable in oxygenated solutions.

However, considering the known products of microbial metabolism of organic substances, including a number of strong complexing agents and the susceptibility of a number of the transuranic elements to complexation, it can be concluded that the transuranic elements, initially immobilized through biological uptake, may be at least as soluble and perhaps more soluble on decomposition.

In preliminary studies (R. E. Wildung and T. R. Garland, unpublished) plutonium-amended soil containing largely undecomposed roots from a previous barley crop was leached with water, and plutonium solubility was compared with a fallow soil containing plutonium at similar levels. The results indicated that soluble plutonium was initially immobilized by incorporation into roots, decreasing by a factor of 10 after root growth. Root decomposition studies are in progress. Previously observed (Romney, Mock, and Larson, 1970) increases in plutonium uptake from soils by plants with increased time, generally attributed to increased root development, may have been due to increased availability through a recycling process on the decomposition of plant roots. The importance of the process will be dependent on transuranic-element availability to different plants and microorganisms, the turnover rate of this tissue in soils under different conditions, and the stability, chemistry, and biological availability of trans-uranic-element metabolites. Studies are presently under way to provide this information as a basis for establishing the long-term effects of recycling processes.

Microbial Influence on the Availability and Form of Transuranic Elements in Plants and Animals

Plants

The results of investigations to physicochemically characterize the mobile forms of plutonium in soils (<0.1% of total plutonium) suggested that mobile plutonium was largely particulate (Garland and Wildung, 1977; previous section). The nonparticulate, or soluble, fraction was present in insufficient quantities, at any single point in time, to separate from soil and chemically characterize with the present methods. However, for plutonium to be available to plants, it must pass through the solution phase. Furthermore, in studies where appropriate measurements have been made, the quantities taken up by plants exceeded the quantity present in the soil solution. Thus plutonium was being resupplied to solution and plants from the solid phase. Since Pu(IV) ions would hydrolyze in solution, precipitating on soil surfaces, it is likely that the plant available fraction was stabilized in soil solution by complexation with inorganic or organic ligands and/or was present in a different, more soluble, valence state.

Circumstantial evidence suggested (Wildung and Garland, 1975; previous section) that inorganic or organic ligands present in limited concentrations in soil stabilized plutonium in soil solution. Furthermore, dissolution of plutonium from the solid phase has been shown to be accelerated by complexing agents (Bondietti, Reynolds, and Shanks, 1976; previous section). Organic ligands, which form the most stable plutonium complexes, are generally derived from microbial processes in soil, and previous studies (Wildung, Garland, and Drucker, 1973; previous section) have shown that plutonium solubility in soil and availability and distribution in plants are influenced by microbial activity, although mechanisms other than complexation may have been responsible. However, a synthetic ligand (DTPA) was shown to maintain plutonium essentially soluble in soil for extended periods. Thus it is likely that organic ligands of microbial origin, which differ markedly in their form, concentration, and stability in soil (Keeney and Wildung, 1977), may play an important role in stabilizing plutonium in solution for subsequent uptake by plants. A key question is whether the relatively low uptake exhibited by plants from plutonium-amended soils (reported concentration ratios of 10^{22} to 10^{-7}, Energy Research and Development Administration, 1976) is due to limited solubility in soil as a result of sorption on particulate surfaces or to discrimination by the plant. If discrimination is not at the plant level, then the potential role of the soil microbiota in increasing plutonium availability from the solid phase (oxide particles from the nuclear fuel cycle or soil particles) becomes very important in influencing the long-term availability of plutonium to plants.

The role played by organic ligands in facilitating plant uptake of ions, particularly hydrolyzable ions, has long been a subject of controversy. The question has been whether the complex simply serves as a means of delivering the ion to the root membrane, supplying the ion to the root by dissociation, or is taken up intact by the plant. Perhaps the most meticulous investigation of this phenomena has been the work of Tiffin and co-workers (summarized by Tiffin, 1972; 1977). At least in the case of iron, the evidence, derived in part from direct analyses of xylem exudates, strongly supports the role of complexors in increasing uptake by plants but indicates that the complex serves mainly to deliver the metal to the root membrane and that the ligand is not taken up by the plant stoichiometrically with the metal. Recently, Malzer and Barber (1976) concluded that less than 16% of several calcium and strontium chelating ligands was removed from nutrient solution by corn *(Zea mays)*, whereas over 90% of the calcium and strontium was taken up by the plant. Both studies used several physicochemical as well as radiochemical measures of chelate concentration in aqueous solution. It should be noted, however, that detailed, exhaustive procedures are required to purify metal chelates, particularly in the case of the transuranic elements (Swanson, Garland, and Wildung, 1975). Without this effort it is possible that studies using only ^{14}C analysis as a measure of chelate concentrations in plants would overestimate uptake since low-molecular impurities containing ^{14}C might account for the ^{14}C present in the plant, particularly where chelate uptake rates are low.

For an aid in distinguishing soil sorption and plant root discrimination and for the evaluation of the role of complexes in the plant uptake of plutonium, hydroponically grown soybeans *(Glycine max)* were placed in micromolar level [238]Pu–DTPA solutions and permitted to accumulate plutonium for up to 48 hr (T. R. Garland, D. A. Cataldo, and R. E. Wildung, unpublished). Concentration ratios (microcuries per gram of plant per microcuries per milliliter of nutrient solution) for shoot tissues were found to be 6×10^{-3} and 0.3 after 1 and 24 hr, respectively. This would suggest that plants do possess the potential to effectively accumulate plutonium, with much of the apparent discrimination in soil–plant studies resulting from the effect of soil sorption in reducing the quantity of plutonium available to the plant. In a preliminary effort to determine the form of mobile plutonium in the plant, several plants were decapitated, xylem exudates were collected in 1-ml aliquots at intervals over a 48-hr period, and the solutions were subjected to thin-layer electrophoresis to resolve plutonium-containing components. The electrophoretic mobilities of plutonium in the medium, in the control exudate spiked with $Pu(NO_3)_4$, and in the exudate collected from decapitated plants grown in $0.1 \mu M$ Pu–DTPA solution over a 24-hr period are illustrated in Fig. 10. The plutonium-containing components in the exudate from plants grown in the absence of Pu–DTPA but spiked with Pu(IV) or Pu(VI) indicated the presence of ligands capable of binding plutonium and forming stable complexes. Similarly, several anionic and cationic plutonium complexes were evident in exudates from plants grown in the presence of Pu–DTPA. A major anionic component with an electrophoretic mobility similar to Pu–DTPA reached maximum concentration in the second aliquot after decapitation and then decreased in concentration with time. The application of several solvent systems to separation subsequently indicated that this component was not the Pu–DTPA complex supplied in the growth media. These data suggest that plants do possess the ability to effectively accumulate soluble plutonium and transport the plutonium to shoots in one or more organic complexes. Furthermore, from the high concentration ratios for plutonium supplied as Pu–DTPA and the lack of uptake of the complex, it can be concluded that much of the apparent discrimination in soil–plant studies results from the effect of soil sorption in reducing the quantity of soluble plutonium available to the plant. The form of plutonium in alfalfa fed to animals was also characterized, and this provided insight into observed differences in gastrointestinal absorption.

Animals

The accepted value for the gastrointestinal transfer ratio of plutonium from food to man is 3×10^{-5} (U. S. Atomic Energy Commission, 1974). This value is based on the gastrointestinal absorption of inorganic Pu(IV) in animals, administered by gavage (Weeks et al., 1956), and the apparent assumption that foodstuffs would contain primarily inorganic Pu(IV). However, the study demonstrated that gavaging Pu(VI) and complexed forms of Pu(IV) resulted in higher absorption rates, e.g., 500 times as great as those for Pu(VI). Until recently it was not possible to test the assumptions because plant tissues with plutonium concentrations sufficient to measure uptake were not available. However, studies with alfalfa indicated that tissues containing up to 400,000 d/min per gram could be obtained under a sequential harvesting regime (T. R. Garland, D. A. Cataldo, and R. E. Wildung, unpublished). These tissues, used in conjunction with a sensitive analytical method (Wessman et al., 1971) for the measurement of plutonium at levels of <1 d/min, allowed preliminary investigations of the availability to animals of plutonium in alfalfa

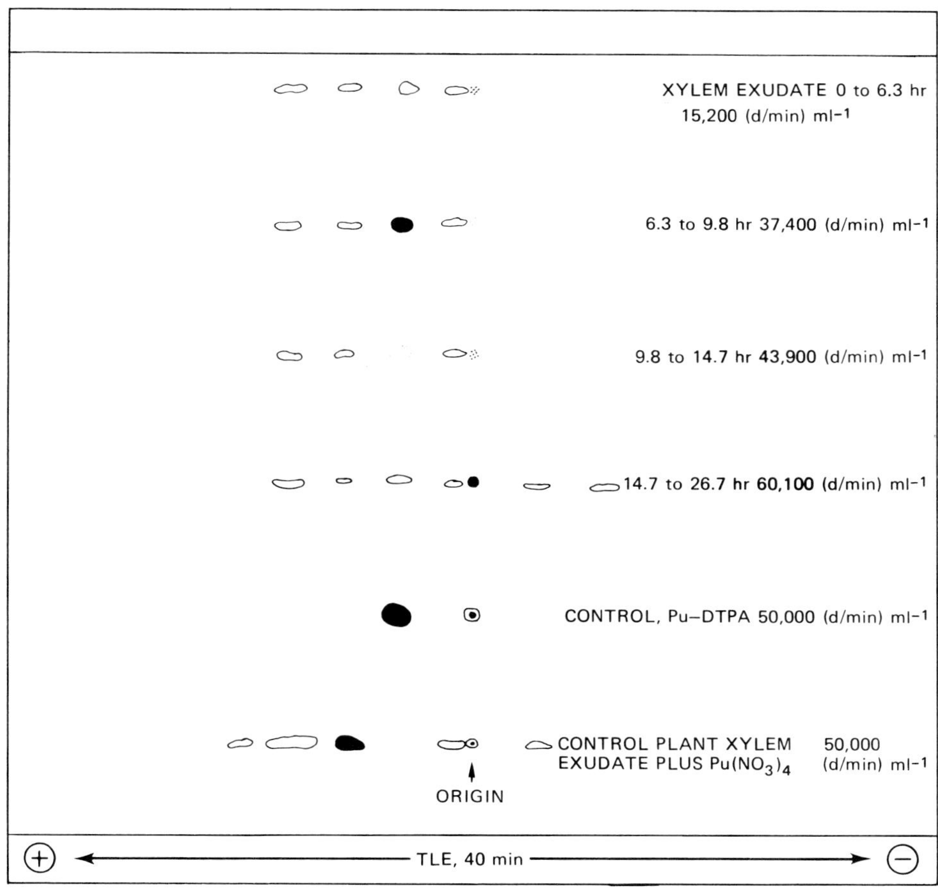

Fig. 10   Thin-layer electrophoretic behavior of plutonium in soybean xylem exudates.
[From Wildung, Drucker, and Au (1977).]

tissues grown on plutonium-containing soil. In these studies gut absorption of plutonium
gavaged in inorganic solution and plutonium fed in alfalfa tissue was compared for rats
(nonherbivorous) and guinea pigs (herbivorous). In both rats and guinea pigs, absorption
through the gastrointestinal tract of plutonium incorporated in alfalfa tissue was greater
(7.7 to 53 times) than that by gavaging inorganic Pu(IV) solutions (Sullivan and Garland,
1977). Uptake by rats fed alfalfa containing the plutonium exceeded that from gavaging
Pu(IV) nitrate or Pu(IV) citrate but was less than that resulting from gavaging Pu(VI)
nitrate (Table 8). Uptake was higher when stems and leaves were fed than when leaves
only were fed. Several variables, including animal species, duration of feeding, and types
of plant tissues fed, were evaluated, but, as a result of the small number of animals that
could be used, the limited quantity of tissues, and variability, further studies are required
to evaluate the statistical significance of individual variables.

Preliminary chemical characterization (T. R. Garland, K. M. McFadden, and R. E.
Wildung, unpublished) of the form of plutonium present in the alfalfa tissue fed to the
rodents indicated that the plutonium was more soluble in stems than in leaves, perhaps by

TABLE 8   Gastrointestinal Transfer of Plutonium in Rats as a
Function of Plutonium Source

Source	Fraction of administered dose in bone plus liver*	Literature cited
Pu(IV) nitrate	0.000012	Weeks et al., 1956
Pu(VI) nitrate	0.0175	Weeks et al., 1956
Pu(IV) citrate	0.00026	Weeks et al., 1956
Plutonium in alfalfa	0.00086	Sullivan and Garland, 1977

*Analyzed 4 days after a single administration.

a factor of 2. Thus the differences in the gastrointestinal absorption between leaves and stems appeared to reflect differences in relative solubility. The results of preliminary studies indicated that more than 90% of the soluble plutonium in either stems or leaves was complexed Pu(IV). These studies are continuing in an effort to evaluate the importance of the ingestion pathway for transuranic-element uptake by man and to determine the factors that influence the rate and extent of biological uptake.

Recommendations for Future Research

A broad range of soil—plant concentration ratios (10^{-2} to 10^{-7}) has been reported for the transuranic elements (Energy Research and Development Administration, 1976). If this variability is attributed to the usual unexplained experimental and environmental parameters, the information would be of little use in predicting the long-term behavior of these elements. However, a closer examination shows that, although the ratios ranged over many orders of magnitude, they encompassed, and were often dependent on, different source terms, soil types, plant species, plant components, climatic conditions, extent of foliar contamination, and kinetic factors. Consideration of these variables and use of rapidly accumulating information on transuranic-element behavior at the chemical, microbiological, and physiological levels should allow reduction of the level of unexplained variability by many orders of magnitude and provide a valuable basis for the prediction of transuranic-element behavior over a broad range of conditions.

There is a need to continue to develop a basic understanding of the processes influencing the fate of transuranic elements in soils and plants over a range of soil types and experimental conditions representing those likely to be encountered in the environment. Research should emphasize those elements which are (1) expected to be present in soil in the highest concentration, (2) most soluble in soil, (3) mobile in the plant, and (4) transported to edible tissues. Source terms receiving future emphasis should be those likely to result from the nuclear fuel cycle as opposed to fallout. From previous discussions of the soil chemistry of the transuranic elements, which illustrate the marked differences in soil behavior and plant availability resulting from different source terms, sources due to fallout and local nuclear testing, useful for initial approximations, cannot be taken as fully representative for validation purposes.

Research should emphasize determination of (1) source-term physicochemical characteristics (composition, mineralogy, particle size, and valence) as a function of source (fallout, reactor operation, reprocessing, and burial leachates); (2) physical redistribution processes (erosion and resuspension); (3) biological redistribution processes (litter incorporation and root decomposition); (4) processes of solubilization and/or transformation of refractory materials entering soil and the factors influencing the form and equilibria between relatively insoluble forms and soluble chemical species (soil type, soil solution composition, pH, temperature, redox conditions, diffusion, and microbial action); (5) the capacity of representative plant species to assimilate soluble chemical species, plant alteration of chemical form, and translocation to edible plant components as a function of plant growth stage, form, and concentration in soil and the presence of competitive ions in the soil solution; and (6) the nature and extent of the above processes on a regional basis, as influenced by soil, plant, and climatic factors and land-use practices. This information is currently being accumulated in laboratory studies, and initial investigations are under way in several geographic regions to selectively validate the findings in the field.

The role of the soil microflora must be viewed as only one contributory factor among a number of highly important physicochemical and biological phenomena influencing the overall behavior of the transuranic elements in terrestrial environments. However, evidence is increasing that organic ligands resulting from microbial activity will play an important part in influencing the behavior and plant availability of hydrolyzable species, such as plutonium, in soils. Future studies in this area should involve a systematic investigation of the major classes of soil organisms exhibiting highest transuranic-element resistance and representing different metabolic types, determination of their ability to alter transuranic-element form in soil, and evaluation of the soil and environmental factors influencing the rate and extent of alteration. In view of the relatively high concentrations of transuranic elements associated with roots and other organic components in soil, particular emphasis should be placed on determining the role of the microflora in recycling and redistribution processes. It is possible that microbial processes responsible for alteration of metal form may function in a like manner for metals exhibiting similar chemical properties, particularly for organisms exhibiting cross resistance to these metals. Thus model systems may be available for those organisms likely to be most responsible for alteration of metal form in soil over the long term, i.e., those organisms capable of growth and reproduction at higher metal concentrations in the immediate vicinity of refractory oxides or soil colloids with surface deposits of the transuranic elements.

A key integrating factor, in all studies of the transuranic elements and in interpretation of environmental phenomena, is a knowledge of the chemical form of the element. With this information studies conducted under broadly different conditions in various substrates and biological media can be compared, and toxicological interpretations can be made on a common basis.

Investigations that are currently under way in several institutions across the country should ultimately provide a realistic evaluation of the role of microbial processes in influencing the long-term behavior of the transuranic elements in soil. Furthermore, the studies should provide a basis for evaluation of the availability of transuranic metabolites to plants and insight into the potential for entrance of these elements into foodstuffs for a broad geographical region over the long term.

Acknowledgments

Appreciation is extended to H. Drucker for his technical contributions and advice during the conduct of the PNL microbiological studies and to F.H.F. Au for review of portions of this manuscript. This chapter is based on work performed under U. S. Department of Energy contract EY-76-C-06-1830.

References

Alexander, M., 1961, *Introduction to Microbiology,* John Wiley & Sons, Inc., New York.

——, 1971, Biochemical Ecology of Microorganisms, *Annu. Rev. Microbiol.,* 25: 361-392.

Balker, E. W., 1969, Porphyrins, in *Organic Geochemistry,* pp. 464-497, G. Eglinton and M. T. Murphy (Eds.), Springer-Verlag, New York.

Beckert, W. F., and F. H. F. Au, 1976, Plutonium Uptake by a Soil Fungus and Transport to Its Spores, in *Transuranium Nuclides in the Environment,* Symposium Proceedings, San Francisco, 1975, pp. 337-345, STI/PUB/410, International Atomic Energy Agency, Vienna.

——, A. A. Moghissi, F. H. F. Au, E. W. Bretthauer, and J. C. McFarlane, 1974, Formation of Methylmercury in a Terrestrial Environment, *Nature,* 249: 674.

Bondietti, E. A., 1974, Adsorption of Plutonium (IV) and Thorium (IV) by Soil Colloids, in *Agronomy Abstracts,* p. 23, American Society of Agronomy, Soil Science Society of America, and Crop Science Society of America Annual Meeting, Chicago, Ill., Nov. 10–15, 1974.

——, and S. A. Reynolds, 1976, Field and Laboratory Observations on Plutonium Oxidation States, in *Proceedings of an Actinide–Sediment Reactions Working Meeting at Seattle, Wash., on Feb. 10–11, 1976,* L. L. Ames (Ed.), USAEC Report BNWL-2117, pp. 505-530, Battelle, Pacific Northwest Laboratories, NTIS.

——, S. A. Reynolds, and M. H. Shanks, 1976, Interaction of Plutonium with Complexing Substances in Soils and Natural Waters, in *Transuranium Nuclides in the Environment,* Symposium Proceedings, San Francisco, 1975, pp. 273-287, STI/PUB/410, International Atomic Energy Agency, Vienna.

Bremner, J. M., 1965, Organic Nitrogen in Soils, in *Soil Nitrogen,* pp. 93-149, W. V. Bartholmew and F. E. Clark (Eds.), American Society of Agronomy, Madison, Wis.

——, 1967, Nitrogenous Compounds, in *Soil Biochemistry,* pp. 19-66, A. D. McLaren and G. H. Peterson (Eds.), Marcel Dekker, New York.

Broadbent, F. E., and G. R. Bradford, 1952, Cation-Exchange Groupings in the Soil Organic Fraction, *Soil Sci.,* 74: 47-457.

Burney, G. A., and R. M. Harbour, 1974, *Radiochemistry of Neptunium,* Report NAS-NS-3060, National Academy of Sciences–National Research Council, NTIS.

Cleveland, J. M., 1970, *The Chemistry of Plutonium,* Gordon & Breach, Science Publishers, New York.

Drucker, H., R. E. Wildung, T. R. Garland, and M. P. Fujihara, 1973, Influence of Seventeen Metals on Microbial Population and Metabolism in Soil, in *Agronomy Abstracts,* p. 90, American Society of Agronomy, Soil Science Society of America, and Crop Science Society of America, Annual Meeting, Las Vegas, Nev., Nov. 11–16, 1973.

Emergy, T., 1974, A Model for Carrier-Mediated Iron Transport, *Biochim. Biophys. Acta,* 363: 19-225.

Energy Research and Development Administration, 1976, Environmental Research for Transuranic Elements, in *Environmental Research for Transuranic Elements, Proceedings of the Workshop,* Battelle Seattle Research Center, Seattle, Wash., Nov. 12–14, 1975, ERDA Report 76/134, NTIS.

Felbeck, G. T., Jr., 1965, Structural Chemistry of Soil Humic Substances, *Adv. Agron.,* 17: 27-368.

Garland, T. R., and R. E. Wildung, 1977, Physiochemical Characterization of Mobile Plutonium Species in Soils, in *Biological Implications of Metals in the Environment,* ERDA Symposium Series, No. 42, Richland, Wash., Sept. 29–Oct. 1, 1975, H. Drucker and R. E. Wildung (Eds.), pp. 254-264, CONF-750929, NTIS.

——, R. E. Wildung, and R. C. Routson, 1976, The Chemistry of Plutonium in Soils. I. Plutonium Solubility, in *Pacific Northwest Laboratory Annual Report for 1975 to the USERDA Division of Biomedical and Environmental Research,* ERDA Report BNWL-2000(Pt. 2), pp. 23-26, Battelle, Pacific Northwest Laboratories, NTIS.

Geering, H. R., and J. F. Hodgson, 1969, Micronutrient Cation Complexes in Soil Solution. III. Characterization of Soil Solution Ligands and Their Complexes with Zn^{2+} and Cu^{2+}, *Soil Sci. Soc. Am., Proc.,* 33: 54-59.

Greenland, D. J., 1965, Interactions Between Clays and Organic Compounds in Soils. Part I. Mechanisms of Interaction Between Clays and Defined Organic Compounds, *Soils Fert.,* 28: 415-425.

Hafez, M. B., 1969, *Spectrophotometric Study of the Complexes of Cerium and Uranides with Diethylenetriaminepentaacetic Acid*, USAEC Report UCRL-Trans.-10366, Lawrence Livermore Laboratory, NTIS.

Hodgson, J. F., 1963, Chemistry of the Micronutrient Elements in Soils, *Adv. Agron.*, 15: 119-159.

Katz, J. J., and G. T. Seaborg, 1957, *The Chemistry of the Actinide Elements*, John Wiley & Sons, Inc., New York.

Keeney, D. R., and R. E. Wildung, 1977, Chemical Properties of Soils, in *Soils for Management of Organic Wastes and Waste Waters*, pp. 75-97, L. Elliott and F. J. Stevenson (Eds.), American Society of Agronomy, Crop Science Society of America, and Soil Science Society of America Monograph Series.

Kononova, M. M., 1966, *Soil Organic Matter*, Pergamon Press, New York.

Lahav, N., and M. Hochberg, 1976, A Simple Technique for Characterizing the Stability of Metal Chelates in the Soil, *Soil Sci.*, 121: 58-59.

Lindsay, W. L., 1972, Inorganic Phase Equilibria of Micronutrients in Soils, in *Micronutrients in Agriculture*, pp. 41-57, J. J. Mortvedt, P. M. Giordano, and W. L. Lindsay (Eds.), Soil Science Society of America, Madison, Wis.

Lloyd, M. H., and R. G. Haire, 1973, in *Proceedings of the XXIVth Congress of the International Union of Pure and Applied Chemistry*, Hamburg, F. R. Germany, Sept. 3–6, 1973, Gesellschaft Deutscher Chemiker, Frankfurt A.M. Main, Germany.

Malzer, G. L., and S. A. Barber, 1976, Calcium and Strontium Absorption by Corn Roots in the Presence of Chelates, *Soil Sci. Soc. Am. J.*, 40: 727-731.

Marks, T. J., 1976, Actinide Organometallic Chemistry, *Acc. Chem. Res.*, 9(6): 223-229.

McBride, B. L., and T. L. Edwards, 1977, Role of Methanogenic Bacteria in the Alkylation of Arsenic and Mercury, in *Biological Implications of Metals in the Environment*, ERDA Symposium Series, No. 42, Richland, Wash., Sept. 29–Oct. 1, 1975, H. Drucker and R. E. Wildung (Eds.), pp. 1-20, CONF-750929, NTIS.

——, and R. S. Wolfe, 1971, Biosynthesis of Dimethylarsine by Nethanobacterim, *Biochemistry*, 10: 4312-4317.

Mitchell, R. L., 1964, Trace Elements in Soils, in *Chemistry of the Soil*, pp. 320-368, F. E. Bear (Ed.), Reinhold Publishing Corp., New York.

——, 1972, Trace Elements in Soils and Factors that Affect Their Availability, *Geol. Soc. Am. Bull.*, 83: 1069-1076.

Mortensen, J. L., 1963, Complexing of Metals by Soil Organic Matter, *Soil Sci. Soc. Am., Proc.*, 27: 179-186.

Norvell, W. A., 1972, Equilibria of Metal Chelates in Soil Solutions, in *Micronutrients in Agriculture*, pp. 115-138, J. J. Mortvedt, P. M. Giordano, and W. L. Lindsay (Eds.), Soil Science Society of America, Madison, Wis.

Ockenden, D. W., and G. A. Welch, 1956, The Preparation and Properties of Some Plutonium Compounds. Part V. Colloidal Quadrivalent Plutonium, *J. Chem. Soc.*, 653: 3358.

Robinson, A. V., T. R. Garland, G. S. Schneiderman, R. E. Wildung, and H. Drucker, 1977, Microbial Transformation of a Soluble Organoplutonium Complex, in *Biological Implications of Metals in the Environment*, ERDA Symposium Series, No. 42, Richland, Wash., Sept. 29–Oct. 1, 1975, H. Drucker and R. E. Wildung (Eds.), pp. 52-63, CONF-750929, NTIS.

Romney, E. M., H. M. Mock, and K. H. Larson, 1970, Persistence of Plutonium in Soil, Plants, and Small Mammals, *Health Phys.*, 18: 487-491.

Routson, R. C., and R. E. Wildung, 1969, Ultimate Disposal of Wastes to Soil, *Chem. Eng. Prog., Symp. Ser.*, 65(97): 19-25.

——, G. Jansen, and A. V. Robinson, 1975, *Sorption of ^{99}Tc, ^{237}Np and ^{241}Am on Two Subsoils from Differing Weathering Intensity Areas*, USAEC Report BNWL-1889, Battelle, Pacific Northwest Laboratories, NTIS.

Schneiderman, G. S., T. R. Garland, H. Drucker, and R. E. Wildung, 1974, Plutonium-Resistant Fungi and Actinomycetes in Soil. I. Mechanisms of Plutonium Toxicity, in *Pacific Northwest Laboratory Annual Report for 1974 to the USAEC Division of Biomedical and Environmental Research*, USAEC Report BNWL-1950(Pt. 2), pp. 25-28, Battelle, Pacific Northwest Laboratories, NTIS.

Schnitzer, M., D. A. Shearer, and J. R. Wright, 1959, A Study in the Infrared of High-Molecular Weight Organic Matter Extracted by Various Reagents from a Podzolic B. Horizon, *Soil Sci.*, 87: 252-257.

Stevenson, F. J., 1967, Organic Acids in Soil, in *Soil Biochemistry,* pp. 119-146, A. D. McLaren and G. H. Peterson (Eds.), Marcel Dekker, New York.

——, and M. S. Ardakani, 1972, Organic Matter Reactions Involving Micronutrients in Soils, in *Micronutrients in Agriculture,* pp. 79-114, J. J. Mortvedt, P. M. Giordano, and W. L. Lindsay (Eds.), Soil Science Society of America, Madison, Wis.

——, and J. H. A. Butler, 1969, Chemistry of Humic Acids and Related Pigments, in *Organic Geochemistry,* pp. 534-557, G. Eglinton and M. T. Murphy (Eds.), Springer-Verlag, New York.

Sullivan, M. F., and T. R. Garland, 1977, Gastrointestinal Absorption of Alfalfa-Bound Plutonium-238 by Rats and Guinea Pigs, in *Pacific Northwest Laboratory Annual Report for 1976 to the ERDA Assistant Administrator for Environment and Safety,* ERDA Report BNWL-2100 (Pt. 1), pp. 137-140, Battelle, Pacific Northwest Laboratories, NTIS.

Swanson, J. L., T. R. Garland, and R. E. Wildung, 1975, Preparation and Chemical Characterization of Plutonium—DTPA Complexes, in *Pacific Northwest Laboratory Annual Report for 1974 to the ERDA Division of Biomedical and Environmental Research,* USAEC Report BNWL-1950 (Pt. 2), Battelle, Pacific Northwest Laboratories, NTIS.

Tait, G. H., 1975, Identification and Biosynthesis of Siderochromes Formed by *Micrococcus dentrificans, Biochem. J.,* 146: 191-204.

Taylor, R. T., and M. L. Hanna, 1977, Methylation of Platinum by Methylcobalamine, in *Biological Implications of Metals in the Environment,* ERDA Symposium Series, No. 42, Richland, Wash., Sept. 29–Oct. 1, 1975, H. Drucker and R. E. Wildung (Eds.), pp. 36-52, CONF-750929, NTIS.

Thompson, S. O., 1965, *Comparative Properties of Plant Lignins and Soil Humic Materials,* Ph.D. Thesis, University of Wisconsin, Madison, Wis.

Tiffin, L. O., 1972, Translocation of Micronutrients in Plants, in *Micronutrients in Agriculture,* pp. 79-114, J. J. Mortvedt, P. M. Giordano, and W. L. Lindsay (Eds.), Soil Science Society of America, Madison, Wis.

——, 1977, The Form and Distribution of Metals in Plants, An Overview, in *Biological Implications of Metals in the Environment,* ERDA Symposium Series, No. 42, Richland, Wash., Sept. 29–Oct. 1, 1975, H. Drucker and R. E. Wildung (Eds.), pp. 315-335, CONF-750929, NTIS.

Vaughan, B. E., R. E. Wildung, and J. J. Fuquay, 1976, Transport of Airborne Effluents to Man Via the Food Chain, in *Controlling Airborne Effluents from Fuel Cycle Plants,* Symposium Proceedings, Sun Valley, Idaho, Aug. 5–6, 1976, American Nuclear Society, Hinsdale, Ill.

U. S. Atomic Energy Commission, 1974, *Proposed Final Environmental Statement for the Liquid Metal Fast Breeder Reactor Program,* Report WASH-1535, Vol. 2, pp. II.G-31, available free from W. H. Pennington, USERDA, Office of Assistant Administrator for Environment and Safety, Washington, D. C.

Weeks, M. H., et al., 1956, Further Studies on the Gastrointestinal Absorption of Plutonium, *Radiat. Res.,* 4: 339.

Wessman, R. A., W. J. Major, K. D. Lee, and L. Leventhal, 1971, Commonality in Water, Soil, Air, Vegetation, and Biological Sample Analysis for Plutonium, in *Proceedings of Environmental Plutonium Symposium,* Los Alamos, N. Mex., Aug. 4–6, 1971, E. B. Fowler, R. W. Henderson, and M. F. Milligan (Eds.), USAEC Report LA-4756, pp. 73-79, Los Alamos Scientific Laboratory, NTIS.

White, M. G., and P. B. Dunaway (Eds.), 1977, *Transuranics in Natural Environments,* Symposium Proceedings, Gatlinburg, Tenn., October 1976, ERDA Report NVO-178, Nevada Operations Office, NTIS.

Wildung, R. E., and T. R. Garland, 1974, Influence of Soil Plutonium Uptake and Distribution in Shoots and Roots of Barley, *J. Agric. Food Chem.,* 22: 836-838.

——, H. Drucker, and F. H. F. Au, 1977, The Relationship of Microbial Processes to the Fate of Transuranic Elements in Soil, in *Transuranics in Natural Environments,* Symposium Proceedings, Gatlinburg, Tenn., October 1976, ERDA Report NVO-178, pp. 127-171, Nevada Operations Office, NTIS.

——, and T. R. Garland, 1975, Relative Solubility of Inorganic and Complexed Forms of Plutonium-238 and Plutonium-239 in Soil, in *Pacific Northwest Laboratory Annual Report for 1974 to the USAEC Division of Biomedical and Environmental Research,* USAEC Report BNWL-1950(Pt. 2), pp. 23-25, Battelle, Pacific Northwest Laboratories, NTIS.

——, T. R. Garland, and R. L. Buschbom, 1975, The Interdependent Effects of Soil Temperature and Water Content on Soil Respiration Rate and Plant Root Decomposition in Arid Grassland Soils, *Soil Biol. Biochem.,* 7: 373-378.

——, T. R. Garland, and H. Drucker, 1973, The Potential Role of the Soil Microbiota in Increasing Plutonium Solubility in Soil and Plutonium Uptake by Plants, in *Agronomy Abstracts,* p. 190, American Society of Agronomy, Soil Science Society of America, and Crop Science Society of America Annual Meeting, Las Vegas, Nev., Nov. 11–16, 1973.

——, T. R. Garland, and H. Drucker, 1974, Potential Role of the Soil Microbiota in the Solubilization of Plutonium in Soil, in *Pacific Northwest Laboratory Annual Report for 1973 to the USAEC Division of Biomedical and Environmental Research,* USAEC Report BNWL-1850 (Pt. 2), pp. 21-22, Battelle, Pacific Northwest Laboratories, NTIS.

——, T. R. Garland, H. Drucker, and G. S. Schneiderman, 1974, Influence of Plutonium on the Soil Microflora, in *Pacific Northwest Laboratory Annual Report for 1973 to the USAEC Division of Biomedical and Environmental Research,* USAEC Report BNWL-1850 (Pt. 2), pp. 19-21, Battelle, Pacific Northwest Laboratories, NTIS.

Wood, J. M., 1974, Biological Cycles for Toxic Elements in the Environment, *Science,* 183: 1049-1052.

——, F. S. Kennedy, and C. G. Rosen, 1968, Synthesis of Methyl-Mercury Compounds by Extracts of a Methanogenic Bacterium, *Nature,* 220: 173-174.

Uptake of Transuranic Nuclides from Soil by Plants Grown Under Controlled Environmental Conditions

D. C. ADRIANO, A. WALLACE, and E. M. ROMNEY

Plant uptake of transuranic nuclides ranges through several orders of magnitude, depending on plant, environmental, and edaphic conditions. Most information presently available concerns root uptake of plutonium and americium. In environments where resuspension prevails, direct deposition on plant foliage may exceed root uptake. Atmospheric deposition is generally short lived, however, and the long-term assessment precludes that root uptake, as in the case of surface land contamination and shallow burial of nuclear wastes, will exceed that obtained from atmospheric deposition. Concentration ratios for ^{241}Am uptake generally ranged from 10^{-4} to 10^{1}; those for ^{239}Pu generally ranged from 10^{-7} to 10^{-3}. Information for curium and neptunium is scarce, but the range appeared to vary from 10^{-4} to 10^{-3} and from 10^{-2} to 10^{-1}, respectively, for these radionuclides.

Studies conducted using soils in pot culture showed that ^{241}Am uptake by crops from southeastern U. S. soils was influenced by clay content and low cation exchange capacity. Lime amendment suppressed ^{241}Am uptake, whereas organic matter amendment appeared to temporarily reduce uptake from these soils. Commonly used agricultural amendments generally were ineffective in altering ^{241}Am and ^{239}Pu uptake from western U. S. desert soils. However, the chelate diethylenetriamine pentaacetic acid markedly and consistently increased root uptake of both plutonium and americium by plants. Chelators and other chemical compounds that enhance complexation reactions with transuranic elements appeared to be most effective in enhancing root uptake from soils. Such compounds, which are usually present in shallow-burial waste-storage areas, may accelerate plant uptake through deep-penetrating root systems.

Numerous studies on the root uptake and translocation of the transuranic elements have been conducted which contributed to the understanding of some aspects of the processes involved. Many investigators have demonstrated that transuranic elements entered plant roots in trace quantities and were transported to aerial parts of plants (Jacobson and Overstreet, 1948; Cline, 1968; Newbould, 1963; Newbould and Mercer, 1961; Rediske and Selders, 1954; Romney, Mork, and Larson, 1970; Romney and Price, 1959; Wilson and Cline, 1966; Rediske, Cline, and Selders, 1955). Adams et al. (1975) found that the availability of plutonium to plants was very low from 100-μm $^{238}PuO_2$ particles [concentration ratio (CR) of 10^{-8} to 10^{-7} in ash]. In general, they found that plant species differed in uptake, with about 25 times more ^{241}Am taken up than ^{239}Pu. Bean seeds contained 200 times less plutonium than bean leaves, but radish roots contained 10 times as much ^{239}Pu as did the tops. Peeling the radish roots, however, removed 99% of the radionuclide, indicating that this radionuclide was mostly contained in or on the peel.

Plant uptake of transuranic elements via the root path has been the subject of recent reviews (Francis, 1973; Price, 1972a; Brown, 1976; Bernhardt and Eadie, 1976; Mueller and Mosley, 1976). There has been partial synthesis of the data, with most of the emphasis on outlining the problems needing further study.

Recently, Adriano et al. (1977), Romney, Mork, and Larson (1970), and Wallace (1969) found that the chelating agent DTPA (diethylenetriamine pentaacetic acid) enhanced plant uptake of both ^{241}Am and ^{239}Pu up to at least one order of magnitude under some conditions. These observations, confirmed by their subsequent studies, are important for two reasons: (1) chelating agents are used in the nuclear industry and are often present in wastes, and (2) chelating agents are also used in plant nutrition to supply micronutrients to plants and therefore are likely to be present in agricultural soils. Agricultural soils are usually treated with various types of amendments to optimize crop production. Such amendments, like lime, organic matter, and fertilizers, change the chemistry of the soil with a concomitant effect on the availability of the elements in question and their subsequent translocation to the plant shoots.

The main objective of this chapter is to evaluate how various soil factors, both indigenous and introduced, affect uptake of the transuranic elements by various plant species grown in potted soils obtained from two distinct regions of the United States: the desert soils of the western United States and soils of the humid southeastern United States.

Materials and Methods

Pot-Culture Experiments with Soils Representing the Humid Environments of the Southeastern United States

The soils used for the Bahia grass and rice experiments were collected from the Savannah River Plant (SRP), near Aiken, S. C. These were uncontaminated soils, highly weathered Ultisols, which were collected from the topsoil in a location that normally receives approximately 130 cm of rainfall annually. These soils are either common in the coastal plain area or similar to the soils found at the burial areas for nuclear wastes at Barnwell, S. C., and at the SRP. The clay is usually reddish owing to ferric oxide and is dominated by kaolin.

Bahia Grass Experiment. Collection, preparation, liming, spiking of soils, and potting of soils (Troup sandy loam and Dothan sandy clay loam) are described in detail in a previous paper (Adriano et al., 1977). Bermuda grass hay was ground to pass a 20-mesh screen and mixed with both the limed and unlimed soils to give 0.0, 1.25, and 5.0% organic matter (OM) by weight. Reagent-grade fertilizers (NH_4NO_3, KH_2PO_4, and KNO_3), lime, and OM were mixed well at the same time with each 2 kg of soil in plastic bags.

A 500-g aliquot of the premixed soils was removed from each pot (top diameter, 15 cm; bottom diameter, 12.5 cm; height, 13.5 cm) and used for spiking. One microcurie of ^{241}Am, dissolved in 1 ml of 0.1N HNO_3, was placed in 125 ml of distilled water; then 10-ml aliquots of this solution were pipeted and added to a thin layer of soil (approximately 40 g) placed on top of the remaining soil. This was repeated until the total 500 g of soil was spiked.

Each treatment of the complete factorial (two soil types × two lime rates × three OM rates) was replicated seven times, but only five replicates were spiked. The two unspiked replicates were used to determine soil chemical changes at various times during the 180-day duration of the study.

After 50 days of equilibration, the pots were transferred into a water bath located in a glasshouse. A 1-cm-thick sheet of glass fiber was placed on top of the soil surface to prevent the soil particles from adhering to lower plant portions during irrigation. Two Bahia grass seeds were placed in each of three holes, cut in the fiber equidistant from each other and the pot perimeter. On the 70th day, the plants were thinned to three per pot, one for each hole, and watered with deionized water as required. The plants were clipped at the 100th, 130th, and 180th day after equilibration. This gave a plant growth time of 50, 30, and 50 days for the first, second, and third clipping periods, respectively. The plants were clipped 2.5 cm from the glass-fiber surface. They were then cut into shorter pieces, placed in paper bags, and dried to constant weights.

Rice Experiments. Only the Dothan soil was used in these experiments. A total of 5 kg of soil was placed in each black plastic pot (top diameter, 22 cm; bottom diameter, 18 cm; height, 20 cm), and 2 μCi of ^{241}Am was added either by the soil-layering technique earlier described or by injection to the ponded water used in the flooded experiment.

In the first experiment ^{241}Am, dissolved in $0.1N$ HNO$_3$, was chelated by adding a ^{241}Am(NO$_3$)$_3$ aliquot to 50 ml of DTPA solution. Chelated or nonchelated ^{241}Am was added to the ponded water at three various stages of growth: booting stage, flowering stage, and dough ripening stage.

In the second experiment DTPA and OM (Bermuda grass hay) were premixed with the whole soil to give 40 ppm DTPA and 5% OM by weight. The nonchelated radionuclide was added to only the top 1 kg of soil by the layering technique. Two rice varieties were grown, one under flooded condition and the other under nonflooded condition.

All pots were supplemented once with reagent-grade NH$_4$NO$_3$, KH$_2$PO$_4$, and KNO$_3$. All treatments were replicated five times. The plants were grown to maturity in a water bath in a glasshouse, harvested, separated into various plant parts, cut into shorter pieces, and dried.

The dried-plant tissues from all experiments were placed in counting tubes and counted for at least 50 min for ^{241}Am with a 7.6- by 7.6-cm NaI well crystal interphased to a multichannel analyzer. Concentration ratios were calculated from the soil and plant tissue radioactivity data.

Pot-Culture Experiments with Soils Representing the Desert Environments of the Western United States

Some areas of the test-range complex in Nevada (NTS) were contaminated by fallout debris during separate high-explosive (nonnuclear) detonations of devices containing plutonium more than 20 yr ago. The ratios of plutonium to americium in soils and vegetation collected in the field indicated that at least americium was taken up via roots from the soil (Romney et al., 1976). Considerable amounts of plutonium and americium have moved to the root zone in the soils involved. The soils in these areas are sandy desert calcareous soils with pH values averaging about 8.0 (Leavitt, 1974). Sampling locations were selected with a portable gamma spectrometer [FIDLER (field instrument for the determination of low-energy radiation)], which measures the 60-keV gamma radiation emitted from ^{241}Am. Radiochemical data of representative soil samples collected from

TABLE 1 ^{239,240}Pu and ^{241}Am Concentrations and
Ratios of Desert Soils from the Nevada Test Site and the
Tonopah Test Range Used in the Soybean Experiment*

	Concentration, nCi/g (dry weight)		
Soil source	^{239,240}Pu	^{241}Am	Pu/Am
Area 11 B	2.8 ± 0.41	0.44 ± 0.06	6.5 ± 0.17
Area 11 C	9.6 ± 1.3	1.8 ± 0.20	5.2 ± 0.17
Area 11 D	4.6 ± 0.24	0.87 ± 0.06	5.2 ± 0.12
Area 13 '	6.0 ± 0.93	1.1 ± 0.11	5.6 ± 0.27
Clean Slate 1	4.3 ± 0.57	0.23 ± 0.03	19.0 ± 0.33
Clean Slate 2	15.0 ± 3.6	0.72 ± 0.13	21.0 ± 1.9
Clean Slate 3	11.0 ± 1.6	0.59 ± 0.07	19.0 ± 0.67
Double Track	5.9 ± 0.79	0.28 ± 0.03	21.0 ± 0.88

*Values are means ±1 standard error.

eight fallout areas at which FIDLER activity readings ranged from 20,000 to 30,000 cpm are given in Table 1. These soils were used for the soybean experiment. The soils used for the alfalfa and barley experiments were collected at a greater distance from ground zero in Area 13 and consequently had much lower ^{239}Pu and ^{241}Am contents.

Soils collected from an intermediate contamination zone in Area 13 were subdivided into twelve 20-kg lots and mixed thoroughly with given amendments in a Patterson— Kelley blender for 1 hr before subdividing the mixture into six 3.2-kg lots for potting. The soil amendments consisted of nitrogen fertilizer (at a rate equal to 200 kg N/ha as NH_4NO_3), 2% agricultural-grade sulfur (to reduce pH from 7.6 to 5.4), and 5% OM (as alfalfa meal). The treatments were done in three separate sets of three replications per treatment and with and without DTPA (72 pots in total). Soil was potted in plastic pots that were sleeved inside plastic buckets and covered with 5 cm of silica sand to prevent soil-particle resuspension. The soil activity levels turned out to be much lower than had been anticipated. Consequently the plant materials from all replicates grown on similar treatments of the three sets had to be combined to produce an adequate sample size for radioassay. Barley plants were grown first. They were harvested in the dough stage by cutting at about 5 cm above the top of the sand layer and were divided into straw and fruit head samples. Alfalfa was grown next. Three successive cuttings of foliage were made in the quarter-bloom stage; then the plants were harvested and combined like the barley plants.

Soils collected from eight different plutonium fallout areas on the NTS and the Tonopah Test Range were used for the pot-culture experiments with soybeans. These soils received only the DTPA chelate (200 ppm) amendment. The eight soils were again arranged in three sets each, containing three replicates, with and without DTPA (144 pots in total). Soil processing and potting and plant harvesting and preparation for radioassay were done as indicated for the barley and alfalfa experiments. The soybean plants were grown to maturity, harvested, and separated into fruit pods and foliage (leaf and stem). All samples were radioassayed for ^{241}Am and ^{239}Pu at LFE, Richmond, Calif. (Majors et al., 1973).

Results and Discussion

Soils from the Humid Southeastern United States

Current field conditions at the SRP do not provide an environment suitable for the study of the incorporation of transuranic elements in plant tissues through root uptake because of a confounding effect from deposition of particles on the foliage following resuspension or stack emission from reprocessing operation. Thus studies were conducted in a glasshouse to evaluate the influence of some common soil amendments on the uptake and translocation of [241]Am by Bahia grass *(Paspalum notatum)* and rice *(Oryza sativa)*. Bahia grass is a common pasture crop grown extensively in the southeast. Rice was included because it is one of the most important food crops in the world and is widely distributed throughout the tropical, subtropical, and temperate zones of all continents (Adair, Miller, and Beachell, 1962; Harlan, 1976).

For brevity, only the CR* values are presented. In studies of this nature, the CR is a convenient method of expressing the availability of an element from the soil and its pattern of translocation to the plant parts. Since the soil [241]Am concentrations are given in the footnotes to the tables of results, the corresponding plant concentrations can be calculated from the CR values.

Bahia Grass Experiment. The effect of soil type on [241]Am uptake in Bahia grass (Table 2) was not so pronounced as that in the bush bean and corn seedlings (Adriano et al., 1977). Nevertheless, the most striking differences in uptake were caused by soil type and lime. On the average, [241]Am concentrations in plant tissues from the unlimed Dothan soil (pH 4.2) were approximately twice as high as those from the unlimed Troup soil (pH 5.0). This wide disparity between these two soils was minimized when both soils were limed. Consequently liming of both soils (pH 7.1 for Dothan soil and pH 6.6 for Troup soil) significantly (p ≤ 0.01) reduced [241]Am availability to Bahia grass. Across the OM treatments, the plant tissues from unlimed Troup soil had 12.0 pCi/g (dry-weight basis), compared to only 1.0 pCi/g from limed soil, on the average. On the average plants from unlimed Dothan soil had 35.3 pCi/g vs. 0 pCi/g (below detection limit) from limed Dothan soil.

The clipping period affected the [241]Am concentration pattern in Bahia grass, particularly in unlimed soils (Table 2). In unlimed Dothan soil, the concentration progressively declined with clipping time. The peak occurred on the first clipping (50th day of growth) and the minimum occurred on the last clipping (130th day of growth), irrespective of OM rate. The concentrations in the first clipping were always significantly higher than those from subsequent clippings. In unlimed Troup soil, the concentrations generally peaked during the second clipping, although at the 1.25% OM rate they were not significantly different from the other clippings. No meaningful pattern can be deduced when either soil was limed, and, in some cases, liming caused plant [241]Am concentrations to be equal to background level.

The addition of OM affected [241]Am plant concentrations to some extent. Plants in unlimed Troup soil in which no OM had been added had the highest concentrations. The mean concentrations for all three clippings were 23.2, 9.5, and 5.6 pCi/g for 0.0, 1.25,

---

$$*CR \text{ (concentration ratio)} = \frac{radioactivity/g \text{ (plant tissue)}}{radioactivity/g \text{ (soil)}}$$

TABLE 2   Influence of Lime and Organic Matter on ^{241}Am Concentration Ratios (Plant Tissue/Soil)*
for Bahia Grass Grown in Two Southeastern U.S. Soils†

Treatment	Dothan soil			Troup soil		
	First clipping	Second clipping	Third clipping	First clipping	Second clipping	Third clipping
Control	$(1.2 \pm 0.18) \times 10^{-1}$	$(8.0 \pm 1.2) \times 10^{-2}$	$(5.9 \pm 1.0) \times 10^{-2}$	$(4.0 \pm 0.54) \times 10^{-2}$	$(7.5 \pm 1.2) \times 10^{-2}$	$(3.0 \pm 0.62) \times 10^{-2}$
Control + lime	$(0.8 \pm 0.54) \times 10^{-2}$	‡	‡	$(3.8 \pm 1.3) \times 10^{-3}$	$(6.2 \pm 1.3) \times 10^{-3}$	$(0.9 \pm 0.62) \times 10^{-3}$
1.25% OM	$(1.50 \pm 0.13) \times 10^{-1}$	$(6.9 \pm 0.67) \times 10^{-2}$	$(3.7 \pm 0.62) \times 10^{-2}$	$(2.0 \pm 0.31) \times 10^{-2}$	$(2.4 \pm 0.3) \times 10^{-2}$	$(1.3 \pm 0.13) \times 10^{-2}$
1.25% OM + lime	$(1.4 \pm 0.62) \times 10^{-3}$	$(1.2 \pm 0.45) \times 10^{-3}$	$(3.3 \pm 1.2) \times 10^{-3}$	$(3.0 \pm 1.2) \times 10^{-3}$	$(5.0 \pm 1.2) \times 10^{-3}$	$(1.3 \pm 0.58) \times 10^{-3}$
5.0% OM	$(5.0 \pm 0.85) \times 10^{-2}$	$(2.8 \pm 0.49) \times 10^{-2}$	$(2.8 \pm 0.36) \times 10^{-2}$	$(9.0 \pm 1.6) \times 10^{-3}$	$(1.5 \pm 0.27) \times 10^{-2}$	$(1.0 \pm 0.10) \times 10^{-2}$
5.0% OM + lime	$(1.7 \pm 0.94) \times 10^{-3}$	$(3.0 \pm 1.43) \times 10^{-3}$	$(0.8 \pm 0.62) \times 10^{-3}$	$(5.6 \pm 3.0) \times 10^{-4}$	$(2.4 \pm 0.62) \times 10^{-3}$	$(1.5 \pm 0.36) \times 10^{-3}$

*Concentrations in plant tissues can be calculated from the CR values and the ^{241}Am concentrations (500 pCi/g soil) of the potted soil.
† Values are means of five replications ±1 standard error.
‡ Activities were below the detection limit.

and 5.0% OM rates, respectively. Similarly, although less pronounced than in the Troup soil, plant concentrations in the unlimed Dothan soil were highest at the 0% OM rate (45.2 pCi/g) and decreased to 36.9 and 14.5 pCi/g for the 1.25 and 5.0% OM rates, respectively. In both unlimed soils, small differences in [241]Am concentrations were obtained from the 1.25% OM rate. However, the 5% OM rate decreased the [241]Am concentrations in both unlimed soils somewhat substantially.

The CR values in Table 2 for limed treatments, in general, are 10 times lower than those for the unlimed treatments and, in some cases, as much as two orders of magnitude lower. The effects of OM rate and clipping period on [241]Am availability can also be easily deduced from Table 2 and further elaborate the effects of these treatments on [241]Am concentrations. The slight reduction in uptake with OM addition was possibly caused by immobilization of [241]Am in the soil microbial biomass and the fixing capacity of OM for metals.

The CR values were calculated on the basis of the total soil mass in the pot. These values can also be calculated by using only the amount of soil that was spiked, in which case the present CR values should be multiplied by a constant factor of 0.25.

Rice Experiment. Results (Table 3) indicate that, in some cases, americium applied in water was detectable in the rice grain. However, these are low radioactivities compared with other plant parts. No radioactivity was detected in the grain when americium was applied to the soil. Americium increased in the following order: unshelled grain < green

TABLE 3 Influence of Chelate DTPA, Time of Spiking, and Method of Placement on [241]Am Concentration Ratios for Rice Grown in a Southeastern U.S. Soil Under Flooded Conditions*†

	Unshelled grain/soil	Green blades/soil	Old (dead) blades/soil
		With DTPA‡	
Applied to water			
Period 1	§	$(3.8 \pm 1.4) \times 10^{-3}$	$(2.7 \pm 1.4) \times 10^{-1}$
Period 2	$(1.1 \pm 0.36) \times 10^{-3}$	$(1.3 \pm 0.18) \times 10^{-2}$	$(6.5 \pm 2.5) \times 10^{-1}$
Period 3	$(1.0 \pm 0.18) \times 10^{-3}$	$(1.7 \pm 0.40) \times 10^{-2}$	$(1.6 \pm 0.54) \times 10^{0}$
Applied to soil¶	§	$(0.2 \pm 0.85) \times 10^{-3}$	$(2.8 \pm 0.27) \times 10^{-2}$
		Without DTPA‡	
Applied to water			
Period 1	§	$(5.3 \pm 2.8) \times 10^{-3}$	$(4.6 \pm 1.3) \times 10^{-2}$
Period 2	$(6.8 \pm 2.3) \times 10^{-4}$	$(5.7 \pm 1.3) \times 10^{-3}$	$(4.8 \pm 3.7) \times 10^{-2}$
Period 3	§	$(1.0 \pm 0.80) \times 10^{-2}$	$(4.2 \pm 3.4) \times 10^{-1}$
Applied to soil¶	§	$(2.6 \pm 0.80) \times 10^{-3}$	$(2.5 \pm 0.22) \times 10^{-2}$

*Concentrations in plant materials can be calculated from the CR values and the assumed concentrations (400 pCi/g dry soil) of the potted soil.

†Values are means of five replicates ±1 standard error.

‡Chelated or nonchelated [241]Am (in 50 ml of 100 ppm DTPA as acid or water) was added to the standing water when the rice plants were at booting (period 1), flowering (period 2), and dough (period 3) stages.

§Activities were below the detection limit.

¶Data taken from Table 4.

blade $\ll$ dead blade. Morphologically, the green blades were located in the top portion of the plant and the dead blades in the lower portion. Thus americium appeared to accumulate in the older leaves to a much greater extent than in the newer leaves, which developed in later stages of growth. The ratio of americium concentrations in dead and green blades ranged from 48 to 140 with DTPA and from 8 to 44 without DTPA. However, most of the americium accumulated in the sheath when this isotope was applied in water and most notably when chelated. The accumulation in the sheath could be attributed to physical absorption of the isotope rather than to physiological assimilation, as in the case for leaf blades. Chelated americium applied to the ponded water was more readily absorbed, as the leaf-blade data suggest, but it was not more readily absorbed when applied to soil.

The relative magnitudes of americium in various parts are easily discernible from the CR values. Plants that received chelated-americium through water application had americium contents 10 times higher than plants supplied with nonchelated americium. Dead blades had americium contents one to two orders of magnitude higher than green blades. The CR values can be used to determine the relative availability of an element from a substrate and the translocation pattern of this element within the plant. The CR values indicate that americium was less available when added to soil and was not readily translocated to younger leaves.

In all CR calculations, 400 pCi/g dry soil was used, taking into account the total soil mass per pot (5 kg). However, if based on only the top 1 kg of spiked soil, 2000 pCi/g should be used. Thus, in the latter case, the reported CR values in Table 3 should be multiplied by a factor of 0.20. With water application it is difficult to assign a conversion factor. It should be pointed out that CR is probably not valid with water application and should be used with caution since traditionally it is used where the radionuclide was applied to the soil and the total soil mass is taken into account in calculating the average radionuclide concentration in the soil. Recently, however, conversion factors have been introduced to consider also the fraction of the soil mass spiked (Lipton and Goldin, 1976). As would be expected, there is some question concerning the use of soil concentration for determining the CR for water application, but the CR would demonstrate the relative translocation or redistribution of ^{241}Am in various rice parts and could serve as a basis for calculating the plant concentration of ^{241}Am.

In the flood variety* the radioactivity in the grain was below the detection limit (Table 4). There was also little translocation from old leaves to green leaves. The chelate DTPA mixed with the soil slightly reduced ^{241}Am uptake. Apparently the chelate level (40 ppm as acid) was harmful to the rice plants, retarding and reducing growth. Organic matter did not have a clear-cut effect, although it tended to suppress the uptake by the nonflood variety. It should be pointed out that the OM retarded growth of the rice seedlings in early stages of growth, presumably because the organic acids inhibited root development (Takijima, 1964). In general, the plant tissues of the nonflood variety had higher ^{241}Am levels.

It appeared that the ^{241}Am content of the grain could be increased slightly by adding chelated ^{241}Am to the standing water. Thus the method of ^{241}Am placement would

---

*The flood variety was a dwarfed "miracle rice" variety from southeast Asia and was ponded with water all the time. The nonflood variety, also from Asia, was not ponded and was taller than the flood variety.

TABLE 4 **Influence of Chelate DTPA and Organic Matter on**
^{241}Am Concentration Ratios for Rice Grown in a Southeastern
U.S. Soil Under Flooded and Nonflooded Conditions*†

	Unshelled grain/soil	Green blades/soil	Old (dead) blades/soil
		Flooded	
Control	‡	$(2.6 \pm 0.80) \times 10^{-3}$	$(2.5 \pm 0.22) \times 10^{-2}$
+ DTPA	‡	$(0.2 \pm 0.85) \times 10^{-3}$	$(2.8 \pm 0.27) \times 10^{-2}$
+ 5% OM	‡	$(7.6 \pm 2.4) \times 10^{-3}$	$(1.5 \pm 0.31) \times 10^{-2}$
		Nonflooded	
Control	$(7.1 \pm 2.0) \times 10^{-4}$	$(2.7 \pm 0.18) \times 10^{-2}$	$(5.9 \pm 1.3) \times 10^{-2}$
+ DTPA	$(2.5 \pm 1.6) \times 10^{-4}$	$(2.9 \pm 0.67) \times 10^{-3}$	$(2.2 \pm 0.76) \times 10^{-2}$
+ 5% OM	$(2.5 \pm 2.1) \times 10^{-4}$	$(1.4 \pm 0.22) \times 10^{-2}$	$(1.5 \pm 0.10) \times 10^{-2}$

*Values are means of five replicates ± standard error.

†Concentrations in plant materials can be calculated from the CR values and the assumed ^{241}Am concentrations (400 pCi/g soil or 435 pCi/g soil + 5% OM) of the potted soil.

‡Activities were below the detection limit.

have an influence on its availability to the rice plants; i.e., application to the standing water is more likely to result in higher uptake.

Rice has a peculiar uptake–translocation physiology (Chandrasekaran and Yoshida, 1973; Myttenaere and Marckwordt, 1967; Myttenaere, Bourdeau, and Masset, 1969). Its only organ of economic importance is the grain. The straw is seldom used for animal feed. The ^{241}Am did not appear to be readily translocated to the grain; therefore its health hazard to man is minimized.

Soils from the Desert Environment of the Western
United States

Barley and Alfalfa Experiment. Results indicate that ^{241}Am generally was taken up by barley two or more times as readily as was ^{239}Pu (Table 5). The exception was for the treatment acidified with sulfur and with DTPA. The reason was a relatively large uptake of ^{239}Pu with DTPA. The americium/plutonium ratios (Table 5) were obtained from the respective CR values to normalize the levels in the soil. The CR values were generally in the 10^{-4} to 10^{-3} range (mean ^{239}Pu = 9.4×10^{-4} and ^{241}Am = 1.4×10^{-3}). Without DTPA they were 1.3×10^{-4} and 3.2×10^{-4}, respectively, for plutonium and americium. Except for the control, which produced poor growth, DTPA enhanced the uptake of both plutonium and americium. However, the increase, which was usually greater than one order of magnitude, was equal for both elements.

The CR values for alfalfa were slightly less than those for barley (Table 5). The mean CR for ^{239}Pu was 1.4×10^{-4} and for ^{241}Am was 9.3×10^{-4}. Without DTPA the CR values were 7.6×10^{-5} and 6.6×10^{-4}, respectively. The preference for americium over plutonium (9.9) was greater in alfalfa than in barley (4.2). The DTPA had much less effect on the uptake of the two elements by alfalfa than by barley (ratio of about 2 for

TABLE 5 Effect of Various Soil Amendments on ^{239,240}Pu and ^{241}Am Concentration Ratios for Barley and Alfalfa Grown on Area 13 Soil from the Nevada Test Site

Soil amendment	Barley			+DTPA/−DTPA		Alfalfa			+DTPA/−DTPA	
	^{239,240}Pu	^{241}Am	Am/Pu	Pu	Am	^{239,240}Pu	^{241}Am	Am/Pu	Pu	Am
Control	1.9×10^{-4}	4.8×10^{-4}	2.5			3.8×10^{-5}	4.8×10^{-4}	12.6		
Control + DTPA	2.3×10^{-4}	1.3×10^{-3}	5.7	1.2	2.7	5.9×10^{-5}	9.2×10^{-4}	15.6	1.6	1.9
Nitrogen	1.8×10^{-5}	1.6×10^{-4}	8.9			6.4×10^{-5}	5.8×10^{-4}	9.0		
Nitrogen + DTPA	2.2×10^{-4}	1.8×10^{-3}	8.2	12.2	11.2	6.5×10^{-5}	9.4×10^{-4}	14.5	1.0	1.6
Sulfur	2.8×10^{-4}	5.8×10^{-4}	2.1			7.2×10^{-5}	1.1×10^{-3}	15.3		
Sulfur + DTPA	5.7×10^{-3}	4.2×10^{-3}	0.7	20.3	7.2	3.6×10^{-4}	1.9×10^{-3}	5.3	5.0	1.7
Organic matter	2.4×10^{-5}	6.0×10^{-5}	2.5			1.3×10^{-4}	4.9×10^{-4}	3.8		
Organic matter + DTPA	8.6×10^{-4}	2.8×10^{-3}	3.3	35.8	46.7	3.4×10^{-4}	1.0×10^{-3}	2.9	2.6	2.0

alfalfa and 17 for barley). This effect may indicate loss of DTPA from the soil with time, possibly by metabolism and degradation.

Soybean Experiment. Results from the barley and alfalfa experiments showed the necessity of using soils containing higher levels of contamination for better accuracy of determination. In addition, the response from the soil amendments indicated a practical influence from only the DTPA. Consequently the soybean experiments were conducted using only chelate treatment of soils collected from eight of the Nevada Applied Ecology Group study areas (Dunaway and White, 1974). Results of radiochemical analyses for ^{239}Pu and ^{241}Am are shown in Table 6 and indicate that the CR values were higher for soybean leaves and stems (Table 6) than for either barley or alfalfa (Table 5). Different soil sources, in part, were involved, but this may not be the major factor. The mean CR values for soybean leaves and stems were 1.4×10^{-3} and 3.7×10^{-2} for plutonium and americium, respectively; with DTPA they were 4.4×10^{-4} and 6.5×10^{-3}, respectively, which are higher than equivalent values for barley and alfalfa.

The mean americium/plutonium ratio was 21.6, which is higher than that for either barley (4.2) or alfalfa (9.9). It was not possible to determine if this ratio for soybeans differed because of variability in soils.

The DTPA enhanced the uptake of both plutonium and americium from the Area 11 and Area 13 soils but only slightly over those from the Tonopah Test Range. For the Area 11 soils, DTPA increased the uptake of americium more than of plutonium (about 2.5 times). This result did not appear to be significant since it was not observed for the other soils described in Table 6 except for one. Different chemical and physical properties of americium and plutonium in the soils from different locations may be involved. The soil plutonium/americium ratios were highest (Table 1) for the soils with least response to DTPA.

The mean CR for fruit pods was 3.7×10^{-5} for ^{239}Pu and 8.0×10^{-4} for ^{241}Am. Without DTPA the values were 1.5×10^{-5} and 3.1×10^{-4}, respectively. The mean americium/plutonium ratios were slightly higher for fruit pods than for vegetative material (32.8 vs. 21.6). The mean americium/plutonium ratio for fruit pods was 23.6 without DTPA and 42.0 with DTPA. The difference, however, was not due to DTPA-induced transport from leaves to fruit because the americium/plutonium ratios with and without DTPA were really not different (Table 6). Also, the mean CR for vegetative parts for plutonium was 9.9 without DTPA and 11.5 with DTPA, which difference was not significant. For americium the means were 15.9 and 10.1, the lower value being with DTPA. It appears that DTPA caused more plutonium and americium to be translocated to the fruit pods because plutonium and americium were higher in the leaves when DTPA was added. This resulted in correlation coefficients of +0.988 for plutonium and +0.983 for americium between these two plant parts. The CR values for fruit pods vs. leaves and stems from the eight soils were calculated for plutonium and americium and show that, on the average, the ratios were 0.035 ± 0.003 SE for ^{239}Pu and 0.058 ± 0.009 SE for ^{241}Am. The americium/plutonium ratio was 1.7 ± 0.24 SE for the eight soils. The ratio of transport with DTPA to that without DTPA was 1.0 ± 0.21 SE for plutonium and 1.3 ± 0.36 SE for americium. This method of calculation indicates that DTPA did not directly increase plutonium and americium transport to fruits, nor did DTPA influence the two elements differentially in transport from shoots to fruits. It appears therefore that there was a mass-action effect for transport from leaves to fruit.

TABLE 6 ^{239,240}Pu and ^{241}Am Concentration Ratios for Soybeans Grown on Soils Containing Aged Fallout Materials from the Nevada Test Site and Tonopah Test Range

Soil source and treatment	Leaf and stem					Fruit pods				
				+DTPA/−DTPA					+DTPA/−DTPA	
	^{239,240}Pu	^{241}Am	Am/Pu	Pu	Am	^{239,240}Pu	^{241}Am	Am/Pu	Pu	Am
Area 11 B	1.5×10^{-4}	1.2×10^{-3}	8.0			7.8×10^{-6}	1.6×10^{-4}	20.5		
Area 11 B + DTPA	1.6×10^{-3}	5.2×10^{-2}	32.5	10.7	43.3	2.6×10^{-5}	1.2×10^{-3}	46.2	3.3	7.5
Area 11 C	1.8×10^{-4}	2.7×10^{-3}	15.0			1.1×10^{-5}	1.9×10^{-4}	17.3		
Area 11 C + DTPA	3.7×10^{-3}	8.9×10^{-2}	24.1	20.5	33.3	1.2×10^{-4}	2.0×10^{-3}	16.7	10.9	10.5
Area 11 D	1.1×10^{-3}	1.7×10^{-2}	15.5			4.2×10^{-5}	7.6×10^{-4}	18.1		
Area 11 D + DTPA	6.6×10^{-3}	3.7×10^{-1}	56.1	6.0	21.8	2.2×10^{-4}	2.3×10^{-2}	104.6	5.2	30.3
Area 13	1.1×10^{-4}	3.3×10^{-3}	30.0			2.4×10^{-6}	8.1×10^{-5}	33.4		
Area 13 + DTPA	3.9×10^{-3}	2.7×10^{-2}	6.9	35.5	8.2	1.5×10^{-4}	1.2×10^{-3}	8.0	65.5	14.8
Clean Slate 1	4.3×10^{-4}	1.4×10^{-2}	32.6			1.7×10^{-5}	2.0×10^{-4}	11.8		
CS 1 + DTPA	7.1×10^{-4}	2.1×10^{-2}	29.6	1.7	1.5	2.2×10^{-5}	1.0×10^{-3}	45.5	1.3	5.0
Clean Slate 2	7.6×10^{-4}	2.5×10^{-2}	3.2			1.4×10^{-5}	1.3×10^{-4}	9.3		
CS 2 + DTPA	1.2×10^{-3}	4.0×10^{-2}	33.3	1.6	16.0	4.4×10^{-5}	1.2×10^{-3}	27.3	3.1	9.2
Clean Slate 3	5.5×10^{-4}	5.5×10^{-3}	10.0			1.8×10^{-5}	3.4×10^{-4}	18.9		
CS 3 + DTPA	7.8×10^{-4}	8.6×10^{-3}	11.00	1.4	1.6	2.7×10^{-5}	5.8×10^{-4}	21.5	1.5	1.7
Double track	2.6×10^{-4}	5.7×10^{-3}	21.9			9.8×10^{-6}	5.8×10^{-4}	59.2		
DT + DTPA	4.2×10^{-4}	6.7×10^{-3}	16.0	1.6	1.2	1.3×10^{-5}	8.6×10^{-4}	66.2	1.3	1.5

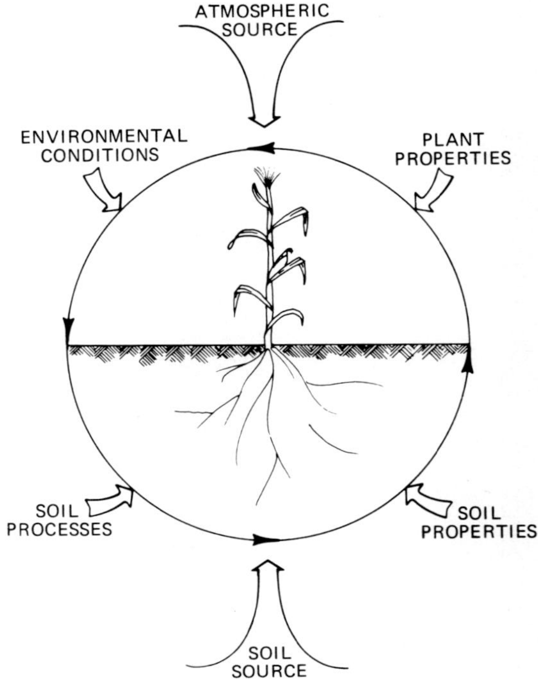

Fig. 1   Schematic diagram of various factors influencing the bioavailability of the transuranic nuclides in the soil–plant system.

Discussion

Several factors influence the availability of the transuranic nuclides to plants. A generalized outline is shown in Fig. 1. Plant uptake is influenced by soil pH, Eh (oxidation state), cation exchange capacity, texture (particularly percent clay), fertilizers and other amendments, and soil OM.

Lime significantly suppressed [241]Am uptake by crops grown in southeastern soils owing either to a lower solubility of [241]Am at higher pH, an increased cation exchange capacity caused by liming (Fiskell, 1970; Helyar and Anderson, 1974), or to calcium–magnesium and [241]Am antagonism. The first two processes could have caused high Kd* values. The latter could have resulted in Ca^{2+} and Mg^{2+} ions suppressing the uptake of [241]Am^{3+} ions. Chelates have been known to make insoluble cations available to plants. Such chelates increase the diffusion and mass flow of cations to roots by replenishing those taken up by the plants. It has been shown that chelates, including DTPA, decreased the Kd values of plutonium; i.e., less plutonium was retained by the soil (Relyea and Brown, 1978). Wallace (1972a; 1972b) observed that [241]Am–DTPA was most stable at about pH 7.7, where plant uptake was greatest.

$$*Kd \text{ (distribution coefficient)} = \frac{\text{concentration of } ^{241}Am/g \text{ (soil)}}{\text{concentration of } ^{241}Am/ml \text{ (solution)}}$$

TABLE 7 (Continued)

Plant species	Range of CR	Conditions and comments	Reference
Barley (shoots)	10^{-4}	Grown in Cinebar (pH 4.5) and Ephrata (pH 7.5) soils using the Neubauer technique; soils spiked with 10 μCi/g of ^{239}Pu as Pu(NO$_3$)$_4$; uptake similar from two soils.	Cline, 1968
Beans (shoots)	10^{-4}	Same as for barley above; seeds mainly used for human food.	Wilson and Cline, 1966
Beans (shoots)	10^{-5}	Grown in standard Hoagland solution spiked with 1.0 μCi/liter of ^{239}Pu as Pu(NO$_3$)$_4$.	Cline, 1968
Clover, Ladino	10^{-5} to 10^{-4}	Grown in 120 kg of NTS soil; high-fired PuO$_2$; 70 nCi/g (soil) of ^{239,240}Pu; CR increased sevenfold in 5 yr; mainly used for livestock feed.	Romney, Mork, and Larson, 1970
Corn			
Grain	10^{-5}	Grown in 5 to 7 kg of soil collected from top layer of a field adjacent to a reprocessing facility at SRP; homogenized soil contained 2 pCi/g total plutonium; used mainly for livestock feed plus some for human food.	Adriano, Corey, and Dahlman, this volume
Leaves	10^{-4}		
Lettuce (plants)	10^{-4}	Same as for alfalfa; mainly for human food.	Adams et al., 1975
Lettuce (plants)	10^{-7} (av.)*	Same as for barley.	Adams et al., 1975
Oats (plants)	10^{-7} (av.)*	Same as for barley; grain mainly used for human food plus ingredient for animal feed; plants used as forage.	Adams et al., 1975
Peas (plants)	10^{-3} to 10^{0}	Grown in about 1.5 kg of sand spiked with ^{239}PuO$_2$ at 20 μCi per container; highest CR caused by chelate; lesser effects by colloid size and placement depth; fruits mainly for human food.	Lipton and Goldin, 1976
Radish			
Bulb	10^{-2}	Same as for alfalfa; mainly for human food, including the shoots as greens.	Adams et al., 1975
Shoots	10^{-3}		

(Table continues on the next page.)

TABLE 7 (Continued)

Plant species	Range of CR	Conditions and comments	Reference
Soybean			
Fruit	10^{-6} to 10^{-4}	Grown in 3 kg of NTS soil;	This study
Forage	10^{-4} to 10^{-2}	5 nCi/g (soil) of ^{239,240}Pu; high-fired PuO$_2$; highest CR caused by chelate treatment.	
Soybean (foliage)	10^{-3}	Same as for corn; fruit used mainly for human food plus ingredient for animal feed.	Adriano, Corey, and Dahlman, this volume
Tomato			
Fruit	10^{-5}	Same as for alfalfa; fruit	Adams et al.,
Plant	10^{-4}	used mainly for human food.	1975
Wheat			
Grain	10^{-8} to 10^{-6}	Grown in 3 kg of soil	Schulz, Tompkins,
Leaf	10^{-6} to 10^{-3}	spiked with ^{239,240}Pu in either the chloride or nitrate form; 10 μCi/g (soil); highest CR occurred when plutonium was added in nitrate form to an alkaline, calcareous soil.	and Babcock, 1976a
Wheat (plants)	10^{-3}	Same as for corn; grain used mainly for human food plus ingredient for animal feed; plants when green can be used for forage.	Adriano, Corey, and Dahlman, this volume
Cheatgrass (plant)	10^{-5} to 10^{-4}	Grown in 1 kg of soil spiked with 50 nCi/g of ^{239}Pu as Pu(NO$_3$)$_4$; high CR values caused by organic acids; native species with no apparent food value to man and livestock animals.	Price, 1972b; 1973
Tumbleweed (plant)	10^{-5} to 10^{-4}	Same as for cheatgrass.	Price, 1973
Crops and vegetables			
Fruit	10^{-5} to 10^{-4}	Bush beans, beets, carrots,	Adriano, Corey,
Foliage	10^{-4} to 10^{-3}	lettuce, millet, potatoes,	and Dahlman,
Subterranean	10^{-4} to 10^{-2}	radishes, soybeans, and tomatoes grown in a flood-plain in Oak Ridge contaminated for 30 yr from weapon development; soil had 25–100 pCi/g total plutonium; nominal surface contamination; peeling the skins of	this volume

TABLE 7 (Continued)

Plant species	Range of CR	Conditions and comments	Reference
Wheat		subterranean crops and vegetables removed most of the plutonium.	
Grain	10^{-3}	Grown on a field adjacent to a reprocessing facility at SRP which aerially released <3 mCi total plutonium per yr; 1 to 3 pCi/g (soil) of total plutonium; 90–97% of contamination was external.	Adriano, Corey, and Dahlman, this volume
Foliage	10^{-2} to 10^{-1}		
Soybean			
Grain	10^{-4}	Grown on a field adjacent to a reprocessing facility at SRP which aerially released <3 mCi total plutonium per yr; 1 to 3 pCi/g (soil) of total plutonium; 90–97% of contamination was external.	Adriano, Corey, and Dahlman, this volume
Foliage	10^{-2}		
Corn			
Grain	10^{-4}	Grown on a field adjacent to a reprocessing facility at SRP which aerially released <3 mCi total plutonium per yr; 1 to 3 pCi/g (soil) of total plutonium; 90–97% of contamination was external.	Adriano, Corey, and Dahlman, this volume
Leaves	10^{-2}		
Corn + cobs	10^{-2}	Corn, potatoes, and peas grown on garden plots in North Eastham, Mass., on Cape Cod; top 30 cm of soil had 6 fCi/g ^{239,240}Pu; crops shielded from direct deposition or resuspension.	Hardy, Bennett, and Alexander, 1977
Potatoes	10^{-2}		
Peas (shelled)	10^{-3}		
Native vegetation	10^{-2} to 10^{0}	Tree foliage, shrubs, and herbaceous samples collected from an area adjacent to a reprocessing plant at SRP.	Unpublished SRP data.
Native vegetation	10^{-4} to 10^{-1}	Leaves of nonedible perennials, messerschmidia, scavola, and pandanas native to Enewetak Atoll; coral sandy soil.	Koranda et al., 1973

(Table continues on the next page.)

TABLE 7 (Continued)

Plant species	Range of CR	Conditions and comments	Reference
Native vegetation	10^{-2} to 10^{0}	Native perennials and shrubs at the NTS environments contaminated by safety shots; resuspension and deposition suspected on hirsute plants.	Romney et al., 1976
Native vegetation	10^{-4} to 10^{-3}	Mixed native grasses grown in contaminated soil blocks at Rocky Flats, Colo.; metallic plutonium in oil at 1 nCi/g (soil); root and foliar surface contamination contributed about equally.	Whicker, 1976
Native vegetation	10^{-3}	Tree foliage and herbaceous samples collected from a floodplain in Oak Ridge contaminated for 30 yr; evidence for plutonium-organic complex; plutonium in monomeric forms.	Energy Research and Development Administration, 1976
Coconut	10^{-3}	Same as for native vegetation; coconut meat used for human food in the tropics.	Koranda et al., 1973
Americium			
Pot culture			
Alfalfa	10^{-4} to 10^{-3}	Grown in 3 kg of NTS soil; 1 nCi/g (soil) of ^{241}Am; chelate increased CR; OM tended to decrease CR.	This study
Alfalfa	10^{-2}	Grown in 1.2 kg of Los Alamos mountain meadow soil spiked with 18 nCi to 0.5 μCi/g (soil) of ^{241}Am.	Adams et al., 1975
Bahia grass	10^{-4} to 10^{-1}	Grown in 2 kg soil from SRP; 0.5 kg of top pot soil spiked with $Am(NO_3)_3$ to give 500 pCi/g (soil); lowest CR caused by lime.	This study
Barley	10^{-1} to 10^{0}	Grown in 500 g of soil spiked with 1 nCi/g of ^{241}Am as $Am(NO_3)_3$;	Wallace et al., 1976

TABLE 7 (Continued)

Plant species	Range of CR	Conditions and comments	Reference
		highest CR caused by chelate applied to a calcareous soil.	
Barley			
Grain	10^{-6}	Same as for barley	Schulz et al.,
Foliage	10^{-4}	plutonium study; 1 to 11 nCi/g (soil) of ^{241}Am.	1976b
Barley			
Grain	10^{-5} to 10^{-3}	Grown in 3 kg of NTS	Energy Research
Foliage	10^{-5} to 10^{-3}	soil; 0.6 nCi/g (soil) of ^{241}Am; highest CR produced by chelate treatment.	and Development Administration, 1976
Barley (plants)	10^{-4}	Grown in 1.2 kg of Los Alamos mountain meadow soil spiked with 18 nCi to 0.5 μCi/g (soil) of ^{241}Am.	Adams et al., 1975
Barley	10^{-3}	Grown in Cinebar (pH 4.5) and Ephrata (pH 7.5) soils using the Neubauer technique; soils spiked with 1.8 μCi/g of ^{241}Am as Am(NO$_3$)$_3$; uptake similar from two soils.	Cline, 1968
Bean, bush (shoots)	10^{-1} to 10^{1}	Same as for barley.	Wallace et al., 1976
Bean, bush (shoots)	10^{-1} to 10^{1}	Grown in 500 g of soil spiked with 2 nCi/g (soil) of ^{241}Am as Am(NO$_3$)$_3$; lowest CR occurred on limed clay soil; highest CR occurred on limed and chelated treatment.	Adriano et al., 1977
Beans	10^{-3}	Grown in standard Hoagland solution spiked with 0.9 μCi/liter of ^{241}Am as Am(NO$_3$)$_3$.	Cline, 1968
Corn (shoots)	10^{-2} to 10^{0}	Same as for barley.	Wallace et al., 1976
Corn (shoots)	10^{-2} to 10^{0}	Same as for beans.	Adriano et al., 1977
Rice			
Grain	BG† to 10^{-3}	Grown under flooded or	This study
Leaves	10^{-3} to 10^{0}	nonflooded conditions in 5 kg of soil from	

(Table continues on the next page.)

TABLE 7 (Continued)

Plant species	Range of CR	Conditions and comments	Reference
		SRP; spiked top 1 kg with $Am(NO_3)_3$ to give 400 pCi/g; highest CR caused by chelate treatment.	
Soybean			
Fruit	10^{-5} to 10^{-2}	Grown in 3 kg of NTS	This study
Foliage	10^{-3} to 10^{-1}	soil; 0.2 to 2 nCi/g (soil); highest CR caused by chelate.	
Wheat (grain)	10^{-7} to 10^{-5}	Same as for wheat plutonium study; [241]Am added as chloride or nitrate to give 45 nCi/g (soil); highest CR obtained when added as chloride to a neutral soil.	Schulz, Tompkins, and Babcock, 1976a
Cheatgrass	10^{-5} to 10^{-4}	Cheatgrass and tumbleweed	Price, 1973
Tumbleweed	10^{-3} to 10^{-2}	grown in 1 kg of soil spiked with 25 μCi/g of [241]Am as $Am(NO_3)_3$; some organics suppressed uptake.	
Field studies			
Corn + cobs	10^{-3}	Same as for plutonium study; 1 fCi/g (soil) of [241]Am	Hardy, Bennett, and Alexander, 1977
Potatoes	10^{-3}	Same as for plutonium study; 1 fCi/g (soil) of [241]Am.	
Peas (shelled)	10^{-3}	Same as for plutonium study; 1 fCi/g (soil) of [241]Am.	
Native vegetation	10^{-2} to 10^{0}	Native perennial shrubs inhabiting the NTS environs; mainly surface contamination.	Energy Research and Development Administration, 1976
Native vegetation	10^{-2} to 10^{0}	Same as for plutonium study.	Koranda et al., 1973

Curium

Plant species	Range of CR	Conditions and comments	Reference
Pot culture			
Cheatgrass	10^{-4}	Cheatgrass and tumbleweed	Price, 1973
Tumbleweed	10^{-3}	grown in 1 kg of soil spiked with 25 nCi/g of [244]Cm as $Cm(NO_3)_3$; uptake not affected by organics.	

TABLE 7 (Continued)

Plant species	Range of CR	Conditions and comments	Reference
		Neptunium	
Pot culture			
Cheatgrass	10^{-2}	Cheatgrass and tumbleweed	Price, 1973
Tumbleweed	10^{-1}	grown in 1 kg of	
		soil spiked with	
		50 nCi/g of ^{237}Np	
		as $Np(NO_3)_5$; organics	
		tended to enhance	
		uptake.	

$$*CR = \frac{\text{radioactivity/g (plant ash)}}{\text{radioactivity/g (soil)}}$$

†Background level of activity.

Summary

Experiments on plant uptake of the transuranic nuclides conducted under controlled environmental conditions using spiked soils from the Savannah River Plant and contaminated soils from the Nevada Test Site revealed the following:

1. In general, the uptake of ^{241}Am by crop plants (Bahia grass and rice) grown on soils from the humid southeastern United States was influenced by soil amendments and indigenous soil factors. Lime generally immobilized ^{241}Am in the soil and decreased uptake by Bahia grass. The DTPA chelate somewhat enhanced the uptake by other crops tested in these acidic soils, but its greatest effect occurred where DTPA was supplied in limed soils.

The addition to soil of up to 5.0% OM appeared to demote ^{241}Am uptake in Bahia grass, but its effect was not so great as that of lime alone. This was possibly caused by temporary immobilization of ^{241}Am in microbial biomass, by an increase in cation-exchange capacity, or production of organic ligands from the OM.

The addition of ^{241}Am to soil resulted in almost no translocation of this radionuclide to the rice grain. However, when ^{241}Am was introduced in a chelated form to the ponded water, it appeared that there was relatively more absorption and translocation to the grain.

2. Crop plants (barley, alfalfa, and soybeans) grown on Nevada Test Site soils had an average CR of 10^{-4} for plutonium in the vegetative parts. Americium appeared more available than plutonium; the average americium/plutonium ratio was about 4 for barley, 10 for alfalfa, and 22 for soybeans.

Of the various soil amendments (nitrogen, sulfur, OM, and DTPA) used, only DTPA markedly and consistently increased both plutonium and americium uptake by plants; the increase with barley and soybeans was usually more than one order of magnitude but only a factor of 2 with alfalfa.

Americium appeared to be more mobile for transport (70% greater) than plutonium from shoots to fruits in soybeans. Concentrations of both plutonium and americium in shoots were highly correlated compared to those in fruits. The chelator DTPA did not differentially influence the transport of plutonium and americium from shoots to fruits.

Acknowledgments

This work was supported by the U. S. Department of Energy through contract Nos. EY-76-C-09-0819 (to Savannah River Ecology Laboratory, University of Georgia) and EY-76-C-03-0012 (to University of California, Los Angeles).

References

Adair, C. R., M. D. Miller, and H. M. Beachell, 1962, Rice Improvement and Culture in the United States, *Adv. Agron.*, 14: 61-108.

Adams, W. H., J. R. Buchholz, C. W. Christenson, G. L. Johnson, and E. B. Fowler, 1975, *Studies of Plutonium, Americium, and Uranium in Environmental Matrices,* ERDA Report LA-5661, Los Alamos Scientific Laboratory, NTIS.

Adriano, D. C., M. S. Delaney, G. D. Hoyt, and D. Paine, 1977, Availability to Plants and Soil Extraction of Americium-241 as Influenced by Chelating Agent, Lime and Soil Type, *Environ. Exp. Bot.*, 17: 69-77.

Bernhardt, D. E., and G. G. Eadie, 1976, *Parameters for Estimating the Uptake of Transuranic Elements by Terrestrial Plants,* Tech. Note ORP/LV-76-2, Office of Radiation Programs, Nevada Operations Office.

Brown, K. W., 1976, *Americium—Its Behavior in Soil-Plant Systems,* Report EPA-600/3-76-005, U. S. Environmental Protection Agency, Las Vegas, Nev.

Cataldo, D. A., R. C. Routson, R. E. Wildung, and T. R. Garland, 1976, *Uptake and Distribution of Plutonium in Soybean Tissues as a Function of Time Following Germination,* ERDA Report BNWL-2000(Pt. 2), Battelle Pacific Northwest Laboratories, NTIS.

Chandrasekaran, W., and T. Yoshida, 1973, Effect of Organic Acid Transformations in Submerged Soils on Growth of the Rice Plant, *Soil Sci. Plant Nutr. (Tokyo)*, 19(1): 39-45.

Cline, J. F., 1968, Uptake of ^{241}Am and ^{239}Pu by Plants, in *Pacific Northwest Laboratory Annual Report for 1977 to the USAEC Division of Biology and Medicine,* USAEC Report BNWL-714, Battelle Pacific Northwest Laboratories, NTIS.

Dunaway, P. B., and M. G. White (Eds.), 1974, *Dynamics of Plutonium in Desert Environments,* USAEC Report NVO-142, Nevada Operations Office, NTIS.

Energy Research and Development Administration, 1976, *Environmental Research for Transuranic Elements,* Proceedings of the Workshop, Battelle Seattle Research Center, Seattle, Wash., Nov. 12–14, 1975, ERDA Report 76/134, NTIS.

Fiskell, J. G. A., 1970, Cation Exchange Capacity and Component Variations of Soils of Southeastern USA, *Soil Sci. Soc. Am., Proc.*, 34: 723-727.

Francis, C. W., 1973, Plutonium Mobility in Soil and Uptake in Plants; A Review, *J. Environ. Qual.*, 2(1): 67-70.

Hardy, E. P., B. G. Bennett, and L. T. Alexander, 1977, *Radionuclide Uptake by Cultivated Crops,* USERDA Environmental Quarterly Report, ERDA Report HASL-321, ERDA Health and Safety Laboratory, NTIS.

Harlan, J. R., 1976, The Plants and Animals that Nourish Man, *Sci. Am.*, 235(3): 89-98.

Helyar, K. R., and A. J. Anderson, 1974, Effects of Calcium Carbonate on the Availability of Nutrients in Acid Soil, *Soil Sci. Soc. Am., Proc.*, 38: 341-346.

Jacobson, L., and R. Overstreet, 1948, The Uptake by Plants of Plutonium and Some Products of Nuclear Fission Adsorbed on Soil Colloids, *Soil Sci.*, 65: 129-134.

Koranda, J. J., J. R. Martin, S. E. Thompson, M. L. Stuart, D. R. McIntyre, and G. Potter, 1973, Terrestrial Biota Survey, in *Enewetak Radiological Survey,* USAEC Report NVO-140, pp. 225-349, Nevada Operations Office, NTIS.

Leavitt, V. D., 1974, *Soil Survey of Five Plutonium Contaminated Areas on the Test Range Complex in Nevada,* USAEC Report NERC-LV-539-Z8, National Environmental Research Center, NTIS.

Lipton, W. V., and A. S. Goldin, 1976, Some Factors Influencing the Uptake of Plutonium-239 by Pea Plants, *Health Phys.*, 31: 425-430.

Majors, W. J., K. D. Lee, R. A. Wessman, and R. Melgard, 1973, Determination of ^{239}Pu and ^{241}Am in Large NAEG Vegetation Samples, in *The Dynamics of Plutonium in Desert Environments,* P. B.

Dunaway and M. G. White (Eds.), USAEC Report NVO-142, pp. 107-118, Nevada Operations Office, NTIS.

Mueller, Anita A., and R. F. Mosley, 1976, *Availability, Uptake and Translocation of Plutonium Within Biological Systems: A Review of the Literature*, Report EPA-600/3-76-943, U. S. Environmental Protection Agency, Las Vegas, Nev.

Myttenaere, C., and U. Marckwordt, 1967, Indirect Contamination of Lowland Rice with Manganese-54, in *Symposium on Radioecology*, Proceedings of the Second National Symposium, Ann Arbor, Mich., May 15–17, 1967, D. J. Nelson and F. C. Evans (Eds.), USAEC Report CONF-670503, pp. 561-570, NTIS.

——, P. Bourdeau, and M. Masset, 1969, Relative Importance of Soil and Water in the Indirect Contamination of Flooded Rice with Radiocesium, *Health Phys.*, 16: 701-707.

Newbould, P., 1963, *Absorption of Plutonium 239 by Plants*, Annual Report on Radiobiology, Agricultural Research Council, p. 86.

——, and E. R. Mercer, 1961, Absorption of Plutonium-239 by Plants, in *Agricultural Research Council Radiobiological Laboratory Annual Report, 1961–1962*, Report ARCRL-8, pp. 81-82, Agricultural Research Council.

Price, K. R., 1972a, The Behavior of Waste Radionuclides in Soil–Plant Systems, in *Pacific Northwest Laboratory Annual Report for 1972 to the USAEC Division of Biomedical and Environmental Research*, USAEC Report BNWL-1750(Pt. 2), Battelle Pacific Northwest Laboratories, NTIS.

——, 1972b, *Uptake of ^{237}Np, ^{239}Pu, ^{241}Am, and ^{244}Cm from Soil by Tumbleweed and Cheatgrass*, USAEC Report BNWL-1688, Battelle Pacific Northwest Laboratories, NTIS.

——, 1973, *Tumbleweed and Cheatgrass Uptake of Transuranium Elements Applied to Soil as Organic Acid Complexes*, USAEC Report BNWL-1755, Battelle Pacific Northwest Laboratories, NTIS.

Rediske, J. H., and A. A. Selders, 1954, The Absorption and Translocation of Plutonium-239 and Cerium-144 by Plants, in *Biology Research—Annual Report for 1953*, USAEC Report HW-30437, Hanford Atomic Products Operation, NTIS.

——, J. F. Cline, and A. A. Selders, 1955, *The Absorption of Fission Products by Plants*, USAEC Report HW-36734, Hanford Atomic Products Operation, NTIS.

Relyea, J. F., and D. A. Brown, 1978, Adsorption and Diffusion of Plutonium in Soil, in *Environmental Chemistry and Cycling Processes*, DOE Symposium Series, Augusta, Ga., Apr. 28–May 1, 1976, D. C. Adriano and I. L. Brisbin (Eds.), pp. 479-494, CONF-760429, NTIS.

Romney, E. M., and K. R. Price, 1959, Availability of Pu-239 to Plants from Contamination, in *Semiannual Progress Report for the Period Ending December 31, 1969*, USAEC Report UCLA-457, University of California at Los Angeles, NTIS.

——, H. M. Mork, and K. H. Larson, 1970, Persistence of Plutonium in Soil, Plants, and Small Mammals, *Health Phys.*, 19: 487-491.

——, A. Wallace, R. O. Gilbert, and J. E. Kinnear, 1976, $^{239-240}Pu$ and ^{241}Am Contamination of Vegetation in Aged Fallout Areas, in *Transuranium Nuclides in the Environment*, Symposium Proceedings, San Francisco, 1975, pp. 470-491, STI/PUB/410, International Atomic Energy Agency, Vienna.

Schulz, R. K., G. A. Tompkins, and K. L. Babcock, 1976a, Uptake of Plutonium and Americium by Plants from Soils: Uptake by Wheat from Various Soils and Effect of Oxidation State of Plutonium Added in Soil, in *Transuranium Nuclides in the Environment*, Symposium Proceedings, San Francisco, 1975, pp. 303-310, STI/PUB/410, International Atomic Energy Agency, Vienna.

——, G. A. Tompkins, L. Leventhal, and K. L. Babcock, 1976b, Uptake of Plutonium and Americium by Barley from Two Contaminated Nevada Test Site Soils, *J. Environ. Qual.*, 5(4): 406-410.

Takijima, Y., 1964, Studies in Organic Acids in Paddy Field Soils with Reference to Their Inhibiting Effects on the Growth of Rice Plants 2: Relations Between Production of Organic Acids in Water-Logged Soils and the Root Growth Inhibition, *Soil Sci. Plant Nutr. (Tokyo)*, 10(5): 22-29.

Wallace, A., 1969, *Behavior of Certain Synthetic Chelating Agents in Biological Soil Systems*, Annual Progress Report for the Fiscal Year, USAEC Report UCLA-34P-51-35, University of California at Los Angeles, NTIS.

——, 1972a, Increased Uptake of ^{241}Am by Plants Caused by the Chelating Agent DTPA, *Health Phys.*, 22: 559-562.

——, 1972b, Effect of Soil pH and Chelating Agent (DTPA) on Uptake by and Distribution of ^{241}Am in Plant Parts of Bush Beans, *Radiat. Bot.*, 12: 433-435.

——, E. M. Romney, R. T. Mueller, and P. M. Patel, 1976, [241]Am Availability to Plants as Influenced by Chelating Agents, in *Radioecology & Energy Resources,* Proceedings of Fourth Symposium on Radioecology, Oregon State University, May 12–14, 1975, pp. 104-107, C. E. Cushing (Ed.), Ecological Society of America Special Publication Series No. 1, Academic Press, Inc., New York.

Whicker, F. W., 1976, *Radioecology of Natural Systems in Colorado,* Fourteenth Annual Progress Report, May 1, 1975–July 31, 1976, USAEC Report COO-1156-84, Colorado State University, NTIS.

Wildung, R. E., and T. R. Garland, 1974, Influence of Soil Plutonium Concentration on Plutonium Uptake and Distribution in Shoots and Roots of Barley, *J. Agric. Food Chem.,* 22: 836-838.

Wilson, D. O., and J. F. Cline, 1966, Removal of Plutonium 239, Tungsten 185, and Lead 210 from Soils, *Nature,* 209: 941-942.

Comparative Uptake and Distribution of Plutonium, Americium, Curium, and Neptunium in Four Plant Species

R. G. SCHRECKHISE and J. F. CLINE

The uptake of the nitrate forms of ^{238}Pu, ^{239}Pu, ^{241}Am, ^{244}Cm, and ^{237}Np from soil into selected parts of four different plant species grown under field conditions was compared. Alfalfa, barley, peas, and cheatgrass were grown outdoors in small weighing lysimeters filled with soil containing these contaminants. The plants were harvested at maturity, divided into selected components, and radiochemically analyzed by alpha-energy analysis. Soil concentration did not appear to affect the plant uptake of ^{238}Pu, ^{239}Pu, ^{241}Am, or ^{244}Cm for the two levels used. The relative uptake values of ^{238}Pu and ^{239}Pu were not significantly different from each other and the ^{241}Am uptake values were not significantly different from the ^{244}Cm values. The relative plant uptake of the four different transuranium elements was $Np > Cm \simeq Am > Pu$. Relative uptake values of neptunium into various plant parts ranged from 2,200 to 45,000 times as great as those of plutonium, whereas americium and curium values were 10 to 20 times as great. The seeds were significantly lower than the rest of the aboveground plant parts for all four transuranics. The legumes accumulated approximately 10 times as much as the grasses. A hypothetical comparison of the radionuclide content of plants grown in soil contaminated with Liquid Metal Fast Breeder Reactor fuels indicates that concentrations of isotopes of americium, curium, and neptunium would exceed ^{239}Pu values.

The release of transuranium nuclides to environmental systems, whether planned or not, poses potential hazards, especially if the biologically toxic materials enter food chains leading to man. Quantitative information on transport parameters is required for an assessment of the potential health hazards from such releases. One parameter that warrants close attention is the plant uptake of transuranics from contaminated soil. It used to be assumed that all transuranium elements behaved like plutonium and were equally discriminated against by plants. However, studies by Cline (1968) and Schulz et al. (1976) indicate a difference in the phytoavailability of ^{241}Am and ^{239}Pu.

Variations in the relative uptake of plutonium by plants, as summarized by the Energy Research and Development Administration (1975), can be explained by differences in plant species or fragments examined. Other factors that affect uptake are edaphic parameters, environmental conditions, and differences in the chemical form or valence state of the plutonium initially added to the soil. This chapter reports the relative plant uptake of ^{238}Pu, ^{239}Pu, ^{241}Am, ^{244}Cm, and ^{237}Np from soil into selected parts of four different plant species grown under field conditions. The data presented here are first-year results of a long-term study of the effects of aging, weathering, and associated biological processes in soil on the phytoavailability of transuranium elements.

Materials and Methods

The radiological safety requirements for experimentally placing transuranic elements in the field are stringent and precise. The radionuclides must be securely contained, readily retrievable, and isolated to eliminate biological and physical transport away from the study site. The plants in this study were grown outdoors in transuranic-contaminated soil contained in small weighing lysimeters. The containers were isolated from biota by a wire-mesh exclosure designed to exclude mammals and birds (Hinds et al., 1976). The exclosure was situated on the Arid Lands Ecology Reserve located on the Department of Energy's Hanford Site in South Central Washington.

The containers were constructed from 1-m lengths of polyvinyl chloride (PVC) pipe measuring 13.2 cm in inside diameter. Metal bale handles were attached to the open end, and the other end was enclosed with a watertight end cap. These containers were placed inside a slightly larger (15.7-cm-diameter) PVC pipe buried vertically in the ground so that the upper end was level with the ground surface. This arrangement facilitated retrieval of the contaminated soil and exposed the soil profile in the containers to realistic outdoor conditions of temperature and precipitation (Hinds, 1975).

Treatment containers were initially filled to within 35 cm from the top with 11.1 kg of oven-dried soil. Nitrate forms of ^{238}Pu, ^{239}Pu, ^{241}Am, ^{244}Cm, and ^{237}Np were individually added to a 3.4-kg aliquot of oven-dried soil, which was then placed in separate containers in a layer 20 cm thick. An additional 1.7 kg (10 cm) of clean soil was added to the top of the contaminated soil. This brought the level of soil to within 5 cm of the top of the container. The surface layer of clean soil was intended to prevent the spread of radionuclides by wind to the surrounding environment or their deposition on the surface of the experimental plants, which would have produced erroneous uptake values.

The soil used for this study was a silt loam of the Ritzville series. The soil has a pH of 6.2 and a cation-exchange capacity of 22.5 meq/100 g at pH 7 (Wildung, 1977). Radionuclides were added to the soil by pipetting 1 ml of the $4M$ HNO_3 solutions directly onto the soil, which had been adjusted to 5% moisture content. The oxidation state of the radionuclides when added to the soil was +4 for the plutonium isotopes, +3 for ^{241}Am and ^{244}Cm, and +5 for the ^{237}Np. Enough $CaCO_3$ was added to the soil to neutralize the HNO_3. The amended soil was stored for 24 hr and then thoroughly mixed in a V-blender before it was transferred to the containers. Two different amounts, 1.0 and 0.1 mCi/3.4 kg soil, of ^{238}Pu, ^{239}Pu, ^{241}Am, and ^{244}Cm were added to the soil. Neptunium-237 was added to the soil only at a concentration of 0.1 mCi/3.4 kg soil.

Control containers containing only uncontaminated soil were also prepared so that the levels of contamination in the treatment vegetation attributable to external deposition or root uptake of radionuclides present in the soil from fallout or other sources could be determined.

Cheatgrass (*Bromus tectorum* L.), an annual grass, was planted in some of the containers. The only water the cheatgrass containers received came from natural precipitation, which averages 16 cm/yr for the study site (Hinds and Thorp, 1971). Peas (*Pisum sativum,* var. Blue Bonnet), barley (*Hordeum vulgare,* var. U. Cal. Briggs), and alfalfa (*Medicago sativa,* var. Ranger) were planted in the spring. The crop plant containers were irrigated and weighed so that the soil moisture content was maintained at about 20% by weight throughout the growing season. The peas, barley, and alfalfa plants were fertilized with NH_4NO_3, at the rate of 300 kg/ha, approximately halfway through the growing seasons. Super phosphate (P_2O_5) was also added to the alfalfa containers at a rate of about 250 kg/ha.

Aboveground plant parts were hand harvested at maturity and divided into selected components. The entire cheatgrass plant, separated from other plant species that had invaded the containers, was analyzed. Barley seeds were analyzed separately from the rest of the plant. Peas, harvested at the dry-seed stage, were divided into seeds, leaves, and stem and pod fragments for analysis. The entire alfalfa plant (three separate harvests) was analyzed. Radiochemical analyses (Major et al., 1973; Wessman et al., 1978) of the plant materials were conducted by the LFE Environmental Analysis Laboratory, Richmond, Calif.

Radiochemical analysis of the control plants was used to determine the net uptake of transuranics by the plants in the amended soil. The radionuclide concentration observed in the treatment plants was corrected by subtracting the corresponding values of the control plant parts.

Results

Results of the radiochemical analyses are summarized in Tables 1 and 2. The data are presented as a ratio of the concentration of the radionuclide in the vegetative part to the total amount of that radionuclide added to the 3.4 kg of soil. The ratio can be used to compare the relative uptake values of the five different radionuclides into the various parts of the four plant species. Since the contaminated soil in this study was covered with a 10-cm layer of clean soil, these values are not to be regarded as concentration ratios (CR). Concentration ratio values are normally calculated by dividing the concentration of the vegetation (activity per unit dry weight) by the concentration of the top 10 to 20 cm of soil (also in activity per unit dry weight).

TABLE 1   Relative Uptake of Transuranium Elements by Cheatgrass and Alfalfa

| | Millicuries per container | $\dfrac{\text{Average (pCi/g dry vegetation)}}{\text{mCi/container}}$ ($\pm$ standard error)* | | | |
| | | | Alfalfa | | |
Isotope		Cheatgrass	1st harvest	2nd harvest	3rd harvest
^{238}Pu	0.108	1.3 ± 0.3	30 ± 7	13 ± 3†	58 ± 42‡
^{238}Pu	1.06	2.7 ± 0.6	16 ± 2‡	22 ± 6‡	34 ± 13‡
^{239}Pu	0.103	12 ± 3‡	80 ± 21	17 ± 5	45 ± 20
^{239}Pu	0.989	5.0 ± 0.8‡	24 ± 7	14 ± 2‡	16 ± 2‡
^{241}Am	0.0946	$(1.4 \pm 0.2) \times 10^2$	$(60 \pm 14) \times 10^2$	$(1.7 \pm 0.2) \times 10^2$	$(2.7 \pm 0.4) \times 10^2$
^{241}Am	0.983	$(0.48 \pm 0.11) \times 10^2$	$(7.0 \pm 1.1) \times 10^2$ ‡	$(1.7 \pm 0.5) \times 10^2$	$(3.5 \pm 0.5) \times 10^2$
^{244}Cm	0.103	$(0.65 \pm 0.25) \times 10^2$ ‡	$(7.8 \pm 1.9) \times 10^2$ ‡	$(2.3 \pm 0.5) \times 10^2$	$(3.8 \pm 0.6) \times 10^2$
^{244}Cm	1.04	$(0.80 \pm 0.17) \times 10^2$	$(3.5 \pm 0.8) \times 10^2$ ‡	$(1.5 \pm 0.3) \times 10^2$	$(2.5 \pm 0.6) \times 10^2$
^{237}Np	0.101	$(1.1 \pm 0.1) \times 10^4$ †	$(21 \pm 7) \times 10^4$	$(5.5 \pm 1.3) \times 10^4$	$(4.7 \pm 1.2) \times 10^4$
Grams of dry tissue per container		3.9 ± 0.3	7.3 ± 0.8	10.0 ± 0.3	6.3 ± 0.2

*n = 5, except as noted.
†n = 3.
‡n = 4.

TABLE 2 Relative Uptake of Transuranium Elements by Barley and Peas

| Isotope | Millicuries per container | Barley | | Peas | | |
| | | $\dfrac{\text{Average (pCi/g dry vegetation)}}{\text{mCi/container}}$ (± standard error)* | | | | |
		Seed	Stem and leaf	Seed	Leaf	Stem and pod
^{238}Pu	0.108	0.21 ± 0.12†	5.9 ± 0.6	0.24 ± 0.06	84 ± 14	24 ± 10
^{238}Pu	1.06	0.10 ± 0.06†	15 ± 10	0.064 ± 0.026	52 ± 14	15 ± 5
^{239}Pu	0.103	0.094 ± 0.013†	4.8 ± 2.0	0.0 ± 0.39	150 ± 56	23 ± 4
^{239}Pu	0.989	0.085 ± 0.017	3.8 ± 1.0	0.089 ± 0.045	100 ± 20	13 ± 4
^{241}Am	0.0946	0.35 ± 0.08	34 ± 6‡	1.8 ± 0.5	$(14 ± 3) \times 10^2$	$(1.3 ± 0.3) \times 10^2$
^{241}Am	0.983	0.16 ± 0.02	9.7 ± 1.0	2.8 ± 0.8	$(20 ± 4) \times 10^2$	$(2.1 ± 0.7) \times 10^2$
^{244}Cm	0.103	0.90 ± 0.28	39 ± 5	3.0 ± 1.1	$(19 ± 5) \times 10^2$	$(1.4 ± 0.4) \times 10^2$
^{244}Cm	1.04	0.16 ± 0.03	10 ± 2	4.7 ± 0.8	$(14 ± 1) \times 10^2$	$(1.3 ± 0.1) \times 10^2$
^{237}Np	0.101	$(0.43 ± 0.17) \times 10^4$	$(2.7 ± 0.3) \times 10^4$	$(0.37 ± 0.04) \times 10^4$	$(35 ± 11) \times 10^4$	$(18 ± 6) \times 10^4$
Grams of dry tissue per container		14.5 ± 0.5	24.6 ± 0.7	12.5 ± 0.7	3.1 ± 0.2	8.0 ± 0.3

*n = 5, except as noted.
†n = 4.
‡n = 3.

As illustrated in Fig. 1, the relative uptake of either plutonium isotope was not statistically different ($\alpha = 0.01$) for the two soil concentrations used (approximately 0.03 and 0.3 μCi/g soil). Soil concentration did not appear to have any effect on the uptake of ^{241}Am or ^{244}Cm, which were also added at the same two levels. Also, as shown in Fig. 1, the relative uptake of ^{238}Pu was not statistically different from ^{239}Pu ($\alpha = 0.01$).

Since soil concentration did not appear to affect phytoavailability or differences in ^{238}Pu vs. ^{239}Pu, the relative plant-uptake data were combined for each element so that further statistical comparisons could be made. Most interesting was a comparison of the uptake values for the four different elements (Fig. 2). The relative uptake of ^{237}Np ranged from 2,200 to 45,000 times as great as that of plutonium, depending on the plant part compared. Neptunium-237 was accumulated 35,000 and 45,000 times as great as plutonium in barley and pea seeds, respectively, and averaged a factor of 4,700 more in the remaining plant tissues. There was no significant difference ($\alpha = 0.01$) in the uptakes of ^{241}Am and ^{244}Cm, which were both 10 to 20 times as great as plutonium.

The relative uptakes of the transuranics by the four different plant species were noticeably different. Generally, the legumes (peas and alfalfa) accumulated approximately 10 times as much as the grasses (cheatgrass and barley). Concentrations of transuranics in various plant tissues examined were also different. The values were considerably lower in the seeds than in other aboveground plant parts. The concentrations in barley seed were lower by a factor of 30 to 50 than those in the entire combined plant parts for plutonium, americium, and curium and about a factor of 5 lower for neptunium. For peas the ratio of concentration in the seeds compared to the rest of the plant was 230, 150, 70, and 30 for plutonium, americium, curium, and neptunium, respectively.

Discussion

Our results showed soil concentration to have no observable effect on the uptake of either plutonium isotope. This differs somewhat from results reported by Wildung and Garland (1974). They observed an increase in the relative uptake of plutonium as the soil concentration decreased. However, they noted little effect on uptake at soil concentrations of 0.5 μCi/g or less, which exceeded the maximum level used in this study (0.3 μCi/g).

Plutonium-238 has been reported to be more available than ^{239}Pu in a grassland ecosystem (Little, 1976) and in the southeastern United States (McLendon et al., 1976). As noted previously, this difference was not observed in this plant uptake study.

The results of this study showed that the relative plant uptake of the four different transuranium elements was Np > Cm $\simeq$ Am > Pu. These trends are consistent with data reported by Price (1972) on the uptake of ^{239}Pu, ^{241}Am, ^{244}Cm, and ^{237}Np by cheatgrass and tumbleweeds *(Salsola kali)*. The same differences in the relative uptake of ^{241}Am and ^{239}Pu have been reported by Cline (1968) and Schulz et al. (1976). A significant aspect of this trend is that, if the CR of plutonium is approximately 0.0001 as summarized by Price (1973) and ^{237}Np was taken up into the entire plant some 3900 times as great as plutonium, one can infer that the CR value for neptunium would be about 0.4; i.e., under usual agronomic conditions, the concentration of vegetation growing on soil contaminated with ^{237}Np in the upper 15 to 20 cm would be equal to approximately one-half the soil concentration on a dry-weight basis. Likewise, the CR values for ^{241}Am and ^{244}Cm would be expected to be about 0.002 since the relative

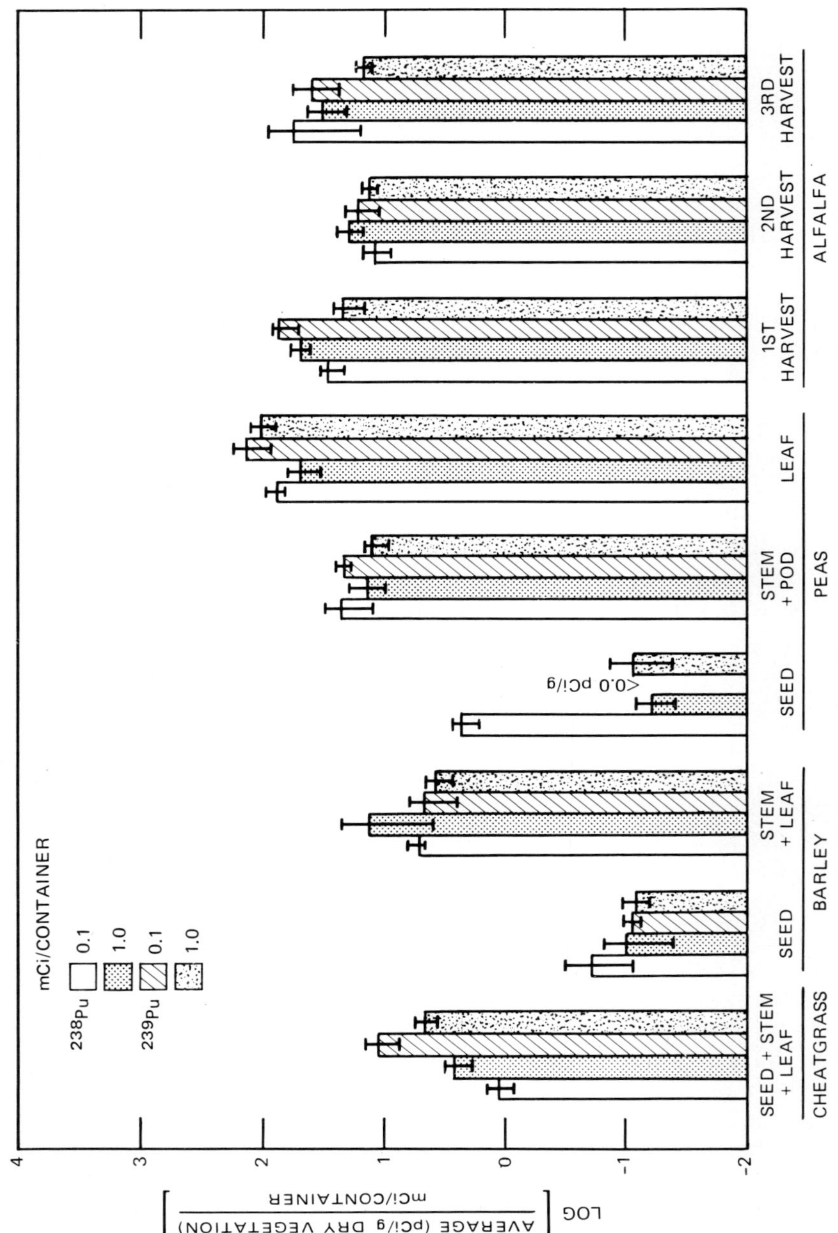

Fig. 1   Relative uptake of ^{238}Pu and ^{239}Pu by plants from soil containing two different concentration levels [0.1 and 1.0 mCi/3400 g (dry soil)]. Bars represent the log transformation of the standard errors.

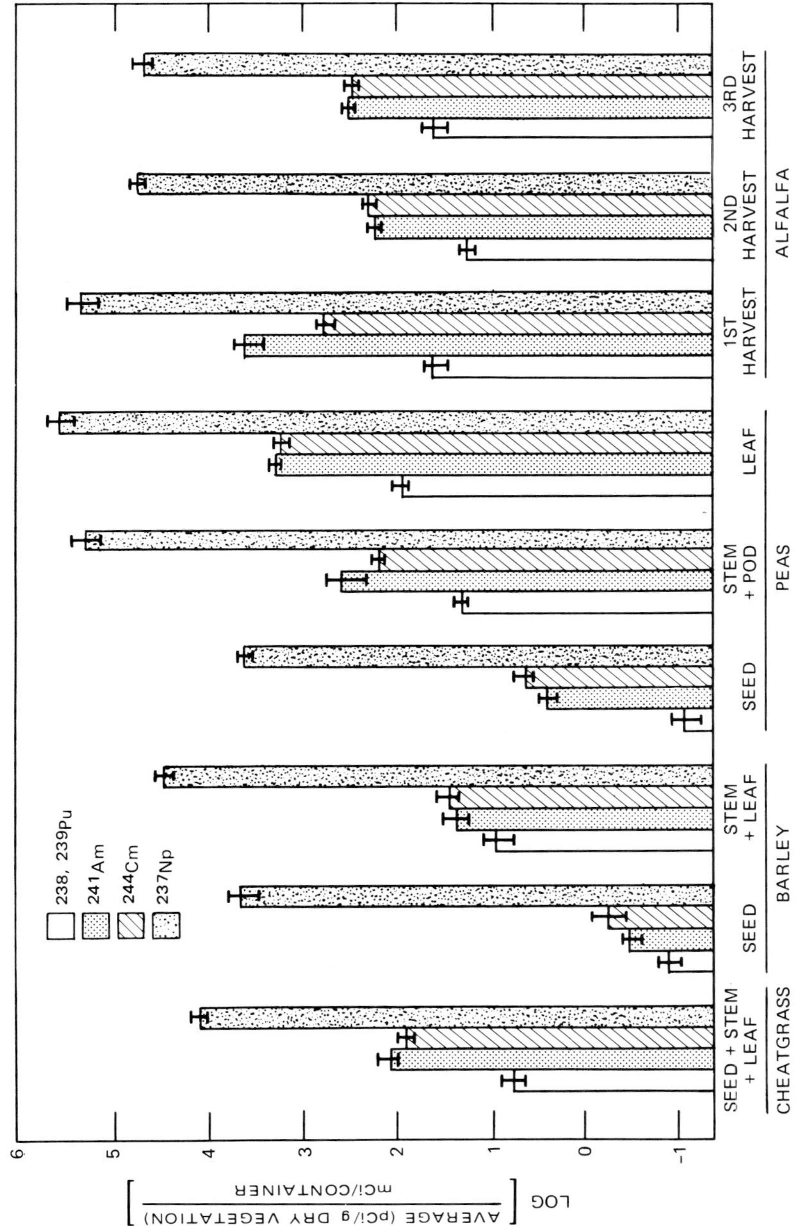

Fig. 2 Relative uptake of plutonium, ^{241}Am, ^{244}Cm, and ^{237}Np into various parts of cheatgrass, barley, peas, and alfalfa plants. Bars represent the log transformation of the standard errors.

uptake of these two transuranics was approximately a factor of 16 times that of plutonium.

The significance of these CR values can be related to transuranics associated with plutonium breeder reactor fuels. Thomas and Healy (1976) reviewed the uptake of neptunium and transplutonium elements by plants. They calculated the relative abundance of a number of long-lived transuranics relative to ^{239}Pu from data reported earlier by Bell (1970) on nuclide levels in spent Liquid Metal Fast Breeder Reactor (LMFBR) fuels. As shown in Table 3, the concentration of ^{241}Am, ^{242}Cm, ^{237}Np, and ^{239}Np in vegetation would be higher than that of ^{239}Pu for various time periods following environmental releases of spent LMFBR fuels. A difference is noted in the relative plant uptake values summarized by Thomas and Healy for americium and neptunium when compared with the values reported in this study. The discrepancy in the americium values is probably due to the fact that they included uptake data in which the soil had been amended with a chelating agent, such as diethylenetriaminepentaacetic acid, which significantly increases plant uptake (Hale and Wallace, 1970; Wallace, 1972). Their relative neptunium uptake value was taken from Price (1972) and was approximately one-tenth of the results presented here. Price noted some toxicity symptoms in the neptunium-contaminated seedlings which may have caused a reduction in neptunium uptake.

Another interesting finding in this study was the low concentrations of transuranics in pea and barley seeds when compared with those of the entire plant. This is important because in many dose-assessment models the CR values used are often calculated from the entire aboveground plant parts. As shown in this study, the levels of plutonium, americium, and curium in barley seeds were lower by a factor of 30 to 50 of those in the entire plant. For neptunium the seeds were one-fifth of the entire plant values. Pea seeds were lower by a factor of 70 to 230 for plutonium, americium, and curium and a factor of 30 lower for neptunium. Differences in plant-part concentrations must be considered when CR values are used in dose-assessment models and can also describe some of the discrepancy in CR values reported in the literature.

Acknowledgments

We thank W. T: Hinds for his technical guidance during the initial stages of this study and H. A. Sweany, M. J. Harris, L. F. Nelson, M. A. Combs, and V. D. Charles for their technical assistance throughout the study. This research was funded by the U.S. Department of Energy, Office of Health and Environmental Research, under contract EY-76-C-06-1830.

References

Bell, M. J., 1970, *Heavy Element Composition of Spent Power Reactor Fuels,* USAEC Report ORNL-TM-2897, Oak Ridge National Laboratory, NTIS.

Cline, J. F., 1968, Uptake of ^{241}Am and ^{239}Pu by Plants, in *Pacific Northwest Laboratory Annual Report for 1967 to the USAEC Division of Biology and Medicine,* USAEC Report BNWL-714, pp. 8.24-8.25, Battelle, Pacific Northwest Laboratories, NTIS.

Energy Research and Development Administration, 1975, *Workshop on Environmental Research for Transuranium Elements,* Proceedings of the Workshop, Battelle Seattle Research Center, Seattle, Wash., Nov. 12–14, 1975, ERDA Report ERDA-76/134, NTIS.

Hale, V. Q., and A. Wallace, 1970, Effect of Chelates on Uptake of Some Heavy Metal Radionuclides from Soil by Bush Beans, *Soil Sci.,* 109: 262-263.

TABLE 3 Relative Importance of Americium, Curium, and Neptunium in Reference to ^{239}Pu for Relating the Uptake of Spent LMFBR Fuels by Plants for 3 and 1000 Yr Following Irradiation

	Data summarized by Thomas and Healy (1976)					Results from this study		
	Abundance in spent LMFBR fuels relative to ^{239}Pu		Plant uptake relative to ^{239}Pu	Importance relative to ^{239}Pu		Plant uptake relative to ^{239}Pu	Importance relative to ^{239}Pu	
Nuclide	3 yr	1000 yr		3 yr	1000 yr		3 yr	1000 yr
^{241}Am	1.5	1.6	800	1200	1300	20	30	33
^{242}Cm	0.28	0.00034	15	4.2	0.0051	20	5.6	0.0068
^{244}Cm	0.40	0	15	6.1	0	20	8.1	0
^{237}Np	1.9×10^{-5}	1.2×10^{-3}	730	0.014	0.89	4000	0.08	4.8
^{239}Np	1.8×10^{-2}	1.7×10^{-2}	730	13	12	4000	72	68

Hinds, W. T., 1975, Energy and Carbon Balances in Cheatgrass: An Essay in Autecology, *Ecol. Monogr.*, 45: 367-388 (1975).

——, and J. M. Thorp, 1971, *Annual Summaries of Microclimatological Data from the Arid Lands Ecology Reserve: 1968 Through 1970,* USAEC Report BNWL-1629, Battelle, Pacific Northwest Laboratories, NTIS.

——, J. F. Cline, H. A. Sweany, and V. D. Charles, 1976, Weather and Aging Effects on Uptake of Transuranics by Plants, in *Pacific Northwest Laboratory Annual Report for 1975 to the USERDA Division of Ecological Science,* ERDA Report BNWL-2000 (Pt.2), pp. 176-177, Battelle, Pacific Northwest Laboratories, NTIS.

Little, C. A., 1976, *Plutonium in a Grassland Ecosystem,* Ph.D. Thesis, Colorado State University, Fort Collins, Colorado.

Major, W. J., K. D. Lee, R. A. Wessman, and R. Melgard, 1973, Determination of Plutonium-239 and Americium-241 in Large NAEG Vegetation Samples, in *Nevada Applied Ecology Group Procedures Handbook for Environmental Transuranics,* M. G. White and P. B. Dunaway (Eds.), pp. 271-282, ERDA Report NVO-166, Nevada Operations Office, NTIS.

McLendon, H. R., O. M. Stewart, A. L. Boni, J. C. Corey, K. W. McLeod, and J. E. Pinder, 1976, Relationships Among Plutonium Contents of Soil, Vegetation and Animals Collected on and Adjacent to an Integrated Nuclear Complex in the Humid Southeastern United States, in *Transuranium Nuclides in the Environment,* Symposium Proceedings, San Francisco, 1975, pp. 347-363, STI/PUB/410, International Atomic Energy Agency, Vienna.

Price, K. R., 1972, *Uptake of ^{237}Np, ^{239}Pu, ^{241}Am and ^{244}Cm from Soil by Tumbleweed and Cheatgrass,* USAEC Report BNWL-1688, Battelle, Pacific Northwest Laboratories, NTIS.

——, 1973, A Review of Transuranic Elements in Soils, Plants, and Animals, *J. Environ. Qual.,* 2(1): 62-66.

Schulz, R. K., G. A. Tompkins, L. Lerenthal, and K. L. Babcock, 1976, Uptake of Plutonium and Americium by Barley from Two Contaminated Nevada Test Site Soils, *J. Environ. Qual.,* 5(4): 406-410.

Thomas, R. L., and J. W. Healy, 1976, *Appraisal of Available Information on Uptake by Plants of Transplutonium Elements and Neptunium,* ERDA Report LA-6460-MS, Los Alamos Scientific Laboratory, NTIS.

Wallace, A., 1972, Increased Uptake of ^{241}Am by Plants Caused by the Chelating Agent DTPA, *Health Phys.,* 22(6): 559-562.

Wessman, R. A., K. D. Lee, B. Curry, and L. Leventhal, 1978, Transuranium Analysis Methodologies for Biological and Environmental Samples, in *Environmental Chemistry and Cycling Processes,* DOE Symposium Series, Augusta, Ga., Apr. 28–May 1, 1976, D. C. Adriano and I. Lehr Brisbin, Jr. (Eds.), pp. 275-289, CONF-760429, NTIS.

Wildung, R. E., 1977, *Soils of the Pacific Northwest Shrub-Steppe, Occurrence and Properties of Soils on the Arid Lands Ecology Reserve, Hanford Reservation,* ERDA Report BNWL-2272, Battelle, Pacific Northwest Laboratories, NTIS.

——, and T. R. Garland, 1974, Influence of Soil Plutonium Concentration on Plutonium Uptake and Distribution in Shoots and Roots of Barley, *J. Agric. Food Chem.,* 22(5): 836-838.

Comparative Distribution of Plutonium in Contaminated Ecosystems at Oak Ridge, Tennessee, and Los Alamos, New Mexico

ROGER C. DAHLMAN, CHARLES T. GARTEN, JR., and THOMAS E. HAKONSON

The distribution of plutonium was compared in portions of forest ecosystems at Oak Ridge, Tenn., and Los Alamos, N. Mex., which were contaminated by liquid effluents. Inventories of plutonium in soil at the two sites were generally similar, but a larger fraction of the plutonium was associated with biota at Los Alamos than at Oak Ridge. Most (99.7 to 99.9%) of the plutonium was present in the soil, and very little (0.1 to 0.3%) was in biotic components. Comparative differences in distributions within the two ecosystems appeared to be related to individual contamination histories and greater physical transport of plutonium in soil to biotic surfaces at Los Alamos.

Currently most of the plutonium present in terrestrial ecosystems of the United States originates from nuclear weapons testing and from the reentry burnup of the SNAP-9A satellite power source (Hanson, 1975). In the future, however, local ecosystems will receive small quantities of plutonium released from nuclear fuel reprocessing and fabrication facilities. The purpose of this chapter is to compare and contrast the distribution of plutonium in two contaminated ecosystems that are representative of humid and semiarid environments of the United States. Current plutonium inventories for ecosystems several decades after initial contamination can help ecologists forecast the fate of plutonium in the environment. One important question is whether the availability of this element to plants and other organisms will change after it has been subjected to weathering and ecological processes of the environment. Potential radiological toxicity and long physical half-lives of plutonium dictate that its behavior in ecosystems be understood.

Although the ecosystems at Oak Ridge, Tenn., and Los Alamos, N. Mex., are dissimilar owing to differences in geology, climate, and ecology, there are certain similar features of these contaminated environments. A forested floodplain at Oak Ridge and a canyon at Los Alamos were contaminated by treated liquid waste effluents which have resulted in detectable levels of plutonium in most ecosystem components. However, the discharge and chemical characteristics of plutonium were not similar at Oak Ridge and Los Alamos; therefore it is not possible to conduct a rigorous intercomparison of plutonium behavior in the different ecosystems. Although many environmental variables are uncontrolled, a comparative analysis of plutonium in both ecosystems provides insight on patterns of plutonium behavior in the environment.

Plutonium in Oak Ridge and Los Alamos Environments

Rates of plutonium releases to the Oak Ridge and Los Alamos environs during the study period and resulting concentrations in abiotic components are summarized in Table 1. The release of plutonium to the atmosphere at Oak Ridge was considerably less than that at Los Alamos, and ambient plutonium in air reflects the higher release rate at Los Alamos. However, these concentrations in air are not easily distinguished from plutonium in air originating from global fallout (Bennett, 1976).

The quantity of plutonium released to surface water is similar at both sites, but the source of plutonium entering the drainage systems is different. At Oak Ridge release from routine operations is negligible; most of the plutonium in surface water originates from contaminated locations in the White Oak Drainage (WOD). At Los Alamos routine release of plutonium in treated liquid waste accounts for nearly all the plutonium entering Mortandad Canyon (Hakonson, Johnson, and Purtymun, 1973).

TABLE 1 Plutonium in the Oak Ridge and Los Alamos Environments*†

	Oak Ridge			Los Alamos		
Component	Floodplain	Plant area	Plant perimeter	Mortandad Canyon	Laboratory area	Off site
Release, mCi/yr						
Atmosphere	NA‡	0.004	NA‡	NA‡	10	NA‡
Surface water	NM§	NM§	20¶	8.8	9	NM§
Air concentration, fCi/m^3	0.11**	0.02	0.01	NM§	0.05	0.05
Soil concentration, pCi/g						
0 to 1 cm	NM§	0.04	0.04	NM§	NM§	NM§
0 to 20 cm	10 to 150	NM§	NM§	9 to 250	0.05	0.02
Surface-water concentration, pCi/liter	ND‡‡	NM§	0.2††	7.7	NM§	0.12

*Total plutonium includes the 238, 239, and 240 nuclides. Prior to 1968 the plutonium released to Mortandad Canyon was ^{239,240}Pu. Since that time ^{238}Pu has been the dominant isotope. The present ^{238}Pu/^{239,240}Pu activity ratio is 3 : 1. The predominant isotope released to the Oak Ridge floodplain is ^{239,240}Pu.

†Data for Oak Ridge obtained from Union Carbide Corporation, Nuclear Division (1976), Oakes and Shank (1977), Dahlman and McLeod (1977), and Bondietti and Sweeton (1977); for Los Alamos from Herceg (1973), and Schiager and Apt (1974). The data matrix is obviously incomplete in terms of measurements for certain components and in terms of estimates of errors. The values presented represent single measurements, averages of a few measurements (accompanied by high variances), and summations or products of several measurements. Accordingly, the inclusion of error estimates was not considered practical. The reader is referred to original data sources where additional information on variability is provided.

‡NA, not applicable.

§NM, not measured.

¶Represents total release from three plant areas to three different surface streams; value is based on total alpha analysis but excludes uranium and thorium and includes other transuranium elements.

**Not detectable in 350-m^3 air samples where the minimum detectable level is 40 fCi per sample; air sampled in the 0- to 10-cm zone contained 0.14 fCi/m^3 under ambient conditions; air samples at 1-m height during soil cultivation contained 26.3 fCi/m^3.

††Water collected from White Oak Lake at the point of discharge, which is approximately 2 km below the floodplain site.

‡‡ND, not determined.

Plutonium in liquid effluents released from Oak Ridge National Laboratory (ORNL) during the Manhattan Project contaminated environments in the White Oak Drainage Basin. The Oak Ridge site, approximately 0.5 km downstream from ORNL, received effluents containing plutonium when it served, for 6 months in 1944, as a temporary settling basin for radioactive wastes. The impoundment drained in late 1944, and since then a forest has developed on the floodplain.

Mortandad Canyon has received liquid waste since 1963, and from 1972 to 1973 the release rate was approximately 9 mCi ^{238}Pu/yr. Soil has become contaminated with both ^{238}Pu and ^{239}Pu; concentrations range from 250 pCi/g at the waste outfall to 9 pCi/g about 2 km down the canyon. Surface water from the outfall completely infiltrates into the alluvium within 1.2 km of the effluent outfall. Downstream transport of plutonium into dry portions of the streambed occurs only during storm runoff events (Hakonson, Nyhan, and Purtymun, 1976).

Chemical and isotopic characteristics of plutonium released from Oak Ridge and Los Alamos are also different. The Oak Ridge method of treating radioactive liquid waste in 1944 involved coprecipitation with carbonate to remove radionuclides, primarily ^{90}Sr. At Los Alamos plutonium may be associated with laundry and laboratory chelating agents [e.g., nitrilotri (methylene phosphoric) acid—ATMP; 1-hydroxyethylidene 1,1 diphosphoric acid—HEDP]. The environmental stability of these complexes is unknown, but plutonium is presently associated with the soil-sediment component of the canyon.

Soil and Biotic Characteristics of the Oak Ridge Floodplain

The floodplain is representative of bottomlands of the East Tennessee valley. The soil profile is azonal because of periodic erosion and deposition of sediments related to flooding. An accumulation of humus is evident from the dark-brown appearance of soil in the 0- to 3-cm zone. Soil texture is a loamy clay (72% silt and 24% clay). The soil reaction is neutral to slightly alkaline (pH = 7.1 to 7.6), which is atypical of regional forest soils. The slightly alkaline condition is attributed to the alkaline coprecipitation of wastes during the Manhattan Project.

The forest ecosystem of the floodplain occupies a 3-ha area. The present successional stage of the forest is dominated by sycamore (*Platanus occidentalis* L.) and white ash (*Fraxinus americana* L.). Ground vegetation is chiefly wild rye grass (*Elymus virginicus* L.), *Microstegium vimineum* [(Trinus) A. Canus], jewelweed *(Impatiens capensis Meerb.)*, and Japanese honeysuckle (*Lonicera japonica* Thunberg). The resident small-mammal population includes the white-footed mouse *(Peromyscus leucopus)*, the rice rat *(Oryzomys palustris)*, and the short-tailed shrew *(Blarina brevicauda)*. Earthworms *(Lumbricus rubellus)* and crayfish (*Cambarus* sp.) are important soil invertebrates.

Biomass [grams (dry weight)] of the major compartments of the floodplain ecosystem is given in Table 2. Biomass for arborescent species was estimated from mensuration data (Van Voris and Dahlman, 1976) and from regression equations (Harris, Goldstein, and Henderson, 1973). The arborescent component (leaf, root, and wood) contains 95% of the total forest biomass, whereas animals comprise only 0.02%. These estimated biomass standing crops compare favorably with average values compiled for different forest stands of the eastern deciduous forest biome (EDFB) (Burgess and O'Neill, 1975) with the exception of litter (550 g/m), which was considerably less than the average for EDFB ($\sim$2000 g/m^2). The low estimate for litter was probably due to (1) the young age of the floodplain forest (approximately 30 yr); (2) the effect of periodic

**TABLE 2 Estimates of Mass of the Major
Components of the Oak Ridge Floodplain and the
Los Alamos Mortandad Canyon**

Component	Oak Ridge floodplain (g/m²)	Los Alamos Mortandad Canyon (g/m²)
Soil*	2.6×10^5	2.4×10^5
Tree wood	10,500†	15,000‡
Tree root	3,000†	3,000‡
Tree leaf	400†	600‡
Litter	550§	1,000
Ground vegetation	110§	Not determined
Grass	Not determined	8
Forb	Not determined	25
Soil fauna	3§	Not determined
Consumer	0.03§	0.07

*Mass is based on a profile depth of 20 cm and on densities of 1.3 and 1.2 g/cm³ for Oak Ridge and Los Alamos, respectively.

†Estimated from mensuration data and regression equations (Harris, Goldstein, and Henderson, 1973).

‡Estimated from mensuration data and regression equations (Wheeler, Smith, and Gallegos, 1977).

§Estimated from field measurements of populations and biomass.

floods on litter accumulation; and (3) moist-mesic conditions, which favor rapid decomposition.

Soil and Biotic Characteristics of Mortandad Canyon

The canyon soil consists of an alluvial deposit (<30 cm) derived from volcanic tuff. The coarse soil is less than 3% by weight silt and clay (Nyhan, Miera, and Peters, 1976a). Cation-exchange capacity is low (2 to 20 meq per 100 g) (Schiager and Apt, 1974). Organic matter ranges from 0.1 to 0.2%. Calcium as high as 3.7% and soil pH up to 9.2 are measurable (Schiager and Apt, 1974), which reflects the carbonate contribution from liquid waste. Uncontaminated soil from the canyon floor has a pH of 5.7.

The dominant arborescent species of the canyon are ponderosa pine *(Pinus ponderosa)*, Douglas fir *(Pseudotsuga menzezii)*, and Gambels oak *(Quercus gambelii)* (Miera et al., 1977). Dominant forb and grass species are wheat grass (*Elymus* sp.), bluegrass *(Poa pratensis)*, wild strawberry *(Fragaria americana)*, and dandelion *(Taraxacum officinale)*. The most common mammal residents are pinon mouse *(Peromyscus truei)*, deer mouse *(P. maniculatus)*, and the least chipmunk *(Eutamias minimus)*.

Standing-crop biomass [grams (dry weight)] for arborescent, herbaceous, and animal components of the floodplain and canyon ecosystems is shown in Table 2. Arborescent species contribute more than 80% of the total mass at the site.

Characteristics of Plutonium in the Floodplain Soil

Concentration of plutonium in the floodplain soil ranges from about 10 to 150 pCi/g over a 3-ha area (Fig. 1). The highest concentrations were found behind the former dike, along

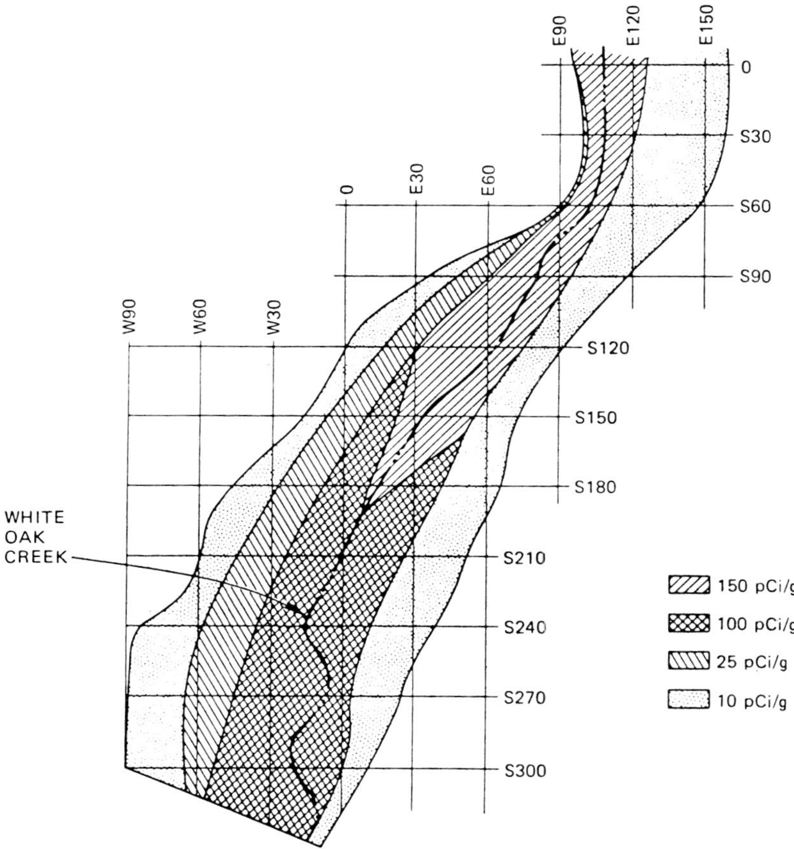

Fig. 1   Approximation of areal distribution of plutonium at the contaminated Oak Ridge floodplain. Grid size is 30 by 30 m for an area of 900 m^2. Distribution is generally estimated from 0- to 10-cm soil samples plus a few samples from the 10- to 20-cm depth.

White Oak Creek, and in the upper part of the floodplain. The maximum concentration of plutonium occurred where the creek is believed to have entered the historic impoundment between coordinates E60-S120 and E120-S30 (Fig. 1). Sediment-borne plutonium probably settled from the water column as the stream velocity decreased on entering the impoundment. Downstream from the site of initial deposition the higher plutonium concentrations tended to follow the watercourse of White Oak Creek (WOC).

Plutonium is not distributed uniformly between the loam and clay fractions of the soil. The loam fraction (>0.002 mm) contains 60% of the plutonium; whereas 24% of the <0.002-mm size class contains 40% of the plutonium. The high affinity of plutonium for colloids may be responsible for plutonium enrichment in clay. The distribution coefficient (K_d), determined by desorption in 0.01M NaHCO$_3$, is 5 × 10^3 (Bondietti and Tamura, this volume).

The highest concentration of plutonium is observed in the 0- to 10-cm zone of the soil profile. Occasionally a higher concentration of plutonium occurs below 10 cm. This atypical distribution is attributed to deposition of sediments over the initial plutonium deposit rather than to movement of the element by biogeochemical processes.

Soil Plutonium Characteristics in the Canyon Ecosystem

Concentrations of plutonium in alluvial soil from the canyon are strongly a function of distance from the waste-effluent outfall (Hakonson and Bostick, 1975; Nyhan, Miera, and Peters, 1976b). Highest concentrations, averaging about 250 pCi/g, occur near the outfall, whereas successively lower concentrations are measured with increasing distance downstream. Downstream transport of waste has occurred to about 3500 m below the outfall where fallout levels of plutonium are measured. Stream-bank soils to 1 m either side of the channel are contaminated to the same level as adjacent channel soils.

Plutonium is rather uniformly distributed within the canyon soil profile to depths of 30 cm, which reflects the effect of turbulent mixing processes during storm runoff events. The mixing of plutonium with depth has been rapid ($<$years) as inferred from the distribution of ^{238}Pu with depth (Hakonson and Bostick, 1975).

Over 85% of the plutonium in canyon alluvium is associated with sand particles of from 0.05 to 23 mm in diameter. About 14% of the plutonium inventory is present in the silt–clay fraction ($<$0.05 mm), which comprises only 2% of the soil mass (Nyhan, Miera, and Peters, 1976a). Significant correlations with solid-phase constituents (i.e., carbonates and the colloidal exchange complex) suggest that plutonium is sorbed to particles and minerals.

Concentration Ratios

The availability of soil plutonium to biota can be inferred from the concentration ratios (CR's) in Table 3. The low CR's for biota at both sites demonstrate the small fraction of plutonium that has moved from soil to biota. Vegetative components (root, litter, and herbaceous species) most intimately associated with the soil exhibited the highest CR's at both sites. Arborescent components generally exhibited the lowest CR's. The fact that all CR's are less than 1 confirms that there is no evidence of biomagnification of plutonium in these terrestrial ecosystems two to three decades after the environments were contaminated.

TABLE 3   **Concentration and Inventory of Plutonium in Major Components of the Oak Ridge Floodplain and the Los Alamos Mortandad Canyon**

Component	Oak Ridge*				Los Alamos†			
	pCi Pu/g	pCi Pu/m^2	CR‡	IR§	pCi Pu/g¶	pCi Pu/m^2	CR‡	IR§
Soil	63	1.6×10^7		0.999	51	1.2×10^7		0.997
Tree wood	0.003	32	5×10^{-5}	2×10^{-6}	0.05	7.5×10^2	1×10^{-3}	6×10^{-5}
Tree root	3.6	1.1×10^4	6×10^{-2}	7×10^{-4}				
Tree leaf	0.003	1.2	5×10^{-5}	8×10^{-8}	0.05	30	1×10^{-3}	3×10^{-6}
Litter	6.0	3.3×10^3	1×10^{-1}	2×10^{-4}	32	3.2×10^4	6×10^{-1}	3×10^{-3}
Soil fauna	1.0	3.0	2×10^{-2}	2×10^{-7}				
Ground vegetation	0.05	5.5	8×10^{-4}	3×10^{-7}				
Grass					60	4.8×10^2	1×10^0	4×10^{-5}
Forb					4.0	1×10^2	8×10^{-2}	8×10^{-6}
Consumer	0.04	0.001	6×10^{-4}	6×10^{-11}	0.27	0.02	5×10^{-3}	2×10^{-9}

*Concentrations and inventories averaged for the 3-ha floodplain.

†Concentrations and inventories based on the 0- to 1500-m segment of Mortandad Canyon.

‡CR = [Pu]$_{component}$/[Pu]$_{soil}$.

§IR = [Pu]$_{receptor}$/[Pu]$_{source}$. The IR's for all components except soil are based on soil as the plutonium source. The soil IR represents the fraction of total plutonium of the ecosystem that is present in soil.

¶Weighted average for intensive study sites I and II.

Plutonium Inventories in the Floodplain and Canyon Ecosystems

Total plutonium present in different ecosystem components is referred to as a static inventory or budget. For each ecosystem the inventory was calculated by multiplying the mass values of Table 2 and the plutonium concentrations (picocuries per gram) in the respective components (Table 3). Results on areal distribution of soil plutonium (Fig. 1) were integrated to provide an estimate of the plutonium inventory for the entire 3-ha floodplain. Total soil plutonium inventory in the top 20 cm, based on the summation of subinventories of four different concentration zones, was 0.5 Ci (8 g). The 100- and 150-pCi/g zones contained 88% of the plutonium, but they occupied only 46% of the area of the 3-ha floodplain. Because only 50% of the soil determinations included both the topsoil (0 to 10 cm) and subsoil (10 to 20 cm) and because the zones of concentration indicated by the isopleths in Fig. 1 are interpolated between sampling stations, the soil inventory is provisional and does not represent high resolution of plutonium distribution in the floodplain.

The inventory of plutonium for the 0- to 20-cm soil depth in Mortandad Canyon was calculated for the 3 × 1500-m segment (intensive study sites I and II) of stream channel immediately below the effluent outfall. Nearly all the plutonium inventory is present in this segment of the canyon (Hakonson and Bostick, 1975). The estimated plutonium inventory was 0.054 Ci as of 1974, which corresponded closely to the estimated input of 0.051 Ci based on treatment-plant release records.

A summation of component inventories shows that, by far, the majority of the plutonium resides in soil. More than 30 yr after the initial deposit, less than 0.1% of the total plutonium is present in living components of the Oak Ridge floodplain. For Mortandad Canyon, the inventory of plutonium in biotic components of the canyon ecosystem was 0.00015 Ci in the 0- to 1500-m segment, or 0.28% of the total plutonium present in the canyon ecosystem.

Inventory Ratios for Plutonium in the Floodplain and Canyon Ecosystems

One approach used to understand the significance of relative distributions of plutonium in ecosystems is to relate the plutonium inventory of the receptor to the plutonium content of the donor or source component in terms of inventory ratios (IR's):

$$IR = \frac{\text{Activity/unit area in receptor}}{\text{Activity/unit area in source}}$$

Contaminated soil is the major source of plutonium for all biotic components of both the floodplain and canyon ecosystems; consequently soil has been used as the denominator to calculate IR's (Table 3). Components other than soil may contribute plutonium to certain other components, e.g., wood is a likely source of plutonium in tree leaves. The reader may choose data from Table 3 to derive other IR's not provided by our analysis.

The IR results (Table 3) demonstrate the importance of soils as the reservoir for environmental plutonium. Over 99% of the plutonium of both study sites was associated with soils. Biota serve as an incidental, although potentially important, receptor for plutonium in the environment, but only a small fraction of the total plutonium is present in the biotic components. Root and litter components of the floodplain had an IR of approximately 10^{-4} compared with a value of 10^{-3} for litter at Los Alamos. Inventory ratios for tree and aboveground vegetative components ranged from 10^{-6} to 10^{-7} on the floodplain. Inventory ratios for grass, forb, and tree components of the canyon ecosystem were about 10^{-5}. The higher IR's of the canyon biota reflect both higher concentrations of plutonium and greater standing crops of biomass.

Discussion

There are limitations in the use of CR and IR data to describe the distribution of plutonium among components of the ecosystem. There is the difficulty of distinguishing between contributions resulting from several simultaneous transfers to a receptor. Sometimes the relationships between sources and receptors are not clear. Another limitation is that IR's and CR's are static indexes of plutonium distribution in the environment. Both indexes are based on an assumption of equilibrium conditions for the ecosystem; i.e., that biomass and concentrations are constant. The equilibrium assumption is not strictly met in terrestrial communities that are dynamic. In spite of these limitations, both CR and IR values can give order-of-magnitude estimates on the availability of plutonium in the environment.

More than 30 yr after deposition in floodplain sediments, plutonium in the Oak Ridge soil appears to be a monomeric species associated with endemic organic and mineral constituents (Bondietti, Reynolds, and Shanks, 1976). There have been no recent amendments of plutonium to the floodplain ecosystem. In contrast, there are annual additions of plutonium in the waste effluents released to Mortandad Canyon. These effluents contain diverse industrial wastes, including chelating agents; consequently plutonium in the canyon environment can be complexed by chelators. Enhanced mobility of plutonium would be expected in the canyon ecosystem if a chelated form of plutonium were a stable and dominant species in the environment. This factor could increase the biochemical assimilation of plutonium by plants and animals; thus, along with surface contamination, it may be partly responsible for the higher CR's and IR's observed for the canyon ecosystem.

Plutonium uptake by the root pathway yields a plant : soil concentration ratio of about 10^{-3} for floodplain species (Table 3). This ratio is about one to two orders of magnitude greater than CR's determined from short-term experiments when plants are grown in soil contaminated with plutonium solutions (Dahlman, Bondietti, and Eyman, 1976). Root uptake is the main mechanism of plutonium incorporation by plants in the floodplain ecosystem because the negligible contribution of plutonium to the Oak Ridge atmosphere from resuspension and industrial release would create minimal contamination of vegetative surfaces. As mentioned previously, external contamination of vegetation is considered an important mechanism of uptake at the canyon.

The fraction of physiologically available plutonium is largely determined by environmental chemistry and reactions of plutonium with soil. The great affinity of plutonium for soil particles results in distribution coefficients of the order of 10^3 to 10^5. Sorption of plutonium to colloids is a surface reaction that occurs predominantly with the clay constituents of soil because this component possesses the greatest specific surface area. Enrichment of plutonium in the clay fraction has been observed in several contaminated environments that contain appreciable clay. The differences in percentage clay in canyon and floodplain soil (2% vs. 24%, respectively) may be responsible for diminished sorption of plutonium to canyon soil, and this could account for the increased incorporation of plutonium into biotic components of the canyon ecosystem. Although plutonium enrichment in soil clay apparently occurs at both sites, and, although incorporation of plutonium by biota of the canyon appears inversely related to clay content, the minimum quantity of clay in soil that is required to sorb plutonium and restrict its movement to biota is unknown.

Comparative studies of the biogeochemical behavior of plutonium in ecosystems can facilitate the application of plutonium data to assessments of future environmental impact. Relative distributions and concentrations of plutonium in components of two different ecosystems confirm that the element is not readily incorporated by biota after it has been in the terrestrial environment for 20 to 30 yr. However, in the absence of data over many decades, it is difficult to forecast with certainty the future biological availability of this element. Yet the small inventory in biota and the absence of any evidence of biomagnification indicate limited environmental mobility of the element.

Currently available indexes of mobility in forest ecosystems 20 to 30 yr after initial contamination suggest that the properties of plutonium have not been modified in a way that would affect its long-term biogeochemical behavior.

Acknowledgments

Research for the ORNL study was sponsored by the U. S. Department of Energy under contract with Union Carbide Corporation, Publication No. 1347. The LASL study was funded under contract No. W-7405-Eng.36 between the U. S. Department of Energy and LASL.

References

Bennett, B. G., 1976, Transuranic Element Pathways to Man, in *Transuranium Elements in the Environment,* Symposium Proceedings, San Francisco, 1975, pp. 367-383, STI/PUB/410, International Atomic Energy Agency, Vienna.

Bondietti, E. A., and F. H. Sweeton, 1977, Transuranic Speciation in the Environment, in *Transuranics in Natural Environments,* Symposium Proceedings, Gatlinburg, Tenn., October 1976, M. G. White and P. B. Dunaway (Eds.), ERDA Report NVO-178, pp. 449-476, Nevada Operations Office, NTIS.

———, S. A. Reynolds, and M. A. Shanks, 1976, Interaction of Plutonium with Complexing Substances in Soils and Natural Waters, in *Transuranium Elements in the Environment,* Symposium Proceedings, San Francisco, 1975, pp. 273-287, STI/PUB/410, International Atomic Energy Agency, Vienna.

Burgess, R. L., and R. V. O'Neill (Eds.), 1975, Eastern Deciduous Forest *Biome Progress Report, Sept. 1—Aug. 31, 1974,* Oak Ridge National Laboratory, NTIS.

Dahlman, R. C., and K. W. McLeod, 1977, Foliar and Root Pathways of Plutonium Contamination of Vegetation, in *Transuranics in Natural Environments,* Symposium Proceedings, Gatlinburg, Tenn., October 1976, M. G. White and P. B. Dunaway (Eds.), ERDA Report NVO-178, pp. 303-320, Nevada Operations Office, NTIS.

———, E. A. Bondietti, and L. D. Eyman, 1976, Biological Pathways and Chemical Behavior of Plutonium and Other Activities in the Environment, in *Actinides in the Environment,* A. M. Friedman (Ed.), ACS Symposium Series, No. 35, pp. 47-80, American Chemical Society.

Hakonson, T. E., and K. V. Bostick, 1975, Cesium-137 and Plutonium in Liquid Waste Discharge Areas at Los Alamos, in *Radioecology and Energy Resources,* Proceedings of the Symposium on Radioecology, Oregon State Univ., May 13—14, 1975, The Ecological Society of America, Special Publications No. 1, C. E. Cushing, Jr. (Ed.), pp. 40-48, Halsted Press, New York.

———, L. J. Johnson, and W. D. Purtymun, 1973, The Distribution of Plutonium in Liquid Waste Disposal Areas at Los Alamos, in *Proceedings of the Third International Radiation Protection Association Meeting,* Washington, D. C., Sept. 9—14, 1973, W. S. Snyder (Ed.), USAEC Report CONF-730901, pp. 248-252, International Radiation Protection Association, NTIS.

——, J. W. Nyhan, and W. D. Purtymun, 1976, Accumulation and Transport of Soil Plutonium in Liquid Waste Discharge Areas at Los Alamos, in *Transuranium Nuclides in the Environment*, Symposium Proceedings, San Francisco, 1975, pp. 175-189, STI/PUB/410, International Atomic Energy Agency, Vienna.

——, J. W. Nyhan, L. J. Johnson, and K. O. Bostick, 1973, *Ecological Investigation of Radioactive Materials in Waste Discharge Areas at Los Alamos,* USAEC Report LA-5282-MS, Los Alamos Scientific Laboratory, NTIS.

Hanson, Wayne C., 1975, Ecological Considerations of the Behavior of Plutonium in the Environment, *Health Phys.,* 28: 529-537.

Harris, W. F., R. A. Goldstein, and G. S. Henderson, 1973, Analysis of Forest Biomass Pools, Annual Primary Production and Turnover of Biomass for a Mixed Deciduous Forest Watershed, in *IUFRO Biomass Studies, Mensuration, Growth and Yield*, pp. 41-64, H. Young (Ed.), Univ. of Maine Press, Orono.

Herceg, Joseph E. (Comp.), 1973, *Environmental Monitoring in the Vicinity of Los Alamos Scientific Laboratory, Calendar Year 1972,* USAEC Report LA-5184, Los Alamos Scientific Laboratory, NTIS.

Miera, F. R., Jr., K. V. Bostick, T. E. Hakonson, and J. W. Nyhan, 1977, *Biotic Survey of Los Alamos Radioactive Liquid-Effluent Receiving Areas,* ERDA Report LA-6503-MS, Los Alamos Scientific Laboratory, NTIS.

Nyhan, J. W., F. R. Miera, Jr., and R. J. Peters, 1976a, Distribution of Plutonium in Soil Particle Size Fractions of Liquid Effluent-Receiving Area at Los Alamos, *J. Environ. Qual.,* 5(1): 50-59.

——, F. R. Miera, Jr., and R. J. Peters, 1976b, The Distribution of Plutonium and Cesium in Alluvial Soils of the Los Alamos Environs, in *Radioecology and Energy Resources,* Proceedings of the Symposium on Radioecology, Oregon State Univ., May 13–14, 1976, The Ecological Society of America, Special Publications No. 1, C. E. Cushing, Jr. (Ed.), pp. 49-57, Halsted Press, New York.

Oakes, T. W., and K. E. Shank, 1977, *Surface Investigations of Energy Systems Research Laboratory Site at Oak Ridge National Laboratory,* ERDA Report ORNL-TM-5695, Oak Ridge National Laboratory, NTIS.

Schiager, Keith J., and Kenneth E. Apt (Comps.), 1974, *Environmental Surveillance at Los Alamos During 1973,* USAEC Report LA-5586, Los Alamos Scientific Laboratory, NTIS.

Union Carbide Corporation, Nuclear Division, 1976, *Environmental Monitoring Report, U. S. Energy Research and Development Administration, Oak Ridge Facilities, Calendar Year 1975,* ERDA Report Y/UB-4, Union Carbide Corporation, NTIS.

Van Voris, P., and R. C. Dahlman, 1976, *Floodplain Data: Ecosystem Characteristics and* ^{137}Cs *Concentrations in Biota and Soil,* USAEC Report ORNL-TM-5526, Oak Ridge National Laboratory, NTIS.

Wheeler, M. L., W. J. Smith, and A. F. Gallegos, 1977, *A Preliminary Evaluation of the Potential for Plutonium Release from Burial Grounds at Los Alamos Scientific Laboratory,* ERDA Report LA-6694-MS, Los Alamos Scientific Laboratory, NTIS.

Plutonium Contents of Field Crops
in the Southeastern United States

D. C. ADRIANO, J. C. COREY, and R. C. DAHLMAN

Agricultural crops were grown at the U. S. Department of Energy Savannah River Plant (SRP) and at Oak Ridge National Laboratory (ORNL) on soils at field sites containing plutonium concentrations above background levels from nuclear weapon tests. Major U. S. grain crops were grown adjacent to a reprocessing facility at SRP, which releases low chronic levels of plutonium through an emission stack. Major vegetable crops were grown at the ORNL White Oak Creek floodplain, which received plutonium effluent wastes in 1944 from the Manhattan Project weapon development.

The plutonium contents of grain crops (wheat, soybeans, and corn) at SRP were affected by distance from the emission stack, plant height, and grain-processing method. In general, vegetative materials growing close to the stack had higher plutonium concentrations than those growing in an adjacent field. Plutonium concentrations of portions of plants, such as wheat and corn, collected highest from the ground level indicate that plutonium contamination of these plant parts from soil resuspendible matter was minimal. The plutonium content of the grain when harvested by combine was elevated because the grain was mixed with extraneous matter and straw, which had relatively higher plutonium concentrations. Results from glasshouse studies using the same field-grown crops indicate that root uptake contributed insignificantly to the total plutonium contents of the field-grown crops.

Plutonium contents of vegetable crops grown at the ORNL White Oak Creek floodplain were influenced by part of plant, stage of maturity, and method of processing for the edible portions of the subterranean crops. Plutonium concentrations of fruits were at least one order of magnitude lower than those of the foliage. The plutonium content of the vegetable foliage was maximum when the foliage biomass was at maximum. Peeling the skins from potatoes and beets removed approximately 99% of the residual plutonium.

In general, the concentration ratios of vegetative parts of crops at SRP were approximately one order of magnitude higher than those at ORNL, which indicates the influence of aerial deposition of plutonium at the SRP site.

Research on transuranic nuclides in the environment has gained momentum in recent years as a result of proposed increases in production and use of plutonium in the nuclear fuel cycle. Because of the toxicity and long half-life of plutonium (specifically ^{239,240}Pu), its health hazard to man is being evaluated. Although inhalation has been considered the major pathway by which plutonium reaches man (Bennett, 1976), ingestion of plutonium-contaminated foodstuff through the soil-to-plant pathway should be critically evaluated because of the long persistence and general immobility of plutonium in the environment (Francis, 1973). Numerous studies have been conducted to determine plant uptake of plutonium under both field and controlled conditions (Francis, 1973; Price, 1972; Hanson, 1975; Bennett, 1976). However, most of these experiments were done in pot culture; experiments done in the field involved only nonagricultural vegetation. This general lack of information on field-grown crops is apparent from the proceedings of the 1976 international symposium on the *Transuranium Nuclides in the Environment* (International Atomic Energy Agency, 1976).

This chapter describes plutonium contents of crops grown on fields at the U. S. Department of Energy Savannah River Plant (SRP) and at Oak Ridge National Laboratory (ORNL) in soils containing plutonium concentrations at levels above those attributable to fallout from nuclear weapon tests. The fields at SRP, near Aiken, S. C., have been receiving plutonium at low chronic levels from the emission stack of a reprocessing facility since 1955; the field plots at ORNL are located on the White Oak Creek floodplain, which received plutonium from Manhattan Project operations in 1944. We compared the plutonium contents of major grain crops (wheat, soybeans, and corn) at SRP, where the major mode of contamination is through deposition from the emission stack, with those of major vegetable crops at Oak Ridge, where the major pathway of contamination is via root uptake.

Description and History of the Study Sites

Savannah River Plant

The SRP is on a reservation of 77,830 ha. Public access to the reservation has been controlled since its acquisition in 1951. The reservation consists of freshwater streams, old fields, and forest; most of the old fields are in the upper Coastal Terraces.

For over 20 yr this integrated nuclear complex has included nuclear reactors (three of the original five are operating at present), two nuclear-fuel reprocessing plants, a fuel fabrication facility, a heavy-water production unit, and nuclear and environmental research laboratories (Fig. 1). It also includes an ecological research laboratory to assess the effects of nuclear technology on the environment and its biota. The reprocessing plants (F and H) and global fallout are the sources of the transuranic elements that enter the SRP environs. Each source releases plutonium of unique isotopic composition: 25 and 95 α % ^{238}Pu* from the F- and H-area reprocessing facilities, respectively, and 10 α % ^{238}Pu from global fallout. These isotopic differences provide a convenient basis for studying the origin and transport of plutonium in various SRP ecosystems.

In 1974 two crop fields were established adjacent to the H-area reprocessing facility. Low-level atmospheric releases of plutonium have occurred from H area since the start of operations in July 1955. Approximately 440 mCi of plutonium had been released from H area before the installation of high-efficiency filters on the exhaust-air systems in December 1955. These releases contained ^{239,240}Pu. From 1956 through 1966, ^{239,240}Pu releases averaged 4 mCi/yr. From 1967 through 1974, normal releases averaged 12 mCi ^{238}Pu/yr and 4 mCi ^{239,240}Pu/yr. An accidental failure of the filtering mechanism in 1969 released an additional 560 mCi of ^{238}Pu and 58 mCi of ^{239,240}Pu. Total releases through 1974 were 640 mCi of ^{238}Pu and 570 mCi of ^{239,240}Pu.

$$*\alpha \ \% \ ^{238}\text{Pu} = \frac{^{238}\text{Pu alpha activity}}{\text{total plutonium alpha activity}} \times 100.$$

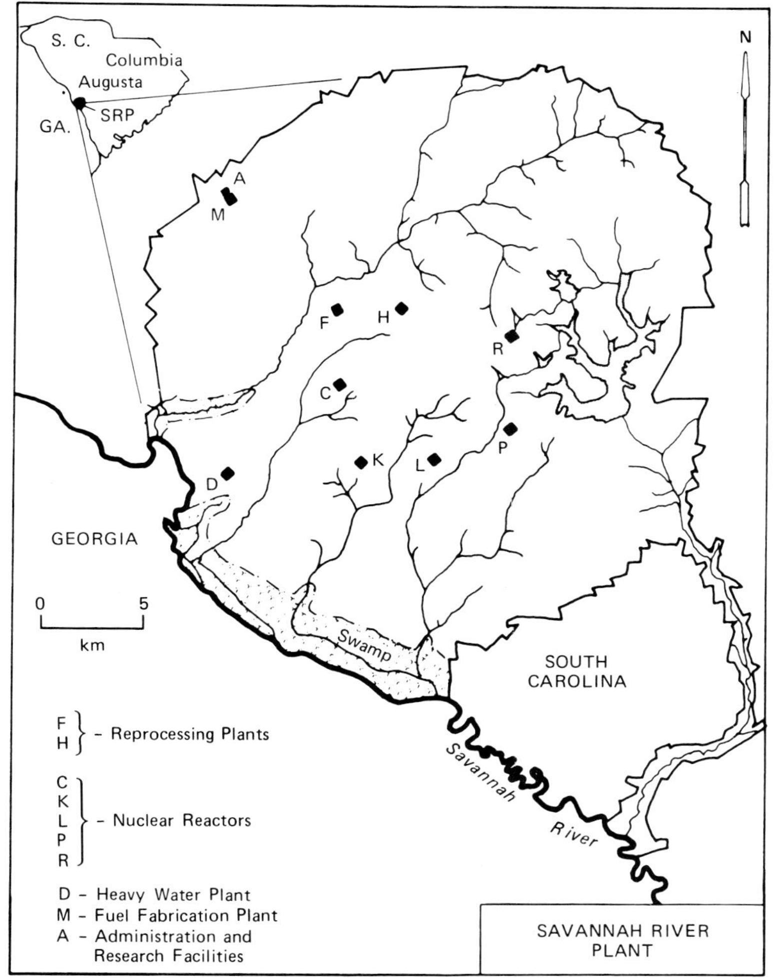

Fig. 1 Locations of the two reprocessing plants (H and F) at the Savannah River Plant. The agricultural study area is near the H area.

The South Field was 145 m by 30 m, and its long axis was oriented to the northwest away from the point of release, a 62-m stack. The North Field shared the same long axis as the South Field but was smaller (105 m by 30 m). The centers of the South and North fields were approximately 230 and 420 m, respectively, from the stack. The A soil horizon and parts of the B soil horizon had been removed from the South Field during construction of the reprocessing facility. In addition, some fill dirt had been deposited on the South Field. The North Field had been disturbed less and had a soil profile that was typical for the area. Before being tilled, both fields supported a herbaceous plant community dominated by *Andropogon* spp., *Lespedeza cuneata* (Dumont) G. Don, *Panicum* spp., and *Smilax* spp. with scattered loblolly pines (*Pinus taeda* L.). The plant communities were typical of those occurring on abandoned fields of similar soil types in the southeastern United States. A 5-m-wide strip along the southwestern margin of each field has been mowed regularly for several years. The soil in both fields is classified as Vaucluse, a highly leached Ultisol characterized by sandy A horizon and sandy clay B horizon. Both fields were acidic with a similar pH ($\sim$4.6).

The climate of the SRP area consists of mild winters and long, warm, humid summers. Temperatures average about 9°C in the winter and 30°C in the summer. The average annual temperature is 18°C. The average annual relative humidity and rainfall are 70% and 120 cm, respectively. The maximum annual precipitation recorded was 187 cm in 1929, and the minimum was 71 cm in 1933 (Langley and Marter, 1973).

White Oak Creek Floodplain, Oak Ridge

Manhattan Project operations in 1944 produced treated wastes containing plutonium, americium, and curium. Following soda-ash treatment to precipitate various ions, the liquid effluents were released to White Oak Creek (WOC). Trace quantities of ^{238}Pu, ^{239}Pu, ^{241}Am, and ^{244}Cm were deposited along the water course, in an intermediate retention pond, and, finally, in White Oak Lake. These elements were deposited in sediments of a retention pond over a 6-month period in 1944. No additional radioactivity is believed to have been deposited at this location since the pond was drained in late 1944. The former retention-pond site currently constitutes the first floodplain terrace of WOC.

The alluvial soil of the floodplain is representative of bottomlands of the Tennessee Valley and Ridge province. Azonal characteristics predominate because of concurrent erosion and deposition of materials during periodic floods. The soil profile remains relatively undeveloped, although the accumulation of humus is evident from a dark-brown coloration in the 0- to 3-cm zone.

Sedimentation during the 6-month impoundment in the temporary holding basin 30 yr ago contributed new sedimentary materials. A Tennessee Valley Authority (1975) survey in 1951 reported that approximately 2100 m^3 of sediment had been deposited at this locality. This volume, distributed uniformly over the research site, represented an increment of approximately 9 cm of new sediment. However, equal deposition was unlikely, and the exact history of sedimentation associated with the 1944 impoundment is unknown.

Textural analyses indicate that soils are silty loam (72% silt and 24% clay) and contain almost no sand or gravel (Tamura, 1976). Although this texture is representative of the floodplain, isolated gravel lenses occur irregularly across the floodplain and within the soil profile.

The soil reaction is mildly alkaline since pH ranges from 7.1 to 7.6. The mild alkalinity is probably caused by the soda-ash method of waste treatment used before effluents were released from laboratory operations. Data have not been obtained on base exchange or soil fertility, although, on the basis of plant growth and crop performance, the site possesses high fertility potential.

The climate at Oak Ridge, Tenn., is characterized by mild winters and warm, humid summers. Average temperatures of the continental climate are 24°C in summer and

5°C in winter; average precipitation is 140 cm, with 10 cm of snowfall. Winter and summer temperature variations tend to be greater at Oak Ridge than at SRP.

Methods

Savannah River Plant

Pine trees were bulldozed from the old fields and bushes and herbaceous vegetation were cut with a tractor-drawn rotary mower to prepare the land for crops. All debris, including tree roots, was removed before any operation. The fields were disked with a bush and bog disk harrow, then with a standard disk harrow, subsoiled, redisked, limed, redisked, fertilized, and, finally, redisked. The disking was done to a depth of about 20 cm. Agricultural lime (consisting of 60% $CaCO_3$, 25% $MgCO_3$, and 15% H_2O by weight) was added at the rates of 908 kg on the South Field and 590 kg on the North Field. Mixed-grade fertilizers were added at rates of 318 kg of 3–9–18* and 227 kg of 5–10–15 for the South and North fields, respectively. The South Field was divided into 18 equal plots, and the North Field, into 12 equal plots.

Wheat (variety Coker 68-19) was sown on the fields by hand in November 1974 at the rate of approximately 53 kg of seed per field. The seeds were then covered by disking the fields at a very shallow depth. Foliage samples were collected in March and April. At harvest in June, plants were separated into grain and straw. Wheat-grain samples were obtained by two techniques. The first technique was to carefully collect wheat heads by hand from each of the 30 plots and separate the grain with a laboratory thrashing machine. This machine was carefully cleaned between samples from each plot to minimize the amount of dust that would adhere to the grain. The second technique was to harvest the grain with a tractor-pulled combine. Grain collected by this method was exposed to the usual soil and dust stirred up by harvest activities. Total biomass for the crop (grain plus hull plus straw) was estimated at 3615 and 3545 kg/ha for the South and North fields, respectively. Wheat was also grown 37 km from the reprocessing plants as a control.

The fields were prepared for soybeans by disking them several times and adding mixed-grade fertilizer (3–9–18) at the rate of 227 kg for each field. Inoculated soybeans (variety Bragg) were planted in July 1975 with a two-row planter; about 27 kg of seeds was used for the two fields. The crop was harvested with a combine in November 1975.

The fertilizer-addition and field-preparation techniques for the corn crop were similar to those for soybeans. Field corn (variety Coker) was planted in May 1976 with a two-row planter and harvested in October with a combine.

The chronic releases of plutonium-bearing particles at low levels from the emission stack make it impossible to determine the plutonium uptake by the crops from the soil through the root pathway. Whatever amount of plutonium is translocated to the plant foliage from the soil would be obscured by external deposition and retention from stack fallout. Therefore glasshouse studies were conducted to determine the amount of plutonium translocated to the aerial portions of the plants from the soil.

---

*3–9–18 refers to 3% elemental nitrogen, 9% P_2O_5, and 18% K_2O, respectively.

Soils used in the glasshouse studies were obtained from the surface layer (0 to 20 cm) of the South Field, where crops were grown. The soil had an average total plutonium content of 1.96 pCi/g (dry weight) with 21 α % ^{238}Pu. The soils ($\sim$8 kg) were potted and fertilized, and the same varieties of wheat, soybeans, and corn were grown to maturity.

In the selection of locations for sampling soil and resuspendible particles, 18 grid points were located in the South Field and 12 in the North Field. Grids were composed of sampling blocks placed 3, 15, and 27 m from the southwestern margin of the fields on transects across the short axis of the fields. The transects originated at 30.4-m intervals on the long axis of each field. Each sampling block was 3 m by 10 m and contained ten 1- by 3-m plots. A randomly chosen plot was permanently marked in each block. Thus there were 18 sampling locations in the South Field and 12 in the North Field.

Soil cores of 3.8-cm diameter were taken with a split-barrel sampler and divided into 0- to 5-cm, 5- to 15-cm, and 15- to 30-cm fractions for plutonium analysis. For unexplained reasons plutonium concentrations in the 0- to 5-cm samples collected before tillage were 50% lower than in samples collected after tillage; therefore these samples were replaced with samples that had been collected at the same depth before tillage with a hand soil auger.

The resuspendible particles on the soil surface from areas where aboveground vegetation had been removed were collected by drawing a nearly laminar flow of air (velocity $\cong$ 6 m/sec) across a 232-cm^2 area under a 1-cm-tall stainless-steel hood. The resuspended materials were collected in the paper bag of a small electric vacuum cleaner. The interior of the plastic compartment holding the paper bag was wiped to collect materials passing through the bag, and these materials were included in the sample.

Samples of soil and of resuspendible particles were collected at three different times: (1) before tillage, (2) after soil preparation for wheat, and (3) after wheat harvest.

All samples were ashed before determination of plutonium contents. Plant, grain, and resuspension samples were ashed at 550°C. Soil samples were ground to a particle size of $\leqslant$500 μm and ashed at 500°C.

Actinide elements were leached from a $\leqslant$10-g aliquot of the sample ash with hot 8M HCl, and valences were adjusted to ensure formation of Pu(IV) and Np(IV). Plutonium, neptunium, and uranium were extracted into 10% triisooctylamine in xylene, and the plutonium was separated from the neptunium and uranium by reducing Pu(IV) to Pu(III) with NH$_4$I and extracting into 8M HCl. This solution was evaporated to dryness and oxidized to destroy residual organic matter. The plutonium was taken up in 8M HNO$_3$, and the valence was adjusted to Pu(IV). Final purification was accomplished by adsorbing the Pu(IV) onto an anion-exchange column and removing any residual iron, uranium, or other contaminants from the column with 4M HNO$_3$. The Pu(IV) was reduced to Pu(III) and eluted from the column with H$_2$SO$_3$. Following purification, the plutonium was electrodeposited on platinum plates, and the amounts of ^{238}Pu and ^{239,240}Pu were determined by alpha spectrometry with low-background high-resolution surface-barrier detectors. Counting times ranged from 2 to 7 days, depending on the plutonium concentration of the sample. An internal standard of ^{236}Pu was used to determine recovery efficiencies.

After the concentrations of ^{238}Pu and ^{239,240}Pu in a sample had been determined, the ratio of the ^{238}Pu concentration to the ^{239,240}Pu concentration was computed.

This ratio was used to evaluate the relative importance of different pathways of plutonium movement.

White Oak Creek Floodplain, Oak Ridge

A successional floodplain forest characterizes the vegetation of the study site (Dahlman and Van Voris, 1976). A small area was cleared of native vegetation to provide adequate sunlight for growing vegetables and field crops on the contaminated site. Soil was cultivated by tilling in the early spring of 1975 and 1976. Common varieties of vegetable and forage seed stock were obtained from local vendors. All plants were grown in parallel rows randomly placed in each of four 5- by 5-m subplots. Each species (except potato and tomato) was grown in two replicated rows per subplot. The entire 100-m^2 plot contained eight row replications of all species except potato and tomato. Single rows of potatoes and tomatoes were grown in each subplot for a total of four rows per 100-m^2 plot. Analysis of variance among subplots showed no significant difference in plutonium concentration for a given species among subplots where sufficient plant material was collected for plutonium analysis.

After the seedlings emerged, black plastic mulch was placed on the soil surface to prevent weed growth and resuspension of soil-borne plutonium. This method proved effective because the ambient air concentration of plutonium 10 cm above the soil surface did not exceed 0.14×10^{-3} pCi/m^3 (Dahlman and McLeod, 1977). Air samples were collected over mulched soil containing 63 ± 0.4 (standard error) pCi Pu/g. Monitoring results for the Oak Ridge area showed that ambient plutonium was 0.016×10^{-3} pCi/m^3 at a height of 1 m above ground (Oakes and Shank, 1977). It was not determined if the higher value at 10 cm represented normal plutonium in ambient air near the floodplain soil or if it represented radioactivity induced by air currents from the high-volume samples (approximately 30 cfm). We suspect the latter caused it.

Vegetative samples were cleaned before radiochemical analysis. Bush bean, soybean, and tomato leaves were washed and rinsed in an ultrasonic cleaning device, where fresh samples (about 5 to 10% by volume) were placed in a 1-liter-capacity cleaning tray containing 700 ml of distilled water. The sonifier was tuned to give an optimal frequency for cleaning each sample for a 2-min period. This was determined by the maximum agitation delivered to the loaded cleaning tray by the wave generator. The cleaned vegetation was carefully separated from residues, which settled in the bottom of the sonifier tray. After the residue had been discarded and the tray rinsed with acid, the vegetation was cleaned a second time with fresh distilled water. Samples were finally removed from the tray, dried, chopped, and prepared for radiochemical analysis.

Other garden vegetables were cleaned according to kitchen food-preparation techniques. Lettuce was cleaned under running tap water by gently rubbing the leaves. The samples were drained, dried, chopped, and prepared for radiochemical analysis. Root samples (radish, carrot, and potato) were vigorously hand scrubbed under running tap water. The samples were dried, chopped, and prepared for radiochemical analysis.

The samples were sent to LFE, Richmond, Calif., for radiochemical analysis. The analytical procedure has been discussed previously by Wessman et al. (1978).

Results and Discussion

Field Crops at SRP

Wheat. Data for wheat are summarized in Table 1. The results for the South and North fields are presented separately because of the differences in proximity to the stack and hence the possibly different deposition patterns. There appears to be no significant change in the plutonium concentrations of wheat foliage with time on either the South or the North Field. The plutonium content of foliage or straw for both fields averaged 4×10^{-2} pCi/g (dry weight).

The isotopic composition of the foliage, as indicated by the $\alpha\%$ ^{238}Pu, changed with time (Table 1). In March the wheat plants were short, and the foliage was not yet dense. At this stage contamination of the foliage appeared to originate primarily from the soil, as in rain splash or resuspendible matter. As the plants grew taller and their foliage became denser, they were able to intercept fallout particles from the stack more efficiently. Also, the dense foliage minimized rainfall energy on impact with the ground. The foliage had much higher ^{238}Pu values in April and June than in March. This discrepancy indicated that contamination originated primarily from fallout particles from the emission stack because the ^{238}Pu values were closer to those of the deposition particles than to those of the soil samples (Dahlman and McLeod, 1977). The plutonium concentration of straw was approximately 300 times greater than that of the laboratory-thrashed grain and 40 times greater than that of the field-combined grain. That the plutonium values for the grain were lower than those for the straw indicates that the grain was possibly shielded from atmospheric deposition. The plutonium concentrations in foliage, straw, and grain samples were slightly higher in the South Field.

The straw from the off-plant control site had only $(8 \pm 2) \times 10^{-4}$ pCi/g (data not shown), about two orders of magnitude lower than SRP field samples. The thrashed grain from the control site, however, had the same plutonium contents as the SRP thrashed samples. However, the high ^{238}Pu percentage of 62 ± 15 for thrashed grain from the control site, which is disparate from the straw value, should be noted. This high value was apparently caused by contamination from thrashing of the SRP samples since all thrashed samples had similar plutonium values.

The concentration ratios (CR) for the June straw were 5×10^{-2} and 1×10^{-1} for the South and North fields, respectively. Similar values (1×10^{-1}) were obtained for samples from the control site. The CR for the grain collected from the combine averaged 2×10^{-3} compared with an average of 2×10^{-4} for the thrashed grain from the two fields.

In Table 2 plutonium contents of field- and glasshouse-grown wheat plants are compared. The glasshouse-grown plants had total plutonium contents one order of magnitude lower than the plants from the South Field. The combined grain from the fields was sifted with a soil screen to evaluate the effects of removing extraneous matter on the plutonium content. Data from Tables 1 and 2 for the unsifted grain harvested with a combine are remarkably similar. The sifted combined grain had a factor of 2 less total plutonium content than the unsifted grain, which indicates that the extraneous matter had higher plutonium contents than the grain. The extraneous matter included unshelled grain and chopped straw.

TABLE 1 Plutonium Concentration and Isotopic Composition of Wheat Grown Near a Reprocessing Facility (H Area) at SRP

	South Field*				North Field*			
	Number of samples	Total Pu, pCi/g	$\alpha\%\,^{238}$Pu	CR	Number of samples	Total Pu, pCi/g	$\alpha\%\,^{238}$Pu	CR
Foliage								
March	6	$(5.2 \pm 1.0) \times 10^{-2}$	38 ± 4	3×10^{-2}	4	$(2.0 \pm 0.60) \times 10^{-2}$	38 ± 2	6×10^{-2}
April	18	$(2.1 \pm 0.42) \times 10^{-2}$	61 ± 3	1×10^{-2}	12	$(9.8 \pm 1.2) \times 10^{-3}$	66 ± 3	3×10^{-2}
Straw, June	18	$(8.8 \pm 1.5) \times 10^{-2}$	58 ± 2	5×10^{-2}	12	$(3.9 \pm 1.8) \times 10^{-2}$	55 ± 3	1×10^{-1}
Grain								
Field combined	8	$(2.0 \pm 0.64) \times 10^{-3}$	56 ± 4	1×10^{-3}	7	$(0.9 \pm 0.19) \times 10^{-3}$	55 ± 5	3×10^{-3}
Laboratory thrashed	18	$(3.0 \pm 0.52) \times 10^{-4}$	60 ± 3	2×10^{-4}	12	$(1.4 \pm 0.43) \times 10^{-4}$	64 ± 6	3×10^{-4}

*Values are means ± 1 standard error.

TABLE 2 Plutonium Concentration and Isotopic Composition of Vegetative Parts and Grain of Crops Grown Near a Reprocessing Facility (H Area) and in a Glasshouse at SRP

	South Field*†				Glasshouse†‡			
	Number of samples	Total Pu, pCi/g	$\alpha \% ^{238}Pu$	CR	Number of samples	Total Pu, pCi/g	$\alpha \% ^{238}Pu$	CR
Wheat								
Straw, June	18	$(8.8 \pm 1.5) \times 10^{-2}$	58 ± 2	6×10^{-2}	10	$(3.0 \pm 0.47) \times 10^{-3}$	41 ± 2	2×10^{-3}
Grain								
Combine, unsifted	7	$(3.0 \pm 0.98) \times 10^{-3}$	50 ± 6	2×10^{-3}				
Combine, sifted§	6	$(1.6 \pm 0.86) \times 10^{-3}$	47 ± 4	1×10^{-3}				
Soybeans								
Vegetative, whole plants	14	$(5.2 \pm 0.45) \times 10^{-2}$	53 ± 2	3×10^{-2}	10	$(5.6 \pm 0.51) \times 10^{-3}$	33 ± 2	3×10^{-3}
Grain, combine	9	$(5.5 \pm 1.3) \times 10^{-4}$	41 ± 3	3×10^{-4}				
Corn								
Leaves								
0- to 1-m height	10	$(2.5 \pm 0.35) \times 10^{-2}$	63 ± 3	2×10^{-2}	6	$(1.1 \pm 0.16) \times 10^{-3}$	55 ± 4	6×10^{-4}
1- to 2-m height	10	$(4.2 \pm 0.57) \times 10^{-2}$	70 ± 1	3×10^{-2}				
Grain								
Combine, unsifted	5	$(1.7 \pm 0.31) \times 10^{-4}$	63 ± 2	1×10^{-4}	6	$(1.2 \pm 0.33) \times 10^{-4}$	76 ± 2	6×10^{-5}
Combine, sifted§	5	$(8.5 \pm 1.1) \times 10^{-5}$	78 ± 4	5×10^{-5}				

*The total plutonium content of 1.61 pCi/g (dry weight) of the 0- to 15-cm soil depth for the South Field for the wheat crop was used. This same value was used for the soybean and corn calculations.

†Values are means ± 1 standard error.

‡The glasshouse soils were from the top layer of the South Field and had the following average contents: Total plutonium, 1.96 pCi/g (dry weight) (n = 9) and 21% ^{238}Pu (n = 9). From the glasshouse the following samples were used: wheat, whole plant without grain; soybeans, whole plant without grain; and corn, leaves from the whole plant and hand-shelled grain.

§The same combined grains were sifted through a soil screen.

Soybeans. Data for soybeans from the field and glasshouse are given in Table 2. Total plutonium concentrations in glasshouse-grown plants were an order of magnitude less than those in field-grown plants. The CR for plants grown in the glasshouse was a factor of 10 lower than that for plants grown in the field. The glasshouse plants (whole vegetative plants) were sampled when the bean pods were still green and before the plants started defoliating. The field-grown plants were sampled approximately 1 month before the field was combined, when the plants had the maximum biomass. The plutonium content of the combined grain was two orders of magnitude lower than that of the vegetative parts.

The bean plants had plutonium concentrations $[10^{-2}$ pCi/g (dry weight)] and ^{238}Pu percentages similar to those of the wheat straw (whole plants without the grain). Consequently the two crops had similar CR values. However, the soybean grain had slightly lower plutonium contents than the wheat grain.

Corn. Plutonium contents of field- and glasshouse-grown corn are shown in Table 2. So that the extent of interception of fallout particles from the stack by foliage could be determined, corn leaves were sampled from the field when the plants had their maximum biomass. The leaves were sampled from two heights: 0 to 1 m from the ground and 1 to 2 m (or 1 m to the top). Plants in the glasshouse were sampled when the ears had matured. The total plutonium content of leaves from glasshouse-grown corn was one order of magnitude lower than that from field-grown corn, and the CR was a factor of 30 lower. Leaves from the 1- to 2-m section of the corn plants had plutonium contents almost a factor of 2 higher than leaves from the lower section (0 to 1 m). This indicates that the upper foliage partially filtered and retained plutonium-bearing particles from atmospheric deposition.

Shelled grain from the glasshouse had slightly lower plutonium concentrations than the field-combined grain (Table 2). However, the CR of the grain from the glasshouse was an order of magnitude lower than that of the field-combined grain. The unsifted combined grain had total plutonium content a factor of 2 higher than that of the sifted. Apparently the extraneous matter separated from the grain contained approximately 50% of the plutonium in the sample.

Relative Contribution of Root Uptake to Plutonium Contents of Field Crops. A comparison (Table 3) of the total plutonium contents of the vegetated material of the crops grown under field and glasshouse conditions indicates that plutonium contamination of the field crops was primarily external in nature. For wheat and corn samples, approximately 97% of the total contamination was external. For soybeans contribution from root uptake appeared greater than 10%. No plausible explanation can be offered at this time for this discrepancy.

Forage and Vegetable Crops at WOC Floodplain, Oak Ridge

Forage and vegetable crops harvested from a field plot located on the contaminated floodplain contained measurable quantities of plutonium. The field experiments were designed to minimize resuspension of dust and subsequent contamination of leaf surfaces by plutonium-contaminated soil from the field experimental plots. Five factors aided this objective: (1) The plastic mulch that covered the plot served as a barrier to the transport

**TABLE 3 Comparison of Total Plutonium Contents of
Vegetative Materials from Crops Grown on the South
Field and in a Glasshouse at SRP***

Vegetative material	Plutonium contents, fCi/g (dry weight)		External contamination,† %
	South Field	Glasshouse	
Wheat straw	88.0	3.0	97
Soybeans, whole plants	52.2	5.6	89
Corn leaves ‡	33.5	1.1	97

*The soils used in the glasshouse were from the top layer of the South Field.

†Calculated from the equation: [(field content − glasshouse content)/field content] × 100.

‡Fiberglass mats were placed on top of the soil in the corn pots in the glasshouse to prevent plant contamination by resuspension.

of particles from the soil surface to the air. (2) The floodplain soil moisture was maintained continuously near field capacity owing to capillary conductivity from a shallow water table. The cohesive force of a thin film of water surrounding the soil prevented particles from becoming airborne. (3) Owing to protection provided by the surrounding forest, the floodplain site was not exposed to wind, and particles did not become airborne by aeolian mechanisms. (4) The plastic mulch used for weed control eliminated the need to cultivate; therefore airborne dust particles were not created by mechanical operations. (5) The surrounding floodplain soil was covered by moist decomposing litter and by a dense cover of herbaceous understory vegetation. The boundary layer afforded by such conditions prevented soil-borne plutonium from becoming airborne. These five factors reduced the likelihood that soil-borne plutonium would become airborne and deposited on vegetation; indeed plutonium was not detectable in 346-m^3 air samples collected next to the test plot at 1-m height. Attempts to measure plutonium in air were unsuccessful because of limited sample size over a 6- to 8-hr collection period, but the air concentration would be less than 0.3×10^{-3} pCi/m^3 on the basis of a minimum detectable level of 0.1 pCi per sample (Dahlman and McLeod, 1977).

Concentrations of ^{239,240}Pu in foliage of bush beans, soybeans, tomatoes, lettuce, radishes, and millet ranged from 0.01 to 0.33 pCi/g (dry weight) (Table 4). Concentrations of plutonium in the fruit of these species were lower by at least an order of magnitude than those in the foliage. Limited observations on root crops (carrots, potatoes, and beets) indicated that the plutonium concentrations of the edible portion were similar to those of the foliage.

The plutonium concentrations of foliage and fruit appeared to represent plutonium assimilated by the root pathway because surface contamination was removed before radiochemical analysis. Samples of bush beans, soybeans, and tomatoes were washed and rinsed in a sonic bath. This procedure effectively removed surface-bound plutonium

TABLE 4 Plutonium Concentration and Concentration Ratio of Field Crops from the Oak Ridge WOC Floodplain

Crop	Number of samples	Total plutonium,* pCi/g	Concentration ratio†
Foliage			
Bush bean	8	$(1.1 \pm 0.3) \times 10^{-1}$	2×10^{-3}
Soybean	24	$(1.0 \pm 0.2) \times 10^{-1}$	2×10^{-3}
Tomato	3	$(3.3 \pm 0.7) \times 10^{-1}$	5×10^{-3}
Lettuce	3	$(8.0 \pm 2.0) \times 10^{-2}$	1×10^{-3}
Radish	7	$(1.0 \pm 0.5) \times 10^{-1}$	2×10^{-3}
Millet	14	$(1.0 \pm 0.3) \times 10^{-2}$	2×10^{-4}
Fruit			
Bush bean			
Whole bean	7	$(7.0 \pm 2.0) \times 10^{-3}$	1×10^{-4}
Shelled bean	5	$(1.0 \pm 0.4) \times 10^{-2}$	2×10^{-4}
Tomato	5	$(4.0 \pm 2.0) \times 10^{-3}$	6×10^{-5}
Soybean			
Whole bean	7	$(1.5 \pm 0.3) \times 10^{-2}$	2×10^{-4}
Shelled bean	3	$(1.7 \pm 1.2) \times 10^{-2}$	3×10^{-4}
Subterranean			
Radish	5	$(4.1 \pm 0.6) \times 10^{-1}$	6×10^{-3}
Carrot	1	3.1×10^{-1}	5×10^{-3}
Irish potato			
Whole	3	$(8.0 \pm 1.0) \times 10^{-2}$	1×10^{-3}
Peeled	2	$(6.0 \pm 1.0) \times 10^{-3}$	1×10^{-4}
Skin only	4	$(5.3 \pm 1.2) \times 10^{-1}$	8×10^{-3}
Beet			
Peeled	3	$(5.0 \pm 1.0) \times 10^{-3}$	8×10^{-5}
Skin only	2	$(1.3 \pm 0.5) \times 10^{0}$	2×10^{-2}

*Values are means ± 1 standard error, which includes analytical error of approximately 10%.

†Concentration ratio values are based on a soil concentration of 63 ± 0.4 (standard error) pCi/g.

because no particles were observed in microscopic examination on any sample except soybean pods where hirsute structures effectively retained surface contaminants.

The plutonium concentration of soybean fruit, which included the pod, was similar to that of the whole bean. Concentrations of whole snap beans and shelled beans were also similar. For both species the bean pod protected bean seeds from possible surface contamination while young bean seeds matured on the vine. Because plutonium concentrations were similar when bean seeds were analyzed separately and when seeds and pods were analyzed together, these observations reinforced the argument that any residual surface soil contaminant was removed from vegetative surfaces by the cleaning process. Thus the plutonium content of these vegetables is attributed to assimilation by the root pathway. Other results in support of root assimilation are discussed elsewhere (Dahlman, Bondietti, and Eyman, 1976; Dahlman and McLeod, 1977).

Concentration of plutonium in root—vegetable crops reflected contamination by residual soil plutonium. Concentrations ranged from 0.08 to 0.41 pCi/g for whole radish, carrot, and potato samples that had been hand scrubbed before radiochemical analysis. These samples were not examined microscopically for the presence of residual particles, as had been done for sonically cleaned vegetation; yet concentrations of plutonium in storage organs of roots were similar to those in cleaned vegetation. Peeling the skins from potatoes and beets removed most of the residual plutonium. Potato skins averaged 0.53 pCi/g, as compared with 0.006 pCi/g for peeled potatoes. This indicates that about 99% of the plutonium was removed by peeling. Beet skins had even greater concentrations of plutonium, 1.3 pCi/g for skins vs. 0.005 pCi/g for peeled beets. Thus peeling the beet removed about 99.5% of the total plutonium in the beet.

TABLE 5 Plutonium Concentration and Concentration Ratio for Soybean Foliage Harvested at Different Stages of Growth at Oak Ridge WOC Floodplain

Growth stage	Number of samples	Total plutonium,* pCi/g	Concentration ratio†
8 weeks	8	$(2.0 \pm 0.4) \times 10^{-2}$	3×10^{-4}
12 weeks	8	$(1.0 \pm 0.3) \times 10^{-1}$	2×10^{-3}
16 weeks	8	$(1.9 \pm 0.4) \times 10^{-1}$	3×10^{-3}

*Values are means ± 1 standard error.

†Concentration ratio values are based on a soil concentration of 63 ± 0.4 (standard error) pCi/g.

The uptake pattern for the soybean plants at the WOC floodplain is shown in Table 5. The plutonium concentration in the foliage after 8 weeks of growth was only 0.02 pCi/g (dry weight). The plutonium concentration had increased five times (0.10 pCi/g) by the 12th week and ten times (0.19 pCi/g) by the 16th week. The foliage biomass was at maximum at around the 16th week. Correspondingly, the CR of the foliage increased by an order of magnitude from the 8th to the 16th week.

Dose to Man from Ingestion of Plutonium-Contaminated Foodstuff

The dose to man from ingesting wheat or other foods (Tables 2 and 4) can be estimated from the nomograms in Fig. 2. For example, if a person ate 2×10^5 g of wheat grain containing 2 fCi Pu/g of dry grain, the 70-yr dose to bone would be ~0.5 mrem. Likewise, if a person ate 2×10^5 g of peeled Irish potatoes (approximately equivalent to 2000 kg of wet potatoes) containing 6 fCi Pu/g (dry weight), the 70-yr dose to bone would be ~1.0 mrem. These doses are low compared with the 130-mrem dose from background radiation (Klement et al., 1972). The basic data used for these calculations came from ICRP publications (International Commission on Radiological Protection, 1959; 1972a; 1972b).

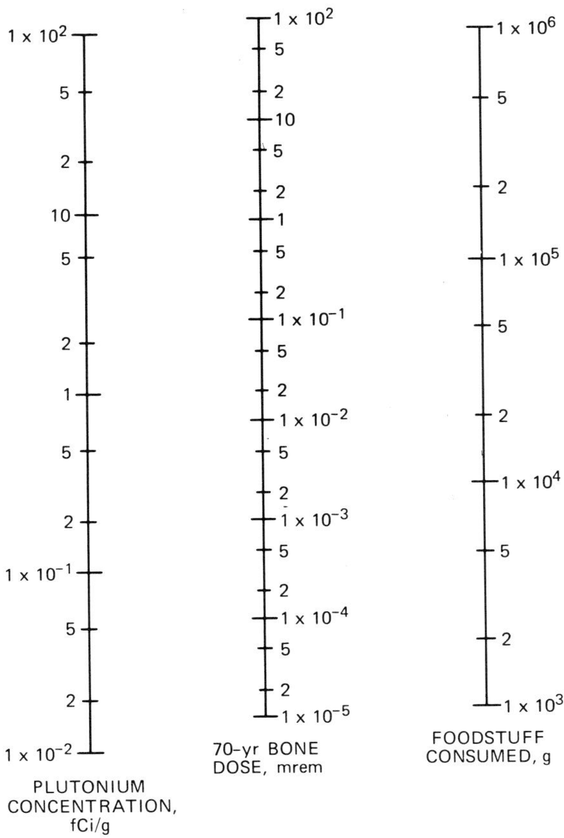

Fig. 2 Nomograph for calculating dose to bone from consumption of plutonium-contaminated foodstuff. The dose is that which will be received by the bone during a 70-yr life-span. The following assumptions from ICRP publications were used: effective absorbed energy per disintegration, 270 MeV (International Commission on Radiological Protection, 1959); fraction from gastrointestinal tract to blood, 3×10^{-5}; fraction from blood to bone, 0.45; half-life in bone, 100 yr (International Commission on Radiological Protection, 1972); mass of bone, 5×10^3 g (International Commission on Radiological Protection, 1972).

Soils at SRP

As with plant materials, the resuspendible matter and soils from the South Field (closer to the stack) had higher plutonium concentrations than those from the North Field (Table 6). The differences in concentrations were more pronounced in the resuspendible matter where the South Field had plutonium levels one order of magnitude higher than the North Field. Soils (0 to 5 cm) from both fields had plutonium levels about three orders of magnitude higher than soils from the control site. Cultivation of the fields resulted in more uniform plutonium concentrations in the soil profile, increasing the plutonium at the deeper depths (5 to 15 cm). Cultivation also altered the isotopic

TABLE 6   Effect of Cultivation on Plutonium Concentration and Isotopic Composition of Soils in an Agricultural Study Area Adjacent to a Reprocessing Facility at SRP

	South Field*			North Field*		
	Number of samples	Total Pu, pCi/g	$\alpha \% {}^{238}Pu$	Number of samples	Total Pu, pCi/g	$\alpha \% {}^{238}Pu$
Before cultivation						
Suspendible matter	18	$(1.3 \pm 0.38) \times 10^{1}$	51 ± 2	12	$(1.6 \pm 0.26) \times 10^{0}$	49 ± 3
0 to 5 cm	21	$(3.2 \pm 0.79) \times 10^{0}$	29 ± 3	14	$(1.0 \pm 0.10) \times 10^{0}$	19 ± 1
6 to 15 cm	18	$(1.2 \pm 0.28) \times 10^{-1}$	30 ± 2	12	$(9.4 \pm 1.8) \times 10^{-2}$	20 ± 2
After 1st cultivation						
Suspendible matter	18	$(4.8 \pm 0.57) \times 10^{0}$	28 ± 1	12	$(8.5 \pm 1.1) \times 10^{-1}$	27 ± 2
0 to 5 cm	21	$(3.0 \pm 0.59) \times 10^{0}$	30 ± 3	14	$(6.3 \pm 0.48) \times 10^{-1}$	29 ± 2
6 to 15 cm	18	$(9.2 \pm 3.4) \times 10^{-1}$	22 ± 2	12	$(1.7 \pm 0.40) \times 10^{-1}$	25 ± 3
After 2nd cultivation						
Suspendible matter	18	$(4.6 \pm 0.69) \times 10^{0}$	34 ± 2	12	$(9.3 \pm 1.4) \times 10^{-1}$	30 ± 1
0 to 5 cm	18	$(4.4 \pm 1.1) \times 10^{0}$	31 ± 3	12	$(7.0 \pm 1.0) \times 10^{-1}$	28 ± 3
6 to 15 cm	18	$(4.7 \pm 1.5) \times 10^{-1}$	30 ± 3	11	$(3.8 \pm 1.0) \times 10^{-1}$	21 ± 3

*Values are means ± 1 standard error.

composition of the resuspendible portion to approach that of the soil values. After the initial cultivation, the ^{238}Pu percentage for the resuspendible matter decreased from an average of 50 to an average of 27 for the two fields.

Soils at WOC Floodplain, Oak Ridge

Concentrations of plutonium in the floodplain soil generally ranged from 10 to 150 pCi/g (dry weight) over a 3-ha area (Fig. 3). The highest concentrations occurred behind the former dike, along WOC, and at the upper end of the floodplain. Concentrations of plutonium were less than 25 pCi/g near the margins of the floodplain. The ^{239}Pu concentration of soil at a garden plot site (W35, S295) was 63 ± 0.4 (standard error) pCi/g.

A typical profile distribution of plutonium showed that the highest concentration of plutonium was observed in the uppermost zone of the profile (Duguid, 1975). Occasionally, however, where higher concentrations of plutonium occurred in the subsoil,

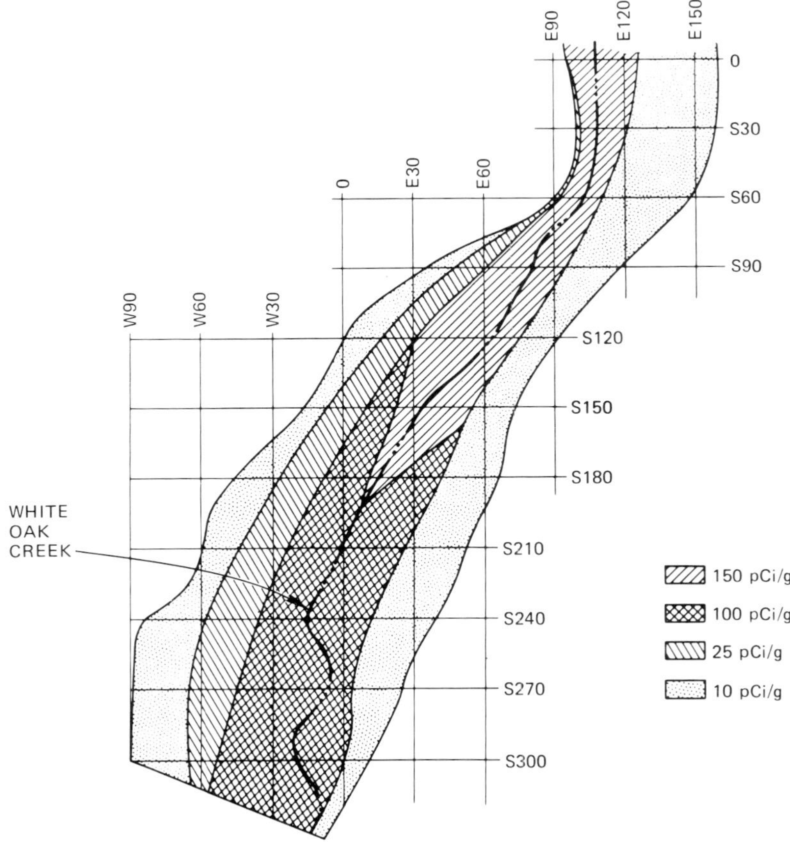

Fig. 3   Areal distribution of plutonium at the contaminated Oak Ridge WOC floodplain. Grid size is 30 m by 30 m (900 m²).

this atypical distribution was attributed to variable deposition of sediments over an initial plutonium deposit rather than to plutonium cycling by leaching processes.

Plutonium was not distributed uniformly between the silt and clay fractions of the soil. Although 72% of the soil was silt, this fraction contained 60% of the plutonium, whereas the 24% clay contained 40% of the plutonium. In this case the affinity of plutonium for colloids, the clay fraction, may be responsible for plutonium enrichment in clay.

The results of various attempts to extract plutonium from the soil indicated that more than 95% of the plutonium can be removed by hot $8M$ HNO_3; milder HNO_3 treatments removed smaller amounts of plutonium (Auerbach, 1975). These treatments indicated that plutonium recovery from the floodplain soil did not require rigorous treatment, such as HNO_3—HF or carbonate fusion. Milder treatment with $0.1M$ citric acid removed 16 to 24% of the soil plutonium. Contact with a CHELEX resin removed 11%. Humic acid solubilized by the presence of Na^+ from a sodium-saturated CHELEX resin contained 5% of the total plutonium after 4 weeks. The carbonate treatment removed 54% of the soil plutonium. Further analysis of the carbonate extract showed that at least 90% of the plutonium behaved as Pu(IV) (Bondietti and Sweeton, 1977; Bondietti, Reynolds, and Shanks, 1976).

General Discussion

The various CR values for the agricultural vegetation at SRP are generally higher than those obtained for indoor-grown plants. Schulz et al. (1976a; 1976b) obtained values for barley on the order of 10^{-5} for vegetative material; grain was a factor of 20 to 100 lower. In wheat grain they obtained CR values ranging from 1.1×10^{-7} to 3.8×10^{-6}. Lipton and Goldin (1976) obtained CR values for pea plants on the order of 10^{-4}. Natural plant species not subject to contamination from atmospheric releases or resuspension were observed to have CR values on the order of 10^{-6} to 10^{-4} (Francis, 1973; Hanson, 1975), much lower than values obtained at SRP. However, in arid, windy environments, dust and soil particles can become airborne and can be deposited and retained in leaves, causing plutonium CR values to approach 10^0 (Romney et al., 1975). Earlier studies at SRP indicated that deposition on the surfaces of the leaves and stems was the principal mechanism of plutonium contamination of natural vegetation (Adriano and Pinder, 1977; McLendon et al., 1976). The plutonium concentrations of all ecosystem components decreased as the distance from reprocessing plants increased (McLendon, 1975; McLendon et al., 1976). Thus considerable external contamination of plants from atmospheric releases and resuspension is a complicating factor in field data interpretation.

The CR values from the glasshouse studies at SRP are on the order of 10^{-4} to 10^{-3} for the vegetative materials and, in general, are similar or slightly higher than the values obtained by Schulz et al. (1976a; 1976b) and Lipton and Goldin (1976). Our glasshouse results suggest that ^{238}Pu is more available than ^{239}Pu, as indicated by relatively higher ^{238}Pu percentage values in the vegetative samples than in the soil. The percent ^{238}Pu ranged from 33 ± 6 for soybeans to 55 ± 11 for corn. The soils used for growing these crops had only 21% ^{238}Pu. This difference in availability between the two radionuclides has been observed also at the Trinity Site ecosystem, where changing ratios of

^{238}Pu/^{239}Pu in soils, vegetation, and animal components were obtained (Hanson, 1975). Although the ^{238}Pu data in the various ecosystem components were not conclusive, earlier studies by McLendon et al. (1976) at SRP support evidence presented in other studies that there was an apparent increase in the bioavailability of ^{238}Pu relative to that of ^{239}Pu in the environment. Hanson (1975) has already discussed the possible mechanisms causing this difference.

In general, plutonium concentrations in vegetative parts of agricultural plants at SRP, where the primary mode of contamination was external in nature, were similar to those in the forage and vegetable foliage cultured at ORNL floodplain ecosystem, where contamination was caused primarily by root assimilation. However, the CR values of vegetative parts of crops from SRP were about one order of magnitude higher than those from ORNL.

Plutonium data on the edible portions of root crops are almost nonexistent. Potatoes grown on soils receiving only global fallout plutonium which had been peeled had a CR of 3×10^{-4} (Bennett, 1976). This is similar to the ORNL data, which indicate that peeling subterranean crops removed most of the plutonium, as high as 99.5% in the case of beets. Whether plutonium in these vegetative tissues occurred as a free ion is still unknown.

Data on plutonium in a variety of species from the ORNL floodplain suggest that CR values in the range of 10^{-4} to 10^{-2} are related to plutonium assimilation by the root pathway. The order of CR values was 2.4×10^{-3} (foliage av.) $> 1.7 \times 10^{-4}$ (fruit av.) $\geqslant 0.9 \times 10^{-4}$ (peeled root). The results for foliage compare favorably with CR values for plants grown in the glasshouse at SRP where aboveground tissues were protected from airborne sources of plutonium. Comparative results described in this chapter and elsewhere (Dahlman, Bondietti, and Eyman, 1976; Dahlman and McLeod, 1977) clearly ascribe high plutonium CR values (10^{-2} to 10^{0}) to contamination of foliage surfaces. Assimilation of plutonium by the root pathway was responsible for CR values in foliage of 10^{-3} or less. There was no evidence that assimilation of plutonium by the root pathway had been especially enhanced as a result of weathering, complexation, or other soil processes in the 30-yr period since the site was contaminated in 1944.

Summary

Agricultural crop experiments conducted adjacent to a reprocessing facility at SRP, which releases low chronic levels of plutonium per year through an emission stack, and at the WOC floodplain at Oak Ridge, where plutonium was deposited in 1944 from the Manhattan Project weapon development, revealed the following:

1. Plutonium concentrations and isotopic compositions in winter wheat grown in 1975 at SRP were affected by distance from the stack, plant height, and method of processing the grain. In general, vegetative materials closer to the stack (South Field) had plutonium concentrations a factor of 2 higher than samples farther away (North Field). The wheat straw had approximately two orders of magnitude higher plutonium content than the straw from the control site. Grain harvested by a combine had an order of magnitude higher plutonium content than laboratory-thrashed grain. When the plants were short, as in March, the sources of contamination were stack emission and resuspendible and soil materials, as indicated by an average 38 α % ^{238}Pu. Later in the

season, as the plants matured and grew taller, emission became the principal source of contamination, as indicated by a higher ^{238}Pu percentage.

2. Plutonium concentrations of the foliage of soybeans and corn grown in the South Field at SRP were similar to those of the June wheat straw [on the order of 10^{-2} pCi/g (dry weight)] collected from the same field. However, the corn foliage had higher α % ^{238}Pu than did the wheat and soybean foliage, which reflects the minimal contribution of resuspension to the corn canopy. Plutonium contents of the combine-harvested grain of the three crops were basically similar. Sifting the combined grain separated about 50% of the plutonium-bearing particles and/or extraneous matter from the grain and resulted in plutonium concentrations a factor of 2 lower in the sifted grain than in the unsifted grain.

3. Preliminary results from the SRP glasshouse studies indicated that only about 3% of the total contamination of field-grown crops adjacent to a reprocessing facility was contributed by root uptake. Also, it appeared from glasshouse results that ^{238}Pu was more available than ^{239}Pu.

4. Cultivation of the fields at SRP before planting wheat caused the plutonium concentrations in the top soil layer to become more uniform and to increase the plutonium concentrations at the deeper depths (5 to 15 cm). It also caused the ^{238}Pu percentage for the resuspendible portion to approach the soil values. Subsequent cultivations caused greater uniformity in plutonium concentration in the top soil.

5. Concentrations of ^{239,240}Pu in the foliage of forage and vegetable crops grown at the WOC floodplain ranged from 0.01 to 0.33 pCi/g. This represented the amounts of plutonium taken up by plants exclusively via the root pathway from soils with plutonium concentrations ranging from 10 to 150 pCi/g and averaging 63 pCi/g. Plutonium concentrations in the fruit of these species, however, were lower by at least an order of magnitude than those in the foliage.

6. Peeling the skins of potatoes and beets grown at the WOC floodplain removed approximately 99% of the residual plutonium.

7. Plutonium contents of soybean foliage were related to the stage of maturity and were maximum when the foliage biomass was maximum.

8. In general, the CR values of vegetative parts of crops at SRP were approximately one order of magnitude higher than those at Oak Ridge, which suggests the influence of aerial deposition of plutonium at the SRP site.

Acknowledgments

This work was supported by the U. S. Department of Energy through contract Nos. EY-76-C-09-0819 (to Savannah River Ecology Laboratory, University of Georgia), AT(07-2)-1 (to Savannah River Laboratory, E. I. du Pont de Nemours and Co.), and W-7405-eng-26 (to Oak Ridge National Laboratory, Union Carbide Corp.).

References

Adriano, D. C., and J. E. Pinder III, 1977, Aerial Deposition of Plutonium in Mixed Forest Stands from Nuclear Fuel Reprocessing, *J. Environ. Qual.*, 6: 303-307.

Auerbach, S. I., 1975, *Environmental Sciences Division Annual Progress Report*, Part I, Plutonium in the Environment, ERDA Report ORNL-5193, pp. 6-7, Oak Ridge National Laboratory, NTIS.

Bennett, B. G., 1976, Transuranic Element Pathways to Man, in *Transuranium Nuclides in the Environment*, Symposium Proceedings, San Francisco, 1975, pp. 367-383, STI/PUB/410, International Atomic Energy Agency, Vienna.

Bondietti, E. A., S. A. Reynolds, and M. H. Shanks, 1976, Interaction of Plutonium with Complexing Substances in Soils and Natural Waters, in *Transuranium Nuclides in the Environment*, Symposium Proceedings, San Francisco, 1975, pp. 273-287, STI/PUB/410, International Atomic Energy Agency, Vienna.

——, and F. H. Sweeton, 1977, Transuranic Speciation Questions, in *Transuranics in Natural Environments*, Symposium Proceedings, Gatlinburg, Tenn., October 1976, M. G. White and P. B. Dunaway (Eds.), ERDA Report NVO-178, Nevada Operations Office, NTIS.

Dahlman, R. C., E. A. Bondietti, and L. D. Eyman, 1976, Biological Pathways and Chemical Behavior of Plutonium and Other Actinides in the Environment, in *Actinides in the Environment*, A. M. Friedman (Ed.), ACS Symposium Series No. 35, pp. 47-80, American Chemical Society.

——, and K. W. McLeod, 1977, Foliar and Root Pathways of Plutonium Contamination of Vegetation, in *Transuranics in Natural Environments*, Symposium Proceedings, Gatlinburg, Tenn., October 1976, M. G. White and P. B. Dunaway (Eds.), ERDA Report NVO-178, pp. 303-320, Nevada Operations Office, NTIS.

——, and P. Van Voris, 1976, Cycling of ^{137}Cs in Soil and Vegetation of a Floodplain 30 Years After Initial Contamination, in *Radioecology & Energy Resources*, Proceedings of the Fourth National Symposium on Radioecology, Oregon State University, May 12–14, 1975, C. E. Cushing (Ed.), pp. 291-298, Ecological Society of America Special Publication Series No. 1, Academic Press, Inc., New York.

Duguid, J. O., 1975, *Status Report on Radioactivity Movement from Burial Grounds in Melton and Bethel Valleys*, ERDA Report ORNL-5017, p. 55, Oak Ridge National Laboratory, NTIS.

Francis, C. W., 1973, Plutonium Mobility in Soil and Uptake in Plants: A Review, *J. Environ. Qual.*, 2: 67-70.

Hanson, W. C., 1975, Ecological Considerations of the Behavior of Plutonium in the Environment, *Health Phys.*, 28: 529-537.

International Atomic Energy Agency, 1976, *Transuranium Nuclides in the Environment*, Symposium Proceedings, San Francisco, 1975, STI/PUB/410.

International Commission on Radiological Protection, 1959, *Report of Committee II on Permissible Dose for Internal Radiation*, ICRP Publication 2, Pergamon Press, Inc., New York.

——, (Task Group of Committee 2), 1972a, *The Metabolism of Compounds of Plutonium and Other Actinides*, ICRP Publication 19, Pergamon Press, Inc., New York.

——, (Task Group of Committee 2), 1972b, *Alkaline Earth Metabolism in Adult Man*, ICRP Publication 20, Pergamon Press, Inc., New York.

Klement, A. W., C. R. Miller, R. P. Minx, and B. Shlelen, 1972, Estimates of Ionizing Radiation Doses in the United States, 1960–2000, Report EPA-ORP/CSD-72-1, U. S. Environmental Protection Agency, Office of Radiation Programs, Washington, D. C.

Langley, T. M., and W. L. Marter, 1973, *The Savannah River Plant Site*, ERDA Report DP-1323, E. I. du Pont de Nemours & Co., Savannah River Laboratory, NTIS.

Lipton, W. V., and A. S. Goldin, 1976, Some Factors Influencing the Uptake of Plutonium-239 by Pea Plants, *Health Phys.*, 31: 425-430.

McLendon, H. R., 1975, Soil Monitoring for Plutonium at the Savannah River Plant, *Health Phys.*, 28: 347-354.

——, O. M. Stewart, A. L. Boni, J. C. Corey, K. W. McLeod, and J. E. Pinder III, 1976, Relationships Among Plutonium Contents of Soil, Vegetation and Animals Collected on and Adjacent to an Integrated Nuclear Complex in the Humid Southeastern United States of America, in *Transuranium Nuclides in the Environment*, Symposium Proceedings, San Francisco, 1975, pp. 347-363, STI/PUB/410, International Atomic Energy Agency, Vienna.

Oakes, T. W., and K. E. Shank, 1977, *Subsurface Investigation of the Energy Systems Research Laboratory Site at Oak Ridge National Laboratory*, ERDA Report ORNL-TM-5695, Oak Ridge National Laboratory, NTIS.

Price, K. R., 1972, *Uptake of ^{237}Np, ^{239}Pu, ^{241}Am, and ^{244}Cm from Soil by Tumbleweed and Cheatgrass*, USAEC Report BNWL-1688, Battelle Pacific Northwest Laboratories, NTIS.

Romney, E. M., A. Wallace, R. O. Gilbert, and J. E. Kinnear, 1975, ^{239,240}Pu and ^{241}Am Contamination of Vegetation in Aged Plutonium Fallout Areas, in *Radioecology of Plutonium and Other Transuranics in Desert Environments*, Nevada Applied Ecology Group Progress Report as of January 1975, M. G. White and P. B. Dunaway (Eds.), ERDA Report NVO-153, pp. 43-88, Nevada Operations Office, NTIS.

Schulz, R. K., G. A. Tomkins, and K. L. Babcock, 1976, Uptake of Plutonium and Americium by Plants from Soils: Uptake by Wheat from Various Soils and Effect of Oxidation State of Plutonium Added to Soil, in *Transuranium Nuclides in the Environment*, Symposium Proceedings, San Francisco, 1975, pp. 303-310, STI/PUB/410, International Atomic Energy Agency, Vienna.

——, G. A. Tomkins, L. Leventhal, and K. L. Babcock, 1976, Uptake of Plutonium and Americium by Barley from Two Contaminated Nevada Test Site Soils, *J. Environ. Qual.*, 5: 406-410.

Tamura, T., 1976, Physical and Chemical Characteristics of Plutonium in Existing Contaminated Soils and Sediments, in *Transuranium Nuclides in the Environment*, Symposium Proceedings, San Francisco, 1975, pp. 218-220, STI/PUB/410, International Atomic Energy Agency, Vienna.

Tennessee Valley Authority, 1975, personal communication from T. C. Bounds to R. C. Dahlman, Sediment Investigation, 1951, White Oak Creek and Lake.

Wessman, R. A., K. D. Lee, B. Curry, and L. Leventhal, 1978, Transuranium Analysis Methodologies for Biological and Environmental Samples, in *Environmental Chemistry and Cycling Processes*, DOE Symposium Series, Augusta, Ga., Apr. 28–May 1, 1976, D. C. Adriano and I. L. Brisbin, Jr. (Eds.), pp. 275-289, CONF-760429, NTIS.

Ecological Relationships of Plutonium in Southwest Ecosystems

T. E. HAKONSON and J. W. NYHAN

A comprehensive summary of results was prepared on plutonium distribution and transport in Los Alamos and Trinity Site study areas. Despite differences in ecosystems and plutonium source, there are several similarities in plutonium distribution between Los Alamos and Trinity Site study areas. First, the soils/sediment component contains virtually all the plutonium inventory, with vegetation and rodents containing less than 0.1% of the total in all cases.

Plutonium has penetrated to considerable soil depths at both locations, although it has occurred much more rapidly and to a greater degree in the alluvial soil at Los Alamos than in the arid terrestrial system at Trinity Site. However, in all cases less than 50% of soil-column plutonium inventories was found in the surface 2.5 cm. The plutonium penetration depth appears to correspond to the moisture penetration depth at Trinity Site. This is probably the governing factor at Los Alamos, although storm runoff and accompanying turbulent mixing processes complicate the process. In Acid–Pueblo Canyon, the bulk of the soil column inventory lies in the lower profiles, an indication of the loss of the plutonium from surface layers due to sediment transport.

Soil plutonium, in most cases, was associated with relatively coarse-size fractions. The silt–clay (<53 µm) fraction contained relatively little (<15%) of the plutonium; this reflects the small amounts of this size fraction in study area soils. An exception was in Area 21 at Trinity, where the <53-µm soil-size fraction contained about 73% of soil plutonium inventories. The importance of these distributional differences was demonstrated for Trinity Site, where Bagnold dust samples from Area 21 contained 54% silt–clay material and samples from Area Ground Zero (GZ) contained less than 10% of this material.

Concentrations in herbaceous vegetation were generally related to those in soils from all sites. Our belief, although unsubstantiated, is that external contamination of the plant surfaces is the major contaminating mechanism in these arid systems. The plutonium concentrations in certain rodent tissues from all study areas were related to corresponding soil concentrations. Over 95% of the plutonium body burden in rodents was associated with pelt and gastrointestinal tract samples, indicating the dominance of physical processes as the contaminating mechanism.

Horizontal transport of soil plutonium is dominated by physical processes. At Los Alamos water governs the downstream transport of soil plutonium, and indications are that wind is a relatively more important transport vector at Trinity Site.

In no case was there evidence for trophic-level increase due to physiological processes as plutonium passes from the soil to vegetation to the rodents, although food habits of rodents are not known sufficiently to preclude a trophic-level increase. We believe, however, that rodents most likely come into contact with environmental plutonium directly from the soil and not through a food-web intermediary.

Several reviews on environmental plutonium distribution and transport indicated a general lack of published field data from representative areas of the United States (Francis, 1973; Price, 1973; Romney, 1977; Hanson, 1975; Hakonson, 1975). Several field studies of plutonium have been initiated in the last few years to address informational needs at a number of locations which encompass a wide spectrum of climatic conditions ranging from deserts to humid forests and contain plutonium from industrial, weapons, or accidental-release sources.

The comparison of plutonium data from two southwest ecosystems in this chapter is one step in the total synthesis of information from various regions of the United States where types of ecosystems and sources of plutonium differ. The southwest United States is an important study locale because of the energy activities that may develop and the lack of understanding of the processes in arid systems which govern distribution and transport of contaminants. In this regard studies on environmental plutonium are useful to develop an understanding of patterns that are applicable to the transport and fate of other materials.

The objective of this chapter is to use existing plutonium contamination in the canyon waste areas at Los Alamos and in the grasslands in the fallout zone at Trinity Site

• To evaluate the role of environmental transport processes in distributing and redistributing surface inputs of plutonium.

• To evaluate the transport of environmental plutonium to the biosphere and the relationships that lead to the potential for human exposure.

• To compare plutonium behavior in these two major southwest ecotypes.

The tasks in this study were to (1) document plutonium inputs where possible, (2) develop an understanding of distributions by inventory of major environmental components, and (3) evaluate transfers as functions of ecological variables. Plutonium, as used in this chapter, denotes ^{238}Pu and/or ^{239,240}Pu.

Site Descriptions

Los Alamos

The canyons at Los Alamos, in north central New Mexico (Fig. 1), are typical of those in the southwest plateau region of New Mexico, Arizona, Colorado, and Utah. They vary from 10 m to over 200 m in depth and were formed by water erosion of the volcanic substrate of the Pajarito Plateau. The area has a semiarid continental mountain climate (Table 1) with annual precipitation ranging from about 20 to 50 cm as elevation increases from 1650 to 2200 m; rainfall accounts for about 75% of the annual precipitation. Drainage from the 113-km^2 Laboratory site is via the many canyons that bisect the plateau. The biotic resources of the canyons are diverse (Miera et al., 1977); total vegetative ground cover is variable but generally high and approaches 100% in some areas owing to the dense overstory, which is partly due to the industrial liquid effluents.

Nearly all the liquid wastes generated by the Laboratory since 1943 have been collected by industrial waste lines, treated (since 1951), and released into one of three canyons (Fig. 1). The results of studies in two of these canyons are emphasized in this chapter since they represent the extremes in temporal use history. The oldest waste-receiving area is Acid–Pueblo Canyon, which was used from 1943 to 1963 and

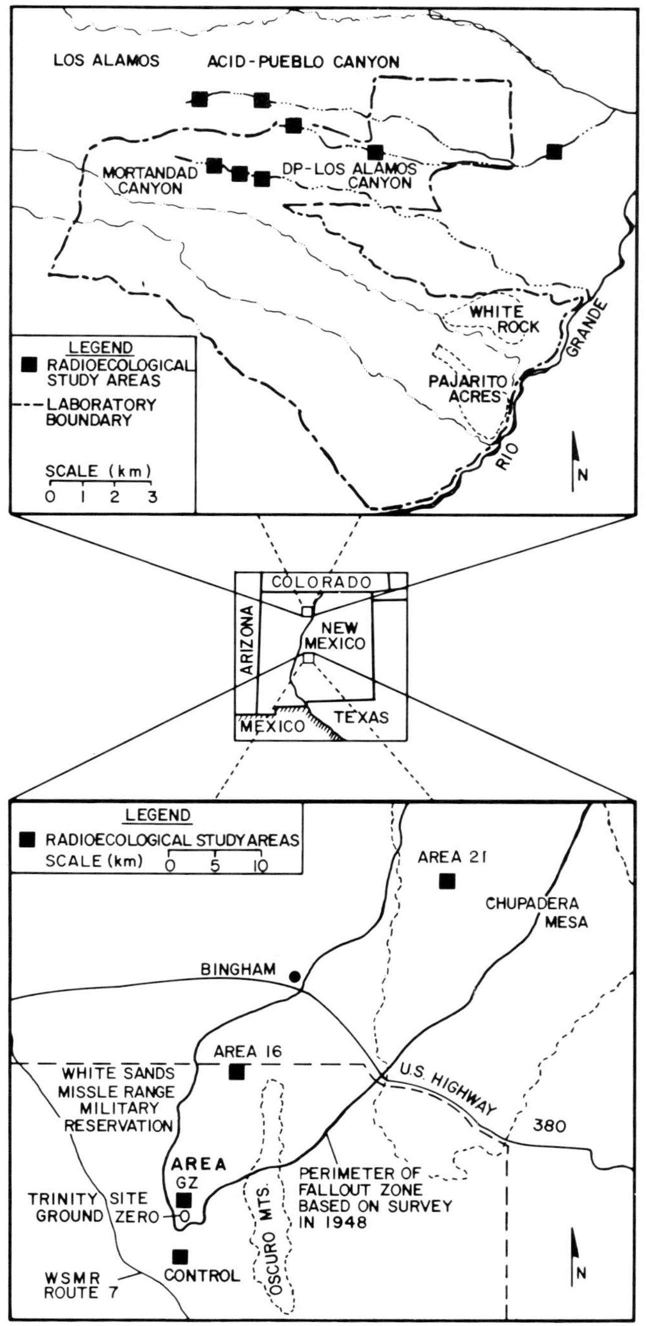

Fig. 1 Plutonium study areas at Los Alamos and Trinity Site, New Mexico.

TABLE 1   Some Characteristics of Plutonium Study Areas at Los Alamos and Trinity, New Mexico

	Mortandad	Acid–Pueblo	Area GZ	Area 21
Annual precipitation, cm	43 to 52	39 to 54	12 to 25	30 to 40
Average annual temperature, °C	7.5	7.1	15	12
Range	−26 to 36	−23 to 31	−5 to 39	−4 to 38
Soil	Sandy alluvium	Sandy alluvium	Sandy loam	Loam
Soil pH	8.6 to 9.2	7.1 to 7.9	7.5 to 8.4	8.2 to 8.4
Soil cation exchange capacity, equivalents/kg soil	0.06 to 0.09	0.05 to 0.10	0.02 to 0.02	0.02 to 0.03
Soil organic carbon, %	0.10 to 0.40	0.04 to 0.54	0.40 to 0.70	0.75 to 1.5
Clay mineralogy	Amorphous	Amorphous	Mixed	Mixed
Plutonium source	Industrial liquid effluent	Industrial liquid effluent	Weapon fallout	Weapon fallout
Soil ^{239,240}Pu/^{238}Pu concentration ratio, (pCi/g)/(pCi/g)	0.35*	100*	12†	21†
Weathering time of plutonium in environment (yr) as of 1973, the year most of the data in this paper were collected	0 to 11	14 to 30	28	28

*See Miera et al. (1977); Nyhan, Miera, and Peters (1976a).
†See Neher and Bailey (1976).

received an estimated 173 mCi of plutonium; Mortandad Canyon has been used for the least amount of time (from 1963 to present) and currently receives most of the Laboratory's liquid waste plutonium. As of 1973 and 1974, the years from which data in this chapter were collected, Mortandad Canyon had received about 61 mCi of plutonium.

Surface water exists in the upper reaches of both canyons as a result of Laboratory effluents and/or domestic sewage-treatment effluent; the lower portions of the canyons are normally dry. Surface water, including the pulse releases of plutonium-contaminated liquid effluent, rapidly percolates into the alluvium and generally disappears about 1 km downstream. Relatively large flows occur in both canyons during storm runoff events. Storm runoff reaches the Rio Grande via Acid–Pueblo and Los Alamos Canyons (Fig. 1), but the runoff water in Mortandad Canyon rapidly soaks into the thick alluvial deposits and seldom reaches postoutfall distances beyond 3 km. Many rainstorms at Los Alamos are intense, of short duration, and result in dramatic flash floods in the canyons.

Trinity

Trinity Site and the associated fallout zone is located in the northern end of the Tularosa basin in south central New Mexico (Fig. 1). The region is characterized (Table 1) by low rainfall (12 to 40 cm), high summer temperatures (commonly greater than 37°C), and severe wind and water erosion on exposed and disturbed ground surfaces. Rainfall accounts for about 90% of the annual precipitation. The area supports a relatively dense vegetation cover, considering the region; total vegetative ground cover ranges from about 15 to 25% (Neher and Bailey, 1976).

On July 16, 1945, a 20-kt atomic bomb was detonated 31 m above the ground surface at Trinity Site during a relatively unstable climatic regime when winds were to the northeast and were accompanied by intermittent thundershowers. Fallout from the cloud deposited in a northeast direction in the general pattern outlined in Fig. 1 (Larson et al., 1951). Relatively high fallout deposition occurred on Chupadera Mesa about 35 to 55 km from the crater. The reasons for the heavy deposition in this zone are unknown but may be related to weather or topographic factors. The elevation increases from about 1500 m at the crater to 2100 m on Chupadera Mesa. The fallout zone within 15 km of the crater is on the White Sands Missile Range, which is under U. S. Army jurisdiction. Beyond 15 km the fallout zone is on mixed private and public (Bureau of Land Management, State, and U. S. Forest Service) lands that are used heavily for domestic livestock grazing.

Plutonium Distribution

General

The chronic release of treated liquid effluents to the Los Alamos canyons has resulted in soil plutonium concentrations that are generally much higher than those at Trinity Site. Concentrations of a few hundred picocuries per gram (dry weight) are found in soils from the canyons, whereas those in Trinity soils average less than 1 pCi/g (Table 2). Worldwide fallout concentrations of ^{239,240}Pu in Los Alamos and Trinity Site soils average about 0.01 pCi/g (Apt and Lee, 1976; Nyhan, Miera, and Neher, 1976b).

TABLE 2   Ranges in Plutonium Concentration and Variability
Estimates in Some Los Alamos and Trinity Ecosystem
Components in 1973 and 1974

Component*	Los Alamos	Trinity
Soil (0 to 15 cm)		
pCi Pu/g	1–290†	0.02–0.32
CV‡	0.32–2	0.52–0.88
nCi Pu/m²	190–80,000	2.8–63
Vegetation		
pCi Pu/g	0.08–76	0.002–0.37
CV‡	0.65–2.2	0.38–1.1
pCi Pu/m²	0.7–600	0.07–5
Rodents		
fCi Pu/g	7–300	3–100
CV‡	0.16–1.3	0.52–1.3
fCi Pu/m²	0.2–20	0.03–2

*Dry-weight concentrations.
†Includes ^{238}Pu and ^{239,240}Pu.
‡Coefficient of variation (CV = standard deviation/mean).

Vegetation at both study locations contains the highest plutonium concentrations of any biotic component yet examined (Hakonson and Bostick, 1976). Plutonium concentrations in native grasses and forbs ranged from 0.08 to 76 pCi/g (dry weight) at Los Alamos and from 0.002 to 0.37 pCi/g in the Trinity fallout zone; levels in vegetation generally do not exceed those in corresponding soil samples. Additionally, the highest plutonium concentrations were associated with plants growing closest to the ground surface; taller growth forms, such as shrubs and trees, contained the lowest concentrations (Hakonson and Bostick, 1976; Hakonson and Johnson, 1974).

Plutonium concentrations in rodents, as representatives of the primary consumer trophic level, reflect the low physiological availability of the element. Pooled samples of internal organs from rodents generally do not contain measurable levels of plutonium, even though habitat soils may contain up to a few hundred picocuries per gram. Plutonium concentrations in whole rodents ranged from analytical detection limits of about 5 fCi/g to a few hundred femtocuries per gram; most of this radioactivity was associated with samples of pelt and gastrointestinal (GI) tract and contents.

Plutonium concentration variability, as characterized by the coefficient of variation in soils, plants, and animals, was uniformly high at all study sites. It commonly varied up to 2.0, with extreme values approaching 3.0 (Hakonson and Bostick, 1976; Nyhan, Miera, and Neher, 1976b). Variability of this magnitude has been observed at several environmental plutonium study sites in the United States (Little, 1976; Gilbert and Eberhardt, 1976) and results in the need for very large sample sizes in field experiments (Gilbert and Eberhardt, 1976; White and Hakonson, 1978).

Soils

Horizontal Distribution.   Horizontal plutonium concentration gradients are evident in both study areas, reflecting the dispersion from point sources of plutonium. Concentrations in the Los Alamos stream channels decrease one to two orders of magnitude in a predictable fashion (Nyhan, Miera, and Peters, 1976a; Hakonson and Bostick, 1976) within 10 km of the effluent sources, whereas similar differences occur over much greater distances at Trinity and do not necessarily decrease with distance. For example, plutonium concentrations in Trinity soils gradually increase from a minimum just outside the crater to a maximum at about 50 km from the crater (Nyhan, Miera, and Neher, 1976b; Larson et al., 1951).

Liquid effluent radionuclides at Los Alamos have been transported laterally into the stream banks as well as to downstream areas. Stream-bank soils accumulate radionuclides to levels equivalent to adjacent channel soil (Anonymous, 1977), and they serve as a long-term source of these materials to stream-bank biota. The stream banks, which are heavily vegetated, retard the downstream movement of radionuclides since they are not subject to the severe erosion encountered in the channel.

Although plutonium concentrations average much higher in the canyons than at Trinity, the extent of the contamination in the canyons is confined to less than 0.1 km^2, whereas the low-level contamination at Trinity Site covers several thousand square kilometers. Consequently the ecosystems at risk at Los Alamos are exposed to higher concentrations of plutonium than those at Trinity; however, the areas involved are smaller, and corrective action could be taken more easily should it ever be necessary.

Vertical Distribution. Some data from Area 21 (see Fig. 1) at Trinity Site indicate that the plutonium originally deposited on those environs in 1945 has been depleted from the soil surface over a 23-yr period (Table 3). Area 21 soils contained about 700 nCi/m^2 in 1950 (Olafson, Nishita, and Larson, 1957) and 18 nCi/m^2 in 1973 (Nyhan, Miera, and Neher, 1976b).

The depletion of plutonium from the soil surface is primarily due to the vertical transport of the element into the soil profile rather than to horizontal transport away from the study site by wind or water. Evidence that plutonium has migrated into the soil profile at the two Trinity Site locations is illustrated in Table 4 and is presented in detail by Nyhan, Miera, and Neher (1976b). In 1973 plutonium was detected at the 28- and 35-cm depths at Areas GZ and 21, respectively, whereas in 1950 plutonium was confined exclusively to the surface 2.5 cm (Olafson, Nishita, and Larson, 1957). The patterns of distribution with depth were typical of those observed in terrestrial soils in that plutonium concentrations decreased with depth.

TABLE 3 Comparison of Plutonium Concentrations in Surface (0 to 2.5 cm) Soils from Chupadera Mesa as a Function of Time After the Atomic Bomb Test at Trinity Site in 1945

Plutonium concentration, nCi/m^2		
1950*	1951*	1973†
746(0.31)‡	341(0.82)‡	18(0.48)‡
n = 6	n = 3	n = 8

*Data for 1950 and 1951 from Larson et al. (1951), and Olafson, Nishita, and Larson (1957).

†Data for 1973 from Nyhan, Miera, and Neher (1976b).

‡Parenthetic value is coefficient of variation (CV = standard deviation/mean).

TABLE 4 Mean Percent Plutonium Inventory in Soil Profiles from Los Alamos and Trinity Site Study Locations in New Mexico

	Trinity Site*			Los Alamos*	
Depth, cm	Area GZ	Area 21	Depth, cm	Mortandad	Acid–Pueblo
0–2.5	29(0.78)†	41(0.46)†	0–2.5	20(0.44)†	4.0(0.76)†
2.5–5.0	18(0.72)	19(0.63)	2.5–7.5	36(0.23)	10(0.48)
5–10	21(0.81)	6.0(0.88)	7.5–12.5	21(0.55)	20(1.3)
10–15	15(0.67)	8.0(0.92)	12.5–30	24(0.79)	67(0.18)
20–25	17(1.3)	16(1.0)			
25–33	ND‡	10(1.2)			

*n = 8 for Trinity Site data; n = 10 for Los Alamos data.

†Parenthetic value is coefficient of variation (CV = standard deviation/mean).

‡Not detectable.

The depth of plutonium transport into channel and bank soil profiles in the Los Alamos canyons is much greater than that at Trinity. In areas where permanent surface water exists (i.e., Mortandad Canyon), elevated plutonium concentrations are found at depths of 100 cm in the channel and at depths of 50 cm in the stream bank. Plutonium concentrations in channel soils do not show any consistent patterns with sampling depth, whereas decreasing concentrations with depth are evident in bank soils. In downstream areas, which are dry except during periods of storm runoff, plutonium occurs at depths of at least 30 cm (Nyhan, Miera, and Peters, 1976a).

The transport of plutonium into the channel alluvium and stream-bank soil has been rapid, as shown by the presence of elevated ^{238}Pu at the lower sampling depths. Elevated ^{238}Pu was observed at soil depths of 30 cm in Mortandad Canyon in 1972, about 4 yr after the first significant release of this element to the canyon (Hakonson and Bostick, 1976). In contrast, fallout ^{239,240}Pu in Trinity soils 5 yr after the bomb test was confined to the upper 2.5 cm of soil (Olafson, Nishita, and Larson, 1957).

A common feature of plutonium distribution in soils from both locations was that in 1974 less than one-half the total plutonium in the soil column was present in the surface 2.5 cm (Table 4) despite differences in soils and source of plutonium. In Acid—Pueblo Canyon 10 yr after the decommissioning of those facilities for waste disposal, an average of 67% of the soil column inventory was below the 12.5-cm depth, which reflects depletion of plutonium from the surface layers by vertical and horizontal transport processes. Previous studies in the canyons have shown that horizontal transport of soil during storm runoff events is an important mechanism in the downstream transport of plutonium (Purtymun, 1974; Hakonson, Nyhan, and Purtymun, 1976).

The depletion of plutonium from the soil surface decreases the probability of horizontal transport by wind and water but may increase the probability of uptake by vegetation during the time that the element is distributed within the plant rooting zone. However, over long periods of time, continued movement of plutonium into the soil profile may remove the element from the biologically active zone of the soil.

Particle Size Relationships.   The highest concentrations of plutonium in soil at the Los Alamos and Area 21 locations were associated with the silt—clay fraction, whereas this fraction at Area GZ, 1 km from the crater, contained the lowest concentrations of plutonium (Table 5) (Nyhan, Miera, and Neher, 1976b; Nyhan, Miera, and Peters, 1976c). At the GZ location, the highest concentrations were measured in the 1- to 2-mm soil particles, which perhaps reflects the physical characteristics of the fallout debris near the detonation site and/or depletion of the plutonium from smaller size fractions by wind or water transport vectors. Decreasing plutonium particle sizes with increasing distance from the crater were also noted at weapons test sites in Nevada (Romney, 1977).

The inventory of plutonium among the various soil size fractions in surface soils at the Los Alamos and Area GZ Trinity study sites was similar in that the silt—clay size fraction (<53 μm) comprised less than 10% of the soil mass and contained less than 15% of the plutonium (Table 5), whereas over 80% of the plutonium was associated with soil particles greater than 53 μm (Nyhan, Miera, and Neher, 1976b; Nyhan, Miera, and Peters, 1976c). The reverse was true at Area 21, Trinity Site, where the <53-μm fraction comprised 36% of the soil mass and contained over 70% of the soil plutonium inventory.

TABLE 5 Comparative Distribution of Plutonium in Surface Soil (0 to 2.5 cm) Size Fractions at the Los Alamos and Trinity Study Areas

	Soil size fraction*					
	<53 μm*	53−105 μm	105−500 μm	500−1000 μm	1−2 mm	2−23 mm
Mortandad Canyon						
pCi Pu/g†	1500.0	1300.0	610.0	310.0	87.0	69.0
Soil weight, %	2.2	1.8	14.0	21.0	26.0	35.0
Pu in fraction, %	14.0	6.0	27.0	21.0	16.0	16.0
Acid−Pueblo Canyon						
pCi Pu/g†	85.0	60.0	25.0	8.8	7.9	25.0
Soil weight, %	3.0	3.0	16.0	26.0	28.0	24.0
Pu in fraction, %	7.0	7.0	31.0	19.0	17.0	19.0
Trinity Site, Area GZ						
pCi Pu/g†	0.07	0.05	0.92	2.1	5.3	0.01
Soil weight, %	8.9	11.0	49.0	23.0	6.1	2.0
Pu in fraction, %	0.78	0.43	36.0	38.0	25.0	0.01
Trinity Site, Area 21						
pCi Pu/g†	3.8	1.7	0.42	0.64	1.6	0.23
Soil weight, %	36.0	18.0	25.0	4.2	2.9	14.0
Pu in fraction, %	73.0	16.0	5.5	1.4	2.4	1.8

*Size fraction data based on composite samples.

†Pu denotes primarily [239]Pu in Mortandad Canyon and [239,240]Pu at all other study locations.

Vegetation

Plant−Soil Relationships. The concentrations of plutonium in the study area vegetation were related to the levels of plutonium in associated soils (Fig. 2). The relationship between plutonium concentrations in vegetation and in soils was predictable over a range of five orders of magnitude in concentrations; this relationship is similar to relationships that were observed in the Rocky Flats environs (Little, 1976).

Plant−soil plutonium concentration ratios (CR = picocuries per gram of vegetation/ picocuries per gram of soil) are a convenient means of estimating the plutonium levels in vegetation growing on contaminated soils. Ratio estimates for native grasses in the Los Alamos and Trinity Site study areas (Table 6) ranged from 0.05 to 1.2, whereas the values for forbs ranged from 0.04 to 1.1. All these ratios are high relative to those derived from experimental studies where root uptake was the contamination mechanism (Romney and Davis, 1972; Wilson and Cline, 1966), which indicates that either plutonium is much more available to plants under field conditions or that mechanisms other than root uptake are responsible for the plutonium measured in plant samples from the field.

The relative amounts of plutonium associated with the internal and external portions of the vegetation are difficult to assess under field conditions, although we contend that

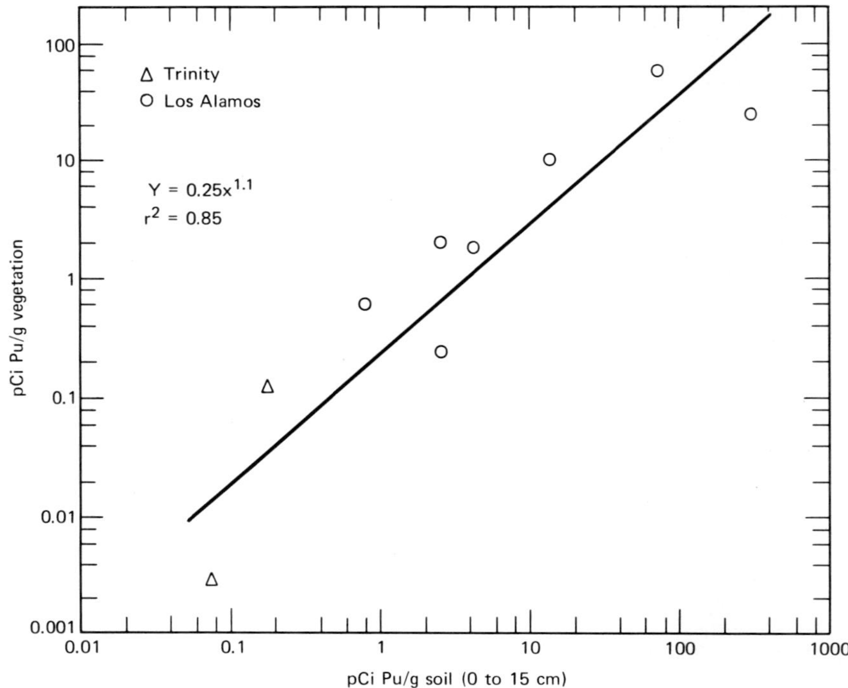

Fig. 2 Relationship of average plutonium concentration in herbaceous vegetation (grasses and forbs combined) and in corresponding soils in Los Alamos and Trinity Site study areas.

TABLE 6 Plutonium Concentration Ratios for Vegetation and Associated Soils from Los Alamos and Trinity Study Sites

		Plutonium concentration ratio*						
		Los Alamos				Trinity Site		
Component	n	Mortandad Canyon	n	Acid–Pueblo Canyon	n	Area GZ	n	Area 21
Grass	24	0.93(0.94)	19	0.13(1.2)	13	0.05(0.54)	16	1.2(0.74)
Forb	16	0.31(1.3)	11	0.23(0.28)	17	0.04(0.97)	21	1.1(0.92)

*Ratio calculated as [pCi Pu/g (dry weight) plant]/[pCi/g (dry weight) soil (0 to 15 cm depth)]. Parenthetic value is coefficient of variation (CV = standard deviation/mean).

most of the plutonium in our study areas is externally deposited on plant surfaces. Information supporting this conclusion includes:

- The high plant/soil plutonium concentration ratios compared to greenhouse studies.
- The obvious presence of soil in vegetation samples.

In addition, other investigators have shown that some of the plutonium associated with native vegetation samples can be removed by a wash treatment (Alldredge, Arthur, and Hiatt, 1977).

Rodents

Rodent–Soil Relationships. Plutonium in internal organs (i.e., liver, bone, and muscle) of rodents sampled within our study areas generally could not be measured. However, concentrations of plutonium in pelt and GI tract samples were readily measured and were

TABLE 7 Inventory of Plutonium in Small Mammal Tissues from Mortandad Canyon

Tissue	Percent of total body weight	Total plutonium,* pCi/g	Percent total plutonium
Pelt	23	0.85	50.0
GI tract	10	1.8	46.0
Lung	2	0.034	0.02
Liver	5	0.035	0.5
Carcass	60	0.018	2.8

*Based on six pooled samples.

directly correlated with levels in the study area soils ($r^2 = 0.90$). Over 95% of the plutonium body burdens in rodents was associated with these two tissues, as shown by the data for Mortandad Canyon in Table 7. Thus we conclude that, in our study areas, physical and biological processes (i.e., contamination of the pelt or ingestion of soil) dominate in the transport of plutonium to rodents.

Plutonium Inventories

The fractional distribution of plutonium in Los Alamos and Trinity ecosystem components (Table 8) is based on quantitative estimates of ecosystem component mass (grams per square meter) and corresponding plutonium concentrations (picocuries per square meter) in those compartments. The distribution of plutonium among five components was generally quite similar between sites in that over 99% of the plutonium was associated with soil and less than 1% with biota. Live vegetation contained 10^{-6} to $10^{-8}\%$ of the plutonium inventory. We conclude that very little of the environmental plutonium present in our study areas has appeared in the biological components of the ecosystem even after 30 yr of exposure. These results are essentially the same as those observed at Rocky Flats and Oak Ridge (Little, 1976; Dahlman, Garten, and Hakonson, this volume).

TABLE 8   Plutonium Inventory Ratios for Some Components of Los Alamos
and Trinity Study Areas in New Mexico

| | | Plutonium inventory ratio* | | | | | | |
| | | Los Alamos | | | | Trinity | | |
Component*	n	Mortandad Canyon	n	Acid–Pueblo Canyon	n	Area GZ	n	Area 21
Grass	24	4.1×10^{-5} (0.90)	20	5.6×10^{-4} (1.6)	13	2.0×10^{-5} (0.99)	16	1.3×10^{-4} (0.76)
Forb	16	4.8×10^{-5} (1.2)	11	1.7×10^{-4} (1.4)	17	1.7×10^{-4} (1.0)	21	3.5×10^{-5} (0.77)
Litter					5	1.6×10^{-4} (2.0)	3	1.1×10^{-4} (0.81)
Rodents	33	1.5×10^{-9} (0.77)	48	4.5×10^{-10} (0.99)	40	3.7×10^{-8} (1.7)	20	2.3×10^{-9} (0.47)
Soil	29	0.99(0.00009)	23	0.99(0.001)	8	0.99(0.0003)	8	0.99(0.00008)

*Inventory ratio = (pCi Pu/m² in component)/(total pCi Pu/m²). All plutonium values are [239,240]Pu except Mortandad Canyon which is [238]Pu; parenthetic value is coefficient of variation (CV = standard deviation/mean).

The relative inventory of plutonium within all our study ecosystems is governed primarily by component mass relationships since differences in mass of the various ecosystem components are much greater than the differences in plutonium concentrations between the same components. The data in Table 9 demonstrate that mass inventory ratios for Mortandad Canyon provide a good approximation of the plutonium inventory ratio.

TABLE 9   Mass and Plutonium Inventory Ratios
in Mortandad Canyon at Los Alamos

Component	Component mass, g/m²	Mass inventory ratio	Plutonium inventory ratio*
Soil (0 to 15 cm)	2.3×10^5	0.999	0.999
Grass	20.0	9.0×10^{-5}	4.1×10^{-5}
Forb	10.0	4.4×10^{-5}	4.8×10^{-5}
Rodents	0.03	1.3×10^{-7}	1.5×10^{-9}

*Data from Table 8.

Plutonium Transport

Soils

Rainstorm runoff in the intermittent streams receiving wastes was identified over 30 yr ago in the environmental transport of plutonium (Kingsley, 1947). Additional studies were begun to determine the relationships of rainfall, runoff, and suspended sediments with radionuclide transport (Purtymun, 1974; Purtymun, Johnson, and John, 1966; Hakonson, Nyhan, and Purtymun, 1976).

Results of these studies demonstrate that runoff from snow melt and summer rainstorms serves as a radionuclide transport vector in Los Alamos intermittent streams and that the magnitude of this transport is highly dependent on the hydrologic characteristics of the watershed and the intensity of runoff flow (Purtymun, 1974; Hakonson, Nyhan, and Purtymun, 1976). The dependency of concentrations of suspended sediments and plutonium in runoff on flow rate is indicated in Fig. 3 for one rainstorm runoff event in Mortandad Canyon. The nonlinearity in the curve is due to the relationship of flow rate with the particle size of resuspended material. At flows less than 0.25 m^3/sec, only the silt—clay size materials were in suspension in the runoff. However, at flows greater than 0.25 m^3/sec, coarser sands containing most of the sediment plutonium inventory (Table 5) were resuspended, which resulted in increased suspended sediment and radionuclide concentrations. High flow rates typically occur during the early phases of runoff events at Los Alamos owing to the intense nature and short duration of area rainstorms. We found that nearly 80% of the sediment and 70% of the radioactivity was transported within the first half of such events.

Additionally, there was a highly significant (P < 0.01) relationship between suspended sediment and radionuclide concentrations in runoff water. About 99% of the radioactivity in runoff was associated with suspended sediments greater than 0.45 μm in diameter, whereas only 1% of the radioactivity in the liquid phase was associated with sediments less than 0.45 μm in diameter.

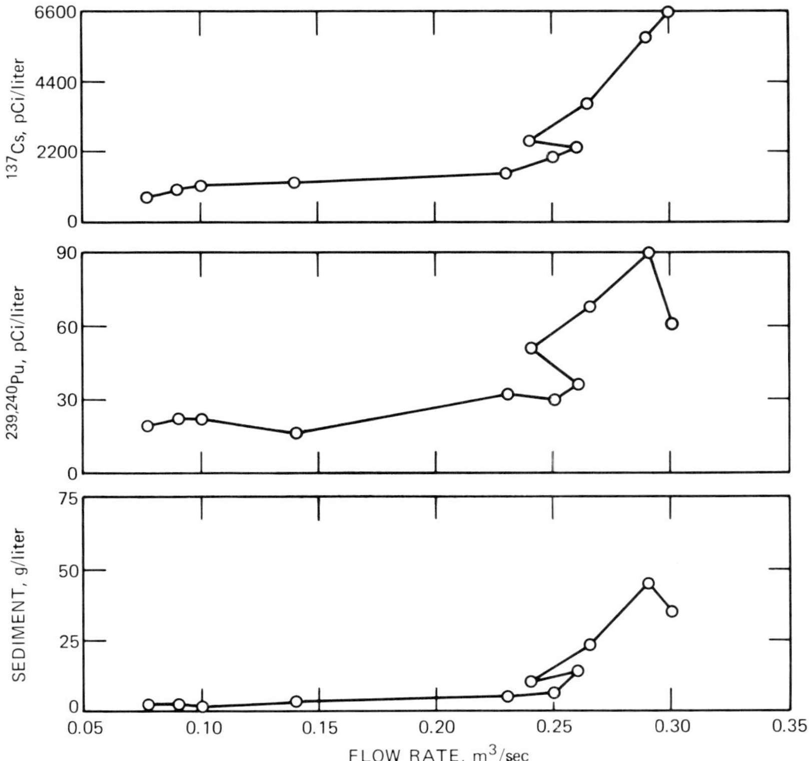

Fig. 3 Concentration of sediment and radioactivity in unfiltered runoff water from Mortandad Canyon as a function of runoff flow rate.

Studies were recently begun on wind transport of plutonium in the Trinity fallout zone, where evidence of wind erosion of soil is readily apparent. Although these studies are not complete, several important observations have been made. First, soil flux by surface creep and saltation processes is highly seasonal and has peaked in the months of July and August for two consecutive years of observation. Second, soil particle size analyses on dust-collector samples show major differences in the amount of silt—clay material between study sites. About half the dust material at Area 21 is in the silt—clay size range, whereas less than 1% of collected dust at Area GZ is in the silt—clay size range. These differences become important when coupled with the plutonium concentrations in the various soil particle size fractions (from Table 5). For example, silt—clay material in dust collectors at Area 21 contains over 200 times as much plutonium as the silt—clay fraction of dust samples at Area GZ.

Summary and Conclusions

Despite differences in ecosystems and plutonium source, there are several similarities in plutonium distribution between the Los Alamos and Trinity study areas. First, the soils—sediment component contains virtually all the plutonium, with vegetation and rodents containing less than 0.1% of the total. Plutonium has penetrated to considerable soil depths at both locations, although it has occurred much more rapidly and to a greater degree in the alluvial soil at Los Alamos than in the arid terrestrial soils at Trinity. At both locations less than 50% of soil column plutonium inventories was found in the surface 2.5 cm.

The plutonium penetration depth appears to correspond to the moisture penetration depth in the Trinity fallout zone. This is probably the governing factor at Los Alamos, although storm runoff and accompanying turbulent mixing complicate the process. In Acid—Pueblo Canyon, the bulk of the soil column inventory lies in the lower profiles, an indication of the loss of plutonium from surface layers due to sediment transport.

The plutonium in most cases was associated with relatively coarse soil size fractions. The silt—clay (<53 μm) fraction contained relatively little ($<15\%$) of the plutonium, a reflection of the small amounts of this size fraction in study area soils. An exception was in Area 21 at Trinity, where the <53-μm soil size fraction contained about 73% of soil plutonium inventories. The importance of these distributional differences stems from the fact that silt—clay soil particles can be transported farther and are more likely to adhere to biological surfaces than larger size fractions.

Concentrations in herbaceous ground vegetation were generally related to those in soils from all sites. Our data strongly indicate that external contamination of plant surfaces is the major soil-to-plant transport mechanism in these arid systems. The plutonium concentrations in pelt and GI tissues were related to corresponding soil concentrations at all sites. Over 95% of the plutonium body burden in rodents was associated with pelt and GI tract samples, an indication of the dominance of physical and/or biological processes as the contaminating mechanism.

Horizontal transport in both areas is dominated by wind- and water-driven processes. At Los Alamos surface runoff water governs the downstream transport of plutonium; indications are that wind is a relatively more important transport vector at Trinity, although splash-up of soil by raindrops may be an important transport mechanism in these arid, sparsely vegetated study locations.

There was no evidence for a trophic-level increase of plutonium from soil to vegetation to rodents. We believe that rodents come into contact with environmental plutonium directly from the soil and to a lesser extent through a food-web intermediary.

Research Needs

The importance of the soils component as a receptor of plutonium released to the Los Alamos and Trinity Site study areas coupled with the direct role these soils play in contamination of biota emphasizes the importance of understanding soil formation and transport processes. Factors governing these processes will be instrumental in determining plutonium distribution and transport as a function of time. Hydrologic and wind transport processes discriminate against certain soil particle sizes; therefore studies on the relationship of plutonium to soil separates will be useful in evaluating the potential importance of a transport pathway and the resulting hazard. We know, for example, that wind transport of silt—clay material at Area GZ, Trinity Site, would represent a relatively smaller inhalation hazard than corresponding transport at Area 21 simply because the silt—clay fraction of Area GZ soil contains very little of the plutonium inventory.

Factors affecting migration of plutonium into the soil profile require understanding since depletion of plutonium from the soil surface will likely reduce the horizontal transport potential and may alter the availability of the element to vegetation.

Field studies should be conducted to quantify the relative importance of the root pathway for contaminating vegetation to serve as a basis for judging changes in physiological availability of environmental plutonium with time. As yet few field studies have been able to show conclusively the relative importance of the two contamination mechanisms.

In our opinion studies should be continued on the availability of plutonium to native animals in our study ecosystems; however, on the basis of present concentrations and the high variability associated with these measurements, we believe that the frequency of sampling should be drastically reduced. Perhaps sampling at intervals of 5 to 10 yr would be adequate to judge whether significant changes in plutonium availability have occurred.

Acknowledgments

We wish to recognize the following individuals for their valuable contributions to this research: J. L. Martinez, G. Trujillo, E. Trujillo, K. Bostick, T. Schofield, G. Martinez, K. Baig, P. Baldwin, R. Peters, W. Schwietzer, and S. Lombard. We also wish to thank R. O. Gilbert, D. Adriano, J. Corey, and G. Matlack for their efforts in reviewing this manuscript. This research was performed under U. S. Department of Energy contract No. W-7405-ENG-36.

References

Alldredge, A. W., W. J. Arthur, and G. S. Hiatt, 1977, Mule Deer as a Plutonium Vector, in *Radioecology of Natural Systems, 15th Annual Progress Report for the Period August 1, 1976–July 31, 1977*, ERDA Report COO-1156-89, Colorado State University, NTIS.

Anonymous, 1977, *Environment Surveillance at Los Alamos During 1976*, ERDA Report LA-6801-MS, Los Alamos Scientific Laboratory, NTIS.

Apt, K. E., and V. J. Lee, 1976, *Environmental Surveillance at Los Alamos During 1975*, USAEC Report LA-6321-MS, Los Alamos Scientific Laboratory, NTIS.

Francis, C. W., 1973, Plutonium Mobility in Soil and Uptake in Plants: A Review, *J. Environ. Qual.*, 2: 67-70.

Gilbert, R. O., and L. L. Eberhardt, 1976, Statistical Analysis of "A Site" Data and Interlaboratory Comparisons for the Nevada Applied Ecology Group, in *Studies of Environmental Plutonium and Other Transuranics in Desert Ecosystems*, M. G. White and P. B. Dunaway (Eds.), USAEC Report NVO-159, pp. 117-154, Nevada Operations Office, NTIS.

Hakonson, T. E., 1975, Environmental Pathways of Plutonium into Terrestrial Plants and Animals, *Health Phys.*, 29: 583.

——, and K. V. Bostick, 1976, Cesium-137 and Plutonium in Liquid Waste Disposal Areas at Los Alamos, in *Radioecology and Energy Resources*, Proceedings of the Symposium on Radioecology, Oregon State Univ., May 13–14, 1975, The Ecological Society of America, Special Publications: No. 1, C. E. Cushing, Jr. (Ed.), pp. 40-48, Halsted Press, New York.

——, and L. J. Johnson, 1974, Distribution of Environmental Plutonium in the Trinity Site Ecosystem After 27 Years, in *Proceedings of the Third International Radiation Protection Association Meeting*, Washington, D. C., Sept. 9–14, 1973, W. S. Snyder (Ed.), USAEC Report CONF-730907-P1, pp. 242-247, International Radiation Protection Association, NTIS.

——, J. W. Nyhan, and W. D. Purtymun, 1976, Accumulation and Transport of Soil Plutonium in Liquid Waste Discharge Areas at Los Alamos, in *Transuranium Nuclides in the Environment*, Symposium Proceedings, San Francisco, 1975, pp. 175-189, STI/PUB/410, International Atomic Energy Agency, Vienna.

Hanson, W. C., 1975, Ecological Considerations of the Behavior of Plutonium in the Environment, *Health Phys.*, 28: 529-537.

Kingsley, W. H., 1947, *Survey of Los Alamos and Pueblo Canyon for Radioactive Contamination and Radioassay Tests Run on Sewer-Water Samples and Water and Soil Samples Taken from Los Alamos and Pueblo Canyons*, USAEC Report LA-516, Los Alamos Scientific Laboratory, NTIS.

Larson, K. H., J. H. Olafson, J. W. Neel, W. F. Dunn, S. H. Gordon, and B. Gillooly, 1951, *The 1949 and 1950 Radiological Survey of Fission Product Contamination and Some Soil-Plant Inter-Relationships of Areas in New Mexico Affected by the First Atomic Bomb Detonation*, USAEC Report UCLA-140, University of California at Los Angeles, NTIS.

Little, C. A., 1976, *Plutonium in a Grassland Ecosystem*, Ph.D. Dissertation, Colorado State University, Fort Collins, Colorado, ERDA Report COO-1156-83.

Miera, F. R., Jr., K. V. Bostick, T. E. Hakonson, and J. W. Nyhan, 1977, *Biotic Survey of Los Alamos Radioactive Liquid-Effluent Receiving Areas*, USAEC Report LA-6503-MS, Los Alamos Scientific Laboratory, NTIS.

Neher, R. E., and O. F. Bailey, 1976, *Soil Survey of White Sands Missile Range, New Mexico*, U. S. Department of Agriculture, Soil Conservation Service, West Regional Technical Service Center, Portland, Oregon.

Nyhan, J. W., F. R. Miera, Jr., and R. J. Peters, 1976a, The Distribution of Plutonium and Cesium in Alluvial Soils of the Los Alamos Environs, in *Radioecology and Energy Resources*, Proceedings of the Symposium on Radioecology, Oregon State Univ., May 13–14, 1975, The Ecological Society of America, Special Publications: No. 1, C. E. Cushing, Jr. (Ed.), pp. 49-57, Halsted Press, New York.

——, F. R. Miera, Jr., and R. E. Neher, 1976b, The Distribution of Plutonium in Trinity Site Soils After 28 Years, *J. Environ. Qual.*, 5: 431-437.

——, F. R. Miera, Jr., and R. J. Peters, 1976c, Distribution of Plutonium in the Soil Particle Size Fractions in Liquid Effluent-Receiving Areas at Los Alamos, *J. Environ. Qual.*, 5: 50-56.

Olafson, J. H., H. Nishita, and K. H. Larson, 1957, *The Distribution of Plutonium in the Soils of Central and Northeastern New Mexico As a Result of the Atomic Bomb Test of July 16, 1945*, USAEC Report UCLA-406, pp. 1-25, University of California at Los Angeles, NTIS.

Price, K. R., 1973, A Review of Transuranic Elements in Soils, Plants and Animals, *J. Environ. Qual.*, 2: 62-66.

Purtymun, W. D., 1974, *Storm Runoff and Transport of Radionuclides in DP Canyon, Los Alamos County, New Mexico*, USAEC Report LA-5744, Los Alamos Scientific Laboratory, NTIS.

——, G. L. Johnson, and E. C. John, 1966, *Distribution of Radioactivity in the Alluvium of a Disposal Area at Los Alamos, New Mexico*, U. S. Geological Survey, Professional Paper 550-D, pp. D250-D252.

Romney, E. M., 1977, Plutonium Contamination of Vegetation in Dusty Field Environments, in *Transuranics in Natural Environments*, M. G. White and P. B. Dunaway (Eds.), USAEC Report NVO-178, pp. 287-302, Nevada Operations Office, NTIS.

——, and J. J. Davis, 1972, Ecological Aspects of Plutonium Dissemination in the Environment, *Health Phys.*, 22: 551.

White, G. C., and T. E. Hakonson, 1978, Statistical Considerations and Survey of Plutonium Concentration Variability in Some Terrestrial Ecosystem Components, *J. Environ. Qual.*, 8: 176-182.

Wilson, D. O., and J. F. Cline, 1966, Removal of Plutonium-239, Tungsten-185, and Lead-210 from Soils, *Nature*, 209: 941-942.

Plutonium in a Grassland Ecosystem

CRAIG A. LITTLE

This chapter is primarily concerned with plutonium contamination of grassland at the U. S. Department of Energy Rocky Flats plant, which is located northwest of Denver, Colo. Major topics include the definition of major plutonium-containing ecosystem compartments; the relative amounts in those compartments; whether or not the predominant isotopes, ^{238}Pu and ^{239}Pu, behaved differently; and what mechanisms might have allowed for the observed patterns of contamination.

Samples of soil, litter, vegetation, arthropods, and small mammals were collected for plutonium analysis and mass determination. Small aliquots (5 g or less) were analyzed by a rapid scintillation technique and by alpha spectrometry.

Of the compartments sampled, greater than 99% of the total plutonium was contained in the soil. The concentrations of plutonium in soil were significantly inversely correlated with distance from the contamination source, depth of sample, and particle size of the sieved soil samples. The soil data suggested that the distribution of contamination largely resulted from physical transport processes.

Concentrations of plutonium in litter and vegetation were inversely correlated to distance from the source and directly correlated to soil concentrations at the same location. Comparatively high concentration ratios of vegetation to soil suggested wind resuspension of contamination as an important transport mechanism.

Arthropod and small-mammal tissue samples were highly skewed, kurtotic, and quite variable. Plutonium concentrations were lower in bone than in other tissues. Hide, gastrointestinal tract, and lung were generally not higher in plutonium concentration than kidney, liver, and muscle. All data tended to indicate that physical transport processes were the most important.

Median isotopic ratios of ^{239}Pu to ^{238}Pu by activity concentration in soil were 40 to 50. Litter and vegetation isotopic ratios were similar to those of soil. Arthropod and small-mammal isotopic ratios were lower than those of soil, which implied that the two isotopes were differentially incorporated into the animal bodies and ^{238}Pu was taken up at a higher rate. However, further investigations suggested that statistical bias may have spuriously contributed to the lower isotopic ratios in small animals.

Most of the world's agriculture occurs on land that, before tilling, was once covered by stands of grasses. An important untilled tract of land contaminated with plutonium* is the grassland immediately adjacent to and contained within the Rocky Flats plutonium processing plant and associated buffer zone about 12 km northwest of Denver, Colo., metropolitan area. Because Rocky Flats is a prime example of plutonium-contaminated

---

*The word "plutonium" indicates $^{239,240}Pu$ in this chapter, unless otherwise noted.

grassland, this chapter will dwell primarily on data from environmental sampling at Rocky Flats.

The Rocky Flats installation uses nearly 30 km^2 as a buffer zone to separate the public from plutonium-handling operations. The climate at the installation is typified by occasional strong WNW winds exceeding 40 m/sec and moderate precipitation, i.e., 40 cm/yr average (Rocky Flats 1975 annual weather summary, unpublished). The Rocky Flats grassland has been modified by the activities of humans and includes plant species typical of short-grass plains (*Bouteloua gracilis* and *Buchloe dactyloides*) as well as tall-grass prairie (*Agropyron* spp. and *Andropogon* spp.) and ponderosa pine (*Pinus ponderosa*) woodland (Weber, Kunkel, and Shultz, 1974). Mule deer (*Odocoileus hemionus*) are found on the site along with grassland species of reptiles, rodents, and birds (Whicker, 1974).

Source of the Contamination

Investigations by Krey and Hardy (1970) of DOE's Environmental Measurements Laboratory (EML, formerly Health and Safety Laboratory) suggested that the most likely contamination source was a storage area of stacked 55-gal barrels that leaked plutonium-polluted oil. Data supporting the conclusion of Krey and Hardy and a description of the nature of the stored oil—plutonium mixture are delineated.

Air-sampling data from Rocky Flats link the barrel storage area to the east-southeast contamination pattern discovered by Krey and Hardy (1970). Air-sampling station S-8, one of many such stations maintained and sampled regularly by Rocky Flats personnel, is located about 75 m east and slightly south of the barrel storage area. Except for a brief period during 1961, monthly averages of daily airborne contamination values have been kept since at least 1960 to the present (Fig. 1).

The S-8 air-sampling data indicated that contamination peaks in the air were associated with dates of perturbation of the contaminated surface (Table 1 and Fig. 1). Except for periods of disturbance, the gross alpha concentrations in ambient air were near 0.01 pCi/m^3. However, during excavation and paving of the barrel storage area, the alpha concentration in air markedly increased (Table 1).

The plutonium-contaminated cutting oil, about as viscous as lightweight motor oil but thinned by carbon tetrachloride, was stored in the 55-gal barrels for periods of up to 7 yr. The interactions between the oil, air, CCl$_4$, and plutonium within the barrels were probably quite important in determining the eventual fate of the element.

The oil was filtered through 2- to 3-μm filters before being placed in the barrels. The discard limit at the time of storing was 1 x 10^{-2} g of plutonium per liter of oil. If the limit and the filtering had been observed and performed faithfully, each of the approximately 3570 plutonium-containing barrels would have had no more than 2.1 g of plutonium (0.13 Ci) (M. A. Thompson, Environmental Sciences, Rocky Flats Plant, and F. J. Miner, Chemical Resources, Rocky Flats Plant, personal communications).

It is difficult to assess what occurred once the oil was inside the barrels. The presence of carbon tetrachloride in the drums allows the possibility that hydrochloric acid was formed, which, in turn, may have reacted with the plutonium metal to form very low concentrations of plutonium chloride, a more-soluble form of the element (J. M. Cleveland, Environmental Studies, Rocky Flats Plant, personal communication). This possibility is given credibility by the work of J. Navratil of Rocky Flats Chemical Research Division, who has studied contaminated cutting oil in recent years.

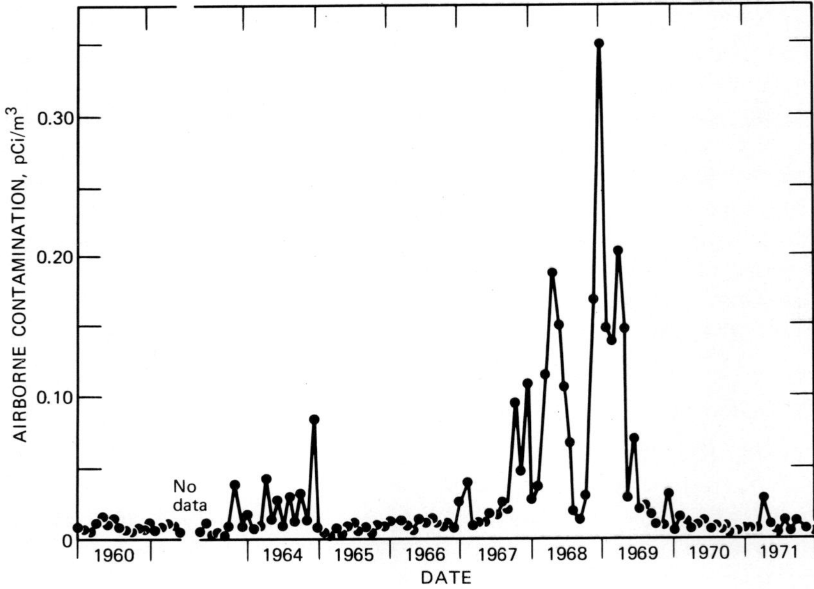

Fig. 1 **Monthly means of daily gross alpha activity in ambient air at station S-8 (75 m east of the oil-barrel storage area). Aliquots of Gelman AE filter material were counted in a gas-flow proportional detector. Data adapted from D. C. Hunt, Environmental Sciences, Rockwell International, personal communication.**

TABLE 1 Total Monthly Gross Alpha Activity in Ambient Air at Station S-8 (75 m East of Oil-Barrel Storage Area) During Disturbances of the Storage-Area Surface*†

Dates	Event	Alpha activity, pCi/m³
7/59–9/63	No large-scale leaking.	0.009
1/64–1/65	Large-scale leaking.	0.025
1/65	Contaminated soil covered with fill.	0.01
1/66	Small building added to filter contaminated oil from leaking to new drums.	0.014
1/67	Drum-removal activity begun.	0.038
6/68	Last drums removed but high winds spread some activity.	0.188
2/69	Weeds burned and area graded for paving.	0.34
9/69	Asphalt pad completed.	0.013
11/69	Four sampling wells dug through pad.	0.033
4/71	Drainage ditch dug on west side of asphalt pad.	0.033

*Air-filter material was counted directly in a gas-flow proportional counter.

†Adapted from D. C. Hunt, Environmental Sciences, Rockwell International, personal communication.

Navratil and Baldwin (1976) found that filtering the contaminated oil through a 0.01-μm filter removed only about 50% of the plutonium. This result strongly suggested that about half the plutonium was in a relatively large particulate form whereas the other half was associated with very small particles. Fission-track analysis of the filtered oil confirmed that the remaining plutonium was monomeric. It is doubtful that the barrels consistently held the above proportions of particulate and nonparticulate plutonium oxide, but each probably contained some plutonium chloride.

J. M. Cleveland (personal communication) also suggested that the filtered 3-μm plutonium particles might combine to form larger aggregates of the metal. Of course, the size and binding tenacity of these conglomerates are unknown.

Methods

Two macroplots were chosen for intensive sampling of plutonium in soil, vegetation, and litter. The locations of these macroplots relative to the supposed plutonium source, the barrel storage area, and the prevailing wind are shown in Fig. 2. A sampling grid was superimposed over each macroplot. The macroplot 1 grid was approximately 0.75 ha

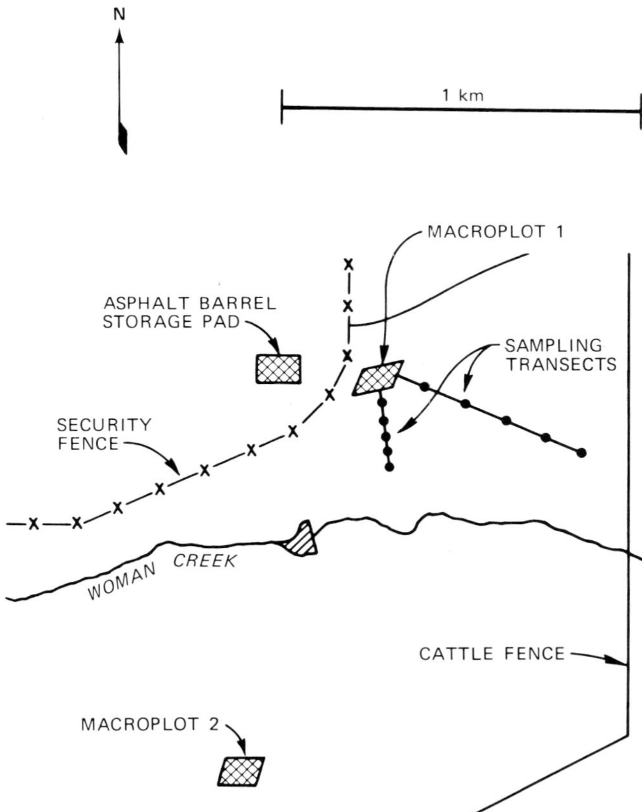

Fig. 2 Schematic representation of the southeast corner of the Rocky Flats Plant indicating the location of study macroplots and sampling transects. The wind rose indicates the relative magnitudes of wind velocities during 1974.

(1 ha = 2.5 acres) and contained 100 sampling markers. Plutonium data in this chapter are from macroplot 1 unless specifically noted otherwise.

Depth-profile soil samples were taken by hand with a trowel. After vegetation and litter had been clipped and bagged separately, four 5- by 5- by 3-cm samples were removed and bagged separately for each of seven 3-cm-depth layers to 21 cm deep. If rocks precluded sampling at a given depth, the column was resumed below the blockage. Soil samples were transported to the laboratory and air-dried. Rocks or debris greater than 0.5 cm in diameter were removed from the sample. After oven-drying and weighing, samples were mechanically shaken on brass soil sieves. The accumulation on each sieve was weighed and placed in a small paper envelope, and the envelope was sealed with tape.

Litter and standing vegetation were sampled from 0.25-m^2 and 0.5-m^2 areas, respectively. Vegetation was clipped and bagged, and litter was gathered by hand and bagged. In the laboratory litter and vegetation samples were air-dried and weighed. Soil was separated from the litter samples by a flotation process (Little, 1976). The net vegetation or the litter dry weight was divided by the microplot size, 0.5 m^2 or 0.25 m^2, respectively, to calculate mass per unit area. For plutonium analysis vegetation and litter samples were ground on a Wiley mill with an 850-μm opening screen, and 5-g aliquots were taken.

Arthropods were sampled by a combination of sweep netting, pitfall trapping, and drop trapping at random sites on established grids. At the laboratory animals were separated into generic groups that were weighed separately. These generic totals were combined for an estimated weight per 0.5-m^2 microplot. Arthropods obtained by the drop-trap method were not analyzed for plutonium owing to fear of cross contamination from soil during the vacuuming process. Samples for plutonium analysis and a species inventory list resulted from the sweep netting and pitfall traps. For plutonium analysis a representative composite was analyzed for each sampling period. Arthropods were not cleaned prior to plutonium analysis.

Small-mammal trapping grids were superimposed over each macroplot in a manner resembling that used by the U. S. International Biological Program Grassland Biome (Packard, 1971). Animals were trapped about six times yearly. Sherman live traps were used for cricetid and sciurid rodents. Geomyid rodents were trapped less regularly in homemade live traps placed in burrows systems. Approximately 15% of the estimated population was removed from each macroplot during each trapping session. These animals were collected by removing dead-in-trap individuals during the regular session and the remainder randomly in one extra trapping night. Small mammals were either dissected immediately or frozen for dissection later. Special precautions were taken during dissection to minimize cross contamination between tissues. Approximately 10 cm^2 of hide was used as an aliquot. Lungs, liver, and gastrointestinal (GI) tract were removed intact. Muscle samples were taken from the legs in most cases. Bone samples consisted of the whole skeleton, which had been cleaned of flesh by a dermestid beetle colony. All samples except bone were placed on tared, ashless filter papers, oven-dried at 50 to 60°C, and weighed. The sample was placed in a snap-cap vial for storage or transport to a commercial laboratory.

Some soil-sample aliquots (5 g) were analyzed for plutonium content by commercial laboratories (LFE, Richmond, Calif., and Eberline Instrument Corp., Albuquerque, N. Mex.). Most soil samples were analyzed in our laboratory, as were most litter and vegetation samples. Small-mammal tissues and arthropods were commercially analyzed. The LFE method used concentrated hydrofluoric acid to dissolve samples (Wessman et al., 1971); Eberline modified a pyrosulfate fusion technique for the same purpose (Sill, 1969). Ion-exchange columns removed interfering elements and isolated plutonium from the samples before alpha spectrometry analysis. Chemical recovery was measured by adding ^{236}Pu tracer to each sample. Agreement between homogenized split samples sent to these laboratories was good (Little, 1976). In our laboratory a procedure was used that incorporated harsh digestion of the sample by nitric and hydrofluoric acids, ion exchange, organic extraction, and liquid scintillation spectrometry (Little, 1976). This method had an estimated minimum detectable activity of 0.42 pCi (P < 0.05). Plutonium data in this chapter are ^{239,240}Pu unless otherwise noted. Plutonium-240 contributed about 20% on the average to the alpha activity of ^{239,240}Pu.

Plutonium Compartmentalization

The inventories of plutonium in the principal compartments of the grassland ecosystem were calculated. Compartments investigated were soil, in 3-cm increments from 0 to 21 cm, litter, standing vegetation, arthropods, and small mammals.

The compartmental inventories of plutonium were calculated by multiplying the mean mass of each compartment (g/m^2) by the mean plutonium concentration of the compartment ($\mu Ci/g$). A total ecosystem inventory was calculated by summing over all compartments. The compartmental fraction (unitless) of the total plutonium inventory was calculated by dividing each compartmental inventory ($\mu Ci/m^2$) by the total inventory ($\mu Ci/m^2$).

The soil compartment had vastly the largest fraction of the total plutonium, 99.69% (Table 2). As expected, the fraction of the total plutonium contained within a soil layer decreased as the depth increased. The litter compartment comprised less than 1% of the total plutonium (53 nCi/m^2) in the study areas, and the vegetation represented only about 0.01% of the total plutonium. By virtue of representing both low biomasses and plutonium concentration, the animal compartments, arthropods and small mammals, had extremely small fractions of the total plutonium, 6.8×10^{-9} and 3.3×10^{-9}, respectively.

In summary, the compartmentalization data indicated that greater than 99% of the plutonium in the study area was located in the soil. At the time of sampling, nearly one-half (49.7%) the total plutonium was in the top 3 cm of soil. In decreasing order, smaller plutonium-inventory fractions were found in litter, vegetation, arthropods, and small mammals. The implication of these results is that, in the present state, transport of plutonium is closely linked to soil movement or erosion. Therefore efforts to prevent plutonium transport off contaminated grasslands should be directed primarily at minimizing soil transport rather than mobilization by biota.

Plutonium in Soil

Analysis of the soil plutonium data suggested that two primary generalizations about plutonium in soil could be stated. First, the plutonium concentrations in the soil samples were highly variable. Second, the plutonium concentration in a soil sample was a function of sample location, sample depth, and the soil particle composition of the sample. Rationales for both of these conclusions are examined in some detail.

Frequency distributions of plutonium in soil samples were positively skewed (P < 0.05) with coefficients of variation (CV = standard deviation ÷ mean) ranging to

**TABLE 2 Distribution of ^{239}Pu in Samples
from the Rocky Flats Study Macroplot***

Compartment	Mean	n†	95% confidence interval‡
Plutonium concentrations, pCi/g			
Soil, 0–3 cm	835	72	554–1120
Soil, 3–21 cm	105	309	69–141
Litter	412	29	314–509
Vegetation	28.6	76	15.7–41.4
Arthropods§	5.48	23	3.13–7.83
Small mammals	6.50	304	2.38–10.6
Fraction of total plutonium			
Soil, 0–3 cm	5.0×10^{-1}		$2.5 \times 10^{-1} - 7.4 \times 10^{-1}$
Soil, 3–21 cm	5.0×10^{-1}		$2.5 \times 10^{-1} - 7.5 \times 10^{-1}$
Litter	2.9×10^{-3}		$1.6 \times 10^{-3} - 4.2 \times 10^{-3}$
Vegetation	1.0×10^{-4}		$4.1 \times 10^{-5} - 1.6 \times 10^{-4}$
Arthropods§	1.2×10^{-8}		$4.6 \times 10^{-9} - 2.0 \times 10^{-8}$
Small mammals	3.3×10^{-9}		$6.6 \times 10^{-10} - 6.0 \times 10^{-9}$

*Compartmental ^{239}Pu inventory (pCi/m^2) equals mean biomass [g(dry)/m^2] times mean concentration [pCi/g(dry)]. Fraction of total equals mean compartmental inventory (pCi/m^2) divided by total inventory.

†Number of samples for which the mean is calculated: For arthropods and vegetation n is the number of groups of individuals analyzed; for small mammals n is the number of tissue samples, not individual animals.

‡95% confidence interval equals mean ± (1.96 standard error of the mean).

§Includes data from Bly (1977).

greater than 2.0. Although positive skewness is a characteristic of lognormal distributions, the natural-log transformation of soil data did not result in normal distributions (Kolmogorov–Smirnov one-sample test, P > 0.05) but did reduce the skewness for the seven depth groups tested.

Three adjacent soil columns (5 by 5 cm) from a 5- by 15-cm area on macroplot 2 exhibited the extreme spatial variability that sometimes occurred in plutonium concentrations in the soil. The mean plutonium concentrations in the 5-g aliquots from each column were 480 (column A), 5.4 (column B), and 0.57 pCi/g (column C) at the 0- to 3-cm depth. Virtually all the plutonium in column A was in the top 3 cm, the other depths (in 3-cm increments to 21 cm) being at or near background. In columns B and C, the majority of the plutonium was found at lower depths. No other cases of such extreme spatial variation in soil plutonium concentrations were detected during the sampling at Rocky Flats.

As expected, surface soil samples (0 to 3 cm) had a higher mean plutonium concentration than subsurface samples (Table 3). This result agreed with data from Rocky Flats soil sampling reported by Krey and Hardy (1970). Plutonium concentrations were also a function of the size range of soil practices comprising the aliquot (Tables 3 and 4).

TABLE 3 Mean Plutonium Concentrations of Soil Samples from Rocky Flats

Soil particle size range, μm	Plutonium concentration, pCi/g						
	0–3 cm	3–6 cm	6–9 cm	9–12 cm	12–15 cm	15–18 cm	18–21 cm
850–2000	740	140	88	27	13	5.4	1.8
425–850	460	120	100	36	29	7.0	5.5
250–425	440	130	89	30	30	13	5.7
180–250	460	120	130	39	30	14	8.9
150–180	770	130	100	35	140	25	6.5
75–150	870	210	100	68	56	44	11
45–75	1400	310	210	100	160	84	35
0–45	1500	180	810	190	220	85	27
0–2000	830	170	200	66	86	35	13

TABLE 4 Regression Parameters of Soil Plutonium Concentration (pCi/g) Adjusted for the Sample Location as a Function of Soil Particle Diameter at Various Depths*

Depth, cm	Intercept (b_0)	Slope (b_1)	Correlation coefficient (r)	Significant at α =	n
0–3	4.8	−0.336	−0.312	0.01	72
3–6	3.6	−0.270	−0.291	0.05	69
6–9	0.89	−0.753	−0.471	0.001	50
9–12	1.6	−0.544	−0.564	0.001	69
12–15	0.67	−0.799	−0.719	0.001	52
15–18	−0.21	−0.775	−0.706	0.001	47
18–21	−0.42	−0.572	−0.358		22

*The model used was $\ln \text{Pu} = b_0 + b_1 \ln \text{D}$.

Least-squares regressions were calculated with linear, exponential, and power-function models of plutonium concentration in surface soil as functions of the distance east or south from the asphalt pad. The power-function model gave the best fits of the data for both curves (Figs. 3 and 4). A t-test indicated that the slope of the distance-south curve (Fig. 4) was steeper (P < 0.05) than that of the distance-east curve (Fig. 3). These results conform to the concept of wind-distributed plutonium; the more effective, stronger winds were to the east, and hence the slope of that curve was smaller.

Several multiple linear-regression models were calculated. The model that accounted for the largest fraction of the total variance (0.868) had the following form:

$$\ln \text{Pu} = 11.15 - 0.0535 \ln \text{E} - 1.628 \ln \text{S}$$

where Pu is the plutonium concentration (pCi/g), E is the distance east of the asphalt pad centerline (m), and S is the distance south of the asphalt pad centerline (m).

With this model plutonium concentrations of samples in the soil depth profile were adjusted to estimate the concentration expected at a common location. The adjusted values were then regressed as a function of sample depth (Fig. 5). As with the distance

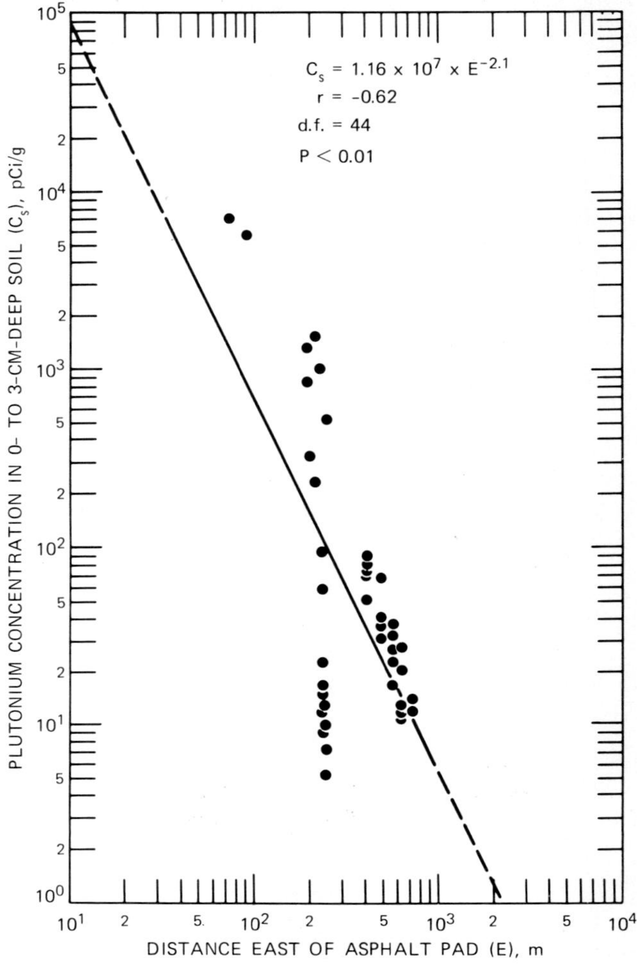

Fig. 3 Plutonium concentration in 0- to 3-cm-deep Rocky Flats macroplot 1 soil as a function of distance east of the center of the asphalt oil-barrel storage pad.

relationships, a power-function regression model had the highest correlation of plutonium concentration with depth of the models attempted and was significant ($P < 0.01$).

The relationship of plutonium concentration in soil as a function of soil particle diameter (as represented by the opening of the final passage sieve) was examined for each depth layer (Table 4). The model resulted in a significant ($P < 0.05$) regression for each depth group except the 18- to 21-cm group. The steepest slope (-0.799), at 12 to 15 cm, was significantly different ($P < 0.05$) from the flattest slope (-0.270), at the 3- to 6-cm level. However, there was no obvious trend in slope vs. soil particle-size curves with depth. Because the amount of surface area represented by the soil particle spheres in a constant mass of soil is inversely related to soil particle diameter, it followed that the plutonium concentration in a soil sample should be inversely related to the surface area of the particles in the sample. A tabulation of the fractions of the total soil-sample mass

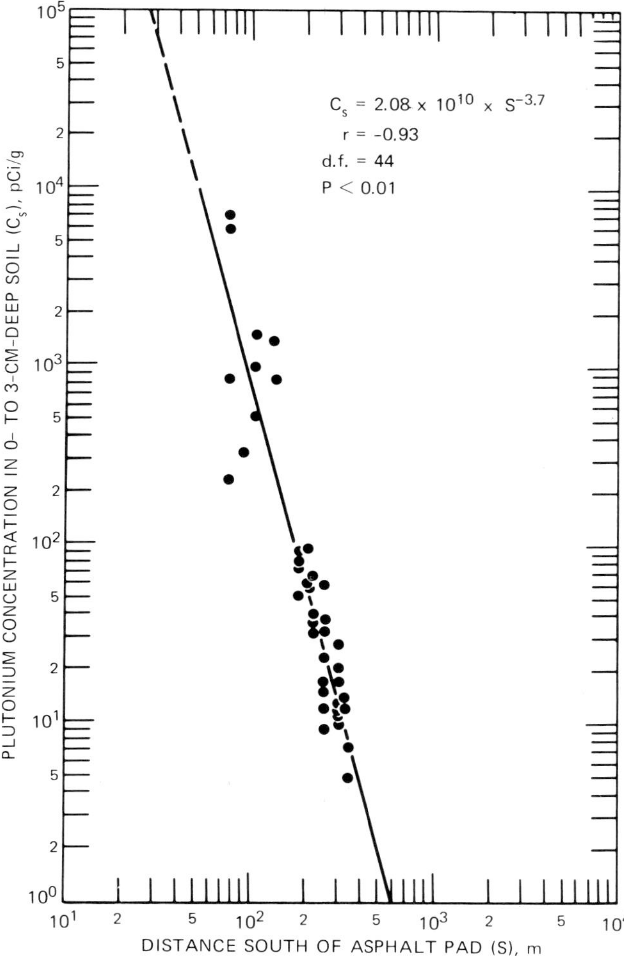

Fig. 4 **Plutonium concentration in 0- to 3-cm-deep Rocky Flats macroplot 1 soil as a function of distance south of the asphalt oil-barrel storage pad.**

represented by each sieve size organized by depths did not produce any obvious patterns with either depth or particle size range. Consequently regressions of the soil mass fraction per sample as a function of depth were not significant for most sieve fractions. However, these last results would not preclude a surface-attachment mechanism.

The data on plutonium in soil at Rocky Flats can be summarized by several statements. First, the variance in the plutonium concentrations of the soil samples was large; CV's within groups of like samples (same depth and particle size) ranged to over 2.0. Frequency distributions for soil samples were positively skewed. Spatial variation was also large; in one instance the plutonium concentrations of aliquots taken less than 15 cm apart varied by nearly three orders of magnitude.

Second, in spite of the large degree of variance in the data, the plutonium concentrations in soil were significantly correlated with the location and soil particle

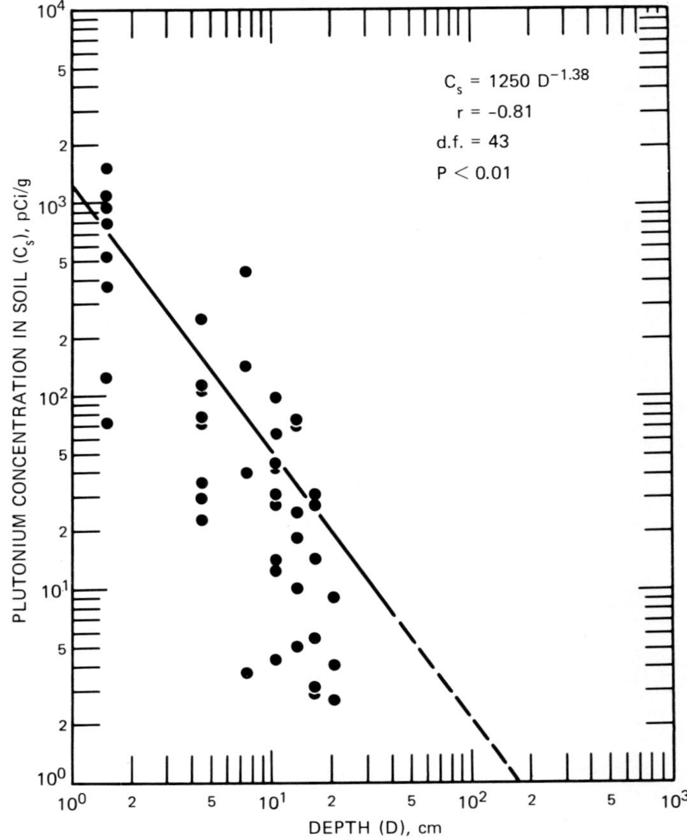

Fig. 5   Plutonium concentration in Rocky Flats macroplot 1 soil as a function of depth of sample. Sample concentrations adjusted for distance east and south of center of asphalt oil-barrel storage pad.

composition of the soil sample. The spatial distribution of plutonium (i.e., more plutonium downwind than downslope) implicated wind as the prime mechanism of plutonium transport onto the studied areas. Such factors as resuspension with or without added mechanical disturbances by humans or fauna undoubtedly contributed to the wind transport of plutonium but to a presently unknown degree. The data also indicated that plutonium was found to a depth of 21 cm in most samples from downwind of the barrel storage area but that about two-thirds of the contamination was in the top 5 cm. The relationship between plutonium concentration and soil particle size suggested a surface-attachment mechanism of plutonium attachment to soil particles. However, the lack of any pattern of soil mass fraction with depth for the various particle sizes probably indicates that plutonium transport with depth is not simply a case of transport of the various plutonium—soil particles downward.

Plutonium in Plant Compartments

The vegetation community of the study area was composed mostly of grasses and members of the sunflower family. Members of the sedge, pea, and mustard families were

**TABLE 5 Plutonium Concentrations
in Rocky Flats Vegetation
and Litter Samples**

| | Plutonium concentration, pCi/g | | |
	Mean	Coefficient of variation	n
Litter	412	0.65	29
Vegetation	28.6	2.02	76

also present but in much lower numbers of individuals. Rather than study numerous plant types individually, two plant-derived compartments were studied, litter and detritus and standing vegetation. Although these compartments accounted for only a small fraction of the total plutonium (about 0.2%), the study of those compartments helped derive some concepts of plutonium transport.

As with the soil, frequency distributions for vegetation samples were positively skewed. Further, the hypothesis that plutonium concentrations in vegetation were lognormally distributed could not be rejected (P > 0.05). Unexpectedly, the hypothesis that plutonium concentrations in litter were normally distributed could not be rejected (P > 0.05).

Mean plutonium concentrations in litter were higher than those in vegetation (Table 5). Concentrations of plutonium in litter and vegetation were each inversely correlated with distance east or south from the asphalt pad (P < 0.01).

The fact that litter had a higher mean concentration of plutonium than standing vegetation is not surprising. This result reinforces the suggestion made above that soil transport was the primary mechanism of plutonium transport.

Plutonium in Animal Compartments

Two animal compartments were studied, arthropods and small mammals. These compartments together contained about 2×10^{-8} of the total plutonium estimated to be in the studied areas. Nevertheless, the mobility of the animals makes them potential transporters of plutonium, albeit relatively small amounts, off the site.

As expected from the soil and vegetative sampling, the frequency distributions of plutonium concentrations in small mammals were positively skewed, as indicated by the histogram in Fig. 6. Not only were there many samples that had plutonium concentrations below the detection limit but also much of the total activity was supplied by relatively few samples. Frequency distributions of plutonium concentrations in arthropods were also positively skewed (Bly, 1977). Bly (1977) further indicated that logarithmic transformations were useful in alleviating the skewness. Therefore the plutonium concentrations of the arthropod samples were probably lognormally distributed.

Concentrations of plutonium in 23 groups of individual arthropods and in small-mammal tissues were of comparable magnitude (Table 6). The small-mammal

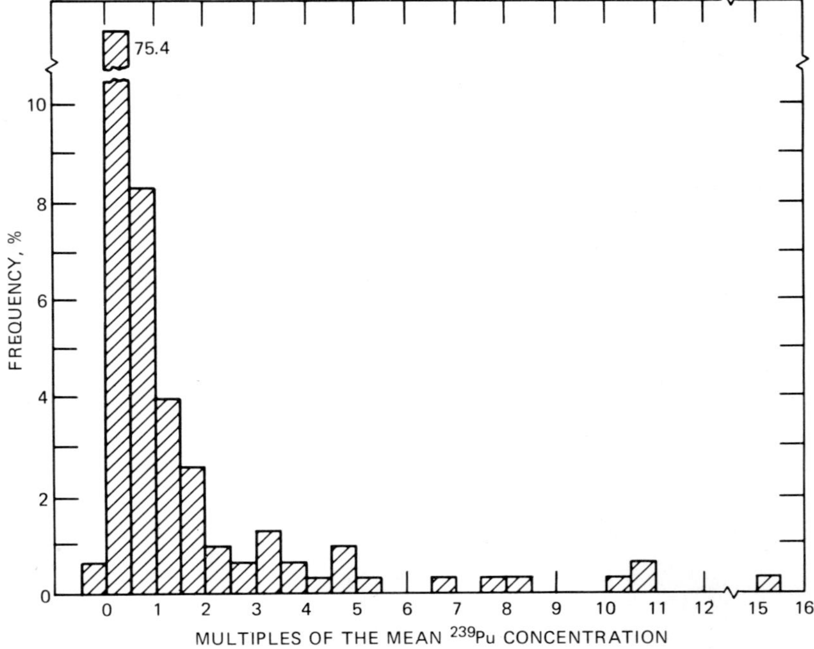

Fig. 6   Representative histogram of small-mammal tissue samples from Rocky Flats.

TABLE 6   Mean Plutonium Concentrations of Arthropods* and Small Mammals Sampled from Rocky Flats Macroplot 1

Sample type	n	Plutonium concentration, pCi/g	
		Mean	Coefficient of variation
Arthropods	23	5.48	1.05
Small mammals			
Bone	28	0.288	2.28
GI tract	40	7.03	2.50
Hide	47	1.51	1.84
Kidney	45	13.6	4.39
Liver	46	8.38	5.45
Lung	47	3.57	1.90
Muscle	50	8.92	5.85
External tissues	134	3.88	2.76
Internal tissues	169	8.59	5.58

*Includes data adapted from Bly (1977).

tissue-sample means ranged from 0.288 pCi/g for bone to 13.6 pCi/g for kidney, and the mean of whole arthropods was 5.48 pCi/g.

The patterns, or rather, lack of patterns, in the small-mammal data were puzzling. The tissues were arbitrarily classed either external or internal, depending on whether or not the tissue had a direct contact with the animal's environment. External tissues included GI tract, hide, and lung; internal tissues included bone, kidney, liver, and muscle. By virtue of the supposed low biological availability of plutonium and the proximity of the external tissues to the contaminated soil, external tissues were expected to have larger plutonium concentrations than internal tissues. Inexplicably, this was not the case. The three highest plutonium concentrations were found in internal tissues, i.e., kidney, muscle, and liver; hide and lung comprised two of the three lowest means. Additionally, the amount of variation in samples within a given tissue was quite high. The minimum tissue variation was in hide samples (CV = 1.84), and the maximum was in muscle (CV = 5.85).

Only two explanations for the high degree of variability are at hand. First, the possibility of cross contamination always exists no matter how carefully one removes tissues during dissection. Second, the extremely small sample mass of a few samples (a dry kidney may be as small as 0.05 g) may have had a tendency to magnify the relative plutonium concentrations. However, a plot of plutonium concentration in small mammals vs. sample mass indicated that about as many samples had large mass and small plutonium concentrations as had small mass and large concentrations. Beyond this, however, the tendency for small-mass samples to skew the distribution has not been investigated.

The nonparametric Kruskal—Wallis technique (Siegel, 1956) was used to test whether or not the seven tissue means were from the same population. The resulting chi-square value of about 44 indicated that the difference between the tissue groups was highly significant (P < 0.001). Although no test was performed, it was intuitively obvious that the mean plutonium concentration of the bone samples (0.29 pCi/g, n = 28) was lower than that of other tissues.

Plutonium Concentration Ratios

The concentration ratio (CR) is a potential indicator of plutonium redistribution by wind, water, or plant uptake. Concentration ratio is defined as the concentration in activity per unit mass or volume divided by the concentration of the same nuclide in the same units in another material. In this section the CR will have 0- to 3-cm-deep soil as the material in the denominator [e.g., CR of litter = (pCi Pu/g litter) ÷ (pCi Pu/g 0- to 3-cm-deep soil)].

The CR's of litter, vegetation, arthropods, and small mammals are listed in Table 7. Litter had the largest CR followed in descending order by vegetation, small mammals, and arthropods. Regressions of litter and vegetation CR's vs. distances east and south of the asphalt pad did not achieve significant correlation coefficients (P > 0.05).

The plutonium concentrations in litter and in vegetation were plotted vs. soil plutonium concentrations from the same locations. Only the litter curve is shown here (Fig. 7). The litter regression was interesting because of its high correlation (r = 0.975) and near-unit slope (1.001). Although the number of samples here was limited, the data comprising Fig. 7 suggested that litter may be an excellent estimator of soil plutonium concentration in the grassland. The regression of plutonium concentration in vegeta-

**TABLE 7   Plutonium Concentration Ratios and
95% Confidence Intervals of Ecosystem Compartments
in Rocky Flats Macroplot 1 with 0- to 3-cm-Deep Soil***

Compartment	Concentration ratio	95% confidence interval
Litter	4.9×10^{-1}	$2.9 \times 10^{-1} - 7.0 \times 10^{-1}$
Vegetation	3.4×10^{-2}	$1.5 \times 10^{-2} - 5.4 \times 10^{-2}$
Arthropods	6.8×10^{-3}	$3.1 \times 10^{-3} - 1.1 \times 10^{-2}$
Small mammals	7.8×10^{-3}	$2.1 \times 10^{-3} - 1.3 \times 10^{-2}$

*Mean plutonium concentration in 0- to 3-cm-deep soil equals 835 pCi/g. Concentration ratio equals mean pCi/g compartment divided by mean pCi/g in 0- to 3-cm-deep soil.

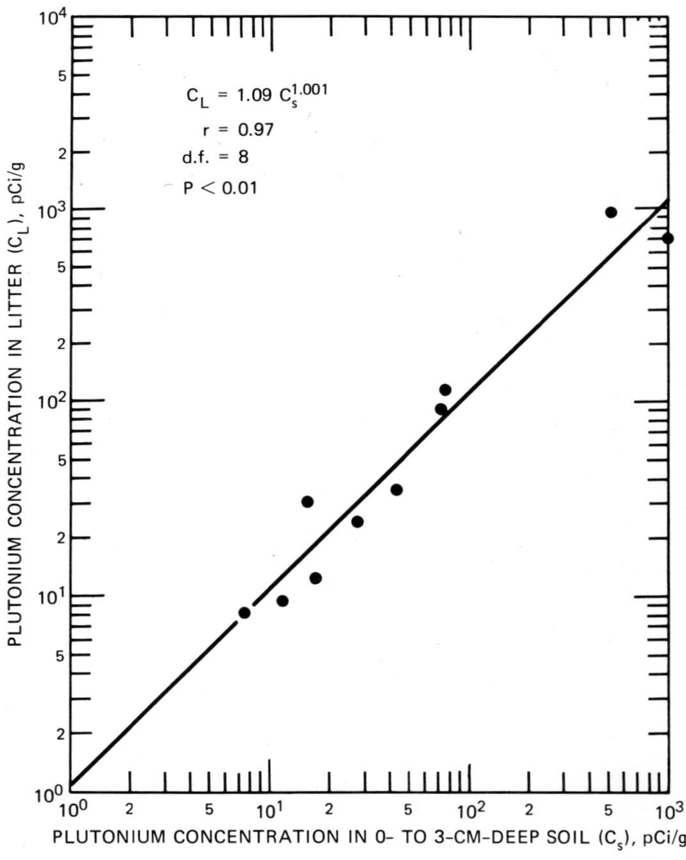

Fig. 7   Plutonium concentration in litter vs. plutonium concentration in soil at the same sample location.

tion vs. plutonium concentration in underlying soil was also statistically significant (P < 0.01) but was less conclusive than the litter vs. soil curve and is not shown here.

The CR's of the vegetation were higher than those produced in greenhouse studies. Typically, uptake of plutonium under laboratory conditions has been on the order of 10^{-6} to 10^{-4} of the soil concentrations (Newbould, 1963; Wilson and Cline, 1966; Romney, Mork, and Larson, 1970; Schulz, Tompkins, and Babcock, 1976). The Rocky Flats CR of 3.4×10^{-2} suggests either increased root uptake by grassland species or another method of contamination, such as aerial deposition of resuspended soil particles. The high surface-to-volume ratio of grasses and the hairy nature of the leaves of many members of the sunflower family would be amenable to a high rate of impaction and attachment of small soil particles. Given wind-redistributed plutonium at Rocky Flats, surficial attachment of contaminated soil particles to plants is the likely mechanism of contaminating the vegetation.

Plutonium Isotopic Ratios

Ratios of plutonium isotopes or ratios of ^{239}Pu and ^{241}Am have been reported from several sites (Emery et al., 1976; Gilbert et al., 1975; Hakonson and Johnson, 1974; Markham, 1976). In the hope that the examination of the isotopic ratios of ^{239}Pu and ^{238}Pu in the grassland would give some insight into the relative ecological availability of these two nuclides, isotopic ratios were calculated for samples analyzed by alpha spectrometry [isotopic ratio (IR) = ^{239}Pu pCi/g of sample ÷ ^{238}Pu pCi/g of same sample]. Ratios were not calculated for samples where either isotope was below the detection limit. Ratios were tabulated according to sample type and tested for goodness of fit to a normal distribution. The distribution of IR in the various soil depths was either lognormal or marginally normal. Small-mammal tissues appeared to be lognormal with respect to the IR.

As suggested by Doctor and Gilbert (1977), the concentration of ^{239}Pu was plotted vs. ^{238}Pu for each of the seventeen sample types. Seventy percent of these groups exhibit a zero intercept, based on a t-test. These results implied that the ratios were constant within the tested sample groups and that $^{239}\overline{Pu}/^{238}\overline{Pu}$ would be an unbiased estimator. However, because of the likelihood that the IR is lognormally distributed within most of the sample groups, median IR's are reported here (Table 8). The R_4 method of calculation discussed by Doctor and Gilbert (1977) was used to calculate these values.

At first glance the median isotopic ratio in soil appeared to decrease as depth increased. However, the overlapping 95% confidence intervals for the listed medians suggested that the ratio is relatively constant. As expected, neither linear, exponential, nor power-function regressions of the raw IR data vs. soil depth were statistically significant (P > 0.05).

Despite the limited number of litter and vegetation samples analyzed for both ^{239}Pu and ^{240}Pu, the median IR's of these two compartments were very similar to IR's of the soil. These results tended to indicate that the litter and vegetation were closely linked to the soil.

The IR's in the animal compartments raised some very interesting questions (Table 8). Only two sample types (GI tract and muscle) exhibited 95% confidence intervals that overlapped with soil IR confidence intervals. Therefore, it appeared that the small-mammal and arthropod compartments had lower IR's than soil. A lower IR would imply

**TABLE 8   Median Isotopic Ratios
(^{239}Pu pCi/g ÷ ^{238}Pu pCi/g) in
Rocky Flats Environmental Samples***

Compartment	Isotopic ratio†		n
	Median	95% confidence interval	
Soil depth, cm			
0–3	54.70	37.04–80.78	10
3–6	43.70	35.09–54.43	7
6–9	46.41	35.23–61.14	8
9–12	46.45	40.67–53.06	6
12–15	48.76	42.81–55.54	6
15–18	46.94	32.27–68.26	3
18–21	38.67	34.16–43.78	2
Litter	55.47	51.62–59.61	5
Vegetation	59.98	39.92–90.12	3
Arthropods	9.88	5.69–17.15	9
Small-mammal tissues			
Bone	7.49	2.99–18.71	9
GI tract	24.82	17.18–35.90	20
Hide	19.94	13.99–28.43	21
Kidney	11.07	3.98–30.80	7
Liver	17.55	11.62–26.50	12
Lung	7.42	3.97–13.87	10
Muscle	13.20	4.66–37.41	9

*Only data in which both ^{239}Pu and ^{238}Pu were above detectable limits were included.

†The median and confidence limits were calculated by method R_4 of Doctor and Gilbert (1977).

relatively enhanced assimilation of ^{238}Pu, compared to ^{239}Pu, into these compartments than into soil.

Obviously, there are some physical reasons for skepticism regarding data which suggest that two isotopes of the same element behave differently in biological systems. The difference in mass between ^{238}Pu and ^{239}Pu is less than that between ^{235}U and ^{238}U, on which millions of dollars have been spent for enrichment. Alpha-recoil energy from ^{238}Pu and ^{239}Pu could displace other atoms from near the surface of a particle of plutonium metal. However, unless the particle is composed of either pure ^{238}Pu or pure ^{239}Pu, there would probably be no preferential displacement of either nuclide relative to their ratio in the original metal. Rocky Flats plutonium metal probably did not contain either pure ^{238}Pu or ^{239}Pu particles.

However, if a particle of pure ^{238}Pu were in some way introduced into an organism, autoradiolysis by this high-specific-activity nuclide might allow relatively fast biological transport compared to ^{239}Pu. This idea is not unprecedented. Rats that inhaled ^{238}PuO$_2$ and ^{239}PuO$_2$ of the same particle size and crystalline form translocated up to seven times as much ^{238}Pu as ^{239}Pu to systemic organs at times up to a year postinhalation (Stuart, 1970). Ballou et al. (1973) allowed rats and beagle dogs to inhale PuO$_2$ aerosols of identical size and preparation. According to these workers, "The much greater translocation of ^{238}Pu ... suggests that solubilization of the ^{238}PuO$_2$ occurs to a significant degree within the dog . . . ."

The previous two paragraphs do little to help explain the animal IR data. A possible explanation may be had in statistical bias that heretofore has gone undetected. Basically, the bias has to do with the fact that both ^{239}Pu and ^{238}Pu are probably lognormally distributed in environmental compartments. Therefore the ratio of ^{239}Pu to ^{238}Pu should also be lognormally distributed (Aitchison and Brown, 1969, p. 11). Unfortunately, the distribution of both ^{238}Pu and ^{239}Pu was censored; i.e., some proportion of the data points was below a detectable limit (Aitchison and Brown, 1969). Shaeffer and Little (1978) have shown that both the mean ratio and the variance of the ratio of two censored lognormal variates will be decreased relative to ratios of uncensored variates if the denominator (^{238}Pu) has a lower magnitude than the numerator (^{239}Pu). The magnitude of the decrease in mean ratio and variance is influenced by the relative closeness to the detection limit of the two variates.

This appears to be essentially the case with the IR data presented herein. The soil, vegetation, and litter compartments had relatively high plutonium concentrations and also relatively large IR's. As the plutonium concentration began to approach the detection limit, e.g., in arthropods and small mammals, the IR also decreased. Therefore, if the censoring is large, an estimate of the mean or median of the uncensored ratios will be in error because of the effect of censoring.

A solution for the problem of ratios of two censored distributions is to try to estimate the population parameters for each distribution and then use method R$_2$, i.e., mean ratio equals mean ^{239}Pu divided by mean ^{238}Pu, as suggested by Doctor and Gilbert (1977). Kushner (1976) discusses two methods of estimating such parameters.

Lognormality was assumed, and the methods of Hald (1949) as modified by Kushner (1976) were used to calculate population parameters. Then, a method of Aitchison and Brown (1969, p. 45) was used to calculate the "minimum variance unbiased estimator" of the arithmetic mean isotopic ratio for hide. The mean ratio of hide by these methods was found to be 37. The median ratio published in this chapter was 20, and the mean ratio calculated by summing all hide ratios and dividing by the number of ratios (method R$_3$ in Doctor and Gilbert, 1977) was 29. Therefore, although no confidence interval was calculated, the mean IR in hide calculated by Kushner's (1976) method would be little different from the mean IR in soil. Unfortunately, some of the small-mammal tissue data are censored to such a degree that some of the functional values are extreme enough that they were not tabulated by Hald (1949), one of Kushner's (1976) prime references. Therefore the parameters of most of the censored small-mammal data cannot be estimated by the methods of Kushner (1976) and Hald (1949).

In summary, the median IR was constant in soil and vegetation compartments. However, the median IR's also suggest that ^{238}Pu is preferentially mobile in animal compartments of the grassland relative to ^{239}Pu and soil. There is reason to believe that the IR data are biased toward lower magnitudes as influenced by their nearness to the detection limit. The mean IR for hide estimated with procedures of Kushner (1976) and Hald (1949) suggested that these data may be similar to soil IR's. Other small-mammal tissues were not compatible with these estimation procedures. Further field sampling to eliminate the censoring difficulties is probably necessary if the question of differential concentration of ^{239}Pu and ^{238}Pu in small mammals is to be resolved.

Summary

The soil to a depth of 21 cm contained more than 99% of the plutonium estimated to be in the studied areas of the Rocky Flats grassland. Litter contained a larger fraction of the total plutonium ($\sim 10^{-3}$) than vegetation ($\sim 10^{-4}$), arthropods ($\sim 10^{-9}$), or small mammals ($\sim 10^{-9}$). These results implied that soil–plutonium relationships and soil-management practices are very important at contaminated sites.

Plutonium-concentration frequency distributions for soil samples were positively skewed and characterized by CV's that were generally greater than 100%. Plutonium concentrations in surface (0 to 3 cm) soil were inversely related to distance from the plutonium source, the former oil-barrel storage area. Soil–plutonium concentrations tended to decrease as depth increased and tended to increase as the soil particle size decreased. This latter result suggested that plutonium–soil interaction was a surface-attachment mechanism.

Mean concentrations of plutonium were higher in litter than in vegetation. Frequency distributions of plutonium concentration were normal in litter and lognormal in vegetation. In a manner similar to soil, plutonium concentration both in litter and in vegetation was also inversely related to distance from the barrel storage area. Plutonium concentrations in plant-derived compartment samples were also significantly correlated to plutonium concentration in surface soil at the same locations.

Plutonium frequency distributions in arthropods and small mammals were also positively skewed. Plutonium concentrations in bone samples were lower than those in the other tissues sampled, namely, GI tract, hide, kidney, liver, lung, and muscle.

Concentration ratios of litter, arthropods, and small mammals relative to soil indicated that litter had the highest value. The other compartments, in descending order, were vegetation (3.4×10^{-2}), small mammals (7.8×10^{-3}), and arthropods (6.8×10^{-3}). The relatively high CR's suggested that most of the contamination of vegetation resulted from surficially attached plutonium–soil particles as opposed to root uptake. All the above data strongly indicate that in the grassland soil is by far the most important compartment insofar as plutonium content and transport are concerned. The primary conclusion is that, if transport of plutonium is to be avoided, then transport of soil should be avoided. Therefore soil stabilization should be promoted by maximizing vegetative cover growth and minimizing mechanical disturbances.

Isotopic ratios of ^{239}Pu to ^{238}Pu were calculated for soil, litter, vegetation, arthropod, and small-mammal samples processed by commercial laboratories. The soil results indicated that the median ratio was about 50. Litter and vegetation IR's were similar to IR's in soil. The IR's of small-mammal tissues and arthropods were likely lower than those of soil.

The meaning of the lower IR's in animal compartments was clouded by the fact that the frequency distributions of the ^{239}Pu and ^{238}Pu concentrations, from which the ratios were formed, were censored. Further, the ^{238}Pu concentration distribution was censored to a much larger degree than was the ^{239}Pu distribution. This situation may have the effect of spuriously decreasing the mean or median ratio if the ratios are formed before the average is calculated. An estimation procedure was used to calculate the mean of both ^{239}Pu and ^{238}Pu by taking into account the degree of censorship. Although most small-mammal compartments may not be amenable to such a procedure, the ratio in hide was calculated to be about 37. This value was within the 95% confidence interval of most of the soil IR's. Without further analysis, the hide data suggested that the IR may not be changing between environmental plutonium compartments, as previously suggested (Little, 1976), but may indeed be constant.

Acknowledgments

Most of the Rocky Flats data presented here were collected by my associates and me in the Department of Radiology and Radiation Biology of Colorado State University while under contract with the Energy Research and Development Administration [now the U. S. Department of Energy (DOE)]. Substantial work was contributed by T. F. Winsor, F. W. Whicker, and J. A. R. Bly.

References

Aitchison, J., and J. A. C. Brown, 1969, *The Lognormal Distribution*, Cambridge University Press, New York.

Ballou, J. E., D. K. Craig, J. F. Park, H. A. Ragan, and C. L. Sanders, 1973, Dose—Effect Studies with Inhaled Plutonium in Rats and Dogs, in *Pacific Northwest Laboratory Annual Report for 1972 to the USAEC Division of Biological and Environmental Research*, USAEC Report BNWL-1750(Pt. 1), pp. 21-27, Battelle, Pacific Northwest Laboratories, NTIS.

Bly, J. A., 1977, *Plutonium Concentrations in Arthropods at a Nuclear Facility*, Colorado State University, unpublished.

Doctor, P. G., and R. O. Gilbert, 1977, Ratio Estimation Techniques in the Analysis of Environmental Transuranic Data, in *Transuranics in Natural Environments*, Symposium Proceedings, Gatlinburg, Tenn., October 1976, M. G. White and P. B. Dunaway (Eds.), ERDA Report NVO-178, pp. 601-619, Nevada Applied Ecology Group, NTIS.

Emery, R. M., D. C. Klopfer, T. R. Garland, and W. C. Weimer, 1976, The Ecological Behavior of Plutonium and Americium in a Freshwater Pond, in *Radioecology and Energy Resources*, Proceedings of the Fourth National Radioecology Symposium, Oregon State University, May 12–14, 1975, C. E. Cushing (Ed.), Ecological Society of America Special Publication Series: No. 1, pp. 74-85, Academic Press, Inc., New York.

Gilbert, R. O., L. L. Eberhardt, E. B. Fowler, E. M. Romney, E. H. Essington, and J. E. Kinnear, 1975, Statistical Analysis of $^{239-240}$Pu and ^{241}Am Contamination of Soil and Vegetation on NAEG Study Sites, in *Radioecology of Plutonium and Other Transuranics in Desert Environments*, Report NVO-153, M. G. White and P. B. Dunaway (Eds.), pp. 339-448, Nevada Operations Office, NTIS.

Hakonson, T. E., and L. J. Johnson, 1974, Distribution of Environmental Plutonium in the Trinity Site Ecosystem After 27 Years, in *Proceedings of the Third International Radiation Protection Association Meeting,* Wash., D. C., Sept. 9–14, 1973, W. S. Snyder (Ed.), USAEC Report CONF-730907-P1, pp. 242-247, International Radiation Protection Association, NTIS.

Hald, A., 1949, Maximum Likelihood Estimation of the Parameters of a Normal Distribution Which Is Truncated at a Known Point, *Skand. Aktuartidskrift*, 32: 119-134.

Krey, P. W., and E. P. Hardy, 1970, *Plutonium in Soil Around the Rocky Flats Plant*, ERDA Report HASL-235, New York Operations Office, NTIS.

Kushner, E. J., 1976, On Determining the Statistical Parameters for Pollution Concentration from a Truncated Data Set, *Atmos. Environ.*, 10: 975-979.

Little, C. A., 1976, *Plutonium in a Grassland Ecosystem*, Ph.D. Thesis, Colorado State University, Ft. Collins, Colo.

Markham, O. D., 1976, Plutonium and Americium in Soils Near the INEL Radioactive Waste Management Complex, in *1975 Progress Report, Idaho National Engineering Laboratory Site Radioecology—Ecology Programs*, ERDA Report IDO-12080, pp. 35-46, Idaho National Engineering Laboratory, NTIS.

Navratil, J. D., and C. E. Baldwin, 1976, *Recovery Studies for Plutonium Machining Oil Coolant*, ERDA Report RFP-2579, Rockwell International, NTIS.

Newbould, P., 1963, Absorption of Plutonium-239 by Plants, in *Annual Report [on Radiobiology]*, Report ARCRL-10, p. 86, Agricultural Research Council.

Packard, R. L., 1971, *Small Mammal Survey on the Jornada and Pantex Sites*, U. S. International Biological Program Grassland Biome Technical Report 114.

Romney, E. M., H. M. Mork, and K. H. Larson, 1970, Persistance of Plutonium in Soil, Plants and Small Mammals, *Health Phys.*, 19: 487.

Schulz, R. K., G. A. Tompkins, and K. L. Babcock, 1976, Uptake of Plutonium and Americium by Plants from Soils, in *Transuranium Nuclides in the Environment*, Symposium Proceedings, San Francisco, 1975, pp. 303-310, STI/PUB/410, International Atomic Energy Agency, Vienna.

Shaeffer, D. L., and C. A. Little, 1978, unpublished data.

Siegel, S., 1956, *Nonparametric Statistics for the Behavioral Sciences*, McGraw-Hill Book Company, New York.

Sill, C. W., 1969, Separation and Radiochemical Determination of Uranium and the Transuranium Elements Using Barium Sulfate, *Health Phys.*, 17: 89.

Stuart, B. O., 1970, Comparative Distribution of ^{238}Pu and ^{239}Pu in Rats Following Inhalation of the Oxide, in *Pacific Northwest Laboratory Annual Report for 1968 to the USAEC Division of Biology and Medicine*, Vol. 1, Life Sciences, USAEC Report BNWL-1050(Pt. 1), pp. 3.19-3.20, Pacific Northwest Laboratory, NTIS.

Weber, W. A., G. Kunkel, and L. Shultz, 1974, *Botanical Inventory of the Rocky Flats AEC Site: Final Report*, USAEC Report COO-2371-2, University of Colorado, NTIS.

Wessman, R. A., W. J. Major, K. D. Lee, and L. Leventhal, 1971, Commonality in Water, Soil, Air, Vegetation, and Biological Sample Analysis for Plutonium, in *Proceedings of Environmental Plutonium Symposium*, Los Alamos Scientific Laboratory, Aug. 4–5, 1971, E. B. Fowler, R. W. Henderson, and M. F. Milligan (Eds.), USAEC Report LA-4756, Los Alamos Scientific Laboratory, NTIS.

Whicker, F. W., 1974, *Radioecology of Some Natural Organisms and Systems in Colorado*, USAEC Report COO-1156-70, Colorado State University, NTIS.

Wilson, D. O., and J. F. Cline, 1966, Removal of Plutonium-239, Tungsten-185 and Lead-210 from Soils, *Nature*, 209: 941.

Transuranic Elements in Arctic Tundra Ecosystems

WAYNE C. HANSON

Concentrations and inventories of worldwide fallout of ^{137}Cs, ^{238}Pu, and $^{239,240}Pu$ in soils, lichens, and animals from northern Alaska and Greenland during the period 1968–1976 are discussed. Cumulative ^{137}Cs fallout deposition at the soil surface at Anaktuvuk Pass, Alaska, during the period 1959–1976 was estimated to be 43 mCi/km², compared to 16 mCi/km² at Thule, Greenland. Measured ^{137}Cs values in surface (top 5 cm) soil were 7.9 mCi/km² at Anaktuvuk Pass and 21.5 mCi/km² at Thule. The discrepancy is presumably due to measuring difficulties and to rapid movement of radionuclides into the soil profiles. An effective half-time of 0.4 to 0.5 yr was estimated for plutonium isotopes in surface soil at Anaktuvuk Pass. Average concentrations and inventories of $^{239,240}Pu$ in uncontaminated Thule lichen communities were, respectively, 0.25 pCi/g and 0.21 nCi/m² in 1968 and 0.33 pCi/g and 0.25 nCi/m² in 1974; however, these values were not significantly different. Inventories of ^{238}Pu and $^{239,240}Pu$ in Alaskan lichen carpets were 0.019 and 0.28 nCi/m², respectively, in 1968 and 0.040 and 0.67 nCi/m², respectively, in 1974. Concentrations of ^{137}Cs, ^{238}Pu, and $^{239,240}Pu$ were significantly higher in the upper 6-cm stratum than in the lower 6-cm stratum of Cladonia–Cetraria lichen carpets at Anaktuvuk Pass; concentrations of ^{90}Sr were less consistent.

Radionuclides in arctic ecosystems have been investigated for nearly 20 yr because of the efficient transfer of worldwide fallout materials through arctic food chains. Initial investigations were concerned primarily with the radiological health aspects of the appreciable body burdens of ^{90}Sr and ^{137}Cs obtained by circumpolar populations that were involved in the lichen–reindeer/caribou–man/carnivore food webs. The discovery of measurable amounts of ^{137}Cs in Nearctic Eskimos and Indians and Palearctic reindeer-herding peoples in 1961 and 1962 coincided with the advent of a second major period of nuclear weapons tests, which resulted in appreciable radioactive fallout deposition and increased the efforts of several investigators.

Several ecological aspects of arctic tundra ecosystems recommended them for study of the transfer of worldwide fallout radionuclides. Although their geographical location, mostly beyond 60°N latitude, is a region of appreciably less fallout deposition than other more-populated areas of the world, the plant and animal communities are so related that ^{137}Cs body burdens of the native peoples in the Arctic regions, for example, are often 100 to 1000 times greater than those of Temperate Zone residents. This has resulted from the effective accumulation of atmospheric materials, radioactive or otherwise, by the lichen communities, which provide a reservoir of such materials at the base of northern food webs. Transfer from this relatively rich source is enhanced by (1) the low concentration of potassium, the chemical analog of cesium, and other nutrients; (2) the arctic climate, which enhances food intake by the native animals; (3) the simple and direct food webs that are clearly defined; and (4) the few interfering factors that expedite evaluation of small changes over a sufficient time span.

This chapter summarizes information on ^{137}Cs and plutonium isotopes which was gathered as a part of intensive studies of worldwide fallout in soils, lichens, and animal samples from northern Alaska during the period 1969 to 1976. Soil and lichen samples from Thule, Greenland, were obtained during participation in Danish expeditions of 1968 and 1974 which investigated ecological consequences of plutonium that had been released to those environs by the nonnuclear explosion of four unarmed nuclear weapons during the crash of a U.S. Air Force B-52 bomber on Jan. 21, 1968 (Aarkrog, 1971a; 1971b; 1977; Hanson, 1971; 1972; 1975). Results from these studies were compared with results from radiological studies of plutonium isotopes in the Scandinavian lichen—reindeer—Lapp food web.

Most of the data are from studies centered at the inland Eskimo village of Anaktuvuk Pass, located in the central Brooks Range (Fig. 1), where annual precipitation is about 20 cm. Soil samples were collected from undisturbed locations at Bettles, some 125 km south of Anaktuvuk Pass, with annual precipitation of about 32 cm, and from Fairbanks, about 480 km southeast of Anaktuvuk Pass, with annual precipitation of 28 cm.

Fig. 1 Map of northern Alaska showing sampling locations.

Methods

Sample Collection and Processing

Soil samples were collected at eight locations near Thule (Fig. 2) during 1974 to estimate the amounts of ^{137}Cs and ^{238,239,240}Pu deposited on the landscapes. Seven of the locations (numbers 5, 7, 13, 14, 14A, 18, and 21A) were chosen in the downwind vector of the debris cloud that drifted from the 1968 crash site, and one location (number 3) was located 20 km upwind from that site. Five 1-dm^2 by 0.5-dm-deep samples were collected at 0.2- to 0.4-km intervals along transects over landscapes selected for

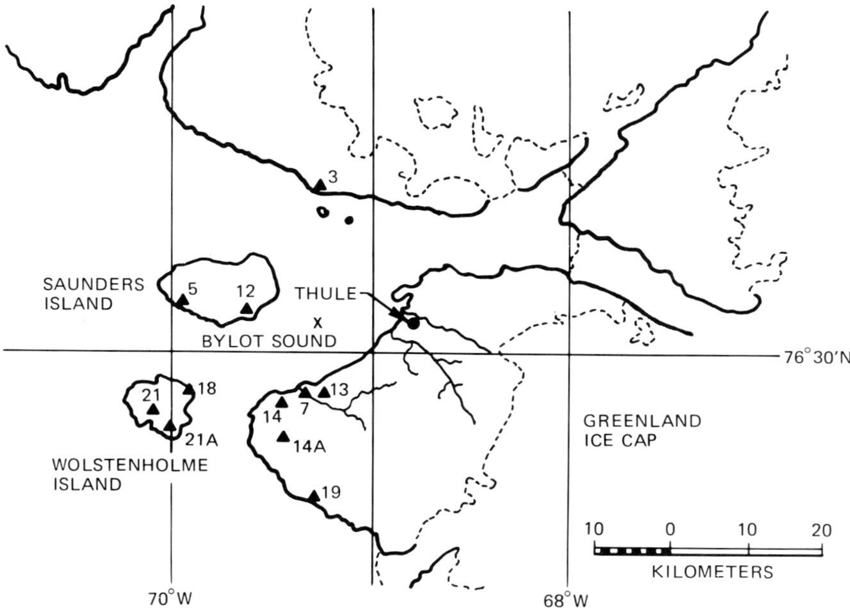

Fig. 2   Map of Thule, Greenland, environs showing 1968 and 1974 sampling sites for soils, alluvium, and lichens.

uniformity of slope, direction, orientation to the crash site and relationship to lichen sampling sites. The variable intervals were selected to best represent the landscape unit to be sampled. Soil samples were taken from sites that were free of vegetation and large rocks. The five samples were composited, yielding 0.05 m^2 of surface area, for inventory of the radionuclide deposition. The method was similar to the template method adopted by the Environmental Measurements Laboratory (Harley, 1972). Alluvium samples were collected in the same manner as soil samples from seasonal streambeds that drained the landscapes across which the soils were collected.

Alaskan soil samples were collected from three locations at Anaktuvuk Pass and at single locations near Bettles and Fairbanks in the same manner as at Greenland locations except that single 1-dm^2 by 0.5-dm-deep samples (0.01 m^2) or three composited samples (0.03 m^2) were analyzed for radionuclides. The different methods were used to better define analytical variability and to remain within the analytical capability of our laboratory.

All soil samples were screened to separate rocks greater than 0.6 cm in diameter from the fine-soil component. Both size fractions were dried, and 100-g aliquots of the fine-size (<0.6 cm) fraction and the entire large-size fraction were leached for about 16 hr with a heated mixture of HCl and HF acids to dissolve the plutonium in the samples.

Lichens were collected on an areal basis whenever possible to provide estimates of biomass, community composition, and compartmental analysis for the various radio-nuclides. Alaskan samples were 0.25-m² blocks cut from the *Cladonia–Cetraria* and *Alectoria–Cladonia–Cetraria* fruticose lichen communities that form carpets on the Anaktuvuk Pass (Fig. 1) landscapes. Greenland samples were obtained from several discontinuous but representative "islands" of each community type which result from microhabitat differences. The term "community" included all lichen or other plant species (populations) within a sample and was designated by the dominant lichen species at a specific location. The various dominant lichen species contributed an average of 90% of the total community biomass in Greenland samples and 80% in the Alaskan samples. Other populations separated from the community samples consisted of subordinate lichen species, vascular plants, lichen and vascular plant debris, and fine soil, which normally comprise a lichen community. The component samples were dried at 100°C for 24 hr to determine dry weights and then dry-ashed at 425°C and dissolved in HCl and HF acids for radiochemical analyses. An average of 1 to 5 g of ash resulted from minor sample components, and up to 100 g of ash from major components of the populations was used for plutonium determination.

Eskimo residents of Anaktuvuk Pass provided the animal samples. Emphasis was placed on sampling caribou *(Rangifer arcticus tarandus)* because of its importance as a food base for the entire carnivore (including human) population in northern Alaska. Major sampling efforts were made in autumn (September–October) and spring (May–June) months when the caribou were intercepted by Eskimo hunters during their migrations to and from wintering grounds and summer ranges. Standard samples consisted of the upper femur and attached muscle. Appreciable numbers of red foxes *(Vulpes fulva)*, tundra wolves *(Canis lupus)*, and wolverines *(Gulo gulo)* and lesser numbers of arctic foxes *(Alopex lagopus)* and lynx *(Lynx canadensis)* were taken each winter by Anaktuvuk Pass Eskimo hunters; entire hindquarters of each animal type were obtained for separation into muscle and bone samples, drying at 100°C, ashing at 425°C, and subsequent radionuclide analyses.

Cesium-137 was measured by counting the 0.661-MeV gamma-ray emissions from dried soil, plant, and animal samples in a calibrated plastic container atop a 7.6- by 7.6-cm sodium iodide crystal connected to a 400-channel analyzer–spectrometer. Most counting times for lichen samples were in the range of 30 to 40 min; counting times were longer for small samples. Spectra were corrected to individual radionuclide amounts by comparison with National Bureau of Standards sources and corrected for background-radiation contributions.

Plutonium-242 tracer was added to sample solutions to determine recovery of the plutonium isotopes, and the mixture was deposited as the nitrate on an anion-exchange resin column. The plutonium was eluted from the column with a nitric acid–ammonium iodide solution and electrodeposited on stainless-steel planchets; the planchets were counted on a silicon surface-barrier alpha spectrometer for 165 to 1330 min. Recovery of plutonium isotopes, as measured by recovery of the ^{242}Pu tracer, was usually in the 60 to 80% range. Isotopic exchange was considered to be uniform within the samples on the basis of standards and interlaboratory comparisons. Counting efficiency for this particular measurement system was 30% with an average background of 3 counts per 1330 min. Counting data were reduced by a computer program that expressed results in picocuries of ^{238}Pu or ^{239,240}Pu per gram of sample with one standard deviation for the counting statistics.

An improved analytical procedure for the determination of ^{241}Am in large (up to 100 g of ash) samples became available during 1977 (Knab, 1977) and yielded the first realistic results for that radionuclide in a limited number of Alaskan samples. This procedure consists of DEHPP [phosphorus pentoxide, bis(2-ethylhexyl) phosphoric acid, and cyclohexane] extraction of both plutonium and americium from the sample residue, separation of plutonium from americium by anion exchange onto a nitric acid prepared column, and purification of americium by ion exchange in methanol—nitric acid and ammonium thiocyanate anion columns. The eluted americium is then electrodeposited on a stainless-steel planchet and counted on the same alpha-spectrometer system used for plutonium. Yields were monitored by ^{243}Am tracer.

Several samples, particularly the animal tissues, which were analyzed for transuranic elements yielded net values that were lower than the minimum detection limits (MDL) of the system; for ^{238}Pu extracted from 10 g of soil samples and counted for 1333 min, the MDL was 0.003 pCi/g, and for ^{239,240}Pu it was 0.002 pCi/g. Values of zero or negative numbers are a common occurrence in environmental sampling owing to statistical fluctuations in the measurements. Although a negative value for a measurement does not represent a physical reality, a valid long-term average of many measurements can be obtained only if very small or negative values are included in the population. The data reported here are often averages of several samples, including those below the minimum detection limit or negative numbers. Zero values were considered to represent ≤0.0061 dpm at the 95% confidence level. This procedure is consistent with data treatment at other laboratories (Harley and Fisenne, 1976). Unless specifically stated, data reported as ^{239}Pu include the minor radioactive contribution of ^{240}Pu.

Results and Discussion

Radionuclide Deposition Estimates

Studies of transuranic elements in ecosystems are greatly aided by relating their behavior to that of fallout ^{137}Cs, which is easily measured and is consistently near a ratio (^{239}Pu/^{137}Cs) of 0.016 (Hardy, 1975). Fallout deposition on the Alaskan and Greenland landscapes was calculated from data published by the U. S. Department of Energy Environmental Measurements Laboratory (Hardy, 1975) with ^{137}Cs deposition estimated from the ratio ^{137}Cs/^{90}Sr = 1.6 ± 0.2 (Hardy and Chu, 1967). Values were directly available for Thule, Greenland, from 1959; however, the fallout deposition at Anaktuvuk Pass was estimated by extrapolating the measured deposition at Fairbanks, some 500 km southeast of Anaktuvuk Pass, to the study area by multiplying by 0.67, the ratio of the annual precipitation rates at the two locations (21 and 32 cm/yr, respectively) (Volchok and Kleinman, 1971). Similarly, fallout deposition at Bettles, about 300 km northwest of Fairbanks, was estimated by multiplying by 1.34, the ratio of their annual precipitation rates (43 and 32 cm/yr, respectively).

The correlation of worldwide fallout deposition with precipitation and the practicality of estimating the integrated fallout deposited in a geographic region by careful soil sampling have been demonstrated by Hardy (1974; 1975) and Hardy and Krey (1971). Our data indicated that ^{137}Cs inventories in Greenland lichen communities were about 60% as great as those in Alaska, or about proportional to their differences in annual precipitation. However, the estimated ^{137}Cs deposition at Thule was only 37% of that estimated for Anaktuvuk Pass, which suggests that "dry fallout" must account for an appreciable part of the worldwide fallout deposited on Thule lichen communities. The close agreement between most of the ^{137}Cs inventories in the soils and adjacent lichen communities of that region suggested that the two components were balanced by some physical transport mechanism, probably the substantial winds of the region.

Fallout collections at both Thule and Fairbanks began near the end of a previous major period of fallout deposition that resulted from an extensive series of atmospheric nuclear weapons tests during 1952 to 1958 by Great Britain, Russia, and the United States. Therefore the ^{137}Cs estimate of 43 mCi/km^2 at Anaktuvuk Pass at this time and of 16 mCi/km^2 at Thule at the time of sample collection in August 1974 represent conservative deposition estimates. Comparison of fallout collection data at New York during periods before 1960 and after 1960 suggests that the preceding deposition estimates for the northern Alaska and Greenland areas should be about doubled to account for the total deposition of fallout since the beginning of nuclear weapons testing.

The estimated deposition of ^{137}Cs from fallout in the Thule environs (annual precipitation, 13 cm/yr) determined from ^{137}Cs concentrations in soil samples (Table 1)

TABLE 1 ^{137}Cs Concentrations and Areal Inventories in Soils* of the Thule, Greenland, Environs During August 1974

	^{137}Cs	
Location/number†	Concentration, pCi/g (dry weight)	Inventory, nCi/m^2
Saunders Island/5	1.05	32.1
Narssarssuk/14	0.29	21.6
Wolstenholme Island/18	0.45	21.4
Narssarssuk/7	0.27	21.0
Cape Abernathy/3	0.22	18.2
Narssarssuk/13	0.22	14.7
Wolstenholme Island/21A	<0.01	
Narssarssuk/14A	<0.01	

*Each value is derived from measurement of a 100-g aliquot of dry soil taken from a composite of five 0.01-m^2 samples taken along a transect of 0.2 to 0.4 km in the various locations.
†Refer to Fig. 2.

ranged from 14.7 to 32.1 nCi/m^2 with an average of 21.5 ± 2.4 (SE). The maximum value occurred on the southwest side of Saunders Island, where large snowdrifts accumulate during winter periods; the minimum values (<0.01 pCi/g) were obtained in two samples from the windswept headlands at the southern edge of the study area. The validity of this estimate was substantiated by a value of 20.6 nCi/m^2 calculated from a ^{239,240}Pu inventory of 0.33 mCi/km^2 measured in a large (1-kg) aliquot of 622 cm^2 of Thule soils during 1970 and 1971 (Hardy, Krey, and Volchok, 1973) and a ^{239,240}Pu/^{137}Cs ratio of 0.016. Fallout collections at Thule indicated that an additional 0.008 mCi of ^{239,240}Pu per square kilometer was deposited between July 1971 and August 1974, bringing the two estimates even closer to agreement.

Radionuclide Concentrations in Soils

Soil samples of the Thule region during 1974 contained an average of 13.0 ± 6.3 (SE) fCi (10^{-15} Ci) of ^{239,240}Pu per gram (dry weight) in those areas considered to be uncontaminated by the January 1968 accident debris (Table 2). The ^{239,240}Pu inventory was estimated to be 0.35 ± 0.10 nCi/m^2 from these samples. Two sampling locations (numbers 7 and 13 in Fig. 2) within an area of about 16 km^2 near the small habitation of Narssarssuk contained 20 to 100 times that inventory, which reflects contamination from the accident. Alluvium samples from seasonal streambeds that drained snowmelt from the landscapes across which the soil transects were taken corroborated the soil measurements. The nature of the variability in aliquots taken from replicate soil samples from the Narssarssuk area indicated that small particles of indeterminate size contributed most of the radioactivity. Plutonium oxide particles with a calculated mass median diameter of 4 μm were determined by nuclear track auto-

TABLE 2 ^{239,240}**Pu Concentrations in Replicate 100-g
Aliquots* of Soil and Alluvium Samples of the Thule,
Greenland, Environs During August 1974**

Location/number†	Sample aliquot (replicates)	^{239,240}Pu concentration	
		fCi/g (dry weight)	nCi/m^2
Soils			
Narssarssuk/7	A	929	72
	B	232	18
Narssarssuk/13	A	217	15
	B	379	26
Narssarssuk/14A	A	50	3.5
	B	3	0.2
Narssarssuk/14	A	16	1.2
	B	14	1.0
Wolstenholme Island/18	A	17	0.8
	B	19	0.9
Saunders Island/5	A	18	0.6
	B	19	0.6
Cape Abernathy/3	A	14	1.2
	B	5	0.5
Wolstenholme Island/21A	A	2	
	B	2	
Alluvium			
Narssarssuk/7	A	622	
	B	197	
Narssarssuk/14	A	130	
	B	6	
Wolstenholme Island/18	A	20	
	B	16	
Cape Abernathy/3	A	18	
	B	10	

*Aliquot values above the dashed lines are considered to be contaminated by the 1968 accident debris.
†Refer to Fig. 2.

radiographic and microscopic studies of debris, and geometric mean particle diameters of 2 to 5.6 μm were reported in snow/ice samples obtained shortly after the accident (Langham, 1970; Gjørup, 1970).

Isotopic ratios in the Thule soil and alluvium samples from uncontaminated sites were in the range of 0.011 to 0.065 for ^{239,240}Pu/^{137}Cs and 0.15 to >1.0 for ^{238}Pu/^{239,240}Pu. The ^{238}Pu/^{239,240}Pu ratios in lichen communities, by comparison, were usually within the range of 0.020 to 0.024, which was reported for global fallout (Harley, 1975).

Radionuclide concentrations in soil samples collected at the Alaskan sites during summer periods of 1975 and 1976 are shown in Table 3. Inventories of ^{137}Cs, expressed as nanocuries per square meter, were generally proportional to average annual rainfall regimes at the three locations. The greatest amounts were in Fairbanks samples, although Fairbanks receives only about three-quarters as much precipitation as Bettles; however, these values were not significantly different (t-test value, 1.54; df, 2; P $\leqslant$ 0.2 to 0.3), nor were the Anaktuvuk Pass and Bettles values (t value, 1.43; df, 6; P $\leqslant$ 0.2). Fairbanks values were significantly different from those for Anaktuvuk Pass (t value, 4.01; df, 6; P $\leqslant$ 0.01). There was a significant difference between the areal inventories of ^{137}Cs measured at Anaktuvuk Pass in 1975 and 1976 (t value, 1.81; df, 7; P $\leqslant$ 0.10) but not between those in July and September 1976 (t value, −0.88; df, 8; P $\leqslant$ 0.40). This was apparently due to slightly greater fallout deposition during 1974 and 1975 after the large atmospheric nuclear weapons tests conducted by the People's Republic of China in 1973 and 1974 (Carter and Moghissi, 1977). The ^{239,240}Pu/^{137}Cs ratio was 0.016 in 1975 and then decreased to 0.004 to 0.007 at all sites in 1976 as greater amounts of ^{137}Cs were apparently deposited.

Concentrations and inventories of ^{239,240}Pu in Alaskan soils during 1975 and 1976 were substantially less than those in Greenland during 1974; in most cases they were

TABLE 3   Radionuclide Concentrations (Mean ± Standard Error) in Soil Samples* Collected at Anaktuvuk Pass, Bettles, and Fairbanks, Alaska, During Summer Periods of 1975 and 1976

			^{137}Cs		^{238}Pu		^{239,240}Pu		^{241}Am	
Location	Date	N†	pCi/g	nCi/m²	pCi/g	nCi/m²	pCi/g	nCi/m²	pCi/g	nCi/m²
Anaktuvuk Pass	7/75	3	0.49 ±0.22	9.6 ±4.3	−0.0018 ±0.0018		0.0078 ±0.0054	0.152 ±0.101	0.0005 ±0.0003	0.0092 ±0.0061
Anaktuvuk Pass	7/76	6	0.26 ±0.16	7.5 ±3.7	0.0004 ±0.0001	0.0185 ±0.0050	0.0012 ±0.0012	0.042 ±0.030		
Bettles	7/76	2	0.47 ±0.27	22.0 ±11.8	0.0004 ±0.0000	0.0287 ±0.0130	0.0016 ±0.0006	0.115 ±0.081		
Fairbanks	7/76	2	1.15 ±0.57	34.1 ±13.2	0.0022	0.0858	0.0057	0.222		
Anaktuvuk Pass	9/76	4	0.36 ±0.20	7.9 ±2.1	0.0003 ±0.0001	0.0040 ±0.0023	0.0010 ±0.0010	0.032 ±0.017		

*Each sample consisted of a 10-g aliquot taken from a 0.1- by 0.1- by 0.05-m core after drying and sieving to remove rocks greater than 6.35 mm in diameter.

†Number of samples.

about one-tenth the Thule values. This difference could be due to the short residence time of ^{239}Pu in the upper 5 cm of soil at both the Alaska and Greenland sites. Plutonium inventories in surface soil at Anaktuvuk Pass decreased between 1975 and 1976 sampling periods at an effective half-time of 0.4 to 0.5 yr. A tentative application of these values to the Greenland situation suggests that the decline in Thule soils is of a similar rate.

Radionuclides in Lichen Communities

The ability of lichens to retain and recycle fallout radionuclides has been observed by several northern investigators. The radiological health aspects of the lichen—caribou—man food web have been the dominant theme of the studies carried out in European nations. Similarly, the ecosystem studies at Thule, Greenland, were mainly oriented toward defining the consequences of the accidentally released plutonium in the marine food webs of that area which were of importance to the local Eskimos (Aarkrog, 1971a; 1971b; 1977). Most of the plutonium contamination (~30 Ci) resulting from the accident was associated with the sea ice and other Bylot Sound marine components. Approximately 1 to 5 Ci of plutonium was estimated to have been contained in the cloud of smoke and debris that drifted west-southwesterly from the crash site and deposited in uncertain amounts on the sea ice and landscape of the area (Langham, 1970). This uncertainty was enhanced by the discontinuous distribution of lichens in the Thule region, the arid climate and light character of the soils, and the appreciable winds that redistributed the plutonium particles that originated from worldwide fallout from nuclear weapons tests, the April 1964 burnup of the SNAP-9A satellite power source, and the aircraft accident.

Lichen samples collected from several Thule locations during 1968 (Hanson, 1972) and 1974 (Table 4) illustrated the highly variable nature of plutonium concentrations compared with more uniform ^{137}Cs concentrations in the lichen communities exposed to the 1968 accident debris. During 1968 most lichen samples from uncontaminated areas contained a mean ^{239,240}Pu concentration of 0.25 ± 0.07 (SE) pCi/g (standard dry weight) and a total inventory of 0.21 nCi/m^2; comparable values during 1974 were 0.33 ± 0.09 pCi/g and 0.25 nCi/m^2, which presumably included the 0.016 nCi of ^{239,240}Pu/m^2 that had been deposited on Thule landscapes in the 6-yr interval between collections. Those two sets of data were not significantly different (t value, -1.47; df, 12; $P \leqslant 0.1$ to 0.2) nor were the ^{238}Pu/^{239,240}Pu ratios of the uncontaminated 1974 lichen samples (last eight values in Table 4) significantly different from those which contained appreciably greater amounts of ^{239,240}Pu that apparently originated from the 1968 accident. However, the ^{239,240}Pu/^{137}Cs ratios of uncontaminated samples varied over a 200-fold range and showed coefficients of variation (CV = standard deviation $\div$ mean) that averaged >2.0. By comparison, the ^{239,240}Pu/^{137}Cs ratios in uncontaminated lichen samples were relatively stable at 0.02 ± 0.01. This greater variation in radionuclide ratios in lichen samples collected near the accident site suggests that the rigorous climatic and edaphic factors of the Thule region probably had a major influence on the redistribution of radionuclides and led to a balancing of concentrations in lichens and soil.

The ^{137}Cs inventory in the *Cladonia—Cetraria* lichen carpet at Anaktuvuk Pass increased steadily from 6.2 nCi/m^2 in the initial sampling in 1962 to maximum values of about 50 nCi/m^2 in 1965 and has subsequently fluctuated near 35 nCi/m^2 (Table 5). The estimated ^{137}Cs deposition and the amount in the lichen carpet were in close agreement, although the lichens had also been exposed to an undetermined amount of fallout during

TABLE 4   Estimated Inventories of ^{238}Pu, ^{239,240}Pu, and ^{137}Cs in Areal Samples* of Greenland Lichen Communities During August 1974

Location/number†	Sample type	Radionuclide inventory, nCi/m^2		
		^{238}Pu	^{239,240}Pu	^{137}Cs
Saunders Island/12	*Cetraria nivalis*	0.068	3.97	46
Saunders Island/12	*Cetraria nivalis*	0.071	3.79	32
Saunders Island/12	*Cetraria delisei*	0.078	5.50	30
Saunders Island/12	*Alectoria ochroleuca*	0.058	0.86	11
Saunders Island/12	*Alectoria ochroleuca*	0.077	3.79	17
Kap Atoll/19	*Cetraria delisei*	0.022	0.68	21
Narssarssuk/7	*Cetraria nivalis*	0.046	9.38	28
Narssarssuk/14	*Cetraria nivalis*	0.154	8.24	7
Narssarssuk/13	*Cetraria nivalis*	1.452	81.61	13
Narssarssuk/13	*Cetraria nivalis*	0.010	0.28	12
Wolstenholme Island/18	*Cetraria nivalis*	0.006	0.12	15
Wolstenholme Island/18	*Cetraria nivalis*	0.009	0.28	15
Wolstenholme Island/18	*Cetraria nivalis*	0.003	0.17	11
Saunders Island/5	*Cetraria nivalis*	0.009	0.21	34
Saunders Island/5	*Cetraria nivalis*	0.006	0.19	8
Saunders Island/5	*Cetraria nivalis*	0.006	0.24	6
Cape Abernathy/3	*Cetraria nivalis*	0.015	0.49	31

*Samples above dashed line are considered contaminated by 1968 accident debris.
†Refer to Fig. 2.

the pre-Health and Safety Laboratory (HASL) measurement period. For example, values in lichens sampled in July 1964 reached 41 ± 2.6 nCi/m^2 compared to a calculated fallout deposition at Anaktuvuk Pass during the period 1960 to 1964 of 26 nCi/m^2. This maintenance of high ^{137}Cs inventory prompted the experiments on effective half-times of radionuclides in lichens and the modeling of Arctic food chains which demonstrated the significant differences in radionuclide behavior within lichen communities and the important implications to Arctic ecosystems (Eberhardt and Hanson, 1969; Hanson and Eberhardt, 1967; Hanson, 1973). A salient feature of those data was the longer effective half-time of ^{137}Cs (>10 yr) compared with that of ^{90}Sr (1.0 to 1.6 yr) in lichens. This was due primarily to the greater mobility and recycling of ^{137}Cs and the impedance of ^{90}Sr translocation by cation-exchange phenomena (Tuominen, 1967; 1968). Similar mechanisms may be operative in the relatively rapid loss ($T_{1/2}$, 6.1 yr) of ^{239,240}Pu from lichen carpets reported from Scandinavia (Holm and Persson, 1975). Similar results were obtained from Alaskan lichen carpets at Anaktuvuk Pass (Table 6 and Fig. 3), which showed (1) a general increase of both ^{238}Pu and ^{239,240}Pu from 1968 to 1971; (2) a decline during 1972 and 1973, periods of low fallout deposition; (3) a sudden increase during 1974 that correlated with increased fallout deposition presumably due to the Chinese nuclear weapons tests of 1973 and 1974; and (4) another decline in 1975 and then an increase in 1976 samples. The lower 1975 values for plutonium are unexplained but also occurred to a lesser degree in the ^{137}Cs values; the decrease was more pronounced when compared with similar decreases that occurred in 1972. The probability that this was due to either sampling or analytical error is considered to be

TABLE 5 Worldwide Fallout ^{137}Cs Inventory*
in the *Cladonia–Cetraria* Lichen Carpet at
Long-Term Sampling Sites Near Anaktuvuk Pass,
Alaska, During the Period 1962 to 1976

Sampling date	N†	Biomass, kg/m²	^{137}Cs inventory,‡ nCi/m²	Deposition,§ nCi/m²
3/7/62	6	0.52	6.2 ± 2.5	5.14
7/7/63	3	0.61	14 ± 1.4	9.82
12/7/64	4	1.53	41 ± 2.6	11.01
26/7/65	10	1.45	48 ± 4.3	10.97
2/8/66	7	1.26	34 ± 1.2	2.13
2/8/67	10	1.20	30 ± 1.2	0.63
30/6/68	10	1.16	30 ± 1.3	0.60
29/7/69	5	2.95	44 ± 1.6	0.66
2/8/70	6	1.66	24 ± 5.3	0.91
9/8/71	1	2.04	42	0.67
25/7/72	2	2.82	44 ± 1.0	0.00
2/7/73	3	2.42	23 ± 0.8	0.06
26/9/74	2	1.74	44 ± 1.3	0.32
21/7/75	2	3.33	33 ± 4.2	0.06
8/7/76	2	2.76	35 ± 3.0	0.05
17/9/76	2	2.78	33 ± 1.0	0.00
				43.03

*Values are based on samples of 0.25-m² to 0.5-m²
replicates from contiguous sampling areas.
† Number of samples.
‡ Mean ± standard error.
§ Cumulative fallout deposition at Anaktuvuk Pass
between successive sampling periods based on 0.67 of
monthly measurement at Fairbanks (Hardy, 1975).

very slight because the decline occurred in both lichen communities sampled and the analytical data have been verified. One-way ANOVA tests performed on untransformed and log-transformed lichen radionuclide data for the years 1974 to 1976 showed significant ($P \leqslant 0.05$) differences between 1974 and 1975 and between 1975 and 1976 which were identified by multicomparison procedures. These differences were confirmed by such nonparametric procedures as the Kruskal–Wallis and Kolmogorov–Smirnov tests (Hollander and Wolfe, 1973). The ^{239,240}Pu/^{137}Cs ratios in Alaskan lichens usually were stable near 0.013, but they decreased to 0.006 during 1975. Statistical analysis (t statistic for two means) of the 1971 to 1976 plutonium concentrations (Table 7) revealed that the 1975 values were significantly lower than the other years and that the ^{238}Pu/^{239,240}Pu ratios in the *Alectoria–Cladonia–Cetraria* lichen community samples during 1976 were significantly greater than those in the *Cladonia–Cetraria* community samples.

During 1969 and 1970 the *Cladonia–Cetraria* lichen carpet samples were fractionated into upper 6-cm (C_u) and lower 6-cm (C_l) components to test the hypothesis that there were no significant differences between their radionuclide concentrations ($H_0 : C_u = C_l$)

TABLE 6   Estimated Inventories of ^{238}Pu, ^{239,240}Pu, ^{241}Am, and ^{137}Cs in Areal Samples of Alaskan Lichen Communities During Summers of 1968–1976

Year	Taxon	N*	Radionuclide inventory, nCi/m^2			
			^{238}Pu	^{239,240}Pu	^{137}Cs	^{241}Am
1968	*Cladonia–Cetraria*	1	0.019	0.28	30.0	
1969	*Cladonia–Cetraria*	5	0.018	0.36	44.5	
1970	*Cladonia–Cetraria*	3	0.027	0.41	50.2	
1971	*Cladonia alpestris*	1	0.030	0.47	42.1	
	Cetraria delisei	1	0.013	0.18	40.4	
	Alectoria ochroleuca	1	0.002	0.04	9.2	
1972	*Cladonia–Cetraria*	2	0.024	0.28	44.0	
	Cladonia alpetris	1	0.029	0.30	27.9	
	Cetraria delisei	2	0.028	0.30	40.2	
1973	*Cladonia–Cetraria*	3	0.024	0.33	23.4	
	Cladonia alpestris	1	0.038	0.30	26.8	
	Cetraria delisei	2	0.007	0.10	27.7	
	Stereocaulon paschale	2	0.069	0.32	57.4	
1974	*Cladonia–Cetraria*	2	0.040	0.67	44.3	
	Alectoria ochroleuca	2	0.023	0.40	38.2	
1975	*Cladonia–Cetraria*	2	0.008	0.14	29.2	
	Alectoria ochroleuca	2	0.009	0.16	34.3	
1976	*Cladonia–Cetraria*	2	0.030	0.43	34.6	0.122
	Alectoria ochroleuca	2	0.034	0.34	21.8	0.078

*Number of samples.

at the $P \leqslant 0.05$ level. Subsequent statistical analyses for four common fallout radionuclides were performed using the untransformed data in a two-tailed t test which allowed separate variance estimates (Nie et al., 1975), and the log-transformed data were tested by Kruskal–Wallis and Kolmogorov–Smirnov procedures (Hollander and Wolfe, 1973) (Table 8). The upper 6 cm usually contained significantly greater concentrations of ^{90}Sr, ^{137}Cs, ^{238}Pu, and ^{239,240}Pu than the lower 6 cm, except for ^{90}Sr during 1970. Cesium-137 showed the greatest differences in concentrations in the layers, apparently owing to its greater mobility and concentration in the more rapidly photosynthesizing upper portion of the lichens (Moser, 1977). These data are consistent with similar studies in Sweden (Holm and Persson, 1975) in which *Cladonia alpestris* carpets were fractionated into several vertical layers. Considering the variation of radionuclide depositions, sampling, and analytical differences, the values reported for the ^{238}Pu and ^{239,240}Pu concentrations in lichen samples from central Sweden are similar to those from northern Alaska.

Concentration ratios of ^{238}Pu, ^{239,240}Pu, and ^{137}Cs in Greenland and Alaska (lichens/soil) (shown in Table 9) were generally consistent; the exception occurred in ^{238}Pu measured in Greenland samples. The values for the plutonium isotopes were considerably higher than the values in the range of 10^{-6} to 10^{-4} reported for most Temperate Zone plants (Francis, 1973). Resuspension of radionuclides from soil to lichens was assumed to be a strong possibility in the Greenland sites and very minor in

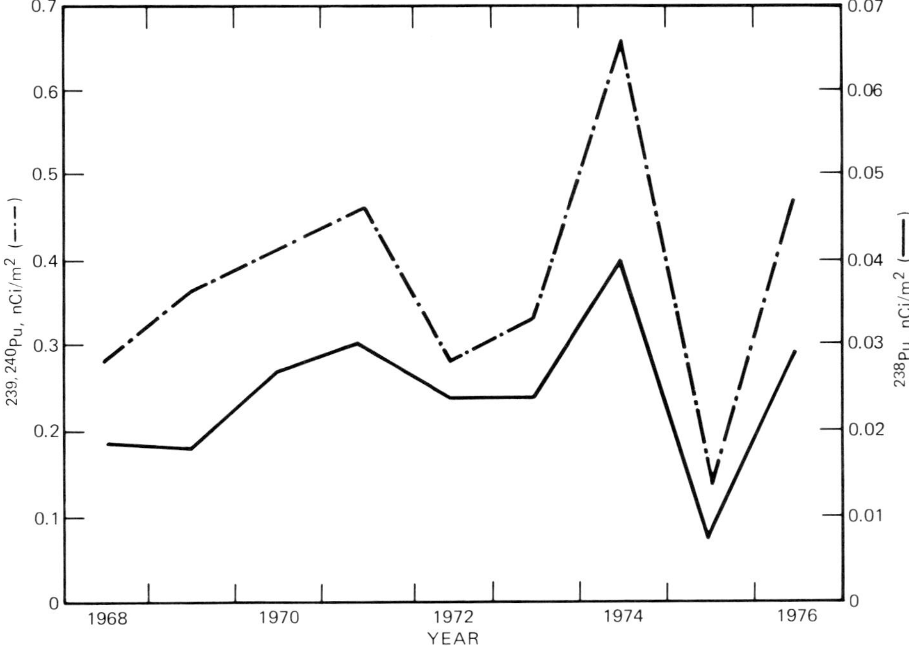

Fig. 3 Inventories of ^{238}Pu and ^{239,240}Pu in the *Cladonia–Cetraria* lichen carpet at Anaktuvuk Pass, Alaska, during the period 1968–1976.

TABLE 7 ^{238}Pu and ^{239,240}Pu Concentrations in
Northern Alaskan Lichen Communities
During 1971–1976

Year	N*	Radionuclide concentration,† pCi/g (dry weight)		
		^{238}Pu	^{239,240}Pu	^{238}Pu/^{239,240}Pu
1971	6	0.012 ± 0.002	0.201 ± 0.100	0.065 ± 0.010
1972	4	0.026 ± 0.003	0.280 ± 0.026	0.091 ± 0.003
1973	8	0.013 ± 0.003	0.146 ± 0.026	0.089 ± 0.010
1974	4	0.017 ± 0.004	0.280 ± 0.006	0.064 ± 0.005
1975	3	0.003 ± 0.001	0.058 ± 0.002	0.050 ± 0.004
1976	6	0.014 ± 0.003	0.176 ± 0.010	0.078 ± 0.012

*Number of samples.
†Mean ± standard error.

Alaska; however, that mechanism did not appear to be of major significance in this study, or possibly it was masked by other parameters.

Correlations between radionuclide isotopic ratios obtained from soil and lichen samples (Table 10) indicated that linear relationships between ^{238}Pu/^{239}Pu and ^{239}Pu/^{137}Cs were more constant in Alaskan samples than in Greenland samples and that ratios in lichen samples were more strongly related than ratios in soil samples. There was

TABLE 8   Statistical Analyses (t Statistic for Two Means) of the Null Hypothesis of Equal Radionuclide Concentrations in Upper and Lower 6-cm Segments of *Cladonia–Cetraria* Lichen Carpet Samples (H_o : C_u = C_l) During 1969 and 1970

Year	Radionuclide	t value	Degrees of freedom	Probability
1969	^{90}Sr	5.23	4	0.002
	^{137}Cs	10.55	4	0.000
	^{238}Pu	3.07	4	0.016
	^{239,240}Pu	3.10	4	0.010
1970	^{90}Sr	0.63	2	0.482
	^{137}Cs	27.45	2	0.000
	^{238}Pu	11.83	2	0.008
	^{239,240}Pu	27.33	2	0.016

TABLE 9   Concentration Ratios of ^{238}Pu, ^{239,240}Pu, and ^{137}Cs in Greenland and Alaska Lichen Samples

Location	Concentration ratio, (pCi/g lichen)/(pCi/g soil)		
	^{238}Pu	^{239,240}Pu	^{137}Cs
Alaska	140	180	36
Greenland	3	140	52

TABLE 10   Correlation Coefficients of ^{238}Pu/^{239}Pu and ^{239}Pu/^{137}Cs Concentration Ratios in Soil and Lichen Samples from Alaska and Greenland

Sample type	^{238}Pu/^{239}Pu			^{239}Pu/^{137}Cs		
	Degrees of freedom	r	Conclusion about hypothesis, $\rho = 0$	Degrees of freedom	r	Conclusion about hypothesis, $\rho = 0$
Alaska soil	7	0.4690	Not rejected	12	0.7280	Rejected at 1%
Alaska lichens	24	0.6481	Rejected at 1%	24	0.5292	Rejected at 1%
Greenland soil	8	0.4369	Not rejected	6	0.6164	Not rejected
Greenland lichens	29	0.9695	Rejected at 1%	22	0.1000	Not rejected

also an indication that sample sizes were often too small to validate the hypothesis that $\rho = 0$, i.e., that the sampled populations were normally distributed about the regression of Y on X.

Radionuclides in Caribou Tissues

Flesh and bone samples of 69 caribou taken by the Anaktuvuk Pass Eskimos during the period 1964 to 1976 contained highly variable concentrations of ^{238}Pu and ^{239,240}Pu.

Highest values were found in samples taken during early 1965; values then declined to 0 to 0.4 fCi ^{239,240}Pu/g (dry weight) and 0 to 0.1 fCi ^{238}Pu/g. A relatively constant isotopic ratio of 1/4 (^{238}Pu/^{239,240}Pu) prevailed throughout the sample series analyzed. There was little or no correlation between plutonium and ^{137}Cs concentrations in flesh on a seasonal basis, and ^{137}Cs concentrations were generally 2.5 x 10^5 times as great as ^{239,240}Pu concentrations. Concentrations of ^{241}Am in 35 of the preceding flesh samples were of the same general values as the plutonium concentrations but were often more variable.

Radionuclides in Carnivore Tissues

A limited number of carnivore tissues have thus far been analyzed for their transuranic nuclide content. The results were so variable and near the minimal detectable amounts of radionuclides that larger sample volumes (up to 100 g of ash) of most mammalian species were analyzed to provide more positive values for meaningful comparison and interpretation. The results (Table 11) showed that (1) only ^{239,240}Pu could be reliably

TABLE 11 Concentrations of ^{238}Pu, ^{239,240}Pu, and ^{241}Am in Large (38- to 100-g ash) Muscle and Bone Samples Composited from Several Animals Collected at Anaktuvuk Pass, Alaska, During April and May 1976

Species sample type	N*	Sample ash weight, g	Radionuclide concentration,† fCi/g ash		
			^{238}Pu	^{239,240}Pu	^{241}Am
Caribou *(Rangifer arcticus)*					
Muscle	9	60.5	−0.19 ± 0.17	0.90 ± 0.20	−0.1 ± 0.3
Bone	3	83.4	0.03 ± 0.07	0.23 ± 0.11	0.7 ± 0.2
Bone	6	100.2	0.33 ± 0.11	0.53 ± 0.14	1.0 ± 0.3
Wolf *(Canis lupus)*					
Muscle	6	43.0	−0.40 ± 0.20	0.80 ± 0.40	0.3 ± 0.4
Bone	3	71.7	−0.09 ± 0.11	0.30 ± 0.13	0.2 ± 0.3
Bone	2	92.2	0.07 ± 0.15	0.05 ± 0.10	0.1 ± 0.2
Fox *(Vulpes fulva)*					
Muscle	9	38.0	−0.30 ± 0.30	0.40 ± 0.40	−0.1 ± 0.5
Bone	4	85.6	−0.07 ± 0.12	0.09 ± 0.13	0.1 ± 0.2

*Number of animals.

†Mean ± standard error (counting statistics).

reported from these composited animal samples, (2) caribou flesh contained slightly higher ^{239,240}Pu concentrations than wolf and fox flesh, and (3) there was no indication of biomagnification in the upper trophic levels of the food web. Concentration ratios of the plutonium isotopes were generally similar; the lichen/soil ratio was 140–180; the caribou/lichen flesh ratio was 0.004–0.005; and the wolf/caribou ratio was 0–0.9.

Conclusions

Two major periods of worldwide fallout deposition on northern Alaskan and Greenland ecosystems have occurred as a result of atmospheric nuclear weapons tests of Great

Britain, Russia, and the United States, the first and most sustained during 1953 to 1959 and the second during 1961 to 1964. Recent additions of fallout have been made by atmospheric tests conducted by France and the People's Republic of China; tests conducted during the summers of 1973 and 1974 by China have apparently made important contributions of radionuclides of interest in this chapter.

Estimates of ^{239,240}Pu fallout inventories in the Arctic landscapes discussed in this chapter began during 1959 and 1960 and have recently been estimated to be 0.33 mCi/km^2 at Thule, Greenland, and 0.40 to 0.60 mCi/km^2 in northern Alaska. If an average ^{239,240}Pu/^{137}Cs ratio of 0.016 is assumed, these values translate to 20.6 mCi ^{137}Cs/km^2 at Thule and 35.6 mCi ^{137}Cs/km^2 at Anaktuvuk Pass. This is reasonably close to the amounts calculated from HASL fallout deposition data for the period after 1959 and 1960, when the measurements began. A large increment of the pre-1959 fallout is therefore unaccounted for in both the soils (to a depth of 5 cm) and lichen communities studied in this chapter. The lichen communities of northern Alaska during 1964 contained 41 nCi ^{137}Cs/m^2 (mCi/km^2), and values since that time have fluctuated near 35 to 40 nCi/m^2. This is equivalent to a predicted level of 0.56 to 0.64 nCi ^{239,240}Pu/m^2 based on the assumption of a ^{239,240}Pu/^{137}Cs ratio of 0.016. However, measurements of ^{238}Pu and ^{239,240}Pu in lichen samples were about one-half the predicted concentration. This lower value was confirmed by consistently lower ^{239,240}Pu/^{137}Cs ratios (near 0.006 to 0.012) in lichens during recent years.

Isotopic ratios in Greenland soil and alluvium samples during 1974 were in the range of 0.011 to 0.014 for ^{239,240}Pu/^{137}Cs and 0.010 to 0.019 for ^{238}Pu/^{239,240}Pu. Although these values do not strongly indicate the presence of plutonium released by the 1968 aircraft accident that deposited an estimated 1 to 5 Ci of ^{239,240}Pu on the Thule landscapes, the ratios are substantially lower than the 0.020 to 0.024 values usually found. The presence of ^{239,240}Pu particles in soils, alluvium, and lichens of the Thule environs in both 1968 and 1974 samples from sites south and southwest of the 1968 crash site was inferred from the extreme variation in sample aliquots from those areas. Uncontaminated areas of Thule showed a more balanced distribution of plutonium concentrations in soil and lichens, probably because of the edaphic and climatic factors of the region which resuspended soils to a considerable degree.

Samples of soils and lichens from northern Alaska contained lower plutonium concentrations in relation to fallout deposition than were noted in Thule samples. Lichen/soil ratios during 1975 and 1976 were 0.92 and 13.0, respectively; this contrasts with the plant/soil ratios that are often of the order of 10^{-4} to 10^{-6} in Temperate Zone environments (Francis, 1973). Caribou/lichen ratios of ^{239,240}Pu concentrations were in the range of 10^{-3} to 10^{-4}, and carnivores contained transuranic nuclide concentrations that were equal to or less than those in caribou, which were undoubtedly their major food source.

Acknowledgments

I am grateful to Mary Ann Hanson for long hours spent in the meticulous separation of the lichen sample components; to Eliza Trujillo, William Goode, and Kenneth Bostick for laboratory assistance; and to Daryl Knab, Richard Peters, and David Curtis for plutonium analyses. Gary White assisted in statistical analyses of the data. Ludi Kupinski and Karen Tallent rendered secretarial and editorial assistance. This work was performed under U.S. Department of Energy contract No. W-7405-ENG-36.

References

Aarkrog, A., 1971a, Radioecological Investigations of Plutonium in an Arctic Marine Environment, *Health Phys.*, 20: 31-47.

——, 1971b, A Re-Examination of Plutonium at Thule, Greenland, in 1970, in *Proceedings of International Symposium on Radioecology Applied to the Protection of Man and His Environment*, Rome, Italy, Sept. 7–10, 1971, Report EUR-4800, Vols. 1 and 2, pp. 1213-1218, European Atomic Energy Community.

——, 1977, Environmental Behavior of Plutonium Accidentally Released at Thule, Greenland, *Health Phys.*, 32: 271-284.

Carter, M. W., and A. A. Moghissi, 1977, Three Decades of Nuclear Testing, *Health Phys.*, 33: 55-71.

Eberhardt, L. L., and W. C. Hanson, 1969, A Simulation Model for an Arctic Food Chain, *Health Phys.*, 17: 793-806.

Francis, C. W., 1973, Plutonium Mobility in Soil and Uptake in Plants: A Review, *J. Environ. Qual.*, 2: 67-70.

Gjørup, H. L., 1970, Investigation and Evaluation of Contamination Levels, *USAF Nucl. Saf.*, 65(2): 57-63 (No. 1, Special Edition, Crested Ice).

Hanson, W. C., 1971, Fallout Radionuclide Distribution in Lichen Communities near Thule, *Arctic*, 24: 269-276.

——, 1972, Plutonium in Lichen Communities of the Thule, Greenland Region During the Summer of 1968, *Health Phys.*, 22: 39-42.

——, 1973, *Fallout Strontium-90 and Cesium-137 in Northern Alaskan Ecosystems During 1959–1970*, Ph.D. Thesis, Colorado State University, Fort Collins, Colo.

——, 1975, Studies of Transuranic Elements in Arctic Ecosystems, in *Radioecology and Energy Resources*, C. E. Cushing, Jr. (Ed.), Ecological Society of America Special Publication Series, No. 1, Proceedings of Symposium on Radioecology, Oregon State University, May 12–14, 1975, pp. 29-39, Academic Press, Inc., New York.

——, and L. L. Eberhardt, 1967, Effective Half-Times of Radionuclides in Alaskan Lichens and Eskimos, in *Symposium on Radioecology*, Proceedings of the Second National Symposium, Ann Arbor, Mich., May 15–17, 1967, D. J. Nelson and F. C. Evans (Eds.), USAEC Report CONF-670503, pp. 627-634, NTIS.

Hardy, E. P., 1974, Depth Distributions of Global Fallout Sr^{90}, Cs^{137}, and $Pu^{239,240}$ in Sandy Loam Soil, in *Fallout Program Quarterly Summary Report*, June 1, 1974–Sept. 1, 1974, USAEC Report HASL-286, pp. I-2 to I-10, Health and Safety Laboratory, NTIS.

——, 1975, Regional Uniformity of Cumulative Radioactive Fallout, in *Fallout Program Quarterly Summary Report*, USAEC Report HASL-288, pp. I-2 to I-9, Health and Safety Laboratory, New York.

——, and N. Chu, 1967, The Ratio of ^{137}Cs to ^{90}Sr in Global Fallout, in *Fallout Program Quarterly Summary Report, Mar. 1, 1967–June 1, 1967*, USAEC Report HASL-182, pp. I-6 to I-9, Health and Safety Laboratory, NTIS.

——, and P. W. Krey, 1971, Determining the Accumulated Deposit of Radionuclides by Soil Sampling and Analysis, in *Proceedings of Environmental Plutonium Symposium*, Los Alamos, N. Mex., Aug. 4–5, 1971, E. B. Fowler, R. W. Henderson, and M. F. Milligan (Eds.), USAEC Report LA-4756, pp. 37-42, Los Alamos Scientific Laboratory, NTIS.

——, P. W. Krey, and H. L. Volchok, 1973, Global Inventory and Distribution of Fallout Plutonium, *Nature*, 241: 444-445.

Harley, J. H. (Ed.), 1972, Soil Sampling Procedure, in *HASL Procedures Manual*, USAEC Report HASL-300, pp. B-05-01 through B-05-11, Health and Safety Laboratory, NTIS.

——, 1975, Transuranium Elements on Land, in *Health and Safety Laboratory Environmental Quarterly*, ERDA Report HASL-291, pp. I-105 to I-109, E. P. Hardy, Jr. (Ed.), Health and Safety Laboratory, NTIS.

Harley, N., and I. M. Fisenne, 1976, Reporting Results of Radioactivity Measurements at Near Zero Levels of Sample Activity and Background, in *Health and Safety Laboratory Environmental Quarterly, March 1–June 1, 1976*, ERDA Report HASL-306, pp. I-43 to I-50, E. P. Hardy, Jr. (Ed.), Health and Safety Laboratory, NTIS.

Hollander, M., and D. A. Wolfe, 1973, *Nonparametric Statistical Methods*, pp. 68-74, 113-114, and 219-228, John Wiley & Sons, Inc., New York.

Holm, E., and R. B. R. Persson, 1975, Fall-Out Plutonium in Swedish Reindeer Lichens, *Health Phys.*, 29: 43-51.

Knab, D., 1977, *A Procedure for the Analysis of Americium in Complex Matrices,* ERDA Report LA-7057, Los Alamos Scientific Laboratory, NTIS.

Langham, W. H., 1970, Technical and Laboratory Support, *USAF Nucl. Saf.*, 65(2): 36-41 (No. 1, Special Edition, Crested Ice).

Moser, T. J., 1977, Cetraria cucullata's *Photosynthetic Patterns and Caribou Grazing Effects on Lichen Biomass,* M.S. Thesis, Arizona State University, Tempe, Ariz.

Nie, N. H., C. H. Hall, J. G. Jenkins, K. Steinbrenner, and D. H. Bent, 1975, *Statistical Package for the Social Sciences,* 2nd ed., McGraw-Hill Book Company, New York.

Tuominen, Y., 1967, Studies on the Strontium Uptake of the *Cladonia alpestris* Thallus, *Ann. Bot. Fenn.*, 4: 1-28.

——, 1968, Studies on the Translocation of Caesium and Strontium Ions in the Thallus of *Cladonia alpestris, Ann. Bot. Fenn.*, 5: 102-111.

Volchok, H. L., and M. T. Kleinman, 1971, Global Sr-90 Fallout and Precipitation: Summary of the Data by 10 Degree Bands of Latitude, in *Fallout Program Quarterly Summary Report, June 1, 1971—Sept. 1, 1971,* USAEC Report HASL-245, pp. I-2 to I-83, Health and Safety Laboratory, NTIS.

Nevada Applied Ecology Group Model for Estimating Plutonium Transport and Dose to Man

W. E. MARTIN and S. G. BLOOM

A Standard Man is assumed to live in and obtain most of his food from a plutonium-contaminated area at the Nevada Test Site (NTS). A plutonium-transport model, based on the results of other Nevada Applied Ecology Group plutonium studies, is used to estimate potential chronic rates of ^{239}Pu inhalation and ingestion as functions of the average concentration of ^{239}Pu (C_S, picocuries per gram) in the surface soil (0- to 5-cm depth) of the reference area. A dose-estimation model, based on parameter values recommended in publications of the International Commission on Radiological Protection (ICRP), is used to estimate organ burdens, accumulated doses, and dose commitments as functions of exposure time. These estimates are combined with ICRP recommendations for allowable public exposure to radiation to arrive at acceptable soil concentrations at NTS.

The plutonium-transport model is based on a relatively simple ecosystem that was used as a preliminary model to guide data-acquisition studies at NTS. The preliminary model provides a framework for developing more detailed dynamic models of the ecosystem, but at present there are insufficient data to implement these dynamic models; so the estimates of inhalation and ingestion rates are based on simpler steady-state models. If we assume the transport system to be in steady state, the estimated inhalation and ingestion rates (picocuries per day) are 0.002 C_S and 0.2 C_S, respectively.

A number of dose-estimation models were examined, and calculations were made for comparison. The results of these calculations indicated that the dose estimates to the most sensitive organs were comparable. The model recommended by the Task Group on Lung Dynamics of ICRP was used for dose estimates at NTS because it is the model most widely accepted. Estimated doses (rem) due to chronic inhalation and ingestion of ^{239}Pu for 50 yr at the rates indicated above are: thoracic lymph nodes, 0.610 C_S; lungs, 0.025 C_S; bone, 0.014 C_S; liver, 0.009 C_S; kidney, 0.003 C_S; total body, 0.0007 C_S; and gastrointestinal tract (lower large intestine), 0.0002 C_S. Inhalation accounts for 100% of the estimated dose to the lungs and thoracic lymph nodes and for about 95% of the estimated dose to bone, liver, kidney, and total body. Ingestion accounts for >99% of the dose to the gastrointestinal tract.

According to the ICRP (International Commission on Radiological Protection, 1966) recommendations for individual members of the public, the dose rate to the lungs after 50 yr exposure should not exceed 1.5 rem/yr. The plutonium-transport and dose-estimation models described in this chapter indicate that the average concentration of ^{239}Pu in the surface (0 to 5 cm) soils of contaminated areas at NTS which could result in a maximum dose rate of 1.5 rem/yr to the lungs is approximately 2.8 nCi/g, or about 140 $\mu Ci/m^2$ for soils weighing 1 g/cm^3.

An important goal of the Nevada Applied Ecology Group (NAEG) plutonium program is to evaluate the potential radiological hazard to man due to the presence of plutonium in various nuclear safety test areas at the Nevada Test Site (NTS). Since the contaminated areas of interest are uninhabited, we have based our analysis on the assumption that a Standard Man resides in and obtains most of his food from a plutonium-contaminated area at NTS.

In this chapter we use information provided by other NAEG studies to develop a plutonium-transport model that attempts to characterize the general behavior of plutonium in a typical NTS ecosystem and to provide a basis for estimating potential rates of plutonium ingestion and inhalation by the hypothetical Standard Man. We discuss the mechanisms involved in the transport processes and, in most cases, include appropriate mathematical expressions for these mechanisms. However, the final form of the transport model is determined by the available data, which often limits us to using only the simplest mathematical expressions.

The estimates of inhalation and ingestion rates provide the input for a dose-estimation model that is used to calculate potential organ burdens, cumulative organ doses, and dose commitments due to chronic inhalation and ingestion of ^{239}Pu. Although several models are considered, the preferred dose-estimation model is based entirely on the recommendations and publications of the International Commission on Radiological Protection (1959; 1964; 1966; 1972).

Finally, a procedure is described whereby the combined results of the transport model and the dose-estimation model can be applied to the practical problem of deciding whether and to what extent environmental decontamination might be required to limit or reduce potential health hazards due to plutonium.

A preliminary model of potential plutonium transport from the environment to man was introduced during the planning stage of the NAEG plutonium program to ensure consideration of laboratory and field studies that would provide the data and parameter estimates required for implementation of more detailed transport and dose-estimation models to be developed later in the program. This model forms the basis for discussing the various transport mechanisms in this chapter. Some of the parameters sought at the outset have proved to be elusive or impossible to measure accurately, and consequently the proposed dynamic model has not been fully implemented. This chapter represents our best effort to judge and interpret the information currently available and to select the best available methods for estimating potential intake rates and doses. The design of the transport and dose-estimation models plus the assumptions and parameter values selected for their implementation comprise what we believe to be a reasonable and conservative working hypothesis that provides a method for evaluating the potential health hazards associated with plutonium-contaminated areas at the NTS. As a working hypothesis, it is subject to continuing reappraisal, and the results or conclusions derived from it are subject to unavoidable uncertainties. To a considerable extent, however, these uncertainties are compensated for by conservative assumptions, which tend to result in overestimates of potential intake rates, organ burdens, and doses rather than underestimates.

Plutonium-Transport Model (Preliminary Model)

Figure 1 is a diagram of the potential transport pathways considered in the preliminary planning model. The large square represents an arbitrary boundary of a contaminated area. Boxes represent the principal ecosystem compartments of interest, and arrows represent net transport via the pathways indicated. Arrows that cross the arbitrary boundary represent net transport out of the system.

The distribution of plutonium in the contaminated areas of principal interest at the NTS has been described by Gilbert et al. (1975). Present levels of soil contamination in the areas of interest range from about 1.0 μCi/m^2 to >6000 μCi/m^2. Because these levels of soil contamination resulted from nuclear safety tests carried out from 1954 through 1963 and because current worldwide fallout rates are insignificant compared with existing levels of contamination, Fig. 1 shows no current plutonium input to the system.

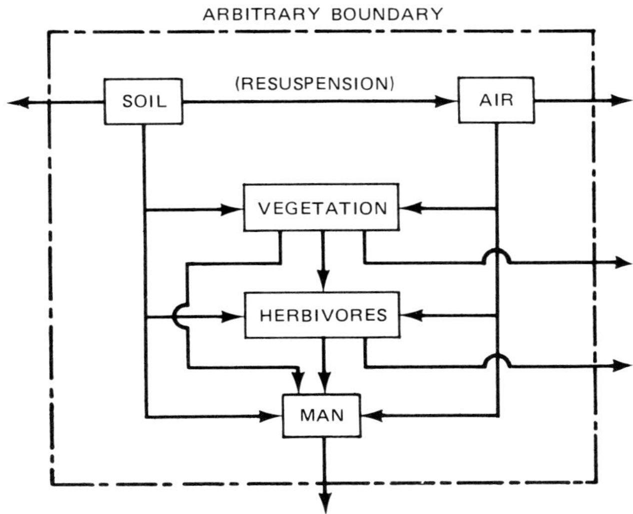

Fig. 1   **Principal pathways of plutonium transport to man.**

Under these conditions the plutonium concentration in soil is the principal factor forcing the transport system. Air is contaminated by resuspension of plutonium-bearing soil particles. Vegetation is contaminated internally by root uptake from soil and externally by deposition of resuspended particles. Plutonium input to herbivores is due to ingestion of soil and vegetation and to inhalation. Plutonium could reach man by inhalation of contaminated air, by accidental ingestion of contaminated soil, by ingestion of contaminated vegetation, and by ingestion of milk or meat (skeletal muscle or internal organs) from animals raised in the contaminated area. Drinking water for herbivores and man is assumed to come from deep wells or from sources outside the contaminated area and to contribute nothing to plutonium intakes by herbivores or by man. Numerous other pathways, most of them trivial and unsubstantiated, could be postulated, but we have tried to limit our consideration to genuinely important pathways.

If it is assumed that (1) the major ecosystem compartments and important transport pathways are as indicated in Fig. 1, (2) the plutonium in each compartment is well mixed with the other contents of the compartment, and (3) the net rate of transfer from one compartment to another is the product of a transfer coefficient and the quantity of plutonium in the transmitting compartment, then the intercompartmental flux of plutonium is represented by a system of linear, first-order, ordinary differential equations, the general formula for which is

$$\frac{dY_j}{dt} = \sum_{\substack{i=1 \\ i \neq j}}^{n} \lambda_{ij} Y_i - Y_j \sum_{\substack{i=1 \\ i \neq j}}^{n} \lambda_{ji} \qquad (j = 1,2,3 \ldots n) \tag{1}$$

where j is the compartment of reference and all other compartments are designated i, Y_j is the amount (picocuries) of plutonium in compartment j at time t (days), and λ_{ij} and λ_{ji} are transfer coefficients (day^{-1}) for flows into and out of compartment j. The positive expression on the right side of Eq. 1 represents the flow rate into compartment j, and the negative expression represents the flow rate out of compartment j. The amount present in a given compartment at a given time, $Y_j(t)$, is therefore dependent on the rates of input and output.

In a general way, Fig. 1 and Eq. 1 identify the principal kinds of information needed to estimate the transport of plutonium to man. The compartments of Fig. 1 indicate the principal ecosystem components, and the arrows indicate the pathways of transport from environment to man via inhalation and ingestion. Equation 1 suggests that intercompartmental flow rates might be expressed as the product of a transfer coefficient and the quantity of plutonium in the transmitting compartment. It was recognized, however, that some parts of the transport system (Fig. 1) might not behave in accordance with the first-order kinetics model suggested by Eq. 1. Consequently the objectives of the NAEG plutonium studies were stated in broader terms. The general objectives related to the estimation of potential human ingestion and inhalation rates were simply to (1) determine plutonium concentrations in ecosystem components and (2) quantify the rates of plutonium transfer among ecosystem components.

In the remainder of this chapter, we discuss in turn each compartment of the preliminary model (Fig. 1), soil, air, vegetation, herbivores (cattle), and man. In the sections on soil, air, vegetation, and cattle, we describe what is known about the compartment and discuss the processes that involve it in the transport of plutonium to man. In the section on man, we provide methods for estimating plutonium inhalation and ingestion rates, based on concentration in soil. These rates are used in the section on Dose-Estimation Models to estimate organ burdens, cumulative doses, and dose commitments by alternative methodologies. In the final section, Practical Applications, we show how the results of the plutonium-transport and dose-estimation models can be used to determine an "acceptable soil concentration."

We wish to emphasize at the outset that this is not a definitive study of the behavior of plutonium in desert ecosystems. It is merely an inquiry that asks how we can best use the theory and data presently available to obtain a reasonable assessment of potential hazards and a credible criterion on which to base preliminary consideration of countermeasures that may or may not be planned and executed in the future. Our study identifies some of the obstacles between present knowledge and a workable cleanup criterion and recommends a pro tem path around these obstacles. In plotting this sometimes tortuous path, we have encountered theory that cannot be applied for lack of data, and we have encountered data that cannot be used because they are too scanty to be fitted into the present theoretical framework. The result is a compromise between knowledge and ignorance. We make use of the knowledge we have, but we are made uneasy by the awareness that there are other paths, perhaps equally defensible, which may lead to far different conclusions. Or, to put the matter more bluntly, the present state of the transport- and dose-modeling art is such that, by careful selection of published parameter values and model equations, one could obtain a preselected result. We have made every effort to avoid doing this, but we feel obliged to offer this comment to warn the reader that such efforts are necessary.

Soil

Plutonium Concentration in Soil.   Various soil surveys have been conducted to delineate highly contaminated areas at NTS, to determine the horizontal and vertical distribution of plutonium in contaminated soils, and for various other purposes (see several papers in reports by Dunaway and White, 1974; White and Dunaway, 1975; 1976; 1977). Inventories of ^{239,240}Pu in the surface soils (0- to 5-cm depth) of NAEG study areas were reported by Gilbert et al. (1975, p. 379) and revised by Gilbert (1977, p. 425).

As mentioned earlier, soil is the principal reservoir for plutonium at NTS, and soil concentration (picocuries per gram) is the factor that drives or forces the transport system. In developing equations to estimate potential plutonium inhalation and ingestion rates for the hypothetical Standard Man, we shall attempt to relate the concentrations in air and foods to the average concentration in soil. Soil concentrations based on data provided by Gilbert et al. (1975) are given in Table 1. The estimated inventories (as revised) are given in Table 2.

TABLE 1   Average Concentrations of ^{239,240}Pu in Surface Soils (0- to 5-cm Depth) of NAEG Study Areas*

Study area	Strata	n	^{239,240}Pu,† nCi/g	Study area	Strata	n	^{239,240}Pu,† nCi/g
13	1	39	0.036 ± 0.0078	Clean Slate 2	1	18	0.086 ± 0.028
	2	31	0.10 ± 0.025		2	12	1.8 ± 0.74
	3	14	0.40 ± 0.075		3	13	6.2 ± 2.5
	4	19	1.1 ± 0.15		4	20	5.4 ± 1.4
	5	20	2.4 ± 0.43	Clean Slate 3	1	28	0.24 ± 0.046
	6	47	14 ± 6.4		2	12	1.2 ± 0.33
5 (GMX)	1	41	0.059 ± 0.013		3	13	4.6 ± 1.3
	2	23	0.73 ± 0.15		4	10	7.9 ± 3.9
	3	13	4.5 ± 1.2	Area 11 sites			
	4	23	7.3 ± 1.6	<5000-cpm region	1	50	0.021 ± 0.0066
	5	13	0.084 ± 0.03	CD overlap	6	6	0.30 ± 0.13
Double Track	1	24	0.12 ± 0.057	B site	2	12	0.73 ± 0.45
	2	10	5.7 ± 4.0		3	14	5.5 ± 1.4
	3	10	2.9 ± 0.97		4	23	33 ± 7.0
	4	9	44 ± 15	C site	2	12	0.85 ± 0.44
Clean Slate 1	1	21	0.36 ± 0.17		3	10	2.2 ± 0.65
	2	13	1.6 ± 0.57		4	19	26 ± 7.5
	3	13	2.7 ± 1.1		5	6	120 ± 52
	4	10	2.9 ± 0.97	D site	2	10	1.0 ± 0.20
					3	12	4.3 ± 1.6
					4	18	18 ± 6.5
					5	14	49 ± 15

*Based on data from Gilbert et al. (1975, pp. 393–395).

†Mean ± standard error (SE). SE = s/(n)$^{1/2}$, where n is the number of samples.

TABLE 2   Estimates of Inventory of ^{239,240}Pu in Surface Soil (0- to 5-cm Depth) of NAEG Study Areas*

Area	Strata	Size of area m²	Percent	n	^{239,240}Pu, μCi/m²	Estimated inventory,† Ci	Percent of total inventory
13	1	1,245,000	31.0	39	1.9 ± 0.34	2.4 ± 0.42	5
	2	2,547,000	63.4	31	5.8 ± 1.4	15 ± 3.6	33
	3	108,000	2.7	14	23 ± 4.3	2.5 ± 0.46	5
	4	74,000	1.8	18	54 ± 8.8	4.0 ± 0.65	9
	5	19,000	0.5	20	110 ± 19	2.1 ± 0.36	5
	6	24,000	0.6	47	820 ± 340	20 ± 8.2	43
Total		4,017,000	100.0	169		46 ± 9.0	100
5 (GMX)	1	111,300	88.8	41	3.1 ± 0.66	0.35 ± 0.073	24
	2	8,400	6.7	22	42 ± 9.1	0.35 ± 0.076	24
	3	800	0.6	12	270 ± 64	0.22 ± 0.051	15
	4	1,000	0.8	23	530 ± 150	0.53 ± 0.15	36
	5	3,800	3.0	13	4.6 ± 1.6	0.02 ± 0.006	1
Total		125,300	99.9	111		1.5 ± 0.19	100
Double Track	1	176,000	98.3	23	6.7 ± 3.5	1.2 ± 0.62	33
	2	1,600	0.9	11	350 ± 250	0.56 ± 0.40	16
	3	800	0.4	10	190 ± 59	0.15 ± 0.047	4
	4	600	0.3	9	2,800 ± 1,000	1.7 ± 0.60	47
Total		179,000	99.9	53		3.6 ± 0.95	100
Clean Slate 1	1	157,000	88.9	21	15 ± 7.0	2.4 ± 1.1	58
	2	10,000	5.7	13	64 ± 22	0.64 ± 0.22	15
	3	8,400	4.5	13	110 ± 35	0.92 ± 0.29	22
	4	1,700	1.0	10	120 ± 39	0.20 ± 0.066	5
Total		177,100	100.1	57		4.2 ± 1.2	100
Clean Slate 2	1	351,000	74.7	18	4.1 ± 1.3	1.4 ± 0.46	8
	2	82,300	17.4	12	73 ± 30	6.0 ± 2.5	34
	3	26,200	5.5	13	270 ± 99	7.1 ± 2.6	41
	4	11,000	2.3	20	260 ± 65	2.9 ± 0.72	17
Total		470,500	99.9	63		17 ± 3.7	100
Clean Slate 3	1	1,615,000	93.2	28	12 ± 2.2	19.4 ± 3.6	52
	2	61,000	3.5	12	58 ± 16	3.5 ± 0.98	9
	3	40,000	2.3	13	210 ± 63	8.4 ± 2.5	23
	4	16,000	0.9	10	370 ± 190	5.9 ± 3.0	16
Total		1,732,000	99.9	63		37 ± 5.4	100
Area 11 sites							
<5000-cpm region	1	4,672,000	96.7	50	0.97 ± 0.30	4.5 ± 1.4	12.4
CD overlap	6	62,200	1.3	6	12 ± 5.2	0.75 ± 0.32	2.1
B site	2	8,200		12	30 ± 18	0.25 ± 0.15	
	3	6,000		14	220 ± 55	1.3 ± 0.33	
	4	3,300		23	1,400 ± 300	4.6 ± 0.99	
Total		17,500	0.4	49		6.2 ± 1.1	17.0

TABLE 2 (Continued)

Area	Strata	Size of area m²	Percent	n	^{239,240}Pu, μCi/m²	Estimated inventory,† Ci	Percent of total inventory
C site	2	16,400		12	34 ± 22	0.56 ± 0.36	
	3	5,600		10	88 ± 25	0.49 ± 0.14	
	4	3,500		18	1,400 ± 390	4.9 ± 1.4	
	5	300		6	6,200 ± 2,800	1.9 ± 0.84	
Total		25,800	0.5	46		7.8 ± 1.7	21.6
D site	2	32,300		10	46 ± 9.5	1.5 ± 0.31	
	3	13,300		12	220 ± 86	2.9 ± 1.1	
	4	4,900		18	990 ± 370	4.9 ± 1.8	
	5	2,900		14	2,700 ± 840	7.8 ± 2.4	
Total		53,400	1.1	54		17.1 ± 3.2	47.0
Area 11 Total		4,830,900	100.0			36 ± 4	100.1

*Based on data from Gilbert (1977, p. 425).
†Mean ± standard error (SE). SE = s/(n)$^{1/2}$ = standard error.

Losses from Soil Compartment. As suggested by Fig. 1, plutonium can be transferred from the soil compartment to compartments representing other ecosystem components. It can also be removed from the soil of a given area by water or wind erosion. Percolation into the profile could remove plutonium from the surface, where it is most susceptible to resuspension, and could, if the soil were plowed and rainfall were plentiful, transport some plutonium below the root zones of crop plants.

Owing to the extreme variability of plutonium concentrations in soil samples taken from the same general area and to the arbitrary nature of soil-compartment boundaries (usually specified by a depth measurement), it would be difficult to design field studies to estimate the overall rate of plutonium loss from the soil compartment. In fact, no such studies have been undertaken in the field or in the laboratory, and we have no basis for assuming that the average concentrations of plutonium in the soils of large contaminated areas will decrease significantly in the next 100 yr or so. Consequently the soil concentrations given in Table 1 will be treated as constants for the areas indicated; i.e., the soil compartment is assumed to be a continuous and constant source for plutonium transfer to other compartments. In the absence of any evidence that the rate of plutonium loss is, in fact, significantly greater than the rate of loss due to radioactive decay, the equation for the soil compartment is

$$C_s = C_s(0) \exp(-\lambda_A t) \tag{2}$$

where C_s = average concentration of plutonium in the surface soil of a contaminated area at time t (pCi/g)

$C_s(0)$ = initial concentration as given in Table 1

λ_A = radioactive decay rate of ^{239}Pu (7.7829 × 10^{-8} day^{-1})

t = time (days)

Air

As indicated in Fig. 1, plutonium contained in surface soils can be resuspended and transported to vegetation via external deposition or to herbivores and man via inhalation, and some of it can be carried by wind and redeposited beyond the arbitrary boundary. In the absence of data to the contrary, we have assumed that deposition and resuspension processes in contaminated areas at NTS are in approximate steady state, although data presented by Anspaugh and Phelps (1974, pp. 292–294) suggest that resuspension may exceed deposition, at least to a small degree. Several methods have been suggested for analyzing and modeling deposition and resuspension processes. These are discussed in the following paragraphs. Of these, the mass-loading approach requires the least information for implementation and was used in the present model owing to the absence of data to implement the other methods at NTS.

Deposition Velocity. The rate at which resuspended plutonium is deposited on soil could be estimated as the product of a deposition velocity (centimeters per day) and concentration in air (microcuries per cubic centimeter) to yield a rate that has dimensions of μCi cm^{-2} day^{-1}. Deposition velocities are functions of meteorological factors and the aerodynamic properties of plutonium-bearing soil particles and soil surfaces.

Deposition velocities measured under field conditions have been reported by Van der Hoven (1968), Sehmel, Sutter, and Dana (1973), and Healy (1974). Measurements under controlled conditions in a wind tunnel have been reported by Sehmel, Sutter, and Dana (1973) and Sehmel (1973; 1975). These data indicate that the deposition velocity increases with increasing air velocity, increases with increasing particle size for sizes greater than about 1 μm, increases with decreasing particle size for sizes less than 0.01 μm, exhibits a minimum somewhere in the range of 0.01 to 1 μm, and is strongly influenced by the type of surface roughness. The wind-tunnel data of Sehmel et al. (1973) for grass surfaces indicate that the deposition velocity is approximately proportioned to both air velocity and particle size in the range of 2 to 12 m/sec and 1 to 100 μm. These grass data appear to correspond closely to field conditions provided that a proper value is assigned to surface roughness.

Tamura (1976) has reported that more than 65% of the plutonium in soil samples from Area 13 is associated with soil particles in the range of 20 to 53 μm. Using the grass data of Sehmel, Sutter, and Dana (1973) at 2.2 m/sec, the corresponding range of deposition velocities is from 3 to 20 cm/sec. Particles on the order of 20 to 50 μm could play an important role with respect to external contamination of vegetation, but particles this large are of little concern with respect to inhalation. Since respirable particles are generally <10 μm, the corresponding deposition velocities suggested by the grass data would be <1 cm/sec.

Deposition Models. Both Healy (1974) and Sehmel (1975) present results of models used to predict deposition velocities. Healy's results indicate that deposition velocity is proportional to air velocity and is strongly dependent on atmospheric stability. Sehmel's results indicate that deposition velocity increases as a nonlinear function of air velocity, exhibits a minimum value as a function of particle size, and is not strongly dependent on atmospheric stability. Both sets of results indicate a strong dependence on surface roughness. To apply either model to field conditions, we must estimate or measure the surface roughness and velocity profile, both of which are variable.

For most applications to NTS, the grass data of Sehmel et al. (1973) appear to be the best analog. The trend of these data, in the range of 2 to 12 m/sec air velocity and 1- to 100-μm particle diameter, is approximately

$$\frac{V_d}{U} = 3 \times 10^{-4} \, d_p \tag{3}$$

where V_d is the deposition velocity (cm/sec), U is the wind velocity (cm/sec), and d_p is the particle diameter (μm). For 10-μm particles, Eq. 3 yields values similar to Healy's results for neutral atmospheric stability.

Resuspension Factor.   The resuspension of plutonium from soil is often expressed as the ratio of air concentration (microcuries per cubic meter) to surface soil concentration (microcuries per square meter). Many such measurements have been made at NTS (Mork, 1970; Anspaugh and Phelps, 1974) and in the vicinity of Rocky Flats, Colo. (Volchok, 1971). The measured magnitudes of this ratio range generally from 10^{-5} to 10^{-11} m^{-1}. To estimate "acceptable soil concentrations," Anspaugh (1974) used a value of 10^{-9} m^{-1} for NTS. These ratios are, to say the least, extremely variable with respect to time and environmental factors, such as wind speed and direction, rainfall, and disturbances affecting aerodynamic properties of soil surfaces. Other factors affecting this ratio are the aerodynamic properties of plutonium-bearing particles and their susceptibility to saltation and resuspension. There is evidence that the ratio tends to decrease with time after fallout contamination of soil (Anspaugh et al., 1973; Anspaugh, 1974; Kathren, 1968). Anspaugh et al. (1975) have proposed a model in which the air/soil ratio decreases as a function of time from a maximum of 10^{-4} to a minimum of 10^{-9} m^{-1}, i.e.,

$$\frac{C_a}{C_{ss}} = 10^{-4} \, \exp\left[-k\,(t)^{\frac{1}{2}}\right] + 10^{-9} \tag{4}$$

where C_a = air concentration (μCi/m^3)
$\quad\quad\; C_{ss}$ = soil surface concentration (μCi/m^2)
$\quad\quad\;\; k = 0.15$ day$^{-\frac{1}{2}}$
$\quad\quad\;\; t$ = time from deposition (days)

This model is consistent with data collected over the years at NTS.

Resuspension Models.   Many attempts have been made to develop mathematical models to simulate resuspension (Amato, 1971; Mills and Olson, 1973; Killough and McKay, 1976). Most of these are based on models of wind erosion developed by Bagnold (1960) and, as a function of wind speed, take the form

$$C_a = K\,(U - U_T)^3 \, \frac{C_{ss}}{U} \tag{5}$$

where U_T is a threshold wind speed (m/sec) and K is a constant (sec/m^3).

Others (Sehmel and Orgill, 1973; Shinn and Anspaugh, 1975) have used a power-law expression of the form

$$C_a = K\,U^n \tag{6}$$

where K and n are empirical constants derived from the data. Sehmel and Orgill (1973) found n = 2.1 when they fit Volchok's (1971) data for plutonium resuspension at Rocky Flats to Eq. 6. Shinn and Anspaugh (1975) also found n = 2.1 for dust flux at NTS. We approximately fit Sehmel's (1975) data for calcium molybdate resuspension at Hanford and also arrived at a value of about 2.1. The only contrary data are Shinn and Anspaugh's results for a plowed field in Texas, which yielded n = 6.4.

The empirical value of about 2 for n when derived for different tracers, different soils, and different climates (provided that the soil is undisturbed) tends to provide indirect confirmation for the theoretically derived form of Eq. 5. However, both K and U_T in Eq. 5 are functions of particle size, soil moisture content, surface roughness, relative humidity, and the time period over which the wind speed is averaged. Some attempts have been made to theoretically include many of these factors (especially particle size), but the theory does not seem to describe adequately the variations in the data. Thus K and U_T must be treated as empirical constants for the present. Consequently there is no practical benefit in using Eq. 5 in preference to the simpler Eq. 6. However, at least one experimental measurement of resuspension and wind speed must be made to set the value of K in Eq. 6 for the particular area.

Mass Loading. In the absence of data to implement Eq. 6 for a given area, Anspaugh (1974) suggests that a mass-loading factor (L_a) of 100 μg(soil)/m³(air) be used for predictive purposes. If we assume that the radioactivity of one square meter is associated with 50 kg of soil (5-cm depth $\times$ 10^4 cm²/m² $\times$ 10^{-3} kg/cm³), a mass-loading factor of 100 μg/m³ is equivalent to a resuspension factor of 2 $\times$ 10^{-9} m⁻¹. The theoretical basis for the mass-loading approach is described by Anspaugh (1974). Anspaugh et al. (1975) provide comparisons showing that predicted air concentrations based on L_a = 100 μg/m³ are in good agreement with measured air concentrations.

We shall use the suggested mass-loading factor to represent average conditions at NTS, but it must be noted that higher than average wind velocities (Shinn and Anspaugh, 1975) or mechanical disturbances, such as plowing (Milham et al., 1976), could cause the mass-loading factor to be temporarily much higher than 100 μg/m³. It should also be noted that some recent work by Sehmel (1977) suggests little if any experimental justification for this approach.

Vegetation

As shown in Fig. 1, vegetation can be contaminated externally by deposition of resuspended material or internally by uptake from soil or by both processes simultaneously. Other mechanisms of external and internal contamination have been identified or postulated, but direct deposition from air and root uptake appear to be the processes most important to consider when attempting to develop a general model.

In the following paragraphs we discuss the mechanisms involved in contaminating vegetation and present mathematical expressions to simulate the dynamics of the contaminating mechanisms. We also discuss the parameters in these expressions and their variations under the influence of different environmental factors. However, we conclude that there are too few data to develop an adequate dynamic model, and we are forced to use a simple steady-state model with a constant vegetation-to-soil contamination factor in the overall transport model.

General Hypothesis. Externally deposited material can be removed from plant surfaces by weathering, i.e., the mechanical action of wind and rain, and it can be diluted by plant growth. Internally deposited material can also be diluted by growth but not by weathering. Processes that remove biomass from vegetation (e.g., grazing, cropping, root decay, and dehiscence of above-ground parts) also remove plutonium. If they exceed growth rates, these processes can reduce the total amount of plutonium in the vegetation compartment of an ecosystem. Different plant species can vary widely with respect to their ability to retain externally deposited plutonium or to assimilate plutonium from foliar deposits or soil, and translocation within the plant can result in large differences regarding plutonium concentrations in different plant organs. In this discussion we do not attempt to distinguish one plant species from another. We assume that plutonium is uniformly distributed in edible plant materials and that processes which remove biomass from the vegetation compartment have no effect on the concentration of plutonium in the remaining biomass.

Differential equations expressing the principal processes described above can be written:

$$\frac{dy_{ve}}{dt} = k_{av}C_a - (\lambda_w + \lambda_g + \lambda_A)\, y_{ve} \tag{7}$$

$$\frac{dy_{vi}}{dt} = k_{sv}C_s - (\lambda_g + \lambda_A)\, y_{vi} \tag{8}$$

where y_{ve} = concentration in vegetation of externally deposited plutonium (pCi/g)
    k_{av} = air-to-vegetation deposition rate coefficient ($m^3/g \cdot day$)
    C_a = concentration of plutonium in air (pCi/m^3)
    λ_w = weathering rate coefficient (day^{-1})
    λ_g = vegetation growth rate coefficient (day^{-1})
    λ_A = radioactive decay rate coefficient for ^{239}Pu (day^{-1})
    y_{vi} = concentration in vegetation of internally deposited plutonium (pCi/g)
    k_{sv} = soil-to-vegetation uptake rate coefficient (day^{-1})
    C_s = concentration of plutonium in soil (pCi/g)

Equations 7 and 8 represent the external and internal components of plutonium in vegetation. The former is due to foliar deposition; the latter, to root uptake. It is assumed that plutonium taken up via roots can be translocated to stems and leaves, but this rate is difficult to estimate. Consumers of vegetation are connected to both compartments simultaneously, and this is the same as summing the two components. Assimilation of externally deposited materials and their translocation to other parts of the plant have been demonstrated experimentally for various kinds of substances applied externally to foliage, but, in the case of plutonium (which is most probably deposited on foliage in the form of insoluble particles), foliar assimilation is assumed to be zero. A recent study (Cataldo, Klepper, and Craig, 1976) has demonstrated that translocation of foliarly deposited plutonium to roots and seeds can occur. However, the accumulation ratios observed in the absence of a solution vector (simulated rainfall) were on the order of 10^{-6} for both fresh and aged PuO_2; i.e., the observed concentrations in leaf tissue were about 200,000 to 500,000 times higher than the observed concentrations in seed and root tissue, respectively.

Although foliar deposition and root uptake of plutonium have been studied separately in a variety of experiments, there is no reliable method for distinguishing between the two components in a plant sample if both are present. If plants are grown in contaminated soil or culture media and only the aerial parts are assayed, we can assume that the activity detected was internally deposited. If the aerial parts of plants are collected from a recently contaminated area that had no plutonium in the soil before the contaminating event, we can assume that all or nearly all the activity detected was externally deposited. If, on the other hand, plant samples are taken from an area that was contaminated years ago, it is likely that most of the plutonium contained therein is due to external contamination by resuspended soil particles and that only a small fraction is due to internal contamination by root uptake.

Recent evidence (Romney et al., 1975; Wildung and Garland, 1974) suggests the possibility that (1) the biological availability of plutonium in contaminated soil may increase with time after environmental release and that successive crops of annuals thus would take up successively greater amounts of plutonium; (2) perennial plants may accumulate plutonium to a much greater extent than previously indicated by short-term uptake experiments; and (3) plant uptake may increase with decreasing plutonium concentration in the soil. Either or all of these factors could have the effect of making biological transport of plutonium to man progressively more important relative to physical transport and inhalation. These considerations are interesting because most assessments of the potential hazards of environmental plutonium attribute little importance to biological transport and ingestion compared with physical transport and inhalation.

Foliar Deposition.   The estimation of deposition velocity, V_d, i.e., the ratio of surface deposition rate to air concentration, was discussed earlier as applied to soil surfaces.

Deposition velocities have also been determined experimentally for different kinds of plants and several kinds of vegetation with respect to various aerosols and particulates. In this study, however, it was more convenient to base estimates of air-to-plant deposition rates on the product of air concentration, soil/air deposition velocity, and a plant (or vegetation) interception factor. The plant interception factor is defined as the amount initially deposited per gram (dry weight) of plant material divided by the amount initially deposited per unit area of soil surface.

Much of the available information concerning the interception of airborne radionuclides by plants has come from studies made in the fallout fields produced by nuclear test detonations at NTS. Romney et al. (1963) summarized early studies in the vicinity of NTS. They found that levels of fallout deposition on plants varied with respect to (1) downwind distance from the detonation point, (2) lateral distance away from the midline of the fallout field, (3) a variety of morphological features associated with different plant species, and (4) the level of fallout deposition on soil surfaces near the contaminated plants. Although there was no constant relationship between fallout concentrations on plants (disintegrations per minute per gram) and on soils (disintegrations per minute per square meter), there was a good correlation between radioactivity in plant samples and in the fraction of fallout samples with particle sizes less than 44 μm. Apparently, the larger particles were deposited close to the detonation point and were not as readily intercepted by plants as were the smaller particles, which were deposited farther downwind; or, if they were intercepted, they were apparently more easily removed by weathering processes. In the laboratory, Romney et al. (1963) found that 50 to 90% of the gross radioactivity on fallout-contaminated plants could be removed by washing with water or a wetting agent, such as versene. These and similar studies demonstrated the possibility of predicting fallout interception by plants, but they did not provide quantitative methods for doing so.

Miller and Lee (1966) carried out extensive studies of fallout interception by plants. The plants were cultivated in gardens near San Jose, Costa Rica, and the fallout was provided by continuing eruptions of Irazú, a nearby volcano. Miller and Lee also developed a comprehensive theoretical model of fallout interception by plants. The model assumes different sets of constants for different fallout particle size classes, different morphological characteristics of foliage, and different meteorological conditions. Unfortunately, their model is practically unworkable, in spite of its elegance, because it requires the use of constants and other parameter values that are rarely, if ever, available for predictive purposes.

The values of the interception factor determined experimentally by Miller and Lee (1966) varied only slightly with respect to the different species of cultivated plants they studied, and the only meteorological condition consistently correlated with large differences in measured values of the interception factor was relative humidity. The particles intercepted by plants were essentially the same sizes as those deposited on adjacent soil surfaces. In both cases the mass median diameters for the volcanic dust deposited as fallout were generally between 50 and 100 μm. The weighted averages of interception factors for all the plant types tested (mostly garden vegetables) were 95.7 ± 66.9 cm^2/g for damp exposure conditions (relative humidity greater than 90%) and 47.4 ± 29.7 cm^2/g for dry exposure conditions.

Interception factors based on nuclear testing experience are about one or two orders of magnitude lower. For detonations involving the incorporation of large quantities of soil material in the initial cloud, estimates range from 1.9 to 11.1 and have a mean of 3.7 cm^2/g. For detonations involving the incorporation of little or no soil material in the initial cloud, the estimated values are about an order of magnitude lower.

Miller and Lee (1966) noted that the "foliar samples obtained at the weapons test experiments were apparently subjected to an unknown degree of weathering before they were taken, while the primary samples [in our studies] were collected at the end of a 12- to 24-hr period of exposure to more or less continuous fallout from Irazú, and the weight of dust deposited on leaves was often greater than the dry weight of the leaves." Heavy dust deposits such as these are easily dislodged by the slightest mechanical disturbance, and moderate rains were observed to remove more than 90% of the material deposited.

Martin (1965) studied the interception and retention of ^{89}Sr and ^{131}I by desert shrubs (primarily *Atriplex confertifolia* and *Artemisia tridentata*). His estimates of the plant interception factor were based on concentrations of ^{89}Sr and ^{131}I in plant samples collected 5 days after fallout deposition and estimates of the theoretical deposition rates for these two radionuclides on unobstructed soil surfaces in the same locations ranging from about 10 to about 100 miles downwind from the detonation point. Estimates for different study areas range from 1.49 to 11.05 cm^2/g, with the higher values occurring in the more distant areas. The overall mean for ^{89}Sr was 4.09 cm^2/g, and the overall mean for ^{131}I was 4.00 cm^2/g, approximately an order of magnitude lower than Miller and Lee's average value for dry deposition conditions. Although the discrepancy may appear to be large, it may be due to the effects of weathering during the 5 days between fallout deposition and the collection of plant samples.

Weathering Rate. To estimate the effective rates of ^{89}Sr and ^{131}I loss from fallout-contaminated plants, Martin (1964) collected additional sets of plant samples at intervals of 10, 15, 30, and 60 days after fallout. Factoring out the loss of ^{131}I due to vaporization and the losses of both radionuclides due to radioactive decay, the data indicated that the weathering (environmental) half-time for these two radionuclides increased with respect to time after fallout. During the interval from 5 to 15 days after fallout, the average weathering half-time was about 20 days. The value obtained for the interval from 15 to 30 days after fallout was about 30 days, and it increased for the interval from 30 to 60 days postdetonation to about 130 days. The D+5-day concentrations can be compared with assumed D+0 concentrations, which would reconcile the difference between Martin's estimate of the plant interception factor and Miller and Lee's estimate of the foliage contamination factor. This procedure suggests an average weathering half-time of approximately 1.4 days during the interval from 0 to 5 days after fallout.

These observations lead us to the hypothesis that the decay-corrected concentration of a radionuclide in fallout-contaminated plant material is a very rapidly declining exponential function of time at times soon after the contaminating event but approaches a lower asymptote. Since this hypothesis appears to be correct, the effective rate at which a radionuclide is removed from surfaces following external deposition cannot be expressed precisely by a single coefficient because the weathering half-time increases as a function of time after contamination. If the initial deposition is a heavy one, a significant fraction of it (perhaps as much as 90%) can be removed by weathering in a matter of hours, or a few days at most. A portion of what remains after this initial period of fast weathering (something in the range of 10 to 60%) is so tightly trapped that it cannot be removed even by vigorous washing (Romney et al., 1963). Presumably, this nonremovable fraction is composed predominantly of particles that are small and mechanically trapped on plant surfaces.

The situation in the plutonium-contaminated area at NTS is one in which foliar deposition of resuspended particles and the loss of these particles from foliage is a more or less continuous process. If the turnover rate is rapid, a foliage/soil steady state would be quickly established.

Plant Growth Rates.   As indicated earlier, the growth of new plant tissue may dilute both the external and the internal concentrations of plutonium or other transuranium elements in plant materials. Since different plant parts may grow at different rates, it is obvious that the growth rate of interest with respect to external contamination is the growth rate of leaves (and other edible parts formed above ground). If we assume that internal plutonium due to root uptake is uniformly distributed to all parts of the plant, the growth rate of interest with respect to dilution of root uptake is the overall growth rate, i.e., the growth rate of leaves plus the growth rate of all other plant parts.

Plant growth is not a continuous process, nor is it the same for all species in a given area or for all the parts of a given plant. In the temperate zone, at least, plant growth is confined to the warm season, and the rate of growth is not uniform throughout the growing season because different plant organs develop at different times. Ignoring the morphogenic aspects of plant growth (i.e., the differentiation and development of structure), growth is most simply conceived as an increase in biomass (i.e., dry weight of tissue per unit area). For annuals, the biomass at the beginning of the growing season consists of seeds; for herbaceous perennials, for which the aboveground parts die back during the winter, it consists mostly of roots and other belowground parts; for woody perennials, it consists of roots and stems (mostly dead tissue) plus, in the case of evergreens, leaves and vestiges of fruits produced the previous growing seasons. In addition to these seasonal and species variations, it is reasonable to suppose that the roots, stems, leaves, fruits, and other organs of a given species grow at different times and at different rates and that additional variations can be expected in response to environmental factors, such as temperature, soil moisture, and availability of nutrients.

To attempt a mathematical description of vegetation growth that includes all factors mentioned above (and others not mentioned) would be a monumental undertaking. What is needed at present is a simple expression of growth rate, as a continuous function, which will provide a reasonable, but conservative, estimate of the potential overall concentration of plutonium in plant materials that have been contaminated externally by airborne deposits and/or internally by root uptake from soil.

To obtain a rough estimate of growth rate, we can define λ_g as follows:

$$\lambda_g = \frac{\ln\,[1 + (P_n/B_0)]}{365} \tag{9}$$

where λ_g is the growth rate coefficient averaged over the year (day^{-1}), P_n is the net gain in biomass during a growing season (g/m^2), and B_0 is the biomass at the beginning of the growing season (g/m^2).

Odum (1971) has estimated that the average gross primary productivity (GPP) of deserts and tundras is about 200 kcal m^{-2} yr^{-1}. Since the fraction of GPP (0.2) used up in respiration does not appear as new tissue, the dilution growth rate is proportional to 0.8 GPP = 160 kcal m^{-2} yr^{-1}. At 4.5 kcal/g (dry weight) (Odum, 1971), this amounts to a net gain of P_n = 36 g/m^2 (approximately). The biomass of desert vegetation varies from place to place. The mean biomass for Area 13 is B_0 = 289 g/m^2 (Wallace and Romney, 1972). Substituting these values of P_n and B_0 in Eq. 9, λ_g = 3 × 10^{-4} day^{-1}. If we assume that internally deposited plutonium is uniformly distributed above and below ground, this would be the value to use in Eq. 8. If we assume that two-thirds of P_n is above ground and one-third is below ground, the dilution growth rate for the external (aboveground) component (Eq. 7) would be λ_g = 2 × 10^{-4} day^{-1}.

Root Uptake and Plant/Soil Concentration Ratio.   For plutonium to enter plants via root uptake, it must first reach the roots. Plowing, of course, accomplishes this "transport" quite rapidly by mixing the soil, but the downward movement of plutonium in an undisturbed soil profile is such a slow process that much of the plutonium deposited on the surface may stay near the surface for many years. To circumvent the variability inherent in these and other soil processes affecting the behavior of plutonium in soils (factors reviewed by Price, 1973a; Francis, 1973), we have made the simplifying and conservative assumption that plutonium deposited on soil is diluted by only the top 5 cm of soil and that root uptake is related to the resulting concentration in surface soil; i.e., the probable concentration of plutonium in the root-zone soil is deliberately overestimated.

Most of the available data (Price, 1973a; Francis, 1973) on plutonium uptake by plants has been derived from short-term greenhouse experiments. Typical values thus derived for the plant/soil concentration ratio range from 10^{-3} to 10^{-6}. Uptake has been shown to be enhanced by the reduction of pH or the addition of chelating agents. There is some evidence that plutonium uptake by plants may increase with time (Romney, Mork, and Larson, 1970) and that the mobility of plutonium, i.e., its ability to move in the soil and its availability for root uptake, may increase with time as a result of chelation due to bacterial decay of soil organic matter.

The evidence that the soil-uptake concentration ratio may increase exponentially as a function of time is scanty but deserves consideration because of its implications. Romney, Mork, and Larson (1970) grew ladino clover (*Trifolium repens L.*) on a plutonium-contaminated soil under greenhouse conditions and cropped it repeatedly over a period of 5 yr. The resulting estimates of C_{RV} were as follows:

Year	C_{RV}, $(d/min)\ g^{-1}$ plants (dry weight) $(d/min)\ g^{-1}$ soil (dry weight)
1	1.91×10^{-5}
2	4.14×10^{-5}
3	4.38×10^{-5}
4	7.10×10^{-5}
5	13.95×10^{-5}

A least-squares fit of these data to an exponential function yields

$$C_{RV} = 1.31 \times 10^{-5} \exp(0.452\ t)$$

where t is time (years) and 0.452 is the apparent growth rate coefficient; i.e., the data indicate that C_{RV} would be expected to double in about 1.5 yr. By extrapolation to 20 yr, the concentration ratio would be 0.11, a value within the range of field observations for perennials (see below) but misleading nonetheless. Romney, Mork, and Larson (1970) attribute the "apparent" increase of C_{RV} to root growth during the time the experiment was being conducted and not to any change in the biological availability of the plutonium contained in the contaminated soil.

In fact, the plant/soil ratios observed under field conditions are generally too high to be explained by root uptake. Romney et al. (1975) collected soil and plant samples from contaminated areas at NTS. The mean plant/soil ratios for different groups of paired samples ranged from 0.004 to 0.44 and were inversely related to the mean soil contamination of each group, which ranged from 5.9×10^{-5} to 0.12 $\mu Ci/g$. The weighted mean ratio for 506 paired samples was 0.096 ± 0.004. The plants in this study were desert shrubs growing in areas that were contaminated with plutonium as a result of nuclear safety tests conducted from 1953 to 1964. Ratios obtained by growing plants in these same soils were on the order of 10^{-3} to 10^{-4}, which indicates that no more than 1% of the plutonium in plant samples from contaminated areas at NTS is likely to be due to root uptake.

Environmental monitoring data from Hanford (Bramson and Corley, 1973; Nees and Corley, 1974; Fix, 1975) indicate ratios in the range of 0.05 to 1.0 for soil concentrations in the range of 2×10^{-9} to 4×10^{-8} $\mu Ci/g$. Similar data from Savannah River (McLendon et al., 1976) indicate ratios from 0.009 to 0.97 for soil concentrations in the range of 1.3×10^{-9} to 1.6×10^{-7} μCi $^{239}Pu/g$ and 2×10^{-10} to 4.6×10^{-8} μCi $^{238}Pu/g$. The higher plant/soil ratios are usually found at lower soil concentrations.

In general, plant/soil ratios for ^{239}Pu, which are based on plant and soil samples collected under field conditions, range from 1 to 10^{-3}, whereas ratios based on laboratory studies, which preclude external contamination, range from 10^{-3} to 10^{-6}. Considering the situation at NTS, we believe it is reasonable to assume that approximately 99% of the plutonium associated with the vegetation compartment is due to external contamination and that no more than 1% is due to root uptake.

Data on the plant/soil ratio for other transuranium elements are very limited. Romney et al. (1975) measured ^{241}Am concentrations of the vegetation from the areas at and near NTS. Grouped according to species and location, the mean plutonium/americium ratio in vegetation ranged from 2.0 to 28.3, with typical values being about 10. Similar groups of soil (Gilbert et al., 1975) ranged from 5.3 to 26, with typical values also being about 10. These analyses indicate that the long-term plant/soil ratio for americium is not significantly different from that for plutonium.* The data on the short-term plant/soil ratio indicate significant differences that may be related to the solubility of the element. Price (1973b) measured the uptake of ^{237}Np, ^{239}Pu, ^{241}Am, and ^{244}Cm by tumbleweed and cheatgrass from various solutions applied to the soil. The americium uptake was about 2 to 30 times as great as that of plutonium, curium uptake was about 2 to 40 times as great as that of plutonium, and neptunium uptake was about 100 to 1000 times as great as that of plutonium. Bennett (1976) summarized much of the short-term data and concluded that americium and curium uptakes were about 10 to 30 times as great as that of plutonium.

Variation of Plant/Soil Ratio. Data presented by Romney et al. (1975) also demonstrate that the mean concentrations of plutonium in soils and plants decrease with increasing distance from ground zero locations, whereas the vegetation/soil ratios within sampling strata show a tendency to increase. Tamura (1976) provides a graph of soil activity vs. distance from ground zero and fits the data to a power curve of the form $y = ax^{-b}$, where y is soil activity and x is distance from ground zero. This is an interesting notion to pursue because we expect both soil concentration and particle size to decrease with increasing distance from ground zero. If foliar retention is greater for small particles, we would expect vegetation/soil ratios to increase as particle size and soil contamination decrease with increasing distance from ground zero. We have no comparable curve for vegetation, but we assume it would be of the same form. On the basis of this assumption, the vegetation/soil ratio could be expressed as a function of distance from ground zero as follows:

$$y_v = a_v x^{-b_v}$$

$$C_s = a_s x^{-b_s}$$

$$\frac{y_v}{C_s} = \frac{a_v}{a_s} x^{(b_s - b_v)}$$

(10)

---

*Gilbert et al. (1975, p. 407), using Double Tracks data, show that the average vegetation/soil ratio for americium is about 1.5 times the average vegetation/soil ratio for plutonium; but, considering the range of ratios contributing to the averages, we are not persuaded that 1.5 is significantly different from 1.0. At other sites the ratios were not different.

where y_v and C_s are plutonium concentrations (pCi/g) of vegetation and soil, respectively, x is distance from ground zero, and a and b are constants derived from least-squares analysis. If $b_s > b_v$, the vegetation/soil ratio will increase with increasing distance from ground zero and decreasing soil concentration.

To obtain an expression for plutonium concentration in vegetation as a function of soil concentration, R. O. Gilbert (personal communication) performed a regression analysis to fit the available data (636 paired samples of vegetation and soil) to the following equation:

$$\ln (y_v) = \ln (a) + b \ln (C_s)$$

or

$$V = A + bS$$

where $V = \ln (y_v)$

$S = \ln (C_s)$

$A = \ln (a)$

$$b = \left[\frac{\Sigma V^2 - (1/n)(\Sigma V)^2}{\Sigma S^2 - (1/n)(\Sigma S)^2} \right]^{1/2}$$

$A = \overline{V} - b\overline{S}$

$\overline{V}$ and $\overline{S}$ = the mean of V and S, respectively.

[This method of calculating b is indicated because measurements of both y_v and C_s are subject to error. See Ricker (1973) for a discussion of this method of calculation.]

The results of this analysis are summarized as follows:

Parameter	Standard deviation
$\overline{V} = -3.4961$	2.3096
$\overline{S} = -0.9322$	3.0382
$b = 0.7602$	0.0221
$A = -2.7871$	0.0458
$r = 0.8084$ (correlation coefficient)	
$n = 636$	

With antilogs, the mean vegetation/soil ratio is

$$C_{RV} = \frac{y_v(nCi/g)}{C_s(nCi/g)}$$

$$= \frac{0.0303}{0.3937}$$

$$= 0.0770$$

which is somewhat lower than the value (0.096) obtained from the grouped data presented by Romney et al. (1975).

The equation for vegetation concentration (nanocuries per gram) as a function of soil concentration (nanocuries per gram) is

$$y_v = 0.0620 \, C_s^{0.76} \tag{11}$$

The corresponding equation for the vegetation/soil ratio is

$$C_{RV} = \frac{y_v}{C_s} = 0.0620 \, C_s^{-0.24} \tag{12}$$

Thus, for areas in which soil concentrations are 10, 1.0, 0.1, and 0.01 nCi/g, the predicted vegetation/soil ratios would be 0.036, 0.062, 0.107, and 0.187, respectively.

Equations 11 and 12 demonstrate the dependence of vegetation concentration on soil concentration and the fact that the vegetation/soil ratio tends to increase as soil concentration decreases. The use of either equation for predictive purposes may be limited by the extreme range of soil and vegetation values and/or by various site specific factors that are not considered in the regression analysis. It might be better to apply the regression analysis to sampling strata means as shown in Table 3.

Except for strata 3 and 4 in Area 13, the measured and predicted values given in Table 3 seem to agree quite well, but the equations for Area 13 and GMX-5 (footnote to Table 3) predict higher values (especially at higher soil concentrations) than would be obtained from Eqs. 11 and 12, which are based on samples from all study areas.

Discussion. Equations 11 and 12 shed some light on how vegetation/soil ratios may be expected to vary with respect to soil concentration, but they do not explain why. To approach this and related questions, we refer back to Eqs. 7 and 8, the proposed differential equations for the external and internal components of plant contamination. For the time being at least, we can dismiss Eq. 8 from further consideration because the greenhouse studies have shown that root uptake cannot account for more than a small fraction of the vegetation/soil ratios observed at NTS.

Equation 7, for external contamination, has the following solution for a constant air concentration (C_a):

$$y_{ve} = \frac{k_{av} C_a}{\lambda_w + \lambda_g + \lambda_A} \{ 1 - \exp [-(\lambda_w + \lambda_g + \lambda_A)t] \} \tag{13}$$

where the parameters are defined following Eq. 7, and

$$k_{av} = V_d F_v \tag{14}$$

$$C_a = L_a C_s \tag{15}$$

where V_d is the deposition velocity on soil (cm/day), F_v is a vegetation interception factor (cm^2/g vegetation), and L_a is a mass-loading factor (g soil/cm^3 air).

If we assume a steady state (large t) between vegetation and soil, the vegetation/soil ratio can be expressed in the parameters of Eqs. 13, 14, and 15 as follows:

$$\frac{y_{ve}}{C_s} = \frac{V_d F_v L_a}{\lambda_v} \tag{16}$$

**TABLE 3   Mean Concentrations of ^{239,240}Pu in Soil and
Vegetation and Vegetation/Soil Ratios by Strata in Area
GMX-5 and Area 13**

| | | ^{239,240}Pu, nCi/g | | | |
| | Soil | Vegetation (means) | | Vegetation/soil ratio | |
Strata	(means)*	Measured*	Predicted†	Measured*	Predicted†
Area GMX-5					
5	0.084	0.0083	0.0099	0.13	0.12
1	0.059	0.0092	0.0075	0.16	0.13
2	0.73	0.064	0.055	0.075	0.075
3	4.5	0.26	0.23	0.052	0.051
4	7.3	0.31	0.34	0.050	0.046
Area 13					
1	0.036	0.0052	0.0067	0.15	0.19
2	0.1	0.013	0.017	0.14	0.17
3	0.4	0.17	0.058	0.44	0.14
4	1.1	0.077	0.14	0.069	0.13
5	2.4	0.28	0.28	0.10	0.12
6	14.0	1.2	1.36	0.078	0.10

*Based on data from Romney et al. (1975).

†$y_v = 0.13\ C_s^{0.79}$ for Area GMX-5. $y_v = 0.07\ C_s^{0.89}$ for Area 13. Both equations are based on the strata means in columns 2 and 3.

where $\lambda_v = \lambda_w + \lambda_g + \lambda_A$ is the effective decay rate coefficient for plutonium-bearing soil particles externally deposited on vegetation.

As noted in the section on air, the deposition velocity is a function of particle size (Eq. 4). In the soils of Area 13 (Tamura, 1976), most of the plutonium is associated with coarse silt (20 to 53 μm), and the estimated deposition velocity (V_d) for particles of 50-μm diameter could be as high as 20 cm/sec, or 1.73×10^6 cm/day.

The plant interception factor (F_v) determined by Miller and Lee (1966) for freshly deposited volcanic dust (50 to 100 μm) was 47.4 cm^2/g for dry exposure conditions.

For predictive purposes Anspaugh (1974) has suggested that a mass-loading factor (L_a) of 100 μg/m^3 (10^{-10} g/cm^3) be used. This is the amount of dust we would expect to find in the GMX area (Shinn and Anspaugh, 1975) when the wind velocity averages about 1.4 m/sec (3 mph).

Substituting these values in Eq. 16, assuming a vegetation/soil ratio of 0.1, and solving for λ_v indicates an effective half-life of about 8.5 days. So

$$\frac{y_{ve}}{C_s} = \frac{(1.73 \times 10^6 \text{ cm/day})\,(47.4\text{ cm}^2/\text{g})\,(10^{-10}\text{ g/cm}^3)}{\ln(2)/8.5}$$

$$= 0.10$$

This exercise proves nothing. It merely demonstrates that Eq. 16 might explain the high vegetation/soil ratios observed at NTS. Shinn and Anspaugh (1975) have demonstrated that mass loading (L_a) increases with wind velocity. The effective half-life may decrease with wind velocity. Sehmel (1975) has shown that deposition velocity (V_d) decreases as the particle size decreases for $d_p > 1$ μm. If small particles are more readily retained by vegetation than larger particles, the plant interception factor and the effective half-life may increase as the particle size decreases. Variations of the factors of Eq. 16 as functions of particle size and wind velocity may account for the range of vegetation/soil ratios implied by Eq. 11, but there are still too many unknowns to develop a descriptive, dynamic model for the vegetation compartment.

For predictive purposes we shall assume that the soil and vegetation compartments are in steady state and that the mean vegetation/soil ratio is 0.1. This ratio is somewhat conservative since it tends to overestimate the plutonium content of vegetation in areas where soil concentrations are greater than 100 pCi/g.

The steady-state assumption is justified to some extent by the fact that movement of contaminated particles from soil to foliage and back to soil is a more or less continuous process. Since the turnover time is apparently short (between 1 and 2 days), a steady state should be established quickly and characterized by a constant vegetation/soil concentration ratio. The choice of 0.1 is in the range of statistical mean (0.096 ± 0.0004) obtained from actual soil and vegetation samples.

Cattle

Transport Pathways. For present purposes the only herbivore assumed to contribute to man's plutonium intake is the cow. Both dairy cattle and beef cattle are considered. The principal plutonium inputs to these herbivores (Fig. 1) are by inhalation and by ingestion of contaminated soil and vegetation. Figure 2 illustrates the assumed pathways of plutonium transport to man via beef cattle and dairy cattle and provides estimates of some of the parameters required to estimate potential concentrations of plutonium in the muscle, liver, and milk of beef and dairy cattle maintained in a contaminated area at NTS.

Formulation of Cow Model. A general equation for the concentration of plutonium in the muscle, liver, or milk compartment of Fig. 2 can be derived as follows:

$$\frac{dy_i}{dt} = \frac{1}{m_i} \left(r_b \, f_{bi} - \lambda_i \, y_i \right) \tag{17}$$

$$y_i(t) = \frac{r_b \, f_{bi}}{m_i} \left(\frac{1 - \exp\left(-\lambda_i t\right)}{\lambda_i} \right) \tag{18}$$

where i = muscle, liver, and milk

y_i = concentration of plutonium in compartment i at time t (pCi/g)

m_i = weight of compartment i (g)

r_b = rate at which plutonium enters blood (pCi/day)

f_{bi} = fraction (dimensionless) transferred from blood to compartment i (Fig. 2: 0.07, 0.12, and 0.007)

λ_i = effective elimination rate coefficient (day^{-1}) for plutonium in compartment i (based on effective half-lives, T, in Fig. 2)

Estimated values for some of the parameters of Eqs. 17 and 18 are given, as indicated above, in Fig. 2. The transfer fractions from the gastrointestinal tract to blood and from blood to muscle, liver, and milk, the weight of muscle and liver, and the effective half-life in milk are based on experimental results reported by Stanley, Bretthauer, and Sutton (1974). The other values were assumed (Martin and Bloom, 1976) for purposes of estimation. Equation 18 ignores the retention of plutonium in the lungs of cattle and

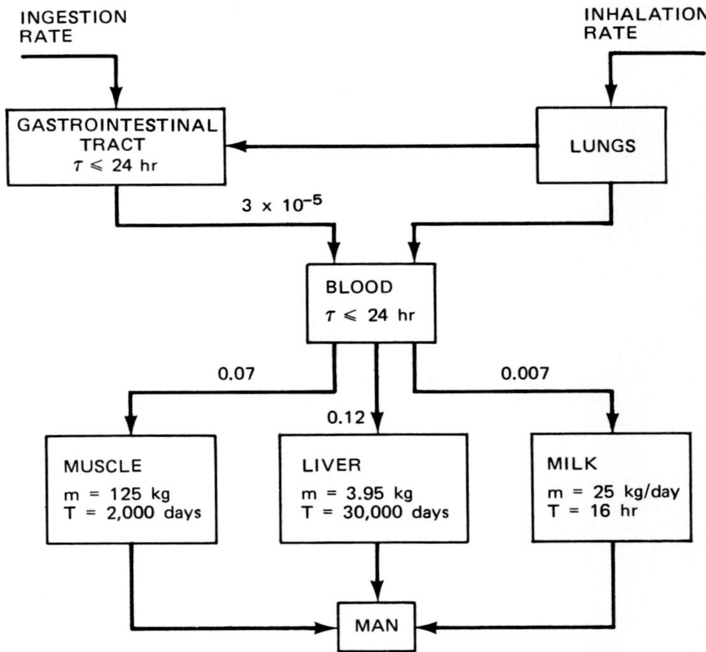

Fig. 2   Pathways of plutonium transport to man via beef cattle and dairy cattle.

assumes that all inhaled plutonium is transferred to the gastrointestinal tract or the blood within 24 hr.

If we assume that the parameter values given in Fig. 2 are reasonably close to the true values, the only parameters remaining to be determined are the plutonium ingestion and inhalation rates and the rate at which plutonium enters the blood (r_b).

Plutonium Ingestion Rates. Kleiber (1961) shows that the basal metabolic rate of mammals (heat production by a fasting animal) is proportional to the three-fourths power of body weight and that the feeding capacity (maximum energy intake) of domestic animals, such as the cow, is about five times the food intake required for basal metabolism. Data given in *The Merck Veterinary Manual* (Siegmund, 1967) for the digestible energy (DE) requirements for maintenance of mature cows are based on

$$DE = 163.5 \ W^{0.73} \tag{19}$$

where DE is the digestible energy required for maintenance (kcal/day) and W is the body weight (kg).

The additional DE requirement for milk production in the range from 20 to 35 kg milk/day at 5% butterfat is 1850 kcal (DE)/kg milk. The additional DE requirement for growth ranges from about 8600 to about 19,800 kcal/kg gained, depending also on body weight. According to McKell (1975), the average digestibility of desert vegetation is about 36% compared with 52% for good alfalfa hay and up to 80% for some concentrates (Siegmund, 1967). The average energy content of most plant materials is about 4.5 kcal/g (dry weight) (Golley, 1961), and the digestible energy content of desert vegetation is about 1.6 kcal/g (dry weight). On the basis of these considerations, the vegetation ingestion rate for a cow grazing desert vegetation can be estimated as follows:

$$V_{ing} = \frac{163.5 \ W^{0.73} \ \text{kcal/day}}{0.36 \times 4.5 \ \text{kcal/g}} \tag{20}$$

where V_{ing} is the vegetation ingestion rate (g/day) for maintenance of a mature cow grazing desert vegetation. For cows that are gaining weight, producing milk, or pregnant, the energy requirement, and thus the vegetation ingestion rate, would be higher than estimated by Eq. 20.

Grazing cattle also ingest soil. In earlier papers (Martin, Bloom, and Yorde, 1974; Martin and Bloom, 1976) we assumed that the soil ingestion rate might be as high as 2000 g/day. Data recently reported by Smith (1977) indicate that this value is probably too high. The amounts of sediment (soil) recovered from the reticulum and rumen of three cows that had been grazing in Area 13 before sacrifice were 8.5, 57.3, and 278 g, respectively. As Smith points out, "These data suggest that the total amount of soil ingested is much less than 2 kg per day, and that a reasonable estimate would be between 0.25 and 0.5 kg."

Smith, Barth, and Patzer (1976) estimate that a 409-kg cow that grazed for 177 days in the inner compound of Area 13 ingested a total of 100 μCi of ^{239,240}Pu, or 0.565 μCi/day. This estimate was based on plutonium concentrations in the rumen contents of fistulated steers allowed to graze in the same compound. Gilbert, Eberhardt, and Smith (1976) made an independent estimate of 0.620 μCi/day based on plutonium concentrations in *Eurotia lanata* (34% of the cow's diet) and *Atriplex canescens* (64% of the cow's diet), as reported for Area 13 by Romney et al. (1975). The average wet weight of vegetation ingested by the fistulated steers was 30 kg/day, and the average dry/wet ratio was about 0.2 (R. O. Gilbert, personal communication). In other words, the cow's vegetation ingestion rate was estimated to be about 6 kg/day. On the basis of Eq. 20, a 409-kg cow would have to ingest about 8 kg/day to meet its energy requirements for maintenance. Neither Smith, Barth, and Patzer (1976) nor Gilbert, Eberhardt, and Smith (1976) include soil ingestion in their estimates of the plutonium ingestion rate for the Area 13 cow.

On the basis of the methods outlined above, we would estimate this cow's plutonium ingestion rate as follows:

$$I_v = 8139 \ \text{g vegetation/day} \times 0.1 \times 5.5 \times 10^{-4} \ \mu\text{Ci/g soil}$$

$$= 0.448 \ \mu\text{Ci/day}$$

$$I_s = 250 \ \text{g soil/day} \times 5.5 \times 10^{-4} \ \mu\text{Ci/g soil}$$

$$= 0.138 \ \mu\text{Ci/day}$$

$$I_v + I_s = 0.585 \ \mu\text{Ci/day}$$

where I_v and I_s are the plutonium ingestion rates through vegetation and soil, respectively. In this calculation, 8139 g/day is the vegetation ingestion rate, which is based on Eq. 20 for a 409-kg cow; 0.1 is the assumed average vegetation/soil ratio; $5.5 \times 10^{-4} \ \mu$Ci/g is the average soil concentration of the Area 13 inner compound (Gilbert, Eberhardt, and Smith, 1976); and 250 g/day is the assumed soil ingestion rate.

The resulting estimate of the cow's total plutonium ingestion rate, 0.585 μCi/day, is comparable to the independent, site-specific estimates of Smith, Barth, and Patzer (1976) (0.565 μCi/day) and of Gilbert, Eberhardt, and Smith (1976) (0.620 μCi/day). Considering the probable variability of the measurements and parameters involved, the agreement of results is remarkably good.

The principal differences between our method and the method of Gilbert et al. are: (1) our estimate of the vegetation ingestion rate, which is based on energy requirements (8 kg/day), is higher than their estimate, which is based on the rumen contents of a fistulated steer (6 kg/day); (2) our estimate of the average concentration of plutonium in vegetation is lower than theirs, 55 pCi/g vs. 103 pCi/g; and (3) we included an input for soil ingestion.

If we assume that the digestibility of Area 13 vegetation is 49% instead of 36%, a ration of 6 kg/day would be adequate to meet the cow's maintenance energy requirements. The weighted mean vegetation/soil ratio for Area 13 is about 0.15, or 50% higher than average. If we use this ratio instead of 0.1 and 6 kg/day instead of 8 kg/day, I_v = 0.495 μCi/day and I_s = 0.565 − 0.495 = 0.07 μCi/day, or 127 g soil/day.

Inhalation Rate.   Standard Man's respiration rate is 20 m^3/day (International Commission on Radiological Protection, 1959), and his digestible energy requirement for maintenance metabolism (no weight gain or loss) is 2600 kcal/day (National Research Council, 1968). We assume that for man and cattle respiration rates are proportional to digestible energy maintenance requirements. The DE requirement (Eq. 19) for a 409-kg cow is 13,185 kcal/day. The cow's estimated respiration rate is therefore 13,185 × 20/2600 = 101 m^3/day. The cow's plutonium inhalation rate (I_a) can be estimated by

$$I_a = \frac{20\,DE}{2600} \times L_a \times C_s \tag{21}$$

where

I_a = plutonium inhalation rate (pCi/day)

DE = digestible energy required (Eq. 19) for maintenance metabolism (kcal/day)

Respiration rate = 20 m^3/day divided by 2600 kcal/day = 7.69 × 10^{-3} m^3/kcal for the Standard Man

L_a = mass-loading factor (100 μg soil/m^3 air) as recommended by Anspaugh (1974)

C_s = average plutonium concentration (pCi/g) in the soil of the area grazed by the cow.

For the 409-kg cow of the Area 13 inner enclosure, where C_s = 550 pCi/g, I_a = 5.56 pCi/day.

Ingestion vs. Inhalation.   It is obvious from the preceding discussion that $I_v \gg I_a$, but the accumulation of plutonium in organs or tissues other than the gastrointestinal tract or lungs and the excretion of plutonium in milk depends (Fig. 2 and Eq. 17) on the rate (r_b) at which plutonium reaches the blood. The rate at which ingested plutonium reaches the blood is simply the plutonium ingestion rate (I_v) multiplied by the fraction transferred from the gastrointestinal tract to the blood. For the Area 13 cow discussed above,

$$\begin{aligned} r_{bv} &= I_v f_{gb} \\ &= 565{,}000 \text{ pCi/day} \times (3 \times 10^{-5}) \\ &= 16.95 \text{ pCi/day} \end{aligned}$$

where r_{bv} is the rate at which ingested plutonium reaches the blood (pCi/day) and f_{gb} is the fraction transferred from the gastrointestinal tract to blood. Since this cow was left in the inner compound of Area 13 for 177 days before sacrifice (Smith, 1977), the total ^{239}Pu expected to have entered the blood via ingestion would be 3000 pCi.

The transfer of inhaled plutonium to blood is considerably more complicated than indicated by Fig. 2. Assuming resuspended particles to have an activity median aerodynamic diameter (AMAD) of 0.5 μm and applying the Task Group on Lung Dynamics model (Fig. 4 and Table 11) to the cow, we obtain the following expression for the rate at which inhaled plutonium would be expected to enter the blood:

$$r_{ba} = 0.0021\, I_a + 0.0833\, \lambda_1 y_1 + \lambda_2 y_2 + 3 \times 10^{-5}\, y_3 \tag{22}$$

$$\frac{dy_1}{dt} = 0.18\, I_a - \lambda_1 y_1 \tag{23}$$

$$\frac{dy_2}{dt} = 0.225\, \lambda_1 y_1 - \lambda_2 y_2 \tag{24}$$

$$\frac{dy_3}{dt} = 0.75\left[I_a\left(0.2079 + \frac{0.12}{\lambda_0}\right) + 0.667\, \lambda_1 y_1 \right] \tag{25}$$

where

r_{ba} = rate (pCi/day) at which inhaled plutonium enters the blood

0.0021 = fraction of inhaled plutonium transferred directly to blood from the upper respiratory tract

0.0833 = fraction of plutonium deposited in the lungs and then transferred to blood

y_1 = amount (pCi) present in the lung at time t

λ_1 = ln (2)/500 days = the lung clearance rate

y_2 = amount (pCi) present in the lymph at time t

λ_2 = ln (2)/1000 days = the lymph clearance rate

y_3 = amount (pCi) present in the gastrointestinal tract at time t

0.18 = fraction deposited in the lungs and cleared with a 500-day half-life

0.225 = fraction transferred from lung to lymph and then cleared to blood with a 1000-day half-life

0.75 (day) = average residence time of plutonium in the gastrointestinal tract

0.2079 = fraction transferred directly from the upper respiratory tract to the gastrointestinal tract

0.12 = fraction cleared from lungs to the gastrointestinal tract with a 1-day half-life

λ_0 = ln (2)/1 day

0.667 = fraction cleared from lungs to the gastrointestinal tract with a 500-day half-life

Note that Eq. 22 consists of four terms. These terms represent transfers of inhaled plutonium to blood from the upper respiratory tract, the lungs, the pulmonary lymph, and the gastrointestinal tract, respectively. Equations 23, 24, and 25 are the differential equations for the long-term components of the lungs and lymph nodes and for the gastrointestinal tract, respectively. The solution to Eq. 22 is

$$r_{ba}(t) = I_a \{0.0021 + 0.015(1 - e^{-\lambda_1 t}) + 5.61 \times 10^{-5} (e^{-\lambda_2 t} - e^{-\lambda_1 t})$$

$$+ 3 \times 10^{-5} [0.381 + 0.12(1 - e^{-\lambda_1 t})] \} \quad (26)$$

Integration of Eq. 26 and evaluation of the integral from $t = 0$ to $t = 177$ days give the amount reaching the blood in 177 days as $0.676\, I_a$. Applying this result to the Area 13 cow, the total ^{239}Pu expected to have entered the blood via inhalation would be 3.76 pCi. Compared with 3000 pCi for ingestion (see above), this is a negligible quantity.

On the basis of this comparison, we shall assume that plutonium concentrations in cow milk, muscle, and liver are due to ingestion only, or, conversely, that the contribution from inhalation is negligible. Even after 10 yr of exposure in Area 13, the contribution from inhalation would be no more than 0.56% of the total.

Comparison of Model Predictions and Field Data.   Plutonium concentrations in lungs, liver, and muscle are given in Table 4 for four cows included in the Area 13 grazing study.

TABLE 4   Concentrations of ^{239,240}Pu in Tissues of Cattle Grazed in Area 13 (NTS)

| | ^{239,240}Pu, pCi/kg | | | | | |
| | Outer compound* | | | | Inner compound† | |
Tissue	1	4	6	Average	2	M
Lungs	74.5	51.4	18.2	48.0	NR‡	NR
Liver	14.5	15.8	10.9	13.7	38.9	6.13 kg
Muscle	0.05	0.195	lost	0.12	0.17	189 kg

*Data extracted from Smith, Barth, and Patzer (1976).
†Data extracted from Smith (1977).
‡Not reported.

Cow 2 (Smith, Barth, and Patzer, 1976) was placed in the inner compound for 177 days. Cows 1, 4, and 6 grazed the outer compound for 433 days (Smith, 1977). Cow 1 weighed 252 kg, and cow 4 weighed 300 kg; the weight of cow 6 is not known. We shall assume that the average weight of the cows from the outer compound was 275 kg. Tissue weights for these cattle were not reported; but we estimate for a 275-kg cow that the average lung weight is about 2.1 kg and the average liver weight is 4.8 kg, based on a study made by Smith and Baldwin (1974). At 45% of body weight (Smith, 1977), the muscle weight for a 275-kg cow would be about 125 kg.

Lungs.   On the basis of these and other considerations, the concentration of ^{239}Pu in the lungs of the cows from the outer compound of Area 13 can be estimated as follows:

$$C_{lung} = \frac{(76\ m^3/day)(10^{-4}\ g/m^3)(215\ pCi/g)(0.18)}{2.1\ kg} \frac{1 - e^{-433\lambda}}{\lambda}$$

$$= 55\ pCi/kg\ (vs.\ 48\ pCi/kg,\ Table\ 4)$$

where   76 m^3/day = estimated respiration rate for a 275-kg cow (Eq. 21)
10^{-4} g(soil)/m^3(air) = estimated mass-loading factor (Eq. 21)
215 pCi/g = average concentration of ^{239}Pu in the soil of Area 13*
0.18 = assumed fraction of inhaled plutonium deposited in the lungs and cleared with a 500-day half-life
2.1 kg = estimated weight of the lungs
λ = ln (2)/500 days

Liver and muscle.   The plutonium ingestion rate for a 275-kg cow can be estimated as follows:

$$I_v = 6158 \text{ g vegetation/day}\dagger \times 0.1 \times 70 \text{ pCi/g soil} + 250 \text{ g soil/day} \times 70 \text{ pCi/g soil}$$

$$= 60,606 \text{ pCi/day}$$

The concentration in liver for cows in the inner compound can now be estimated as follows:

$$C_{liver} = \frac{(60,606 \text{ g/day})(3 \times 10^{-5})(0.12)}{4.8 \text{ kg}} \frac{1 - e^{-433\lambda}}{\lambda}$$

$$= 19.6 \text{ pCi/kg (vs. 13.7 pCi/kg, Table 4)}$$

where 3 × 10^{-5} = fraction transferred from the gastrointestinal tract to blood (Fig. 2)
0.12 = fraction transferred from the blood to the liver (Fig. 2)
4.8 kg = estimated weight of the liver
λ = ln (2)/30,000 days (Fig. 2)

Similar calculations were made for the other cases given in Table 4. The observed values from Table 4 and the estimated values [pCi/kg (wet weight)] are compared below:

| | pCi/kg (wet weight) | | | |
| | Outer compound | | Inner compound | |
	Observed	Estimated	Observed	Estimated
Lungs	48.0	55.0	NR	61.0
Liver	13.7	19.6	38.9	60.7
Muscle	0.12	0.4	0.17	1.12

These comparisons suggest that the model for beef cattle may be somewhat conservative, but the order-of-magnitude agreement between observed and estimated values appears to be good, better than might be expected, as a matter of fact. However, partial data for

---

*The average soil concentration in the outer compound is 70 pCi/g, but some of the resuspended material in the air of the outer compound is assumed to come from the soil of the inner compound.
† Based on Eq. 20 for a 275-kg cow.

four cows, two areas, and two grazing times are hardly an adequate basis for model validation. The best we can conclude from these comparisons is that they provide no basis for rejecting the model. The discrepancy between the experimental values and values predicted by the model is less than an order of magnitude, and we have little reason to expect better than an order of magnitude accuracy.

The milk cow.   The model milk cow is assumed to weigh 650 kg and to produce milk at a rate of 25 kg/day. Such an animal would require a digestible energy intake of 64,750 kcal/day, i.e., 18,500 kcal/day for maintenance plus 25 kg/day × 1850 kcal/kg for milk production (Siegmund, 1967). To meet this high energy requirement and, at the same time, to provide a conservatively high estimate of plutonium transport to man via milk, we shall assume that the model milk cow consumes 10 kg of desert vegetation per day and 15 kg of alfalfa hay grown in the same contaminated area per day. The remainder of the diet consists of commercial concentrates containing no plutonium. For the model milk cow, the plutonium ingestion rate is estimated as follows:

$$I_v = C_s \, (250 \text{ g soil/day} + 0.1 \times 10,000 \text{ g vegetation/day} + 0.017 \times 15,000 \text{ g alfalfa/day})$$

$$= 1505 \, C_s \, (\text{pCi/day})$$

where C_s is the soil concentration (pCi/g) and 0.017 is the alfalfa/soil ratio, which is assumed to be one-sixth the desert vegetation/soil ratio due to plowing and mixing of the soil to a depth of 30 cm. The equation for estimating the concentration in milk is

$$C_{milk} = \frac{(1505)(3 \times 10^{-5})(0.007)}{25 \, \lambda_{milk}} \, C_s \qquad (27)$$

$$= 1.37 \times 10^{-5} C_s \, (\text{pCi/kg})$$

where $\lambda_{milk} = \ln (2)/0.75$ (Fig. 2).

Man

In the preceding discussion we have considered the dynamics of the plutonium transport system (Fig. 1) and have attempted to establish mathematical relationships between compartments. Our present knowledge of the food-chain kinetics of plutonium in contaminated areas at NTS is not adequate for modeling the dynamic aspects of all parts of the transport system. To simplify estimation of the plutonium inhalation and ingestion rates for herbivores (cattle), we assumed a steady-state system and constant intake rates. We now apply the same simplifying assumptions to estimate potential plutonium inhalation and ingestion rates for the hypothetical Standard Man.

Inhalation Rate.   The plutonium inhalation rate (A_m) is defined as the product of the respiration rate (B_m) and the concentration of plutonium in air. The concentration of plutonium in air is, of course, quite variable, but, since it is due to resuspension of contaminated soil, it can be related to the average concentration in surface soil (C_s).

For predictive purposes, Anspaugh (1974) has suggested the use of a mean mass-loading factor of 100 μg soil/m^3 air. We combine this factor with the further assumption that the specific activity of plutonium in resuspended materials is the same as that in the associated soil and estimate A_m as follows:

$$A_m = B_m L_a C_s = 0.002\ C_s \qquad (28)$$

where A_m = plutonium inhalation rate for man (pCi/day)
  B_m = respiration rate ($20\ \text{m}^3/\text{day}$)
  L_a = mass-loading factor ($100\ \mu\text{g/m}^3$)
  C_s = average concentration of plutonium (pCi/g) in the soil of the contaminated area

The observed mass-loading factor during cascade impactor runs at NTS was 70 μg/m^3 (Anspaugh, 1974), and the specific activity of particles recovered from the impactors was about one-third as high as that of surface-soil samples from the same locations (Phelps and Anspaugh, 1974). Compared with these observations, the estimate of A_m provided by Eq. 28 may be conservatively high by a factor of about 4 under average conditions. High winds or mechanical disturbances, such as vehicular traffic, plowing, etc., could cause the mass-loading factor to increase temporarily to very high levels. However, a comparison of observed and predicted air concentrations based on L_a = 100 μg/m^3 showed very good agreement (Anspaugh et al., 1975).

Ingestion Rate. The plutonium ingestion rate is defined as the sum of products of the rates at which different kinds of contaminated materials are ingested and the concentration of plutonium in each kind of material. The formula used for estimating a probable ingestion rate for use in this study was

$$H_m = C_s \sum_{i=1}^{n=6} I_i D_i \qquad (29)$$

where H_m is the plutonium ingestion rate for man (pCi/day), I_i is the ingestion rate for substance i (g/day), and D_i is the discrimination ratio (dimensionless) for substance i.

The kinds of materials considered, their assumed ingestion rates (I_i), and associated discrimination factors (D_i) are listed, together with their products and sum, in Table 5. The methods, experimental data, and assumptions used to estimate the discrimination factors (D_i) are explained in the following text.

Soil. The assumption that the Standard Man of the model accidentally ingests soil at an average rate of 0.01 g/day is purely speculative but not unreasonable considering the amount of dust that can be raised in desert environments by activities that disturb the soil surface.

Vegetation. To estimate the plutonium concentration in native vegetation, we assume an average vegetation/soil ratio of 0.1. As explained earlier, this ratio should tend to overestimate the concentration of plutonium growing in areas of relatively high soil concentration at NTS. To distinguish between native vegetation and cultivated plants (alfalfa hay), we assumed a sixfold dilution of soil concentration due to plowing to a depth of 30 cm; i.e., the plant/soil ratio for cultivated plants is 0.017 instead of 0.1. In preparing Table 1, we assumed that 90% of the external contamination of "leafy vegetables" and that 99% of the contamination associated with "other food plants" would be removed by washing, peeling, etc., during preparation for consumption. In spite

TABLE 5   Estimation of Standard Man's Plutonium Ingestion Rate

i	Substance suggested	I_i*	D_i*	I_iD_i	Percent
1	Soil	0.01†	1.0	1.0×10^{-2}	5.151
2	Leafy vegetables	81‡	1.7×10^{-3}	1.4×10^{-1}	70.935
3	Other food plants	222‡	1.7×10^{-4}	3.8×10^{-2}	19.441
4	Beef muscle	273‡	9.4×10^{-6}	2.6×10^{-3}	1.322
5	Beef liver	13§	4.7×10^{-4}	6.1×10^{-3}	3.148
6	Cow milk	436‡	1.4×10^{-8}	6.1×10^{-6}	0.003
				$\Sigma\ I_iD_i = 1.9 \times 10^{-1}$	100.000

*See Eq. 29 and explanation in text.
†Assumed accidental ingestion rate.
‡From U. S. Department of Agriculture (1973).
§From Organization for Economic Cooperation and Development (1970).

of this assumed reduction, leafy vegetables and other food plants account for 90% of Standard Man's estimated plutonium ingestion rate (Table 5).

Muscle, liver, and milk.   The model beef cow weighs about 275 kg. Its plutonium ingestion rate (I_v), owing to ingestion of 6.2 kg native vegetation/day and 0.25 kg soil/day, is about 870 Cs (pCi/day). Given this ingestion rate and the parameters noted in Fig. 2, the discrimination ratios for muscle and liver (Table 5) were estimated as follows:

$$\frac{C_{muscle}}{C_s} = \frac{(870)(3 \times 10^{-5})(0.07)}{125,000} \left(\frac{1 - \exp\{-730\ [\ln(2)/2000]\}}{\ln(2)/2000} \right)$$

$$= 9.4 \times 10^{-6}$$

$$\frac{C_{liver}}{C_s} = \frac{(870)(3 \times 10^{-5})(0.12)}{4800} \left(\frac{1 - \exp\{-730\ [\ln(2)/30,000]\}}{\ln(2)/30,000} \right)$$

$$= 4.7 \times 10^{-4}$$

In these examples t was set equal to 730 days (2 yr), and this is the assumed average age of beef cattle at the time of slaughter.

The method of estimating the discrimination factor for milk was described earlier in the description of the model milk cow (see Eq. 27).

Discussion. Models for calculating organ burdens and cumulative organ doses due to ingestion and inhalation of ^{239}Pu are discussed in considerable detail later in this chapter. On the basis of the ICRP-II model for ingestion and the Task Group on Lung Dynamics model for inhalation, radiation doses to the respiratory system would be entirely due to inhalation, and doses to the gastrointestinal tract would be primarily due to ingestion. Doses to organs receiving the radionuclide from blood (bone, liver, kidney, etc.) would be due to both ingestion and inhalation. The relative importance of inhalation vs. ingestion can be compared by comparing the two components of organ burden after a period of chronic exposure. Such a comparison is provided in Table 6.

TABLE 6   Fractions of ^{239}Pu in Bone, Liver, or Kidney Due to Chronic Ingestion and Inhalation for a Period of 50 yr*

Ingestion/inhalation	Fraction due to ingestion	Fraction due to inhalation
1	0.0005	0.9995
10	0.0053	0.9947
100	0.0506	0.9494
200	0.0964	0.9036
400	0.1758	0.8242
1000	0.3478	0.6522

*Estimated burdens based on ICRP Publications 2 and 19 (International Commission on Radiological Protection, 1959; 1972).

On the basis of our estimates for the hypothetical Standard Man at NTS [inhalation = 0.002 C_s (pCi/day) and ingestion = 0.2 C_s (pCi/day)], the ingestion/inhalation ratio would be 100, and ingestion would contribute about 5% of the 50-yr bone burden. As indicated by Table 6, the relative importance of ingestion vs. inhalation increases as the ingestion/inhalation ratio increases. Any factor tending to increase the transfer from the gastrointestinal tract to blood would have the same effect as an increase in the ingestion/inhalation ratio. A factor tending to decrease the inhalation rate would also increase the ingestion/inhalation ratio. The point of Table 6 is that, to have a significant effect on internal organ burden, dose, or dose commitment, the ingestion rate must exceed the inhalation rate by a factor of 100 or more.

Dose-Estimation Models

Plutonium reaches man by ingestion of contaminated food and water or by inhalation of contaminated air. Part of this plutonium is distributed throughout the body where it may remain for some time. While it remains within the body, organs that retain the plutonium will receive a radiation dose that depends on the weight of the organ, the amount of plutonium retained, and the time that the plutonium is retained. The several models we have used to estimate plutonium distribution in man are discussed in this section.

Dose and Dose Commitment

In all the dose-estimation models, the formula for estimating the radiation dose to a critical organ of man, i.e., one of those organs which tends to receive the highest radiation dose, is

$$\frac{dD}{dt} = \frac{E}{m} y \tag{30}$$

where t = time (days)

    D = dose to the reference organ (rem)

    E = 51.2159ϵ, a dose-rate factor (g $\cdot$ rem μCi^{-1} day^{-1})

    ϵ = effective energy absorbed in the reference organ per disintegration of radionuclide (MeV/dis)

    y = plutonium burden in the organ (μCi)

    m = either the mass of the organ if the organ is not part of the gastrointestinal tract or twice the mass of the contents if the organ is part of the gastrointestinal tract (g)

The values of the parameters in Eq. 30 for ^{239}Pu and other transuranium elements are given in Table 7. Most of these values were reported by the International Commission on Radiological Protection (1959; 1964). The masses of deep lung and other portions of the respiratory tract are the values used by Snyder (1967) and Kotrappa (1968; 1969). The mass of thoracic lymph nodes was assumed to be the value (15 g) reported by Pochin (1966). The mass of abdominal lymph nodes was assumed to be less than the mass of thoracic lymph nodes and was arbitrarily set at 10 g. The dose accumulated in the organ from the beginning of the exposure period (t = 0) to some later time (t = T_D) is given by

$$D = \frac{E}{m} \int_0^{T_D} y \, dt \tag{31}$$

If ingestion and inhalation of plutonium were halted at time T_D and the individual were to live to some later time T_L, each organ would accumulate an additional dose from the plutonium already within the body at time T_D. The dose commitment is the sum of the dose accumulated to T_D plus the additional dose, or

$$D_c = D + D_a \tag{32}$$

where D_a is the additional dose (rem) and D_c is the dose commitment (rem).

ICRP Committee II Model

The report of the ICRP Committee II (International Commission on Radiological Protection, 1959) contains a model and data that were used to estimate maximum permissible concentrations (MPC's) of radionuclides in air and water. The model, as

TABLE 7 Parameters for Calculating Radiation Doses from Transuranium Radionuclides

Radionuclide	τ_A†	Organ: Mass, g:	Effective energy,* MeV/disintegration					
			GIT 150‡	Lung§ 500	Bone 7,000	Liver 1,700	Kidney 300	TB 70,000
^{237}Np	8×10^8		0.62	49	250	49	49	49
^{239}Np	2.33		0.14	0.16	0.63	0.16	0.15	0.22
^{238}Pu	33000		0.55	57	284	57	57	57
^{239}Pu	8.9×10^6		0.52	53	266	53	53	53
^{240}Pu	2.4×10^6		0.52	53	266	53	53	53
^{241}Pu	4800		0.010	0.013	0.048	0.013	0.012	0.014
^{242}Pu	1.4×10^8		0.49	51	253	51	51	51
^{243}Pu	0.208		0.18	0.18	0.88	0.18	0.18	0.19
^{244}Pu	2.8×10^{10}		1.14	59	292	59	59	59
^{241}Am	1.7×10^5		0.56	57	283	57	57	57
^{242m}Am	5.6×10^4		0.745	53.2	266	53.2	53.1	53.2
^{242}Am	0.667		0.734	53.1	266	53.1	53.1	53.1
^{243}Am	2.9×10^6		0.79	54.2	273	54.2	54.2	54.2
^{244}Am	0.0181		0.52	0.52	2.6	0.52	0.52	0.52
^{242}Cm	162.5		0.61	63	315	63	63	63
^{243}Cm	13000		0.61	60	299	60	60	60
^{244}Cm	6700		0.59	60	299	60	60	60
^{245}Cm	7.3×10^6		0.55	55	277	55	55	55
^{246}Cm	2.4×10^6		0.54	56	278	56	56	56
^{247}Cm	3.3×10^{10}		0.54	56	278	56	56	56
^{248}Cm	1.7×10^8		11.5	453	2244	453	453	453
^{249}Cm	0.044			0.31	27			5.2
^{249}Bk	290		0.026	0.026	0.13			0.026
^{250}Bk	0.134		0.41	0.52	1.5			0.83
^{249}Cf	1.7×10^5		0.63	60	301			60
^{250}Cf	3700		0.61	62	311			62
^{251}Cf	2.9×10^5		0.59	59	295			59
^{252}Cf	803		2.1	210	1100			210
^{253}Cf	18.0		0.78	68	343			68
^{254}Cf	56.0		120	3800	18900			3800

*Includes energy from daughter products with half-times less than 1 yr.

†τ_A is radioactive half-life of radionuclide, days.

‡GIT is gastrointestinal tract (principally, lower large intestine). Mass of contents of lower large intestine is 150 g.

§ICRP-II used 1000 g for mass of lungs. All other models use 500 g. Effective energy to lymph nodes and all portions of the respiratory tract are assumed equal to lung. Masses for these organs are: nasopharyngeal region, 1.35 g; tracheobronchial region, 400 g; abdominal lymph nodes, 10 g.

applied to plutonium or other transuranic elements, is shown in Fig. 3. This model distinguishes between insoluble and readily soluble compounds that are inhaled. In the absence of other data, it is assumed that 25% of a soluble compound is exhaled, 50% is swallowed and reaches the gastrointestinal tract (GIT) almost immediately, and 25% is immediately transferred directly to the blood.

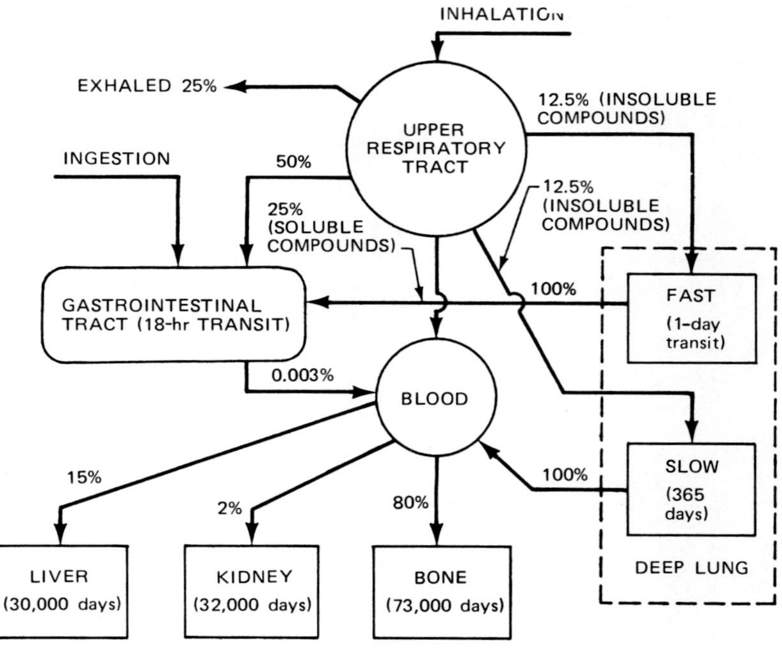

Fig. 3   International Commission on Radiological Protection Committee II model for plutonium. Fraction from blood to total body is 100%. Biological half-time in total body is 65,000 days.

None of the soluble compound resides in the lung for any significant length of time and therefore does not contribute a radiation dose to the lung. For insoluble compounds, in the absence of other data, it is assumed that 25% is exhaled, 50% is swallowed and reaches the GIT immediately, and 25% is immediately transferred directly to tissues deep in the lung. Of the amount transferred to the deep lung, half (12.5% of that inhaled) is coughed up within 24 hr and is swallowed (reaches the GIT). The remainder (12.5% of that inhaled) resides in the deep lung with a half-time of 365 days and is then transferred to the blood. To avoid underestimating radiation doses, we use the insoluble parameters for calculating the transfer to the lung and GIT but use the soluble parameters for the transfer to all other organs.

On the basis of Fig. 3 and the above discussion, the equations for the ICRP II model are

$$r_{GIT} = f_{URG} \, A_m + f_{URDLF} \, A_m + H_m \qquad (33)$$

$$r_B = f_{URB} \, A_m + f_{GITB} \, r_{GIT} \qquad (34)$$

$$y_{GIT} = r_{GIT} \, T_{GIT} \qquad (35)$$

$$\frac{dy_{DLS}}{dt} = f_{URDLS} \, A_m - (\lambda_A + \lambda_{DLS}) \, y_{DLS} \qquad (36)$$

$$y_{DL} = y_{DLS} + f_{URDLF} A_m T_{DLF} \tag{37}$$

$$\frac{dy_{liver}}{dt} = f_{BL}\, r_B - (\lambda_A + \lambda_L)\, y_{liver} \tag{38}$$

$$\frac{dy_K}{dt} = f_{BK}\, r_B - (\lambda_A + \lambda_K)\, y_K \tag{39}$$

$$\frac{dy_{bone}}{dt} = f_{BBN}\, r_B - (\lambda_A + \lambda_B)\, y_{bone} \tag{40}$$

$$\frac{dy_{TB}}{dt} = f_{BTB}\, r_B - (\lambda_A + \lambda_{TB})\, y_{TB} \tag{41}$$

where r_B, r_{GIT} = rates that plutonium reaches the blood and gastrointestinal tract (GIT), respectively (μCi/day)

A_m, H_m = plutonium inhalation and ingestion rates, respectively (μCi/day)

f = fraction of plutonium transferred from one location to another within the body with the subscript notation as follows: URG is upper respiratory tract (URT) to GIT; URB is URT to blood; GITB is GIT to blood; URDLF and URDLS are URT to the fast and slow portions, respectively, of the deep lung; BL, BK, BBN, and BTB are blood to liver, kidney, bone, and total body, respectively

T_{DLF}, T_{GIT} = transit times for the fast portion of the deep lung and the GIT (lower large intestine), respectively

λ = biological elimination rate constant with the subscript notation as follows: DLS is deep lung (slow portion), TB is total body, L is liver, K is kidney, and B is bone (day^{-1})

$\lambda = \ln(2)/\tau$, where τ is the biological half-time for the organ (days)

y = plutonium burden in the organ with the subscript notation either self-evident or identical to that for λ except DLF is the fast portion of the deep lung and DL is the total deep lung (μCi)

Values of the parameters in Eqs. 33 through 41 can be obtained from Fig. 3 for plutonium. The values for other transuranic elements, if different from plutonium, are given in Table 8. The radioactive half-times for each transuranic nuclide are given in Table 7.

Task Group on Lung Dynamics Model

The ICRP Task Group on Lung Dynamics (Morrow et al., 1966) provided a more detailed description of the inhalation pathway and arrived at the model indicated in Fig. 4. A subsequent report [ICRP Publication 19 (International Commission on Radiological Protection, 1972)] provided specific data for applying this model to plutonium and other transuranic elements. The Task Group model treats the respiratory tract as a series of compartments in which the amount initially deposited depends on the particle size and the clearance rate (biological half-time) depends on the type of compound.

**TABLE 8   Parameter Values Applicable to the ICRP II Model
for Transuranic Elements Other than Plutonium***

	Element				
	Neptunium	Americium	Curium	Berkelium	Californium
Biological Half-Time, days					
Deep lung (slow portion)	120	120	120	120	120
Liver	54,000	3,480	3,000		
Kidney	64,000	27,000	24,000		
Total body	39,000	20,000	24,000	65,000	65,000
Fraction from Blood to Organ					
Liver	0.05	0.35	0.40		
Kidney	0.03	0.03	0.02		
Bone	0.45	0.25	0.30	0.80	0.80

*All other parameter values are identical to those of plutonium.

On the basis of Fig. 4, the equations for the compartments of the respiratory tract
and lymph nodes plus the transfer to the GIT and blood are

$$r_{GIT} = \lambda_b y_{NP_b} + \lambda_d y_{TB_d} + \lambda_f y_{P_f} + \lambda_g y_{P_g} + H_m \tag{42}$$

$$r_B = \lambda_a y_{NP_a} + \lambda_c y_{TB_c} + \lambda_e y_{P_e} + \lambda_i y_{LM_i} + f_j r_{GIT} \tag{43}$$

$$\frac{dy_{NP_a}}{dt} = f_a D_3 A_m - (\lambda_A + \lambda_a) y_{NP_a} \tag{44}$$

$$\frac{dy_{NP_b}}{dt} = f_b D_3 A_m - (\lambda_A + \lambda_b) y_{NP_b} \tag{45}$$

$$y_{NP} = y_{NP_a} + y_{NP_b} \tag{46}$$

$$\frac{dy_{TB_c}}{dt} = f_c D_4 A_m - (\lambda_A + \lambda_c) y_{TB_c} \tag{47}$$

$$\frac{dy_{TB_d}}{dt} = f_d D_4 A_m - (\lambda_A + \lambda_d) y_{TB_d} \tag{48}$$

$$y_{TB_{f,g}} = (\lambda_f y_{P_f} + \lambda_g y_{P_g}) T_{TB_{f,g}} \tag{49}$$

$$y_{TB} = y_{TB_c} + y_{TB_d} + y_{TB_{f,g}} \tag{50}$$

$$\frac{dy_{P_e}}{dt} = f_e D_5 A_m - (\lambda_A + \lambda_e) y_{P_e} \tag{51}$$

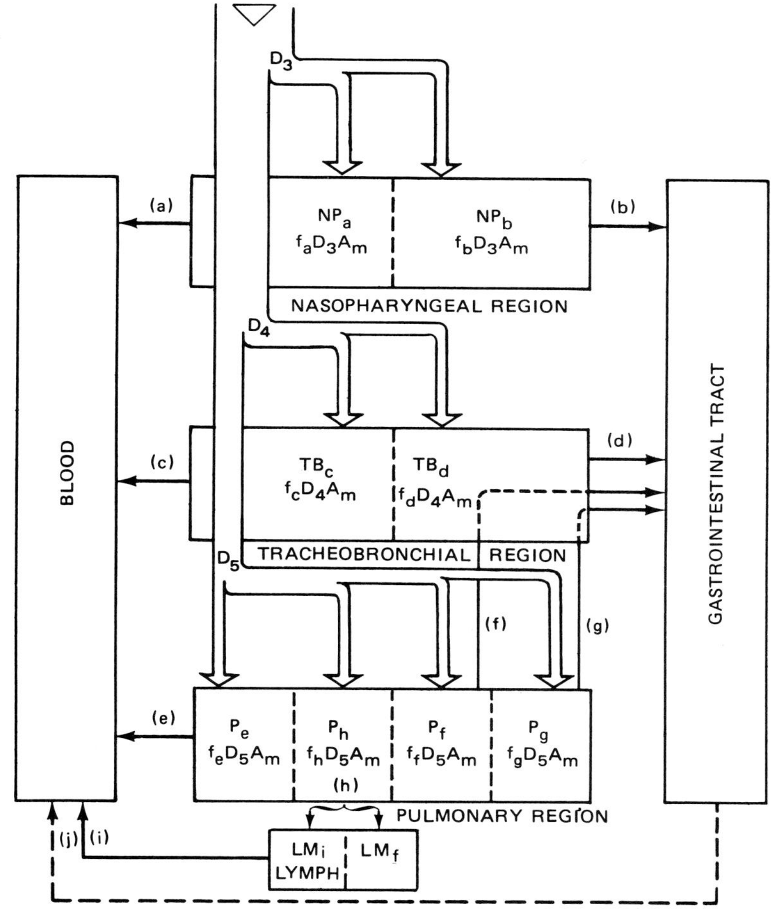

Fig. 4 Schematic diagram of the Task Group on Lung Dynamics model. [From Houston, Strenge, and Watson (1975).]

$$\frac{dy_{P_f}}{dt} = f_f D_5 A_m - (\lambda_A + \lambda_f)\, y_{P_f} \tag{52}$$

$$\frac{dy_{P_g}}{dt} = f_g D_5 A_m - (\lambda_A + \lambda_g)\, y_{P_g} \tag{53}$$

$$\frac{dy_{P_h}}{dt} = f_h D_5 A_m - (\lambda_A + \lambda_h)\, y_{P_h} \tag{54}$$

$$y_P = y_{P_e} + y_{P_f} + y_{P_g} + y_{P_h} \tag{55}$$

$$\frac{dy_{LM_i}}{dt} = f_i \lambda_h y_{P_h} - (\lambda_A + \lambda_i)\, y_{LM_i} \tag{56}$$

$$\frac{dy_{LM_f}}{dt} = (1 - f_i)\, \lambda_h y_{P_h} - \lambda_A y_{LM_f} \qquad (57)$$

$$y_{LM} = y_{LM_i} + y_{LM_f} \qquad (58)$$

where r, λ, y, H_m, f, and A_m are as defined in the ICRP II model and the subscripts, where different, refer to the compartments and the pathways listed in Fig. 4 and Table 9; $T_{TB_{f,g}}$ is the residence time of material following pathways f and g in the tracheobronchial region (days); and D_3, D_4, and D_5 are the fractions of inhaled material that are deposited in the nasopharyngeal, tracheobronchial, and pulmonary regions, respectively, of the respiratory tract (see Fig. 4 and Table 9).

Values of most of the parameters in the above equations are given in Table 9 (U. S. Nuclear Regulatory Commission, 1976). $T_{TB_{f,g}}$ is assumed to be 1 hr, or $\frac{1}{24}$ day (Snyder, 1967; Kotrappa, 1968; 1969), and f_j is identical to f_{GITB} (0.003%) in the

TABLE 9 Task Group Lung Model Parameter Values

Fraction of Inhaled Particles Deposited in the Respiratory System vs. Particle Diameter*

| | Fraction of inhaled quantity retained | | |
Particle size (AMAD), μm	Nasopharyngeal region (D_3)	Tracheobronchial region (D_4)	Pulmonary region (D_5)
0.05	0.001	0.08	0.59
0.1	0.008	0.08	0.50
0.3	0.063	0.08	0.36
0.5	0.13	0.08	0.31
1.0	0.29	0.08	0.23
2.0	0.50	0.08	0.17
5.0	0.77	0.08	0.11

Clearance Parameter Values†

| | | Translocation class | | | | | |
| | | Days | | Weeks | | Years | |
Compartment	k‡	T_k^b‡	f_k‡	T_k^b‡	f_k‡	T_k^b‡	f_k‡
NP	a	0.01	0.50	0.01	0.10	0.01	0.01
	b	0.01	0.50	0.40	0.90	0.40	0.99
TB	c	0.01	0.95	0.01	0.50	0.01	0.01
	d	0.20	0.05	0.20	0.50	0.20	0.99
P	e	0.50	0.80	50	0.15	500	0.05
	f	NA	NA	1	0.40	1	0.40
	g	NA	NA	50	0.40	500	0.40
	h	0.50	0.20	50	0.05	500	0.15
LM	i	0.50	1.00	50	1.00	1000	0.90

*Estimated from data of Morrow et al. (1966).

†As amended by ICRP Publication 19 [International Commission on Radiological Protection (1972)].

‡k is clearance pathway (see Fig. 4); T_k^b is biological half-time (days) for pathway k; f_k is fraction cleared by pathway k.

ICRP II model. The equations for GIT, liver, kidney, bone, and total body are identical to the ICRP II model and have identical parameter values except for the following values, which apply to all transuranic elements: $f_{BL} = 0.45$, $f_{BBN} = 0.45$, $\tau_L = 40$ yr, and $\tau_B = 100$ yr.

Stuart, Dionne, and Bair (SDB) Model

Stuart, Dionne, and Bair (1968) developed models to describe the distribution and retention of plutonium in the body following a single inhalation. These models were based on the results of several studies with dogs, and these results were extrapolated to where they might apply to man. The short-term model is shown in Fig. 5, and the long-term form is shown in Fig. 6. Stuart, Dionne, and Bair (1971) revised the long-term model, and these revisions are incorporated in Fig. 6. Stuart et al. (1971) combined the nasopharyngeal and tracheobronchial regions of the Task Group model into one compartment but expanded the pulmonary region into two compartments, one with a constant biological half-time of 3 yr and another with a variable half-time.

They also added compartments for abdominal lymph nodes and treated the transfers from the pulmonary region to lymph nodes in a slightly different manner than the Task

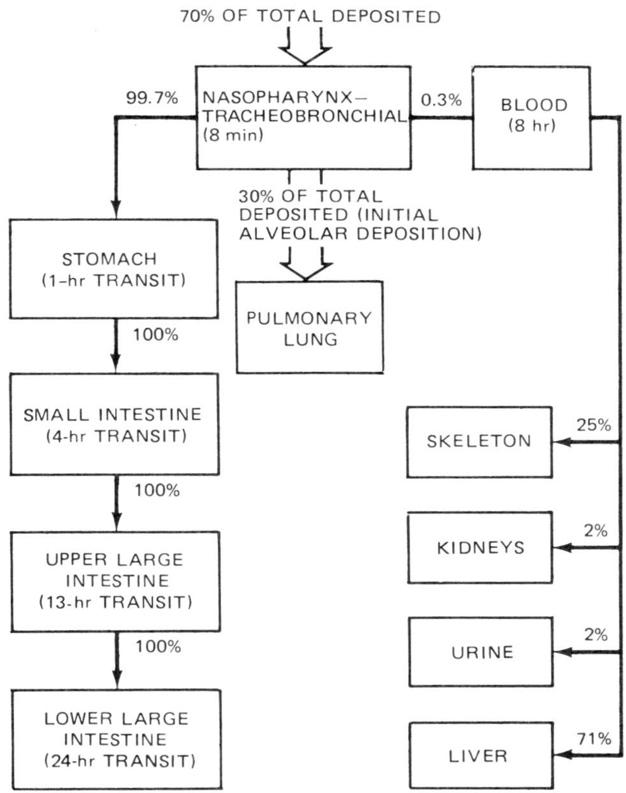

Fig. 5 Block diagram of the short-term form of the Stuart, Dionne, and Bair (SDB) model.

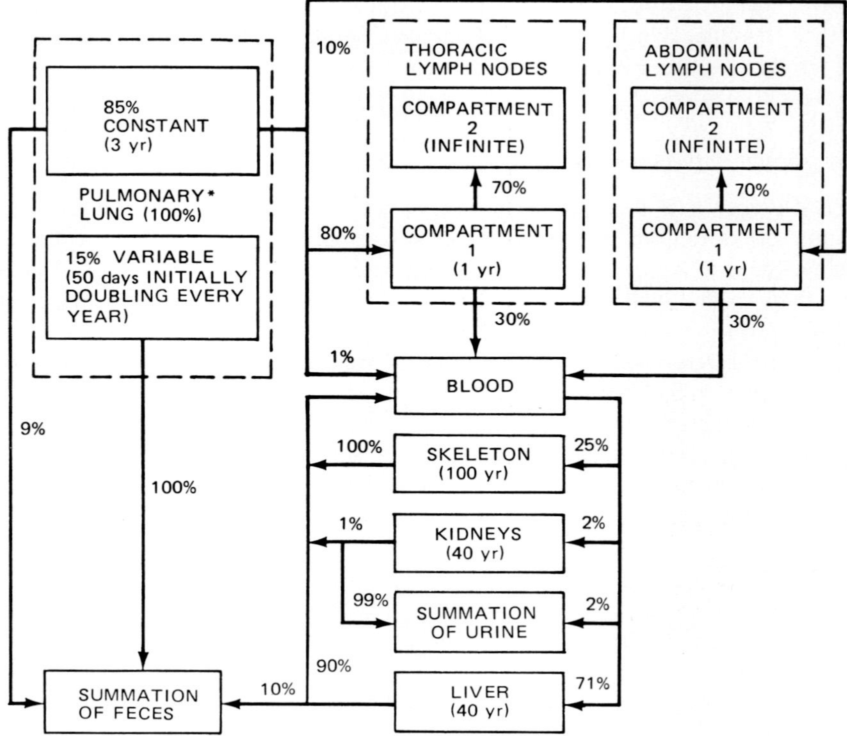

*BASED ON 100% OF THE INITIAL ALVEOLAR DEPOSITION.

Fig. 6 Block diagram of the long-term form of the Stuart, Dionne, and Bair (SDB) model.

Group model. In addition, they included feedback pathways for liver to the gastrointestinal tract (GIT) and for liver, kidneys, and skeleton to blood. Since they applied this model to inhalation only, they did not consider the transfer from GIT to blood since the fraction transferred is so small (less than 0.01%).

On the basis of Figs. 5 and 6, the equations for the compartments of the respiratory tract and lymph nodes plus the transfer to GIT and blood are

$$r_{GIT} = f_{URG}\lambda_{NPTB}y_{NPTB} + f_{LCG}\lambda_{LC}y_{LC} + \lambda_{LV}y_{LV} + f_{LG}\lambda_{LY}_{liver} \tag{59}$$

$$r_B = f_{URB}\lambda_{NPTB}y_{NPTB} + f_{LCB}\lambda_{LC}y_{LC} + f_{TLB}\lambda_{TL}y_{TLNM}$$

$$+ f_{ALB}\lambda_{AL}y_{ALNM} + f_{LB}\lambda_{LY}_{liver} + f_{KB}\lambda_{K}y_{K} + \lambda_{BY}_{bone} \tag{60}$$

$$\frac{dy_{NPTB}}{dt} = (0.7)(D_3 + D_4 + D_5)A_m - (\lambda_A + \lambda_{NPTB})y_{NPTB} \tag{61}$$

$$\frac{dy_{LC}}{dt} = (0.3)(0.85)(D_3 + D_4 + D_5)A_m - (\lambda_A + \lambda_{LC})y_{LC} \tag{62}$$

$$\frac{dy_{ALNR}}{dt} = f_{ALAL}\lambda_{AL}y_{ALNM} - \lambda_A y_{ALNR} \tag{76}$$

$$\frac{dy_{liver}}{dt} = f_{BL}r_B - (\lambda_A + \lambda_L)y_{liver} \tag{77}$$

$$\frac{dy_K}{dt} = f_{BK}r_B - (\lambda_A + \lambda_K)y_K \tag{78}$$

$$\frac{dy_{bone}}{dt} = f_{BBN}r_B - (\lambda_A + \lambda_B)y_{bone} \tag{79}$$

$$\frac{dy_{TB}}{dt} = f_{BTB}r_B - (\lambda_A + \lambda_{TB})y_{TB} \tag{80}$$

where r, λ, y, f, D, A_m, and H_m are as defined in the ICRP II, Task Group, and SDB models. The subscript URDL refers to transfer from URT to DL, BDL is from blood to DL, DLTL is from DL to TLNM, BTL is from blood to TLNM, DLTR is from DL to TLNR, DLAL is from DL to ALNM, and BAL is from blood to ALNM.

Parameters for the above equations can be obtained from Fig. 7. Most of these values are identical to those of the SDB model (Stuart, Dionne, and Bair, 1971). Our modifications are designed to make the model more applicable to chronic inhalation and to include ingestion and transfer of plutonium from blood to DL and from blood to the lymph nodes. The value for the transfer from GIT to blood is identical to the ICRP II value. The values for the other parameters are based on data reported by Ballou, Park, and Morrow (1972) for the translocation of a soluble form of plutonium (plutonium citrate) in dogs. Stuart, Dionne, and Bair (1971) recommended a variable half-time for part of the transfer rate from the lungs to blood, but it was felt that a constant 3-yr half-time for the entire lung was an adequate representation of their dog data. It was also much easier to use the constant half-time for the mathematical description of chronic inhalation.

Like the SDB model, we assumed that 30% of the material initially deposited in URT is rapidly transferred to DL. Since the short-term form of the SDB model transfers 99.7% of the material in URT to GIT in a short time (8 min), the effective transfer from URT to GIT is 69.79% (0.997 × 0.70). The corresponding transfer from URT to blood is 0.21% (0.003 × 0.70).

Model Comparisons

The preceding models were compared by computing organ burdens and radiation doses resulting from unit intakes by ingestion or inhalation. The units were either 1 μCi as a single intake or 1 μCi/day continuous intake, and the calculations were carried to 50 yr. The resulting organ burdens and doses after 50 yr are given in Tables 10 and 11.

Stuart, Dionne, and Bair (1971) (SDB) applied their model to single inhalations, and the variable half-time in the pulmonary lung did not present any difficulty. However, there is some doubt as to the interpretation of the variable half-time for the chronic case. A strict interpretation would imply that a significant fraction (15%) of material continuously deposited in the pulmonary lung is eliminated with a half-time that exceeds

TABLE 10 Comparison of Calculations for 1-μCi Single Intake* Using Various Models

Model[†]	Path	URT[‡]	GI tract	Lung	Lymph nodes Thoracic	Lymph nodes Abdominal	Liver	Kidney	Bone	Total body
					50-yr Organ Burdens, μCi					
ICRP II	Ingestion						2.95×10^{-6}	4.03×10^{-7}	2.01×10^{-5}	2.47×10^{-5}
Task Group	Ingestion						5.67×10^{-6}	4.03×10^{-7}	9.53×10^{-6}	2.47×10^{-5}
SDB	Ingestion		6.37×10^{-11}				1.74×10^{-5}	4.91×10^{-7}	9.10×10^{-6}	2.70×10^{-5}
Modified	Ingestion		6.37×10^{-11}	1.47×10^{-8}	2.90×10^{-7}	9.54×10^{-8}	1.74×10^{-5}	4.91×10^{-7}	9.10×10^{-6}	4.11×10^{-5}
ICRP II	Inhalation[§]			1.11×10^{-16}			2.49×10^{-2}	3.37×10^{-3}	1.68×10^{-1}	2.05×10^{-1}
Task Group	Inhalation	7.26×10^{-17}	1.31×10^{-15}	1.89×10^{-12}	4.64×10^{-3}		1.22×10^{-2}	8.30×10^{-4}	1.95×10^{-2}	4.98×10^{-2}
SDB	Inhalation		8.30×10^{-8}	1.70×10^{-5}	7.45×10^{-2}	9.38×10^{-3}	2.27×10^{-2}	6.39×10^{-4}	1.14×10^{-2}	1.19×10^{-1}
Modified	Inhalation		8.30×10^{-8}	2.06×10^{-5}	7.45×10^{-2}	9.38×10^{-3}	2.27×10^{-2}	6.39×10^{-4}	1.14×10^{-2}	5.10×10^{-2}
					50-yr Cumulative Dose to Organs, rem					
ICRP II	Ingestion		6.66×10^{-2}				1.07×10^{-1}	8.18×10^{-2}	7.94×10^{-1}	1.93×10^{-2}
Task Group	Ingestion		2.50×10^{-2}				2.63×10^{-1}	8.18×10^{-2}	4.05×10^{-1}	1.93×10^{-2}
SDB	Ingestion		6.66×10^{-2}				5.62×10^{-1}	8.96×10^{-2}	2.99×10^{-1}	2.02×10^{-2}
Modified	Ingestion		6.66×10^{-2}	4.00×10^{-3}	7.48×10^{-1}	3.82×10^{-1}	5.62×10^{-1}	8.96×10^{-2}	2.99×10^{-1}	2.55×10^{-2}
ICRP II	Inhalation[§]		4.16×10^{-2}	1.79×10^{2}			9.04×10^{2}	6.82×10^{2}	6.61×10^{3}	1.61×10^{2}
Task Group	Inhalation	7.27×10^{-1}	3.04×10^{-2}	7.29×10^{2}	2.59×10^{4}		5.03×10^{2}	1.54×10^{2}	7.59×10^{2}	3.50×10^{1}
SDB	Inhalation	1.97×10^{-2}	2.67×10^{-2}	1.13×10^{3}	2.28×10^{5}	4.30×10^{4}	6.42×10^{2}	1.03×10^{2}	3.36×10^{2}	8.73×10^{1}
Modified	Inhalation		2.67×10^{-2}	1.34×10^{3}	2.28×10^{5}	4.30×10^{4}	6.42×10^{2}	1.03×10^{2}	3.36×10^{2}	2.84×10^{1}

*Particle size assumed to be 0.5 μm (activity mean aerodynamic diameter).

†ICRP is International Commission on Radiological Protection, Report of Committee II (1959); Task Group is Task Group on Lung Dynamics (Morrow et al., 1966) as modified in ICRP Publication 19 (International Commission on Radiological Protection, 1972); SDB is Stuart, Dionne, and Bair (1968, 1971); Modified is our modifications to the SDB model (Bloom and Martin, 1976).

‡URT is upper respiratory tract, which includes the nasopharyngeal and tracheobronchial regions.

§Particles assumed insoluble for lung and GI tract and soluble for other organs.

TABLE 11 Comparison of Calculations for 1-μCi/day Intake* Using Various Models

Model[†]	Path	URT[‡]	GI tract	Lung	Lymph nodes		Liver	Kidney	Bone	Total body
					Thoracic	Abdominal				
					50-yr Organ Burdens, μCi					
ICRP II	Ingestion		7.50×10^{-1}				6.70×10^{-2}	9.04×10^{-3}	4.02×10^{-1}	4.97×10^{-1}
Task Group	Ingestion		7.50×10^{-1}				1.65×10^{-1}	9.04×10^{-3}	2.08×10^{-1}	4.97×10^{-1}
SDB	Ingestion		7.50×10^{-1}				3.50×10^{-1}	9.90×10^{-3}	1.52×10^{-1}	5.12×10^{-1}
Modified	Ingestion		7.50×10^{-1}	7.37×10^{-4}	3.41×10^{-3}	1.41×10^{-3}	3.52×10^{-1}	9.91×10^{-3}	1.54×10^{-1}	6.57×10^{-1}
ICRP II	Inhalation §		4.69×10^{-1}	6.58×10^{1}			5.66×10^{2}	7.54×10^{1}	3.55×10^{3}	4.14×10^{3}
Task Group	Inhalation	1.07×10^{-1}	3.42×10^{-1}	1.35×10^{2}	1.42×10^{2}		3.15×10^{2}	1.70×10^{1}	3.90×10^{2}	9.26×10^{2}
SDB	Inhalation	2.92×10^{-3}	3.00×10^{-1}	2.48×10^{2}	1.26×10^{3}	1.58×10^{2}	4.02×10^{2}	1.13×10^{1}	1.72×10^{2}	2.25×10^{3}
Modified	Inhalation		3.00×10^{-1}	2.48×10^{2}	1.26×10^{3}	1.58×10^{2}	4.02×10^{2}	1.13×10^{1}	1.72×10^{2}	7.38×10^{2}
					50-yr Cumulative Dose to Organs, rem					
ICRP II	Ingestion		1.22×10^{3}				1.04×10^{3}	7.96×10^{2}	7.45×10^{3}	1.82×10^{2}
Task Group	Ingestion		1.22×10^{3}				2.74×10^{3}	7.96×10^{2}	3.97×10^{3}	1.82×10^{2}
SDB	Ingestion		1.22×10^{3}				5.28×10^{3}	8.48×10^{2}	2.62×10^{3}	1.87×10^{2}
Modified	Ingestion		1.22×10^{3}	5.52×10^{1}	6.03×10^{3}	3.19×10^{3}	5.30×10^{3}	8.46×10^{2}	2.64×10^{3}	2.21×10^{2}
ICRP II	Inhalation §		7.59×10^{2}	3.17×10^{6}			8.82×10^{6}	6.63×10^{6}	6.21×10^{7}	1.51×10^{6}
Task Group	Inhalation	1.33×10^{4}	5.48×10^{2}	1.28×10^{7}	3.05×10^{8}		4.74×10^{6}	1.36×10^{6}	6.79×10^{6}	3.09×10^{5}
SDB	Inhalation	3.60×10^{2}	4.82×10^{2}	2.24×10^{7}	1.94×10^{9}	3.66×10^{8}	5.43×10^{6}	8.66×10^{5}	2.67×10^{6}	8.17×10^{5}
Modified	Inhalation		4.82×10^{2}	2.24×10^{7}	1.94×10^{9}	3.66×10^{8}	5.43×10^{6}	8.66×10^{5}	2.67×10^{6}	2.22×10^{5}

*Particle size assumed to be 0.5 μm (activity mean aerodynamic diameter).

†ICRP is International Commission on Radiological Protection, Report of Committee II (1959); Task Group is Task Group on Lung Dynamics (Morrow et al., 1966) as modified in ICRP Publication 19 (International Commission on Radiological Protection, 1972); SDB is Stuart, Dionne, and Bair (1968; 1971); Modified is our modifications to the SDB model (Bloom and Martin, 1976).

‡URT is upper respiratory tract, which includes the nasopharyngeal and tracheobronchial regions.

§Particles assumed insoluble for lung and GI tract and soluble for other organs.

140 yr after 10 yr of chronic inhalation. The lung burden could thus reach very high levels. A fraction may indeed be retained with a long half-time, but we find it difficult to believe that this fraction could be so high (15%). We therefore examined the long-term behavior of the burden in this portion of the lung resulting from a single inhalation. Equations 63 and 64 can be solved to yield

$$y_{LV} = 0.15 \, y_P(0) \exp\left(-\lambda_A t\right) \exp\left\{-7.3[1 - \exp\left(-0.001899t\right)]\right\} \tag{81}$$

where $y_P(0)$ is the initial amount deposited in the pulmonary lung.

If we neglect radioactive decay ($\lambda_A = 0$), the initial fraction (15%) is reduced to 0.4% after 1 yr, 0.013% after 5 yr, and asymptotically approaches a constant value of about 0.01%. On the basis of these results, we assume that this fraction (0.01%) is retained indefinitely and the remaining 14.99% is removed with a 3-yr biological half-time.

As shown in Tables 10 and 11, inhalation is the critical pathway for plutonium to all organs except GIT. The organ burdens and radiation doses from inhalation are generally 1,000 to 10,000 times as great as the corresponding burdens and doses from ingesting the same amount of plutonium. This is due to the relatively large fraction (0.2 to 25%) that reaches the blood directly from inhalation vs. the relatively small fraction (0.003%) from ingestion. For ingestion bone is the critical organ for the ICRP II and Task Group models, whereas liver is the critical organ for the SDB and modified models. This difference is explained by the fraction transferred from blood to the organ, which is 71% to the liver and 25% to bone for the SDB and modified models, whereas the corresponding values for the ICRP II model are 15 and 80%, respectively, and those for the Task Group model are 45 and 45%, respectively. Where the fractions are equal, the bone has the larger burden and dose because it has the larger biological half-time.

For inhalation the lung is the critical organ for all models except ICRP II. The dose to lymph nodes is actually higher, but ICRP (International Commission for Radiological Protection, 1959) does not recognize lymph nodes as critical organs. For the ICRP II model the bone is the critical organ because this model has the highest fraction of inhaled material that reaches the blood immediately (25%) and the shortest biological half-time in lung (365 days).

In spite of differences in translocation pathways and biological half-times, the radiation doses to critical organs are surprisingly similar for a given intake situation. This leads us to use the Task Group model because it is recognized by ICRP [ICRP Publication 19 (International Commission on Radiological Protection, 1972)], and the results using this model are not too different from the more elaborate SDB and modified models. Although still the official model of ICRP, the ICRP II model is generally considered to be outdated. The Task Group model was used to calculate the accumulated doses and dose commitments (to 70 yr) due to constant intake rates [$A_m = 0.002 \, C_s$ (pCi/day) and $H_m = 0.19 \, C_s$ (pCi/day)], and the results are shown in Figs. 8 and 9.

Practical Applications

Our purpose in this discussion is to show how the results of a transport- and dose-estimation model can be applied to the practical problem of deciding whether and to what extent environmental decontamination might be required to limit or reduce potential health hazards. The procedure suggested for this purpose and outlined below is analogous to the procedure followed by ICRP in calculating maximum permissible

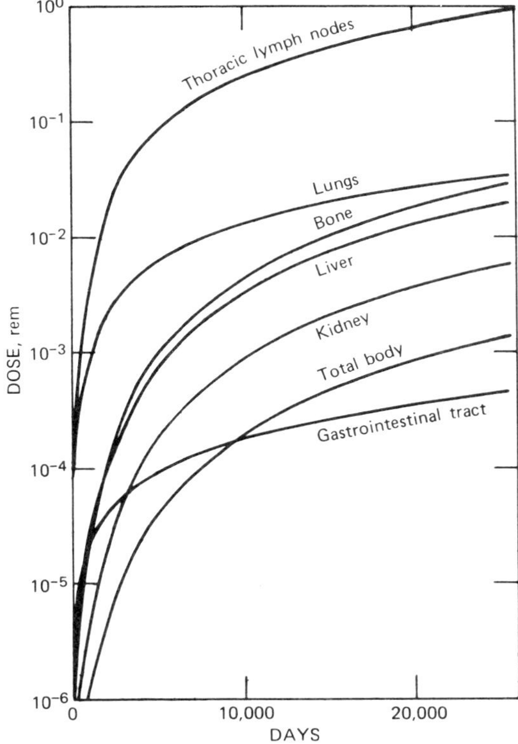

Fig. 8 Predicted cumulative doses due to ^{239}Pu in different organs of Standard Man.

concentrations (MPC's) of radionuclides in air and water. The principal steps involved are (1) identification of the critical exposure pathway, (2) identification of the critical organ or organs, (3) selection of maximum-permissible-dose criteria, (4) calculation of the corresponding MPC of plutonium in soil $(MPC)_s$, and (5) comparison of the $(MPC)_s$ with estimated inventories of plutonium in the surface soils of contaminated areas at NTS.

Critical Pathway

The estimated plutonium ingestion rate for a hypothetical Standard Man living in a contaminated area at NTS is about 100 times the estimated inhalation rate, but, owing to the very small fraction of plutonium transferred from the GIT to blood (3×10^{-5}), the GIT is the only organ that receives a significant dose from ingested plutonium. The preferred dose-estimation model, based on ICRP recommendations (the Task Group model, Fig. 4), shows that inhalation accounts for 100% of the plutonium deposited in the lungs and thoracic lymph nodes, and, for an ingestion/inhalation ratio of 100, inhalation accounts for about 95% of the plutonium in bone, liver, and kidney after 50 yr of chronic exposure (Table 6). Clearly, inhalation is the critical pathway.

Critical Organ

According to the Task Group model (Fig. 4), thoracic lymph nodes receive the highest dose (Figs. 8 and 9), but the critical organs recognized by ICRP (International

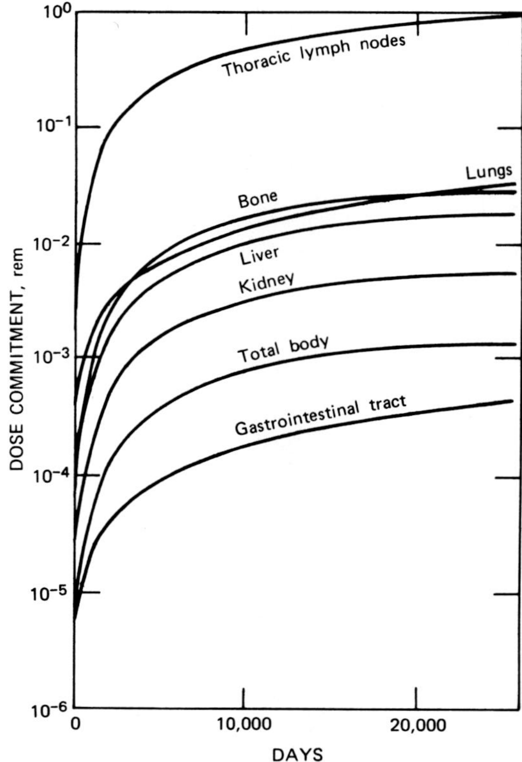

Fig. 9   Predicted dose commitments due to ^{239}Pu in different organs of Standard Man.

Commission on Radiological Protection, 1959) are bone, if the plutonium is "soluble," and lung if the plutonium is "insoluble." For relatively short exposure times, the model, which makes no distinction between soluble and insoluble, predicts that the cumulative dose to lungs would be considerably higher than that to bone (Fig. 8), but the cumulative doses to 70 yr are about the same. For exposure times longer than 70 yr, the dose to bone would be higher than the dose to lungs because of the relatively short biological half-life of plutonium in lungs (500 days) compared with that of bone (36,500 days). Since the estimated dose to lungs is higher than that to bone and the exposure periods (<70 yr) and the permissible dose to lungs are lower (see below), the lung (i.e., the pulmonary region of the respiratory tract) is the critical organ.

Permissible Dose Criteria

Current ICRP recommendations (International Commission on Radiological Protection, 1966) concerning "dose limits for individual members of the public" indicate that the dose to lungs should not exceed "1.5 rems in a year." Annual dose rates to a given organ can be estimated on the basis of predicted organ burdens as a function of exposure time. For present purposes we shall consider only the equilibrium lung burden, which, for practical purposes, is constant for chronic exposure times in excess of about 10 yr.

Estimation of Acceptable Soil Concentration

The acceptable soil concentration (ASC) is defined as the average concentration of [239]Pu in the soil of an area occupied by the hypothetical Standard Man which would result in a dose to the lungs equal to or less than the permissible dose. This value is estimated as follows:

$$ASC = \frac{(1.5/365)}{(E\ Y_{eq}/m)}$$

$$= 2817\ pCi\ ^{239}Pu/g\ soil$$

where $(1.5/365)$ = "permissible dose rate" (rem/day)
  E for [239]Pu = $51.2159 \times 53 = 2714$ (rem/day)/μCi, or 0.0027144 (rem/day)/pCi
    m = 500 g for lung
  $Y_{eq} = 0.002\ C_s y$
    y = lung burden after 50 yr = 134.34 μCi per μCi inhaled per day (see Table 11)
  $0.002\ C_s$ = plutonium inhalation rate (pCi/day), i.e., 20 m^3 air/day $\times\ 10^{-4}$ g soil/m^3 air, and C_s is the average soil concentration (pCi/g)

Comparison of ASC and Soil Inventory Data

Tables 1 and 2 summarize the mean soil concentrations and the estimated inventories of [239,240]Pu in the surface soils (0- to 5-cm depth) at NTS. In each contaminated study area, soil sampling was stratified according to contour intervals (strata) previously established by field instrument for the determination of low-energy radiation (FIDLER) surveys. The pertinent results for Area 13 are given in Table 12.

Only stratum 6 exceeds ASC = 2800 pCi/g, but stratum 5 is close enough to be included in the contaminated region. Complete decontamination of strata 5 and 6 (1.1%

**TABLE 12   Estimated Inventory of [239,240]Pu in
Surface Soil (0- to 5-cm Depth) in Area 13**

Strata*	Area,* m^2	[239,240]Pu,*† μCi/m^2	Soil concentration,‡ pCi/g
1	1,245,000	1.9 ± 0.34	36 ± 7.8
2	2,547,000	5.8 ± 1.4	100 ± 25
3	108,000	23 ± 4.3	400 ± 75
4	74,000	54 ± 8.8	1100 ± 150
5	19,000	110 ± 19	2400 ± 430
6	24,000	220 ± 340	14000 ± 6400
	4,017,000		
		Area weighted mean	201

*From Gilbert (1977, Table 1).
†From Gilbert (1977, Table 1). Mean ± standard error.
‡From Gilbert et al (1975, p. 393). Mean ± standard error.

of Area 13) would remove about 48% of the total plutonium in Area 13 and reduce the average soil concentration from 201 to 106 pCi/g. If it were decided to decontaminate all areas at NTS in which the average soil concentration exceeds 2 nCi/g, decontamination would be required for about 0.2 km^2 (about 50 acres) of the 11.5 km^2 (about 2850 acres) included in the soil inventory study (Table 1). If the decontamination criteria were further reduced to 1 nCi/g, the area requiring decontamination would be about 0.5 km^2 (111 acres). In other words, the plutonium contamination at NTS is so concentrated in areas near ground zero sites that decontamination of from 2 to 4% of the total soil inventory area would reduce average soil concentrations by 40 to 50%.

It should be noted that possible decontamination activities of these sites are complicated by potential damage to the desert ecosystem. Further information on this subject can be found in reports by Wallace and Romney (1975) and Rhoads (1976).

Discussion

On the basis of the preceding results (ASC = 2800 pCi/g) and the mass-loading factor of 100 μg soil/m^3 air, the expected air concentration would be 2.8 × 10^{-13} μCi/cm^3. The maximum permissible concentration in air (MPC)$_a$ indicated by ICRP Publication 2 (10^{-12} μCi/cm^3) is higher than this by a factor of about 3.6. Using (MPC)$_a$ = 10^{-12} μCi/cm^3 and a mass-loading factor of 100 μg/m^3, we would find the acceptable soil concentration to be 10 nCi/g instead of 2.8 nCi/g, which would be equivalent to assuming a mass-loading factor of 355 instead of 100 μg soil/m^3 air.

Another conservative factor in our estimate of ASC is that the lung deposition factor (D_5 = 0.31) is based on the assumption that the mean size of resuspended soil particles is 0.5 μm (AMAD). The value obtained from cascade impactor studies in the GMX area was 3 μm (AMAD), which would indicate $D_5 \leqslant 0.2$. Changing only this parameter would increase the estimate of ASC by a factor of 1.55 to 4266 pCi/g.

The least conservative factor involved in arriving at ASC = 2.8 nCi/g is the assumed mass-loading factor of 100 μg soil/m^3 air. As demonstrated by Shinn and Anspaugh (1975) and Anspaugh et al. (1975), this estimate appears to be adequate for undisturbed areas and normal winds, but high winds or mechanical disturbances, such as vehicular traffic, plowing, excavation, etc., might increase the mass-loading factor to several milligrams per cubic meter. If we assume, for example, that the hypothetical Standard Man at NTS were exposed, for one reason or another, to mass-loading factors of 5000 μg/m^3 during 30 days each year, the average mass-loading factor would increase to about 500 μg/m^3, and our estimate of ASC would decrease to about 560 pCi/g.

The point of this discussion is that the notion of an "acceptable soil concentration" is not fixed but is very much dependent on how man plans to use a contaminated area. Under present conditions the ASC for contaminated areas at NTS is 2.8 nCi/g. If these same areas were to be used for agricultural purposes or for any other purpose that would tend to increase the average mass-loading factor, a lower ASC would be indicated. In other words, the notion of an "acceptable soil concentration" is an attractive one. It implies the existence of a numerical criterion that can be used to make important determinations concerning the need for or effectiveness of countermeasures to ensure safety. How to determine an ASC value is a different matter, and how to determine whether a particular ASC value is entirely appropriate under a given set of physical, social, and political circumstances is a far different matter.

References

Amato, A. J., 1971, Mathematical Analysis of the Effects of Wind on Redistribution of Surface Contamination, USAEC Report WASH-1187, GPO.

Anspaugh, L. R., 1974, Resuspension Element Status Report, Appendix X. The Use of NTIS Data and Experience to Predict Air Concentrations of Plutonium due to Resuspension on the Enewetak Atoll, in *The Dynamics of Plutonium in Desert Environments*, P. B. Dunaway and M. G. White (Eds.), USAEC Report NVO-142, pp. 299-310, Nevada Operations Office, NTIS.

——, and P. L. Phelps, 1974, Resuspension Element Status Report, Results and Data Analysis, in *The Dynamics of Plutonium in Desert Environments*, P. B. Dunaway and M. G. White (Eds.), USAEC Report NVO-142, pp. 265-298, Nevada Operations Office, NTIS.

——, P. L. Phelps, N. C. Kennedy, and H. G. Booth, 1973, Wind-Driven Redistribution of Surface-Deposited Radioactivity, in *Environmental Behavior of Radionuclides Released in the Nuclear Industry*, Symposium Proceedings, Aix-en-Provence, 1973, STI/PUB/345, International Atomic Energy Agency, Vienna.

——, J. H. Shinn, P. L. Phelps, and N. C. Kennedy, 1975, Resuspension and Redistribution of Plutonium in Soils, *Health Phys.*, 29: 571-582.

Bagnold, R. A., 1960, *The Physics of Blown Sand and Desert Dunes*, Methuen, Inc., New York.

Ballou, J. E., J. F. Park, and W. G. Morrow, 1972, On the Metabolic Equivalence of Ingested, Injected, and Inhaled 239-Pu Citrate, *Health Phys.*, 22: 857-862.

Bennett, B. G., 1976, Transuranic Element Pathways to Man, in *Transuranium Nuclides in the Environment*, Symposium Proceedings, San Francisco, 1975, pp. 367-383, STI/PUB/410, International Atomic Energy Agency, Vienna.

Bloom, S. G., and W. E. Martin, 1976, A Model to Predict the Environmental Impact of the Release of Long-Lived Radionuclides, Battelle Columbus Laboratories, Report to U. S. Environmental Protection Agency.

Bramson, P. E., and J. P. Corley, 1973, *Environmental Surveillance at Hanford for CY-1972*, USAEC Report BNWL-1727, Battelle, Pacific Northwest Laboratories, NTIS.

Cataldo, D. A., E. L. Klepper, and D. K. Craig, 1976, Fate of Plutonium Intercepted by Leaf Surfaces: Leachability and Translocation to Seed and Root Tissues, in *Transuranium Nuclides in the Environment*, Symposium Proceedings, San Francisco, 1975, pp. 291–301, STI/PUB/410, International Atomic Energy Agency, Vienna.

Dunaway, P. B., and M. G. White (Eds.), 1974, *The Dynamics of Plutonium in Desert Environments*, USAEC Report NVO-142, Nevada Operations Office, NTIS.

Francis, C. W., 1973, Plutonium Mobility in Soil and Uptake in Plants: A Review, *J. Environ. Qual.*, 2: 67-70.

Fix, J. J., 1975, *Environmental Surveillance at Hanford for CY-1974*, USAEC Report BNWL-1910, Battelle, Pacific Northwest Laboratories, NTIS.

Gilbert, R. O., 1977, Revised Total Amounts of ^{239,240}Pu in Surface Soil at Safety-Shot Sites, in *Transuranics in Desert Ecosystems*, M. G. White, P. B. Dunaway, and D. L. Wireman (Eds.), USDOE Report NVO-181, pp. 423-429, Nevada Operations Office, NTIS.

——, L. L. Eberhardt, and D. D. Smith, 1976, *Initial Synthesis of Area 13 ^{239}Pu Data and Other Statistical Analyses*, USAEC Report BNWL-SA-5667, Battelle, Pacific Northwest Laboratories, NTIS.

——, L. L. Eberhardt, E. B. Fowler, E. M. Romney, E. H. Essington, and J. E. Kinnear, 1975, Statistical Analysis of 239-240-Pu and 241-Am Contamination of Soil and Vegetation on NAEG Study Sites, in *The Radioecology of Plutonium and Other Transuranics in Desert Environments*, M. G. White and P. B. Dunaway (Eds.), ERDA Report NVO-153, pp. 339-448, Nevada Operations Office, NTIS.

Golley, F. B., 1961, Energy Values of Ecological Materials, *Ecology*, 42(3): 581-584.

Healy, J. W., 1974, *A Proposed Interim Standard for Plutonium in Soils*, USAEC Report LA-5483-MS, Los Alamos Scientific Laboratory, NTIS.

Houston, J. R., Strenge, D. L., and Watson, E. C., 1975, *DACRIN—A Computer Program for Calculating Organ Dose from Acute or Chronic Radionuclide Inhalation*, USAEC Report BNWL-B-389, Battelle, Pacific Northwest Laboratories, NTIS.

International Commission on Radiological Protection, 1959, *Recommendations Report of Committee II on Permissible Dose for Internal Radiation*, ICRP Publication 2, Pergamon Press, Inc., New York.

——, 1964, *Recommendations of the International Commission on Radiological Protection* (as amended 1959 and revised 1962), ICRP Publication 6, Pergamon Press, Inc., New York.

——, 1966, *Recommendations of the ICRP* (Adopted Sept. 17, 1965), ICRP Publication 9, Pergamon Press, Inc., New York.

——, 1972, *The Metabolism of Compounds of Plutonium and Other Actinides,* ICRP Publication 19, Pergamon Press, Inc., New York.

Kathren, R. L., 1968, Towards Interim Acceptable Surface Contamination Levels for Environmental PuO_2, in *Radiological Protection of the Public in a Nuclear Mass Disaster,* Symposium Proceedings, Interlakes, Switzerland, May 27–June 1, 1968, USAEC Report CONF-680507, NTIS.

Killough, G. G., and L. R. McKay, 1976, *Methodology for Calculating Radiation Doses from Radioactivity Released to the Environment,* USAEC Report ORNL-4992, Oak Ridge National Laboratory, NTIS.

Kleiber, M., 1961, *The Fire of Life: An Introduction to Animal Energetics,* John Wiley & Sons, Inc., New York.

Kotrappa, P., 1968, *Dose to Respiratory Tract from Continuous Inhalation of Radioactive Aerosol,* USAEC Report UR-49-964, University of Rochester, NTIS.

——, 1969, Calculation of the Burden and Dose to the Respiratory Tract from Continuous Inhalation of a Radioactive Aerosol, *Health Phys.,* 17(3): 429-432.

Martin, W. E., 1964, Losses of 90-Sr, 89-Sr, and 131-I from Fallout-Contaminated Plants, *Radiat. Bot.,* 4: 275-284.

——, 1965, Interception and Retention of Fallout by Desert Shrubs, *Health Phys.,* 11: 1341-1354.

——, and S. G. Bloom, 1976, Plutonium Transport and Dose Estimation Model, in *Transuranium Nuclides in the Environment,* Symposium Proceedings, San Francisco, 1975, pp. 385–400, STI/PUB/410, International Atomic Energy Agency, Vienna.

——, S. G. Bloom, and R. J. Yorde, Jr., 1974, Plutonium Transport and Dose Estimation Model, in *The Dynamics of Plutonium in Desert Environments,* P. B. Dunaway and M. G. White (Eds.), USAEC Report NVO-142, pp. 331–360, Nevada Operations Office, NTIS.

McKell, C. M., 1975, Shrubs—A Neglected Resource of Arid Lands, *Science,* 187: 803-809.

McLendon, H. R., et al., 1976, Relationships Among Plutonium Contents of Soil, Vegetation, and Animals Collected on and Adjacent to an Integrated Nuclear Complex in the Humid Southeastern United States of America, in *Transuranium Nuclides in the Environment,* Symposium Proceedings, San Francisco, 1975, pp. 347–363, STI/PUB/410, International Atomic Energy Agency, Vienna.

Milham, R. C., J. F. Schubert, J. R. Watts, A. L. Boni, and J. C. Corey, 1976, Measured Plutonium Resuspension and Resulting Dose from Agricultural Operations on an Old Field at the Savannah River Plant in the Southeastern United States of America, in *Transuranium Nuclides in the Environment,* Symposium Proceedings, San Francisco, 1975, pp. 409–421, STI/PUB/410, International Atomic Energy Agency, Vienna.

Miller, C. F., and H. Lee, 1966, Operation Ceniza-Arena: The Retention of Fallout Particles from Volcano Irazu (Costa Rica) by Plants and People, Pt. 1, Report AD-637313, Stanford Research Institute.

Mills, M. T., and J. S. Olson, 1973, A Model for Suspension and Atmospheric Transport of Dust, paper presented at the fall annual meeting of the American Geophysical Union, San Francisco, Dec. 10–14, 1973, American Geophysical Union.

Mork, H. M., 1970, *Redistribution of Plutonium in the Environs of the Nevada Test Site,* USAEC Report UCLA-12-590, University of California, NTIS.

Morrow, P. E., D. V. Bates, B. R. Fish, T. F. Hatch, and T. T. Mercer, 1966, Deposition and Retention Models for Internal Dosimetry of the Human Respiratory Tract, *Health Phys.,* 12: 173-207.

National Research Council, 1968, *Recommended Dietary Allowances,* National Academy of Sciences, Publication 1964.

Nees, W. L., and J. P. Corley, 1974, *Environmental Surveillance at Hanford for CY-1973*, USAEC Report BNWL-1811, Battelle, Pacific Northwest Laboratories, NTIS.

Odum, E. P., 1971, *Fundamentals of Ecology*, 3rd ed., W. B. Saunders Company, Philadelphia.

Organization for Economic Cooperation and Development, 1970, *Food Consumption Statistics 1960–1968*.

Phelps, P. L., and L. R. Anspaugh, 1974, Resuspension Element Status Report, Introduction, in *The Dynamics of Plutonium in Desert Environments*, P. B. Dunaway and M. G. White (Eds.), USAEC Report NVO-142, pp. 223-233, Nevada Operations Office, NTIS.

Pochin, E. E., 1966, The Mass of the Tracheobronchial Lymph Glands, *Health Phys.*, 12: 563-564.

Price, K. R., 1973a, A Review of Transuranic Elements in Soils, Plants, and Animals, *J. Environ. Qual.*, 2: 62-66.

——, 1973b, *Tumbleweed and Cheatgrass Uptake of Transuranium Elements Applied to Soil as Organic Acid Complexes*, USAEC Report BNWL-1755, Battelle, Pacific Northwest Laboratories, NTIS.

Rhoads, W. A., 1976, A Wise Resolution of the Plutonium Contaminated Lands Problem in Certain Parts of NTS May Be To Isolate and Maintain Them Without Further Disturbance: A Position Paper, in *Studies of Environmental Plutonium and Other Transuranics in Desert Ecosystems*, M. G. White and P. B. Dunaway (Eds.), USAEC Report NVO-159, pp. 165–180, Nevada Operations Office, NTIS.

Ricker, W. E., 1973, Linear Regression in Fishery Research, *J. Fish. Res. Board Can.*, 30: 409-434.

Romney, E. M., R. G. Lindberg, H. A. Hawthorne, B. G. Bystorm, and K. H. Larson, 1963, Contamination of Plant Foliage by Radioactive Fallout, *Ecology*, 44: 343-349.

——, H. M. Mork, and K. H. Larson, 1970, Persistence of Plutonium in Soils, Plants and Small Mammals, *Health Phys.*, 19: 487-491.

——, A. Wallace, R. O. Gilbert, and J. E. Kinnear, 1975, 239-240-Pu and 241-Am Contamination of Vegetation in Aged Plutonium Fallout Areas, in *The Radioecology of Plutonium and Other Transuranics in Desert Environments*, M. G. White and P. B. Dunaway (Eds.), ERDA Report NVO-153, pp. 43-87, Nevada Operations Office, NTIS.

Sehmel, G. A., 1973, Particle Eddy Diffusivities and Deposition Velocities for Isothermal Flow and Smooth Surfaces, *Aerosol Sci.*, 4: 125-138.

——, 1975, *Experimental Measurements and Predictions of Particle Deposition and Resuspension Rates*, USAEC Report BNWL-SA-5228, Battelle, Pacific Northwest Laboratories, NTIS.

——, 1977, *Radioactive Particle Resuspension Research Experiments on the Hanford Reservation*, ERDA Report BNWL-2081, Battelle, Pacific Northwest Laboratories, NTIS.

——, and M. M. Orgill, 1973, Resuspension by Wind at Rocky Flats, in *Pacific Northwest Laboratory Annual Report for 1972 to the USAEC Division of Biomedical and Environmental Research*, Vol. II, Physical Sciences, Pt. 1, Atmospheric Sciences, USAEC Report BNWL-1751(Pt.1), Battelle, Pacific Northwest Laboratories, NTIS.

——, S. L. Sutter, and M. T. Dana, 1973, Dry Deposition Processes, in *Pacific Northwest Laboratory Annual Report for 1972 to the USAEC Division of Biomedical and Environmental Research*, Vol. II, Physical Sciences, Pt. 1, Atmospheric Sciences, USAEC Report BNWL-1751(Pt. 1), Battelle, Pacific Northwest Laboratories, NTIS.

Shinn, J. H., and L. R. Anspaugh, 1975, Resuspension—New Results in Predicting the Vertical Dust Flux, in *The Radioecology of Plutonium and Other Transuranics in Desert Environments*, M. G. White and P. B. Dunaway (Eds.), ERDA Report NVO-153, pp. 207–215, Nevada Operations Office, NTIS.

Siegmund, O. H. (Ed.), 1967, *The Merck Veterinary Manual*, 3rd ed., Merck & Co., Inc., Rahway, N. J.

Smith, D. D., 1977, Grazing Studies on a Contaminated Range of the Nevada Test Site, in *Environmental Plutonium on the Nevada Test Site and Environs*, ERDA Report NVO-171, Nevada Operations Office, NTIS.

——, J. Barth, and R. G. Patzer, 1976, Grazing Studies on a Plutonium-Contaminated Range of the Nevada Test Site, in *Transuranium Nuclides in the Environment*, Symposium Proceedings, San Francisco, 1975, pp. 325–336, STI/PUB/410, International Atomic Energy Agency, Vienna.

Smith, N. E., and R. L. Baldwin, 1974, Effects of Breed, Pregnancy, and Lactation on Weight of Organs and Tissues in Dairy Cattle, *J. Dairy Sci.*, 57(9): 1055-1060.

Snyder, W. S., 1967, The Use of the Lung Model for Estimation of Dose, in *Proceedings of the 12th Annual Bioassay and Analytical Chemistry Meeting*, Gatlinburg, Tenn., Oct. 13–14, 1966, C. M. West (Comp.), USAEC Report CONF-661018, NTIS.

Stanley, R. E., E. W. Bretthauer, and W. W. Sutton, 1974, Absorption, Distribution, and Excretion of Plutonium by Dairy Cattle, in *The Dynamics of Plutonium in Desert Environments*, P. B. Dunaway and M. G. White (Eds.), USAEC Report NVO-142, pp. 163–185, Nevada Operations Office, NTIS.

Stuart, B. W., P. J. Dionne, and W. J. Bair, 1968, A Dynamic Simulation of the Retention and Translocation of Inhaled Plutonium Oxide in Beagle Dogs, in *Proceedings of the Eleventh AEC Air Cleaning Conference*, Richland, Wash., Aug. 31–Sept. 3, 1970, M. W. First and J. M. Morgan, Jr. (Eds.), USAEC Report CONF-700816, NTIS.

——, P. J. Dionne, and W. J. Bair, 1971, Computer Simulation of the Retention and Translocation of Inhaled 239-PuO_2 in Beagle Dogs, in *Pacific Northwest Laboratory Annual Report for 1970 to the USAEC Division of Biology and Medicine*, Vol. 1, Life Sciences, Pt. 1, Biological Sciences, USAEC Report BNWL-1550(Pt. 1), Battelle, Pacific Northwest Laboratories, NTIS.

Tamura, T., 1976, Plutonium in Surface Soils of Area 13: A Progress Report, in *Studies of Environmental Plutonium and Other Transuranics in Desert Ecosystems*, M. G. White and P. B. Dunaway (Eds.), USAEC Report NVO-159, pp. 5-16, Nevada Operations Office, NTIS.

U. S. Department of Agriculture, 1973, *Agricultural Statistics 1973*, GPO.

U. S. Nuclear Regulatory Commission, 1976, *Calculated Doses from Inhaled Transuranium Radionuclides and Potential Risk Equivalence to Whole-Body Radiation*, internal working paper, Office of Nuclear Regulatory Research, Division of Safeguards, Fuel Cycle and Environmental Research, Report NR-SFE-001.

Van der Hoven, I., 1968, Deposition of Particles and Gases, in *Meteorology and Atomic Energy*, D. H. Slade (Ed.), USAEC Report TID-24190, Chap. 5-3, Environmental Science Services Administration, NTIS.

Volchok, H. L., 1971, Resuspension of Plutonium 239 in the Vicinity of Rocky Flats, in *Proceedings of Environmental Plutonium Symposium*, Los Alamos, Aug. 4–5, 1971, E. B. Fowler, R. W. Henderson, and M. F. Milligan (Eds.), USAEC Report LA-4756, Los Alamos Scientific Laboratory, NTIS.

Wallace, A., and E. M. Romney, 1972, *Radioecology and Ecophysiology of Desert Plants at the Nevada Test Site*, USAEC Report TID-25954, University of California, NTIS.

——, and E. M. Romney, 1975, Feasibility and Alternate Procedures for Decontamination and Post-Treatment Management of Pu-Contaminated Areas in Nevada, in *The Radioecology of Plutonium and Other Transuranics in Desert Environments*, M. G. White and P. B. Dunaway (Eds.), USAEC Report NVO-153, pp. 251–337, Nevada Operations Office, NTIS.

White, M. G., and P. B. Dunaway (Eds.), 1975, *The Radioecology of Plutonium and Other Transuranics in Desert Environments*, USAEC Report NVO-153, pp. 251-337, Nevada Operations Office, NTIS.

——, and P. B. Dunaway (Eds.), 1976, *Studies of Environmental Plutonium and Other Transuranics in Desert Ecosystems*, ERDA Report NVO-159, Nevada Operations Office, NTIS.

——, and P. B. Dunaway (Eds.), 1977, *Environmental Plutonium on the Nevada Test Site and Environs*, ERDA Report NVO-171, Nevada Operations Office, NTIS.

Wildung, R. E., and T. R. Garland, 1974, Influence of Soil Plutonium Concentration on Plutonium Uptake and Distribution in Shoots and Roots of Barley, *J. Agric. Food Chem.*, 22: 836-838.

A Model of Plutonium Dynamics in a Deciduous Forest Ecosystem

CHARLES T. GARTEN, JR., ROBERT H. GARDNER, and ROGER C. DAHLMAN*

A linear compartment model with donor-controlled flows between compartments was designed to describe and simulate the behavior of plutonium (^{239,240}Pu) in a contaminated forest ecosystem at Oak Ridge, Tenn. At steady states predicted by the model, less than 0.25% of the plutonium in the ecosystem resides in biota. Soil is the major repository of plutonium in the forest, and exchanges of plutonium between soil and litter or soil and tree roots were dominant transfers affecting the ecosystem distribution of plutonium. Variation in predicted steady-state amounts of plutonium in the forest, given variability in the model parameters, indicates that our ability to develop models of plutonium transport in ecosystems should improve with greater precision in data from natural environments and a better understanding of sources of variation in plutonium data.

Systems analysis techniques have been useful in simulating the fate and dynamics of a variety of substances in ecosystems, including radionuclides (Olson, 1965; Wheeler, Smith, and Gallegos, 1977), pesticides (Webb, Schroeder, and Norris, 1975), and stable elements (Shugart et al., 1976). Both descriptive and predictive purposes are considered in the building of these models. Past applications of ecosystem modeling of radionuclide behavior in the environment have included (1) projection of the time-dependent distribution of material within the system and (2) manipulation of the model system to determine the sensitivity of various components to variation in transfer coefficients. The latter exercise allows identification of critical pathways affecting radionuclide distribution in the system.

This chapter describes an ecosystem model of plutonium (^{239,240}Pu) behavior in a Tennessee forest. In ecosystem models the complexities of community structure and ecological processes are often simplified to an abstract compartmental system linked by linear differential equations (Hudetz, 1973). The model described was formulated on the basis of data from International Biological Program (IBP) studies on deciduous forests and investigations at a plutonium-contaminated forest in eastern Tennessee. A primary objective of the model was to holistically describe plutonium behavior in the forest. To this end Monte Carlo simulations of the transfer of plutonium from soil to biota and a sensitivity analysis of the model forest were performed.

---

*Present address: U. S. Department of Energy, Washington, D. C.

513

Design of the Model

The model was conceived to describe plutonium cycling in a deciduous forest. Studies from a plutonium-contaminated forest adjacent to White Oak Creek on the U. S. Department of Energy's Oak Ridge reservation were used as much as possible to design the model and establish its parameter values. In 1944, during the Manhattan Project, the White Oak Creek floodplain was contaminated with ^{239,240}Pu and mixed fission products. After 33 yr a deciduous forest dominated by white ash (*Fraxinus americana,* eight trees per 100 m^2) and sycamore (*Plantanus occidentalis,* three trees per 100 m^2) has developed on the site (Van Voris and Dahlman, 1976).

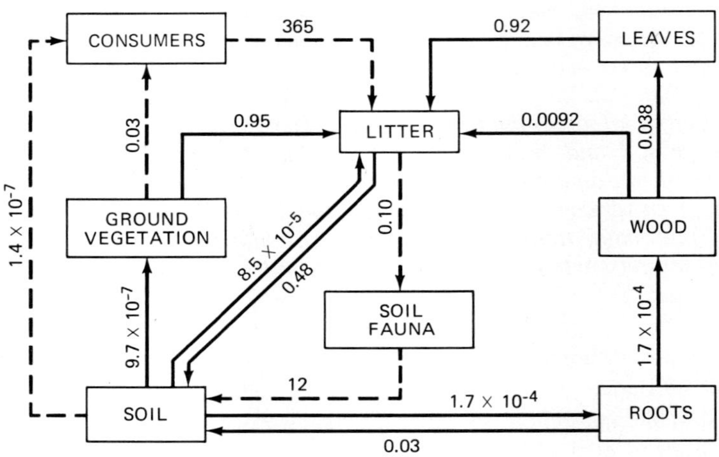

Fig. 1 Diagrammatic model of plutonium transfers in a deciduous forest ecosystem showing abstracted compartments and annual transfer coefficients. The basic model included compartments linked by solid lines. Dashed lines indicate transfers and compartments coupled after calibration of the basic model.

Initially, a six-compartment model with 10 transfers was set up to represent plutonium dynamics in soil and vegetative components of the ecosystem (Fig. 1). Average annual biomass values (grams of dry weight per square meter) and plutonium concentrations (picocuries per gram of dry weight) were multiplied to arrive at the amount of plutonium (picocuries per square meter) in each compartment of the forest (Table 1). A majority of the transfers in the model were calculated on the basis of biomass flux (grams per square meter per year) from the donor compartment. In lieu of site-specific data, fluxes were derived from data collected for eastern deciduous forests in the Oak Ridge area during the IBP (Harris, Goldstein, and Henderson, 1973; Sollins, Reichle, and Olson, 1973; Harris et al., 1975). Values for annual transfer coefficients and their derivation are given in Table 2. Parameter values for the six-compartment model were arrived at independent of model performance. Later, some parameter adjustment (Table 2) was necessary to calibrate the predicted amount of plutonium (picocuries per square meter) in the forest after a 30-yr computer simulation; the calculated inventory was based on field data.

TABLE 1   Standing-Crop Biomass, Plutonium Concentrations, and Areal Amounts of Plutonium in a 30-yr-old Contaminated Deciduous Forest at Oak Ridge, Tennessee

Component	Biomass, g/m^2	Plutonium concentration, pCi/g	^{239,240}Pu content, pCi/m^2
Soil*	2.6 × 10^5	65	1.7 × 10^7
Tree roots†	3,000	4	12,000
Litter‡	500	6	3,000
Tree wood†	10,500	0.003	32
Ground vegetation‡	110	0.15	17
Tree leaves†	400	0.003	1.2

*Mass is based on a 20-cm soil depth and a soil density of 1.3 g/cm^3; concentration is the floodplain average.

†Biomass is estimated from mensuration data (Van Voris and Dahlman, 1976) and regression equations (Harris, Goldstein, and Henderson, 1973); concentrations are based on field measurements.

‡Biomass and concentration data are based on field measurements.

After the six-compartment model had been calibrated, plutonium transfers to animal components of the ecosystem were simulated. The complexities associated with transfers of plutonium to consumers and soil fauna in the model necessitated our simplifying assumptions to arrive at parameter estimates. For example, pathways for resuspension of plutonium-contaminated soil to atmosphere and subsequent inhalation by animals were not represented in the model. Resuspension factors (Anspaugh et al., 1975) range from 10^{-10} to 10^{-11} m^{-1} for the forest, based on plutonium concentrations measured in soil and air (Dahlman and McLeod, 1977). Since this resuspension factor is in the lower range of values measured in natural environments [i.e., 10^{-2} to 10^{-11} m^{-1} (Hanson, 1975)], food-chain transfer to animals is modeled as the chief transport pathway.

The forest is modeled as a closed system since inputs of plutonium (e.g., fallout resulting in a cumulative plutonium concentration in soil of 0.02 pCi/g) are negligible relative to the existing soil contamination (plutonium in soil ranges from ~25 to ~150 pCi/g). Forest outputs include surface runoff, erosion, and groundwater seepage, but these processes are beyond the scope of this model. We believe that solution-phase transport is negligible as an output from the ecosystem because plutonium is strongly sorbed to soil (Bondietti, Reynolds, and Shanks, 1976). In addition, the downward movement of plutonium in soil is not considered important over the time frame of the simulations. Although there is evidence of a downward movement of plutonium in soil over time (Bennett, 1976; Jakubick, 1976), soil fauna could promote redistribution of plutonium from the subsoil to the soil surface. Reichle et al. (1973) calculated a 44-yr turnover time for the top 25 cm of forest soil due solely to earthworm activity.

Since the purpose of the model is descriptive, we have not made millennium predictions of plutonium dynamics in the forest. Loucks (1970) points out that there is a tendency in forests toward perturbation at time intervals ranging from decades to centuries. Therefore simulations with the model were limited to time spans of less than 500 yr. Over this period of time, decreasing plutonium concentrations from radioactive decay have a negligible effect on model predictions.

Modeling Methods

The model was formulated as a linear compartment model with donor-controlled flows between compartments. The behavior of the state variables (X_i) was described by simple differential equations of the form

$$\frac{dX_i}{dt} = (\text{inputs to } X_i) - (\text{losses from } X_i)$$

TABLE 2   Values and Derivation of Annual Transfer Coefficients for Plutonium Dynamics in a Contaminated Mixed Deciduous Forest*

Transfer	Derivation	Value, yr^{-1}
Soil to ground vegetation	[110 g (ground vegetation produced) $\text{m}^{-2}\ \text{yr}^{-1}$] x (0.15 pCi Pu/g)/[$1.7 \times 10^7$ pCi Pu/m^2 (soil)]	9.7×10^{-7}
Ground vegetation to litter	95% lost to litter via plant mortality (105 g $\text{m}^{-2}\ \text{yr}^{-1}$)/(110 g/$\text{m}^2$)	0.95
Soil to roots	The initial value, [750 g (root production) $\text{m}^{-2}\ \text{yr}^{-1}$] x (4 pCi Pu/g)/[$1.7 \times 10^7$ pCi Pu/m^2 (soil)], was adjusted downward to make predictions of plutonium in roots at 30 yr match the calculated inventory in Table 1	1.7×10^{-4}
Roots to wood	The initial value, [460 g (wood production) $\text{m}^{-2}\ \text{yr}^{-1}$] x (0.003 pCi Pu/g)/[3000 g (roots)/m^2 x 4 pCi Pu/g], was adjusted upward to make predictions of plutonium in wood at 30 yr match the inventory in Table 1	1.7×10^{-4}
Wood to leaves	[400 g (leaf production) $\text{m}^{-2}\ \text{yr}^{-1}$]/[10,500 g (standing wood)/m^2]	0.038
Leaves to litter	92% of the forest canopy is returned in leaf fall to the litter each year (368 g $\text{m}^{-2}\ \text{yr}^{-1}$)/(400 g/$\text{m}^2$)	0.92
Wood to litter	Based on average values of branch-bole mortality in deciduous forests (97 g $\text{m}^{-2}\ \text{yr}^{-1}$)/[10,500 g (wood)/$\text{m}^2$]	0.0092
Roots to soil	Based on measurements of root mortality in deciduous forests (700 g $\text{m}^{-2}\ \text{yr}^{-1}$)/[3000 g (roots)/$\text{m}^2$]	0.23
Litter to soil	Average decomposition rate for *Fraxinus* and *Plantanus* litter (240 g $\text{m}^{-2}\ \text{yr}^{-1}$)/[500 g (litter)/$\text{m}^2$]	0.48
Soil to litter	Parameter fitting; represents physical resuspension of soil to litter and not biological uptake	8.5×10^{-5}
Vegetation to consumers	Small mammals are assumed to consume 33% of their biomass each day [3.5 g (consumed) $\text{m}^{-2}\ \text{yr}^{-1}$]/[110 g (ground vegetation)/m^2]	0.03
Consumer to litter	Assumes most of the plutonium in small mammals is present in the gut and has a turnover time of 1 day (1 day/365 days)$^{-1}$	365
Litter to soil fauna	Assumes 14% of the annual leaf fall is processed by soil fauna (50 g $\text{m}^{-2}\ \text{yr}^{-1}$)/[500 g (litter)/$\text{m}^2$]	0.10
Soil fauna to soil	Assumes a 30-day turnover time for soil fauna (30 days/365 days)$^{-1}$	12
Soil to consumer	Assumes ~1% of the amount ingested is soil [0.035 g (soil)/yr] x (65 pCi Pu/g)/[1.7×10^7 pCi Pu/m^2 (soil)]	1.4×10^{-7}

*The diagrammatic model is shown in Fig. 1. Biomass data for deciduous forests were obtained from Harris, Goldstein, and Henderson, 1973; Harris et al., 1975; and Sollins, Reichle, and Olson, 1973. If plutonium concentrations are not shown, they cancel out of the calculation.

The model uses time-invariant transfer coefficients. Since this is an annual model seasonal variations in transfers are not represented. The system was modeled interactively on a computer with a differential-equation modeling program (Rust and Mankin, 1976). The average steady-state value of all compartments and their variabilities was determined with COMEX (Gardner, Mankin, and Shugart, 1976). COMEX is a computer program that uses Monte Carlo methods for analyses of donor-controlled linear compartment models. Its features are: (1) selection of transfer coefficients for each simulation from a multivariate normal random-number generator (the distribution is determined by specifying means and variances for all transfers); (2) solution of the system by matrix calculus; and (3) output of results to SAS, a statistical analysis package (Barr et al., 1976), for analyses of state variables and relationships between state variables and model transfers. From COMEX it is possible to evaluate the sensitivities of model compartments to changes in transfer coefficients and provide estimates of variation in model predictions, given variation in the model parameters.

Results and Analyses

First, the model was used to simulate plutonium dynamics in vegetative components of a forest on the basis of the data set for the forest at Oak Ridge (basic model). Second, simulations including the animal components of the forest were performed (expanded model). Third, the variability in state variables under equilibrium conditions was calculated for both cases, given variability in the model transfer coefficients. Last, a correlation analysis was performed on the expanded model to identify important plutonium transfers.

The accumulation of plutonium in vegetative components of the forest (basic model) was simulated, starting from an initial condition of 1.7×10^7 pCi Pu/m^2 (soil). The results are shown in Fig. 2. The amount of plutonium in litter, ground vegetation, roots, wood, and leaves reached steady state in approximately 120 yr. After this time, less than 0.1% of the total soil plutonium had transferred to aboveground components. Tree roots and litter were the principal biological reservoirs of plutonium in the forest. Expansion of the model by coupling it to animal compartments resulted in only a slight alteration in model performance. At steady state the consumer and soil-fauna components contained <0.01 and 21 pCi Pu/m^2, respectively. The addition of soil fauna lowered the steady-state amount of plutonium in litter from $\approx$3000 to $\approx$2500 pCi/m^2. All other components were negligibly affected.

The selection of values of transfer-coefficient variability for the COMEX simulations proved difficult. An accurate assignment of variation to each parameter would require information on the form of the probability density function of each transfer coefficient (e.g., a frequency distribution of litter decay rates measured throughout the forest). Presently, such statistical information is rarely available for a single parameter much less an entire ecosystem. It is not uncommon for coefficients of variation (CV = standard deviation/mean) to range from 0.5 to 3.0 for field measurements of biomass and plutonium concentrations. Nevertheless, we have assigned CV values of 0.2 to transfer coefficients on the assumption that variation is inversely related to sample size, and, with a sufficient number of measurements from the forest, transfers could be quantified with standard deviations much smaller than those normally encountered. We also assume that in large samples these parameters will be normally distributed. Therefore the question posed was, "Given that transfer coefficients can be measured with CV values equal to

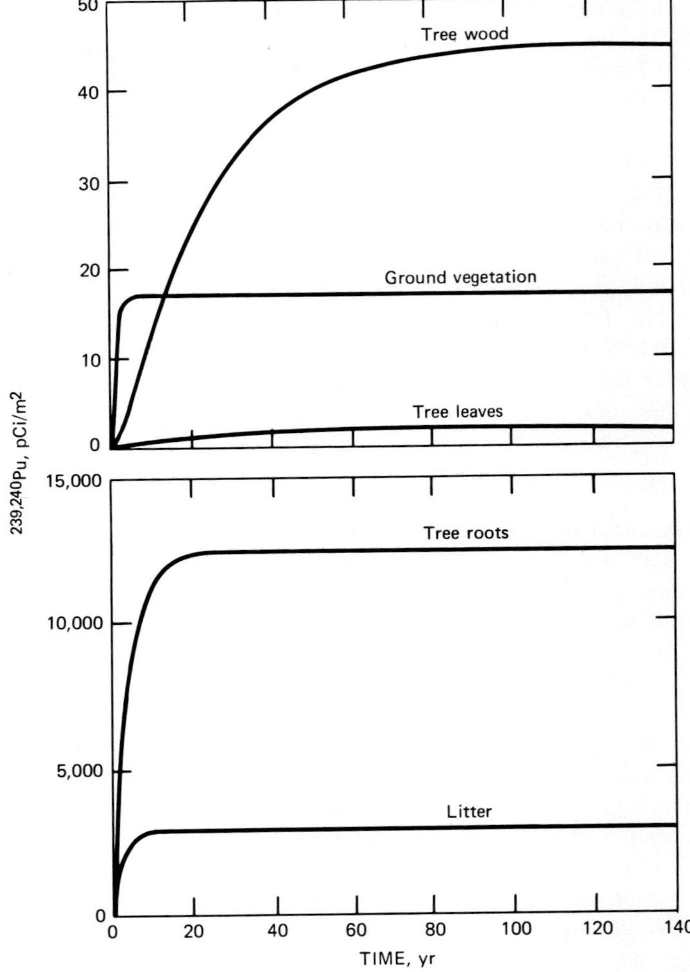

Fig. 2 Simulated uptake of plutonium from contaminated soil (65 pCi/g) by major biotic components of a deciduous forest.

0.2 and that the normally distributed model parameters are permitted random excursions about their mean values, what is the variation in predicted amounts of plutonium in the forest at steady state?''

Coefficients of variation in predicted values for plutonium in the biotic components of the basic and expanded forest model ranged from 0.28 to 0.46 when CV values were set at 0.2 for all transfer coefficients. Considering the basic model (i.e., vegetation only), the predicted amount of plutonium in the forest exhibited considerable variation (e.g., the range for litter was 1300 to 8900 pCi/m²). Similar variation was observed in the expanded model (Table 3). Consequently, even if parameters for plutonium transport in the forest could be measured with CV values approaching 0.2, which is unlikely, the variation in predicted plutonium would be greater than the variation in model transfer coefficients.

**TABLE 3   Mean, Minimum, and Maximum Predicted Amounts of
Plutonium and Coefficients of Variation (CV) for Ecosystem
Components of the Model Deciduous Forest***

Compartment	Mean, pCi/m²	Minimum, pCi/m²	Maximum, pCi/m²	CV†
Soil‡	1.698×10^7	1.696×10^7	1.699×10^7	0.025
Tree roots	13,300	4,920	34,100	0.325
Litter§	2,600	1,230	8,480	0.278
Tree wood	50	16	145	0.438
Soil fauna	23	7.8	109	0.460
Ground vegetation	18	7.5	35	0.301
Tree leaves	2.0	0.6	5.5	0.425
Consumers	0.0084	0.0039	0.022	0.309

*Statistics are based on 300 deterministic simulations to steady state with
varying transfer coefficients having CV values of 0.2.
†CV = standard deviation/mean.
‡A 20-cm soil depth.
§Litter contaminated with some soil plutonium.

Correlations between transfers and the predicted steady-state values of plutonium in
each compartment of the model are given in Table 4. The magnitude of the correlations
between transfers and state variables indicates that the amount of plutonium in the
modeled forest was most sensitive to changes in transfers of plutonium from soil to other
components, especially roots and litter. The influence of these transfers is related to the
large pool of soil plutonium and its central location within the complex of model
pathways (Fig. 1).

Discussion

Three questions emerge from the experience of building this model:
  • What does the model reveal about the behavior of plutonium in forest ecosystems?
  • What are the possible sources of variation in model predictions?
  • What does the uncertainty in model predictions tell us about our present ability to
develop ecosystem-scale models of plutonium behavior in the environment?

Plutonium is expected to accumulate in forest components that are characterized by
large biomass and long turnover times. For example, among the biotic components,
wood, roots, and litter made up 96% of the biomass and contained 99% of the plutonium
inventory. Nevertheless, the amount of plutonium in biota at steady state was always
<0.25% of the inventory in soil; most of the ecosystem's mass is in the soil, which has a
longer turnover time than any biotic component. Therefore the fractional transfer of
plutonium from soil to forest biomass is extremely small and is expected to remain so
indefinitely in the absence of major changes in forest structure or existing environmental
conditions.

The model transfers having the greatest effect on the amount of plutonium in soil,
tree roots, wood, and litter include (1) reciprocal exchanges between soil and tree roots,
(2) reciprocal exchanges between soil and forest litter, and (3) transfers from roots to
wood and wood to leaves. Because of the potential importance of these transfers to the
distribution of plutonium among forest components, more research is needed to

accurately quantify them before other model parameters are considered. We know the least about critical transfers to and from soil because they are the most difficult to quantify. For example, in this model three of the preceding transfers required some parameter fitting to calibrate the performance of the model against field measurements. Although potentially important to animal components because of the radiotoxicity of plutonium, the addition of plutonium transfers to animals had a minor influence on the major plutonium reservoirs in the forest.

Sources of variation in model predictions include spatial variation, temporal variation, and measurement error in transfer coefficients. These sources are independent of any bias due solely to model structure or coupling. COMEX simulations of plutonium behavior in the model forest exclude temporal (annual and seasonal) variation in transfers because the model transfers are considered time invariant. In this sense our simulations differ from the time variant or stochastic modeling of plutonium behavior in pinon—juniper forests in New Mexico (Wheeler, Smith, and Gallegos, 1977), where ecosystem behavior in response to random variations in climate was considered. Variation in model predictions from the COMEX simulations can be interpreted as a consequence of spatial variation in transfers, measurement error, or both, depending on perspective. Intraforest variation in transfer coefficients resulting from different edaphic or microclimatic conditions within a single forest will produce local differences in the amount of plutonium (picocuries per square meter) in biota. Intraforest variation, however, could be negligible relative to differences in transfer coefficients between distinct forest stands (e.g., in different counties). Therefore the average predicted amount of plutonium in each forest, given a uniform soil contamination, could vary, depending on site conditions and forest species composition.

An example of how such variation between forests could bear on the assessment of the environmental impact of plutonium is provided by considerations of fire. Assume that

TABLE 4   Correlations Between Varying Transfer Coefficients and Predicted Steady-State Values for Each Forest Model Compartment*

| | Compartment | | | | | | | |
Transfer	Soil	Ground vegetation	Consumers	Tree roots	Tree wood	Tree leaves	Forest litter	Soil fauna
Soil to tree roots	−0.67			0.68	0.54	0.51		
Soil to litter							0.68	0.45
Tree roots to soil	0.69			−0.69	−0.49	−0.52		
Soil to ground vegetation		0.71						
Litter to soil							−0.65	−0.39
Soil fauna to soil								−0.57
Ground vegetation to litter		−0.74						
Tree roots to wood					0.40	0.35		
Consumers to litter			−0.77					
Tree wood to leaves					−0.50			
Litter to soil fauna								0.34
Tree leaves to litter						−0.48		
Soil to consumers			0.52					

*Data are based on 300 independent simulations. Transfers were varied simultaneously before each simulation, using a Monte Carlo random-number generator. Only correlations greater than 0.30 are reported.

a regional assessment involves the impact of a fire that burns litter, ground vegetation, tree wood, and leaves. From field experiments transfer coefficients for plutonium have been measured in forest stands over the region of interest with a precision such that CV values are nearly 0.2 for all model parameters. Given that the soil is contaminated with 1.7×10^7 pCi Pu/m^2, the predicted amount of plutonium in the forest, at steady state, at risk of release by fire from an average forest is 26.7 μCi/ha. This amount could range from 12 to 87 μCi/ha, however, depending on which forest was contaminated.

Attempts to model plutonium dynamics in ecosystems are hampered by uncertainties in predictions arising from problems with system identification, quality of data, and lack of validation. System identification, which involves determining transfers and model structure in a way that model performance fits real-world data, is a problematic area in ecosystem modeling because of the variability in ecological data and our lack of control over the natural environment (Halfon, 1975). The recommendation of O'Neill (1973) and of Shelley (1976) to "build the simplest model appropriate to achieving the objective" was followed in designing the present plutonium forest model. In simple models the effects of measurement error on predictions are reduced, but inaccuracies arising from systematic bias are increased (O'Neill, 1973).

Even when an optimal model structure can be found (i.e., one that simultaneously minimizes inaccuracies due to systematic bias and measurement error), problems remain because of natural variation in plutonium data from ecosystems. By simultaneously varying all transfers, COMEX is a statistically based simulation technique that permits an assessment of the variation in predicted amounts of plutonium in the forest, recognizing that variance in transfers exists. Even with small amounts of variation about the model parameters (CV = 0.2), there are typically eightfold differences in model predictions. As Shelley (1976) points out, we cannot expect the variability in predicted values from ecosystem models to be less than the variation in the data used to calibrate the model. The COMEX simulations of plutonium in forest biota support this argument. Given that CV values on plutonium data from field studies range from 0.5 to 3.0, we conclude that the ability to adequately model plutonium transport in ecosystems is strongly dependent on better data from natural environments and on an understanding of the causes of variation in the data.

Data on plutonium in the White Oak Creek floodplain forest were used to calibrate this model. Calibration data cannot be used for model validation (Shelley, 1976), and criteria for validity (Mankin et al., 1977) are difficult to define. The model cannot be judged valid, but it has been useful for the identification of areas where research is needed to better our understanding of plutonium dynamics in forests and thereby develop more precise models. Future field studies should provide the data necessary for systems analysis and comparison of plutonium dynamics in forests and other ecosystems. It is unlikely, however, that these models will be validated over the full time frame of some simulations (e.g., >100 yr) before advances in ecosystem analysis make the models obsolete. Nevertheless, the long radioactive half-life of ^{239}Pu (24,400 yr) and its potential for accumulation in the biosphere, necessitate some predictions in lieu of none at all. The model reported here represents an hypothesis that presents testable predictions about the dynamics and distribution of plutonium in deciduous forests.

Acknowledgments

We thank S. I. Auerbach, W. F. Harris, C. A. Little, D. E. Reichle, H. H. Shugart, Jr., and anonymous technical reviewers for their helpful comments and suggestions. The research was sponsored by the U. S. Department of Energy (formerly Energy Research and Development Administration) under contract with Union Carbide Corporation. Publication No. 1334, Environmental Sciences Division, Oak Ridge National Laboratory.

References

Anspaugh, L. R., J. H. Shinn, P. L. Phelps, and N. C. Kennedy, 1975, Resuspension and Redistribution of Plutonium in Soils, *Health Phys.*, 29: 571.

Barr, A. J., J. H. Goodnight, J. P. Sall, and J. T. Helwig, 1976, *A User's Guide to SAS 76*, Sparks Press, Raleigh, N. C.

Bennett, B. G., 1976, Transuranic Element Pathways to Man, in *Transuranium Nuclides in the Environment*, Symposium Proceedings, San Francisco, 1975, pp. 367-383, STI/PUB/410, International Atomic Energy Agency, Vienna.

Bondietti, E. A., S. A. Reynolds, and M. H. Shanks, 1976, Interaction of Plutonium with Complexing Substances in Soils and Natural Waters, in *Transuranium Nuclides in the Environment*, Symposium Proceedings, San Francisco, 1975, pp. 273-287, STI/PUB/410, International Atomic Energy Agency, Vienna.

Dahlman, R. C., and K. W. McLeod, 1977, Foliar and Root Pathways of Plutonium Contamination of Vegetation, in *Transuranics in Natural Environments*, Symposium Proceedings, Gatlinburg, Tenn., Oct. 5–7, 1976, M. G. White and P. B. Dunaway (Eds.), USDOE Report NVO-178, Nevada Operations Office, NTIS.

Gardner, R. H., J. B. Mankin, and H. H. Shugart, Jr., 1976, *The COMEX Computer Code Documentation and Description*, ERDA Report EDFB-IBP-76-4, Oak Ridge National Laboratory, NTIS.

Halfon, E., 1975, The Systems Identification Problem and the Development of Ecosystem Models, *Simulation*, 25: center section.

Hanson, W. C., 1975, Ecological Considerations of the Behavior of Plutonium in the Environment, *Health Phys.*, 28: 529.

Harris, W. F., R. A. Goldstein, and G. S. Henderson, 1973, Analysis of Forest Biomass Pools, Annual Primary Production and Turnover of Biomass for a Mixed Deciduous Forest Watershed, in *Proceedings of the Working Party on Forest Biomass of IUFRO*, H. Young (Ed.), pp. 41-64, University of Maine Press, Orono, Maine.

——, P. Sollins, N. T. Edwards, B. E. Dinger, and H. H. Shugart, 1975, Analysis of Carbon Flow and Productivity in a Temperate Deciduous Forest Ecosystem, in *Productivity of World Ecosystems*, D. E. Reichle, J. F. Franklin, and D. W. Goodall (Eds.), pp. 116-122, National Academy of Sciences, Washington, D. C.

Hudetz, W. J., 1973, Ecosystem Models and Simulation, *Simulation*, 21: 81.

Jakubick, A. T., 1976, Migration of Plutonium in Natural Soils, in *Transuranium Nuclides in the Environment*, Symposium Proceedings, San Francisco, 1975, pp. 47-62, STI/PUB/410, International Atomic Energy Agency, Vienna.

Loucks, O. L., 1970, Evolution of Diversity, Efficiency, and Community Stability, *Am. Zool.*, 10: 17.

Mankin, J. B., R. V. O'Neill, H. H. Shugart, and B. W. Rust, 1977, The Importance of Validation in Ecosystem Analysis, in *New Directions in the Analysis of Ecological Systems*, Part 1, pp. 63-72, G. S. Innis (Ed.), Simulation Councils Proceedings Series, Simulation Councils, Inc.

Olson, J. S., 1965, Equations for Cesium Transfer in a *Liriodendron* Forest, *Health Phys.*, 11: 1385.

O'Neill, R. V., 1973, Error Analysis of Ecological Models, in *Radionuclides in Ecosystems*, Proceedings of the Third National Symposium on Radioecology, Oak Ridge, Tenn., May 10–12, 1971, D. J. Nelson (Ed.), pp. 898-908, USAEC Report CONF-710501, Oak Ridge National Laboratory, NTIS.

Reichle, D. E., D. A. Crossley, Jr., C. A. Edwards, J. F. McBrayer, and P. Sollins, 1973, Organic Matter and ^{137}Cs Turnover in Forest Soil by Earthworm Populations: Application of Bioenergetic Models to Radionuclide Transport, in *Radionuclides in Ecosystems*, Proceedings of the Third National Symposium on Radioecology, Oak Ridge, Tenn., May 10–12, 1971, D. J. Nelson (Ed), pp. 240-246, USAEC Report CONF-710501, Oak Ridge National Laboratory, NTIS.

Rust, B. W., and J. B. Mankin, 1976, Interactive Differential Equations Modeling Program, ERDA Report EDFB-IBP-74-2, Oak Ridge National Laboratory, NTIS.

Shelley, P. E., 1976, Field Data for Environmental Modeling—Adjacent or Integral? in *Environmental Modeling and Simulation,* W. R. Ott (Ed.), Report EPA-600/9-76-016, pp. 583-585, U. S. Environmental Protection Agency, Washington, D. C.

Shugart, H. H., D. E. Reichle, N. T. Edwards, and J. R. Kercher, 1976, A Model of Calcium Cycling in an East Tennessee *Liriodendron* Forest: Model Structure, Parameters, and Frequency Response Analysis, *Ecology,* 57: 99.

Sollins, P., D. E. Reichle, and J. S. Olson, 1973, Organic Matter Budget and Model for a Southern Appalachian *Liriodendron* Forest, USAEC Report EDFB-IBP-73-2, Oak Ridge National Laboratory, NTIS.

Van Voris, P., and R. C. Dahlman, 1976, Floodplain Data: Ecosystem Characteristics and ^{137}Cs Concentrations in Biota and Soil, ERDA Report ORNL/TM-5526, Oak Ridge National Laboratory, NTIS.

Webb, W. L., H. J. Schroeder, Jr., and L. A. Norris, 1975, Pesticide Residue Dynamics in a Forest Ecosystem: A Compartment Model, *Simulation,* 24: 161.

Wheeler, M. L., W. J. Smith, and A. F. Gallegos, 1977, A Preliminary Evaluation of the Potential for Plutonium Release from Burial Grounds at Los Alamos Scientific Laboratory, ERDA Report LA-6694-MS, Los Alamos Scientific Laboratory, NTIS.

A Review of Biokinetic and Biological Transport of Transuranic Radionuclides in the Marine Environment

T. M. BEASLEY and F. A. CROSS

Present understanding of the uptake, retention, and loss of transuranic radionuclides by marine biota is limited. Laboratory experiments have demonstrated that for certain species assimilation of plutonium and americium from labeled food is an efficient process and that direct uptake from seawater is important in the bioaccumulation of all transuranic radionuclides studied to date. Organisms appear to play an important role in the vertical transport of these radioelements from the surface layers of the ocean to greater depths.

A discussion of the biokinetic behavior of transuranic radionuclides in marine organisms should address the rates at which these radioelements are ingested, assimilated, and egested as well as the rates at which they are lost from the organism's tissues over time (e.g., turnover time and biological half-life). In terms of oceanic processes, such information is of little value, however, unless it can be used in predicting the ultimate fate of transuranics released into the marine environment. We have chosen, therefore, to address both these aspects of transuranic behavior and believe that this chapter, along with the chapters by Noshkin (this volume) and by Eyman and Trabalka (this volume), will provide a comprehensive summary of the behavior of transuranics in aquatic ecosystems.

Background

By far the greatest amount of information to date dealing with transuranic radionuclides in marine organisms has been confined to the determination of absolute amounts of these radioelements in both whole animals and selected tissues with the subsequent computation of a concentration factor to indicate the degree of biomagnification between the organism and its environment. The use of this approach will be discussed later.

The review articles by Noshkin (1972), Cherry and Shannon (1974), and Eyman and Trabalka (this volume) describe the general features observed to date in aquatic ecosystems relative to the accumulation of plutonium in aquatic organisms. Although these data do not give information as to the rates of uptake and loss of the transuranics in aquatic organisms, they are useful in clarifying one unfortunate error in terminology that requires rectification. In the truest sense of the word, aquatic organisms do not discriminate against transuranic radionuclides; if they did, the radioelements could not be measured in the organisms deriving these entities from their labeled environments. The data in hand simply indicate that uptake by organisms either by assimilation or by surface adsorption is greater for certain species than for others.

Biokinetics of Transuranics in Marine Organisms

The published information dealing with the uptake, assimilation, and loss of the elements plutonium, americium, curium, and neptunium in marine organisms is limited. Apart from the general lack of concern regarding transuranics as potential marine pollutants until the late 1960s, there are other reasons why progress in this field has been retarded. First, the number of laboratories having access to fresh flowing seawater and the culture facilities necessary to undertake such research are limited. Second, the extensive radiation protection measures required to conduct even modest tank experiments with these alpha-emitting radionuclides coupled with the analytical task of making large numbers of low-level alpha measurements have discouraged most investigators. Finally, those laboratories which have been involved in marine transuranic measurements are reluctant to house even small amounts of these radioelements so as to preclude the possibility of sample contamination, which would compromise their low-level determinations.

It is surprising that as early as 1966 Todd (1968) and Todd and Logan (1966) had demonstrated the feasibility of using ^{237}Pu ($T_{1/2}$ = 45.6 days), which decays by electron capture, as a tracer for metabolic studies. However, it was not until 1974 that Bair et al. (1974) used this isotope in a dual-labeling experiment with ^{239,240}Pu in a comparative study of the distribution and excretion of the element in beagle dogs, and only recently Fowler, Heyraud, and Beasley (1975) used the isotope to perform metabolic studies with marine organisms. Because of its high specific activity (curies per gram), it is possible to approach low atom concentrations in labeling solutions more comparable to those found in environments contaminated by the lower specific-activity isotopes ^{239,240}Pu and ^{238}Pu. In addition, the 100-keV X ray emitted in the deexcitation of ^{237}Np permits easy detection by NaI(Tl) scintillation techniques and therefore permits whole-body counting techniques to be used with small marine organisms. Therefore, for plutonium many of the obstacles for laboratory tank experiments can be minimized, even though current production costs of ^{237}Pu are high ($500 per microcurie) and small amounts of ^{236}Pu and ^{238}Pu are present in the purified ^{237}Pu.

Although laboratory experiments can be performed with care using ^{241}Am as a tracer and counting its 60-keV X ray by scintillation techniques, relatively high activity levels must be used and small experimental animals must be used to preclude serious geometry problems associated with the absorption of the weak X ray in the organisms. For curium and neptunium, there are no isotopes of long enough half-life and decay characteristics to permit in vivo measurements; thus one must use the more demanding techniques of total alpha counting by thick source measurements (Cherry, 1964; Guary and Fowler, 1977) or chemical isolation of the isotopes and alpha spectrometry. It is not an exaggeration, therefore, to say that much time and effort will be expended before sufficient data are in hand to present a reasonable picture of the biokinetic behavior of transuranics in aquatic organisms.

Plutonium

Perhaps the first open-literature publication dealing with the uptake and tissue distribution of plutonium in a marine organism was the work of Ward (1966). Using the lobster *Homarus vulgaris,* she demonstrated that direct uptake of ^{239,240}Pu from labeled seawater did occur. Near equilibrium was reached in the exoskeleton and gills after 50 days of exposure. At 220 days the gut and hepatopancreas concentration factors approached that of the shell and gills ($\simeq$100) but gave evidence of still being far from equilibrium. Flesh at day 220 showed concentration factors of $\simeq$3, but the shape of the uptake curve suggests that higher values would have been reached had the experiment been continued. Approximately 90% of the total plutonium taken up by the lobster was found in the exoskeleton; some 4.6 and 1.2% was retained by the hepatopancreas and flesh, respectively. As would be expected from these results, the major portion of the plutonium accumulated by the lobster is lost during molting.

Between 1966 and 1975 very little information was published in the open literature dealing with the actual rates of uptake and loss of plutonium in marine organisms. Zlobin (1966; 1971), Zlobin and Mokanu (1970), and Zlobin and Perlyuk (1971) presented data on the uptake of ^{239,240}Pu in marine algae, principally the brown algae *Ascophyllum nodosum,* which suggested that the element was actually assimilated by the plant rather than simply adsorbed to it. However, the subsequent work of Wong, Hodge, and Folsom (1972), Hodge, Hoffman, and Folsom (1974), Folsom, Hodge, and Gurney (1975), and Folsom and Hodge (1975) using other macroalgae suggests that adsorption is the more likely mechanism for accumulation and that plutonium might be attached to large macromolecules or micelles, which have slow diffusivities but great affinity for a variety of surfaces. In any case evidence now exists that marine plants (phytoplankton and free-floating and rooted algae) do accumulate transuranics to a relatively high level and that the rate of accumulation is rapid. This process applies to both fallout-derived plutonium (Noshkin, 1972) and that introduced from fuel reprocessing plants (Hetherington, Jefferies, and Lovett, 1975; Hetherington et al., 1976; Fraizier and Guary, 1976; 1977). For phytoplankton equilibrium between the algae and water can be established in as little as 5 to 10 days (Gromov, 1976).

The first laboratory experiments using ^{237}Pu to determine plutonium biokinetics in marine organisms appear to be those of Fowler, Heyraud, and Beasley (1975). Using mussels *(Mytilus galloprovincialis),* shrimp *(Lysmata seticaudata),* and marine worms *(Nereis diversicolor),* they followed both uptake and loss of plutonium in the organisms after direct uptake from seawater and, in the case of mussels and shrimp, from labeled food as well. The valence state of ^{237}Pu tracer was chemically adjusted to either the quadrivalent (+4) or hexavalent (+6) state before the isotope was introduced into the experimental aquaria. No subsequent attempt was made to determine the valence state of the isotope during the course of the experiment. In all three organisms direct uptake from seawater occurred quite readily. For mussels exposed to filtered seawater containing Pu(+6), concentration factors ranged from 20 to 60 after 26 days of exposure, and a large percentage of the plutonium taken up resided in the shell and byssus threads. In those cases where the byssus was removed from the mussel, greater than 80% of the activity was associated with the shell. The activity was firmly bound to the shell material, and the shell showed only minor losses even when rinsed for as long as 8 hr in 0.1N HCl solution. Mussels that accumulated Pu(+6) directly from seawater showed a two-component loss when placed in unlabeled seawater; the biological half-life ($T_{b\frac{1}{2}}$) for the fast pool containing 35% of the total plutonium was 7 days; that for the slower turnover pool, which contained 65% of the total plutonium, was 776 days (>2 yr). Mussels that had accumulated Pu(+4) from both food and water showed more rapid turnover owing both to a shorter labeling time and presumably to a more rapid clearance of labeled material voided as feces. Direct uptake of Pu(+4) by mussels from filtered seawater was not investigated.

For shrimp direct uptake from Pu(+6)-labeled seawater was slow and was strongly influenced by molting. A single individual that did not molt during a 25-day exposure period reached a concentration factor of only 19. Three individuals that molted during the first 18 days of exposure lost between 92 and 100% of their total body content of plutonium, a value that may be artificially high owing to the capacity of such material to further adsorb plutonium from labeled solutions once they have been cast. Animals that molted twice during the loss period showed virtually no plutonium in these second exuviae, which indicated that their whole-body content was indeed the result of systemically deposited plutonium. Excretion in *L. seticaudata* following a single feeding of Pu(+6)-labeled brine shrimp *(Artemia)* was rapid during the first 3 days but then decreased sharply to an exponential rate for 1 month until only 1% of the initial burden remained. Shrimp fed daily rations of labeled *Artemia* for 15 days did not accumulate higher levels of plutonium than those fed a single ration of labeled food. Although shrimp that were starved after a single feeding of labeled *Artemia* retained a significant fraction of the initial dose up to day 8 (40%), they quickly eliminated this material once feeding was resumed. It is therefore likely that accumulation in tissues other than the exoskeleton in the shrimp would be a slow process.

Interestingly, the marked decrease in the rate of excretion after a single feeding of labeled food and the gut clearance of this material suggested that the assimilation efficiency in the shrimp greatly exceeded the tenths to hundredths of 1% assimilation efficiencies reported for terrestrial mammals (Thompson, 1967). This appears to be the case whether the plutonium is derived from food or directly from water. This was perhaps one of the first indications that marine invertebrates were capable of retaining a substantial portion of the plutonium they derived from these two major routes.

For worms direct uptake of plutonium from water for either the +4 or +6 valence state was both rapid and efficient since, after 15 days of exposure, concentration factors approached 200 for both valences. Eight days following uptake [Pu(+6)], worms placed in unlabeled seawater rapidly lost some 30% of their plutonium in 4 days; thereafter the rate of loss slowed dramatically, giving a $T_{b\frac{1}{2}}$ of 79 days (computed between days 4 and 35). Once again a surprisingly high percentage of the initial plutonium body burden appeared to have been retained by the organism, but it was not determined whether the plutonium had been systematically incorporated into tissue or sequestered by the external mucus. Moreover, it was clearly shown that the exometabolites excreted by worms into seawater can render the plutonium less available to fresh worms introduced into this conditioned water.

Not only do these experiments give interesting insight into the rates of accumulation and loss of the plutonium in the animals used but they also confirm the general tissue distributions found in similar species that accumulate plutonium from fallout and in those at Thule, Greenland (Aarkrog, 1971; 1977). Crustacea contain large amounts of plutonium in their exoskeletons, molluscs retain the majority of their plutonium in the shell and byssus threads, and polychaetes efficiently accumulate plutonium and are expected to evidence high levels of the element when exposed to contaminated water. Finally, if one were to assess the relative importance of the pathways by which the element is accumulated by marine organisms, direct uptake from water may be significant and in some cases more significant than uptake from labeled food. By contrast, for molluscs, which derive their plutonium from reprocessing wastes, tissue concentrations often exceed those found in the shell, and concentration factors for benthic fishes appear higher as well (Hetherington, Jefferies, and Lovett, 1975; Hetherington et al., 1976; Pillai and Mathew, 1976; Guary and Fraizier, 1977; Guary, Masson, and Fraizier, 1976). This discrepancy may result from differences in (1) bioavailability of physicochemical form, (2) absolute levels in the environment, (3) duration of exposure, and (4) environmental processes in various ecosystems.

Data on the biokinetic behavior of plutonium in marine fishes are scarce. Pentreath (personal communication, 1977) has studied the direct uptake of ^{237}Pu(+6) from seawater by the plaice *(Pleuronectes platessa),* as well as assimilation from *Nereis* sp. injected with the isotope. Sixty-three days following direct uptake from water, the whole-body concentration factors were < 1. Complete dissections of the fish showed that ^{237}Pu had concentrated in all fish in the stomach (concentration factor, 1 to 2), upper gut (11 to 26), and lower gut (2 to 10). Concentration factors for skin and gill were 1 and 2 to 3, respectively. Plutonium was detectable in the livers of six of the seven fish used in the experiment (0.8 to 1.7), and only traces were measurable in blood cells, plasma, and bone of two fish.

The retention of ^{237}Pu by plaice fed injected *Nereis* sp. [both Pu(+4) and Pu(+6)] and subsequently fed unlabeled *Nereis* sp. as maintenance rations ranged between 0.4 and 3.0% 5 days after exposure. The measured $T_{b\frac{1}{2}}$ values of ^{237}Pu for 10 fish in the experiment ranged from 9 to 49 days. By contrast, the retention of parenterally administered ^{237}Pu(+4) in five fish that had been injected in the right dorsal muscle was relatively high; $T_{b\frac{1}{2}}$ values ranged between 642 and 877 days. Similar results were obtained from fish that had been injected directly in the body cavity ($T_{b\frac{1}{2}}$, 282 to 1100 days). Redistribution of the isotope within the fish 158 days postinjection was marked; the highest accumulations occurred in the liver, kidney, and spleen. Similar distributions were observed for both injection sites.

Unlike the plaice the thornback ray *(Raja clavata)* appears to assimilate more ^{237}Pu from *Nereis* sp. when the isotope is injected into the worm. Measurable amounts of the isotope are detected in the liver at dissection (Pentreath, personal communication, 1977). Crab digestive gland, which was incubated with ^{237}Pu and subsequently fed to both plaice and rays, produced results similar to those from experiments in which *Nereis* sp. was used. Clearly, the digestive physiologies of the two fish are sufficiently different that enhanced plutonium uptake occurs in the ray.

The most recent evidence suggesting efficient assimilation of plutonium from labeled food by invertebrates is the work of Fowler and Guary (1977a) in which crabs (*Cancer maenas* and *Cancer pagurus*) were fed ^{237}Pu-labeled *N. diversicolor*. Remarkably high assimilation efficiencies ranging from 20 to 60% were observed. Of the ^{237}Pu absorbed across the gut wall, 43 to 85% was found in the hepatopancreas, 8 to 43% in the shell, and 5 to 10% in the gill. It appears to make no difference whether the initial plutonium labeling solution contains Pu(+6) or Pu(+4). Initial results from experiments in which mussel *(M. galloprovincialis)* tissue labeled by ^{237}Pu uptake from water and phytoplankton was fed to starfish indicate similarly high assimilation efficiencies (Fowler and Guary, 1977b). These latter experiments would tend to support the earlier contention of Noshkin et al. (1971) that food-chain magnification in the simple food chain mussel— starfish can occur. That crabs efficiently take up plutonium from labeled environments has equally been demonstrated by the work of Guary, Masson, and Fraizier (1976) and Guary and Fraizier (1977).

It would thus appear that assimilation of plutonium in invertebrates does not constitute an inefficient process nor does it parallel the low gastrointestinal adsorption of plutonium by terrestrial vertebrates where the plutonium has been administered by gavage (Thompson, 1967). What role digestive physiology or the manner in which the label is administered plays in these strikingly dissimilar results remains to be elucidated. If biochemically-bound plutonium is more readily assimilated across gut walls of experimental animals than plutonium in either ionic or chelated forms (citrate or tartrate), then it remains to be demonstrated whether substantial differences exist in either the distribution or the elimination of the plutonium retained by the animal. Should such differences be observed, the implications for revising current radiation protection standards are obvious.

Americium

As far as we are able to determine, the only published data dealing with the biokinetic behavior of americium in marine organisms are those of Fowler and Heyraud (1974). Using the brine shrimp *(Artemia)* and the euphausiid *(Meganyctiphanes norvegica),* Fowler and Heyraud studied ^{241}Am uptake directly from seawater (brine shrimp and euphausiid) and a combined food—seawater pathway (brine shrimp) in a short-term experiment. *Artemia* accumulated ^{241}Am efficiently from water; concentration factors of 1700 were attained in as little as 48 hr. By contrast, concentration factors from these animals in a labeled phytoplankton suspension reached 400 in the same period of time. Filtration of the seawater showed that some 81% of the ^{241}Am was associated with the algal cells; 2.0% was retained by 0.45-μm filters, and some 17% remained in the filtrate. Recalculation of the concentration factor based on the activity in water and the 0.45-μm filterable fraction alone gave concentration factors virtually identical to those found when accumulation was from water only. Clearly, little ^{241}Am (if any) was assimilated by the food pathway in this experiment. When placed in fresh seawater, both groups of *Artemia* retained less than 1% of their accumulated ^{241}Am after 3 hr, which suggested that very little of the ^{241}Am was incorporated metabolically.

Euphausiids appeared to accumulate ^{241}Am less efficiently than the brine shrimp and at a somewhat slower rate. After the euphausiids had been exposed for 64 hr to labeled seawater, concentration factors had reached a value of 125, and there was an indication that equilibrium was being approached. When placed in unlabeled seawater, a single euphausiid lost only 40% of its ^{241}Am burden during the first 8 days. At molting, however, virtually all the ^{241}Am was lost with the cast molt.

Aside from these observations, we have little data concerning the uptake rate, assimilation, and loss of ^{241}Am in marine organisms. However, Fowler and Guary (1977b) have observed relatively high assimilation efficiencies for starfish fed ^{241}Am-labeled mussel tissue, as observed for ^{237}Pu, and retention times appear to be long. Starfish fed a single ration of ^{241}Am-labeled mussel tissue retained 85 to 95% of the ingested dose 5 weeks postexposure. At dissection, $\simeq$90% of the retained ^{241}Am was found in the pyloric caeca.

The relationships between the behavior of americium and plutonium in the marine environment have recently been summarized by Bowen (1975) and by Livingston and Bowen (1976a). For marine organisms there appears to be no consistent trend regarding their bioavailability, i.e., that one is preferentially taken up by biota in favor of the other. Virtually all the comparisons to date rest with measurements of these isotopes in biota deriving the elements either from fallout or from fuel reprocessing wastes. However, Pentreath and Lovett (1976) and Beasley and Fowler (1976b) have presented additional data contrasting the relative amounts of the two isotopes in plaice collected from the Irish Sea and the relative uptake of the two elements by polychaetes exposed to labeled sediments. In plaice *(Pleuronectes platessa)* there was an indication that ^{241}Am was taken up in preference to ^{239,240}Pu, although the variable discharge rates of the two isotopes from the Windscale effluent before sampling and the mobility of the fish themselves in the discharge area make a definitive statement regarding differential uptake problematical (Pentreath and Lovett, 1976). For these reasons, and others raised by Fowler and Beasley (1977), accurate estimations of the concentration factors for the fish could not be made. For the polychaetes *(Nereis diversicolor),* the case is somewhat clearer. Whether exposed to labeled Windscale sediment or to that collected from the Bikini Atoll, uptake of plutonium was greater than that of americium (Beasley and Fowler, 1976b).

Curium

At this time (early 1978), we are unable to find any reference to biokinetic experiments, either field or laboratory, which deal specifically with the element curium. Moreover, until proven otherwise, it would not be prudent to extrapolate experimental findings derived from ^{241}Am studies to unequivocally predict curium behavior in marine organisms. We base this statement on the previously published data of Sugihara and Bowen (1962) and Bowen and Sugihara (1965), in which a distinct difference was noted in the behavior of fallout ^{144}Ce and ^{147}Pm with respect to their uptake on particulate matter in the oceans. In addition, there continue to be unsettling reports in the literature which suggest a differential behavior between isotopes of the same element due principally to differences in their specific activities or to potential differences due to their existence in different physicochemical forms as a result of "hot-atom chemistry" when formed by decay of a parent radionuclide (Volchok et al., 1975; Hakonson and Johnson, 1973; Emery, Klopfer, and Weimer, 1974; Emery and Garland, 1974; Bowen and Livingston, 1975). In the case of plutonium isotopes (Beasley and Fowler, 1976a), polychaetes exposed to labeled sediments from Windscale and Bikini Atoll and to spiked Mediterranean sediments showed no preferential uptake of the isotopes measured (^{238}Pu and ^{239,240}Pu). There have been no comparable experiments, however, dealing with ^{241}Am or ^{242,244}Cm. The rather large differences in specific activities between ^{241}Am (3.45 Ci/g), ^{242}Cm (6.83 × 10³ Ci/g), and to a lesser degree ^{244}Cm (82.8 Ci/g) still leave open the possibility of differential behavior between the isotopes, particularly if concentration effects become operative at high activity levels, as have been noted between mammalian experiments using ^{237}Pu and ^{239,240}Pu (Bair et al., 1974).

Neptunium

The recent paper by Guary and Fowler (1977) on the biokinetic behavior of ^{237}Np in mussels and shrimp appears to be the only published paper on this subject. Direct uptake of ^{237}Np from water by both the mussel *(M. galloprovincialis)* and the shrimp *(L. seticaudata)* appears to be much less than that observed from plutonium. Whole-body concentration factors of 15 to 20 for both species were observed after exposure for 3 months. Tissue distributions for the mussel followed patterns previously seen for plutonium, i.e., high concentrations in the shell and lesser amounts in flesh. The exoskeleton of the shrimp accumulates the major portion of neptunium, and loss rates are strongly influenced by molting. When the shrimp were placed in unlabeled seawater, the loss was biphasic; the largest part of the ^{237}Np initially present was lost with a $T_{b\frac{1}{2}}$ of 4 days, and about 3% of the original activity exhibited a $T_{b\frac{1}{2}}$ of 252 days. Mussels held in the laboratory had $T_{b\frac{1}{2}}$ values of between 180 and 226 days; faster turnover was observed in animals held at temperatures of 25°C than in those held at 13°C. Mussels placed in the sea showed whole-body $T_{b\frac{1}{2}}$ values of some 81 days; the faster turnover rates were attributed to active growth of the organism.

As yet there is no information regarding the assimilation of ^{237}Np by marine organisms fed labeled food and subsequent turnover rates of assimilated material.

Extrapolation of Laboratory-Derived Information to Natural Conditions

The entire subject of biokinetics of radioactive and stable isotopes of elements and the extrapolation of these laboratory-derived data to the real world has been a point of discussion among aquatic scientists for a number of years. It is extremely difficult to design laboratory experiments that are short term and simplistic relative to oceanic processes which will provide predictive information on the accumulation and redistribution of radionuclides by marine organisms. In addition, field verification of laboratory-derived information often is not possible. Inherent problems in experimental designs for studying the transfer of radionuclides in marine food chains are discussed in a recent review by Cross, Renfro, and Gilat (1975).

Laboratory experiments must be designed which will present the radionuclide to the test organism in a manner similar to that which occurs in nature; i.e., it must be allowed to distribute between particulate and dissolved fractions realistically and must be of a similar specific activity. In addition, feeding rates of the test organisms must reflect natural conditions, as should population densities in the experimental aquaria. These conditions require a combination of laboratory and field observations and basic information on the general ecology of the test organism, which often is not available. Realistic estimates of uptake rates, assimilation efficiencies, and turnover rates can only be obtained by carefully designed experiments (Cross, Willis, and Baptist, 1971; Cross et al., 1975; Cross, Renfro, and Gilat, 1975; Willis and Jones, 1977). Perhaps one of the major problems in radiotracer experiments is that of incomplete labeling (Willis and Jones, 1977).

Another important aspect of biokinetics of radionuclides which warrants discussion here is the use (or misuse) of concentration factors. At this time we do not know what fraction of an element in nature is bioavailable relative to concentration factors, and we obviously are not in agreement (Lowman, Rice, and Richards, 1971; Cross, Renfro, and Gilat, 1975; Fowler and Beasley, 1977). Examples can be found in the literature which base concentration factors on (1) dissolved concentrations in the water, (2) total concentrations in the water, (3) concentrations in food organisms, and (4) concentrations in sediment. The use of any of these four fractions will depend on the author's prejudice relative to the source of the element to the organisms under study. Presently, we do not know what physicochemical forms of elements are most bioavailable to marine organisms. In fact, we have not even developed adequate experimental designs to determine the relative importance of food and water in conveying elements to aquatic animals (Cross and Sunda, 1979). In reality, bioavailable fractions of elements probably consist of a complex interaction of food, water, and sediment, depending on trophic level and feeding type.

Another problem in using concentration factors in field studies is that concentrations of elements in water or particulate materials are highly variable and organisms under study usually are constantly moving both horizontally and vertically in the ocean. Concentrations of elements in organisms represent an integration of environmental levels to which the organisms have been exposed, and environmental levels at the point of capture may not actually reflect the history of exposure. Therefore there is no known bioavailable fraction or constant environmental level that can serve as a realistic value on which we can base concentration factors. For these reasons we believe that the use of concentration factors in aquatic studies should be approached more cautiously in the future than in the past.

Biological Transport of Transuranics in the Ocean

Lowman, Rice, and Richards (1971) have discussed the relative importance of biological vs. physical processes in the distribution of radionuclides and trace metals in the ocean and have included an excellent bibliography on the subject. It has become increasingly apparent that the biological processes operative in the mixed layer of the ocean greatly influence the vertical transport of materials from it and that biological activity near the sediment—water interface may be a key mechanism for transport of recently deposited materials to depth in the sediment. Although we have chosen to address only the data relating to transuranics in this regard, the bulk of the evidence for vertical biological transport of materials from the mixed layer to depth has come from measurements of trace metals (Kuenzler, 1969; Fowler et al., 1973; Small and Fowler, 1973; Small, Fowler, and Keckes, 1973), natural radionuclides (Broecker, Kaufman, and Trier, 1973; Cherry et al., 1975; Beasley et al., 1977), artificial radionuclides (Osterberg, Carey, and Curl, 1963), and synthetic organic compounds (Elder and Fowler, 1977). Vertical mixing of sediments by biological activity has been well established (Davidson, 1891; Dapples, 1942; Emery, 1953; Gordon, 1966; Glass, 1969; Rhoads, 1974), and this process obviously will affect the vertical distribution of radionuclides associated with sediments.

Bowen, Wong, and Noshkin (1971) first demonstrated that plutonium subsurface maxima occur in the upper 1000 m of the ocean and attributed its removal from the mixed layer to biogenic particle fluxes. Subsequently Noshkin and Bowen (1973) proposed a model to explain both the vertical distributions observed in the water column and the small but measurable amounts of plutonium found in deep-sea sediments as a function of fallout delivery. This heuristic model invokes a mixed-particle population, 30% sinking at 392 m/yr, 40% sinking at 140 m/yr, and 30% sinking at 70 m/yr; no assumptions concerning the exact nature of the particulate matter involved are required.

Direct evidence for the association of plutonium and americium with particulate matter in the upper mixed layer of the oceans has been confirmed by several investigators. Livingston and Bowen (1976b) have found that as much as 70% of fallout plutonium can be removed by Millipore filtration of open North Atlantic surface seawater (presumably 0.45 μm); in coastal waters near Woods Hole, Mass., something over 90% of the plutonium is associated with the particulate phase. Evidence is accruing which would suggest that americium is also associated with particulate matter, although the number of analyses are fewer than those for plutonium. Silker (1974) has measured both the soluble and particulate plutonium in Pacific surface waters and has found that some 55 ± 7% of the plutonium is being retained by alumina; the remainder is fixed to glass-fiber prefilters with an effective pore size of 0.3 μm. More recently, Holm et al. (1977) measured the particulate plutonium and americium retained by 0.45-μm Millipore filters from open ocean surface Mediterranean waters (Tables 1 and 2). Substantially lower fractions of plutonium were found on the filters than was reported by Livingston and Bowen or by Silker. The average percent of plutonium retained for 14 samples (1700 to 7680 liters) was only 3.8 ± 0.2%. By contrast, the average percent of the ^{241}Am retained by the filters was 10 ± 1%.

That fractionation of ^{241}Am and plutonium isotopes is occurring in the water column is inferred from the profile data of Livingston and Bowen (1976a) by the frequency with which the ^{241}Am/^{239,240}Pu ratio in deep waters from the Atlantic exceeds those ratios observed either on land (Krey et al., 1976) or in coastal sediments (Livingston and Bowen, 1976a). This ratio at present is calculated to be 0.22, which is in good agreement with undisturbed soil measurements (0.22 to 0.25) and shallow coastal sediments (0.20) (Krey et al., 1976; Livingston and Bowen, 1976a). Deep-water samples from the North Atlantic, however, appear to have ^{241}Am/^{239,240}Pu ratios significantly different (higher) from this range of values. Profile data from the North Pacific are fewer in number but do not appear to evidence the same trend. Although the actual mechanism for this apparent fractionation is yet to be proven, it is conceivable, as suggested by Livingston, Bowen, and Burke (1976), that ^{241}Am is preferentially more associated with inorganic particles than are plutonium isotopes. Certainly the data from the Mediterranean make the hypothesis an attractive one. Fukai, Ballestra, and Holm (1976) have shown that ^{241}Am is depleted in Mediterranean surface waters relative to ^{239,240}Pu. The mean ratio was 0.055 ± 0.007 from nine stations throughout the Mediterranean (July to September 1975), a value significantly different from 0.22. From core samples taken from the Mediterranean by Livingston, Bowen, and Burke (1976), the surface-sediment ratios averaged 1.2 ± 0.4 and are equaled only by values from the northeast South American slope sediments off the Guiana coast (0.7 to 1.2). The Millipore filter data of Holm et al. (1977) shown in Tables 1 and 2 evidence a distinct enrichment of ^{241}Am/^{239,240}Pu; this ratio in unfiltered seawater is 0.055 ± 0.007, whereas the same ratio on 0.45-μm filters is 0.13 ± 0.05 (1σ level).

It is well known that the Mediterranean has a high proportion of terrigenous detritus in its waters and sediments (Emelyanov and Shimkus, 1972), and biological productivity is known to be low (Brouardel and Rink, 1956). These facts, coupled to the data already in hand from the Mediterranean basin, appear to us to be increasingly compelling evidence that, in fact, fractionation of ^{241}Am and plutonium isotopes can occur, depending on the nature of the particulate matter in the water column. Slightly damaging to the argument, however, is the fact that ^{241}Am/^{239,240}Pu ratios in open-ocean plankton appear to be very nearly those observed for the mixed-layer waters (Livingston and Bowen, 1976b). The same appears to be true for plankton samples taken from Lake Michigan (Wahlgren et al., 1976). One piece of critical information that is still missing, however, is the measurement of ^{241}Am/^{239,240}Pu ratios in zooplankton metabolic particulate products (molts and fecal pellets), which would be invaluable in helping to clarify this question.

That molts and fecal pellets from zooplankton can transport plutonium to depth has been clearly shown by the recent laboratory data of Higgo et al. (1977), in which ^{239,240}Pu measurements on the molts and fecal pellets of the euphausiid *Meganycti-*

TABLE 1   Plutonium and Americium in Mediterranean Surface Waters Retained by 0.45-μm Filters

Station No.	Position	Date of collection	Volume of seawater filtered, liters	$^{239+240}$Pu		^{238}Pu		^{241}Am	
				aCi/liter*	Percent on filter	aCi/liter*	Percent on filter	aCi/liter*	Percent on filter
C	43°11'N, 06°32'E	Sept. 14, 1975	4540	43 ± 4	4.5	1.5 ± 0.3	1.5	4.2 ± 0.8	12
1	42°30'N, 06°30'E	Sept. 14, 1975	3790	44 ± 5	4.3†	1.8 ± 0.3	2.6†	4.8 ± 0.8	8.4†
2	41°30'N, 06°30'E	Sept. 15, 1975	4540	27 ± 2	2.6†	1.3 ± 0.2	1.9†	4.8 ± 0.9	8.4†
3	40°30'N, 06°30'E	Sept. 16, 1975	7000	37 ± 3	3.6†	1.6 ± 0.2	2.4†	3.4 ± 1.0	6.0†
7	37°30'N, 11°00'E	Sept. 20, 1975	3440	39 ± 3	3.5	1.4 ± 0.3	2.5	5.7 ± 0.8	10†
8	38°40'N, 12°00'E	Sept. 20, 1975	2350	35 ± 2	3.4	1.4 ± 0.3	2.8	7.2 ± 0.8	14
9	40°40'N, 11°40'E	Sept. 20, 1975	7680	41 ± 3	3.8	1.7 ± 0.3	3.4	3.4 ± 0.7	4.9
13	42°47'N, 09°25'E	Sept. 21, 1975	4150	40 ± 5	3.6	1.9 ± 0.5	3.2	4.4 ± 1.1	15
14	43°55'N, 09°00'E	Sept. 22, 1975	1700	53 ± 5	4.8	2.0 ± 0.7	2.0		
			Average‡	40 ± 2	3.8 ± 0.2	1.6 ± 0.1	2.5 ± 0.2	4.7 ± 0.4	10 ± 1

*In terms of attocuries (10^{-18} Ci) per liter with 1σ propagated errors.

†Values for unfiltered seawater were not available. Calculated on the basis of average concentrations in unfiltered water ($^{239+240}$Pu, 1.03 ± 0.05 fCi/liter; ^{238}Pu, 0.068 ± 0.007 fCi/liter; ^{241}Am, 0.057 ± 0.007 fCi/liter).

‡Errors are given in standard errors.

TABLE 2 ^{238}Pu/$^{239+240}$Pu and ^{241}Am/$^{239+240}$Pu Activity
Ratios in Mediterranean Surface Waters

Station No.	(^{238}Pu/$^{239+240}$Pu) × 100		(^{241}Am/$^{239+240}$Pu) × 100	
	In water	In particulate	In water	In particulate
C	11 ± 1	3.5 ± 0.8	3 ± 1	10 ± 2
1		4.1 ± 0.8		11 ± 2
2		4.8 ± 0.8		18 ± 4
3		4.3 ± 0.6		9 ± 3
7	5 ± 1	3.6 ± 0.8		15 ± 2
8	5 ± 1	4.8 ± 0.8	4 ± 2	21 ± 3
9	5 ± 1	4.1 ± 0.8	6 ± 2	8 ± 2
13	6 ± 2	4.8 ± 1.4	3 ± 1	11 ± 3
14	9 ± 2	3.8 ± 1.4	4 ± 1	
Average*	7 ± 3	4.2 ± 0.5	4 ± 1	13 ± 5

*Errors are given in standard deviations.

phanes norvegica were combined with soluble plutonium excretion experiments to arrive at the flux rates of plutonium through the organism. Essentially 99% of the plutonium taken up by *M. norvegica* is excreted by fecal pellets. A crude estimate of the removal time of plutonium from the mixed layer by fecal-pellet production alone is 3.6 yr. Although the uncertainties of such an estimate were willingly admitted by the authors, it is interesting to note the approximate order-of-magnitude agreement between this estimate and that of Hodge, Folsom, and Young (1973) and Folsom (1975), which ranged from half removal times of 3.5 yr to complete removal times of 1 yr from the surface layers of the ocean.

Although evidence is accumulating on vertical transport mechanisms for the transuranics as a result of planktonic biological processes (at a painfully slow rate!), we are unaware of any published data on the redistribution of these elements either horizontally or vertically by other marine organisms except reports that attribute the redistribution of plutonium and americium from surface to deeper sediments to bioturbation (Livingston and Bowen, 1976a; Livingston, Bowen, and Burke, 1976).

Conclusions

The amount of information now in hand concerning the biokinetic behavior of transuranic radionuclides in marine organisms is astonishingly small and, as discussed earlier, is plagued with difficulties inherent with experiments of this type. Even so there is now evidence that certain marine organisms exhibit relatively high assimilation efficiencies for the transuranics which are quite unlike those seen for terrestrial vertebrates. There is increasing evidence to suggest that, despite similar aqueous chemistries, fractionations among transuranics in the ocean can occur and are mediated by the nature of the particulate matter to which they are adsorbed. Removal rates to deep water and sediments would thus be quite different through the world oceans. A single attempt at numerically estimating the importance of zooplankton metabolic products as a transport mechanism of plutonium to depth has shown that such particulates are likely a key mode of removal. Yet this evidence comes from investigations using only one species of a rather large macroplankton whose particulate products (molts, fecal pellets, and carcasses) would be expected to reach depths at fairly rapid rates. The microzooplankton, which constitutes by far the larger portion of the biomass at trophic level II (herbivores), most certainly contributes to this process, yet the basic biological data concerning fecal-pellet and molt production rates under varying food conditions comparable to data obtained for selected macrozooplankton species have yet to be formulated in a way that could be used to refine vertical transport estimates for the transuranics. Finally, the importance of the food vs. water pathway for uptake of transuranics by marine biota has been established for only a very few species, and even fewer definitive experiments have investigated direct uptake from labeled sediment. Until substantial progress is made in each of these areas, our understanding of the behavior of transuranics in the biotic component of the marine environment and its attendant influence on the movement of these elements in the ocean, including transport back to man, will continue to be inadequate.

Acknowledgments

We have attempted to select information that would give the reader a general overview of the field and the directions that are being followed by researchers in the discipline. There are likely omissions of some of the work of our colleagues, and for this we apologize. The interested reader is urged to consult the articles cited to gain further insight into the complexities of the subject as well as a full appreciation of the arduous tasks ahead. Support for this work has been received from the U. S. Department of Energy under contract Ey-76-5-06-2227, Task Agreement No. 30, and from a cooperative agreement between the U. S. Department of Energy and the National Marine Fisheries Service.

References

Aarkrog, A., 1971, Radioecological Investigations of Plutonium in an Arctic Marine Environment, *Health Phys.,* 20: 31-47.

——, 1977, Environmental Behavior of Plutonium Accidentally Released at Thule, Greenland, *Health Phys.,* 32: 271-284.

Bair, W. I., E. H. Willard, I. C. Nelson, and A. C. Case, 1974, Comparative Distribution and Excretion of ^{237}Pu and ^{239}Pu Nitrates in Beagle Dogs, *Health Phys.,* 27: 392-396.

Beasley, T. M., and S. W. Fowler, 1976a, Plutonium Isotope Ratios in Polychaete Worms, *Nature,* 262: 813-814.

——, and S. W. Fowler, 1976b, Plutonium and Americium: Uptake from Contaminated Sediments by the Polychaete *Nereis diversicolor, Mar. Biol.,* 38: 95-100.

——, M. Heyraud, J. J. W. Higgo, R. D. Cherry, and S. W. Fowler, 1977, ^{210}Po and ^{210}Pb in Zooplankton Fecal Pellets, *Mar. Biol.,* 44: 325-328.

Bowen, V. T., 1975, *Transuranic Elements in Marine Environments,* ERDA Report HASL-291, Health and Safety Laboratory, NTIS.

——, and H. D. Livingston, 1975, Americium-242*m* in Nuclear Test Debris, *Nature,* 256: 482.

——, and T. T. Sugihara, 1965, Oceanographic Implications of Radioactive Fallout Distributions in the Atlantic Ocean: From 20°N to 25°S from 1957 to 1961, *J. Mar. Res.,* 23: 123-146.

——, K. M. Wong, and V. E. Noshkin, 1971, Plutonium-239 in and over the Atlantic Ocean, *J. Mar. Res.,* 29: 1-10.

Broecker, W. S., A. Kaufman, and R. M. Trier, 1973, The Residence Time of Thorium in Surface Seawater and Its Implications Regarding the Fate of Reactive Pollutants, *Earth Planet. Sci. Lett.,* 20: 35-44.

Brouardel, J., and E. Rink, 1956, Détermination de la Production de Matiére Organique en Méditerranée a l'aide du ^{14}C, *C. R. Acad. Sci.,* 243: 1797-1800.

Cherry, R. D., 1964, Alpha-Radioactivity in Plankton, *Nature,* 203: 139-143.

——, and L. V. Shannon, 1974, The Alpha Radioactivity of Marine Organisms, *At. Energy Rev.,* 12: 3-45.

——, S. W. Fowler, T. M. Beasley, and M. Heyraud, 1975, Polonium 210: Its Vertical Oceanic Transport by Zooplankton Metabolic Activity, *Mar. Chem.,* 3: 105-110.

Cross, F. A., W. C. Renfro, and E. Gilat, 1975, A Review of Methodology for Studying the Transfer of Radionuclides in Marine Food Chain, in *Design of Radiotracer Experiments in Marine Biological Systems,* pp. 185-210, STI/DOC/10/167, International Atomic Energy Agency, Vienna.

——, and W. A. Sunda, 1979, Bioavailability of Metals in Estuaries, *Estuarine Res. Fed.* (in press).

——, J. N. Willis, and J. P. Baptist, 1971, Distribution of Radioactive and Stable Zinc in an Experimental Marine Ecosystem, *J. Fish. Res. Board Can.,* 28: 1783-1788.

——, J. N. Willis, L. H. Hardy, N. Y. Jones, and J. M. Lewis, 1975, Role of Juvenile Fish in Cycling of Mn, Fe, Cu and Zn in a Coastal-Plain Estuary, *Estuarine Res.,* 1: 45-65.

Dapples, E. C., 1942, The Effect of Macro-Organisms upon Nearshore Marine Sediments, *J. Sediment. Petrol.,* 12: 118-126.

Davidson, C., 1891, On the Amount of Sand Brought up by Lobworms to the Surface, *Geol. Mag.,* 8: 489-493.

Elder, D. L., and S. W. Fowler, 1977, Polychlorinated Biphenyls: Penetration into the Deep Ocean by Zooplankton Fecal Pellet Transport, *Science,* 197: 459-461.

Emelyanov, E. M., and K. M. Shimkus, 1972, Suspended Matter in the Mediterranean Sea, in *The Mediterranean Sea: A Natural Sedimentation Laboratory,* Symposium Proceedings, Heidelberg, Germany, Aug. 30–Sept. 1, 1971, D. J. Stanley (Ed.), pp. 417-439, Dowden, Hutchinson and Ross, Inc., Stroudsburg, Pa.

Emery, K. O., 1953, Some Surface Features of Marine Sediments made by Animals, *J. Sediment. Petrol.,* 23: 202-204.

Emery, R. M., and T. R. Garland, 1974, *Ecological Behavior of Plutonium and Americium in a Freshwater Ecosystem,* ERDA Report BNWL-1879, pp. 1-29, Battelle Pacific Northwest Laboratories, NTIS.

——, D. C. Klopfer, and W. C. Weimer, 1974, *Ecological Behavior of Plutonium and Americium in a Freshwater Ecosystem,* ERDA Report BNWL-1867, pp. 1-82, Battelle Pacific Northwest Laboratories, NTIS.

Folsom, T. R., 1975, Comments During Discussions of Papers, in Proceedings of the Second Los Alamos Life Sciences Symposium, *Health Phys.,* 29: 597-611.

——, and V. F. Hodge, 1975, Experiments Suggesting some First Steps in the Dispersal and Disposal of Plutonium and Other Alpha Emitters in the Ocean, *Marine Sci. Commun.,* 1: 213-247.

——, V. F. Hodge, and M. E. Gurney, 1975, Plutonium Observed on Algal Surfaces in the Ocean, *Marine Sci. Commun.,* 1: 39-49.

Fowler, S. W., and T. M. Beasley, 1977, Plutonium and Americium in Fish, *Nature,* 265: 384.

——, and J. C. Guary, 1977a, High Absorption Efficiency for Ingested Plutonium in Crabs, *Nature,* 266: 827-828.

——, and J. C. Guary, 1977b, International Laboratory for Marine Radioactivity, Musée Océanographique, Principality of Monaco, personal communication.

——, and M. Heyraud, 1974, Accumulation and Retention of ^{241}Am in Marine Plankton, in *Activities of the International Laboratory of Marine Radioactivity, 1974 Report,* Technical Report, IAEA-163, pp. 28-32, International Atomic Energy Agency, Vienna.

——, M. Heyraud, and T. M. Beasley, 1975, Experimental Studies on Plutonium Kinetics in Marine Biota, in *Impacts of Nuclear Releases into the Aquatic Environment,* Symposium Proceedings, Otaniemi, Finland, 1975, pp. 157-177, STI/PUB/406, International Atomic Energy Agency, Vienna.

——, M. Heyraud, L. F. Small, and G. Benayoun, 1973, Flux of ^{141}Ce Through a Euphausiid Crustacean, *Marine Biol.,* 21: 317-325.

Fraizier, A., and J. C. Guary, 1976, Recherche d'Indicateurs Biologiques Appropriés au Contrôle de la Contamination du Littoral par le Plutonium, in *Transuranium Nuclides in the Environment,* Symposium Proceedings, San Francisco, 1975, pp. 679-689, STI/PUB/410, International Atomic Energy Agency, Vienna.

——, and J. C. Guary, 1977, *Diffusion du Plutonium en Milieu Marin: Etude Quantitative Effectuee sur des Especes Marines du Littoral de la Manche de Brest (Pointe St. Mathieu) a Honfleur,* Report

CEA-R-4822, Centre de la Hague, Laboratoire de Radioécologie Marine, Service de Documentation, Gif-sur-Yvette, France.

Fukai, R., S. Ballestra, and E. Holm, 1976, [241]Am in Mediterranean Surface Waters, *Nature,* 264: 739-740.

Glass, B. P., 1969, Reworking of Deep-Sea Sediments as Indicated by the Vertical Dispersion of the Australasian and Ivory Coast Microtektite Horizons, *Earth Planet. Sci. Lett.,* 6: 409-415.

Gordon, D. C., Jr., 1966, The Effects of the Deposit Feeding Polychaete, *Pectinaria gouldii,* on the Intertidal Sediments of Barnstable Harbor, *Limnol. Oceanogr.,* 11: 327-332.

Gromov, V. V., 1976, Uptake of Plutonium and Other Nuclear Waste by Plankton, *Mar. Sci. Commun.,* 2: 227-247.

Guary, J. C., and S. W. Fowler, 1977, Biokinetics of Neptunium-237 in Mussels and Shrimp, *Mar. Sci. Commun.,* 3: 211-229.

——, and A. Fraizier, 1977, Influence of Trophic Level and Calcification on the Uptake of Plutonium Observed, In Situ, in Marine Organisms, *Health Phys.,* 32: 21-28.

——, M. Masson, and A. Fraizier, 1976, Etude Preliminarie, In Situ de la Distribution du Plutonium dans Different Tissus et Organes de *Cancer pagurus* (Crustacea: Decapoda) et de *Pleuronectes platessa* (Pisces: Pleuronectidae), *Mar. Biol.,* 36: 13-17.

Hakonson, T. E., and L. J. Johnson, 1973, *Distribution of Environmental Plutonium in the Trinity Site Ecosystem After 27 Years,* AEC Report LA-UR-73-1291, Los Alamos Scientific Laboratory, NTIS.

Hetherington, J. A., D. F. Jefferies, and M. B. Lovett, 1975, Some Investigations into the Behavior of Plutonium in the Marine Environment, in *Impacts of Nuclear Releases into the Aquatic Environment,* Symposium Proceedings, Otaniemi, Finland, 1975, pp. 193-212, STI/PUB/406, International Atomic Energy Agency, Vienna.

——, D. F. Jefferies, N. T. Mitchell, R. J. Pentreath, and D. J. Woodhead, 1976, Environmental and Public Health Consequences of the Controlled Disposal of Transuranic Elements to the Marine Environment, in *Transuranium Nuclides in the Environment,* Symposium Proceedings, San Francisco, 1975, pp. 139-154, STI/PUB/410, International Atomic Energy Agency, Vienna.

Higgo, J. J. W., R. D. Cherry, M. Heyraud, and S. W. Fowler, 1977, Rapid Removal of Plutonium from the Ocean Surface Layer by Zooplankton Fecal Pellets, *Nature,* 266: 623-624.

Hodge, V. F., T. R. Folsom, and D. R. Young, 1973, Retention of Fall-Out Constituents in Upper Layers of the Pacific Ocean as Estimated from Studies of a Tuna Population, in *Radioactive Contamination of the Marine Environment,* Symposium Proceedings, Seattle, 1972, pp. 263-276, STI/PUB/313, International Atomic Energy Agency, Vienna.

——, F. L. Hoffman, and T. R. Folsom, 1974, Rapid Accumulation of Plutonium and Polonium on Giant Brown Algae, *Health Phys.,* 27: 29-35.

Holm, E., S. Ballestra, R. Fukai, and T. M. Beasley, 1977, *Plutonium Isotopes in the Environment: A Radioecological Study,* Ph.D. Thesis, University of Lund, Lund, Sweden.

Krey, P. W., E. P. Hardy, C. Pachucki, F. Rourke, J. Coluzza, and W. K. Benson, 1976, Mass Isotopic Composition of Global Fall-Out Plutonium in Soil, in *Transuranium Nuclides in the Environment,* Symposium Proceedings, San Francisco, 1976, pp. 671-678, STI/PUB/410, International Atomic Energy Agency, Vienna.

Kuenzler, E. J., 1969, Elimination of Iodine, Cobalt, Iron and Zinc by Marine Zooplankton, in *Symposium on Radioecology,* Proceedings of the Second National Symposium, Ann Arbor, Mich., May 15–17, 1967, D. J. Nelson and F. C. Evans (Eds.), USAEC Report CONF-670503, pp. 462-473, NTIS.

Livingston, H. D., and V. T. Bowen, 1976a, Americium in the Marine Environment: Relationships to Plutonium, in *Environmental Toxicity of Aquatic Radionuclides: Model and Mechanisms,* Conference Proceedings, Rochester, N. Y., June 2–4, 1975, pp. 107-130, Ann Arbor Science Publishers, Ann Arbor, Mich.

——, and V. T. Bowen, 1976b, *Contrasts Between the Marine and Freshwater Biological Interactions of Plutonium and Americium,* ERDA Report HASL-315, Health and Safety Laboratory, NTIS.

——, V. T. Bowen, and J. C. Burke, 1976, Fallout Radionuclides in Mediterranean Sediments, in *XXVth Congress and Plenary Assembly of ICSEM,* Conference Proceedings, Split, Yugoslavia, Oct. 25–30, 1976 (to be published).

Lowman, F. G., T. R. Rice, and F. A. Richards, 1971, Accumulation and Redistribution of Radionuclides by Marine Organisms, in *Radioactivity in the Marine Environment*, pp. 161-199, National Academy of Sciences, Washington, D. C.

Noshkin, V. E., 1972, Ecological Aspects of Plutonium Dissemination in Aquatic Environments, *Health Phys.*, 22: 537-549.

——, and V. T. Bowen, 1973, Concentrations and Distributions of Long-Lived Fallout Radionuclides in Open Ocean Sediments, in *Radioactive Contamination of the Marine Environment*, Symposium Proceedings, Seattle, 1972, pp. 671-686, STI/PUB/313, International Atomic Energy Agency, Vienna.

——, V. T. Bowen, K. M. Wong, and J. C. Burke, 1971, Plutonium in North Atlantic Ocean Organisms: Ecological Relationships, in *Radionuclides in Ecosystems*, Proceedings of the Third National Symposium on Radioecology, Oak Ridge, Tenn., May 10–12, 1971, D. J. Nelson (Ed.), USAEC Report CONF-710501, pp. 691-699, Oak Ridge National Laboratory, NTIS.

Osterberg, C. A., A. J. Carey, Jr., and H. Curl, Jr., 1963, Acceleration of Sinking Rates of Radionuclides in the Ocean, *Nature*, 200: 1276-1277.

Pentreath, R. J., 1977, Fisheries Radiobiological Laboratory, Lowestoft, United Kingdom, private communication.

——, and M. B. Lovett, 1976, Occurrence of Plutonium and Americium in Plaice from the North-Eastern Irish Sea, *Nature*, 262: 814-816.

Pillai, K. C., and E. Mathew, 1976, Plutonium in the Aquatic Environment: Its Behavior, Distribution and Significance, in *Transuranium Nuclides in the Environment*, Symposium Proceedings, San Francisco, 1975, pp. 25-45, STI/PUB/410, International Atomic Energy Agency, Vienna.

Rhoads, D. C., 1974, Organism-Sediment Relations on the Muddy Sea Floor, in *Oceanography and Marine Biology: Annual Review*, H. Barnes (Ed.), Vol. 12, pp. 263-300, Hafner Press, New York.

Silker, W. B., 1974, *Plutonium Concentrations in the Pacific Ocean*, ERDA Report BNWL-1950, Battelle Pacific Northwest Laboratories, NTIS.

Small, L. F., and S. W. Fowler, 1973, Turnover and Vertical Transport of Zinc by the Euphausiid, *Meganyctiphanes norvegica*, in the Ligurian Sea, *Mar. Biol.*, 18: 284-290.

——, S. W. Fowler, and S. Keckes, 1973, Flux of Zinc Through a Macroplankton Crustacean, in *Radioactive Contamination of the Marine Environment*, Symposium Proceedings, Seattle, 1972, pp. 437-452, STI/PUB/313, International Atomic Energy Agency, Vienna.

Sugihara, T. T., and V. T. Bowen, 1962, Radioactive Rare Earths from Fallout for Study of Particle Movement in the Sea, in *Radioisotopes in the Physical Sciences and Industry*, Symposium Proceedings, Copenhagen, Denmark, 1960, pp. 57-65, STI/PUB/20, International Atomic Energy Agency, Vienna.

Thompson, R. C., 1967, Biological Factors, in *Plutonium Handbook*, Vol. II, pp. 785-829, Gordon and Breach Science Publishers, New York.

Todd, R., 1968, Plutonium-237 for Metabolic Studies: Its Preparation and Use, *Int. J. Appl. Radiat. Isot.*, 19: 141-145.

——, and R. Logan, 1966, Plutonium-237 Labelling: A New Technique for Plutonium Metabolic Studies in Animals, *Int. J. Appl. Radiat. Isot.*, 17: 253-255.

Volchok, H. L., et al., 1975, Transuranic Elements, in *Assessing Potential Ocean Pollutants*, pp. 27-63, National Academy of Sciences, Washington, D. C.

Wahlgren, M. A., J. J. Alberts, D. M. Nelson, and K. A. Orlandini, 1976, Study of the Behavior of Transuranics and Possible Chemical Homologues in Lake Michigan Water and Biota, in *Transuranium Nuclides in the Environment*, Symposium Proceedings, San Francisco, 1975, pp. 9-24, STI/PUB/410, International Atomic Energy Agency, Vienna.

Ward, E. E., 1966, Uptake of Plutonium by the Lobster, *Homarus vulgaris*, *Nature*, 209: 625-626.

Willis, J. N., and N. Y. Jones, 1977, The Use of Uniform Labelling with Zinc-65 to Measure Stable Zinc Turnover in the Mosquito Fish, *Gambusia affinis*. I. Retention, *Health Phys.*, 32: 381-387.

Wong, K. M., V. F. Hodge, and T. R. Folsom, 1972, Plutonium and Polonium Inside Giant Brown Algae, *Nature*, 237: 460-462.

Zlobin, V. S., 1966, Accumulation of Uranium and Plutonium by Marine Algae [in Russian], *Radiobiologiya*, 6(4): 613-617; English translation, pp. 219-227.

——, 1971, Active Phase Assimilation of ^{239}Pu by the Marine Algae *Ascophyllum nodosum*, USAEC Report AEC-tr-7418, pp. 207-217, translated from *Tr., Polyarn. Nauchno.-Issled. Proektn. Inst. Morsk. Rybn. Khoz. Okeanogr.*, 29: 169-175.

——, and O. V. Mokanu, 1970, Mechanisms of Accumulation of ^{239}Pu and ^{210}Po by the Brown Algae *Ascophyllum nodosum* and Marine Phytoplankton, USAEC Report AEC-tr-7205, pp. 160-169, translated from *Radiobiologiya*, 10: 584-589.

——, and M. F. Perlyuk, 1971, Photosynthesis and the Mechanism of the Action of Cyanide on Cell Respiration and ^{239}Pu Accumulation by Marine Algae, USAEC Report AEC-tr-7418, pp. 195-206, translated from *Tr., Polyarn. Nauchno.-Issled. Proektn. Inst. Morsk. Rybn. Khoz. Okeanogr.*, 29: 159-168.

Geochemistry of Transuranic Elements at Bikini Atoll

W. R. SCHELL, F. G. LOWMAN, and R. P. MARSHALL

The distribution of transuranic and other radionuclides in the marine environment at Bikini Atoll was studied to better understand the geochemical cycling of radionuclides produced by nuclear testing between 1946 and 1958. The reef areas, which are washed continually by clean ocean water, have low levels of radionuclide concentrations. Radionuclides are contained in fallout particles of pulverized coral. In the water these particles may dissolve, be transported by currents within the Atoll, or enter the North Equatorial Current by tidal exchange of water in the lagoon. The transuranic elements are distributed widely in sediments over the northwest quadrant of the Atoll, which suggests that this area serves as a settling basin for particles. The distribution of plutonium in the water column indicates that plutonium in the sediments is released to the bottom waters and then is transported and diluted by the prevailing currents. Upon interaction with the lagoon environment, plutonium occurs in several physicochemical states. Laboratory tests and field studies at Bikini show that approximately 15% of the plutonium is associated with the colloidal fraction. Different $^{238}Pu/^{239,240}Pu$ ratios found in sediments, suspended particulates, and soluble fractions suggest that ^{238}Pu may be more "soluble" than $^{239,240}Pu$. Different isotope ratios for the physicochemical states of plutonium radionuclides may be due to differences in decay rates and/or the mode of formation.

Bikini Atoll was one of the sites used for nuclear weapons testing between 1946 and 1958. In the 19 yr since cessation of testing, physical decay and environmental processes have removed or reduced significantly many of the radionuclides that resulted. However, several fission and neutron-induced radionuclides, such as ^{90}Sr, ^{137}Cs, ^{60}Co, ^{55}Fe, ^{155}Eu, and ^{207}Bi, which have half-lives of 2 to 30 yr, can still be measured easily in sediments, soils, and some biota. In addition, unburned fissile and device materials of uranium and plutonium, as well as many of the neutron-induced transuranium radionuclides, such as americium and neptunium, which have half-lives of 10^2 to 10^9 yr, still remain in the Atoll ecosystem. These transuranic elements generally decay by alpha-particle emission, and their measurement requires detailed chemical analysis of samples.

In 1946 the Marshallese living on Bikini were evacuated from the Atoll during the U. S. nuclear testing program. Today at Bikini Atoll a potential health hazard from these long-lived radionuclides may still exist to the returning Marshallese people. The potential release of transuranic elements to coastal marine environments other than Bikini is indicated by the projected increase in the global use of plutonium in power reactors by more than 10^3 times between 1971 and 2000 (Shapley, 1971). This exposure to the

population from plutonium will come from power-reactor waste products and/or accidents.

In the marine environment the distribution of transuranium elements occurs through biogeochemical processes, which effect a transfer of materials between the sediments, waters, and biota of the ecosystem. Surprisingly few data have been published on the redistribution of radioactivity between the sediments and waters of the lagoon in the contaminated environment of Bikini Atoll. This has been partly due to the secondary importance placed on sediment studies compared with studies of the uptake of radioactivity by flora and fauna and with studies of the radiation exposure to inhabitants of the Atolls. The sediment environment at Bikini has been disturbed significantly at the sites of the 23 nuclear detonations, and a significant time period may have been required to achieve quasi-steady-state concentrations of sediments and radionuclides. Recent studies on the concentrations of long-lived radionuclides remaining in the lagoon environment have indicated that nearly steady-state processes may now exist (Noshkin et al., 1974). Although the problems of data interpretation presented by the complex sources of the introduction of radioactivity into the lagoon still remain, the present situation at Bikini offers unique opportunities and advantages for the study of the physical and biogeochemical processes, which have governed and will continue to govern the fate of radionuclides in this marine ecosystem. The data and interpretations at Bikini will help in assessing the general hazards of plutonium in the marine environment. The objective of this chapter, then, is to review the current status of studies on the processes and mechanisms that control the distribution of transuranic elements in the Bikini lagoon ecosystem.

Sources in the Bikini Ecosystem

The formation of transuranic elements at Bikini resulted from the detonation of fission and fusion devices of different sizes using different fissile materials (^{235}U, ^{239}Pu, and ^{238}U). The largest test was the Bravo event of the Castle series (1954)—15 Mt equivalent TNT. This device consisted of the fission–fusion–fission process [^{235}U (^{239}Pu)–LiD(T)–^{238}U]. The transuranic elements now present in the lagoon environment are from unburned fissile material, energetic particle-induced activation products, and decay products which were incorporated in or on coral material. Recently, information has been obtained that ^{242}Cm was used as a fallout tracer of the transuranic elements in several nuclear devices. Significant amounts of ^{238}Pu would now be present in debris from the alpha-particle decay of ^{242}Cm ($t_{1/2}$, 162 days). The formation of the coral fallout particles resulted from interaction of vaporized device and soil materials in the fireball with the environmental materials that were swept into the expanding fireball and cloud at later times (Adams, Farlow, and Schell, 1960).

The locations of the 23 detonations reported at Bikini are shown in Fig. 1, and the detonation parameters are given in Table 1. The yields of the largest detonations reported were: Bravo, 15 Mt in 1954 at location B; Zuni, 3.53 Mt in 1956 at location C; and Tewa, 5.01 Mt in 1956 at location G. There was also a "several megaton" airburst detonation in 1956 which probably resulted in relatively minor contamination of lagoon sediments. Typically, two types of sites were used for testing nuclear devices at Bikini, and each probably gave rise to fallout particles of distinctly different compositions and structures.

The first was for devices exploded over water deep enough to prevent the incorporation of large quantities of soil in the ensuing fireball and cloud (sites A, F, D,

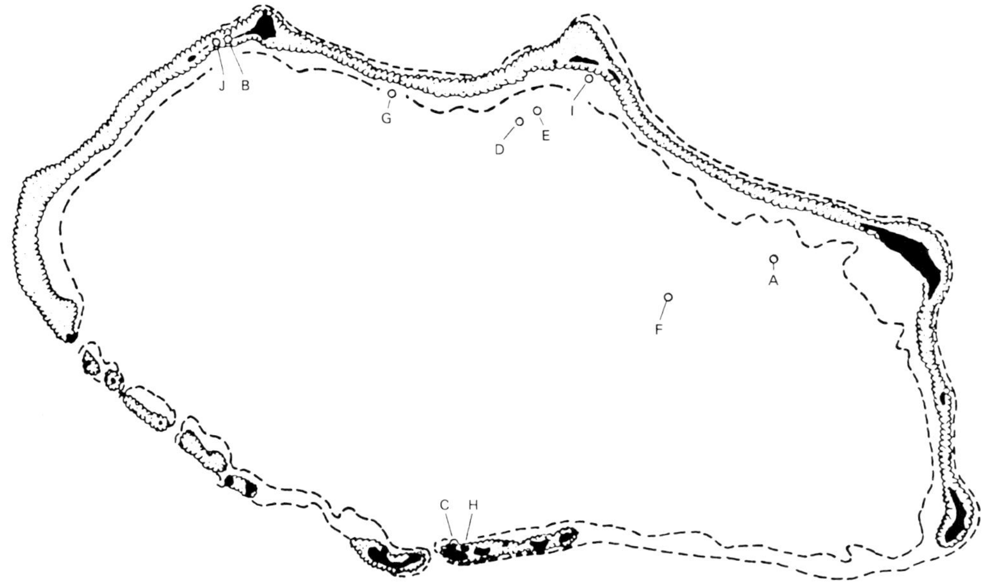

Fig. 1   Approximate locations of nuclear tests at Bikini Atoll.

and E in Fig. 1). Devices detonated on barges at Bikini under these conditions contained large quantities of iron and coral which were used as barge ballast (Adams, Farlow, and Schell, 1960). Spherical particles ($<1\ \mu$m) of "dicalcium ferrite" ($2\ CaO \times Fe_2O_3$) formed from vaporization of the barge and ballast contained about 85% of the radioactivity in the fallout droplets (Schell, 1959); the saturated sodium chloride (sea salt) droplets, in which these insoluble solids were suspended, contained the remaining 15% of the measured radioactivity (Farlow and Schell, 1957).

The second common site was the shallow water or island environments where the largest tests were conducted (sites B, J, G, I, C, and H in Fig. 1). From explosions of this type, Adams, Farlow, and Schell (1960) found that condensation of the vaporized materials typically occurred as impurities into and on the surfaces of the coral soils swept into the fireball, which produced two distinct types of fallout particles, spherical and angular. The spherical particles consisted of CaO, which was partially hydrated to $Ca(OH)_2$. A surface coating of $Ca(OH)_2$ and/or $CaCO_3$ was present owing to the reaction of the particles with water vapor and atmospheric CO_2 during the fallout. These particles were formed by high-temperature ($>2570°$C) vaporization of coral with subsequent condensation of the oxide as spherical particles, which lost their normal porosity. The radioactivity was almost uniformly distributed throughout the particles. The angular particles consisted of $Ca(OH)_2$ with a thin outer coating of $CaCO_3$. Some of these particles contained unmelted coralline sand fragments as the central core; the bulk of the radioactivity was in the outer carbonate shell. The angular shape of these particles, the lack of incorporated radioactivity, and the presence of occasional unmodified sand grains suggested that these particles were formed from nonvolatilized coral that had been heated enough to melt and decarbonate (800 to 900°C) while incorporating only an outer surface of condensing radionuclides. Occasionally 10-μm and smaller oxide spherical

TABLE 1 **Announced Nuclear Detonations at Bikini Atoll**

Operation and event	Date	Height, ft	Location	Yield	Map ref. (Fig. 1)
Crossroads					
Able	6/30/1946	+520	Air	Nominal	A
Baker	7/24/1946	−90	Water	Nominal	A
Castle					
Bravo	2/28/1954		Surface	15 Mt	B
Romeo	3/26/1954		Barge		B
Koon	4/6/1954		Surface	110 kt	C
Union	4/25/1954		Barge		D
Yankee	5/4/1954		Barge		D
Redwing					
Cherokee	5/20/1956	4320	Air	Several megatons	E
Zuni	5/27/1956		Surface	3.53 Mt	C
Flathead	6/11/1956		Barge		F
Dakota	6/25/1956		Barge		F
Navajo	7/10/1956		Barge		D
Tewa	7/20/1956		Barge	5.01 Mt	G
Hardtack Phase I					
Fir	5/11/1958		Barge		B
Nutmeg	5/21/1958		Barge		H
Sycamore	5/31/1958		Barge		B
Maple	6/10/1958		Barge		I
Aspen	6/11/1958		Barge		B
Redwood	6/27/1958		Barge		I
Hickory	6/29/1958		Barge		H
Cedar	7/2/1958		Barge		B
Poplar	7/12/1958		Barge		J
Juniper	7/22/1958		Barge		H

particles were observed adhering to these particles. Because of the dense property of the $CaO/Ca(OH)_2$ particles, their atmospheric hydration was dependent on the aqueous environment encountered during fallout and sedimentation. Complete hydration, which was observed in laboratory tests over several weeks' time, was found to be accompanied by a 100% increase in particle volume and in the development of a crumbly, fluffy structure. The $CaO/Ca(OH)_2$ particles began to dissolve slowly when wet with seawater. The freed calcium ions reacted with sulfate ions in the seawater to form calcium sulfate-dihydrate (gypsum), whereas the hydroxyl ions reacted to form insoluble $Mg(OH)_2$. A hard shell of $Mg(OH)_2$ formed around the particle, which, during the period of observation, apparently stopped any further reaction with seawater; a region of $Ca(OH)_2$ remained on the inner surfaces of the spherical particles. The remaining radioactivity was associated with the $Ca(OH)_2$ in the center of the sphere. Some of the freed calcium ions in the spheres also formed $CaCO_3$ by reaction with bicarbonate ions in seawater. The time history of the distribution and redistribution of transuranic elements has been intimately associated with these particles and with their redistribution by the physical circulation system of the lagoon.

Distribution in Surface Sediments

The size of the detonation craters and the extent of the impact on the reef ecosystem are shown in Figs. 2 and 3. The Bravo crater, a dish in the reef, is approximately 550 m in diameter and about 40 m deep. The several large craters in the reef are evident from the photographs. Finely divided coral particles, which resulted from the detonations, appear on the reef flat near Bravo crater and on the lagoon terrace extending south toward the islands of Bokdrolulu and Bokaetoktok. The sampling stations for the biogeochemical survey trip* in 1972 are shown in Fig. 4.

A thin-source survey method for alpha radioactivity was developed to initially scan the surface sediments collected in 1972 (Marshall, 1975). The results of this rapid total alpha analysis are shown in Fig. 5. The highest total alpha radioactivity is shown not to be in the bomb craters but to be distributed widely over the northwestern quadrant of the lagoon. Thus the principal source of transuranic elements to the water is a large area in the lagoon; the maximum concentrations are near the Namu Island—Bravo crater area. The plutonium and americium concentrations were determined in the surface sediments, and the results of these analyses are shown in Figs. 6 and 7, respectively, and in Table 2 (Marshall, 1975; Nevissi and Schell, 1975; Schell and Watters, 1975). The general distribution pattern of plutonium and americium in the isopleths of Figs. 6 and 7 is the same as that shown previously for the results obtained by the total alpha method of analysis of sediments, which indicates that most of the alpha radioactivity is derived from plutonium and americium.

Many of the sediment samples collected for analysis in the study by Marshall (1975) consisted predominantly of coralline particles, which were much smaller in size than natural Marshall Island Atoll sediments, as described by Emery, Tracey, and Ladd (1954) and Anikouchine (1961). The sediments were probably pulverized by the detonations and were distributed in the lagoon; the finely divided particles contained the highest concentrations of radioactivity. The proportion of the finely divided material ($<16\ \mu$m) in each sample was estimated visually.

Surface sediments collected from stations C-1, C-3, C-4 (Bravo crater), B-2, and B-20 (lagoon) consisted entirely of fine-grain material. Surface sediments collected from stations C-7, C-8, B-21, and B-30 contained 45 to 95% fine-grain material. Sediments collected from stations B-18 and B-19 contained approximately 20 to 40% pulverized material. All other sediments contained widely varying portions of fine material but generally less than approximately 10 to 15% by volume (Marshall, 1975).

Two observations were made regarding the distribution of pulverized sediments and the distribution of ^{239,240}Pu, for example. Sediments collected at stations C-5, C-10, and C-11 (S-16), which had much lower concentrations of radionuclides than did sediments collected at the nearby stations, C-1, C-2, C-3, C-4, C-6, C-8, and C-11 (S-19), respectively, also contained lower proportions of fine-grain material. Although a similar relationship held for most of the sediments collected, there were three obvious exceptions. These exceptions occurred for sediments collected at stations B-21, B-22, and

---

*This sampling trip was initiated by the Energy Research and Development Administration. The Puerto Rico Nuclear Center vessel *R. V. Palumbo* was used for the trip, and the chief scientists were Frank Lowman, Puerto Rico Nuclear Center, Victor Noshkin, Lawrence Livermore Laboratory, and William Schell, University of Washington.

(Text continues on page 551.)

(a)

(b)

Fig. 2 Aerial photograph of northwestern Bikini Atoll. (a) Tewa crater area, including Namu Island. (b) Bravo crater area and the western reef.

(a)

(b)

Fig. 3　Aerial photograph of Bikini Atoll. (a) Southwestern reef toward the deep passes at Bokdrolulu and Bokaetoktok Islands. (b) Zuni crater area.

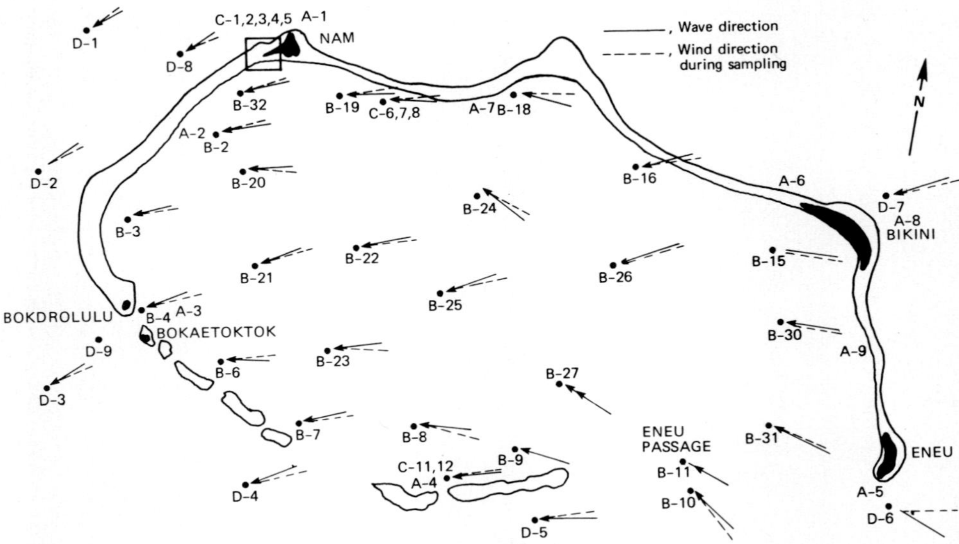

Fig. 4 Sampling stations and wind and wave directions measured during the joint sampling program in 1972 (Adapted from Noshkin, 1974).

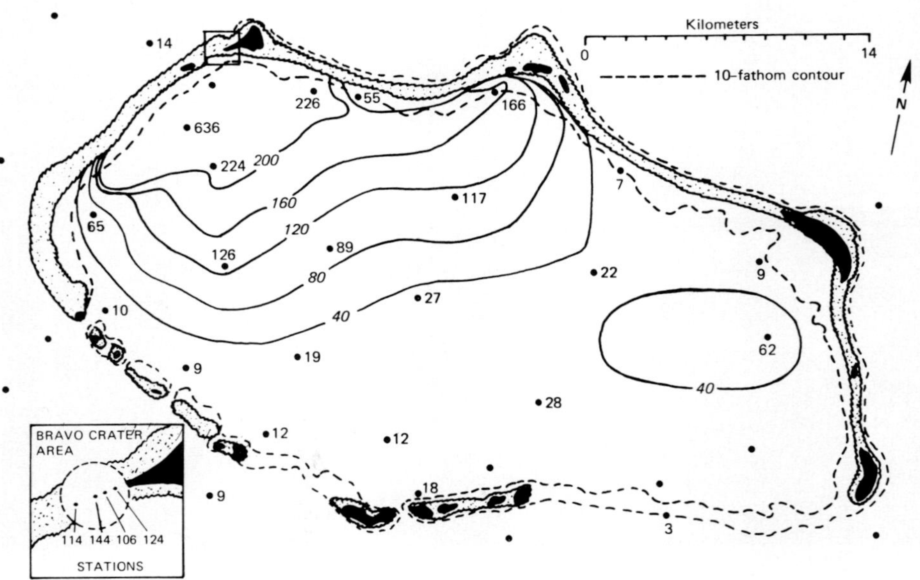

Fig. 5 Distribution of total alpha concentrations at Bikini Atoll lagoon. Concentrations in picocuries per gram of sediment.

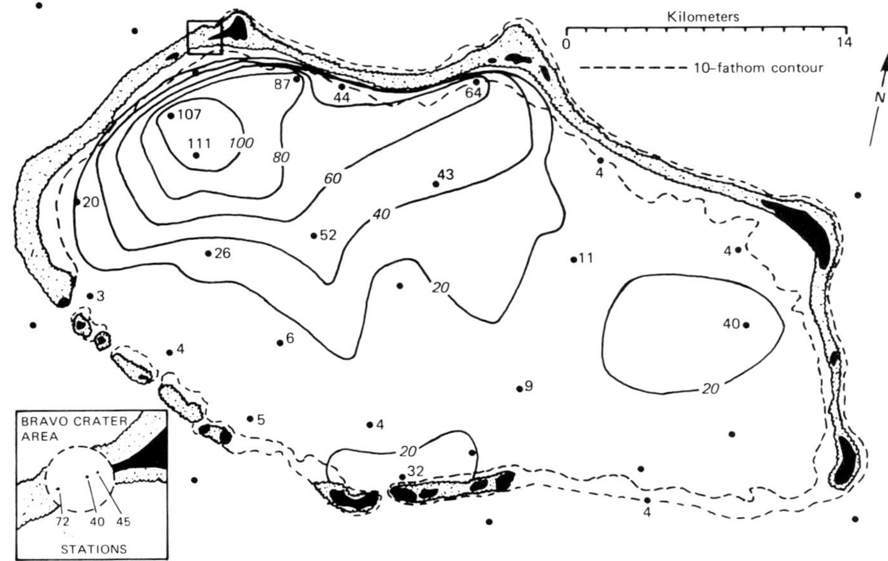

Fig. 6 Distribution of ^{239,240}Pu concentrations at Bikini Atoll lagoon. Concentrations in picocuries per gram of sediment.

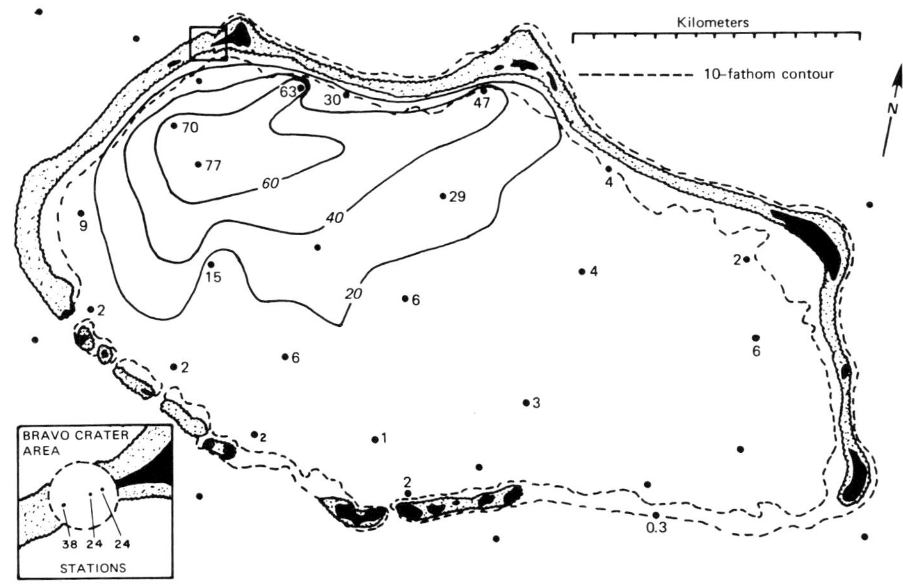

Fig. 7 Distribution of ^{241}Am concentrations at Bikini Atoll lagoon. Concentrations in picocuries per gram of sediment.

TABLE 2   Distribution of ^{239,240}Pu, ^{238}Pu, and ^{241}Am and the ^{238}Pu/^{239,240}Pu Ratio in Surface Sediments

Station location	Sample location	Concentration,* pCi/g (dry weight)			
		^{239,240}Pu	^{238}Pu	^{238}Pu/^{239,240}Pu	^{241}Am
B-2	S-20	107.0 ± 11	1.11 ± 0.41	0.010 ± 0.004	69.5 ± 2.6
B-3	S-23	19.7 ± 1.3	0.80 ± 0.18	0.041 ± 0.009	9.35 ± 0.52
B-4	S-21	3.17 ± 0.65	0.27 ± 0.15	0.087 ± 0.018	1.74 ± 0.16
B-6	S-14	3.44 ± 0.78			1.52 ± 0.06
B-7	S-18	5.16 ± 0.51	0.54 ± 0.12	0.104 ± 0.025	2.41 ± 0.18
B-8	S-12	4.11 ± 0.47	0.234 ± 0.078	0.057 ± 0.02	1.09 ± 0.14
B-10	S-5	0.381 ± 0.083	0.015 ± 0.022		0.272 ± 0.087
B-15	S-1	3.71 ± 0.38	0.085 ± 0.046	0.022 ± 0.012	1.88 ± 0.17
B-16	S-8	3.13 ± 0.27	0.050 ± 0.019	0.016 ± 0.006	1.80 ± 0.15
B-16	S-7	5.73 ± 0.43	0.067 ± 0.045	0.012 ± 0.008	2.83 ± 0.27
B-18	S-9	64.2 ± 1.7	2.11 ± 0.22	0.033 ± 0.004	46.5 ± 0.8
B-19	S-24	87.0 ± 6.2	0.70 ± 0.18	0.008 ± 0.002	62.6 ± 1.6
B-20	S-22	121.0 ± 7.0	0.53 ± 0.13	0.004 ± 0.001	77.3 ± 1.4
B-21	S-15	27.4 ± 2.4	0.42 ± 0.31	0.015 ± 0.012	15.4 ± 0.4
B-22	S-11	52.1 ± 2.2	0.27 ± 0.07	0.005 ± 0.001	33.0 ± 1.1
B-23	S-17	6.39 ± 0.36	0.199 ± 0.043	0.031 ± 0.008	3.06 ± 0.17
B-24	S-10	43.0 ± 3.4	0.21 ± 0.07	0.005 ± 0.002	28.9 ± 0.5
B-25	S-13	9.27 ± 0.45	0.134 ± 0.048	0.014 ± 0.004	5.87 ± 0.28
B-26	S-6	10.6 ± 0.5	0.218 ± 0.052	0.020 ± 0.006	4.34 ± 0.42
B-27	S-4	8.52 ± 0.41	0.073 ± 0.021	0.009 ± 0.003	3.28 ± 0.18
B-30	S-3	42.5 ± 2.1	1.59 ± 0.24	0.037 ± 0.006	6.77 ± 0.26
B-30	S-2	38.4 ± 2.1	1.21 ± 0.20	0.032 ± 0.006	5.65 ± 0.24
C-1	S-30(K)†	71.6 ± 3.8	4.16 ± 0.54	0.058 ± 0.008	37.6 ± 1.9
C-4	S-29	40.3 ± 1.6	1.72 ± 0.17	0.043 ± 0.004	24.1 ± 0.9
C-5	S-28‡	45.7 ± 2.2	0.67 ± 0.23	0.015 ± 0.004	24.7 ± 0.8
C-6	S-33	46.6 ± 1.4	0.33 ± 0.14	0.007 ± 0.002	33.5 ± 1.0
C-8	S-32	43.3 ± 2.2	0.69 ± 0.27	0.016 ± 0.006	31.7 ± 2.0
C-8	S-31	41.5 ± 1.9	0.550 ± 0.097	0.013 ± 0.002	28.8 ± 1.5
C-10	S-34				12.5 ± 0.4
C-11	S-19	28.9 ± 1.1	15.0 ± 0.9	0.518 ± 0.002	3.46 ± 0.41
C-11	S-16	5.36 ± 0.36	1.32 ± 0.19	0.246 ± 0.030	0.71 ± 0.11
D-4	S-26	2.20 ± 0.19	0.206 ± 0.053	0.093 ± 0.022	0.72 ± 0.11
D-8	S-27	30.6 ± 4.5	0.51 ± 0.23	0.017 ± 0.008	

*Mean ± 2 standard deviations.

†Surface sediment C-1 S-30 (n) was not analyzed for plutonium. Sediment C-1 S-30 (K) is a portion of the remaining grab sample.

‡Contained sediment from upper several centimeters (see Table 4).

B-24, which were located to the south and east of the area of the highest radionuclide concentrations measured at stations B-2 and B-20. Station B-21 was located at the extreme southern end of the region of high radionuclide concentration, and the finely divided sediments found there were similar in appearance to those collected at stations B-2 and B-20. Even though both stations B-21 and B-20 had similar proportions of fine sediments, only about 23% of the ^{239,240}Pu measured at station B-20 was found at station B-21. In contrast, stations B-22 and B-24, which were located some distance downstream and to the east of the area of high ^{239,240}Pu concentrations (i.e., stations B-2 and B-20), contained low proportions of finely divided material (less than ~15%) but contained 43% and 39%, respectively, of the total ^{239,240}Pu measured at station B-20.

These observations can be explained by two processes: the first is by dilution of the concentration of radioactive particles deposited at station B-21 by material of a lower concentration (resulting from biological activity or erosion of the reef); the second is by physical or chemical fractionation of the radioactivity in, or from, debris particles that are transported in suspension. (Physical fractionation could arise from differences in the concentrations and rates of dissolution of different-size particles.) The plutonium concentrations of relatively larger size particles deposited at station B-21, for instance, may have been lower than those of smaller size particles deposited farther downstream at stations B-22 and B-24. Second, since chemical fractionation of the radionuclides may be a function of the length of time the particles remained in suspension, particles deposited at station B-21 may also have lost a higher proportion of their surface-associated radioactivity than those deposited at stations B-2 and B-20. The relatively high concentrations of the sediments collected at stations B-22 and B-24 would be consistent with the deposition of finely divided material of a high specific activity.

Distribution in Sediment Cores

Measurements of the concentration distribution of elements in the sediment column are fundamental to the study of the exchange of materials across the sediment—water interface. In the Bikini lagoon measurements of both the transuranic and fission-product radionuclides with depth were considered to be particularly informative since debris from several detonations have been added to the lagoon at different times.

Nine sediment cores were collected from various locations in the lagoon. Three types of profiles of the radionuclide concentration with depth were observed. These occurred in (1) crater sediments (stations C-3 and C-12), which had either relatively homogeneous or constant distributions of most radionuclides with depth; (2) northwest quadrant lagoon sediments (stations B-2, B-20, and B-21), which had large proportions of finely pulverized material and which had radionuclide concentrations that changed regularly with depth; and (3) central and eastern lagoon sediments (stations B-15, B-16, B-27, and B-30), which had variable radionuclide concentrations with depth (Marshall, 1975).

Crater Cores

The distribution of radionuclides measured in the sediment core collected from the center of Zuni crater (station C-12) showed approximately constant transuranic and fission-product concentrations with depth (Table 3). No appreciable portion of the sediments in the Zuni crater core was finely pulverized. A unique concentration sequence of the following order was found: ^{239,240}Pu $>$ ^{155}Eu $>$ ^{238}Pu $>$ ^{60}Co $>$ ^{137}Cs $>$ ^{241}Am $>$ ^{207}Bi.

TABLE 3 Distribution of ^{241}Am, ^{239,240}Pu, ^{238}Pu, and ^{210}Po and the ^{238}Pu/^{239,240}Pu Ratio with Depth in Sediment Cores

Core depth, cm	Concentration,* pCi/g (dry weight)				
	^{241}Am	^{239,240}Pu	^{238}Pu	^{238}Pu/^{239,240}Pu	^{210}Po
Core C-12					
0 to 2	5.7 ± 1.5	35.5 ± 1.9	19.0 ± 1.1	0.56 ± 0.04	0.860 ± 0.054
2 to 4	6.33 ± 0.70				
4 to 6	6.04 ± 0.87				
6 to 8	6.27 ± 1.0	38.8 ± 1.4	18.8 ± 0.8	0.48 ± 0.03	
8 to 10	6.04 ± 0.97				
10 to 12	6.4 ± 1.1				
12 to 15	6.12 ± 0.61	36.4 ± 1.4	18.6 ± 0.8	0.51 ± 0.03	
Core C-3					
0 to 2	29.8 ± 1.0	49.4 ± 4.0	2.98 ± 0.62	0.063 ± 0.013	1.18 ± 0.13
2 to 4	34.4 ± 2.1				
4 to 6	43.2 ± 2.2				
6 to 8	67.7 ± 2.4	98.7 ± 5.6	8.27 ± 0.96	0.083 ± 0.011	2.99 ± 0.16
8 to 10	26.1 ± 0.8				
10 to 12	17.5 ± 0.9				
26 to 28	13.6 ± 0.5				
28 to 30	13.6 ± 0.6				
30 to 32	8.32 ± 0.28	6.24 ± 0.46	0.163 ± 0.064	0.026 ± 0.010	0.712 ± 0.045
32 to 34	1.44 ± 0.06				
48 to 50	19.0 ± 0.6				
50 to 52	23.1 ± 0.5	35.4 ± 2.2	1.28 ± 0.26	0.036 ± 0.007	1.08 ± 0.07
54 to 56	26.1 ± 0.9				
Core B-2					
0 to 2	103.0 ± 1.0	107.0 ± 4.0	1.30 ± 0.22	0.012 ± 0.002	Lost
2 to 4	85.7 ± 1.4				
4 to 6	80.2 ± 0.8	102.0 ± 4.0	1.32 ± 0.22	0.013 ± 0.002	0.677 ± 0.072
6 to 8	75.4 ± 1.1	97.2 ± 3.6	1.28 ± 0.22	0.013 ± 0.002	0.572 ± 0.072
6 to 10	56.2 ± 1.0				
10 to 12	48.7 ± 1.6	16.4 ± 1.5	0.21 ± 0.14	0.013 ± 0.008	0.505 ± 0.032
12 to 14	20.2 ± 0.4				
14 to 16	14.4 ± 0.5				
16 to 18	8.27 ± 0.21				
18 to 20	5.17 ± 0.27	6.76 ± 0.58	0.320 ± 0.088	0.047 ± 0.013	0.194 ± 0.018
20 to 22	2.66 ± 0.20				
22 to 24	1.65 ± 0.16				
24 to 26	0.73 ± 0.14	0.919 ± 0.028	0.01 ± 0.002	0.011 ± 0.002	Lost
26 to 28	0.178 ± 0.073				
28 to 30	0.24 ± 0.13				
30 to 32	0.124 ± 0.084				
32 to 34	0.203 ± 0.058				
34 to 36		0.165 ± 0.008	0.002 ± 0.001	0.010 ± 0.007	0.032 ± 0.018
36 to 38	0.070 ± 0.050				
38 to 40	0.92 ± 0.18	1.63 ± 0.08	0.011 ± 0.007	0.007 ± 0.004	0.090 ± 0.010

TABLE 3 (Continued)

Core depth, cm	Concentration,* pCi/g (dry weight)				
	^{241}Am	^{239,240}Pu	^{238}Pu	^{238}Pu/^{239,240}Pu	^{210}Po
Core B-20					
0 to 2	81.7 ± 1.6	101.0 ± 3.0	0.303 ± 0.088	0.003 ± 0.001	1.88 ± 0.10
2 to 4	57.4 ± 1.7				
4 to 6	53.3 ± 1.2				
6 to 8	61.3 ± 1.8	71.8 ± 4.6	0.41 ± 0.22	0.006 ± 0.003	1.72 ± 0.09
8 to 11	40.1 ± 1.3				
11 to 12	25.8 ± 1.0	34.7 ± 1.4	0.126 ± 0.060	0.004 ± 0.002	0.901 ± 0.054
Core B-21					
0 to 2	16.7 ± 0.5	23.6 ± 2.1	0.58 ± 0.20	0.024 ± 0.009	0.473 ± 0.036
2 to 4	19.0 ± 0.4				
4 to 6	18.5 ± 0.7	29.8 ± 1.0	0.604 ± 0.062	0.020 ± 0.002	
6 to 8	13.2 ± 0.8				
8 to 10	12.8 ± 0.7	19.1 ± 1.2	0.38 ± 0.10	0.020 ± 0.005	
10 to 12	4.34 ± 0.20				
12 to 14	2.47 ± 0.22	4.34 ± 0.38	0.075 ± 0.032	0.017 ± 0.008	0.225 ± 0.036
14 to 16	7.02 ± 0.35				
Core B-15					
0 to 2	2.86 ± 0.19	3.73 ± 0.48	0.067 ± 0.011	<0.034	2.21 ± 0.11
2 to 4	1.72 ± 0.15				
4 to 6	1.80 ± 0.21				
6 to 8	1.75 ± 0.25	4.19 ± 0.20	0.085 ± 0.022	0.020 ± 0.005	
8 to 10	2.56 ± 0.27				
10 to 12	3.75 ± 0.31	3.32 ± 0.18	0.056 ± 0.012	0.017 ± 0.004	
12 to 14	2.89 ± 0.49				
14 to 16	2.74 ± 0.29	6.99 ± 0.34	0.130 ± 0.036	0.018 ± 0.005	

*Mean ± 2 standard deviations.

A long core (56 cm) of entirely pulverized sediment was collected from the center of the Bravo crater (station C-3). Three segments of this core (the 0- to 12-, 26- to 34-, and 48- to 56-cm regions) were cut into 2-cm sections for the radionuclide measurements. The concentrations of radionuclides (Fig. 8 and Table 3) measured in the two lower regions of the core were similar to the uniform concentrations measured in the Zuni crater core. In the surface 12 cm, however, a well-defined layer of high-radionuclide concentrations was centered at the 6- to 8-cm depth. Elevated concentrations of all radionuclides were measured in this section, which contained the highest concentrations of ^{238}Pu (8.3 pCi/g), ^{207}Bi (432 pCi/g), and ^{60}Co (306 pCi/g) measured in any Bikini sediments except for the one higher ^{238}Pu concentration (19.0 pCi/g) that was found in Zuni crater (station C-12) sediments. The ordering sequence of radionuclide concentrations in different regions in the core differed greatly. This ordering sequence can be compared with that for surface sediments shown in Fig. 9. The sequence in the 0- to 2-cm section of the core differed from that in lower sections and from that found in the three other surface grab samples collected across the crater; these grab samples also differed from each other. In the 2- to 12-cm region of the core, the order in the sections was Bi > Co >

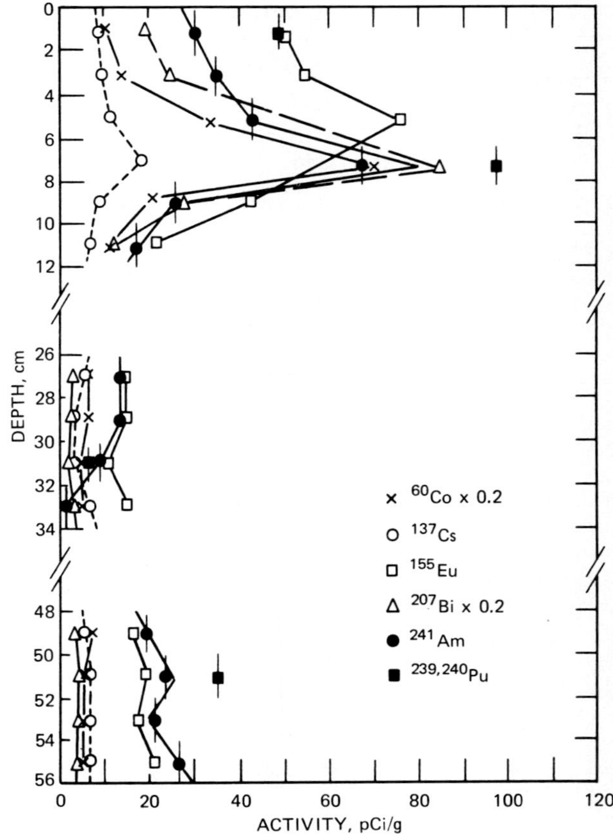

Fig. 8   Distribution of radionuclides in the sediment core collected at station C-3.

Eu > Am > Cs. In the 26- to 34-cm region of the core, three or four sections had the order Co > Bi > Eu > Am > Cs, and in the 48- to 56-cm region of the core, the sequence was Co > Am > Eu > Bi > Cs or Co > Am > Bi > Eu > Cs. Since plutonium was measured in only one section from each region, its placement is not included as characteristic of the larger regions. The high ^{238}Pu/^{239,240}Pu ratios found in the upper 12 cm of this core, however, showed that the origin of these radionuclides was different from those found in the lower regions. The radionuclides measured in the uppermost sections may be remnants from one of the smaller post-Bravo tests conducted in this area. The ^{238}Pu/^{239,240}Pu ratios of 0.036 and 0.026 found in the two deeper segments of the core were similar to several other ratios found in surface sediments of the lagoon that were collected away from the region south of the Tewa crater (station C-8).

Northwest Lagoon Cores

Three sediment cores were collected from this region of the lagoon (stations B-2, B-20, and B-21). Pulverized sediments were found at all three stations, although the radionuclide concentrations were significantly lower at station B-21 than at stations B-2 and B-20.

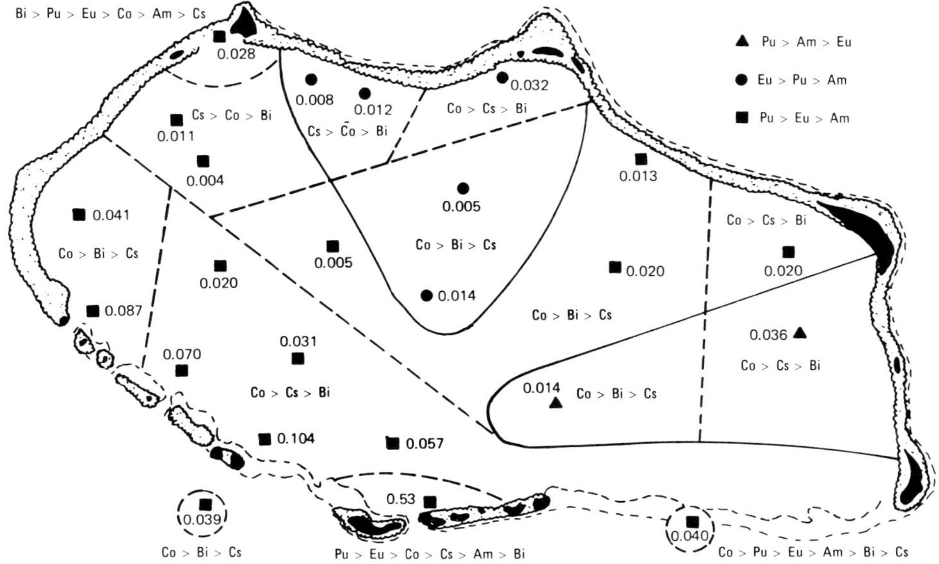

Fig. 9 Distribution of the ordering sequence of radionuclide concentrations in the surface sediments of Bikini Atoll lagoon. The ^{238}Pu/^{239,240}Pu ratios at each station for the year 1972 are also shown.

Station B-2.   The distribution of ^{239,240}Pu, ^{241}Am, ^{207}Bi, ^{155}Eu, ^{137}Cs, and ^{60}Co in the sediment core collected at station B-2 is shown in Fig. 10, and that of ^{241}Am, ^{239,240}Pu, ^{238}Pu, and ^{210}Po is shown in Table 3. Several features of this long core were similar to features in other sediment cores collected from the northwest quadrant. The sediments in this core consisted of mixtures of *Halimeda* and pulverized fine coral material; *Halimeda* predominated at the 8- to 10-cm section. The distribution of ^{241}Am, ^{239,240}Pu, ^{155}Eu, and ^{137}Cs concentrations at station B-2 was similar to that at station B-20 except that (1) the absolute concentrations of ^{137}Cs measured were lower than the ^{241}Am and ^{155}Eu concentrations by a factor of 10 and (2) the ^{239,240}Pu concentrations measured were slightly more irregular with depth than the concentrations measured for ^{241}Am or ^{155}Eu. In the top 11 cm of the core, the concentrations of ^{239,240}Pu, ^{241}Am, ^{137}Eu, and ^{137}Cs decreased regularly with depth to 50% of their respective concentrations, which were measured in the surface layer. In the 12- to 26-cm region of the core, the concentrations of ^{239,240}Pu, ^{241}Am, ^{155}Eu, and ^{137}Cs decreased nearly logarithmically with depth. In the 28- to 38-cm region of the core, the concentrations again decreased slowly with depth.

The distribution of ^{60}Co and ^{207}Bi concentrations in the core was unusual in that decreasing concentrations (with increasing depth) were not found in the upper 10 cm of the sediment core. Although the concentration of ^{60}Co was relatively constant in the upper 12 cm of the core, the concentration of ^{207}Bi increased 50% between the 2- to 4- and 8- to 10-cm sections. Below the 8- to 10-cm section in the core, the decrease in concentration of ^{207}Bi was similar to that of ^{241}Am, ^{155}Eu, and ^{137}Cs; however, the concentration of ^{60}Co was almost constant with depth. The distribution of ^{238}Pu/^{239,240}Pu ratios measured in different sections of the core was separated by the value of

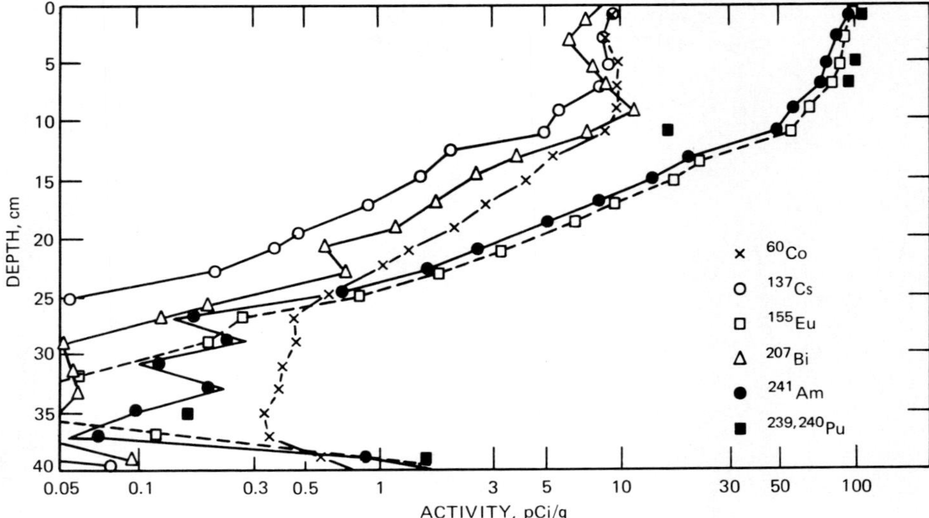

Fig. 10 Distribution of radionuclides in the sediment core collected at station B-2 in 1972.

0.047, which was found in the 18- to 20-cm section. Below this 18- to 20-cm section, the ratios decreased with depth from 0.0108 to 0.0069, whereas above the 18- to 20-cm section, the ratios ranged from 0.0138 to 0.0113.

The ordering sequence of the radionuclide concentrations measured in this core varied with depth. Below the 0- to 2-cm section, the order changed from that shown in Fig. 9 to Pu > Eu > Am > Co > Cs > Bi in the 2- to 6-cm region, to Pu > Eu > Am > Co > Bi > Cs in the 6- to 26-cm region, to Co > Pu > Eu > Am > Bi > Cs in the 26- to 38-cm region, and to Pu > Eu > Am > Co > Bi > Cs in the 38- to 40-cm region.

Station B-20. The distribution of ^{239,240}Pu, ^{241}Am, and ^{155}Eu concentrations in this core (Fig. 11 and Table 3) was again quite similar and decreased with depth by 50% at about the 9-cm section. At 11 cm in the core, a sharp break occurred between the finely divided material in overlying sections to coarse sand. Considering the range and distribution of ^{238}Pu/^{239,240}Pu ratios measured in surface sediments across the lagoon, the ratios found in the three sections of this core were uniquely low, which possibly indicates a common source(s) for the majority of the plutonium contamination in the sediment column collected at this station. Only in the 6- to 8-cm section of this core does the radionuclide ordering sequence differ from that found in surface sediments (Fig. 9). In this section the ordering of ^{241}Am and ^{155}Eu concentrations was reversed from those in other sections. This order, Pu > Am > Eu, was found in sediments only in the four sections from the bottom of the Bravo crater core and from the eastern lagoon.

Bismuth-207 concentrations were below the limit of detection in most sections of the core. However, the concentration of ^{207}Bi in the 0- to 2-cm section was at least four to five times as high as that in any lower section.

The concentrations of ^{60}Co and ^{137}Cs decreased, respectively, to 50% of their largest concentration at the 9- and 11-cm levels in the core. However, neither of these

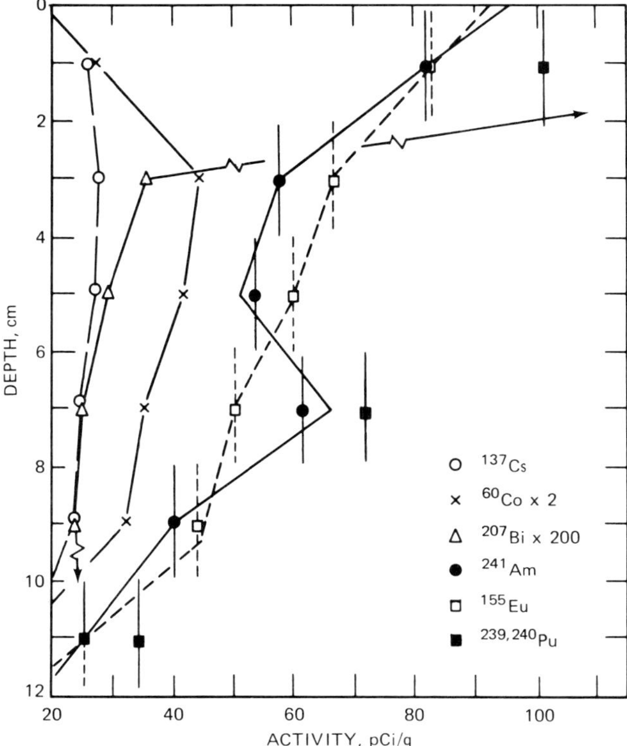

Fig. 11   **Distribution of radionuclides in the sediment core collected at station B-20 in 1972.**

radionuclides showed steadily decreasing concentrations in the upper layers. The concentration of ^{60}Co in the 0- to 2-cm section of the core was significantly lower than that in lower sections. The concentration of ^{137}Cs at the 0- to 9-cm level of the core showed no appreciable change with depth.

Station B-21. The concentration profiles of all the radionuclides measured in this sediment core are roughly similar to those in station B-20 in that the concentrations increased to a maximum at between 5 and 7 cm (except at 3 to 5 cm for ^{241}Am) and then decreased to 50% of their highest measured concentrations at depths of 10 cm, after which increasing proportions of *Halimeda* began to appear. As in the B-2 core, increased concentrations of radionuclide distributions were measured in the lowest section of this core. As in both of the other cores from this region of the lagoon, the ^{238}Pu/^{239,240}Pu ratios measured with depth in the core (Table 3) showed only a slight decrease with depth.

In the 8- to 16-cm region of the core, the ordering sequence of radionuclide concentrations changed from those found in surface sediments (Fig. 9) to the order Pu > Eu > Am > Co > Bi > Cs. This sequence was the same as that observed below the 6-cm section at station B-2 and in surface sediments at the far western region of the Atoll.

Central and Eastern Lagoon Cores

The four sediment cores collected from the central and eastern regions of the Atoll (stations B-27, B-16, B-15, and B-30) were similar in three respects: First, there was no significant net increase or decrease in the concentration of radionuclides measured between the upper and lower sections in any of these four cores; second, the distribution profiles of ^{239,240}Pu, ^{241}Am, ^{155}Eu, ^{137}Cs, ^{207}Bi, and ^{60}Co concentrations were roughly similar with depth in the individual cores; and third, the ^{238}Pu/^{239,240}Pu ratios measured in all but the lower section of the cores from station B-27 were similar to that in the surface-sediment section. Because of the very short (6 cm) length of the cores at stations B-16 and B-30, no further interpretation of the observed radionuclide profiles was warranted. Except for the ordering of ^{241}Am in one section and the ^{239,240}Pu concentration in the 10- to 12-cm section immediately below, the ordering sequence of radionuclide concentrations in the 16-cm core from station B-15 was the same as that in the surface sediments (Fig. 9). In the 10-cm core collected from station B-27, the sequence of ^{60}Co, ^{207}Bi, and ^{137}Cs concentrations measured did not change with depth from that shown in Fig. 9. However, in the 4- to 8-cm region, ^{241}Am concentration was higher than ^{239,240}Pu, and in the 8- to 10-cm section, the ordering sequence was the same as that at lagoon stations B-16, B-26, and B-22 to the west.

The constancy of the concentrations of all radionuclides measured to depths of 10 cm (core B-27) and 16 cm (core B-15, Table 3) showed that a considerable penetration of radionuclides occurred in these sediments, which appeared physically to be normal lagoon deposits. Assuming a negligible natural sedimentation rate, the penetration of radionuclides into these sediments is significantly greater than that observed by Held (E. Held, University of Washington, unpublished results) in Rongelap Atoll sediments. However, these two sediment cores were the longest obtained from any station in the Atoll having unpulverized sediments, which suggests that these sediments may have been significantly less consolidated than average. This could explain both the length of the core collected and the radionuclide concentrations measured with depth.

Sedimentation Rates at Station B-2

Measurements of the concentrations of ^{210}Pb and ^{226}Ra with depth in core B-2 were used to determine the effective sedimentation rates based on the ^{210}Pb age dating technique (Goldberg, 1963; Koide, Soutar, and Goldberg, 1971). The average ^{226}Ra concentration of 0.131 pCi/g, which was measured by gamma counting, was used to determine the concentrations of unsupported ^{210}Pb. The unsupported ^{210}Pb concentrations measured in the 0- to 2-, 4- to 6-, 6- to 8-, and 10- to 12-cm sections decreased logarithmically with depth, which indicates a constant sedimentation rate for the upper layers. Below 12 cm the ^{210}Pb concentrations were not significantly different from the ^{226}Ra concentrations measured, which indicates no unsupported ^{210}Pb. The effective sedimentation rate was determined by calculating a linear regression of the unsupported ^{210}Pb concentrations in the upper 11 cm of sediment. A sedimentation rate of 0.58 cm/yr (correlation 0.98) was calculated for the upper 11 cm of sediment. Thus the approximate date calculated for the deposition of the 11-cm section was 1953, a date consistent with the period of nuclear testing at Bikini.

These data indicate that two different processes were responsible for the deposition of the 40 cm of sediment sampled at this station (B-2): (1) slow accumulation of sediment occurred in the upper layers (11 cm) and (2) at some point below 11 cm in the core, rapid accumulation of the sediments containing no unsupported ^{210}Pb predominated. Figure 10 shows that both ^{60}Co and ^{207}Bi have concentration profiles that are markedly different from those of other radionuclides in the core above and below about 11 cm. This may indicate that not only the process of deposition but also the source of contaminated debris may have differed for the two depth regions in the sediment column at station B-2.

Given the dynamic hydrological environment at Bikini, the most significant contamination of the sediment environment a priori would arise from the large surface bursts (such as Bravo, Koon, and Zuni) whose fireballs strongly interacted with the soil or sediment and from similar interactions of deep lagoon or barge bursts (such as the Baker and Tewa tests). At Bikini the initial introduction of highly contaminated debris to the lagoon from detonations of this type can be described as fallout deposition of a large mass of chemically altered coralline soils reduced in size and containing the condensed radionuclides. A large mass of crushed coralline material of a relatively low specific activity must also have been ejected by the detonations. The areal distribution of two different types of materials (altered and unaltered coral) would overlap at progressively greater distances away from the detonation craters, and mixed particles would descend at rates depending on their sizes and shapes. In the aqueous environment the particles would be transported a distance that would be determined by their settling velocities, sizes, densities, and the speed of the prevailing lagoon currents. The result of these physical forces with time would be to yield a concentration of fine particles in the surface deposits. The net result of the radionuclide concentrations would be to yield sediment concentrations (picocuries per gram) that were progressively more dilute (by natural sediments) at increasing distances downstream.

The sedimentation rate measured in the upper 11 cm of sediment collected at the station near the Bravo crater (station B-2) showed that the material was deposited at a constant rate between the 1950s and 1972. Although the initial source for the material deposited at these locations was the detonation craters, the present location of the source(s) supplying the material for redistribution at these lagoon stations is not known. The importance of this point should not be underestimated because the location and extent of the source of these fine sediments may determine the continued availability of the radionuclides for redistribution and uptake by biota.

It is clear, from the large size of the Bravo, Tewa, and Zuni detonation craters, that a huge quantity of pulverized sediment was removed from the reef immediately after the detonations. However, as noted by Welander et al. (1966), lagoon currents were capable of maintaining a large flow of the finely divided sediment out of certain craters at Enewetak long (>1 yr) after the testing stopped. It is quite likely that much of this material at Bikini was deposited outside the detonation craters and was the source for part of the material redistributed in the lagoon. The ^{238}Pu/^{239,240}Pu ratios measured in the craters and at various stations in the northwest quadrant suggest three possibilities for the source of the redistributed material deposited at station B-2: (1) from locations between station B-2 and the Bravo crater; (2) from (1) above and from the area between station B-2 and the northern reef (near station B-19); or (3) from (1) or (2) above and from within the detonation craters. The reason for making these hypotheses is that the ^{238}Pu/^{239,240}Pu ratios in the top 11 cm of redistributed sediments at station B-2 are about 0.0125, whereas the ratios measured in the fine surface sediments collected in Bravo crater are about 0.05, and the ratios at station B-19 are about 0.008. Thus a mixture of sediments from different sources may be deposited at station B-2.

Concentrations and Physicochemical States in the Water Column

At each of the sediment sampling stations, plutonium and americium were measured in water collected at 2 m from the bottom and at 1 to 2 m from the surface. During the collections in November 1972, the wind and wave conditions were typical of the winter season—ESE and ENE as given in Fig. 4 (Noshkin et al., 1974). During the collections in July 1976, the wind direction and velocity were essentially the same as those in November 1972. The physical circulation of the water in the lagoon may be traced by using the plutonium concentrations and distributions measured at lagoon stations away from the northwestern quadrant.

Distribution of $^{239,240}Pu$ *Concentrations*

With the use of all total plutonium concentration values obtained in 1972 by F. G. Lowman, Puerto Rico Nuclear Center, V. Noshkin, Lawrence Livermore Laboratory, and W. R. Schell, University of Washington (unpublished data) and the few additional values obtained in 1976 for the lagoon water and by averaging the values when duplicate collections were available for the same station, isopleths of the plutonium concentrations for the surface and deep water of the lagoon have been constructed and are shown in Figs. 12 and 13, respectively. These isopleths show clearly the distribution patterns of plutonium, caused by the lagoon circulation, from its main source in the sediments of the northwestern quadrant of the lagoon. With the use of $^{239,240}Pu$ concentrations as the tracer, the transport and circulation of the water in the lagoon have been estimated. The surface water appears to be diluted by incoming ocean water through the wide pass near Eneu and by oceanic water over the northeastern reef. The outlet of lagoon water is

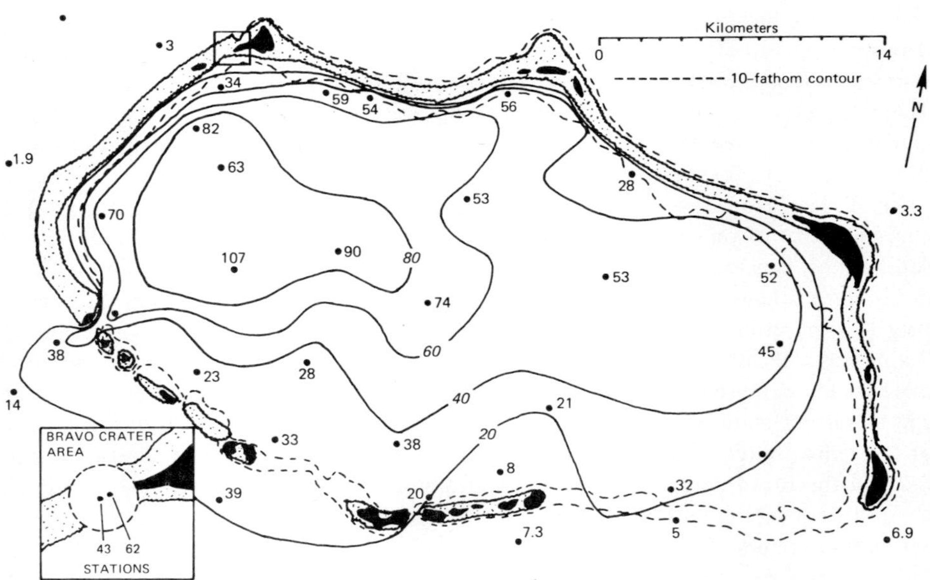

Fig. 12   Distribution of $^{239,240}Pu$ concentrations in surface water (2 m) at Bikini Atoll. Concentrations in picocuries per cubic meter.

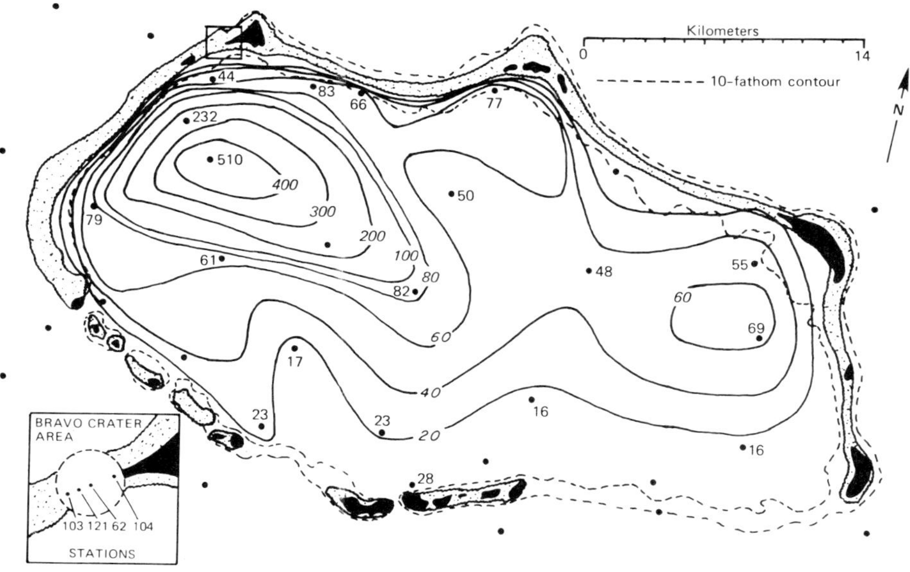

Fig. 13 Distribution of ^{239,240}Pu concentrations in deep water (2 m above bottom) at Bikini Atoll. Concentrations in picocuries per cubic meter.

through the deep passes in the southwestern part of the Atoll. It is here that lagoon water exits into the North Equatorial Current. The pattern for the transport of deep water appears to be that of oceanic water entering the lagoon through the Eneu passage and moving as far as the deep passes at the southwestern part of the lagoon. Oceanic water also either enters through or over the reef at the northwestern part of the lagoon and dilutes the deep water in this region. As shown in Fig. 6, the highest ^{239,240}Pu concentrations in sediments are near the northwest reef. The highest ^{239,240}Pu concentrations in the deep water appear to be displaced southward slightly from the higher concentrations that were measured in the sediments and to be transported in an easterly direction. For such a complex water circulation pattern in the lagoon, water must pile up and descend near the western reef and upwell near the eastern reef. This process was identified by Von Arx (1954). Water must flow in opposite directions at the southcentral part of the lagoon. The tongue of oceanic water at the bottom in the southern part of the lagoon has a lower ^{239,240}Pu concentration than the surface water that is leaving the lagoon through the deep passes. The same general circulation pattern was inferred from ^{55}Fe measurements in the same samples except that the source was more concentrated near the northern reef (Schell, 1976).

The distribution of plutonium in the lagoon water indicates that the near-reef areas contain low concentrations of plutonium; this is probably due to dilution by oceanic water that enters the lagoon from over the reef. Thus the organisms that inhabit the lagoon terrace or seaward reef areas are exposed to very low levels of plutonium even though much higher levels exist inside the lagoon. The predominant source of fish for the returning Marshallese will be those reef fish, which should contain low levels of transuranic radionuclides. The radionuclide levels are probably not much higher than those found at Atolls that have not been contaminated by fallout (Noshkin, Eagle, and Wong, 1976).

Physicochemical States of $^{239,240}Pu$ and ^{238}Pu

In a model of the marine environment that can be used to predict effects, the water is the central component; the water receives the inputs of metals and provides the transport medium for uptake into the biosphere or loss to the sediments. Because of this central role, the behavior and physicochemical states of the transuranic elements in seawater must be known if a useful model of the system is to be constructed. Once the speciation and behavior of these elements are known, their relative hazards can be predicted. High concentrations of salts in seawater interfere with the methods used to measure the physicochemical states and must be removed before the analysis for the transuranic elements. Consequently separation and preconcentration from the saltwater matrix are required at some stage of the analysis. Ideally the best method would be direct in situ measurement of the elements in the field. Unfortunately, with present technology direct field measurement is not feasible. The next best thing is to extract and concentrate the elements in the field. This would eliminate some of the problems resulting from contamination, losses during transport and storage of samples, and changes in chemical speciation on storage in containers.

The Battelle large volume water sampler (BLVWS) is a sampling system that can be used to concentrate low levels of trace metals or radionuclides from natural waters in the field (Silker, Perkins, and Rieck, 1971). This method effectively eliminates the need for preservation and storage of water samples, extraction in the field, and evaporation or ion exchange in the laboratory to concentrate the elements to a level sufficient for analysis. The advantage of the BLVWS technique over the "conventional" techniques is that both the total concentration and the physicochemical-state concentrations of the particulate and soluble fractions can be measured. Collection efficiencies of the soluble fraction are determined individually for each element during collection. In addition, much larger volumes of water permit lower concentrations to be measured.

An evaluation of a new sampling and measurement technique for transuranic elements requires detailed studies of the precision and accuracy of the technique in both controlled and natural environments. It also requires simultaneous measurements of samples that have been collected by the more conventional methods. For the past several years, we have attempted to set up experiments that would test the validity of the BLVWS technique for plutonium measurements in both laboratory and field studies (Huntamer, 1976; Nevissi and Schell, 1975; Schell, Nevissi, and Huntamer, 1978).

Description of the Sampler.   The BLVWS is a field collector that can process as much as 4000 liters of water in 3 hr with the large sampler (28-cm diameter) and about 800 liters of water with the small sampler (13-cm diameter), depending on the particulate loading. The filtering section of the BLVWS normally consists of eight filters arranged in parallel. The number of filters used can be expanded or reduced by removing or adding plates to the BLVWS. The water, after passing through one of the filters, is then channeled through the sorption beds. The sorption beds generally consist of two to four 0.6-cm-thick sections (Fig. 14). The use of individual sorption beds rather than one thick bed permits the calculation of the collection efficiency for individual elements and permits easy variation of the sorption-bed thickness. It also allows for the use of a mixture of different sorption beds if desired.

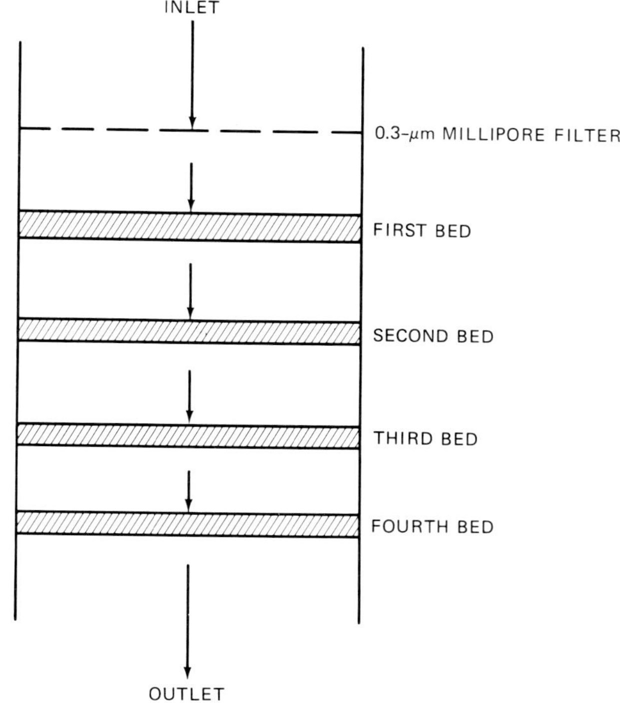

Fig. 14 Schematic representation of the BLVWS showing the flow of water through the filters and sorption beds.

Water is forced through the BLVWS with an electric pump. The smaller BLVWS requires a lower flow rate than the larger samples; so the flow is controlled by the use of a valve and a water "bypass." The entire pumping system is shown schematically in Fig. 15. The volume of the water sampled is measured with a recording water meter.

Collection Efficiencies. Most of the sampling procedures and techniques used in seawater analysis assume 100% efficiency for the collection and measurement or use radioactive tracers to determine the chemical yield of the samples. One problem with the use of tracers is that they are usually not added to the samples in the same chemical form as the element in the sample. If the chemical states of the sample and tracer element are not identical, the isotopic equilibrium might not be reached for a long time, and the chemical yield may be in error.

Because of the short column lengths used in the BLVWS, collection efficiencies seldom are 100% except for a few elements that are totally retained on the first sorption beds, such as ^{241}Am, ^{207}Bi, ^{55}Fe, and ^{155}Eu (Schell, Nevissi, and Huntamer, 1978). Elements that are not collected quantitatively can be determined by the differences in the amounts collected on successive sorption beds, as outlined by Held (1971), Schell, Jokela, and Eagle (1973), and Schell, Nevissi, and Huntamer (1978).

This method of determining collection efficiencies, referred to as the "BLVWS echnique," is an empirical method which assumes that a constant fraction of the

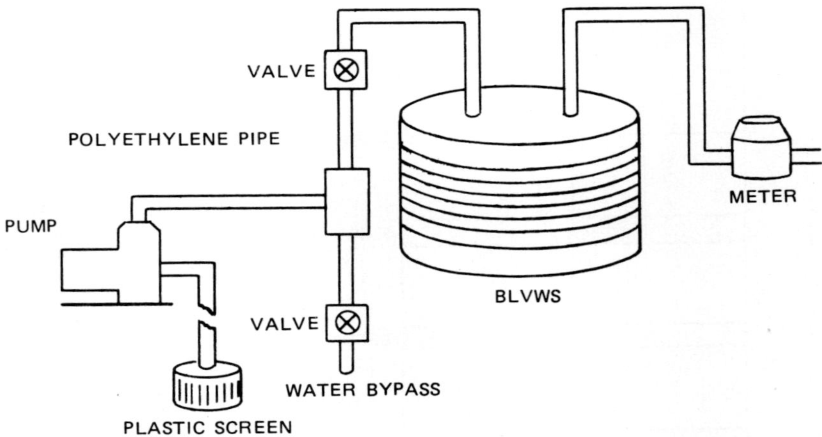

Fig. 15 Schematic representation of the BLVWS.

available solute is removed by each sorption bed. When this approach is used to measure the amount of solute, N, on the individual sorption beds, N_m and N_{m+1}, the collection efficiency, E, between beds m and m + 1 can be determined from Eq. 1 (Schell, Nevissi, and Huntamer, 1978).

$$E_{(m, m+1)} = \frac{N_m - N_{m+1}}{N_m} \tag{1}$$

this efficiency can then be used to obtain the concentration of the solute in the soluble phase, C_s.

$$C_s = \frac{N_m}{E_{(m, m+1)}} + \sum_{m}^{m-1} N_{m-1} \tag{2}$$

The total concentration, C_t, is found by adding the contribution of the particulates, C_p, to the soluble.

$$C_t = C_s + C_p \tag{3}$$

where E = collection efficiency between m and m + 1 beds
    N_m = concentration of solute retained on the m*th* sorption bed
  N_{m+1} = concentration of solute retained on the m + 1 sorption bed
    C_s = concentration of solute in the soluble fraction of the water
    C_p = concentration of solute in the particulate fraction retained on the filters
    C_t = total concentration of solute in the water volume sampled

Tank Experiments with the BLVWS. The BLVWS has been evaluated at different salinities in the laboratory by a series of tank experiments. The experimental procedures are discussed in detail in an M.S. thesis by Huntamer (1976) and in papers by Schell, Nevissi, and Huntamer (1978) and Nevissi and Schell (1976). Six elements, americium, bismuth, cobalt, europium, iron, and plutonium, have been evaluated in the tank experiments by using two different sorption beds, Al_2O_3 and Chelex-100. These elements are of interest in that they are either products of nuclear weapons testing or nuclear reactor operations.

The results show that the BLVWS can be used to measure the concentration of the above elements in marine waters over a wide range of salinities. The precision of the BLVWS method has been evaluated by comparing duplicate samples taken with the BLVWS. For example, the mean percent variation between duplicate samples (collected simultaneously from the same tank) in experiments using Al_2O_3 was $\pm10\%$ for plutonium.

The results of the tank experiments indicate that, with the use of either Al_2O_3 or Chelex-100 sorption beds, the BLVWS is a suitable sampling method for some elements. In addition, the behavior of individual elements on the sorption beds provides qualitative information on the physicochemical speciation of the elements. Table 4 gives a summary of the physicochemical states observed at high salinity ($>31\ \%_0$) with the BLVWS in both the tank and field studies at Bikini lagoon. The results are compared with the chemical species predicted by equilibrium calculations or determined by the other measurement methods. Americium was found to be 80 to 100% particulate (i.e., $>0.3\ \mu m$) and consequently was collected efficiently on the Millipore filters and/or first Al_2O_3 bed at all salinities. Plutonium was collected by the BLVWS technique, using

TABLE 4 Physicochemical States of the Trace Metals in Marine Waters Estimated by the BLVWS Method Compared with the Physicochemical States Measured and Predicted with Other Methods

Elements	Samples	Particulate, %	Colloidal,* %	Soluble, %	Predicted and observed species
Americium	Tank	76 to 87	13 to 24	0	Particulates
	Field	30 to 100	0 to 70	0	
Bismuth	Tank	69 to 90	10 to 31		$Bi(OH)^{2+}$, $Bi(OH)^{6+}_{12}$,[†] insoluble
	Field	0 to 22	78 to 100	0	alkaline solution[‡]
Cobalt	Tank	0 to 2		98 to 100	$CoCl^+$, Co^{2+}, $(CoSO_4^0)$,[§][¶]
	Field	10		90	organics[**]
Europium	Tank	29 to 71	29 to 71		Particulates (freshwater)[††]
	Field	11 to 100	0 to 89		
Iron	Tank	72 to 95	2 to 28		$Fe(OH)_2^+$, $Fe(OH)_4^-$,[§]
	Field	31 to 60	40 to 69		particulates[¶]
Polonium	Tank	100			Particulates[‡]
	Field	95 to 100	0 to 5		
Plutonium	Tank	69 to 93	4 to 28	3	Colloidal,[‡‡] $PuO_2(CO_3)_2^{2-}$,
	Field	2 to 60		40 to 98	Pu^{3+}, Pu_2OH^+[‡‡]

*Colloidal species based on the complete retention of the fraction passing through a 0.3-μm Millipore filter on the first Al_2O_3 bed.

[†]Stumm (1967).

[‡]Nozaki and Tsunogai (1973).

[§]Stumm and Morgan (1970).

[¶]Sibley and Morgan (1976).

[**]Lowman and Ting (1973).

[††]Robertson et al. (1973).

[‡‡]Andelman and Ruzzell (1970).

Al_2O_3 beds, at all the salinities tested. The efficiencies derived from the concentrations on each sorption bed can be used to obtain the total concentration in the water.

The tank experiments, in addition to testing the BLVWS technique, also indicated that chemical speciation beyond particulate and soluble fractions can be made. There appeared to be evidence of a "colloidal" form of plutonium that passed through a 0.3-μm Millipore filter but was efficiently sorbed on the first Al_2O_3 bed. The colloidal plutonium, first discussed by Nevissi and Schell (1975), appeared at the higher salinities, 31.6 ‰, and was identified by a much greater collection efficiency on the first Al_2O_3 bed compared with the second and third beds, as shown in Fig. 16. An abrupt change in

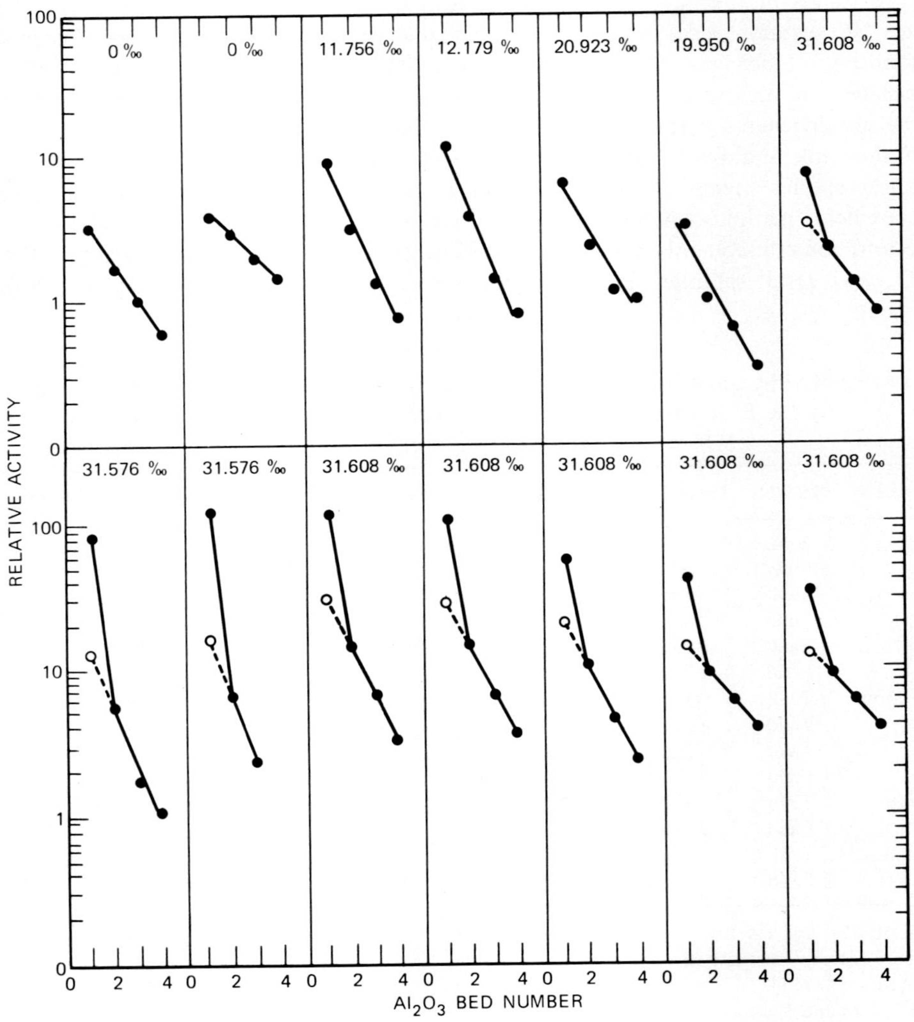

Fig. 16 Relative concentrations of plutonium sorbed on Al_2O_3 beds at different salinities using prefiltered water (0.3 μm) that had been aged for 2 to 3 weeks before sampling with the BLVWS. ○, extrapolated values for soluble activity. ●, activity measurements.

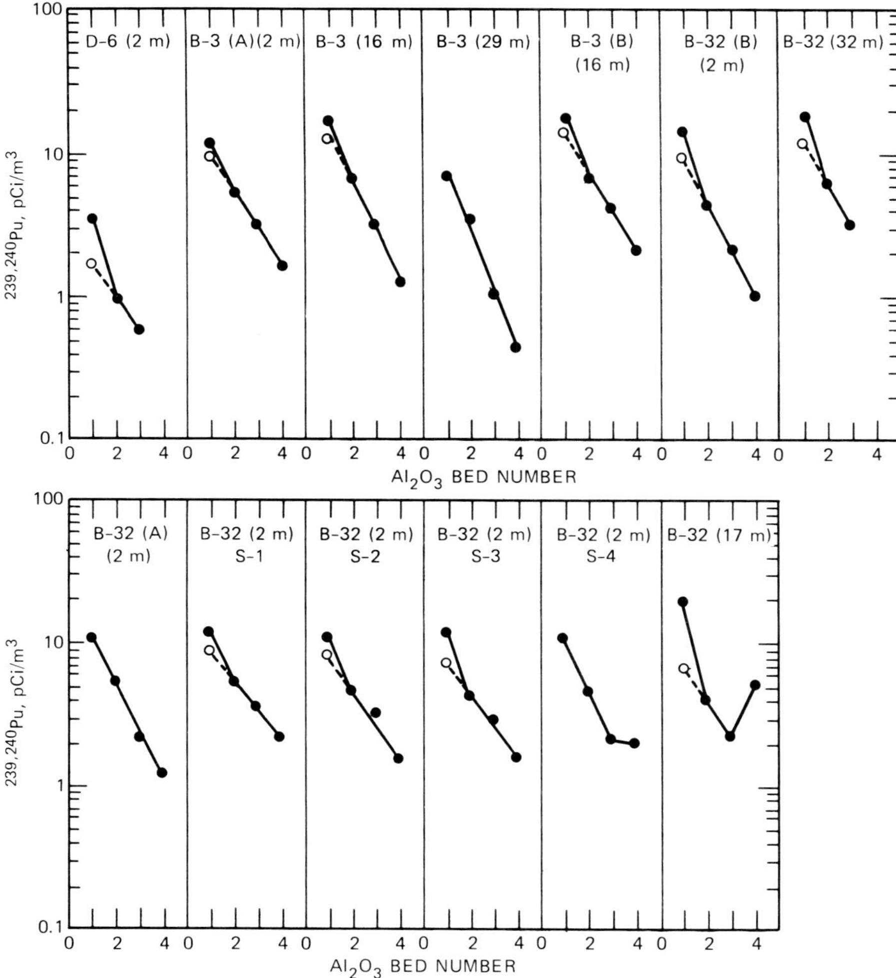

Fig. 17 Concentrations of plutonium sorbed on Al_2O_3 beds from BLVWS samples collected at Bikini Atoll in 1976. ○, extrapolated values for soluble activity. ●, activity measurements.

slope occurred in bed 2 for the high-salinity samples. This effect was not observed at the lower salinities. By extrapolation back to the first bed, the colloidal fraction was estimated. For plutonium the colloidal fraction of the total averaged $15 \pm 3\%$ with a range of 11 to 17%.

The BLVWS collections of ^{239,240}Pu in water samples collected in July 1976 at Bikini Atoll have been evaluated to determine the fraction of the total concentration present in the colloidal state. The colloidal fraction has been determined, as before, by extrapolating the amounts of ^{239,240}Pu collected on the second, third, and fourth bed to the first bed, as shown in Fig. 17. The least-squares regression line with its error through these data extrapolated to bed 1 gives the soluble fraction in bed 1. The difference between this extrapolated value (soluble) and the total concentration measured in bed 1 is defined as the amount present in the colloidal state. Table 5 gives the amount of ^{239,240}Pu present in the total sample, i.e., soluble, colloidal, and percent colloidal. The average of the colloidal concentration of ^{239,240}Pu in the 22 lagoon samples measured from Bikini Atoll was $12 \pm 8\%$ (SD). This average can be compared with the tank studies at 31.6 $\permil$ salinity in which $15 \pm 3\%$ (SD) was found in the colloidal state.

Since only two Al_2O_3 beds in the BLVWS were used for the collections at Bikini Atoll in 1972, no colloidal fraction could be determined, and a systematic error may have been present in the total plutonium concentration data. The 1976 results have been evaluated with only two beds to observe if significant systematic errors were present in the 1972 data. Of the total plutonium present in 14 samples tested, 91% (range, 78 to 100%) would have been measured if only the first two beds had been used in the calculations of total concentrations. Thus the 1972 data for individual samples measured by the BLVWS technique should be reliable to within about 20%.

Speciation of ^{238}Pu and ^{239,240}Pu into Particulate and Soluble Fractions.   The isotope ratios of ^{238}Pu/^{239,240}Pu in surface sediments and in the overlying batch water samples are shown in Figs. 18 and 19, respectively. The range of ^{238}Pu/^{239,240}Pu ratios in the surface sediments is 0.004 to 0.52; the average value determined from 28 measurements is 0.029 ± 0.027 (SD). The ^{238}Pu/^{239,240}Pu ratios in the water samples collected at 2 m above the bottom show that at several stations the ^{238}Pu/^{239,240}Pu ratios in water are similar to those of underlying sediments. However, the ^{238}Pu/^{239,240}Pu ratios in water, which range from <0.02 to 1.11, are considerably more variable than those in the sediments. It appears that there may be enrichment in ^{238}Pu/^{239,240}Pu ratios in the overlying water compared with those in sediment.

The ^{238}Pu/^{239,240}Pu ratios for the BLVWS collection of plutonium in surface- and deep-water samples with 0.3-μm Millipore filters and two Al_2O_3 sorption beds are shown in Figs. 20 and 21, respectively. The ^{238}Pu/^{239,240}Pu ratios were not usually constant for the three fractions of the same surface- or deep-water samples. This unexpected finding was at first questioned, and the analysis and counting of the samples were repeated. Possible contamination by ^{241}Am and ^{228}Th, which decay by alpha particles of nearly the same energy as ^{238}Pu, was rechecked. The results showed that the original values were real and that the particulate fraction had ^{238}Pu/^{239,240}Pu ratios that were significantly different from the two soluble (<0.3 μm) fractions. In fact, the two soluble fractions (first Al_2O_3 bed and second Al_2O_3 bed) also had different ^{238}Pu/^{239,240}Pu ratios. The source of plutonium isotopes to the water column is the contaminated lagoon sediments, but only a few water samples have ^{238}Pu/^{239,240}Pu ratios equal to those of the sediment. In the surface waters of the lagoon, the ranges of ^{238}Pu/^{239,240}Pu ratios were: Millipore filters, 0.014 to 0.57; first Al_2O_3 bed, 0.06 to 0.64; second Al_2O_3 bed, 0.09 to 0.46. The ^{238}Pu/^{239,240}Pu ratios in surface-water samples from the craters and outside the lagoon were even more variable. In fact, the ^{238}Pu/^{239,240}Pu ratios in a few collections made outside the lagoon were greater than 1; i.e., more ^{238}Pu than ^{239,240}Pu was present in the samples where the total concentration was less than 3 pCi/m^3.

An external source of ^{238}Pu, such as SNAP-9A, which burned up in the atmosphere over the Indian Ocean in 1964 (Volchok, 1969), may contribute to the higher ^{238}Pu/^{239,240}Pu ratios in the water column at Bikini. However, the amount of this material at any location would be small since 17 kCi of ^{238}Pu was dispersed throughout the earth's atmosphere and hydrosphere. Thus, on the basis of the above findings, the conclusions must be that different properties of the plutonium isotopes exist when the isotopes interact with various components of the marine environment at Bikini Atoll.

The ^{238}Pu appears to be more soluble than the ^{239,240}Pu in lagoon samples, as evidenced by the higher ratios found in the soluble fractions than in the particulate fractions from the BLVWS collections; this preferential solubility is also illustrated by the fact that the ^{238}Pu/^{239,240}Pu ratio is higher in many "batch" samples from the water column than in samples from the surface sediments. A source of ^{238}Pu that is different from bomb plutonium is indicated; most bomb debris would have much greater ^{239,240}Pu than ^{238}Pu concentrations except for those devices which used ^{242}Cm as a tracer. Since both ^{239}Pu and ^{240}Pu were measured together by alpha spectroscopy, some of the differences in the ^{238}Pu/^{239,240}Pu ratios possibly could be ascribed to variability in the ^{240}Pu isotope in samples. However, an evaluation of this radionuclide would require a more detailed study using mass spectrometry to measure the ^{240}Pu concentrations.

Plutonium in seawater at a pH 8.0 to 8.2 forms oxy–hydroxy–carbonatoplutonyl complexes. The size of the aggregates of the plutonyl complexes would depend on the number of plutonium atoms available and on the charge field surrounding the aggregates or clusters (the cluster hypothesis). At Bikini Atoll the coralline particles that experienced the effects of the fireball contain the plutonium isotopes. The release of plutonium into the water column from these particles may depend on recoil from the alpha decay of the plutonium isotopes; this decay would break the bonds between plutonium clusters and the coral matrix. The ^{238}Pu clusters would have a higher probability of being released from the coral particles than ^{239}Pu because of the differences in alpha-decay half-lives (86 yr for ^{238}Pu and 24,400 yr for ^{239}Pu) and possibly by ^{238}Pu formation from the decay of ^{242}Cm ($t_{\frac{1}{2}}$ of 162.5 days); thus it is reasonable to assume that ^{238}Pu could be more soluble than ^{239}Pu. However, the magnitude of this preferential solubility has not yet been determined.

If the clusters containing ^{238}Pu are smaller (i.e., in effect, more soluble) than those containing ^{239,240}Pu, then the results of the measurements made at Bikini lagoon and deep ocean areas could be explained. The larger clusters of ^{239,240}Pu could attach to the natural particles and could be removed from the water column at a more rapid rate than the more-soluble ^{238}Pu clusters. The availability of these different physicochemical states of plutonium may help decide the potential hazards of transuranic elements in the aquatic food chain to man. A concentrated effort is needed to collect additional data and to interpret further these preliminary findings.

Conclusions

The measurements of the radionuclides in Bikini lagoon sediments show that bomb craters are only one of the sources for the transuranic elements in the ecosystem. Sediments in the northwest quadrant of the lagoon contribute significantly to the concentrations of the radionuclides found in the water and biota. Coral particles that have been altered by the bomb and the environment contain the radionuclides. These particles must be transported and subsequently deposited at different locations; this is indicated by the high sediment rate found at station B-2 (0.58 cm/yr) and by the changes observed in the radionuclide concentrations found in the sediment-core profile.

(Text continues on page 576.)

TABLE 5   Measurements of the Particulate, Soluble, Colloidal, and Total Concentration of ^{239,240}Pu in Samples Collected by the Battelle Large Volume Water Sampler and Batch Methods at Bikini Atoll in July 1976

Station location	Depth, m	Volume,* m³	Battelle Large Volume Water Sampler						Batch‡ total,† pCi/m³
			Particulate,† pCi/m³	Particulate, %	Soluble,† pCi/m³	Colloidal, %	Total,† pCi/m³		
Station B-3									
N 11° 37.2′; E 165° 13.2′	2	1.624	16.7 ± 1.0	39	26 ± 2	7 ± 3	43 ± 2		55 ± 7
	2	0.445	29.0 ± 3.0		62 ± 47		91 ± 47		
	16	0.284	7 ± 2.3	19	31 ± 2	6 ± 3	39 ± 3		
	16	0.314	11.0 ± 1.2	24	35 ± 4	10 ± 3	46 ± 4		
	29	0.260	50 ± 3.6	80	13 ± 1		63 ± 4		72 ± 8
1930 hr	2								86 ± 29
2130 hr	2								<8
2330 hr	2								79 ± 8
0130 hr	2								42 ± 6
0330 hr	2								73 ± 13
0530 hr	2								69 ± 9
Station B-22									
N 11° 37.4′; E 165° 19.1′	2								94 ± 11
	2								98 ± 13
	2								78 ± 12
Station B-25									
N 11° 34.3′; E 165° 22′	2§	1.157	3.2 ± 0.2	11	27 ± 4	23 ± 5	30 ± 4		77 ± 10
	2§	0.343	7.3 ± 1.8	5	137 ± 10		144 ± 10		
	40§	0.204	7.5 ± 0.7	3	288 ± 61	3 ± 1	296 ± 62		90 ± 21
	40§	0.215	<1	<1	260 ± 40		260 ± 40		84 ± 10
	40§	0.281	8.2 ± 1.6	4	185 ± 31		193 ± 31		

Station B-8

N 11° 31.7′; E 165° 21′	2	0.284	<0.42	<1	85 ± 12	5 ± 1	85 ± 12	42 ± 10
	17	0.350	3.1 ± 0.3	3	100 ± 22		103 ± 22	33 ± 6
	40	0.280	3.2 ± 0.2	4	69 ± 11	3 ± 1	72 ± 11	28 ± 4

Station B-15

N 11° 36.6′; E 165° 30.9′	2	1.049	2.3 ± 0.3	9	23 ± 3		25 ± 4	61 ± 22
	2	0.986	2.7 ± 0.2	7	36 ± 4	19 ± 5	39 ± 4	
	17	0.235	2.6 ± 0.3	2	139 ± 35		142 ± 35	36 ± 5
	37	0.394	3.2 ± 0.2	3	91 ± 10	22 ± 3	93 ± 10	44 ± 9

Station B-18

N 11° 40.2′; E 165° 23.5′	2	0.553	17.5 ± 1.4	24	54 ± 11	10 ± 3	72 ± 12	
	15	0.319	21.8 ± 1.7	11	175 ± 75		197 ± 76	
	40	0.407	282 ± 21	49	294 ± 21	7 ± 2	576 ± 55	

Station B-32

N 11° 40′; E 165° 17′								
A	2§	2.337	6.6 ± 0.4	23	22 ± 2	4 ± 2	28 ± 2	41 ± 9
B	2§	2.354	6.1 ± 0.5	20	24 ± 1	20 ± 4	30 ± 1	
1	2¶	0.441	5.0 ± 0.7	15	29 ± 3	14 ± 2	34 ± 3	
2	2¶	0.263	6.2 ± 0.5	20	24 ± 3	8 ± 4	30 ± 3	
3	2¶	0.239	7.0 ± 1.8	22	26 ± 4	20 ± 4	33 ± 4	
4	2¶	0.280	13.5 ± 2.6	42	19 ± 2	5 ± 2	32 ± 3	
	17	0.242	5.0 ± 0.6	14	30 ± 3	34 ± 4	35 ± 3	46 ± 5
	33	0.301	10.2 ± 1.6	24	32 ± 3	16 ± 3	42 ± 3	45 ± 6
Chelex	2	0.276	10.0 ± 0.8	5	209 ± 24		219 ± 24	
Chelex	32	0.280	56.4 ± 4.1	53	50 ± 10		106 ± 11	
Chelex	32	0.250	63.2 ± 4.6	39	97 ± 15		160 ± 16	

(Table continues on the next page.)

TABLE 5 (Continued)

Station location	Depth, m	Volume,* m³	Battelle Large Volume Water Sampler					Batch‡ total,† pCi/m³
			Particulate,† pCi/m³	Particulate, %	Soluble,† pCi/m³	Colloidal, %	Total,† pCi/m³	
Station C-3								
N 11° 41.4′; E 165° 16.2′	2	0.424	8.6 ± 0.8	18	39 ± 7	10 ± 3	48 ± 7	39 ± 4
	2	0.401	9.4 ± 0.9	16	50 ± 10	6 ± 2	59 ± 10	
	41	0.227	81 ± 6	38	134 ± 21	8 ± 1	215 ± 22	86 ± 16
Station D-6								
N 11° 29.3′; E 165° 33.4′	2	0.361	5.2 ± 0.3	47	5.83 ± 0.71	16 ± 6	11.0 ± 0.8	6.8 ± 4
	100	0.343	<0.23	<1	1.39 ± 0.34	100	2.0	6.7 ± 2
	150	0.237	<0.29	<1	<1.20		<1.5	<3.8
Station D-9								
N 11° 33.3′; E 165° 12.1′	2	0.305	0.3 ± 0.1					<10
	67	0.341	0.5 ± 0.2	14	2.9 ± 0.4	85 ± 10	3.4 ± 0.5	12 ± 2
	148	0.319	<0.3	<1	0.8 ± 0.6		<1.0	
Station B-2								
N 11° 39.2′; E 165° 15.2′	2							45 ± 8
	26							110 ± 13
	42							332 ± 5

Station B-20

N 11° 38.5'; E 165° 16.4'   2   41 ± 5

Station B-21

N 11° 36'; E 165° 15.8'   2   243 ± 57

---

*Two sizes of BLVWS were used, 142 mm and 293 mm; the large sampler processed 1 to 3 m^3 of water whereas the smaller sampler processed 0.1 to 0.45 m^3 of water.

†The concentrations measured in picocuries per square meter ± 1 SD single-sample counting error.

‡The batch samples consisted of 90 liters of water which had been coprecipitated with MnO_2 using the method of Wong et al. (1975).

§Duplicate sample collections.

¶Four sequential samples collected during the times the duplicate samples were collected to measure the precision in collecting large and small volumes of water with the BLVWS.

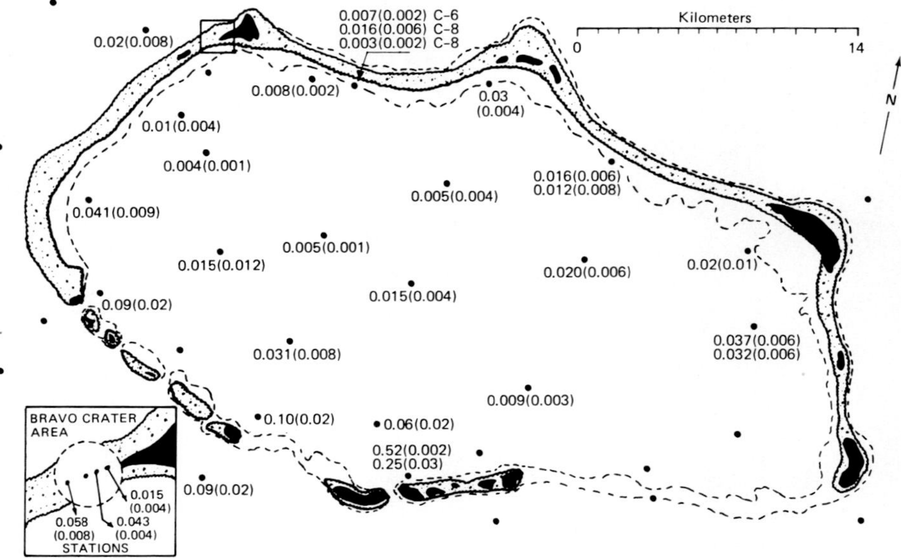

Fig. 18   Ratios of ^{238}Pu/^{239,240}Pu concentration in surface sediments collected in 1972.

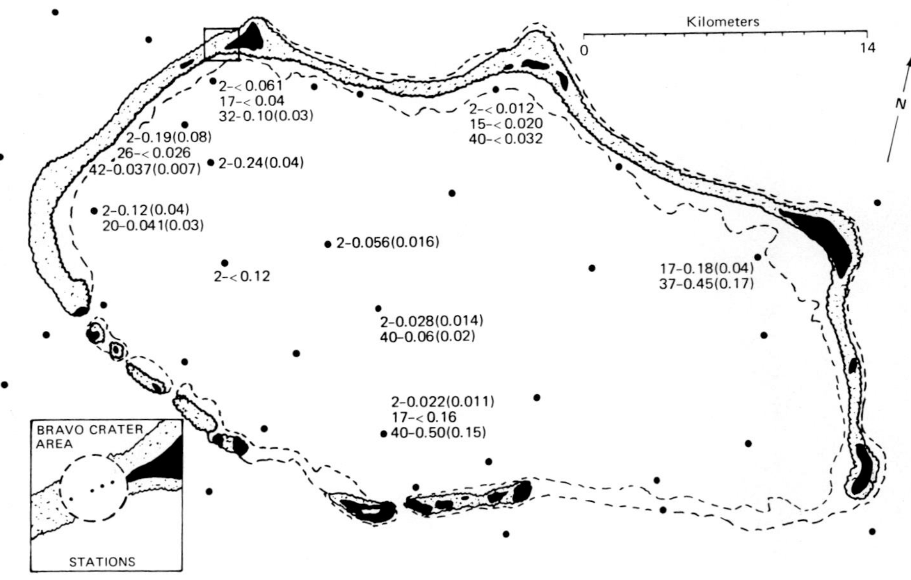

Fig. 19   Ratios of ^{238}Pu/^{239,240}Pu concentration in 90-liter batch samples collected in July 1976 at Bikini Atoll and coprecipitated with MnO_2.

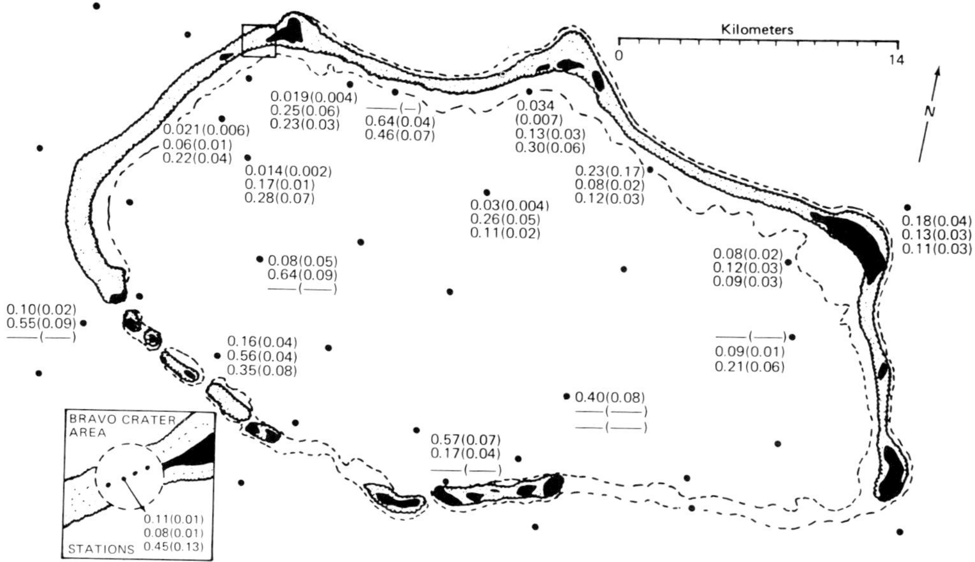

Fig. 20 Ratios of ^{238}Pu/^{239,240}Pu concentration in surface-water fractions from BLVWS collections in 1972: Top number, particulates (error); middle number, first Al_2O_3 bed (error); bottom number, second Al_2O_3 bed (error). — (–), no data available.

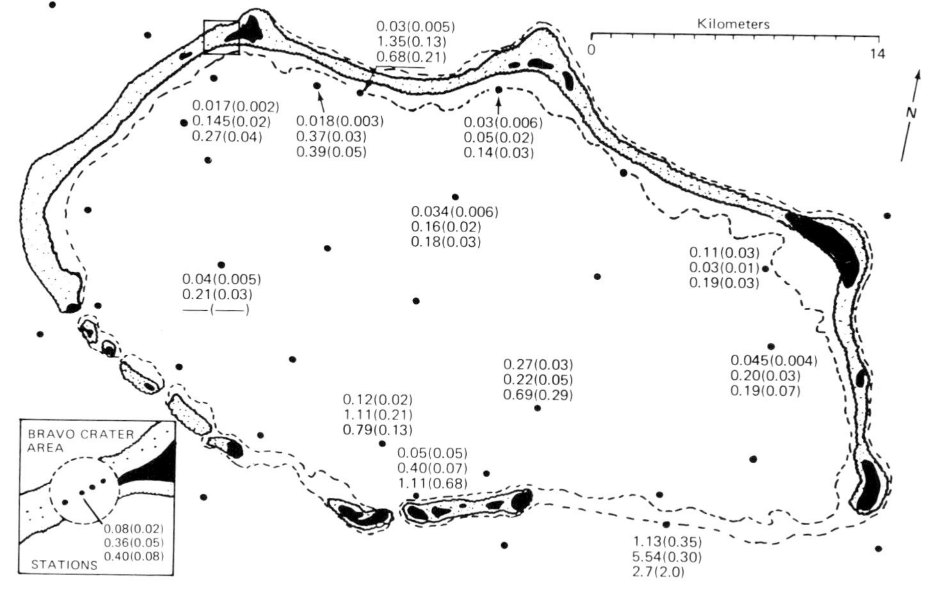

Fig. 21 Ratios of ^{238}Pu/^{239,240}Pu concentration in deep-water fractions from BLVWS collections in 1972: Top number, particulates (error); middle number, first Al_2O_3 bed (error); bottom number, second Al_2O_3 bed (error). — (–), no data available.

The distribution of plutonium concentrations in the water column follows the general lagoon circulation pattern with the highest concentration in the northwest quadrant and with decreasing concentration gradually toward the east and south. The outlet of the radioactive lagoon water is through the deep passes in the southwestern part of the Atoll where the exit into the North Equatorial Current occurs. The physicochemical state studies of plutonium show that approximately 15% of the total concentration is present in the colloidal fraction; varying amounts are found in the soluble and particulate fractions, depending on location. The americium is found primarily associated with the particulate fraction (>0.3 μm). Measurements of the isotope ratios of $^{238}Pu/^{239,240}Pu$ in the particulate, first Al_2O_3, and second Al_2O_3 fractions of a water sample are not constant. The ratios in these water fractions also vary with location in the lagoon. The ^{238}Pu appears to be more soluble than the $^{239,240}Pu$ in lagoon samples since the ratios are higher in the soluble fractions (first Al_2O_3 and second Al_2O_3) than in the particulate fraction. In an attempt to explain these observations, it is speculated that the ^{238}Pu forms smaller clusters than the $^{239,240}Pu$ owing to possible differences in specific ionization. This could be a result of the differences in decay half-lives (^{238}Pu, 86 yr; ^{239}Pu, 24,400 yr).

References

Adams, C. E., N. H. Farlow, and W. R. Schell, 1960, The Compositions, Structures and Origins of Radioactive Fallout Particles, *Geochim. Cosmochim. Acta,* 18: 42-56.

Andelman, J. B., and T. B. Ruzzell, 1970, Plutonium in the Water Environment, in *Radionuclides in the Environment,* R. F. Gould (Ed.), Advances in Chemistry Series, American Chemical Society, Washington, D. C.

Anikouchine, W. A., 1961, *Bottom Sediments of Rongelap Lagoon, Marshall Islands,* M.S. Thesis, University of Washington, Seattle, Wash.

Emery, K. O., J. I. Tracey, Jr., and H. S. Ladd, 1954, *Bikini and Nearby Atolls:* Part 1, Geology, Professional Paper No. 260-A, U. S. Geological Survey.

Farlow, N. H., and W. R. Schell, 1957, *Physical, Chemical and Radiological Properties of Slurry Particulate Fallout Collected During Operation Redwing,* Report USNRDL-TR-170, U. S. Naval Radiological Defense Laboratory, NTIS.

Goldberg, E. D., 1963, Geochronology with Pb-210, in *Radioactive Dating,* Symposium Proceedings, Athens, November 1962, pp. 121-131, STI/PUB/68, International Atomic Energy Agency, Vienna.

Held, E., 1971, *Amchitka Radiobiological Program Progress Report, July 1970–April 1971,* USAEC Report NVO-269-11, University of Washington, College of Fisheries, NTIS.

Huntamer, D. D., 1976, *An Evaluation of the Battelle Large Volume Water Sampler for Measuring Concentrations and Physico-Chemical States of Some Trace Elements in Marine Waters,* M.S. Thesis, University of Washington, Seattle, Wash.

Koide, M., A. Soutar, and E. D. Goldberg, 1971, Marine Geochronology with Pb-210, Earth and Planet, *Sci. Letters,* 14: 442-446.

Lowman, F. G., and R. Y. Ting, 1973, The State of Cobalt in Sea Water and Its Uptake by Marine Organisms and Sediments, in *Radioactive Contamination of the Marine Environment,* Symposium Proceedings, Seattle, July 10–14, 1972, STI/PUB/313, International Atomic Energy Agency, Vienna.

Marshall, R. P., 1975, *Concentrations and Redistributions of Plutonium, Americium and Other Radionuclides on Sediments at Bikini Atoll Lagoon,* M.S. Thesis, University of Washington, Seattle, Wash.

Nevissi, A., and W. R. Schell, 1975, Distribution of Plutonium and Americium in Bikini Atoll Lagoon, *Health Phys.,* 28: 539-547.

——, and W. R. Schell, 1976, Efficiency of a Large Volume Water Sampler for Some Radionuclides in Salt and Fresh Water, in *Radioecology and Energy Resources,* Proceedings of the Fourth National Symposium on Radioecology, Oregon State University, May 12–14, 1975, pp. 277-282, C. E.

Cushing, Jr. (Ed.), Ecological Society of America Special Publication Series, No. 1, Dowden, Hutchinson and Ross, Inc., Stroudsburg, Pa.

Noshkin, V. E., R. J. Eagle, and K. M. Wong, 1976, Plutonium Levels in Kwajalein Lagoon, *Nature,* 262: 745-748.

——, K. M. Wong, R. J. Eagle, and C. Gatrousis, 1974, *Transuranics at Pacific Atolls.* I. Concentrations in the Waters at Enewetak and Bikini, USAEC Report UCRL-51612, Lawrence Livermore Laboratory, NTIS.

Nozaki, Y., and S. Tsunogai, 1973, A Simultaneous Determination of Lead-210 and Polonium-210 in Sea Water, *Anal. Chem. Acta.,* 64: 209-216.

Robertson, D. E., W. B. Silker, J. C. Langford, M. R. Peterson, and R. W. Perkins, 1973, Transport and Depletion of Radionuclides in the Columbia River, in *Radioactive Contamination of the Marine Environment,* Symposium Proceedings, Seattle, July 10–14, 1972, STI/PUB/313, International Atomic Energy Agency, Vienna.

Schell, W. R., 1959, *Identification of Micron-Sized, Insoluble-Solids Fallout Particles Collected During Operation Redwing,* Report USNRDL-TR-364, U. S. Naval Radiological Defense Laboratory.

——, 1976, Concentrations and Physical–Chemical States of ^{55}Fe in Bikini Atoll Lagoon, in *Radioecology and Energy Resources,* Proceedings of the Fourth National Symposium on Radioecology, Oregon State University, May 12–14, 1975, pp. 271-276, C. E. Cushing, Jr. (Ed.), Ecological Society of America Special Publication Series, No. 1, Dowden, Hutchinson and Ross, Inc., Stroudsburg, Pa.

——, 1977, Concentrations, Physico-Chemical States and Mean Residence Times of ^{210}Pb and ^{210}Po in Marine and Estuarine Waters, *Geochim. Cosmochim. Acta,* 41: 1019-1031.

——, T. Jokela, and R. Eagle, 1973, Natural ^{210}Pb and ^{210}Po in a Marine Environment, in *Radioactive Contamination of the Marine Environment,* Symposium Proceedings, Seattle, July 10–14, 1972, pp. 701-724, STI/PUB/313, International Atomic Energy Agency, Vienna.

——, A. Nevissi, and D. D. Huntamer, 1978, Sampling and Analysis of Am and Pu in Natural Waters, *Mar. Chem.,* 6: 143-153.

——, and R. L. Watters, 1975, Plutonium in Aqueous Systems, *Health Phys.,* 29: 589-597.

Shapley, D., 1971, Plutonium: Reactor Proliferation Threatens a Nuclear Black Market, *Science,* 172: 143-146.

Sibley, T. H., and J. J. Morgan, 1976, *Equilibrium Speciation of Trace Metals in Freshwater. Sea Water Mixtures,* California Institute of Technology, unpublished.

Silker, W. R., R. W. Perkins, and H. C. Rieck, 1971, A Sampler for Concentrating Radionuclides from Natural Waters, *Ocean Eng.,* 2: 49-55.

Stumm, W., 1967, Metal Ions in Aqueous Solutions, in *Principles and Applications of Water Chemistry,* S. D. Faust and J. V. Hunter (Eds.), John Wiley & Sons, Inc., New York.

——, and J. J. Morgan, 1970, *Aquatic Chemistry: An Introduction Emphasizing Chemical Equilibria in Natural Waters,* John Wiley & Sons, Inc., New York.

Volchok, H. L., 1969, *Fallout of Pu-238 from the SNAP-9A Burnup-IV,* in USAEC Report HASL-207, pp. 1-5 to 1-13, Health and Safety Laboratory, NTIS.

Von Arx, U. S., 1954, *Circulation Systems of Bikini and Rongelap Lagoons,* in Professional Paper No. 260-B, pp. 254-273, U. S. Geological Survey.

Welander, A. D., et al., 1966, *Bikini–Eniwetok Studies, 1964.* Part I. Ecological Observations, USAEC Report UWFL-93(Pt. 1), University of Washington, College of Fisheries, Laboratory of Radiation Ecology, NTIS.

Transuranium Radionuclides
in Components of the Benthic Environment
of Enewetak Atoll

V. E. NOSHKIN

Data on the concentrations and distributions of transuranium radionuclides in the marine environment of Enewetak Atoll are reviewed. The distributions of the transuranics in the lagoon are very heterogeneous. The quantities of transuranics generated during the nuclear-test years at the Atoll and now associated with various sediment components are discussed. Whenever possible, concentrations of ^{241}Am and $^{239+240}Pu$ are compared. The lagoon is the largest reservoir of transuranics at the Atoll, and radionuclides are remobilized continuously to the hydrosphere from the solid source terms and are cycled with components of the biosphere. Although $^{239+240}Pu$ is associated with filterable material in the water column, the amount that is relocated and redeposited to different areas in the lagoon is small. Barring catastrophic events, little alteration in the present distribution of transuranics in the sediment is anticipated during the next few decades. The Atoll seems to have reached a chemical steady state in the partitioning of $^{239+240}Pu$ between soluble and insoluble phases of the environment. The amount of dissolved radionuclides predicted, with an experimentally determined K_d for $^{239+240}Pu$, to be in equilibrium with concentrations in the sediment agrees well with recently measured average concentrations in the water at both Enewetak and Bikini atolls. The remobilized $^{239+240}Pu$ has solute-like characteristics. It passes readily and rapidly through dialysis membranes and can be traced as a solute for considerable distances in the water. It is estimated that 50% of the present inventory of $^{239+240}Pu$ in sediment will be remobilized in solution and discharged to the North Equatorial Pacific over the next 250 yr.

Large inventories of several transuranium radionuclides (U. S. Atomic Energy Commission, 1973) persist in the marine environment of Enewetak Atoll. Forty-three nuclear weapons tests were conducted by the United States at Enewetak between 1948 and 1958. The testing produced close-in fallout debris which was contaminated with transuranics and which entered the aquatic environment of the Atoll. More transuranics were transported westward to Enewetak in airborne debris and water contaminated from nuclear testing at Bikini Atoll. Global fallout deposited a small additional amount of transuranics on the Atoll. Presently, the largest inventory of transuranics introduced from these source terms is associated with components of the benthic environment.

Because of the high level of deposition, the Atoll is now its own transuranic source term. Plutonium, for example, is not permanently fixed with the carbonates and other material with which it was originally deposited in the lagoon and on the reef during nuclear testing. Small amounts of plutonium are now remobilized, resuspended, assimilated, and transferred continuously within the Atoll environment by physical, chemical, and biological processes.

More than half the U. S. nuclear tests in the Pacific were conducted at Enewetak Atoll. Surface and tower shots left craters and contaminated scrap on land and generated radioactive debris that was redistributed to the adjacent reef and lagoon. Megaton tests that left underwater craters and barge shots in the lagoon contributed significantly to the present transuranic inventory.

The impact of nuclear testing and the fate of the residual radioactive materials introduced to the aquatic environment at both Enewetak and Bikini atolls are the subjects of reports too numerous to list herein. Not until late 1972, however, when a radiological resurvey of Enewetak Atoll was conducted to gather data for the development of cleanup and rehabilitation procedures for the resettlement of the Enewetak people to their homeland, did extensive measurements of transuranics in the Atoll environment begin. The information was published in a survey report (U. S. Atomic Energy Commission, 1973), which contains data on most long-lived residual radionuclides, including plutonium and americium, in components of the marine environment. The survey was followed by other more-extensive investigations, which concentrated on the measurement of transuranics to better assess the impact of these radionuclides on the environment and inhabitants of the Atoll and to increase our understanding of the mobilization, reconcentration, and redistribution processes from sources within the environment.

This chapter contains a summary of data related to the concentrations of the transuranium elements in components of the benthic and pelagic environment of the Atoll lagoon. Data from the survey report (U. S. Atomic Energy Commission, 1973), more-recent publications, and unpublished results from this laboratory are discussed. Some published and unpublished data from Lawrence Livermore Laboratory (LLL) studies at Bikini Atoll are presented when necessary for comparison with Enewetak data and, in the absence of Enewetak data, for the clarification of characteristics of transuranic radionuclide concentrations at the Atolls. Whenever possible, the Atoll data are compared with those from other marine ecosystems.

Geography and Atoll Test History

Enewetak Atoll, with U. S.-assigned and native names and several landmarks, including the locations of craters formed by nuclear tests, is shown in Fig. 1. The U. S.-assigned island names are used throughout this chapter.

The Atoll originally consisted of a ring of 42 low islands arranged on a roughly elliptical reef, 40.2 by 32.2 km (Emery, Tracy, and Ladd, 1954), with the elongated axis in the northwesterly direction. Nuclear testing completely destroyed the islands of Gene and Flora, and only a sandbar now remains to distinguish the island of Helen. Only 39 of the original 42 islands of the Atoll remain; these islands make up a total land area of approximately 6.9 km^2, which is situated on the reef which has an area of 84 km^2. The average depth of the lagoon is 47.4 m; the maximum depth is 60 m. The lagoon area is 933 km^2. The sedimentary components in Enewetak lagoon were studied extensively during the late 1940s (Emery, Tracy, and Ladd, 1954). The main components in the lagoon sediments included foraminifera, coral, *Halimeda* remains, shells of mollusks, and fine material. Material finer than 0.5 mm in diameter was too fine to identify and was classified as fine debris. Distributions and average abundance of the sedimentary components were described (Emery, Tracy, and Ladd, 1954). Fine debris made up 57% of the lagoon sediments and was abundant throughout the lagoon to within a few hundred feet from the shore.

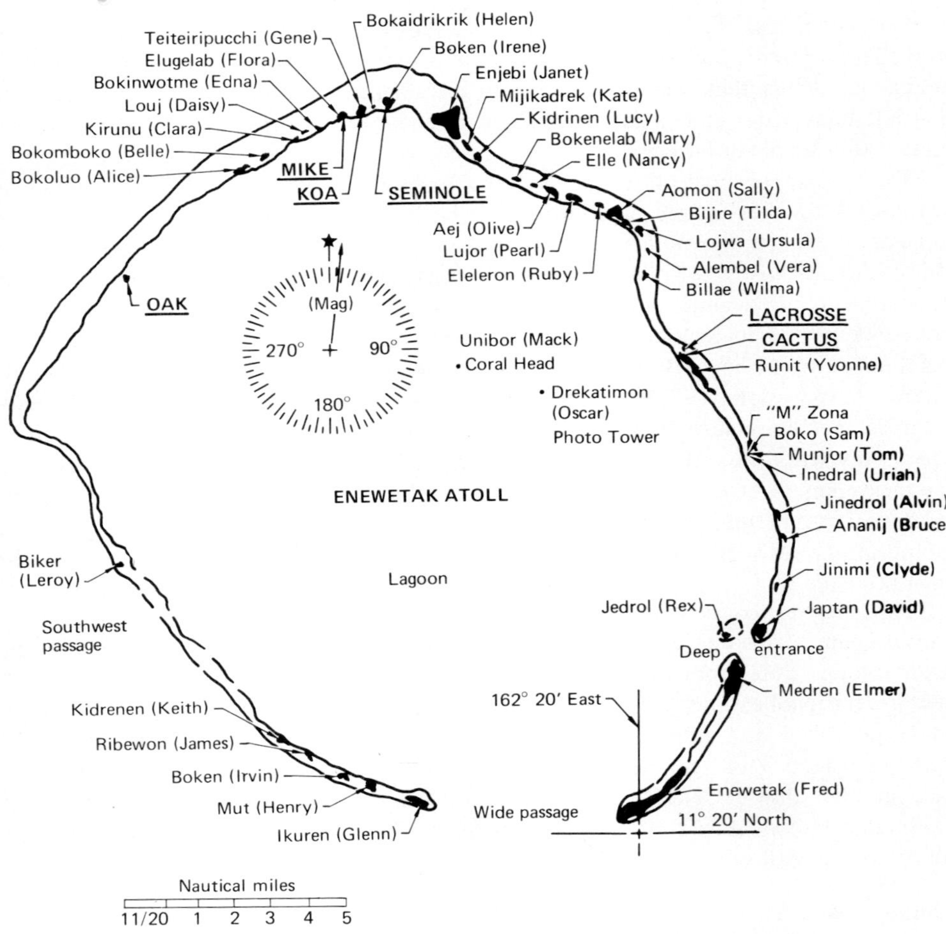

Fig. 1 Map of Enewetak Atoll with names and locations of the islands and the six nuclear craters.

A detailed description of the forms, living habits, populations, and specific relationships of the aquatic biological components at the Atoll is beyond the scope of this report. A significant number of articles published between 1955 and 1974 describing the research conducted at the Enewetak Marine Biological Laboratory were compiled recently in a three-volume report (Mid-Pacific Marine Laboratory, 1976). The individual reports dealing with specific ecological studies at the Atoll are too numerous to list. The reader is referred to the compilation for descriptions of the biology and ecology of the Atoll.

The most severe radiological impact on the aquatic environment of Enewetak occurred during the nuclear-test years between 1948 and 1958. The types of nuclear events, shot frequencies, geographical locations, yields, generated particles, conditions after the tests, and other factors determined the resulting distributions of transuranics and influenced the physical and chemical forms of the elements deposited in the benthic environment. A brief historical review of testing at Enewetak, abstracted from several unclassified documents (U. S. Atomic Energy Commission, 1973; Circeo and Nordyke, 1964; Hines, 1962), explains a few conditions responsible for the transuranic distributions and inventories at the Atoll.

The test series at Enewetak began in 1948 (Operation Sandstone) when 37-, 49-, and 18-kt devices were detonated from 200-ft towers on the islands of Janet, Sally, and Yvonne between April 14 and May 14. In 1951, testing was resumed (Operation Greenhouse), and four tower shots were conducted during a 47-day interval. The island of Janet was again the location of two ground zeros. In 1952, the first thermonuclear device (Mike) destroyed the island of Flora on the northwest reef. The Mike event, a 10.4-Mt surface detonation, occurred on October 31. Water surging from the point of the explosion sent a wave over adjacent islands, including Janet, the site of three previous ground zeros. The original crater where Flora had once been had an irregular outline and was more than 1 mile in diameter. Before the crater was partially refilled by the returning rush of coral sediment, it was almost 200 ft deep; it is presently 90 ft deep. The 1952 series of tests concluded with the King event, a high-yield airdrop over Yvonne Island. In 1954, a single device, Nectar, was detonated on a barge located over Mike crater. Not only did this test greatly disturb the radionuclides already deposited in the crater sediments but it also again sent a surge of contaminated water over adjacent islands, including Janet. In 1956, the Redwing series began with a tower detonation on Yvonne and included two additional cratering events, LaCross and Seminole. LaCross was a 39.5-kt device detonated on an earth-filled causeway built on the reef off the north end of Yvonne. Seminole, detonated on the island of Irene, was first placed in a 15-ft-diameter tank that was itself then placed in a 50-ft-diameter tank filled with water before it was fired. During 1958, the final year of testing at Enewetak, 22 tests of various types were held at different Atoll locations during an 82-day period. The series opened with an 86,000-ft balloon shot over the Atoll on April 28. On May 5, an 18-kt device produced Cactus crater on the northwest end of Yvonne and west of LaCross crater. During May 11 and 12, one of three tests conducted was the Koa event, a 1.37-Mt nuclear device housed in a tank of water and detonated on the east end of the Gene—Helen island complex. A sizable crater was produced, which connected with Mike crater. On June 8, the Umbrella device was detonated on the floor of the lagoon. Twenty days later, the 8.9-Mt Oak device was fired on a barge 4 miles southwest of Alice off the edge of the reef. The test left a crater that breached to the lagoon. The Quince event on Yvonne Island failed to produce a fission yield; so the plutonium within the device was dispersed by a high explosive. Subsequently another nuclear device was successfully detonated over the same area and undoubtedly further dispersed the nonnuclear-generated plutonium. In addition to the nuclear tests, radionuclides were dispersed by plowing on many of the islands during the test years. Unfortunately, none of the radiological safety reports during these operations provided details to determine the eventual fate of the radioactive debris, e.g., location and quantity of the disposal (U. S. Atomic Energy Commission, 1973).

From this brief summary, we can assume safely that the transuranic elements were introduced to the aquatic environment not only as complicated carbonate particles fused or condensed with other material from the environment or with devices and associated structures but also as soluble and particulate species of transuranium oxide.

Despite the complexities in the formation processes, much of the behavior of the transuranics is similar to that determined from investigations of fallout and other aquatic pathways. The results from the Atoll studies therefore have great value in predicting transuranic behavior and fate on a global aquatic scale.

Transuranic Elements Identified at the Pacific Test Site Atolls Since 1972

Neptunium

Concentrations of ^{237}Np in several 1972 samples of unfiltered lagoon and crater water from Enewetak were determined by mass spectrometry (Noshkin et al., 1974). The average concentration in six samples from the lagoon was 0.058 ± 0.013 fCi/liter. Water samples from Mike and Koa craters averaged 0.45 ± 0.22 fCi/liter. Outside the lagoon and to the east of the Atoll, concentrations in water samples from the open ocean surface averaged 0.013 ± 0.003 fCi/liter. This comparison shows, as do results for all other transuranics, that Atoll sources contribute the major fraction of the transuranic inventory in the water column of the lagoon. The ^{237}Np concentrations in the lagoon and crater water samples were less than 0.2% of the measured $^{239+240}$Pu concentrations in those samples.

Plutonium

Many types of samples from the Atoll contain ^{238}Pu, ^{239}Pu, ^{240}Pu, and ^{241}Pu. Most reported values are the sum of ^{239}Pu and ^{240}Pu activities determined by alpha spectrometry. These radionuclides are distributed widely throughout the Atoll and have been detected in nearly every type of marine and terrestrial sample analyzed to date. Atoll water samples, sedimentary components (including fine unidentifiable carbonate sands, coral fragments, *Halimeda* debris, foraminifera, and mollusk shells), living algae, benthic invertebrate tissues, planktonic species, and marine vertebrate tissue all contain $^{239+240}$Pu.

The distribution of ^{238}Pu is as wide among components in the marine environment as is $^{239+240}$Pu but at lower concentrations. The ^{238}Pu/$^{239+240}$Pu ratio determined in a variety of aquatic samples from different regions of the lagoon ranges from less than 0.04 to greater than 0.50.

A few activity ratios of ^{240}Pu/^{239}Pu were determined by mass spectrometry. The ratios in two water samples collected from the lagoon during 1972 were 0.432 and 0.289 (Noshkin et al., 1974). Samples of mackeral bone and of viscera collected in 1972 near the island of Glenn had ^{240}Pu/^{239}Pu activity ratios of 1.15 ± 0.25 and 1.27 ± 0.26, respectively; goatfish viscera and tridacna tissue from nearby David had ratios of 0.68 + 0.07 and 0.66 ± 0.19 (Gatrousis, 1975), respectively. The activity ratios in 56 soil samples from seven islands ranged from 0.066 to 1.42 and averaged 0.84 ± 0.37 (Gatrousis, 1975), and the average ratio in seven marine water and biota samples was 0.66 ± 0.40. Neither average value determined in the environmental samples differed greatly from the average of 0.65 ± 0.05 for global fallout debris (Krey et al., 1976). The similar isotopic ratio in mackeral tissue shows no obvious discrimination in uptake of isotopes by tissues of organisms in the Atoll if feeding and living are restricted to specific regions of the Atoll.

The average ^{240}Pu/^{239}Pu ratio in the yearly growth sections of a live sample of *Favites virens* coral collected from the western basin in Bikini lagoon was 0.77 ± 0.07 (Noshkin et al., 1975). This value is similar to the isotopic ratio in Enewetak samples. However, the mean isotopic concentration ratio in soil and vegetation of Bikini and Eneu islands is 1.15 (Mount et al., 1976), which is somewhat higher than the average in the Bikini coral sample.

Since the ^{240}Pu/^{239}Pu activity ratio in some environmental samples exceeds 1, it seems inappropriate to use the shorthand notation, ^{239}Pu, when referring to the sum of

^{239}Pu and ^{240}Pu activities as has so often been done in the literature. Throughout this report, $^{239+240}$Pu will refer to the sum of the activities of the two radionuclides, and ^{239}Pu will refer to only that isotope.

In two Enewetak lagoon water samples collected during 1972 (Noshkin, 1974), ^{241}Pu was measured by mass spectrometry. The ^{241}Pu/$^{239+240}$Pu activity ratios as of December 1972 were 1.14 and 2.56. In the 1972 growth section of the previously mentioned live coral from Bikini, the ^{241}Pu/$^{239+240}$Pu activity ratio was 11.7 and the ^{241}Pu/^{239}Pu ratio was 21.0 ± 1.1 (Noshkin et al., 1975). As of Jan. 1, 1975, the ^{241}Pu/^{239}Pu ratio in soil samples from Bikini and Eneu islands and in Bikini Island vegetation averaged 22.0 ± 3.3 (Mount et al., 1976). Correcting the ^{241}Pu in the November 1972 coral growth section for decay to Jan. 1, 1975, yields a ^{241}Pu/^{239}Pu ratio of 18.9 ± 1.1. Bikini and Eneu islands and the sedimentary environment from which the coral was obtained were contaminated principally with radioactive debris from the 1954 Bravo event. The good agreement between the ratios determined in the terrestrial and marine samples indicates a lack of discrimination between ^{241}Pu and ^{239}Pu isotopes in processes in these environments. Bikini and Enewetak have very different isotopic ratios and therefore different inventories of plutonium isotopes. The amount of ^{241}Pu in the environment regulates the projected inventory of ^{241}Am through growth and beta decay of the parent radionuclide. The amount of ^{241}Am that will be generated at Bikini from ^{241}Pu decay will exceed the amount produced by this source at Enewetak.

Americium

The distribution of ^{241}Am is also wide spread in the aquatic environment of the Atoll. Although the highest concentrations of plutonium and americium are in the same areas of the lagoon at Enewetak, the two transuranics are distributed differently. The ^{241}Am/$^{239+240}$Pu ratio in sediments collected from the lagoon during 1972 ranged from 0.06 to 0.93. Significant errors, therefore, can be introduced if one transuranic is used to predict the levels of others at any given location in the lagoon. No other americium isotopes were detected in the aquatic environment of either Enewetak or Bikini.

Curium

No strenuous effort has been made to obtain an inventory of ^{242}Cm or ^{244}Cm in Enewetak by alpha spectrometry. Curium activities were separated and measured in several lagoon water samples; ^{242}Cm activities were less than 0.2 fCi/liter. No ^{242}Cm or ^{244}Cm was detected in sediment samples from the Bravo crater at Bikini Atoll (Beasley, 1976).

Higher Transuranic Elements

No information is available to my knowledge on either berkelium or californium in marine samples from Enewetak or Bikini.

Transuranic Elements in the Benthic Environment of Enewetak Atoll

Surface-Sediment Distributions and Inventories

The distributions of $^{239+240}$Pu and ^{241}Am activities measured in December 1972 in the 2.5-cm-thick surface layer of sediment from the lagoon floor are shown in Figs. 2 and 3,

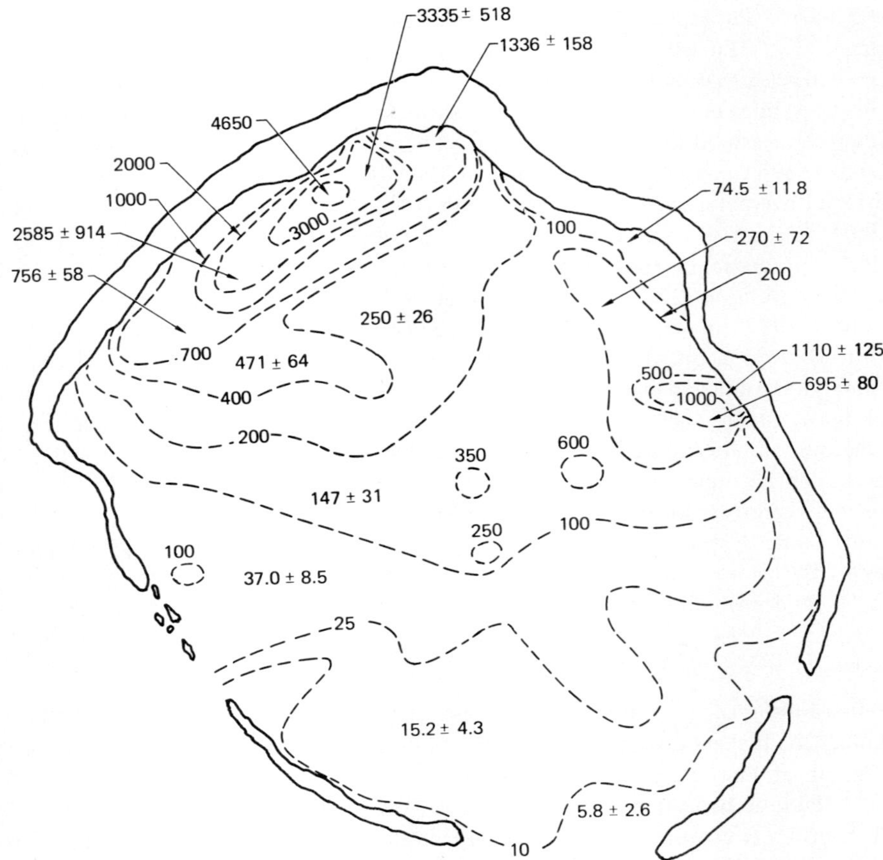

Fig. 2   Activities of $^{239+240}$**Pu (millicuries per square kilometer) associated with the sediment components in the top 2.5-cm layer of Enewetak lagoon.**

respectively. Isolines were constructed to distinguish regions of the lagoon having similar concentrations. The mean transuranic inventory in the surface layer and the range of concentrations within the defined areas are shown in the two figures. Figures 4 and 5 show regions of the surface layer of sediments with similar ^{238}Pu/$^{239+240}$Pu and ^{241}Am/$^{239+240}$Pu ratios, respectively.

The transuranic concentrations in the surface layer of sediments were determined in over 150 ball-milled surface samples of known thickness and in 20 core samples obtained throughout the lagoon. The lagoon was divided into a grid consisting of a series of 6-km^2 regions; at least one sediment sample was obtained from each region to provide radiological data for areal distributions. All sediments are composed of different quantities of fine- and coarse-grained carbonate material, shells, coral fragments, and *Halimeda* debris. No attempt was made in assessing the sediment inventory to distinguish concentration levels in specific sedimentary components. Figures 2 through 5 illustrate the main features of the transuranic distributions in the surface layer of the lagoon sediment. Isolated regions of relatively high concentrations of $^{239+240}$Pu are evident in some lesser contaminated areas of the lagoon; other small regions of high surface

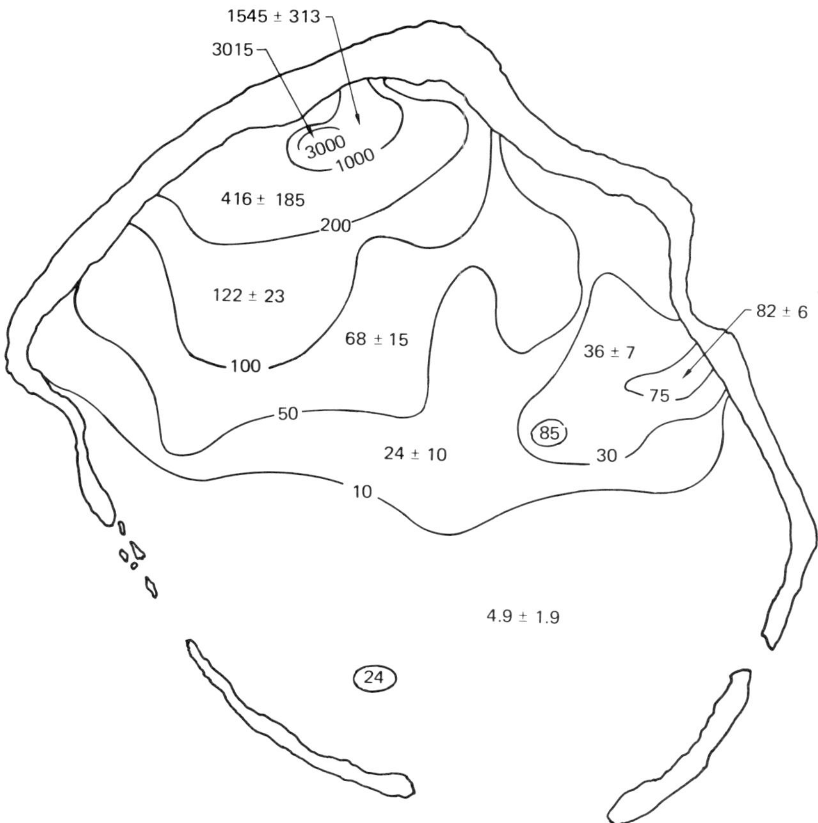

Fig. 3 Activities of [241]Am (millicuries per square kilometer) associated with the sediment components in the top 2.5-cm layer of Enewetak lagoon.

radioactivity might have escaped detection. The areal distributions are based on available data from the samples that were collected and analyzed.

The transuranics are distributed nonuniformly over the lagoon floor. Highest surface concentrations are associated with the sediments near, but not necessarily adjacent to, the locations of larger or more numerous nuclear tests. Highest plutonium concentrations are associated with the sediments from the northwest quadrant in a north- and south-oriented elliptical area that is roughly 2 to 3 km east of the islands of Alice and Belle and several kilometers southwest of Mike and Koa craters. A second region of relatively high concentration is in sediments off the shore of Yvonne Island. The activity in this region is lower than that in sediments in the northwest. Most of the transuranic inventory in the surface sediments can be separated roughly from the lesser contaminated deposits by a line extending from the Southwest Passage to the island of Tom (Munjor), which is south of Yvonne on the eastern reef. The surface $^{239+240}$Pu concentrations north of this line range between 2 and 170 pCi/g (dry weight); those south of this line are less than 2 pCi/g. All surface-sediment samples obtained during and since 1972 contained $^{239+240}$Pu. The inventory (mCi/km^2) in only the top 2.5-cm layer of sediment exceeds the activity deposited to the earth's surface as worldwide fallout in any latitude band in the northern or southern hemisphere (Hardy, Krey, and Volchok, 1973).

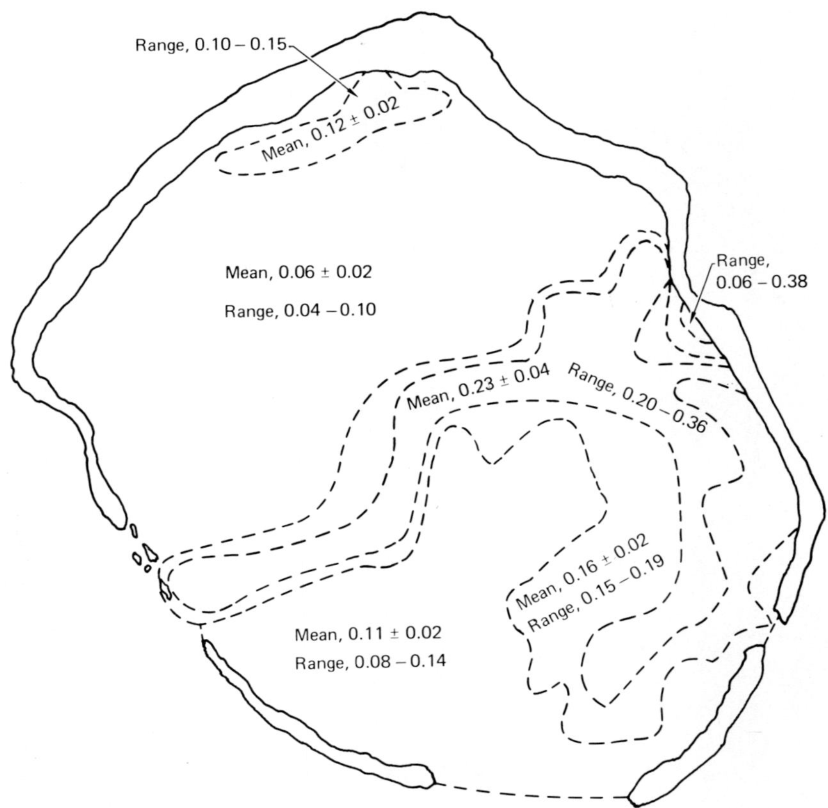

Fig. 4 Activity ratios of ^{238}Pu/$^{239+240}$Pu in the surface sediments of Enewetak lagoon.

Although the surface distribution of ^{241}Am in the sediments appears similar to that of $^{239+240}$Pu, the ratios of ^{241}Am/$^{239+240}$Pu activities (Fig. 5) show that the radionuclides are not well mixed throughout the surface deposits. The ratios in the sediments range from 0.06 to 0.93. The mean ratio, however, determined by averaging ^{241}Am/$^{239+240}$Pu ratios from all surface-sediment samples, is 0.29 ± 0.17, and the ratio determined from the mean surface concentrations (Table 1) is 0.30 ± 0.06. The average ratio is similar to that found in central Pacific and northeast Atlantic sediments (Livingston and Bowen, 1976), which receive only worldwide fallout deposition but have, in contrast, one-half the average concentration ratio of surface sediments at Bikini (Nevissi and Schell, 1975).

The ratios of ^{238}Pu/$^{239+240}$Pu activities in the surface sediments (Fig. 4) demonstrate the nonuniformity among plutonium isotopes in components of the sediment in the Atoll environment. There are, however, large geographical regions of the lagoon with similar isotopic ratios in the sediment. On the other hand, small areas of the lagoon, such as a 600-m strip on the lagoon side of Yvonne Island, contain plutonium with isotopic ratios ranging from 0.05 to 0.38 (U. S. Atomic Energy Commission, 1973). In Cactus crater, at the northern end of Yvonne, the isotopic ratio of 0.55 in the sediments is one of the highest at the Atoll. The average concentration ratio in the lagoon sediments,

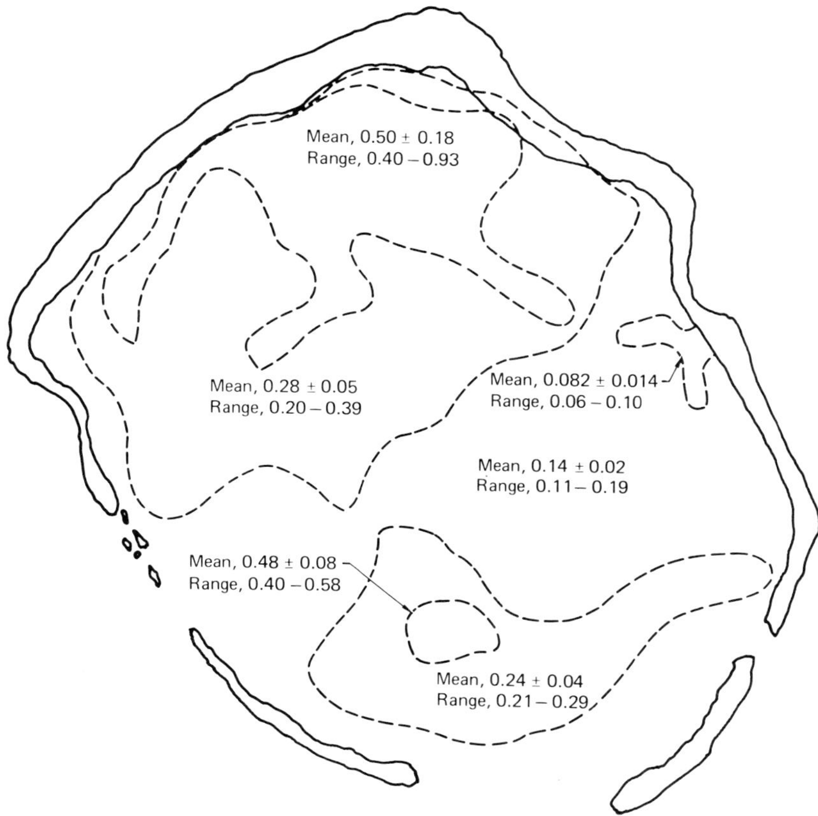

Fig. 5 Activity ratios of ^{241}Am/$^{239+240}$Pu in the surface sediments of Enewetak lagoon.

TABLE 1 Estimated Transuranic Sediment Inventory, Enewetak
and Bikini Atolls, Jan. 1, 1973

	$^{239+240}$Pu	^{238}Pu	^{239}Pu	^{240}Pu	^{241}Pu	^{241}Am
Enewetak Atoll (area, 933 km²)						
Areal activity to 2.5-cm depth, mCi/km²	267	38	145	122	493	81
Total radioactivity to 2.5-cm depth, Ci	249	35	135	114	460	76
Total radioactivity to 16-cm depth, Ci	1,185	167	642	543	2,190	475
Bikini Atoll (area, 629 km²)						
Areal activity to 2.5-cm depth, mCi/km²	492	16	229	263	4,809	289
Total radioactivity to 2.5-cm depth, Ci	309	10	144	165	3,025	182
Total radioactivity to 16-cm depth, Ci	1,470	76	686	786	14,405	1,140
Total radioactivity due to global fallout from weapons testing, January 1971, kCi	319	22*	192	127	3,010	72

*Weapons, 8.6 kCi; fallout debris from SNAP 9A, 13.4 kCi.

determined from the mean surface concentrations, is 0.14. The average ratio determined in the lagoon water samples during 1972, 1974, and 1976 is identical to the sediment ratio. A steady-state condition is reached where plutonium isotopes are remobilized to the aqueous phase in proportion to their concentrations in different regions of the sediments and reef environments. One region of the lagoon sediments with lower or higher isotopic ratios, for example, is not the dominant source term supplying plutonium isotopes to the water column.

In 1977, several core samples were obtained from the lagoon basin near stations that had been sampled in 1972. The ^{241}Am concentrations in surface-sediment layers sampled in 1972 and 1977 were nearly identical, which showed that there was little change in the surface concentrations of transuranics at many lagoon locations during those years. Only small quantities of the transuranics were remobilized or reworked to greater depths in the sediment column during these years. Little resuspended material from other areas of the lagoon having different concentrations of transuranics was transported and deposited to the areas that were resampled.

The largest inventory of transuranics at Enewetak Atoll is associated with the components of the lagoon sediment. The estimated lagoon sediment inventories given in Table 1 were determined from Figs. 2 and 3 by summing the products of the areas in the lagoon by the average inventory of the transuranics present there. Approximately 250 Ci of $^{239+240}$Pu and 75 Ci of ^{241}Am are unevenly distributed throughout the 2.5-cm-thick surface-sediment layer of the lagoon. The total $^{239+240}$Pu inventory in island soils, sampled to depths of 35 to 150 cm, is estimated from available data (U. S. Atomic Energy Commission, 1973; Noshkin et al., 1976) at <25 Ci. Transuranic distributions in surface sediment at Bikini Atoll were constructed, and inventories were estimated from published (Nevissi and Schell, 1975; Noshkin et al., 1975) and unpublished (Noshkin et al., 1978a) data. Bikini sediment inventories were estimated from substantially fewer data than were available from Enewetak. Future results from Bikini might change the present estimates of transuranic inventories given in Table 1. Analysis of 25 cores (12 to 21 cm deep) from different locations in Bikini and Enewetak lagoons showed that only $21 \pm 11\%$ of the $^{239+240}$Pu and $16 \pm 6\%$ of the ^{241}Am (Noshkin et al., 1978a) in the sediment column are associated with components in the top 2.5-cm layer. If the average $^{239+240}$Pu inventory in the surface sediment is only 21% of the total inventory to a mean depth of 16 cm for the entire lagoon, then the estimated $^{239+240}$Pu inventories in the sediment column to a 16-cm depth at Enewetak and Bikini are 1.2 and 1.5 kCi, respectively. However, in a few deeper cores, which are difficult to obtain from carbonate deposits, $^{239+240}$Pu and ^{241}Am were detected at depths below 20 cm. The inventories computed to a depth of 16 cm then can be assumed only to represent lower limits. With the average isotope ratios from samples from the Atoll environment (discussed earlier), an estimate of the concentration for each plutonium isotope and ^{241}Am in the Atoll sediments is made (see Table 1). Transuranic isotopes deposited from global fallout of weapons debris are estimated from available data (Krey et al., 1976; Hardy, Krey, and Volchok, 1973) and are also given in Table 1.

The inventory at the Atolls of transuranics produced by weapons is only a small fraction of the total quantity deposited to the earth's surface from global fallout debris. Some specific marine environments were contaminated with substantial quantities of transuranics from other source terms. These lagoon sediments, however, are the most contaminated aquatic regions in the world that received transuranic inputs only from nuclear weapons. The estimated ^{239}Pu, ^{240}Pu, ^{241}Pu, and ^{241}Am inventories at Bikini exceed the respective isotopic inventories at Enewetak, but ^{238}Pu is higher in Enewetak lagoon. Inventories of ^{241}Am at Bikini will increase by 25% from ^{241}Pu decay, but only a 10% increase over present ^{241}Am levels is expected at Enewetak from ^{241}Pu decay.

Transuranic Elements Associated with Components in the Sediment Column

In line with the definition of Emery, Tracy, and Ladd (1954) for classifying fines as material less than 0.5 mm in diameter, 23 surface samples and several cores were separated into fine and coarse fractions. The dry weight of the fines ranged from 25 to 80% of the total dry weight of the surface volume (Noshkin et al., 1978a). A similar range of fine material was found in Bikini Atoll sediments. At least 93% of the sediment weight in Mike and Koa crater deposits was fine material. In over 98% of the sediment samples from Enewetak lagoon, the ^{241}Am and $^{239+240}$Pu concentrations (pCi/g) associated with the fine sediment components were greater than or equal to the concentrations associated with the coarse fraction. The activity of ^{241}Am and $^{239+240}$Pu in the fines was 0.6 to more than 10 times that in the coarse fraction. These distributions between size fractions are very unlike those encountered for fallout of $^{239+240}$Pu in sediments in Buzzards Bay, Mass., where the $^{239+240}$Pu was not preferentially associated with the fine fractions of sedimentary deposits (Bowen, Livingston, and Burke, 1976). This difference is perhaps not unexpected because most of the transuranic inventory deposited to the lagoon environment was probably associated with small particulate carbonates. During the years since nuclear testing, some plutonium has exchanged slowly as a result of chemical reactions with exposed surfaces of the larger sedimentary components.

The transuranic inventory at lagoon locations, however, is dependent on the local abundance of the fine and coarse materials. Table 2 shows the ^{241}Am concentrations in the fine and coarse components of two core samples from midlagoon locations at Enewetak. The fraction of the coarse components in the sediment column of core 6 decreases with depth and in core 1 increases with depth. The ^{241}Am concentration associated with the fine fraction in the surface 2-cm section is 2 to 5 times the concentration associated with the coarse fraction; but, because the fine material in the surface layer of core 6 accounts for only 25% of the total dry weight of the sediment volume, 58% of the ^{241}Am in the surface 2-cm layer is associated with the coarse fraction. In core 1, on the other hand, 95% of the total ^{241}Am in the surface 2-cm layer is associated with the fine fraction. Although the ^{241}Am concentrations (pCi/g) associated with the fine material at various depths in the sediment column exceed the concentrations associated with the coarse components in both cores, the inventory of the radionuclide (pCi/cm^3) within any depth interval associated with the fine and coarse components can be variable throughout the sediment column. Areal transuranic distributions like those shown in Figs. 2 and 3 but associated with only the fine or only the coarse component of sediment would differ.

The vertical distributions of the transuranics in the lagoon sediment are very complex. No generalization about the shape of the concentration profile in any region can be made. Table 2, for example, shows a ^{241}Am peak associated with the fine components of core 6 at depths of 25 to 30 cm with little ^{241}Am associated with the coarse components at these depths. In core 1, the highest ^{241}Am concentrations are associated with the fine components at depths of between 8 and 10 cm in the sediment column. The ^{241}Am concentrations associated with the coarse component in both cores generally decrease gradually with depth. Transuranic concentrations increase, decrease, or remain constant

TABLE 2 Americium-241 Associated with Components in
Core Samples of Sediment

| Depth, cm | Concentration, pCi/g (dry weight) | | Inventory, pCi/cm³ | | Radioactivity associated with coarse component, % | Relative amount of coarse fraction, % (dry weight) |
	Fine fraction*	Coarse fraction*	Fine fraction	Coarse fraction		
Core 6						
0–2	5.97(23)	2.84(28)	0.57	0.78	57.8	74.6
2–4	4.99(7)	1.81(11)	1.54	0.99	39.1	64.0
4–6	6.44(5)	2.08(12)	2.68	0.95	26.1	52.4
6–8	5.51(6)	1.40(11)	2.88	0.63	17.9	46.4
8–10	3.10(6)	0.96(19)	1.74	0.39	18.3	41.6
10–15	0.72(18)	0.20(40)	0.38	0.09	19.1	45.9
15–20	<0.09	<0.09	<0.05	<0.05		48.6
20–25	5.83(7)	<0.09	3.75	<0.03	8.0	37.0
25–30	11.1(7)	0.16(42)	6.91	0.06	0.9	39.0
30–35	0.06	<0.08	<0.02	<0.02		48.3
35–40	0.06	<0.04	<0.04	<0.03		36.6
Core 1						
0–2	42.5(3)	12.9(16)	43.4	2.2	4.8	14.6
2–4	30.4(6)	6.31(5)	24.7	3.7	13.0	41.8
4–6	34.8(5)	4.90(13)	26.1	2.3	8.1	38.0
6–8	41.1(3)	4.91(20)	33.7	2.4	6.6	37.2
8–10	51.0(3)	4.07(14)	37.6	2.4	6.0	44.4
10–15	25.2(3)	0.98(17)	10.5	0.7	6.3	62.8
15–19	Lost	0.36(17)		0.2		93.0
19–25	3.55(18)	0.23(34)	0.2	0.01	4.7	90.7

*The values in parentheses are the 1σ counting errors expressed as the percent of the value listed.

with depth in sediment cores from other lagoon locations (Noshkin et al., 1978a). The concentrations of $^{239+240}$Pu and ^{241}Am associated with the carbonate components in four cores taken along a 1.5-km transect across Mike and Koa craters are shown in Fig. 6. The concentrations in the sediments from the Atolls' largest craters are surprisingly nonhomogeneous. Turbulence and large-scale mixing of the sediments during and after testing should have produced a much more uniform distribution than that found. The $^{239+240}$Pu concentration in the sediment column at station 17E is fairly uniform to a depth of 50 cm. At station 16E, the concentration increases with depth to 35 cm. The ^{241}Am concentration in the sediment column at station 16E decreases with depth. No correlation is obvious between the ^{241}Am and $^{239+240}$Pu concentrations associated with the components of these crater sediments. The craters should act as natural sediment traps, but little sedimentation in the Mike and Koa craters has occurred since the bottom depths were redetermined in 1964. In 1964 the maximum bottom depth of Mike crater was 27.4 m below sea level (U. S. Atomic Energy Commission, 1973). We have found no measurable change in the depth of the crater bottom during the period 1972 to 1977. Only small quantities of resuspended or reef-generated particulate material are then transported in the water masses to the western reef. Very little sedimentary material therefore escapes from the lagoon, and any resuspended bottom material probably settles out again on the lagoon floor close to its origin. The complex areal and vertical patterns

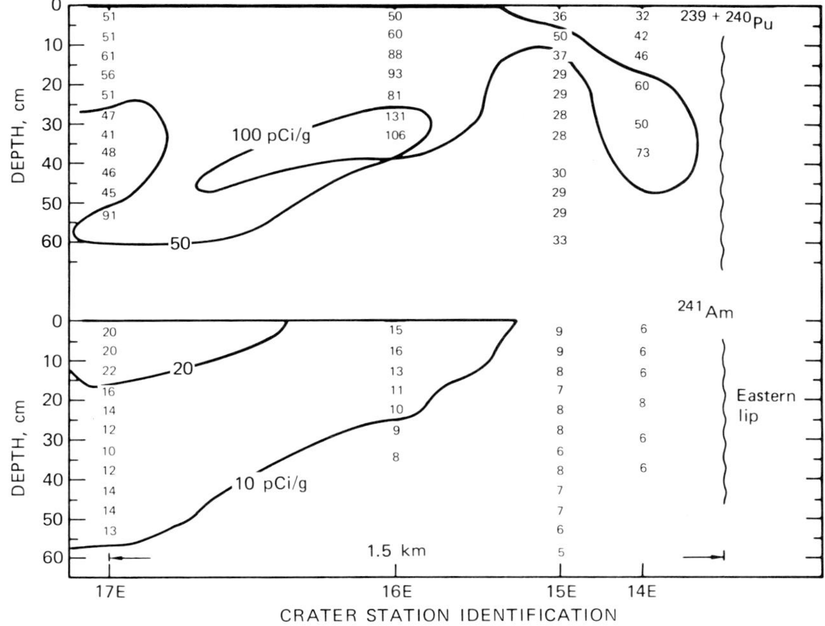

Fig. 6 Vertical and areal distributions of $^{239+240}$Pu and ^{241}Am activities in sediments in Mike and Koa craters.

of transuranics detected in this relatively small region of the lagoon where the distributions are expected to be more uniform are but examples of the complex patterns in the lagoon.

Halimeda, shells, coral, and foraminifera fragments were sorted from the coarse fraction of several sediment samples by hand. Table 3 shows the $^{239+240}$Pu concentrations associated with each component in the surface layer from two locations in Enewetak lagoon and at various depths in a core from Bikini lagoon. The $^{239+240}$Pu concentration associated with *Halimeda* fragments at station 40C only slightly exceeds that in fragments from station 3D. The concentrations associated with the separated foraminifera and coral fragments from station 40C are, however, at least 2.5 times higher than those associated with their respective components at station 3D. The distribution of $^{239+240}$Pu is different among components in the sediment from different regions of the lagoon. The fine fraction at these locations contained the highest concentration of $^{239+240}$Pu. To within our analytical precision, the ^{238}Pu/$^{239+240}$Pu concentration ratio is identical in the components from both stations.

In the sediment column from Bikini station B3, $^{239+240}$Pu was associated with all components that were separated. At all depths in the sediment column, the highest concentration of $^{239+240}$Pu in the coarse components was associated with *Halimeda* fragments. Sedimentation of labeled material to the lagoon occurs at a rate that is too slow to account for the buried activities below a few millimeters in the sediment column. Although the age of the *Halimeda* fragments, coral, and other components at depths greater than a few centimeters must therefore predate the test years, $^{239+240}$Pu is associated with these components.

**TABLE 3   Plutonium Concentrations Associated with
Sediment Components [pCi/g (dry weight)]**

Plutonium concentration of $^{239+240}$Pu [pCi/g(dry weight)] and
ratio of ^{238}Pu/$^{239+240}$Pu at Enewetak Atoll (surface 2.5-cm layer)

	Station 3D		Station 40C	
	$^{239+240}$Pu	^{238}Pu/$^{239+240}$Pu	$^{239+240}$Pu	^{238}Pu/$^{239+240}$Pu
Mollusk shells	0.64 ± 0.06	0.09 ± 0.03	Absent in sample	
Dead *Halimeda* fragments	4.8 ± 0.5	0.08 ± 0.01	6.0 ± 0.4	0.08 ± 0.01
Coral fragments	1.3 ± 0.1	0.08 ± 0.03	10 ± 5	0.13 ± 0.07
Fines (<0.5 mm)	6.85 ± 0.05	0.07 ± 0.01	23.5 ± 0.2	0.10 ± 0.01
Foraminifera	1.1 ± 0.1	0.07 ± 0.03	2.7 ± 0.2	0.10 ± 0.02

Plutonium concentrations at Bikini core Station B3,
pCi/g(dry weight)

Depth in sediment column, cm	*Halimeda* fragments	Foraminifera	Mollusk shells	Coral fragments
0–5	1.15 ± 0.06	0.324 ± 0.013	0.265 ± 0.013	0.101 ± 0.004
5–10	0.95 ± 0.03	0.154 ± 0.008	0.090 ± 0.004	0.130 ± 0.004
10–15	0.62 ± 0.01	0.093 ± 0.004	0.038 ± 0.003	0.063 ± 0.004
15–20	0.33 ± 0.01	0.024 ± 0.004	0.032 ± 0.002	0.073 ± 0.004
20–25	0.177 ± 0.003	0.012 ± 0.002	0.018 ± 0.002	0.007 ± 0.002
25–30	0.013 ± 0.001	0.002 ± 0.001	0.006 ± 0.001	0.001 ± 0.000
30–35	0.009 ± 0.001	0.001 ± 0.000	0.003 ± 0.001	0.008 ± 0.003

The possibility that subsurface remains labeled during testing were buried later in the sediment column by large-scale turbulence can be discounted. Coral or *Halimeda* fragments directly subject to a nuclear explosion probably would not retain their identity. In recent yearly growth increments of a living sample of *Favites virens* coral from station B3, the $^{239+240}$Pu concentrations averaged 104 ± 12 pCi/kg (Noshkin et al., 1975). This value agrees well with the $^{239+240}$Pu concentrations in dead coral remains in surface layers at station B3. In no yearly growth increment from this coral since 1954 was the $^{239+240}$Pu concentration below 104 ± 12 pCi/kg. Lower concentrations are associated with coral remains deeper in the sediment column. From the radiological record retained in the skeletal matrix of the *Favites virens,* coral labeled during 1954 and 1958, for example, should have $^{239+240}$Pu concentrations of 39 x 10^3 and 4.5 x 10^3 pCi/kg, respectively (Noshkin et al., 1975). These concentrations are orders of magnitude larger than those in any subsurface coral remains. These data, therefore, do not support the translocation of labeled coral material deeper into the sediment column by physical processes during or after testing. Burrowing organisms could redistribute some fraction of labeled sedimentary components to depths in the sediment column. However, when the $^{239+240}$Pu activities associated with each component at various depths are compared to the activity in the corresponding component at the surface, the $^{239+240}$Pu activities differ. For example, between 5 and 10 cm, the $^{239+240}$Pu concentrations associated with the coral, *Halimeda,* foraminifera, and shells are, respectively, 1.29, 0.83, 0.43, and 0.34 times the concentrations associated with those components in the surface layer. Burrowing and mixing processes by organisms are not likely to move specific components selectively down through the sediment column.

The data indicate that all plutonium does not remain associated with the sedimentary material with which it was originally deposited. Small quantities of plutonium are remobilized continuously from the sediments to the lagoon water column by surface-exchange mechanisms. Plutonium is also detected in the interstitial water extracted in situ from sediments (Noshkin et al., 1978b) at concentrations higher than those in the overlying bottom water at Enewetak Atoll. By equilibration, small quantities of $^{239+240}$Pu from the sediments are exchanged and released. Vertical diffusion moves the radionuclides to the sediment water interface where the plutonium mixes with the lagoon water mass. Remobilized plutonium can then be concentrated by members of the marine food chain. Vertical diffusion can also move the exchanged plutonium in the interstitial fluid deeper into the sediment column. Exchange of plutonium with exposed carbonate surfaces might account for the concentrations associated with material deeper in the sediment column.

Transuranic Elements Associated with the Calcareous Algae *Halimeda*

Debris from the calcareous algae *Halimeda* is the second most abundant component of Enewetak lagoon sediments and covers an estimated 26% of the lagoon floor (Emery, Tracy, and Ladd, 1954). Live species have recently been collected by divers and during dredging operations from numerous locations at both Enewetak and Bikini. Because algae were shown previously to concentrate plutonium (Noshkin, 1972), the role of this benthic algae in recycling the transuranic elements at the Atoll should be assessed.

The mean concentration factor for $^{239+240}$Pu associated with algae species from both atolls is 6×10^4 and ranges from 1×10^4 to 32×10^4 (Noshkin et al., 1978a). To within the precision of our measurements, the concentration factors for plutonium at the two atolls and of different *Halimeda* species from both atolls do not differ (Noshkin et al., 1978a). Concentrations of $^{239+240}$Pu associated with the live algae ranged from 0.4 to 22 pCi/g (wet weight), and the concentrations in the water where the algae were obtained ranged from 10 to 116 fCi/liter. Surface-sediment concentrations at the stations (Noshkin et al., 1978a; Nevissi and Schell, 1975) were compared to the algae concentrations at these sites. The average ratio of the $^{239+240}$Pu concentrations associated with the *Halimeda* species [pCi/g (dry weight)] to that in the top 2.5-cm sediment layer [pCi/g (dry weight)] was 0.24 ± 0.13, and the ^{241}Am concentration ratio was 0.32 ± 0.24. Concentrations of $^{239+240}$Pu and ^{241}Am in the sediment ranged from 9 to 82 pCi/g and from 1.1 to 67 pCi/g, respectively. On an equivalent weight basis, the live benthic algae have lower $^{239+240}$Pu and ^{241}Am levels than sediments in the immediate environment. The average plant/sediment concentration ratios of $^{239+240}$Pu and ^{241}Am are not statistically different. Thus there is no discrimination between $^{239+240}$Pu and ^{241}Am in processes beginning with remobilization of the transuranics from the environment and ending with concentration by the algae.

Table 4 summarizes data on transuranic concentrations in algae, water, and sediment from Cactus crater at Enewetak. The data show that the ^{238}Pu/$^{239+240}$Pu ratios in plants, water, and sediment are identical. In this crater ecosystem marine algae do not discriminate among the plutonium isotopes in the environment. The plant/sediment concentration ratios of $^{239+240}$Pu and ^{241}Am are nearly identical, which again shows that the processes of environmental release and plant uptake of the two transuranics are similar.

TABLE 4 **Concentrations of Transuranium Radionuclides in**
Halimeda **Algae, Sediment, and Water from Cactus Crater**

Radionuclides	*Halimeda monile**	Crater sediment	Crater water
[241]Am, pCi/g (dry weight)	2.53 ± 0.30	9.2 ± 2.0	
[239+240]Pu, pCi/g (dry weight)	18.68 ± 0.033	82 ± 2	116 ± 62 fCi/liter†
[238]Pu/[239+240]Pu	0.54 ± 0.03	0.54 ± 0.02	0.53 ± 0.02
[241]Am/[239+240]Pu	0.14 ± 0.02	0.11 ± 0.02	

[241]Am (*Halimeda*/sediment)‡ 0.28
[239+240]Pu (*Halimeda*/sediment)‡ 0.23
Concentration factor for [239+240]Pu = 7.5 × 10^4 [pCi/kg (wet weight) per pCi/kg H$_2$O]

*Wet weight/dry weight = 2.13.
†Average concentration in eight water samples from the crater bottom collected between 1974 and 1977.
‡Values are expressed as pCi/g (dry weight) *Halimeda*/pCi/g (dry weight) sediment.

The mean surface-sediment inventory of [239+240]Pu at Enewetak is 249 Ci (Table 1). The lagoon is 933 km^2 in area, and the average specific gravity of the *Halimeda* and other sediment components is 1.8 g/cm^3 (Emery, Tracy, and Ladd, 1954). Activities of [239+240]Pu associated with the algae are related to the activity in the surface sediment. The mean wet/dry ratio of the *Halimeda* species is 2.3, and the average wet weight of the plants, without holdfast, is 6.4 ± 3.8 g (Noshkin et al., 1978a). Therefore the average [239+240]Pu concentration associated with the live *Halimeda* species at Enewetak is 0.62 pCi/g (wet weight). Approximately 4.0 pCi is associated with each plant. If the number of *Halimeda* plants were known, the mean plutonium inventory associated with the living *Halimeda* reservoir could be computed. Unfortunately no estimates of *Halimeda* biomass at Enewetak are available. During the late 1940s, the mean sedimentation rate of *Halimeda* at Bikini was estimated at 3.8 mm/yr (Emery, Tracy, and Ladd, 1954). If this sedimentation rate is applicable to Enewetak Atoll, approximately 1 Ci of [239+240]Pu is deposited annually in the sediments in association with *Halimeda* detritus. This quantity represents only 0.4% of the surface-sediment inventory and a smaller yet fraction of the total inventory in the sediment column. If, however, the life-span of each plant is 1 yr, for example, a quantity of [239+240]Pu equivalent to half the present sediment inventory, or 125 Ci, could be recycled with the algae in approximately 175 yr. Spies, Marsh, and Colsher (1978) demonstrated that, when live *Halimeda* from Enewetak were cleaned and treated with 1N acetic acid, the acid-soluble fraction, or the carbonate material, contained 58% of the total [239+240]Pu, and 42% remained bound to the plant tissue. As the plant decomposes after death, the organic material and associated radioelements are released to the environment, leaving the skeletal carbonate matrix and its associated transuranics in the sedimentary deposits. The transuranics associated with the organic fraction released during decomposition are recycled to the benthic or pelagic environments. Over the long term the algae could play a key role in cycling the transuranics between the sediments and the aqueous environment.

Plutonium Concentrations in the Lagoon Seawater

A considerable number of lagoon water samples have been collected for plutonium analysis by this laboratory since 1972 (U. S. Atomic Energy Commission, 1973; Noshkin et al., 1974; Noshkin et al., 1976; Noshkin et al., 1978b). Several studies are in progress at the Atoll which require data on concentrations in lagoon water; so the number of samples and the locations sampled are predicated by the requirements of the current program. Contours of $^{239+240}$Pu concentrations in the water show complex distribution patterns (Noshkin et al., 1974) in various regions of the lagoon. The spatial patterns of surface and bottom $^{239+240}$Pu concentrations in solution and in association with filterable material are very different, as are the ^{238}Pu/$^{239+240}$Pu ratios in the water mass. A detailed discussion of the plutonium levels in the pelagic environment of the lagoon is in preparation (Noshkin et al., 1978b). Instead of relating all results from the analysis of lagoon water samples collected since 1972 with hydrological, seasonal, or spatial factors, I will summarize some of the data that are related to remobilization and redistribution of plutonium.

In 1972, 1974, and 1976, a sufficient number of water samples from the lagoon were analyzed for $^{239+240}$Pu to permit an estimate of mean concentrations in the lagoon. The mean concentrations are summarized in Table 5. In 1972, the average $^{239+240}$Pu concentration in the lagoon was determined for 34 unfiltered surface and bottom samples. A more-detailed water sampling program was conducted in 1974. In 1976, a smaller number of water samples were collected around the perimeter of the lagoon 2 km off the shore of the reef. Water samples collected during 1974 and 1976 were filtered through 1-μm filters. In the discussion to follow, the estimated average soluble $^{239+240}$Pu concentrations shown in Table 5 refer to material passing through a 1-μm filter.

During July 1974, the soluble $^{239+240}$Pu in the lagoon water ranged in concentration from 2 to 75 fCi/liter. The percentage of the total activity associated with the filterable material in the water samples during 1974 and 1976 ranged from 2% to 54% and from 12% to 94%, respectively. The concentrations of plutonium radionuclides in solution above fallout background concentrations in the lagoon water are direct evidence of the remobilization of transuranics from the solid phases of the environment. Dissolved plutonium released from the sediments of Cactus crater was traced for considerable distances along the reef by a plutonium radionuclide balance, which involved the change in the ^{238}Pu/$^{239+240}$Pu ratio in the water, and dyes to trace the crater water (Noshkin et al., 1978c). The dissolved plutonium moves in solution apparently without interacting with the sediment deposits during transport. The dissolved plutonium passes readily through dialysis membranes (Noshkin et al., 1978c). Equilibration between dissolved plutonium in the crater seawater and low-activity seawater contained in dialysis bags is achieved in 3 days (Noshkin et al., 1978c). These characteristics suggest that the plutonium remobilized to the environmental waters has very solute-like characteristics. It is tempting to suggest, considering the environment, that the remobilized chemical species is some form of carbonate complex.

The average concentration of total $^{239+240}$Pu in the water was essentially the same in 1972 and 1974, but a marked decrease was noted during 1976. In 1976, the average concentration associated with the filterable material in the lagoon doubled over the mean 1974 level, and the mean soluble concentration was reduced to half. Forty percent fewer samples were collected in 1976 than in 1974. During the 1974 program, samples were taken at stations throughout the lagoon, whereas the 1976 samplings were restricted to locations only 2 km from the reef. Similar $^{239+240}$Pu concentrations were found in water samples from the few 1974 locations resampled in 1976, which suggests that any computed mean concentration in the lagoon is contingent on the number and location of samples. The mean ^{238}Pu/$^{239+240}$Pu ratios in the lagoon water samples in 1972, 1974, and 1976 were virtually the same. Differences in sample ratios between soluble and particulate phases were noted sometimes, but the average ratios associated with the two phases from all stations were not significantly different.

During 1974 and 1976, 1.5 and 0.7 Ci of $^{239+240}$Pu, respectively, were found in solution, and 0.27 and 0.53 Ci, respectively, were associated with particulate material. These latter quantities represent less than 0.2% of the plutonium inventory in the surface sediment and less than 0.04% of the inventory estimated to a 16-cm depth in the sediment column (Table 1). The average quantities of soluble plutonium in the water are also small fractions of the sediment inventory. Therefore in recent years only small fractions of the Atoll plutonium inventory are either remobilized to the solution phase or resuspended to the water column.

During 1976, zooplankton samples contained less than 1% of the $^{239+240}$Pu activity in the total material filtered from an equivalent volume of water (Noshkin et al., 1978b).

TABLE 5 Plutonium Concentrations in the Water Column at Enewetak Atoll

		Sampling time		
		December 1972	July 1974	April 1976
Mean concentrations in lagoon water, fCi/liter				
Soluble (<1 μm)			35	16
Particulate (>1 μm)			6	12
	Total	39	41	28
Mean ^{238}Pu/$^{239+240}$Pu		0.12	0.13	0.13
Water-column inventory, mCi/km^2				
Soluble			1.65	0.76
Particulate			0.29	0.57
	Total	1.84	1.94	1.33
Lagoon inventory, Ci				
Soluble			1.54	0.71
Particulate			0.27	0.53
	Total	1.70	1.81	1.24
Water inventory, % of sediment inventory				
Water inventory compared to top 2.5 cm of sediment-surface inventory, %				
Soluble			0.62	0.29
Particulate			0.11	0.21
	Total	0.68	0.73	0.50
Water inventory compared to top 16 cm of sediment inventory, %				
Soluble			0.13	0.056
Particulate			0.023	0.045
	Total	0.14	0.15	0.10

The remaining $^{239+240}Pu$ in the particulate material is therefore associated with other forms of suspended matter. Between 1960 and 1963, Johannes (1967) investigated the composition of the suspended particles in the lagoon. Progressing from the eastern reef toward the lagoon, suspended benthic algae and sediment particles became less abundant with depth of the water as they settled to the bottom, and suspended macroscopic organic aggregates, consisting largely of mucus released by coral, increased progressively in size and number (Johannes, 1967). Often calcareous grains resuspended near the reef; microorganisms, copepod fecal pellets, and other undifferentiated material were incorporated with the aggregates. These materials and other particles produced in the pelagic environment are the most important food components for lagoon zooplankton and certain plankton-feeding fish (Gerber and Marshall, 1974). The small quantities of plutonium ingested with this particulate debris are dispersed over the lagoon by these organisms. Herbivorous fish play a role in the generation of particles in the water column (Smith, 1973). These fish are not efficient assimilators; while satisfying their energetic requirements, they disturb large quantities of material and release large amounts of unassimilated material containing plutonium in their feces. Moriarty (1976) estimates that a 200-g mullet, a species common to Enewetak, which feeds by scooping up bottom material to sift and remove small algae, will pass 50 g of dry sediment through its gut per day.

Bottom particles from the northwest quadrant of the lagoon, where highest plutonium concentrations in sediment are found, usually have high plutonium concentrations, which indicates that a fraction of the plutonium in the particulate phase may originate from turbulent resuspension of the sediment components in deep (60 m) water. The resuspended material and associated plutonium are probably not transported for any distance in the lagoon. Previous results indicate that the material is redeposited in the same general area of its origin. Only a few of the variety of active processes capable of generating and moving particulate plutonium in the water mass have been considered. It is remarkable that these and other processes resuspend so little of the plutonium inventory. Barring catastrophic events, the present distribution and inventory of plutonium in the sediments will be only slightly altered during the years by relocation of labeled material from other regions in the lagoon.

Laboratory studies with contaminated sediments and soils from Enewetak show that plutonium is rapidly partitioned between the solid phase and solution and reaches equilibrium after several days with an average distribution coefficient for plutonium of 1.8×10^5. Table 6 shows this and some recent determinations of the distribution coefficient for plutonium in laboratory and field experiments using a variety of sediments. Considering the difference in the types of environmental samples represented in Table 6, it is striking that the K_d for plutonium differs so little.

Table 1 shows that the mean plutonium inventory associated with the sediment components in the top 2.5 cm at Enewetak is 249 Ci. The lagoon sediment has an average density of 1.8 g/cm^3 (Emery, Tracy, and Ladd, 1954) and occupies an area of 933 km^2. The mean depth of the lagoon is 47.4 m. With these data and the K_d for plutonium of 1.8×10^5, one can construct a simple model to predict the average concentration expected in the lagoon water by assuming that the plutonium in solution is in equilibrium with that in the sediments. At any time the amount of plutonium in solution is limited by the saturation of the solution under equilibrium conditions. The rate at which water and its dissolved plutonium is flushed from the lagoon is balanced by input of uncontaminated ocean water, which is rapidly saturated with remobilized plutonium from the

**TABLE 6   Recent Determinations of the Distribution
Coefficient for Plutonium**

Sediment type	K_d		Reference
	Range	Average*	
Enewetak, coral soil and sediment (laboratory desorption study)	$4 \times 10^4 - 3 \times 10^5$	1.8×10^5	Unpublished
Enewetak, groundwater particulates	$1.4 \times 10^4 - 10^6$	2.5×10^5	Noshkin et al., 1976
Trombay Harbour, suspended silt	$4.8 \times 10^4 - 1.3 \times 10^5$	0.9×10^5	Pillai and Mathew, 1976
Bikini Tewa crater sediment (laboratory desorption study, oxic−anoxic conditions)	$4 \times 10^4 - 4 \times 10^5$	2.2×10^5	Mo and Lowman, 1976
Windscale area, 5% clay, 50% silt, 45% sand	$0.6 \times 10^4 - 22 \times 10^4$	0.5×10^5	Hetherington, Jefferies, and Lovett, 1975
Humboldt Bay, Calif., suspended clay−silt particulates	$4.7 \times 10^4 - 11.8 \times 10^4$	0.8×10^5	Unpublished data, this laboratory
Lake Michigan, suspended particulates		3.0×10^5	Wahlgren et al., 1976
Mediterranean sediment (laboratory sorption study)	$1.3 \times 10^4 - 9.4 \times 10^4$	0.5×10^5	Duursma and Parsi, 1974
	Mean	$1.4 \times 10^5 \ cm^3/g$	

*Quantity of $^{239+240}Pu$ bound to the sediment per unit dry weight of sediment divided by the amount of $^{239+240}Pu$ in water per cubic centimeter.

Atoll source terms. If plutonium is cycled through an intermediate host, such as the *Halimeda,* the rate at which it is released from decaying plants must be balanced by uptake in the new growth to maintain a steady state condition. Given that steady state conditions exist, the mean plutonium inventory in the lagoon water and the concentration expected in solution computed from the basic equation relating K_d to water and sediment concentrations are 1.4 Ci and 32 fCi/liter, respectively. There is general agreement between the average quantity of $^{239+240}Pu$ predicted and that measured in solution (see Table 5). In 1976, the computed value differed from the measured mean soluble concentration by a factor of 2. Although this is not a large discrepancy, the average concentration, as was mentioned previously, probably does not represent the real mean for the lagoon at the time sampled. From the appropriate dimensions for the Bikini lagoon, the sediment data in Table 1, and the K_d for $^{239+240}Pu$, the average inventory in the water column and the concentration at Bikini are computed to be 1.7 Ci and 60 fCi/liter, respectively. During December 1972, the mean soluble $^{239+240}Pu$ inventory and the concentration in the lagoon water were 1.2 Ci and 42 ± 21 fCi/liter, respectively (Noshkin et al., 1974), and in January 1977 the respective values were 1.4 Ci and 49 ± 21 fCi/liter (Noshkin et al., 1978b). These average values also are consistent with the amounts predicted.

For many reasons it may be argued that some of this agreement is fortuitous. Nevertheless, the general agreement found between computed and twice-measured average concentrations in both lagoons between 1972 and 1977 demonstrates the general usefulness of this simple equilibrium model in predicting long-term average concentrations in lagoon water. From radiological records retained in yearly growth of coral sections (Noshkin et al., 1975; Noshkin et al., 1978a), Bikini and Enewetak lagoon water along with dissolved species are estimated to be exchanged approximately twice per year. At this rate of exchange under steady-state conditions, slightly more than 250 yr will be required to reduce the plutonium inventory in the sediment by 50%. The rates of the mobilization and migration processes of plutonium away from the Atoll to the equatorial Pacific waters are much faster than the rate of radioactive decay. These figures and results should be considered when the consequences of disposal methods for transuranic wastes to the oceans are discussed.

Some massive corals collected from the atolls contain well-defined growth bands dating from the collection time to the early 1950s. Each yearly growth concentrates plutonium in proportion to the levels in the environment (Noshkin et al., 1975; Noshkin et al., 1978a). Concentrations of $^{239+240}$Pu associated with growth increments dated since 1965 in three Enewetak corals from different locations in the lagoon and one Bikini lagoon sample are given in Table 7. The average amount of plutonium concentrated by the coral from 1965 until the year of collection has been computed and is shown in Table 7.

The average absolute concentrations in the corals are different, as expected, and reflect the local environmental concentrations in the region. In only a few growth sections are the $^{239+240}$Pu concentrations different from the mean by more than a factor of 2, and only corals 1 and 2 show this magnitude of variation. Corals 1 and 2 were obtained in the water on the lagoon side of the eastern reef. The patterns of current in

TABLE 7 Concentrations of $^{239+240}$Pu in Yearly Growth Sections of Enewetak and Bikini Coral

| | $^{239+240}$Pu concentrations, fCi/g (dry weight) | | | |
| | Enewetak | | | Bikini, |
	Coral 1 *(Favia pallida)*	Coral 2 *(Gonrastrea retiformis)*	Coral 3 *(Favia pallida)*	Coral 4 *(Favites virens)*
Year of growth section				
1974			7.4(13)*	
1973	3.9(12)*	19.6(5)*	5.2(14)	
1972	2.3(9)	35.5(4)	6.7(13)	130(7)*
1971	0.9(25)	13.5(6)	7.0(10)	130(7)
1970	1.3(14)	5.0(12)	6.3(12)	100(5)
1969	4.9(10)	41.7(4)	5.9(11)	100(4)
1968	7.9(8)	10.4(6)	6.9(10)	100(2)
1967	3.6(6)	12.9(7)	Lost	100(5)
1966	5.1(11)	11.2(7)	3.7(13)	90(5)
1965	3.2(9)	10.5(6)	6.0(8)	110(5)
Average concentration (1965 to year of collection)	3.7 ± 2.1	17.8 ± 12.5	6.1 ± 1.1	108 ± 15
Date of collection	October 73	April 74	August 74	November 72

*Values in parentheses are 1σ counting errors expressed as the percent of the value listed.

this region of the lagoon are variable, and the windward reef community contributes a significant detrital load with associated plutonium to the lagoon. Since growing coral is a point source in the environment, small changes in even the local circulation, to name one of many factors, will greatly alter the plutonium concentration in the vicinity of the coral. It is rather more surprising that, for the most part, the $^{239+240}$Pu levels associated with the last 9 yr of growth are nearly constant. This shows that the dissolved $^{239+240}$Pu levels available to the corals in a specific region have also been similar during the last 9 yr.

These results from coral and other studies demonstrate that Enewetak lagoon has attained a chemical steady-state condition with respect to plutonium remobilization from solid components to solution. Not only will the simple equilibrium model explain average concentrations in lagoon water but also it can be used to estimate local concentrations expected in the waters from areas of the Atoll with different levels of contamination. By using appropriate concentration factors for plutonium, one can estimate the quantities accumulated by marine organisms anywhere in the lagoon. The data on biotic concentration can be used to estimate the potential dose to man if part or all of the Atoll were to supply his marine food requirements.

Acknowledgments

I wish to express my appreciation to several coworkers, K. Wong, R. Eagle, R. Spies, K. Marsh, T. Jokela, J. Brunk, G. Holladay, and L. Nelson, who provided much of the previously unreported data and without whose efforts in the field our program in the Marshall Islands could not be carried out. This work is supported by the Division of Biology and Environmental Research of the U. S. Department of Energy, contract No. W-7405-ENG-48.

References

Beasley, T. M., 1976, Analysis of Windscale and Bikini Atoll Sediments for ^{242}Am, in *Activities of the International Laboratory of Marine Radioactivity 1976 Report,* Report IAEA-187, International Atomic Energy Agency, Vienna.

Bowen, V. T., H. D. Livingston, and J. C. Burke, 1976, Distributions of Transuranium Nuclides in Sediment and Biota of the North Atlantic Ocean, in *Transuranium Nuclides in the Environment,* Symposium Proceedings, San Francisco, 1975, p. 107, STI/PUB/410, International Atomic Energy Agency, Vienna.

Circeo, L. J., Jr., and M. D. Nordyke, 1964, *Nuclear Cratering Experience at the Pacific Proving Grounds,* USAEC Report UCRL-12172, Lawrence Livermore Laboratory, NTIS.

Duursma, E. K., and P. Parsi, 1974, Distribution Coefficient of Plutonium Between Sediment and Seawater, in *Activities of the International Laboratory of Marine Radioactivity 1974 Report,* Report IAEA-163, p. 94, International Atomic Energy Agency, Vienna.

Emery, K. O., J. I. Tracy, and H. S. Ladd, 1954, *Geology of Bikini and Nearby Atolls,* Professional Paper 260-A, U. S. Geological Survey.

Gatrousis, C., 1975, Lawrence Livermore Laboratory, private communication.

Gerber, R. P., and N. Marshall, 1974, Ingestion of Detritus by the Lagoon Pelagic Community at Enewetak Atoll, *Limnol. Oceanogr., 19:* 815.

Hardy, E. P., P. W. Krey, and H. L. Volchok, 1973, Global Inventory and Distribution of Fallout Plutonium, *Nature, 241:* 444.

Hetherington, J. A., D. F. Jefferies, and M. B. Lovett, 1975, Some Investigations into the Behaviour of Plutonium in the Marine Environment, in *Impacts of Nuclear Releases into the Aquatic Environment,* Symposium Proceedings, Otaniemi, Finland, 1975, p. 193, STI/PUB/410, International Atomic Energy Agency, Vienna.

Hines, N. O., 1962, *Proving Grounds, An Account of the Radiobiological Studies in the Pacific, 1946–1961,* University of Washington Press, Seattle, Wash.

Johannes, R. E., 1967, Ecology of Organic Aggregates in the Vicinity of a Coral Reef, *Limnol. Oceanogr.*, 12: 189.

Krey, P. W., E. P. Hardy, C. Pachucki, F. Rourke, J. Coluzza, and W. K. Benson, 1976, Mass Isotopic Composition of Global Fallout Plutonium in Soil, in *Transuranium Nuclides in the Environment*, Symposium Proceedings, San Francisco, 1975, p. 671, STI/PUB/410, International Atomic Energy Agency, Vienna.

Livingston, H. D., and V. T. Bowen, 1976, Americium in the Marine Environment—Relationships to Plutonium, in *Environmental Toxicity of Aquatic Radionuclides; Models and Mechanisms*, M. W. Miller and J. Newell Stannard (Eds.), p. 107, Ann Arbor Science Publishers, Inc., Ann Arbor, Mich.

Mid-Pacific Marine Laboratory, Enewetak Marine Biological Laboratory, 1976, *Contributions 1955–1974*, Report NVO-628-1, Vols. I-III.

Mo, T., and F. G. Lowman, 1976, Laboratory Experiments on the Transfer of Plutonium from Marine Sediments to Seawater and to Marine Organisms, in *Radioecology and Energy Resources*, Proceedings of the Symposium on Radioecology, Oregon State Univ., May 13–14, 1975, The Ecological Society of America Special Publications: No. 1, C. E. Cushing, Jr. (Ed.), p. 86, Halsted Press, New York.

Moriarty, D. W., 1976, Quantitative Studies on Bacteria and Algae in the Food of the Mullet, *Mugil Cephalus L.*, and the Prawn, *Metapenaeus Bennettae, J. Exp. Mar. Biol. Ecol.*, 22: 131.

Mount, M. E., W. L. Robison, S. E. Thompson, K. O. Hamby, A. L. Prindle, and H. B. Levy, 1976, *Analytical Program—1975 Bikini Radiological Survey*, USAEC Report UCRL-51879, Part 2, Lawrence Livermore Laboratory, NTIS.

Nevissi, A., and W. R. Schell, 1975, Distribution of Plutonium and Americium in Bikini Atoll Lagoon, *Health Phys.*, 28: 539.

Noshkin, V. E., 1972, Ecological Aspects of Plutonium Dissemination in the Aquatic Environment, *Health Phys.*, 22: 537.

——, K. M. Wong, R. J. Eagle, and C. Gatrousis, 1974, *Transuranics at Pacific Atolls. I. Concentrations in the Waters at Enewetak and Bikini*, USAEC Report UCRL-51612, Lawrence Livermore Laboratory, NTIS.

——, K. M. Wong, R. J. Eagle, and C. Gatrousis, 1975, Transuranics and Other Radionuclides in Bikini Lagoon: Concentration Data Retrieved from Aged Coral Sections, *Limnol. Oceanogr.*, 20: 729.

——, K. M. Wong, K. Marsh, R. Eagle, G. Holladay, and R. W. Buddemeier, 1976, Plutonium Radionuclides in the Groundwater at Enewetak Atoll, in *Transuranium Nuclides in the Environment*, Symposium Proceedings, San Francisco, 1975, p. 517, STI/PUB/410, International Atomic Energy Agency, Vienna.

——, K. M. Wong, R. Spies, R. Eagle, and K. Marsh, 1978a, *Transuranics at Pacific Atolls— Concentrations in Sediment Components and Benthic Organisms at Enewetak and Bikini*, Lawrence Livermore Laboratory, in preparation.

——, K. M. Wong, R. J. Eagle, T. Jokela, J. Brunk, and K. V. Marsh, 1978b, *Transuranics in Seawater and Pelagic Components at Enewetak and Bikini*, Lawrence Livermore Laboratory, in preparation.

——, R. J. Eagle, K. M. Wong, and K. V. Marsh, 1978c, *Remobilization of Plutonium Radionuclides from Cactus Crater Sediment at Enewetak*, in preparation.

Pillai, K. C., and E. Mathew, 1976, Plutonium in the Aquatic Environment, Its Behaviour, Distribution and Significance, in *Transuranium Nuclides in the Environment*, Symposium Proceedings, San Francisco, 1975, p. 25, STI/PUB/410, International Atomic Energy Agency, Vienna.

Smith, S. V., 1973, Carbon Dioxide Dynamics: A Record of Organic Carbon Production, Respiration and Calcification in the Eniwetok Reef Flat Community, *Limnol. Oceanogr.*, 18: 106.

Spies, R. B., K. V. Marsh, and J. Colsher, 1978, Dynamics of Radionuclide Exchange in the Calcareous Algae, *Halimeda, Limnol. Oceanogr.* (in preparation).

U. S. Atomic Energy Commission, 1973, *Enewetak, Radiological Survey*, USAEC Report NVO-140, Vols. I-III, Nevada Operations Office, NTIS.

Wahlgren, M. A., J. J. Alberts, D. M. Nelson, and K. A. Orlandini, 1976, Study of the Behaviour of Transuranics and Possible Chemical Homologues in Lake Michigan Water and Biota, in *Transuranium Nuclides in the Environment*, Symposium Proceedings, San Francisco, 1975, p. 9, STI/PUB/410, International Atomic Energy Agency, Vienna.

Plutonium and Americium Behavior
in the Savannah River Marine Environment

D. W. HAYES and J. H. HORTON

The $^{239,240}Pu$ and ^{241}Am concentrations in the Savannah River are about the same as those measured in other U. S. rivers (0.25 fCi/liter and 0.05 ± 0.05 fCi/liter, respectively). If $^{239,240}Pu$ in the Savannah River originated from the watershed $^{239,240}Pu$ inventory, then the net annual removal from the watershed represents 0.005% per year. This indicates that thousands of years will likely pass before all the plutonium is displaced to the river. The plutonium and americium concentrations [highest, ~50 fCi/g (dry weight)] in sediment cores from the tidal freshwater and near the mouth of the Savannah River estuary were comparable and are not greatly different from those of other locations in which the only source is nuclear fallout from nuclear weapons testing. The transuranic activity in these sediments represents less than 10% of the gross alpha activity from the natural radionuclides that are present. Current concentrations of plutonium and americium in seafoods make only a very minor contribution to the overall dose commitment to humans.

Thirteen power reactors, two fuel fabrication plants, and a U. S. Department of Energy nuclear production complex, the Savannah River Plant (SRP), are operating on rivers or in coastal regions of the southeastern United States. Rivers and estuaries are a major geographic feature of this region and can represent both transport paths and sinks for transuranics. Studies are in progress to establish the distribution and transport properties of transuranic elements in the rivers and estuaries of this region. Of particular interest is the Savannah River and its estuary because located on the watershed are the Savannah River Plant and three commercial power reactors. These facilities make the Savannah River watershed one of the most intensively developed nuclear watersheds in the United States. The SRP consists of three production reactors, two fuel separation plants, a fuel fabrication facility, and a heavy water plant. The SRP has been in operation since 1952, whereas the three power reactors located at the headwaters of the Savannah River have operated for less than 10 yr. Included in this chapter are estimates of watershed loss rates for plutonium, measurements of plutonium and americium concentrations in the water and sediments of the Savannah River and its estuary, estimates of plutonium concentrations in seafood, and dose-rate estimates for seafood ingestion.

Savannah River Basin and Estuary

The Savannah River basin has a surface area of 27,400 km^2. It can be divided into three physiographic provinces that transect the basin (Fig. 1). The Blue Ridge Mountains include portions of North Carolina, South Carolina, and Georgia and range in elevation

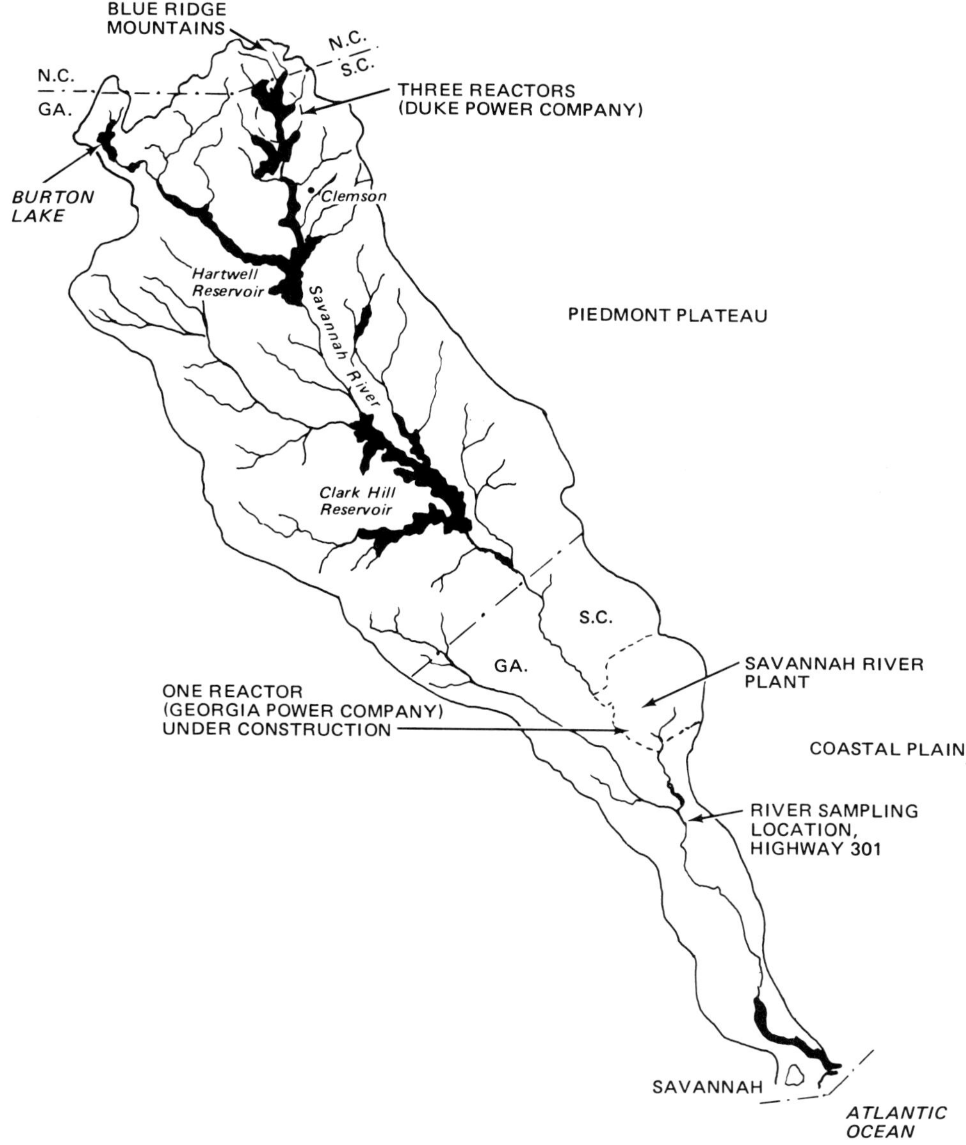

Fig. 1 Map of the Savannah River basin in the southeastern United States showing physiographic provinces, major reservoirs, nuclear power plants, and sampling points.

from 5500 ft at the headwaters to about 1000 ft at the Piedmont plateau. The hilly plateau descends from 1000 ft to about 200 ft near Augusta, Ga. The gently rolling (upper) to nearly level (lower) Atlantic Coastal Plains extend from Augusta, Ga., to the ocean.

Of the 16 rivers in the southeastern United States with flows greater than 30 m³/sec (1000 cfs), the Savannah River is fourth in volume flow with an average of about 340 m³/sec (12,000 cfs). The flow is regulated by two large impoundments, the Clark

Hill and Hartwell reservoirs, located in the Piedmont region (Fig. 1). Each reservoir has a capacity exceeding 3 km^3 and can contain the equivalent of one-half to one year's flow for the region of the river where it is located.

The Savannah River estuary is relatively narrow (about 0.5 km) and is maintained at a minimum depth of 11 m throughout its length of 35 km to accommodate shipping. To maintain the depth of 11 m in the harbor requires practically continual dredging, and the dredge materials are dumped on adjacent areas to the north of the harbor. The harbor region has a tidal range of 2.1 to 2.4 m. The estuary classification is that of a moderately stratified one. The estuary has been heavily polluted with raw sewage and industrial waste, but these pollutants have been reduced considerably in the last few years.

Sampling

Water, sediment, and seafood samples from the Savannah River and its estuary were collected and analyzed to permit transport, inventory, and dose-to-man calculations. The location of the sampling station for plutonium transport in the Savannah River watershed is shown in Fig. 1. This location was chosen because the sampler could be easily located to take water samples near midstream. Monthly composite samples were collected. Sample bias was avoided by taking four samples per day, about 300 cm^3 each, from a depth of 1 m with an automatic compositing sampler.

Estuary water and sediment sampling locations are shown in Fig. 2. Sediment was collected in marshes where vegetation lended stability to the sediments. The sediments were collected by inserting a 3.6-cm-diameter core barrel into the sediments. The cores were then extruded, sectioned, and bottled. Estuary water samples were 50-liter grab samples from a depth of 1 m.

Oysters and crab meat were obtained from a local wholesaler, who obtained the oysters from Wassaw Island and the crabs from crab pots located in Wassaw Sound (Fig. 2). Clams were collected from Port Royal Sound, which is about 32 km north of the mouth of the Savannah River estuary. Shad were netted in the Savannah River, and mullet and speckled trout were obtained from a local wholesaler whose boats work in the Savannah River estuary and nearby waters.

Analytical Methods

The procedure developed by Wong, Brown, and Noshkin (1978) for concentrating plutonium from large volumes of water was adapted for use on these water samples for both plutonium and americium analyses. In the modified method the 50 liters of water in the drum was adjusted to pH 2 with hydrochloric acid. Plutonium-236 and americium-243 spikes were added, and the sample was equilibrated for 7 to 10 days. At the end of the equilibrium period, 40 cm^3 of saturated potassium permanganate was added and the pH was adjusted to 8 with sodium hydroxide. The potassium permanganate was reduced by using a slight excess of sodium bisulfite. The hydrated manganese oxide was collected on a 1-μm cotton filter by continually recirculating the sample through the filter at 12 liters/min for 25 min. Recirculation had the advantage of keeping the water vigorously stirred as it was continually passed through the manganese oxide bed being collected on the filter. The samples were ashed while wet to avoid rapid combustion. The plutonium analyses were performed according to a procedure developed by Butler (1965) and Sanders and Leidt (1961) or by the LFE Laboratories, Richmond, Calif. All americium analyses were done by the LFE Laboratories.

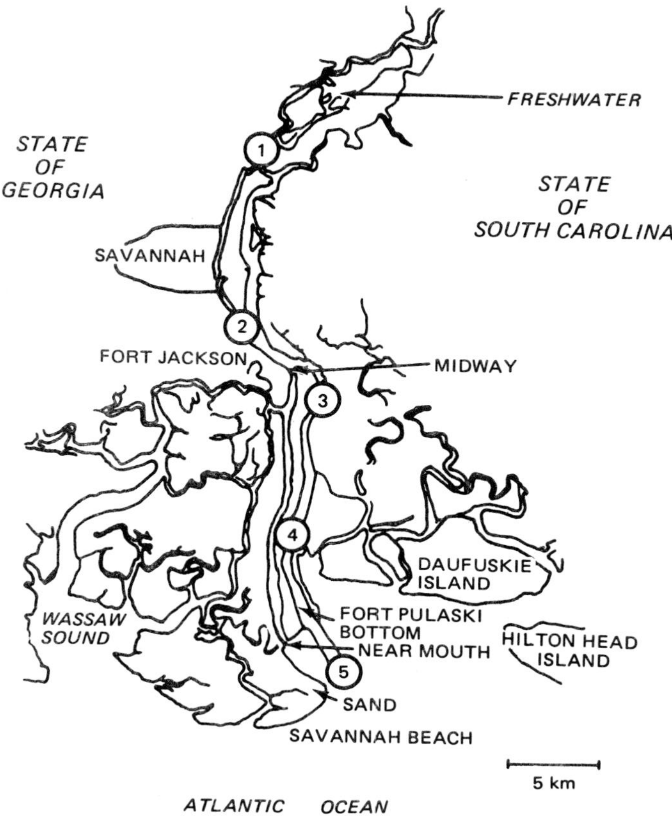

Fig. 2 Map of the Savannah River estuary showing water and sediment sampling locations.

Sources of Transuranics

The Savannah River receives transuranics by direct deposition of fallout from nuclear weapons tests, watershed runoff, and discharges from nuclear facilities. In addition to receiving the transuranics by deposition of fallout from nuclear weapons tests, watershed runoff, and discharges from nuclear facilities, the estuary also receives transuranics from the ocean via the movement of salt water some 35 km up the estuary.

Estimates of the total amount of transuranics deposited on the watershed from nuclear weapons tests are based on analyses of soil cores in the southeastern United States. These estimates range from 1.5 to about 2.2 mCi/km^2 with ^{238}Pu/^{239,240}Pu ratios of about 0.04 to 0.18 (McLendon, 1975). If fallout deposition is uniform over the Savannah River watershed, then the inventory is approximately 55 Ci, of which about 1.5 Ci was deposited on the water impoundments. No americium data are available for estimating its inventory in the southeastern United States.

Data on plutonium releases from the SRP, located 256 km from the mouth of the Savannah River, are available (Ashley and Zeigler, 1975). Atmospheric releases have totaled about 3.7 Ci since fuel reprocessing operations began in 1954. Of the 3.7 Ci, about 2 Ci was released in 1955, which was before the installation of high-efficiency filters on the air exhaust system, and about 0.8 Ci in 1969 when a sand filter failed. Currently, atmospheric releases average about 10 mCi/yr. Most of the plutonium from the SRP operations is probably on site because analyses of soil cores from the plant perimeter and off-site soils have about the same concentration, 1.96 ± 0.7 mCi/km² at the plant perimeter and 1.81 ± 0.58 mCi/km² at 160 km; other values at the same latitude are about 2 mCi/km².

Savannah River Plant plutonium releases to surface water were estimated to be about 0.3 Ci for the 20-yr period from 1954 to 1974 and were fairly consistent over this interval (Hayes, LeRoy, and Cross, 1976). Until 1971 plutonium releases were estimated by measuring gross alpha only and by assuming that all alpha activity was from plutonium. Since then plutonium has been analyzed for specifically. The water after release is subjected to cleanup by on-site streams (about 16 km in length), an on-site swamp, and the Savannah River before it reaches the estuary.

The fate of ^{137}Cs released to surface water by the SRP has been extensively studied and can be used to estimate the fate of plutonium released to surface waters. Plutonium and cesium have similar transport properties in most environmental systems owing to their strong association with the very fine suspended solids in stream water and with stream-bed sediments (Simpson et al., 1976). Some 500 Ci of ^{137}Cs has been discharged to effluent streams, and only 90 Ci (about 18%) has been measured at Highway 301 (Fig. 1). About 58% of the 500 Ci of ^{137}Cs that has been discharged is estimated to have been deposited in the SRP streams before reaching the on-site swamp, and the swamp is estimated to contain about 120 Ci, or about 24%, of cesium that has been discharged (Marter, 1974). If these ^{137}Cs data are extrapolated to plutonium, about 0.054 Ci of plutonium is estimated to have left the SRP site since start-up.

From the above data, the total amount of plutonium on the watershed is estimated to be about 59 Ci.

Results and Discussion

Savannah River

Plutonium concentrations in the Savannah River water are lower than would be predicted by considering the concentration in other freshwaters. Concentrations measured for 3 months in the Savannah River at Highway 301 (Fig. 1) varied from 0.13 to 0.32 fCi/liter. In comparison, Lake Michigan contains 0.6 fCi/liter (Edgington et al., 1976), the Great Miami River in Ohio contains about 1 fCi/liter, and the Neuse and Newport rivers in North Carolina contain about 1.2 and 1.7 fCi/liter, respectively (Hayes, LeRoy, and Cross, 1976). Concentrations in the Savannah River appear to be greatly influenced by reservoirs. Sedimentation in the two large reservoirs on the Savannah River should remove all except some of the very small plutonium-bearing particles. About 73% of the Savannah River flow originates above Clark Hill Dam. Erosion is greater in the hilly Piedmont region above the reservoirs than in the coastal plains; so removal of plutonium-bearing particles from the watershed below the reservoirs would be less rapid than that above the reservoirs. In contrast, the Great Miami River of Ohio and the Neuse and Newport rivers of North Carolina do not have large water impoundments on them. Also, the percentage of the Savannah River watershed that is under cultivation is only one-third as large as that of the Great Miami River.

The calculated rate of plutonium removal from the Savannah River watershed is about 10% of that from the Great Miami River. On the basis of the 3-month average plutonium concentration and measured flow rates (see Table 1), the estimated plutonium transport in the Savannah River at Highway 301 is 0.22 mCi/month, or 2.6 mCi/yr. The area of the Savannah River watershed above Highway 301 is 81% of the total watershed. So the amount of plutonium in the watershed above the sampling point is 0.81 × 55 Ci from nuclear weapons fallout plus 4 Ci released by the SRP, or a total of 48.6 Ci. Thus the annual removal rate is approximately 0.005%. The value reported for the 1400-km^2 watershed of the Great Miami River is 0.05% (Sprugel and Bartelt, 1978).

TABLE 1   Plutonium Transport in the Savannah River

Sampling period (1976)	River flow, liters/period	^{239,240}Pu,* fCi/liter	^{239,240}Pu transport, mCi/sampling period
6/8 to 7/7	1.25 × 10^{12}	0.13 ± 0.09	0.16
7/7 to 8/4	1.20 × 10^{12}	0.27 ± 0.08	0.32
8/5 to 9/7	7.31 × 10^{11}	0.26 ± 0.11	0.19
Average	1.06 × 10^{12}	0.22	0.22

*Mean ± standard error.

Information concerning the transport and fate of americium in rivers and estuaries is limited. The concentration of ^{241}Am in Savannah River water has not been accurately determined; a few samples collected at Highway 301 indicate that it is about 0.05 fCi/liter, as compared with ^{239,240}Pu concentration of 0.25 fCi/liter. The ^{241}Am concentration is the same as that in the Mediterranean Sea (Fukai, Bullestra, and Holm, 1976) and Lake Michigan (Wahlgren et al., 1976) water, where ^{241}Am concentration is 3 to 5% of the ^{239,240}Pu concentration. If the same percentage existed in Savannah River water, the concentration of ^{241}Am would be only about 0.01 fCi/liter.

Savannah River Estuary

The plutonium concentrations in the Savannah River estuary are not much different from those in other estuaries in the southeastern United States; in fact, the concentrations in this estuary are lower than those in some others. Water concentrations of ^{239,240}Pu were determined in the Neuse and Newport river estuaries of North Carolina for comparison with concentrations in the Savannah River estuary. The results (Fig. 3) show that the concentrations in the Newport River estuary are about three times as great as those in the Neuse River or Savannah River estuaries, which are about equal.

The three estuaries and the rivers supplying them are quite different. The Neuse and Savannah rivers flow through both the Piedmont plateau and Atlantic Coastal Plains. Only the Savannah River has its flow regulated by reservoirs. The volume of flow in the Savannah River is about twice that of the Neuse River and 10 times that of the Newport River. The Newport River estuary is extremely small and shallow with depths of less than 1 m at mean low water as compared with at least 4 m in the other two. Suspended solid (5-μm fraction) concentrations in the Newport River estuary are about one and one-half

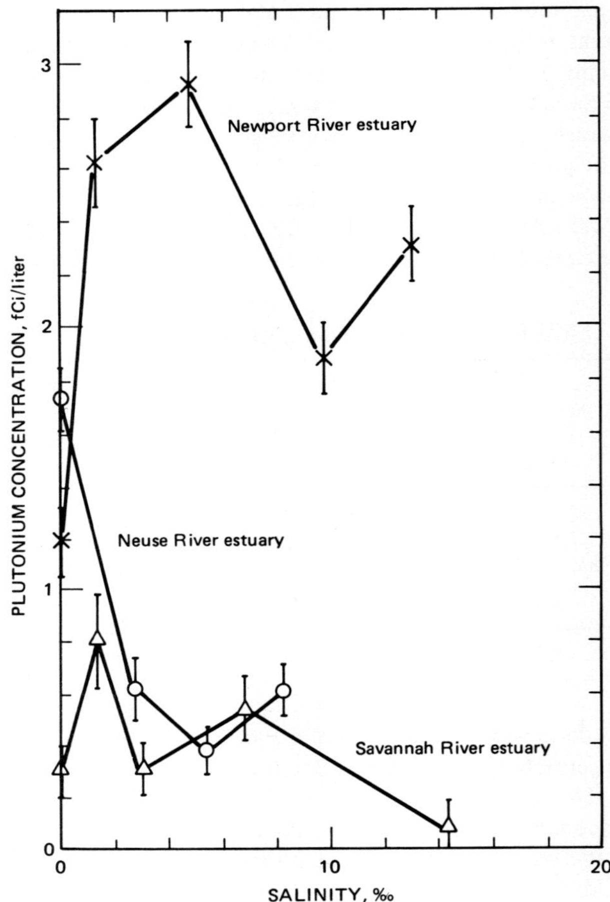

Fig. 3 Total plutonium concentrations in the waters of three southeastern U. S. estuaries.

times as great as those in the Savannah River or Neuse River estuaries (Hayes, LeRoy, and Cross, 1976); this may be due to shallow water in the Newport River estuary, which could resuspend bottom sediments throughout its depth. These sediments are likely to be very fine since the Newport River flows entirely in the coastal plains where the slope is small. The higher plutonium concentration in the Newport River estuary could be due to the larger quantity of suspended solids.

Within the Savannah River estuary, the plutonium concentrations in the sediment from the tidal freshwater region and near the mouth of the estuary were comparable (Table 2). The values are not greatly different from those of other locations that received transuranic input from nuclear weapons fallout only. Plutonium concentrations up to 200 fCi/g have been reported for the Great Lakes sediments (Edgington et al., 1976); about 60 to 70 fCi/g for Atlantic coastal waters, e.g., Buzzards Bay (Livingston and Bowen, 1975); and about 10 to 30 fCi/g for the Savannah River system (Hayes, LeRoy, and Cross, 1976). Fallout ^{238}Pu/^{239,240}Pu ratios are generally less than 0.1. Ratios

TABLE 2 Plutonium, Americium, and Gross Alpha Activities (dry-weight basis) in the Savannah River Estuary Sediments

Location	Core depth interval, cm	^{233}Pu,* fCi/g	^{239,240}Pu,* fCi/g	$\frac{^{238}\text{Pu}}{^{239,240}\text{Pu}}$ ratio*	^{241}Am,* fCi/g	$\frac{^{241}\text{Am}}{^{239,240}\text{Pu}}$ ratio*	Gross alpha, fCi/g
Tidal	0–5	4.3 ± 1.3	27.2 ± 3.3	0.16 ± 0.05	11.5 ± 4.1	0.42 ± 0.16	24,000
freshwater	5–15	6.1 ± 1.2	35.5 ± 2.8	0.17 ± 0.04	4.0 ± 1.9	0.11 ± 0.05	20,000
	15–30	2.8 ± 1.3	30.9 ± 2.8	0.09 ± 0.04	5.4 ± 1.9	0.17 ± 0.06	20,000
	30–50	0.4 ± 0.4	10.6 ± 1.0	0.04 ± 0.04	2.0 ± 0.8	0.19 ± 0.08	18,000
	50–70	0.05 ± 0.10	0.05 ± 0.05		0.23 ± 0.23		18,000
Mouth of	0–5	3.2 ± 1.1	50.6 ± 4.1	0.06 ± 0.02	11.1 ± 1.7	0.22 ± 0.04	10,000
estuary	5–15	1.7 ± 1.0	21.6 ± 2.2	0.08 ± 0.05	3.0 ± 2.3	0.14 ± 0.11	13,000
	15–25	0.2 ± 0.2	2.9 ± 0.5	0.07 ± 0.07	1.8 ± 0.6	0.62 ± 0.23	12,000
	45–65		0.5 ± 0.5				13,000

*Mean ± standard error.

greater than this are usually indicative of other sources of plutonium in the system. The ratios for the Savannah River estuary cores are reported in Table 2, and only in the freshwater core in the upper 0 to 15 cm were ratios found to be different from those from fallout. These ratios were about a factor of 2 greater than fallout ratios and presumably resulted from the SRP releases to the river system.

If americium dynamics in estuaries are different from those in freshwater or seawater systems, this difference would be evidenced by the ^{241}Am/^{239,240}Pu ratios. The average value for such ratios in shallow near-shore sediments (Livingston and Bowen, 1975) and in Lake Michigan sediments (Edgington et al., 1976) varies from 0.14 to 0.34, with an average of 0.22, and no fractionation between americium and plutonium has been found in these sediments, even when the radionuclides are being lost from the sediment following upward migration (Livingston and Bowen, 1975). Except for one value of 0.62, the ^{241}Am/^{239,240}Pu ratios for two sediment cores reported in Table 2 are not significantly different from those quoted in the literature. The indication is that the chemistries of americium and plutonium are similar in this estuarine system and that the ^{241}Am has grown in from ^{241}Pu.

The transuranic alpha activity in these cores represents less than 1% of the gross alpha activity from the natural radionuclides that are present. Indeed, modern civilization's impact on the alpha activity of these cores is small compared with the natural background.

Seafoods

At present plutonium levels, seafoods make a very minor contribution to the overall radiation-dose commitment to the populations in the southeast. Seafood samples that represent all trophic levels consumed by people in the southeastern United States were collected in and near the Savannah River estuary and analyzed for plutonium. The plutonium concentrations decreased as expected from the molluscs to the fish, with the oyster having the highest concentration, 0.12 pCi/kg, compared with 0.001 pCi/kg for shad (Table 3). The 50-yr bone-dose commitment from consuming 5.9 kg of oysters per

**TABLE 3 Southeastern Seafood Plutonium
Dose Commitments**

	pCi/kg (wet weight)*	50-yr bone-dose commitment,* mrem
Oysters	0.12	5.5×10^{-4}
Clams	0.05	2.3×10^{-4}
Crabs	0.007	2.4×10^{-5}
Mullet	0.005	5×10^{-5}
Speckled trout	0.004	3.9×10^{-5}
Shad	0.001	1.3×10^{-5}

*Consumption assumed for the dose calculation,
5.9 kg/yr molluscs, 11.8 kg/yr fish.

year is less than 0.0004% of the annual background radiation dose of about 120 mrem which is received by man in the southeastern United States.

Conclusions

Nuclear facilities operating on the Savannah River watershed have contributed less than 10% as much plutonium to the watershed as has nuclear weapons fallout. Transport of plutonium from the watershed to the estuary is very slow and appears to be influenced by two large reservoirs that serve as sinks for suspended plutonium-bearing particles. Consequently plutonium concentrations in Savannah River water and estuary sediments are no higher—and in some cases are much lower—than plutonium concentrations in other rivers and estuaries on which there are no nuclear facilities.

At present plutonium levels, seafoods make a very minor contribution to the overall dose commitment to the population in the southeastern United States.

Acknowledgment

The information contained in this chapter was developed during the course of work under contract No. AT(07-2)-1 with the U. S. Department of Energy.

References

Ashley, C. A., and C. Zeigler, 1975, *Releases of Radioactivity at the Savannah River Plant, 1954 Through 1975,* ERDA Report DPSPU-75-25-1, E. I. du Pont de Nemours & Company, NTIS.

Butler, F. E., 1965, Determination of Uranium and Americium–Curium in Urine by Liquid Ion Exchange, *Anal. Chem.,* 37: 340.

Edgington, D. N., J. J. Alberts, M. A. Wahlgren, J. O. Karttunen, and C. A. Reeve, 1976, Plutonium and Americium in Lake Michigan Sediments, in *Transuranium Nuclides in the Environment,* Symposium Proceedings, San Francisco, 1975, p. 493, STI/PUB/410, International Atomic Energy Agency, Vienna.

Fukai, R., S. Bullestra, and E. Holm, 1976, [241]Am in Mediterranean Surface Waters, *Nature,* 264: 739.

Hayes, D. W., J. H. LeRoy, and F. A. Cross, 1976, Plutonium in Atlantic Coastal Estuaries in the Southeastern United States, in *Transuranium Nuclides in the Environment,* Symposium Proceedings, San Francisco, 1975, p. 79, STI/PUB/410, International Atomic Energy Agency, Vienna.

Livingston, H. D., and V. T. Bowen, 1975, Americium in the Marine Environment—Relationships to Plutonium, in *Environmental Toxicity of Aquatic Radionuclides: Models and Mechanisms,* Proceedings of the International Conference on Environmental Toxicology, Rochester, N. Y., June

2–4, 1975, M. W. Miller and J. W. Stannard (Eds.), Ann Arbor Science Publishers, Ann Arbor, Mich.

Marter, W. L., 1974, *Radioactivity from SRP Operations in a Downstream Savannah River Swamp*, USAEC Report DP-1370, E. I. du Pont de Nemours & Company, NTIS.

McLendon, H. R., 1975, Soil Monitoring for Plutonium at the Savannah River Plant, *Health Phys.*, 28: 347.

Sanders, S. M., and S. C. Leidt, 1961, A New Procedure for Plutonium Analysis, *Health Phys.*, 6: 189.

Simpson, H. J., C. R. Olsen, R. M. Trier, and S. C. Williams, 1976, Man-Made Radionuclides and Sedimentation in the Hudson River Estuary, *Science*, 194: 179.

Sprugel, D. G., and G. E. Bartelt, 1978, Erosional Removal of Fallout Plutonium from a Large Midwestern Watershed, *J. Environ. Qual.*, 7: 175.

Wahlgren, M. A., J. J. Alberts, D. M. Nelson, and K. A. Orlandini, 1976, Study of the Behavior of Transuranics and Possible Chemical Homologues in Lake Michigan Water and Biota, in *Transuranium Nuclides in the Environment*, Symposium Proceedings, San Francisco, 1975, p. 9, STI/PUB/410, International Atomic Energy Agency, Vienna.

Wong, K. M., G. S. Brown, and V. E. Noshkin, 1978, A Rapid Procedure for Plutonium Separation in Large Volumes of Fresh and Saline Water by Manganese Dioxide Coprecipitation, *J. Radioanal. Chem.*, 42(1): 7-15.

Patterns of Transuranic Uptake by Aquatic Organisms: Consequences and Implications

L. D. EYMAN and J. R. TRABALKA

Literature on the behavior of plutonium and transuranic elements in aquatic organisms is reviewed. The commonality of observed distribution coefficients over a wide array of aquatic environments (both freshwater and marine) and the lack of biomagnification in aquatic food chains from these environments are demonstrated. These findings lead to the conclusion that physical processes dominate in the transfer of transuranic elements from aquatic environments to man. The question of the nature of the association of plutonium with aquatic biota (surface sorption vs. biological incorporation) is discussed as well as the importance of short food chains in the transfer of plutonium to man.

For years plutonium and the transuranic elements were considered to be unimportant in ecological transfers and food chains because of their low solubility and uptake when ingested by mammals. It is true that the mobility and availability of plutonium are limited compared with cesium or strontium, the major concerns for years. However, sufficient information has been developed in the past 5 yr to indicate that plutonium and americium are available to a greater extent than was shown in earlier studies. Current information demonstrates, even with the greater uptake by aquatic biota, that the transuranic elements are not enriched in aquatic food chains but rather are discriminated against. Data on plutonium and americium concentration in aquatic organisms are difficult to interpret owing to differing degrees of surface contamination and/or gut loading. Thus true biological accumulation is often masked by these contributions, which frequently indicate a higher degree of assimilation than is actually the case.

To assess the potential hazards of transuranic materials released to aquatic environments, such as in low-level waste effluents from fuel-cycle processes and burial grounds, some measure of their environmental behavior is needed, particularly as it relates to the accumulation of these isotopes in man. The following review of the literature on the behavior of plutonium and americium in aquatic environments is intended to provide information on which such an assessment can be based.

Literature Review

Emery et al. (Emery, Klopfer, and Weimer, 1974; Emery and Farland, 1974; Emery et al., 1975; Emery et al., 1976) have described the behavior of plutonium and americium in a pond at the Hanford plant. The pond is fed by a plutonium processing plant and laundry wastes with a flow of about 10 m³ of water per minute. Percolation accounts for about 95% of the water loss from the pond. The pond is described as an ultraeutrophic system with most of the plant nutrients supplied by laundry wastes. Analyses of sediment from trenches leading to the pond indicate that most of the historic transuranic releases were removed in the sediments of the trenches and thus did not reach the pond. This raises the possibility that the plutonium and americium deposited years earlier may constitute a supply of possibly solubilized radionuclides to the pond which could influence the measured values.

Sediment contains more than 95% of the total plutonium pool in the pond. The potential availability of ^{238}Pu, ^{239,240}Pu, and ^{241}Am from sediments was estimated from a series of extractions using sodium chloride, oxalate, and ethylene-diaminetetraacetic acid (EDTA). Emery et al. (Emery, Klopfer, and Weimer, 1974; Emery and Farland, 1974; Emery et al., 1975; Emery et al., 1976) corrected the concentration ratios for pond biota to account for the estimated available fraction from the sediments or water. However, for purposes of this presentation, we have chosen to present concentration ratios (CR, ratio of sample plutonium concentration to source plutonium concentration) for plutonium and americium related to the total concentrations reported by Emery et al. in sediments and water. Table 1 clearly demonstrates the difference in calculated CR values for aquatic biota when different sources are assumed.

The ratios of either ^{241}Am or ^{238}Pu to ^{239,240}Pu in organisms divided by those same ratios in various potential source compartments (sediment, interstitial water, and overlying water) are given in Table 2. A value of 1.0 would indicate that the acceptor compartment (biota) contains exactly the same isotopic ratios as the proposed source compartment, i.e., that the ^{241}Am/^{239,240}Pu ratios or ^{238}Pu/^{239,240}Pu ratios are the same in both biota and source. Therefore, if we assume that there are no significant differences in metabolism of these isotopes by aquatic biota, the compartment that exhibits ratios closest to 1.0 for both isotopes can be considered the prime source of transuranic elements to the biota in this system. As shown in Table 2, interstitial water from sediments most closely meets this requirement.

Marshall, Waller, and Yaguchi (1974) provided an earlier assessment of the role of organisms in the removal of plutonium from Great Lakes waters and the potential for plutonium to reach man via the food chain. Concentration ratios for ^{239,240}Pu are

**TABLE 1   Concentration Ratios for Plutonium and Americium
in Aquatic Biota from U-Pond Calculated Using Different Sources***

	Sediment		Interstitial water		Water	
	Pu	Am	Pu	Am	Pu	Am
Algae floc	5	3	219	185	2×10^8	2×10^5
Snail	0.2	1	6	62	5×10^6	8×10^4
Submergent cattail	0.2	0.6	7	34	5×10^6	5×10^4
Algae	0.1		6		5×10^6	
Native goldfish, without gut	0.05	0.1	2	7	2×10^6	8×10^3
Emergent cattail	0.02	0.05	0.8	3	5×10^5	3×10^3
Beetle *(Coleoptera)*	0.009	0.01	0.3	0.7	3×10^5	8×10^2
Submergent bulrush	0.003		0.1		1×10^5	
Emergent bulrush	0.003	0.001	0.1	0.2	1×10^5	3×10^2
Native goldfish muscle	0.002	0.005	0.06	0.3	5×10^4	4×10^2

*Data from Emery, Klopfer, and Weimer, 1974; Emery and Farland, 1974; Emery et al., 1975; Emery et al., 1976.

**TABLE 2 Quotients of Isotopic Ratios in Biota
and Potential Source Compartments**

| | Ratios in biota/ratio in source | | |
| | Sources | | |
Compared isotopic ratio	Sediment	Interstitial water	Overlying water
^{238}Pu/^{239,240}Pu	1.7–1.9	0.74–0.84	0.41–0.46
^{241}Am/^{239,240}Pu	2.3–2.8	1.7–2.0	0.002–0.003

**TABLE 3 Concentration Ratios for ^{239}Pu in Great Lakes
Biota, June 1972 to November 1973 (Wet Weight)**

| | | | Concentration ratios over water | |
Lake	Sample	Number	Mean ± standard error	Range
Superior	Mixed plankton	3	4000 ± 900	2680–5730
	Zooplankton	1	630	
	Smelt	1	6	
Michigan	Mixed plankton	22	5700 ± 800	620–15,300
	Cladophora sp.	16	3800 ± 500	1060–6930
	Zooplankton	9	350 ± 60	122–653
	Mysis relicta	7	760 ± 60	587–989
	Pontoporeia affinis	2	1600	1450–1830
	Slimy sculpin	7	250 ± 60	128–560
	Chub	8	37 ± 3	21–50
	Alewife	7	25 ± 2	17–30
	Smelt	6	20 ± 4	6–33
	Perch	2	16	4–29
	Whitefish	2	14	5–23
	Coho	1	7	
	Chinook	1	4	
	Lake trout	2	1	1–2
Huron	Mixed plankton	2	4460	3340–5680
	Alewife	2	165	25–305
	Smelt	1	13	
	Perch	1	24	
Erie	Zooplankton	3	500 ± 150	316–788
	Smelt	1	235	
	Perch	1	10	
Ontario	Mixed plankton	3	2420 ± 200	2030–2670
	Alewife	1	176	

reported for biota of the five Great Lakes, with emphasis on samples from Lake Michigan
(Table 3). High CR's in mixed plankton compared with zooplankton are thought to be
due, at least partially, to plutonium associated with phytoplankton, which predominate in
mixed plankton samples. It is further suggested that much of the ^{239}Pu in zooplankton
samples is due to phytoplankton in their digestive tracts. Food-chain relationships
between most of the species analyzed are mentioned in this chapter and are listed in
Table 4.

TABLE 4 Primary Food Sources for Various Great Lakes Species

Species	Primary food sources
Zooplankton (general)	Phytoplankton
Mysis, Pontoporeia (crustaceans)	Periphyton near surficial sediment layer
Sculpin	*Pontoporeia*
Chub, alewife, smelt, and whitefish	Zooplankton or mixture of zooplankton and benthic invertebrates
Perch	Mixture of invertebrates and fish
Coho salmon, chinook salmon, and lake trout	Smaller fish

As summarized by Marshall, Waller, and Yaguchi (1974), "The results clearly indicate that although the concentration of plutonium in phytoplankton is several thousand times that in the water, it decreases by an order of magnitude in each successive link in the food chain leading to man."

Wahlgren and Marshall (1974) studied the distribution of residual fallout plutonium in Lake Michigan between the water and the various trophic levels of the food chain. They found that the CR value for plutonium in phytoplankton compared with that in water was about 5000. A reduction in concentration by a ratio of about 10 was observed at each trophic level consisting of zooplankton, planktivorous fish, and piscivorous fish. The top predators had concentrations only slightly greater than the lake water. However, the concentration in benthic (bottom-feeding) fish was considerably higher than that in the planktivorous fish. It should be noted that the fish were not dissected but were analyzed in their entirety (including GI tract). From Fig. 1 we estimate the planktivorous fish to have a CR of about 20 over water and the piscivorous fish a CR of about 2, whereas benthic fish exhibit a CR of approximately 300.

The major role of phytoplankton in plutonium kinetics in aquatic systems has been postulated to be one of the removal of a significant fraction of plutonium from the water column (Wahlgren et al., 1976; Hetherington, 1976). However, collection techniques are such that phytoplankton cannot readily be separated from inorganic suspended particulate matter. The CR values for algae reported in Table 3, in fact, may be high owing to the inclusion of suspended inorganic particulate matter, which would have a CR value of approximately 10^5 (see below). The correlation between percent silicon content and plutonium concentrations in phytoplankton samples (Yaguchi, Nelson, and Marshall, 1974) was attributed to the predominance of diatom frustules in the samples analyzed. Wahlgren et al. (1976) reported a correlation between percent ash weight and plutonium concentrations. They also concluded that the plutonium was associated with diatom frustules in the plankton samples. However, in neither case was the contribution of associated suspended inorganic particulate matter quantified. Wahlgren et al. (1976) reported a distribution coefficient

$$K_d = \frac{\text{Concentration on solid phase (g/g)}}{\text{Concentration in liquid phase (g/ml)}}$$

of plutonium for suspended sediment materials of about 3×10^5. The inclusion of a small amount of those materials in the ash residue of phytoplankton samples is a plausible alternative interpretation of the observed correlations.

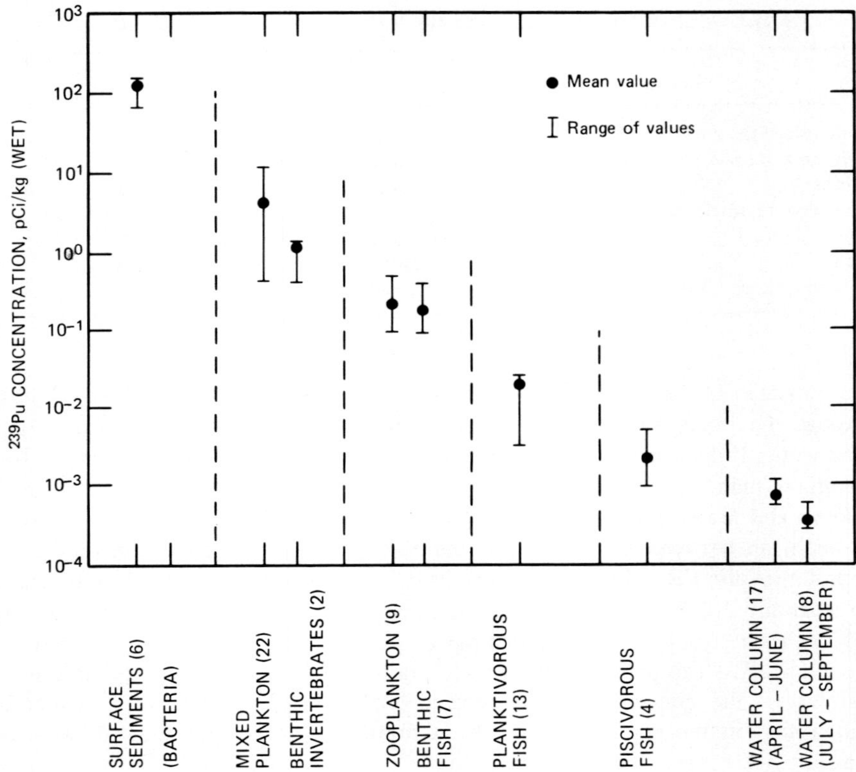

Fig. 1 Plutonium concentrations in various Lake Michigan compartments. (Data from Wahlgren and Marshall, 1974.)

Edgington et al. (1976) measured the plutonium and americium concentrations in Lake Michigan sediments and identified some of their characteristics. They calculated that approximately 97% of the plutonium that has entered Lake Michigan now resides in the sediments. Surface sediments in Lake Michigan now contain between 140 and 400 fCi/g of dry sediment, whereas the concentration in the water column is less than 1 fCi/liter. Earlier, Edgington et al. (1976) had indicated a negligible contribution to the input of plutonium to Lake Michigan from runoff via tributary rivers and streams. Their data suggest a significant redistribution of sedimentary material in the lake with a rapid movement of the radioactivity from its site of deposition on the sediment surface to a final site of deposition in the sediments. Apparently there are large areas of Lake Michigan where no significant sedimentation occurs. These areas have a layer of floc, containing somewhat higher concentrations of plutonium, overlying the glacial till or sand. Since no significant accumulations of sediment have occurred in these areas, it is likely that these 1- to 2-cm-thick deposits are transitory and that the material is readily resuspended.

The ^{239,240}Pu CR values for shoreline plants in Lake Ontario found by Bowen (1974) were similar to those reported by Marshall, Waller, and Yaguchi (1974). However, the plankton CR values ranged from 10 to 300 (vs. 600 to 15,000 as reported by Marshall, Waller, and Yaguchi, 1974). Further, Marshall, Waller, and Yaguchi reported CR values for Lake Ontario which ranged from 2000 to 2700. Observed differences in CR values for plankton reported in these two studies have not been resolved. Bowen (1974) and Marshall, Waller, and Yaguchi (1974) reported that benthic feeders accumulate plutonium to higher levels than limnetic feeders. Some predator species (i.e., largemouth bass, rock bass, white perch, and coho salmon) deposit most of the plutonium in the bone, whereas other forms (i.e., northern pike, yellow perch, and freshwater drum) deposit plutonium primarily in the liver. These apparent differences in tissue distributions between species have not been satisfactorily explained.

Dahlman, Bondietti, and Eyman (1976) reported on the behavior of plutonium in the biotic components of White Oak Lake. Results from analyses of various components of the lake system support the finding by Marshall, Waller, and Yaguchi (1974) in Lake Michigan, i.e., decreased concentrations of plutonium at higher trophic levels (Table 5).

TABLE 5 Concentrations of ^{239,240}Pu and Concentration Ratios for Fishes from White Oak Lake

Species	Plutonium content				Concentration ratio[†]	
	Carcass*		GI tract			
	pCi/g	Standard error	pCi/g	Standard error	CR	Standard error
Largemouth bass	2×10^{-4}	6×10^{-5}	6×10^{-3}	4×10^{-4}	0.04	0.2
Bluegill	1×10^{-3}	8×10^{-4}	4×10^{-4}	5×10^{-5}	3	2
Goldfish	1×10^{-3}	1×10^{-3}	8×10^{-2}	3×10^{-2}	3	3
Shad	2×10^{-3}	3×10^{-5}	4×10^{-1}	7×10^{-2}	4	0.1

*Total fish minus GI tract.

†Concentration ratio (CR) is defined as [plutonium] in organisms (wet weight)/[plutonium] in water. Water concentration of plutonium used in the calculation of CR values is 4×10^{-4} pCi/g. The CR values are for carcass.

Organisms living in or on the bottom of sediments in the White Oak Lake (filamentous algae and benthic invertebrates) have plutonium concentrations that are two or three orders of magnitude higher than those in predatory fish, such as largemouth bass and bluegill. Filamentous algae associated with sediments in shallow areas of White Oak Lake had the highest concentration of plutonium of any biotic component measured. Gastrointestinal contents of goldfish, gizzard shad, and bluegill had plutonium concentrations intermediate between those of sediment and food organisms, which indicates the importance of sediment as a dietary source of plutonium. Further evidence of sediment ingestion by these fish was the fact that the ^{238}Pu/^{239}Pu ratio of gut contents was similar to that measured in sediments.

Long-term, chemically and/or biologically mediated transformations of plutonium compounds may be expected to occur in aquatic systems. These transformations may result in plutonium being complexed by naturally occurring chelating agents, such as carboxylic acids (citrate), fulvic acids, or proteins. Several laboratory experiments were carried out at Oak Ridge (Eyman, Trabalka, and Case, 1976; Eyman and Trabalka, 1977; Trabalka and Eyman, 1976) to determine the uptake and distribution of monomeric plutonium (IV) in chelated forms both in an aquatic vertebrate, the channel catfish *Ictalurus punctatus,* and in a littoral aquatic microecosystem. A primary finding was that the gastrointestinal intake by catfish was significantly higher than that reported for mammals. The highest observed retention (whole body) at 63 days was 3.8% of ingested dose for ^{237}Pu citrate, whereas retention of the fulvate was 0.6%. Reduced uptake of the fulvate complex is thought to be due to either its high molecular weight (>10,000) or its stability in metabolic systems. Increased uptake of ^{237}Pu citrate is attributable to both high gut permeability and instability of the complex in metabolic systems. Chelation can either enhance or reduce the uptake of ingested plutonium relative to plutonium hydroxide (monomer) in channel catfish. Approximately two-thirds of the initial body burden of plutonium (administered as citrate) was lost within 63 days after gut clearance. These results suggest that the observed retention pattern of plutonium in channel catfish was due to plutonium labeling of the gut followed by subsequent turnover by cell-renewal processes. This suggestion is based on the observed slow gut-cell renewal times in fish (20 to 30 days) (Hyodo-Taguchi, 1968) compared with those in mammals (48 hr in mouse) (Lesher, Wahlburg, and Sacher, 1964).

Tissue-distribution studies in channel catfish revealed that relatively little (≤10%) of the intracardially injected plutonium citrate was excreted. Blood clearance rates were similar to those found in small mammals, the plutonium being associated primarily with the plasma protein transferrin. The fractional body burdens in bone, liver, and kidney 17 days after injection were 31, 24, and 9% of the injected dose, respectively. High kidney burdens relative to mammals are expected since the kidney functions as the major site of hemopoiesis in teleosts. The absence of significant excretion of plutonium reinforces the previous suggestion that a short half-life component of elimination following gut clearance in gavage studies is due to plutonium uptake by, and subsequent turnover of, cells in the gut wall.

A distribution coefficient of 9×10^4 was observed for sediment in a year-old aquatic microcosm spiked with ^{237}Pu nitrate. A materials balance at 90 days postspike provided the following estimates of plutonium distribution: 0.001% in water, 0.04% in biota, and over 99.9% in sediments. Concentrations of ^{237}Pu in whole animals, including fish, were surprisingly uniform (within a factor of 10, 1.2 to 9.9% of mean sediment concentration). This was related to gut loading of sediments and/or surface contamination. The uptake by emergent macrophytes not exposed to surface contamination was quite small: ≤0.03 to 0.1% of the sediment concentration.

On the basis of this set of laboratory experiments, sorption to plant surfaces, on gut walls, and on exoskeletons appears to provide the dominant sites for plutonium deposition in or on submerged components of aquatic systems. Interestingly, the sediment distribution coefficient observed in the laboratory microcosm study was well within the range of values reported from a wide variety of laboratory and field studies (Table 6).

In a study of crayfish from the Great Miami River, Wayman, Bartelt, and Groves (1976) reported that most of the plutonium was concentrated in soft tissues rather than in the sclerotized shell. Similar results were reported earlier by Nelson and Noshkin (1973) for the *Tridacna* clam and lobster in marine studies at Enewetak Atoll. Noshkin (1972) reported higher concentrations of ^{239,240}Pu in the shell than in the body of scallops, whelks, and moonshells collected off Cape Cod. Ward (1976) also observed that the calcified shell appeared to accumulate ^{239,240}Pu at a rapid rate. A very high proportion (89.5%) of the total plutonium was in the skeleton, which accounted for about 43% of the total weight of the lobster. Concentration factors for shells were on the order of 200, and gills were about 100. The flesh, which comprised about 28.7% of the total body weight, contained only 1.2% of the plutonium present in the entire body. The

**TABLE 6   Distribution Coefficients ($\check{K}_d$) for
Plutonium Isotopes in Aquatic Systems**

K_d*	Isotope	Environment†	Reference
9.0×10^4	237	L,F	Trabalka and Eyman (1976)
$1.3 - 9.4 \times 10^4$	237	L,M	Duursma and Parsi (1974)
$1.2 - 7.9 \times 10^4$	239	L,F	Trabalka and Frank (1976)
$4.8 \times 10^4 -$ 1.3×10^5	239	M	Pillai and Mathew (1976)
5×10^4	239	F	Bowen (1974)
3×10^5	239	F	Wahlgren et al. (1975)
5×10^4	239	M	Hetherington et al. (1976)

$$*K_d = \frac{\text{Concentration on solid phase (grams per gram)}}{\text{Concentration in liquid phase (grams per milliliter)}}.$$

† F, freshwater; M, marine; L, laboratory.

content of the cast shell was approximately twice that of the shell of the intermoult lobster. Whether the differences reported are due to the chemical species of plutonium the organisms were exposed to at the various sites or to interspecific physiological differences is open to question at this time.

Some authors (Emery et al., 1975; Bair et al., 1974; Hakonson and Johnson, 1973; Morin, Nenot, and LaFuma, 1972) have suggested that plutonium isotopes can behave differently in biological systems. In environmental studies the discrepancies in behavior can be explained by different physicochemical forms of the isotopes in the original source or by different sources of uptake. In laboratory experiments differences in metabolic behavior can be attributed to differences in concentrations of the isotopes tested. However, Fowler, Heyraud, and Beasley (1975) found that, if the different plutonium isotopes (^{236}Pu, ^{237}Pu, ^{238}Pu, and ^{239,240}Pu) were present in the same physicochemical form, aquatic organisms were unable to discriminate between them in either accumulation or excretion studies. They also concluded that the suggestion of Moghissi and Carter (1975), reinforced by the work of Eyman, Trabalka, and Case (1976), is correct, "that in environmental studies, ^{237}Pu is most likely the best tracer for measuring plutonium kinetics in biological systems."

Discussion

This chapter is not intended to be definitive, but, rather, we have attempted to evaluate the most pertinent data sets. Although there appears to be a commonality (i.e., no biomagnification, highest concentrations in benthic organisms and phytoplankton) among all environmental data sets, whether marine or freshwater, there are also significant discrepancies. The most interesting common factor we were able to extract from a wide range of studies, both laboratory and field, is the similarity in the observed distribution coefficient for plutonium in sediment (Table 6). Inconsistencies involve differences in CR values for freshwater algae and some tissue distributions observed in both freshwater and marine studies. These discrepancies are probably explainable on the basis of differences in physicochemical form to which organisms were exposed at different sites.

From a review of the data available for both freshwater and marine environments, there appear to be relatively few significant differences in the patterns of accumulation or in the observed CR's.

There is a lack of any definitive information on the other three elements of interest, americium, neptunium, and curium. One study of neptunium in the Columbia River (Davis et al., 1958) involved an isotope with a half-life of only 2.3 days (vs. 2×10^6 days for ^{237}Np); so the radioactive decay limited the quantities of this isotope in the organism. Data from Lake Michigan indicate increased CR's for ^{241}Am over plutonium in the lower trophic levels by factors of 1.5 to 5 (Wahlgren et al., 1976). However, americium CR values for fish are less than or equal to 10 times those for plutonium: the data were not adequate to calculate specific values. Data from the discharge of waste from Windscale (Hetherington et al., 1976) indicate little difference between ^{239}Pu and ^{241}Am concentrations in fish within 10 km of the discharge for equal discharge rates of the two isotopes.

Concentration ratios chosen for use in both freshwater and marine environments are given in Table 7. For americium, neptunium, and curium (^{241}Am, ^{237}Np, ^{242}Cm, and ^{244}Cm), we have applied a factor of 10 over the plutonium values. These values were

TABLE 7   Recommended Concentration Ratios

Material	Plutonium	Americium, curium, and neptunium
Sediment	100,000	100,000
Plankton	5,000	50,000
Benthic algae and macrophytes	5,000	50,000
Benthic invertebrates	1,000	10,000
Fish		
Bottom feeders	250	2,500
Plankton feeders	25	250
Piscivorous (fish eaters)	5	50

chosen because these isotopes have a greater availability than plutonium in terrestrial systems (Dahlman, Bondietti, and Eyman, 1976), and it is expected that, when adequate data are available, the same may be true in aquatic systems. Limited data from Lake Michigan and Windscale indicate that the factor for ^{241}Am may be less than 10; however, owing to the incomplete nature of the data, we have chosen a more conservative figure as the factor for biota.

The very high affinity of plutonium for particulate matter in aquatic ecosystems (distribution coefficient, $\sim 10^5$) suggests that it may not be appropriate to use the traditional expression of CR to estimate the concentration of this element in biota. Rather, we feel that the observed concentrations of plutonium in aquatic biota should be related to the primary abiotic source in the system, sediment (both suspended and bottom). To express this relationship, the term Trophic Transfer Factor (TTF) has been used by various researchers (Lipke, 1971; Trabalka and Eyman, 1976; Elwood, Hildebrand, and Beauchamp, 1976). The concentration of an element in sediment or food is substituted for the concentration in water, which is normally used in the calculation of a CR. The underlying assumption is that, owing to the high distribution coefficients (K_d's) observed, element accumulation in tissues of higher trophic levels will be dominated by gut absorption rather than by direct uptake from water. Trophic transfer of plutonium by aquatic animals is comprised of three fractions: exterior surface sorption, gut labeling, and absorption from the gut. Again we stress that external contamination with sedimentary particulate matter and gut loading are not considered to represent true uptake and should be considered separately. The TTF, then, serves as a realistic measure of plutonium discrimination in food chains. It should be noted that the CR values for the Hanford waste pond fall into line with other data sets when considered on this basis (Table 1).

Some of the variation in TTF values observed can be explained by the relative trophic position of the organisms analyzed. The number of intervening food-chain transfers between the organism analyzed and the abiotic source of plutonium should be inversely related to the observed TTF value.

Conclusions and Recommendations

To assess the potential transfer of plutonium to man from aquatic ecosystems, we must concentrate on those food sources most closely linked to sediment as a measure of maximum plutonium in human food. One could postulate, as an extreme, although improbable, case, the direct consumption of sediment by man. Owing to the high K_d of plutonium in sediments, direct ingestion could expose humans to plutonium concentrations up to several hundred thousand times those found in water. A more probable projection, however, would be exposure to plutonium via a food chain involving a single trophic transfer from sediment to an organism that is consumed by man. Such short, single trophic transfer food chains should result in the highest plutonium concentrations in human food derived from aquatic ecosystems (Critical Exposure Pathway). Some examples would include bottom-feeding fishes, shellfish, and rooted macrophytes, such as rice. Although we could find no data on the accumulation of plutonium in rice, this information seems critical since it is representative of a single trophic transfer from sediment to man and is a major dietary component of a large segment of the world population. Further, both marine (Pillai and Mathew, 1976) and freshwater (Wahlgren and Marshall, 1974) organisms associated with the sediment—water interface (i.e., benthos) contain plutonium burdens that are one hundred times as high as those of free-swimming forms.

Present National Committee on Radiation Protection and Measurement (NCRP) guidelines (Title 10, Part 20) for plutonium in food are derived by inference from standards based on drinking water consumption at a fixed rate (International Commission on Radiological Protection, 1959; National Bureau of Standards, 1959). The total radioactivity ingested as food and/or water cannot exceed the product of the Maximum Permissible Concentration (MPC) times the consumption rate for water. Therefore, if an individual is ingesting water contaminated at the MPC level, he or she cannot be exposed to plutonium from any other dietary source. Weights of water and food ingested are approximately equal for the Standard Man. It is apparent, therefore, that knowledge of the expected dietary plutonium contribution from food is as important as that of the contribution from water.

The MPC for plutonium in drinking water was derived by using a fractional gut transfer factor of 3×10^{-5} for ingested plutonium (International Commission on Radiological Protection, 1959). The value of the human gut transfer factor is based on studies whereby laboratory mammals were fed plutonium in a variety of chemical forms. The value of 3×10^{-5} is based on plutonium administered in the nitrate form. Actual gut transfer factors reported in these studies ranged from 1×10^{-5} to 3×10^{-3}, depending on a number of factors (chemical form, species of test animal, age of test animal, etc.) (Hodge, Stannard, and Hursh, 1973). To our knowledge no studies have been published which determine the fractional gut transfer factor for plutonium incorporated in actual food materials. On the basis of the variability in data reported above, one cannot rule out the possibility that this factor may be significantly greater than the values used by the NCRP. In addition, plutonium releases in low-level effluent streams from fuel-cycle processes or from burial grounds may be in a more available chelated form, which is either due to release in a chelated form or to long-term environmental transformation to the chelated form. These potential routes need to be quantified, and the dominant mobile forms need to be identified.

Even greater uncertainty arises when trying to predict probable plutonium concentrations in an aquatic food source for purposes of dosimetric calculations. This predicted concentration is the product of three values, plutonium concentration in water, K_d for plutonium in sediment, and the TTF for a given single transfer food chain (Critical Exposure Pathway). Since the K_d is known to be very high ($\simeq 10^5$), if the TTF value is greater than 10^{-5}, the potential dose from food intake could be higher than the value for which the standards (based on water intake) were designed, notwithstanding the uncertainties cited above. Since TTF values of 5×10^{-3} to 5×10^{-2} appear reasonable, based on data presented in this chapter [laboratory results (Eyman and Trabalka, 1977; Trabalka and Eyman, 1976; Trabalka and Frank, 1978; Beasley and Fowler, 1976a; 1976b) and apparent CR values in benthic biota of 500 to 5000], our hypothetical individual consuming *such aquatic food organisms exclusively* (obtained from a water source contaminated at MPC levels) could receive radiation doses significantly higher than intended by the standards. *Although the radiation doses projected above obviously represent a purely hypothetical case,* the conditions sufficient to produce such doses cannot be excluded because of the present uncertainty associated with parameters used to derive the dose estimates. Further, it should be recognized that certain cases involving the assessment of the plutonium contribution to human diet require that the TTF be separated into its components (i.e., for animals that are not to be consumed whole). It is not our purpose to "single out" plutonium or any other actinide as an unusual hazard. Many isotopes released to the environment from various nuclear-fuel-cycle processes have been subjected to similar scrutiny in attempting to assess transport to man. Several (radionuclides of cesium, strontium, and cobalt) will undoubtedly contribute significantly higher doses to man than expected for plutonium (Blaylock and Witherspoon, 1978). To demonstrate that the conservative assumptions stated above are unwarranted, we must develop a data base on the environmental behavior of actinides reasonably comparable to existing information on cesium, strontium, and cobalt.

Future research on the transport of plutonium to man from aquatic ecosystems should concentrate on those food chains which have the lowest number of trophic transfers between abiotic sources in the system and man. Data generated from such research will provide critical information necessary for the evaluation of present standards by determining plutonium concentrations in critical aquatic organisms that serve as food sources to man. Additional research on the relationship between the chemical characteristics of plutonium in abiotic components of the system and observed concentrations in edible aquatic foods will strengthen our ability to predict potential transfer of plutonium to man from aquatic ecosystems.

References

Bair, W. J., D. H. Willard, I. C. Nelson, and A. C. Case, 1974, Comparative Distribution and Excretion of ^{237}Pu and ^{239}Pu Nitrates in Beagle Dogs, *Health Phys.*, 27: 392.

Beasley, T. M., and S. W. Fowler, 1976a, Plutonium and Americium: Uptake from Contaminated Sediments by the Polychaete (*Nereis diversicolor*), *Mar. Biol.*, 38: 95.

——, and S. W. Fowler, 1976b, Plutonium and Americium Uptake From Sediment by Polychaete Worms, in *Activities of the International Laboratory of Marine Radioactivity, 1976 Report*, Report IAEA-187, International Atomic Energy Agency, Vienna.

Blaylock, B. G., and J. P. Witherspoon, 1978, Evaluation of Radionuclides Released from the Light Water Reactor Nuclear Fuel Cycle to the Aquatic Environment, in *Environmental Chemistry and Cycling Processes*, DOE Symposium Series, No. 38, Augusta, Ga., Apr. 28–May 1, 1976, D. C. Adriano and I. Lehr Brisbin, Jr. (Eds.), pp. 851-865, CONF-760429, NTIS.

Bowen, V. T., 1974, *Plutonium and Americium Concentration Along Freshwater Food Chains of the Great Lakes, U.S.A.*, General Summary of Progress 1973–1974, USAEC Report COO-3568-4, Woods Hole Oceanographic Institution.

Dahlman, R. C., E. A. Bondietti, and L. D. Eyman, 1976, Biological Pathways and Chemical Behavior of Plutonium and Other Actinides in the Environment, in *Actinides in the Environment*, A. M. Friedman (Ed.), American Chemical Society, Symposium Series 35, American Chemical Society, Washington, D. C.

Davis, J. J., R. W. Perkins, R. F. Palmer, W. C. Hanson, and J. F. Cline, 1958, Radioactive Materials in Aquatic and Terrestrial Organisms Exposed to Reactor Effluent Water, in *Proceedings of the Second United Nations International Conference on the Peaceful Uses of Atomic Energy, Geneva, 1958*, Vol. 18, pp. 423-428, United Nations, New York.

Duursma, E. K., and P. Parsi, 1974, Distribution Coefficient of Plutonium Between Sediment and Sea Water, in *Activities of the International Laboratory of Marine Radioactivity, 1974 Report*, Report IAEA-163, pp. 94-96, International Atomic Energy Agency, Vienna.

Edgington, D. N., J. J. Alberts, M. A. Wahlgren, J. O. Karttunen, and C. A. Reeve, 1976, Plutonium and Americium in Lake Michigan Sediments, in *Transuranium Nuclides in the Environment*, Symposium Proceedings, San Francisco, 1975, pp. 493-516, STI/PUB/410, International Atomic Energy Agency, Vienna.

Elwood, J. W., S. G. Hildebrand, and J. J. Beauchamp, 1976, Contribution of Gut Contents to the Concentration and Body Burden of Elements in *Tipula* spp. from a Spring-Fed Stream, *J. Fish. Res. Board Can.*, 33: 1930-1938.

Emery, R. M., and T. R. Farland, 1974, *Ecological Behavior of Plutonium and Americium in a Freshwater Ecosystem. Phase II. Implications of Differences in Transuranic Isotopic Ratios*, USAEC Report BNWL-1879, Battelle, Pacific Northwest Laboratories, NTIS.

——, D. C. Klopfer, and W. C. Weimer, 1974, *Ecological Behavior of Plutonium and Americium in a Freshwater Ecosystem. Phase I. Limnological Characterization and Isotopic Distribution*, USAEC Report BNWL-1867, Battelle, Pacific Northwest Laboratories, NTIS.

——, D. C. Klopfer, T. R. Farland, and W. C. Weimer, 1975, *Ecological Behavior of Plutonium and Americium in a Freshwater Pond*, USAEC Report BNWL-SA-5346, Battelle, Pacific Northwest Laboratories, NTIS.

——, D. C. Klopfer, T. R. Farland, and W. C. Weimer, 1976, The Ecological Behavior of Plutonium and Americium in a Freshwater Pond, in *Radioecology and Energy Resources*, C. E. Cushing, Jr. (Ed.), Ecological Society of America, Special Publication No. 1, Dowden, Hutchinson & Ross, Inc., Stroudsburg, Pa.

Eyman, L. D., and J. R. Trabalka, 1977, Plutonium-237: Comparative Uptake in Chelated and Non-Chelated Forms by Channel Catfish (*Ictalurus punctatus*), *Health Phys.*, 32: 475.

——, J. R. Trabalka, and F. N. Case, 1976, Plutonium-237 and 246: Their Production and Use as Gamma Tracers in Research on Plutonium Kinetics in an Aquatic Consumer, in *Environmental Toxicity of Radionuclides: Models and Mechanisms*, pp. 193-206, M. W. Miller and J. N. Stannard (Eds.), Ann Arbor Science Publishers, Inc., Ann Arbor, Mich.

Fowler, S., M. Heyraud, and T. M. Beasley, 1975, Experimental Studies on Plutonium Kinetics in Marine Biota, in *Impacts of Nuclear Releases into the Aquatic Environment*, Symposium Proceedings, Otaniemi, Finland, STI/PUB/406, International Atomic Energy Agency, Vienna.

Hakonson, T. E., and L. J. Johnson, 1973, *Distribution of Environmental Plutonium in the Trinity*

Site Ecosystem After 27 Years, USAEC Report LA-UR-73-1291, Los Alamos Scientific Laboratory, NTIS.

Hetherington, J. A., 1976, The Behavior of Plutonium Nuclides in the Irish Sea, in *Environmental Toxicity of Aquatic Radionuclides: Models and Mechanisms,* M. W. Miller and J. N. Stannard (Eds.), Ann Arbor Science Publishers, Inc., Ann Arbor, Mich.

——, D. F. Jefferies, N. T. Mitchell, R. J. Pentreath, and D. S. Woodhead, 1976, Environmental and Public Health Consequences of the Controlled Disposal of Transuranic Elements to the Marine Environment, in *Transuranium Nuclides in the Environment,* Symposium Proceedings, San Francisco, 1975, pp. 139-154, STI/PUB/410, International Atomic Energy Agency, Vienna.

Hodge, H. C., J. N. Stannard, and J. B. Hursh (Eds.), 1973, *Uranium Plutonium Transplutonic Elements,* Springer-Verlag, New York.

Hyodo-Taguchi, Y., 1968, Rate of Development of Intestinal Damage in the Goldfish After X-Irradiation and Mucosal Cell Kinetics at Different Temperatures, in *Gastrointestinal Radiation Injury,* pp. 120-126, M. F. Sullivan (Ed.), Excerpta Medica Foundation, Amsterdam.

International Commission on Radiological Protection, 1959, *Recommendations of the International Commission on Radiological Protection,* Publication 2, Pergamon Press, New York.

Lesher, S., H. W. Wahlburg, and G. A. Sacher, 1964, Generation Cycle in the Duodenal Crypt Cells of Germ-Free and Conventional Mice, *Nature,* 202(4935): 884-886.

Lipke, E. J., Jr., 1971, *Effects of Environmental Parameters on the Uptake of Radioisotopes in Freshwater Fish,* Ph.D. Dissertation, University of Michigan, Ann Arbor.

Marshall, J. S., B. J. Waller, and E. M. Yaguchi, 1974, Plutonium in the Laurentian Great Lakes: Food Chain Relationships, in *Proceedings of the XIX Congress of Association of Limnology,* Winnipeg, Manitoba, Canada, Aug. 23–28, 1974.

Moghissi, A. A., and M. W. Carter, 1975, Comments on "Comparative Distribution and Excretion of ^{237}Pu and ^{239}Pu Nitrates in Beagle Dogs," *Health Phys.,* 28: 825.

Morin, M., J. C. Nenot, and J. LaFuma, 1972, Metabolic and Therapeutic Study Following Administration to Rats of ^{238}Pu, *Health Phys.,* 23: 475.

National Bureau of Standards, 1959, *Maximum Permissible Body Burdens and Maximum Permissible Concentrations of Radionuclides in Air and in Water for Occupational Exposure,* Handbook 69, GPO.

Nelson, V., and V. E. Noshkin, 1973, *Eniwetok Radiological Survey,* USAEC Report NVO-140, Nevada Operations Office, NTIS.

Noshkin, V. E., 1972, Ecological Aspects of Plutonium Dissemination in Aquatic Environments, *Health Phys.,* 22: 537-549.

Pillai, K. C., and E. Mathew, 1976, Plutonium in the Aquatic Environment: Its Behaviour, Distribution and Significance, in *Transuranium Nuclides in the Environment,* Symposium Proceedings, San Francisco, 1975, pp. 25-45, STI/PUB/410, International Atomic Energy Agency, Vienna.

Trabalka, J. R., and L. D. Eyman, 1976, Distribution of Plutonium-237 in a Littoral Freshwater Microcosm, *Health Phys.,* 31: 390.

——, and M. L. Frank, 1978, Trophic Transfer by Chironomids and Distribution of Plutonium-239 in Simple Aquatic Microcosms, *Health Phys.,* 35: 492.

Wahlgren, M. A., and J. S. Marshall, 1974, Distribution Studies of Plutonium in the Great Lakes, in *Proceedings of the 2nd International Conference on Nuclear Methods in Environmental Research,* University of Missouri, Columbia, Mo., July 29–31, 1974, CONF-740701, NTIS.

——, J. J. Alberts, D. M. Nelson, and K. A. Orlandini, 1976, Study of the Behavior of Transuranic and Possible Chemical Homologues in Lake Michigan Water and Biota, in *Transuranium Nuclides in the Environment,* Symposium Proceedings, San Francisco, 1975, pp. 9-24, STI/PUB/410, International Atomic Energy Agency, Vienna.

Ward, E. E., 1976, Uptake of Plutonium by the Lobster *Homarus vulgaris, Nature,* 209(5023): 625-626.

Wayman, C. W., G. E. Bartelt, and S. E. Groves, 1976, Further Investigations of Plutonium in Aquatic Biota of the Great Miami River Watershed, Including the Canal and Ponds in Miamisburg, Ohio, in *Radiological and Environmental Research Division Annual Report, January–December 1975,* USAEC Report ANL-75-60 (Pt. 3), Argonne National Laboratory, NTIS.

Yaguchi, E. M., D. M. Nelson, and J. S. Marshall, 1974, Plutonium in Lake Michigan Plankton and Benthos, in *Radiological and Environmental Research Division Annual Report, July 1972–June 1973,* USAEC Report ANL-8060, Argonne National Laboratory, NTIS.

The Migration of Plutonium
from a Freshwater Ecosystem at Hanford

RICHARD M. EMERY, DONALD C. KLOPFER, and M. COLLEEN McSHANE

A reprocessing waste pond at Hanford has been inventoried to determine quantities of plutonium that have accumulated since its formation in 1944. Expressions of export were developed from these inventory data and from informed assumptions about the vectors that act to mobilize material containing plutonium. This 14-acre pond provides a realistic illustration of the mobility of plutonium in a lentic ecosystem. The ecological behavior of plutonium in this pond is similar to that in other contaminated aquatic systems with widely differing limnological characteristics. Since its creation this pond has received about 1 Ci of ^{239,240}Pu and ^{238}Pu, most of which has been retained by its sediments. Submerged plants, mainly diatoms and Potamogeton, *accumulate more than 95% of the plutonium contained in biota. Emergent insects are the only direct biological route of export, mobilizing about 5×10^3 nCi of plutonium annually, which is also the estimated maximum quantity of the plutonium exported by waterfowl, birds, and mammals collectively. There is no apparent significant export by wind, and it is not likely that plutonium has migrated to the groundwater below U-Pond via percolation. Although this pond has a rapid flushing rate, a eutrophic nutrient supply with a diverse biotic profile, and interacts with an active terrestrial environment, it appears to effectively bind plutonium and prevent it from entering pathways to man and other life.*

The dissemination of plutonium in our environment continues to be a major issue centering around the development and application of nuclear energy. In addressing this problem, investigators have made efforts to inventory the worldwide plutonium burden in terms of fallout and point-source deposition (Electric Power Research Institute, 1976). A number of locations have been identified as having above-background plutonium concentrations, such as weapons-testing sites, sites of accidental releases where plutonium has escaped its container, and sites of controlled releases associated with waste management areas. For some of these areas, such as the Eniwetak and Bikini Atolls and the Mortandad Canyon leading to the Rio Grande, plutonium inventories are being investigated in attempts to estimate quantities that migrate away from these sites over time (Schell and Watters, 1975; Hakonson, Nyhan, and Purtymun, 1976). The export of plutonium away from any contaminated aquatic site has not been sufficiently studied to provide a quantitative example of the environmental mobility of this element. In this regard waste ponds can serve as useful study sites since they often have diverse ecological profiles, receive additions of plutonium over extended periods of time, and have frequent contact with terrestrial forces that can mobilize plutonium.

Waste ponds likely constitute the most probable and greatest percentage of freshwater environments that become contaminated with plutonium from local sources, whether from controlled low-level releases or by accident. The few that exist in this country are located at Rocky Flats, Colo. (Johnson, Svalberg, and Paine, 1974); Oak Ridge, Tenn. (Dahlman, Bondietti, and Eastwood, 1975); the Savannah River plant near Aiken, S. C. (D. Paine, 1977, Battelle, Pacific Northwest Laboratories, private communication); Idaho Falls, Idaho (D. Markham, 1976, Rockwell Hanford Operations, Richland, Wash., private communication), and Hanford near Richland, Wash. (Emery, Klopfer, and Weimer, 1976). These ponds are managed in association with fuel separation, reprocessing, and reactor testing operations. The three unmanaged freshwater systems reported to have received small amounts of plutonium include the Miami River near Miamisburg, Ohio (Bartelt, Wayman, and Edgington, 1975), Sawmill Creek at Argonne, Ill. (Singh and Marshall, 1977), and streams leading to the Rio Grande River near Los Alamos, N. Mex. (Hakonson, Nyhan, and Purtymun, 1976). Although amounts of plutonium released to these aquatic systems are usually quite small, concentrations accumulated in waste ponds are often significantly above background levels.

Information about the ecological transport of plutonium from any of these systems would be of special interest since they represent the results of actual contamination events as they exist today. The distribution and fate of plutonium in a waste pond at Hanford, specifically called U-Pond (Fig. 1), have been studied since 1973 (Emery, Klopfer, and Weimer, 1974; Emery et al., 1976). The results appear to provide a good example of the behavior of plutonium in a freshwater environment. U-Pond has an established ecosystem and has been exposed to plutonium longer than any other aquatic environment. Since 1944 plutonium has reached this 14-acre pond via waste ditches, which have received occasional pulses of transuranic elements from the clean-up of minor accidental spills of low-level contaminants within the reprocessing laboratories.

One of the major goals of the study at U-Pond has been to obtain sufficient information about the pond's ecosystem and the distribution of plutonium within it so that plutonium export routes can be assessed quantitatively. Although it is often difficult to measure with reasonable certainty the parameters necessary for describing these export routes, the purpose of this work is to formulate the best expressions of export given the conditions that limit this process. The objectives are to determine ranges of quantities of plutonium in the pond's ecosystem and assess the amount of plutonium being exported in relation to this inventory. To accomplish this task, we estimated the pond's plutonium inventory quantities on a basis of minimum, mean, and maximum values for each ecosystem compartment to postulate the amount of these inventories that is exported yearly.

Methods and Materials

To examine the pond's inventory and export conditions, we separated the ecosystem into two categories, the aquatic system and the contacting terrestrial system. The entire inventory and export scenario is shown in Fig. 2. The aquatic system is divided into 10 compartments:

1. Nonfilamentous algae (including sestonic diatoms).
2. Filamentous algae.
3. Submerged macrophytes.
4. Emergent macrophytes.
5. Lower invertebrates (excluding insects and gastropods).
6. Resident insects (those with aquatic adult stages).

7. Emergent insects (those with emerging adult stages).
8. Gastropods.
9. Goldfish (anthropogenically introduced).
10. Sediments (down to 10 cm and including organic floc generated from decomposing plant material).

Water itself is not considered as a compartment, but the particulate contents contained by the water mass are accounted for in other compartments. Suspended particles greater than 0.1 μm (seston) are fractioned by weight into inert particles and algae on the basis of microscopic inspection and dry weight—ash weight comparisons. Weights of the sestonic algae, which are mostly diatoms, are added to the nonfilamentous algae compartment, and weights of inert particles are placed with the sediments. Particles

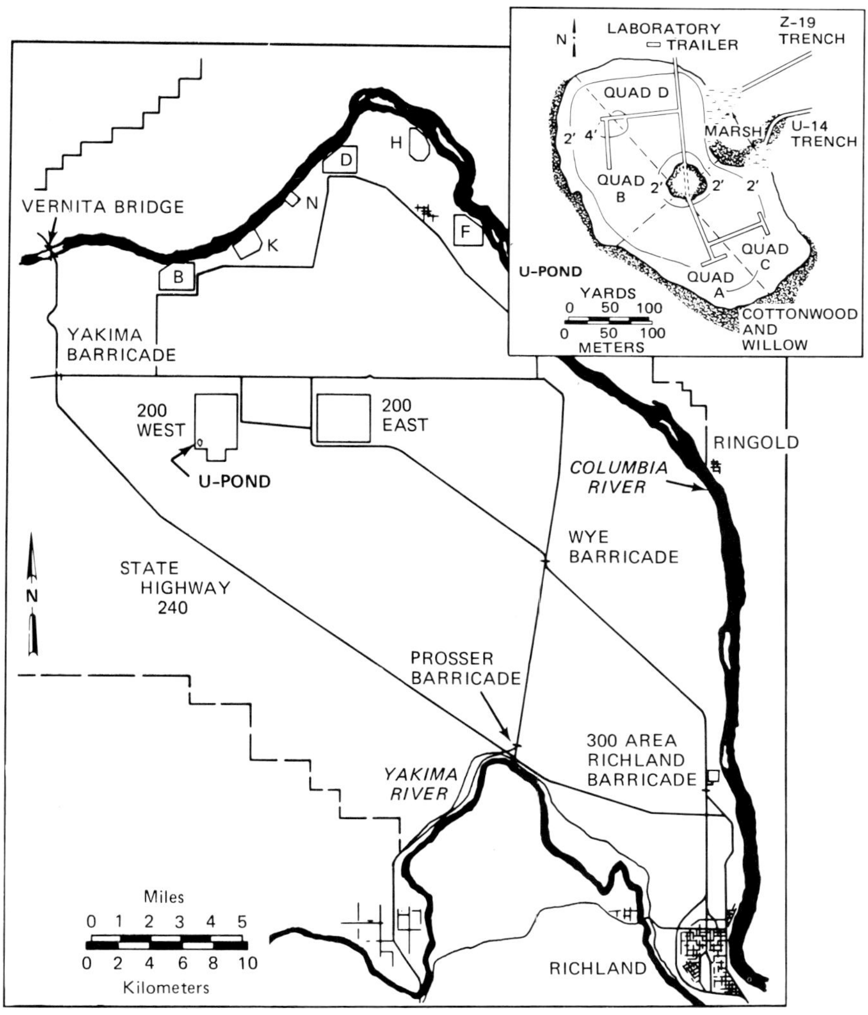

Fig. 1 Map of Hanford site showing location and detail of U-Pond.

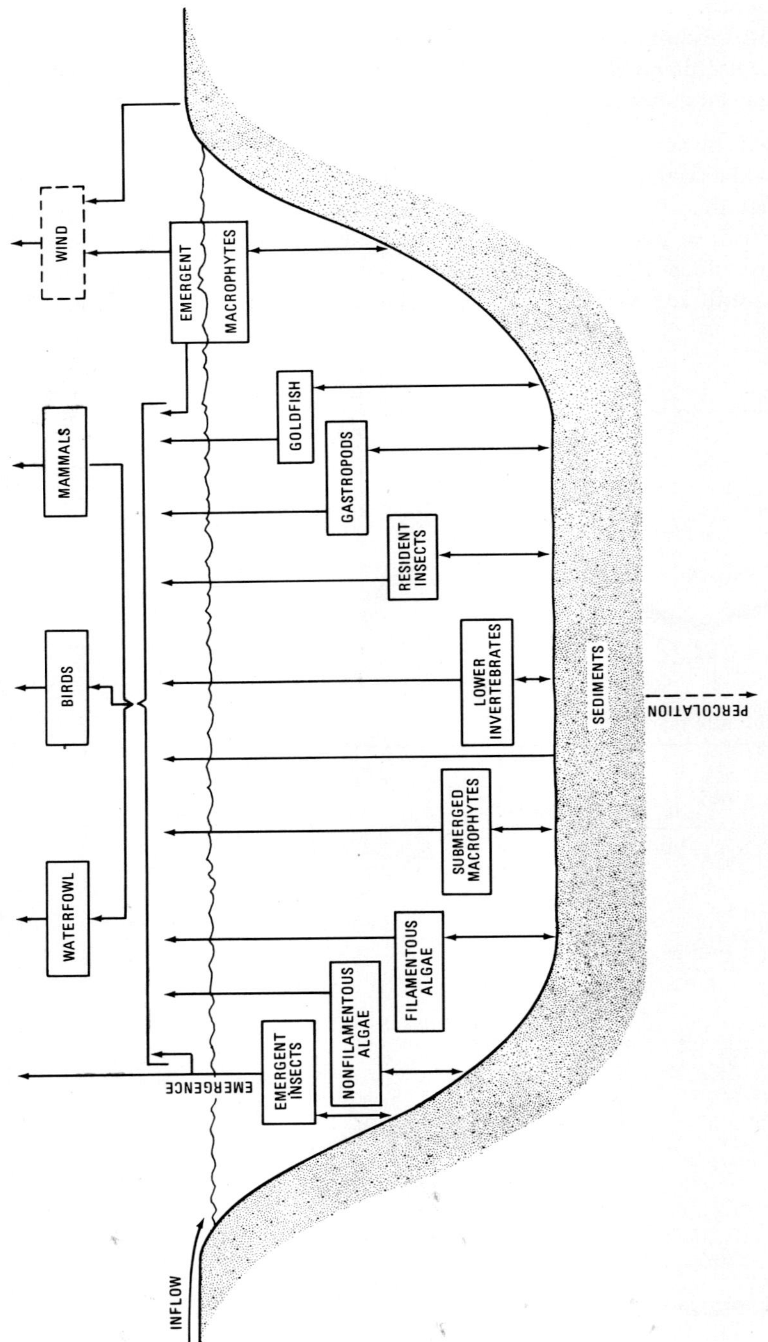

Fig. 2 Schematic representation of ecosystem compartments and plutonium export routes from U-Pond via biological mobilization, wind, and percolation.

of plutonium less than 0.1 μm and soluble plutonium are in very dilute concentrations ($\ll 1$ pCi/liter), which are lower than standards that regulate plutonium concentrations in drinking water. It is assumed that this small fraction of plutonium remains in the pond water until it enters any one of the ecosystem compartments that could provide a route of export. Dissolved and suspended materials in the pond have a short duration since nearly all the water leaves the pond by percolation after a mean residence time of 40 hr (Emery, Klopfer, and Weimer, 1974). The pond has no surface outflow.

The contacting terrestrial system has four compartments:

1. Waterfowl.
2. Birds (other than waterfowl).
3. Mammals.
4. Airborne particulates.

Since these compartments have a transient association with the pond, they also serve as routes of plutonium export. In addition, the transient insect population in the pond, along with the emergent macrophytes, provides means for plutonium to leave the pond. The emergent macrophytes would require assistance from one of the other export routes to release any of their plutonium content to adjacent areas. This is also true for the plutonium that resides in the shoreline sediments. Thus the only export vectors through which plutonium can leave U-Pond are:

- Percolation.
- Emergent insects.
- Waterfowl.
- Birds.
- Mammals.
- Wind (containing airborne particles).

Methods of sample preparation and plutonium analysis of pond samples, which involve drying, ashing, chemical separation, and electrodeposition, are described by Emery, Klopfer, and Weimer (1974).

Several techniques were used in the measurements of annual production of biomass in the aquatic system. For all compartments except sediments, the annual production is expressed as the quantity of biomass that is generated in 1 yr. The material quantity of the pond's sediments is expressed as the dry weight of sediments to a depth of 10 cm. The inventory of plutonium in the sediments is then the amount of plutonium in the upper 10 cm. The dry weight of the upper 10-cm layer of sediments is 3.4×10^6 kg, which extends over an area of 5.6×10^4 m^2.

Nonfilamentous algae are composed mainly of sestonic diatoms (not always a true phytoplankton population) and, to a much lesser extent, *Tetraspora,* which rests on the bottom in loose globular masses. The annual production of sestonic diatoms is estimated by using the weight of the average fraction of seston that is diatoms. This was done by microscopic examination of concentrated seston samples to determine the mean percentage of the total number of the particles that is diatoms (28%). Seston concentrations were then multiplied by 0.28 to obtain an estimate of the concentration of diatoms (in milligrams per liter). This concentration was proportioned to the volume of the pond to estimate an instantaneous standing crop of diatoms in the pond. Sestonic diatom standing crops were sampled seasonally to establish an annual mean. Since the mean residence time of this water mass is 40 hr, this mean standing-crop value is multiplied by the number of 40-hr intervals in a year (219) to estimate the annual production. The production of *Tetraspora* was estimated by direct observations of appearance in sampling grids made periodically during the growing season. The mean dry weight of a volumetric quantity was established and then proportioned to the volume of *Tetraspora* observed to occur in U-Pond.

Inventories of filamentous algae and submerged macrophytes are based on periodic measurements of primary productivity and standing crop [see Emery, Klopfer, and Weimer (1974) for a detailed description]. A method described by Verduin (1964) was used to measure pond-wide primary production, which accounts for the photosynthesis of all plant life in the pond inclusive of phytoplankton, macroalgae, and submerged macrophytes. Results of these measurements express the net accumulation of plant mass per unit time. Hence a summation of the monthly rates of net productivity for the entire year provides one estimate for the annual quantity of submerged plant biomass that may accumulate plutonium. For verification of this estimate, submerged plant standing crop was measured at the beginning and end of the growing season by the areal sampling methods described by Emery, Klopfer, and Weimer (1974). Areal sampling of submerged plant biomass was also done periodically to provide a direct accounting of changes in standing crop. So that standing crop could be measured, in this way plant material was harvested from known surface areas, dried, weighed, and projected on a weight basis to the area of the pond observed to be covered by these algae and macrophytes. Results of both methods of estimating the pond's production of submerged plant biomass were in reasonable agreement with each other (less than an order of magnitude difference).

Emergent macrophytes represent 10 to 15% of the pond's biomass at the peak of the growing season. The annual production of emergent plants was estimated by the same method used for submerged plants.

Pond invertebrates live mainly in association with rooted macrophytes and to a lesser extent in the organic floc covering much of the pond's bottom. A 0.25-ft^2 Ekman dredge was used to estimate the area concentrations of invertebrates living in the organic floc. The more densely populated surfaces of macrophytes were quantitatively sampled by direct collection and enumeration of invertebrates from the plants (mainly *Potamogeton*), for which dry weights were also determined. Dry weights of the various types of invertebrates were assessed per unit weight of plant material and then projected to the pond-wide weight of the specific plant types to estimate an invertebrate standing crop. All invertebrate life was assumed to appear in the pond on an annual basis; hence the maximum standing crop served to represent the weight of invertebrates produced annually by U-Pond.

The goldfish population in U-Pond is largest during the summer months. During the colder months this population appears to decrease by several orders of magnitude, and few goldfish are observed during the winter. Since this goldfish population appears to reproduce annually and undergo substantial depletion in the colder months, the only valid expression of annual production can be based on the summer standing crop. For this reason the standing crop observed in August 1974, the only time when goldfish were counted, will serve as the expression of annual production.

The standing crop of goldfish was estimated by first establishing a weight for an individual and then counting the numbers of individuals appearing in 125, 9-m^2 grids placed randomly about the pond. The mean number of individuals was converted to mean weight per unit area and proportioned to the entire pond area.

The biomasses of terrestrial life contacting the pond annually were estimated by using mean weights for designated taxonomic groups and observations of frequency of contact by these groups. Although it was possible to obtain reliable measurements of mean weights for these organisms, the measurements of their contact frequencies were much less precise. It would be desirable to have a more precise understanding of the intensity of these export activities; however, it is possible to suggest a range within which these vectors operate based on the ranges of plutonium concentrations found in these organisms.

Means and 95% confidence limits of plutonium concentrations in pond compartments were determined on an arithmetic basis. Concentrations, inventories, percentages, and exported quantities of plutonium are expressed in scientific notation rather than in a decimal format to draw attention to the order of magnitude rather than to emphasize the exact quantity as a primary matter of consideration.

Results and Discussion

The Inventory

The scope of this study and the resources available for it placed limitations on the resolution of compartment-size determination. Although it was possible to determine with accuracy the weights of sediments down to 10 cm and of emergent macrophytes, the remaining compartment sizes were evaluated with a variety of estimation procedures. Since much of these data are derived in this way, we feel that they represent only a reasonable approximation of compartment sizes. Efforts to examine these data statistically were not redeeming, and it was concluded that statistical confidence intervals and central tendencies are not appropriate expressions for these results. Instead, these results are intended to suggest best approximations without indicating ranges within which the compartment sizes fluctuate.

The biota in U-Pond contain about 1% of the total mass of the pond, including sediments down to 10 cm (Table 1). Concentrations of plutonium isotopes in the pond's ecosystem compartments are shown in Table 2. Nonfilamentous algae and sediments show the highest mean plutonium concentrations of 2.8×10^2 and 5.0×10^2 pCi/g, respectively. This similarly reflects the close association between the two compartments. Submerged macrophytes and gastropods also have mean plutonium concentrations exceeding 1×10^2 pCi/g (1.6×10^2 and 2.4×10^2 pCi Pu/g, respectively). Filamentous algae and emergent insects show mean plutonium concentrations of 8.6×10^1 and 4.6×10^1 pCi/g, respectively, whereas the remaining compartments have mean plutonium concentrations ranging from 1 to 2×10^1 pCi/g.

U-Pond's eutrophic condition is reflected by its high rate of primary production, which occurs as high as 42 kg C ha^{-1} day^{-1}. This rate of productivity can also be expressed as 440 μg C $liter^{-1}$ hr^{-1}. Verduin (1964) found primary productivity rates in two Pennsylvania ponds to range from 120 to 760 μg C $liter^{-1}$ hr^{-1}. Hence U-Pond's primary productivity resembles that in ponds not associated with nuclear facilities. Its rate of carbon assimilation also approaches that of a highly productive terrestrial community, a cornfield, which has an average assimilation rate of 63 kg C ha^{-1} day^{-1} (Robbins, Weier, and Stocking, 1957).

Submerged plant life has most of the total plutonium inventory in pond biomass, more than 95% (Tables 3 and 4). Submerged flora are composed mainly of algae,

**TABLE 1   Estimated Quantities of Annual Production
of Biomass in U-Pond
(Weights of existing sediments are also included)**

	Estimated annual production	
	Weight, kg (dry)	Percent of total
Nonfilamentous algae	8.5×10^3	2.4×10^{-1}
Filamentous algae	5.1×10^3	1.5×10^{-1}
Submerged macrophytes	1.8×10^4	5.2×10^{-1}
Emergent macrophytes	9.4×10^3	2.7×10^{-1}
Lower invertebrates	6.8×10^1	2.0×10^{-3}
Resident insects	2.2×10^1	6.3×10^{-4}
Emergent insects	1.1×10^2	3.2×10^{-3}
Gastropods	7.6×10^1	2.2×10^{-3}
Goldfish	2.1×10^2	6.1×10^{-3}
Biota subtotal	4.1×10^4	1.2×10^0
Sediments	3.4×10^6	9.9×10^1
Total	3.4×10^6	1.0×10^2

TABLE 2   Means and Confidence Limits (95%) of Plutonium Concentrations in the U-Pond System

	Concentrations of plutonium isotopes, pCi/g (dry)					
	Lower limit		Mean		Upper limit	
	^{239,240}Pu	^{238}Pu	^{239,240}Pu	^{238}Pu	^{239,240}Pu	^{238}Pu
Nonfilamentous algae (26)*	5.2×10^1	7.5×10^1	1.2×10^2	1.6×10^2	1.9×10^2	2.5×10^2
Filamentous algae (14)	1.3×10^1	2.3×10^1	3.4×10^1	5.2×10^1	5.5×10^1	8.1×10^1
Submerged macrophytes (21)	7.2×10^0	4.5×10^0	6.3×10^1	9.7×10^1	1.2×10^2	1.9×10^2
Emergent macrophytes† (18)	2.0×10^{-1}	2.0×10^{-1}	8.3×10^0	9.9×10^0	1.8×10^1	2.2×10^1
Lower invertebrates (8)	9.0×10^{-1}	7.2×10^{-1}	8.3×10^0	8.5×10^0	1.6×10^1	1.6×10^1
Resident insects (12)	2.7×10^0	3.4×10^0	4.8×10^0	5.6×10^0	7.0×10^0	7.8×10^0
Emergent insects (84)	1.3×10^1	1.9×10^1	1.8×10^1	2.8×10^1	2.4×10^1	3.7×10^1
Gastropods† (11)	1.2×10^0	2.0×10^0	9.1×10^1	1.5×10^2	2.6×10^2	4.3×10^2
Goldfish (8)	5.3×10^0	7.7×10^0	7.8×10^0	1.2×10^1	1.0×10^1	1.6×10^1
Sediments (123)	1.9×10^2	2.0×10^2	2.3×10^2	2.7×10^2	2.7×10^2	3.3×10^2

*Sample numbers are shown in parentheses.

†Lowest observed concentration replaces lower 95% confidence limit since the latter concentration is a negative number.

including diatoms, *Cladophora, Hydrodictyon,* and *Tetraspora,* and the macrophyte *Potamogeton.* Among these the diatoms and *Potamogeton* are the principal components where plutonium is accumulated, which contain more than 99% of the pond's plutonium inventory accumulated by plants. This suggests that an approximate plutonium inventory for the biota of a similar ecosystem could be rapidly determined by estimating the inventories in only a few populations of flora.

Of the invertebrate life emergent insects and gastropods have the highest mean plutonium inventories (Tables 3 and 4). Gastropods show an inventory of 1.8×10^4 nCi, whereas emergent insects account for 5.1×10^3 nCi of plutonium. Quantities of plutonium in emergent insects are of particular interest since they provide the only direct route of biological mobilization from the pond. However, they do not contain more than $4 \times 10^{-4}\%$ of the plutonium in the pond or more than $1 \times 10^{-1}\%$ of the inventory in pond biota.

Estimates of goldfish production in U-Pond fall within production ranges for suckers and carp reported by Carlander (1955) for a number of North American lakes and reservoirs. This goldfish population appears to contain about 4×10^3 nCi of plutonium, less than $1 \times 10^{-1}\%$ of the plutonium in all the pond biota and less than $1 \times 10^{-3}\%$ of the entire pond inventory. Goldfish are occasionally eaten by herons, coyotes, and waterfowl.

The bulk of material in the pond is, of course, in the sediments (Table 1). They contain about 99% of the entire mass of the pond's ecosystem (excluding water).

TABLE 3   Ranges and Means of the ^{239,240}Pu Inventory in U-Pond

| | Inventory of ^{239,240}Pu | | | | | |
| | Lower limit | | Mean | | Upper limit | |
	Activity, nCi	Percent of total	Activity, nCi	Percent of total	Activity, nCi	Percent of total
Nonfilamentous algae	4.4×10^5	6.8×10^{-2}	1.0×10^6	1.3×10^{-1}	1.6×10^6	1.7×10^{-1}
Filamentous algae	6.6×10^4	1.0×10^{-2}	1.7×10^5	2.2×10^{-2}	2.8×10^5	3.0×10^{-2}
Submerged macrophytes	1.3×10^5	2.0×10^{-2}	1.1×10^6	1.4×10^{-1}	2.2×10^6	2.4×10^{-1}
Emergent macrophytes	1.9×10^3	2.9×10^{-4}	7.8×10^4	1.0×10^{-2}	1.7×10^5	1.8×10^{-2}
Lower invertebrates	6.1×10^1	9.4×10^{-6}	5.6×10^2	7.2×10^{-5}	1.1×10^3	1.2×10^{-4}
Resident insects	5.9×10^1	9.1×10^{-6}	1.1×10^2	1.4×10^{-5}	1.5×10^2	1.6×10^{-5}
Emergent insects	1.4×10^3	2.2×10^{-4}	2.0×10^3	2.6×10^{-4}	2.6×10^3	2.8×10^{-4}
Gastropods	9.1×10^1	1.4×10^{-5}	6.9×10^3	8.8×10^{-4}	2.0×10^4	2.2×10^{-3}
Goldfish	1.1×10^3	1.7×10^{-4}	1.6×10^3	2.0×10^{-4}	2.1×10^3	2.3×10^{-4}
Biota subtotal	6.4×10^5	9.8×10^{-2}	2.4×10^6	3.0×10^{-1}	4.3×10^6	4.6×10^{-1}
Sediments	6.5×10^8	$>9.9 \times 10^1$	7.8×10^8	$>9.9 \times 10^1$	9.2×10^8	$>9.9 \times 10^1$
Total	6.5×10^8	1.0×10^2	7.8×10^8	1.0×10^2	9.2×10^8	1.0×10^2

TABLE 4   Ranges and Means of the ^{238}Pu Inventory in U-Pond

| | Inventory of ^{238}Pu | | | | | |
| | Lower limit | | Mean | | Upper limit | |
	Activity, nCi	Percent ot total	Activity, nCi	Percent of total	Activity, nCi	Percent of total
Nonfilamentous algae	6.4×10^5	9.4×10^{-2}	1.4×10^6	1.5×10^{-1}	2.1×10^6	1.9×10^{-1}
Filamentous algae	1.2×10^5	1.8×10^{-2}	2.7×10^5	2.9×10^{-2}	4.1×10^5	3.7×10^{-2}
Submerged macrophytes	8.1×10^4	1.2×10^{-2}	1.7×10^6	1.8×10^{-1}	3.4×10^6	3.1×10^{-1}
Emergent macrophytes	1.9×10^3	2.8×10^{-4}	9.0×10^4	1.0×10^{-2}	2.1×10^5	1.9×10^{-2}
Lower invertebrates	4.9×10^1	7.2×10^{-6}	5.8×10^2	6.3×10^{-5}	1.1×10^3	9.9×10^{-5}
Resident insects	7.5×10^1	1.1×10^{-5}	1.2×10^2	1.3×10^{-5}	1.7×10^2	1.5×10^{-6}
Emergent insects	2.1×10^3	3.1×10^{-4}	3.1×10^3	3.4×10^{-4}	4.1×10^3	3.7×10^{-4}
Gastropods	1.5×10^2	2.2×10^{-5}	1.1×10^4	1.2×10^{-3}	3.3×10^4	3.0×10^{-3}
Goldfish	1.6×10^3	2.3×10^{-4}	2.5×10^3	2.7×10^{-4}	3.4×10^3	3.1×10^{-4}
Biota subtotal	8.5×10^5	1.2×10^{-1}	3.5×10^6	3.7×10^{-1}	6.2×10^6	5.6×10^{-1}
Sediments	6.8×10^8	$>9.9 \times 10^1$	9.2×10^8	$>9.9 \times 10^1$	1.1×10^9	$>9.9 \times 10^1$
Total	6.8×10^8	1.0×10^2	9.2×10^8	1.0×10^2	1.1×10^9	1.0×10^2

Sediments also have the highest concentrations of plutonium in the pond (Table 2). The 95% confidence interval of plutonium in sediment samples extended from 1.9 to 2.7 x 10^2 pCi/g for ^{239,240}Pu and from 2.0 to 3.3 x 10^2 pCi/g for ^{238}Pu. Hence the inventory of plutonium in the sediments is more than 99% of the inventory for the entire system (Tables 3 and 4).

Other studies of plutonium in aquatic systems show sediments playing the dominating role in the plutonium inventory of their respective ecosystems (Johnson, Svalberg, and Paine, 1974; Patterson et al., 1976; Trabalka and Eyman, 1976). In nearly all studies that have reported inventories of plutonium in freshwater systems, the sediments contain at least 99% of the total plutonium burden. This commonality among ecosystems having widely different limnological characteristics suggests that the accumulation, retention, and transport of plutonium is strongly associated with sediment and sedimenting particles. This idea is supported by results of numerous studies in which particulate plutonium concentrations were measured apart from those dissolved (or suspended) in water and interstitial water (Bartelt, Wayman, and Edgington, 1975; Dahlman, Bondietti, and Eastwood, 1975; Emery, Klopfer, and Weimer, 1974; Hakonson, Nyhan, and Purtymun, 1976; Johnson, Svalberg, and Paine, 1974; Magno, Reaney, and Apidianakis, 1970; Noshkin, 1972; Singh and Marshall, 1977; Trabalka and Eyman, 1976). These studies show that plutonium associated with particulates makes up more than 80% of the total plutonium concentrations in water. This characteristic of plutonium distribution and transport in freshwaters is the most significant aspect of its environmental behavior.

The pond is highly enriched with nutrients coming from laundry effluents via U-14 ditch (Fig. 1, Emery, Klopfer, and Weimer, 1974). This nutrient supply supports luxuriant growths of algae and macrophytes (Table 1) which eventually settle to the bottom and decompose. The result of this process is the formation of a layer of organic floc that rests on the surface of older floc and sediments. Although this material is sedimentary, it has several special characteristics. The density of floc approaches that of water, which causes it to be loosely compacted and easily resuspended. The large quantity of floc generated each year serves as a source of food for many animal populations in and around the pond. Perhaps most important, the floc is the primary concentrator of plutonium in the ecosystem.

Hydrologic considerations of the pond provide additional significance to the functional role of the organic floc in the ecosystem. Since the pond has no surface outflow and a short retention time (40 hr), there is a rapid deposition of suspended material (seston). The sedimentation rate is approximately 1 kg m^{-2} yr^{-1} (dry). This means that about 5.6 x 10^4 kg of seston is deposited each year in this loosely compacted floc/sediment.

If U-Pond has sustained a continuous annual sedimentation rate of 5.6 x 10^4 kg since its formation in 1944, about 1.8 x 10^6 kg of sestonic sediments has been deposited. The total organic production of biomass in U-Pond is about 5 x 10^4 kg/yr (Table 1), which is approximately equal to the annual deposition of suspended matter. If losses of organic matter in the sediments caused by decomposition are ignored, all sources of sedimentary materials have deposited 3.5 x 10^6 kg since 1944. The weight of U-Pond sediments down to 10 cm is 3.4 x 10^6 kg (dry, Table 1). This suggests that U-Pond has not deposited more than about a 10-cm layer of sediments since its creation. The actual thickness of deposition would probably be smaller because of the decomposition of organic matter. This does not account for wind-blown dust accumulated in the pond.

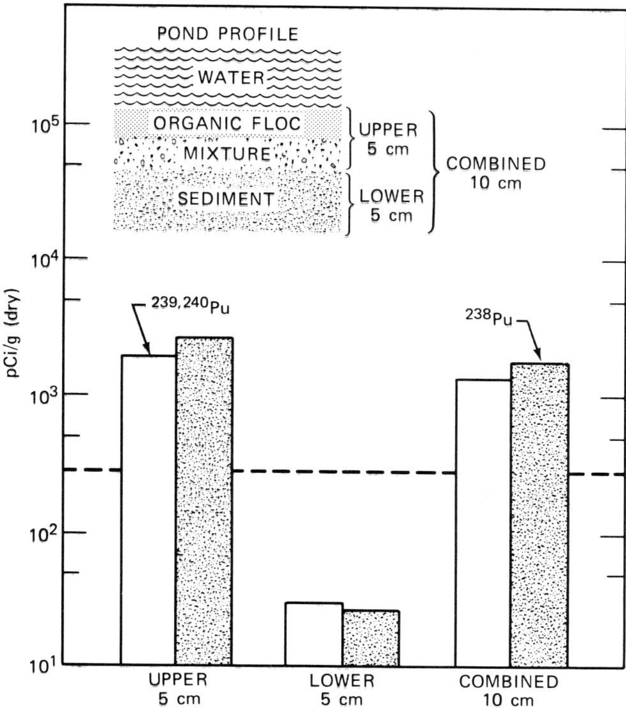

Fig. 3 Distribution of plutonium isotopes in the upper 10 cm of U-Pond sediments. The broken horizontal line indicates the pond-wide mean for both plutonium isotopes in a 10-cm core.

Horizontal distributions of plutonium in the sediments are spatially and temporally random, but several vertical profiles indicate that plutonium is most heavily concentrated in the upper 5 cm (Emery et al., 1976). When three sediment cores were analyzed for plutonium concentrations in the upper 5 cm, lower 5 cm, and combined 10 cm, it was found that most of the plutonium was located in the upper layer (Fig. 3). In these samples the plutonium concentration in the upper 5 cm was about 4×10^3 pCi/g, whereas that in the lower 5 cm was only about 5×10^1 pCi/g. The combined 10-cm core showed a plutonium concentration of about 2×10^3 pCi/g. It should be noted, however, that the concentrations in these samples did not resemble the mean plutonium concentrations in samples of 10-cm sediment cores. The difference of plutonium concentrations between the upper and lower sections of the sediment core in Fig. 3 appears to be exaggerated beyond the normal range, but a vertical reduction of plutonium in the top 10 cm of sediments is indicated.

This vertical distribution of plutonium appears to be largely the result of the rapid accumulation of sedimenting seston discussed earlier. Seston has shown the highest concentrations of plutonium in any subcompartment sample, often greater than 1×10^1 nCi/g. However, this material settles to the bottom and is captured by a layer of floc. This accumulation of seston, decomposing plant material and a mixture of older sediments, contains the largest portion of the pond's inventory.

The existing mean inventory of plutonium in the sediments (to 10 cm) is 1.7×10^9 nCi (Tables 3 and 4). This amount of plutonium represents the total accumulation from all sources minus the losses via the various routes of export. The validity of this can be examined by calculating a theoretical accumulation of plutonium in the sediments using annual rates of plutonium deposition via sedimenting seston and annual accumulation of plutonium by pond biota.

The mean concentration of plutonium in U-Pond seston measured over the study period is 5.6×10^0 nCi/m^3. Thus the pond's water mass (2.27×10^4 m^3) contains an average of 1.3×10^5 nCi of plutonium. This mass passes through the pond's basin at an average rate of 219 times per year (i.e., 40-hr retention time). If we assume that the flushing rate and plutonium content of U-Pond water remain constant, then 2.8×10^7 nCi of sestonic plutonium is deposited as sediments each year. Proceeding with the same assumptions, the 33-yr total theoretical accumulation of plutonium in the sediments is 9.2×10^8 nCi. The mean annual accumulation of plutonium by pond biota is 2.4×10^6 nCi (Tables 3 and 4), which suggests a historic total deposition of 7.9×10^7 nCi if we assume that each year biotic accumulation of plutonium is the same. This supply increases the theoretical accumulation of plutonium in the pond's sediments to 1.0×10^9 nCi. This quantity is also a theoretical expression of the historic supply of the plutonium to the pond.

The Export

Percolation. No experiments were undertaken to measure the percolative loss of plutonium from U-Pond, and definitive conclusions about the movement of plutonium from the sediments into the ground below the pond cannot be made. However, there are indications that nearly all the plutonium that has reached the pond has been retained by its sediments. In the above discussions, we concluded that the pond has deposited about 10 cm of sediments since its formation in 1944. This was based on present-day measurements of sedimentation processes occurring in the pond. It was also theorized that about 1 Ci of plutonium has reached these sediments during the lifetime of the pond. Intensive sampling of the pond's sediments to a depth of 10 cm has shown that about 1.7 Ci of plutonium presently resides there. This agreement between theoretical and observed accumulation of plutonium in U-Pond sediments suggests that the pond has received about 1 Ci of plutonium and that most of it has been retained by its sediments.

The downward migration of plutonium in Hanford soils has been studied by several workers to assess the seepage of reprocessing wastes from crib sites (Crawley, 1969; Ames, 1974; Price and Ames, 1976). Their findings indicate a vertical reduction of plutonium concentrations over the upper 10 m of the vadose zone (i.e., soil lying above the water table). Price and Ames (1976) found that particulate plutonium (>2 μm) was deposited within the upper 1 m of the vadose zone. The nonparticulate plutonium (<2 μm), which was less than 0.5% (by weight) of the plutonium entering the vadose zone, showed deeper penetration and was eventually deposited in association with the silicate hydrolysis of sediment particles. Brown (1967), studying the vertical migration of other long-lived radionuclides below disposal facilities, found that more than 99.9% (by activity) of these materials was deposited within the upper 10 m of the vadose zone. In addition to this, Myers, Fix, and Raymond (1977) indicate that plutonium concentrations in the groundwater below Hanford (at $\simeq$50 m) are not significantly different from those of other areas.

Although we have not determined how much plutonium has percolated out of U-Pond since its formation in 1944, the available evidence points to the retention of virtually all of it in the sediments—probably in the upper several centimeters. Furthermore, we find no evidence to indicate that plutonium has migrated from U-Pond into the groundwater below Hanford. Thus we have no reason to believe that percolation is a significant route of plutonium export from U-Pond.

Emergent Insects. Insects emerging from U-Pond constitute the only direct route of biological export. However, if the life cycles of these insects are considered, it appears that emergence alone does not account for the export of the entire plutonium inventory contained in this compartment. The cast exoskeletons left in the pond at the final ecdysial stage prior to emergence may contain a substantial portion of the plutonium burden of the insects. It is also possible that some of these insects complete their life cycles without leaving the pond's ecosystem, and their plutonium burdens may ultimately be returned to the pond. We will not attempt to estimate the fraction of plutonium that is left in the pond by these processes, but instead we will assume that the entire inventory of this compartment leaves the pond when these insects emerge.

With the foregoing considerations taken into account, the mean annual export of plutonium by emerging insects is about 5×10^3 nCi. This quantity is about $9 \times 10^{-2}\%$ of the plutonium inventory of the biota and $6 \times 10^{-4}\%$ of the total pond inventory (Tables 3 and 4).

Waterfowl. The waterfowl that contact the U-Pond ecosystem are mostly mallards and an assortment of other ducks and coots. Some of these waterfowl nest along the shoreline of the pond. Since it is unlikely that these waterfowl contact other locations where they may be exposed to above-background levels of plutonium, it is assumed that most of the plutonium found in their gut and tissues came from U-Pond.

Examination of crop and gut contents of ducks collected from U-Pond indicates that they feed most heavily on the organically rich floc that covers the pond's sediments and, to a lesser extent, on goldfish and other material. Recall that floc contains most of the plutonium in the pond's ecosystem.

Concentrations of plutonium in whole bodies (including gut and contents) of four wild ducks *(Anas)* ranged from 3×10^{-2} to 3×10^0 pCi/g, with a mean of 4×10^{-1} pCi/g. These ducks were in contact with the pond when they were sampled, and most of their plutonium burdens were contained in the gut. Less than 5% of their entire plutonium burdens was contained in the body tissue.

Knowledge of the relationship between the length of time a duck spends in the U-Pond ecosystem and the amount of plutonium accumulated would be useful. Information about the contact frequency and duration is not available for waterfowl sampled from the pond; thus there is no basis to establish this relationship. However, a short experiment was performed to determine the amount of plutonium accumulated in ducks *(Anas)* held on the pond in large cages for 5 days and fed a continuous diet of organic floc (Emery and Klopfer, 1977). These experimental conditions represent the highest potential for the accumulation of plutonium by a duck on a short-term basis.

Results of this experiment suggest that ducks could accumulate about 6×10^0 pCi Pu/g (whole duck) in 2 to 5 days of continuous contact with U-Pond (Fig. 4). Under these conditions the accumulation of plutonium in the gut could be around 7×10^1 pCi/g, and the body tissue may concentrate about 3×10^0 pCi Pu/g. It is evident that in the experimental ducks the gut contained most of the plutonium burden (>95%), which

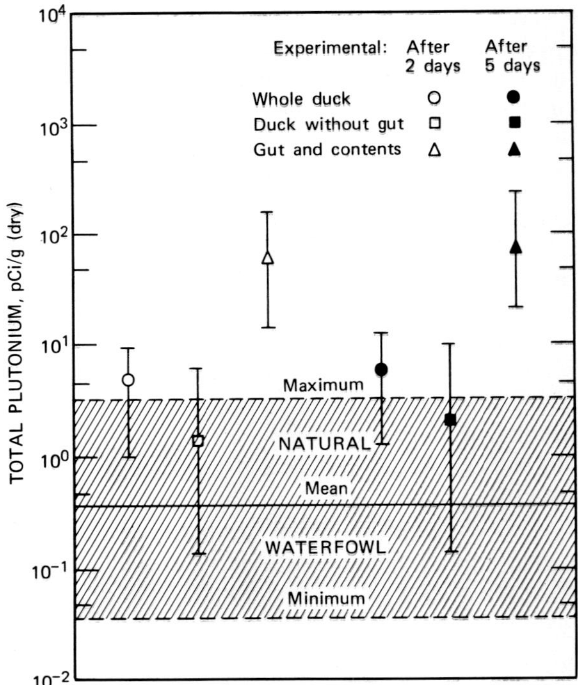

Fig. 4 Total plutonium accumulated by experimental ducks after 2 and 5 days of continuous contact with U-Pond compared with minimum, maximum, and mean concentrations of plutonium in whole waterfowl occurring naturally on U-Pond. Mean plutonium concentrations in experimental ducks are shown by location of appropriate symbol within the ranges of data depicted by horizontal lines.

indicates that most of the plutonium accumulated while the ducks were in contact with the pond would be lost soon after they flew away.

The mean plutonium concentrations of whole experimental ducks after 2 to 5 days of continuous contact with U-Pond ($\sim 6 \times 10^0$ pCi/g) were more than an order of magnitude greater than the mean plutonium concentrations in whole wild duck samples ($\sim 4 \times 10^{-1}$ pCi/g, Fig. 4). However, the maximum plutonium concentrations in whole wild ducks were higher than the minimum concentrations in whole experimental ducks. This comparison suggests an upper limit of contact duration, and it is concluded that wild ducks have a plutonium burden that is less than that obtained from 2 days of continuous contact with U-Pond.

The estimated total weight of waterfowl contacting U-Pond annually is 1.2×10^3 kg. Since the range of plutonium concentrations in whole wild ducks was from 3×10^{-2} to 3×10^0 pCi/g, the amount of plutonium exported by waterfowl annually could be from approximately 4×10^1 nCi to 4×10^3 nCi. The mean plutonium concentration in whole wild ducks of 4×10^{-1} pCi/g suggests a mean annual export of 5×10^2 nCi of plutonium. These export quantities are approximately four orders of magnitude lower than the total plutonium inventory for U-Pond.

Birds.   Birds observed around U-Pond are mainly sparrows, swallows, blackbirds, doves, and shorebirds (Emery, Klopfer, and Weimer, 1974). Estimates of contact frequency along with a mean weight for each taxon of bird suggest that approximately 5×10^2 kg of bird biomass moves through the air mass around U-Pond each year. Greater resolution of this annual biomass quantity is limited by the lack of information on the magnitude of bird activity in the pond region.

Samples of sparrow, swallows, and killdeer (total of 6) were analyzed for whole-body content of plutonium (including gut and contents). Plutonium concentrations ranged from less than 1×10^{-3} to 5×10^0 pCi/g, with swallows showing the highest concentrations. This may be associated with their mud-gathering activities involved in nest construction. The mean of these samples is approximately 2×10^{-1} pCi/g.

An estimation of plutonium exported by birds suggests that approximately 1×10^2 nCi is removed from the pond each year. The maximum export is approximately 2×10^3 nCi of plutonium, whereas a minimum annual export is less than detectable. These export quantities are below the estimated amounts of plutonium exported by waterfowl. They also represent less than $1 \times 10^{-4}\%$ of the total plutonium inventory for U-Pond.

Mammals. Mammals (other than man) that contact U-Pond are mice, rabbits, coyotes, and deer. Only mice were sampled from the mammal population and were collected within 10 m of the pond's shoreline.

The mouse population density around U-Pond is estimated to be not greater than 1 per $10\ m^2$, or $\leqslant 6 \times 10^3$ mice in a region around U-Pond that is equivalent to the pond's surface area ($\sim 6 \times 10^4\ m^2$). Since one mouse weighs approximately 2 g (dry), the mass of this population is about 1×10^1 kg. Whole-body samples of seven mice analyzed for total plutonium showed a range of concentrations from 1×10^{-3} to about 1×10^0 pCi/g (mean, $\sim 5 \times 10^{-1}$ pCi/g). This suggests that the mouse population may contain a maximum of 1×10^{-1} nCi of plutonium, or a mean of about 5×10^0 nCi. It is not known if all the plutonium found in mice came from U-Pond since there are regions adjacent to the pond that have plutonium concentrations above background levels.

The coyote population on the Hanford Site ($600\ mi^2$) has been estimated to be around 300 following spring breeding (Rickard et al., 1977). Occasionally coyotes are observed feeding on goldfish at U-Pond. A contact frequency by coyotes is estimated to be not greater than about 1 per day, or about 4×10^2 coyote visits per year. If we assume that each coyote removes 1 goldfish (3 g, dry weight) with each visit, the dry weight of goldfish removed by coyotes each year is about 1×10^3 g. The maximum concentration of plutonium in goldfish is about 3×10^1 pCi/g (Table 2), which suggests that as much as 3×10^1 nCi of plutonium might be exported by coyotes each year.

Rabbits and deer are seldom observed at U-Pond; less than 50 sightings of either mammal have been made by the study team in over 3 yr. It is likely that these mammals visit U-Pond to drink water only; therefore we will consider their annual export of plutonium to be negligible.

It is interesting to note that the largest plutonium export route among mammals is the researcher. During a normal study year, approximately 2×10^2 kg of samples are taken from the pond, which contains about 2×10^2 pCi Pu/g. This "export" quantity of 4×10^4 nCi/yr as research samples appears to be greater than that caused by all other mammals combined.

Wind. Attempts to quantify the export of airborne particulate plutonium from U-Pond via wind were not made. It would be desirable to have data that express the rate of plutonium movement away from the pond and its shore as airborne particles, but there are many limitations to this assessment. Sehmel (1977) reports that airborne plutonium concentrations in samples taken at resuspension sites in the U-Pond region have been significantly higher than fallout levels in other areas (but still less than maximum permissible concentrations). This region has other sources of airborne plutonium besides U-Pond, and it was not possible to determine if the plutonium detected in these samples came from U-Pond, fallout, or some adjacent source.

However, it appears that U-Pond does not contribute significantly to the plutonium concentrations in the air downwind from the chemical processing areas. This is indicated when plutonium concentrations in the air downwind from these areas are compared with the concentrations in the air of distant upwind and downwind perimeter communities. Differences between the plutonium concentrations of the two air masses were not significant. In 1972, for example, the average plutonium concentration in the air of distant communities was 1.8×10^{-5} pCi/m^3, whereas the mean air concentration of plutonium downwind from the chemical processing areas was 1.9×10^{-5} pCi/m^3 (Energy Research and Development Administration, 1975).

The premise that U-Pond does not release appreciable amounts of plutonium particles through wind action is strengthened by an additional consideration. Between 1944 and 1955 U-Pond was occasionally flooded to the limits of its basin and subsequently overflowed into an auxiliary basin. Since 1955 the pond has remained within its original shoreline. Plutonium deposited as sediments while these areas were flooded is now exposed to the movements of air. Although this exposed area is larger than the present surface area of U-Pond, the plutonium concentrations of downwind air are not significantly elevated by its presence.

Summary and Conclusions

In its 34-yr history, U-Pond has received an estimated 1 Ci of plutonium. Since the same quantity presently resides in the sediments, it appears that U-Pond has retained nearly all the plutonium that has been discharged into it.

In relative terms, sediments, submerged plants, and gastropods have the highest concentrations of plutonium, ranging from 3.2×10^0 to 6.9×10^2 pCi/g. Plutonium concentrations of emergent plants and the remaining fauna range from 4.0×10^{-1} to 6.1×10^1 pCi/g. Emerging insects had the highest plutonium concentrations of the latter group, ranging from 3.2×10^1 to 6.1×10^1 pCi/g.

The mean plutonium inventory of the sediment is 1.7×10^9 nCi, ranging from 1.3×10^9 to 2.0×10^9 nCi of plutonium (Fig. 5). This essentially represents the total pond inventory since more than 99% of the plutonium in the pond is found in the sediments. The mean plutonium inventory for the biota is 6×10^6 nCi, ranging from 1×10^6 to 1×10^7 nCi (Fig. 5). Among these, biota plant life contains more than 95% of the plutonium. Diatoms and pondweed *(Potamogeton)* alone account for more than 99% of the plutonium in plants. Emergent insects contain less than 1×10^{-1}% of the plutonium in biota and less than 1×10^{-3}% of the plutonium in the pond. The inventory of this compartment has particular relevance since it is the only direct biological route of export from the pond. Remaining pond biota contain less than 1×10^{-2}% of the total plutonium inventory in the pond and can leave the pond only by the forces of external export vectors.

If all emergent insects successfully leave the pond, they could export from 3.5×10^3 to 7×10^3 nCi of plutonium. These quantities are more than five orders of magnitude lower than the total pond plutonium inventory (Fig. 5). Estimated quantities of

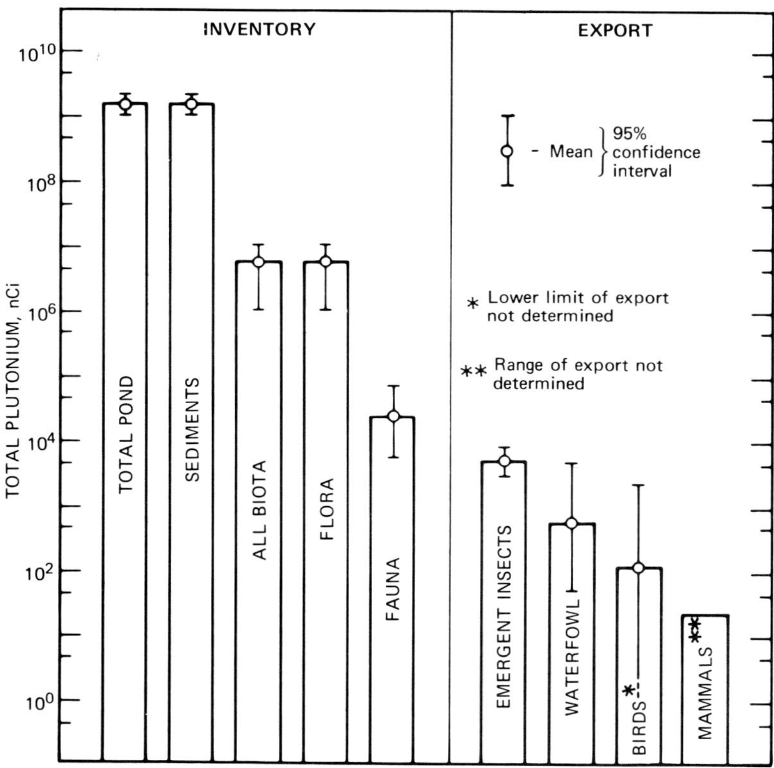

Fig. 5 Plutonium in ecological compartments of U-Pond compared with estimated quantities that are exported from the pond annually. Export of plutonium by percolation or wind does not appear to be significant.

plutonium annually exported by waterfowl range from 4×10^1 to 4×10^3 nCi, with a mean annual export of 5×10^2 nCi of plutonium (Fig. 5). Other birds appear to export about 1×10^2 nCi of plutonium each year, with a maximum of 2×10^3 nCi (Fig. 5). These export quantities are about six orders of magnitude lower than the total inventory of plutonium in the pond. Mammals are estimated to export a maximum of 3×10^1 nCi of plutonium from the pond (Fig. 5) annually, which is at least five orders of magnitude lower than the minimum total plutonium inventory of the pond. There is no apparent significant export of plutonium from the pond via wind or percolation.

In conclusion, U-Pond has been exposed to plutonium longer than any other aquatic system and has received about 1 Ci of ^{239,240}Pu and ^{238}Pu. This 14-acre pond provides a realistic illustration of the mobility of plutonium in a lentic or nonflowing ecosystem. Although this pond has a rapid flushing rate, is highly enriched with plant nutrients, is ecologically well established with a natural complexity of populations and diversity of communities, and has continuous interaction with associated terrestrial life, it appears to effectively bind the plutonium discharged into it and prevent it from moving significantly into routes leading to man and other remote life. Furthermore, the environmental behavior of plutonium in U-Pond appears to be quite similar to that of other aquatic systems having vastly different ecological character. So long as this pond remains in its present condition, the likelihood of its releasing hazardous quantities of plutonium to man and his environment is very small.

Acknowledgments

Advice in the interpretation of data provided by Battelle Scientists R. F. Foster, D. G. Watson, J. M. Thomas, L. L. Eberhardt, K. A. Gano, R. E. Fitzner, and R. J. Serne is greatly appreciated. We also wish to express thanks to the Atlantic Richfield Hanford Company for providing assistance in the funding of radioanalysis. This research was funded by the U. S. Department of Energy, Division of Biomedical and Environmental Research, under contract No. EX-76C-06-1830.

References

Ames, L. L., Jr., 1974, *Characterization of Actinide Bearing Soils: Top Sixty Centimeters of 216-Z-9 Enclosed Trench,* USAEC Report BNWL-1812, Battelle, Pacific Northwest Laboratories, NTIS.

Bartelt, G. E., C. W. Wayman, and D. N. Edgington, 1975, Plutonium Concentrations in Water and Suspended Sediment from the Miami River Watershed, Ohio, in *Radiological and Environmental Research Division Annual Report, January–December, 1974,* USAEC Report ANL-75-3 (Pt. 3), pp. 72-77, Argonne National Laboratory, NTIS.

Brown, D. J., 1967, Migration Characteristics of Radionuclides Through Sediments Underlying the Hanford Reservation, in *Disposal of Radioactive Wastes into the Ground,* Symposium Proceedings, Vienna, May 29–June 2, 1967, pp. 215-228, STI/PUB/156, International Atomic Energy Agency, Vienna.

Carlander, K. D., 1955, The Standing Crop of Fish in Lakes, *J. Fish. Res. Board Can.,* 12: 543-570.

Crawley, D. T., 1969, *Plutonium–Americium Soil Penetration at 234-5 Building Crib Sites,* Report ARH-1278, Atlantic Richfield Hanford Company.

Dahlman, R. C., E. A. Bondietti, and E. R. Eastwood, 1975, Plutonium in Aquatic and Terrestrial Environments, in *Environmental Sciences Division Annual Progress Report for Period Ending Sept. 30, 1974,* USAEC Report ORNL-5016, pp. 1-8, Oak Ridge National Laboratory, NTIS.

Electric Power Research Institute, 1976, *Plutonium: Facts and Inferences,* C. L. Comar (Ed.), Report EPRI-EA-43-SR, Electric Power Research Institute.

Emery, R. M., and D. C. Klopfer, 1977, Ecological Distribution and Fate of Plutonium and Americium in a Processing Waste Pond on the Hanford Reservation, in *Pacific Northwest Laboratory Annual Report for 1976 to the ERDA Assistant Administrator for Environment and Safety,* B. E. Vaughan (Ed.), ERDA Report BNWL-2100 (Pt. 2), p. 4.15, Battelle, Pacific Northwest Laboratories, NTIS.

——, D. C. Klopfer, and W. C. Weimer, 1974, *Ecological Behavior of Plutonium and Americium in a Freshwater Ecosystem. Phase I. Limnological Characterization and Isotopic Distribution,* USAEC Report BNWL-1867, Battelle, Pacific Northwest Laboratories, NTIS.

——, D. C. Klopfer, T. R. Garland, and W. C. Weimer, 1976, The Ecological Behavior of Plutonium and Americium in a Freshwater Pond, in *Radioecology and Energy Resources,* Fourth National Radioecology Symposium, Corvallis, Ore., May 12–14, 1975, C. E. Cushing, Jr. (Ed.), pp. 74-85, Dowden, Hutchinson and Ross, Inc., Stroudsburg, Pa.

Energy Research and Development Administration, 1975, Final Environmental Statement, *Waste Management Operations, Hanford Reservation, Richland, Washington,* ERDA Report ERDA-1538 (Vol. 1), NTIS.

Hakonson, T. E., J. W. Nyhan, and W. D. Purtymun, 1976, Accumulation and Transport of Soil Plutonium in Liquid Waste Discharge Areas at Los Alamos, in *Transuranium Nuclides in the Environment,* Symposium Proceedings, San Francisco, 1975, pp. 175-189, STI/PUB/410, International Atomic Energy Agency, Vienna.

Johnson, J. E., S. Svalberg, and D. Paine, 1974, *The Study of Plutonium in Aquatic Systems of the Rocky Flats Environs,* Final Technical Report, Department of Radiation Biology and the Department of Animal Sciences, Colorado State University.

Magno, P., T. Reaney, and J. Apidianakis, 1970, Liquid Waste Effluents from a Nuclear Fuel Reprocessing Plant, in *Health Physics Aspects of Nuclear Facility Siting,* Proceedings of the 5th Annual Health Physics Society Midyear Topical Symposium, Idaho Falls, Nov. 3–6, 1970, pp. 208-220, Health Physics Society.

Myers, D. A., J. J. Fix, and J. R. Raymond, 1977, *Environmental Monitoring Report on the Status of Ground Water Beneath The Hanford Site, January–December, 1976,* ERDA Report BNWL-2199, Battelle, Pacific Northwest Laboratories, NTIS.

Noshkin, V. E., 1972, Ecological Aspects of Plutonium Dissemination in Aquatic Environments, *Health Phys.,* 22: 537-549.

Patterson, J. H., G. B. Nelson, G. M. Matlack, and G. R. Waterbury, 1976, Interaction of ^{238}PuO$_2$ Heat Sources with Terrestrial and Aquatic Environment, in *Transuranium Nuclides in the Environment,* Symposium Proceedings, San Francisco, 1975, pp. 63-78, STI/PUB/410, International Atomic Energy Agency, Vienna.

Price, S. M., and L. L. Ames, 1976, Characterization of Actinide-Bearing Sediments Underlying Liquid Waste Disposal Facilities at Hanford, in *Transuranium Nuclides in the Environment,* Symposium Proceedings, San Francisco, 1975, pp. 191-211, STI/PUB/410, International Atomic Energy Agency, Vienna.

Rickard, W. H., et al., 1977, Densities of Large and Medium-Sized Mammals on the Hanford Reservation, in *Pacific Northwest Laboratory Annual Report for 1976 to the ERDA Division of Biomedical and Environmental Research,* B. E. Vaughan (Ed.), ERDA Report BNWL-2100 (Pt. 2), p. 4.33, Battelle, Pacific Northwest Laboratories, NTIS.

Robbins, W. W., T. E. Weier, and D. R. Stocking, 1957, *Botany: An Introduction to Plant Science,* p. 204, John Wiley & Sons, Inc., New York.

Schell, W. R., and R. L. Watters, 1975, Plutonium in Aqueous Systems, *Health Phys.,* 29: 589-597.

Sehmel, G. A., 1977, *Radioactive Particle Resuspension Research Experiments on the Hanford Reservation,* ERDA Report BNWL-2081, Battelle, Pacific Northwest Laboratories, NTIS.

Singh, H., and J. S. Marshall, 1977, A Preliminary Assessment of ^{239,240}Pu Concentrations in a Stream near Argonne National Laboratory, *Health Phys.,* 32: 195-198.

Trabalka, J. R., and L. D. Eyman, 1976, Distribution of Plutonium-237 in a Littoral Freshwater Microcosm, *Health Phys.,* 31: 390-393.

Verduin, J., 1964, Principles of Primary Productivity: Photosynthesis Under Completely Natural Conditions, in *Algae and Man,* D. F. Jackson (Ed.), Proceedings of the NATO Advanced Study Institute, Louisville, Ky., July 22–Aug. 11, 1962, pp. 221-228, Plenum Press, Inc., New York.

Plutonium in Rocky Flats Freshwater Systems

D. PAINE

This study was initiated to determine the behavior of plutonium in the freshwater aquatic environs at the Rocky Flats Dow Chemical plutonium fabrication plant, Golden, Colo. The principal study area included four holding ponds for waste solutions generated at the plant complex.

Samples of biotic and abiotic components were collected from the spring of 1971 through the summer of 1973. These components consisted of sediment, water, seston, zooplankton, fish, vegetation, and small mammals in close proximity to the aquatic systems. Laboratory experiments were performed to quantify field results. Owing to the high variability of plutonium concentrations in the environment, numerous samples were collected and analyzed by a modified solvent-extraction liquid-scintillation counting procedure.

Sediments were the major site of $^{239,240}Pu$ deposition. Coring analysis revealed the largest concentrations at subsurface-sediment depths, and thus depth-profile data were used in calculating total inventory. A retention function determined in the laboratory demonstrated a rapid transfer of plutonium from water to sediment. Pond reconstruction during the study period resulted in significant increases in mean-surface(top 5 cm)-sediment concentrations.

Seston contained 30 to 80% of the $^{239,240}Pu$ in an unfiltered water sample. Concentration ratios in seston, ranging from 10^4 to 10^5, were higher than those found in marine studies. No vertical distribution of $^{239,240}Pu$ was noted in pond water. Laboratory experiments suggested active uptake by algae rather than by simple surface adsorption. Zooplankton showed a discrimination against plutonium concentration along the simple phytoplankton-to-zooplankton food chain. Fish flesh and bone showed no levels above minimum detectable activity (MDA, 0.03 d/min per 10-g sample for a 100-min count). Vegetation associated with pond sediments contained higher concentration ratios from sediment to aerial portions of plants than previously observed, ranging from 10^{-2} to 10^{-1}.

Although plutonium in the biosphere presently exists at very low concentrations, trophic biomagnification and possible localized contamination may result in increased plutonium concentrations in organisms of higher trophic levels. Cycling processes and biological uptake of plutonium must be understood before environmental releases so that rational assessment of its potential hazard can be performed. The major concern with plutonium is its potential hazard to man. Plutonium could enter man either directly through inhalation of atmospherically suspended material or indirectly through incorporation into his food chain. The inhalation route is considered the most hazardous mode of entry to man (Taylor, 1973). However, the concentration of plutonium in sediments or in aquatic

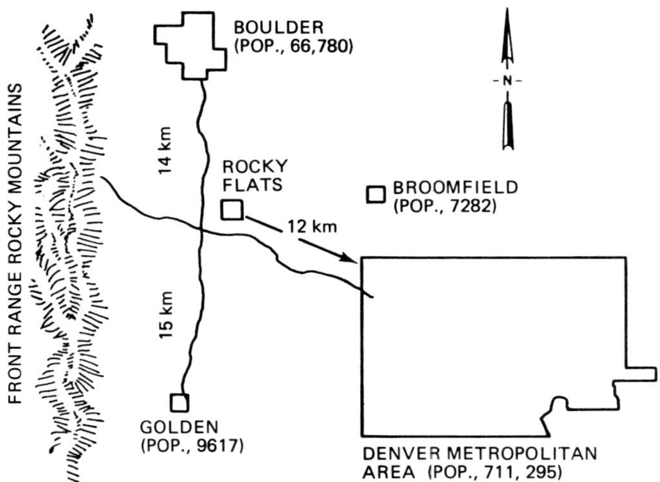

Fig. 1   Rocky Flats installation relative to nearby population centers.

organisms frequently exceeds concentrations in surrounding waters by orders of magnitude. This concentration process may pose unexpected hazards when considering food-chain transport.

Before this investigation little information concerning plutonium movement in aquatic systems was available (Stannard, 1973; Noshkin, 1972). Fallout and marine studies comprised the bulk of this environmental data, with average concentrations in the femtocurie range (Pillai, Smith, and Folsom, 1964; Aarkrog, 1971; Noshkin et al., 1971). In general, all freshwater studies have concurred that sediments appear to be the major reservoir for ultimate plutonium deposition and that relatively insignificant transport of plutonium through biotic systems to man exists (Emery and Klopfer, 1976; Hakonson, Nyham, and Purtymun, 1976).

The purpose of this investigation was to determine the behavior of plutonium in freshwater systems at the Rocky Flats Dow Chemical plutonium fabrication plant, Golden, Colo. The objectives were to (1) investigate the distribution patterns of plutonium in the biotic and abiotic components of the Rocky Flats freshwater systems, (2) determine any concentrating processes that were occurring, and (3) determine if any biological mobilization processes existed. It was the first attempt of its kind at delineating the cycling processes of plutonium using a holistic systems approach.

Methods and Materials

Figure 1 shows the location of the Rocky Flats area relative to the larger surrounding metropolitan areas. The plant site itself covers approximately 10 km^2.

Figure 2 shows the general sampling area at Rocky Flats. The principal study area included the four holding ponds (B-series ponds) for waste solutions generated at the plant complex. These ponds were drained by Walnut Creek, which flowed into Great Western reservoir, the City of Broomfield's municipal water supply. Great Western's water sources were provided by Walnut Creek (2%), Coal Creek (8%), and Clear Creek watershed (90%) (Hammond, 1971). The A-series and C$_1$ ponds were monitoring ponds that did not receive routine releases of plutonium waste. Pond A$_1$ had received low-level

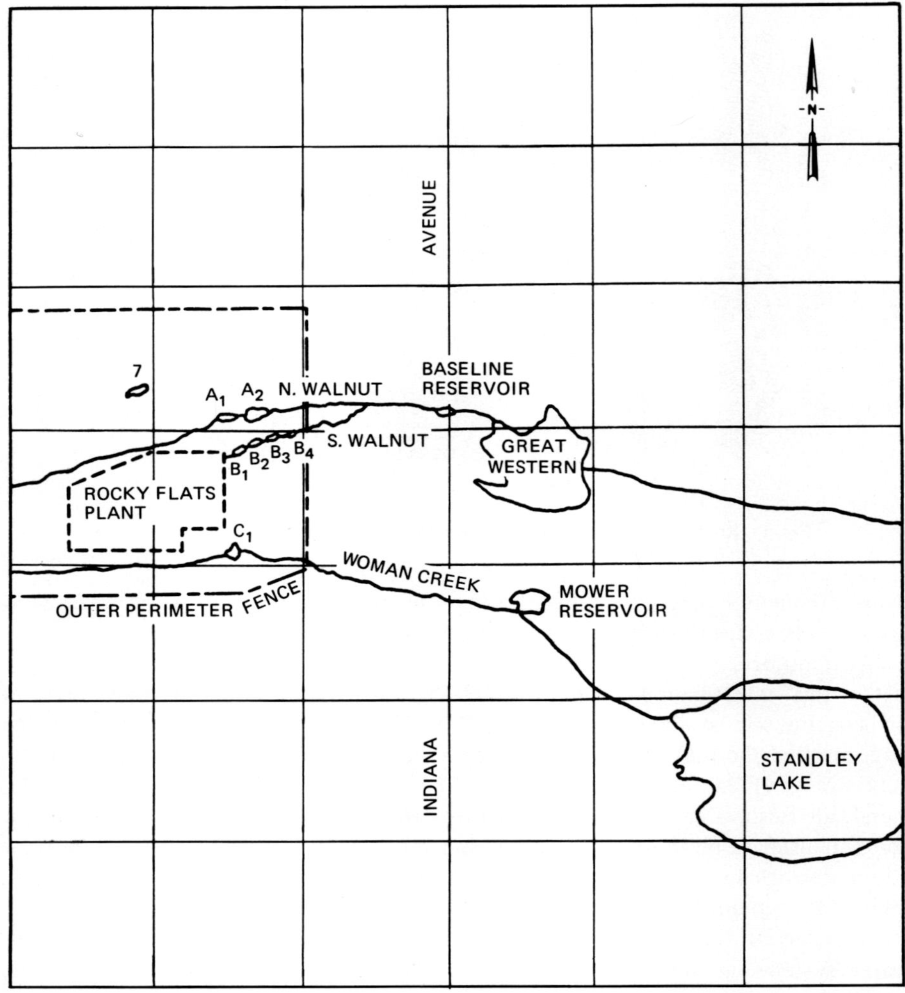

Fig. 2   Study area of Rocky Flats environs showing ponds, streams, and reservoirs. Flow on Woman and Walnut Creeks is from west to east.

plutonium contamination from past nonroutine releases. After the completion of this study, pond A_2 was constructed to handle excessive water runoff. This pond was not investigated in this study. Pond C_1 received runoff from the "pad" area located several hundred yards due northeast. This area was previously contaminated by plutonium from leaking 50-gal drums and was subsequently covered with an asphalt pad. The southernmost pond (pond C_1) drained into Woman Creek, which flowed into Standley Lake, an irrigation reservoir as well as the municipal water supply for Westminster, Colo. Pond 7, located several miles northeast of the study area, was used as a control pond for this study. Samples were periodically collected from pond 7 and used as background correction.

Primary plutonium waste discharged to the pond complex included laundry wastes and process waste solutions generated by various phases of the plant's operation. Owing to the high variability of plutonium concentrations in the environment, numerous samples were collected and analyzed by a modified solvent-extraction liquid-scintillation counting procedure (Keough and Powers, 1970). Counting yield was 96%. Overall chemical recovery was 90%. The minimum detectable activity (MDA) was 0.30 d/min per sample for a 100-min count. Modifications and additional analysis information are presented in a report by Johnson, Svalberg, and Paine (1974). Unless otherwise stated, all references to plutonium in this chapter include both ^{239}Pu and ^{240}Pu plutonium isotopes because the analytical procedure did not discriminate between plutonium isotopes.

From 20 to 30 surface-sediment cores (approximately the top 5 cm) were obtained during each sampling period (~1 per month) from each pond. The procedure followed that outlined by Hakonson (1972). Additionally, core samples were extracted from the sediment beds of the pond, and the procedure defined by Johnson, Svalberg, and Paine (1974) was used to determine vertical distribution of plutonium within the pond sediments. Incremental samples were composited for analysis. The coefficient of variation determined from composited sediment samples was approximately 30%.

Surface-water samples were initially taken at each sediment sampling location from each pond. Later samples were taken not only at the surface but also at 0.5-m increments to the sediment—water interface. A mechanical water sampler was used to collect the subsurface water samples.

Water samples were filtered in a Millipore filtering apparatus that was modified by adding a brass screen with a pore size of 250 μm to the top of the water intake funnel. The screen removed most of the zooplankton and large organic material from the water sample. The water was pulled through a glass-fiber filter (Whatman GF/A, 4.7-cm diameter) with a vacuum pump to remove the remaining suspended material. The residue was analyzed as a separate component called seston, which included primarily phytoplankton, detritus, and other suspended solids. Seston as defined here does include some small zooplankton. The filtration process was usually carried out within 24 to 48 hr after the collection. Water samples were kept in darkness to inhibit growth until filtration could be accomplished.

The term "zooplankton" was used collectively for all small planktonic animals trapped in a number 10 plankton net (160-μm mesh size). These samples contained seston as well as large aquatic insects. The insects were separated from the samples. It was assumed that most of the sestonic material was probably smaller than the 160-μm mesh size and would pass through the net since the zooplankton sample was rinsed in the collecting net by repeated dunkings in the pond. Zooplankton were identified to species, but biomass estimates were not determined. A 12.7-cm-diameter Clark—Bumpus plankton sampler was towed behind a boat in an attempt to sample organisms.

Bass (*Ictiobus bubalus*) and carp (*Cyprinus carpio*) were collected by seining and angling. No fish were present in the B-series ponds, but minnows (*Hybosis* sp.) were collected in pond C$_1$ and pond A$_1$ with a large collecting net. The fish were too small for angling, and seining would have disturbed the bottom sediments. The maximum fish length observed in the ponds was approximately 6 cm. Vegetation (primarily *Juncus balticus, Rumex crispus*, and *Typha latifolia*) was collected in and around the ponds, streams, and reservoirs throughout the study period. Generally, the aerial portions of the plant samples were clipped with grass shears, and the roots were extracted separately owing to excessive sediment—soil contamination.

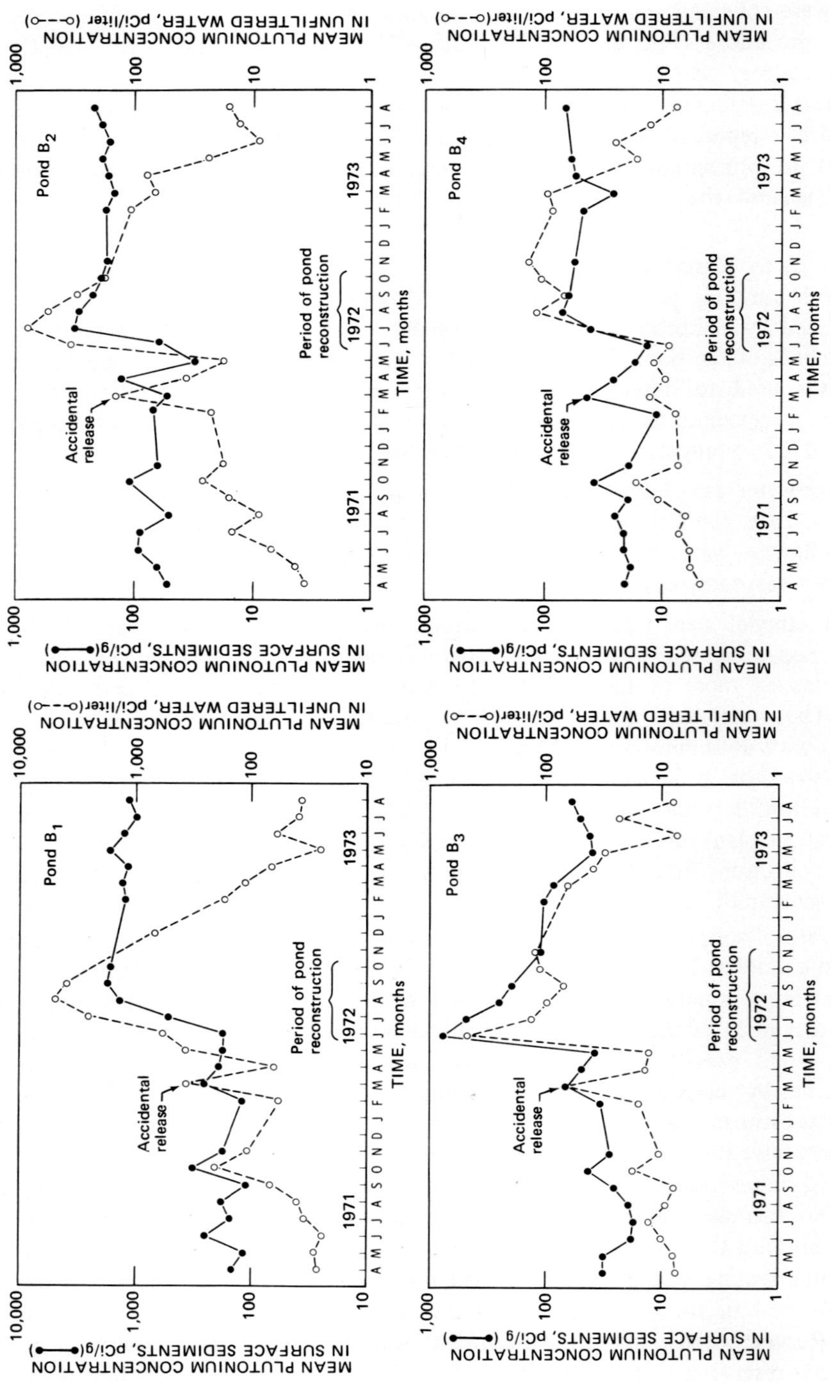
MEAN PLUTONIUM CONCENTRATION
IN UNFILTERED WATER, pCi/liter (o——o)
MEAN PLUTONIUM CONCENTRATION
IN SURFACE SEDIMENTS, pCi/g (●——●)
Pond B₁
Pond B₂
Pond B₃
Pond B₄
Accidental release
Period of pond reconstruction
TIME, months
1971
1972
1973
A M J J A S O N D J F M A M J J A S O N D J F M A M J J A

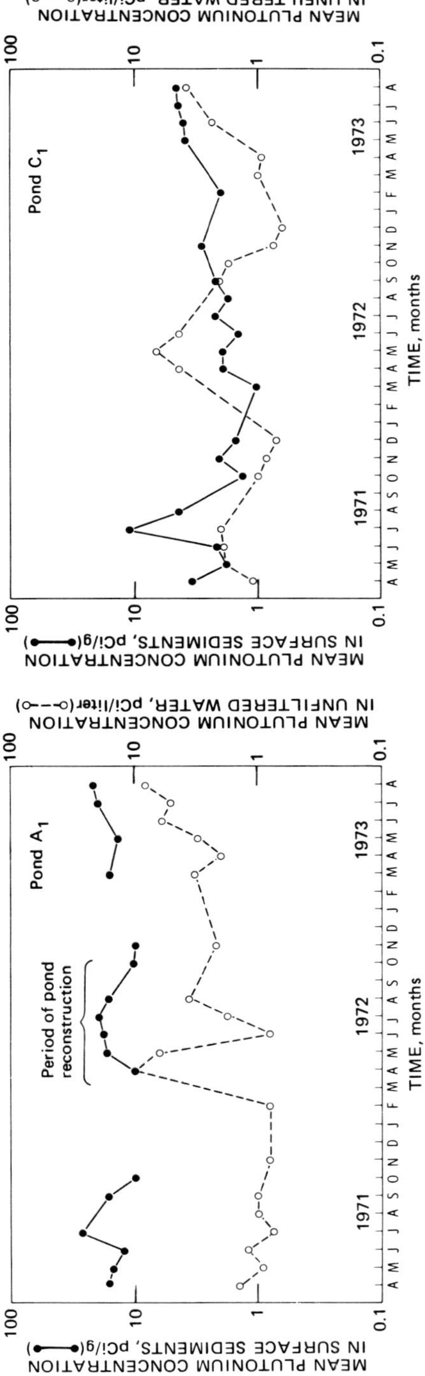

Fig. 3 Mean plutonium concentrations in surface sediments (pCi/g) and mean plutonium concentrations in unfiltered water (pCi/liter) for ponds B_1, B_2, B_3, B_4, A_1, and C_1. The sampling period was from April 1971 through August 1973. The B-series ponds received routine discharges containing low-level plutonium waste.

Small mammals (*Microtus pennsylvanicus modestus* and *Peromyscus nasutus*) were kill-trapped throughout the study period. Mule deer (*Odocoileus hemionus*) samples were collected from road kills.

Results

Sediment

Figure 3 shows the average ^{239,240}Pu concentrations in water and surface sediments over the entire study period. The accidental release in March 1971 resulted when process waste solutions low in ^{239,240}Pu content, due to be pumped to solar evaporation ponds for concentration, were accidentally released to the sanitary waste system. The resultant elevation over ambient conditions is readily apparent in Figs. 3 and 4. Reconstruction of the pond area had a marked effect on the mean sediment concentrations. Concentration levels increased significantly during construction and remained high during the rest of the study period for most ponds (Table 1). Plutonium concentrations also increased in the sediment sampled at Walnut Creek at Indiana Avenue (baseline reservoir), which indicated that considerable activity escaped the holding-pond system during the period of reconstruction (Fig. 4).

Pond B_1 showed the highest surface concentrations throughout the study period. A slight decrease in mean sediment concentrations was noted in ponds B_2 and B_4 following peak levels. However, pond B_3, which was the first pond to undergo reconstruction and underwent the most extensive remodeling, showed a marked decrease in sediment activity after the period of maximum values. This can be attributed to the deposition of appreciable soil that contained lower concentrations of plutonium during and after the dam and pond remodeling.

Subsurface sediments were probably mixed when the ponds were refilled. Core samples contained highest plutonium concentrations at 20- to 30-cm depths. Some minor construction modifications were made in the effluent bypass system which could have caused redistribution of high-level plutonium sediments from this area. Sewage-treatment modifications in May 1972, before reconstruction, could also have resulted in high-activity flocculate being released to the holding ponds.

Plutonium-239,240 concentrations in pond B_3 sediment peaked in late June 1972; those in pond B_2 peaked in July, and those in ponds B_1 and B_4 peaked in August. This suggests that pond reconstruction played a major role in the redistribution of plutonium since this is the order in which remodeling occurred. In any case it is readily apparent that mean-surface-sediment values increased markedly during the period of pond reconstruction and remained at higher levels except in pond B_3.

The clay sediments showed an extremely high affinity for plutonium, and, if left undisturbed, they appear to be an excellent reservoir for plutonium in an aquatic system.

Water

The mean concentrations of plutonium in unfiltered water samples during the course of this study (Fig. 3) showed that construction played a major role in the redistribution of plutonium from pond to pond. The increase in plutonium concentrations was also detected downstream at the Walnut Creek at Indiana Avenue sampling station (Fig. 4).

The majority of the plutonium in the water component was usually associated with the filterable fraction ($>0.45\,\mu$m) (Table 2). However, ponds A_1 and C_1, the monitoring

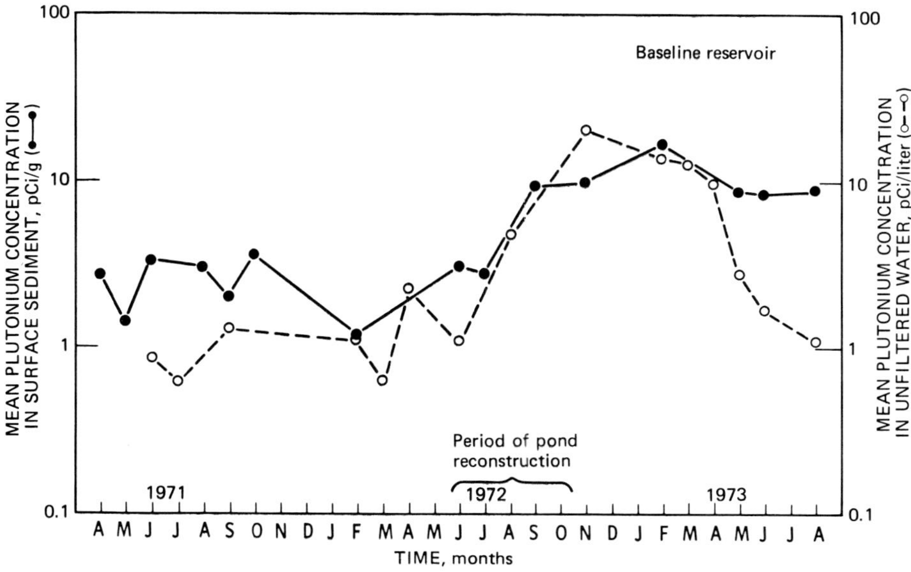

Fig. 4 Mean plutonium concentrations in surface sediment (pCi/g) and mean plutonium concentrations in unfiltered water (pCi/liter) for baseline reservoir. This sampling station is located where Walnut Creek crosses under Indiana Avenue.

TABLE 1 Mean-Surface(Top 5 cm)-Sediment ^{239,240}Pu
Concentrations During Preconstruction
and Postconstruction Periods

	Preconstruction		Postconstruction	
Pond	n*	pCi/g†	n*	pCi/g†
B₁	13	200 ± 70	11	1300 ± 350
B₂	12	80 ± 30	11	200 ± 60
B₃	13	30 ± 10	11	200 ± 200
B₄	14	20 ± 10	10	55 ± 15
C₁	12	3 ± 3	9	3 ± 1
A₁	9	15 ± 5	8	15 ± 4
Baseline reservoir	8	3 ± 1	7	10 ± 4
Bypass dam	2‡	750 ± 150		

*n = number of sampling periods.
†Mean ± standard error.
‡Mean of two samples taken in June 1971.

ponds that were separate from the holding-pond chain, received little contaminated plant effluent and contained a larger fraction of nonfilterable plutonium. Less suspended material, including phytoplankton, in ponds A₁ and C₁ is probably the explanation for this phenomenon. Because of the shallow nature of the pond systems, no apparent vertical distribution of plutonium could be determined.

**TABLE 2   Percent of Plutonium Isotopes
Associated with Filterable Fraction of Water
Samples from Rocky Flats Ponds**

Pond	Filterable fraction*
B_1	90 ± 6
B_2	80 ± 12
B_3	80 ± 8
B_4	70 ± 12
C_1	30 ± 30
A_1	35 ± 20

*Mean ± standard error.

Laboratory experiments were performed to study the transfer of plutonium from water to sediment as a function of time. The function appeared to consist of two exponential terms and was described by the equation

$$C(t) = C_0 \left(0.75\,e^{-0.92t} + 0.25\,e^{-0.58t}\right) \tag{1}$$

where $C(t)$ is the concentration of plutonium in water at any time t, t is time (in days), and C_0 is initial concentration of plutonium in water.

This experimental finding fits remarkably well with actual pond limnological data. The average S. Walnut Creek flow into pond B_1 was measured to be 480 m³/day during 1971. The water volume of pond B_1 was calculated to be 1500 m³. Therefore the mean lifetime of any parcel of water in pond B_1, if mixing is uniform, can be calculated as follows:

$$t = \frac{1500 \text{ m}^3}{480 \text{ m}^3/\text{day}} = 3.1 \text{ days}$$

According to Eq. 1, 75% of the plutonium in water delivered to pond B_1 should be exchanged to sediment in an average residence time of 3.1 days.

Plutonium concentration as a function of sediment depth for the ponds is illustrated in Table 3. These data were plotted and integrated by a planimeter to determine the area (picocuries-centimeter per gram). This value was then divided by the mean sediment depth to give the mean sediment concentration of plutonium. When multiplied by the estimated sediment volume, these data yielded plutonium inventories for the sediment of the ponds. The same sediment characteristics were assumed for each pond. The variation could be due to shunting of water past ponds at unknown times. The calculated value also assumed that no plutonium was being transferred from pond to pond by suspended materials. The agreement between the calculated and measured inventories is shown for the holding ponds in Table 4.

Seston

Seston was defined as primarily phytoplankton, some detritus, and some small zooplankton. Planktonic algae constituted by far the majority of aquatic plant material found in the holding-pond chain on S. Walnut Creek (B ponds).

TABLE 3   Distribution of Plutonium Concentrations in Depth Profiles of Ponds B_1, B_2, B_3, B_4, and C_1

Depth, cm	Plutonium concentration, $(d/min)\,g^{-1}$				
	B_1	B_2	B_3	B_4	C_1
0 − 5	2,000	260	40	430	<1
5 − 10	2,200	80	170	190	<1
10 − 15	10,900	40	370	20	<1
15 − 20	32,100	60	330	4	2
20 − 25	8,800	230	20	7	4
25 − 30	1,100	190	3	7	<1
30 − 35	880	390	3	<1	
35 − 40	900	340	380		
40 − 45	190	100	6		
45 − 50	30	20	6		
50 − 55	100	10	3		
55 − 60	70	9	2		
60 − 65	5	4	2		
70 − 75	7	3			
75 − 80		2			
80 − 85		12			
85 − 90		50			
90 − 95		30			
95 − 100		3			
100 − 105		7			
105 − 110		40			
110 − 115		12			
115 − 120					

TABLE 4   Calculated and Measured Inventories of $^{239,240}Pu$ in Holding Ponds During 1971

Pond	Measured		Calculated,* %
	mCi	%	
B_1	84.5	62.4	74.8
B_2	27.0	19.9	21.9
B_3	19.4	14.3	2.9
B_4	4.6	3.4	0.4

*Calculated values are from the retention function obtained from laboratory experimentation.

**TABLE 5 Concentration Ratios (CR's) in Seston,
Zooplankton, and Crayfish Relative to Filtered Pond Water**

Pond	No. of sampling periods	CR*
	Seston	
A_1	13	$7.5 \pm 1.1 \ (10^4)$
B_1	18	$100 \pm 20 \ (10^4)$
B_2	19	$1.7 \pm 2.8 \ (10^4)$
B_3	19	$2.5 \pm 5.7 \ (10^4)$
B_4	21	$16 \pm 3.7 \ (10^4)$
C_1	12	$12 \pm 3.4 \ (10^4)$
	Zooplankton	
B_2	2	$0.14 \pm 0.02 \ (10^4)$
B_4	2	$0.17 \pm 0.02 \ (10^4)$
	Whole crayfish	
B_3	2	$0.13 \pm 0.002 \ (10^4)$
B_4	4	$0.07 \pm 0.019 \ (10^4)$
C_1	2	$0.06 \pm 0.003 \ (10^4)$

*pCi ^{239,240}Pu/g acceptor $\div$ pCi ^{239,240}Pu/ml water. Mean ±
standard error.

The transfer of plutonium from water to seston was extremely high (Table 5). The
concentration ratios (CR's) relative to filtered water were of the order of 10^4 to 10^5.
Concentration ratio is defined as

$$CR = \frac{\text{picocuries per gram seston (dry weight)}}{\text{picocuries per milliliter water (filtered)}}$$

These CR's were higher than those previously observed in marine systems. Laboratory
experiments revealed that the mechanisms involved were more than simple surface
sorption (Johnson, Svalberg, and Paine, 1974).

Zooplankton

Although several species of cladocerans, copepods, and amphipods were collected,
sufficient biomasses for analysis were never obtained at any one sampling period. This
necessitated a pooling of the samples over several months. This was especially true for the
B-series ponds, which contained almost no zooplankton throughout the study. Zooplankton showed CR's relative to filtered water in the 10^4 range (Table 5). These CR's
are similar to those reported in marine studies. If ingestion is the primary route of
transfer in these organisms, then higher concentration factors would be expected from
the simple phytoplankton-to-zooplankton food chain. Since an increase in trophic-level
concentration of plutonium did not occur, there appears to be a selective mechanism that
discriminates against plutonium at this level. This would result in a decreased potential
hazard when considering the transfer of plutonium through ingestion routes.

TABLE 6 Concentration of ^{239,240}Pu in Fish Inhabiting Rocky Flats Environs

Sample type	Location	n	Sample	Concentration,[†] pCi/g
Minnow (*Hybosis* sp.)	C_1	5	Whole	1.7 ± 0.2
Minnow (*Hybosis* sp.)	A_1	8	Whole	5.1 ± 1.8
Carp (*Cyprinus carpio*)	Great Western	6	Whole and dissected	<0.02
Bass (*Ictiobus bubalus*)	Pond 7	6	Whole and dissected	<0.02
Minnow (*Hybosis* sp.)	C_1	3*	GI tract	0.6 ± 0.7
			Flesh	<0.02
			Head	0.9 ± 0.9
			Skin	2.3 ± 0.4
			Bone	<0.02
Minnow (*Hybosis* sp.)	A_1	3*	GI tract	0.9 ± 0.9
			Flesh	<0.02
			Head	2.3 ± 2.2
			Skin	4.6 ± 4.2
			Bone	<0.02

*Number of composite fish samples analyzed (5 fish/composite).
[†]Mean ± standard error.

Crayfish

Crayfish, a large invertebrate common to the pond system, showed CR's relative to unfiltered water in the range of 320 to 1290 with a mean value of 830 (Table 5). These values are similar to those found in other studies. Seventy-seven percent of the plutonium in crayfish was associated with the exoskeleton, even though the crayfish were scrubbed extensively. The benthic origin of these organisms probably explains the high plutonium concentrations associated with the exoskeleton.

Fish

Fish flesh and bone from ponds A and C and reservoirs were never above MDA (0.30 d/min per sample) even when several samples were composited (Table 6). Whole fish, however, contained measurable amounts of plutonium in the gut contents, the head, and the outer skin. This suggests that plutonium is being discriminated against at this trophic level.

Flora

No true aquatic vascular plants and relatively few emergent species existed in the pond systems at Rocky Flats. Bulrush (*Juncus balticus*) rooted sporadically within the ponds, and cattail (*Typha latifolia*) frequently grew with its roots submerged. Dock (*Rumex crispus*) was abundant in the riparian area. Concentration ratios for plants associated closely with pond sediments confirmed the observation that the transfer of plutonium from sediments to aerial portions through roots is higher than that previously reported in laboratory experiments (Romney, Mork, and Larson, 1970) (Table 7). Concentration ratios were in the 10^{-2} to 10^{-1} range. This could suggest that the plutonium associated

TABLE 7   Concentrations of ^{239,240}Pu in Vegetation
Samples (*Juncus balticus, Rumex crispus,* and *Typha
latifolia*) Associated with the Rocky Flats Pond System

| | Concentration, pCi/g (dry weight) | | | | |
	Mean	Min.	Max.	CV	n*
Total roots	11.2	0.31	93.2	2.03	51
Total standing vegetation	5.1	0.01	44.3	2.00	52
A_1 roots	1.69	0.31	4.77	0.76	12
A_1 standing vegetation	1.47	0.28	3.68	0.80	12
B_1 roots	45.4	2.46	93.2	0.76	9
B_1 standing vegetation	18.9	1.23	44.3	0.80	9
B_2 roots	2.45	0.67	4.39	0.55	9
B_2 standing vegetation	2.04	0.01	8.27	1.26	9
B_3 roots	1.16	1.91	1.41	0.30	2
B_3 standing vegetation	0.45	0.33	0.56	0.36	2
B_4 roots	3.89	1.49	6.22	0.50	4
B_4 standing vegetation	1.28	0.21	3.94	1.06	4
C_1 roots	2.84	0.55	7.38	0.55	15
C_1 standing vegetation	1.83	0.12	5.66	0.93	16

*Number of samples.

with the ponds is of a more biologically available form. This appears contradictory to
laboratory experiments which do not include a variety of environmental factors that
could contribute to an increased uptake of plutonium, such as surface contamination.

Fauna

A variety of small and large mammals were opportunistically captured during the course
of this study. The data associated with this compartment were too few except to draw
tentative conclusions. However, it would appear that fauna associated with the Rocky
Flats area, in general, maintained a relatively low systemic body burden of plutonium
(Table 8).

Conclusions

The results obtained in this study were of a very preliminary nature because of the more
general systems approach to the study and to the use of an analytical technique that
provided no isotopic discrimination. Owing to the cost of sophisticated sample analyses, a
majority of activity levels near fallout background and/or analytical detectability, and the
overall complexity of a systems approach, only tentative conclusions can usually be
ascertained for transuranic elements in the environment. However, the tentative
conclusions drawn from this study and others are, in general, the same.

Although the various components of the aquatic system at Rocky Flats are
concentrating plutonium to a relatively high degree, there appears to be no direct
evidence that concentrations of plutonium observed will result in a biological hazard to

**TABLE 8 Plutonium Concentrations in Some
Animals Collected at Rocky Flats**

Odocoileus hemionus (Rocky Mountain mule deer)

Sample	Concentration, pCi/g (dry weight)				
	Mean	Min.	Max.	CV	n*
Spleen	0.03	0.02	0.05	0.42	3
Kidney	0.08	0.05	0.10	0.41	2
Lung	0.03	0.01	0.10	0.90	7
Broncheoles	0.07				1
Bronchus	0.08				1
Liver	0.03	0.01	0.09	1.43	3
Heart	0.01				1
Hide	0.06	0.03	0.16	0.74	7
Lymph nodes (broncheolar)	0.33				1
Esophagus	0.08				1
Rumen contents	0.05	0.01	0.15	1.08	5
Blood	0.02	0.01	0.02	0.30	5
Muscle	0.33	<0.01	1.80	1.05	6

	Mean	SD	n
Internal =	0.12	± 0.15	7
External =	0.06	± 0.02	6

Microtus pennsylvanicus modestus (meadow mouse)

Sample	pCi/g (dry weight)
Liver	0.58
Lungs	5.10
GI tract	0.17
Bone	0.06

Peromyscus nasutus (white-footed deer mouse)

Sample	pCi/g (dry weight)
Liver	0.99
Lungs	40.10
Flesh	0.07
Bone	0.55

Rana pipiens (leopard frog)

Sample	pCi/g (dry weight)
Liver	0.01
Lungs	14.40
Flesh	0.17
Bone	0.31

*Number of samples.

man through ingestion routes. This was concluded on the basis that (1) the majority of plutonium in the system was associated with sediments; (2) plutonium in unfiltered water leaving the Rocky Flats plant site averaged <10 pCi/liter, even during pond reconstruction, which was below accepted maximum permissible concentration (1600 pCi/liter, International Commission on Radiological Protection); and (3) plutonium concentrations did not increase along simple trophic-level routes to any significant extent.

References

Aarkrog, A., 1971, Radioecological Investigations of Plutonium in an Arctic Marine Environment, *Health Phys.*, 20: 31-47.

Emery, R. M., and D. C. Klopfer, 1976, The Distribution of Transuranic Elements in a Freshwater Pond Ecosystem, in *Environmental Toxicity of Aquatic Radionuclides: Models and Mechanisms*, pp. 269-285, M. W. Miller and J. N. Stannard (Eds.), Ann Arbor Science Publishers, Ann Arbor, Mich.

Hakonson, T. E., 1972, *Cesium Kinetics in a Montane Lake System*, Ph.D. Dissertation, Colorado State University, Fort Collins, Colo.

——, J. W. Nyham, and W. D. Purtymun, 1976, Accumulation and Transport of Soil Plutonium in Liquid Waste Discharge Areas at Los Alamos, in *Transuranium Nuclides in the Environment*, Symposium Proceedings, San Francisco, Nov. 17–21, 1975, pp. 175-189, STI/PUB/410, International Atomic Energy Agency, Vienna.

Hammond, S. E., 1971, Industrial-Type Operations as a Source of Environmental Plutonium, in *Proceedings of Environmental Plutonium Symposium*, Los Alamos, N. M., Aug. 4–6, 1971, E. B. Fowler, R. W. Henderson, and M. F. Milligan (Coordinators), USAEC Report LA-4756, pp. 25-35, Los Alamos Scientific Laboratory, NTIS.

Johnson, J. E., S. Svalberg, and D. Paine, 1974, *Study of Plutonium in Aquatic Systems of the Rocky Flats Environs*, Final Technical Report, Dept. of Radiology and Radiation Biology and the Dept. of Animal Sciences, Colorado State University.

Keough, R. F., and G. J. Powers, 1970, Determination of Plutonium in Biological Materials by Extraction and Liquid Scintillation Counting, *Anal. Chem.*, 42: 419-421.

Noshkin, V. E., 1972, Ecological Aspects of Plutonium Dissemination in Aquatic Environments, *Health Phys.*, 22: 537-549.

——, V. T. Bowen, K. M. Wond, and J. C. Burke, 1971, Plutonium in North Atlantic Ocean Organisms Ecological Relationships, in *Radionuclides in Ecosystems*, Proceedings of the Third National Symposium on Radioecology, Oak Ridge, Tenn., May 10–12, 1971, D. J. Nelson (Ed.), USAEC Report CONF-710501, pp. 681-688, Oak Ridge National Laboratory, NTIS.

Pillai, K. C., R. C. Smith, and T. R. Folsom, 1964, Plutonium in the Marine Environment, *Nature (London)*, 203: 568.

Romney, E. M., H. M. Mork, and K. H. Larson, 1970, Persistence of Plutonium in Soil, Plants, and Small Animals, *Health Phys.*, 19: 487-491.

Stannard, J. N., 1973, Chemical and Physical Properties of Plutonium, in *Uranium and Plutonium Transplutonic Elements*, pp. 670-686, H. C. Hodge, J. N. Stannard, and J. B. Hursh (Eds.), Springer-Verlag, New York.

Taylor, D. M., 1973, Chemical and Physical Properties of Plutonium, in *Uranium and Plutonium Transplutonic Elements*, pp. 323-347, H. C. Hodge, J. N. Stannard, and J. B. Hursh (Eds.), Springer-Verlag, New York.

Plutonium in the Great Lakes

M. A. WAHLGREN, J. A. ROBBINS, and D. N. EDGINGTON

Since 1971 plutonium concentrations have been measured annually in Lake Michigan and Lake Ontario and at less frequent intervals in the other Great Lakes. The concentrations of plutonium in the water column have decreased only slightly during the 7 yr of measurement. The residence times for plutonium in the lakes have been estimated by simple time–concentration models. The apparent sinking rates for plutonium have been found to be essentially constant in all the Great Lakes, which suggest that the basic processes that control the concentrations of dissolved plutonium are similar despite considerable differences in chemical, biological, and physical characteristics of the lakes. Analyses of plutonium in water, suspended solids, material from sediment traps, and sediment cores show that considerable resuspension of previously sedimented material into the hypolimnion occurs throughout a major part of the year. A mechanism is proposed to account for the seasonal cycling of plutonium in the epilimnion of Lake Michigan. Recent studies show that plutonium in Lake Michigan (and in the Irish Sea) exists primarily in the water column as Pu(VI) and on the sediments as Pu(IV). For a better understanding of the long-term geochemical and biological behaviors of plutonium in aquatic environments, further study of the limnological factors that control the chemical forms of plutonium is required.

Approximately 40% of the population of the United States lives in states bordering the Laurentian Great Lakes (Fig. 1). The economic advantages to the electrical generating industry of using these lakes for once-through cooling have long been recognized in both the United States and Canada. The advent of large multiple-unit nuclear plants has led to the operation of 8 such reactors on Lake Michigan and a total of 16 on the four lower Great Lakes. The rapid growth of the nuclear power industry has generated considerable public concern about possible environmental effects of radioactive discharges, whether routine or accidental, and this concern has been directed primarily toward plutonium.

Very little, if any, plutonium from nuclear power plants has entered the lakes. The source of plutonium in the Great Lakes is almost entirely stratospheric fallout as a result of nuclear weapons testing. Because of the very low concentration and consequent analytical difficulties of plutonium, neither concentrations nor inventories of ^{239,240}Pu or ^{238}Pu were measured during the period of maximum fallout. However, excellent records exist for the deposition of ^{90}Sr from the worldwide fallout monitoring program (Environmental Measurements Laboratory, 1978). Therefore, if a general ^{239,240}Pu/^{90}Sr ratio can be assigned, it is possible to obtain a reasonable estimate of the annual deposition of plutonium. Measurements of plutonium and strontium in the atmosphere since 1965 have been summarized by Krey, Schonberg, and Toonkel (1974), who found

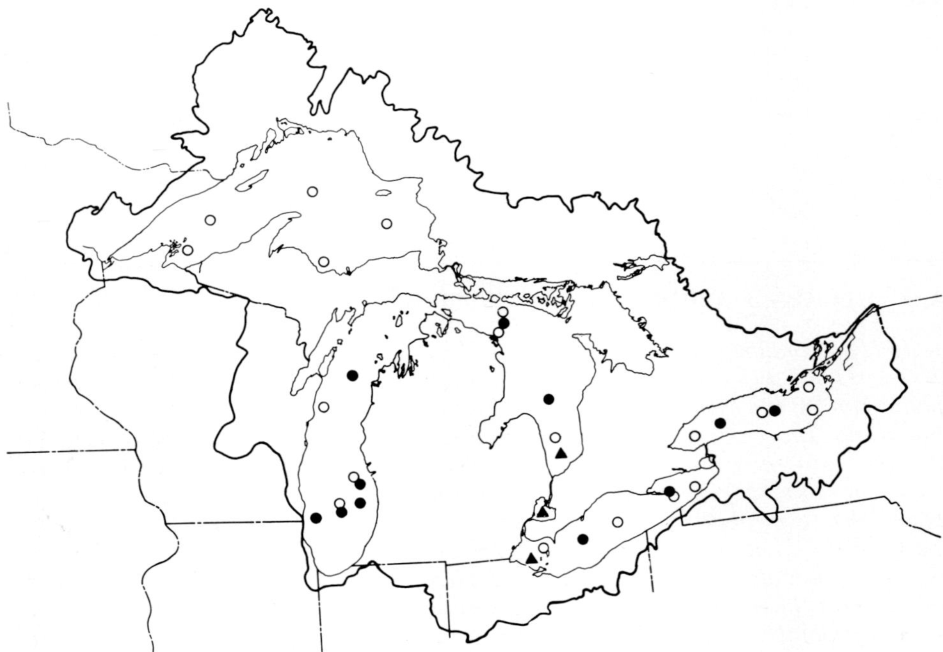

Fig. 1 Laurentian Great Lakes, with water sampling sites for comparative studies. ○, May 10 to June 6, 1973. ▲, July 1 to 3, 1974. ●, June 30 to July 6, 1976.

that a ^{239,240}Pu/^{90}Sr ratio of 0.017 should give a good estimate of plutonium deposition. The annual inputs of plutonium to Lake Michigan, which lies almost wholly within the 40°N latitude band, are shown in Fig. 2. A knowledge of this source term is important for the interpretation of residence times in the water column and concentration profiles in the sediments. Measurements of the cumulative deposition of plutonium in soil at Argonne ($\sim$2.2 mCi/km^2; Golchert, Duffy, and Sedlet, 1978) are in agreement with the integral of all the inputs shown in Fig. 2 and the worldwide fallout data summarized by Hardy, Krey, and Volchok (1973). On this basis the total content of plutonium in the Great Lakes basin is 1660 Ci (or 25 kg), of which 540 Ci was deposited directly onto the surface of the lakes. The plutonium content in each lake is summarized in Table 1. Sprugel and Bartelt (1976) have measured the loss of plutonium from a typical midwestern watershed to be about 0.05% of the total accumulated deposition per year.

The presence in the Great Lakes waters of these low concentrations of long-lived radioactivity from nuclear fallout provides the opportunity to characterize the environmental behavior of these isotopes and to study the biogeochemical and geophysical processes that determine the residence times of radioactive and stable trace materials entering the lakes. For plutonium it is of particular importance to determine (1) the potential pathways to man (food chains and drinking water); (2) a radiological baseline data set for the Great Lakes; and (3) the likely distribution of possible future inputs between various compartments of the lake, including the final sinks, if any.

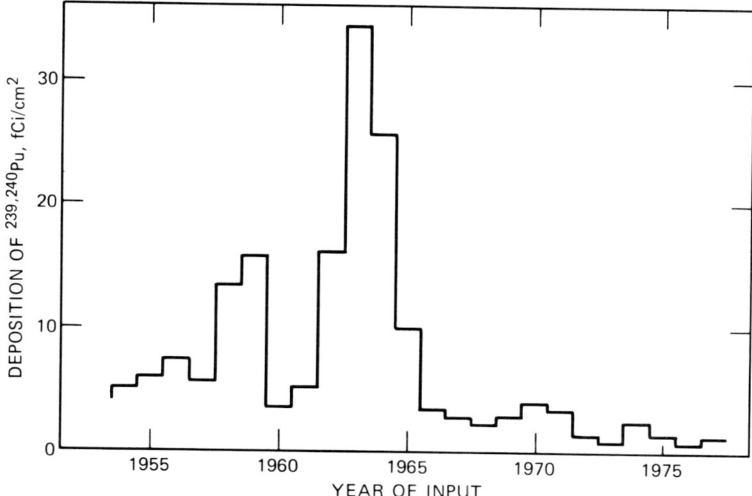

Fig. 2 Estimates of the annual deposition of ^{239,240}Pu per unit area over Lake Michigan on the basis of monitoring ^{90}Sr in rainwater at Argonne, Ill., and Green Bay, Wis., the average rainfall, and the monitoring of ^{239,240}Pu in atmospheric fallout at Argonne.

TABLE 1 Inventory of Plutonium
in the Great Lakes as of 1977

	Plutonium, Ci			
Lake	Watershed	Deposited on lake surface	Water	Sediments
Superior	290	180	5	175
Michigan	270	130	2	128
Huron	290	130	2	128
Erie	130	60	<1	45
Ontario	140	40	<1	50*

*The amount stored in the sediments is greater than that deposited on the lake surface because of the ^{239}Pu exported from Lake Erie down the Niagara River.

The purpose of this chapter is to describe mainly the biogeochemical and physical processes that appear to determine the behavior of plutonium and other transuranic elements in the Great Lakes. The roles of recent atmospheric and watershed erosion inputs, sedimentation and resuspension, and export by outflow as controls of the longer term availability of plutonium are discussed. Rapid sedimentation from the epilimnion by association with autochthonous particulate material is shown to account for the seasonal cycling of plutonium in Lake Michigan. The methods for measuring the transuranic elements in the lakes have been adequately described elsewhere (Golchert and Sedlet, 1972; Nelson et al., 1974; Wahlgren et al., 1976).

Long-Term Behavior of Plutonium in the Great Lakes

The earliest measurements of plutonium in any of the Great Lakes were made in 1971 (Maletskos, 1972). Since that time measurements of plutonium have been made every year in Lake Michigan and Lake Ontario (Bowen and Noshkin, 1972; 1973; Bowen, 1974; 1975; 1976; 1977; Wahlgren and Nelson, 1973; 1974a; 1974b; 1975; 1977a; Wahlgren and Marshall, 1976; Wahlgren, Nelson, and Kucera, 1977, unpublished) and occasionally in the other lakes. The annual data for plutonium for the years 1971 to 1977 are summarized in Table 2. Since it was found in 1972 that the fallout radionuclides ^{239}Pu, ^{137}Cs, and ^{90}Sr in Lake Michigan have a homogeneous distribution throughout the water column following the winter mixing period (Wahlgren and Nelson, 1973), the concentrations in early spring or from the hypolimnion during early summer may be taken to represent the mean for the lake. Therefore the numbers in the table are the best estimate of the total concentration of each nuclide remaining in the water column each year. From these data it is possible, knowing the volume of the lake, to calculate the inventory of plutonium in the water column. As indicated in Table 1, the fraction of the total amount in each of the lakes in the water column at the present time is very small. Since very little plutonium can be lost by outflow from the upper Great Lakes (Michigan, Superior, and Huron), it follows that there must be a very efficient transfer of this element to the sediments. In the lower lakes (Erie and Ontario) losses due to outflow and gains from the upper lakes must be considered as well. The mean ^{239}Pu concentrations for the three upper lakes (about 0.6 fCi/liter) are significantly different from those for Lake Erie and Lake Ontario (about 0.2 fCi/liter). At first this would appear to be due solely to the greater outflow in the last two lakes. However, the situation is more complicated and interesting.

To understand the long-term behavior of plutonium in the Great Lakes, one must consider not only the inputs and losses but also the volume and area of each lake and the efficiency of the scavenging of plutonium from the water column. To the extent that each lake is well mixed, the change in concentration of plutonium in the water column is

$$\frac{dC}{dt} = \frac{1}{V}(A\phi + \alpha W + I - S - O) \qquad (1)$$

where V = volume of the lake

    A = area of the lake

    ϕ = plutonium flux to the lake surface, femtocuries per square centimeter per year

    W = amount stored in the watershed, femtocuries

    α = annual fraction lost to the lake

Since the lake is assumed to be well mixed, the amount lost by outflow, O, is given by QC, where Q is the mean annual outflow from the lake (cubic centimeters per year). In addition, I is the input (femtocuries per year) from the next higher lake, where I = QC, and S is the amount lost to the sediments (femtocuries per year). Introducing subscripts i = 1 to 5, which refer, respectively, to Lakes Superior, Michigan, Huron, Erie, and Ontario, the change in concentration in each lake is given by

$$\frac{dC_i}{dt} = \frac{1}{V_i}(A_i\phi_i + \alpha_1 W_i + Q_{i-1}\,C_{i-1}) - \frac{Q_i}{V_i}C_i - \frac{S_i}{V_i} \qquad (2)$$

TABLE 2 Concentrations* of ^{239,240}Pu in Unfiltered Great Lakes Water

Lake							

Concentrations of ^{239,240}Pu, fCi/liter — Year of sampling

Lake	1971	1972	1973	1974	1975	1976	1977
Superior			0.57 ± 0.11(3)				
Michigan†	0.97 ± 0.16(6)	0.80 ± 0.10(14)	0.77 ± 0.11(5)	0.73 ± 0.10(4)	0.62 ± 0.10(4)	0.63 ± 0.12(9)	0.58 ± 0.06(6)
ANL-5					0.74 ± 0.12(12)	0.63 ± 0.20(10)	0.56 ± 0.15(8)
Huron			0.62 ± 0.02(2)	0.51 ± 0.06		0.54 ± 0.14(4)	
St. Clair				0.39 ± 0.06			
Erie			0.21 ± 0.07(3)	0.20 ± 0.05		0.11 ± 0.04(4)	
Ontario							
ANL			0.26 ± 0.04(3)			0.35 ± 0.04(2)	
Woods Hole Oceano- graphic Institution	0.90 ± 0.12(8)‡	0.57 ± 0.30(9)§	0.36 ± 0.36(21)‡				

*Number in parentheses denotes mean of (n) samples taken at various depths during the spring convective mixing period or from the hypolimnion during early summer.

†All offshore stations.

‡August.

§September.

The ratio V_i/Q_i is the residence time of plutonium in the lake with respect to outflow (T_{R_i}). If the rate of loss by sedimentation is proportional to the concentration in the water column, then the term S_i/V_i can be replaced by C_i/T'_{R_i}, where T'_{R_i} is the residence time of plutonium in the water column with respect to losses via sedimentation. In this case Eq. 2 can be written as

$$\frac{dC_i}{dt} = \frac{A_i\phi_i + \alpha W_i + Q_{i-1}\,C_{i-1}}{V_i} - \left(\frac{1}{T_{R_i}} + \frac{1}{T'_{R_i}}\right)C_i \tag{3}$$

The mean residence time $\overline{T_{R_i}}$ of plutonium in each lake is given by

$$\frac{1}{\overline{T_{R_i}}} = \frac{1}{T_{R_i}} + \frac{1}{T'_{R_i}} \tag{4}$$

For Lake Superior and Lake Michigan, there is no inflow from the other lakes, i.e., $I_1 = I_2 = 0$ (Fig. 1); for Lake Huron, however, $I_3 = Q_2 C_2 + Q_1 C_1$. The above system of equations (Eq. 3) is equivalent to the concentration–time model described by Lerman (1972) to describe the behavior of ^{90}Sr in the Great Lakes. For plutonium, losses by radioactive decay are, of course, negligible. Since the values of $A_i\phi_i$, V_i, and Q_i are known (Table 3) and α can be assumed as a first approximation to be zero, the only undetermined parameter in the model is T'_R.

TABLE 3 Physical and Hydrological Data for the Great Lakes

| | Area, 10^4 km^2 | | | |
Lake	Drainage basin	Lake surface	Volume, 10^3 km^3	Outflow, km^3 yr
Superior	12.8	8.2	12	65
Michigan	11.8	5.8	4.9	49
Huron	13.1	6.0	4.6	157
Erie	5.9	2.6	0.48	175
Ontario	6.0	2.0	1.6	209

The results of evaluating Eq. 3 are shown in Fig. 3. The monthly values of ϕ_i used were taken from Lerman (1972). For deposition from 1973 to 1977, values measured at Argonne National Laboratory were used for each lake (Golchert, Duffy, and Sedlet, 1978). Values of T'_R were chosen to reproduce the earliest available measured concentration in each lake. This approach was taken to evaluate subsequent changes in the value of T'_R. In all lakes but Lake Erie, the mean residence time is around 2 to 3 yr. Under such conditions short-term (monthly) fluctuations in deposition are averaged out in the water column, and the calculated time–dependence of plutonium levels shows a smooth variation over the past 25 yr or so. Since plutonium in waters of the Great Lakes has been measured at the "tail-end" of the concentration–time record, reconstruction of previous levels in the water column is an exercise in extrapolation. However, the choice of residence times, T'_R, summarized in Table 2, gives a very reasonable prediction of recent levels in all the lakes as well. In other words, values of T'_R, which by definition correctly

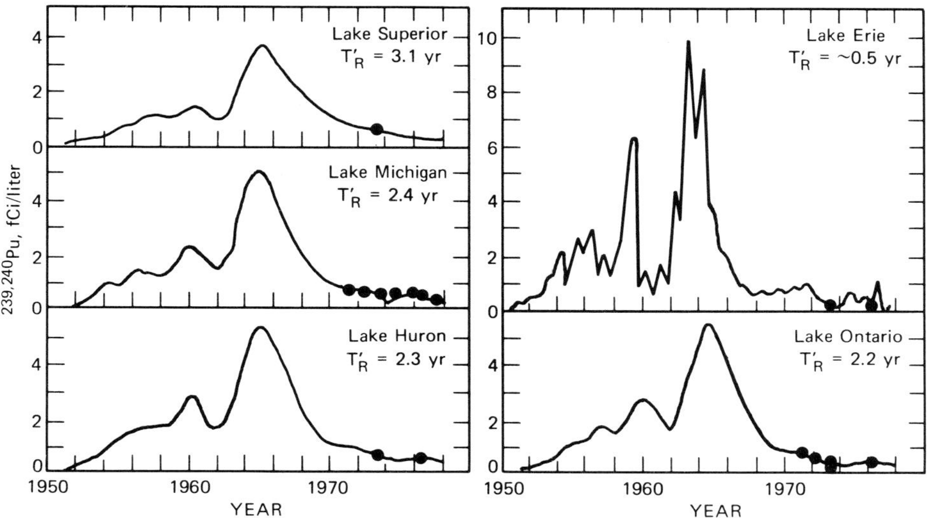

Fig. 3 Comparison of concentrations of ^{239,240}Pu in the water column predicted by the coupled-lakes model with available experimental data points (●). The best estimate of the residence time for deposition in the sediments is given for each lake.

predict the concentration of plutonium in the early 1970s, give an adequate account of levels in each lake (except Lake Superior, for which there is only one data point) 3 to 7 yr later. For Lake Michigan, where plutonium has been measured each year since 1971, the model appears to slightly underestimate concentrations from 1973 onward. Plutonium levels in the water of Lake Erie show comparatively strong fluctuations over the past two decades because of the very short mean residence time resulting from rapid losses to sediments.

The small differences between observed and calculated mean plutonium concentrations in Lake Michigan after 1973 could be due to a combination of small effects because levels were so low in the 1970s. For example, a 20% increase in the T'_R gives a better least-squares fit to all the mean concentration data. Thus the lower value of T'_R resulting from use of earliest (1971) concentration values alone could be an artifact of the approach or reflect uncertainties in the estimate of the mean concentration in 1971. Alternatively, the slightly higher recent value of T'_R could reflect an increasing importance of sediment—water exchange or watershed erosion in the regulation of the very low plutonium concentrations in Lake Michigan. Unfortunately there are insufficient data to discuss the other lakes in these terms.

Little is presently known about the inputs of plutonium from watersheds of the Great Lakes. The few measurements of the concentration of plutonium in the Grand River, one of the major tributaries to Lake Michigan, suggest total concentrations of 0.5 to 1.0 fCi/liter. If this range is representative of average concentrations, tributary rivers would contribute up to ~0.1 Ci/yr at the present time compared with ~0.7 Ci/yr from direct fallout to the surface of Lake Michigan.

The recent results of Sprugel and Bartelt (1976) suggest that watershed contributions may be more important than previously supposed. They found that 0.05% of the total plutonium stored on a typical midwestern watershed is lost by erosion each year. If this erosion rate applies to the large watersheds of the Great Lakes, the annual input to Lake Michigan, for example (see Table 1), should be ~0.14 Ci/yr (assuming that there is 270 Ci of plutonium deposited on the watershed). The possible effect of such a 0.05%/yr erosion rate on plutonium concentrations in Lake Michigan since 1973 is illustrated in Fig. 4. Equation 3 is evaluated on a monthly time interval starting in 1973, with initial conditions determined by observed concentrations in the lake, and T'_R = 2.4 yr. Monthly inputs from the rivers are estimated by the crude assumptions: (1) The total inflow of water to Lake Michigan is proportional to that from the Grand River (U. S. Geological Survey, 1973–1976; U. S. Geological Survey, personal communication, for 1977 data); and (2) the concentration of plutonium in river water is constant. Clearly the addition of plutonium from the watershed at a rate of 0.05%/yr (~3 fCi/liter) better reproduces the data. In years like 1976, when there is little new atmospheric deposition, levels do not decrease significantly because of continued inputs from the tributaries.

Although there may be many ways to account for the minor variations in mean concentration of plutonium in the lakes each year, the overall behavior is adequately

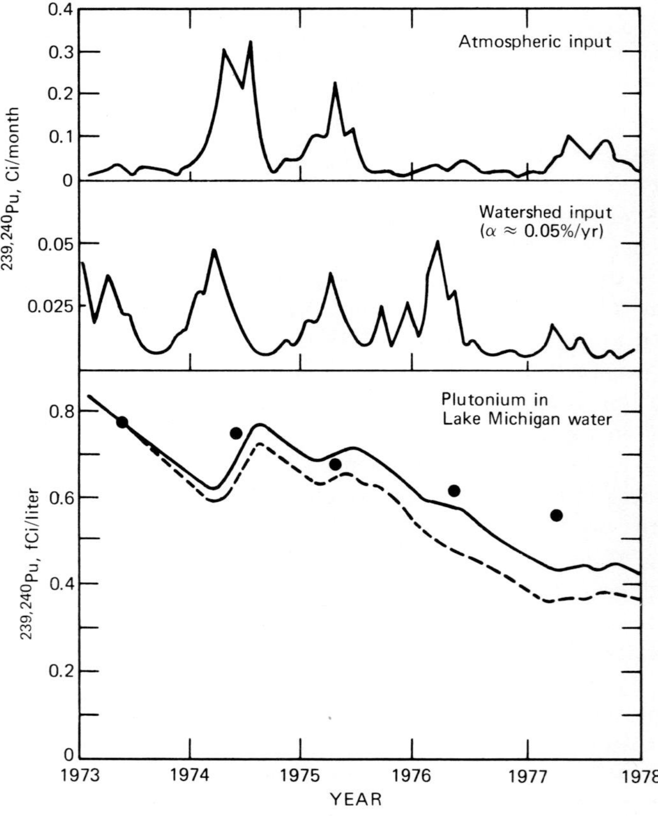

Fig. 4 Evaluation of the possible role of watershed erosion in maintaining the recent levels of plutonium in Lake Michigan waters. ——, predicted for unstratified, well-mixed lake, α = 0.05%. – – –, predicted for unstratified well-mixed lake, α = 0. ●, measured values of mean concentration.

described in terms of a single parameter, T'_R. The preceding discussion suggests that uncertainty in the estimated value for Lake Michigan is around 20% and somewhat higher ($\sim$30%) for the other lakes. Thus the variation in the values of T'_R from lake to lake by a factor of 6 is real.

Although the variation of values of T'_R is large, the apparent settling rate ($\bar{L}/T'_R$, where $\bar{L}$ is the mean lake depth) is essentially independent of the lake. These values are given in Table 4. For all lakes except Lake Superior, the apparent settling rate is 35 ± 2 m/yr; for Lake Superior the value is 48 m/yr. However, in view of the uncertainties in the average plutonium concentration for each lake and therefore in calculating T'_R, this value is not significantly different, and the mean apparent settling rate for the five lakes is 37 ± 3 m/yr[1]. This result is consistent with the observed loss rate, corrected for

TABLE 4 Residence Time of Plutonium in the Great Lakes

| | | Residence time, yr | | | Apparent settling |
Lake	Depth $(\bar{L})$, m	Mean $(\overline{T_R})$	Outflow (T_R)	Sedimentation (T'_R)	rate $(\bar{L}/T'_R)$, m/yr
Superior	149	3.1	190	3.1	48
Michigan	84	2.4	100	2.4	35
Huron	77	2.1	30	2.3	34
Erie	17	0.44	3	0.52	33
Ontario	86	1.8	8	2.3	37

degradation, of DDT (Bierman and Swain, 1978) in Lake Michigan and Lake Superior. It is also similar to the apparent settling rate for total phosphorus, 10 m/yr[1] (S. C. Chapra, Great Lakes Environmental Research Laboratory, personal communication), and for detrital particles, 36 m/yr[1] (D. M. DiToro, Manhattan College, personal communication).

This rate of 37 m/yr[1] is comparable to mean particle settling rates inferred from trap and sedimentation studies. The rate of accumulation of particles in traps placed in southern Lake Michigan shows both a strong seasonal dependence and a marked increase with increasing water depth (Wahlgren and Nelson, 1977b). However, for most of the year, the net accumulation rate in near-surface waters is about 0.02 mg cm^{-2} day^{-1}. Increases in flux with increasing depth must result either from resuspension or from transient effects associated with earlier particle production in surface waters. The low average downward particle flux from surface waters is comparable to average sedimentation in the southern basin of Lake Michigan (7 mg cm^{-2} yr^{-1} = 0.02 mg cm^{-2} day^{-1}) estimated by Edgington and Robbins (1976) on the basis of ^{210}Pb and ^{137}Cs profiles in a large number of sediment cores. Since the mean concentration of particles in the water column is about 1 to 2 mg/liter (Wahlgren and Nelson, 1974b; additional unpublished data for the period 1973 to 1977), the mean particle settling rate is about 40 m/yr (0.02 mg cm^{-2} day^{-1} × 2 mg/liter). That plutonium carrying particles has a net settling rate that is comparable to the mean rate indicates that plutonium is not selectively scavenged by an atypical suite of particles in the water column. This idea is also supported by studies on the distribution of plutonium between particles and water discussed further in the following text.

That the apparent sinking rate is essentially invariant from lake to lake is surprising considering the diversity of limnological characteristics of these lakes. Evidently, the processes determining the long-term removal of plutonium from the water column must be similar for each of the Great Lakes, and the rate of removal from the water column to underlying sediments is determined by the net rate at which the particles scavenging plutonium sink. Furthermore, this result suggests that the availability of suitable particles for scavenging plutonium and kinetics of exchange between dissolved and particulate phases are not presently determining the long-term removal rate.

It must be emphasized that the apparent settling rate could be very different from the actual net downward motion of particles. Vertical turbulence and resuspension of bottom sediments can significantly alter the apparent particle settling rate. The role of resuspension will be discussed further in the following text.

It is clear that, since the apparent settling rate is constant from lake to lake, the processes controlling the removal of plutonium from the water column are similar for each lake, and, therefore, considering the large differences in productivity, it is unlikely that association with autochthonous organic matter is the rate determining step. However, although this generalization is true for the water column as a whole, it will become evident, from the experimental data discussed in the next section, that association with biogenic material can explain, at least in part, the more rapid removal of plutonium from the epilimnion (at least for Lake Michigan) during the period of stratification.

Seasonal Cycling of Plutonium in Lake Michigan

The seasonal cycling of plutonium from surface waters was first observed (Wahlgren and Nelson, 1974b) at an offshore station 40 km west of Grand Haven, Mich. Since 1973 onward the cycle has been followed in detail in Lake Michigan (Wahlgren and Nelson, 1977a; Wahlgren, Nelson, and Kucera, 1977, unpublished data). This seasonal cycling of plutonium occurs at other stations in Lake Michigan and probably is common to all the Great Lakes (Bowen, 1975; Alberts, Wahlgren, and Nelson, 1977) except Lake Superior. More than 75% of the total (dissolved and suspended) plutonium is lost from the epilimnion at station ANL-5 in Lake Michigan during the summer months and returned during the fall and winter mixing period each year. The results, summarized in Fig. 5, contrast the strong seasonal cycle in surface waters with mean levels expected from the concentration—time model discussed in the preceding text. The data shown are primarily from a single sampling station, ANL-5 (12 km southwest of Grand Haven, Mich.), but the same trends have been observed at other stations farther offshore, including EPA-18, which is in the middle of the southern basin. Farther offshore the onset of stratification occurs later in the spring season, and the initial removal of plutonium from surface waters is delayed correspondingly. A limited number of cross-lake transects during 1976 show that the degree of removal from surface water by September was comparable across the whole of the southern half of the lake.

The change in concentration of plutonium in the well-mixed epilimnion of depth is given by

$$\frac{dC}{dt} = \frac{\phi}{L_E} - \frac{C}{T'_{R_E}} \tag{5}$$

provided that there is no upward transfer across the thermocline. (The effect of outflow can be ignored.) The results of evaluating Eq. 5 with $\phi = 0$ as well as with values of ϕ

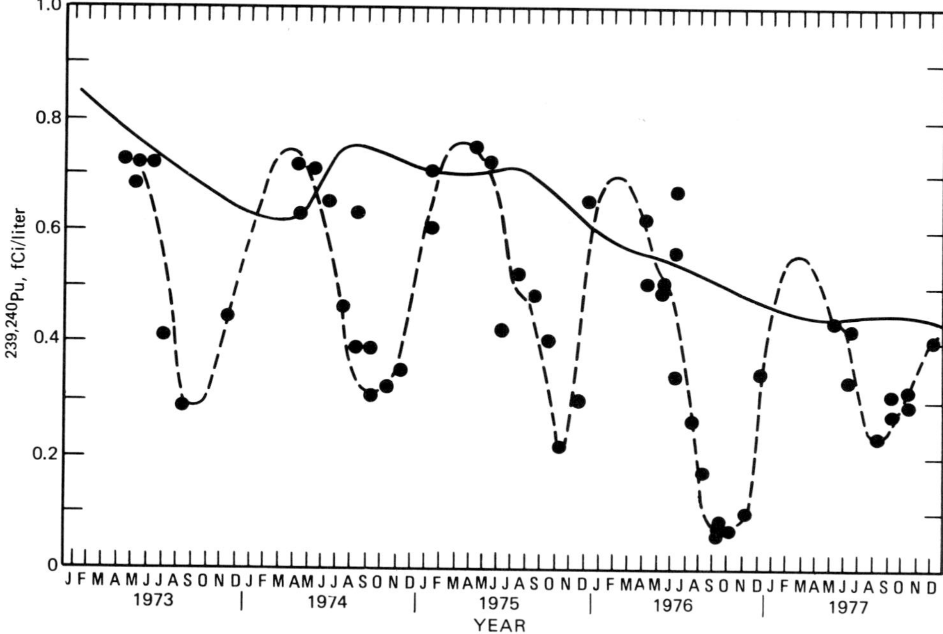

Fig. 5 Comparison of concentrations of ^{239,240}Pu in the water column predicted by the coupled-lakes model with experimental measurements of surface-water concentrations in Lake Michigan. The annual cycling of plutonium is clearly evident in this comparison. ———, surface-water values predicted by the model. – – –, surface-water values measured.

determined from monthly measurements of fallout (Environmental Measurements Laboratory, 1978) are summarized in Table 5. The results of these calculations indicate that, for the years 1973 to 1977, there is a slightly greater variation in the calculated residence time in the epilimnion when new inputs are ignored (i.e., $\phi = 0$). For example, the effect of new fallout on the calculated values was most important in 1977; the residence time for plutonium was reduced from 0.31 to 0.21 yr. When new inputs are included, the residence time is almost constant from year to year. The mean residence time $\overline{T_{R_E}} = 0.22$ yr corresponds to an apparent particle settling rate ($\overline{\omega}_E$) of about 90 m/yr. The effect of an additional input of plutonium from the underlying waters would be to further decrease the estimate of T'_{R_E} and increase the calculated value of $\overline{\omega}_E$.

The losses of plutonium from the epilimnion are more rapid than expected on the basis of the average residence time of this radionuclide in the lake. To the extent that the epilimnion is isolated from underlying waters and the T'_R scales with water depth, the residence time of plutonium in the epilimnion would be $T'_{R_E} = T'_R \times (\overline{L_E}/\overline{L})$, where $\overline{L_E}$ is the mean depth of the epilimnion (~20 m). Thus $T'_{R_E} \sim 2.4 \times (20/84) \simeq 0.6$ yr. Thus the observed value T'_{R_E} is at least 2.5 times lower than that expected from the average residence time of plutonium in the lake. It is therefore clear that the removal of plutonium from the epilimnion is not solely due to its isolation from underlying waters.

This increased efficiency of the removal of plutonium from surface waters during the period just before and during stratification of the lake is probably due to intensified

**TABLE 5   Parameters Describing the Removal of Plutonium
from the Epilimnion**

Year	$\overline{C}$,* fCi/liter	$\overline{\delta C/\delta t}$,† fCi liter^{-1} yr^{-1}	T^{*}_{RE},‡ yr	$\phi/\overline{L}_E$,‡ fCi liter^{-1} yr^{-1}	T'_{RE},§ yr	$\overline{\omega}_E$,‡ m/yr
1973	0.45	2.0	0.27	0.19	0.24	83
1974	0.52	2.3	0.23	0.38	0.22	91
1975	0.39	1.3	0.28	0.16	0.25	80
1976	0.32	1.5	0.21	0.16	0.19	105
1977	0.31	0.90	0.31	0.64	0.21	95
Average			0.27		0.22	90

*$\overline{C}$ is the mean plutonium concentration in the epilimnion, May through September.

†$\overline{\delta C/\delta t}$ is the mean rate of change in plutonium concentration, May through September.

‡T^{*}_{RE} is uncorrected for atmospheric inputs (i.e., $\phi = 0$ in Eq. 5); $\overline{\omega}_E \equiv \overline{L}_E/T'_{RE}$.

§$T'_{RE} = \overline{C}/\overline{\delta C/\delta t}$.

scavenging by particles produced in the epilimnion. Limnological factors affecting the cycling of plutonium in the Great Lakes are summarized in Fig. 6. The major inputs of particles from external sources (allochthonous) occur during the very early spring and late fall months and tend to be rapidly distributed throughout the length of the water column. During the late spring and through the summer and fall, two principal types of particles, diatoms and calcite, are produced in surface waters. The onset of the decrease in plutonium levels occurs just before stratification in June (Fig. 5), which coincides with the end of the major plankton bloom. The reduction is largely completed during the period of in situ calcite formation during August and September (Fig. 5). If it is assumed that the thermocline averages 15 m deep over the whole season, then the total clearance of plutonium is about 1 fCi/cm^2.

The initial decrease in plutonium parallels, in time and extent, that reported for soluble silicon in offshore Lake Michigan waters (Holland and Beeton, 1972). One possible removal mechanism would therefore appear to be the accumulation of plutonium by phytoplankton (primarily diatoms) and the subsequent settling of phytodetritus and zooplankton fecal pellets from the epilimnion. From a knowledge of the concurrent decrease in the silica content of the epilimnion and the uptake of plutonium by net plankton, it is possible to estimate the removal of plutonium from the epilimnion due to the production and settling of diatoms in May and June.

As shown in Fig. 7, after the spring diatom bloom there is a reduction in the concentration of particles in the 8- to 80-μm size range through June and July followed by a dramatic increase in the concentration of particles due primarily to the appearance of a large number in the 3- to 8-μm size range. Since the particulate matter collected in August and September is predominantly calcium carbonate (up to 75%), it is taken that these particles result from the in situ production of calcite particles, which agrees with observations made elsewhere (Brunskill, 1969).

Thus the formation of calcite may also be an important mechanism for efficient clearance of plutonium from the epilimnion. From July to September the concentration of calcium in the epilimnion decreases by about 1 to 2 mg/liter. If it is taken that at this time the epilimnion is 20 m deep, then the formation of calcite could clear from 5 to 10

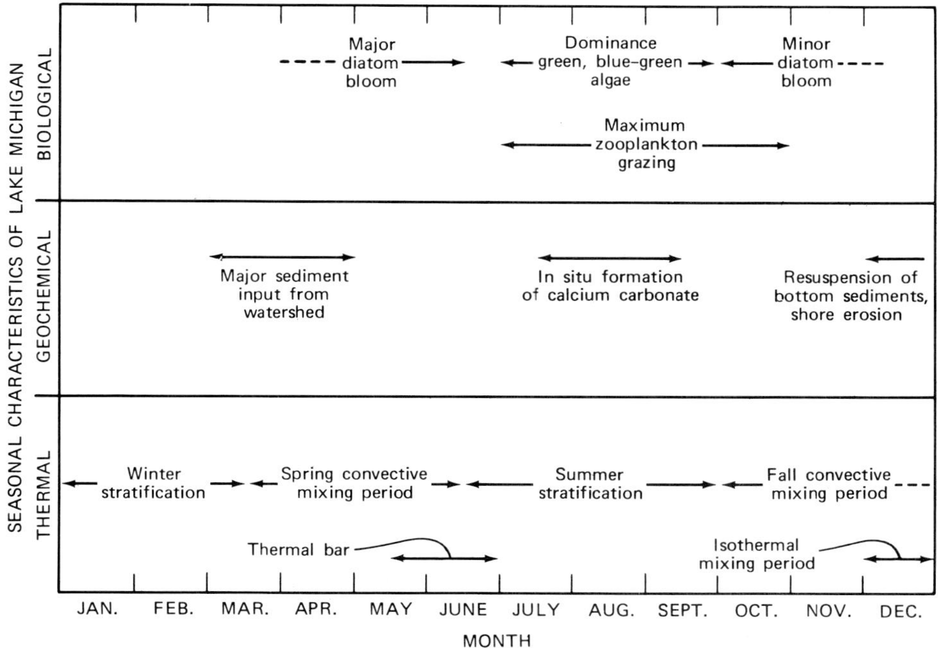

Fig. 6 Seasonal processes in Lake Michigan.

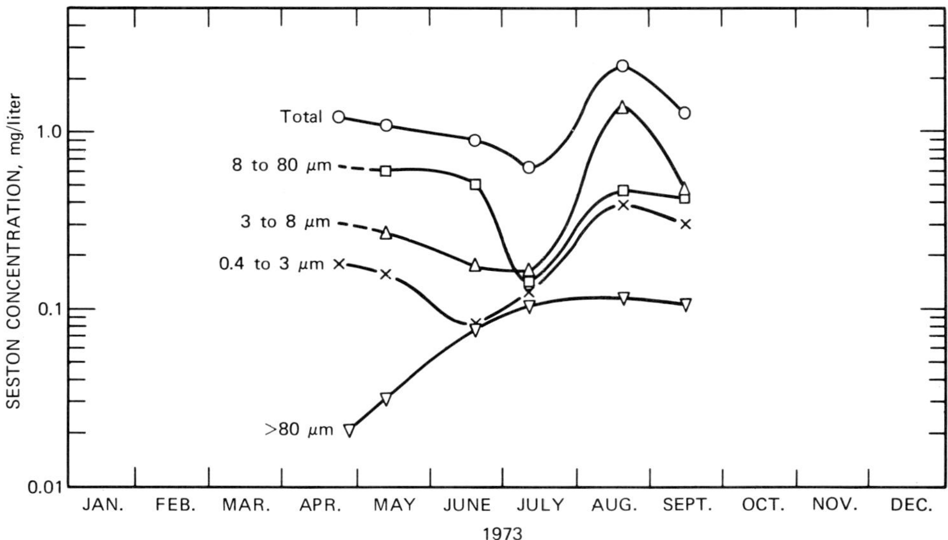

Fig. 7 Seasonal dependence of the concentration and size distribution of particulate matter in the surface waters of Lake Michigan (ANL-73-3, 40 km west of Grand Haven, Mich.).

**TABLE 6   Comparison of the Ability of Authigenic Silica and Calcite
to Remove Plutonium from the Epilimnion**

Analysis	Net plankton, May and June	Suspended particles	
		May and June	August and September
Plutonium, fCi/g ash	55	165	90 to 100
Percent SiO_2 ash	~95	22	~3
Percent $CaCO_3$ ash		~10	~75
Average plutonium, fCi/g SiO_2	50	750	
Average plutonium, fCi/g $CaCO_3$			95
Clearance from epilimnion,* pCi/cm²			
In relation to loss of SiO_2 (May and June)	0.06	0.2	
In relation to loss of $CaCO_3$ (August and September)			0.6 to 1.2

*See text.

mg $CaCO_3/cm^2$. From the data in Table 6 and assuming that all the plutonium is carried on calcite, the clearance from the epilimnion could range from 0.6 to 1.2 fCi/cm². Thus the combined scavenging ability of diatoms and calcite could account for the removal of all the plutonium from the epilimnion. It must be made clear that in these calculations it is assumed that plutonium is carried exclusively by biogenic silica or calcite. No direct measurements have been made as yet.

The results of analyses for plutonium and silica in samples of phytoplankton from net tows are compared with those for suspended particles filtered from the epilimnion and are summarized in Table 6. Detailed measurements of the concentration of SiO_2 in the water column have shown that during the phytoplankton bloom its concentration in the epilimnion decreases by about 1 mg/liter. If it is taken that the early epilimnion is about 10 m deep, then the total clearance of SiO_2 is equivalent to about 1 mg/cm², and, therefore, since the concentration of plutonium expressed in terms of SiO_2 content is 58 fCi/g SiO_2, about 0.06 fCi/cm² of plutonium will be carried with the sinking diatoms. That is less than 10% of the total removed from the epilimnion during the whole summer. The concentration of plutonium in the ashed material collected on filters is far higher than that in ashed phytoplankton. It is not clear whether this additional plutonium is associated in any way with planktonic detrital silica. However, if plutonium is, in fact, removed in association with this silica, then about 0.2 fCi/cm², or about 20%, would be lost from the epilimnion.

The enhanced removal of plutonium from the epilimnion is probably accomplished by scavenging particles with a short lifetime. An extremely small fraction (<5%) of the diatoms produced annually is incorporated into permanent sediments (Parker and Edgington, 1976), and most of the silica tied up in their frustules is redissolved in the water column (Parker, Conway, and Yaguchi, 1977), whereas the remainder is resolubilized within a few weeks after reaching the benthic zone. The concentration of plutonium in surface water starts to increase again with the breakdown of the thermocline but does not attain its maximum value until sometime after early December. In contrast, reactive silica, which also undergoes an annual cycle in the water column, returns to nearly its spring value as early as the end of September. Thus the release of plutonium from dissolving frustules is delayed in relation to the silicon (and calcite) cycles. The production of calcite later in the year may account for this delay, but the lifetime of calcite particles in the water column is presently unknown for the Great Lakes. Alternatively, this delay could indicate that, if plutonium is released from dissolving frustules or calcite particles, it is transferred to other particles in hypolimnion or in the vicinity of the sediment—water interface. Alberts, Wahlgren, and Nelson (1977) have shown that almost all the plutonium in floc collected at the sediment—water interface is associated with reducible hydrated oxides, such as ferric and manganese oxides, and not with carbonates or silica.

Since epilimnetic plutonium losses are mainly due to scavenging by particles that dissolve rapidly, this transient process may not result in appreciable net transfer to sediments during most of a year. In 1975 the mean concentration of total (dissolved and particulate) plutonium at station ANL-5 (from the surface to 60 m) shows a seasonal variation that is consistent with its average residence time in the lake (Fig. 8; the dashed line in the figure corresponds to $T'_R = 2.4$ yr; $\alpha = 0.05\%/\text{yr}$). The net loss to the sediments during each month of this year is predicted by the coupled-lakes model. Since

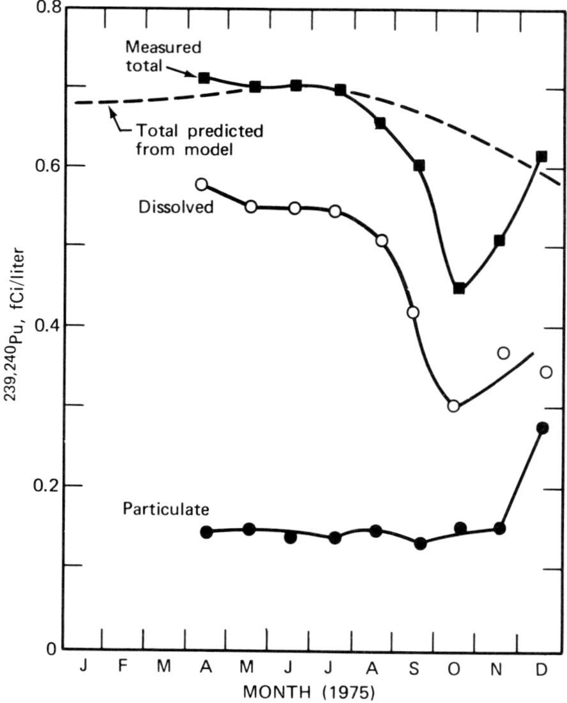

Fig. 8 Mean concentrations of plutonium in whole water, in solution, and on particles in the upper 60 m of the water column of Lake Michigan (ANL-5; 1975; water depth, 67 m). The concentrations expected for a mean residence time (T'_R) of 2.4 yr and watershed erosion (α) of 0.05%/yr are indicated by the dashed curve.

no enhanced transfer to the sediments is expected during the period of diatom and calcite production, there appears to be a dramatic exchange of plutonium between dissolved and particulate phases and temporary loss to the benthic layer (between 60 m and the sediment—water interface). This complementary behavior can be understood in terms of the constancy of the distribution coefficient K_D (Wahlgren and Nelson, 1977c). Since K_D is essentially independent of particle type ($\sim 2 \times 10^5$), the proportion of plutonium tied up with particles should reflect the amount of particulate matter in the water column. From a comparison of the total concentration of particles in surface water each month (Fig. 7) with the amount of plutonium in the water column above the benthic layer (Fig. 8), it can be seen that the proportion of plutonium above the benthic layer reflects the time—dependence of particle concentrations in surface waters.

The extent to which plutonium is removed from the dissolved phase before fall overturn is illustrated in Fig. 9. The seasonal variation in the profiles of dissolved plutonium from the years 1975 to 1977 shows that the removal from the soluble phase extends progressively deeper into the water column as the season advances.

With the onset of fall overturn in October, appreciable losses of plutonium from solution extend essentially to the bottom at this station (ANL-5). The effect was most pronounced in 1976. On the basis of a series of samples collected from sediment traps moored at all depths across the southern basin, it appears that this behavior is characteristic of the deeper waters of the lake as well.* During November 1976 there was a significant reduction in the total amount of plutonium in the water column at this station. Provided that horizontal transport is ignored, the subsequent regeneration of plutonium in the water column would require transfers of plutonium from sources below the deepest sampling point, which is 7 m off the bottom.

The composition of particles reaching this depth illustrates the complementary behavior of silica and calcite in regulating the cycling of plutonium. It can be seen (Fig. 10) that the proportion of silica in the particulate material reflects diatom productivity, with peak values in May and a secondary maximum in November. In contrast, the percentage of calcium carbonate (calcite) remains low through the spring and early summer but rises dramatically in September. There is approximately a 1-month delay between the onset of the plankton blooms or formation of calcite and the appearance of increased concentrations of SiO_2 or $CaCO_3$ in particulate material 7 m above the bottom. Since it is possible to distinguish seasonal variations at this depth, it strongly suggests that there is either little resuspension of bottom sediment at 7 m above the bottom or that the only material that is resuspended is newly deposited detritus. The sharp drop in proportion of SiO_2 and $CaCO_3$ in November is probably associated with the input of terrigenic material washed into the lake as a result of shoreline erosion from severe early winter storms.† Beyond November both the silica and calcite contents of the seston decline dramatically. The total (dissolved and particulate) plutonium per unit weight of particulate material exhibits a monthly variation that reflects the combined effects of silica and calcite scavenging. The total plutonium concentration is high in the spring, goes through a minimum in June, and then steadily rises to a maximum with the addition of calcite to near-bottom seston. Considerable plutonium, like silica and calcite, is lost from the seston and regenerated in the water column during the period from

---

*Earth Resources Technology Satellite satellite photographs show that the "whiting" of the surface waters of Lake Michigan is a lake-wide process.

†It has been shown that up to 50% of the total annual erosion can occur in November (E. Siebel, University of Michigan, personal communication).

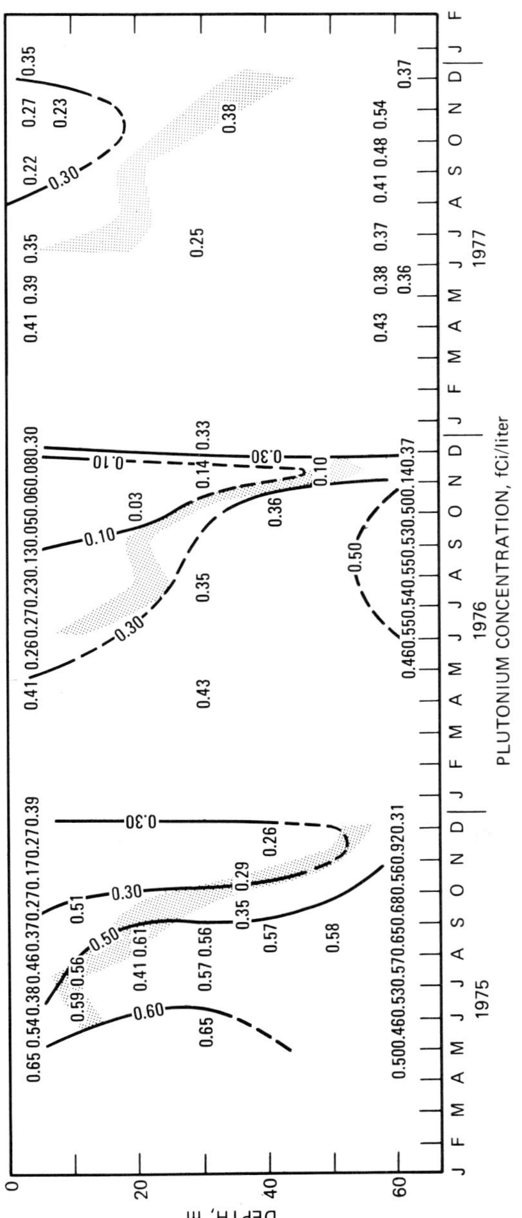

Fig. 9   Seasonal variations in the concentration of dissolved plutonium in the water column of Lake Michigan (ANL-5, 1975 to 1977). The shaded regions indicate the development and breakdown of thermoclines.

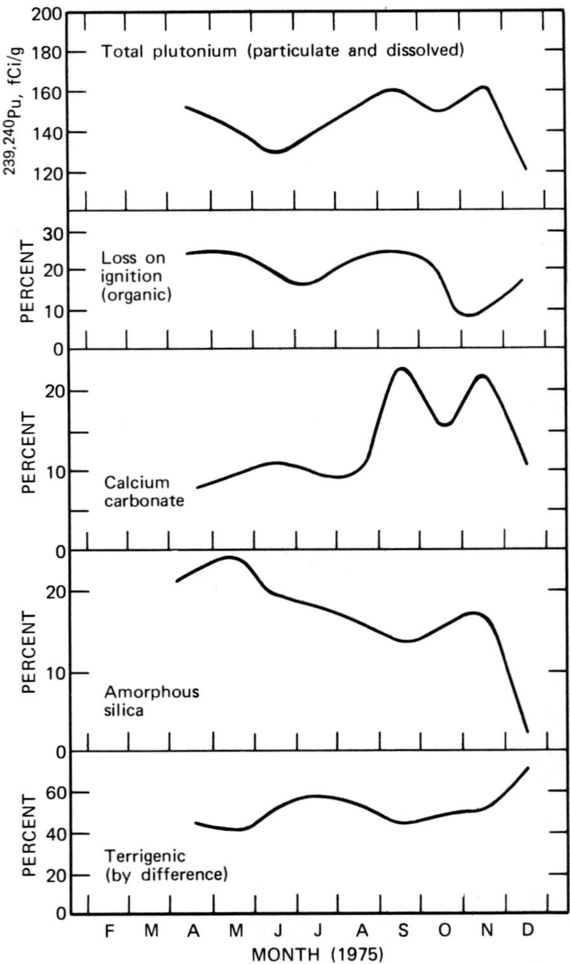

Fig. 10 Total (particulate and dissolved) plutonium and principal constituents of particulate material from filtration (ANL-5, 7 m above bottom).

November to December. The excellent correlation between total plutonium and the combination of SiO_2 and $CaCO_3$ is shown in Fig. 11. To the extent that a sample of particles and water constitute a closed system, it is the total plutonium that should correlate with this combination of variables if plutonium is primarily derived from the dissolution of either silica or calcite particles.

Regeneration of plutonium in the water column would thus appear to result from the dissolution of calcite and silica, especially during November and December. The role that resuspension plays in this event is uncertain. It has been suggested (Wahlgren et al., 1976) that resuspension of bottom sediments and redistribution of plutonium between particles and water must occur to account for the reappearance of plutonium in surface waters and for the apparent plateau leveling-off of average plutonium concentrations in the lake from

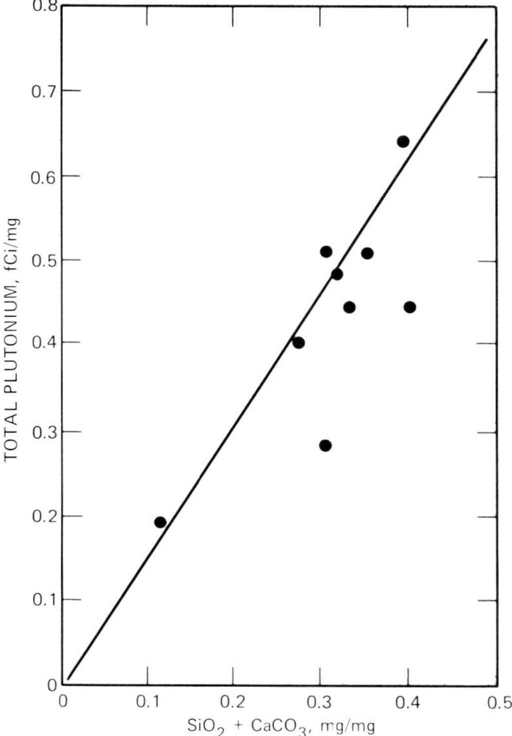

Fig. 11 Relation between total (in solution and on particles) plutonium and the content of SiO_2 and $CaCO_3$ in particles from filtration (ANL-5, 7 m above bottom).

1973 onward. The results of the concentration–time models presented above show that it is unnecessary to assume that present levels reflect equilibrium between sediments and water. The preceding arguments indicate that plutonium is released from dissolving particles that may be localized in the benthic zone, depending on lake dynamics. Redistribution of soluble plutonium in the water column can readily occur without particle resuspension ($\omega_s = 0$).

The study of Chase and Tisue (1977) indicates that the material most available to be resuspended (benthic floc residing above consolidated sediments) is probably hydro-dynamically unsuited to appreciable upward movement in the water column. The benthic floc consists primarily of organic-mineral aggregates, typically a few tens of microns in diameter having bulk densities of about 1.05 g/cm^3. The aggregates consist of diatom frustules, calcite and other minerals, and other unidentified detritus. Particles with such properties ($\bar{r} \sim 20$ μm) remain aggregates on resuspension and have Stokes' settling velocities ω_s of roughly 10^{-3} cm/sec, or about 500 m/yr (Lerman, Lal, and Dacey, 1974). Under normal conditions (excluding fall overturn) the vertical eddy diffusivity K_V in the hypolimnion is on the order of 1 cm^2/sec, or $\sim 3 \times 10^3$ m^2/yr (Kullenberg, Murthy, and Westerberg, 1973). Thus the scale length for the resuspended material flux in the water column under steady-state conditions is $K_V/\omega_s \sim 5$ m. The flux of resuspended material is given by $J = J_0 \, e^{-\omega_s h/K_V}$, where h is the height above the bottom and J_0

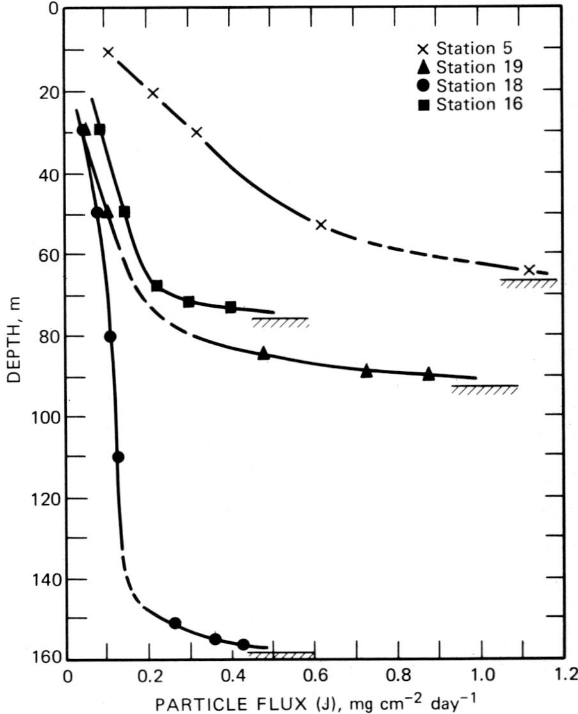

Fig. 12 Particle-settling-rate profiles from sediment-trap studies at several stations in the southern basin of Lake Michigan (1976).

(milligrams per square centimeter per year) is the downward particle flux at the sediment—water interface.

This scale length is confirmed by sediment-trap studies conducted at a series of locations in Lake Michigan. Each of the particle-settling-rate profiles (J in milligrams per square centimeter per day) shown in Fig. 12 has a "foot" that extends from the bottom upward to about 10 to 15 m. The rate of decrease in J over the interval is very nearly exponential and has a scale of about 5 m, which is essentially the same at each location except at station 5. Therefore it is likely that the effects of resuspension are confined to the bottom 10 to 15 m in deep waters (over 80 m). It is possible that limited numbers of smaller particles of freshly formed aggregates of lower densities in the benthic floc may be resuspended to greater average heights above the bottom.

However, the monthly series of settling-rate profiles from station ANL-5 (Fig. 13) suggest that there is a nonconstant settling rate above 20 m that is probably due to seasonal variations in particle production in the epilimnion. In addition, these profiles suggest that resuspension may account for most of the settling material at relatively shallow stations similar to this.* During the period in August when the nonash

---

*This presents a problem in that, although the water depth of ANL-5 is representative of the mean depth of the lake (~84 m), it is close enough to shore (~12 km) to make it highly susceptible to contributions from shoreline erosion and tributary inputs.

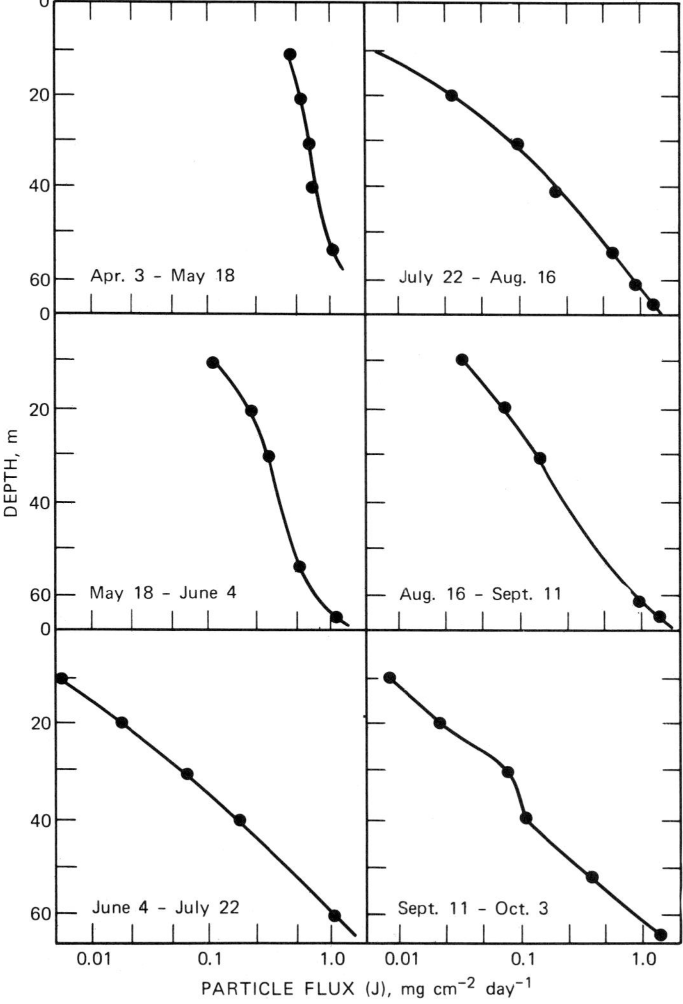

Fig. 13 Seasonal variation in particle-settling-rate profiles from sediment-trap studies in the southern basin of Lake Michigan (ANL-5, 1976).

component (plankton remains) is low, the particle flux profile is very nearly exponential from depths of 65 up to 10 m. This may indicate vertical mixing by eddy diffusion, at least below the thermocline. The apparently high sedimentation rate ($\sim$365 mg cm^{-2} yr^{-1}) in traps near the bottom must reflect both the effect of resuspension and deposition of transient particles, such as diatom frustules and calcite. The mass sedimentation rate for consolidated sediments in this region is only 30 mg cm^{-2} yr^{-1} (Edgington and Robbins, 1976). In view of these results, resuspension during isothermal mixing could be of some importance for the return of plutonium to the whole water column at this inshore station. The effect of resuspension in shallower waters could also influence plutonium levels in offshore areas, depending on the efficiency of horizontal

**TABLE 7   Seston Concentrations, Particle Fluxes,
and Apparent Settling Rates in the Upper 10 m of Water
(ANL-5, 1976)**

Month	Seston concentration (C),* mg/liter	Particle flux (J), $mg\ cm^{-2}\ day^{-1}$	Apparent settling rate (V = J/C), m/yr
April	2.1	0.35	610
May	2.1	0.12	220
June	1.5	0.005	12
July	2.0	0.006	11
August	2.7	0.03	45
September	1.7	0.01	21

*Average of samples taken at beginning and end of sediment-trap collection interval.

transport of resuspended particles. The fact that regeneration of plutonium proceeds concurrently at offshore stations during the winter indicates that either inshore resuspension is intense and horizontal transport rapid or that dissolution rather than resuspension is more important in recycling plutonium lost to particle phases during the earlier sequence of diatom and calcite production.

The particle flux profiles (Fig. 13) provide a crude measure of the seasonal variation in apparent particle settling rates. Although the lower portions of the profile must result, at least partially, from resuspension, the particle flux measured above the thermocline is likely to be unaffected by resuspension. Under conditions of a stable thermocline (July and August), the upward particle flux into the epilimnion will be negligible. The ratio of this flux to the estimated concentration of particles in surface waters (Fig. 7) is the apparent particle settling rate given in Table 7. Apart from extremely high rates during the months of the spring convection mixing period and diatom bloom, the rates during the succeeding months are comparable (22 m/yr; June to September average) to the rate inferred from the concentration—time model. Thus plutonium is apparently scavenged by particles that, on the average, are settling at the same apparent rate as most particles in the water column. Further, it seems likely that resuspension does not have an appreciable effect on this apparent rate of settling.

Conclusions

The results of extensive measurements of plutonium in the Great Lakes have shown that for each lake the concentration of plutonium in the water column at the end of the spring convective mixing period can be described by a simple time—concentration model with only one variable for each lake, viz., the residence time of plutonium. This retention time for the loss of plutonium in each lake is controlled by two processes, the outflow of water to the next lower lake (T_R) and transfer to the sediment on settling particles (T'_R). With the use of a plutonium source term based on the fallout ^{90}Sr monitoring values from the Environmental Measurements Laboratory, values of T'_R that gave best fit to the experimental data were obtained from the model. The values of T'_R were found to be proportional to the mean depth of the lakes, which implies that the apparent settling rate of the particles carrying plutonium is the same in each lake. Considering the large differences in their limnological properties, such as primary productivity, this result is rather surprising in that it might have been expected that a large proportion of plutonium would be carried with organic detritus.

The rate of clearance of plutonium from the epilimnion of Lake Michigan between June and September is far faster than that from the whole water column of the lake and is constant from year to year. Enhanced removal of plutonium from the epilimnion results from intensified particle production during the spring and summer months. Most likely, plutonium is scavenged by diatoms and calcite particles, which subsequently redissolve.

Although it is possible to develop an understanding of what appears to be an extremely complex system, such as the behavior of plutonium in the Great Lakes, in terms of a simple model, there are still many unanswered questions. For example, the uptake of plutonium on biogenic silica or autochthonous calcium carbonate in the epilimnion must be a transient process because it is clear that almost all this material redissolved in the hypolimnion. Some redissolved plutonium may be taken up by other particulate material. Presently it is impossible to balance the downward flux of plutonium on SiO_2 and $CaCO_3$, measured near the bottom in sediment traps, with that apparently deposited in the surface sediments each year because of the subsequent horizontal redistribution of older sediment containing plutonium in the lake.

Since the lakes have very similar general chemical properties, it is possible that the exchange of plutonium between the water column and sediments is controlled by chemical reactions. In fact, it has been suggested that the concentration of plutonium in the water column is largely controlled by chemical equilibrium between specific species in the water column and the sediments. If this equilibrium is a major factor in controlling the concentration of plutonium in the lake, the value of T'_R should have increased significantly during the period of major deposition in the sediments (1963–1970). The data at present demonstrate that little change has occurred in the past 7 yr since the present concentration in the water column can be described by the value of T'_R calculated for 1971.

The situation is complicated by the very recent observations that (1) plutonium in Lake Michigan (and the Irish Sea) exists primarily in the water column in the VI oxidation state and on the sediments as the IV oxidation state and (2) Lake Michigan water can readily extract sorbed plutonium from high-activity pond sediments. Since the distribution coefficients of Pu(VI) and Pu(IV) with sediment are very different, a critical step in the clearance of plutonium from the water column may be the reduction of Pu(VI) to Pu(IV) either in the water or at the sediment surfaces. It is clear, however, that, if there is to be a complete understanding of the long-term behavior of plutonium, especially from other source terms in aquatic environments, more attention must be paid to determining its chemical forms.

Acknowledgments

This work was performed under the auspices of the U. S. Department of Energy.

We gratefully acknowledge the contributions of D. M. Nelson, K. A. Orlandini, and other members of the Ecological Sciences Section to various aspects of the field sampling program that made this chapter possible.

References

Alberts, J. J., M. A. Wahlgren, and D. M. Nelson, 1977, Distribution of Fallout Plutonium in the Waters of the Lower Great Lakes, in *Radiological and Environmental Research Division Annual Report, January–December 1976,* ERDA Report ANL-76-88(Pt.3), pp. 50-52, Argonne National Laboratory, NTIS.

Bierman, V. T., Jr., and W. R. Swain, 1978, DDT Losses in Low-Lipid Fishes of the Great Lakes, in *Abstracts of the 21st Conference on Great Lakes Research,* Windsor, Ontario, May 9–11, 1978, International Association for Great Lakes Research.

Bowen, V. T., 1974, *Plutonium and Americium Concentrations Along Fresh Water Food Chains of the Great Lakes, U.S.A.,* General Summary of Progress, USAEC Report COO-3568-4, Woods Hole Oceanographic Institution, NTIS.

——, 1975, *Plutonium and Americium Concentration Along Fresh Water Food Chains of the Great Lakes, U.S.A.,* General Summary of Progress, USAEC Report COO-3568-8, Woods Hole Oceanographic Institution, NTIS.

——, 1976, *Plutonium and Americium Concentration Along Fresh Water Food Chains of the Great Lakes, U.S.A.,* General Summary of Progress, USAEC Report COO-3568-13, Woods Hole Oceanographic Institution, NTIS.

——, 1977, *Plutonium and Americium Concentration Along Fresh Water Food Chains of the Great Lakes, U.S.A.,* General Summary of Progress, ERDA Report COO-3568-18, Woods Hole Oceanographic Institution, NTIS.

——, and V. E. Noshkin, 1972, *Plutonium Concentrations Along Fresh Water Food Chains of the Great Lakes, U. S. A.,* General Summary of Progress, USAEC Report COO-3568-1, Woods Hole Oceanographic Institution, NTIS.

——, and V. E. Noshkin, 1973, *Plutonium Concentrations Along Fresh Water Food Chains of the Great Lakes, U.S.A.,* General Summary of Progress, USAEC Report COO-3568-3, Woods Hole Oceanographic Institution, NTIS.

Brunskill, G. J., 1969, Fayetteville Green Lake, New York. II. Precipitation and Sedimentation of Calcite in a Meromictic Lake with Laminated Sediments, *Limnol. Oceanogr.,* 14: 830-847.

Chase, R. R. P., and G. T. Tisue, 1977, Settling Behavior of Lacustrine Organic Mineral Aggregates, in *Radiological and Environmental Research Division Annual Report, January–December 1976,* ERDA Report ANL-76-88(Pt.3), pp. 99-102, Argonne National Laboratory, NTIS.

Conway, H. L., J. I. Parker, E. M. Yaguchi, and D. L. Mellinger, 1977, Biological Utilization and Regeneration of Silicon in Lake Michigan, *J. Fish. Res. Board Can.,* 34: 537-544.

Edgington, D. N., and J. A. Robbins, 1976, Records of Lead Deposition in Lake Michigan Sediments Since 1800, *Environ. Sci. Technol.,* 10: 266-274.

Environmental Measurements Laboratory, 1978, *Environmental Quarterly,* Appendix, USDOE Report EML-342 Appendix, NTIS.

Golchert, N. W., and J. Sedlet, 1972, Radiochemical Determination of Plutonium in Environmental Water Samples, *Radiochem. Radioanal. Lett.,* 12: 215-221.

——, T. L. Duffy, and J. Sedlet, 1978, *Environmental Monitoring at Argonne National Laboratory, Annual Report for 1977,* USDOE Report ANL-78-26, Argonne National Laboratory, NTIS.

Hardy, E. P., P. W. Krey, and H. L. Volchok, 1973, Global Inventory and Distribution of Fallout Plutonium, *Nature,* 241: 444-445.

Holland, R. E., and A. M. Beeton, 1972, Significance to Eutrophication of Spatial Differences in Nutrients and Diatoms in Lake Michigan, *Limnol. Oceanogr.,* 17: 88-96.

Krey, P. W., M. Schonberg, and L. Toonkel, 1974, *Updating Stratospheric Inventories to January 1973, Fallout Program Quarterly Summary Report,* USAEC Report HASL-281, Health and Safety Laboratory, NTIS.

Kullenberg, G., C. R. Murthy, and H. Westerberg, 1973, An Experimental Study of Diffusion Characteristics in the Thermocline and Hypolimnion Regions of Lake Ontario, in *Proceedings of the Sixteenth Conference on Great Lakes Research,* Huron, Ohio, Apr. 16–18, 1973, pp. 774-790, International Association for Great Lakes Research.

Lerman, A., 1972, Strontium-90 in the Great Lakes: Concentration–Time Model, *J. Geophys. Res.,* 77: 3256-3264.

——, D. Lal, and M. F. Dacey, 1974, Stokes' Settling and Chemical Reactivity of Suspended Particles in Natural Waters, in *Suspended Solids in Water,* R. J. Gibbs (Ed.), Plenum Press, New York.

Maletskos, C. J., 1972, *Intercalibration and Intercomparison Study of Radionuclide Concentrations in Lake Michigan Samples,* Progress Report, Oct. 1, 1971–Mar. 31, 1972, USAEC Report COO-3371-1, New England Deaconess Hospital, NTIS.

Nelson, D. M., E. M. Yaguchi, B. J. Waller, and M. A. Wahlgren, 1974, Radiochemical Methods, in *Radiological and Environmental Research Division Annual Report, January–December 1973,* USAEC Report ANL-8060(Pt.3), pp. 6–17, Argonne National Laboratory, NTIS.

Parker, J. I., H. L. Conway, and E. M. Yaguchi, 1977, Seasonal Periodicity of Diatoms, and Silicon Limitation in Offshore Lake Michigan, *J. Fish. Res. Board Can.,* 34: 552-558.

——, and D. N. Edgington, 1976, Concentration of Diatom Frustules in Lake Michigan Sediment Cores, *Limnol. Oceanogr.,* 21: 887-893.

Sprugel, D. G., and G. E. Bartelt, 1976, Preliminary Mass Balance of Plutonium in a Watershed Near Sidney, Ohio, in *Radiological and Environmental Research Division Annual Report, January–December 1975,* USAEC Report ANL-75-60(Pt.3), pp. 20-22, Argonne National Laboratory, NTIS.

U. S. Geological Survey, 1973–1976, *Surface Water Records for Michigan.*

Wahlgren, M. A., J. J. Alberts, D. M. Nelson, and K. A. Orlandini, 1976, Study of the Behavior of Transuranics and Possible Chemical Homologues in Lake Michigan Water and Biota, in *Transuranium Nuclides in the Environment,* Symposium Proceedings, San Francisco, 1975, pp. 9-24, STI/PUB/410, International Atomic Energy Agency, Vienna.

——, and J. S. Marshall, 1976, The Behavior of Plutonium and Other Long-Lived Radionuclides in Lake Michigan. I. Biological Transport, Seasonal Cycling, and Residence Times in the Water Column, in *Impacts of Nuclear Releases into the Aquatic Environment,* Symposium Proceedings, Otaniemi, Finland, 1975, pp. 227-243, STI/PUB/406, International Atomic Energy Agency, Vienna.

——, and D. M. Nelson, 1973, Plutonium in Lake Michigan Water, in *Radiological and Environmental Research Division Annual Report, January–December 1972,* USAEC Report ANL-7960(Pt.3), pp. 7-14, Argonne National Laboratory, NTIS.

——, and D. M. Nelson, 1974a, Residence Times for ^{239}Pu and ^{137}Cs in Lake Michigan Water, in *Radiological and Environmental Research Division Annual Report, January–December 1973,* USAEC Report ANL-8060(Pt.3), pp. 85-92, Argonne National Laboratory, NTIS.

——, and D. M. Nelson, 1974b, Studies of Plutonium Cycling and Sedimentation in Lake Michigan, in *Proceedings of the Seventeenth Conference on Great Lakes Research,* Ontario, Canada, Aug. 12–14, 1974, pp. 212-218, International Association for Great Lakes Research.

——, and D. M. Nelson, 1975, Plutonium in the Laurentian Great Lakes: Comparison of Surface Waters, *Verh. Int. Ver. Theor. Angew. Limnol.,* 19: 317-322.

——, and D. M. Nelson, 1977a, Seasonal Cycling of Plutonium in Lake Michigan, in *Radiological and Environmental Research Division Annual Report, January–December 1976,* ERDA Report ANL-76-88(Pt.3), pp. 53-55, Argonne National Laboratory, NTIS.

——, and D. M. Nelson, 1977b, Lake Michigan Sediment Trap Study: Preliminary Assessment of Results, in *Radiological and Environmental Research Division Annual Report, January–December 1976,* ERDA Report ANL-76-88(Pt.3), pp. 107-110, Argonne National Laboratory, NTIS.

——, and D. M. Nelson, 1977c, A Comparison of the Distribution Coefficients of Plutonium and Other Radionuclides to Those in Other Systems, in *Radiological and Environmental Research Division Annual Report, January–December 1976,* ERDA Report ANL-76-88(Pt.3), pp. 56-60, Argonne National Laboratory, NTIS.

Transport of Plutonium by Rivers

H. J. SIMPSON, R. M. TRIER, and C. R. OLSEN

A number of nuclear facilities are located on rivers and estuaries, and thus it is important to understand the primary transport pathways of transuranic elements in such systems. Relatively few field studies of point-source releases of plutonium to river systems have been made up to now. Information from research on the behavior of fallout plutonium in rivers can, however, provide some useful insights. The range of variation of soluble-phase fallout, $^{239,240}Pu$, in freshwaters and estuaries is relatively small (0.3 ± 0.2 fCi/liter) and appears to be "buffered" to some extent by the large reservoir of fallout $^{239,240}Pu$ in soils and the relative uniformity of the specific activity on soil particles (~20 pCi/kg). The Hudson River, Hudson estuary, New York City tap water, New York bight, and Great Lakes all have reasonably similar concentrations of soluble-phase $^{239,240}Pu$ despite the large range of chemical and other characteristics. The distribution of fallout $^{239,240}Pu$ between soluble phases and particles in rivers can be approximated by a partition coefficient of about 10^{-5}. For suspended particle loads of about 10 mg/liter, which are reasonably typical of low-flow summer conditions for rivers in the northeastern United States, $^{239,240}Pu$ is transported by both soluble phases and particles in approximately equal amounts. For higher suspended loads, typical of northeastern rivers during greater freshwater discharge and of most other large, nontropical rivers, the transport of fallout $^{239,240}Pu$ is clearly dominated by particles (by about an order of magnitude). For point-source addition of plutonium to a river, the most important transport pathway appears to be binding to the suspended load and the mobile portions of the fine-grain sediments and subsequent downstream movement with the fine particles. Since the kinetics and downstream transport pathways of fine particles of a particular river depend on a number of factors peculiar to each system, the most direct approach would be to exploit the presence of "tracers" already present to define the parameters of most relevance to transuranic-element transport over various time scales. Nuclear facilities often release sufficient quantities of fission and activation products during normal operations which can be used as indicators of fine-particle transport pathways. The behavior of these radionuclides cannot be expected to be identical to transuranic elements in river systems, but those elements with strong particle-phase associations can provide very useful information for sites of primary interest for transuranic-element transport assessments.

A number of nuclear power plants are now located on rivers and estuaries, and many more probably will be in the future. The only major reprocessing facility currently operating in the United States is located on a small tributary of the Savannah River. Thus knowledge of the transport pathways of transuranic elements in rivers is essential for proper monitoring of the routine operations of these facilities and for developing plans for dealing with any abnormally large releases of transuranic elements that might occur.

In principle, rivers can carry plutonium and other transuranic elements either in solution or as part of the suspended load. These two transport pathways are probably strongly coupled by some type of quasi-equilibrium partitioning between the two phases and thus cannot really be considered separately. As with many elements that are reactive in natural waters, the classifications of "dissolved" and "particulate" plutonium are based largely on operational procedures, such as whether or not material will pass through a filter of a certain nominal pore size. The actual species distribution of plutonium in natural waters is probably some kind of continuum from small-molecular-weight complexes through silt- or sand-size particles. To further complicate matters, particles can be transported in suspension or as bed load in a stream or accumulated in depositional environments and either buried or resuspended at a later time.

There have been relatively few field studies of point-source releases of plutonium to river systems. Three areas in the eastern United States that have received such attention are the Savannah River and its tributary downstream of the reprocessing facility in South Carolina (Hayes and Horton, this volume), the Miami River (a tributary of the Ohio River) downstream of Mound Laboratory in Ohio (Sprugel and Bartelt, 1978) and streams near Oak Ridge, Tenn. These river systems are the focus of ongoing research programs which should provide considerable information about the transport by rivers of plutonium derived from point sources. This chapter discusses the distribution of fallout plutonium in a few natural systems, including the Hudson River and estuary, and attempts to derive some first-order principles by which the transport pathways of plutonium in other river systems can be predicted. The Hudson estuary is now the site of three nuclear reactors, and at least half a dozen other units which are planned for this estuary in the next two decades.

Plutonium in the Hudson River Estuary

The Hudson River discharges into one of the large estuarine systems that dominate much of the coastal environment of the northeastern United States (Fig. 1). The Hudson has an unusually long, narrow reach of tidal water (>250 km), most of which is usually fresh. Saline water intrudes only about 40 km from the coastline during seasonal high freshwater discharge and reaches as far inland as 120 km during summer and early fall months of drought years. The near-surface suspended load of the Hudson is relatively low (10 to 20 mg/liter), as it is for nearly all the larger rivers in the northeastern United States, except during maximum spring runoff and following major storms.

From studies of the distribution of fallout nuclides and gamma-emitting nuclides released from Indian Point, the patterns of suspended particle transport and recent sediment accumulation in the Hudson estuary have been described (Simpson et al., 1976; 1978; Olsen et al., 1978). Much of the estuary has relatively little net accumulation of fine particles, whereas a few areas, such as marginal coves and especially New York harbor, account for a major fraction of the total deposition of fine particles containing fallout and reactor nuclides. The zone of major sediment accumulation is more than 60 km downstream from the reactor site, and the time scale of transport of fine particles labeled with reactor nuclides from the release area to burial in the harbor sediments varies from probably less than a month to years. At present there is no evidence in the Hudson sediments, including New York harbor, of releases of reactor

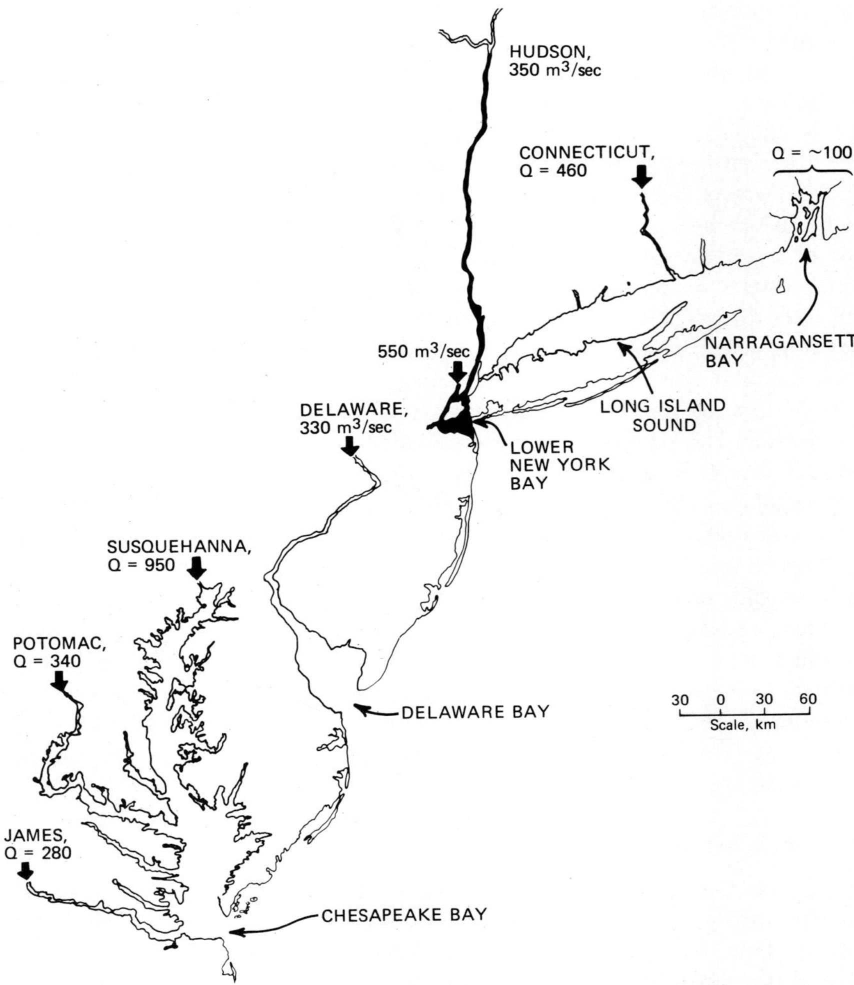

Fig. 1   Location map of the major estuarine systems of the northeastern United States. Q is mean annual freshwater flow.

^{239,240}Pu which are resolvable in the presence of the burden of fallout ^{239,240}Pu. Thus the current distribution of ^{239,240}Pu in the Hudson appears to be governed primarily by the delivery of global fallout to the drainage basin mostly more than a decade ago and the transport processes that have occurred since delivery.

Table 1 shows concentrations of ^{239,240}Pu in large-volume Hudson water samples that have had the suspended load removed by settling for 24 to 48 hr or by passing through a continuous-flow centrifuge followed by a 0.45-μm filter. The range of observed values for samples collected in 1975 and 1976 was 0.12 to 0.88 fCi/liter; the median value was about 0.3 fCi/liter. The current annual transport of ^{239,240}Pu in the "dissolved" phase in the Hudson can be estimated to be about 5×10^{-3} Ci if we assume a concentration of 0.3 fCi/liter and a mean annual river discharge of 550 m^3/sec. This represents somewhat less than 0.01% of the fallout burden of ^{239,240}Pu ($\sim$80 Ci) in the

TABLE 1   Dissolved ^{239,240}Pu in Continental Waters

Location*	fCi/liter†	Sample volume, liters
Hudson River (mp 61) (S)	0.32 ± 0.01	660
Hudson estuary (mp 19) (S)	0.88 ± 0.07	625
Hudson estuary (mp 18) (F)	0.47 ± 0.03	490
Hudson River (mp 47) (S)	0.27 ± 0.02	570
Hudson estuary (mp 19) (S)	0.12 ± 0.02	570
Hudson estuary (mp 8) (S)	0.15 ± 0.02	570
Hudson estuary (mp 24) (F)	0.30 ± 0.03	1500
New York bight (S)	0.25 ± 0.03	380
New York bight (U)	0.59 ± 0.09	660
New York bight (U)	0.68 ± 0.05	660
New York bight (U)	0.68 ± 0.09	660
New York bight (U)	0.68 ± 0.09	660
New York bight (U)	0.91 ± 0.14	660
New York bight (U)	0.95 ± 0.14	660
New York bight (U)	1.18 ± 0.14	660
New York City tap water (1973 to 1975)‡	~0.3	
Lake Ontario (1973)§	~0.3	
Great Lakes (1972 to 1973)¶	~0.5	

*mp indicates "mile point" upstream of the mouth of the Hudson, defined as the southern tip of Manhattan Island. The pretreatment procedure of the large-volume samples is indicated by one of the following three letters: U, unfiltered; F, filtered after passing through a continuous-flow centrifuge; S, suspended particles allowed to settle, usually for 24 to 48 hr, before the clarified water was transferred to another tank for processing.

†Mean ± standard error.

‡Data from Bennett (1976).

§Data from Farmer et al. (1973).

¶Data from Wahlgren and Marshall (1975).

soils of the Hudson drainage basin (~3.5 × 10⁴ km²). Soluble-phase release of fallout ^{239,240}Pu from Hudson soils thus has a half-time of the order of 10^4 yr and supplies an insignificant amount of dissolved ^{239,240}Pu to the coastal ocean compared with that transported onto the shelf from surface waters of the deep ocean.

The suspended-load activity of ^{239,240}Pu for which dissolved plutonium concentrations are listed in Table 1 averaged about 20 pCi/kg (18.9 ± 0.9 and 23.4 ± 1.0 pCi/kg). The distribution coefficient (K_d) of ^{239,240}Pu between the dissolved phase and suspended particles for those two samples was about 1.5×10^{-5}. Thus the transport of ^{239,240}Pu by suspended particles equals that in the dissolved phase when the concentration of suspended particles is about 15 mg/liter, a value that is reasonably typical of moderate and low freshwater flow periods in the Hudson. During periods of higher suspended load, the transport of ^{239,240}Pu in the Hudson is predominantly on particles.

From the quantities of material dredged annually from New York harbor (~2 × 10⁶ tons), the downstream transport of particles by the Hudson must be about a factor of 4 higher than indicated by multiplying typical near-surface suspended-load concentrations times mean annual flow. The "extra" transport of particles is probably accomplished by some combination of very high suspended loads coinciding with the highest freshwater flow rates and bed-load transport, which in the Hudson appears to consist largely of resuspension and deposition of fine particles in the lowest meter of the water column on the time scale of a tidal cycle. Thus in the Hudson the total downstream transport of ^{239,240}Pu is approximately a factor of 4 greater than that in the dissolved phase, which indicates a half-time for removal of fallout ^{239,240}Pu from the drainage basin, largely on particles, of about 10^3 years. Similar calculations for the Savannah River (Hayes and Horton, this volume) and the Greater Miami River (Sprugel and Bartelt, 1978) suggest drainage-basin removal times of about 2×10^4 and 2×10^3 yr, respectively. Again this supply term to the coastal ocean is not significant relative to advection of deep-ocean fallout ^{239,240}Pu onto the shelf. For the Hudson most of the delivery of ^{239,240}Pu on particles to the coastal ocean is accomplished by the dumping of dredge spoils rather than by estuarine discharge of suspended particles.

Plutonium in the New York Bight

The concentrations of dissolved ^{239,240}Pu in the coastal waters off the New York City area are two to three times those in the Hudson (Table 1). The suspended loads in the New York bight are almost two orders of magnitude lower than those in the Hudson, and fine-grain sediments in the bight have activities of ^{239,240}Pu comparable to those in the Hudson. Thus the transport of ^{239,240}Pu in the shelf environment appears to be largely in the dissolved phase, in contrast to the situation in the Hudson River and estuary.

Plutonium in Other Freshwaters

Data for the concentration of ^{239,240}Pu in New York City tap water (Bennett, 1976) are available for the period 1973 to 1975 (Table 1). The water supply for New York City is derived from tributaries of the Hudson and Delaware rivers. The activities ranged from 0.08 to 0.60 fCi/liter, with a mean value of about 0.3 fCi/liter (about 2% of the average rain activities during the same period). The range and mean value of the tap water ^{239,240}Pu concentrations are almost identical with the values observed for the Hudson River and estuary.

Farmer et al. (1973) have reported ^{239,240}Pu activities in Lake Ontario (Table 1) that are in the same range as the data discussed here for the Hudson River and New York bight. During the period 1971 to 1973, the average ^{239,240}Pu activity for the entire lake declined from about 0.8 fCi/liter to about 0.3 fCi/liter. The average ^{239,240}Pu activity in all five Great Lakes (Wahlgren and Marshall, 1975; Wahlgren et al., 1976) during 1972 and 1973 was about 0.5 fCi/liter (Table 1).

Transport of Fallout Plutonium to the Oceans

The data available indicate that the range of variation of soluble-phase ^{239,240}Pu in freshwaters is relatively small. The transport by rivers of fallout ^{239,240}Pu in "solution" can thus be estimated relatively easily solely on the basis of the rate of freshwater discharge. The concentrations in freshwaters appear to be "buffered" to some extent by the large reservoir of fallout ^{239,240}Pu in soils and the relative uniformity of the specific activity of ^{239,240}Pu on soil particles and river suspended particles ($\sim$20 pCi/kg). The distribution of fallout ^{239,240}Pu between soluble phases and particles in rivers can probably be approximated by a partition coefficient of about 10^{-5}. The total delivery of dissolved fallout ^{239,240}Pu to the oceans by rivers is probably about 10 Ci/yr, if we assume a discharge rate for all rivers of about 10^6 m^3/sec and a concentration of about 0.3 fCi/liter. Since the global average of suspended load in rivers is about 600 mg/liter, the transport of fallout ^{239,240}Pu by rivers will clearly be dominated by particles. If we assume that the specific activity of all river suspended matter is similar to that of surface soils, the total delivery of fallout ^{239,240}Pu to the ocean by rivers is about 5 × 10^2 Ci/yr, about 50 times the soluble-phase delivery. The specific activity of particles in rivers with very high suspended loads is probably somewhat lower owing to the presence of more large silt- and sand-size particles; so a more reasonable estimate for the total annual delivery of fallout ^{239,240}Pu to the ocean by rivers is probably 1 to 5 × 10^2 Ci.

Transport of Plutonium by Rivers Added at Point Sources

The distribution of fallout ^{239,240}Pu provides information about the partitioning of plutonium between soluble- and suspended-particle phases in rivers and about the processes by which transuranic-element transport occurs in rivers. For point-source addition of plutonium to a river, the most important transport pathway appears to be binding to the suspended load and the mobile portions of the fine-grain sediments and downstream movement with the fine particles. Since the effective concentrations of suspended particles, including the upper few centimeters of fine-grain sediment, in a river will be far greater than 10 to 15 mg/liter, the dominant transport of plutonium would be in association with particles. The kinetics and downstream transport pathways of a particular river system will depend on many factors, such as the frequency and duration of deposition and resuspension episodes for the suspended particles. In the tidal reach of the Hudson, the downstream movement of fine particles tagged with reactor nuclides is distributed such that some particles require several years to move 50 km whereas others probably require considerably less than a few months. In other rivers, such as the Columbia, which is above tidal influence, the downstream transport of some portions of the suspended load is probably similar to the rate of water transport, whereas other portions of the suspended particles are trapped for long periods, perhaps indefinitely, behind dams.

The distribution of fallout nuclides can provide valuable information about which areas of the bottom in a river system are actively scoured and which portions accumulate fine-grain sediments rapidly but probably cannot provide a very detailed picture of the kinetics of downstream transport of fine particles. A tracer added relatively uniformly to the earth's surface, as was weapons-testing fallout, is not very powerful for providing such information. Fortunately the river systems for which the kinetics of fine-particle movement are most important to understand for predicting transport of transuranic elements are also the ones for which point-source tracers are available. Many nuclear power plants and reprocessing facilities release sufficient quantities of fission or activation products during normal operations which can be used as indicators of fine-particle transport pathways. The behavior of these radioactive tracers cannot be expected to be identical to that of transuranic elements in river systems, but some of these tracers are associated with particles sufficiently to provide very valuable information about the patterns and kinetics of movement and accumulation of fine particles of most importance for evaluating the transport pathways of point-source releases of transuranic elements.

References

Bennett, B. G., 1976, *Fallout* ^{239,240}*Pu in Diet,* USAEC Report HASL-306, pp. 115-125, Health and Safety Laboratory, NTIS.

Farmer, J. G., V. T. Bowen, V. E. Noshkin, and M. B. Gavini, 1973, *Long-Lived Artificial Radionuclides in Lake Ontario. I. Supply from Fallout, and Concentrations in Lake Water, of Plutonium, Americium, Strontium-90 and Cesium-137,* unpublished.

Olsen, C. R., H. J. Simpson, R. F. Bopp, S. C. Williams, T. H. Peng, and B. L. Deck, 1978, A Geochemical Analysis of the Sediments and Sedimentation in the Hudson Estuary, *J. Sediment. Petrol.,* 48: 401-418.

Simpson, H. J., C. R. Olsen, R. M. Trier, and S. C. Williams, 1976, Man-Made Radionuclides and Sedimentation in the Hudson River Estuary, *Science,* 194: 179-183.

——, R. F. Bopp, C. R. Olsen, R. M. Trier, and S. C. Williams, 1978, Cesium-137 as a Tracer for Reactive Pollutants in Estuarine Sediments, in *First American—Soviet Symposium on Chemical Pollution of the Marine Environment,* L. L. Turekian and A. I. Simonov (Eds.), Report EPA-600/9-78-038, pp. 102-111, U. S. Environmental Protection Agency.

Sprugel, D. G., and G. E. Bartelt, 1978, Erosional Removal of Fallout Plutonium from a Large Midwestern Watershed, *J. Environ. Qual.,* 7: 175.

Wahlgren, M. A., and J. S. Marshall, 1975, The Behavior of Plutonium and Other Long-Lived Radionuclides in Lake Michigan. I. Biological Transport, Seasonal Cycling and Residence Times in the Water Column, in *Impacts of Nuclear Releases into the Aquatic Environment,* Symposium Proceedings, Otaniemi, Finland, 1975, pp. 227-243, STI/PUB/406, International Atomic Energy Agency, Vienna.

——, J. J. Alberts, D. M. Nelson, and K. A. Orlandini, 1976, Study of the Behavior of Transuranics and Possible Chemical Homologues in Lake Michigan Water and Biota, in *Transuranium Nuclides in the Environment,* Symposium Proceedings, San Francisco, 1975, pp. 9-24, STI/PUB/410, International Atomic Energy Agency, Vienna.

Biological Effects of Transuranic Elements in the Environment: Human Effects and Risk Estimates

ROY C. THOMPSON and BRUCE W. WACHHOLZ

The potential for human effects from environmentally dispersed transuranic elements is briefly reviewed. Inhalation of transuranics suspended in air and ingestion of transuranics deposited on or incorporated in foodstuffs are the significant routes of entry. Inhalation is probably the more important of these routes because gastrointestinal absorption of ingested transuranics is so inefficient. Major uncertainties are those concerned with substantially enhanced absorption by the very young and the possibility of increased availability as transuranics become incorporated in biological food chains.

Our knowledge of plutonium distribution and retention in the human is based on human autopsy data and on the extrapolation of a large body of experimental animal data. These data are undoubtedly more precise than our knowledge of the environmental exposure pathways that may lead to such deposition and more precise than our knowledge of the health consequences that may result from this deposition.

There is no positive information on the effects of transuranic elements in either man or experimental animals at the very low exposure levels with which we are concerned. Various approaches to the evaluation of this problem are discussed. We can conclude with some certainty that effects from present fallout levels will never be detected as a perturbation on normal cancer death rates. The possibility of no cancer deaths from fallout plutonium cannot be precluded.

The principal focus of this book is the environment, exclusive of man. Our ultimate concern, however, is for the effect of this environment on man. This chapter, therefore, reviews briefly the routes by which man can interact with transuranics in the environment and the possible consequences to man of such interaction. The level of treatment in this chapter is less detailed than in other chapters. The reader interested more specifically in effects on man and in the animal studies bearing on that problem can find such detail in several recent compilations: Hodge, Stannard, and Hursh, 1973; Bair, 1974; Thompson and Bair, 1972; Jee, 1976; and Wachholz, 1974.

Routes of Exposure

Opportunities for exposure of man to transuranics are of two quite different types: those resulting from employment in the nuclear industry and those resulting from the general dispersal of transuranics throughout the environment. It is the latter type which concerns us in the context of this book. Routes of exposure will differ for the two types, but effects are presumed to be similar in both cases.

691

Two routes of exposure, ingestion and inhalation, are of significance for general environmental dispersal. Of these, inhalation is the better understood because the mechanisms and kinetics of deposition and retention in the lung are primarily determined by physical factors, which have been studied for many substances in man. Such data can be extrapolated with some confidence in predicting the behavior of transuranics in the lung. The absorption of ingested transuranics from the intestinal tract is of more critical uncertainty, however, because such absorption is primarily determined by chemical factors unique to each individual transuranic and because only that fraction which is absorbed is of primary hazard concern. Unabsorbed alpha-emitting transuranics, while passing through the gastrointestinal tract, are conservatively assumed by the International Commission on Radiological Protection (ICRP) to deliver 1% of their decay energy to sensitive cells of the intestinal wall. On this basis the intestine becomes the critical organ for ingestion of insoluble transuranics (International Commission on Radiological Protection, 1960), although it is unlikely that significant damage actually occurs by this mechanism.

Inhalation

The model usually employed to describe the kinetics of inhaled transuranics in man is that of the ICRP Task Group on Lung Dynamics (1966) as modified by the ICRP Task Group on Plutonium and Other Actinides (1972). This model adjusts for three classes of particle solubility and for a wide range of particle sizes. To illustrate the general case of environmental transuranics, we will assume insoluble particles (class Y) of 0.4-μm activity median aerodynamic diameter (AMAD), which were assumed by Bennett (1976) to be typical of airborne fallout. The model predicts that 32% of such an inhaled aerosol will be deposited in the pulmonary region of the lung; the remainder will be immediately exhaled or rapidly cleared from the nasopharynx or tracheobronchial region. Of the 32% deposited in the pulmonary region, 40% is cleared with a half-life of 1 day; the remaining 60%, which is equivalent to about 20% of the aerosol initially inhaled, is retained in the lung with a half-life of 500 days—this fraction is responsible for essentially the total irradiation of the lung. About 6% of the quantity initially inhaled eventually reaches the bloodstream and is distributed among the systemic organs. Although these fractions will vary with chemical form, particle size, and other exposure variables, a sizable fraction of inhaled transuranic will, under any circumstance, be tenaciously retained by man. This is an efficient route of entry, as compared to ingestion, and atmospheric transport is a correspondingly hazardous environmental pathway.

Ingestion

Because of the general insolubility of transuranic oxides and hydroxides and the propensity for more soluble compounds to hydrolyze at physiological pH, one anticipates little absorption of these elements from the gastrointestinal tract. The ICRP assumes a fraction absorbed of 3×10^{-5} for plutonium and 10^{-4} for americium and curium (International Commission on Radiological Protection, 1960); a value of 10^{-6} is suggested as appropriate for plutonium oxide (International Commission on Radiological Protection, 1972). These values are based on the results of animal studies. Several recent investigations have indicated somewhat higher absorption than that assumed by the ICRP and a typically large variability from experiment to experiment (Durbin, 1973; Sullivan and Crosby, 1975; 1976). Because of this variability and because there is no direct measurement of gastrointestinal absorption in man, one must be cautious in applying the animal data to populations ingesting very low levels of transuranics in unknown chemical forms. Certain complexed forms of plutonium and hexavalent plutonium compounds are known to be more readily absorbed than other plutonium compounds (Durbin, 1973).

Under present conditions of recent fallout deposition, most ingested transuranics are likely to be either swallowed following inhalation or consumed as external contaminants on food (Bennett, 1976). With the passage of time, however, biologically incorporated transuranics may become a more important factor relative to other forms of ingested transuranics and relative also to inhaled transuranics. One might expect biologically incorporated transuranics to be more readily absorbed than inorganic forms, and there are limited animal data on milk (Finkel and Kisieleski, 1976), meat (Sullivan and Crosby, 1976), and alfalfa (Sullivan and Garland, 1977) which suggest that this may indeed be true for plutonium. There are also data that suggest the opposite conclusion for neptunium (Sullivan and Crosby, 1976). The extent to which transuranics may become biologically incorporated in foods and the gastrointestinal absorbability of such material are uncertain factors in any evaluation of the impact of environmental transuranics on man.

Another uncertainty relating to ingestion is the question of enhanced absorption in the very young. There is now an abundance of data attesting to the fact that neonatal rats (Sullivan and Crosby, 1975; 1976; Ballou, 1958; Sikov and Mahlum, 1972a; Sullivan, 1978), cats (Finkel and Kisieleski, 1976), and swine (Sullivan, 1978; Buldakov et al., 1969) absorb a very much larger fraction of ingested plutonium than do adult animals. Such anomalous absorption by the infant has been reported for many other normally nonabsorbed substances in many species, including man (Koldovsky, 1969; Sikov and Mahlum, 1972b). How one should extrapolate these animal data on transuranics to man is not clear either with regard to the magnitude of increased absorption or to the duration of this effect. In rats absorption drops to near-adult levels by the age of weaning (3 weeks) (Ballou, 1958). The life-style of the infant may protect it from plutonium ingestion, as compared with the adult, and plutonium deposited at an early age will be subsequently diluted as a consequence of growth. It is possible, however, that the infant is more sensitive to the production of deleterious effects from deposited plutonium, although preliminary reports of studies in rats (Mahlum and Sikov, 1974) and dogs (Stevens et al., 1978) suggest that this is not the case.

Distribution and Retention of Transuranics in Man

Although few health consequences and no fatalities have been observed to result from transuranic deposition in man, the distribution and retention of these elements in man can be measured. Such distribution and retention data permit the calculation of radiation doses in human tissues which can be compared with tissue doses from other forms of radiation known to produce effects in man. The similarity of human distribution and retention data to that measured in experimental animals also lends confidence to the extrapolation of health-effects data obtained in experimental animals.

Most directly relevant to environmental transuranics are the deposition data for fallout plutonium in man. An example of such data is shown in Table 1, which lists 50th percentile values (50% of individual values are lower) for tissues from more than 170 U.S. autopsies (McInroy et al., 1977). Also shown in Table 1 are computed estimates of plutonium concentration in tissues (Bennett, 1976), which are based on measured

TABLE 1   Concentration of Fallout Plutonium in Man

| | Plutonium concentration, pCi/g | | |
| | Autopsy samples* (50th percentile values) | Computed New York estimates† | |
Organ		1964	1974
Bone (vertebra)	0.50	0.08	0.20
Liver	0.58	0.23	0.54
Lung	0.24	2.48	0.12
Lymph nodes	2.42	43	27
Kidney	0.06	0.03	0.06
Gonads	0.13	0.02‡	0.12‡

*Data from McInroy et al. (1977).

†Data from Bennett (1976).

‡Estimated on the assumption that 0.05% of the total body burden is present in 10 g of ovaries.

airborne plutonium concentrations in the New York area and on the assumption that all intake by the human is via inhalation according to the ICRP lung model and ICRP assumptions of tissue distribution and retention (International Commission on Radiological Protection, 1972). The autopsies, on material collected from many states, were performed over a period extending from 1959 to 1976; the estimates are calculated for New York cumulative exposure to 1964 and 1974. Considering the uncertainties of time and place, the agreement between measured and computed organ burdens is quite reasonable for bone, liver, and lung, the organs of principal hazard concern as deduced from animal toxicity studies. Kidney and gonads also show excellent agreement. This agreement lends confidence to the ICRP assumptions regarding deposition and subsequent redistribution. The lower than predicted lymph node measurements may, at least partially, reflect the considerable difficulties of sampling pulmonary lymph nodes.

Transuranic distribution data in man are also available from autopsies on occupationally exposed persons (McInroy, 1976; Norwood and Newton, 1976) and on intentionally injected patients considered to be suffering from terminal illness (Durbin, 1972; Rowland and Durbin, 1976). In these cases tissue levels are much higher, and a larger number of organs can be analyzed with higher precision. These autopsy data are generally in accord with the more extensive animal data and with the ICRP assumption that 45% of a transuranic reaching the blood will deposit in bone and 45% in liver (International Commission on Radiological Protection, 1972).

Of particular interest are recent human data relating to the distribution of plutonium within organs. Limited data from sectioned lungs of occupationally exposed persons suggest that initial distribution of inhaled plutonium is relatively uniform (McInroy et al., 1976) but that, at long time periods following exposure, plutonium concentration is higher in the periphery of the lung (McInroy et al., 1976; Nelson et al., 1972). The distribution of plutonium among different bones (Larsen, Toohey, and Ilcewicz, 1976) and the microscopic distribution within bone (Schlenker, Oltman, and Cummins, 1976) have been studied in autopsy material from a patient who died 17 months after plutonium injection. This patient was suffering from Cushing's syndrome, and bone metabolism was not normal; nevertheless, the general distribution pattern was encouragingly similar to that which would have been predicted from animal studies.

The retention of plutonium in the various organs of man must be known if radiation doses are to be calculated. We have already noted the assumptions regarding retention in the lung as postulated in the ICRP lung model. Retention in the systemic organs is known to be prolonged, as deduced from human plutonium excretion data following intentional or accidental administration and from much data showing long-term retention in a variety of animal species (Durbin, 1972). The ICRP has assumed a biological half-life of 100 yr for transuranics in bone and 40 yr in liver; 90% of the plutonium deposited in lymph nodes is assumed to be retained with a biological half-life of 1000 days, and the remaining 10% is assumed to be retained without loss (International Commission on Radiological Protection, 1972). On the basis of a thorough review of the pertinent data, Durbin (1972) concluded that human bone plutonium might exhibit a shorter retention half-life than liver plutonium, and, on the basis of recently acquired nonhuman primate data, Durbin and Jeung (1976) have suggested shorter half-lives for both bone and liver plutonium.

In summary, it would seem fair to conclude that our knowledge of plutonium distribution and retention within the human, although uncertain in many details, is considerably more precise than our knowledge of the environmental and exposure pathways that lead to this deposition and more precise than our knowledge of the health consequences that may result from this deposition.

Effects of Transuranics in Man

An unevaluatable uncertainty attaches to any prediction of specific health effects from the exposure of humans to transuranic elements at levels contemplated for environmental dispersal. This uncertainty is due to the absence of any positive information on the effects of these elements in either man or experimental animals at the exposure levels of concern. Data are available at much higher exposure levels on the effects of transuranics in experimental animals and on the effects of certain other forms of radiation in man. The extrapolation of these data is made difficult by our lack of understanding of the mechanisms by which these effects occur. In the absence of such understanding, it has been common practice to extrapolate from the high-dose data by assuming a linear relationship between radiation dose and biological effect. Such a practice is endorsed by the Advisory Committee on the Biological Effects of Ionizing Radiations (BEIR) of the National Academy of Sciences—National Research Council (1972) as "warrant[ing] use in determining public policy on radiation protection"; in the same sentence they caution that "explicit explanation and qualification of the assumptions and procedures involved in such risk estimates are called for to prevent their acceptance as scientific dogma." Although in this chapter we have used the linear dose—effect assumption in estimates of the consequences of human exposure to environmental transuranics, we must emphasize that these estimated effects, if they occur at all, will be difficult to detect over the background of indistinguishable effects from other causes.

Experience with Transuranics in Animals

Direct information on the toxicity of transuranic elements is available only from studies in experimental animals. The radiobiological literature suggests that the effects observed in such animal experiments will at least qualitatively approximate those which would occur in man if he were exposed under the same conditions. On the basis of extensive data from several animal species, it is concluded that the most probable serious effects of

TABLE 2 Comparison of Transuranic Health Risk Estimates

	Cancer deaths or genetic defects per 10^6 organ-rem					
	Risk estimates based on data from humans					Risks estimates based on data from animals
	BEIR*					
	High†	Low‡	UNSCEAR§	MRC¶	Mays**	
Lung tumors	100	16	25–50	25	20	60–200††
Bone tumors	17	2	2–5‡‡	5	4	10–100§§
Liver tumors			10	20	10	
Genetic defects (in all subsequent	500¶¶	50¶¶	140¶¶			
quent generations)	1000***	10***	45***			

*National Academy of Sciences–National Research Council, 1972.

†Relative-risk model with lifetime plateau (U. S. Atomic Energy Commission, 1974).

‡Absolute-risk model with 30-year plateau (U. S. Atomic Energy Commission, 1974).

§United Nations, New York, 1977.

¶Medical Research Council, 1975.

**Mays, 1976.

††Data from Bair and Thomas, 1976.

‡‡Expressed by UNSCEAR as risk per 10^6 rads of low-LET radiation to endosteal cells, which should be roughly equivalent to risk per 10^6 rem of plutonium alpha radiation averaged throughout bone.

§§Data from Mays, 1976.

¶¶Specific genetic defects.

***Defects with complex etiology.

long-term low-level exposure to transuranics are lung, bone, and possibly liver cancers. Most of these data are from experiments with plutonium, but they can probably be applied to other transuranics with less error than is involved in many other unavoidable assumptions. Although quantitative extrapolation from animal to man involves considerable uncertainty, the animal data suggest cancer risks per 10^6 organ-rem of 60 to 200 for lung (Bair and Thomas, 1976) and 10 to 100 for bone (Bair, 1974; Mays et al., 1976). These estimates are compared with others in Table 2.

Experience with Transuranics in Man

It is clearly impossible to relate specific observed biological effects in man to the exposure of man at present levels of environmental plutonium. Some conclusions have been drawn from the absence of observed effects in the substantial numbers of persons occupationally exposed to very much higher levels of plutonium. Cave and Freedman (1976), investigating the adequacy of present plutonium exposure limits, conclude that, "total exposure represented by the available human data is not yet large enough to substantiate fully, on a statistical basis, the value of 0.016 μCi for the maximum permissible lung burden. However, regarded as a 'best estimate' this value should not be too high by a factor of more than 15 or by a factor of more than 40 at the 95% upper confidence level." On the basis of the long-term survival without bone tumors of eight "terminal" patients injected with plutonium, Rowland and Durbin (1976) conclude that, "the bone-tumor risk from plutonium is no greater than that from radium, and might be less." Certainly it would seem clear by now that occupational exposure to plutonium has not resulted in the kind of tragedy visited on the radium dial painters or the uranium miners.

Experience with Natural Radiation in Man

Alpha-emitting elements are a natural part of man's environment. He has lived with these internally deposited radioelements and with radiation from other natural sources throughout the history of the species. It is of some relevance to note that inhaled naturally occurring alpha-emitting radionuclides contribute an average annual dose of about 100 mrem to the lung and that naturally occurring alpha emitters in bone contribute an average annual dose at bone surfaces of about 40 mrem (National Council on Radiation Protection and Measurements, 1975). Although these doses cannot be related to any measure of specific effects, they have been at least tolerable on the evolutionary scale, and therefore slight increases would not be expected to have catastrophic effects.

Experience with Other Types of Radiation in Man

Inferences concerning the effects of transuranic elements in man may be drawn from information available on the effects of other forms of ionizing radiation in man; e.g., data derived from medical, occupational, accidental, or wartime exposure of humans to different radiation sources, including external X radiation, atomic bomb gamma and neutron radiation, and radium, radon, and radon daughters. Such information was summarized by the BEIR Committee (National Academy of Sciences–National Research Council, 1972) and, most recently, by the United Nations Scientific Committee on the Effects of Atomic Radiation (UNSCEAR) (1977). Both groups arrived at comparable risk estimates. England's Medical Research Council (MRC) (1975), considering much the same information covered in the BEIR and UNSCEAR reports, derived risk estimates specifically applicable to plutonium.

Of particular relevance are recently accumulated data on the carcinogenicity of ^{224}Ra in human bone (Spiess and Mays, 1970; 1973); ^{224}Ra has a very short half-life (3.62 days) and, because of this, irradiates only the surface layer of bone in much the same manner as transuranics. From these ^{224}Ra data, Mays et al. (1976) have estimated human bone cancer risks from plutonium; Mays (1976) has also estimated liver cancer risks, which are based largely on experience with Thorotrast, and lung cancer risks, which are based largely on data from the Japanese atomic bomb survivors.

Concluding Comments

Table 2 compares estimates of cancer risk from several sources previously discussed. Also included in Table 2 are estimates of genetic risk as derived in the BEIR and UNSCEAR reports.

Although of dubious quantitative applicability to the problems of environmental exposure because of the extrapolation uncertainties discussed previously, the kind of data presented in Table 2 will inevitably be used to estimate health effects from such exposure. As an example of such an exercise, in Table 3 we have derived an estimate of the human health consequences of the environmental dispersal of bomb-test plutonium in

**TABLE 3   An Estimate of Cancer Deaths in the United States
Due to Fallout Plutonium**

Organ	Individual dose,* mrem	Popula- tion dose,† 10^6 organ-rem	Risk factors,‡ cancer deaths/ 10^6 organ-rem	Cancer deaths §
Lung	16	3.2	20	64
Bone	34	6.8	4	27
Liver	17	3.4	10	34
			Total	125

*Cumulative organ dose to year 2000 based on inhalation exposure from 1954 to 1973. Data from Bennett (1976).

†Product of individual dose and U.S. population of 2×10^8.

‡Risk factors suggested by Mays (1976).

§Product of population dose and risk factor. Cancer death estimate is uncorrected for prior death from other causes.

the United States. On the basis of New York air samples, the ICRP lung model, and other metabolic parameters previously described, Bennett (1976) has calculated cumulative organ dose rates to the year 2000 for an individual exposed from 1954 through 1973. If we multiply these doses by 200 million people, we have an estimate of the total man-rem exposure resulting from plutonium fallout in the United States—an estimate that has obvious limitations but is probably more accurate than many other factors that go into the health-effects estimate. Multiplying this population dose by the cancer risk factors of Mays (1976), we arrive at an estimate of 125 cancer deaths. Because of the difficulty in defining a genetic effect and uncertainties in regard to the genetically effective dose from transuranics, we did not attempt an estimate of genetic effects in Table 3; it is generally agreed that such effects are probably "a minor part of the total" (Medical Research Council, 1975).

Mays' (1976) risk factors were used in Table 3 as the "best guesses" in our opinion. Use of the most pessimistic estimates of Table 2 would have led to a maximum cancer death count about four times higher. Neither estimate would constitute a detectable perturbation on normal cancer death rates; the possibility of no cancer deaths from fallout plutonium is not precluded.

Acknowledgment

This work was supported by the U. S. Department of Energy under contract EY-76-C-06-1830.

References

Bair, W. J., 1974, Toxicology of Plutonium, *Adv. Radiat. Biol.,* 4: 255-315.

——, and J. M. Thomas, 1976, Prediction of the Health Effects of Inhaled Transuranium Elements from Experimental Animal Data, in *Transuranium Nuclides in the Environment,* Symposium Proceedings, San Francisco, 1975, pp. 569-585, STI/PUB/410, International Atomic Energy Agency, Vienna.

Ballou, J. E., 1958, Effects of Age and Mode of Ingestion on Absorption of Plutonium, *Proc. Soc. Exp. Biol. Med.,* 98: 726-727.

Bennett, B. G., 1976, Transuranic Element Pathways to Man, in *Transuranium Nuclides in the Environment,* Symposium Proceedings, San Francisco, 1975, pp. 367-383, STI/PUB/410, International Atomic Energy Agency, Vienna.

Buldakov, L. A., E. R. Lyubchanskii, Y. I. Moskalev, and A. P. Nifatov, 1969, *Problems of Plutonium Toxicology,* USAEC Report LF-tr-41, translated by A. A. Horvath from *Problemy Toksikologie Plutoniya,* NTIS.

Cave, L., and L. Freedman, 1976, A Statistical Evaluation of the Radiotoxicity of Inhaled Plutonium in Insoluble Form, in *Transuranium Nuclides in the Environment,* Symposium Proceedings, San Francisco, 1975, pp. 547-567, STI/PUB/410, International Atomic Energy Agency, Vienna.

Durbin, P. W., 1972, Plutonium in Man: A New Look at the Old Data, in *Radiobiology of Plutonium,* pp. 469-530, B. J. Stover and W. S. S. Jee (Eds.), J. W. Press, Salt Lake City, Utah.

——, 1973, Metabolism and Biological Effects of the Transplutonium Elements, in *Uranium–Plutonium–Transplutonic Elements,* pp. 739-896, H. C. Hodge, J. N. Stannard, and J. B. Hursh (Eds.), Springer-Verlag, New York.

——, and N. Jeung, 1976, Reassessment of Distribution of Plutonium in the Human Body Based on Experiments with Non-human Primates, in *The Health Effects of Plutonium and Radium,* pp. 297-313, W. S. S. Jee (Ed.), J. W. Press, Salt Lake City, Utah.

Finkel, M. P., and W. E. Kisieleski, 1976, Plutonium Incorporation Through Ingestion by Young Animals, in *The Health Effects of Plutonium and Radium,* pp. 57-69, W. S. S. Jee (Ed.), J. W. Press, Salt Lake City, Utah.

Hodge, H. C., J. N. Stannard, and J. B. Hursh (Eds.), 1973, Uranium–Plutonium–Transplutonic Elements, in *Handbook of Experimental Pharmacology,* Vol. 36, Springer-Verlag, New York.

International Commission on Radiological Protection, 1960, *Permissible Dose for Internal Radiation,* ICRP Publication 2, Pergamon Press, Inc., New York.

International Commission on Radiological Protection—Task Group on Lung Dynamics, 1966, Deposition and Retention Models for Internal Dosimetry of the Human Respiratory Tract, *Health Phys.,* 12: 173-207.

International Commission on Radiological Protection (Task Group of Committee 2), 1972, *The Metabolism of Compounds of Plutonium and Other Actinides,* ICRP Publication 19, Pergamon Press, Inc., New York.

Jee, W. S. S. (Ed.), 1976, *The Health Effects of Plutonium and Radium,* J. W. Press, Salt Lake City, Utah.

Koldovsky, O., 1969, *Development of the Functions of the Small Intestine in Mammals and Man,* S. Karger, Basel.

Larsen, R. P., R. E. Toohey, and F. H. Ilcewicz, 1976, Macro-distribution of Plutonium in Selected Bones from an Abnormal Skeleton, in *The Health Effects of Plutonium and Radium,* pp. 315-320, W. S. S. Jee (Ed.), J. W. Press, Salt Lake City, Utah.

McInroy, J. F., 1976, The Los Alamos Scientific Laboratory's Human Autopsy Tissue Analysis Study, in *The Health Effects of Plutonium and Radium,* pp. 249-270, W. S. S. Jee (Ed.), J. W. Press, Salt Lake City, Utah.

——, H. A. Boyd, B. C. Eutsler, M. W. Stewart, and G. L. Tietjen, 1977, Determination of Plutonium in Man, in *Biomedical and Environmental Research Program of the LASL Health Division, January–December 1976,* ERDA Report LA-6898-PR, pp. 42-50, Los Alamos Scientific Laboratory, NTIS.

——, H. A. Boyd, M. W. Stewart, B. C. Eutsler, and G. L. Tietjen, 1976, Determination of Plutonium in Man, in *Biomedical and Environmental Research Program of the LASL Health Division, January–December 1975,* ERDA Report LA-6313-PR, pp. 19-22, Los Alamos Scientific Laboratory, NTIS.

Mahlum, D. D., and M. R. Sikov, 1974, Influence of Age on the Late Effects of Plutonium-239 in Rats, *Radiat. Res.,* 59: 171.

Mays, C. W., 1976, Estimated Risk from [239]Pu to Human Bone, Liver and Lung, in *Biological and Environmental Effects of Low-Level Radiation,* Symposium Proceedings, Chicago, 1975, Vol. 2, pp. 373-384, STI/PUB/409, International Atomic Energy Agency, Vienna.

——, H. Spiess, G. N. Taylor, R. D. Lloyd, W. S. S. Jee, S. S. McFarland, D. H. Taysum, T. W. Brammer, D. Brammer, and T. A. Pollard, 1976, Estimated Risk to Human Bone from [239]Pu, in *The Health Effects of Plutonium and Radium,* pp. 343-362, W. S. S. Jee (Ed.), J. W. Press, Salt Lake City, Utah.

Medical Research Council, London, 1975, *The Toxicity of Plutonium*, Her Majesty's Stationery Office, London.

National Academy of Sciences–National Research Council, 1972, *The Effects on Populations of Exposure to Low Levels of Ionizing Radiation*, Advisory Committee on the Biological Effects of Ionizing Radiations.

National Council on Radiation Protection and Measurements, 1975, *Natural Background Radiation in the United States*, NCRP Report No. 45.

Nelson, I. C., K. R. Heid, P. A. Fuqua, and T. D. Mahony, 1972, Plutonium in Autopsy Tissue Samples, *Health Phys.*, 22: 925-930.

Norwood, W. D., and C. E. Newton, Jr., 1976, U. S. Transuranium Registry Study of Thirty Autopsies, *Health Phys.*, 28: 669-675.

Rowland, R. E., and P. W. Durbin, 1976, Survival, Causes of Death, and Estimated Tissue Doses in a Group of Human Beings Injected with Plutonium, in *The Health Effects of Plutonium and Radium*, pp. 329-341, W. S. S. Jee (Ed.), J. W. Press, Salt Lake City, Utah.

Schlenker, R. A., B. G. Oltman, and H. T. Cummins, 1976, Microscopic Distribution of ^{239}Pu Deposited in Bone from a Human Injection Case, in *The Health Effects of Plutonium and Radium*, pp. 321-328, W. S. S. Jee (Ed.), J. W. Press, Salt Lake City, Utah.

Sikov, M. R., and D. D. Mahlum, 1972a, Age-Dependence of ^{239}Pu Metabolism and Effect in the Rat, in *Radiobiology of Plutonium*, pp. 261-272, B. J. Stover and W. S. S. Jee (Eds.), J. W. Press, Salt Lake City, Utah.

——, and D. D. Mahlum, 1972b, Plutonium in the Developing Animal, *Health Phys.*, 22: 707-712.

Spiess, H., and C. W. Mays, 1970, Bone Cancers Induced by ^{224}Ra(ThX) in Children and Adults, *Health Phys.*, 19: 713-729; also Addendum, *Health Phys.*, 20: 543-545 (1971).

——, and C. W. Mays, 1973, Protraction Effect on Bone Sarcoma Induction of ^{224}Ra in Children and Adults, in *Radionuclide Carcinogenesis*, AEC Symposium Series, Richland, Wash., May 10–12, 1972, C. L. Sanders, R. H. Busch, J. E. Ballou, and D. D. Mahlum (Eds.), pp. 437-450, CONF-720505, NTIS.

Stevens, W., D. R. Atherton, F. W. Bruenger, and B. J. Stover, 1978, Retention, Distribution and Toxicity of ^{239}Pu in Beagles Injected at 3 Months of Age, in *Abstracts of Papers for the Twenty-Sixth Annual Meeting of Radiation Research Society*, Toronto, Canada, May 14–18, 1978, *Radiat. Res.*, 74: 592.

Sullivan, M. F., 1978, Gastrointestinal Absorption and Retention of Plutonium-238 in Neonatal Rats and Swine, in *Pacific Northwest Laboratory Annual Report for 1977 to the DOE Assistant Secretary for Environment*, DOE Report PNL-2500(Pt.1), pp. 3.91-3.92, Battelle Pacific Northwest Laboratories, NTIS.

——, and A. L. Crosby, 1975, Absorption of ^{233}U, ^{237}Np, ^{238}Pu, ^{241}Am, ^{244}Cm and ^{253}Es from the Gastrointestinal Tract of Newborn and Adult Rats, in *Pacific Northwest Laboratory Annual Report for 1974 to the ERDA Division of Biomedical and Environmental Research*, ERDA Report BNWL-1950(Pt.1), pp. 105-108, Battelle Pacific Northwest Laboratories, NTIS.

——, and A. L. Crosby, 1976, Absorption of Transuranic Elements from Rat Gut, in *Pacific Northwest Laboratory Annual Report for 1975 to the ERDA Division of Biomedical and Environmental Research*, ERDA Report BNWL-2000(Pt.1), pp. 91-93, Battelle Pacific Northwest Laboratories, NTIS.

——, and T. R. Garland, 1977, Gastrointestinal Absorption of Alfalfa-Bound Plutonium-238 by Rats and Guinea Pigs, in *Pacific Northwest Laboratory Annual Report for 1976 to the ERDA Assistant Administrator for Environment and Safety*, ERDA Report BNWL-2100(Pt.1), pp. 137-139, Battelle Pacific Northwest Laboratories, NTIS.

Thompson, R. C., and W. J. Bair (Eds.), 1972, Biological Implications of the Transuranium Elements, *Health Phys.*, 22: 553-954.

United Nations Scientific Committee on the Effects of Atomic Radiation, 1977, *Sources and Effects of Ionizing Radiation*, United Nations, New York.

U. S. Atomic Energy Commission, 1974, *Liquid Metal Fast Breeder Reactor Program*. Vol. 2, p. G-61. Available free from W. H. Pennington, USERDA, Office of Assistant Administrator for Environment and Safety, Washington, D. C.

Wachholz, B. W. (Comp.), 1974, *Plutonium and Other Transuranium Elements: Sources, Environmental Distribution and Biomedical Effects*, USAEC Report WASH-1359, NTIS.

Ecological Effects of Transuranics in the Terrestrial Environment

---

F. W. WHICKER

This chapter explores the ecological effects of transuranium radionuclides in terrestrial environments. No direct studies that relate the level of transuranic contamination to specific changes in structure or function of ecological systems have been carried out. The only alternative approach presently available is to infer such relationships from observations of biota in contaminated environments and models. Advantages and shortcomings of these observations as well as those of the direct experimental approach are discussed. Searches for ecological effects of plutonium contamination in terrestrial ecosystems adjacent to the Rocky Flats plant (Colorado) and at the Nevada Test Site have not positively demonstrated plutonium-induced perturbations. These studies were carried out in areas containing of the order of 10 to 1000 μCi $^{239}Pu/m^2$ in the upper 3 cm of soil. Simple calculations suggest that ^{239}Pu applications on the order of 1 Ci/m^2 may be required to cause significant mortality in plant populations. Models and calculations indicate that over 1 mCi $^{239}Pu/m^2$ would likely be required to produce subacute mortality in mammals. Additional research applicable to ecological effects is suggested.

To grasp the ecological implications of transuranium elements in the environment, we must understand their chemical, physical, and biological behavior through time. We must also understand the effects on biological systems of these elements when they are dispersed into the environment. Knowledge of the biological effects is particularly lacking. This may seem surprising in view of the large research efforts that have been devoted to the biological effects of plutonium and other transuranics. The lack of quantitative understanding in the area of ecological effects is not so surprising, however, when the complexities of the problem are considered. Such complexities include the environmental behavior of transuranics, which is dependent on the physical and chemical form of the nuclides as well as on the nature of the ecosystem. Of major importance is the dose to certain tissues, but dose distribution is especially complex for relatively insoluble alpha emitters. A high-level application of transuranic may have little radiation effect if energy is not deposited in critical cells.

Although we know a great deal about the effects of plutonium on experimental mammals (Bair and Thompson, 1974), we know very little about its effects on the other classes of animals that have important functions in natural systems and even less about its effects on plants. Also, very little is known about the general biological effects of the other transuranics. The effects of X- and gamma radiation on major plant and animal phyla have been studied in depth, but the extrapolation of X- and gamma radiation information to insoluble alpha emitters is seriously complicated by dosimetry and the relative biological effectiveness (RBE) of alpha particles for most actinides. Recent reviews on plutonium and other actinides in the environment have very little to say about ecological effects; rather, they dwell primarily on distribution and behavior (Romney and Davis, 1972; Martell, 1975; Hakonson, 1975; Hanson, 1975).

There are three basic approaches to the study of ecological effects of transuranics: (1) direct experiments in which radionuclides are applied at various levels to study systems, (2) observations of populations that occupy contaminated areas, and (3) modeling and extrapolation from applicable research data. Each approach has inherent advantages and shortcomings. The direct experimental approach might enjoy a relatively high degree of credibility and accuracy, but it has not been used with transuranics for reasons of safety and lack of public acceptance. Examination of contaminated areas is quite feasible and has been done at such places as the Nevada Test Site, Enewetak, and Rocky Flats (in Colorado). This approach is less than ideal, however, because of the usual lack of good experimental control and the common presence of more than one potentially toxic substance, which lead to uncertainties in data interpretation. The third approach can be used when needed with existing data, but accuracy may be poor because of the complexity and uncertainty associated with parameter values.

Ecosystems can probably tolerate higher levels of radioactivity from most, if not all, of the transuranics than from the more biologically mobile fission products, such as ^{90}Sr and ^{137}Cs. Low solubility, lack of essential nutrient analogues, and the virtual lack of penetrating radiations for most transuranics form the basis for this opinion. However, critical experiments to make this comparison have not been done; there is some concern that biological incorporation of long-lived transuranics in the environment may slowly increase with time, and it is known that very low levels may be carcinogenic.

Direct Experiments

The literature dealing with effects of ionizing radiation on plants and animals is massive. Important reviews and bibliographies include the BEIR report (National Academy of Sciences–National Research Council, 1972), the UNSCEAR report (United Nations, 1972), and the bibliography by Sparrow, Binnington, and Pond (1958). The vast majority of this literature, however, is based on laboratory studies with X- or gamma radiation. A far smaller body of literature exists on radiation effects on natural populations. Whicker and Fraley (1974) reviewed field studies dealing with the effects of ionizing radiation on terrestrial plant communities, and Turner (1975) prepared a similar review for native animal populations. This literature also is restricted primarily to X- and gamma radiation, but it provides a substantial basis for understanding dose–effect relationships.

A major problem in applying this information to transuranics is that of determining the equivalent dose to critical tissues which would result from a given level of contamination. Of the 17 transuranic nuclides listed as being of some importance in the nuclear industry to the year 2000 (Energy Research and Development Administration, 1976), 13 are alpha emitters with generally infrequent emission of weak (mostly <0.07 MeV) photons. The other 4 are beta emitters with accompanying weak photon emissions. Alpha–weak-photon emitters include the particularly important nuclides ^{238}Pu, ^{239}Pu, ^{241}Am, ^{242}Cm, and ^{244}Cm. Alphas from these nuclides have energies of 5 to 6 MeV and ranges in air and biological tissue of roughly 4 cm and 40 μm, respectively (Walsh, 1970), which lend considerable complexity to the problem of dosimetry.

From field studies in plutonium-contaminated areas, most of the plutonium associated with vegetation appears to be surficial and not incorporated within tissues. Therefore critical tissues (meristem for growth and flower bud for reproduction) may receive a widely variable dose from surface contamination, depending on the location of the material and the thickness of epidermal tissue layers. I am not aware of any studies designed to show the detailed histological distribution of transuranics in and on plant tissues in contaminated-field environments.

The effective dose to animal tissues is equally difficult to determine. The dose from inhalation and ingestion of transuranics is subject to many variables. Absorption, translocation, deposition, and retention are affected by the physical and chemical forms of the nuclide and physiology of the animal (International Commission on Radiological Protection, 1972). The environmental chemistry of plutonium is extremely complex (Wildung et al., 1977), and our overall understanding is inadequate (Dahlman, Bondietti, and Eyman, 1976).

A few studies have been conducted in which simulated fallout particles containing beta and beta—gamma emitters were administered to field plots. The studies by Murphy and McCormick (1973) and Dahlman, Beauchamp, and Tanaka (1973) come closer to the kind needed for transuranics in that the problems of dosimetry are circumvented by simply relating effects to the level of fallout simulant applied. Murphy and McCormick applied ^{90}Y-coated albite particles to experimental granite outcrop plant communities. The effects on the reproductive potential of *Viguiera porteri* treated with 0, 205, and 526 mCi/m^2 were measured. Dahlman, Beauchamp, and Tanaka applied ^{137}Cs fused to silica sand particles to 100-m^2 plots in a fescue meadow. The levels applied (22 mCi/m^2) caused measurable decreases in seed production of *Festuca arundinacea*. A similar study using ^{90}Y-tagged sand grains to produce effects on crop plants was conducted by Schulz (1971). Fallout simulants containing ^{137}Cs were also applied to field plots at Oak Ridge, Tenn., to study the effects on arthropods and small mammals (Auerbach and Dunaway, 1970).

For research findings to be integrated and understood, however, it is highly desirable to estimate the dose to critical tissues from the levels of simulants applied. In the studies cited, beta-particle doses were estimated by thermoluminescent dosimeters and various computations. The Stanford Research Institute developed fallout-particle simulants for the field studies and measurement and computational techniques for beta dosimetry (Lane, 1971; Brown, 1965; Mackin, Brown, and Lane, 1971). Similar technology could probably be applied to alpha emitters for their use in field studies.

I am not aware of any studies in which physically and/or chemically characterized transuranics have been experimentally applied to field plots at levels sufficient to cause measurable ecological effects. The safe conduct of such studies would require an area remote from human habitation and stringent health physics practice and cleanup. Such a study would be expensive, possibly hazardous, and difficult to justify. A greenhouse study involving plants growing on soil that has been heavily contaminated with transuranics is being conducted by A. Wallace and E. M. Romney at the University of California at Los Angeles. One of the objectives of this study will be the effects of alpha particles.

Another investigation that bears on the problem of biological effects of transuranics in the environment is under way at Battelle—Pacific Northwest Laboratories under the direction of R. E. Wildung. Early results indicate radiation toxicity from ^{238}Pu and ^{239}Pu to some strains of soil actinomycetes and fungi at levels of 0.7 μCi/g (soil)

($\sim$2.5 $\times$ 10^4 μCi/m^2). Such toxicity was expressed as a decline in microbial numbers. Since microbes perform functions in soil that are important to plant growth, indirect effects to higher plants and animals could be elicited through microbial perturbations from plutonium in soil.

It appears that large quantities of a transuranic nuclide would be required in the field to cause obvious ecological effects. Two very crude calculations, one for higher plants and one for animals, illustrate the approximate levels of ^{239}Pu required to produce, for instance, detectable mortality.

Plant Communities

Assumptions: A grassland plant community requires a dose rate of about 40 rad/day to show measurable changes in diversity (Whicker and Fraley, 1974); the effective decay energy for ^{239}Pu is 53 MeV/d, considering an RBE of 10 (International Commission on Radiological Protection, 1960); a concentration ratio (CR = activity per gram of plant $\div$ activity per gram of soil) of 10^{-4} is assumed (Energy Research and Development Administration, 1976); the ^{239}Pu is assumed to be uniformly distributed within plant tissues and uniformly distributed in the upper 3 cm of soil, which has a bulk density of 1.2 g/cm^3. Surficial contamination is neglected.

Calculations:

Required ^{239}Pu concentration in plant tissue

$$= \frac{(40 \text{ rad/day})(6.25 \times 10^7 \text{ MeV/g-rad})}{(53 \text{ MeV/d})(3.2 \times 10^9 \text{ d/day-}\mu\text{Ci})} = 1.5 \times 10^{-2} \ \mu\text{Ci/g}$$

Required ^{239}Pu concentration in soil

$$= \frac{1.5 \times 10^{-2} \ \mu\text{Ci/g}}{10^{-4}} = 150 \ \mu\text{Ci/g}$$

Required ^{239}Pu application to soil

$$= (150 \ \mu\text{Ci/g})(1.2 \text{ g/cm}^3)(3 \text{ cm}^3/\text{cm}^2)(10^4 \text{ cm}^2/\text{m}^2) = 5.4 \times 10^6 \ \mu\text{Ci/m}^2$$

This value is within an order of magnitude of the soil plutonium levels that appear to evoke some toxic effects in plants under greenhouse conditions (R. E. Wildung and T. R. Garland, Battelle–Pacific Northwest Laboratories, personal communication).

Animals

Assumptions: Inhalation of suspended soil is considered the critical route of entry; human and experimental animal data and standards are used; the maximum permissible ^{239}Pu human lung burden of 1.6 $\times$ 10^{-5} μCi/g is achieved with a mean air concentration of 10^{-5} μCi/m^3 (International Commission on Radiological Protection, 1960); the critical concentration of ^{239}Pu in the lung required for subacute death is 1 $\times$ 10^{-2} μCi/g (Bair, 1974); and a mean resuspension factor of 10^{-5}/m is assumed.

Calculations:

Required air concentration

$$= \frac{(1 \times 10^{-2} \ \mu Ci/g)(10^{-5} \ \mu Ci/m^3)}{1.6 \times 10^{-5} \ \mu Ci/g} = 6.3 \times 10^{-3} \ \mu Ci/m^3$$

Required ^{239}Pu application to soil

$$= \frac{6.3 \times 10^{-3} \ \mu Ci/m^3}{10^{-5}/m} = 630 \ \mu Ci/m^2$$

If these calculations approach reality, it is clear that very large applications of ^{239}Pu would be required to produce measurable ecological changes, especially in plant communities. Nevertheless, such studies, if done, would carry more credibility than crude extrapolations and simplified calculations.

Contaminated Environments

The approach of examining ecosystems that have been accidentally contaminated with transuranics is feasible and probably desirable. Because of the lack of direct experimental data and the inherent complexity and uncertainty in computational models, we should look at areas that have been contaminated to ^{239}Pu activity levels that significantly exceed worldwide fallout levels. Several such areas exist or have existed in the past. These include Rocky Flats, Trinity, several areas at the Nevada Test Site, and various sites on Enewetak atoll (in the Pacific). In addition, plutonium releases to the environment have occurred from nuclear facilities at Oak Ridge, Hanford, Mound Laboratory, Los Alamos, Savannah River, Idaho National Engineering Laboratory, and from bomber crashes in local areas in Greenland and Spain.

If sufficiently careful searches for ecological changes in contaminated areas prove to be negative, then it probably can be concluded that the observed levels had no detectable consequence. Such data should be examined in the light of laboratory information for additional assurance. If biological perturbations are discovered in contaminated areas, then it may or may not be possible to assign causal factors. In many contaminated sites, more than one toxic substance may be present, or other factors may be responsible for changes. It may be possible to offset these problems if a proper control area is available, but this should be determined before the initiation of any search for effects.

A comparison of various biological measurements between two ecologically similar study areas of substantially differing ^{239}Pu levels at Rocky Flats was conducted by T. F. Winsor* and C. A. Little† of Colorado State University and G. E. Dagle of Battelle—Pacific Northwest Laboratories (Whicker, 1976; Little, 1976). The ^{239}Pu readings from soil in the principal study areas ranged from 100 to over 20,000 d/min per gram in the upper 3 cm (2 to 400 $\mu Ci/m^2$). In addition, comparative data were obtained from control areas containing only worldwide fallout plutonium of the order of 0.1 d/min per gram

---

*Present address: Rockwell International, Rocky Flats Plant, Golden, Colo.

†Present address: Division of Health and Safety Research, Oak Ridge National Laboratory, Oak Ridge, Tenn.

(0.002 μCi/m^2). Biological measurements were made, including vegetation-community structure and biomass; litter mass; arthropod community structure and biomass; and small mammal species occurrence, population density, biomass, reproduction, and whole carcass and organ masses. In addition, small mammals were examined by X ray for skeletal sarcomas, microscopy for lung tumors, and necropsy for general pathology and parasite occurrence.

Although minor differences in certain biological attributes between study areas were observed, none could be related to plutonium levels. Pathological conditions and parasites were found in some rodents, but occurrence frequencies between control and contaminated areas were similar. No evidence of cancers or other radiogenic disease was found. These observations and measurements, combined with intensive field observations over a period of 5 years, led to the conclusion that plutonium contamination at Rocky Flats has not produced demonstrable ecological changes. Furthermore, the levels of plutonium observed in tissues of plants and animals in contaminated areas were insufficient to produce the doses that would be required to produce obvious biological changes.

Subcellular biological changes, such as chromosome aberrations, cannot be ruled out at Rocky Flats. Even if chromosome aberration frequencies were increased in the more highly contaminated areas, however, population-level changes would likely not persist because of the surrounding reservoir of normal genetic information. The possibility of long-term biological effects cannot be discounted either, although this would appear highly unlikely, nor can we conclude that a similar level of plutonium contamination spread over a much larger area would be without ecological consequence. The latter idea, discussed by Odum (1963), stems from the fact that population effects in a highly localized area can be readily masked by immigration of unaffected individuals and propagules from the surrounding area, emigration of affected individuals, and gene pool mixing between the contaminated and surrounding areas. Such masking might not operate, at least to the same degree, for a large area. The validity of any future studies of animals in small-size contaminated areas might be increased if a barrier were erected to prevent the animals from moving into or out of the study area.

Extensive searches for ecological changes in contaminated areas have been carried out at the Nevada Test Site (Wallace and Romney, 1972; Allred, Beck, and Jorgensen, 1965; and Rhoads and Platt, 1971). In the majority of these studies, however, the contamination consisted principally of mixed fission products, and, except for the work reported by Rhoads and Platt, the more dramatic ecological effects were generally attributed to nonradiological perturbations. The best opportunities for searching for ecological effects from plutonium alone exist in a number of areas on or adjacent to the Nevada Test Site which have been used for "safety shot" tests. These tests involved detonation by conventional explosives of plutonium in various containment configurations. Some 300 acres containing on the order of 10 μCi Pu/m^2 exist, and a few acres have levels exceeding 6000 μCi Pu/m^2 (Wallace and Romney, 1975; Martin and Bloom, 1976). Studies of small mammals and grazing cattle in these areas have failed to discover any evidence of radiogenic pathology (Romney and Davis, 1972; Smith, Barth, and Patzer, 1976). Varney and Rhoads (1977) have examined shrubs in areas assumed to have been contaminated primarily with plutonium. Although their data implied that such shrubs had increased frequencies of chromosomal aberrations in comparison to controls, the evidence was not conclusive.

Although, as mentioned, other sites in the world have been contaminated with plutonium, I am not aware of any specific searches for ecological effects at these sites. Competent ecologists have conducted studies on plutonium distribution and behavior within many of these sites, however, and any readily apparent ecological changes would likely have been reported. I am also not aware of any sites at which other transuranics have been released at levels greater than existing plutonium levels.

Another approach to the study of contaminated environments is to examine areas containing above-normal amounts of the naturally occurring radionuclides. Many areas contain substantial quantities of natural uranium and thorium. These primordial radionuclides and many of their progeny are alpha emitters. Possibly some inferences to the transuranics could be made from studies in areas of high natural alpha activity. For instance, the rodents on Morro do Ferro in Minas Gerais, Brazil, which receive an estimated lung dose of 10^3 to 10^4 rem/yr, might provide a good study opportunity (Drew and Eisenbud, 1966). Major problems with such an approach include differences in radiation schemes and chemical properties between the naturally occurring and transuranium radionuclides. We know something about the relative toxicities of ^{239}Pu and ^{226}Ra (Thompson, 1976) but very little about the relative ecological and physiological behavior and toxicities of transuranics with other naturally occurring alpha emitters. Pochin (1976) points out some other difficulties inherent in trying to quantify biological effects of environmental radioactivity at low levels.

On the basis of data summarized by the United Nations (1972), I calculate that the upper 3 cm of soil in the United States averages roughly 0.3 μCi of natural alpha activity per square meter. A similar calculation applied to atypically high natural radiation background areas yielded alpha activities of 7 μCi/m^2 in the upper 3 cm near Central City, Colo. (Mericle and Mericle, 1965), 50 μCi/m^2 in local areas in Brazil (Eisenbud et al., 1964), and 200 μCi/m^2 in the USSR (Maslov, Maslova, and Verkhovskaya, 1967).

A number of genetic and ecological studies have been done in some of these and similar areas. Rats occupying a monazite sand area in Kerala, India, had no discernible phenotypic effects (Gruneberg et al., 1966). There is, however, suggestion of radiation-induced genetic damage leading to mental retardation of humans who occupy the same region in India (Kochupillai et al., 1976). Furthermore, Gopal-Ayengar et al. (1977) report genetic alterations in plants indigenous to the monazite belt in Kerala. Cullen (1968) reported preliminary findings of a human cytogenetic study in Guarapari, Brazil, in which an apparently increased incidence of somatic chromosome aberrations in comparison to a control area was found. A high incidence of multiple-break aberrations was noted which was thought to be compatible with the presence of internal alpha emitters. These findings were apparently corroborated more recently by Barcinski et al. (1975).

Osburn (1961) observed an increased incidence of morphological anomalies and flower abortion in *Penstemon virens* growing on the more radioactive sites near Central City, Colo. However, the chemical toxicity of thorium and possibly other factors cannot be ruled out as causal. In the USSR, Maslov, Maslova, and Verkhovskaya (1967) reported various deleterious effects on reproduction, parasite infestation, and the general condition of small mammals in areas of high natural radiation. Although radiation was implied as the cause of such effects, it was not the only variable between experimental and control populations. I am not convinced from these studies that naturally occurring alpha emitters, even in the unusually high natural background regions of the world, cause demonstrable ecological consequences. Potential genetic changes in local areas would likely be disadvantageous and selected out of populations (National Academy of Sciences–National Research Council, 1972; Muller, 1950).

Application of Existing Data

Existing data can be used to predict the magnitude of human or ecological hazard from a given level of transuranic contamination. As mentioned, however, computational models that reasonably simulate actual environmental and physiological processes require many parameter values which in themselves vary with circumstances and site. Existing knowledge of appropriate parameter values for plutonium behavior in extensively studied areas, such as the Nevada Test Site, the White Oak Creek drainage at Oak Ridge, and Rocky Flats, appears sufficient to develop models with reasonable credibility. Data that could be applied to most other terrestrial environments, however, are essentially lacking. This is especially true for transuranics other than plutonium.

Complexities involved in computational models have been discussed in considerable detail by Healy (1974), Anspaugh et al. (1975), and Martin and Bloom (1976). Healy (1974) undertook the difficult task of calculating the levels of plutonium in soil which might be considered standard or guideline levels for humans residing on and deriving sustenance from such soils. The standard levels calculated could conceivably result in the attainment of maximum permissible doses for members of the public. The computations were general in application and used available experimental data and conservative assumptions. The conceptual model considered surface soil to be the major reservoir and source of plutonium and considered processes by which the material might reach the critical organs of man. These processes included resuspension, atmospheric dispersion, cloud depletion, deposition, inhalation, ingestion of soil and contaminated foods, skin absorption, and metabolic behavior following intake. The calculations suggested that 4×10^{-4} μCi ^{239}Pu/g or 25 μCi ^{239}Pu/m^2 in the top 3 cm of soil was probably a conservative standard.

Using a similar approach but with site-specific data from the Nevada Test Site, Martin and Bloom (1976) calculated that 3 nCi ^{239}Pu/g (soil) (170 μCi/m^2) could result in the nonoccupational maximum permissible dose to the lung (1.5 rem/yr) of a standard man living over and obtaining food from the soil in question. This model was presented in a lucid and practical way, and the relative degree of confidence that can be placed on each parameter used in the model was made clear. The basic approach relates intake rates for ingestion and inhalation to surface soil concentrations; human metabolic and dose calculations are based on International Commission on Radiological Protection (ICRP) models and recommended parameter values.

In the ecological context, it seems important to consider the concept of the "critical organism." Although our primary concern is focused on man, the general welfare of the human population cannot be separated from environmental quality. Legal, moral, and scientific justification exists for ensuring the protection of species other than man from environmental contaminants. Indigenous species of plants and animals, by virtue of proximity and life habitats, will receive substantially higher radiation dose rates than man at many sites likely to receive transuranic contamination. On the other hand, many wild species, because of shorter normal life-spans, may not live long enough to develop serious pathology from chronic low-level exposures. In addition, for wildlife, society is generally concerned about performance of the population, whereas for humans, we are concerned about the more limiting case of individuals (Auerbach, 1971).

Historically, assessments of radionuclides in the environment have considered man to be the critical organism. The assumption has often been made that, if adequate protection for man is assured, we need not worry about ecological effects. Auerbach (1971) has addressed this question with the conclusion, "Present knowledge based on these and similar studies of the ecological effects of low-level chronic doses, such as could result from routine reactor releases under current standards, guidelines, and operational experience, indicates that any possible biological effects would be undetectable." Although this philosophy generally appears defensible, especially for reactor effluents as stated, I hope that we do not blindly adopt it for all situations. For example, nuclear-waste disposal could present unanticipated ecological problems in the future, possibly without causing hazardous doses to humans.

Present Status and Directions for Future Work

To add clarity, before discussing research needs and possible directions for future work, I will recapitulate what I think is the status of our knowledge on biological responses to alpha emitters in the environment. Apparently, transuranics have not been experimentally applied to study plots in the field. On the basis of limited observations of terrestrial environments accidentally or inadvertently contaminated with plutonium in the range of 10 to 1000 μCi/m^2, no clear-cut ecological effects attributable to plutonium have been found. A few investigations have shown biological differences between areas containing natural alpha radioactivity in the range of 5 to 200 μCi/m^2 in the top 3 cm of soil and nearby control areas. It is not clear, however, that the differences are caused by variations in radiation dose. Simulation models and available data imply that humans should be able to occupy and derive sustenance from land areas containing of the order of 20 to 200 μCi ^{239}Pu/m^2 in the top 3 cm of soil without exceeding the nonoccupational maximum permissible dose to critical organs as recognized by the ICRP. Simplified calculations suggest that ^{239}Pu applications of roughly 1 Ci/m^2 may be required in grassland areas to cause significant mortality in plant populations. I am not aware of computational models relating ecological effects to the level of application of transuranics other than plutonium.

The general lack of confidence in the accuracy of our predictive capability at present appears to justify substantial research efforts in this area. The shortcomings of the three general approaches have been discussed; yet I see no other approaches to the problem. Therefore it seems that enhanced efforts in each area are called for with continual integration of findings from each.

For direct measurements of the relationship between levels of transuranic application and ecological effects, such applications would need to be made under controlled experimental designs. The use of shorter lived transuranics and engineered barriers to prevent unwanted dispersal of the radioactive material would reduce the risks from such an experiment. If such experiments ever become feasible, remote, controlled areas, such as the Hanford Reservation, the Idaho National Engineering Laboratory, and the Nevada Test Site, might be considered. In addition, the application of effect-inducing quantities of transuranics to terrestrial microcosms might be considered. Although direct-application experiments seem needed from a scientific viewpoint, I do not necessarily advocate them.

Areas presently contaminated with substantial quantities of transuranics should be investigated for suitability for long-term study. Areas in which higher levels of transuranics occur without a previous history of contamination with other materials, such as fission products, and for which good control areas exist would seem particularly valuable for study. Such areas have existed in the past (e.g., Rocky Flats), but, as a result of public concern, cleanup operations were judged more expedient than biological studies. Cleanup decisions are deserving of greater scientific input because in some cases the operation itself may expose the public to greater risk than leaving the protected material in place.

Areas that contain notably high levels of naturally occurring alpha emitters seem deserving of further study, particularly if it can be shown how results might be integrated with current knowledge of transuranic behavior and effects. Potentially valuable study areas exist in Brazil, Colorado, Wyoming, and the USSR.

In terms of theoretical efforts, it seems clear that more generally applicable models are needed. This will require more data from a greater diversity of environments, however, and a much better understanding of basic transport mechanisms. For example, we need to know how climate, vegetation, soil, and other ecosystem attributes affect model parameters that describe such processes as erosion, resuspension, assimilation, and retention. The substantial quantities of data on the environmental behavior of plutonium in the Nevada desert or in Colorado grasslands have only limited applicability to ecosystems in regions of higher precipitation. Resuspension seems to be a particularly critical process affecting the hazard of deposited transuranics, especially in arid regions. As a final point, our knowledge of the effects of pure alpha emitters on plants is far less than our knowledge on animals and is grossly inadequate. Since plants provide stability and the food base of ecosystems, this deficiency should be corrected.

From a scientific viewpoint, it is clear that additional and redirected research can be justified for transuranium elements in the environment. Social tolerance of environmental contamination with radioactive materials, however, appears to be far lower than biological tolerance. In other words, the level of contamination that appears in many cases to prompt cleanup efforts is considerably lower than that which might be expected to elicit obvious biological change. This argument might be used against continued funding for environmental transuranic research. If this is to be the case, scientists in the field may need to provide stronger justification for their work in the future.

Acknowledgments

Preparation of this manuscript was made possible through support from the U. S. Energy Research and Development Administration under Contract No. EY-76-S-02-1156 with Colorado State University. I am indebted to a number of colleagues who assisted in the development of the manuscript. In particular, I wish to cite the exceptional help of A. W. Alldredge, R. O. Gilbert, D. C. Hunt, C. A. Little, W. A. Rhoads, R. C. Thompson, T. F. Winsor, and M. R. Zelle.

References

Allred, D. M., D. E. Beck, and C. D. Jorgensen, 1965, A Summary of the Ecological Effects of Nuclear Testing on Native Animals at the Nevada Test Site, *Utah Acad. Sci., Arts Lett., Proc.,* 42(2): 252-260.

Anspaugh, L. R., J. H. Shinn, P. L. Phelps, and N. C. Kennedy, 1975, Resuspension and Redistribution of Plutonium in Soils, *Health Phys.,* 29(4): 571-582.

Auerbach, S. I., 1971, Ecological Considerations in Siting Nuclear Power Plants: The Long-Term Biotic Effects Problem, *Nucl. Saf.,* 12(1): 25-34.

——, and P. B. Dunaway (Eds.), 1970, *Progress Report in Postattack Ecology,* USAEC Report ORNL-TM-2983, Oak Ridge National Laboratory, NTIS.

Bair, W. J., 1974, The Biomedical Effects of Transuranium Elements in Experimental Animals, in *Plutonium and Other Transuranium Elements: Sources, Environmental Distribution and Biomedical Effects,* USAEC Report WASH-1359, p. 229, NTIS.

——, and R. C. Thompson, 1974, Plutonium: Biomedical Research, *Science,* 183: 715-722.

Barcinski, M. A., et al., 1975, Cytogenetic Investigation in a Brazilian Population Living in an Area of High Natural Radioactivity, *Am. J. Hum. Genet.,* 27(6): 802-806.

Brown, S. L., 1965, *Disintegration Rate Multipliers in Beta-Emitter Dose Calculations,* USAEC file number NP-15714, Stanford Research Institute.

Cullen, T. L., 1968, Dosimetric Measurements in Brazilian Regions of High Natural Radioactivity Leading to Cytogenetic Studies, in *Radiation Protection,* pp. 371-377, W. S. Snyder (Ed.), Pergamon Press, Inc., New York.

Dahlman, R. C., J. J. Beauchamp, and Y. Tanaka, 1973, Effects of Simulated Fallout Radiation on Reproductive Capacity of Fescue, in *Radionuclides in Ecosystems,* Proceedings of the Third National Symposium on Radioecology, Oak Ridge, Tenn., May 10–12, 1971, D. J. Nelson (Ed.), USAEC Report CONF-710501-P2, pp. 988-998, Oak Ridge National Laboratory, NTIS.

——, E. A. Bondietti, and L. D. Eyman, 1976, Biological Pathways and Chemical Behavior of Plutonium and Other Activities in the Environment, in *Actinides in the Environment,* A. M. Friedman (Ed.), ACS Symposium Series, No. 35, pp. 47-80, American Chemical Society.

Drew, R. T., and M. Eisenbud, 1966, The Natural Radiation Dose to Indigenous Rodents on the Morro do Ferro, Brazil, *Health Phys.,* 12(9): 1267-1274.

Eisenbud, M., H. Petrow, R. T. Drew, F. X. Roser, G. Kegel, and T. L. Cullen, 1964, Naturally Occurring Radionuclides in Foods and Waters from the Brazilian Areas of High Radioactivity, in *The Natural Radiation Environment,* pp. 837-854, J. A. S. Adams and W. M. Lowder (Eds.), University of Chicago Press, Chicago.

Energy Research and Development Administration, 1976, *Workshop on Environmental Research for Transuranic Elements,* Proceedings of the Workshop, Battelle Seattle Research Center, Seattle, Wash., Nov. 12–14, 1975, ERDA Report ERDA-76/134, NTIS.

Gopal-Ayengar, A. R., K. Sundaram, K. B. Mistry, and K. P. George, 1977, Current Status of Investigations on Biological Effects of High Background Radioactivity in the Monazite Bearing Areas of Kerala Coast in South-West India, in *International Symposium on Areas of High Natural Radioactivity,* pp. 19-27, Academia Brasileira de Ciencias, Rio de Janeiro.

Gruneberg, H., G. S. Bains, R. J. Berry, L. Riles, C. A. B. Smith, and R. A. Weiss, 1966, *A Search for Genetic Effects of High Natural Radioactivity in South India,* Medical Research Council, Land. Report 307, Her Majesty's Stationery Office, London.

Hakonson, T. E., 1975, Environmental Pathways of Plutonium into Terrestrial Plants and Animals, *Health Phys.,* 29(4): 583-588.

Hanson, W. C., 1975, Ecological Considerations of the Behavior of Plutonium in the Environment, *Health Phys.,* 28: 529-537.

Healy, J. W., 1974, *Proposed Interim Standard for Plutonium in Soils,* USAEC Report LA-5483-MS, Los Alamos Scientific Laboratory, NTIS.

International Commission on Radiological Protection, 1960, Report of Committee II on Permissible Dose for Internal Radiation, *Health Phys.,* 3: 1-380.

——, 1972, *The Metabolism of Compounds of Plutonium and Other Actinides,* ICRP Publication 19, Pergamon Press, Inc., New York.

Kochupillai, N., I. C. Verma, M. S. Grewal, and V. Ramalingaswami, 1976, Down's Syndrome and Related Abnormalities in an Area of High Background Radiation in Coastal Kerala, *Nature,* 262: 60-61.

Lane, W. B., 1971, Preparation and Use of Fallout Simulants in Biological Experiments, in *Survival of Food Crops and Livestock in the Event of Nuclear War,* AEC Symposium Series, Upton, Long Island, New York, Sept. 15–18, 1970, D. W. Benson and A. H. Sparrow (Eds.), pp. 107-114, CONF-700909, NTIS.

Little, C. A., 1976, *Plutonium in a Grassland Ecosystem,* Ph.D. Dissertation, Colorado State University, Ft. Collins, Colo.

Mackin, J., S. Brown, and W. Lane, 1971, Measurement and Computational Techniques in Beta Dosimetry, in *Survival of Food Crops and Livestock in the Event of Nuclear War,* AEC Symposium Series, Upton, Long Island, New York, Sept. 15–18, 1970, D. W. Benson and A. H. Sparrow (Eds.), pp. 51-55, CONF-700909, NTIS.

Martell, E. A., 1975, *Actinides in the Environment and Their Uptake by Man,* Report NCAR-TN/STR-110, National Center for Atmospheric Research.

Martin, W. E., and S. G. Bloom, 1976, Plutonium Transport and Dose Estimation Model, in *Transuranium Nuclides in the Environment,* Symposium Proceedings, San Francisco, 1975, pp. 385-400, STI/PUB/410, International Atomic Energy Agency, Vienna.

Maslov, V. I., K. I. Maslova, and I. N. Verkhovskaya, 1967, Characteristics of the Radioecological Groups of Mammals and Birds of Biocoenoses with High Natural Radiation, in *Radioecological Concentration Processes,* pp. 561-571, B. Aberg and F. P. Hungate (Eds.), Pergamon Press, Inc., New York.

Mericle, L. W., and R. P. Mericle, 1965, Biological Discrimination of Differences in Natural Background Radiation Level, *Radiat. Bot.,* 5: 475-492.

Muller, H. J., 1950, Our Load of Mutations, *Am. J. Hum. Genet.,* 2(2): 111-176.

Murphy, P. G., and J. F. McCormick, 1973, Effects of Beta Radiation from Simulated Fallout on the Reproductive Potential of a Dominant Plant Species in a Natural Community, in *Radionuclides in Ecosystems,* Proceedings of the Third National Symposium on Radioecology, Oak Ridge, Tenn., May 10–12, 1971, D. J. Nelson (Ed.), USAEC Report CONF-710501-P2, pp. 983-987, Oak Ridge National Laboratory, NTIS.

National Academy of Sciences–National Research Council, 1972, *The Effects on Populations of Exposure to Low Levels of Ionizing Radiation,* Report of Advisory Committee on Biological Effects of Ionizing Radiation.

Odum, E. P., 1963, Summary, in *Ecological Effects of Nuclear War,* G. M. Woodwell (Ed.), USAEC Report BNL-917, pp. 69-72, Brookhaven National Laboratory, NTIS.

Osburn, W. S., 1961, Variation in Clones of Penstemon Growing in Natural Areas of Differing Radioactivity, *Science,* 134: 342-343.

Pochin, E. E., 1976, Problems Involved in Detecting Increased Malignancy Rates in Areas of High Natural Radiation Background, *Health Phys.,* 31(2): 148-151.

Rhoads, W. A., and R. B. Platt, 1971, Beta Radiation Damage to Vegetation from Close-In Fallout from Two Nuclear Detonations, *Bioscience,* 21(22): 1121-1125.

Romney, E. M., and J. J. Davis, 1972, Ecological Aspects of Plutonium Dissemination in Terrestrial Environments, *Health Phys.,* 22: 551-557.

Schulz, R. K., 1971, Survival and Yield of Crop Plants Following Beta Irradiation, in *Survival of Food Crops and Livestock in the Event of Nuclear War,* AEC Symposium Series, Upton, Long Island, New York, Sept. 15–18, 1970, D. W. Benson and A. H. Sparrow (Eds.), pp. 370-395, CONF-700909, NTIS.

Smith, D. D., J. Barth, and R. G. Patzer, 1976, Grazing Studies on a Plutonium-Contaminated Range of the Nevada Test Site, in *Transuranium Nuclides in the Environment,* Symposium Proceedings, San Francisco, 1975, pp. 325-336, STI/PUB/410, International Atomic Energy Agency, Vienna.

Sparrow, A. H., J. P. Binnington, and V. Pond, 1958, *Bibliography on the Effects of Ionizing Radiations on Plants, 1896–1955,* USAEC Report BNL-504, Brookhaven National Laboratory, NTIS.

Thompson, R. C., 1976, Panel Discussion, in *Transuranium Nuclides in the Environment,* Symposium Proceedings, San Francisco, 1975, p. 631, STI/PUB/410, International Atomic Energy Agency, Vienna.

Turner, F. B., 1975, Effects of Continuous Irradiation on Animal Populations, in *Advances in Radiation Biology,* Vol. 5, pp. 83-144, J. T. Lett and H. Adler (Eds.), Academic Press, Inc., New York.

United Nations, New York, 1972, *Ionizing Radiation: Levels and Effects,* A Report of the United Nations Scientific Committee on the Effects of Atomic Radiation.

Varney, D. M., and W. A. Rhoads, 1977, Chronic Radiation Induced Chromosomal Aberrations in Native Shrubs at Nevada Test Site, in *Transuranics in Natural Environments,* Symposium Proceedings, Gatlinburg, Tenn., Oct. 5–7, 1976, M. G. White and P. B. Dunaway (Eds.), USDOE Report NVO-178, Nevada Operations Office, NTIS.

Wallace, A., and E. M. Romney, 1972, *Radioecology and Ecophysiology of Desert Plants at the Nevada Test Site,* USAEC Report TID-25954, NTIS.

——, and E. M. Romney, 1975, Feasibility and Alternate Procedures for Decontamination and Post-Treatment Management of Pu-Contaminated Areas in Nevada, in *The Radioecology of*

Plutonium and Other Transuranics in Desert Environments, M. G. White and P. B. Dunaway (Eds.), ERDA Report NVO-153, pp. 251-337, Nevada Operations Office, NTIS.

Walsh, P. J., 1970, Stopping Power and Range of Alpha Particles, *Health Phys.,* 19(2): 312-316.

Whicker, F. W., 1976, *Radioecology of Natural Systems in Colorado,* Fourteenth Annual Progress Report, May 1, 1975–July 31, 1976, ERDA Report COO-1156-84, Colorado State University, NTIS.

——, and L. Fraley, Jr., 1974, Effects of Ionizing Radiation on Terrestrial Plant Communities, in *Advances in Radiation Biology,* Vol. 4, pp. 317-366, J. T. Lett, H. Adler, and M. R. Zelle (Eds.), Academic Press, Inc., New York.

Wildung, R. E., et al., 1977, Transuranic Complexation in Soil and Uptake by Plants, in *Pacific Northwest Laboratory Annual Report for 1976 to the ERDA Assistant Administrator for Environment and Safety,* B. E. Vaughan (Ed.), ERDA Report BNWL-2100(Pt. 2), pp. 4.6–4.10, Battelle, Pacific Northwest Laboratories, NTIS.

Dosimetry and Ecological Effects
of Transuranics in the Marine Environment

WILLIAM L. TEMPLETON

Radiation doses received by aquatic organisms as a result of exposure to transuranics in the environment are comparable to those received from natural radionuclides even in known contaminated areas. At these levels it is doubtful whether experimental studies in the field, or in the laboratory at similar levels, could reasonably be conducted which would offer some degree of success in determining radiological effects on individuals, populations, or ecosystems. Some of the mechanisms of recruitment to exploited fish populations are considered, and these mechanisms suggest that any radiation-induced effects would probably be compensated for by density-dependent responses in highly fecund species. In species with low fecundity, increased stress would clearly increase the chances of diminishing these populations; however, from the dose-rate estimates, present levels of radiation are unlikely to provide any additional stress in comparison to exploitation of some of these species by man. Although little quantitative genetic information is available for aquatic populations, it appears unlikely, from estimated mutation rates, that significant deleterious genetic effects due to radiation would be produced at the present low levels in the environment.

Over the past quarter of a century, the open oceans and coastal waters of the world have received quantities of artificially produced transuranic elements. These elements have been distributed globally in the atmosphere as a result of nuclear weapons testing conducted by the United States, the United Kingdom, the Union of Soviet Socialist Republics, France, India, and China and the burnup of a U. S. space satellite (SNAP-9A) in 1964. Estimates are that about 325 kCi of ^{239,240}Pu and about 8 kCi of ^{238}Pu have been deposited over the globe by weapons testing and about 17 kCi of ^{238}Pu by the SNAP failure (Hardy, Krey, and Volchok, 1973). On a more local scale, transuranic elements have been, and continue to be, introduced to the marine environment at nuclear weapons testing grounds, at nuclear-fuel reprocessing plants, and, to a much lesser degree, by nuclear power facilities and occasional nuclear-device accidents. In the Pacific the total inventory of transuranic elements in the Bikini and Enewetak atolls at the Pacific Proving Grounds probably is as high as 10 kCi, with a reported net flux from the Bikini lagoon to the North Equatorial Current of about 6 Ci of ^{239,240}Pu per year and 3 Ci of ^{241}Am per year (Nevissi and Schell, 1975). During the period 1960 to 1974 (Hetherington et al., 1975; 1976), the nuclear-fuel reprocessing plant at Windscale (United Kingdom) discharged approximately 10 kCi of ^{238,239,240}Pu to the northeast Irish Sea. In recent years the average transport out of the Irish Sea has been 40 Ci/yr.

The major world inventories of transuranics are, of course, contained in reactors, weapons stockpiles, reprocessing plants, and waste-storage systems, and only a very small fraction of the transuranics produced has been or will be released or disposed of to the environment. We still, however, have only a scant knowledge of the processes that control the behavior and fate of the transuranics in the environment, and hence our ability to predict and assess the potential effects is hampered.

In the past, and this is still partly true today, the primary consideration has been anthropocentric, and research priorities have been particularly directed to the assumed primary pathway of inhalation by man. Less attention has been given to secondary pathways, which result in chronic long-term exposures to man through the food web and to biota in the ecosystem. Although the primary short-term hazard to man from an atmospheric release may be via the inhalation pathway, it behooves us to give increased consideration to the long-term hazard potentials because of the potential time lag in transfer and long-term persistence in natural reservoirs.

The behavior and fate of transuranics in the marine environment were given very little attention before the mid-1960s. One reason for this was the lack of methodology and instrumentation to determine the ultralow levels that existed in the aquatic environment. In fact, it was not until Pillai, Smith, and Folsom (1964) determined the levels of ^{239,240}Pu in marine organisms from weapons tests that any data were published other than total-alpha measurements from some selected sites. Since that time data have become available for plutonium isotopes and americium from weapons tests, SNAP-9A, and some reprocessing plants in particular in air, freshwater, seawater, sediment, and biological materials. Much of the available published data was reviewed by Noshkin (1972). Although intensive monitoring and research studies have been conducted more recently at the Pacific test sites (U. S. Atomic Energy Commission, 1973), in the northeast Irish Sea (Hetherington et al., 1976), at La Hague in France (Frazier and Guary, 1976; Guary and Frazier, 1977a; 1977b), and in Lake Michigan (Wahlgren et al., 1976), by far the greatest amount of data on plutonium isotopes generated in the late 1960s was applicable only to the determination of the residence times of these materials in the oceans. Laboratory and field studies on plutonium kinetics in marine ecosystems have been very limited until recently, and even today little research has been conducted on transuranic elements other than plutonium (International Atomic Energy Agency, 1976). Data for americium, neptunium, and curium are sparse, and none have been published for berkelium and californium. This chapter discusses one aspect of transuranics in the marine environment: the potential effects of radiation from these materials on organisms in the marine ecosystem.

One way to assess whether the present levels of transuranics, more particularly plutonium, in the aquatic environment have a potential to result in somatic or genetic damage to aquatic organisms is to compare the radiation dose rate from plutonium, both from weapons-test fallout and from selected sites contaminated to a relatively higher level, with that from natural radiation to which organisms, populations, communities, and ecosystems have been exposed for near-geological time for their life-spans. These calculated dose rates can also be compared with experimentally determined data on effects. Finally, where the availability of data for the latter are lacking, we need to assess, and perhaps hypothesize, on the basis of other related data on radiation effects, ecological interactions, and population dynamics, whether or not these dose rates in the environment could result in somatic and genetic effects that would be detrimental to the maintenance of aquatic populations. This chapter draws heavily on a recent review and assessment of the ecological effects of radiation in the marine environment (International

Atomic Energy Agency, 1976) and reviews some recent experimental data on the effects of plutonium on aquatic organisms.

Dose Rates from Transuranics and Natural Radionuclides in Natural Waters

A few recent papers have dealt with dose rates received by aquatic organisms exposed to plutonium in natural waters, and they are compared here with dose rates from the natural radionuclide ^{210}Po.

Till, Kaye, and Trebalka (1976), Till and Franks (1977), and Till (1978) have estimated the dose rate during embryogenesis for six species of marine and freshwater fishes exposed to plutonium in a variety of natural waters (northeast Irish Sea; White Oak Lake, Oak Ridge National Laboratory; U-Pond, Hanford; Enewetak lagoon; and Lake Michigan). They estimated that the dose rates ranged from 7×10^{-2} μrad of plutonium per hour for plaice *(Pleuronectes platessa)* eggs exposed at 1 pCi of plutonium per liter of water in the Irish Sea to 3×10^{-5} μrad/hr for carp *(Cyprinus carpio)* eggs exposed at 1×10^{-3} pCi/liter in Lake Michigan. These dose rates are less than those from the natural radionuclides ^{210}Pb and ^{210}Po (Woodhead et al., 1976).

Hetherington et al. (1976) estimated the dose rates to the embryos of plaice, adult plaice, and other organisms from the northeast Irish Sea where plutonium concentrations in water are of the order of 1 pCi/liter and the concentrations of ^{238,239,240}Pu in the sediment out to 10 km from the discharge point are of the order of 40 pCi of plutonium per gram of sediment. The ratio of ^{241}Am to ^{238,239,240}Pu for eight stations had a mean value of 1.3 ± 0.2. The estimated dose rates to the developing plaice eggs that had been exposed to 1 pCi/liter were in the range of 0.09 to 0.47 μrad/hr, which is somewhat less than the 0.7 μrad/hr from natural ^{40}K and less than the dose received by zooplankton from ^{210}Po (Woodhead et al., 1976). Dose rates from ^{238}Pu, ^{239,240}Pu, and ^{241}Am have also been estimated for young (Group III) plaice, crab *(Cancer pagurus)*, and mussels *(Mytilus edulis)*. The total dose rates (Table 1) are of the same order as those from the natural background from ^{210}Po except for the soft tissues of mussels and gills of crab. In these exceptions there is a possibility that the tissues were contaminated by sediment.

Effects Studies

There is a paucity of data in the literature on the effects of transuranics on aquatic biota. Studies of the effects of ^{234}Pu on the eggs of carp have been reported by Auerbach, Nelson, and Struxness (1974). No observable effects were seen either in the rate of hatching or in the frequency of abnormalities when eggs were exposed to concentrations ranging from 5×10^5 to 5×10^7 pCi/liter, which is many orders of magnitude above concentrations known to exist in the natural environment. Till, Kaye, and Trebalka (1976) have reported on the doses that produced effects on hatching, survival, and abnormalities in carp and fathead minnow *(Pimephales promelas)* eggs (Table 2). Accumulated doses that, over the period of embryogenesis, first produced a significant effect on hatching and survival were in excess of 2000 rad for both species. Abnormalities were first produced when the accumulated dose exceeded 750 rad. These doses would be significantly greater in dose-equivalent units, where for alpha radioactivity the dose (in rad) is multiplied by a quality factor. These data indicate that these fish eggs are relatively insensitive to the effects of alpha radiation from ^{238}Pu. The data also infer that concentrations of 1 μCi of

TABLE 1 Total Dose Rates (μrad/hr) to Biota in the Northeast
Irish Sea from Internal Plutonium, Americium, and Polonium*

Species	Tissue	^{238}Pu	^{239,240}Pu	^{241}Am	Total	^{210}Po
			Dose rate, μrad/hr			
M. edulis	Visceral mass	4.7	17	44	66	Mollusca: 4.5 to 12.0
C. pagurus	Muscle	0.35	1.1	6.9	8.4	Crustacea: Whole, 4.5 to 18
	Gill	0.9	31	90	130	Hepatopancreas, 140
	Digestive gland	0.58	2.2	17	20	
P. platessa	Skin	9.4×10^{-3}	3.5×10^{-2}	0.23	0.27	Fish: - - - -
	Bone	4.4×10^{-2}	0.15	1.3	1.5	0.2 to 2.5
	Gill	- - - -	0.14	2.4	2.5	- - - -
	Gut	4.3×10^{-2}	0.16	1.8	2.0	2.2 to 2.9
	Muscle	1.2×10^{-3}	4.4×10^{-3}	2.3×10^{-2}	2.8×10^{-2}	5×10^{-3} to 1.5
	Liver	8.7×10^{-2}	0.32	6.7	7.1	2.2 to 10
	Kidney	0.5	0.96	6.1	6.1	- - - -

*Based on data from Hetherington et al. (1976) and Woodhead et al. (1976).

TABLE 2 Summary of Estimated* Doses of ^{238}Pu to *C. carpio* and
P. promelas Eggs Which Produced Effects on Hatching, Survival,
and Abnormalities†

Egg‡	Hatching§	Survival	Abnormalities	Number abnormal / Sample size
		^{238}Pu dose, rad		
C. carpio	1.57×10^4 (8.17×10^3)	8.19×10^3	4.27×10^3	5/377
P. promelas	9.71×10^3 (5.60×10^3)	1.94×10^3	5.68×10^2	15/454

*Except where noted, the dose is for the concentration that first produced a
significant effect on hatching, survival, and abnormalities.

†Based on data from Till et al. (1976).

‡Period of embryogenesis (*C. carpio*, 3 days; *P. promelas*, 7 days).

§Number in parentheses is the dose at which little or no effect was observed
on hatching and may be considered as an estimate of the threshold dose to affect
hatching.

^{238}Pu per liter of water may result in a synergistic effect between chemical and
radiological toxicity because of the high concentration of plutonium mass present.

These doses are within the range of experimental dose rates reported in the
literature that result in damage and are clearly many orders of magnitude above
those experienced even in the most highly contaminated areas (International Atomic
Energy Agency, 1976).

Potential Impact of Ionizing Radiation on Aquatic Resources

The available data on the effects of ionizing radiation on individual organisms indicate that, when experimental and field exposure dose rates are less than 1 rad/day, it is difficult to demonstrate effects that are not already within the inherent variation already present. It is pertinent, however, to discuss some of the concerns and potential problems to provide an improved basis for the assessment, understanding, and acceptance of the degree of risk that aquatic populations are exposed to as a result of the introduction of radionuclides into the aquatic environment. A valid perspective can be developed by comparing potential effects with losses caused by natural mortality and fishing. This aspect has been discussed in some depth by an international panel (International Atomic Energy Agency, 1976).

The available evidence suggests that (1) fish are the most radiosensitive component; (2) developing gametes, fertilized eggs, and larvae are the sensitive stages; and (3) any damage that might occur to a fishery resource would most likely arise from the direct effects of radiation on the fish rather than from effects from disturbances or changes in the food web. Some of the earlier literature from laboratory experimentation observed that effects, seen as the number of abnormal larvae hatched, resulted from exposure to radionuclides at extremely low concentrations. These findings were extrapolated to suggest that, if fish were exposed to these concentrations, the yield from commercial fisheries would be adversely affected. However, consideration needs to be given to (1) the nature of the stock and recruitment relationships for highly fecund fish where large numbers of eggs are produced at each spawning (10^3 to 10^6 per female); (2) the small number of eggs required to survive to maintain stock at equilibrium; (3) the high mortality that occurs during the larval stages such that, say, only 1 in 10,000 survives; and (4) the fact that a direct relationship does not necessarily exist between spawning stock size and recruitment to the reproductive stock. These factors suggest that, even if mortality at this stage of development is being enhanced by the low dose rates presently existing in the aquatic environment, recruitment in highly fecund species of marine fish would unlikely be adversely affected unless these stocks were already at risk because of severe overexploitation. The mechanisms controlling recruitment in invertebrates appear similar, except that environmental factors probably play a more important role.

For species with low fecundity (rays, sharks, dogfish, and marine mammals), most of whom produce live young, recruitment is related to adult stock size, and, for the marine mammals, the relationship may be almost direct. Although it is reported that the fecundity of baleen whales (Laws, 1962) and elasmobranch stocks (Holden, 1973) has increased as a result of exploitation, there is an obvious upper limit; available data would suggest that this upper limit has been reached by the exploited stocks. Further stress would clearly increase the chances of extinguishing the populations. However, although there are no data on the radiosensitivity of these low-fecund species, the estimates of the dose rates received by aquatic organisms in the natural waters of the world are rarely of the same order as, and generally less than, the limits recommended for humans. Hence they are unlikely to provide any additional stress in comparison to man's continued exploitation of some of these stocks.

The possible increase in mortality rates due to radiation in exploited fish stocks can also be compared with mortality rates experienced by those fisheries presently exploited by man. Rates of exploitation as high as 50% a year on all classes recruited to a fishery are common. In addition, mortality due to natural causes occurs; thus a heavily exploited stock may survive even though it is subjected to a total mortality rate of 60 to 70% per year. Any additional mortality, say, due to radiation, would reduce the stock but would not necessarily affect the ability of the stock to replace itself. Additionally, any mortality caused by radiation would probably not be detectable as such. There are only a very few studies on natural populations that have been subjected to low-level irradiation; but, in those reported (Donaldson and Bonham, 1964; 1970; Bonham and Donaldson, 1966; Blaylock, 1961; Templeton et al., 1976), there is no evidence that low levels of radiation have any adverse effects on these populations.

There is virtually no published data on the genetic effects of irradiation on fish populations. Although studies have indicated that chromosomal abnormalities can occur in irradiated eggs and larvae of aquatic species (Ophel et al., 1976), there is no evidence to indicate that these abnormalities have been detrimental to the population. Predictions therefore can be based only on studies conducted with other species, e.g., *Drosophila*. It can be argued from those data that modest increases in mutation rates with concomitant enhancement in the genetic variability may even lead to improved fitness (Neel, 1972). Additionally, the large amount of genetic variability revealed by recent biochemical techniques may challenge the consensus that mutations are always detrimental in nature and emphasize the importance of understanding the dynamics of newly introduced mutants in the gene pool and selective processes. These developments have increased the difficulty of assessing the potential long-term genetic implications of the irradiation of natural populations (Neel, 1972; International Atomic Energy Agency, 1976). In this connection, Woodhead (1974) has conservatively estimated, on the basis of very limited data, that, if all mutations are dominant lethals resulting in nonviable zygotes, then less than 1 of every 1000 fish embryos would be eliminated as the result of an integrated dose of 0.5 rad received by each of the parents.

Research Needs

Many assumptions used in considering the somatic and genetic effects of radiation on populations in the aquatic environment are to some degree speculative. Since this is also true for assessments of the effects of any energy-related pollutants that enter the aquatic environment, the recommendations for future research made by the International Atomic Energy Agency (1976) could equally be applied to pollutants other than radiation.

It is recommended that prime consideration be given to:

● Comprehensive studies on a sufficient spatial and temporal scale to determine the significance of changes in populations, communities, and ecosystems resulting from low-level chronic exposure to pollutants, singly and in combination. Emphasis should be given to determining the rates of change, the rates of recovery from various degrees of damage, and the rates of repopulation in decimated areas.

● Comparative studies of mutation rates induced by pollutants, singly and in concert, on a wide range of marine organisms, including species with both high and low fecundities. Emphasis should be given to both genetic damage (gene mutation, chromosomal aberrations, recombination, etc.) and effects on population size, biomass, fecundity, and fitness components.

● Studies designed to provide an understanding of the role of genetic variation, expressed as discrete polymorphisms and quantitative variations of individual species, in the maintenance of aquatic communities. Emphasis should be given to clarifying the significance of the response of these polymorphisms to varying physical, chemical, and biological parameters.

Summary

Since the dose rates received by aquatic organisms as a result of exposure to transuranics are comparable to those received from natural radionuclides, even in known contaminated areas, it is apparent that there are few experimental field studies that reasonably could be conducted which would determine whether radiological effects are occurring in the environment as a result of present levels of radionuclides. The comparisons drawn here between the estimated doses from plutonium and americium on the one hand and naturally occurring polonium on the other would be more accentuated if the total dose rates from all natural radionuclides were computed.

Consideration of some of the mechanisms of recruitment to exploited fish populations would suggest that any effects as a result of chronic exposure to low-level ionizing radiations would probably be compensated for by density-dependent responses in highly fecund species. Effects due to radiation therefore would not likely be distinguishable from those due to natural fluctuations in aquatic populations. Although little quantitative genetic information is available for aquatic populations, it is unlikely, on the basis of predicted mutation rates, that significant deleterious genetic effects due to radiation at the low levels present in the environment today would be produced in aquatic populations.

References

Auerbach, S. I., D. J. Nelson, and E. G. Struxness, 1974, *Environmental Sciences Division Annual Report,* C. I. Taylor (Ed.), USAEC Report ORNL-5016, p. 10, Oak Ridge National Laboratory, NTIS.

Blaylock, B. G., 1961, The Fecundity of *Gambusia affinis affinis* Population Exposed to Chronic Environmental Contamination, *Radiat. Res.,* 37: 108.

Bonham, K., and L. R. Donaldson, 1966, Low Level Chronic Irradiation of Salmon Eggs and Alewives, in *Disposal of Radioactive Wastes into Seas, Oceans, and Surface Waters,* Symposium Proceedings, Vienna, 1966, p. 869, STI/PUB/126, International Atomic Energy Agency, Vienna.

Donaldson, L. R., and K. Bonham, 1964, Effects of Low Level Chronic Irradiation of Chinook and Coho Salmon Eggs and Alevins, *Trans. Am. Fish. Soc.,* 93: 333.

——, and K. Bonham, 1970, Effects of Chronic Exposure to Chinook Salmon Eggs and Alevins to Gamma Irradiation, *Trans. Am. Fish. Soc.,* 99: 112.

Frazier, A., and J. C. Guary, 1976, Recherche d'indicateurs biologiques appropriés au controle de la contamination du littoral par le plutonium, in *Transuranium Nuclides in the Environment,* Symposium Proceedings, San Francisco, 1975, pp. 679-689, STI/PUB/410, International Atomic Energy Agency, Vienna.

Guary, J. C., and A. Frazier, 1977a, Influence of Trophic Levels and Calcification on the Uptake of Plutonium Observed In Situ in Marine Organisms, *Health Phys.,* 32: 21.

——, and A. Frazier, 1977b, Etude comparée de teneurs en plutonium chez divers mollusques de quelques sites littoraux francais, *Mar. Biol.,* 41: 263.

Hardy, E. P., P. W. Krey, and H. L. Volchok, 1973, Global Inventory and Distribution of Fallout from Plutonium, *Nature,* 241: 444.

Hetherington, J. A., D. F. Jeffries, and M. B. Lovett, 1975, Some Investigations into the Behavior of Plutonium in the Marine Environment, in *Impacts of Nuclear Releases into the Aquatic Environment,* Symposium Proceedings, Otaniemi, 1975, p. 193, STI/PUB/406, International Atomic Energy Agency, Vienna.

——, D. F. Jeffries, N. T. Mitchell, R. J. Pentreath, and D. S. Woodhead, 1976, Environmental and Public Health Consequences of the Controlled Disposal of Transuranic Elements to the Marine Environment, in *Transuranium Nuclides in the Environment,* Symposium Proceedings, San Francisco, 1975, pp. 139-154, STI/PUB/410, International Atomic Energy Agency, Vienna.

Holden, M. J., 1973, Are Long-Term Sustainable Fisheries for Elasmobranchs Possible? *Rapp. P.-V. Reun. Cons. Int. Explor. Mer.,* 164: 360.

International Atomic Energy Agency, 1976, *Effects of Ionizing Radiation on Aquatic Organisms and Ecosystems,* Technical Reports Series No. 172, STI/DOC/10/172, International Atomic Energy Agency, Vienna.

Laws, R. M., 1962, Some Effects of Whaling on the Southern Stocks of Baleen Whales, in *The Exploitation of National Animal Populations,* E. C. LeCueu and M. W. Holdgate (Eds.), p. 137, Oxford University Press.

Neel, J. V., 1972, The Detection of Increased Mutation Rates in Human Populations, in *Mutagenic Effects of Environmental Contaminants,* H. E. Sutton and M. I. Harris (Eds.), p. 39, Academic Press, Inc., New York.

Nevissi, A., and W. R. Schell, 1975, Distribution of Plutonium and Americium in Bikini Atoll Lagoon, *Health Phys.,* 28: 539.

Noshkin, V. E., 1972, Ecological Aspects of Plutonium Dissemination in Aquatic Environments, *Health Phys.,* 22: 537.

Ophel, I., et al., 1976, Effects of Ionizing Radiation on Aquatic Organisms, in *Effects of Ionizing Radiation on Aquatic Organisms and Ecosystems,* Technical Reports Series No. 172, Chap. 2, p. 57, STI/DOC/10/172, International Atomic Energy Agency, Vienna.

Pillai, K. C., R. C. Smith, and T. R. Folsom, 1964, Plutonium in the Marine Environment, *Nature,* 203: 568.

Templeton, W. L., et al., 1976, Effects of Ionizing Radiation on Aquatic Populations and Ecosystems, in *Effects of Ionizing Radiation on Aquatic Organisms and Ecosystems,* Technical Reports Series No. 172, Chap. 3, STI/DOC/10/172, International Atomic Energy Agency, Vienna.

Till, J. E., 1978, The Effect of Chronic Exposure to ^{238}Pu(IV) Nitrate on the Embryonic Development of Carp and Fathead Minnow Eggs, *Health Phys.,* 34: 333-343.

——, and M. L. Franks, 1977, Bioaccumulation, Distribution and Dose of ^{241}Am, ^{244}Am and ^{238}Pu in Developing Fish Embryos, in *Proceedings of the Fourth International Congress of International Radiation Protection Association,* Paris, Vol. 2, pp. 645-648.

——, S. V. Kaye, and J. R. Trebalka, 1976, *The Toxicity of Uranium and Plutonium to the Developing Embryos of Fish,* ERDA Report ORNL-5160, Oak Ridge National Laboratory, NTIS.

U. S. Atomic Energy Commission, 1973, *Enewetak Radiological Survey,* USAEC Report NVO-140, Vols. I-III, Nevada Operations Office, NTIS.

Wahlgren, M. A., J. J. Alberts, D. M. Nelson, and K. A. Orlandi, 1976, Study of Behavior of Transuranics and Possible Chemical Homologues in Lake Michigan Water and Biota, in *Transuranium Nuclides in the Environment,* Symposium Proceedings, San Francisco, 1975, pp. 9-24, STI/PUB/410, International Atomic Energy Agency, Vienna.

Woodhead, D. S., 1974, The Estimation of Radiation Dose Rates to Fish in Contaminated Environments, and the Assessment of the Possible Consequences, in *Population Dose Evaluation and Standards for Man and His Environment,* Symposium Proceedings, Portoróz, 1974, STI/PUB/375, International Atomic Energy Agency, Vienna.

——, et al., 1976, Concentrations of Radionuclides in Aquatic Environments and the Resultant Radiation Dose Rates Received by Aquatic Organisms, in *Effects of Ionizing Radiation on Aquatic Organisms and Ecosystems,* Technical Reports Series No. 172, Chap. 1, STI/DOC/10/172, International Atomic Energy Agency, Vienna.

Index

DISCLAIMER

FUNDERBURG LIBRARY
MANCHESTER COLLEGE